Second Edition

MODERN DIFFERENTIAL GEOMETRY
of Curves and Surfaces with
MATHEMATICA®

Alfred Gray

Second Edition

MODERN DIFFERENTIAL GEOMETRY
of Curves and Surfaces with
MATHEMATICA®

CRC Press
Boca Raton   Boston   London   New York   Washington, D.C.

Library of Congress Cataloging-in-Publication Data

Gray, Alfred, 1939–
    Modern Differential Geometry of Curves and Surfaces with Mathematica / by Alfred Gray.
        p.   cm. -- (Studies in advanced mathematics)
    Includes bibliographical references and indexes.
    ISBN 0-8493-7164-3   (alk. paper)
    1. Geometry, Differential--Data processing. 2. Mathematica (Computer file) I. Title. II. Series.
QA641.G72 1997
516.3′6—dc21
    for Library of Congress   97-17161
                                                                                                CIP

This book contains information obtained from authentic and highly regarded sources. Reprinted material is quoted with permission, and sources are indicated. A wide variety of references are listed. Reasonable efforts have been made to publish reliable data and information, but the author and the publisher cannot assume responsibility for the validity of all materials or for the consequences of their use.

Neither this book nor any part may be reproduced or transmitted in any form or by any means, electronic or mechanical, including photocopying, microfilming, and recording, or by any information storage or retrieval system, without prior permission in writing from the publisher.

The consent of CRC Press LLC does not extend to copying for general distribution, for promotion, for creating new works, or for resale. Specific permission must be obtained in writing from CRC Press LLC for such copying.

Direct all inquiries to CRC Press LLC, 2000 Corporate Blvd., N.W., Boca Raton, Florida 33431.

Trademark Notice: Product or corporate names may be trademarks or registered trademarks, and are used only for identification and explanation, without intent to infringe.

© 1998 by CRC Press LLC

No claim to original U.S. Government works
International Standard Book Number 0-8493-7164-3
Library of Congress Card Number 97-17161
Printed in the United States of America   2 3 4 5 6 7 8 9 0
Printed on acid-free paper

PREFACE

Modern Differential Geometry of Curves and Surfaces is a traditional text, but it uses the symbolic manipulation program *Mathematica*. This important computer program, available on PCs, Macs, NeXTs, Suns, Silicon Graphics Workstations and many other computers, can be used very effectively for plotting and computing. The book presents standard material about curves and surfaces, together with accurate interesting pictures, *Mathematica* instructions for making the pictures and *Mathematica* programs for computing functions such as curvature and torsion.

Although **Curves and Surfaces** makes use of *Mathematica*, the book should also be useful for those with no access to *Mathematica*. All calculations mentioned in the book can in theory be done by hand, but some of the longer calculations might be just as tedious as they were for differential geometers in the 19th century. Furthermore, the pictures (most of which were done with *Mathematica*) elucidate concepts, whether or not *Mathematica* is used by the reader.

The main prerequisite for the book is a course in calculus, both single variable and multi-variable. In addition, some knowledge of linear algebra and a few basic concepts of point set topology are needed. These can easily be obtained from standard sources. **No computer knowledge is presumed.** In fact, the book provides a good introduction to *Mathematica*; the book is compatible with both versions 2.2 and 3.0. For those who want to use **Curves and Surfaces** to learn *Mathematica*, it is advisable to have access to Wolfram's book **Mathematica** for reference. (In version 3.0 of *Mathematica*, Wolfram's book is available through the help menus.)

Curves and Surfaces is designed for a traditional course in differential geometry. At an American university such a course would probably be taught at the junior-senior level. When I taught a one-year course based on **Curves and Surfaces** at the University of Maryland, some of my students had computer experience, others had not. All of them had acquired sufficient knowledge of *Mathematica* after one week. I chose not to have computers in my classroom because I needed the classroom time to explain concepts. I assigned all of the problems at the end of each chapter. The students used workstations, PCs and Macs to do those problems that required *Mathematica*. They either gave me a printed version of each assignment, or they sent the assignment to me by electronic mail.

Symbolic manipulation programs such as *Mathematica* are very useful tools for differential geometry. Computations that are very complicated to do by hand can frequently be performed with ease in *Mathematica*. However, they are no substitute for the theoretical aspects of differential geometry. So **Curves and Surfaces** presents theory and uses *Mathematica* programs in a complementary way.

Some of the aims of the book are the following.

- To show how to use *Mathematica* to plot many interesting curves and surfaces, more than in the standard texts. Using the techniques described in **Curves and Surfaces**, students can understand concepts geometrically by plotting curves and surfaces on a monitor and then printing them. The effect of changes in parameters can be strikingly portrayed.

- The presentation of pictures of curves and surfaces that are informative, interesting and accurate. The book contains over 400 illustrations.

- The inclusion of as many topics of the classical differential geometry and surfaces as possible. In particular, the book contains many examples to illustrate important theorems.

- Alleviation of the drudgery of computing things such as the curvature and torsion of a curve in space. When the curvature and torsion become too complicated to compute, they can be graphed instead. There are more than 175 miniprograms for computing various geometric objects and plotting them.

- The introduction of techniques from numerical analysis into differential geometry. *Mathematica* programs for numerical computation and drawing of geodesics on an arbitrary surface are given. Curves can be found numerically when their torsion and curvature are specified. See the pictures in Section 6.4.

- To place the material in perspective through informative historical notes. There are capsule biographies with portraits of over 75 mathematicians and scientists.

- To introduce interesting topics that, in spite of their simplicity, deserve to be better known. I mention triply orthogonal systems of surfaces (Chapter 29), Björling's formula for constructing a minimal surface containing a given plane curve as a geodesic (Chapter 33) and canal surfaces and cylides of Dupin as Maxwell discussed them (Chapter 35).

- To develop a dialect of *Mathematica* for handling functions that facilitates the construction of new curves and surfaces from old. For example, there is a simple program to generates a surface of revolution from a plane curve.

- To provide explicit definitions of curves and surfaces. Over 300 *Mathematica* definitions of curves and surfaces can be used for further study.

The approach of **Curves and Surfaces** is admittedly more computational than is usual for a book on the subject. For example, Brioschi's formula for the Gaussian curvature in terms of the first fundamental form can be too complicated for use in hand calculations, but *Mathematica* handles it easily, either through computations or through graphing the curvature. Another part of *Mathematica* that can be used effectively in differential geometry is its special function library. For example, nonstandard spaces of constant curvature can be defined in terms of elliptic functions and then plotted.

Frequently, I have been asked if new mathematical results can be obtained by means of computers. Although the answer is generally no, it is certainly the case that computers can be an effective supplement to pure thought, because they allow experimentation and the graphs provide insights into complex relationships. I hope that many research mathematicians will find **Curves and Surfaces** useful for that purpose. Two results that I found with the aid of *Mathematica* are the interpretation of torsion in terms of tube twisting in Chapter 9 and the construction of a conjugate minimal surface without integration in Chapter 31. I have not seen these results in the literature, but they may not be new.

The programs in the book, as well as some descriptive *Mathematica* notebooks, will eventually be available on the web. In the meantime the programs and notebooks can be down-loaded via anonymous ftp from **bianchi.umd.edu**, which is the computer in my office. I have also made images of many curves and surfaces that are displayable using the programs acrospin and geomview, as described in Appendix D. These files are also available from **bianchi.umd.edu**. On the ftp site *Maple* definitions of curves and surfaces are given as well. My email address is **gray@bianchi.umd.edu**.

Sample Course Outlines

There is ample time to cover the whole book in three semesters at the undergraduate level or two semesters at the graduate level. Here are suggestions for courses of shorter length.

- **One semester undergraduate differential geometry course:** Chapters 1, 2, 8, 11—17, 20, 22, 25, 26, 27.

- **Two semester undergraduate differential geometry course:** Chapters 1, 2, 4, 6—27.

- **One semester graduate differential geometry course:** Chapters 1, 2, 8—17, 20—27, 30—33.

- **One semester course on curves and their history:** Chapters 1–10 and Section 36.2.

- **One semester course on *Mathematica* and graphics** Chapters 1, 2, 4, 8—14, 20, 21, 29—36.

I have tried to include more details than are usually found in mathematics books. This, plus the fact that the *Mathematica* programs can be used to elucidate theoretical concepts, makes the book easy to use for independent study.

Curves and Surfaces is an ongoing project. In the course of writing this book, I have become aware of the vast amount of material that was well-known over a hundred years ago, but now is not as popular as it should be. So I plan to develop a web site, and to write a problem book to accompany the present text. Spanish, German, Japanese and Italian versions of **Curves and Surfaces** are already available.

Graphics

Although *Mathematica* graphics are very good, and can be used to create Quick-Time movies, the reader may also want to consider the following additional display methods:

- **Acrospin** is an inexpensive easy-to-use program that works on even the humblest PC. See page 994 for details of how to use Acrospin with *Mathematica*. Acrospin allows any surface or space curve to be spun.

- **Geomview** is program for interactive display of curves and surfaces. It works on high-end computers, including NeXTs, Suns, Silicon Graphics (best), and PCs running Linux. See page 999 for details. Geomview displays have the added advantage of smooth shading of polygons.

- **Live** is an add-on program to *Mathematica* that allows interactive display. Contact Wolfram Research or True-D Software, whose email address is `info@true-d.co.uk`.

- **AVS** programs have been developed by David McNabb for spectacular stereo three-dimensional images for the surfaces described in this book. These are viewable using special glasses on a Silicon Graphics Workstation using AVS. For more information contact `mcnabb@avl.umd.edu`.

A Perspective

Mathematical trends come and go. R. Osserman in his article ("The Geometry Renaissance in America: 1938-1988" in **A Century of Mathematics in America**, volume 2, American Mathematical Society, Providence, 1988) makes the point that in the 1950s when he was a student at Harvard, algebra dominated mathematics, the attention given to analysis was small, and the interest in differential geometry was converging to zero.

It was not always that way. In the last half of the 19th century surface theory was a very important area of mathematics, both in research and teaching. Brill, then Schilling, made an extensive number of plaster models available to the mathematical public. Darboux's **Leçons sur la Théorie Générale des Surfaces** and Bianchi's **Lezioni di Geometria Differenziale** were studied intensely. I attribute the decline of differential geometry, especially in the United States, to the rise of tensor analysis. Instead of drawing pictures it became fashionable to raise and lower indices.

I strongly feel that pictures need to be much more stressed in differential geometry than is presently the case. It is unfortunate that the great differential geometers of the past did not share their extraordinary intuitions with others by means of pictures. I hope that the present book contributes in some way to returning the differential geometry of curves and surfaces to its proper place in the mathematics curriculum.

I wish to thank Elsa Abbena, James Anderson, Thomas Banchoff, Marcel Berger, Michel Berry, Nancy Blachman, William Bruce, Renzo Caddeo, Eugenio Calabi, Thomas Cecil, Luis A. Cordero, Al Currier, Luis C. de Andrés, Mirjana

Djorić, Franco Fava, Helaman Fergason, Marisa Fernández, Frank Flaherty, Anatoly Fomenko, V.E. Fomin, David Fowler, George Francis, Ben Friedman, Thomas Friedrick, Pedro M. Gadea, Sergio Garbiero, Laura Geatti, Peter Giblin, Vladislav Goldberg, William M. Goldman, Hubert Gollek, Mary Gray, Joe Grohens, Garry Helzer, A.O. Ivanov, Gary Jensen, Alfredo Jiménez, Raj Jakkumpudi, Gary Jensen, David Johannsen, Joe Kaiping, Ben Kedem, Robert Kragler, Steve Krantz, Henning Leidecker, Stuart Levy, Mats Liljedahl, Lee Lorch, Sanchez Santiago Lopez de Medrano, Roman Maeder, Steen Markvorsen, Mikhail A. Malakhaltsev, Armando Machado, David McNabb, José J. Mencía, Michael Mezzino, Vicente Miquel Molina, Deanne Montgomery, Tamara Munzner, Emilio Musso, John Novak, Barrett O'Neill, Richard Palais, Mark Phillips, Lori Pickert, David Pierce, Mark Pinsky, Paola Piu, Andreas Igleias Prieto, Emma Previato, Valeri Pouchnia, Lilia del Riego, Patrick Ryan, Giacomo Saban, George Sadler, Isabel Salavessa, Simon Salamon, Jason P. Schultz, Walter Seaman, B.N. Shapukov, V.V. Shurygin, E.P. Shustova, Sonya Šimek, Cameron Smith, Dirk Struik, Rolf Sulanke, John Sullivan, Daniel Tanrè, C. Terng, A.A. Tuzhilin, Lieven Vanhecke, Gus Vlahacos, Tom Wickam-Jones and Stephen Wolfram for valuable suggestions.

CONTENTS

1. Curves in the Plane 1
    1.1 Euclidean Spaces 2
    1.2 Curves in $\mathbb{R}^n$ 5
    1.3 The Length of a Curve 7
    1.4 Vector Fields along Curves 12
    1.5 Curvature of Curves in the Plane 14
    1.6 Angle Functions Between Plane Curves 17
    1.7 The Turning Angle 19
    1.8 The Semicubical Parabola 21
    1.9 Exercises 22

2. Studying Curves in the Plane with Mathematica 25
    2.1 Computing Curvature of Curves in the Plane 29
    2.2 Computing Lengths of Curves 33
    2.3 Filling Curves 34
    2.4 Examples of Curves in $\mathbb{R}^2$ 35
    2.5 Plotting Piecewise Defined Curves 42
    2.6 Exercises 45

2.7　Animations of Definitions of Curves 47

3. **Famous Plane Curves** **49**
　　　3.1　Cycloids 50
　　　3.2　Lemniscates of Bernoulli 52
　　　3.3　Cardioids 54
　　　3.4　The Catenary 55
　　　3.5　The Cissoid of Diocles 57
　　　3.6　The Tractrix 61
　　　3.7　Clothoids 64
　　　3.8　Pursuit Curves 66
　　　3.9　Exercises 70
　　　3.10　Animation of a Prolate Cycloid 74

4. **Alternate Methods for Plotting Plane Curves** **75**
　　　4.1　Implicitly Defined Curves in $\mathbb{R}^2$ 75
　　　4.2　Cassinian Ovals 82
　　　4.3　Plane Curves in Polar Coordinates 86
　　　4.4　Exercises 92
　　　4.5　Animation of Fermat's Spiral 96

5. **New Curves from Old** **97**
　　　5.1　Evolutes 98
　　　5.2　Iterated Evolutes 101
　　　5.3　The Evolute of a Tractrix is a Catenary 102
　　　5.4　Involutes 103

	5.5	Tangent and Normal Lines to Plane Curves 108
	5.6	Osculating Circles to Plane Curves 111
	5.7	Parallel Curves 115
	5.8	Pedal Curves 117
	5.9	Exercises 120
	5.10	Animation of a Cycloidial Pendulum 126

6. Determining a Plane Curve from its Curvature 127

	6.1	Euclidean Motions 128
	6.2	Curves and Euclidean Motions 134
	6.3	Intrinsic Equations for Plane Curves 136
	6.4	Drawing Plane Curves with Assigned Curvature 140
	6.5	Exercises 146
	6.6	Animation of Epicycloids and Hypocycloids 152

7. Global Properties of Plane Curves 153

	7.1	The Total Signed Curvature of a Plane Curve 154
	7.2	The Rotation Index of a Closed Plane Curve 159
	7.3	Convex Plane Curves 163
	7.4	The Four Vertex Theorem 165
	7.5	Curves of Constant Width 168
	7.6	Envelopes of Curves 172
	7.7	The Support Function of an Oval 173
	7.8	Reuleaux Polygons 176
	7.9	The Involute Construction of Curves of Constant Width 177
	7.10	Exercises 178

8. Curves in Space 181

    8.1 The Vector Cross Product on $\mathbb{R}^3$ 182

    8.2 Curvature and Torsion of Unit-Speed Curves in $\mathbb{R}^3$ 183

    8.3 Curvature and Torsion of Arbitrary-Speed Curves in $\mathbb{R}^3$ 189

    8.4 Computing Curvature and Torsion with Mathematica 194

    8.5 The Helix and its Generalizations 198

    8.6 Viviani's Curve 201

    8.7 Exercises 202

    8.8 Animation of the Frenet Frame of Viviani's Curve 206

9. Tubes and Knots 207

    9.1 Tubes about Curves 207

    9.2 Torus Knots 209

    9.3 Exercises 215

10. Construction of Space Curves 217

    10.1 The Fundamental Theorem of Space Curves 217

    10.2 Drawing Space Curves with Assigned Curvature 222

    10.3 Contact 225

    10.4 Space Curves that Lie on a Sphere 231

    10.5 Curves of Constant Slope 234

    10.6 Loxodromes on Spheres 238

    10.7 Exercises 240

    10.8 Animation of Curves on a Sphere 243

11. Calculus on Euclidean Space 245

- 11.1 Tangent Vectors to $\mathbb{R}^n$ 246
- 11.2 Tangent Vectors as Directional Derivatives 248
- 11.3 Tangent Maps 250
- 11.4 Vector Fields on $\mathbb{R}^n$ 255
- 11.5 Derivatives of Vector Fields on $\mathbb{R}^n$ 258
- 11.6 Curves Revisited 265
- 11.7 Exercises 266

12. Surfaces in Euclidean Space 269

- 12.1 Patches in $\mathbb{R}^n$ 269
- 12.2 Patches in $\mathbb{R}^3$ 277
- 12.3 The Local Gauss Map 279
- 12.4 The Definition of a Regular Surface in $\mathbb{R}^n$ 281
- 12.5 Tangent Vectors to Regular Surfaces in $\mathbb{R}^n$ 286
- 12.6 Surface Mappings 288
- 12.7 Level Surfaces in $\mathbb{R}^3$ 291
- 12.8 Exercises 293

13. Examples of Surfaces 295

- 13.1 The Graph of a Function of Two Variables 296
- 13.2 The Ellipsoid 301
- 13.3 The Stereographic Ellipsoid 302
- 13.4 Tori 304
- 13.5 The Paraboloid 307
- 13.6 Sea Shells 308

13.7	Patches with Singularities 309
13.8	Implicit Plots of Surfaces 311
13.9	Exercises 312

14. Nonorientable Surfaces 317

14.1	Orientability of Surfaces 317
14.2	Nonorientable Surfaces Described by Identifications 322
14.3	The Möbius Strip 325
14.4	The Klein Bottle 327
14.5	A Different Klein Bottle 329
14.6	Realizations of the Real Projective Plane 330
14.7	Coloring Surfaces with Mathematica 335
14.8	Exercises 337
14.9	Animation of Steiner's Roman Surface 340

15. Metrics on Surfaces 341

15.1	The Intuitive Idea of Distance on a Surface 341
15.2	Isometries and Conformal Maps of Surfaces 346
15.3	The Intuitive Idea of Area on a Surface 351
15.4	Programs for Computing Metrics and Areas on a Surface 353
15.5	Examples of Metrics 354
15.6	Exercises 356

16. Surfaces in 3-Dimensional Space 359

| 16.1 | The Shape Operator ....360 |
| 16.2 | Normal Curvature ....363 |

16.3 Calculation of the Shape Operator 367
16.4 The Eigenvalues of the Shape Operator 371
16.5 The Gaussian and Mean Curvatures 373
16.6 The Three Fundamental Forms 380
16.7 Examples of Curvature Calculations by Hand 382
16.8 A Global Curvature Theorem 386
16.9 Exercises... 387

17. Surfaces in 3-Dimensional Space via Mathematica ... 391

17.1 Programs for Computing the Shape Operator and Curvature ... 392
17.2 Examples of Curvature Calculations with Mathematica 395
17.3 Principal Curvatures via Mathematica 402
17.4 The Gauss Map via Mathematica 403
17.5 The Curvature of Nonparametrically Defined Surfaces 409
17.6 Exercises 415

18. Asymptotic Curves on Surfaces 417

18.1 Asymptotic Curves 418
18.2 Examples of Asymptotic Curves 422
18.3 Using Mathematica to Find Asymptotic Curves 426
18.4 Exercises 429

19. Ruled Surfaces 431

19.1 Examples of Ruled Surfaces 432
19.2 Flat Ruled Surfaces 439
19.3 Tangent Developables 441

19.4	Noncylindrical Ruled Surfaces 445
19.5	Examples of Striction Curves of Noncylindrical Ruled Surfaces . 449
19.6	A Program for Ruled Surfaces 450
19.7	Other Examples of Ruled Surfaces 452
19.8	Exercises 454

20. Surfaces of Revolution 457

20.1	Principal Curves 459
20.2	The Curvature of a Surface of Revolution 461
20.3	Generating a Surface of Revolution with Mathematica 465
20.4	The Catenoid 467
20.5	The Hyperboloid of Revolution 470
20.6	Surfaces of Revolution of Curves with Specified Curvature 471
20.7	Surfaces of Revolution Generated by Data 472
20.8	Generalized Helicoids 475
20.9	Exercises 478

21. Surfaces of Constant Gaussian Curvature 481

21.1	The Elliptic Integral of the Second Kind 482
21.2	Surfaces of Revolution of Constant Positive Curvature 482
21.3	Surfaces of Revolution of Constant Negative Curvature 486
21.4	Flat Generalized Helicoids 490
21.5	Dini's Surface 493
21.6	Kuen's Surface 496
21.7	Exercises 497

22. Intrinsic Surface Geometry 501
- 22.1 Intrinsic Formulas for the Gaussian Curvature 502
- 22.2 Gauss's Theorema Egregium 507
- 22.3 Christoffel Symbols 509
- 22.4 Geodesic Curvature and Torsion 513
- 22.5 Exercises 519

23. Differentiable Manifolds 521
- 23.1 The Definition of Differentiable Manifold 522
- 23.2 Differentiable Functions on Differentiable Manifolds 526
- 23.3 Tangent Vectors on Differentiable Manifolds 532
- 23.4 Induced Maps 540
- 23.5 Vector Fields on Differentiable Manifolds 545
- 23.6 Tensor Fields on Differentiable Manifolds 550
- 23.7 Exercises 554

24. Riemannian Manifolds 557
- 24.1 Covariant Derivatives 558
- 24.2 Indefinite Riemannian Metrics 564
- 24.3 The Classical Treatment of Metrics 567
- 24.4 Exercises 572

25. Abstract Surfaces 573
- 25.1 Metrics and Christoffel Symbols for Abstract Surfaces 575
- 25.2 Examples of Metrics on Abstract Surfaces 578
- 25.3 Computing Curvature of Metrics on Abstract Surfaces 580

- 25.4 Orientability of an Abstract Surface 582
- 25.5 Geodesic Curvature for Abstract Surfaces 583
- 25.6 The Mean Curvature Vector Field 589
- 25.7 Exercises 591

26. Geodesics on Surfaces 595
- 26.1 The Geodesic Equations 596
- 26.2 Clairaut Patches 604
- 26.3 Finding Geodesics Numerically with Mathematica 613
- 26.4 The Exponential Map and the Gauss Lemma 617
- 26.5 Length Minimizing Properties of Geodesics 621
- 26.6 An Abstract Surface as a Metric Space 623
- 26.7 Exercises 625

27. The Gauss-Bonnet Theorem 627
- 27.1 The Local Gauss-Bonnet Theorem 628
- 27.2 Topology of Surfaces 634
- 27.3 The Global Gauss-Bonnet Theorem 636
- 27.4 Applications of the Gauss-Bonnet Theorem 638
- 27.5 Exercises 640

28. Principal Curves and Umbilic Points 641
- 28.1 The Differential Equation for the Principal Curves of a Surface .. 642
- 28.2 Umbilic Points 645
- 28.3 The Peterson-Mainardi-Codazzi Equations 649
- 28.4 Hilbert's Lemma and Liebmann's Theorem 652

| | 28.5 | The Fundamental Theorem of Surfaces .... *654* |
| | 28.6 | Exercises .... *659* |

29. Triply Orthogonal Systems of Surfaces *661*

	29.1	Examples of Triply Orthogonal Systems *662*
	29.2	Curvilinear Patches and Dupin's Theorem *665*
	29.3	Elliptic Coordinates *669*
	29.4	Parabolic Coordinates *676*
	29.5	Exercises *678*

30. Minimal Surfaces *681*

	30.1	Normal Variation *681*
	30.2	Examples of Minimal Surfaces *684*
	30.3	The Gauss Map of a Minimal Surface *694*
	30.4	Exercises *696*
	30.5	Animations of Minimal Surfaces *699*

31. Minimal Surfaces and Complex Variables *701*

	31.1	Isothermal Coordinates *702*
	31.2	Isometric Deformations of Minimal Surfaces *705*
	31.3	Complex Derivatives *710*
	31.4	Elementary Complex Vector Algebra *714*
	31.5	Minimal Curves *716*
	31.6	Finding Conjugate Minimal Surfaces *721*
	31.7	Enneper's Surface of Degree n *728*
	31.8	Exercises *732*

31.9 Animations of Minimal Surfaces 734

32. Minimal Surfaces via the Weierstrass Representation 735

32.1 The Weierstrass Representation 736

32.2 Weierstrass Patches via Mathematica 740

32.3 Examples of Weierstrass Patches 741

32.4 Minimal Surfaces with One Planar End 744

32.5 Costa's Minimal Surface 747

32.6 Exercises 760

33. Minimal Surfaces via Björling's Formula 761

33.1 Björling's Formula 761

33.2 Minimal Surfaces from Plane Curves 764

33.3 Examples of Minimal Surfaces Constructed from Plane Curves . 766

33.4 Exercises 772

34. Construction of Surfaces 773

34.1 Parallel Surfaces 773

34.2 Parallel Surfaces to a Möbius Strip 777

34.3 The Shape Operator of a Parallel Surface 779

34.4 A General Construction of a Triply Orthogonal System 782

34.5 Pedal Surfaces 784

34.6 Twisted Surfaces 786

34.7 Exercises 788

35. Canal Surfaces and Cyclides of Dupin 789
  35.1 Surfaces Whose Focal Sets are 2-Dimensional 791
  35.2 Canal Surfaces 798
  35.3 Cyclides of Dupin via Focal Sets 809
  35.4 Exercises 817

36. Inversions of Curves and Surfaces 819
  36.1 The Definition of Inversion 819
  36.2 Inversion of Curves 822
  36.3 Inversion of Surfaces 825
  36.4 Cyclides of Dupin via Inversions 830
  36.5 Liouville's Theorem 832
  36.6 Exercises 834

Appendices 837
  A General Programs 837
  B Curves 891
      Parametrically Defined Plane Curves 891
      Implicitly Defined Plane Curves 916
      Polar Defined Plane Curves 923
      Parametrically Defined Space Curves 927
  C Surfaces 935
      Parametrically Defined Surfaces 935
      Implicitly Defined Surfaces 969
      Drum Plots 974
      Minimal Curves 976
      Surface Metrics 984

D Plotting Programs 987
   Miscellaneous Plotting Programs 987
   Mathematica to Acrospin Programs 994
   Mathematica to Geomview Programs 999

Bibliography 1005
Index 1021
Name Index 1041
Miniprogram and Mathematica Command Index 1045

1

CURVES IN THE PLANE

Geometry before calculus involves only the simplest curves: straight lines, broken lines, arcs of circles, ellipses, hyperbolas and parabolas. These are the curves that form the basis for Greek geometry. Although other curves (such as the cissoid of Diocles and the spiral of Archimedes) were known in antiquity, the general theory of plane curves began to be developed only after the invention of Cartesian coordinates by Descartes[1] in the early 1600s.

Interesting plane curves arise in every day life. Examples include the orbit of a planet (an ellipse), the path of a projectile (a parabola), the form of a suspension bridge (a catenary) and the path of a point on the wheel of a car (a cycloid). In this chapter we get underway with the study of the local properties of curves that has developed over the last few centuries.

We begin in Section 1.1 by recalling some standard operations on Euclidean space $\mathbb{R}^n$, and in Section 1.2 we define the notion of curve in $\mathbb{R}^n$. Arc length for curves in $\mathbb{R}^n$ is defined and discussed in Section 1.3, and vector fields along curves are discussed in Section 1.4. In Section 1.5 we define the important notion of signed curvature of a curve in the plane $\mathbb{R}^2$ and show in Section 1.7 that it is the derivative of the turning angle. Section 1.6 is devoted to the problem of defining an angle function between plane curves. The semicubical parabola is used to illustrate some

---

[1]

René du Perron Descartes (1596-1650). French mathematician and philosopher. Descartes developed algebraic techniques to solve geometric problems, thus establishing the foundations of analytic geometry. Although most widely known as a philosopher, he also made important contributions to physiology, optics and mechanics.

of these concepts in Section 1.8.

Our basic approach in Chapters 1–10 is to study curves by means of their parametric representations (but implicitly defined curves in $\mathbb{R}^2$ are discussed in Section 4.1). This means we must show that any geometric invariant depending only on the point set traced out by the curve is independent of the parametrization, at least up to sign. The most important geometric quantities associated with a curve are of two types: (1) those that are independent of parametrization, and (2) those that do not change under a positive reparametrization, but change sign under a negative reparametrization. For example, we show in Section 1.3 that the length of a curve is independent of the parametrization chosen. The curvature of a plane curve is a more subtle invariant, because it can change sign if the curve is traversed in the opposite direction, but is otherwise independent of the parametrization. This fact we prove in Section 1.5.

1.1 Euclidean Spaces

Since we shall be studying curves and surfaces in a Euclidean space, we summarize some of the algebraic properties of Euclidean space in this section.

Definition. *Euclidean n-space $\mathbb{R}^n$ consists of the set of all real n-tuples:*

$$\mathbb{R}^n = \{\, (p_1, \ldots, p_n) \mid p_j \text{ is a real number for } j = 1, \ldots, n \,\}.$$

The elements of $\mathbb{R}^n$ are called **vectors**. *We write $\mathbb{R}$ for $\mathbb{R}^1$; it is simply the set of all real numbers. $\mathbb{R}^2$ is frequently called the* **plane**.

$\mathbb{R}^n$ is an n-dimensional vector space; this means that the operations of **addition** and **scalar multiplication** are defined. Thus if

$$\mathbf{p} = (p_1, \ldots, p_n) \quad \text{and} \quad \mathbf{q} = (q_1, \ldots, q_n),$$

then $\mathbf{p} + \mathbf{q}$ is the element of $\mathbb{R}^n$ given by

$$\mathbf{p} + \mathbf{q} = (p_1 + q_1, \ldots, p_n + q_n).$$

Similarly, for $\lambda \in \mathbb{R}$ the vector $\lambda \mathbf{p}$ is defined by

$$\lambda \mathbf{p} = (\lambda p_1, \ldots, \lambda p_n).$$

1.1 Euclidean Spaces

Furthermore, we shall denote by $\cdot$ the **dot product** (or **scalar product**) of $\mathbb{R}^n$. It is an operation that assigns to each pair of vectors $\mathbf{p} = (p_1, \ldots, p_n)$ and $\mathbf{q} = (q_1, \ldots, q_n)$ the real number

$$\mathbf{p} \cdot \mathbf{q} = \sum_{j=1}^{n} p_j q_j.$$

The **norm** and **distance** functions of $\mathbb{R}^n$ are defined by

$$\|\mathbf{p}\| = \sqrt{\mathbf{p} \cdot \mathbf{p}}, \quad \text{and} \quad \text{distance}(\mathbf{p}, \mathbf{q}) = \|\mathbf{p} - \mathbf{q}\|$$

for $\mathbf{p}, \mathbf{q} \in \mathbb{R}^n$. These functions have the following properties:

$$\mathbf{p} \cdot \mathbf{q} = \mathbf{q} \cdot \mathbf{p}, \qquad (\mathbf{p} + \mathbf{r}) \cdot \mathbf{q} = \mathbf{p} \cdot \mathbf{q} + \mathbf{r} \cdot \mathbf{q},$$

$$(\lambda \mathbf{p}) \cdot \mathbf{q} = \lambda(\mathbf{p} \cdot \mathbf{q}) = \mathbf{p} \cdot (\lambda \mathbf{q}), \qquad \|\lambda \mathbf{p}\| = |\lambda| \|\mathbf{p}\|,$$

for $\lambda \in \mathbb{R}$ and $\mathbf{p}, \mathbf{q}, \mathbf{r} \in \mathbb{R}^n$. Furthermore, the **Cauchy-Schwarz** and **triangle** inequalities state that for all $\mathbf{p}, \mathbf{q} \in \mathbb{R}^n$ we have

$$|\mathbf{p} \cdot \mathbf{q}| \leq \|\mathbf{p}\| \|\mathbf{q}\| \quad \text{and} \quad \|\mathbf{p} + \mathbf{q}\| \leq \|\mathbf{p}\| + \|\mathbf{q}\|.$$

(See, for example, [MS, pages 22–24].)

We shall also need the notion of the **angle** θ between nonzero vectors $\mathbf{p}, \mathbf{q} \in \mathbb{R}^n$. It is defined by the conditions:

$$\cos \theta = \frac{\mathbf{p} \cdot \mathbf{q}}{\|\mathbf{p}\| \|\mathbf{q}\|}, \qquad 0 \leq \theta \leq \pi.$$

Notice that the Cauchy-Schwarz inequality implies that

$$-1 \leq \frac{\mathbf{p} \cdot \mathbf{q}}{\|\mathbf{p}\| \|\mathbf{q}\|} \leq 1$$

for nonzero $\mathbf{p}, \mathbf{q} \in \mathbb{R}^n$, so that the definition of angle makes sense. Occasionally, it will be necessary to relax the requirement that $0 \leq \theta \leq \pi$.

A **linear map** of $\mathbb{R}^n$ into $\mathbb{R}^m$ is a function $A: \mathbb{R}^n \longrightarrow \mathbb{R}^m$ such that

$$A(\lambda \mathbf{p} + \mu \mathbf{q}) = \lambda A \mathbf{p} + \mu A \mathbf{q}$$

for $\lambda, \mu \in \mathbb{R}$ and $\mathbf{p}, \mathbf{q} \in \mathbb{R}^n$.

So far we have been dealing with $\mathbb{R}^n$ for general n. When n is 2 or 3, the vector space $\mathbb{R}^n$ has special structures that are useful for describing curves and surfaces.

In Chapters 1–7 we shall be studying curves in the plane, so we discuss $\mathbb{R}^2$ now. A discussion of the additional algebraic properties of $\mathbb{R}^3$ is postponed until Section 8.1.

For the differential geometry of curves in the plane an essential tool is the **complex structure** of $\mathbb{R}^2$; it is the linear map $J\colon \mathbb{R}^2 \longrightarrow \mathbb{R}^2$ given by

$$J(p_1, p_2) = (-p_2, p_1).$$

Geometrically, J is rotation by $\pi/2$ in a counterclockwise direction. It is easy to show that the complex structure J has the following properties:

$$J^2 = -I,$$
$$(J\mathbf{p}) \cdot (J\mathbf{q}) = \mathbf{p} \cdot \mathbf{q},$$
$$(J\mathbf{p}) \cdot \mathbf{p} = 0,$$

for $\mathbf{p}, \mathbf{q} \in \mathbb{R}^2$ (where $I\colon \mathbb{R}^2 \longrightarrow \mathbb{R}^2$ is the identity linear map). We shall use the complex structure J of $\mathbb{R}^2$ to define the signed curvature of a plane curve.

A point in the plane $\mathbb{R}^2$ can be considered a complex number via the canonical isomorphism

(1.1) $$\mathbf{p} = (p_1, p_2) \longleftrightarrow p_1 + i\, p_2 = \mathfrak{Re}(\mathbf{p}) + i\,\mathfrak{Im}(\mathbf{p}),$$

where $\mathfrak{Re}(\mathbf{p})$ and $\mathfrak{Im}(\mathbf{p})$ denote the real and imaginary parts of $\mathbf{p}$. We shall need descriptions of the dot product $\cdot$ and the complex structure J in terms of complex numbers. Recall that the **complex conjugate** and **absolute value** of a complex number $\mathbf{p}$ are defined by

$$\overline{\mathbf{p}} = \mathfrak{Re}(\mathbf{p}) - i\,\mathfrak{Im}(\mathbf{p}) \quad \text{and} \quad |\mathbf{p}| = \sqrt{\mathbf{p}\overline{\mathbf{p}}}.$$

The proof of the following lemma is elementary.

Lemma 1.1. *Identify the plane $\mathbb{R}^2$ with the set of complex numbers $\mathbb{C}$, and let $\mathbf{p}, \mathbf{q} \in \mathbb{R}^2 = \mathbb{C}$. Then*

(1.2) $$J\mathbf{p} = i\mathbf{p}, \qquad |\mathbf{p}| = \|\mathbf{p}\|, \quad \text{and} \quad \mathbf{p}\,\overline{\mathbf{q}} = \mathbf{p}\cdot\mathbf{q} + i(\mathbf{p}\cdot J\mathbf{q}).$$

The angle between vectors in $\mathbb{R}^n$ does not distinguish between the order of the vectors, but there is a refined notion of angle between vectors in $\mathbb{R}^2$ that makes this distinction.

Lemma 1.2. *Let $\mathbf{p}$ and $\mathbf{q}$ be nonzero vectors in $\mathbb{R}^2$. There exists a unique θ with the following properties:*

(1.3) $$\cos\theta = \frac{\mathbf{p}\cdot\mathbf{q}}{\|\mathbf{p}\|\,\|\mathbf{q}\|}, \qquad \sin\theta = \frac{\mathbf{p}\cdot J\mathbf{q}}{\|\mathbf{p}\|\,\|\mathbf{q}\|}, \qquad 0 \leq \theta < 2\pi.$$

*We call θ the **oriented angle** from $\mathbf{p}$ to $\mathbf{q}$.*

1.2 Curves in $\mathbb{R}^n$

Proof. Since $(\mathbf{p}\,\overline{\mathbf{q}})/(|\mathbf{p}|\,|\mathbf{q}|)$ is a complex number of absolute value 1, it lies on the unit circle in $\mathbb{C}$; thus there exists a unique θ with $0 \leq \theta < 2\pi$ such that

$$\frac{\mathbf{p}\,\overline{\mathbf{q}}}{|\mathbf{p}|\,|\mathbf{q}|} = e^{i\theta}. \tag{1.4}$$

Then using the expressions for $\mathbf{p}\,\overline{\mathbf{q}}$ in (1.2) and (1.4), we find that

$$\mathbf{p}\cdot\mathbf{q} + i(\mathbf{p}\cdot J\mathbf{q}) = e^{i\theta}|\mathbf{p}|\,|\mathbf{q}| = |\mathbf{p}|\,|\mathbf{q}|\cos\theta + i\,|\mathbf{p}|\,|\mathbf{q}|\sin\theta. \tag{1.5}$$

When we take the real and imaginary parts in (1.5), we get (1.3). ∎

1.2 Curves in $\mathbb{R}^n$

In the previous section we reviewed some of the algebraic properties of $\mathbb{R}^n$. We need to go one step further and study differentiation. In this chapter, and generally throughout the rest of the book, we shall use the word "differentiable" to mean "possessing derivatives or partial derivatives of all orders". We begin by studying $\mathbb{R}^n$-valued functions of one variable.

Definition. *Let $\boldsymbol{\alpha}\colon(a,b) \longrightarrow \mathbb{R}^n$ be a function, where (a,b) is an open interval in $\mathbb{R}$. We write*

$$\boldsymbol{\alpha}(t) = \big(a_1(t),\ldots,a_n(t)\big),$$

where each a_j is an ordinary real-valued function of a real variable. We say that $\boldsymbol{\alpha}$ is **differentiable** *provided a_j is differentiable for $j = 1,\ldots,n$. Similarly, $\boldsymbol{\alpha}$ is* **piecewise-differentiable** *provided a_j is piecewise-differentiable for $j = 1,\ldots,n$.*

Definition. *A (parametrized)* **curve** *in $\mathbb{R}^n$ is a piecewise-differentiable function*

$$\boldsymbol{\alpha}\colon(a,b) \longrightarrow \mathbb{R}^n,$$

where (a,b) is an open interval in $\mathbb{R}$. We allow the interval to be finite, infinite or half-infinite. If I is any other subset of $\mathbb{R}$, we say that

$$\boldsymbol{\alpha}\colon I \longrightarrow \mathbb{R}^n$$

is a **curve** *provided there is an open interval (a,b) containing I such that $\boldsymbol{\alpha}$ can be extended as a piecewise-differentiable function from (a,b) into $\mathbb{R}^n$.*

It is important to distinguish a curve α, which is a function, from the set of points traced out by α, which we call the **trace** of α. Note that the trace of α is $\alpha\big((a,b)\big)$ (or more generally $\alpha(I)$). We say that a subset S of $\mathbb{R}^n$ is **parametrized by** α provided there is a subset $I \subseteq \mathbb{R}$ such that $\alpha: I \longrightarrow \mathbb{R}^n$ is a curve for which $\alpha(I) = S$.

Definition. *Let $\alpha: (a,b) \longrightarrow \mathbb{R}^n$ be a curve with $\alpha(t) = \big(a_1(t), \ldots, a_n(t)\big)$. Then the* **velocity** *of α is the function $\alpha': (a,b) \longrightarrow \mathbb{R}^n$ given by*

$$\alpha'(t) = \big(a_1'(t), \ldots, a_n'(t)\big).$$

The function v defined by $v(t) = \|\alpha'(t)\|$ is called the **speed** *of α. The* **acceleration** *of α is α''.*

Notice that $\alpha'(t)$ is defined for those t for which $a_1'(t), \ldots, a_n'(t)$ are defined.

Some curves are better than others.

Definition. *A curve $\alpha: (a,b) \longrightarrow \mathbb{R}^n$ is said to be* **regular** *if it is differentiable and its velocity is everywhere defined and nonzero. If $\|\alpha'(t)\| = 1$ for $a < t < b$, then α is said to have* **unit-speed**.

The simplest example of a parametrized curve in $\mathbb{R}^n$ is a **straight line** containing distinct points $\mathbf{p}, \mathbf{q} \in \mathbb{R}^n$; it is most naturally parametrized by

(1.6) $$\tilde{\beta}(t) = (1-t)\mathbf{p} + t\mathbf{q}, \qquad -\infty < t < \infty.$$

The second simplest curve in $\mathbb{R}^2$ is a **circle** of radius a centered at $(p_1, p_2) \in \mathbb{R}^2$; it has the parametrization:

$$\tilde{\gamma}(t) = (p_1 + a\cos t, p_2 + a\sin t), \qquad 0 \leq t < 2\pi.$$

Straight lines and circles have more complicated unit-speed parametrizations:

$$\beta(s) = \frac{(\|\mathbf{p}-\mathbf{q}\| - s)\mathbf{p} + s\mathbf{q}}{\|\mathbf{p}-\mathbf{q}\|} \quad \text{and} \quad \gamma(s) = \left(p_1 + a\cos\left(\frac{s}{a}\right), p_2 + a\sin\left(\frac{s}{a}\right)\right).$$

Our definition of a curve in $\mathbb{R}^n$ is the **parametric form** of a curve. In contrast, the **nonparametric form** of a curve consists of a system of $n-1$ equations

$$F_1(p_1, \ldots, p_n) = \cdots = F_{n-1}(p_1, \ldots, p_n) = 0.$$

We study the nonparametric form of curves in $\mathbb{R}^2$ in Section 4.1. It is usually easier to work with the parametric form of a curve, but there is one disadvantage: distinct parametrizations may trace out the same point set. Therefore, it is important to know when two curves are the same, except for a change of variables.

Definition. Let $\alpha\colon (a,b) \longrightarrow \mathbb{R}^n$ and $\beta\colon (c,d) \longrightarrow \mathbb{R}^n$ be differentiable curves. Then β is said to be a **positive reparametrization** of α provided there exists a differentiable function $h\colon (c,d) \longrightarrow (a,b)$ such that $h'(t) > 0$ for all $c < t < d$ and $\beta = \alpha \circ h$.

Similarly, β is called a **negative reparametrization** of α provided there exists a differentiable function $h\colon (c,d) \longrightarrow (a,b)$ such that $h'(t) < 0$ for all $c < t < d$ and $\beta = \alpha \circ h$.

We say that β is a **reparametrization** of α if it is either a positive or negative reparametrization of α.

Next, we determine the relation between the velocity of a curve α and the velocity of a reparametrization of α.

Lemma 1.3. (The chain rule for curves.) *Suppose that β is a reparametrization of α. Write $\beta = \alpha \circ h$, where $h\colon (c,d) \longrightarrow (a,b)$ is differentiable. Then*

(1.7) $$\beta'(u) = h'(u)\alpha'\bigl(h(u)\bigr)$$

for $c < u < d$.

Proof. Write $\alpha(t) = \bigl(a_1(t),\ldots,a_n(t)\bigr)$ and $\beta(u) = \bigl(b_1(u),\ldots,b_n(u)\bigr)$. Then we have $b_j(u) = a_j\bigl(h(u)\bigr)$ for $j = 1,\ldots,n$. The ordinary chain rule implies that $b_j'(u) = a_j'\bigl(h(u)\bigr)h'(u)$ for $j = 1,\ldots,n$, and so we get (1.7). ∎

1.3 The Length of a Curve

One of the simplest and most important geometric quantities associated with a curve is its length. Everyone has a natural idea of what is meant by length, but this idea must be converted into an exact definition. For our purposes the simplest definition is as follows:

Definition. Let $\alpha\colon (a,b) \longrightarrow \mathbb{R}^n$ be a curve. Assume that α is defined on a slightly larger interval containing (a,b), so that α is defined and differentiable at a and b. Then the **length** of α over the interval $[a,b]$ is given by

$$\mathbf{Length}[a,b][\alpha] = \mathbf{Length}[\alpha] = \int_a^b \|\alpha'(t)\|\,dt.$$

It is clear intuitively that length should not depend on the parametrization of the curve. We now prove this.

Theorem 1.4. *Let β be a reparametrization of α. Then*
$$\text{Length}[\alpha] = \text{Length}[\beta].$$

Proof. We do the positive reparametrization case first. Let $\beta = \alpha \circ h$, where $h: (c,d) \longrightarrow (a,b)$ and $h'(t) > 0$ for $c < t < d$. Then by Lemma 1.3 we have
$$\|\beta'(t)\| = \|\alpha'(h(t))h'(t)\| = \|\alpha'(h(t))\| \, |h'(t)| = \|\alpha'(h(t))\| h'(t).$$

Hence using the change of variables formula for integrals, we compute
$$\begin{aligned}\text{Length}[\alpha] &= \int_a^b \|\alpha'(t)\| dt = \int_c^d \|\alpha'(h(u))\| h'(u) du \\ &= \int_c^d \|\alpha'(h(u))h'(u)\| du = \int_c^d \|\beta'(u)\| du = \text{Length}[\beta].\end{aligned}$$

In the negative reparametrization case we have
$$\lim_{t \to c} h(t) = b \quad \text{and} \quad \lim_{t \to d} h(t) = a.$$

Also,
$$\|\beta'(t)\| = \|\alpha'(h(t))h'(t)\| = \|\alpha'(h(t))\| \, |h'(t)| = -\|\alpha'(h(t))\| h'(t).$$

Again, using the change of variables formula for integrals, we compute
$$\begin{aligned}\text{Length}[\alpha] &= \int_a^b \|\alpha'(t)\| dt = \int_d^c \|\alpha'(h(u))\| h'(u) du \\ &= -\int_c^d \|\alpha'(h(u))\| h'(u) du = \int_c^d \|\beta'(u)\| du = \text{Length}[\beta]. \blacksquare\end{aligned}$$

The trace of a curve α can be approximated by a series of line segments connecting a sequence of points on the trace of α. Intuitively, the length of this approximation to α should tend to the length of α as the individual line segments become smaller and smaller. This is in fact true. To explain exactly what happens, let $\alpha: (c,d) \longrightarrow \mathbb{R}^n$ be a curve and $[a,b] \subset (c,d)$ a closed finite interval. For every partition
$$P = \{a = t_0 < t_1 < \cdots < t_N = b\}$$
of $[a,b]$, let
$$|P| = \max_{1 \leq i \leq N}(t_i - t_{i-1}) \quad \text{and} \quad \ell(\alpha, P) = \sum_{i=1}^N \|\alpha(t_i) - \alpha(t_{i-1})\|.$$

1.3 The Length of a Curve

Geometrically, $\ell(\alpha, P)$ is the length of the polygon in $\mathbb{R}^n$ whose vertices are $\alpha(t_i)$, $i = 1, \ldots, n$. Then **Length**$[\alpha]$ is the limit of inscribed polygons in the sense of the following theorem.

Theorem 1.5. *Let $\alpha\colon (c,d) \longrightarrow \mathbb{R}^n$ be a curve, and let **Length**$[\alpha]$ denote the length of the restriction of α to a closed subinterval $[a,b]$. Given $\epsilon > 0$, there exists $\delta > 0$ such that*

(1.8) $$|P| < \delta \quad \text{implies} \quad \big|\mathbf{Length}[\alpha] - \ell(\alpha, P)\big| < \epsilon.$$

Proof. Let $P = \{a = t_0 < t_1 < \cdots < t_N = b\}$ be a partition of $[a,b]$ and write $\alpha = (a_1, \ldots, a_n)$. For each i the mean value theorem from calculus implies that for $1 \leq j \leq N$ there exists $t_j^{(i)}$ with $t_{j-1} < t_j^{(i)} < t_j$ such that

$$a_i(t_j) - a_i(t_{j-1}) = a_i'(t_j^{(i)})(t_j - t_{j-1}).$$

Then

$$\|\alpha(t_j) - \alpha(t_{j-1})\|^2 = \sum_{i=1}^N \big(a_i(t_j) - a_i(t_{j-1})\big)^2 = \sum_{i=1}^N a_i'(t_j^{(i)})^2 (t_j - t_{j-1})^2$$

$$= (t_j - t_{j-1})^2 \sum_{i=1}^N a_i'(t_j^{(i)})^2.$$

Hence

$$\|\alpha(t_j) - \alpha(t_{j-1})\| = (t_j - t_{j-1})\sqrt{\sum_{i=1}^N a_i'(t_j^{(i)})^2} = (t_j - t_{j-1})(A_j + B_j),$$

where

$$A_j = \sqrt{\sum_{i=1}^N a_i'(t_j)^2} = \|\alpha'(t_j)\| \quad \text{and} \quad B_j = \sqrt{\sum_{i=1}^N a_i'(t_j^{(i)})^2} - \sqrt{\sum_{i=1}^N a_i'(t_j)^2}.$$

The triangle inequality implies that

$$|B_j| = \Big|\|(a_1'(t_j^{(1)}), \ldots, a_n'(t_j^{(n)}))\| - \|\alpha'(t_j)\|\Big|$$

$$\leq \|(a_1'(t_j^{(1)}), \ldots, a_1'(t_j^{(n)})) - \alpha'(t_j)\| = \left(\sum_{i=1}^N (a_i'(t_j^{(i)}) - a_i'(t_j))^2\right)^{1/2}$$

$$\leq \sum_{i=1}^N |a_i'(t_j^{(i)}) - a_i'(t_j)|.$$

Let $\epsilon > 0$. Since $t \longmapsto \boldsymbol{\alpha}'(t)$ is continuous on the closed interval $[a, b]$, there exists $\delta_1 > 0$ such that

$$|u - v| < \delta_1 \quad \text{implies} \quad |a_i'(u) - a_i'(v)| < \frac{\epsilon}{2N(b-a)}$$

for $i = 1, \ldots, n$. Then $|P| < \delta_1$ implies that

$$|B_j| < \frac{\epsilon}{2(b-a)},$$

for each j, so that

$$\left| \sum_{j=1}^{N} (t_j - t_{j-1}) B_j \right| \leq \sum_{j=1}^{N} (t_j - t_{j-1}) \frac{\epsilon}{2(b-a)} = \frac{\epsilon}{2}.$$

Hence

$$|\mathbf{Length}[\boldsymbol{\alpha}] - \ell(\boldsymbol{\alpha}, P)| \leq \left| \mathbf{Length}[\boldsymbol{\alpha}] - \sum_{j=1}^{N} \|\boldsymbol{\alpha}(t_j) - \boldsymbol{\alpha}(t_{j-1})\| \right|$$

$$= \left| \mathbf{Length}[\boldsymbol{\alpha}] - \sum_{j=1}^{N} (A_j + B_j)(t_j - t_{j-1}) \right|$$

$$\leq \left| \mathbf{Length}[\boldsymbol{\alpha}] - \sum_{j=1}^{N} A_j (t_j - t_{j-1}) \right| + \left| \sum_{j=1}^{N} B_j (t_j - t_{j-1}) \right|$$

$$\leq \left| \int_a^b \|\boldsymbol{\alpha}'(t)\| \, dt - \sum_{j=1}^{N} \|\boldsymbol{\alpha}'(t_j)(t_j - t_{j-1})\| \right| + \frac{\epsilon}{2}.$$

But

$$\sum_{j=1}^{N} \|\boldsymbol{\alpha}'(t_j)(t_j - t_{j-1})\|$$

is a Riemann sum approximation to

$$\int_a^b \|\boldsymbol{\alpha}'(t)\| \, dt.$$

1.3 The Length of a Curve

Hence there exists $\delta_2 > 0$ such that $|P| < \delta_2$ implies that

$$\left| \sum_{j=1}^{N} \|\alpha'(t_j)(t_j - t_{j-1})\| - \int_a^b \|\alpha'(t)\| dt \right| < \frac{\epsilon}{2}.$$

Thus if we take $\delta = \min(\delta_1, \delta_2)$, we get (1.8). ∎

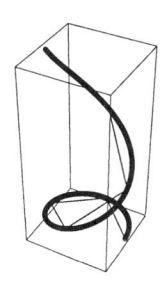

A polygonal approximation to the curve
$t \longmapsto (\cos t, \sin t, t^{2.7})$

We need to know how the length of a curve varies when we change the limits of integration.

Definition. *Fix a number c with $a < c < b$. The **arc length function** s_α starting at c of a curve $\alpha \colon (a, b) \longrightarrow \mathbb{R}^n$ is defined by*

$$s_\alpha(t) = \int_c^t \|\alpha'(u)\| du,$$

for $c \leq t \leq b$. When there is no danger of confusion, we simplify s_α to s.

Note that

$$\mathbf{Length}[d, f][\alpha] = s_\alpha(f) - s_\alpha(d).$$

Theorem 1.6. *Let $\alpha \colon (a, b) \longrightarrow \mathbb{R}^n$ be a regular curve. Then there exists a unit-speed reparametrization β of α.*

Proof. By the fundamental theorem of calculus, any arc length function s of α satisfies

$$\frac{ds}{dt}(t) = s'(t) = \|\alpha'(t)\|.$$

Since α is regular, by definition $\alpha'(t)$ is never zero; hence ds/dt is always positive. The inverse function theorem from calculus implies that $t \longmapsto s(t)$ has an inverse $s \longmapsto t(s)$, and that

$$\left. \frac{dt}{ds} \right|_{s(t)} = \frac{1}{\left. \frac{ds}{dt} \right|_{t(s)}}.$$

Now define β by $\beta(s) = \alpha(t(s))$. By Lemma 1.3 we have $\beta'(s) = (dt/ds)\alpha'(t(s))$. Hence

$$\|\beta'(s)\| = \left\|\frac{dt}{ds}\alpha'(t(s))\right\| = \frac{dt}{ds}\|\alpha'(t(s))\| = \frac{dt}{ds}(s)\frac{ds}{dt}(t(s)) = 1. \blacksquare$$

The arc length function of any unit-speed curve $\beta\colon (c,d) \longrightarrow \mathbb{R}^n$ starting at c satisfies

$$s = s(t) = \int_c^t \|\beta'(u)\|\,du = t - c.$$

Thus the function s actually measures length along β. This is the reason why unit-speed curves are said to be **parametrized by arc length**.

We conclude this section by showing that unit-speed curves are unique up to starting point and direction.

Lemma 1.7. *Let $\alpha\colon (a,b) \longrightarrow \mathbb{R}^n$ be a unit-speed curve, and let $\beta\colon (c,d) \longrightarrow \mathbb{R}^n$ be a reparametrization of α such that β also has unit-speed. Then*

$$\beta(t) = \alpha(\pm t + t_0)$$

for some constant real number t_0.

Proof. By hypothesis, there exists a differentiable function $h\colon (c,d) \longrightarrow (a,b)$ such that $\beta = \alpha \circ h$ and $h'(t) \neq 0$ for $c < t < d$. Then Lemma 1.3 implies that

$$1 = \|\beta'(t)\| = \|\alpha'(h(t))\|\,|h'(t)| = |h'(t)|.$$

Therefore, $h'(t) = \pm 1$ for all t. Since h' is continuous on a connected open set, its sign is constant. Thus there is a constant t_0 such that $h(t) = \pm t + t_0$ for $c < t < d$. $\blacksquare$

1.4 Vector Fields along Curves

We have already seen examples of vector fields along a curve: α', α'', as well as the higher derivatives of α. We now give the general definition.

Definition. Let $\alpha\colon (a,b) \longrightarrow \mathbb{R}^n$ *be a curve. A* **vector field** *along α is a function* $\mathbf{Y}$ *that assigns to each t with $a < t < b$ a vector $\mathbf{Y}(t)$ at the point $\alpha(t)$.*

1.4 Vector Fields along Curves

At this beginning stage we shall not distinguish between a vector at $\alpha(t)$ and the vector parallel to it at the origin. This means a vector field **Y** along a curve α is really an n-tuple of functions:

$$\mathbf{Y}(t) = \bigl(y_1(t), \ldots, y_n(t)\bigr).$$

Thus differentiability of **Y** means that each of the functions $y_1, \ldots, y_n$ is differentiable; piecewise-differentiability is defined similarly. If **Y** is a differentiable vector field along a curve, then the derivative **Y**$'$ is defined by

$$\mathbf{Y}'(t) = \bigl(y_1'(t), \ldots, y_n'(t)\bigr).$$

(Occasionally, we do need to distinguish between a vector at $\alpha(t)$ and the vector parallel to it at the origin. See Section 11.6.)

Addition, scalar multiplication and dot product of vector fields along a curve $\alpha: (a, b) \longrightarrow \mathbb{R}^n$ are defined in the obvious way. Furthermore, we can multiply a vector field **Y** along α by a function $f: (a,b) \longrightarrow \mathbb{R}$: we define the vector field $f\mathbf{Y}$ by $(f\mathbf{Y})(t) = f(t)\mathbf{Y}(t)$. Finally, if $n = 2$ and **X** is a vector field along a curve with $\mathbf{X}(t) = \bigl(x(t), y(t)\bigr)$, then another vector field $J\mathbf{X}$ can be defined by $J\mathbf{X}(t) = \bigl(-y(t), x(t)\bigr)$. Taking the derivative is related to these operations in the obvious way:

Lemma 1.8. *Let* **X** *and* **Y** *be differentiable vector fields along a curve* $\alpha: (a,b) \longrightarrow \mathbb{R}^n$, *and let* $f: (a,b) \longrightarrow \mathbb{R}$ *be differentiable. Then*

(i) $(f\mathbf{Y})' = f'\mathbf{Y} + f\mathbf{Y}'$;

(ii) $(\mathbf{X} + \mathbf{Y})' = \mathbf{X}' + \mathbf{Y}'$;

(iii) $(\mathbf{X} \cdot \mathbf{Y})' = \mathbf{X}' \cdot \mathbf{Y} + \mathbf{X} \cdot \mathbf{Y}'$;

(iv) $(J\mathbf{Y})' = J\mathbf{Y}'$ *(for $n = 2$).*

Proof. For example, we prove (i):

$$\begin{aligned}(f\mathbf{Y})' &= \bigl((fy_1)', \ldots, (fy_n)'\bigr) = (f'y_1 + fy_1', \ldots, f'y_n + fy_n') \\ &= (y_1, \ldots, y_n)f' + (y_1', \ldots, y_n')f = f'\mathbf{Y} + f\mathbf{Y}'. \ \blacksquare\end{aligned}$$

1.5 Curvature of Curves in the Plane

Intuitively, the curvature of a curve measures the failure of a curve to be a straight line. The faster the velocity α' turns along a curve α the larger the curvature.[2]

In Chapter 8 we shall define the notion of curvature $\kappa[\alpha]$ of a curve α in $\mathbb{R}^n$ for arbitrary n. In the special case that $n = 2$ there is a refined version $\kappa 2$ of κ, which we now define. We first give the definition that is the easiest to use in computations. In Section 1.7 we show that the curvature can be interpreted as the derivative of a turning angle.

Definition. Let $\alpha\colon (a,b) \longrightarrow \mathbb{R}^2$ be a regular curve. The **curvature** $\kappa 2[\alpha]$ of α is given by the formula

$$(1.9) \qquad \kappa 2[\alpha](t) = \frac{\alpha''(t) \cdot J\alpha'(t)}{\|\alpha'(t)\|^3}.$$

The positive function $1/|\kappa 2[\alpha]|$ is called the **radius of curvature** of α.

Notice that the curvature can assume both positive and negative values. Sometimes $\kappa 2$ is called the **signed curvature** in order to distinguish it from the curvature κ that we shall define in Chapter 8. Of course, we can speak of the curvature of a plane curve α that is only piecewise-differentiable. When $\alpha'(t) = 0$ or $\alpha'(t)$ does not exist, $\kappa 2[\alpha](t)$ remains undefined. Similarly, the radius of curvature $|1/\kappa 2[\alpha]|$ is undefined at those t for which either of $\alpha'(t)$ or $\alpha''(t)$ vanishes or is undefined.

Whenever the acceleration of a curve is zero, the curvature also vanishes. In particular, the curvature of the straight line parametrized by (1.6) vanishes identically.

Next, we derive the formula for curvature that can be found in most calculus books, and also a formula for curvature using complex numbers.

Lemma 1.9. *If $\alpha\colon (a,b) \longrightarrow \mathbb{R}^2$ is a regular curve with $\alpha(t) = \bigl(x(t), y(t)\bigr)$, then the curvature $\kappa 2[\alpha]$ of α is given by*

$$(1.10) \qquad \kappa 2[\alpha](t) = \frac{x'(t)y''(t) - x''(t)y'(t)}{\bigl(x'^2(t) + y'^2(t)\bigr)^{3/2}},$$

$$(1.11) \qquad\qquad\quad = \frac{\mathfrak{Re}\bigl(\alpha''(t)\,\overline{i\alpha'(t)}\bigr)}{\|\alpha'(t)\|^3}.$$

---

[2] The notion of curvature of a plane curve first appears implicitly in the work of Apollonius of Perga (262-180 B.C.). Newton was the first to study the curvature of plane curves explicitly. In particular, he found formula (1.10).

1.5 Curvature of Curves in the Plane 15

Proof. We have $\alpha''(t) = \bigl(x''(t), y''(t)\bigr)$ and $J\alpha'(t) = \bigl(-y'(t), x'(t)\bigr)$, so that

$$\kappa 2[\alpha](t) = \frac{\bigl(x''(t), y''(t)\bigr) \cdot \bigl(-y'(t), x'(t)\bigr)}{\bigl(x'^2(t) + y'^2(t)\bigr)^{3/2}} = \frac{x'(t)y''(t) - x''(t)y'(t)}{\bigl(x'^2(t) + y'^2(t)\bigr)^{3/2}},$$

proving 1.10.

Equation (1.11) is immediate from the definition of $\kappa 2$ and Lemma 1.1. ∎

It is useful to have some pictures to understand the meaning of positive and negative curvature for plane curves. There are four cases; we illustrate each by a parabola.

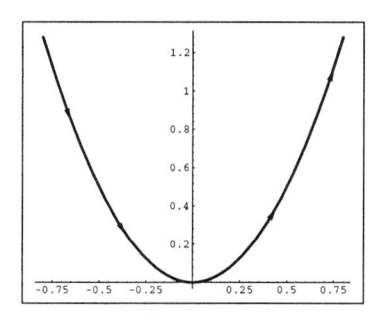

The parabola $t \longmapsto (t, 2t^2)$
$\kappa 2 > 0$

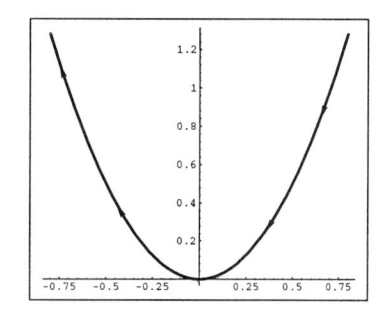

The parabola $t \longmapsto (-t, 2t^2)$
$\kappa 2 < 0$

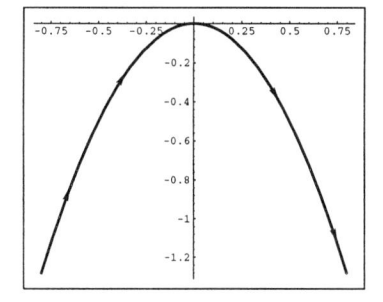

The parabola $t \longmapsto (t, -2t^2)$
$\kappa 2 < 0$

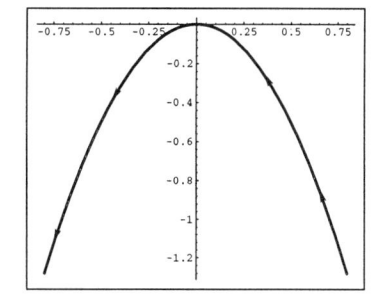

The parabola $t \longmapsto (-t, -2t^2)$
$\kappa 2 > 0$

Like the function **Length**$[\alpha]$ the curvature $\kappa 2[\alpha]$ is a geometric quantity associated with a curve α. As such, it should be independent of the parametrization of the curve α. We now show that this is almost true. Recall that if a is a nonzero real number, then $\mathrm{sign}\, a = a/|a|$.

Theorem 1.10. *Let $\alpha\colon(a,b) \longrightarrow \mathbb{R}^2$ be a regular curve, and let $\beta\colon(c,d) \longrightarrow \mathbb{R}^2$ be a reparametrization of α. Write $\beta = \alpha \circ h$, where $h\colon(c,d) \longrightarrow (a,b)$ is differentiable. Then*

$$\kappa 2[\beta](t) = \bigl(\operatorname{sign} h'(t)\bigr)\kappa 2[\alpha]\bigl(h(t)\bigr), \tag{1.12}$$

wherever $\operatorname{sign} h'(t)$ is defined. Thus the curvature of a plane curve is up to sign independent of the parametrization.

Proof. We have $\beta' = (\alpha' \circ h)h'$, so that $J\beta' = \bigl(J(\alpha' \circ h)\bigr)h'$, and

$$\beta'' = (\alpha'' \circ h)h'^2 + (\alpha' \circ h)h''. \tag{1.13}$$

Thus

$$\kappa 2[\beta] = \frac{\bigl((\alpha'' \circ h)h'^2 + (\alpha' \circ h)h''\bigr) \cdot \bigl(J(\alpha' \circ h)\bigr)h'}{\|(\alpha' \circ h)h'\|^3}$$

$$= \left(\frac{h'^3}{|h'|^3}\right)\frac{(\alpha'' \circ h) \cdot J(\alpha' \circ h)}{\|(\alpha' \circ h)\|^3}.$$

Hence we get (1.12). ∎

The formula for $\kappa 2$ simplifies for a unit-speed curve.

Lemma 1.11. *Let β be a unit-speed curve in the plane. Then*

$$\beta'' = \kappa 2[\beta] J\beta'. \tag{1.14}$$

Proof. Differentiating $\beta' \cdot \beta' = 1$, we obtain $\beta'' \cdot \beta' = 0$. Thus β'' must be a multiple of $J\beta'$. In fact, it follows easily from (1.9) that this multiple must be $\kappa 2[\beta]$, so we obtain (1.14). ∎

Finally, we give simple characterizations of straight lines and circles by means of curvature.

Theorem 1.12. *Let $\alpha\colon(a,b) \longrightarrow \mathbb{R}^2$ be a regular curve.*

(i) *α is part of a straight line if and only if $\kappa 2[\alpha](t) \equiv 0$.*

(ii) *α is part of a circle of radius $a > 0$ if and only if $|\kappa 2[\alpha]|(t) \equiv 1/a$.*

1.6 Angle Functions Between Plane Curves

Proof. It is easy to show that the curvature of the straight line is zero, and that the curvature of a circle of radius a is $+1/a$ for a counterclockwise circle and $-1/a$ for a clockwise circle (see exercise 3).

For the converse, we can assume without loss of generality that α is a unit-speed curve. Suppose $\kappa 2[\alpha](t) \equiv 0$. Lemma 1.11 implies that $\alpha''(t) = 0$ for $a < t < b$. Hence there exist constant vectors $\mathbf{p}$ and $\mathbf{q}$ such that

$$\alpha(t) = \mathbf{p}\,t + \mathbf{q}$$

for $a < t < b$; thus α is part of a straight line.

Next, assume that $\kappa 2[\alpha](t) \equiv 1/a$. Without loss of generality, α has unit-speed. We define a curve $\gamma \colon (a,b) \longrightarrow \mathbb{R}^2$ by

$$\gamma(t) = \alpha(t) + a\, J\alpha'(t).$$

Lemma 1.11 implies that $\gamma'(t) = 0$ for $a < t < b$. Hence there exists $\mathbf{q}$ such that $\gamma(t) = \mathbf{q}$ for $a < t < b$. Then $\|\alpha(t) - \mathbf{q}\| = \|a\,J\alpha'(t)\| = a$. Hence $\alpha(t)$ lies on a circle of radius a centered at $\mathbf{q}$.

Similarly, $\kappa 2[\alpha](t) \equiv -1/a$ also implies that α is part of a circle of radius a (see exercise 4). ∎

1.6 Angle Functions Between Plane Curves

In Section 1.1 we defined the notion of oriented angle between vectors in $\mathbb{R}^2$; now we wish to define a similar notion for curves. Clearly, we can compute the oriented angle between corresponding velocity vectors. This oriented angle θ satisfies $0 \leq \theta < 2\pi$, but such a restriction is not desirable for curves. The problem is that the two curves can twist and turn so that eventually the angle between them lies outside the interval $0 \leq \theta < 2\pi$. To arrive at the proper definition, we need the following useful lemma of O'Neill [ON1].

Lemma 1.13. *Let $f, g \colon (a, b) \longrightarrow \mathbb{R}$ be differentiable functions with $f^2 + g^2 = 1$. Fix t_0 with $a < t_0 < b$ and suppose θ_0 is such that $f(t_0) = \cos\theta_0$ and $g(t_0) = \sin\theta_0$. Then there exists a unique function $\vartheta \colon (a, b) \longrightarrow \mathbb{R}$ such that*

(1.15) $\qquad \vartheta(t_0) = \theta_0, \qquad f(t) = \cos\vartheta(t) \qquad \text{and} \qquad g(t) = \sin\vartheta(t)$

for $a < t < b$.

Proof. Let $h = f + ig$; then $h\,\overline{h} = 1$. Define

(1.16) $$\vartheta(t) = \theta_0 - i \int_{t_0}^{t} h'(s)\overline{h(s)}\,ds.$$

Then
$$\frac{d}{dt}(h\,e^{-i\vartheta}) = e^{-i\vartheta}(h' - i\,h\,\vartheta') = e^{-i\vartheta}(h' - h\,h'\,\overline{h}) = 0.$$

Thus $h\,e^{-i\vartheta} = c$ for some constant c. Since $h(t_0) = e^{i\theta_0}$, it follows that
$$c = h(t_0)e^{-i\vartheta(t_0)} = 1.$$

Hence $h(t) = e^{i\vartheta(t)}$, and so we get (1.15).

Let $\hat{\vartheta}$ be another continuous function such that
$$\hat{\vartheta}(t_0) = \theta_0, \quad f(t) = \cos\hat{\vartheta}(t) \quad \text{and} \quad g(t) = \sin\hat{\vartheta}(t)$$

for $a < t < b$. Then $e^{i\vartheta(t)} = e^{i\hat{\vartheta}(t)}$ for $a < t < b$. Since both ϑ and $\hat{\vartheta}$ are continuous, there is an integer n such that
$$\vartheta(t) - \hat{\vartheta}(t) = 2\pi n$$

for $a < t < b$. But $\vartheta(t_0) = \hat{\vartheta}(t_0)$, so that $n = 0$. Hence ϑ and $\hat{\vartheta}$ coincide. ∎

We now apply this lemma to deduce the existence and uniqueness of a differentiable angle function between curves in $\mathbb{R}^2$.

Corollary 1.14. *Let α and β be regular curves in $\mathbb{R}^2$ defined on the same interval (a, b), and let $a < t_0 < b$. Choose θ_0 such that*
$$\frac{\alpha'(t_0) \cdot \beta'(t_0)}{\|\alpha'(t_0)\|\,\|\beta'(t_0)\|} = \cos\theta_0 \quad \text{and} \quad \frac{\alpha'(t_0) \cdot J\beta'(t_0)}{\|\alpha'(t_0)\|\,\|\beta'(t_0)\|} = \sin\theta_0.$$

Then there exists a unique differentiable function $\vartheta\colon (a, b) \longrightarrow \mathbb{R}$ such that
$$\vartheta(t_0) = \theta_0, \quad \frac{\alpha'(t) \cdot \beta'(t)}{\|\alpha'(t)\|\,\|\beta'(t)\|} = \cos\vartheta(t) \quad \text{and} \quad \frac{\alpha'(t) \cdot J\beta'(t)}{\|\alpha'(t)\|\,\|\beta'(t)\|} = \sin\vartheta(t)$$

*for $a < t < b$. We call ϑ the **angle function** between α and β determined by θ_0.*

Proof. We take
$$f(t) = \frac{\alpha'(t) \cdot \beta'(t)}{\|\alpha'(t)\|\,\|\beta'(t)\|} \quad \text{and} \quad g(t) = \frac{\alpha'(t) \cdot J\beta'(t)}{\|\alpha'(t)\|\,\|\beta'(t)\|}$$

in Lemma 1.13. ∎

1.7 The Turning Angle

Intuitively, it is clear that the velocity vector of a highly curved plane curve changes rapidly. This idea can be made precise mathematically.

Lemma 1.15. *Let $\alpha\colon (a,b) \longrightarrow \mathbb{R}^2$ be a regular curve and fix t_0 with $a < t_0 < b$. Let θ_0 be a number such that*

$$\frac{\alpha'(t_0)}{\|\alpha'(t_0)\|} = (\cos\theta_0, \sin\theta_0).$$

Then there exists a unique differentiable function $\theta[\alpha]\colon (a,b) \longrightarrow \mathbb{R}$ such that $\theta[\alpha](t_0) = \theta_0$ and

$$(1.17) \qquad \frac{\alpha'(t)}{\|\alpha'(t)\|} = \bigl(\cos\theta[\alpha](t), \sin\theta[\alpha](t)\bigr) = e^{i\theta[\alpha](t)}.$$

*for $a < t < b$. We call $\theta[\alpha]$ the **turning angle** determined by θ_0.*

Proof. Let $\beta(t) = (t, 0)$; then β parametrizes a horizontal straight line and $\beta'(t) = (1, 0)$ for all t. Write $\alpha(t) = \bigl(a_1(t), a_2(t)\bigr)$; then

$$\alpha'(t) \cdot \beta'(t) = a_1'(t) \qquad \text{and} \qquad \alpha'(t) \cdot J\beta'(t) = a_2'(t).$$

Corollary 1.14 implies the existence of a unique function $\theta[\alpha]\colon (a,b) \longrightarrow \mathbb{R}$ such that $\theta[\alpha](t_0) = \theta_0$ and

$$(1.18) \qquad \begin{cases} \cos\theta[\alpha](t) = \dfrac{\alpha'(t) \cdot \beta'(t)}{\|\alpha'(t)\| \, \|\beta'(t)\|} = \dfrac{a_1'(t)}{\|\alpha'(t)\|}, \\[2mm] \sin\theta[\alpha](t) = \dfrac{\alpha'(t) \cdot J\beta'(t)}{\|\alpha'(t)\| \, \|\beta'(t)\|} = \dfrac{a_2'(t)}{\|\alpha'(t)\|}. \end{cases}$$

Then (1.18) is equivalent to (1.17). ∎

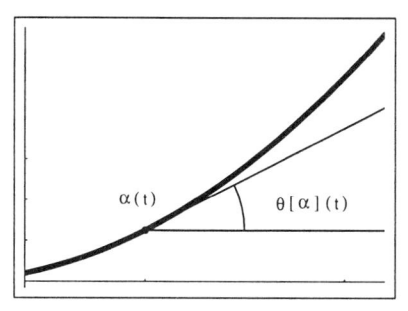

The turning angle

Geometrically, the turning angle $\theta[\alpha](t)$ is the angle between the horizontal and $\alpha'(t)$. Next, we derive a relation (first observed by Kästner[3]) between the turning angle and the curvature of a plane curve.

Lemma 1.16. *The turning angle and curvature of a regular curve α in the plane are related by*

(1.19) $$\theta[\alpha]'(t) = \|\alpha'(t)\|\kappa 2[\alpha](t).$$

Proof. The derivative of the left-hand side of (1.17) is

(1.20) $$\frac{\alpha''(t)}{\|\alpha'(t)\|} + \alpha'(t)\frac{d}{dt}\left(\frac{1}{\|\alpha'(t)\|}\right).$$

Furthermore, by the chain rule (Lemma 1.3) it follows that the derivative of the right-hand side of (1.17) is

(1.21) $$\theta[\alpha]'(t)\big(-\sin\theta[\alpha](t), \cos\theta[\alpha](t)\big) = \theta[\alpha]'(t)\frac{J\alpha'(t)}{\|J\alpha'(t)\|}.$$

Setting (1.21) equal to (1.20) and taking the dot product with $J\alpha'(t)$, we obtain

(1.22) $$\theta[\alpha]'(t)\|\alpha'(t)\| = \frac{\alpha''(t)\cdot J\alpha'(t)}{\|\alpha'(t)\|} = \|\alpha'(t)\|^2\kappa 2[\alpha](t).$$

Then (1.19) is immediate from (1.22). ∎

Corollary 1.17. *The turning angle and curvature of a unit-speed curve β in the plane are related by*

$$\kappa 2[\beta](s) = \frac{d\theta[\beta](s)}{ds}.$$

Thus changes in the turning angle of a curve are measured by its curvature.

---

[3] Abraham Gotthelf Kästner (1719-1800). German mathematician, professor at Leipzig and Göttingen. Although Kästner's German contemporaries ranked his mathematical and expository skills very high, Gauss found his lectures too elementary to attend. Gauss declared that Kästner was first mathematician among poets and first poet among mathematicians.

1.8 The Semicubical Parabola

In Chapter 2 we shall show how to use **Mathematica** to compute length and curvature of plane curves. However, it is informative to calculate these geometric quantities by hand in a simple case. We choose the **semicubical parabola** defined parametrically by
$$\mathbf{sc}(t) = (t^2, t^3).$$
The nonparametric version of **sc** is obviously $x^3 = y^2$.

The semicubical parabola has a special place in the history of curves, because it was the first algebraic curve (other than a straight line) to be found whose arc length function is algebraic. See [Lock, pages 3–11] for an extensive discussion.

We compute easily
$$\mathbf{sc}'(t) = (2t, 3t^2), \qquad J\mathbf{sc}'(t) = (-3t^2, 2t), \qquad \mathbf{sc}''(t) = (2, 6t),$$
so that
$$\kappa 2[\mathbf{sc}](t) = \frac{(2, 6t) \cdot (-3t^2, 2t)}{\|(2t, 3t^2)\|^3} = \frac{6}{|t|(4+9t^2)^{3/2}}.$$
Also, the length of **sc** over the interval $[0, t]$ is
$$s_{\mathbf{sc}}(t) = \int_0^t \|\mathbf{sc}'(u)\| \, du = \int_0^t u\sqrt{4+9u^2} \, du = \frac{1}{27}(4+9t^2)^{3/2} - \frac{8}{27}.$$

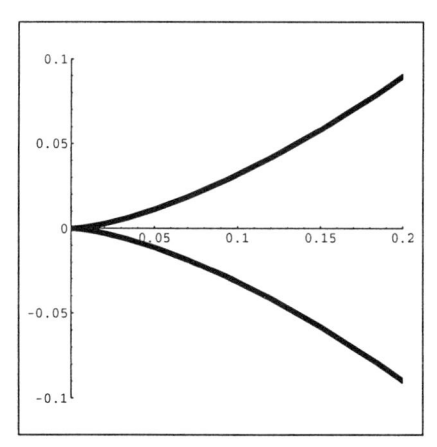

The semicubical parabola $t \longmapsto (t^2, t^3)$

Let $\boldsymbol{\alpha}\colon (a,b) \longrightarrow \mathbb{R}^n$ be a curve, and let $a < t_0 < b$. There are two ways that $\boldsymbol{\alpha}$ can fail to be regular at t_0. On the one hand, regularity can fail because

Let $\alpha\colon (a,b) \longrightarrow \mathbb{R}^n$ be a curve, and let $a < t_0 < b$. There are two ways that α can fail to be regular at t_0. On the one hand, regularity can fail because of the particular parametrization. For example, a horizontal straight line, when parametrized as $t \longmapsto (t^3, 0)$, is not regular at 0. This kind of nonregularity can be avoided by changing the parametrization. A more fundamental kind of nonregularity occurs when it is impossible to find a regular reparametrization of a curve near a point.

Definition. *We say that a curve $\alpha\colon (a,b) \longrightarrow \mathbb{R}^n$ is **regular** at t_0 provided that it is possible to extend the function*

$$t \longmapsto \frac{\alpha'(t)}{\|\alpha'(t)\|}$$

*to be a differentiable function at t_0. Otherwise α is said to be **singular** at t_0.*

It is easy to see that the definitions of regular and singular at a t_0 are independent of the parametrization of α. The notion of regular curve $\alpha\colon (a,b) \longrightarrow \mathbb{R}^n$, as defined on page 6, implies regularity at each t_0 for $a < t_0 < b$, but is stronger, because $\alpha'(t_0)$ must be nonzero.

It is not difficult to prove:

Lemma 1.18. *A curve α is regular at t_0 if and only if there exists a unit-speed parametrization of α near t_0.*

For example, a computation shows that

$$\frac{\mathbf{sc}'(t)}{\|\mathbf{sc}'(t)\|} = \frac{(2t, 3t^2)}{\sqrt{t^2(4+9t^2)}},$$

which is discontinuous at the origin because

$$(1,0) = \lim_{\substack{t \to 0 \\ t > 0}} \frac{(2t, 3t^2)}{\sqrt{t^2(4+9t^2)}} \neq \lim_{\substack{t \to 0 \\ t < 0}} \frac{(2t, 3t^2)}{\sqrt{t^2(4+9t^2)}} = (-1, 0).$$

This explains the singularity of $t \longmapsto \mathbf{sc}(t)$ at 0.

1.9 Exercises

1. Find the arc length function and the curvature for each of the following curves, where $a > 0$.

- a. $t \longmapsto a(\cos t + t\sin t, \sin t - t\cos t)$
- c. $t \longmapsto a(\cos^3 t, \sin^3 t)$
- b. $t \longmapsto (a\cosh(t/a), t)$
- d. $t \longmapsto a(2t, t^2)$

2. Show that the velocity α' of a differentiable curve $\alpha: (a, b) \longrightarrow \mathbb{R}^2$ is given by

$$\lim_{\Delta t \to 0} \alpha'(t) = \frac{\alpha(t + \Delta t) - \alpha(t)}{\Delta t}.$$

3. Let $\beta: (a, b) \longrightarrow \mathbb{R}^2$ be a circle of radius $a > 0$ centered at $\mathbf{q} \in \mathbb{R}^2$. Using the fact that $\|\beta(t) - \mathbf{q}\| = a$ for all t, show that $|\kappa 2[\beta](t)| = a^{-1}$ for all t.

4. Let $\beta: (a, b) \longrightarrow \mathbb{R}^2$ be a unit-speed curve whose curvature is given by $|\kappa 2[\beta](s)| = a^{-1}$ for all s, where $a > 0$ is a constant. Show that β is part of a circle of radius a centered at some point $\mathbf{q} \in \mathbb{R}^2$.

5. Prove Lemma 1.1.

6. Show that the curvature of a curve $\alpha: (a, b) \longrightarrow \mathbb{C}$ is given by

$$\kappa 2[\alpha](t) = \frac{\mathfrak{Im}(\alpha''(t)\overline{\alpha'(t)})}{|\alpha'(t)|^3}.$$

7. Let $\alpha: (a, b) \longrightarrow \mathbb{R}^2$ be a regular curve which does not pass through the origin. Fix t_0 and choose ϕ_0 such that

$$\frac{\alpha'(t_0) \cdot \alpha(t_0)}{\|\alpha'(t_0)\| \|\alpha(t_0)\|} = \cos \phi_0 \quad \text{and} \quad \frac{\alpha'(t_0) \cdot J\alpha(t_0)}{\|\alpha'(t_0)\| \|\alpha(t_0)\|} = \sin \phi_0.$$

Show that there exists a unique differentiable function $\varphi[\alpha]: (a, b) \longrightarrow \mathbb{R}$ such that $\varphi[\alpha](t_0) = \phi_0$,

$$\frac{\alpha'(t) \cdot \alpha(t)}{\|\alpha'(t)\| \|\alpha(t)\|} = \cos(\varphi[\alpha](t)) \quad \text{and} \quad \frac{\alpha'(t) \cdot J\alpha(t)}{\|\alpha'(t)\| \|\alpha(t)\|} = \sin(\varphi[\alpha](t))$$

for $a < t < b$. Geometrically, $\varphi[\alpha](t)$ represents the angle between the radius vector $\alpha(t)$ and the tangent vector $\alpha'(t)$. We call $\varphi[\alpha]$ the **tangent-radius angle** of α.

8. Find a unit-speed parametrization of the semicubical parabola $t \longmapsto (t^2, t^3)$, valid for $t > 0$.

9. Prove Lemma 1.18.

10. Let $\gamma_a \colon (-\infty, \infty) \longrightarrow \mathbb{R}^2$ be the curve defined by

$$\gamma_a(t) = \begin{cases} \left(t, t^a \sin\left(\dfrac{1}{t}\right)\right) & \text{if } t \neq 0, \\ (0,0) & \text{if } t = 0. \end{cases}$$

The curve $\gamma_{3/2}$

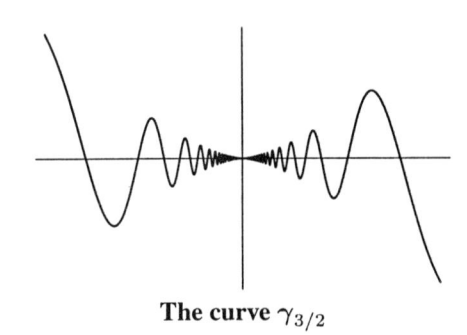

(i) Show that γ_a is continuous if $a > 0$, but discontinuous if $a \leq 0$.

(ii) Show that γ_a is differentiable if $1 < a < 2$, but that the curve γ_a' is discontinuous.

2

STUDYING CURVES IN THE PLANE WITH MATHEMATICA

We explain in this chapter how to use *Mathematica* to study curves in the plane. This will permit us easy access to a wide variety of curves for further study and experimentation. In the past, plane curves were drawn by ingenious but laborious methods that one can find, for example, in Frost's **Curve Tracing** ([Frost]). But now *Mathematica* will allow us to reproduce any curve quickly and accurately.

Not only can we draw curves with *Mathematica*, but we can also study them analytically. *Mathematica* programs for computing the curvatures and lengths of plane curves are described in Sections 2.1 and 2.2. A miniprogram for filling closed curves is described in Section 2.3. In Section 2.4 we give the simplest examples of curves in the plane, and we show how to graph these curves. Although we can compute the curvature of these curves symbolically, it is usually more informative to graph the curvature. The study of more complicated plane curves is postponed to Chapter 3. Piecewise-differentiable curves are studied in Section 2.5.

Before beginning, however, let us make some important remarks on how *Mathematica* syntax differs from ordinary mathematical syntax.

Mathematica *Notation versus Ordinary Mathematical Notation*

1. Parentheses, as used in ordinary mathematics, have at least four distinct

meanings according to the context. One of these is the notation for an open interval:
$$(a, b) = \{\, t \in \mathbb{R} \mid a < t < b \,\};$$
we shall use this formalism, but *Mathematica* does not need it. *Mathematica* employs different notations for two of the three other uses of parentheses.

Use of parenthesis	Ordinary mathematical notation	*Mathematica* notation
Grouping	$a(b+c)$	`a(b+c)`
Point in $\mathbb{R}^2$	(p, q)	`{p,q}`
Function f applied to a variable t	$f(t)$	`f[t]`

2. *Mathematica* uses `:=` instead of `=` in function definitions. Thus to define in *Mathematica* a function `f` that assigns one real number to another we use

   `f[t_] :=` some expression in `t`.

   Here the underscore after the `t` is an important part of the syntax. Whereas `t` denotes the symbol t, `t_` means generic t.

3. *Mathematica* also has different symbols for equality depending on the usage. Here are some examples.

Use of equality	Ordinary mathematical notation	*Mathematica* notation
Assignment	$a = 5$	`a=5`
Defining a function	$f(x) = e^x \sin x$	`f[x_] := E^x Sin[x]`
An equation to be solved	$2x + 3y = 5$	`2x + 3y == 5`

4. Multiplication is denoted in *Mathematica* by either an asterisk `*` or a space. For esthetic reasons we usually prefer to denote multiplication by a space. However, it is sometimes necessary to use an asterisk for multiplication at the end of an intermediate line in the middle of a multi-line expression. Sometimes it is possible to omit the multiplication symbol entirely. For example, `2*E^x*Sin[x]*Cos[x]` can be abbreviated to `2E^x Sin[x]Cos[x]`.

5. A list consisting of three elements **a1, a2, a3** is denoted in *Mathematica* by **{a1,a2,a3}**. Thus a list of n elements can be thought of as a point in $\mathbb{R}^n$.

6. The dot product is denoted by **.** in *Mathematica*; although it is written lower on the line, it behaves like its counterpart · in ordinary mathematical notation. To take the dot product of **{a1,a2,a3}** and **{b1,b2,b3}** we use

    {a1,a2,a3}.{b1,b2,b3}

    to get

    ```
    a1 b1 + a2 b2 + a3 b3
    ```

7. Functions in *Mathematica* are quite general. They can be list-valued, or even plot-valued. Thus a curve in $\mathbb{R}^n$ is just a list-valued function of 1 variable of length n.

Plotting Commands

The main command in *Mathematica* for plotting plane curves is **ParametricPlot**; many examples that illustrate its use and options are given below. On the other hand, ordinary functions are plotted in *Mathematica* with **Plot**; we shall use this command to plot the curvature of a curve. *Mathematica* has many more specialized plotting commands, for example, **ImplicitPlot** and **PolarPlot**, which we explain in Sections 4.1 and 4.3.

Functions in Mathematica

It will be convenient to make use of the shorthand described on pages 207–209 of [Wm] to represent functions. *Mathematica*, like ordinary mathematics, distinguishes functions and the values of functions. **Exp** is a function whose value on **z** (which can be either a number or a symbol) is **Exp[z]**. There are only a few built-in functions in *Mathematica* of this sort, such as **Cos**, **Sinh**, **ArcSin**. One way to generate others is to use *Mathematica's* command **Function**. Thus, for example, the function that assigns z^2 to z is **Function[z,z^2]**.

Mathematica has another way of representing functions which requires less writing. In *Mathematica* **#** is a place holder in a function and **&** terminates the function. So another way to write **Function[z,z^2]** is **#^2&**. For more details see [Wm, pages 207–209].

Common Mistakes

Error	What to Do
Mathematica hangs up.	• First, try "COMMAND-." to abort the command. (On systems without a COMMAND key try "ALT-.".) • If this does not work, try quitting from the kernel. Then reevaluate the commands, carefully eliminating syntax errors. • If this does not work, try to save the current file and quit *Mathematica*. Then open the file again in *Mathematica* and reevaluate all commands, carefully eliminating syntax errors. • If this does not work, try to kill the *Mathematica* process using system commands. • If this does not work, reboot the computer.
Failure to clear values of a symbol. For example, writing **y=5** assigns to **y** the number **5**, so that **y** can no longer be used with commands such as **Solve** or **DSolve**.	Usually it is best to use **Remove** to wipe out *Mathematica's* knowledge of the symbol. A weaker version of **Remove** is **Clear**, which removes any assignments associated with the symbol.
Use of "**=**" instead of "**==**".	"**=**" is used to assign values, whereas "**==**" is used to solve an algebraic or differential equation. Sometimes the cure is to replace "**=**" by "**==**", but more often than not values need to be cleared with **Clear** or **Remove**.

Error	What to Do
Incorrect use of parentheses ( ), brackets [ ] and braces { }	Use parentheses for grouping, brackets for function definitions and evaluation and braces for lists. Examples: **a(b+c)**, **Sin[x]** and **{1,2,3}**
Improper omission of spaces. For example, **ab** is one expression in *Mathematica*, but **a b** or **a*b** is **a** *times* **b**.	Check the syntax carefully. Note that **2b** is interpreted as **2** *times* **b**, but **b2** is interpreted as a single expression.
Using a command from a package before that package has been loaded into *Mathematica*	Usually the cure is to use **Remove** on the offending command; then the command can be used normally. If this fails, quit from the kernel and reload the package before the command is used.

2.1 Computing Curvature of Curves in the Plane

A curve **alpha** in the plane can be defined in *Mathematica* by

```
alpha[t_]:= {a1[t],a2[t]}
```

where **a1** and **a2** are functions of **t**. We can also define **J**, which is the *Mathematica* version of the complex structure J of $\mathbb{R}^2$ that we defined on page 4, by

```
J[{p1_,p2_}]:= {-p2,p1}
```

The command to take the derivative of an expression in *Mathematica* is **D**. For example:

```
D[Sin[a x],x]
```

yields

```
           a Cos[a x]
```

More generally, to calculate the nth derivative of $x \longmapsto \sin(a\,x)$ we use

`D[Sin[a x],{x,n}]`

To find the value of a derivative at a specific number, we must first carry out the differentiation using a dummy variable and then substitute the number for the dummy variable. For example, let us find the value of the derivative of $x \longmapsto \sin x$ at $\pi/4$ using *Mathematica*:

`D[Sin[x],x] /. x->Pi/4`

```
      1
   -------
   Sqrt[2]
```

Here, "`/.`" means "such that". The expression "`x -> Pi/4`" should be read as "`x` goes to `Pi/4`".

We can also use **`D`** to compute the velocity of a curve **`alpha`**:

`D[alpha[t],t]`

An equivalent command to find the velocity of **`alpha`** is

`alpha'[t]`

Curvature is really a function that assigns to each curve **`alpha`** and each number **`t`** another number. We call the function **`kappa2`**, because in Chapter 8 we shall define a slightly different curvature for space curves. The definition of **`kappa2`** is almost an exact translation of equation (1.9) into *Mathematica*:

```
kappa2[alpha_][t_]:=
    D[alpha[tt],{tt,2}].J[D[alpha[tt],tt]]/
    Simplify[D[alpha[tt],tt].
        D[alpha[tt],tt]]^(3/2) /. tt->t
```

For example, if after typing in this definition we now type in

`kappa2[alpha][t]`

2.1 Computing Curvature of Curves in the Plane

we get the response

```
alpha''[t] . J[alpha'[t]]
─────────────────────────
                      3/2
   alpha'[t] . alpha'[t]
```

which is *Mathematica's* way of writing (1.9). For the convenience of the reader, the commands such as **kappa2** and **J** are available in the file **CSPROGS.m**. It can be read into a *Mathematica* session by typing in the following command:

<<CSPROGS.m

In fact, all programs described in this book (as well as many additional programs) can be read into a *Mathematica* session by typing

<<CAS.m

Let us clarify some things about *Mathematica's* definition of **kappa2**.

1. **Simplify** is the most common command in *Mathematica* for simplifying expressions. Version 3.0 of *Mathematica* also has a command **FullSimplify** that takes longer, but uses more transformations.

2. We have used **tt** throughout as a local variable for the purposes of symbolic differentiation. At the very end of the definition of **kappa2** we use "**tt -> t**" to replace **tt** by **t**.

3. The reason we need to use a local variable is that we want to be able to substitute a number for **t**. So the differentiation has to be carried out using a dummy variable, in this case **tt**. Then we replace **tt** by **t**. If we want to, we can then replace **t** either by another symbol, or by a number.

Let us use *Mathematica* to compute the curvature of the **figure eight curve** parametrized by

(2.1) $$\text{eight}(t) = (\sin t, \sin t \cos t).$$

(See [Law, page 124] and [Gomes, volume 1, page 272], where the figure eight curve is also called the **lemniscate of Gerono.**[1]) First, we define **J** and **kappa2** by the above commands, and then we define the figure eight curve by

eight[t_] := {Sin[t],Sin[t]Cos[t]}

Mathematica's response to **kappa2[eight][t]** is

---
[1] Camille Christophe Gerono (1799-1891). French mathematician.

$$\frac{-4\,\text{Cos}[t]^2\,\text{Sin}[t] - \text{Sin}[t]\,(-\text{Cos}[t]^2 + \text{Sin}[t]^2)}{(\text{Cos}[t]^2 + \text{Cos}[2\,t]^2)^{3/2}}$$

which in ordinary mathematical notation would be

$$\frac{-4\cos^2 t\,\sin t + \sin t\,(\cos^2 t - \sin^2 t)}{(\cos^2 t + (\cos^2 t - \sin^2 t)^2)^{3/2}}.$$

Thus the resemblance between *Mathematica* output and ordinary mathematical notation is sufficiently close that no further explanation is required.

There is a potential problem with computing curvature in this way. For mildly complicated curves *Mathematica* can give quite complicated symbolic expressions for the curvature. Frequently, it is unclear how to simplify these complicated expressions or whether they can be simplified at all.

On the other hand, a formula is probably not the best way to understand the curvature of a curve. More often than not, a graph of the curvature is more informative. Fortunately, *Mathematica* does not need the curvature to be simplified in order to graph it. Here is a plot of the figure eight curve and a graph of its curvature.

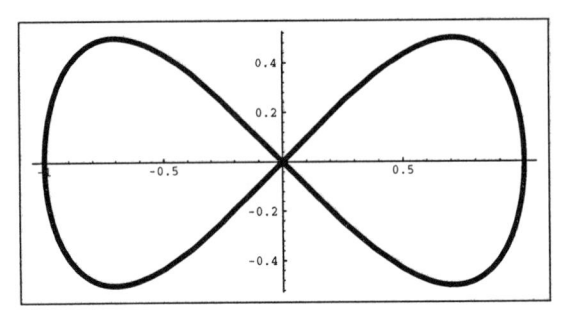

Figure eight curve

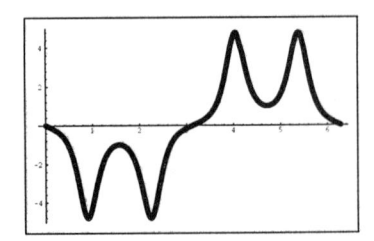

Curvature of the figure eight curve

The commands to create these graphs are:

```
ParametricPlot[
eight[t],{t,0,2Pi},
AspectRatio->Automatic];
```

and

```
Plot[
kappa2[eight][t],
{t,0,2Pi}];
```

The semicolon at the end of a graphics command prevents *Mathematica* from printing out "Graphics" after it does the graph. *Mathematica* uses `AspectRatio` to set the

2.2 Computing Lengths of Curves

height-to-width ratio in various plotting commands. For example, the default for **AspectRatio** for **ParametricPlot** and **Plot** is about 0.618. This number is the reciprocal of the so-called golden ratio, defined as $(1 + \sqrt{5})/2$. On the other hand, the option **AspectRatio->Automatic** will ensure that one unit in the vertical direction is the same as one unit in the horizontal direction.

2.2 Computing Lengths of Curves

Mathematica can be used to compute the length of a curve **alpha**. It is more efficient to compute the derivative of the arc length function before computing the arc length function itself:

```
arclengthprime[alpha_][t_]:=
    Sqrt[Simplify[D[alpha[tt],tt].
        D[alpha[tt],tt]]] /. tt->t
```

Then the arc length function itself is the indefinite integral of **arclengthprime**:

```
arclength[alpha_][t_]:=
    Integrate[arclengthprime[alpha][tt],tt]  /. tt->t
```

Similarly, the length function is the definite integral of **arclength**:

```
length[a_,b_][alpha_]:=
    Integrate[arclengthprime[alpha][u],{u,a,b}]
```

The commands **arclength[alpha]** and **length[alpha]** attempt to integrate **arclengthprime[alpha]** symbolically. (Both use *Mathematica's* symbolic integrator **Integrate**.) They may fail if **arclengthprime[alpha]** is too complicated. However, a numerical value for the length of a curve can always be obtained using the following command, which makes use of *Mathematica's* numerical integrator **NIntegrate**:

```
lengthn[a_,b_][alpha_]:=
    NIntegrate[arclengthprime[alpha][u],{u,a,b}]
```

(For **length** we can take **a** and **b** to be symbols, but for **lengthn** they must be numbers.)

For example, the semicubical parabola **sc** given on page 21 can be defined in *Mathematica* by

```
sc[t_] := {t^2,t^3}
```

To compute the length of **sc** with *Mathematica*, we can use

```
Simplify[PowerExpand[
length[a,b][sc] /. Sqrt[x_]:>Sqrt[Factor[x]]]]
```

to which *Mathematica* responds

$$\frac{-(4 + 9a^2)^{3/2} + (4 + 9b^2)^{3/2}}{27}$$

(*Mathematica*'s command **PowerExpand** converts $(ab)^s$ into $a^s b^s$; sometimes this is correct mathematically, but sometimes it is not. For example, $\sqrt{x^2}$ is not equal to x if $x < 0$. See page 60 for another example of an erroneous result using **PowerExpand**. Thus **PowerExpand** must be used with great care.)

In general, numerical values of lengths can be found using *Mathematica*'s command **N**. For example, either of the commands

```
lengthn[1,2][sc]    N[length[1,2][sc]]
```

tells us that a string stretched along the curve from **sc**(1) to **sc**(2) will have length

7.63371

2.3 Filling Curves

The command **ParametricPlot** carries out its work by drawing a large number of very short lines in such a way that the resulting picture is indistinguishable from the actual curve to the naked eye. In this section we describe a miniprogram **filledcurve**, which has the same syntax as **ParametricPlot**. Instead of drawing a closed curve, the command **filledcurve** will fill it in with a desired gray level or color.

2.4 Examples of Curves in $\mathbb{R}^2$

```
filledcurve[curve_,{u_,u0_,u1_},opts___]:=
    Module[{plottmp,grtmp},
        plottmp = ParametricPlot[
        curve//Evaluate,{u,u0,u1},
        AspectRatio->Automatic,
        DisplayFunction->Identity,opts];
        grtmp = plottmp /.(Line[pts_]:> Polygon[pts]);
    Show[grtmp,DisplayFunction->$DisplayFunction]]
```

A complete description of **filledcurve** would be too much of a diversion at this stage. Suffice it to say that descriptions of the commands used in the definition of **filledcurve** can be understood by reading [Wm] and short summaries in [Blach]. Also, **filledcurve** is similar to the command **wireframe** that is explained on page 404. The specialized plotting commands described in this book, including **filledcurve**, are available in the file **PLTPROGS.m**. It can be read into a *Mathematica* session by typing in the following command:

<<PLTPROGS.m

The file **CURVES.m**, which contains the parametrizations of the plane curves used in this book, can be read in the same way. We illustrate the use of **filledcurve** by filling the plot of the figure eight curve on page 32:

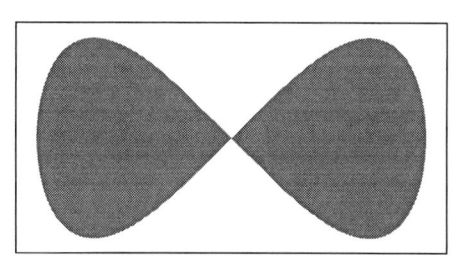

```
filledcurve[
eight[t]//Evaluate,
{t,0,2Pi},
Axes->None,
AspectRatio->
Automatic,
Prolog->
GrayLevel[0.5]];
```

Filled figure eight curve

2.4 Examples of Curves in $\mathbb{R}^2$

In this section we give several examples of curves in $\mathbb{R}^2$. In addition to showing how to define and plot each curve using *Mathematica*, we also graph its curvature, when appropriate.

Circles

A circle of radius a with center $(0,0)$ is usually described by the equation

$$x^2 + y^2 = a^2.$$

This is the nonparametric or implicit form of the circle. For more details on nonparametric forms of curves, see Section 4.1. The simplest parametrization of the circle is

$$\alpha(t) = (a\cos t, a\sin t) \qquad (0 \leq t < 2\pi);$$

another parametrization is

$$\beta(t) = (a\cos 2t, a\sin 2t) \qquad (0 \leq t < \pi).$$

Either of these parametrizations will trace out a circle. However, let us use a standard definition of a parametrization of a circle of radius a, namely,

$$\mathrm{circle}[a](t) = (a\cos t, a\sin t).$$

In *Mathematica* this definition becomes

```
circle[a_][t_] := {a Cos[t], a Sin[t]}
```

To plot a circle of radius 2 with center at the origin, we use

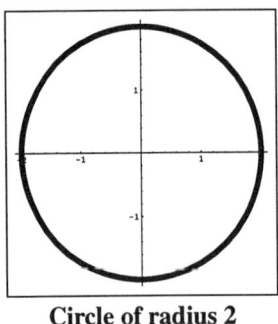

```
ParametricPlot[
circle[2][t],{t,0,2Pi},
AspectRatio->Automatic];
```

Circle of radius 2

Since the curvature of a circle is easily seen to be constant, it is unnecessary to graph the curvature in this simple case.

Let us make two remarks on how *Mathematica* handles circles.

1. Although we have explained how to define and graph a circle centered at the origin, it is easy to plot one or more circles with centers at other points. Here is an example of a plot[2] of two circles centered at $(1,2)$ and $(2,0)$:

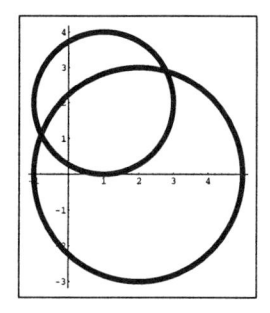

    ```
    ParametricPlot[Evaluate[{
    circle[2][t] + {1,2},
    circle[3][t] + {2,0}}],
    {t,0,2Pi},
    AspectRatio->Automatic];
    ```

    Plots of translates of other curves can be done in a similar fashion. A more general scheme for translating, rotating and reflecting curves with *Mathematica* is described in Chapter 6.

2. *Mathematica* has an internal command `Circle` for drawing circles. For an example of its use see Section 5.6. We have defined `circle[a]` as a function so that we can perform on it the standard operations of calculus such as differentiation.

Ellipses

Perhaps the next simplest curves after the circle are the ellipses. The name "ellipse" (which means "falling short") is due to Apollonius.[3] The nonparametric form of the **ellipse** centered at the origin with axes of lengths a and b is

$$\frac{x^2}{a^2} + \frac{y^2}{b^2} = 1,$$

and the standard parametrization is

(2.2) $\qquad$ **ellipse**$[a, b](t) = (a\cos t, b\sin t) \qquad (0 \le t < 2\pi),$

---

[2] In version 1.2 the command `Release` should be used instead of `Evaluate`. Sometimes `Release` can be omitted. In version 2.0 and greater `Evaluate` makes graphics commands perform faster when used with user-defined functions. Frequently, it is easier to type it in after the command to which it applies using the syntax `//Evaluate`.

[3] Apollonius of Perga (262-180 B.C.). His eight volume treatise on conic sections is the standard ancient source of information about ellipses, hyperbolas and parabolas.

or in *Mathematica*

```
ellipse[a_,b_][t_]:= {a Cos[t],b Sin[t]}
```

For example, let us compute the curvature of the ellipse **ellipse[a, b]**. *Mathematica's* response to **kappa2[ellipse[a,b]][t]** is

$$\frac{a\,b\,\cos[t]^2 + a\,b\,\sin[t]^2}{(b^2\,\cos[t]^2 + a^2\,\sin[t]^2)^{3/2}}$$

The ellipse **ellipse[2,1]** is drawn with the *Mathematica* command

```
ParametricPlot[
ellipse[2,1][t]//Evaluate,{t,0,2Pi},
AspectRatio->Automatic];
```

and the graph of the curvature of the ellipse is drawn with the command

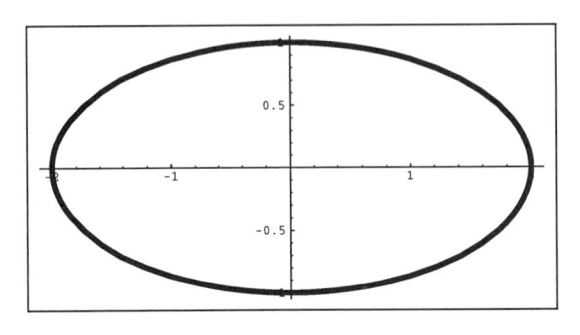

```
Plot[kappa2[ellipse[2,1]][t]//Evaluate,{t,0,2Pi}];
```

The ellipse
$$\frac{x^2}{4} + y^2 = 1$$

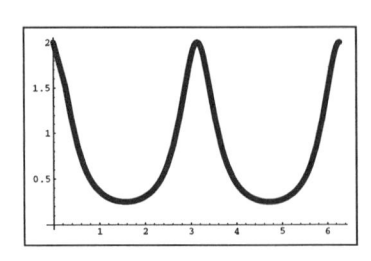

Curvature of the ellipse
$$\frac{x^2}{4} + y^2 = 1$$

An ellipse is traditionally defined as the locus of points such that the sum of the distances from two fixed points $\mathbf{F}_-$ and $\mathbf{F}_+$ is constant. The points $\mathbf{F}_-$ and $\mathbf{F}_+$ are called the **foci** of the ellipse. In the case of **ellipse[a, b]** with $a > b$ the foci are $\left(\pm\sqrt{a^2 - b^2}, 0\right)$:

2.4 Examples of Curves in $\mathbb{R}^2$

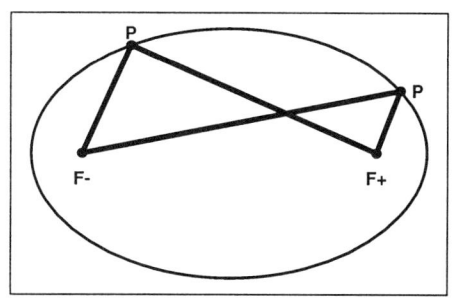

Definition of the ellipse:

The locus of points P such that
distance($\mathbf{F}_-$,P) + distance($\mathbf{F}_+$,P)
is constant.

Theorem 2.1. *Let E be an ellipse with foci $\mathbf{F}_-$ and $\mathbf{F}_+$. Then:*

(i) *for any point $\mathbf{P}$ on E the sum*

$$\text{distance}(\mathbf{F}_-,\mathbf{P}) + \text{distance}(\mathbf{F}_+,\mathbf{P})$$

is constant.

(ii) *the tangent line to E at $\mathbf{P}$ makes equal angles with the line segments connecting $\mathbf{P}$ to $\mathbf{F}_-$ and $\mathbf{F}_+$.*

Proof. To establish these two properties of an ellipse, it is highly advantageous to use complex numbers. Clearly, the parametrization (2.2) is equivalent to

(2.3) $$\mathbf{z}(t) = \frac{1}{2}(a+b)e^{it} + \frac{1}{2}(a-b)e^{-it} \qquad (0 \le t < 2\pi).$$

The vector from $\mathbf{P}$ to $\mathbf{F}_{\mp}$ is given by

(2.4) $$\mathbf{z}_{\pm}(t) = \frac{1}{2}(a+b)e^{it} + \frac{1}{2}(a-b)e^{-it} \pm \sqrt{a^2 - b^2}.$$

It is remarkable that the right-hand side of (2.4) can be written as a perfect square:

(2.5) $$\mathbf{z}_{\pm}(t) = \frac{1}{2}\left(\sqrt{a+b}\,e^{it/2} \pm \sqrt{a-b}\,e^{-it/2}\right)^2.$$

An easy calculation using (2.5) shows that

$$|\mathbf{z}_{\pm}(t)| = a \pm \sqrt{a^2 - b^2}\,\cos t.$$

Consequently, $|\mathbf{z}_+(t)| + |\mathbf{z}_-(t)| = 2a$; this proves (i).

To prove (ii), we first use (2.3) to compute

$$(2.6) \quad \mathbf{z}'(t) = \frac{i}{2}(a+b)e^{it} - \frac{i}{2}(a-b)e^{-it}$$

$$= \frac{i}{2}\left(\sqrt{a+b}\,e^{it/2} + \sqrt{a+b}\,e^{-it/2}\right)\left(\sqrt{a+b}\,e^{it/2} - \sqrt{a+b}\,e^{-it/2}\right).$$

It follows from (2.5) and (2.6) that

$$\frac{\mathbf{z}_+(t)}{\mathbf{z}'(t)} = \frac{1}{i}\frac{\sqrt{a+b}\,e^{it/2} + \sqrt{a+b}\,e^{-it/2}}{\sqrt{a+b}\,e^{it/2} - \sqrt{a+b}\,e^{-it/2}} = -\frac{\mathbf{z}'(t)}{\mathbf{z}_-(t)},$$

which is equivalent to (ii). ∎

Logarithmic Spirals

A **logarithmic spiral** is parametrized by

$$\mathbf{logspiral}[a, b](t) = a(e^{bt}\cos t, e^{bt}\sin t),$$

or in *Mathematica*

```
logspiral[a_,b_][t_] := a{E^(b t)Cos[t],E^(b t)Sin[t]}
```

A small value of b makes the curve spiral tightly. Here is how to draw the logarithmic spiral when $b = 0.08$.

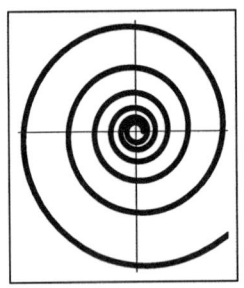

```
ParametricPlot[
  logspiral[1,0.08][t]
  //Evaluate,
  {t,0,12Pi},
  AspectRatio->Automatic,
  Ticks->None,
  PlotPoints->80];
```

The logarithmic spiral
$t \longmapsto (e^{0.08t}\cos t, e^{0.08t}\sin t)$

The option **Ticks->None** removes all the ticks along the axes, and the option **PlotPoints->80** plots the curve more accurately than *Mathematica's* default (which is **PlotPoints->25**).

2.4   Examples of Curves in $\mathbb{R}^2$

If we had used $b = -0.08$ instead of $b = 0.08$, the spiral would have unwound the other way. See the pictures on page 135.

It is evident from the plot of the logspiral that the curvature decreases as t increases. We check this by graphing the curvature of the logarithmic spiral:

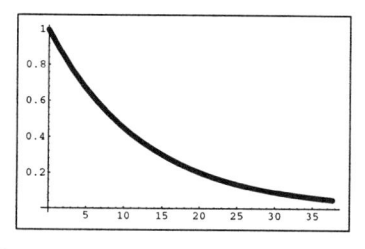

```
Plot[kappa2[
logspiral[1,0.08]][t]
//Evaluate,
{t,0,12Pi}];
```

Curvature of the logarithmic spiral
$$t \longmapsto (e^{0.08t} \cos t, e^{0.08t} \sin t)$$

Also, *Mathematica* tells us the formula for the curvature:

```
PowerExpand[Simplify[kappa2[logspiral[a,b]][t]]]
              1
     ─────────────────
              2    b t
     a Sqrt[1 + b ] E
```

Thus $\kappa 2[\text{logspiral}[a,b]](t) = \dfrac{1}{a\, e^{bt}\sqrt{1+b^2}}$.

Logarithmic spirals were first discussed by Descartes in 1638 in connection with a problem from dynamics. He was interested in determining all plane curves α with the property that the angle between $\alpha(t)$ and the tangent vector $\alpha'(t)$ is constant. Such curves turn out to be precisely the logarithmic spirals, as we now prove.

Let $\alpha\colon (a,b) \longrightarrow \mathbb{R}^2$ be a curve which does not pass through the origin. Recall (see page 23) that the tangent-radius angle of α is a function $\varphi[\alpha]\colon (a,b) \longrightarrow \mathbb{R}^2$ such that

$$\frac{\alpha'(t) \cdot \alpha(t)}{\|\alpha'(t)\|\,\|\alpha(t)\|} = \cos(\varphi[\alpha](t)) \quad \text{and} \quad \frac{\alpha'(t) \cdot J\alpha(t)}{\|\alpha'(t)\|\,\|\alpha(t)\|} = \sin(\varphi[\alpha](t))$$

for $a < t < b$. (The existence of $\varphi[\alpha]$ follows from Lemma 1.13. See exercise 7 of Chapter 1.)

Lemma 2.2. *Let* $\alpha\colon (a,b) \longrightarrow \mathbb{R}^2$ *be a curve which does not pass through the origin. The following conditions are equivalent:*

(i) *The tangent-radius angle* $\varphi[\alpha]$ *is constant;*

(ii) α is a reparametrization of a logarithmic spiral.

Proof. We write $\alpha(t) = r e^{i\theta}$; then $\alpha'(t) = e^{i\theta}(r' + i r \theta')$, so that

$$|\alpha(t)| = r, \quad |\alpha'(t)| = \sqrt{r'^2 + r^2 \theta'^2} \quad \text{and} \quad \overline{\alpha(t)}\alpha'(t) = r(r' + i r \theta').$$

Assume that (i) holds and let γ be the constant value of $\varphi[\alpha](t)$. Lemma 1.2 implies that

(2.7) $$e^{i\gamma} = \frac{\alpha'(t)\overline{\alpha(t)}}{|\alpha'(t)||\alpha(t)|} = \frac{r' + i r \theta'}{\sqrt{r'^2 + r^2 \theta'^2}}.$$

From (2.7) we get

$$\cos\gamma = \frac{r'}{\sqrt{r'^2 + r^2 \theta'^2}} \quad \text{and} \quad \sin\gamma = \frac{r \theta'}{\sqrt{r'^2 + r^2 \theta'^2}},$$

so that

$$\frac{r'}{r} = \theta' \cot\gamma.$$

The solution of this differential equation is easily found to be

(2.8) $$r = a \exp\bigl((\cot\gamma)\theta\bigr),$$

where a is a constant. From (2.8) we obtain

(2.9) $$\alpha(t) = a \exp\bigl(((\cot\gamma) + i)\theta(t)\bigr).$$

Hence α is a reparametrization of a logspiral.

Conversely, it is easy to check that the tangent-radius angle of any logspiral is a constant γ. Under a reparametrization, the tangent-radius angle has the constant value $\pm\gamma$. ∎

2.5 Plotting Piecewise Defined Curves

An **equilateral triangle** may be parametrized by

(2.10) $$\text{triangle}(t) = \begin{cases} (t, 0) & \text{if } 0 \leq t < 1, \\ (1, 0) + (t - 1)(-1/2, \sqrt{3}/2) & \text{if } 1 \leq t < 2, \\ (1/2, \sqrt{3}/2) + (t - 2)(-1/2, -\sqrt{3}/2) & \text{if } 2 \leq t < 3. \end{cases}$$

2.5 Plotting Piecewise Defined Curves

This curve can be defined in *Mathematica* using the command **If**, which takes 3 arguments. The first argument is a condition (an inequality in our case). The second argument specifies what happens if the condition is true and the third argument specifies what happens if the condition is false. For a function whose definition requires several conditions, we can nest **If** several times. Definition (2.10) in *Mathematica* becomes:

```
triangle[t_]:= If[t<1,t{1,0},
    If[t<2,{1,0} + (t - 1){-1/2,Sqrt[3]/2},
    {1/2,Sqrt[3]/2} + (t - 2){-1/2,-Sqrt[3]/2}]]
```

Furthermore, *Mathematica* has no difficulty plotting the triangle with the command

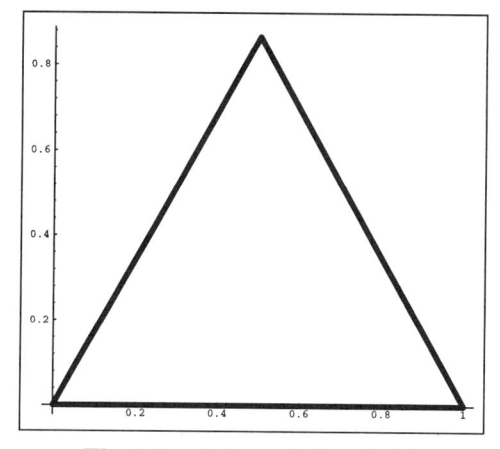

```
ParametricPlot[
triangle[t]
//Evaluate,
{t,0,3},
AspectRatio->
Automatic];
```

The triangle $t \longmapsto \text{triangle}(t)$

See exercise 9 for a miniprogram **ngon** that parametrizes any regular polygon.

It is sometimes possible to define piecewise-linear functions using the fact that $x \longmapsto |x|$ can be represented in *Mathematica* as $x \longmapsto \sqrt{x^2}$, provided x is real. For example, let us define

$$\mathbf{diamond}[\mathbf{n}, \mathbf{a}, \mathbf{b}](t) = \left(a \left(\sqrt{(\cos t)^2} \right)^{n-1} \cos t, \; b \left(\sqrt{(\sin t)^2} \right)^{n-1} \sin t \right).$$

or

```
diamond[n_,a_,b_][t_]:=
    {a Sqrt[Cos[t]^2]^(n - 1)Cos[t],
     b Sqrt[Sin[t]^2]^(n - 1)Sin[t]}
```

Then **diamond[2, 1, 1]** has the shape of a diamond. Here are more examples:

```
ParametricPlot[
Table[
diamond[n,1,1][t],
{n,1,8}]//Evaluate,
{t,0.1,2Pi + 0.1},
Ticks->None,
AspectRatio->
Automatic];
```

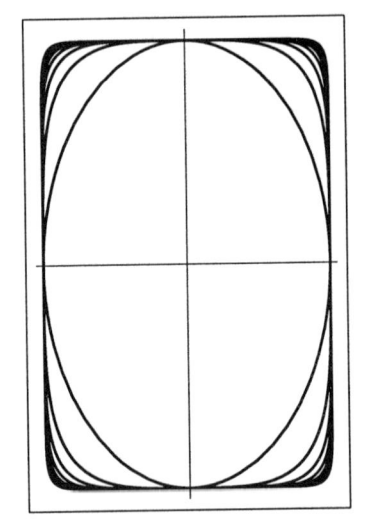

$t \longmapsto \text{diamond}[\mathbf{n}, \mathbf{1}, \mathbf{1}](t)$
$1 \leq n \leq 8$

```
ParametricPlot[
Table[
diamond[1/n,1,1.5][t],
{n,1,8}]//Evaluate,
{t,0.1,2Pi + 0.1},
AspectRatio->
Automatic];
```

$t \longmapsto \text{diamond}[\mathbf{1/n}, \mathbf{1}, \mathbf{1.5}](t)$
$1 \leq n \leq 8$

2.6 Exercises

1. The **parabola[a]** is defined by
$$\mathbf{parabola}[a](t), = (2at, at^2)$$
or

```
parabola[a_][t_] := {2a t,a t^2}
```

The trace of a parabola is the set of points **p** such that the distance of **p** from a point **F** (called the **focus** of the parabola) equals the distance from **p** to a line **L** (called the **directrix** of the parabola). The **vertex V** of a parabola is the point closest to the directrix. In the case of the parabola $t \longmapsto (2at, at^2)$, the focus is $(0, a)$, the directrix is the line $t \longmapsto (t, -a)$ and the vertex is the origin.

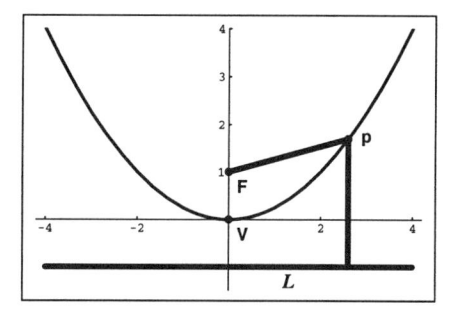

Definition of the parabola:

The locus of points p such that distance(p,F) = distance(p,L).

Plot several parabolas as one graph and their curvatures as another graph.

2. The **hyperbola[a, b]** is defined by
$$\mathbf{hyperbola}[a, b](t) = (a \cosh t, b \sinh t)$$
or

```
hyperbola[a_,b_][t_] := {a Cosh[t],b Sinh[t]}
```

Plot several hyperbolas as one graph and their curvatures as another graph.

3. The **astroid** is defined by
$$\text{astroid}[n, a, b](t) = (a\cos^n t, b\sin^n t),$$

or

```
astroid[n_,a_,b_][t_]:= {a Cos[t]^n,b Sin[t]^n}
```

Find the formula for the arc length and curvature of **astroid**$[3, a, a]$. Plot **astroid**$[3, 1, 1]$ and its curvature.

4. Find formulas for the lengths of the following curves using *Mathematica*:
   (a) The involute of a circle defined by
   $$\text{circleinvolute}[a](t) = a(\cos t + t\sin t, \sin t - t\cos t).$$
   (b) The graph of the logarithm function $t \longmapsto (t, \log t)$.
   (c) The graph of the exponential function $t \longmapsto (t, \exp t)$.

5. Find numerical values for the lengths of the following curves using *Mathematica*:
   (a) The cubic parabola $t \longmapsto (t, t^3)$ from -1 to 1.
   (b) The graph of $t \longmapsto (t, \sin t)$ from 0 to π.
   (c) The ellipse $t \longmapsto (\cos t, 2\sin t)$ from 0 to 2π.

6. *Mathematica's* commands **Epilog** and **Line** can be used to draw a line on the plot of a curve. For example, to draw a line from the focus of a parabola to a point on the parabola, we can use

```
ParametricPlot[
parabola[1][t]
//Evaluate,
{t,-1.0,1.0},
Epilog->
Line[{{0,1},
parabola[1][0.8]}],
AspectRatio->
Automatic];
```

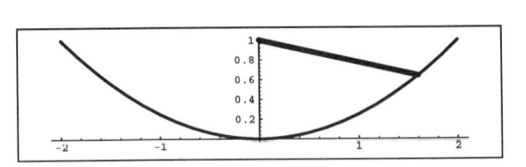

Use this model to draw an ellipse with a line connecting a focus to points on the ellipse. Do the same thing for a hyperbola.

2.7 Animations of Definitions of Curves

7. Plot the following curves (the *Mathematica* definitions can be found in the appendix): conchoid, epitrochoid, hypotrochoid, piriform, strophoid, nephroid.

8. Use *Mathematica* to draw the picture on page 39 that defines the ellipse.

9. The following command is a parametrization of a regular polygon with n sides.

```
ngon[n_,a_][t_]:=
  Which@@Flatten[Table[{(k - 1)/n<=t<=k/n,
    a*{Cos[2*Pi*t],Sin[2*Pi*t]}/Cos[2*Pi*t-(2*k - 1)*Pi/n]},
     {k,1,n}],1]
```

Explain it, and use it to draw a pentagon.

2.7 Animations of Definitions of Curves

The following commands animate the definitions of an ellipse:

```
Do[ParametricPlot[Evaluate[
ellipse[1.5,1][t]],{t,0,2Pi},
PlotStyle->{{AbsoluteThickness[1],RGBColor[0,0,1]}},
Axes->None,Epilog->{AbsoluteThickness[2],
Line[{{Sqrt[1.5^2 - 1],0},ellipse[1.5,1][a],
     {-Sqrt[1.5^2 - 1],0}}],
AbsolutePointSize[6],Point[{Sqrt[1.5^2 - 1],0}],
Point[ellipse[1.5,1][a]],Point[{-Sqrt[1.5^2 - 1],0}]}],
{a,0,2Pi,Pi/100}]
```

The command **Do** creates a succession of pictures, each of which is created by **ParametricPlot**. Similarly, the definition of a parabola can be created with

```
Do[ParametricPlot[Evaluate[
parabola[1][t]],{t,-2,2},
PlotStyle->{{AbsoluteThickness[1]}},
AspectRatio->Automatic,Axes->None,
PlotRange->{Automatic,{-1.5,4.1}},
Ticks->{{-4,-2,2,4},{1,2,3,4}},
Epilog->{AbsoluteThickness[2],Line[{{-4,-1},{4,-1}}],
Line[{{0,1},parabola[1][a]}],
Line[{{parabola[1][a][[1]],-1},parabola[1][a]}],
AbsolutePointSize[6],Point[{0,0}],Point[{0,1}],
Point[parabola[1][a]]}],{a,-2,2,0.04}]
```

3
FAMOUS PLANE CURVES

Plane curves have been a subject of much interest beginning with the Greeks. Both physical and geometric problems frequently lead to curves other than ellipses, parabolas and hyperbolas. The literature on plane curves is extensive. Diocles[1] studied the *cissoid* in connection with the classic problem of doubling the cube. Newton[2] gave a classification of cubic curves (see [Newton] or [BrKn]). Mathematicians from Fermat to Cayley frequently had curves named after them. In this

---

[1] Diocles (180 B.C.).

[2] Sir Isaac Newton (1642–1727). English mathematician, physicist, and astronomer. Newton's contributions to mathematics encompass not only his fundamental work on calculus and his discovery of the binomial theorem for negative and fractional exponents, but also substantial work in algebra, number theory, classical and analytic geometry, methods of computation and approximation, and probability. His classification of cubic curves was published as an appendix to his book on optics; his work in analytic geometry included the introduction of polar coordinates.

As Lucasian professor at Cambridge, Newton was required to lecture once a week, but his lectures were so abstruse that he frequently had no auditors. Twice elected as Member of Parliament for the University, Newton was appointed warden of the mint; his knighthood was awarded primarily for his services in that role. In **Philosophiæ Naturalis Principia Mathematica**, Newton set forth fundamental mathematical principles and then applied them to the development of a world system. This is the basis of the Newtonian physics that determined how the universe was perceived until the twentieth century work of Einstein.

chapter we choose a number of historically interesting plane curves and show how *Mathematica* can be used to study them.

There are many books on plane curves. Four excellent classical books are those of Cesàro [Ces], Gomes Teixeira[3] [Gomes], Loria[4] [Loria1], and Wieleitner [Wiel2]. In addition, Struik's[5] classic book [Stru2] contains much useful information, both theoretical and historical. Modern books on curves include [Arg], [BrGi], [Law], [Lock], [Sav], [Shikin], [vonSeg], [Yates] and [Zwi].

Cycloids are discussed in Section 3.1, *lemniscates of Bernoulli* in Section 3.2 and *cardioids* in Section 3.3. Then in Section 3.4 we derive the differential equation for a *catenary*. The *cissoid of Diocles* is studied in Section 3.5, the *tractrix* in Section 3.6 and *clothoids* in Section 3.7. Finally, *pursuit curves* are discussed in Section 3.8.

3.1 Cycloids

Let us define the **cycloid** by

$$\mathbf{cycloid}[a, b](t) = (a\,t - b\sin t, a - b\cos t);$$

In *Mathematica* the corresponding definition is

```
cycloid[a_,b_][t_] := {a t - b Sin[t], a - b Cos[t]}
```

Then **cycloid**$[a, a]$ is the locus of points traced by a point on a circle of radius a which rolls without slipping along a straight line. Here is a typical plot made with

```
ParametricPlot[cycloid[1,1][t]//Evaluate,
    {t,-Pi/2,5Pi/2},AspectRatio->Automatic];
```

[3] Francisco Gomes Teixeira (1851-1933). A leading Portuguese mathematician of the last half of the 19th century. There is a statue of him in Oporto.

[4] Gino Loria (1862-1954). Italian mathematician, professor at the University of Genoa. In addition to his books on curve theory, [Loria1] and [Loria2], Loria wrote several books on the history of mathematics.

[5]Dirk Jan Struik celebrated his 100th birthday on Sept. 30, 1994.

3.1 Cycloids

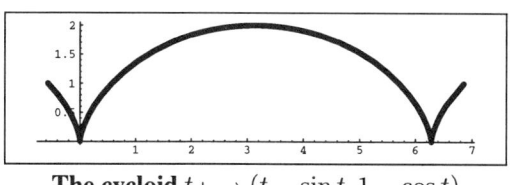

The cycloid $t \longmapsto (t - \sin t, 1 - \cos t)$

Next, we plot the graph of the curvature of **cycloid[1,1][t]** over the range $0 \leq t \leq 2\pi$:

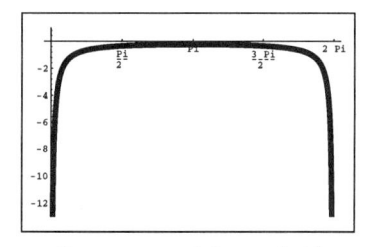

```
Plot[kappa2[
cycloid[1,1]][t]
//Evaluate,
{t,0,2Pi},Ticks->
{Range[0,2Pi,Pi/2],
Automatic}];
```

Curvature of the cycloid
$t \longmapsto (t - \sin t, 1 - \cos t)$

More generally, the curve traced by a point on a circle of radius b when a concentric circle of radius a rolls without slipping along a straight line is **cycloid[a, b]**. The cycloid is called a **prolate cycloid** if $b > a$ and a **curtate cycloid** if $a > b$.

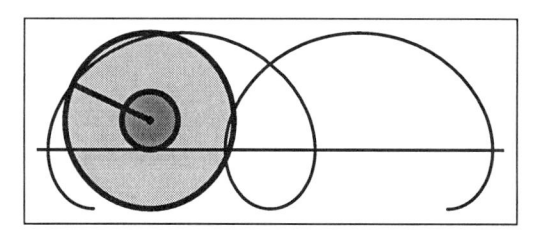

The prolate cycloid $t \longmapsto (t - 3 \sin t, 1 - 3 \cos t)$

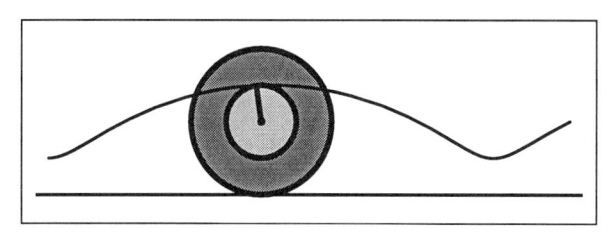

The curtate cycloid $t \longmapsto (2t - \sin t, 2 - \cos t)$

See exercise 1 for graphing of prolate and curtate cycloids. A discussion of cycloids is given in [Lem, Chapter 4] and [Wagon, Chapter 2]. Instructions for making cycloid movies are given in Section 3.10.

3.2 Lemniscates of Bernoulli

Frequently, it is useful to plot a family of curves on the same graph. For example, consider the family

(3.1) $$\text{lemniscate}[a](t) = \left(\frac{a \cos t}{1 + \sin^2 t}, \frac{a \sin t \cos t}{1 + \sin^2 t} \right),$$

or

```
lemniscate[a_][t_] := a{Cos[t],Sin[t]Cos[t]}/
                          (1 + Sin[t]^2)
```

Each curve in this family is called a **lemniscate of Bernoulli**.[6] Like an ellipse, a lemniscate has foci $\mathbf{F}_1$ and $\mathbf{F}_2$, but the lemniscate is the locus S of points $\mathbf{P}$ for which the *product* of distances from $\mathbf{F}_1$ and $\mathbf{F}_2$ is a certain constant f^2. More precisely,

$$S = \{ (x,y) \mid \text{distance}((x,y), \mathbf{F}_1) \text{distance}((x,y), \mathbf{F}_2) = f^2 \},$$

where **distance**$(\mathbf{F}_1, \mathbf{F}_2) = 2f$. Thus the constant f is chosen so that the midpoint of the segment connecting $\mathbf{F}_1$ with $\mathbf{F}_2$ lies on the curve S.

Let us derive (3.1). Let the two foci be $(\pm f, 0)$ and let S be a set of points containing $(0,0)$ such that the product of the distances from $\mathbf{F}_1 = (-f, 0)$ and $\mathbf{F}_2 = (f, 0)$ is the same for all $\mathbf{P} \in S$. Write $\mathbf{P} = (x, y)$. Then the condition that $\mathbf{P}$ lie on S is

(3.2) $$((x-f)^2 + y^2)((x+f)^2 + y^2) = f^4,$$

or equivalently

(3.3) $$(x^2 + y^2)^2 = 2f^2(x^2 - y^2).$$

---

[6] Jakob Bernoulli (1654-1705). Jakob and his brother Johann were the first of a Swiss mathematical dynasty. The work of the Bernoullis was instrumental in establishing the dominance of Leibniz's methods of exposition. Jakob Bernoulli laid basic foundations for the development of the calculus of variations, as well as working in probability, astronomy and mathematical physics. In 1694 Bernoulli studied the lemniscate named after him.

3.2 Lemniscates of Bernoulli

The substitutions $y = x \sin t$, $f = a/\sqrt{2}$ transform (3.3) into (3.1).

The command in *Mathematica* to draw the curves **lemniscate[a]** for $a = 2, 4, 6, 8$ simultaneously is:

```
ParametricPlot[Evaluate[Table[lemniscate[a][t],{a,2,8,2}]],
    {t,0,2Pi},AspectRatio->Automatic];
```

yielding

Lemniscates

Table creates a table of expressions for each of the values of a that we want, and then **ParametricPlot** displays them simultaneously. Thus

```
Table[lemniscate[a][t],{a,2,8,2}]
```

means "print out the expression **lemniscate[a][t]** as **a** goes from **2** to **8** with step **2**".

Here is the graph of the curvature of one of the lemniscates:

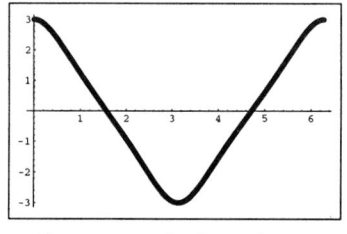

```
Plot[kappa2[
    lemniscate[1]][t]
    //Evaluate,
    {t,0,2Pi}];
```

Curvature of a lemniscate

Although a lemniscate of Bernoulli and an eight curve both resemble a figure eight (see page 32), a comparison of the graphs of their respective curvatures shows the difference between the two curves: the curvature of a lemniscate of Bernoulli has only one maximum and one minimum in the range $0 \leq t < 2\pi$, whereas an eight curve has three maxima, three minima, and two inflection points.

3.3 Cardioids

A **cardioid** is the locus of points traced out by a point on a circle of radius a which rolls without slipping on another circle of radius a; the parametric equation of a cardioid is

$$\mathbf{cardioid}[a](t) = \big(2a\cos t(1+\cos t), 2a\sin t(1+\cos t)\big),$$

or in *Mathematica*

```
cardioid[a_][t_]:=2a(1 + Cos[t]){Cos[t],Sin[t]}
```

The following command draws four cardioids:

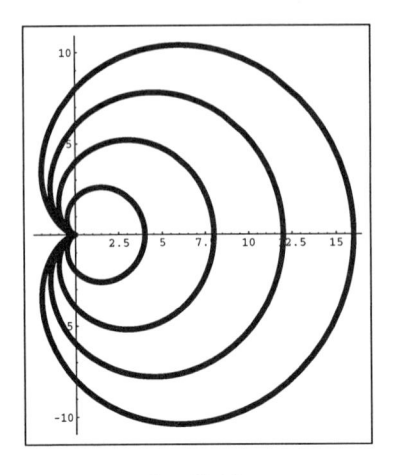

```
ParametricPlot[
Table[cardioid[a][t],
{a,1,4,1}]//Evaluate,
{t,0,2Pi},
AspectRatio->Automatic];
```

Cardioids

The command to graph the curvature of the cardioid with $a = 1$ is

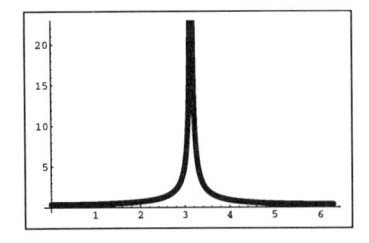

```
Plot[kappa2[
cardioid[1]][t]
//Evaluate,
{t,0,2Pi}];
```

Curvature of a cardioid

Mathematica computes the formula for the curvature of the cardioid as follows:

```
kappa2[cardioid[a]][t]//Simplify
```

$$\frac{3\,a^2\,\text{Cos}[\frac{t}{2}]^2}{8\,(a^2\,\text{Cos}[\frac{t}{2}]^2)^{3/2}}$$

We can simplify this formula by hand to get

$$\kappa 2[\text{cardioid}[a]](t) = \frac{3}{8\left|a\cos\left(\dfrac{t}{2}\right)\right|}.$$

3.4 The Catenary

In 1691 Jakob Bernoulli gave a solution to the problem of finding the curve assumed by a flexible inextensible cord hung freely from two fixed points; Leibniz has called such a curve a **catenary** (which stems from the Latin word *catena*, meaning chain). The solution is based on the differential equation

(3.4) $$\frac{d^2y}{dx^2} = \frac{1}{a}\sqrt{1 + \left(\frac{dy}{dx}\right)^2}.$$

To derive (3.4), we consider a portion $\overline{pq}$ of the cable between the lowest point **p** and an arbitrary point **q**. Three forces act on the cable: the weight of the portion $\overline{pq}$ and the tensions **T** and **U** at **p** and **q**. If w is the linear density and s is the length of **p** and **q**, then the weight of the portion $\overline{pq}$ is ws.

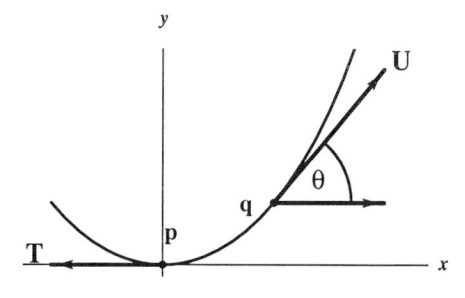

Definition of a Catenary

Let $|\mathbf{T}|$ and $|\mathbf{U}|$ denote the magnitudes of the forces $\mathbf{T}$ and $\mathbf{U}$, and write

$$\mathbf{U} = (|\mathbf{U}|\cos\theta, |\mathbf{U}|\sin\theta).$$

Because of equilibrium we have

(3.5) $\qquad |\mathbf{T}| = |\mathbf{U}|\cos\theta \quad \text{and} \quad ws = |\mathbf{U}|\sin\theta.$

Let $\mathbf{q} = (x,y)$, where x and y are functions of s. From (3.5) we obtain

(3.6) $\qquad \dfrac{dy}{dx} = \tan\theta = \dfrac{ws}{|\mathbf{T}|}.$

Since the length of $\overline{\mathbf{pq}}$ is

$$s = \int_0^x \sqrt{1 + \left(\dfrac{dy}{dx}\right)^2}\, dx,$$

the fundamental theorem of calculus tells us that

(3.7) $\qquad \dfrac{ds}{dx} = \sqrt{1 + \left(\dfrac{dy}{dx}\right)^2}.$

When we differentiate (3.6) with respect to x and use (3.7), we get (3.4).

Although at first glance the catenary looks like a parabola, it is in fact the graph of the hyperbolic cosine.

A solution of (3.4) is given by

$$y = a\cosh(x/a).$$

We rotate this graph and define

$$\mathbf{catenary}[a](t) = \left(a\cosh\left(\dfrac{t}{a}\right), t\right),$$

where without loss of generality, we assume that $a > 0$. The corresponding *Mathematica* definition is

```
catenary[a_][t_] := {a Cosh[t/a],t}
```

A catenary is one of the few curves for which it is easy compute the arc length function. To find it, we use

```
arclength[catenary[a]][s]//PowerExpand
```

obtaining

```
         s
a Sinh[ - ]
         a
```

Thus a unit-speed parametrization of a catenary is given by

```
catenaryunitspeed[a_][s_]:=
    a{Sqrt[1 + s^2/a^2],ArcSinh[s/a]}
```

In Chapters 20 and 30 we shall use the catenary to construct an important minimal surface called the catenoid.

3.5 The Cissoid of Diocles

The **cissoid of Diocles** is the curve defined nonparametrically by

(3.8) $$x^3 + xy^2 - 2ay^2 = 0.$$

To find a parametrization of the cissoid, we substitute $y = xt$ in (3.8) and obtain

$$x = \frac{2at^2}{1+t^2}, \qquad y = \frac{2at^3}{1+t^2}.$$

Thus we define

$$\mathbf{cissoid}[a](t) = \left(\frac{2at^2}{1+t^2}, \frac{2at^3}{1+t^2}\right),$$

or in *Mathematica*

```
cissoid[a_][t_] := 2a(1 + t^2){t^2,t^3}
```

The Greeks used the cissoid of Diocles to try to find solutions to the problems of doubling a cube and trisecting an angle. For more historical information and the definitions used by the Greeks and Newton see [BrKn, pages 9–12], [Gomes, volume 1, pages 1–25] and [Lock, pages 130–133]. Cissoid means "ivy-shaped".

Notice that **cissoid**$[a]'(0) = 0$ so that **cissoid** is not regular at 0. In fact, a cissoid has a cusp at 0, as can be seen by drawing it with the command

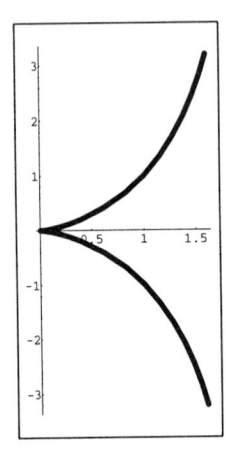

```
ParametricPlot[
  cissoid[1][t]//Evaluate,
  {t,-2,2},
  AspectRatio->2,
  Ticks->
   {Range[0,2,0.5],Automatic},
  PlotRange->All];
```

The cissoid

(The option `PlotRange->All` prevents *Mathematica* from clipping part of the plot.)

The definition of the cissoid used by the Greeks and by Newton can best be explained by considering a generalization of the cissoid. Let ξ and η be two curves and **A** a fixed point. Draw a line L through **A** cutting ξ and η at points **Q** and **R** respectively, and find a point **P** on L such that the distance from **A** to **P** equals the distance from **Q** to **R**. The locus of such points **P** is called the **cissoid of ξ and η with respect to A**.

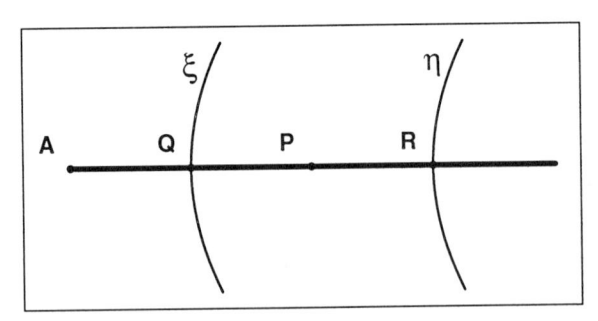

The cissoid of ξ and η with respect to the point A

Then $t \longmapsto \text{cissoid}[a](t)$ is precisely the cissoid of a circle of radius $\sqrt{a}/2$ and one of its tangent lines with respect to the point diametrically opposite to the tangent line, as the following picture shows:

3.5 The Cissoid of Diocles

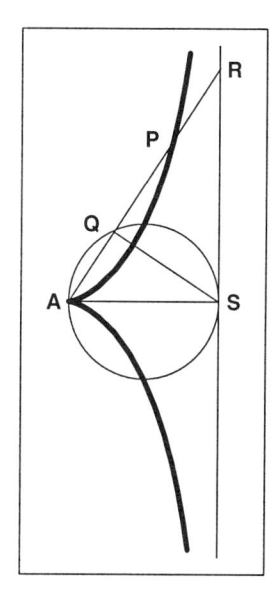

Definition of the cissoid:
As the line AR moves so that
distance(A,P) = distance(Q,R),
the point P traces out a cissoid.

Let us derive (3.8). Consider a circle of radius a centered at $(a,0)$. Let (x,y) be the coordinates of a point **P** on the cissoid. Then $2a = $ **distance(A,S)**, so by the Pythagorean theorem we have

(3.9) $\quad$ **distance(Q,S)**^{2} = **distance(A,S)**^{2} − **distance(A,Q)**^{2}

$\qquad\qquad\quad = 4a^2 - \bigl(\textbf{distance(A,R)} - \textbf{distance(Q,R)}\bigr)^2.$

The definition of the cissoid says that

$$\textbf{distance(Q,R)} = \textbf{distance(A,P)} = \sqrt{x^2 + y^2}.$$

Also, using similar triangles, we see that

$$\frac{\textbf{distance(A,R)}}{2a} = \frac{\textbf{distance(A,P)}}{x}.$$

Therefore, (3.9) becomes

(3.10) $\quad$ **distance(Q,S)**$^2 = 4a^2 - \left(\dfrac{2a}{x} - 1\right)^2 (x^2 + y^2).$

On the other hand,

(3.11) $$\text{distance}(Q,S)^2 = \text{distance}(R,S)^2 - \text{distance}(Q,R)^2$$
$$= \text{distance}(R,S)^2 - \text{distance}(A,P)^2$$
$$= \left(\frac{2ay}{x}\right)^2 - (x^2 + y^2),$$

since by similar triangles, $\text{distance}(R,S)/(2a) = y/x$. We equate the right-hand sides of (3.10) and (3.11). The resulting equation is easy enough to solve by hand, but we can also use *Mathematica's* command `Solve`. To the query

```
Solve[4a^2 - (2a/x-1)^2(x^2+y^2)==(y 2a/x)^2 - (x^2+y^2),y]
```

Mathematica answers

```
                3/2                   3/2
               x                     x
{{y ->  ---------------}, {y -> -(---------------)}}
         Sqrt[2 a - x]              Sqrt[2 a - x]
```

In both cases we obtain (3.8).

Mathematica tells us that the curvature of a cissoid is given by

```
Simplify[kappa2[cissoid[a]][t]]
```

```
               2  2
            3 a  t
------------------------------------
          2  2        2
       2 3  a  t  (4 + t ) 3/2
(1 + t )  (---------------------)
                    2 2
                (1 + t )
```

The *Mathematica* command `PowerExpand` can be used to cancel (incorrectly) all of the factors in this last expression:

```
PowerExpand[Simplify[kappa2[cissoid[a]][t]]]
```

```
         3
------------------
      2 3/2
  a t (4 + t )
```

In fact, the curvature of the cissoid is actually

$$\kappa 2[\text{cissoid}[a]](t) = \frac{3}{a|t|(4+t^2)^{3/2}}.$$

That the curvature of the cissoid is always positive is clear from the expression that *Mathematica* gave us before we used `PowerExpand` and is confirmed visually when we plot it with the command

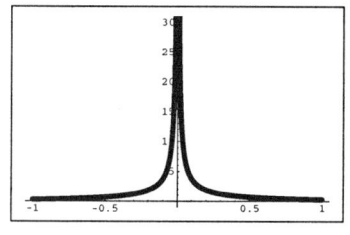

```
Plot[
kappa2[
cissoid[1]][t]
//Evaluate,
{t,-1,1}];
```

Curvature of the cissoid

3.6 The Tractrix

A **tractrix** is a curve α passing through the point $A = (a, 0)$ on the horizontal axis with the property that the length of the segment of the tangent line from any point on the curve to the vertical axis is constant.

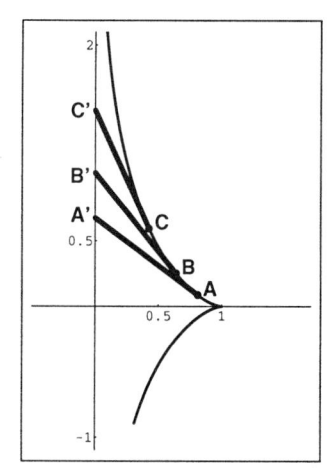

Definition of the tractrix: tangent segments AA', BB', CC' ... have equal lengths

The German word for tractrix is the more descriptive **Hundekurve**. It is the path that an obstinate dog takes when his master walks along a north-south path.

One way to parametrize the tractrix is:

$$(3.12) \qquad \mathbf{tractrix}[a](t) = a\left(\sin t, \cos t + \log\left(\tan\left(\frac{t}{2}\right)\right)\right),$$

or in *Mathematica*

```
tractrix[a_][t_]:= a{Sin[t],Cos[t] + Log[Tan[t/2]]}
```

It approaches the vertical axis asymptotically as $t \longmapsto 0$ or $t \longmapsto \pi$, and has a cusp at $t = \pi/2$.

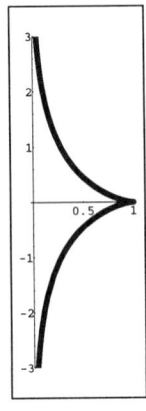

```
ParametricPlot[
 tractrix[1][t]
 //Evaluate,
 {t,0.01,0.99Pi},
 AspectRatio->3,
 PlotPoints->50,
 PlotRange->
  {Automatic,{-3.0,3.0}},
 Ticks->{Range[0,1,0.5],
 Automatic}];
```

The tractrix

To find the differential equation of a tractrix, write **tractrix**$[a](t) = (x(t), y(t))$. Then dy/dx is the slope of the curve, which we recognize as

$$-\frac{\sqrt{a^2 - x^2}}{x}.$$

Thus the differential equation is

$$\frac{dy}{dx} = -\frac{\sqrt{a^2 - x^2}}{x}.$$

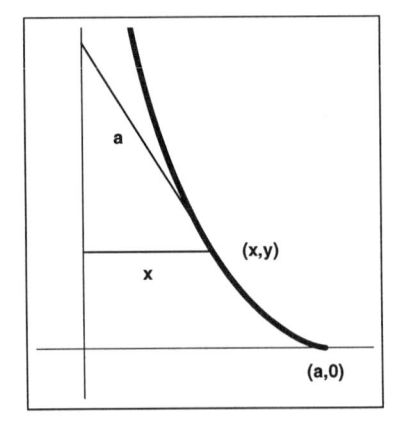

It is easy to check that this differential equation is satisfied by $x(t) = a \sin t$ and $y(t) = a(\cos t + \log(\tan(t/2)))$.

Just as with the cissoid, *Mathematica* almost gives the correct formula for the curvature of the tractrix; the only thing wrong is the sign. The correct formula for the curvature is

$$\kappa 2[\text{tractrix}[1]](t) = -|\tan t|.$$

3.6 The Tractrix

The following plot shows that for the tractrix the curvature is $-\infty$ at $t = 0$.

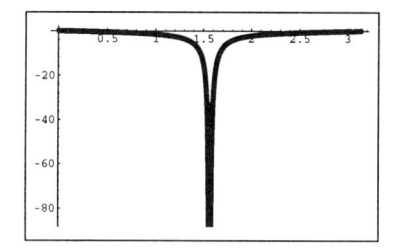

```
Plot[Evaluate[
  kappa2[tractrix[1]][t]],
  {t,0,Pi}];
```

Curvature of a tractrix

For future use let us find a unit-speed parametrization of the tractrix.

Lemma 3.1. *A unit-speed parametrization of the tractrix is given by*

$$(3.13) \quad \alpha(v) = \begin{cases} \left(ae^{-v/a}, \int_0^v \sqrt{1 - e^{-2t/a}}\, dt\right) & \text{for } 0 \leq v < \infty, \\ \left(ae^{v/a}, \int_0^v \sqrt{1 - e^{2t/a}}\, dt\right) & \text{for } -\infty < v \leq 0. \end{cases}$$

(*Note that*

$$\int_0^v \sqrt{1 - e^{-2t/a}}\, dt = a\,\text{arctanh}\left(\sqrt{1 - e^{-2v/a}}\right) - a\sqrt{1 - e^{-2v/a}}.)$$

Proof. First, let us note that

$$(3.14) \quad \mathbf{tractrix}[a]'(\phi) = a\left(\cos\phi, -\sin\phi + \frac{\sec^2(\phi/2)}{2\tan(\phi/2)}\right)$$

$$= a\left(\cos\phi, -\sin\phi + \frac{1}{\sin\phi}\right).$$

For $0 \leq v < \infty$ we define $\phi(v) = \pi - \arcsin(e^{-v/a})$. Then $\sin\phi(v) = e^{-v/a}$; furthermore, $0 \leq v < \infty$ implies $\pi/2 \leq \phi(v) < \pi$, so that $\cos\phi(v) = -\sqrt{1 - e^{-2v/a}}$. Hence

$$(3.15) \quad \phi'(v) = \frac{e^{-v/a}}{a\sqrt{1 - e^{-2v/a}}} = -\frac{\sin\phi(v)}{a\cos\phi(v)}.$$

Therefore, if we define a curve β by $\beta(v) = \mathbf{tractrix}[a](\phi(v))$, it follows from (3.14) and (3.15) that

$$\beta'(v) = \mathbf{tractrix}[a]'(\phi(v))\phi'(v)$$

$$= a\left(\cos\phi(v), -\sin\phi(v) + \frac{1}{\sin\phi(v)}\right)\left(-\frac{\sin\phi(v)}{a\cos\phi(v)}\right)$$

$$= \left(-\sin\phi(v), -\cos\phi(v)\right) = \left(-e^{-v/a}, \sqrt{1-e^{-2v/a}}\right) = \alpha'(v).$$

Also, $\beta(0) = (a,0) = \alpha(0)$. Thus α and β coincide for $0 \leq v < \infty$, so that α is a reparametrization of a tractrix in that range. The proof that α is a reparametrization of **tractrix[a]** for $-\infty < v \leq 0$ is similar. Finally, an easy calculation shows that α has unit-speed. ∎

When dealing with real-valued functions in *Mathematica*, it is frequently advantageous to use **Sqrt[x^2]** to represent $|x|$, instead of **Abs[x]**. The reason is that *Mathematica* knows how to differentiate **Sqrt[x^2]**, but not **Abs[x]**. For example, a unit-speed parametrization of a tractrix can be defined in *Mathematica* by

```
tractrixunitspeed[a_][s_]:= {a E^(-Sqrt[s^2]/a),
     a ArcTanh[Sqrt[1 - E^(-2Sqrt[s^2]/a)]] -
     a Sqrt[1 - E^(-2Sqrt[s^2]/a)]}
```

3.7 Clothoids

One of the most elegant of all plane curves is the **clothoid** or **spiral of Cornu**.[7] We give a generalization of the clothoid by defining

$$\mathbf{clothoid}[n, a](t) = a\left(\int_0^t \sin\left(\frac{u^{n+1}}{n+1}\right)du, \int_0^t \cos\left(\frac{u^{n+1}}{n+1}\right)du\right).$$

Clothoids are important curves used in freeway and railroad construction (see [Higg] and [Roth]). For example, a clothoid is needed to make the gradual transition from a highway, which has zero curvature, to the midpoint of a freeway exit, which has nonzero curvature. A clothoid is clearly preferable to a path consisting of straight lines and circles, for which the curvature is discontinuous. The standard clothoid is the first clothoid **clothoid[1, a]**. The integrals

$$\int_0^t \sin(\pi u^2/2)du \quad \text{and} \quad \int_0^t \cos(\pi u^2/2)du,$$

---

[7] Marie Alfred Cornu (1841-1902). French scientist, who studied the clothoid in connection with diffraction. The clothoid was also known to Euler and Jakob Bernoulli. See [Gomes, volume 2, page 102–107] and [Law, page 190].

3.7 Clothoids

are called **Fresnel**[8] **integrals**; clothoid[1, a] is expressible in terms of these functions. In *Mathematica* the clothoid is given by

```
clothoidprime[n_,a_][t_]:=
    a{Sin[t^(n + 1)/(n + 1)],Cos[t^(n + 1)/(n + 1)]}

clothoid[n_,a_][t_]:=
    Integrate[clothoidprime[n,a][tt],{tt,0,t}]
```

Mathematica has internal commands **FresnelS** and **FresnelC** that define the Fresnel integrals. In fact, **clothoid[1,a]** will be converted automatically by *Mathematica* into an equivalent expression using these two functions:

clothoid[1,a][t]

$$\{a\ \text{Sqrt[Pi] FresnelS}[\frac{t}{\text{Sqrt[Pi]}}],\ a\ \text{Sqrt[Pi] FresnelC}[\frac{t}{\text{Sqrt[Pi]}}]\}$$

Here is the plot of the first clothoid:

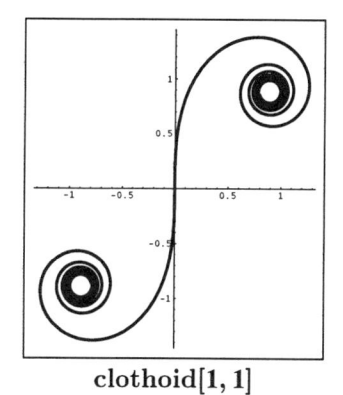

```
ParametricPlot[
clothoid[1,1][t]
//Evaluate,
{t,-10,10},
AspectRatio->
Automatic];
```

clothoid[1, 1]

Since

$$\int_0^{\pm\infty} \sin(u^2/2)du = \int_0^{\pm\infty} \cos(u^2/2)du = \pm\frac{\sqrt{\pi}}{2}$$

---

[8] Augustin Jean Fresnel (1788-1827). French physicist, one of the founders of the wave theory of light.

(as is easily checked with **Mathematica**), the ends of the first clothoid curl around the points $\pm(a\sqrt{\pi}/2, a\sqrt{\pi}/2)$.

The first clothoid is symmetric with respect to the origin, but the second clothoid is symmetric with respect to the horizontal axis:

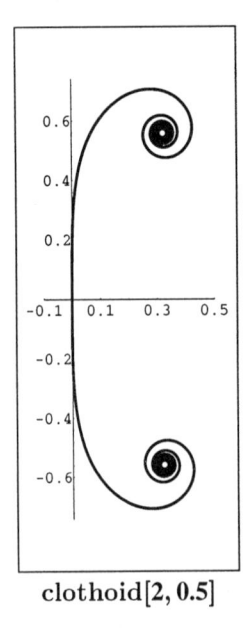

```
ParametricPlot[
Evaluate[
clothoid[2,0.5][t]],
{t,-6,6},
AspectRatio->
Automatic,
Ticks->
{Range[-0.1,0.5,0.2],
Automatic},
PlotRange->
{{-0.1,0.5},
Automatic}]
```

clothoid[2, 0.5]

The odd clothoids have shapes similar to **clothoid[1, a]**, while the even clothoids have shapes similar to **clothoid[2, a]**. Although the definition of **clothoid[n, a]** is quite complicated, its curvature is simple. *Mathematica* can be used to check that

$$\kappa 2[\text{clothoid}[n,a]](t) = -\frac{t^n}{a}.$$

In Chapter 6 we show how to define the clothoid as a numerical solution to a differential equation. See also exercise 9.

3.8 Pursuit Curves

The problem of pursuit probably originated with Leonardo da Vinci. It is to find the curve by which a vessel moves while pursuing another vessel, supposing that the speeds of the two vessels are always in the same ratio. Let us formulate this problem mathematically.

3.8 Pursuit Curves

Definition. *Let α and β be plane curves parametrized on an interval $a < t < b$. We say that α is a **pursuit curve** of β provided that*

(i) *the velocity vector $\alpha'(t)$ points towards the point $\beta(t)$ for $a < t < b$; that is, $\alpha'(t)$ is a multiple of $\alpha(t) - \beta(t)$;*

(ii) *the speeds of α and β are related by $\|\alpha'\| = k\|\beta'\|$, where k is a positive constant. We call k the **speed ratio**.*

A **capture point** *is a point $\mathbf{p}$ for which $\mathbf{p} = \alpha(t_1) = \beta(t_1)$ for some t_1.*

In the diagram below α is the curve of the pursuer, and β is the curve of the pursued.

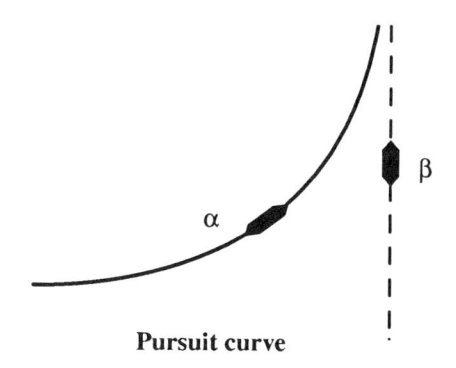

Pursuit curve

When the speed ratio k is larger than 1, the pursuer travels faster than the pursued. Although this would usually be the case in a physical situation, it is not a necessary assumption for the mathematical analysis of the problem.

We derive differential equations for pursuit curves in terms of coordinates.

Lemma 3.2. *Write $\alpha = (x, y)$ and $\beta = (f, g)$, and assume that α is a pursuit curve of β. Then*

(3.16) $$x'^2 + y'^2 = k^2(f'^2 + g'^2)$$

and

(3.17) $$x'(y - g) - y'(x - f) = 0.$$

Proof. Equation (3.16) is the same as $\|\alpha'\| = k\|\beta'\|$. To prove (3.17), we observe that $\alpha(t) - \beta(t) = \big(x(t) - f(t), y(t) - g(t)\big)$ and $\alpha'(t) = \big(x'(t), y'(t)\big)$. Note that the vector $\big(-y(t) + g(t), x(t) - f(t)\big)$ is perpendicular to $\alpha(t) - \beta(t)$. The condition that $\boldsymbol{\alpha}'(t)$ is a multiple of $\boldsymbol{\alpha}(t) - \boldsymbol{\beta}(t)$ is conveniently expressed by saying that $\boldsymbol{\alpha}'(t)$ is perpendicular to $\bigl(-y(t) + g(t), x(t) - f(t)\bigr)$, that is,

$$\begin{aligned}
(3.18) \quad 0 &= \bigl(x'(t), y'(t)\bigr) \cdot \bigl(-y(t) + g(t), x(t) - f(t)\bigr) \\
&= x'(t)\bigl(-y(t) + g(t)\bigr) + y'(t)\bigl(x(t) - f(t)\bigr).
\end{aligned}$$

Then (3.18) is equivalent to (3.17). ∎

Next, we specialize to the case when the curve of the pursued is a straight line. Assume that the curve $\boldsymbol{\beta}$ of the pursued is a vertical straight line passing through the point $(a, 0)$, and that the speed ratio k is larger than 1. We want to find the curve $\boldsymbol{\alpha}$ of the pursuer, assuming the initial conditions $\boldsymbol{\alpha}(0) = (0,0)$ and $\boldsymbol{\alpha}'(0) = (1,0)$.

We can parameterize $\boldsymbol{\beta}$ as

$$\boldsymbol{\beta}(t) = \bigl(a, g(t)\bigr).$$

Furthermore, the curve $\boldsymbol{\alpha}$ of the pursuer can be parametrized as

$$\boldsymbol{\alpha}(t) = \bigl(t, y(t)\bigr).$$

The condition (3.16) becomes

$$(3.19) \qquad 1 + y'^2 = k^2 g'^2,$$

and (3.17) reduces to

$$(3.20) \qquad (y - g) - y'(t - a) = 0.$$

Differentiation of (3.20) with respect to t yields

$$(3.21) \qquad -y''(t - a) = g'.$$

From (3.19) and (3.21) we get

$$(3.22) \qquad 1 + y'^2 = k^2 (a - t)^2 y''^2.$$

Let $p = y'$; then (3.22) can be rewritten as

$$(3.23) \qquad \frac{k\, dp}{\sqrt{1 + p^2}} = \frac{dt}{a - t}.$$

This separable first-order equation has the solution

$$(3.24) \qquad \operatorname{arcsinh}(p) = -\frac{1}{k} \log\left(\frac{a - t}{a}\right) = \log\left(\left(\frac{a-t}{a}\right)^{-1/k}\right),$$

3.8 Pursuit Curves

when we make use of the initial condition $y'(0) = 0$. Then (3.24) can be rewritten as

$$y' = p = \sinh(\operatorname{arcsinh}(p)) = \frac{1}{2}\left(e^{\operatorname{arcsinh}(p)} - e^{-\operatorname{arcsinh}(p)}\right)$$

(3.25)
$$= \frac{1}{2}\left(\left(\frac{a-t}{a}\right)^{-1/k} - \left(\frac{a-t}{a}\right)^{1/k}\right).$$

Integration of (3.25), making use of the initial condition $y(0) = 0$, yields

(3.26) $$y = \frac{ak}{k^2 - 1} + \frac{1}{2}\left(\frac{ak}{k+1}\left(\frac{a-t}{a}\right)^{1+1/k} - \frac{ak}{k-1}\left(\frac{a-t}{a}\right)^{1-1/k}\right).$$

The curve of the pursuer is then $\boldsymbol{\alpha}(t) = (t, y(t))$, where y is given by (3.26). Since $\boldsymbol{\alpha}(t_1) = \boldsymbol{\beta}(t_1)$ if and only if $t_1 = a$, the capture point is

(3.27) $$\mathbf{p} = \left(a, \frac{ak}{k^2 - 1}\right). \blacksquare$$

The graph below depicts the case when $a = 1$ and k has the values 1.5, 2.0, 2.5, 3.0 and 3.5. As the speed ratio k becomes smaller and smaller, the capture point goes higher and higher.

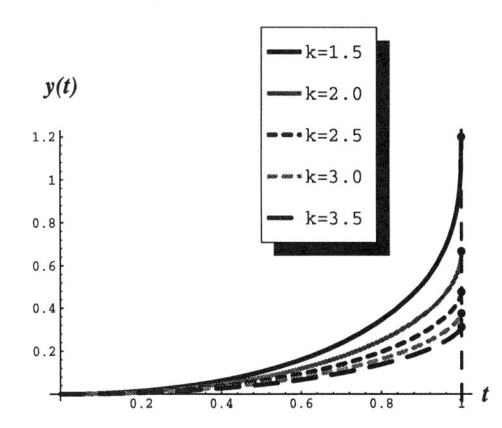

Pursuit curves for the case when the pursued moves in a straight line

3.9 Exercises

1. Plot as one graph the cycloids **cycloid[a, b]** with $a = 1$ and $b = 0.5, 1, 2$; also graph the curvatures of these cycloids. Use *Mathematica* to find the formula for the curvature $\kappa 2$ of the general cycloid **cycloid[a, b]**. Then use *Mathematica* to define and draw ordinary, prolate and curtate cycloids together with the defining circle such as those on page 51.

2. A **deltoid** is defined by

$$\text{deltoid}[a](t) = \big(2a\cos t(1 + \cos t) - a, 2a\sin t(1 - \cos t)\big),$$

or in *Mathematica* by

```
deltoid[a_][t_] := {2a Cos[t](1 + Cos[t]) - a,
                    2a Sin[t](1 - Cos[t])}
```

The curve is so named because it resembles a Greek capital delta. Plot as one graph the deltoids **deltoid[a]** for $a = 1, 2, 3, 4$. Graph the curvature of the first deltoid.

3. The **Lissajous**[9] or **Bowditch**[10] **curve** is defined by

$$\text{lissajous}[n, d, a, b](t) = (a\sin(n\,t + d), b\sin t)$$

or in *Mathematica* by

[9] Jules Antoine Lissajous (1822-1880). French physicist, who studied similar curves in 1857 in connection with his optical method for studying vibrations.

[10] Nathaniel Bowditch (1773-1838). American mathematician and astronomer. His **New American Practical Navigator**, written in 1802, was highly successful. Bowditch also translated and annotated **Laplace's Mécanique Céleste**. His study of pendulums in 1815 included the figures named after him. Preferring his post as president of the Essex Fire and Marine Insurance Company from 1804 to 1823, Bowditch refused chairs of mathematics at several universities.

3.9 Exercises

```
lissajous[n_,d_,a_,b_][t_]:= {a Sin[n t + d],b Sin[t]}
```

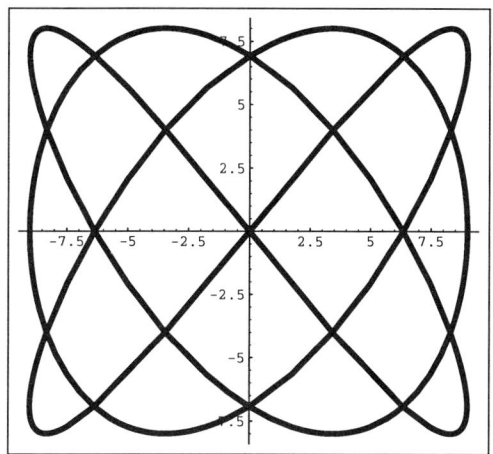

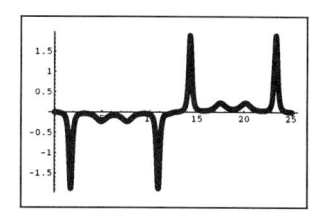

Lissajous curve

Curvature of a Lissajous curve

Draw several of these curves and plot their curvatures.

4. The **limaçon**, sometimes called **Pascal's snail**, named after Étienne Pascal, father of Blaise Pascal[11], is a generalization of the cardioid. It is defined by

$$\text{limacon}[a, b](t) = (2a\cos t + b)(\cos t, \sin t).$$

A limaçon is defined in *Mathematica* by

```
limacon[a_,b_][t_]:= (2a Cos[t] + b){Cos[t],Sin[t]}
```

---

[11] Blaise Pascal (1623-1662). French mathematician, philosopher and inventor. Pascal was an early investigator in projective geometry and invented the first mechanical device for performing addition and subtraction.

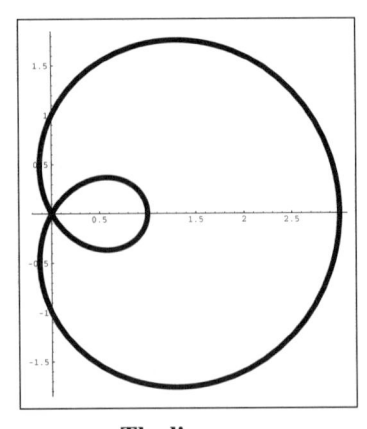

The limaçon

$$t \longmapsto (2\cos t + 1)(\cos t, \sin t)$$

Find the formula for the curvature of **limacon[a, b]** and plot several limaçons.

5. Consider a circle with center $\mathbf{C} = (0, a)$ and radius a. Let $\boldsymbol{L}$ be the line tangent to the circle at $(0, 2a)$. A line from the origin $\mathbf{O} = (0, 0)$ intersecting $\boldsymbol{L}$ at a point $\mathbf{A}$ intersects the circle at a point $\mathbf{Q}$. Let x be the first coordinate of $\mathbf{A}$ and y the second coordinate of $\mathbf{Q}$, and put $\mathbf{P} = (x, y)$. As $\mathbf{A}$ varies along $\boldsymbol{L}$ the point $\mathbf{P}$ traces out a curve called in Italian **versiera**, and misnamed in English as the **witch of Agnesi**.[12]

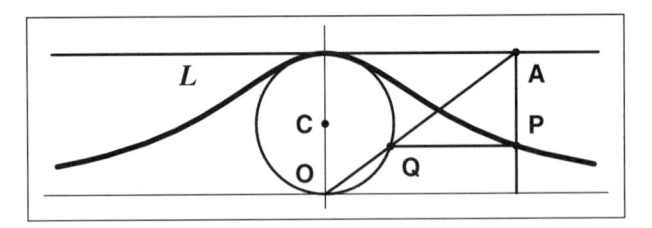

Definition of the witch of Agnesi

---

[12] Maria Gaetana Agnesi (1718-1799). Professor at the University of Bologna. She was the first woman to occupy a chair of mathematics. Her widely used calculus book **Instituzioni Analitiche** was translated into French and English.

3.9 Exercises

6. Use *Mathematica* to draw the picture on page 59 that defines the cissoid.

7. Use *Mathematica* to draw the picture on page 61 that defines the tractrix.

8. Define a curve in *Mathematica* by

```
tschirnhausen[n_,a_][t_]:={a Cos[t]/Cos[t/3]^n,
                          a Sin[t]/Cos[t/3]^n}
```

When $n = 1$, this curve is called **Tschirnhausen's**[13] **cubic**. Find the formula for the curvature of **tschirnhausen[n,a][t]** and make a simultaneous plot of the curves for $1 \leq n \leq 8$.

9. The plotting of **clothoid** can be extremely slow because *Mathematica* uses numerical integration. If we use a different definition of the clothoid (which calls *Mathematica's* **NDSolve**), the curve is plotted much more quickly. We define **clothoidnds** by

```
clothoidnds[smin_,smax_][n_,a_][s_]:=
    Module[{tmp,ss},
       tmp=NDSolve[{x'[ss]==clothoidprime[n,a][ss][[1]],
           y'[ss]==clothoidprime[n,a][ss][[2]],
           x[0]==0,y[0]==0},{x,y},{ss,smin,smax}];
       First[{x[s],y[s]} /. tmp]]
```

Since numerical solution of a differential equation requires the specification of an interval, **clothoidnds** will only be defined on the interval **{smin, smax}**.

Plot **clothoidnds[-6,6][2,0.5]** and **clothoidnds[-4,4][3,1]**.

10. Use *Mathematica* to find the points around which the second clothoid curls.

---

[13]Ehrenfried Walter Tschirnhausen (1651-1708). German mathematician, who tried to solve equations of any degree by removing all terms except the first and last. He contributed to the rediscovery of the process for making hard-paste porcelain. Sometimes the name is written von Tschirnhaus.

11. In the case that the speed ratio is 1, show that the equation for the pursuit curve reduces to

$$y = \frac{a}{4}\left(\left(\frac{a-t}{a}\right)^2 - 1 - 2\log\left(\frac{a-t}{a}\right)\right),$$

and that the pursuer never catches the pursued.

12. Equation (3.26) can be used to define a function in *Mathematica* by

```
linearpursuit[a_,k_][t_]:= (a k)/(k^2 - 1) +
    (k(a - t)(a - t)^k^(-1))/(2a^(1/k)(1 + k)) -
    (a^(1/k)k(a - t)^(1 - 1/k))/(2(k - 1))
```

Use this function to plot a pursuit curve with $a = 1$ and $k = 1.2$.

3.10 Animation of a Prolate Cycloid

To create a rolling prolate cycloid, we first define a blue wheel at position t with

```
wheel[t_]:={{RGBColor[0.8,0.8,1],Disk[{t,1},3]},
{RGBColor[0.5,0.5,1],Disk[{t,1},1]},
{AbsolutePointSize[3],Point[{t,1}]},
{AbsoluteThickness[2],Circle[{t,1},1],
Circle[{t,1},3],
Line[{{t,1},cycloid[1,3][t]}]},
{AbsoluteThickness[1],Line[{{-3Pi,0},{3Pi,0}}]}}
```

Next, we define a plot-valued function which draws the prolate cycloid together with the wheel at position a.

```
proplot[a_]:=ParametricPlot[cycloid[1,3][t]//Evaluate,
{t,-3Pi,3Pi},PlotStyle->{{AbsoluteThickness[1]}},
AspectRatio->Automatic,
PlotRange->{{-4Pi,4Pi},Automatic},Axes->None,
Prolog->wheel[a]]
```

The command to create an animation of the wheel generating a prolate cycloid is:

```
Do[proplot[a],{a,-3Pi,3Pi,Pi/32}]
```

4

ALTERNATE METHODS FOR PLOTTING PLANE CURVES

So far we have been using `ParametricPlot` to plot plane curves, but in order to plot a curve when we know its nonparametric form, we must first find a parametrization of the curve. In Section 4.1 we discuss implicitly defined curves and contrast them with parametrically defined curves. Then we show how to use the *Mathematica* command `ImplicitPlot` to plot nonparametrically defined curves without first parametrizing them. Cassinian ovals, which we consider in Section 4.2, are easily plotted with `ImplicitPlot`.

We show how to use polar coordinates to plot curves and compute their curvatures in Section 4.3. An animation of Fermat's spiral is described in Section 4.5.

4.1 Implicitly Defined Curves in $\mathbb{R}^2$

In this section we consider an alternative to parametrically defined curves.

Definition. *Let* $F: \mathbb{R}^2 \longrightarrow \mathbb{R}$ *be a not necessarily continuous function. The* **set of zeros** *of* F *is the set*

$$F^{-1}(0) = \{\, \mathbf{p} \in \mathbb{R}^2 \mid F(\mathbf{p}) = 0 \,\}.$$

If no restrictions are placed on F, then not much can be said about the set of zeros of F, so we make an additional assumption.

Definition. *An* **implicitly defined curve** *in $\mathbb{R}^2$ is the set of zeros of a differentiable function $F\colon \mathbb{R}^2 \longrightarrow \mathbb{R}$. Frequently, we refer to the set of zeros as $F^{-1}(0)$ or "the curve $F(x,y) = 0$".*

Even when F is assumed to be differentiable, the set of zeros of F may have cusps, and hence appear to be nondifferentiable. Moreover, a theorem of Whitney[1] states that any closed subset of $\mathbb{R}^2$ is the set of zeros of some differentiable function (see [BrLa, page 56]). However, there is an important case when it is possible to find a parametrized curve whose trace is the set of zeros of F.

Theorem 4.1. *Let $F\colon \mathbb{R}^2 \longrightarrow \mathbb{R}$ be a differentiable function, and let $\mathbf{q} = (q_1, q_2)$ be a point such that $F(\mathbf{q}) = 0$. Assume that at least one of the partial derivatives F_u, F_v is nonzero at $\mathbf{q}$. Then there is a small neighborhood $\mathcal{U}$ of $\mathbf{q}$ in $\mathbb{R}^2$ and a parametrized curve $\boldsymbol{\alpha}\colon (a,b) \longrightarrow \mathbb{R}^2$ such that the trace of $\boldsymbol{\alpha}$ is precisely*

$$\{\,\mathbf{p} \in \mathcal{U} \mid F(\mathbf{p}) = 0\,\}.$$

Proof. Suppose, for example, that $F_v(\mathbf{q}) \neq 0$. The implicit function theorem from calculus states that there is a differentiable real-valued function g defined on a small neighborhood of q_1 in $\mathbb{R}$ such that $g(q_1) = q_2$ and $t \longmapsto F(t, g(t))$ vanishes identically. Then we define $\boldsymbol{\alpha}(t) = (t, g(t))$. ∎

Definition. *Let S be a subset of $\mathbb{R}^2$. If $F\colon \mathbb{R}^2 \longrightarrow \mathbb{R}$ is a differentiable function whose set of zeros is S, we say that the equation $F(x, y) = 0$ is a* **nonparametric form** *of S. If $\boldsymbol{\alpha}\colon (a, b) \longrightarrow \mathbb{R}^2$ is a curve whose trace is S, we say that $t \longmapsto \boldsymbol{\alpha}(t)$ is a* **parametrization** *or* **parametric form** *of S.*

For example, let $\mathbf{p}, \mathbf{v} \in \mathbb{R}^2$ with $\mathbf{v} \neq 0$, and define functions $F[\mathbf{p}, \mathbf{v}], G[\mathbf{p}, \mathbf{v}]\colon \mathbb{R}^2 \longrightarrow \mathbb{R}$ by

$$F[\mathbf{p}, \mathbf{v}](\mathbf{q}) = (\mathbf{q} - \mathbf{p}) \cdot \mathbf{v} \quad \text{and} \quad G[\mathbf{p}, \mathbf{v}](\mathbf{q}) = \|\mathbf{q} - \mathbf{v}\|^2 - \|\mathbf{p} - \mathbf{v}\|^2.$$

---

[1] Hassler Whitney (1907-1989). Influential American differential-topologist.

4.1 Implicitly Defined Curves in $\mathbb{R}^2$

Then $F[\mathbf{p}, \mathbf{v}]^{-1}(0)$ is the straight line through $\mathbf{p}$ perpendicular to $\mathbf{v}$ and $G[\mathbf{p}, \mathbf{v}]^{-1}(0)$ is the circle with center $\mathbf{v}$ that contains $\mathbf{p}$.

The following *Mathematica* command will attempt to convert the parametric form of a curve α to the implicit form. The command will usually fail if the components of α are not rational functions.

```
paramtoimp[alpha_,{x_,y_}]:=
    Eliminate[{x,y} == alpha[t],t]
```

We have used *Mathematica's* command **Eliminate** to eliminate variables between a set of simultaneous equations.

For example, let us find the nonparametric form of the cissoid. The command

```
paramtoimp[cissoid[a],{x,y}]
```

yields

$$x^3 + x y^2 == 2 a y^2$$

Thus the nonparametric equation of the cissoid is, as expected, $x^3 = (2a - x)y^2$.

It can happen that a disconnected set is the set of zeros of a differentiable function $F: \mathbb{R}^2 \longrightarrow \mathbb{R}$. On the other hand, the trace of a curve $\alpha: (a, b) \longrightarrow \mathbb{R}^2$ is always connected. (This is a special case of the theorem from topology that states that the image under a continuous map of a connected set remains connected.) So the notions of parametric form and nonparametric form are not the same.

For example, consider the family F_ε of functions defined by

$$F_\varepsilon(x, y) = x^3 + y^3 - 3xy - \varepsilon = 0.$$

Then the set of zeros of F_0 is the **folium of Descartes**:[2]

---

[2] It was Huygens who first drew the curve correctly. See [Still1, pages 67–68].

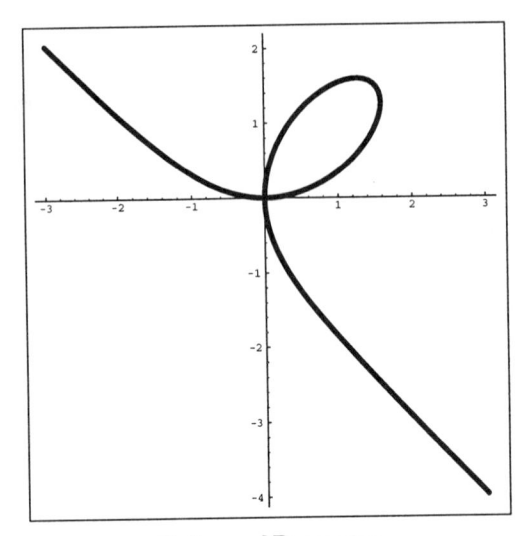

Folium of Descartes
$$x^3 + y^3 = 3xy$$

But the set of zeros of $F_{-0.1}$ is disconnected:

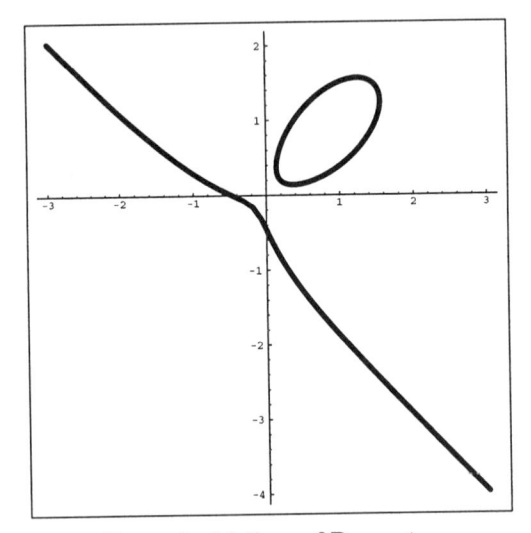

Perturbed folium of Descartes

Versions of *Mathematica* 2.0 and greater have a package for plotting the sets of zeros of many functions $F \colon \mathbb{R}^2 \longrightarrow \mathbb{R}$; it may be read into a *Mathematica* session with

4.1 Implicitly Defined Curves in $\mathbb{R}^2$

the command

Needs["Graphics`ImplicitPlot`"]

There are two ways that the command **ImplicitPlot** inside this package can be used. The first method attempts to plot the curve $F(x, y) = a$ by using *Mathematica's* command **Solve** to find solutions of $F(x, y) = a$ at each point in the x range. Because some equations may be too complicated symbolically for *Mathematica*, this version of **ImplicitPlot** does not always work. But there are many interesting curves that **ImplicitPlot** handles very nicely using this method, making it a very useful command. The perturbed folium was drawn with the command

ImplicitPlot[x^3 + y^3 - 3 x y == -0.1, {x,-3,3}];

This syntax is typical: the equation, an x-range and possibly some plotting options need to be specified. Any variable **z** (other than **x**) can be used for **y**, provided **z** has not already been assigned a value. This first version of **ImplicitPlot** produces a type of graphics called simply **Graphics**.[3]

The second version of **ImplicitPlot** produces a different type of graphics, called **ContourGraphics**. In contrast to the first version of **ImplicitPlot**, this second version always works. However, the resulting plot may be more jagged than a plot produced using the first version. For example:

**ImplicitPlot[(y - x^2)(y - 2x^2) == 0,{x,-1,1},{y,-1,1},
    PlotPoints->100];**

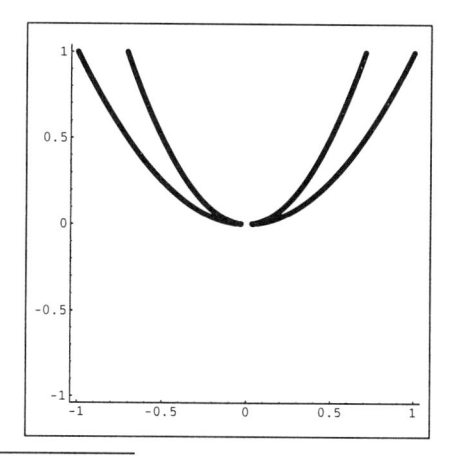

---

[3] *Mathematica's* graphics programs produce several different kinds of graphics. In addition to **Graphics** and **ContourGraphics**, we shall come across **SurfaceGraphics** and **Graphics3D**.

The syntax for the second version of **ImplicitPlot** is the same as the first version, except that a *y*-range is also specified.

Since we have introduced the folium of Descartes, let us study it in a little more detail. It is easy to find a parametrization of the folium of Descartes:

$$\mathbf{folium}(t) = \left(\frac{3t}{1+t^3}, \frac{3t^2}{1+t^3}\right),$$

or in *Mathematica*

```
folium[t_]:= {3 t/(1 + t^3),3 t^2/(1 + t^3)}
```

The *Mathematica* command to draw the parametric version of the folium of Descartes does it incorrectly (the asymptote is not part of the curve):

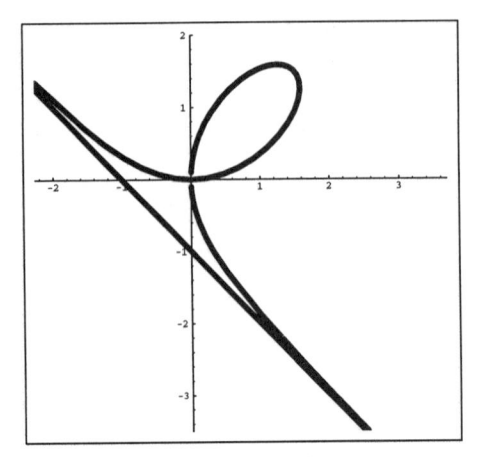

Folium of Descartes with asymptote plotted parametrically

```
foliumgraph=ParametricPlot[folium[t]//Evaluate,
{t,-30,30},PlotRange->{Automatic,{-3.5,2}},
AspectRatio->Automatic];
```

To get rid of the asymptote, we use[4]

```
foliumna=foliumgraph //. {x___,{a_,b_},{c_,d_},y___} /;
Negative[b d] -> {x,Null,Null,y};Show[foliumna];
```

---

[4]This trick is due to Michael Mezzino.

4.1 Implicitly Defined Curves in $\mathbb{R}^2$

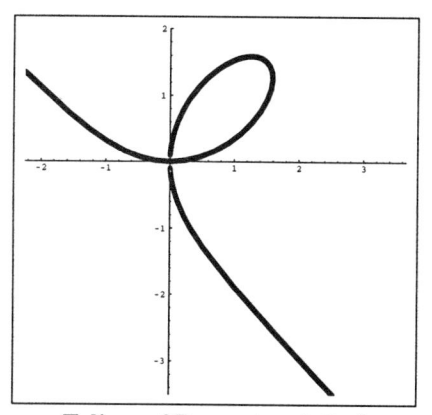

**Folium of Descartes plotted
parametrically without asymptote**

To understand the curvature of the folium of Descartes, we plot it first over the range $-30 < t < 30$ and then over the range $-3 < t < 3$.

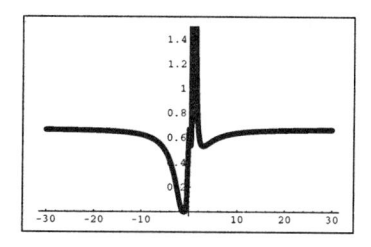

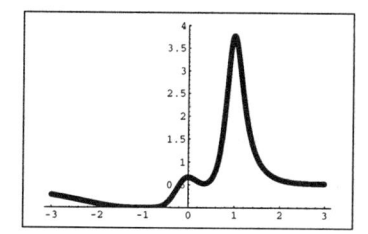

The formula for the curvature, produced by the command

```
Simplify[PowerExpand[kappa2[folium][t]]]
```

is

$$\frac{2(1+t^3)^4}{3(1+4t^2-4t^3-4t^5+4t^6+t^8)^{3/2}}$$

The graphs of the curvature of the folium show several local maxima and minima. The values of t that give rise to these local maxima and minima can be found with the command

```
Select[t /. NSolve[kappa2[folium]'[t]==0,t],Im[#]==0&]
```

yielding

```
{-1., 0., 0.343056, 1., 2.91498}
```

4.2 Cassinian Ovals

A **Cassinian**[5] oval is a generalization of the lemniscate of Bernoulli that we studied in Section 3.2. It is the locus $S(a,b)$ of points for which the product of the distances from two fixed points $\mathbf{F}_1$ and $\mathbf{F}_2$ is a constant b^2, where $\mathbf{distance}(\mathbf{F}_1, \mathbf{F}_2) = 2a$. Thus

$$S(a,b) = \{\,(x,y) \mid \mathbf{distance}((x,y), \mathbf{F}_1)\mathbf{distance}((x,y), \mathbf{F}_2) = b^2\,\}.$$

To study Cassinian ovals, we first define

$$\mathbf{cassiniimplicit}[a,b](x,y) = (x^2 + y^2 + a^2)^2 - b^4 - 4a^2x^2,$$

or

```
cassiniimplicit[a_,b_][x_,y_]:=
    (x^2 + y^2 + a^2)^2 - b^4 - 4 a^2 x^2
```

Generalizing the computations of Section 3.2, we can prove that $S(a,b)$ is the zero set of **cassiniimplicit**$[a,b]$; more explicitly:

Lemma 4.2. *An oval of Cassini with $(\pm a, 0)$ as foci is the implicitly defined curve*

$$(x^2 + y^2)^2 + 2a^2(y^2 - x^2) = b^4 - a^4.$$

For a proof see exercise 2. Note that **cassiniimplicit**$[b,b]$ is a lemniscate of Bernoulli, and **cassiniimplicit**$[0,b]$ is a circle.

ImplicitPlot can be used to plot Cassinian ovals. Here are some examples:

---

[5] Gian Domenico Cassini (1625-1712). Italian astronomer, who did his most important work in France. He proposed the fourth degree curves now called ovals of Cassini to describe planetary motion.

4.2 Cassinian Ovals

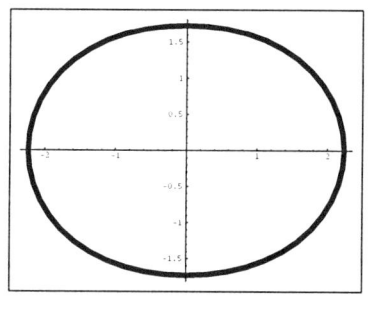

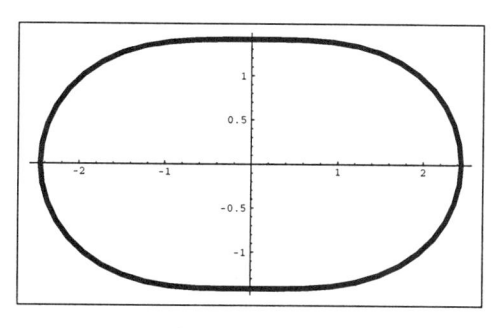

$(1 + x^2 + y^2)^2 = 4x^2 + 16$

$(2 + x^2 + y^2)^2 = 8x^2 + 16$

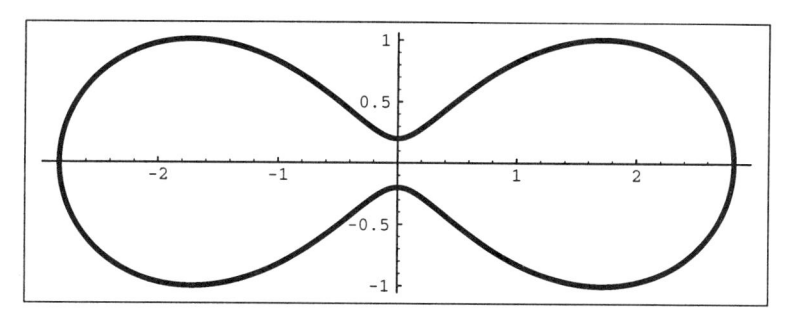

$(3.96 + x^2 + y^2)^2 = 15.84x^2 + 16$

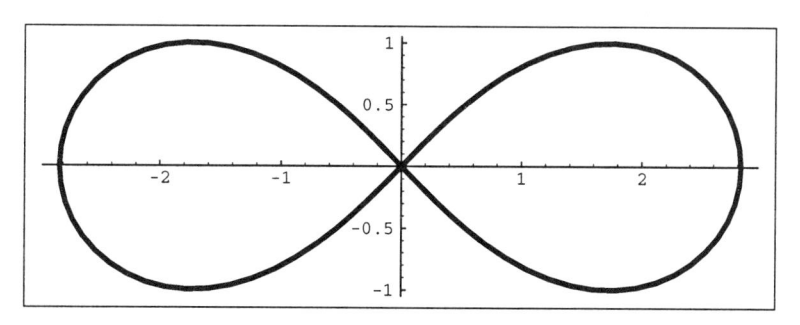

$(4 + x^2 + y^2)^2 = 16x^2 + 16$

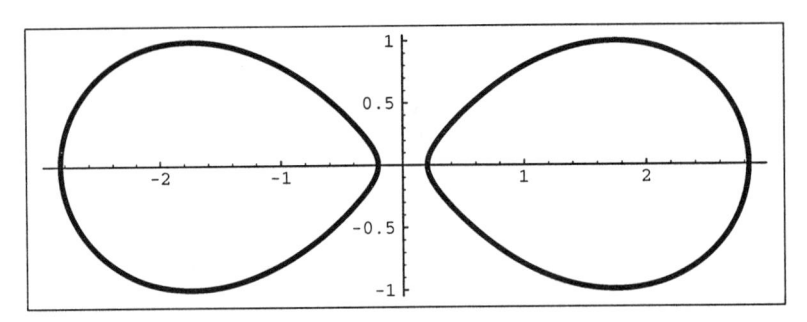

$$(4.04 + x^2 + y^2)^2 = 16.16x^2 + 16$$

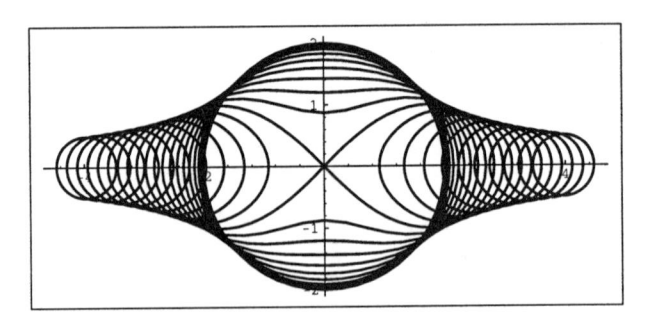

cassiniimplicit[a, 2], $0 \le a \le 4$.

The plotting command for **cassiniimplicit**[a, 2], $0 \le a \le 4$ is

```
ImplicitPlot[Table[cassiniimplicit[a,2][x,y] == 0,
    {a,0,4,1/5}]//Evaluate,{x,-5,5}];
```

Solve can be used to find the intercepts of a curve (that is, where the curve meets the horizontal and vertical axes). For example,

```
Solve[cassiniimplicit[a,b][x,0] == 0,x]

                2    2                    2    2
{{x -> Sqrt[a  + b ]}, {x -> -Sqrt[a  + b ]},

              2    2                    2    2
 {x -> Sqrt[a  - b ]}, {x -> -Sqrt[a  - b ]}}
```

4.2 Cassinian Ovals

tells us that the horizontal intercepts of **cassiniimplicit**[a, b] are

$$\left(\pm\sqrt{a^2 \pm b^2}, 0\right).$$

When $a^2 > b^2$, there are four horizontal intercepts, but when $a^2 < b^2$ there are only two horizontal intercepts, since we must exclude imaginary solutions. Similarly,

```
Solve[cassiniimplicit[a,b][0,y] == 0,y]
```

```
                  2   2                     2   2
{{y -> Sqrt[-a  + b ]}, {y -> -Sqrt[-a  + b ]},
             2   2                    2   2
 {y -> Sqrt[-a  - b ]}, {y -> -Sqrt[-a  - b ]}}
```

tells us that **cassiniimplicit**[a, b] has no vertical intercepts when $a^2 > b^2$; when $a^2 < b^2$ the vertical intercepts are

$$\left(0, \pm\sqrt{b^2 - a^2}\right).$$

Mathematica's command **StackGraphics** can be used to good effect to display Cassinian ovals in 3 dimensions. To make this command available we read in the relevant package with the command **Needs["Graphics`Graphics3D`"]**. Then we create, but do not display, the Cassinian ovals with the command

```
cassinis=
Table[ImplicitPlot[cassiniimplicit[a,2][x,y] == 0,{x,-5,5},
DisplayFunction->Identity],{a,0,4,1/5}]
```

The actual 3-dimensional display is then accomplished with

```
Show[StackGraphics[cassinis],
ViewPoint->{0.617,-1.087,3.144},
Axes->False]
```

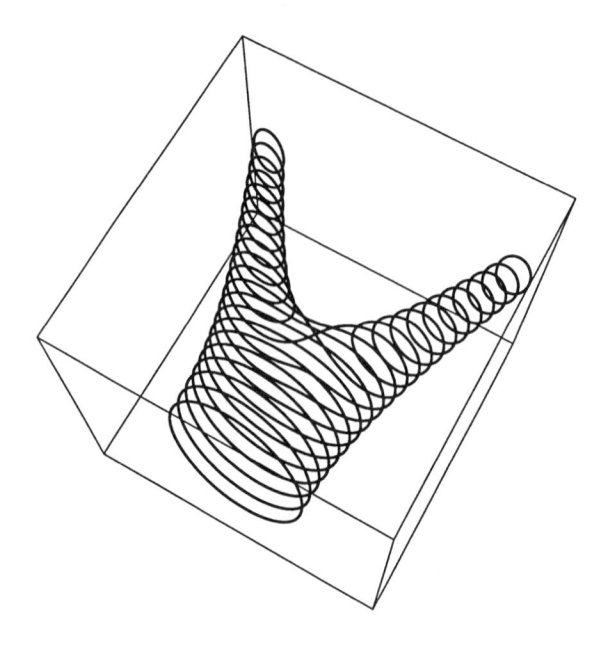

86   Chapter 4   Alternate Methods for Plotting Plane Curves

cassiniimplicit[a, 2], $0 \le a \le 4$.

4.3 Plane Curves in Polar Coordinates

In this section we show how to plot and compute the length and curvature of a plane curve using polar coordinates.

Definition. *A* **polar parametrization** *is a curve* $\gamma\colon (a,b) \longrightarrow \mathbb{R}^2$ *of the form*

(4.1) $$\gamma(\theta) = \mathbf{r}[\gamma](\theta)(\cos\theta, \sin\theta),$$

where $\mathbf{r}[\gamma](\theta) \ge 0$ *for* $a < \theta < b$. *We call* $\mathbf{r}[\gamma]$ *the* **radius function** *of the curve* γ, *and abbreviate it to* $\mathbf{r}$ *when there is no danger of confusion.*

The radius function completely determines a polar parametrization γ, so usually a curve is described in polar coordinates simply by giving the definition of $\mathbf{r}[\gamma]$.

4.3 Plane Curves in Polar Coordinates

A polar parametrization of a plane curve is frequently a very simple parametrization. Its description is simplified further by writing just the definition of the radius function of the curve. We shall see shortly that there are formulas for the arc length and curvature of a polar parametrization in terms of the radius function alone.

A good example of a polar parametrization is the logarithmic spiral. The radius function of the logarithmic spiral is given by

$$\text{logspiralpolar}[a, b](\theta) = a\, e^{b\theta}$$

or

```
logspiralpolar[a_,b_][theta_]:= a E^(b theta)
```

There is a *Mathematica* package for doing a variety of specialized plots including polar plots; it can be read in with the command

```
Needs["Graphics`Graphics`"]
```

The command **PolarPlot** contained in **Graphics/Graphics.m** plots a curve using the radius function. In fact, **PolarPlot** is a *Mathematica* command to convert the radius function of a polar parametrization to the form required for **ParametricPlot**. Here is a slightly simplified version:

```
PolarPlot[r_, {t_,tmin_,tmax_},opts___]:=
ParametricPlot[{r Cos[t],r Sin[t]},{t,tmin,tmax},
    opts,AspectRatio->Automatic]
```

(The **opts___** (with a triple underscore) allows the insertion of zero or more options of the type that one would use with **ParametricPlot**.) It is easy to use **PolarPlot** to graph the logarithmic spiral.

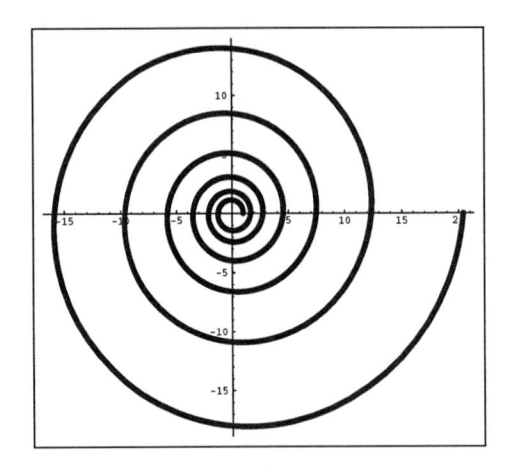

```
PolarPlot[
logspiralpolar[1,0.08][
theta]//Evaluate,
{theta,0,12Pi}];
```

Let us give a generalization of the cardioid and limaçon (see exercise 4 of Chapter 3) by defining

$$\mathbf{limaconpolar}[n, a, b](\theta) = 2a \cos n\theta + b,$$

or

```
limaconpolar[n_,a_,b_][theta_]:= 2a Cos[n theta] + b
```

Then $\theta \longmapsto \mathbf{limaconpolar}[1, a, 2a](\theta) = 2(\cos\theta + 1)$ is the radius function of a standard cardioid.

As an example we do the polar plot of **limaconpolar[3, 1/2, 2]**:

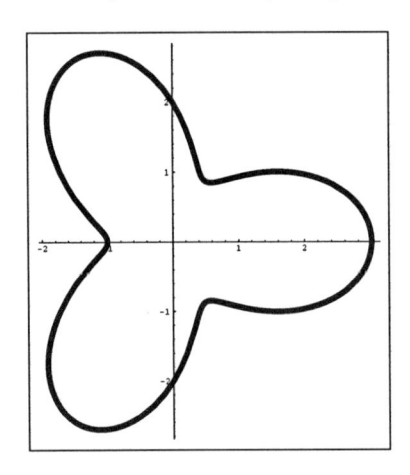

```
PolarPlot[
limaconpolar[3,1/2,2][
theta]//Evaluate,
{theta,0,2Pi}];
```

4.3 Plane Curves in Polar Coordinates

Next, we derive the polar coordinate formulas for the arc length and curvature.

Lemma 4.3. *The length and curvature of a polar parametrization $\gamma\colon (a,b) \longrightarrow \mathbb{R}^2$ are given in terms of the radius function $\mathbf{r}$ of γ by the formulas*

$$(4.2) \qquad \mathbf{Length}[\gamma] = \int_a^b \sqrt{\mathbf{r}'(\theta)^2 + \mathbf{r}(\theta)^2}\, d\theta,$$

$$(4.3) \qquad \kappa 2[\gamma] = \frac{-\mathbf{r}''\mathbf{r} + 2\mathbf{r}'^2 + \mathbf{r}^2}{(\mathbf{r}'^2 + \mathbf{r}^2)^{3/2}}.$$

Proof. The calculations can be carried out most easily using complex numbers. Equation (4.1) can be written more succinctly as

$$(4.4) \qquad \gamma(\theta) = \mathbf{r}(\theta)e^{i\theta}.$$

From (4.4) we get

$$\gamma'(\theta) = \bigl(\mathbf{r}'(\theta) + i\mathbf{r}(\theta)\bigr)e^{i\theta} \quad \text{and} \quad \gamma''(\theta) = \bigl(\mathbf{r}''(\theta) + 2i\mathbf{r}'(\theta) - \mathbf{r}(\theta)\bigr)e^{i\theta}.$$

Therefore

$$(4.5) \qquad \|\gamma'(\theta)\|^2 = \mathbf{r}'(\theta)^2 + \mathbf{r}(\theta)^2;$$

then (4.2) follows immediately from (4.5) and the definition of length. Furthermore, using Lemma 1.1 we find that

$$(4.6) \quad \gamma''(\theta) \cdot J\gamma'(\theta) = \mathfrak{Re}\bigl((\mathbf{r}''(\theta) + 2i\mathbf{r}'(\theta) - \mathbf{r}(\theta))e^{i\theta}\overline{(i(\mathbf{r}'(\theta) + i\mathbf{r}(\theta))e^{i\theta})}\bigr)$$

$$= \mathfrak{Re}\bigl((\mathbf{r}''(\theta) + 2i\mathbf{r}'(\theta) - \mathbf{r}(\theta))(-i\mathbf{r}'(\theta) - \mathbf{r}(\theta))\bigr)$$

$$= -\mathbf{r}''(\theta)\mathbf{r}(\theta) + 2\mathbf{r}'(\theta)^2 + \mathbf{r}(\theta)^2.$$

Then (4.3) follows from (4.5), (4.6) and (1.9). ∎

The *Mathematica* formula for the curvature of a polar parametrization (equation (4.3)) is

```
kappa2polar[rho_][t_]:= (-D[rho[tt],{tt,2}]rho[tt] +
                        2D[rho[tt],tt]^2 + rho[tt]^2)/
    (D[rho[tt],tt]^2 + rho[tt]^2)^(3/2) /. tt->t
```

Let us use this formula to compute the curvature of the curve whose radius function is given by

$$\mathbf{r}^n = a^n \theta.$$

Chapter 4 Alternate Methods for Plotting Plane Curves

In *Mathematica* we define

```
archimedesspiralpolar[n_,a_][theta_] := a theta^(1/n)
```

Mathematica tells us the curvature:

**kappa2polar[archimedesspiralpolar[n,a]
][theta]//Simplify//PowerExpand//Simplify**

$$\frac{n \, \text{theta}^{1 - 1/n} \, (1 + n^2 + n^2 \, \text{theta}^2)}{a \, (1 + n^2 \, \text{theta}^2)^{3/2}}$$

Several of these spirals have been given names:

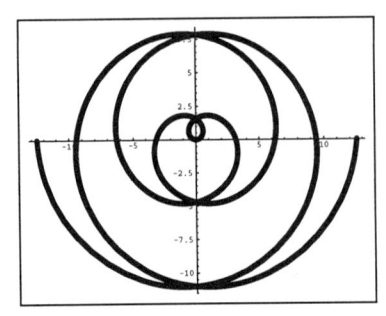

Spiral of Archimedes: $r = \theta$

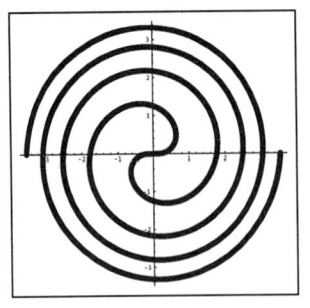

```
PolarPlot[t,
{t,-4Pi,4Pi}];
```

Fermat's spiral: $r^2 = \theta$

```
PolarPlot[
{t^(1/2),-t^(1/2)},
{t,0,4Pi}];
```

4.3 Plane Curves in Polar Coordinates

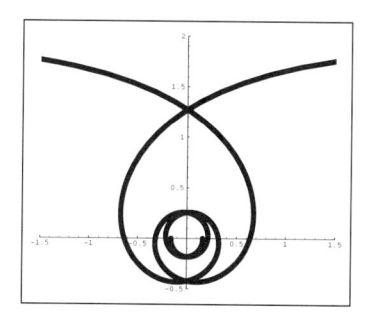

Hyperbolic spiral: $r = 2/\theta$

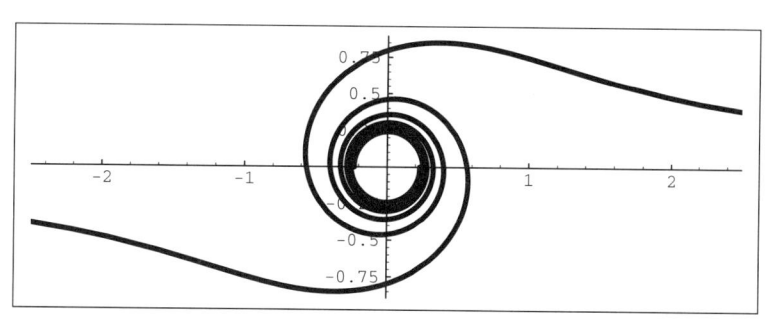

```
Show[PolarPlot[2/t,
{t,0.6,4Pi},
PlotRange->
{{-1.5,1.5},{-0.5,2}},
DisplayFunction->
Identity],
PolarPlot[2/t,
{t,-4Pi,-0.6},
PlotRange->
{{-1.5,1.5},{-0.5,2}},
DisplayFunction->
Identity],
DisplayFunction->
$DisplayFunction];
```

Lituus: $r^2 = 1/\theta$

```
PolarPlot[{t^(-1/2),-t^(-1/2)},{t,0.05,6Pi},
PlotRange->{{-2.5,2.5},Automatic}];
```

Each of these spirals has two branches, which are oppositely oriented. **PolarPlot** plots both branches of the spiral of Archimedes[6] by allowing the radius function to

---

[6] Archimedes of Syracuse (287-212 B.C.). Archimedes is credited with the creation of many mechanical devices such as compound pulleys, water clocks, catapults and burning mirrors. Legend has it that he was killed by a Roman soldier as he traced mathematical figures in the sand during the siege of Syracuse, at that time a Greek colony on what is now Sicily. His mathematical work included finding the area of a circle and the area under a parabola using the method of exhaustion.

be negative. For the spiral of Fermat[7] and the lituus, however, the two branches must be considered separate curves, but **PolarPlot** will still plot them simultaneously. We must plot the hyperbolic spiral twice and then combine plots.

4.4 Exercises

1. Find the nonparametric form of the following curves:

    (a) $\alpha(t) = (t^{15}, t^6)$.

    (b) The hyperbola-like curve $\gamma(t) = (t^3, t^{-4})$.

    (c) The strophoid defined by
    $$\text{strophoid}[a](t) = a\left(\frac{t^2-1}{t^2+1}, \frac{t(t^2-1)}{t^2+1}\right), \quad \text{or}$$

    ```
    strophoid[a_][t_] := a(t^2 - 1)/(t^2 + 1){1,t}
    ```

2. Prove Lemma 4.2.

3. The **devil's curve** is defined nonparametrically as the zeros of the function
    $$\text{devilimplicit}[a,b](x,y) = y^2(y^2 - b^2) - x^2(x^2 - a^2), \quad \text{or}$$

    ```
    devilimplicit[a_,b_][x_,y_] := y^2(y^2 - b^2)
                                   - x^2(x^2 - a^2)
    ```

    Plot the devil's curve using **ImplicitPlot** and find its intercepts.

---

[7] Pierre de Fermat (1601-1665). Fermat, like his contemporary Descartes, was trained as a lawyer. Fermat generalized the work of Archimedes on spirals. Like Archimedes he had the germs of the ideas of both differential and integral calculus—using the new techniques of analytic geometry—but failed to see the connections. Fermat is most remembered for his scribble in a margin purporting to have a proof for what is now known as Fermat's Last Theorem: $x^n + y^n = z^n$ has no solution in integers for $n > 2$. Only recently has the conjecture been proved by Wiles.

4.4 Exercises

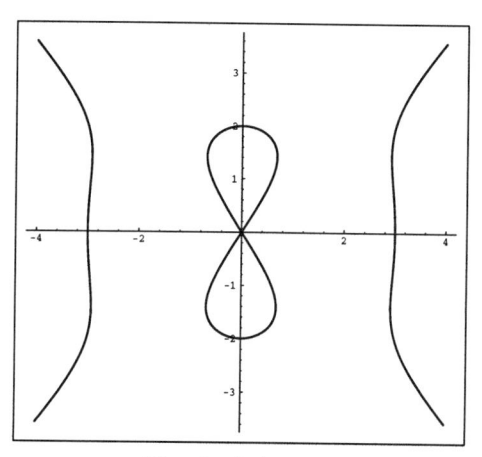

The devils' curve

4. Kepler's[8] **folium** is the curve defined nonparametrically as the set of zeros of the function

$$\text{keplerimplicit}[a, b](x, y) = \left((x-b)^2 + y^2\right)\left(x(x-b) + y^2\right) - 4a(x-b)y^2, \quad \text{or}$$

```
keplerimplicit[a_,b_][x_,y_]:=
    ((x - b)^2 + y^2)(x(x - b) + y^2) - 4a(x - b)y^2
```

Plot Kepler's folium and some of its perturbations using **ImplicitPlot**.

5. Use **ImplicitPlot** to plot the following curves (see [Walk, page 56]) with interesting singularities:

---

[8] Johannes Kepler (1571-1630). German astronomer, who lived in Prague. By empirical observations Kepler showed that a planet moves around the sun in an elliptical orbit having the sun at one of its two foci, and that a line joining the planet to the sun sweeps out equal areas in equal times as the planet moves along its orbit.

(a) $x^3 - x^2 + y^2 = 0$,
(b) $x^4 + x^2y^2 - 2x^2y - xy^2 + y^2 = 0$,
(c) $2x^4 - 3x^2y + y^2 - 2y^3 - y^4 = 0$,
(d) $x^3 - y^2 = 0$,
(e) $(x^2 + y^2)^2 + 3x^2y - y^3 = 0$,
(f) $(x^2 + y^2)^3 - 4x^2y^2 = 0$.

6. The polar equation for the **trisectrix of Maclaurin**[9] is

$$\text{polartrisectrix}[a](\theta) = a \sec\left(\frac{\theta}{3}\right),$$

or

```
trisectrixpolar[a_][theta_] := a/Cos[theta/3]
```

Plot it using **PolarPlot**.

7. Define the function **pacman** by

$$\text{pacman}[n](\theta) = \cos^n \theta + 1,$$

or

```
pacman[n_][theta_] := Cos[theta]^n + 1
```

Plot $\theta \longmapsto \text{pacman}[999](\theta)$ using **PolarPlot**, showing that the curve has finite curvature at $\theta = 0$ but infinite curvature at $\theta = \pi$.

---

[9] Colin Maclaurin (1698-1746). Scottish mathematician. His **Geometrica organica, sive discriptio linearum curvarum universalis** dealt with general properties of conics and higher plane curves. In addition to Maclaurin's own results, this book contained the proofs of many important theorems that Newton had given without proofs. Maclaurin is also known for his work on power series and the defense of his book **Theory of Fluxions** against the religious attacks of Bishop George Berkeley.

4.4 Exercises

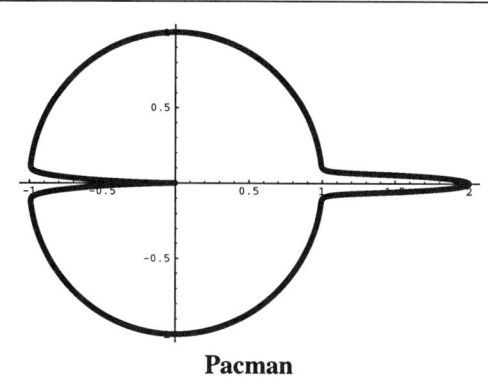

Pacman

8. The general semicubical parabola is defined implicitly in *Mathematica* by

```
semicubicimplicit[a_,b_,c_,d_][x_,y_]:=
            a x^3 + b x^2 + c x + d - y^2
```

Experiment with **ImplicitPlot** to find appropriate values of the **a**, **b**, **c** and **d** to make the following plots:

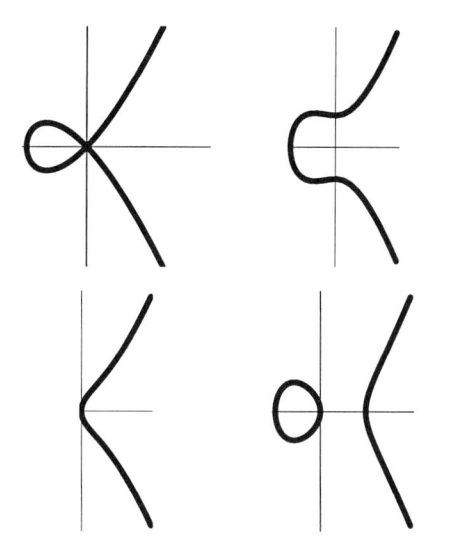

Semicubical curves of the form $y^2 = ax^3 + bx^2 + cx + d$

9. Use **paramtoimp** to find implicit equations for the following curves:

(a) sc
(b) cubicparabola[a,b,c,d]
(c) parabola[a]
(d) serpentine[a,b]

10. Explain the use on page 82 of *Mathematica's* commands **NSolve** and **Select**.

4.5 Animation of Fermat's Spiral

The following command creates an increasing spiral of Fermat.

```
Do[ParametricPlot[Evaluate[
{fermatspiral[1][t],-fermatspiral[1][t]}],
{t,0,a Pi},AspectRatio->Automatic,
PlotRange->{{-4,4},{-4,4}}],{a,0.02,4,0.02}]
```

5
NEW CURVES FROM OLD

Any plane curve gives rise to other plane curves through a variety of general constructions. Each such construction can be thought of as a function which assigns one curve to another. *Mathematica* can be used to define and carry out these constructions. What needs to be done in each case is to use a function to write down carefully what the construction does. The translation of the definition of the function from ordinary mathematics to *Mathematica* is usually straightforward.

Four classic examples of constructing one plane curve from another are studied in the present chapter: *evolutes* in Sections 5.1–5.3, *involutes* in Section 5.4, *parallel curves* in Section 5.7 and *pedal curves* in Section 5.8. The *Mathematica* definitions of evolute, parallel curve and pedal curve can then be used either to draw a new curve or to study the new curve analytically, for example, by computing its curvature.

In addition, we show how to construct normal and tangent lines to a curve in Section 5.5 and osculating circles to curves in Section 5.6. In Section 5.6 we also show that the circle through three points on a curve tends to the *osculating circle* as the three points converge to a common point.

An animation of a *cycloidial pendulum* is given in Section 5.10.

5.1 Evolutes

A point $\mathbf{p} \in \mathbb{R}^2$ is called a **center of curvature** at $\mathbf{q}$ of a curve $\alpha \colon (a,b) \longrightarrow \mathbb{R}^2$, provided that there is a circle γ with center $\mathbf{p}$ which is tangent to α at $\mathbf{q}$ such that the curvatures of α and γ are the same at $\mathbf{q}$. This implies that there is a line L from $\mathbf{p}$ to α which meets α perpendicularly at $\mathbf{q}$, and the distance from $\mathbf{p}$ to $\mathbf{q}$ is the radius of curvature of α at $\mathbf{q}$.

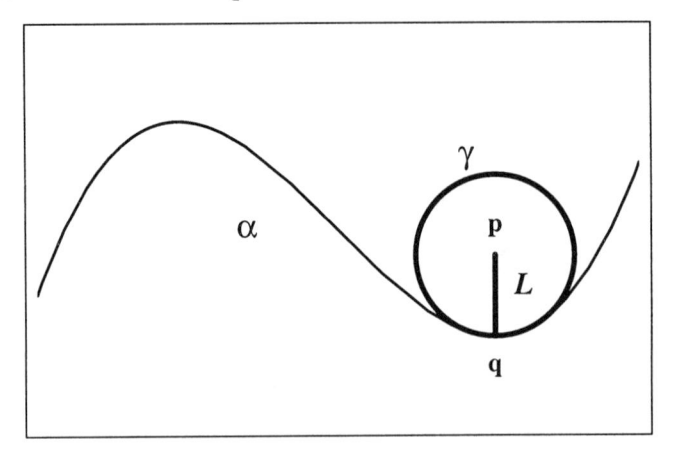

A center of curvature of a cubic parabola

The centers of curvature form a new curve, called the evolute of α, whose precise definition is as follows.

Definition. *The* **evolute** *of a regular plane curve α is the plane curve given by*

$$(5.1) \qquad \mathbf{evolute}[\alpha](t) = \alpha(t) + \left(\frac{1}{\kappa 2[\alpha](t)}\right) \frac{J\alpha'(t)}{\|\alpha'(t)\|}.$$

It turns out that a circle with center $\mathbf{evolute}[\alpha](t)$ and radius $1/|\kappa 2[\alpha](t)|$ will be tangent to the plane curve α at $\alpha(t)$; this is the circle, called the **osculating circle**, that best approximates α near $\alpha(t)$. See Section 5.6.

Using formula (1.9), page 14, we see that the formula for the evolute can be written more succinctly as

$$(5.2) \qquad \mathbf{evolute}[\alpha] = \alpha + \frac{\|\alpha'\|^2}{\alpha'' \cdot J(\alpha')} J\alpha'.$$

An easy consequence of (5.1) and (1.12) is the following important fact:

5.1 Evolutes

Lemma 5.1. *The definition of evolute of a curve α is independent of parametrization; that is,*
$$\text{evolute}[\alpha \circ h] = \text{evolute}[\alpha] \circ h,$$
for any differentiable function $h\colon (c,d) \longrightarrow (a,b)$.

The evolute of a circle consists of a single point. The evolute of any plane curve γ can be described physically. Imagine light rays starting at all points of the trace of γ and propagating down the normals of γ. In the case of a circle, these rays focus perfectly at the center, so for γ the focusing occurs along the centers of best fitting circles, that is, along the evolute of γ.

The *Mathematica* version of (5.2) is

```
evolute[alpha_][t_]:= alpha[tt] +
    D[alpha[tt],tt].D[alpha[tt],tt]/
    (D[alpha[tt],{tt,2}].
    J[D[alpha[tt],tt]]) J[D[alpha[tt],tt]] /. {tt->t}
```

It is easily verified using *Mathematica* that, as expected, the evolute of a circle is a point. A more interesting example is the evolute of an ellipse. The command

```
Simplify[evolute[ellipse[a,b]][t]]
```

yields

$$\left\{ \frac{(a^2 - b^2)\,\text{Cos}[t]^3}{a},\ \frac{(-a^2 + b^2)\,\text{Sin}[t]^3}{b} \right\}$$

Hence the evolute of the ellipse $x^2/a^2 + y^2/b^2 = 1$ is the curve γ defined by

(5.3) $$\gamma(t) = \left(\frac{(a^2 - b^2)(\cos(t))^3}{a}, \frac{(a^2 - b^2)(\sin(t))^3}{a} \right).$$

This is the parametrization of an astroid.

The command to draw ellipse **ellipse[1.5,1]** and its evolute simultaneously is

```
ParametricPlot[Evaluate[{
ellipse[1.5,1][t],
evolute[ellipse[1.5,1]][t]}],{t,0,2Pi},
AspectRatio->Automatic];
```

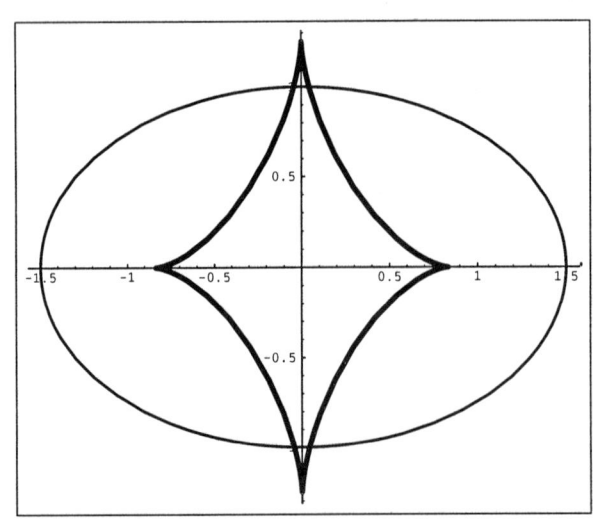

The ellipse $t \longmapsto \text{ellipse}[1.5,1](t)$
and its evolute

The notions of tangent line and normal line to a curve are clear intuitively; here is the mathematical definition.

Definition. The **tangent line** *and* **normal line** *to a curve* $\alpha: (a,b) \longrightarrow \mathbb{R}^2$ *at* $\alpha(t)$ *are the straight lines passing through* $\alpha(t)$ *with velocity vectors equal to* $\alpha'(t)$ *and* $J\alpha'(t)$, *respectively.*

Next, we obtain a characterization of the evolute of a curve in terms of tangent lines and normal lines, and also determine the singular points of the evolute.

Theorem 5.2. *Let* $\beta: (a,b) \longrightarrow \mathbb{R}^2$ *be a unit-speed curve. Then:*

(i) *the evolute of* β *is the unique curve of the form* $\gamma = \beta + f J\beta'$ *for some function* f *for which the tangent line to* γ *at each point* $\gamma(s)$ *coincides with the normal line to* β *at* $\beta(s)$.

(ii) *The singular points of the evolute of* β *are precisely those values of* s *for which* $\kappa 2[\beta]'(s) = 0$.

Proof. When we differentiate (5.1) and use Lemma 1.11, page 16, we obtain

$$(5.4) \qquad \text{evolute}[\beta]' = -\frac{\kappa 2[\beta]'}{\left(\kappa 2[\beta]\right)^2} J\beta'.$$

5.2 Iterated Evolutes

Hence the tangent line to **evolute[β]** at **evolute[β]**(s) coincides with the normal line to β at $\beta(s)$.

Conversely, suppose $\gamma = \beta + f\,J\beta'$. Again, using Lemma 1.11, we compute

$$\gamma' = \left(1 - f\kappa 2[\beta]\right)\beta' + f'J\beta'.$$

If the tangent line to γ at each point $\gamma(s)$ coincides with the normal line to β at $\beta(s)$, then $f = 1/\kappa 2[\beta]$, and so γ is the evolute of β.

This proves (i); then (ii) is a consequence of (5.4). ∎

For example, using *Mathematica* we compute for **ellipse[a,b]** that

$$\kappa 2'(t) = \frac{3a\,b(b^2 - a^2)\sin(2t)}{2(b^2\cos^2 t + \sin^2 t)^{5/2}},$$

so that the evolute of an ellipse is singular at 0, $\pi/2$, π and $3\pi/2$. This is confirmed by (5.3) and the plot of the evolute.

5.2 Iterated Evolutes

Mathematica can also be used to find the evolute of an evolute of a curve. Let us modify the definition of **evolute** by defining

```
evolute[1,alpha_][t_]:= Simplify[alpha[tt] +
    D[alpha[tt],tt].D[alpha[tt],tt]/
    (D[alpha[tt],{tt,2}].J[D[alpha[tt],tt]])*
    J[D[alpha[tt],tt]] /. {tt->t}]

evolute[n_,alpha_][t_]:= evolute[n,alpha][t]=
    evolute[1,evolute[n - 1,alpha]][t]//Simplify
```

This modified version of **evolute** is a function of three variables instead of two, and so it will not conflict with the original version **evolute**. It is clear that **evolute[1,alpha][t]** and **evolute[alpha][t]** coincide as functions of **t**.

Since the formulas for the iterated evolutes of a curve α might be complicated, it is best to find them symbolically and simplify them. Then new curves can be defined to be equal to the resulting expressions. Here are the first three evolutes of a cissoid:

```
evolute[1,cissoid[a_]][t_]:=
{-(a*t^2*(6 + t^2))/3, (8*a*t)/3}

evolute[2,cissoid[a_]][t_]:=
{-(a*(16 + 38*t^2 + 7*t^4))/9,
 (-2*a*t^3*(19 + 8*t^2 + t^4))/9}

evolute[3,cissoid[a_]][t_]:=
{a*(-48 + 114*t^2 + 424*t^4 + 285*t^6 + 75*t^8 + 7*t^10)/27,
 (-4*a*t*(38 + 90*t^2 + 39*t^4 + 5*t^6))/27}
```

If we had not performed these symbolic simplifications, it would have taken a long time for *Mathematica* to draw the iterated evolutes of a cissoid. However, now that *Mathematica* knows the explicit definitions of **cissoid[a][t]** and its first three evolutes, the four curves can be plotted quickly:

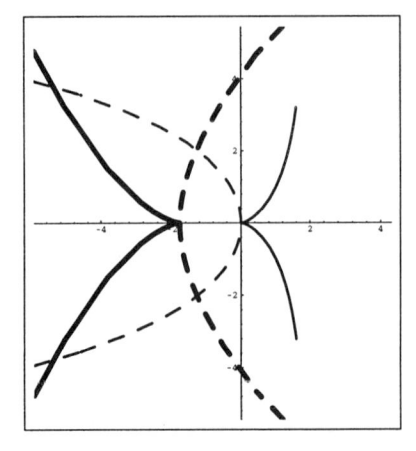

```
ParametricPlot[
 Evaluate[{
  cissoid[1][t],
  evolute[1,cissoid[1]][t],
  evolute[2,cissoid[1]][t],
  evolute[3,cissoid[1]][t]}],
 {t,-2,2},
 AspectRatio->Automatic];
```

A cissoid and its first three evolutes

Note that in spite of the fact that the cissoid has a cusp, its first and third evolutes do not.

5.3 The Evolute of a Tractrix is a Catenary

Another example of a curve with a cusp whose evolute has no cusp is the tractrix. Let us compute the evolute of **tractrix[a]** with *Mathematica*:

```
evolute[tractrix[a]][t]//Simplify
```

$$\left\{\frac{a\,\text{Csc}[\frac{t}{2}]\,\text{Sec}[\frac{t}{2}]}{2},\ a\,\text{Log}[\text{Tan}[\frac{t}{2}]]\right\}$$

So the evolute of the tractrix is the curve $t \longmapsto a\bigl(1/\sin t, \log \tan(t/2)\bigr)$. To see that this curve is actually a catenary, let $u = \log \tan(t/2)$. An easy hand calculation shows that $e^u + e^{-u} = 2/\sin t$. Thus the evolute of the tractrix can be reparametrized as $u \longmapsto a(\cosh u, u)$, which is indeed a catenary. Here is the plot of the tractrix together with its evolute, which is a catenary.

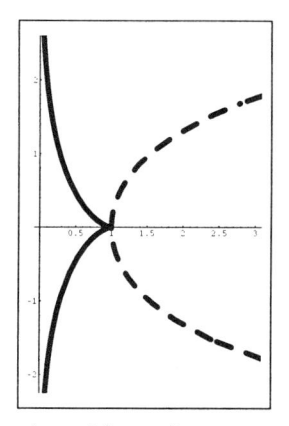

```
ParametricPlot[
{tractrix[1][t],
catenary[1][Log[Tan[t/2]]]}
//Evaluate,
{t,0.05,Pi-0.05},
AspectRatio->Automatic];
```

A tractrix and its evolute: a catenary

5.4 Involutes

The involute is a geometrically important operation that is the inverse to the operation $\alpha \longmapsto \text{evolute}[\alpha]$. In fact, the evolute is related to the involute in the same way that differentiation is related to indefinite integration. Just like indefinite integration, the operation of taking the involute involves an arbitrary constant. Furthermore, we shall prove (see Theorem 5.6) that the evolute of the involute of a curve γ is again γ; this corresponds to the fact that the derivative of the indefinite integral of a function f is again f.

We first give the definition of the involute of a unit-speed curve.

Definition. Let $\beta \colon (a, b) \longrightarrow \mathbb{R}^2$ be a unit-speed curve, and let $a < c < b$. The *involute* of β starting at $\beta(c)$ is the curve given by

(5.5) $\qquad \text{involute}[\beta, c](s) = \beta(s) + (c - s)\beta'(s).$

Whereas the evolute of a plane curve β is a linear combination of β and $J\beta'$, an involute of β is a linear combination of β and β'. Note that although we use s as the arc length parameter of β, it is not necessarily an arc length parameter for the involute of β.

The formula for the involute of an arbitrary-speed curve needs the arc length function:

Lemma 5.3. *Let* $\alpha\colon (a,b) \longrightarrow \mathbb{R}^2$ *be a regular arbitrary-speed curve. Then the involute of* α *starting at* c *(where* $a < c < b$*) is given by*

$$(5.6) \qquad \mathbf{involute}[\alpha, c](t) = \alpha(t) + \bigl(s_\alpha(c) - s_\alpha(t)\bigr)\frac{\alpha'(t)}{\|\alpha'(t)\|},$$

where $t \longmapsto s_\alpha(t)$ *denotes the arc length function of* α.

The *Mathematica* version of (5.6) is

```
involute[alpha_,c_][t_]:= alpha[tt] +
    (arclength[alpha][c] -
    arclength[alpha][tt])*D[alpha[tt],tt]/
    Sqrt[Factor[D[alpha[tt],tt].
    D[alpha[tt],tt]]] /. tt->t
```

We also put

```
involute[alpha_][t_]:= involute[alpha,0][t]
```

The involute of a curve can be described geometrically.

Theorem 5.4. *An involute of a regular plane curve* β *is formed by unwinding a taut string which has been wrapped around* β.

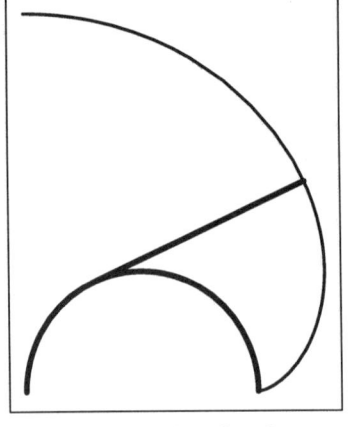

Definition of an involute

5.4 Involutes

Proof. Without loss of generality, we may suppose that β has unit-speed. Then

$$\text{involute}[\beta, c](s) - \beta(s) = (c - s)\beta'(s),$$

so that

(5.7) $$\|\text{involute}[\beta, c](s) - \beta(s)\| = |s - c|.$$

Here $|s - c|$ is the distance from $\beta(s)$ to $\beta(c)$ measured along the curve β, while $\|\text{involute}[\beta, c](s) - \beta(s)\|$ is the distance from $\text{involute}[\beta, c](s)$ to $\beta(s)$ measured along the tangent line to β emanating from $\beta(s)$. Then (5.7) says that these quantities are equal. ∎

Here are involutes of a circle and a figure eight:

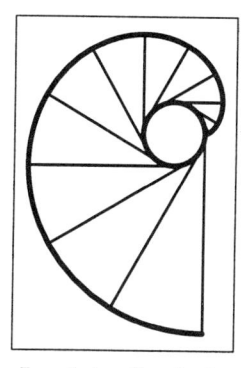

Involute of a circle 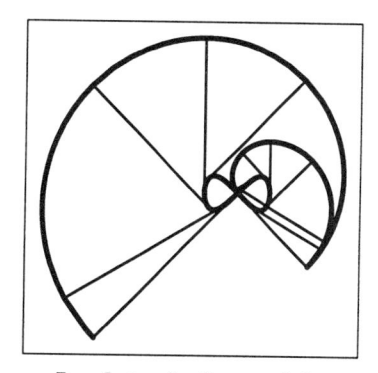   **Involute of a figure eight**

To find the formula for an involute of a circle with *Mathematica* we use

```
Simplify[PowerExpand[involute[circle[a],b][t]]]
```

to get

{a (Cos[t] - b Sin[t] + t Sin[t]), a (b Cos[t] - t Cos[t] + Sin[t])}

Thus

$$\text{involute}[\text{circle}[a], b](t) = a\bigl(\cos t + (-b + t)\sin t, (b - t)\cos t + \sin t\bigr).$$

The formula for the involute of a figure eight involves a complicated integral, so we omit it. The picture of the involute of a figure eight is more informative.

Next, we find the relation between the curvature of a curve and the curvature of its involute.

Lemma 5.5. Let $\beta\colon (a,b) \longrightarrow \mathbb{R}^2$ be a unit-speed curve, and let γ be the involute of β starting at c, where $a < c < b$. Then the curvature of γ is given by the formula

$$\kappa 2[\gamma](t) = \frac{\operatorname{sign}\left(\kappa 2[\beta](t)\right)}{|c-t|}. \tag{5.8}$$

Proof. First, we compute

$$\gamma'(t) = (c-t)\beta''(t) = (c-t)\kappa 2[\beta](t)J\beta'(t), \tag{5.9}$$

and

$$\begin{aligned}\gamma''(t) &= -\kappa 2[\beta](t)J\beta'(t) + (c-t)\kappa 2[\beta]'(t)J\beta'(t) \\ &\quad + (c-t)\kappa 2[\beta](t)J\beta''(t) \\ &= \bigl(-\kappa 2[\beta](t) + (c-t)\kappa 2[\beta]'(t)\bigr)J\beta'(t) \\ &\quad - (c-t)\bigl(\kappa 2[\beta](t)\bigr)^2\beta'(t).\end{aligned} \tag{5.10}$$

From (5.9) and (5.10) we get

$$\gamma''(t) \cdot J\gamma'(t) = (c-t)^2\bigl(\kappa 2[\beta](t)\bigr)^3. \tag{5.11}$$

Now (5.8) follows from (5.9), (5.11) and the definition of $\kappa 2[\gamma]$. ∎

Lemma 5.5 says that the absolute value of the curvature of the involute of a curve is always decreasing for $t \geq c$. This can be seen clearly in the pictures of the involutes of a circle and a figure eight.

Next, we show that the involute is the inverse of the evolute in the same way that integration is the inverse of differentiation.

Theorem 5.6. Let $\beta\colon (a,b) \longrightarrow \mathbb{R}^2$ be a unit-speed curve and let γ be the involute of β starting at c, where $a < c < b$. Then the evolute of γ is β.

Proof. By definition the evolute of γ is the curve ζ given by

$$\zeta(t) = \gamma(t) + \frac{1}{\kappa 2[\gamma](t)} \frac{J\gamma'(t)}{\|\gamma'(t)\|}. \tag{5.12}$$

When we substitute (5.5), (5.8) and (5.9) into (5.12), we get

$$\zeta(t) = \beta(t) + (c-t)\beta'(t) + \frac{|c-t|}{\operatorname{sign}\left(\kappa 2[\beta](t)\right)} \frac{(c-t)\kappa 2[\beta](t)J^2\beta'(t)}{\|(c-t)\kappa 2[\beta](t)J\beta'(t)\|} = \beta(t).$$

5.4 Involutes

Thus β and ζ coincide. ∎

Involutes, like evolutes, can be iterated. One approach to iteration in *Mathematica* was given in Section 5.2. Here is another approach using the command **Nest**. The syntax of **Nest** is

```
Nest[the function,expression,number of iterations]
```

Thus **Nest** applies **function** to **expression** n times, where n is the number of iterations. To apply the function $\alpha \longmapsto \text{involute}[\alpha, 0]$ three times to a circle of radius a centered at the origin, we can use

```
Nest[involute[#,0]&,circle[a],3][t]//Simplify
```

$$\left\{ \frac{a\,(6\,\cos[t] - 3\,t^2\,\cos[t] + 6\,t\,\sin[t] - t^3\,\sin[t])}{6}, \right.$$

$$\left. \frac{a\,(-6\,t\,\cos[t] + t^3\,\cos[t] + 6\,\sin[t] - 3\,t^2\,\sin[t])}{6} \right\}$$

(We are using the shorthand notation **involute[#,0]&** to denote the function that assigns to a curve **alpha** its involute **involute[alpha,0]** starting at **0**. See page 27.) Using **Nest**, we plot a circle of radius 1 and its first three involutes:

```
ParametricPlot[
 {circle[1][t],
  involute[circle[1],0][t],
  Nest[involute[#,0]&,
  circle[1],2][t],
  Nest[involute[#,0]&,
  circle[1],3][t]}
 //Evaluate,
 {t,0,2Pi},
 AspectRatio->Automatic,
 Axes->None,
 PlotRange->All];
```

Circle and three of its involutes

5.5 Tangent and Normal Lines to Plane Curves

The tangent line to a plane curve at a point **p** on the curve is the best linear approximation to the curve at **p**, and the normal line is the tangent line rotated by $\pi/2$. We now explain how to draw these lines using *Mathematica*. We put

```
tangentline[alpha_][s_,t_] := Line[{alpha[tt],alpha[tt] +
    s D[alpha[tt],tt]/Sqrt[Factor[D[alpha[tt],tt].
    D[alpha[tt],tt]]]}] /. tt->t
```

and

```
normalline[alpha_][s_,t_] :=
    Line[{alpha[tt],alpha[tt] + s J[D[alpha[tt],tt]]/
    (D[alpha[tt],tt].D[alpha[tt],tt])^(1/2)}] /. tt->t
```

Neither of these commands by itself will display the line, because they are defined in terms of the *Mathematica* graphics primitive **Line**. But when combined with **Show** and **Graphics** the commands **tangentline** and **normalline** will display lines of length **s**. As an example we draw some tangent lines to an ellipse.

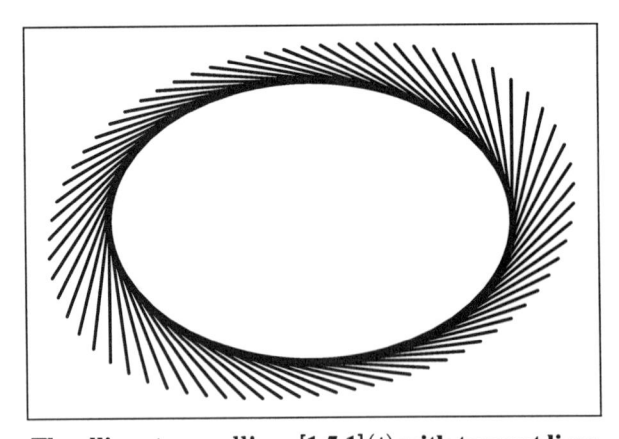

The ellipse $t \longmapsto$ ellipse[1.5,1](t) with tangent lines

The miniprogram to draw this picture illustrates the use of **tangentline**:

5.5 Tangent and Normal Lines to Plane Curves

```
Show[Graphics[{AbsoluteThickness[1],
    Evaluate[Table[tangentline[ellipse[1.5,1]][1,t],
    {t,0,2Pi,Pi/36}]]},DisplayFunction->Identity],
    ParametricPlot[Evaluate[ellipse[1.5,1][t]],{t,0,2Pi},
    PlotStyle->{{AbsoluteThickness[2]}},
    DisplayFunction->Identity],
    AspectRatio->Automatic,Axes->None,PlotRange->All,
    DisplayFunction->$DisplayFunction];
```

This miniprogram uses the low level commands **Show** and **Graphics** in order to control more precisely the graphics output. The command **Show** tells *Mathematica* to display graphics commands. In the above example these graphics commands are divided into three groups. The part

```
Graphics[AbsoluteThickness[1],
    Evaluate[Table[
        tangentline[ellipse[1.5,1]][1,t],
        {t,0,2Pi,Pi/36}]],
    DisplayFunction->Identity]
```

tells *Mathematica* to compute the tangent lines, but the option **DisplayFunction->Identity** prevents *Mathematica* from actually displaying the lines until other parts of the picture have been computed. Next,

```
ParametricPlot[Evaluate[ellipse[1.5,1][t]],
    {t,0,2Pi},PlotStyle->AbsoluteThickness[2],
    DisplayFunction->Identity]
```

computes the ellipse but does not display it. Notice that we have told *Mathematica* to make the ellipse twice as thick as the tangent lines. Having completed the two main parts of our picture, we give **Show** three display options:

```
AspectRatio->Automatic,
Axes->None,PlotRange->All
```

Finally,

```
DisplayFunction->$DisplayFunction
```

tells *Mathematica* to display the results of the previous commands.

Next, we draw some short normal lines to the same ellipse.

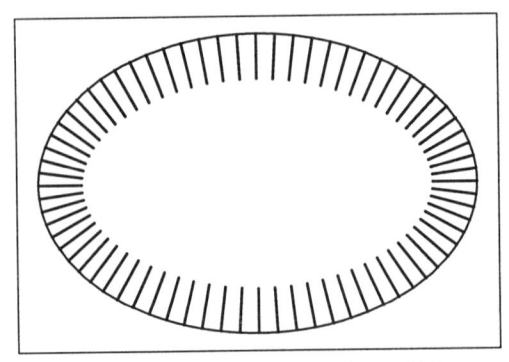

The ellipse $t \longmapsto$ **ellipse[1.5,1]**(t)
with normal lines

```
Show[Graphics[{AbsoluteThickness[1],
    Evaluate[Table[normalline[ellipse[1.5,1]][0.3,t],
    {t,0,2Pi,Pi/36}]]},
    DisplayFunction->Identity],
    ParametricPlot[Evaluate[
    ellipse[1.5,1][t]],{t,0,2Pi},
    Prolog->Thickness[0.01],
    DisplayFunction->Identity],
    AspectRatio->Automatic,
    Axes->None,PlotRange->All,
    DisplayFunction->$DisplayFunction];
```

Longer normal lines may intersect one another, as we see from the next picture.

The ellipse $t \longmapsto$ **ellipse[1.5,1]**(t)
with intersecting normal lines

```
Show[Graphics[{AbsoluteThickness[1],
        Evaluate[Table[normalline[ellipse[1.5,1]][2.5,t],
        {t,0,2Pi,Pi/36}]]},DisplayFunction->Identity],
        ParametricPlot[Evaluate[ellipse[1.5,1][t]],{t,0,2Pi},
        Prolog->AbsoluteThickness[3],DisplayFunction->Identity],
        AspectRatio->Automatic,Axes->None,PlotRange->All,
        DisplayFunction->$DisplayFunction];
```

5.6 Osculating Circles to Plane Curves

Just as the tangent line is the best line that approximates a curve at a point **p** on the curve, the osculating circle is the best circle that approximates the curve at **p**.

Definition. Let α be a regular plane curve defined on an interval (a, b), and let $a < t < b$ be such that $\kappa 2[\alpha](t) \neq 0$. Then the **osculating circle** to α at $\alpha(t)$ is the circle of

$$\text{radius} \quad \frac{1}{|\kappa 2[\alpha](t)|} \quad \text{and center} \quad \alpha(t) + \frac{1}{\kappa 2[\alpha](t)} \frac{J\alpha'(t)}{\|\alpha'(t)\|}.$$

The dictionary definition of "osculating" is kissing. In fact, the osculating circle at a point **p** on a curve approximates the curve much more closely than the tangent line. Not only do α and its osculating circle at $\alpha(t)$ have the same tangent line and normal line, but also the same curvature. The *Mathematica* function for the osculating circle is as follows.

```
osculatingcircle[alpha_][t_]:=
    Circle[alpha[tt] + D[alpha[tt],tt].D[alpha[tt],tt]/
    (D[alpha[tt],{tt,2}].
    J[D[alpha[tt],tt]])J[D[alpha[tt],tt]],
    (D[alpha[tt],tt].D[alpha[tt],tt])^(3/2)/
    (D[alpha[tt],{tt,2}].J[D[alpha[tt],
        tt]])] /. tt->t
```

It is easy to see from equation (5.1) that

Lemma 5.7. *The centers of the osculating circles to a curve form the evolute to the curve.*

The osculating circles to a logarithmic spiral are a good example of close approximation to the curve. We draw the osculating circles without the logarithmic spiral itself.

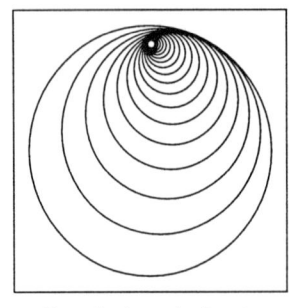

Osculating circles to
$t \longmapsto \text{logspiral}[1,-1.5](t)$

```
Show[
Graphics[Evaluate[
Table[
osculatingcircle[
logspiral[1,-1.5]][t],
{t,0,3Pi/4,Pi/24}]
]]];
```

An osculating circle to a cubic parabola is shown on page 98.

Next, we show that an osculating circle to a plane curve is the limit of circles passing through three points of the curve as the points tend to the point of contact of the osculating circle.

Theorem 5.8. *Let α be a plane curve defined on an interval (a,b), and let $a < t_1 < t_2 < t_3$. Denote by $\mathbf{C}(t_1, t_2, t_3)$ the circle passing through the points $\alpha(t_1)$, $\alpha(t_2)$, $\alpha(t_3)$ provided these points are distinct and do not lie on the same straight line. Assume that $\kappa_2[\alpha](t_0) \neq 0$. Then the osculating circle to α at $\alpha(t_0)$ is the circle*

$$\mathbf{C} = \lim_{\substack{t_1 \to t_0 \\ t_2 \to t_0 \\ t_3 \to t_0}} \mathbf{C}(t_1, t_2, t_3).$$

Proof. Denote by $\mathbf{p}(t_1, t_2, t_3)$ the center of $\mathbf{C}(t_1, t_2, t_3)$, and define $f \colon (a,b) \longrightarrow \mathbb{R}$ by

$$f(t) = \|\alpha(t) - \mathbf{p}(t_1, t_2, t_3)\|^2.$$

Then

(5.13) $$\begin{cases} f'(t) = 2\alpha'(t) \cdot (\alpha(t) - \mathbf{p}(t_1, t_2, t_3)), \\ f''(t) = 2\alpha''(t) \cdot (\alpha(t) - \mathbf{p}(t_1, t_2, t_3)) + 2\|\alpha'(t)\|^2. \end{cases}$$

Since f is differentiable and $f(t_1) = f(t_2) = f(t_3)$, there exist u_1 and u_2 with $t_1 < u_1 < t_2 < u_2 < t_3$ such that

(5.14) $$f'(u_1) = f'(u_2) = 0.$$

Similarly, there exists v with $u_1 < v < u_2$ such that

(5.15) $$f''(v) = 0.$$

5.6 Osculating Circles to Plane Curves

((5.14) and (5.15) follow from Rolle's[1] Theorem. See for example, [Buck, page 90].) Clearly, as t_1, t_2, t_3 tend to t_0, so do u_1, u_2 and v. Hence (5.13)—(5.15) imply that

(5.16)
$$\begin{cases} \boldsymbol{\alpha}'(t_0) \cdot (\boldsymbol{\alpha}(t_0) - \mathbf{p}) = 0, \\ \boldsymbol{\alpha}''(t_0) \cdot (\boldsymbol{\alpha}(t_0) - \mathbf{p}) = -\|\boldsymbol{\alpha}'(t_0)\|^2, \end{cases}$$

where

$$\mathbf{p} = \lim_{\substack{t_1 \to t_0 \\ t_2 \to t_0 \\ t_3 \to t_0}} \mathbf{p}(t_1, t_2, t_3).$$

It follows from (5.16) and the definition of $\kappa 2$ that

$$\boldsymbol{\alpha}(t_0) - \mathbf{p} = \frac{-1}{\kappa 2[\boldsymbol{\alpha}](t_0)} \frac{J\boldsymbol{\alpha}'(t_0)}{\|\boldsymbol{\alpha}'(t_0)\|}.$$

Thus, by definition, **C** is the osculating circle to $\boldsymbol{\alpha}$ at $\boldsymbol{\alpha}(t)$. ∎

Mathematica can be used to illustrate how circles through three points on a curve tend to the osculating circle at **p** as the three points approach **p**. First, we need a miniprogram, of interest in itself, to construct the circle passing through three noncollinear points. To define the miniprogram we proceed as follows. We can find the coordinates of the center {centx,centy} and the square rsqr of the radius of the circle passing through points {p1,q1}, {p2,q2} and {p3,q3} by means of Solve:

```
bb=Solve[{(p1 - centx)^2 + (q1 - centy)^2==rsqr,
          (p2 - centx)^2 + (q2 - centy)^2==rsqr,
          (p3 - centx)^2 + (q3 - centy)^2==rsqr},
         {centx,centy,rsqr}]//Simplify;
```

We put a semicolon at the end to suppress the rather complicated output. The radius, up to sign, is then given by the command

```
Sqrt[rsqr /. bb]//PowerExpand
```

---
[1] Michel Rolle (1652-1719). French mathematician, who resisted the infinitesimal techniques of calculus.

$$\{(\text{Sqrt}[p1^2 - 2\ p1\ p2 + p2^2 + q1^2 - 2\ q1\ q2 + q2^2]$$
$$\text{Sqrt}[p1^2 - 2\ p1\ p3 + p3^2 + q1^2 - 2\ q1\ q3 + q3^2]$$
$$\text{Sqrt}[p2^2 - 2\ p2\ p3 + p3^2 + q2^2 - 2\ q2\ q3 + q3^2])\ /$$
$$(2\ (-(p2\ q1) + p3\ q1 + p1\ q2 - p3\ q2 - p1\ q3 + p2\ q3))\}$$

Similarly, the coordinates of the center are given by

{centx, centy} /. bb

$$\{\{(p2^2\ q1 - p3^2\ q1 - p1^2\ q2 + p3^2\ q2 - q1^2\ q2 + q1\ q2^2 +$$
$$p1^2\ q3 - p2^2\ q3 + q1^2\ q3 - q2^2\ q3 - q1\ q3^2 + q2\ q3^2)$$
$$/\ (2\ (p2\ q1 - p3\ q1 - p1\ q2 + p3\ q2 + p1\ q3 - p2\ q3)),$$
$$(p1^2\ p2 - p1\ p2^2 - p1^2\ p3 + p2^2\ p3 +$$
$$p1\ p3^2 - p2\ p3^2 + p2\ q1^2 - p3\ q1^2 - p1\ q2^2 +$$
$$p3\ q2^2 + p1\ q3^2 - p2\ q3^2)\ /$$
$$(2\ (p2\ q1 - p3\ q1 - p1\ q2 + p3\ q2 + p1\ q3 - p2\ q3))\}\}$$

Armed with this information, we define the function we need:

```
circle3point[{p1_,q1_},{p2_,q2_},{p3_,q3_}]:=
   Circle[{(p2^2*q1 - p3^2*q1 - p1^2*q2 + p3^2*q2 -
   q1^2*q2 + q1*q2^2 + p1^2*q3 - p2^2*q3 + q1^2*q3 -
   q2^2*q3 - q1*q3^2 + q2*q3^2)/
   (2*(p2*q1 - p3*q1 - p1*q2 + p3*q2 + p1*q3 - p2*q3)),
   (p1^2*p2 - p1*p2^2 - p1^2*p3 + p2^2*p3 + p1*p3^2 -
   p2*p3^2 + p2*q1^2 - p3*q1^2 - p1*q2^2 + p3*q2^2 +
   p1*q3^2 - p2*q3^2)/
   (2*(p2*q1 - p3*q1 - p1*q2 + p3*q2 + p1*q3 - p2*q3))},
   Sqrt[(p1^2 - 2*p1*p2 + p2^2 + q1^2 - 2*q1*q2 + q2^2)*
   (p1^2 - 2*p1*p3 + p3^2 + q1^2 - 2*q1*q3 + q3^2)*
   (p2^2 - 2*p2*p3 + p3^2 + q2^2 - 2*q2*q3 + q3^2)]/Abs[
   2*(-p2*q1 + p3*q1 + p1*q2 - p3*q2 - p1*q3 + p2*q3)]]
```

Now we can use **circle3point** to construct the circle we need. This is obviously much faster than using **Solve** each time to find a circle passing through three points.

Next, we use *Mathematica* to draw a circle through three points on the curve $t \longmapsto \mathbf{parabola[1,1]}(t)$. A typical drawing command is

```
Show[ParametricPlot[parabola[1][t]//Evaluate,
{t,-2,2},AspectRatio->Automatic,Axes->None,
DisplayFunction->Identity],
Graphics[circle3point[parabola[1][-1.5],
parabola[1][-0.5],parabola[1][2]]],
PlotRange->{{-4,4.1},{-0.1,4}},
DisplayFunction->$DisplayFunction];
```

As the three points converge to the vertex of the parabola, the circle through the three points converges to the osculating circle at the vertex. Since the curvature of the parabola at the vertex is 1/2, the osculating circle at the vertex has radius 2.

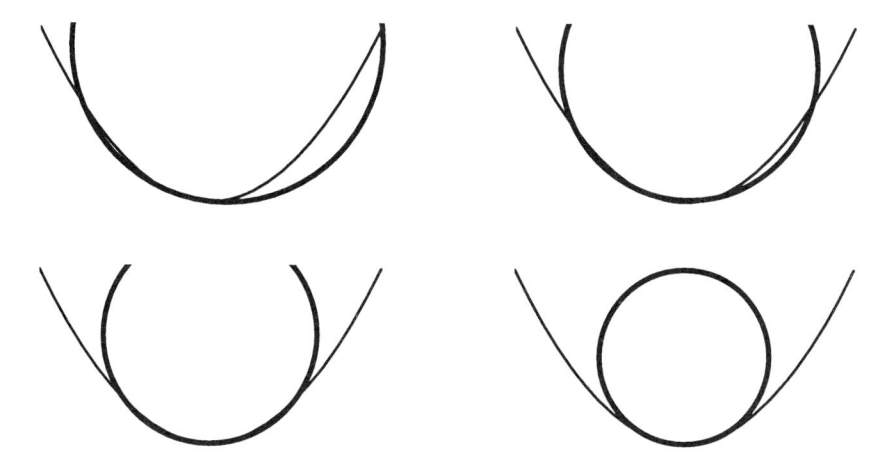

Circles converging to an osculating circle of a parabola

5.7 Parallel Curves

In this section we show how to draw a curve γ at a fixed distance $s > 0$ from a given curve α, where s is not too large. Let α and γ be defined on an interval (a, b); then

$$\|\gamma(t) - \alpha(t)\| = s \quad \text{and} \quad \bigl(\gamma(t) - \alpha(t)\bigr) \cdot \alpha'(t) = 0$$

for $a < t < b$. This leads us to make the following definition.

Definition. *A* **parallel curve** *to a regular plane curve* α *at a distance s is the plane curve given by*

$$(5.17) \qquad \text{parcurve}[\alpha][s](t) = \alpha(t) + \frac{s\, J\alpha'(t)}{\|\alpha'(t)\|}.$$

We allow s to be either positive or negative in (5.17) in order to obtain parallel curves on either side of α. The definition of parallel curve does not depend on the choice of positive reparametrization. In fact, it is not hard to prove:

Lemma 5.9. *Let* $\alpha \colon (a,b) \longrightarrow \mathbb{R}^2$ *be a plane curve, and let* $h \colon (c,d) \longrightarrow (a,b)$ *be differentiable. Then*

$$\text{parcurve}[\alpha \circ h][s](t) = \text{parcurve}[\alpha][(\text{sign } h')s](h(t)).$$

The *Mathematica* version of (5.17) is an operator **parcurve** given by

```
parcurve[alpha_][s_][t_]:= alpha[tt] +
    s J[D[alpha[tt],tt]]/Sqrt[Simplify[
        D[alpha[tt],tt].D[alpha[tt],tt]]] /. tt->t
```

As an example, let us draw some parallel curves to an ellipse.

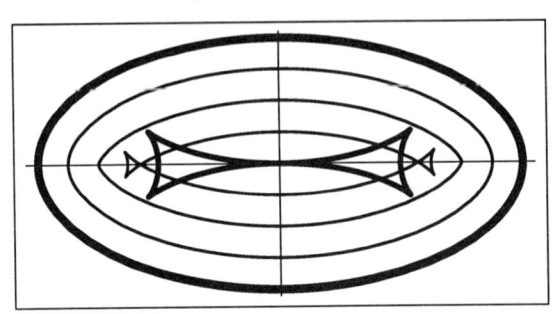

Four parallel curves to $t \longmapsto \text{ellipse}[2,1](t)$

The command to draw this picture is:

```
ParametricPlot[Evaluate[Table[parcurve[ellipse[2,1]][s][t],
    {s,0,1,0.25}]],{t,0,2Pi},AspectRatio->Automatic];
```

From the picture it is clear that when s is too large, a parallel curve may intersect itself and points on it will not necessarily be at a distance $|s|$ from the original curve. Furthermore, we can estimate when this first happens and also compute the curvature of the parallel curve.

Lemma 5.10. *Let α be a regular plane curve. Then the parallel curve* **parcurve**$[\alpha][s]$ *is a regular at those t for which $1 - s\kappa 2[\alpha](t) \neq 0$. Furthermore, its curvature is given by*

$$\kappa 2[\text{parcurve}[\alpha][s]](t) = \frac{\kappa 2[\alpha](t)}{\left|1 - s\kappa 2[\alpha](t)\right|}.$$

Proof. By Theorem 1.10, page 16, and Lemma 5.9 we can assume that α has unit-speed. Write $\beta(t) = \text{parcurve}[\alpha][s](t)$. Then $\beta = \alpha + sJ\alpha'$ so that by Lemma 1.11 we have

$$\beta' = \alpha' + sJ\alpha'' = \alpha' + sJ^2\kappa 2[\alpha]\alpha' = (1 - s\kappa 2[\alpha])\alpha'.$$

Hence the regularity statement follows. Also, we compute

$$\beta'' = (1 - s\kappa 2[\alpha])\kappa 2[\alpha]J\alpha' - s\kappa 2[\alpha]'\alpha'.$$

Hence

$$\kappa 2[\beta] = \frac{\beta'' \cdot J\beta'}{\|\beta'\|^3} = \frac{(1 - s\kappa 2[\alpha])^2 \kappa 2[\alpha]}{\left|1 - s\kappa 2[\alpha]\right|^3} = \frac{\kappa 2[\alpha]}{\left|1 - s\kappa 2[\alpha]\right|}. \blacksquare$$

5.8 Pedal Curves

Let α be a curve in the plane, and let $\mathbf{p} \in \mathbb{R}^2$. The locus of bases of perpendicular lines let down from a point $\mathbf{p}$ to a variable tangent line to α is called the **pedal curve** to α.

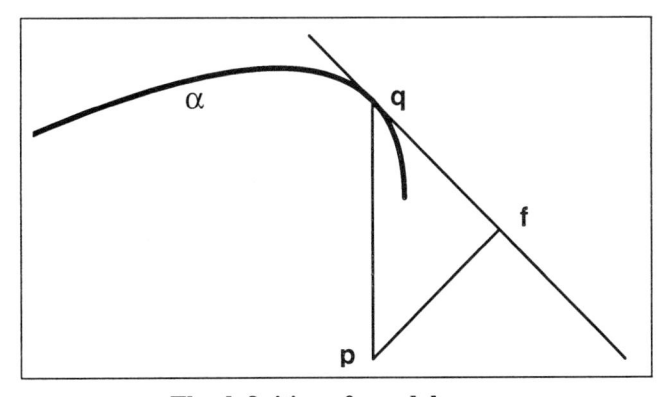

The definition of a pedal curve

Let β be the pedal curve of α with respect to a point $\mathbf{p}$, and write $\alpha(t) = \mathbf{q}$, $\beta(t) = \mathbf{f}$. Then $\beta(t) - \mathbf{p}$ must be the projection of $\alpha(t) - \mathbf{p}$ in the $J\alpha'(t)$ direction. This leads to the following definition of the pedal curve.

Definition. *The* **pedal curve** *of a regular curve* $\alpha\colon (a,b) \longrightarrow \mathbb{R}^2$ *with respect to a point* $\mathbf{p} \in \mathbb{R}^2$ *is defined by*

$$\mathbf{pedal}[\mathbf{p}, \alpha](t) = \mathbf{p} + \frac{(\alpha(t) - \mathbf{p}) \cdot J\alpha'(t)}{\|\alpha'(t)\|^2} J\alpha'(t).$$

The corresponding *Mathematica* definition is

```
pedal[{a_,b_},alpha_][t_]:= {a,b} +
    ((alpha[tt]-{a,b}).J[D[alpha[tt],tt]]/
    (D[alpha[tt],tt].D[alpha[tt],tt]))J[
    D[alpha[tt],tt]] /. {tt->t}
```

It is useful to have a simplified version of **pedal** in the case that `{a,b}={0,0}`:

```
pedal[alpha_][t_]:= pedal[{0,0},alpha][t]
```

The point $\mathbf{pedal}[\mathbf{p}, \alpha](t)$ is the foot of the perpendicular from $\mathbf{p}$ to the tangent line to α at t.

The proof of the following lemma follows the model of Theorem 1.10; see exercise 25.

Lemma 5.11. *The definition of pedal curve of α is independent of the parametrization of α; that is,*

$$\mathbf{pedal}[\mathbf{p}, \alpha \circ h] = \mathbf{pedal}[\mathbf{p}, \alpha] \circ h.$$

For example, we draw the pedal curve of the curve $t \longmapsto (t, t^3/3 + 1/2)$ discussed by Banchoff, Gaffney, and McCrory (see [BGM, page 5]). It is defined in *Mathematica* by `alpha[t_]:= {t,t^3/3 + 1/2}`

5.8 Pedal Curves

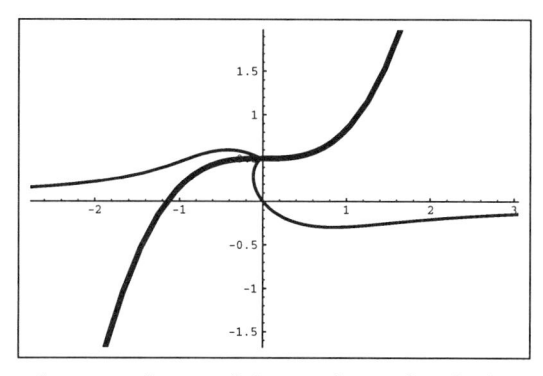

```
ParametricPlot[
{alpha[t],
pedal[alpha][t]
}//Evaluate,
{t,-5,5},
AspectRatio->
Automatic];
```

A curve whose pedal curve has a singularity

Examples of pedal curves are:

(i) The pedal curve of a parabola with respect to its vertex is a cissoid.

(ii) The pedal curve of a circle with respect to any point **p** other than the center of the circle is a limaçon. If **p** lies on the circumference of the circle, the limaçon reduces to a cardioid. The pedal curve of a circle with respect to its center is the circle itself.

(iii) The pedal curve of cardioid with respect to its cusp point is called **Cayley's**[2] **sextic**. We use *Mathematica* to compute `pedal[{0,0},cardioid[a]][t]` and then define a new curve by putting `cayleysextic[a_][t_]` equal to the resulting expression. In other words, we define

```
cayleysextic[a_][t_]:=
    {4 a Cos[t/2]^4(-1 + 2 Cos[t]),
     4 a Cos[t/2]^3Sin[(3 t)/2]}
```

To draw the cardioid and Cayley's sextic simultaneously, we use

---

[2] Arthur Cayley (1821-1895). Cayley was one of the leading English mathematicians of the 19th century; his complete works fill many volumes. He is particularly known for his work on matrices, elliptic functions and nonassociative algebras. His first 14 professional years were spent as a lawyer; during that time he published over 250 papers. In 1863 Cayley was appointed Sadleirian professor of mathematics at Cambridge with a greatly reduced salary.

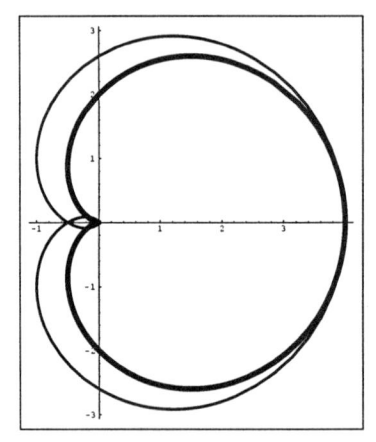

```
ParametricPlot[
{cardioid[1][t],
pedal[
cardioid[1]][t]
}//Evaluate,
{t,0,2Pi},
AspectRatio->
Automatic];
```

Cardioid and its pedal curve: Cayley's sextic

For more information about pedal curves see [Law, pages 46–49], [Lock, pages 153–160] and [Zwi, pages 150–158].

5.9 Exercises

1. Use *Mathematica* to draw evolutes of a cycloid, a cardioid and a logarithmic spiral. Show that the evolute of a cardioid is another cardioid, the evolute of a cycloid is another cycloid and the evolute of a logarithmic spiral is another logarithmic spiral.

2. Prove Lemma 5.1.

3. Draw normal lines to a cardioid, making the lines sufficiently long so that they intersect.

4. Draw several osculating circles to a parabola.

5. Draw as one graph four parallel curves to a lemniscate. Do the same for a cardioid and a deltoid.

6. Define a *Mathematica* function for a line that meets a curve at a given point at an arbitrary angle θ, so that for $\theta = 0$ and $\theta = \pi$ the function reduces to **tangentline** and **normalline**, respectively.

5.9 Exercises

7. Generalize **osculatingcircle** so that it draws a portion of the osculating circle at each point.

8. Find and draw the pedal curve of an ellipse with respect to its center.

9. Find and draw the pedal curve of a catenary with respect to the origin.

10. The general cubic parabola can be defined in *Mathematica* with the command

```
cubicparabola[a_,b_,c_,d_][t_]:=
            {t,a t^3 + b t^2 + c t + d}
```

Use *Mathematica* to draw the picture on page 98 of an osculating circle to the cubic parabola $t \longmapsto$ **cubicparabola**$[1, 0, -1, 1](t)$.

11. Use *Mathematica* to draw the picture on page 117 that describes the definition of pedal point.

12. Use **Nest** to find the second and third involutes of a circle. Define each involute as a curve and compute its signed curvature.

13. Show that the curve defined by

$$\alpha[n,a](t) = a\, e^{it} \sum_{k=0}^{n} \frac{(-it)^k}{k!}$$

is the nth involute starting at $(a, 0)$ of a circle of radius a.

14. A **strophoid** of a curve α with respect to a point $\mathbf{p} \in \mathbb{R}^2$ is a curve γ such that
$$\|\alpha(t) - \gamma(t)\| = \|\alpha(t) - \mathbf{p}\| \qquad \text{and} \qquad \gamma(t) = s\,\alpha(t)$$
for some s. Show that α has two strophoids and find the equations for them. Define the corresponding *Mathematica* function.

15. Show that the involute of a catenary is a tractrix. Draw the following picture, which illustrates this fact.

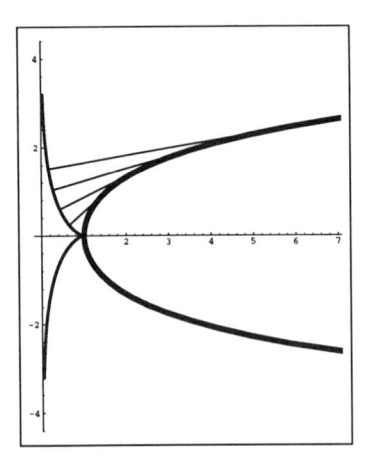

Unwinding a string from a catenary

16. In general it is difficult or impossible for *Mathematica* to find the involute of a curve symbolically, because an integration is required; this integration is equivalent to finding the solution of a differential equation. On the other hand, *Mathematica* can solve a differential equation numerically using *Mathematica's* command **NDSolve**, which is discussed in Chapter 6. Here is a miniprogram that uses **NDSolve** to find the involute of a curve:

```
involutends[a_,b_][alpha_][t_]:=
    Module[{aaa,bbb,s,tt},
        aaa=NDSolve[{s'[tt]==
            arclengthprime[alpha][tt],
            s[a]==0},s,{tt,a,b}];
        bbb=alpha[tt] +
            (s[a] - s[tt])D[alpha[tt],tt]/
            Sqrt[D[alpha[tt],tt].
            D[alpha[tt],tt]]   /.aaa;
        bbb /. tt->t]
```

Use this miniprogram to draw the picture on page 105 of a figure eight and its involute.

17. Draw a picture of the cubic parabola $t \longmapsto$ **cubicparabola[1, 0, -1, 1]**(t) and its involute; use **cubicparabola[1, 0, -1, 1]**(-1) as a starting point.

5.9 Exercises

18. An ordinary pendulum swings back and forth in a circular arc. The oscillations are not isochronous. To phrase it differently, the time it takes for the (circular) pendulum to go from its staring point to its lowest point will be almost, but not quite independent, of the height from which the pendulum is released. In 1680 Huygens[3] observed two important facts: (a) a pendulum that swings back and forth in an inverted cycloidial arc is isochronous, and (b) the involute of a cycloid is a cycloid. Combined, these facts show that the involute of an inverted cycloidial arc can be used to constrain a pendulum so that it moves in a cycloidial arc. Use *Mathematica* to show (b). Then use *Mathematica* to draw the following picture, which shows an inverted cycloid, its involute, and the pendulum, proving (a).

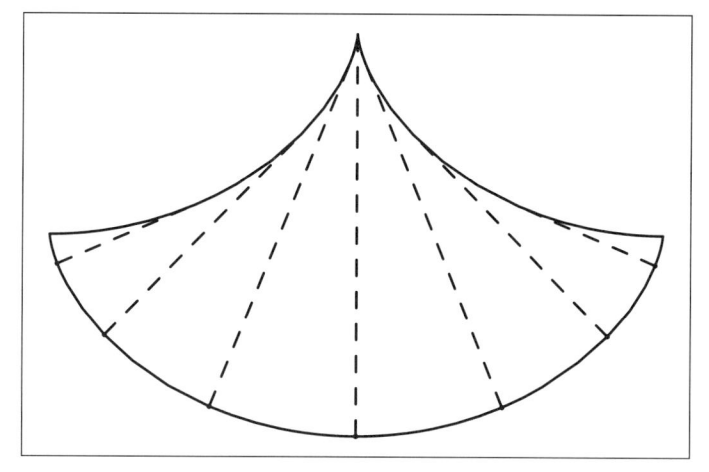

Cycloidial pendulum

19. A **point of inflection** of a plane curve α is a point $\alpha(t_0)$ for which $\kappa 2[\alpha](t_0) = 0$; a **strong point of inflection** is a point $\alpha(t_0)$ for which there exists $\varepsilon > 0$ such that $\kappa 2[\alpha](t)$ is negative for $t_0 - \varepsilon < t < t_0$ and positive for $t_0 < t < t_0 + \varepsilon$, or vice versa.

---

[3] Christiaan Huygens (1629-1695). A leading Dutch scientist of the 17th century. As a mathematician he was a major precursor of Leibniz and Newton. His astronomical contributions include the discovery of the rings of Saturn. In 1656 Huygens patented the first pendulum clock.

(a) Show that if $\alpha(t_0)$ is a strong inflection point of α, and if $t \longmapsto \kappa 2[\alpha](t)$ is continuous at t_0, then $\alpha(t_0)$ is an inflection point of α.

(b) For the curve $t \longmapsto (t^3, t^5)$ show that $(0,0)$ is a strong inflection point which is not an inflection point. Plot the curve.

20. Let $\alpha(t_0)$ be a strong inflection point of a curve $\alpha \colon (a,b) \longrightarrow \mathbb{R}^2$. Show that any involute of α must have discontinuous curvature at t_0. This accounts for the cusps on the involutes of a figure eight and of a cubic parabola.

21. Find the parametric form of the pedal curve of the parabola $t \longmapsto (2at, t^2)$ with respect to the point $(0, b)$. Show that the nonparametric form of the pedal curve is

$$(x^2 + y^2)y + (a-b)x^2 + 2ay^2 - a^2y = 0.$$

22. Show that the formula for the pedal curve of α with respect to the origin can be written as

$$\mathbf{pedal}[0, \alpha] = \frac{\alpha\overline{\alpha'} - \overline{\alpha}\alpha'}{2\overline{\alpha'}}.$$

(See [Zwi, page 150].)

23. The **contrapedal** of a plane curve α is defined analogously to the pedal of α. It is the locus of bases of perpendicular lines let down from a point $\mathbf{p}$ to a variable normal line to α. The exact formula is

$$\mathbf{contrapedal}[\mathbf{p}, \alpha](t) = \mathbf{p} + \frac{(\alpha(t) - \mathbf{p}) \cdot J\alpha'(t)}{\|\alpha'(t)\|^2} J\alpha'(t).$$

Show that the formula for the contrapedal curve of α with respect to the origin can be written as

$$\mathbf{contrapedal}[0, \alpha] = \frac{\alpha\overline{\alpha'} + \overline{\alpha}\alpha'}{2\overline{\alpha'}}.$$

24. Show that the pedal curve of a circle with respect to a point on the circle is a cardioid. Plot pedal curves of a circle with respect to points inside the circle, and then with respect to points outside the circle.

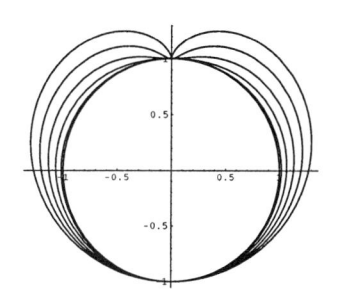

Pedal curves of a circle with respect to points inside the circle

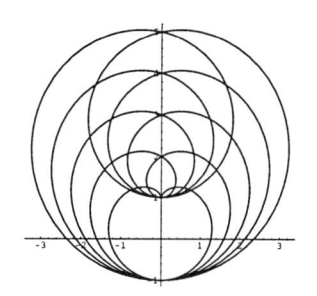

Pedal curves of a circle with respect to points outside the circle

25. Prove Lemma 5.11.

5.10 Animation of a Cycloidial Pendulum

First, we define the evolute of a cycloid.

```
evolute[cycloid[a_,a_]][t_]:=
    {a(t + Sin[t]),-2a Sin[t/2]^2}
```

Then the following command animates a cycloidial pendulum.

```
Do[ParametricPlot[Evaluate[
{cycloid[-1,-1][t],evolute[cycloid[-1,-1]][t]}],
{t,-0.001,2Pi + 0.001},
AspectRatio->Automatic,Axes->None,
Epilog->{
Line[{cycloid[-1,-1][a],
evolute[cycloid[-1,-1]][a]}],
AbsolutePointSize[8],Point[cycloid[-1,-1][a]]}],
{a,-0.001,2Pi + 0.001,Pi/30}]
```

6

DETERMINING A PLANE CURVE FROM ITS CURVATURE

In this chapter we confront the following question: *to what extent does the curvature determine a plane curve?* This question has two parts: *when do two curves have the same curvature, and when can a curve be determined from its curvature?*

To answer the first part, we begin by asking: which transformations of $\mathbb{R}^2$ preserve curvature? We have seen (Theorem 1.10, page 16) that the curvature of a plane curve is independent up to sign of the parametrization. Therefore, without loss of generality, we can assume in this chapter that all curves have unit-speed.

It is intuitively clear that the image of a plane curve α under a rotation or translation of $\mathbb{R}^2$ has the same curvature as α. Rotations and translations are examples of Euclidean motions of $\mathbb{R}^2$, those maps of $\mathbb{R}^2$ into itself which do not distort distances. We first discuss Euclidean motions in general in Section 6.1. *Mathematica* functions for translating, rotating and reflecting curves are given in Section 6.2.

The invariance of curvature under Euclidean motions is established in Section 6.3. In Section 6.3 we also prove the *Fundamental Theorem of Plane Curves*; this important theorem states that two unit-speed curves in $\mathbb{R}^2$ that have the same curvature differ only by a Euclidean motion of $\mathbb{R}^2$.

The second part of our question now becomes: can a unit-speed curve in $\mathbb{R}^2$ be determined up to a Euclidean motion from its curvature? In fact, in Section 6.3

we give an explicit system of differential equations for determining a plane curve from its curvature. In simple cases this system can be solved explicitly, but in the general case only a numerical solution is possible. So the answer to the second part of the question is yes.

Mathematica can be used effectively to determine numerically a unit-speed plane curve with a given curvature and then to graph it; we give a miniprogram in Section 6.4. Many strange interesting curves can be constructed numerically using the miniprogram.

In Section 6.6 we describe animations of epicycloids and epitrocoids.

6.1 Euclidean Motions

In order to discuss the invariance of the curvature of plane curves under rotations and translations, we need to make precise what we mean by invariance. This requires discussing various kinds of maps of $\mathbb{R}^2$ into itself. We begin by defining several classes of maps of $\mathbb{R}^n$ into itself before specializing to $\mathbb{R}^2$.

First, we define several important types of linear maps. The linearity property $(A(a\mathbf{p} + b\mathbf{q}) = a\,A\mathbf{p} + b\,A\mathbf{q})$ implies that each of these maps takes the origin into the origin.

Definition. *Let* $A\colon \mathbb{R}^n \longrightarrow \mathbb{R}^n$ *be a nonsingular linear map.*

(i) *We say that A is* **orientation-preserving** *if* $\det(A)$ *is positive, or* **orientation-reversing** *if* $\det(A)$ *is negative.*

(ii) *A is called an* **orthogonal transformation** *if*

$$A\mathbf{p} \cdot A\mathbf{q} = \mathbf{p} \cdot \mathbf{q}$$

for all $\mathbf{p}, \mathbf{q} \in \mathbb{R}^n$.

(iii) *A* **rotation** *of $\mathbb{R}^n$ is an orientation-preserving orthogonal transformation.*

(iv) *A* **reflection** *of $\mathbb{R}^n$ is an orthogonal transformation of the form $S_\mathbf{q}$, where $\mathbf{q} \neq \mathbf{0}$ and*

(6.1) $$S_\mathbf{q}(\mathbf{p}) = \mathbf{p} - \frac{2(\mathbf{p} \cdot \mathbf{q})}{\|\mathbf{q}\|^2}\mathbf{q}$$

for all $\mathbf{p} \in \mathbb{R}^n$.

6.1 Euclidean Motions

Lemma 6.1. *Let* $B: \mathbb{R}^n \longrightarrow \mathbb{R}^n$ *be an orthogonal transformation. Then*

$$\det(B) = \pm 1.$$

Proof. If $B: \mathbb{R}^n \longrightarrow \mathbb{R}^n$ is any linear map, let ${}^tB$ denote the transpose of B. By definition $({}^tB\mathbf{p}) \cdot \mathbf{q} = \mathbf{p} \cdot (B\mathbf{q})$ for all $\mathbf{p}$ and $\mathbf{q}$. The matrix of ${}^tB$ with respect to any orthonormal basis of $\mathbb{R}^n$ is (a_{ji}), where (a_{ij}) is the matrix of B; consequently, $\det({}^tB) = \det(B)$. Now assume that B is orthogonal. Then ${}^tBB$ is the identity transformation I. Hence

$$1 = \det I = \det({}^tBB) = \det({}^tB)\det(B) = \det(B)^2. \quad \blacksquare$$

We can characterize reflections geometrically.

Lemma 6.2. *Let* $S_{\mathbf{q}}: \mathbb{R}^n \longrightarrow \mathbb{R}^n$ *be a reflection, where* $\mathbf{q} \neq 0$. *Then* $\det(S_{\mathbf{q}}) = -1$. *Furthermore, $S_{\mathbf{q}}$ restricts to the identity map on a hyperplane P (that is, a subspace of $\mathbb{R}^n$ of dimension $n - 1$) and reverses directions on the orthogonal complement of P. Conversely, any hyperplane gives rise to a reflection $S_{\mathbf{q}}$ which restricts to the identity map on P and reverses directions on the orthogonal complement of P.*

Proof. Let P be the orthogonal complement of $\mathbf{q}$; that is,

$$P = \{\mathbf{p} \in \mathbb{R}^n \mid \mathbf{p} \cdot \mathbf{q} = 0\}.$$

Clearly, $\dim(P) = n - 1$, and it is easy to check that $S_{\mathbf{q}}(\mathbf{p}) = \mathbf{p}$ for all $\mathbf{p} \in P$. Since $S_{\mathbf{q}}(\mathbf{q}) = -\mathbf{q}$, it follows that $S_{\mathbf{q}}$ reverses directions on the orthogonal complement of P.

To show that $\det(S_{\mathbf{q}}) = -1$, we choose a basis $\{\mathbf{f}_1, \ldots, \mathbf{f}_{n-1}\}$ of P; then it is easy to check that $\{\mathbf{q}, \mathbf{f}_1, \ldots, \mathbf{f}_{n-1}\}$ is a basis of $\mathbb{R}^n$. Since the matrix of $S_{\mathbf{q}}$ with respect to this basis is

$$\begin{pmatrix} -1 & 0 & \cdots & 0 \\ 0 & 1 & \cdots & 0 \\ \vdots & \vdots & \ddots & \vdots \\ 0 & 0 & \cdots & 1 \end{pmatrix},$$

it follows that $\det(S_{\mathbf{q}}) = -1$.

Conversely, given a hyperplane P let $\mathbf{q}$ be a nonzero vector perpendicular to P. Then the reflection $S_{\mathbf{q}}$ (defined by (6.1)) has the required properties. $\blacksquare$

Next, we define several kinds of maps of $\mathbb{R}^n$ into itself that may not take the origin into the origin.

Definition. Let $\mathbf{q} \in \mathbb{R}^n$.

(i) An **affine transformation** *of $\mathbb{R}^n$ is a map* $F: \mathbb{R}^n \longrightarrow \mathbb{R}^n$ *of the form*

$$F(\mathbf{p}) = A\mathbf{p} + \mathbf{q}$$

for all $\mathbf{p} \in \mathbb{R}^n$, *where A is a linear transformation of $\mathbb{R}^n$. We call A the* **linear part** *of the affine transformation F. An affine transformation F is* **orientation-preserving** *if* $\det(A)$ *is positive, or* **orientation-reversing** *if* $\det(A)$ *is negative.*

(ii) *A* **translation** *of $\mathbb{R}^n$ is an affine map* $T_\mathbf{q}: \mathbb{R}^n \longrightarrow \mathbb{R}^n$ *of the form*

$$T_\mathbf{q}(\mathbf{p}) = \mathbf{p} + \mathbf{q}$$

for all $\mathbf{p} \in \mathbb{R}^n$.

(iii) *A* **Euclidean motion** *of $\mathbb{R}^n$ is an affine transformation whose linear part is an orthogonal transformation.*

(iv) *An* **isometry** *of $\mathbb{R}^n$ is a map* $F: \mathbb{R}^n \longrightarrow \mathbb{R}^n$ *that preserves distance; that is,*

$$\|F(\mathbf{p}_1) - F(\mathbf{p}_2)\| = \|\mathbf{p}_1 - \mathbf{p}_2\|$$

for all $\mathbf{p}_1, \mathbf{p}_2 \in \mathbb{R}^n$. *It is easy to see that an isometry must be a bijective map, that is, a one-to-one map that is onto.*

To put these notions in their proper context, let us recall the following important notions from abstract algebra.

Definition. *A* **group** *consists of a nonempty set $\mathcal{G}$ and a multiplication*

$$\circ: \mathcal{G} \times \mathcal{G} \longrightarrow \mathcal{G}$$

with the following properties.

(i) *There exists an* **identity element** $\mathbf{1} \in \mathcal{G}$, *that is, an element $\mathbf{1}$ such that*

$$a \circ \mathbf{1} = \mathbf{1} \circ a = a$$

for all $a \in \mathcal{G}$.

(ii) *Multiplication is* **associative**; *that is,*

$$(a \circ b) \circ c = a \circ (b \circ c).$$

for all $a, b, c \in \mathcal{G}$.

6.1 Euclidean Motions

(iii) *Every element $a \in \mathcal{G}$ has a unique inverse; that is, for each $a \in \mathcal{G}$ there exists a unique element $a^{-1} \in \mathcal{G}$ such that*

$$a \circ a^{-1} = a^{-1} \circ a = \mathbf{1}.$$

Definition. *A* **subgroup** *of a group $\mathcal{G}$ is a nonempty subset $\mathcal{S}$ of $\mathcal{G}$ closed under multiplication and inverses; that is,*

$$a \circ b \in \mathcal{S} \quad \text{and} \quad a^{-1} \in \mathcal{S}$$

for all $a, b \in \mathcal{S}$.

It is easy to check that a subgroup of a group is itself a group. See any book on abstract algebra for more details on groups, for example [Scott] or [NiSh].

Many sets of maps form groups. Recall that if $F: \mathbb{R}^m \longrightarrow \mathbb{R}^n$ and $G: \mathbb{R}^n \longrightarrow \mathbb{R}^p$ are maps, the **composition** $(G \circ F): \mathbb{R}^m \longrightarrow \mathbb{R}^p$ is defined by

$$(G \circ F)(\mathbf{p}) = G(F(\mathbf{p}))$$

for $\mathbf{p} \in \mathbb{R}^m$. Denote by $\mathbf{GL}(n)$ the set of all nonsingular linear maps of $\mathbb{R}^n$ into itself. It is easy to verify that the composition makes $\mathbf{GL}(n)$ into a group. Similarly, we may speak of the following groups:

- the group $\mathbf{Affine}(\mathbb{R}^n)$ of affine maps of $\mathbb{R}^n$;
- the group $\mathbf{O}(n)$ of orthogonal transformations of $\mathbb{R}^n$;
- the group $\mathbf{Translation}(\mathbb{R}^n)$ of translations of $\mathbb{R}^n$;
- the group $\mathbf{Isometry}(\mathbb{R}^n)$ of isometries of $\mathbb{R}^n$;
- the group $\mathbf{Euclidean}(\mathbb{R}^n)$ of Euclidean motions of $\mathbb{R}^n$;

The verification that each is a group is easy. Moreover:

Theorem 6.3. *A map $F: \mathbb{R}^n \longrightarrow \mathbb{R}^n$ is an isometry of $\mathbb{R}^n$ if and only if it is a composition of a translation and an orthogonal transformation of $\mathbb{R}^n$. Thus the group of Euclidean motions $\mathbf{Euclidean}(\mathbb{R}^n)$ coincides with the group $\mathbf{Isometry}(\mathbb{R}^n)$ of isometries of $\mathbb{R}^n$.*

Proof. It is easy to check that any orthogonal transformation or translation of $\mathbb{R}^n$ preserves the distance function. Hence the composition of an orthogonal transformation and a translation also preserves the distance function. This shows that **Euclidean**$(\mathbb{R}^n) \subseteq$ **Isometry**$(\mathbb{R}^n)$.

Conversely, let $F\colon \mathbb{R}^n \longrightarrow \mathbb{R}^n$ be an isometry, and define $G\colon \mathbb{R}^n \longrightarrow \mathbb{R}^n$ by $G(\mathbf{p}) = F(\mathbf{p}) - F(\mathbf{0})$. Then G is an isometry and $G(\mathbf{0}) = \mathbf{0}$. To show that G preserves the scalar product of $\mathbb{R}^n$, we calculate

$$-2G(\mathbf{p}) \cdot G(\mathbf{q}) = \|G(\mathbf{p}) - G(\mathbf{q})\|^2 - \|G(\mathbf{p}) - G(\mathbf{0})\|^2 - \|G(\mathbf{q}) - G(\mathbf{0})\|^2$$

$$= \|\mathbf{p} - \mathbf{q}\|^2 - \|\mathbf{p} - \mathbf{0}\|^2 - \|\mathbf{q} - \mathbf{0}\|^2$$

$$= -2\mathbf{p} \cdot \mathbf{q}.$$

To show that G is linear, let $\{\mathbf{e}_1, \ldots, \mathbf{e}_n\}$ be an orthonormal basis of $\mathbb{R}^n$. By definition each $\mathbf{e}_i$ has unit length and is perpendicular to every other $\mathbf{e}_j$. Then $\{G(\mathbf{e}_1), \ldots, G(\mathbf{e}_n)\}$ is also an orthonormal basis of $\mathbb{R}^n$. Therefore, for any $\mathbf{p} \in \mathbb{R}^n$ we have

$$G(\mathbf{p}) = \sum_{j=1}^{n} \big(G(\mathbf{p}) \cdot G(\mathbf{e}_j)\big) G(\mathbf{e}_j) = \sum_{j=1}^{n} (\mathbf{p} \cdot \mathbf{e}_j) G(\mathbf{e}_j).$$

(See exercise 1.) Since each $\mathbf{p} \longmapsto (\mathbf{p} \cdot \mathbf{e}_j)$ is linear, so is G. Hence G is an orthogonal transformation of $\mathbb{R}^n$, so that F is the composition of the orthogonal transformation G and the translation $T_{F(\mathbf{0})}$. We have proved that **Isometry**$(\mathbb{R}^n) \subseteq$ **Euclidean**$(\mathbb{R}^n)$. Since we have already proved the reverse inclusion, it follows that **Isometry**$(\mathbb{R}^n)$ and **Euclidean**$(\mathbb{R}^n)$ coincide. ∎

Next, we specialize to $\mathbb{R}^2$. First, we give a characterization of the orientation-preserving and reversing transformations of $\mathbb{R}^2$ in terms of the complex structure J of $\mathbb{R}^2$, defined on page 4.

Lemma 6.4. *Let A be an orthogonal transformation of $\mathbb{R}^2$. Then*

(6.2) $$AJ = \varepsilon J A,$$

(6.3) $$\det(A) = \varepsilon,$$

where

$$\varepsilon = \begin{cases} +1 & \text{if } A \text{ is orientation-preserving,} \\ -1 & \text{if } A \text{ is orientation-reversing.} \end{cases}$$

Furthermore, $\det(A) = -1$ if and only if A is a reflection.

6.1 Euclidean Motions

Proof. Let $\mathbf{p} \in \mathbb{R}^2$ be a nonzero vector. Then $JA\mathbf{p} \cdot A\mathbf{p} = 0$; since A is orthogonal,

$$AJ\mathbf{p} \cdot A\mathbf{p} = J\mathbf{p} \cdot \mathbf{p} = 0.$$

Thus both $AJ\mathbf{p}$ and $JA\mathbf{p}$ are perpendicular to $A\mathbf{p}$, so that for some $\lambda \in \mathbb{R}$ we have $AJ\mathbf{p} = \lambda J A\mathbf{p}$. Since both A and J are orthogonal transformations, it is easy to see that $|\lambda| = 1$.

Let us compute the matrix of A with respect to the basis $\{\mathbf{p}, J\mathbf{p}\}$ of $\mathbb{R}^2$. First, we write

$$A\mathbf{p} = a_{11}\mathbf{p} + a_{12}J\mathbf{p}.$$

Then

$$AJ\mathbf{p} = \lambda J A\mathbf{p} = \lambda J(a_{11}\mathbf{p} + a_{12}J\mathbf{p}) = \lambda(-a_{12}\mathbf{p} + a_{11}J\mathbf{p}),$$

so that the matrix of A is

$$\begin{pmatrix} a_{11} & a_{12} \\ -\lambda a_{12} & \lambda a_{11} \end{pmatrix}.$$

Thus $\varepsilon = \det(A) = \lambda(a_{11}^2 + a_{12}^2)$. Also, for any nonzero $\mathbf{p} \in \mathbb{R}^2$ we have

$$\|\mathbf{p}\|^2(a_{11}^2 + a_{12}^2) = \|A\mathbf{p}\|^2 = \|\mathbf{p}\|^2,$$

so that $(a_{11}^2 + a_{12}^2) = 1$. Hence $\lambda = \varepsilon$, and we obtain both (6.2) and (6.3).

Lemma 6.2 implies that the determinant of any reflection of $\mathbb{R}^2$ is -1. Conversely, an orthogonal transformation $A: \mathbb{R}^2 \longrightarrow \mathbb{R}^2$ with $\det(A) = -1$ must have 1 and -1 as its eigenvalues. Let $\mathbf{q} \in \mathbb{R}^2$ be such that $A(\mathbf{q}) = -\mathbf{q}$ and $\|\mathbf{q}\| = 1$. Then $A = S_\mathbf{q}$. ∎

The following theorem is due to Chasles.[1]

Theorem 6.5. *Every isometry of $\mathbb{R}^2$ is the composition of translations, reflections and rotations.*

Proof. Theorem 6.3 implies that an isometry F of $\mathbb{R}^2$ is a Euclidean motion. There is an orthogonal transformation $A: \mathbb{R}^2 \longrightarrow \mathbb{R}^2$ and $\mathbf{q} \in \mathbb{R}^2$ such that $F(\mathbf{p}) = A(\mathbf{p}) + \mathbf{q}$. Otherwise said,

$$F = T_\mathbf{q} \circ A.$$

---

[1] Michel Chasles (1793-1880). French geometer and mathematical historian. He worked on algebraic and projective geometry.

From Lemma 6.4 we know that A is a rotation if $\det(A) = 1$ or a reflection if $\det(A) = -1$. ∎

The proof of the next lemma is an easy calculation.

Lemma 6.6. *Let A be a rotation of $\mathbb{R}^2$. Then*

$$\mathbf{p} \longmapsto \frac{A\mathbf{p} \cdot \mathbf{p}}{\|\mathbf{p}\|^2} \quad \text{and} \quad \mathbf{p} \longmapsto \frac{AJ\mathbf{p} \cdot \mathbf{p}}{\|\mathbf{p}\|^2}$$

are constant functions.

Definition. *Let A be an orientation-preserving orthogonal transformation of $\mathbb{R}^2$. The* **rotation angle** *of A is the unique number θ such that $0 \leq \theta < 2\pi$ and*

$$\cos\theta = \frac{A\mathbf{p} \cdot \mathbf{p}}{\|\mathbf{p}\|^2}, \quad \sin\theta = \frac{AJ\mathbf{p} \cdot \mathbf{p}}{\|\mathbf{p}\|^2}$$

for any nonzero $\mathbf{p} \in \mathbb{R}^2$. Hence using the identification (1.1), page 4, of $\mathbb{C}$ with $\mathbb{R}^2$, we can write

$$A\mathbf{p} = \cos\theta\,\mathbf{p} + \sin\theta\,J\mathbf{p} = e^{i\theta}\mathbf{p}$$

for all $\mathbf{p} \in \mathbb{R}^2$.

6.2 Curves and Euclidean Motions

It is clear intuitively that any mapping of $\mathbb{R}^2$ into itself maps a curve into another curve. Here is the exact definition.

Definition. *Let $F\colon \mathbb{R}^2 \longrightarrow \mathbb{R}^2$ be a map, and let $\boldsymbol{\alpha}\colon (a,b) \longrightarrow \mathbb{R}^2$ be a curve. The* **image** *of $\boldsymbol{\alpha}$ under F is the curve $F \circ \boldsymbol{\alpha}$.*

It is easy to translate, rotate or reflect plane curves using *Mathematica* with functions `tl` (for translation), `rt` (for rotation) and `rf` (for reflection) defined as follows:

6.2 Curves and Euclidean Motions

```
tl[p_,alpha_][t_]:= p + alpha[t]

rt[theta_,alpha_][t_]:=
    {Cos[theta]alpha[t][[1]] - Sin[theta]alpha[t][[2]],
     Sin[theta]alpha[t][[1]] + Cos[theta]alpha[t][[2]]}

rf[alpha_][t_]:= {alpha[t][[1]],-alpha[t][[2]]}
```

For example, we can use **rt** to rotate an ellipse by an angle of $\pi/12$:

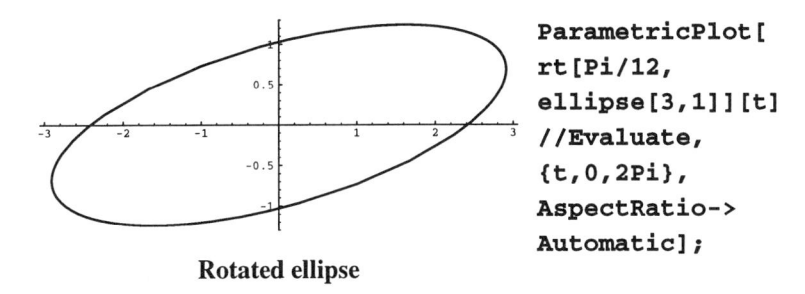

```
ParametricPlot[
rt[Pi/12,
ellipse[3,1]][t]
//Evaluate,
{t,0,2Pi},
AspectRatio->
Automatic];
```

Rotated ellipse

A reflected logspiral spirals the other way:

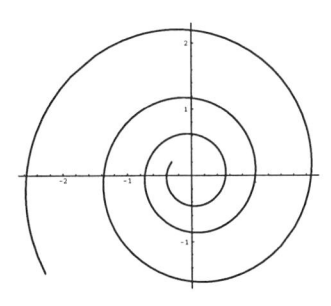

Logspiral

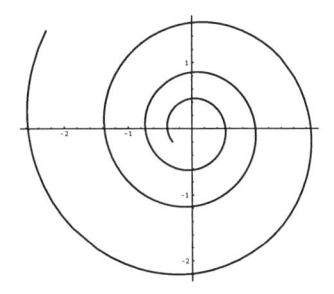

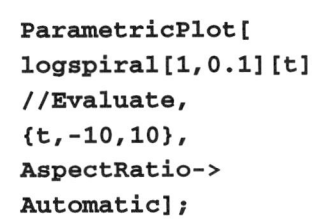

Reflected logspiral

```
ParametricPlot[
logspiral[1,0.1][t]
//Evaluate,
{t,-10,10},
AspectRatio->
Automatic];
```

```
ParametricPlot[
rf[logspiral[1,0.1]][t]
//Evaluate,
{t,-10,10},
AspectRatio->
Automatic];
```

6.3 Intrinsic Equations for Plane Curves

Since a Euclidean motion F of $\mathbb{R}^2$ preserves distance, it cannot stretch or otherwise distort a plane curve. We now show that F preserves the curvature up to sign.

Theorem 6.7. *The absolute value of the curvature and the derivative of arc length of a curve are invariant under Euclidean motions of $\mathbb{R}^2$. The curvature $\kappa 2$ is preserved by an orientation-preserving Euclidean motion of $\mathbb{R}^2$ and changes sign under an orientation-reversing Euclidean motion.*

Proof. Let $\alpha\colon (a,b) \longrightarrow \mathbb{R}^2$ be a curve, and let $F\colon \mathbb{R}^2 \longrightarrow \mathbb{R}^2$ be a Euclidean motion. Let A denote the linear part of F, so that for all $\mathbf{p} \in \mathbb{R}^2$ we have $F(\mathbf{p}) = A\mathbf{p} + F(0)$. Define a curve $\gamma\colon (a,b) \longrightarrow \mathbb{R}^2$ by $\gamma = F \circ \alpha$; then for $a < t < b$ we have

$$\gamma(t) = A\alpha(t) + F(0).$$

Hence $\gamma'(t) = A\alpha'(t)$ and $\gamma''(t) = A\alpha''(t)$. Let s_α and s_γ denote the arc length functions with respect to α and γ. Since A is an orthogonal transformation, we have

$$s'_\gamma(t) = \|\gamma'(t)\| = \|A\alpha'(t)\| = s'_\alpha(t).$$

We compute the curvature of γ as follows:

$$\kappa 2[\gamma](t) = \frac{\gamma''(t) \cdot J\gamma'(t)}{\|\gamma'(t)\|^3} = \frac{A\alpha''(t) \cdot JA\alpha'(t)}{\|A\alpha'(t)\|^3}$$

$$= \varepsilon \frac{A\alpha''(t) \cdot AJ\alpha'(t)}{\|A\alpha'(t)\|^3} = \varepsilon \frac{\alpha''(t) \cdot J\alpha'(t)}{\|\alpha'(t)\|^3} = \varepsilon\, \kappa 2[\alpha](t).\ \blacksquare$$

Next, we prove the converse of Theorem 6.7.

Theorem 6.8. (**Fundamental Theorem of Plane Curves, Uniqueness**) *Let α and γ be unit-speed regular curves in $\mathbb{R}^2$ defined on the same interval (a,b). Assume that α and γ have the same signed curvature. Then there is an orientation-preserving Euclidean motion F of $\mathbb{R}^2$ mapping α into γ.*

Proof. Fix $s_0 \in (a,b)$. Clearly, there exists a translation of $\mathbb{R}^2$ taking $\alpha(s_0)$ into $\gamma(s_0)$. Moreover, we can find a rotation of $\mathbb{R}^2$ that maps $\alpha'(s_0)$ into $\gamma'(s_0)$. Thus there exists an orientation-preserving Euclidean motion F of $\mathbb{R}^2$ such that

$$F(\alpha(s_0)) = \gamma(s_0) \quad \text{and} \quad F(\alpha'(s_0)) = \gamma'(s_0).$$

6.3 Intrinsic Equations for Plane Curves

To show that $F \circ \alpha$ coincides with γ, we define a real-valued function f by

$$f(s) = \|(F \circ \alpha)'(s) - \gamma'(s)\|^2$$

for $a < s < b$. The derivative of f is easily computed to be

(6.4) $\quad f'(s) = 2\big((F \circ \alpha)''(s) - \gamma''(s)\big) \cdot \big((F \circ \alpha)'(s) - \gamma'(s)\big)$

$\qquad \quad = 2\big((F \circ \alpha)''(s) \cdot (F \circ \alpha)'(s) + \gamma''(s) \cdot \gamma'(s)$

$\qquad \qquad - (F \circ \alpha)''(s) \cdot \gamma'(s) - (F \circ \alpha)'(s) \cdot \gamma''(s)\big).$

Since both $F \circ \alpha$ and γ have unit-speed, it follows that

$$(F \circ \alpha)''(s) \cdot (F \circ \alpha)'(s) = \gamma''(s) \cdot \gamma'(s) = 0.$$

Hence (6.4) reduces to

(6.5) $\quad f'(s) = -2\big((F \circ \alpha)''(s) \cdot \gamma'(s) + (F \circ \alpha)'(s) \cdot \gamma''(s)\big).$

Let A denote the linear part of F. Since $(F \circ \alpha)'(s) = A\alpha'(s)$ and $(F \circ \alpha)''(s) = A\alpha''(s)$, we can rewrite (6.5) as

(6.6) $\quad f'(s) = -2\big(A\alpha''(s) \cdot \gamma'(s) + A\alpha'(s) \cdot \gamma''(s)\big).$

Now we use the assumption that $\kappa 2[\alpha] = \kappa 2[\gamma]$ and (6.6) to get

$$f'(s) = -2\big(\kappa 2[\alpha](s) A J\alpha'(s) \cdot \gamma'(s) + A\alpha'(s) \cdot \kappa 2[\gamma](s) J\gamma'(s)\big)$$

$$= -2\kappa 2[\alpha](s)\big(J A\alpha'(s) \cdot \gamma'(s) + A\alpha'(s) \cdot J\gamma'(s)\big) = 0.$$

Since $f(s_0) = 0$, we conclude that $f(s) = 0$ for all s. Hence $(F \circ \alpha)'(s) = \gamma'(s)$ for all s, and so there exists $\mathbf{q} \in \mathbb{R}^2$ such that $(F \circ \alpha)(s) = \gamma(s) - \mathbf{q}$ for all s. In fact, $\mathbf{q} = 0$ because $F(\alpha(s_0)) = \gamma(s_0)$. Thus the Euclidean motion F maps α into γ. ∎

Next, we turn to the problem of explicitly determining a plane curve from its curvature.

Theorem 6.9. (**Fundamental Theorem of Space Curves, Existence**) *A unit-speed curve $\beta\colon (a,b) \longrightarrow \mathbb{R}^2$ whose curvature is a given piecewise-continuous function $k\colon (a,b) \longrightarrow \mathbb{R}$ is parametrized by*

(6.7) $\quad \begin{cases} \beta(s) = \left(\int \cos \theta(s)ds + c, \int \sin \theta(s)ds + d\right), \\ \theta(s) = \int k(s)ds + \theta_0, \end{cases}$

where c, d, θ_0 are constants of integration.

Proof. We define β and θ by (6.7); it follows that

(6.8)
$$\begin{cases} \beta'(s) = (\cos\theta(s), \sin\theta(s)), \\ \theta'(s) = k(s). \end{cases}$$

Thus β has unit-speed, so it follows from Corollary 1.17, page 20, that the curvature of β is k. ∎

One classical way to describe a plane curve α is by means of a **natural equation**, which is an equation of the form

(6.9)
$$F(\kappa 2[\alpha], s) = 0,$$

where s denotes the arc length function of α. Although such an equation may not be the most useful equation for calculational purposes, it shows clearly how curvature changes with arc length and is obviously invariant under translations and rotations. To find the natural equation of a curve $t \mapsto \alpha(t)$, one first computes the curvature and arc length as functions of t and then eliminates t. Here are some examples:

Curve	Natural equation
Straight line	$\kappa 2 = 0$
Circle of radius a	$\kappa 2 = \dfrac{1}{a}$
catenary[a]	$\kappa 2 = -\dfrac{a}{a^2 + s^2}$
clothoid[n, a]	$\kappa 2 = -\dfrac{s^n}{a^{n+1}}$
logspiral[a, b]	$\kappa 2 = \dfrac{1}{bs}$
involute[circle[a]]	$\kappa 2 = \dfrac{1}{\sqrt{2as}}$

Conversely, the ordinary equation can be found (at least in principle) by solving (6.9) and using Theorem 6.9.

6.3 Intrinsic Equations for Plane Curves

Natural equations arose as a reaction against the use of Cartesian and polar coordinates, which, despite their utility, were considered arbitrary. Instead Euler[2] in 1736 ([Euler1]) proposed the use of the arc length s and the radius of curvature $1/|\kappa \mathbf{2}|$. Natural equations were also studied by Lacroix[3] and Hill[4]. Excellent discussions of natural equations are given in [Melz, volume 2, pages 33–41] and [Ces]. The book [Ces] of Cesàro[5] uses natural equations as its starting point; it contains many plots (quite remarkable for a book published at the end of the 19th century) of plane curves given by means of natural equations.

For example, let us find the natural equation of the catenary defined by

$$\textbf{catenary}[a](t) = \left(a\cosh\left(\frac{t}{a}\right), t\right). \tag{6.10}$$

Mathematica tells us that the curvature and arc length are given by

$$\kappa\mathbf{2}(t) = -\frac{\operatorname{sech}(t/a)^2}{a} \quad \text{and} \quad s(t) = a\sinh(t/a). \tag{6.11}$$

Then (6.11) implies that

$$(s^2 + a^2)\kappa\mathbf{2} = -a. \tag{6.12}$$

Although (6.10) and (6.12) both describe a catenary, (6.12) is considered more natural because it does not depend on the choice of parametrization of the catenary.

---

[2] Leonhard Euler (1707-1783). Swiss mathematician. Euler was a geometer in the broad sense in which the term was used during his time. Not only did he contribute greatly to the evolution and systematization of analysis–in particular to the founding of the calculus of variation and the theories of differential equations, functions of complex variables, and special functions–but he also laid the foundations of number theory as a rigorous discipline. Moreover, he concerned himself with applications of mathematics to fields as diverse as lotteries, hydraulic systems, shipbuilding and navigation, actuarial science, demography, fluid mechanics, astronomy, and ballistics.

[3] Sylvestre Francois Lacroix (1765-1843). French writer of mathematical texts who was a student of Monge.

[4] Thomas Hill (1818-1891). American mathematician who became president of Harvard University.

[5] Ernesto Cesàro (1859-1906). Italian mathematician, born in Naples, professor in Palermo and Rome. Although his most important contribution was his monograph **Lezioni di Geometria Intrinseca**, he is also remembered for his work on divergent series.

6.4 Drawing Plane Curves with Assigned Curvature

Mathematica's command **NDSolve** can be used to find numerical solutions to ordinary differential equations or systems of ordinary differential equations. The output of **NDSolve** is a new type of object in *Mathematica* called an interpolating function. Interpolating functions can then be manipulated symbolically much like ordinary functions.

A curve β parametrized by arc length whose curvature is a given function k can be found by solving the system (6.8). To solve (6.8) using **NDSolve**, we can use the following miniprogram:

```
intrinsic[fun_,a_:0,{c_:0,d_:0,theta0_:0},optsnd___,
    {smin_:-10,smax_:10}][t_]:=
Flatten[Module[{x,y,theta},
{x[t],y[t]} /.
NDSolve[{x'[ss]==Cos[theta[ss]],
    y'[ss]==Sin[theta[ss]],
    theta'[ss]==fun[ss],
    x[a]==c,y[a]==d,theta[a]==theta0},
    {x,y,theta},{ss,smin,smax},optsnd]]
```

If **fun** is a known function, then the output of the miniprogram will be the curve $t \longmapsto \bigl(x(t), y(t)\bigr)$ such that **x[a]=c** and **y[a]=d**. The command **NDSolve** needs a range over which it finds the numerical solution. In the miniprogram this range is specified by **smin** and **smax**.

Since we are most often interested in the plot of a curve whose curvature is a given function, we give a miniprogram which solves (6.8) and plots the result.

6.4 Drawing Plane Curves with Assigned Curvature

```
plotintrinsic[fun_,a_:0,{c_:0,d_:0,theta0_:0},optsnd___,
            {smin_:-10,smax_:10},optspp___]:=
ParametricPlot[Module[{x,y,theta},
{x[t],y[t]} /.
NDSolve[{x'[ss]==Cos[theta[ss]],
        y'[ss]==Sin[theta[ss]],
        theta'[ss]==fun[ss],
        x[a]==c,y[a]==d,theta[a]==theta0},
        {x,y,theta},{ss,smin,smax},optsnd]]//Evaluate,
    {t,smin,smax},AspectRatio->Automatic,optspp];
```

Quite simple curvature functions can be used to produce very interesting curves. We have already seen in Section 3.7 that the curve **clothoid[1, 1]** has curvature given by $\kappa 2[\text{clothoid}[1,1]](s) = -s$, where s denotes arc length. Here is a modified clothoid, a curve whose curvature is $s \longmapsto s + \sin s$. The spiraling behavior of the modified clothoid is in the opposite direction and more complicated than the ordinary clothoid:

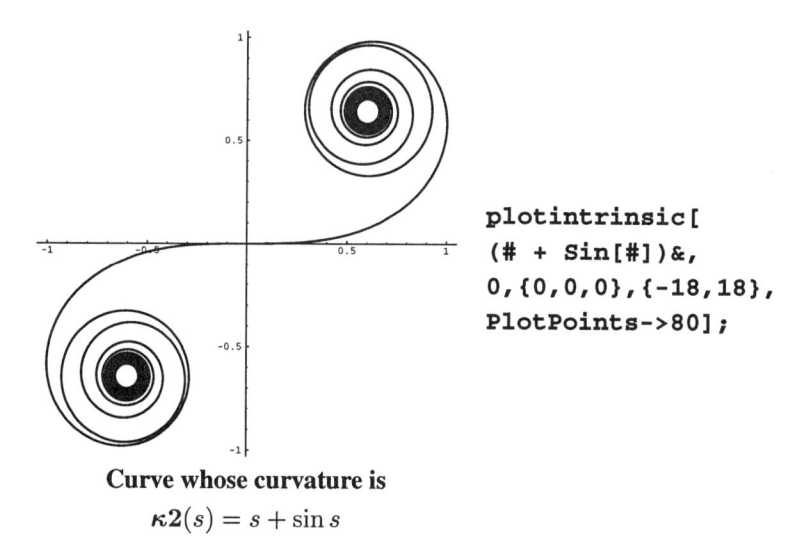

```
plotintrinsic[
(# + Sin[#])&,
0,{0,0,0},{-18,18},
PlotPoints->80];
```

Curve whose curvature is
$\kappa 2(s) = s + \sin s$

The fact that the curvature of this curve tends to infinity as $s \longrightarrow \infty$ accounts for the spiraling.

If, on the other hand, the curvature is just proportional to $\sin s$, the resulting curve is quite different:

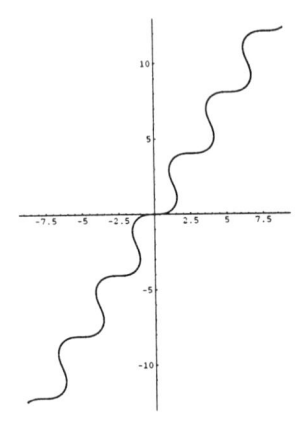

Curve whose curvature is

$$\kappa 2(s) = \sin s$$

```
plotintrinsic[
Sin[#]&,
0,{0,0,0},{-20,20},
PlotPoints->80];
```

Next, we consider several functions that oscillate wildly as $s \longrightarrow \infty$. The resulting curves are strange.

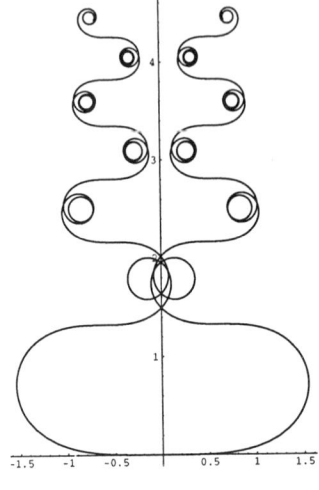

```
plotintrinsic[
(#Sin[#])&,
0,{0,0,0},{-20,20},
PlotPoints->80];
```

Curve whose curvature is

$$\kappa 2(s) = s \sin s$$

Similarly, the command
```
plotintrinsic[(#^2Sin[#])&,
    0,{0,0,0},{-8,8},PlotPoints->80];
```

6.4 Drawing Plane Curves with Assigned Curvature

generates the picture

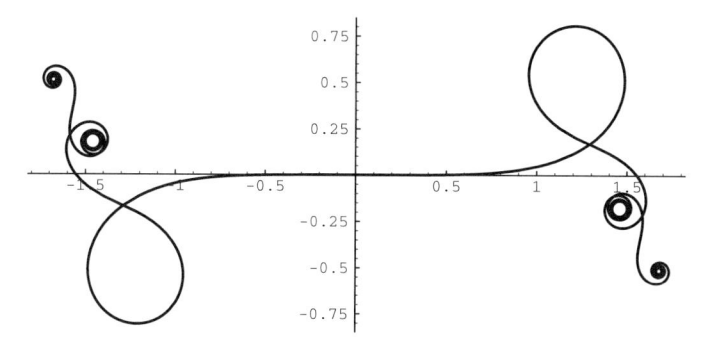

Curve whose curvature is
$$\kappa 2(s) = s^2 \sin s$$

And finally:

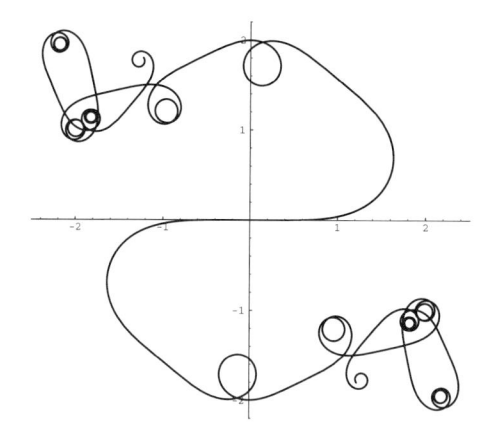

```
plotintrinsic[
(#Sin[#]^2)&,
0,{0,0,0},{-20,20},
PlotPoints->80];
```

Curve whose curvature is
$$\kappa 2(s) = s(\sin s)^2$$

Next, let us find the curves whose curvatures have the form $s \longmapsto s\,f(s)$, where f is one of the Bessel[6] functions, J_0, J_1, J_2. Although f has a wildly oscillatory

---

[6] Bessel (1784-1846). German astronomer and friend of Gauss. In 1800 Bessel was appointed director of the observatory at Königsberg, where he remained for the rest of his life. Bessel functions were introduced in 1817 in connection with a problem of Kepler of determining the motion of three bodies moving under mutual gravitation.

behavior, it is not as wild as $s \longmapsto s\sin(s)$, so the shapes are somewhat different.

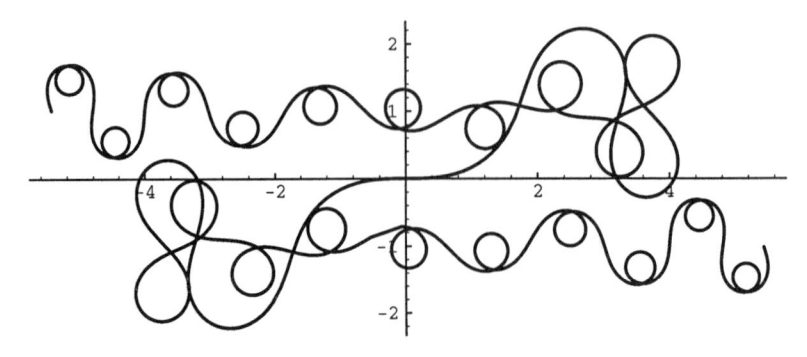

Curve whose curvature is
$$\kappa 2(s) = s\, J_0(s)$$

```
plotintrinsic[(#BesselJ[0,#])&,
0,{0,0,0},{-40,40},PlotPoints->80];
```

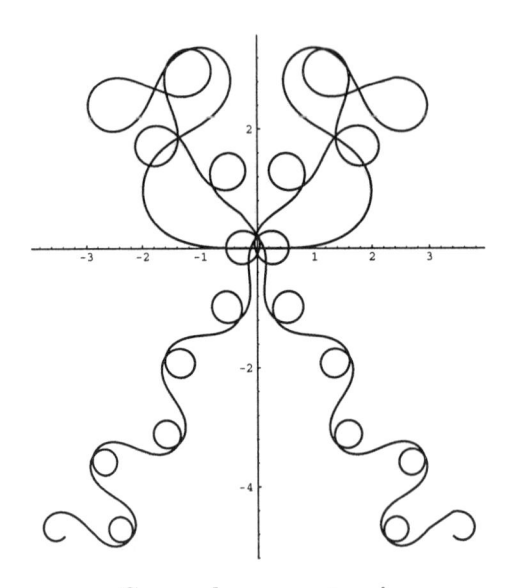

```
plotintrinsic[
(#BesselJ[1,#])&,
0,{0,0,0},{-40,40},
PlotPoints->80];
```

Curve whose curvature is
$$\kappa 2(s) = s\, J_1(s)$$

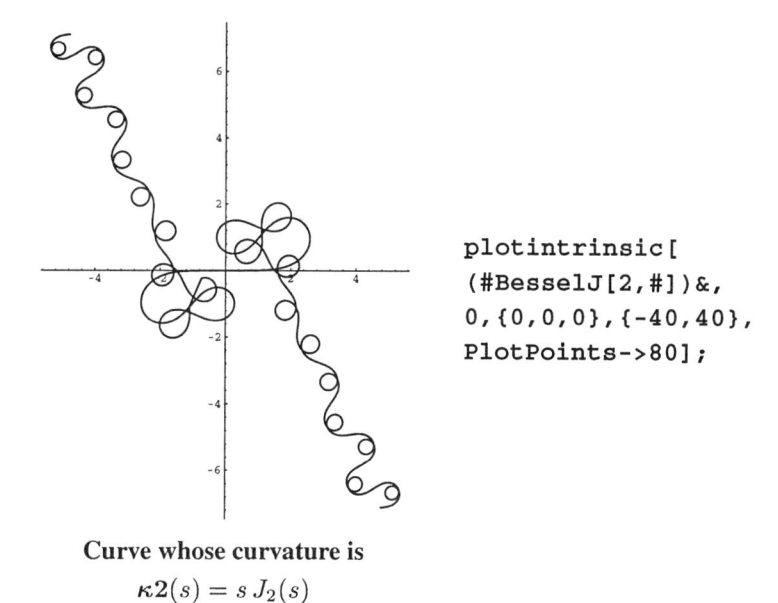

6.4 Drawing Plane Curves with Assigned Curvature

```
plotintrinsic[
(#BesselJ[2,#])&,
0,{0,0,0},{-40,40},
PlotPoints->80];
```

Curve whose curvature is
$\kappa 2(s) = s\, J_2(s)$

Finally, we draw a clothoid through the origin, together with two rotated clothoids:

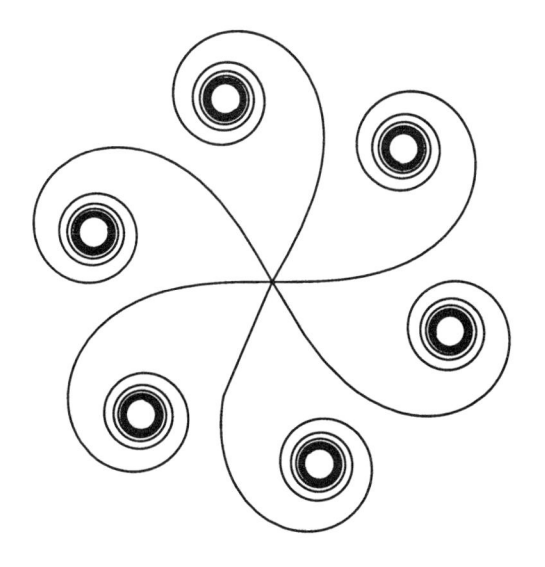

```
ParametricPlot[
Table[
intrinsic[
#&,0,{0,0,a},
{-10,10}][t],
{a,0,2Pi/3,Pi/3}]
//Evaluate,
{t,-10,10},
AspectRatio->
Automatic];
```

Bouquet of Clothoids

6.5 Exercises

1. Let $\{e_1, \ldots, e_n\}$ be an orthonormal basis of $\mathbb{R}^n$. Show that any vector $\mathbf{v} \in \mathbb{R}^n$ can be written as
$$\mathbf{v} = (\mathbf{v} \cdot \mathbf{e}_1)\mathbf{e}_1 + \cdots + (\mathbf{v} \cdot \mathbf{e}_n)\mathbf{e}_n.$$

2. A **homothety** of $\mathbb{R}^n$ is a map $F: \mathbb{R}^n \longrightarrow \mathbb{R}^n$ for which there exists a nonzero constant $\lambda \neq 0$ such that
$$\|F(\mathbf{p}_1) - F(\mathbf{p}_2)\| = \lambda \|\mathbf{p}_1 - \mathbf{p}_2\|$$
for all $\mathbf{p}_1, \mathbf{p}_2 \in \mathbb{R}^n$. Show that a homothety is a linear bijective map.

3. Show that a Euclidean motion $F: \mathbb{R}^2 \longrightarrow \mathbb{R}^2$ preserves evolutes, involutes, pedal curves and parallel curves. More precisely, if $\boldsymbol{\alpha}: (a, b) \longrightarrow \mathbb{R}^2$ is a curve, show that the following formulas hold for $a < t < b$:
$$\begin{cases} \mathbf{evolute}[F \circ \boldsymbol{\alpha}](t) = (F \circ \mathbf{evolute}[\boldsymbol{\alpha}])(t), \\ \mathbf{involute}[F \circ \boldsymbol{\alpha}, c](t) = (F \circ \mathbf{involute}[\boldsymbol{\alpha}, c])(t), \\ \mathbf{pedal}[F(\mathbf{p}), F \circ \boldsymbol{\alpha}](t) = (F \circ \mathbf{pedal}[\mathbf{p}, \boldsymbol{\alpha}])(t), \end{cases}$$
and
$$\mathbf{parcurve}[F \circ \boldsymbol{\alpha}][s](t) = \begin{cases} (F \circ \mathbf{parcurve}[\boldsymbol{\alpha}][s])(t) \\ \qquad \text{if } F \text{ is orientation-preserving,} \\ (F \circ \mathbf{parcurve}[\boldsymbol{\alpha}][-s])(t) \\ \qquad \text{if } F \text{ is orientation-reversing.} \end{cases}$$

4. An **algebraic curve** is an implicitly defined curve of the form $P(x, y) = 0$, where P is a polynomial in x and y. The **order** of an algebraic curve is the order of the polynomial P. Show that an affine transformation preserves the order of an algebraic curve.

5. Find the natural equations of the first three involutes of a circle.

6. **Nielsen's**[7] **spiral** is defined by

---
[7] Niels Nielsen (1865-1931). Danish mathematician.

6.5 Exercises

```
nielsenspiral[a_][t_] := a{CosIntegral[t],SinIntegral[t]}
```

where `SinIntegral[t]` and `CosIntegral[t]` are the integrals

$$\int_0^t \frac{\sin u}{u}\,du \quad \text{and} \quad \int_0^t \frac{(\cos u - 1)}{u}\,du + \log t + \gamma.$$

(γ = Euler's constant $\approx 0.577216\ldots$). Show that the natural equation of Nielsen's spiral is

$$\kappa 2(s) = \frac{e^{s/a}}{a}.$$

Nielsen's Spiral

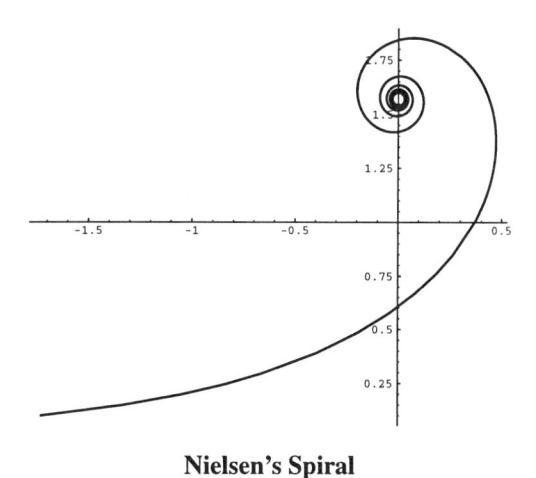

7. The following miniprogram sometimes can be used to find the natural equations of a plane curve:

```
nateq[alpha_][k2_,s_] := ToRules[Eliminate[
    {s==arclength[alpha][t]//PowerExpand,
     k2==kappa2[alpha][t]//PowerExpand//Simplify},t]]
```

Use it to check the formulas in the table on page 138.

8. Plot curves whose curvature is $ns \sin s$ for $n = \pm 2, \pm 2.5, \pm 3$.

9. The **Gamma function** is defined by

$$\Gamma(z) = \int_0^\infty t^{z-1} e^{-t} dt.$$

The corresponding *Mathematica* function is denoted by **Gamma**.

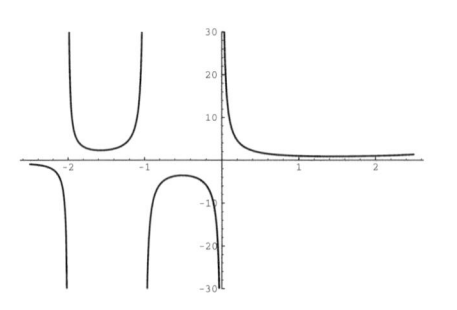

```
Gammagraph=
Plot[Gamma[t],
{t,-2.5,2.5}];
Gammagrapha=
Gammagraph //.
{x__,{a_,b_},
{c_,d_},y__} /;
Negative[b d]->
{x,Null,Null,y};
Show[Gammagrapha];
```

The Gamma function

Plot from -5 to 5 a curve whose curvature is $s \longmapsto \Gamma(s + 0.5)$.

10. The **Airy**[8] **functions**, denoted in *Mathematica* as **AiryAi** and **AiryBi** are linearly independent solutions of the differential equation $y''(t) - t\, y(t) = 0$.

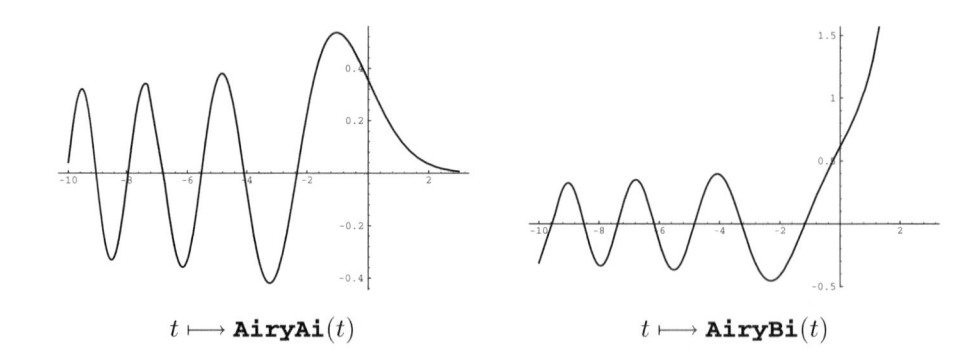

$t \longmapsto$ **AiryAi**(t)

$t \longmapsto$ **AiryBi**(t)

---

[8] Sir George Biddel Airy (1801-1892). Royal astronomer of England.

6.5 Exercises

Plot a curve whose signed curvature is $t \longmapsto$ **AiryAi**(t). Do the same for $t \longmapsto$ **AiryBi**(t).

11. *Mathematica* has a two-argument version of **ArcTan**; it can be used to define a sawtooth function:

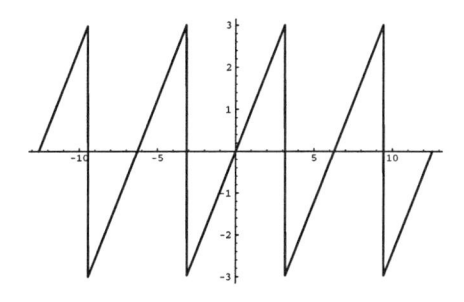

```
sawtooth[a_,b_,c_][t_]:= (c/Pi)ArcTan[a Cos[t],b Sin[t]]
```

```
Plot[
sawtooth[1,1,3][t]
//Evaluate,
{t,-4Pi,4Pi}];
```

A sawtooth wave

Plot a curve whose curvature is **sawtooth[1,1,4]**.

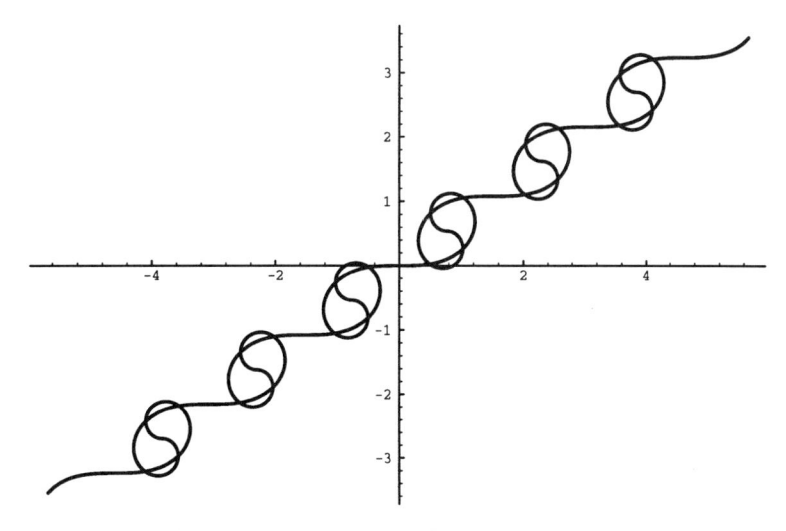

Curve whose curvature is a sawtooth wave

150   Chapter 6   Determining a Plane Curve from its Curvature

12. A square wave function can be defined in *Mathematica* by

```
squarewave[a_,b_][t_]:= b Sin[a t]/Sqrt[Sin[a t]^2]
```

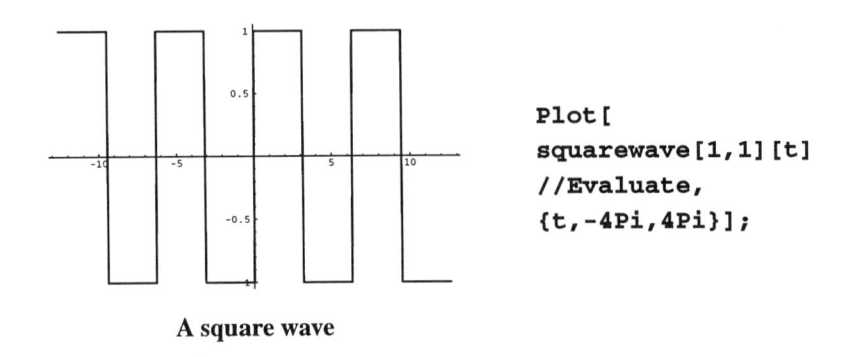

```
Plot[
squarewave[1,1][t]
//Evaluate,
{t,-4Pi,4Pi}];
```

A square wave

Plot a curve whose curvature is **squarewave[1,1]**.

13. An **epicycloid** is the curve that is traced out by a point P on the circumference of a circle (of radius b) rolling *outside* another circle (of radius a). Similarly, a **hypocycloid** is the curve that is traced out by a point P on the circumference of a circle (of radius b) rolling *inside* another circle (of radius a).

An epicycloid with 11 cusps 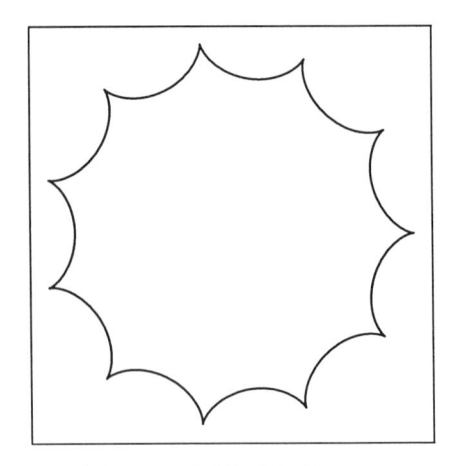   **A hypocycloid with 11 cusps**

6.5 Exercises

These curves are parametrized as follows:

$$\text{epicycloid}[a,b](t) = \left((a+b)\cos t - b\cos\left(\frac{(a+b)t}{b}\right),\right.$$
$$\left.(a+b)\sin t - b\sin\left(\frac{(a+b)t}{b}\right)\right),$$

and

$$\text{hypocycloid}[a,b](t) = \left((a-b)\cos t + b\cos\left(\frac{(a-b)t}{b}\right),\right.$$
$$\left.(a-b)\sin t - b\sin\left(\frac{(a-b)t}{b}\right)\right).$$

Use *Mathematica* to show that the natural equation of an epicycloid is

(6.13) $$a^2 s^2 + \left(\frac{a+2b}{\kappa 2}\right)^2 = 16b^2(a+b)^2,$$

and that the natural equation of a hypocycloid is

$$a^2 s^2 + \left(\frac{a-2b}{\kappa 2}\right)^2 = 16b^2(a-b)^2.$$

14. Show that the natural equation of a tractrix is

$$\kappa 2 = -\frac{1}{a\sqrt{e^{2s/a}-1}}.$$

15. Let $\alpha_n \colon (-\infty, \infty) \longrightarrow \mathbb{R}^2$ be a curve whose curvature is $n(\sin s)^2$, where n is an integer and s denotes arc length. Show that α_n is a closed curve if and only if n is not divisible by 4. Plot α_n for $1 \le n \le 12$.

6.6 Animation of Epicycloids and Hypocycloids

Typical animations of one circle rolling on another to form an epicycloid and a hypocycloid are given by the following commands:

```
Do[ParametricPlot[Evaluate[
epicycloid[3,1][t]],
{t,0,2Pi},PlotRange->{{-5.1,5.1},{-5.1,5.1}},
PlotStyle->{{RGBColor[1,0,0],AbsoluteThickness[2]}},
AspectRatio->Automatic,Axes->None,
Epilog->{Circle[{0,0},3],
{RGBColor[0,1,0],Circle[circle[4][a],1]},
AbsolutePointSize[6],Point[epicycloid[3,1][a]]}],
{a,0,2Pi,Pi/108}]

Do[ParametricPlot[Evaluate[
hypocycloid[3,1][t]],
{t,0,2Pi},PlotRange->{{-3.1,3.1},{-3.1,3.1}},
PlotStyle->{{RGBColor[1,0,0],AbsoluteThickness[2]}},
AspectRatio->Automatic,Axes->None,
Epilog->{Circle[{0,0},3],
{RGBColor[0,1,0],Circle[circle[2][a],1]},
AbsolutePointSize[6],Point[hypocycloid[3,1][a]]}],
{a,0,2Pi,Pi/72}]
```

7

GLOBAL PROPERTIES OF PLANE CURVES

The geometry of plane curves that we have been studying in Chapters 1—6 has been local in nature. For example, the curvature of a plane curve describes the bending of a curve, point by point. In this chapter we consider global properties of plane curves.

In Section 7.1 we define the notion of *total signed curvature* of a plane curve α; it measures the curvature of α as a whole. A closely related quantity is an integer called the *turning number* of a closed curve. **Mathematica** is a useful tool to illustrate this concept. Section 7.2 is devoted to the *rotation index* of a closed curve, which is defined topologically; we show that it coincides with the turning number.

Convex plane curves are considered in Section 7.3, where it is proved that a closed plane curve is convex if and only if its signed curvature $\kappa 2$ does not change sign. We prove the four vertex theorem in Section 7.4.

Basic facts about *curves of constant width* are established in Section 7.5. We briefly discuss *envelopes* of curves in Section 7.6 so that in Section 7.7 we can derive the formula for an *oval* in terms of its *support function*. This allows us in Sections 7.8 and 7.9 to give two constructions for curves of constant width.

7.1 The Total Signed Curvature of a Plane Curve

The signed curvature $\kappa 2$ of a plane curve α measures the bending of the curve at each of its points. A measure of the total bending of α is given by an integral involving $\kappa 2$.

Definition. *The* **total signed curvature** *of a curve* $\alpha\colon (a,b) \longrightarrow \mathbb{R}^2$ *is*

$$\mathbf{TSC}[\alpha] = \int_a^b \kappa 2[\alpha](t) \|\alpha'(t)\| dt,$$

where $\kappa 2[\alpha]$ *denotes the signed curvature of* α.

First, let us check that the total signed curvature is a geometric concept.

Lemma 7.1. *The total signed curvature of a plane curve remains unchanged under a positive reparametrization, but changes sign under a negative reparametrization.*

Proof. Let $\gamma = \alpha \circ h$ where $\gamma\colon (c,d) \longrightarrow \mathbb{R}^2$ and $\alpha\colon (a,b) \longrightarrow \mathbb{R}^2$ are curves. We do the case when $h'(t) > 0$ for all t. Using Theorem 1.10, page 16, and the formula from calculus for the change of variables in an integral, we compute

$$\begin{aligned}\mathbf{TSC}[\alpha] &= \int_a^b \kappa 2[\alpha](t)\|\alpha'(t)\|dt = \int_c^d \kappa 2[\alpha](h(u))\|\alpha'(h(u))\|h'(u)du \\ &= \int_c^d \kappa 2[\alpha](h(u))\|\alpha'(h(u))h'(u)\|du = \int_c^d \kappa 2[\gamma](u)\|\gamma'(u)\|du \\ &= \mathbf{TSC}[\gamma].\end{aligned}$$

The proof that $\mathbf{TSC}[\gamma] = -\mathbf{TSC}[\alpha]$ when γ is a negative reparametrization of α is similar. ∎

There is a simple relation linking the total signed curvature and the turning angle of a curve defined in Section 1.7.

Lemma 7.2. *The total signed curvature can be expressed in terms of the turning angle* $\theta[\alpha]$ *of* α *by*
(7.1) $$\mathbf{TSC}[\alpha] = \theta[\alpha](b) - \theta[\alpha](a).$$

7.1 The Total Signed Curvature of a Plane Curve

Proof. Equation (7.1) results when (1.19), page 20, is integrated. ∎

The total signed curvature of a closed curve is especially important. First, we define carefully the notion of closed curve.

Definition. *A regular curve $\alpha\colon (a,b) \longrightarrow \mathbb{R}^n$ is* **closed** *provided there is a constant $c > 0$ such that*
$$(7.2) \qquad \alpha(t+c) = \alpha(t)$$
for all t. The least such number c is called the **period** *of α.*

See exercise 1 to see why closed curves are assumed to be regular.

Clearly, we can use (7.2) to define $\alpha(t)$ for all t; that is, the domain of definition of closed curve can be extended from (a,b) to $\mathbb{R}$. Just as in the case of a circle, the trace C of a regular closed curve α is covered over and over again by α. Intuitively, when we speak of the length of a closed curve, we mean the length of the trace C. Therefore, in the case of a closed curve we can use either $\mathbb{R}$ or any closed interval $[a,b]$ for the domain of definition of the curve, where $b - a$ is the period. Furthermore, we modify the definition of length given on page 7 as follows:

Definition. *Let $\alpha\colon \mathbb{R} \longrightarrow \mathbb{R}^n$ be a regular closed curve with period c. By the* **length** *of α we mean the length of the restriction of α to $[0,c]$, namely,*
$$\int_0^c \|\alpha'(t)\|\, dt.$$

Next, we find the relation between the period and length of a closed curve.

Lemma 7.3. *Let $\alpha\colon \mathbb{R} \longrightarrow \mathbb{R}^n$ be a closed curve with period c, and let $\beta\colon \mathbb{R} \longrightarrow \mathbb{R}^n$ be a unit-speed reparametrization of α. Then β is also closed; the period of β is L, where L is the length of α.*

Proof. Let s denote the arc length function of α starting at 0; we can assume (by Lemma 1.7, page 12) that $\alpha(t) = \beta(s(t))$. We compute

$$s(t+c) = \int_0^{c+t} \|\alpha'(u)\|\, du = \int_0^c \|\alpha'(u)\|\, du + \int_c^{c+t} \|\alpha'(u)\|\, du$$

$$= L + \int_c^{c+t} \|\alpha'(u)\|\, du = L + \int_0^t \|\alpha'(u)\|\, du = L + s(t).$$

Thus
$$\beta(s+L) = \beta(s(t)+L) = \beta(s(t+c)) = \alpha(t+c) = \alpha(t) = \beta(s(t)) = \beta(s)$$

for all s; it follows that β is closed. Furthermore, since c is the least positive number such that $\alpha(t+c) = \alpha(c)$ for all t, it must be the case that L is the least positive number such that $\beta(s+L) = \beta(s)$ for all s. Hence L is the period of β. ∎

Next, we specialize to plane curves, that is, to the case $n=2$. If $\alpha\colon\mathbb{R} \longrightarrow \mathbb{R}^2$ is a regular closed curve with trace C, then the mapping

$$t \longmapsto \frac{\alpha'(t)}{\|\alpha'(t)\|}$$

gives rise to a mapping Φ from C to the unit circle $S^1(1)$ of $\mathbb{R}^2$. Note that $\Phi(\mathbf{p})$ is just the end point of the unit tangent vector to α at $\mathbf{p}$. It is intuitively clear that when a point $\mathbf{p}$ goes around C once, the image point $\Phi(\mathbf{p})$ goes around the unit circle $S^1(1)$ an integral number of times. There are two completely different ways to define this number; the first is in terms of the total signed curvature.

Definition. *The* **turning number** *of a closed curve* $\alpha\colon\mathbb{R} \longrightarrow \mathbb{R}^2$ *is*

$$\mathbf{Turn}[\alpha] = \frac{1}{2\pi}\int_0^c \kappa 2[\alpha](t)\|\alpha'(t)\|dt,$$

where c denotes the period of α.

Thus the turning number of a closed curve α is just the total signed curvature of α divided by 2π.

Closed plane curves have the remarkable property that their turning numbers are integers, as we shall see in Section 7.2. For example, let $\alpha[n,a]$ be the curve that covers a circle n times; it is defined by

$$\alpha[n,a](t) = a(\cos t, \sin t), \qquad 0 \le t \le 2n\pi,$$

where n is an integer. It is easy to compute that

$$\kappa 2(\alpha[n,a]) = \frac{1}{a} \quad \text{and} \quad \|\alpha[n,a]'(t)\| = a.$$

Hence

$$\mathrm{TSC}[\alpha[n,a]] = 2n\pi \quad \text{and} \quad \mathbf{Turn}[\alpha[n,a]] = n.$$

The following *Mathematica* miniprogram will compute the turning number of a closed plane curve, provided a parametrization of the curve is known.

7.1 The Total Signed Curvature of a Plane Curve

```
turningnumber[alpha_,{a_,b_},opts___]:=
    Module[{tt,tmp1,tmp2},
        tmp1 = kappa2[alpha][tt]Sqrt[alpha'[tt].
            alpha'[tt]];
        tmp2 = NIntegrate[Evaluate[tmp1],{tt,a,b},
            AccuracyGoal->2,opts];
        Round[tmp2/N[2Pi]]]
```

In this miniprogram **NIntegrate** is used to integrate the signed curvature of a plane curve **alpha**. Then **Round** finds the nearest integer to the computed integral.

The **epitrochoid** and **hypotrochoid** are good curves to illustrate the turning number; they are defined by

$$\text{epitrochoid}[a,b,h](t) = \left((a+b)\cos t - h\cos\left(\frac{(a+b)t}{b}\right),\right.$$
$$\left.(a+b)\sin t - h\sin\left(\frac{(a+b)t}{b}\right)\right)$$

and

$$\text{hypotrochoid}[a,b,h](t) = \left((a-b)\cos t + h\cos\left(\frac{(a-b)t}{b}\right),\right.$$
$$\left.(a-b)\sin t - h\sin\left(\frac{(a-b)t}{b}\right)\right).$$

Note that **hypotrochoid**$[a,b,h]$ = **epitrochoid**$[a,-b,-h]$ (see exercise 7). The *Mathematica* versions of these two curves are

```
epitrochoid[a_,b_,h_][t_]:=
    {(a + b)Cos[t] - h Cos[((a + b)t)/b],
     (a + b)Sin[t] - h Sin[((a + b)t)/b]}
```

and

```
hypotrochoid[a_,b_,h_][t_]:=
    {(a - b)Cos[t] + h Cos[((a - b)t)/b],
     (a - b)Sin[t] - h Sin[((a - b)t)/b]}
```

We can describe **epitrochoid**[a, b, h] as the parametrized curve that is traced out by a point **p** on one circle (of radius b) rolling *outside* another circle (of radius a). Here h denotes the distance from **p** to the center of the fixed circle. In the case that $h = b$, the loops degenerate into points, and the resulting curve is called an **epicycloid**. Similarly, **hypotrochoid**[a, b, h] is the parametrized curve that is traced out by a point **p** on one circle (of radius b) rolling *inside* another circle (of radius a). A **hypocycloid** is a hypotrochoid for which the loops degenerate to points; this is the case $h = b$. Both **epitrochoid**[a, b, h] and **hypotrochoid**[a, b, h] have a/b loops provided a/b is an integer, and each curve is contained in a circle of radius $a + b + h$.

As an example let us draw an epitrochoid with 9 inner loops. When we type in

```
epitrochoid[18,2,6][t]
```

Mathematica gives us the explicit parametrization:

```
{20 Cos[t] - 6 Cos[10 t], 20 Sin[t] - 6 Sin[10 t]}
```

The plots of **epitrochoid**[**18, 2, 6**] and its curvature show more clearly that the curvature is always positive:

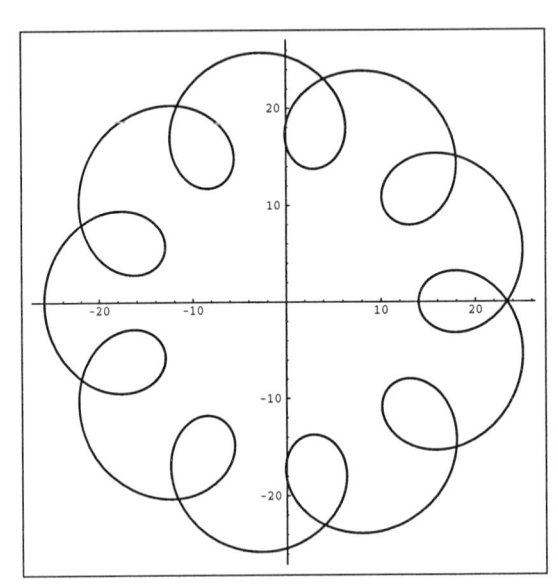

```
ParametricPlot[
epitrochoid[
18,2,6][t]
//Evaluate,
{t,0,2Pi},
AspectRatio->
Automatic];
```

The epitrochoid
$t \longrightarrow (20 \cos t - 6 \cos 10t, 20 \sin t - 6 \sin 10t)$

7.2 The Rotation Index of a Closed Plane Curve

A glance at the picture shows that the turning number of **epitrochoid[18, 2, 6]** is 10. To check this with *Mathematica* we use the command

```
turningnumber[epitrochoid[18,2,6],{0,2Pi}]
```

obtaining

```
10
```

Here is the plot of the curvature:

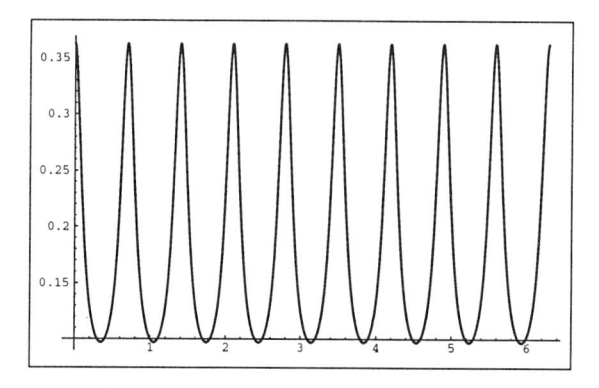

The curvature of the epitrochoid
$t \longrightarrow (20\cos t - 6\cos 10t, 20\sin t - 6\sin 10t)$

7.2 The Rotation Index of a Closed Plane Curve

In Section 7.1 we gave the definition of turning number in terms of the total signed curvature. In this section we define turning number in a completely different way. First, we need an auxiliary definition.

Definition. *Let* $\phi\colon S^1(1) \longrightarrow S^1(1)$ *be a continuous function, where* $S^1(1)$ *denotes a circle of radius 1 in the plane, and let* $\widetilde{\phi}\colon S^1(1) \longrightarrow \mathbb{R}$ *be a function such that*

$$\phi(\cos t, \sin t) = \bigl(\cos(\widetilde{\phi}(t)), \sin(\widetilde{\phi}(t))\bigr)$$

for all t. The **degree** *of* ϕ *is the integer n such that*

$$\widetilde{\phi}(2\pi) - \widetilde{\phi}(0) = 2\pi n.$$

Lemma 7.4. *The definition of degree is independent of the choice of $\widetilde{\phi}$.*

Proof. Let $\widehat{\phi}\colon S^1(1) \longrightarrow \mathbb{R}$ be another function satisfying the same conditions as $\widetilde{\phi}$. Then we have
$$\widehat{\phi}(t) - \widetilde{\phi}(t) = 2\pi\, n(t)$$
where $n(t)$ is an integer. Since $n(t)$ is continuous, it must be constant. Thus $\widetilde{\phi}(2\pi) - \widetilde{\phi}(0) = \widehat{\phi}(2\pi) - \widehat{\phi}(0)$. ∎

We use the notion of degree of a map to define the rotation index of a curve.

Definition. *Let $\alpha\colon \mathbb{R} \longrightarrow \mathbb{R}$ be a regular closed curve. The **rotation index** of α is the degree of the map $\varphi[\alpha]\colon \mathbb{R} \longrightarrow S^1(1)$ defined by*
$$\varphi[\alpha](t) = \frac{\alpha'(t)}{\|\alpha'(t)\|}.$$

Notice that the rotation index of a curve is defined topologically; the turning number, in contrast, is defined as an integral. It is remarkable, but easy to prove, that these integers are the same.

Theorem 7.5. *The rotation index of a regular closed curve α coincides with the turning number of α.*

Proof. Without loss of generality, we can assume that α has unit-speed. Then by (6.8), page 138, we have
$$\varphi[\alpha](s) = \alpha'(s) = \bigl(\cos\theta(s), \sin\theta(s)\bigr),$$
where θ denotes the turning angle of α. Therefore, we can choose $\widetilde{\varphi[\alpha]} = \theta$. It follows from Lemma 7.2 and Corollary 1.17, page 20, that
$$\operatorname{degree}(\varphi[\alpha]) = \frac{1}{2\pi}\bigl(\theta(2\pi) - \theta(0)\bigr) = \frac{1}{2\pi}\int_0^{2\pi}\frac{d\theta(s)}{ds}ds$$
$$= \frac{1}{2\pi}\int_0^{2\pi} \kappa 2[\alpha](s)\,ds = \mathbf{Turn}[\alpha]. \blacksquare$$

Those closed curves which do not cross themselves form an important subclass.

Definition. *A regular closed curve $\alpha\colon(a,b) \longrightarrow \mathbb{R}^n$ with period c is **simple** provided $\alpha(t_1) = \alpha(t_2)$ if and only if $t_1 = t_2 = nc$ for some integer n.*

7.2 The Rotation Index of a Closed Plane Curve

For example, an ellipse is a simple closed curve, but the epitrochoid on page 159 is not.

It is clear intuitively that the rotation index of a simple closed curve is ±1. To prove this rigorously, we need the important topological notion of homotopy. Even though we give the definition for a general topological space X, we shall use it in this chapter only for the case $X = \mathbb{R}^2$.

Definition. *Let X be a topological space, and let α and γ be continuous curves reparametrized so that both are maps*

$$\alpha, \gamma \colon [a, b] \longrightarrow X.$$

*We say that α and γ are **homotopic** provided there exists a map*

$$F \colon [0,1] \times [a,b] \longrightarrow X$$

with the following properties:

(i) $F_0 = \alpha$,

(ii) $F_1 = \gamma$,

(iii) $t \longrightarrow F_u(t)$, *is a continuous curve in X for $0 \leq u \leq 1$.*

*The map F is called a **homotopy** between α and γ.*

The proof of the following theorem is due to H. Hopf ([Hopf1]).

Theorem 7.6. (**H. Hopf**[1]) *The turning number of a simple closed plane curve α is ±1.*

Proof. The idea of the proof is to construct a homotopy between the unit tangent map $t \longmapsto \alpha'(t)/\|\alpha'(t)\|$ (whose degree is **Turn**[α]), and the so-called *secant map* (whose degree is ±1). The details are somewhat lengthy.

Denote by L the length of α and put

$$\mathcal{T} = \{\, (t_1, t_2) \mid 0 \leq t_1 \leq t_2 \leq L\,\};$$

---

[1] Heinz Hopf (1894-1971). Professor at the Eidgenössische Technische Hochschule in Zürich. The greater part of his work was in algebraic topology, motivated by an exceptional geometric intuition. In 1931 Hopf studied homotopy classes of maps between the spheres S^3 to S^2 and defined what is now known as the Hopf invariant.

then $\mathcal{T}$ is a triangular region in the plane. Let β be a reparametrization of α with $\beta(0) = \mathbf{p}$. The secant map $\psi\colon \mathcal{T} \longrightarrow S^1(1)$ is defined by

$$\psi(t_1, t_2) = \begin{cases} \dfrac{\beta(t_2) - \beta(t_1)}{\|\beta(t_2) - \beta(t_1)\|} & \text{for } t_1 \neq 0 \text{ or } t_1 \neq L, \text{ and } t_1 \neq t_2, \\ \dfrac{\beta'(t)}{\|\beta'(t)\|} & \text{for } t_1 = t_2 = t, \\ -\dfrac{\beta'(0)}{\|\beta'(0)\|} & \text{for } t_1 = 0 \text{ and } t_2 = L. \end{cases}$$

Since β is regular, ψ is continuous. Let $A = (0,0)$, $B = (0,L)$ and $C = (L,L)$ be the vertices of the triangular region $\mathcal{T}$. Because the restriction of ψ to the side AC is $\beta'/\|\beta'\|$, the degree of this restriction is the turning number of β. Thus, by construction, $\beta'/\|\beta'\|$ is homotopic to the restriction of ψ to the path consisting of the sides AB and BC. We must show that the degree of the latter map is ± 1.

Assume that the β is oriented with respect to $\mathbb{R}^2$, so that the angle from $\beta'(0)$ to $-\beta'(0)$ is π. Then the restriction of ψ to AB covers one half of $S^1(1)$, while the restriction of ψ to BC covers the other half of $S^1(1)$. Hence the degree of ψ restricted to AB and BC is $+1$. Reversing the orientation, we obtain -1 for the degree. This completes the proof. ∎

Corollary 7.7. *If β is a simple closed unit-speed curve of period L, then the map $s \longmapsto \beta'(s)$ maps the interval $[0, L]$ onto all of the unit circle $S^1(1)$.*

There is a far-reaching generalization of Theorem 7.6:

Theorem 7.8. (Whitney-Graustein) *Two curves that have the same turning number are homotopic.*

For a proof of this theorem see [BeGo, page 325].

7.3 Convex Plane Curves

Any straight line L divides $\mathbb{R}^2$ into two half-planes H_1 and H_2 such that

$$H_1 \cup H_2 = \mathbb{R}^2 \quad \text{and} \quad H_1 \cap H_2 = L.$$

We say that a curve C **lies on one side of** L provided either C is completely contained in H_1 or C is completely contained in H_2.

Definition. *A plane curve is* **convex** *if it lies on one side of each of its tangent lines.*

Clearly, any ellipse is a convex curve; we shall encounter many other examples in this chapter. For a characterization of convex curves in terms of curvature, we need the notions of monotone increasing and monotone decreasing functions.

Definition. *Let $f:(a,b) \longrightarrow \mathbb{R}$ be a function, not necessarily differentiable or even continuous. We say that f is* **monotone increasing** *provided that $s \leq t$ implies $f(s) \leq f(t)$, and* **monotone decreasing** *provided that $s \leq t$ implies $f(s) \geq f(t)$. If f is either monotone decreasing or monotone increasing, we say that f is* **monotone**.

It is easy to find examples of noncontinuous monotone functions, and also of continuous monotone functions that are not differentiable. In the differentiable case we have the following result:

Lemma 7.9. *A function $f:(a,b) \longrightarrow \mathbb{R}$ is monotone if and only if the derivative f' has constant sign on (a,b). More precisely, $f' \geq 0$ implies monotone increasing and $f' \leq 0$ implies monotone decreasing.*

A glance at any simple closed convex curve C convinces us that the unsigned curvature of C does not change sign. We now prove this rigorously.

Theorem 7.10. *A simple closed regular plane curve C is convex if and only if its curvature $\kappa 2$ has constant sign; that is, $\kappa 2$ is either always nonpositive or always nonnegative.*

Proof. We parametrize C by a unit-speed curve β whose turning angle is $\theta[\beta]$ (see Section 1.7). Since $\theta[\beta]' = \kappa 2[\beta]$, we must show that $\theta[\beta]$ is monotone if and only if β is convex, and then use Lemma 7.9.

Suppose that $\theta[\beta]$ is monotone, but that β is not convex. Then there is a point **p** on β for which β lies on both sides of the tangent line L to β at **p**. Since β is closed, there are points $\mathbf{q}_1$ and $\mathbf{q}_2$ on opposite sides of L that are farthest from L.

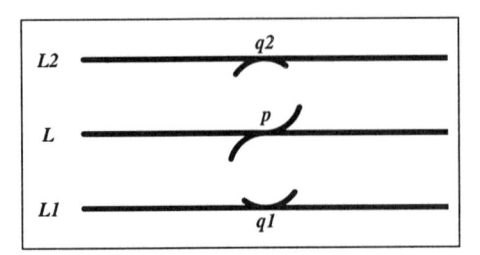

The tangent lines L_1 at $\mathbf{q}_1$ and L_2 at $\mathbf{q}_2$ must be parallel to L. If this were not the case, we could construct a line $\tilde{L}$ through $\mathbf{q}_1$ (or $\mathbf{q}_2$) parallel to L. Since $\tilde{L}$ would pass through $\mathbf{q}_1$ (or $\mathbf{q}_2$) but would not be tangent to β, there would be points on β on both sides of $\tilde{L}$. There would then be points on β more distant from L than $\mathbf{q}_1$ (or $\mathbf{q}_2$).

Two of the three points $\{\mathbf{p}, \mathbf{q}_1, \mathbf{q}_2\}$ must have tangents pointing in the same direction. In other words, if $\mathbf{p} = \beta(s_0)$, $\mathbf{q}_1 = \beta(s_1)$ and $\mathbf{q}_2 = \beta(s_2)$, then there exist s_i and s_j with $s_i < s_j$ such that

$$\beta'(s_i) = \beta'(s_j) \quad \text{and} \quad \theta[\beta](s_j) = \theta[\beta](s_i) + 2n\pi,$$

for some integer n. Since $\theta[\beta]$ is monotone, Theorem 7.6 implies that $n = -1$, $n = 0$ or $n = 1$. If $n = 0$, then $\theta[\beta](s_i) = \theta[\beta](s_j)$, and the monotonicity of $\theta[\beta]$ implies that $\theta[\beta]$ is constant on the interval $[s_i, s_j]$. If $n = \pm 1$, then $\theta[\beta]$ is constant on the intervals $[0, s_i]$ and $[s_j, L]$. In either case, one of the segments of β between $\beta(s_i)$ and $\beta(s_j)$ is a straight line.

Hence the tangent lines at $\beta(s_i)$ and $\beta(s_j)$ coincide. But L, L_1 and L_2 are distinct. Thus we reach a contradiction, and so β must be convex.

To prove the converse, assume that β is convex, but that the turning angle $\theta[\beta]$ is not monotone. Then we can find s_0, s_1 and s_2 such that $s_1 < s_0 < s_2$ with $\theta[\beta](s_1) = \theta[\beta](s_2) \neq \theta[\beta](s_0)$. Corollary 7.7 says that $s \longmapsto \beta'(s)$ maps the interval $[0, L]$ onto all of the unit circle $S^1(1)$; hence there is s_3 such that $\beta'(s_3) = -\beta'(s_1)$. If the tangent lines at $\beta(s_1)$, $\beta(s_2)$ and $\beta(s_3)$ are distinct, they are parallel, and one lies between the other two. This cannot be the case, since β is convex. Thus two of the tangent lines coincide, and there are points **p** and **q** of β lying on the same tangent line. We show that the curve β is a straight line connecting **p** and **q**. Let $L(\mathbf{p}, \mathbf{q})$ denote the straight line segment from **p** to

q. Suppose that some point **r** of $L(\mathbf{p},\mathbf{q})$ is not on β. Let $\widehat{L}$ be the straight line perpendicular to $L(\mathbf{p},\mathbf{q})$ at **r**. Since β is convex, $\widehat{L}$ is nowhere tangent to β. Thus $\widehat{L}$ intersects β in at least two points $\mathbf{r}_1$ and $\mathbf{r}_2$. If $\mathbf{r}_1$ denotes the point closer to **r**, then the tangent line to β at $\mathbf{r}_1$ has at least one of the points **p**, **q**, **r** on each side, contradicting the assumption that β is convex.

Hence **r** cannot exist, and so the straight line segment $L(\mathbf{p},\mathbf{q})$ is contained in the trace of β. Thus **p** and **q** are $\beta(s_1)$ and $\beta(s_2)$, so that the restriction of β to the interval $[s_1, s_2]$ is a straight line. Therefore, $\theta[\beta]$ is constant on $[s_1, s_2]$, contradicting the assumption that $\theta[\beta]$ was not monotone. Hence the assumption that $\theta[\beta]$ was not monotone leads to a contradiction. It follows that $\theta[\beta]$ is monotone, and consequently $\kappa 2[\beta]$ has constant sign. ∎

The proof of Theorem 7.10 contains the proof of the following result.

Corollary 7.11. *Let α be a regular simple closed curve with angle function $\theta[\alpha]$. If $\theta[\alpha](t_1) = \theta[\alpha](t_2)$ with $t_1 < t_2$, then the restriction of α to the interval $[t_1, t_2]$ is a straight line.*

7.4 The Four Vertex Theorem

In this section we prove a celebrated global theorem about plane curves. To understand the result, let us first consider an example. The curve **sinoval** is defined by

```
sinoval[n_,a_][t_]:= {a Cos[t],a Nest[Sin,t,n]}
```

Clearly, **sinoval[1,a]** is a circle of radius **a**. Then **sinoval[2,a]** is the curve $t \longmapsto \bigl(a\cos t, a\sin(\sin t)\bigr)$, and so forth. As n increases, the top and bottom of **sinoval[n,a]** are pushed together more and more. Here is typical plot of a sinoval:

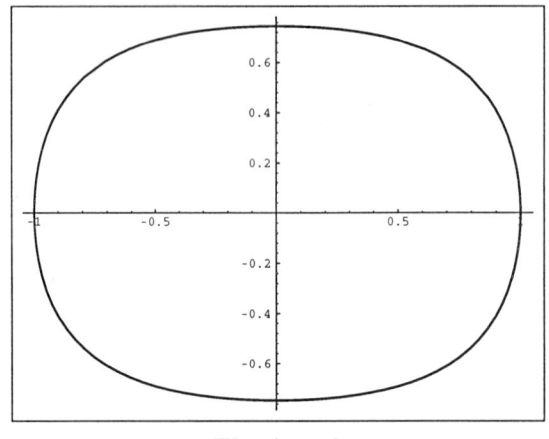

```
ParametricPlot[
Evaluate[
sinoval[3,1][t]],
{t,0,2Pi},
AspectRatio->
Automatic]
```

The sinoval

$$t \longrightarrow \bigl(\cos t, \sin(\sin(\sin t))\bigr)$$

It is clear from this picture that the curvature of the sinoval has four maxima and four minima. More precise information is obtained when we graph the curvature:

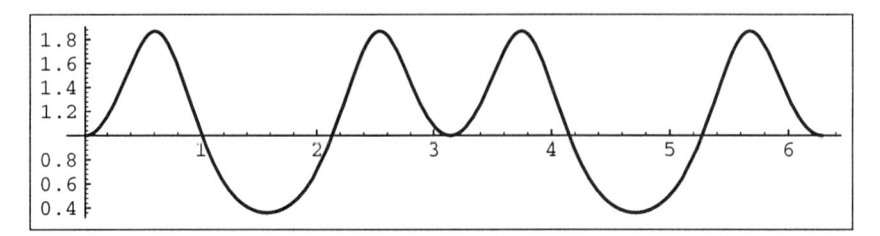

Curvature of the sinoval

$$t \longrightarrow \bigl(\cos t, \sin(\sin(\sin t))\bigr)$$

An even simpler example is an ellipse. The curvature of an ellipse has exactly 2 maxima and 2 minima (see exercise 8). These and other simple examples lead naturally to the conjecture that the signed curvature of any simple closed convex curve has at least two maxima and two minima. To prove this conjecture, we first make the following definition.

Definition. *A* **vertex** *of a regular plane curve is a point where the signed curvature has a relative maximum or minimum.*

Thus to find the vertices of a simple closed convex curve, we must determine those points where the derivative of the curvature vanishes.

7.4 The Four Vertex Theorem

Lemma 7.12. *The definition of vertex is independent of the choice of regular parametrization.*

Proof. The conclusion follows when we differentiate equation (1.12), page 16. ∎

We need the following elementary lemma.

Lemma 7.13. *Let L be a line in the plane. Then there exist constant vectors $\mathbf{a}, \mathbf{c} \in \mathbb{R}^2$ with $\mathbf{c} \neq 0$ such that $\mathbf{z} \in L$ if and only if*

$$(\mathbf{z} - \mathbf{a}) \cdot \mathbf{c} = 0.$$

Proof. If we parametrize L by $\alpha(t) = \mathbf{p} + t\mathbf{q}$, we can take $\mathbf{a} = \mathbf{p}$ and $\mathbf{c} = J\mathbf{q}$. ∎

Finally, we are ready to prove the conjecture.

Theorem 7.14. (Four vertex theorem) *A simple closed convex curve α has at least four vertices.*

Proof. Clearly, the derivative $\kappa 2'$ vanishes at each vertex of α. If $\kappa 2$ is constant on any segment of α, then every point on the segment is a vertex, and we are done. We can therefore assume that α contains neither circular arcs nor straight line segments. Moreover, we can suppose that α has at least two distinct vertices $\mathbf{p}, \mathbf{q} \in \mathbb{R}^2$; without loss of generality, $\alpha(0) = \mathbf{p}$. We now show that the assumption that $\mathbf{p}$ and $\mathbf{q}$ are the only vertices leads to a contradiction. Because vertices come in pairs, this will complete the proof of the theorem.

Let L be the straight line joining $\mathbf{p}$ and $\mathbf{q}$; then L divides α into two segments. Since we have assumed that there are exactly two vertices, it must be the case that $\kappa 2'$ is positive on one segment of α and negative on the other.

Lemma 7.13 says that there are constant vectors $\mathbf{a}$ and $\mathbf{c} \neq 0$ such that $\mathbf{z} \in L$ if and only if $(\mathbf{z} - \mathbf{a}) \cdot \mathbf{c} = 0$. Because α is convex, $(\mathbf{z} - \mathbf{a}) \cdot \mathbf{c}$ is positive on one segment of α and negative on the other.

It can be checked case by case that $\kappa 2'(s)(\alpha(s) - \mathbf{a}) \cdot \mathbf{c}$ does not change sign on α. Hence there must be an s_0 for which $\kappa 2'(s_0)(\alpha(s_0) - \mathbf{a}) \cdot \mathbf{c} > 0$, and so the integral of $\kappa 2'(s)(\alpha(s_0) - \mathbf{a}) \cdot \mathbf{c}$ from 0 to L is nonzero, where L is the length of C. Integrating by parts, we obtain

$$(7.3) \qquad 0 \neq \int_0^L \kappa 2'(s)(\alpha(s) - \mathbf{a}) \cdot \mathbf{c} \, ds$$

$$= \kappa 2(s)(\alpha(s) - \mathbf{a}) \cdot \mathbf{c} \Big|_0^L - \int_0^L \kappa 2(s)(\alpha'(s) \cdot \mathbf{c}) ds = \int_0^L \bigl(-\kappa 2(s)\alpha'(s) \cdot \mathbf{c}\bigr) ds.$$

Lemma 1.11, page 16, implies that $J\alpha''(s) = -\kappa_2(s)\alpha'(s)$, and so the last integral on the right-hand side of (7.3) can be written as

$$(7.4) \qquad \int_0^L \bigl(-\kappa_2(s)\alpha'(s)\cdot \mathbf{c}\bigr)ds = \int_0^L \bigl(\alpha''(s)\cdot \mathbf{c}\bigr)ds = \alpha'(s)\cdot \mathbf{c}\Big|_0^L.$$

But $\alpha'(L) = \alpha'(0)$, so from (7.3) and (7.4) we reach the contradiction that $0 \neq 0$. It follows that the assumption that C had only two vertices is false, and therefore C must have at least four vertices. ∎

It turns out that any simple closed curve, convex or not, has at least four vertices; this is the result of Mukhopadhyaya [Muk]. On the other hand, there exist nonsimple closed curves with only two vertices; see exercise 9.

7.5 Curves of Constant Width

Why is a manhole cover (at least in the U.S.) round? Probably a square manhole cover would be easier to manufacture. But a square manhole cover, when rotated, could slip through the manhole; however, a circular manhole cover can never slip underground. The reason is that a circular manhole cover has constant width, but the width of a square manhole cover varies between a and $a\sqrt{2}$, where a is the length of a side.

Are there other curves of constant width? This question was answered affirmatively by Euler [Euler4] over two hundred years ago. Thus a city governed by a mathematician might want to use manhole covers in the shape of a Reuleaux triangle or the involute of a deltoid; see Sections 7.8 and 7.9.

In this section we give a few details of the theory of curves of constant width. The literature on this subject is large: see, for example, [Bar], [Bieb, pages 27-29], [Dark], [Euler4], [Fischer, chapter 4], [HC-V, page 216], [MiPa, pages 66-71], [RaTo1, pages 137-150], [Strub, volume 1, pages 120-124] and [Stru2, pages 47-51].

Definition. *An* **oval** *is a simple closed plane curve for which the signed curvature* κ_2 *is always positive or always negative.*

By Theorem 7.10 an oval is convex. If α is an oval whose signed curvature is always negative, we can consider instead the curve $t \longmapsto \alpha(-t)$; thus the signed curvature of an oval can be assumed to be positive.

7.5 Curves of Constant Width

Let $\beta: \mathbb{R} \longrightarrow \mathbb{R}^2$ be a unit-speed oval and let $\beta(s)$ be a point on β. Corollary 7.7 implies that there is a point $\beta(\tilde{s})$ on β for which $\beta'(\tilde{s}) = -\beta'(s)$, and the reasoning in the proof of Theorem 7.10 implies that $\beta(\tilde{s})$ is unique. We call $\beta(\tilde{s})$ the **point opposite to** $\beta(s)$. Let $\widehat{\beta}: \mathbb{R} \longrightarrow \mathbb{R}^2$ be the curve such that $\widehat{\beta}(s)$ is the point opposite to $\beta(s)$ for all s. Let us write $\mathbf{T}(s) = \beta'(s)$. Since $\mathbf{T}(s)$ and $J\mathbf{T}(s)$ are linearly independent, we can write

(7.5) $$\widehat{\beta}(s) - \beta(s) = \lambda(s)\mathbf{T}(s) + \mu(s)J\mathbf{T}(s),$$

where $\lambda, \mu: \mathbb{R} \longrightarrow \mathbb{R}$ are piecewise-differentiable functions. Note that even though β has unit-speed, $\widehat{\beta}$ may not have unit-speed. We put $\widehat{\mathbf{T}}(s) = \widehat{\beta}'(s)/\|\widehat{\beta}'(s)\|$.

Clearly, $\lambda(s)$ represents the distance between the tangent lines at opposite points $\beta(s)$ and $\beta(\tilde{s})$. This leads to the following definitions.

Definition. *The* **width** *and* **spread** *of a unit-speed oval* $\beta: \mathbb{R} \longrightarrow \mathbb{R}^2$ *are the functions* λ *and* μ. *We say that an oval has* **constant width** *provided its width function* λ *is constant.*

Note that $\sqrt{\lambda(s)^2 + \mu(s)^2}$ represents the distance between the points $\beta(s)$ and $\beta(\tilde{s})$.

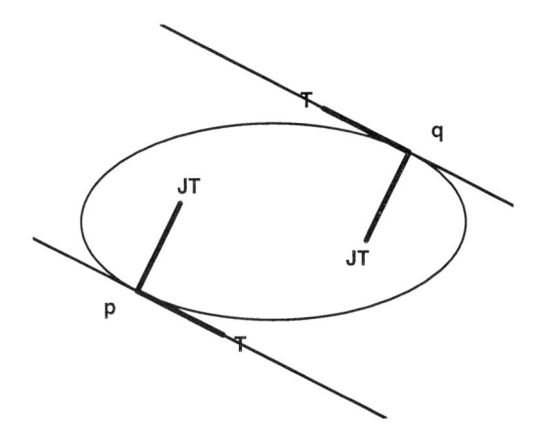

Opposite points on an oval

Next, we prove a fundamental theorem about constant width ovals.

Theorem 7.15. (Barbier[2]) *An oval of constant width w has length πw.*

---

[2] Joseph Émile Barbier (1839-1889). French mathematician. After a career in which he wrote many excellent papers on differential geometry, number theory and probability, he became insane.

Proof. We parametrize the oval by a unit-speed curve $\beta\colon \mathbb{R} \longrightarrow \mathbb{R}^2$. From (7.5) and Lemma 1.11 it follows that

$$\widehat{\beta}' = \left(1 + \frac{d\lambda}{ds} - \mu\kappa 2\right)\mathbf{T} + \left(\lambda\kappa 2 + \frac{d\mu}{ds}\right)J\mathbf{T}. \tag{7.6}$$

On the other hand, if $\widehat{s}$ denotes the arc length function of $\widehat{\beta}$ we have

$$\frac{\widehat{\beta}'}{\|\widehat{\beta}'\|} = -\mathbf{T} \quad \text{and} \quad \frac{d\widehat{s}}{ds} = \|\widehat{\beta}'\|,$$

so that

$$\widehat{\beta}' = -\frac{d\widehat{s}}{ds}\mathbf{T}, \tag{7.7}$$

From (7.6) and (7.7) we obtain

$$\left(1 + \frac{d\widehat{s}}{ds} + \frac{d\lambda}{ds} - \mu\kappa 2\right)\mathbf{T} + \left(\lambda\kappa 2 + \frac{d\mu}{ds}\right)J\mathbf{T} = 0. \tag{7.8}$$

Let ϑ denote the turning angle of β; from Corollary 1.17 we know that

$$\kappa 2 = \frac{d\vartheta}{ds}. \tag{7.9}$$

We can use (7.9) to rewrite (7.8) as

$$\left(\frac{d(s+\widehat{s})}{ds} + \frac{d\lambda}{ds} - \mu\frac{d\vartheta}{ds}\right)\mathbf{T} + \left(\lambda\frac{d\vartheta}{ds} + \frac{d\mu}{ds}\right)J\mathbf{T} = 0,$$

from which we conclude that

$$\frac{d(s+\widehat{s})}{ds} + \frac{d\lambda}{ds} - \mu\frac{d\vartheta}{ds} = 0 \quad \text{and} \quad \lambda\frac{d\vartheta}{ds} + \frac{d\mu}{ds} = 0. \tag{7.10}$$

Now suppose that μ has the constant value w. Since the curvature of β is always positive, the second equation of (7.10) tells us that $\lambda = 0$, and so the first equation of (7.10) reduces to

$$\frac{d(s+\widehat{s})}{ds} - w\frac{d\vartheta}{ds} = 0. \tag{7.11}$$

Fix a point $\mathbf{p}$ on the oval, and let s_0 and s_1 be such that $\beta(s_0) = \mathbf{p}$ and $\beta(s_1) = \widehat{\mathbf{p}}$, where $\widehat{\mathbf{p}}$ is the point on the oval opposite to $\mathbf{p}$. If L denotes the length of the oval, we have from (7.11) that

$$L = \int_{s_0}^{s_1} d(s+\widehat{s}) = \int_{s_0}^{s_1} \frac{d(s+\widehat{s})}{ds}ds = \int_{s_0}^{s_1} w\frac{d\vartheta}{ds}ds = \int_0^\pi w\, d\vartheta = w\,\pi. \blacksquare$$

7.5 Curves of Constant Width

An elegant generalization of Barbier's theorem, giving explicit formulas for the width and spread of a general oval, has been proved by Mellish.[3]

Theorem 7.16. (Mellish) *The width μ of an oval parametrized by a unit-speed curve β, when regarded as a function of ϑ, is a solution of the linear second-order differential equation*

(7.12) $$\frac{d^2\mu}{d\vartheta^2} + \mu = f(\vartheta),$$

where

$$f(\vartheta) = \frac{1}{\kappa 2(\vartheta)} + \frac{1}{\kappa 2(\vartheta + \pi)}.$$

Moreover, let

(7.13) $$\begin{cases} U(\vartheta) = \int_0^\vartheta f(t) \cos t\, dt - \frac{1}{2}\int_0^\pi f(t)\cos t\, dt, \\ V(\vartheta) = \int_0^\vartheta f(t) \sin t\, dt - \frac{1}{2}\int_0^\pi f(t)\sin t\, dt. \end{cases}$$

Then

(7.14) $$\begin{cases} \mu(\vartheta) = U(\vartheta)\sin\vartheta - V(\vartheta)\cos\vartheta, \\ \lambda(\vartheta) = -U(\vartheta)\cos\vartheta - V(\vartheta)\sin\vartheta. \end{cases}$$

Proof. We can rewrite (7.10) in terms of differentials:

(7.15) $$ds + d\widehat{s} + d\lambda - \mu\, d\vartheta = 0 \quad \text{and} \quad \lambda\, d\vartheta + d\mu = 0.$$

Clearly, (7.15) implies that

(7.16) $$\frac{ds}{d\vartheta} + \frac{d\widehat{s}}{d\vartheta} + \frac{d\lambda}{d\vartheta} - \mu = 0 \quad \text{and} \quad \lambda + \frac{d\mu}{d\vartheta} = 0.$$

But

$$\frac{ds}{d\vartheta} = \frac{1}{\kappa 2(\vartheta)} \quad \text{and} \quad \frac{d\widehat{s}}{d\vartheta} = \frac{1}{\kappa 2(\vartheta+\pi)},$$

so that (7.16) can be rewritten as

(7.17) $$f(\vartheta) + \frac{d\lambda}{d\vartheta} - \mu = 0 \quad \text{and} \quad \lambda + \frac{d\mu}{d\vartheta} = 0.$$

Elimination of λ in (7.17) yields (7.12).

---
[3] Arthur Preston Mellish. (1905-1930). Canadian mathematician.

The general solution of the differential equation (7.12) is

$$(7.18) \quad \mu(\vartheta) = \sin\vartheta \left(\int_0^\vartheta f(t) \cos t\, dt + C_1 \right) - \cos\vartheta \left(\int_0^\vartheta f(t) \sin t\, dt + C_2 \right).$$

To determine the arbitrary constants C_1 and C_2 in (7.18), we observe that the functions $\mu(\vartheta)$ and $\lambda(\vartheta)$ and also $f(\vartheta)$ are periodic in ϑ with period π. Hence (7.12) and (7.17) imply that

$$(7.19) \quad \mu(\pi) = \mu(0) \quad \text{and} \quad \frac{d\mu}{d\vartheta}(\pi) = \frac{d\mu}{d\vartheta}(0).$$

From (7.18) and (7.19) we get

$$(7.20) \quad C_1 = -\frac{1}{2} \int_0^\pi f(t) \cos t\, dt \quad \text{and} \quad C_2 = -\frac{1}{2} \int_0^\pi f(t) \sin t\, dt.$$

Now (7.14) follows from (7.13) and (7.20). ∎

7.6 Envelopes of Curves

Let $F(x, y, \alpha)$ be a function of the three variables x, y and α, for which the implicit equation

$$(7.21) \quad F(x, y, \alpha) = 0$$

defines a curve for each α. The totality of curves represented by (7.21) is called a **curve family**. When a given α is replaced by $\alpha + \Delta\alpha$, we obtain a new curve implicitly defined by

$$(7.22) \quad F(x, y, \alpha + \Delta\alpha) = 0.$$

The set of points that belong to both curves satisfies

$$(7.23) \quad \frac{F(x, y, \alpha + \Delta\alpha) - F(x, y, \alpha)}{\Delta\alpha} = 0.$$

Suppose that F is differentiable; when we take the limit as $\Delta\alpha$ tends to zero in (7.23), we obtain

$$(7.24) \quad \frac{\partial F(x, y, \alpha)}{\partial \alpha} = 0.$$

7.7 The Support Function of an Oval

Definition. *If the variable α can be eliminated from (7.21) and (7.24) resulting in*

(7.25) $$f(x,y) = 0,$$

*we call the curve implicitly defined by (7.25) the **envelope** of the curve family represented by (7.21).*

In imprecise but descriptive language we say that the envelope consists of those points which belong to each pair of infinitely near curves in the family (7.21).

To clarify these ideas, let us look at a simple example. Fix a and consider the curve family defined by

(7.26) $$(x - \alpha)^2 + y^2 - a^2 = 0,$$

where α is a parameter. Differentiation of (7.26) with respect to α yields

(7.27) $$x - \alpha = 0.$$

When we eliminate α from (7.26) and (7.27), we obtain

(7.28) $$y^2 - a^2 = 0.$$

Clearly, as α varies between $-\infty$ and ∞, the curve family given by (7.26) consists of all circles of radius a with centers on the x axis. The envelope of the family consists of the two horizontal lines $y = \pm a$.

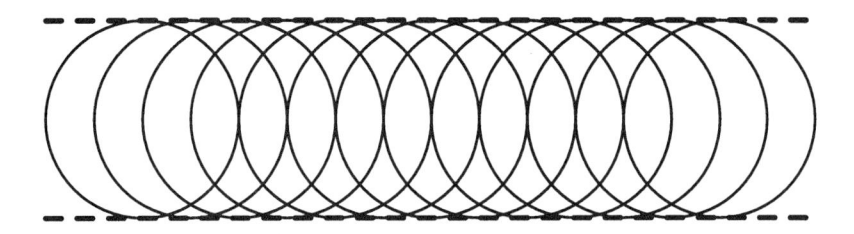

A family of circles with centers on the x-axis together with its envelope

7.7 The Support Function of an Oval

Consider the tangent line M to a plane curve C at a point $\mathbf{q}$ on C. Let H denote the line from a point $\mathbf{p}$ (taken to be the origin) passing through $\mathbf{p}$ and $\mathbf{q}$, and let L denote the line through $\mathbf{p}$ meeting M perpendicularly at a point $\mathbf{f}$. Denote by p

L denote the line through **p** meeting M perpendicularly at a point **f**. Denote by p the distance from M to **p**, and let ψ be the angle between L and the x-axis. The point **q** is given in polar coordinates as $r\,e^{i\theta}$, where r is the distance from **p** to **q** and θ is the angle between H and the x-axis. Then $p = r\cos(\theta - \psi)$, so that the line M is given by

(7.29) $$p = r\cos(\theta - \psi) = x\cos(\psi) + y\sin(\psi).$$

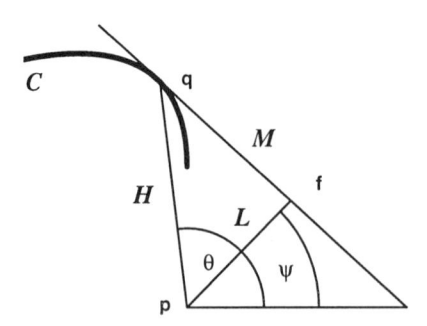

The pedal angle ψ of a curve C

Conversely, let $p(\psi)$ be a given function of ψ; for each value of ψ, (7.29) defines a straight line, and so we obtain a family of lines. This family is the zero set of the function F defined by

(7.30) $$F(x, y, \psi) = p(\psi) - x\cos(\psi) - y\sin(\psi).$$

To find the envelope of this family of lines, we must solve the system of equations

$$F(x, y, \psi) = 0 \quad \text{and} \quad \frac{\partial F(x, y, \psi)}{\partial \psi} = 0,$$

that is, the system

(7.31) $$\begin{cases} x\cos(\psi) + y\sin(\psi) = p(\psi), \\ -x\sin(\psi) + y\cos(\psi) = p'(\psi). \end{cases}$$

The solution of (7.31) is

(7.32) $$\begin{cases} x = p(\psi)\cos\psi - p'(\psi)\sin\psi, \\ y = p(\psi)\sin\psi + p'(\psi)\cos\psi. \end{cases}$$

The parameter ψ can be used to parametrize C.

7.7 The Support Function of an Oval

Definition. *Let C be a plane curve and $\mathbf{p}$ a point. The* **support function** *of C with respect to $\mathbf{p}$ is the function p defined by (7.29). The* **pedal parametrization** *of C with respect to $\mathbf{p}$ is*

(7.33) $\quad \mathbf{oval}[p](\psi) = \bigl(p(\psi)\cos\psi - p'(\psi)\sin\psi, p(\psi)\sin\psi + p'(\psi)\cos\psi\bigr).$

The *Mathematica* version is

```
oval[p_][psi_]:={p[psi]Cos[psi] - p'[psi]Sin[psi],
                 p[psi]Sin[psi] + p'[psi]Cos[psi]}
```

Note that (7.33) defines a curve for any function p; such a curve will be an oval if and only if it is a simple closed curve and has positive curvature $\kappa 2$.

Lemma 7.17. *Let $p \colon \mathbb{R} \longrightarrow \mathbb{R}$ be a differentiable function. The curvature of the plane curve $\mathbf{oval}[p]$ is given by*

$$(7.34) \qquad \kappa 2(\psi) = \frac{1}{p(\psi) + p''(\psi)}.$$

Hence $\mathbf{oval}[p]$ is an oval if and only if it is a simple closed curve and $p(\psi) + p''(\psi) > 0$ for all ψ.

Proof. (7.34) is an easy calculation from (7.33) and the definition of $\kappa 2$. ∎

(7.34) can also be checked with *Mathematica*; see exercise 10.

There is a formula relating the width and support functions of an oval.

Lemma 7.18. *Let p be the support function of an oval C with respect to some point $\mathbf{q}$ inside C, and let λ denote the width function of C. Then*

$$(7.35) \qquad \lambda(\psi) = p(\psi) + p(\psi + \pi)$$

for $0 \le \psi \le 2\pi$.

Proof. Let $\mathbf{p}$ be a point on C, and let $\widehat{\mathbf{p}}$ denote the point on C opposite to $\mathbf{p}$. By definition the ray $\overline{\mathbf{qp}}$ from $\mathbf{q}$ to $\mathbf{p}$ meets the tangent line to C at $\mathbf{p}$ perpendicularly, and similarly for $\widehat{\mathbf{p}}$. Hence the rays $\overline{\mathbf{qp}}$ and $\overline{\mathbf{q}\widehat{\mathbf{p}}}$ are part of a line segment L that meets each of the two tangent lines perpendicularly. The length of L is the width of the oval measured at $\mathbf{p}$ or at $\widehat{\mathbf{p}}$. Since the length of $\overline{\mathbf{qp}}$ is $p(\psi)$ and the length of $\overline{\mathbf{q}\widehat{\mathbf{p}}}$ is $p(\psi + \pi)$ we obtain (7.35). ∎

7.8 Reuleaux Polygons

Particularly simple examples of plane curves of constant width can be constructed from regular polygons with odd numbers of sides. Let $P[n,a]$ be a regular polygon with $2n+1$ sides, where a denotes the length of any side. Corresponding to each vertex $\mathbf{p}$, there is a side of $P[n,a]$ that is most distant from $\mathbf{p}$. Let $\mathbf{p}_1$ and $\mathbf{p}_2$ be vertices of $P[n,a]$, and let $\widehat{\mathbf{p}_1\mathbf{p}_2}$ be the arc of the circle with center $\mathbf{p}$ connecting $\mathbf{p}_1$ and $\mathbf{p}_2$.

Definition. *The* **Reuleaux**[4] *polygon is the curve $R[n,a]$ made up of all circular arcs $\widehat{\mathbf{p}_1\mathbf{p}_2}$ formed when $\mathbf{p}$ ranges over the vertices of $P[n,a]$.*

It is clear that the width of the convex curve $P[n,a]$ is constant and equal to a.

Here is how to parametrize a Reuleaux polygon in *Mathematica*.

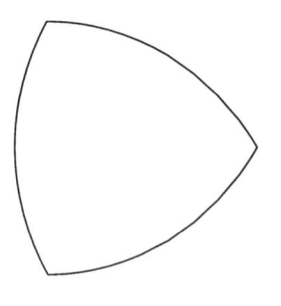

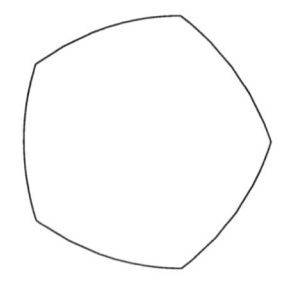

```
reuleauxpolygon[n_,a_][t_]:=Which@Flatten[
    Table[{k/(2n + 1)<=t<=(k + 1)/(2n + 1),
        a{Re[E^(2Pi I k/(2n + 1)) +
        2I Sin[Pi n/(2n + 1)]*
        E^(Pi I((k + n)/(2n + 1) + t))],
        Im[E^(2Pi I k/(2n + 1)) +
        2I Sin[Pi n/(2n + 1)]*
        E^(Pi I((k + n)/(2n + 1) + t))]}},
    {k,0,2n}],1]
```

Reuleaux triangle   **Reuleaux pentagon**

---
[4]Franz Reuleaux, (1829-1905). German professor of machine design.

A Reuleaux triangle is the model for a cross section of the rotor in the Wankel[5] engine.

7.9 The Involute Construction of Curves of Constant Width

Consider a deltoid (see exercise 2 of Chapter 3) with vertices **a**, **b** and **c**, together with a flexible cord attached to the curved side $\widehat{\mathbf{bc}}$. Keep the cord attached to a point **p** on $\widehat{\mathbf{bc}}$, and let both ends of the cord unwind.

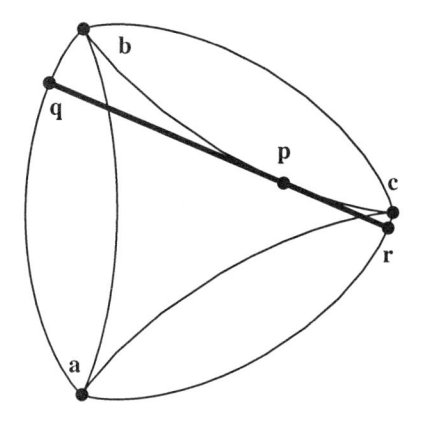

A deltoid and its involute

The end of the cord that was originally attached to **b** traces out a curve, part of an involute of a deltoid, and similarly for the end of the cord attached to **c**. Let **b** move to **q** and **c** to **r**, and denote by $\overline{\mathbf{pq}}$ and $\overline{\mathbf{pr}}$ the line segments from **p** to **q** and

---

[5] Felix Heinrch Wankel (1902-1988). German engineer. The Wankel engine differs greatly from conventional engines. It retains the familiar intake, compression, power, and exhaust cycle but uses a rotor in the shape of a Reuleaux triangle, instead of a piston, cylinder, and mechanical valves. The Wankel engine has 40 percent fewer parts and roughly one third the bulk and weight of a comparable reciprocating engine. Within the Wankel, three chambers are formed by the sides of the rotor and the wall of the housing. The shape, size, and position of these chambers are constantly altered by the rotor's clockwise rotation. The engine is unique in that the power impulse is spread over approximately 270 degrees of crank shaft rotation, as compared to 180 degrees for the conventional reciprocating two-stroke engine.

from **p** to **r**. By definition of the involute,

$$\text{Length}(\overline{pq}) = \text{Length}(\widehat{pb}) \quad \text{and} \quad \text{Length}(\overline{pr}) = \text{Length}(\widehat{pc}),$$

where $\widehat{pb}$ and $\widehat{pc}$ denote arcs of the deltoid. Since the line segments $\overline{pq}$ and $\overline{pr}$ are both tangent to the deltoid at **p**, they are part of the straight line segment $\overline{qr}$ connecting **q** to **r**. Consequently,

$$\text{Length}(\overline{qr}) = \text{Length along the deltoid of } \widehat{bc}.$$

Therefore, $\overline{qr}$ has the same length, no matter the position of **p**. It follows that the involute of a deltoid has constant width.

This construction was originally carried out by Euler [Euler4] in more generality.

The parametrization of the involute of a deltoid can be found with the command

```
Simplify[PowerExpand[Simplify[
    involute[deltoid[a],c][t]]] /. c->0]
```

Therefore, an involute of a deltoid is defined by

```
deltoidinvolute[a_][t_]:=
    {(a/3)(8Cos[t/2] + 2Cos[t] - Cos[2t]),
     (a/3)(-8Sin[t/2] + 2Sin[t] + Sin[2t])}
```

7.10 Exercises

1. Consider the **piriform** defined by

```
piriform[a_,b_][t_]:=
    {a(1 + Sin[t]),b Cos[t](1 + Sin[t])}
```

7.10 Exercises

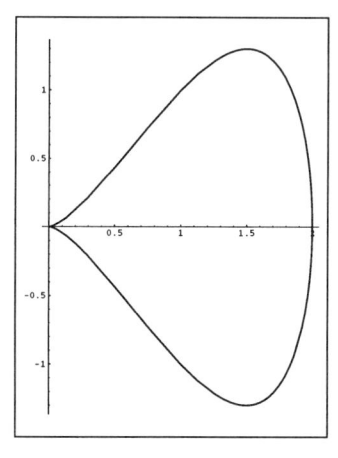

The piriform
$$t \longrightarrow (1 + \sin t)(1, \cos t)$$

```
ParametricPlot[Evaluate[
piriform[1,1][t]],
{t,0,2Pi},
AspectRatio->Automatic]
```

Show that the piriform is *not* a closed curve by finding a point where the curve fails to be regular.

2. Finish the proof of Lemma 7.1.

3. Show that the figure eight curve (see page 31)

$$t \longrightarrow (\sin t, \sin t \cos t), \qquad (0 \le t \le 2\pi)$$

and the lemniscate (see page 52)

$$t \longrightarrow \left(\frac{a \cos t}{1 + \sin^2 t}, \frac{a \sin t \cos t}{1 + \sin^2 t} \right), \qquad (0 \le t \le 2\pi)$$

are both closed curves with total signed curvature and turning number equal to zero. Use *Mathematica* to check the results.

4. Show that the limaçon (see exercise 4 of Chapter 3)

$$t \longrightarrow (2a \cos t + b)(\cos t, \sin t). \qquad (0 \le t \le 2\pi)$$

has turning number equal to 2. Use *Mathematica* to check the result. Why is the turning number of a limaçon different from that of a figure eight curve or that of a lemniscate?

5. Plot **hypotrochoid[18, 2, 6]** and compute its turning number.

6. Use *Mathematica* to find explicit formulas for the signed curvature of a general epitrochoid.

7. Use the command

    ```
    epitrochoid[a,b,h][t]===hypotrochoid[a,-b,-h][t]
    ```

    to show that the two curves coincide.

8. Verify that a noncircular ellipse has exactly 4 vertices and plot its curvature.

9. Verify that the `limacon[1,1]` is a nonsimple closed curve that has exactly 2 vertices, and plot its curvature.

10. Prove (7.34) using *Mathematica*.

11. Show that a parallel curve B to a closed curve C of constant width also has constant width, provided C is interior to B.

8

CURVES IN SPACE

In Chapters 1–7 we saw that the curvature $\kappa 2[\alpha]$ of a plane curve α measured the failure of α to be a straight line. In this chapter we define a similar curvature $\kappa[\alpha]$ for a curve in $\mathbb{R}^n$; it also measures the failure of a curve to be a straight line. $\kappa[\alpha]$ reduces to the absolute value of $\kappa 2[\alpha]$ when $n = 2$. For curves in $\mathbb{R}^3$ we can also measure the failure of a curve to lie in a plane by means of another function called the *torsion* $\tau[\alpha]$.

We shall need the Gibbs[1] vector cross product to study curves in $\mathbb{R}^3$, just as we needed a complex structure to study curves in $\mathbb{R}^2$. We recall the definition and some of the properties of the vector cross product on $\mathbb{R}^3$ in Section 8.1. Curvature and torsion are defined in Section 8.2. In Section 8.2 we also define three orthogonal unit vector fields $\{\mathbf{T}, \mathbf{N}, \mathbf{B}\}$ along a space curve; they constitute the *Frenet frame field* of the curve. In Chapters 1–6 we made frequent use of the frame field $\{\mathbf{T}, J\mathbf{T}\}$ to study the geometry of a plane curve. The Frenet frame $\{\mathbf{T}, \mathbf{N}, \mathbf{B}\}$ plays the same role for space curves, but the algebra is somewhat different. The twisting and turning of the Frenet frame field can be measured by curvature and torsion. In fact, the Frenet formulas use curvature and torsion to express the derivatives of the three vector fields of the Frenet frame field in terms of the vector fields themselves. We

---

[1] Josiah Willard Gibbs (1839-1903). American physicist. When he tried to make use of Hamilton's quaternions, he found it more useful to split quaternion multiplication into a scalar part and a vector part, and to regard them as separate multiplications. Although this caused great consternation among some of Hamilton's followers, the formalism of Gibbs eventually prevailed; engineering and physics books using it started to appear in the early 1900s. Gibbs also played a large role in reviving Grassman's vector calculus.

establish the Frenet formulas for unit-speed space curves in Section 8.2, and for arbitrary-speed space curves in Section 8.3.

Mathematica programs for computing the curvature, torsion and Frenet frame field of a space curve are described in Section 8.4. Also, *Mathematica* is used in Section 8.5 to construct a helical curve over any plane curve. Viviani's curve is discussed in Section 8.6, and an animation of the Frenet Frame moving on Viviani's curve is given in Section 8.8.

8.1 The Vector Cross Product on $\mathbb{R}^3$

Let us recall the notion of vector cross product on $\mathbb{R}^3$. First, let us agree to use the notation

$$\mathbf{i} = (1,0,0), \quad \mathbf{j} = (0,1,0), \quad \mathbf{k} = (0,0,1).$$

For $\mathbf{a} \in \mathbb{R}^3$ we can write either $\mathbf{a} = (a_1, a_2, a_3)$ or $\mathbf{a} = a_1\mathbf{i} + a_2\mathbf{j} + a_3\mathbf{k}$.

Definition. *Let* $\mathbf{a} = (a_1, a_2, a_3)$ *and* $\mathbf{b} = (b_1, b_2, b_3)$ *be vectors in* $\mathbb{R}^3$. *Then the* **vector cross product** *of* $\mathbf{a}$ *and* $\mathbf{b}$ *is given by*

$$\mathbf{a} \times \mathbf{b} = \det \begin{pmatrix} \mathbf{i} & \mathbf{j} & \mathbf{k} \\ a_1 & a_2 & a_3 \\ b_1 & b_2 & b_3 \end{pmatrix},$$

or more explicitly,

$$\mathbf{a} \times \mathbf{b} = \det \begin{pmatrix} a_2 & a_3 \\ b_2 & b_3 \end{pmatrix} \mathbf{i} - \det \begin{pmatrix} a_1 & a_3 \\ b_1 & b_3 \end{pmatrix} \mathbf{j} + \det \begin{pmatrix} a_1 & a_2 \\ b_1 & b_2 \end{pmatrix} \mathbf{k}.$$

The vector cross product on $\mathbb{R}^3$ has the following well-known properties:

$$\mathbf{a} \times \mathbf{c} = -\mathbf{c} \times \mathbf{a},$$

$$(\mathbf{a} + \mathbf{b}) \times \mathbf{c} = \mathbf{a} \times \mathbf{c} + \mathbf{b} \times \mathbf{c},$$

$$(\lambda \mathbf{a}) \times \mathbf{c} = \lambda(\mathbf{a} \times \mathbf{c}) = \mathbf{a} \times (\lambda \mathbf{c}),$$

(8.1) $\qquad (\mathbf{a} \times \mathbf{b}) \cdot (\mathbf{c} \times \mathbf{d}) = (\mathbf{a} \cdot \mathbf{c})(\mathbf{b} \cdot \mathbf{d}) - (\mathbf{a} \cdot \mathbf{d})(\mathbf{b} \cdot \mathbf{c}),$

(8.2) $$\|\mathbf{a} \times \mathbf{b}\|^2 = \|\mathbf{a}\|^2 \|\mathbf{b}\|^2 - (\mathbf{a} \cdot \mathbf{b})^2,$$

(8.3) $$\mathbf{a} \times (\mathbf{b} \times \mathbf{c}) = (\mathbf{a} \cdot \mathbf{c})\mathbf{b} - (\mathbf{a} \cdot \mathbf{b})\mathbf{c},$$

(8.4) $$(\mathbf{a} \times \mathbf{b}) \cdot \mathbf{c} = \mathbf{a} \cdot (\mathbf{b} \times \mathbf{c}),$$

for $\lambda \in \mathbb{R}$ and $\mathbf{a}, \mathbf{b}, \mathbf{c}, \mathbf{d} \in \mathbb{R}^3$. Frequently, (8.1) is referred to as the **Lagrange**[2] **identity**. For $\mathbf{a}, \mathbf{b}, \mathbf{c} \in \mathbb{R}^3$ we define the **vector triple product (abc)** by

$$(\mathbf{abc}) = \det \begin{pmatrix} \mathbf{a} \\ \mathbf{b} \\ \mathbf{c} \end{pmatrix} = \det \begin{pmatrix} a_1 & a_2 & a_3 \\ b_1 & b_2 & b_3 \\ c_1 & c_2 & c_3 \end{pmatrix},$$

where $\mathbf{a} = (a_1, a_2, a_3)$, and so forth. The vector triple product is related to the dot product and the cross product by the formulas

(8.5) $$(\mathbf{abc}) = (\mathbf{a} \times \mathbf{b}) \cdot \mathbf{c} = \mathbf{a} \cdot (\mathbf{b} \times \mathbf{c}).$$

Finally, we note the following consequence of (8.2):

(8.6) $$\|\mathbf{a} \times \mathbf{b}\| = \sin \theta \|\mathbf{a}\| \|\mathbf{b}\|,$$

where θ is the positive angle between $\mathbf{a}$ and $\mathbf{b}$. As a trivial consequence $\mathbf{a} \times \mathbf{a} = \mathbf{0}$ for any $\mathbf{a} \in \mathbb{R}^3$. Equation (8.6) can be interpreted geometrically; it says that $\|\mathbf{a} \times \mathbf{b}\|$ equals the area of the parallelogram spanned by $\mathbf{a}$ and $\mathbf{b}$.

8.2 Curvature and Torsion of Unit-Speed Curves in $\mathbb{R}^3$

We first define the notion of curvature of a unit-speed curve in $\mathbb{R}^n$ for general n. (Arbitrary-speed curves will be considered in Section 8.3.)

---

[2] Joseph Louis Lagrange (1736-1813). Born in Turin, Italy, Lagrange succeeded Euler as director of mathematics of the Berlin Academy of Science in 1766. In 1787 he left Berlin to become a member of the Paris Academy of Science, where he remained for the rest of his life. He made important contributions to mechanics, the calculus of variations and differential equations, in addition to differential geometry. During the 1790s he worked on the metric system and advocated a decimal base.

Definition. Let $\beta\colon (c,d) \longrightarrow \mathbb{R}^n$ be a unit-speed curve. Write

$$\kappa[\beta](s) = \|\beta''(s)\|.$$

Then the function $\kappa[\beta]\colon (c,d) \longrightarrow \mathbb{R}$ is called the **curvature** of β.

We abbreviate $\kappa[\beta]$ to κ when there is no danger of confusion.

Intuitively, curvature measures the failure of a curve to be a straight line. More precisely, a straight line in $\mathbb{R}^n$ is characterized by the fact that its curvature vanishes, as we show in the following lemma.

Lemma 8.1. Let $\beta\colon (c,d) \longrightarrow \mathbb{R}^n$ be a unit-speed curve. The following conditions are equivalent:

(i) $\kappa \equiv 0$;

(ii) $\beta'' \equiv 0$;

(iii) β is a straight line.

Proof. It is clear from the definition that (i) and (ii) are equivalent. To show that (ii) and (iii) are equivalent, suppose that $\beta'' \equiv 0$. Integration of this n-tuple of differential equations yields

(8.7) $$\beta(s) = \mathbf{u}s + \mathbf{v},$$

where $\mathbf{u}, \mathbf{v} \in \mathbb{R}^n$ are constant vectors with $\|\mathbf{u}\| = 1$. Then (8.7) is a unit-speed parametrization of a straight line in $\mathbb{R}^n$. Conversely, any straight line in $\mathbb{R}^n$ has a parametrization of the form (8.7), and from (8.7) it is clear that $\beta'' \equiv 0$. ∎

The curvature κ of a curve in the plane is related in a simple way to the curvature $\kappa 2$ defined in Chapter 1.

Lemma 8.2. Let $\beta\colon (a,b) \longrightarrow \mathbb{R}^2$ be a unit-speed curve. Then

$$\kappa[\beta] = |\kappa 2[\beta]|.$$

Proof. From Lemma 1.11 we have

$$\kappa[\beta] = \|\beta''\| = \|\kappa 2[\beta] J\beta'\| = |\kappa 2[\beta]|\,\|\beta'\| = |\kappa 2[\beta]|. \quad \blacksquare$$

In contrast to the signed curvature $\kappa 2$, the curvature κ is always nonnegative.

8.2 Curvature and Torsion of Unit-Speed Curves in $\mathbb{R}^3$

Definition. *Let $\beta\colon (c,d) \longrightarrow \mathbb{R}^n$ be a unit-speed curve. Then*

$$\mathbf{T} = \beta'$$

is called the **unit tangent vector field** *of β.*

We need a general fact about a vector field along a curve. (The definition of vector field along a curve is given in Section 1.4.)

Lemma 8.3. *If $\mathbf{F}$ is a vector field of unit length along a curve $\alpha\colon (a,b) \longrightarrow \mathbb{R}^n$, then $\mathbf{F}' \cdot \mathbf{F} = 0$.*

Proof. We have
(8.8) $$\mathbf{F}(t) \cdot \mathbf{F}(t) = 1$$
for all t. When we differentiate (8.8), we get $2\mathbf{F}(t) \cdot \mathbf{F}'(t) = 0$. ∎

Taking $\mathbf{F} = \mathbf{T}$ in Lemma 8.3, we see that $\mathbf{T}' \cdot \mathbf{T} = 0$.

We now restrict our attention to unit-speed curves in $\mathbb{R}^3$ whose curvature is strictly positive. This implies that $\mathbf{T}' = \beta''$ never vanishes. Now we can define the torsion, as well as the vector fields $\mathbf{N}$ and $\mathbf{B}$.

Definition. *Let $\beta\colon (c,d) \longrightarrow \mathbb{R}^3$ be a unit-speed curve with $\kappa(s) > 0$ for $c < s < d$. The vector field*

$$\mathbf{N} = \frac{\mathbf{T}'}{\kappa}$$

is called the **principal normal vector field** *and $\mathbf{B} = \mathbf{T} \times \mathbf{N}$ is called the* **binormal vector field***. The triple $\{\mathbf{T}, \mathbf{N}, \mathbf{B}\}$ is called the* **Frenet**[3] **frame field** *on β.*

We are now ready to establish the Frenet formulas for space curves; they form one of the basic tools for the differential geometry of space curves.

Theorem 8.4. *Let $\beta\colon (c,d) \longrightarrow \mathbb{R}^3$ be a unit-speed curve with $\kappa(s) > 0$ for $c < s < d$. Then:*

(i) *$\|\mathbf{T}\| = \|\mathbf{N}\| = \|\mathbf{B}\| = 1$ and $\mathbf{T} \cdot \mathbf{N} = \mathbf{N} \cdot \mathbf{B} = \mathbf{B} \cdot \mathbf{T} = 0$.*

(ii) *Any vector field $\mathbf{F}$ along β can be expanded as*

(8.9) $$\mathbf{F} = (\mathbf{F} \cdot \mathbf{T})\mathbf{T} + (\mathbf{F} \cdot \mathbf{N})\mathbf{N} + (\mathbf{F} \cdot \mathbf{B})\mathbf{B}.$$

---
[3] Jean Frédéric Frenet (1816-1900). French mathematician. Professor at Toulouse and Lyon.

(iii) *The* **Frenet Formulas** *hold*:

(8.10)
$$\begin{cases} \mathbf{T}' = & \kappa \mathbf{N}, \\ \mathbf{N}' = -\kappa \mathbf{T} & +\tau \mathbf{B}, \\ \mathbf{B}' = & -\tau \mathbf{N}. \end{cases}$$

Here $\tau = \tau[\beta]$ is called the **torsion** *of the curve β.*

Proof. By definition $\|\mathbf{T}\| = 1$; furthermore,

$$\|\mathbf{N}\| = \left\|\frac{1}{\kappa}\mathbf{T}'\right\| = \frac{\|\beta''\|}{|\kappa|} = 1,$$

and by Lemma 8.3 we have $\mathbf{T} \cdot \mathbf{N} = 0$. Therefore, the Lagrange identity (8.1) implies that

$$\|\mathbf{B}\|^2 = \|\mathbf{T} \times \mathbf{N}\|^2 = \|\mathbf{T}\|^2\|\mathbf{N}\|^2 - (\mathbf{T}\cdot\mathbf{N})^2 = \|\mathbf{T}\|^2\|\mathbf{N}\|^2 = 1.$$

Finally, (8.5) implies that $\mathbf{B} \cdot \mathbf{T} = \mathbf{B} \cdot \mathbf{N} = 0$. This completes the proof of (i).

Since $\mathbf{T}$, $\mathbf{N}$ and $\mathbf{B}$ are mutually orthogonal, they form a basis for the vector fields along β. Hence there exist functions λ, μ and ν such that

(8.11)
$$\mathbf{F} = \lambda \mathbf{T} + \mu \mathbf{N} + \nu \mathbf{B}.$$

When we take the dot product of both sides of (8.11) with $\mathbf{T}$ and use (i), we find that $\lambda = \mathbf{F} \cdot \mathbf{T}$. Similarly, $\mu = \mathbf{F} \cdot \mathbf{N}$ and $\nu = \mathbf{F} \cdot \mathbf{B}$, proving (ii).

For (iii) we first observe that the first equation of (8.10) holds by definition of $\mathbf{N}$. To prove the third equation of (8.10), we differentiate $\mathbf{B} \cdot \mathbf{T} = 0$, obtaining $\mathbf{B}' \cdot \mathbf{T} + \mathbf{B} \cdot \mathbf{T}' = 0$; then

$$\mathbf{B}' \cdot \mathbf{T} = -\mathbf{B} \cdot \mathbf{T}' = -\mathbf{B} \cdot \kappa \mathbf{N} = 0.$$

Lemma 8.3 implies $\mathbf{B}' \cdot \mathbf{B} = 0$. Since $\mathbf{B}'$ is perpendicular to $\mathbf{T}$ and $\mathbf{B}$, it must be a multiple of $\mathbf{N}$ by part (ii). This means that we can define the torsion τ of β by the equation

$$\mathbf{B}' = -\tau \mathbf{N}.$$

We have established the first and third of the Frenet formulas. To prove the second, we use the orthonormal expansion of $\mathbf{N}'$ in terms of $\mathbf{T}$, $\mathbf{N}$, and $\mathbf{B}$ given by (ii), namely,

(8.12)
$$\mathbf{N}' = (\mathbf{N}' \cdot \mathbf{T})\mathbf{T} + (\mathbf{N}' \cdot \mathbf{N})\mathbf{N} + (\mathbf{N}' \cdot \mathbf{B})\mathbf{B}.$$

The coefficients in (8.12) are easy to find. First, differentiating $\mathbf{N} \cdot \mathbf{T} = 0$ we get

$$\mathbf{N}' \cdot \mathbf{T} = -\mathbf{N} \cdot \mathbf{T}' = -\mathbf{N} \cdot \kappa \mathbf{N} = -\kappa.$$

8.2 Curvature and Torsion of Unit-Speed Curves in $\mathbb{R}^3$

That $\mathbf{N}' \cdot \mathbf{N} = 0$ follows from Lemma 8.3. Finally,

$$\mathbf{N}' \cdot \mathbf{B} = -\mathbf{N} \cdot \mathbf{B}' = -\mathbf{N} \cdot (-\tau \mathbf{N}) = \tau.$$

Hence (8.12) reduces to the second Frenet Formula. ∎

The first book containing a systematic treatment of space curves is Clairaut's[4] **Recherches sur les courbes à double courbure**. After that the term "courbe à double courbure" became a technical term for a space curve. The theory of space curves became much simpler after Frenet discovered the formulas named after him in 1847. Serret[5] found the formulas independently in 1851, and for this reason they are sometimes called the Frenet-Serret formulas (see [Frenet] and [Serret1]). In spite of their simplicity and usefulness, many years passed before they gained wide acceptance. Finally, the Frenet formulas were actually discovered for the first time in 1831 by Senff[6] and his teacher Bartels.[7] Needless to say, the isolation of Senff and Bartels in Dorpat, then a part of Russia, prevented their work from becoming widely known. (See [Reich] for details.)

There is a version of the Frenet formulas (8.10) for a straight line parametrized as a unit-speed curve β. In that case $\mathbf{T}$ is a constant vector field, and the curvature of β vanishes. We take $\mathbf{N}$ and $\mathbf{B}$ to be arbitrary constant vector fields along β, and we define the torsion of β to be zero. Then the formulas (8.10) remain valid. (See exercise 5 for details.)

The following lemma shows that the torsion measures the failure of a curve to lie in a plane.

---

[4] Alexis Claude Clairaut (1713-1765). French mathematician and astronomer, who at the age of 18 was elected to the French Academy of Sciences for his work on curve theory. In 1736-1737 he took part in an expedition to Lapland led by Maupertuis, the purpose of which was to verify Newton's theoretical proof that the earth is an oblate spheroid. Clairaut's precise calculations lead to a near perfect prediction of the arrival of Halley's comet in 1759.

[5] Joseph Alfred Serret (1819-1885). French mathematician. Serret with other Paris mathematicians greatly advanced differential calculus in the period 1840-1865. Serret also worked in number theory and mechanics. Another Serret (Paul Joseph (1827-1898)) also wrote a book **Théorie nouvelle géométrique et mécanique des lignes à double courbure** emphasizing space curves.

[6] Karl Eduard Senff (1810-1849). German professor at the University of Dorpat (now Tartu) in Estonia.

[7] Johann Martin Bartels (1769-1836). German professor at the University of Dorpat (now Tartu) in Estonia. Bartels was the first teacher of Gauss (in Brunswick); he and Gauss kept in contact over the years.

Lemma 8.5. Let $\beta\colon (c,d) \longrightarrow \mathbb{R}^3$ be a unit-speed curve with $\kappa(s) > 0$ for $c < s < d$. The following conditions are equivalent:

(i) β is a plane curve;

(ii) $\tau \equiv 0$.

When (i) and (ii) hold,
$$\tag{8.13} \mathbf{N} = \pm J\mathbf{T},$$
and the binormal $\mathbf{B}$ is perpendicular to the plane containing β.

Proof. The condition that a curve β lie in a plane Π can be expressed analytically as
$$\tag{8.14} (\beta(s) - \mathbf{p}) \cdot \mathbf{q} \equiv 0,$$
where $\mathbf{q}$ is a nonzero vector perpendicular to Π. Differentiation of (8.14) yields
$$\beta'(s) \cdot \mathbf{q} \equiv \beta''(s) \cdot \mathbf{q} \equiv 0.$$

Thus both $\mathbf{T}$ and $\mathbf{N}$ are perpendicular to $\mathbf{q}$. Since $\mathbf{B}$ is also perpendicular to $\mathbf{T}$ and $\mathbf{N}$, it follows that $\mathbf{B}(s) = \pm \mathbf{q}/\|\mathbf{q}\|$ for all s. Therefore $\mathbf{B}' \equiv 0$, and from the definition of torsion, it follows that $\tau \equiv 0$.

Conversely, suppose that $\tau \equiv 0$; then (8.10) implies that $\mathbf{B}' \equiv -\tau \mathbf{N} \equiv 0$. Thus $s \longmapsto \mathbf{B}(s)$ is a constant curve; that is, there exists a unit vector $\mathbf{u} \in \mathbb{R}^3$ such that $\mathbf{B} \equiv \mathbf{u}$. Choose t_0 with $c < t_0 < d$ and consider the real-valued function $f\colon (c,d) \longrightarrow \mathbb{R}^3$ given by
$$f(s) = \bigl(\beta(s) - \beta(t_0)\bigr) \cdot \mathbf{u}.$$

Then $f(t_0) = 0$ and $f'(s) \equiv \beta'(s) \cdot \mathbf{u} \equiv \mathbf{T} \cdot \mathbf{B} \equiv 0$, so that f is identically zero. Thus $\bigl(\beta(s) - \beta(t_0)\bigr) \cdot \mathbf{u} \equiv 0$, and it follows that β lies in the plane perpendicular to $\mathbf{u}$ that passes through $\beta(t_0)$.

Equation (8.13) follows from the fact that both $\mathbf{N}$ and $J\mathbf{T}$ are perpendicular to $\mathbf{T}$ and lie in the same plane Π. For the same reason $\mathbf{B}$ is perpendicular to Π. ∎

Example: The Curvature and Torsion of a Helix

We shall give a more detailed description of the helix on page 198. We mention it here because it is one of the few curves for which a unit-speed parametrization is easy to find. In fact, a unit-speed parametrization of the helix is given by
$$\beta(s) = \left(a\cos\left(\frac{s}{\sqrt{a^2+b^2}}\right), a\sin\left(\frac{s}{\sqrt{a^2+b^2}}\right), \frac{bs}{\sqrt{a^2+b^2}} \right).$$

We use Theorem 8.4 to compute κ, τ, $\mathbf{T}$, $\mathbf{N}$ and $\mathbf{B}$ for β. For convenience, we assume $a > 0$. Then we compute

$$\mathbf{T}(s) = \beta'(s) = \frac{1}{\sqrt{a^2+b^2}}\left(-a\sin\left(\frac{s}{\sqrt{a^2+b^2}}\right), a\cos\left(\frac{s}{\sqrt{a^2+b^2}}\right), b\right),$$

and

$$\mathbf{T}'(s) = \frac{1}{a^2+b^2}\left(-a\cos\left(\frac{s}{\sqrt{a^2+b^2}}\right), -a\sin\left(\frac{s}{\sqrt{a^2+b^2}}\right), 0\right).$$

Thus $\|\mathbf{T}(s)\| = 1$, so that β is indeed a unit-speed curve. Furthermore,

$$\kappa(s) = \|\mathbf{T}'(s)\| = \frac{a}{a^2+b^2},$$

and

$$\mathbf{N}(s) = \frac{\mathbf{T}'(s)}{\|\mathbf{T}'(s)\|} = \left(-\cos\left(\frac{s}{\sqrt{a^2+b^2}}\right), -\sin\left(\frac{s}{\sqrt{a^2+b^2}}\right), 0\right).$$

From the formula $\mathbf{B} = \mathbf{T} \times \mathbf{N}$ we get

$$\mathbf{B}(s) = \frac{1}{\sqrt{a^2+b^2}}\left(b\sin\left(\frac{s}{\sqrt{a^2+b^2}}\right), -b\cos\left(\frac{s}{\sqrt{a^2+b^2}}\right), a\right).$$

Finally, to compute the torsion we first note that

$$\mathbf{B}'(s) = \frac{1}{a^2+b^2}\left(b\cos\left(\frac{s}{\sqrt{a^2+b^2}}\right), b\sin\left(\frac{s}{\sqrt{a^2+b^2}}\right), 0\right).$$

When we compare the formulas for $\mathbf{B}'(s)$ and $\mathbf{N}(s)$ and use the Frenet formula $\mathbf{B}' = -\tau \mathbf{N}$, we see that

$$\tau(s) = \frac{b}{a^2+b^2}.$$

Thus both the curvature and torsion of a helix are constant.

8.3 Curvature and Torsion of Arbitrary-Speed Curves in $\mathbb{R}^3$

Although we know that every regular curve has a unit-speed parametrization, in practice it is very difficult to find it. For example, consider the **twisted cubic** defined by

$$\mathbf{twicubic}(t) = (t, t^2, t^3).$$

or

```
twicubic[t_] := {t,t^2,t^3}
```

The arc length function of this curve is

$$s(t) = \int_0^t \sqrt{1 + 4u^2 + 9u^4}\, du.$$

The inverse of $s(t)$, which is needed to find the unit-speed parametrization, is an elliptic function that too complicated to be of much use. Here is the plot of **twicubic** together with a combined plot of the curvature and torsion of **twicubic**; shortly we shall show how to compute these quantities.

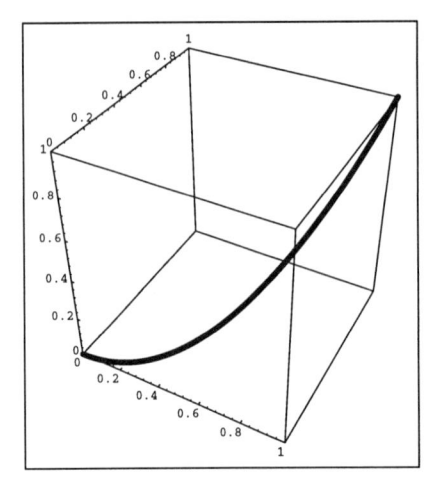

The twisted cubic $t \longmapsto (t, t^2, t^3)$

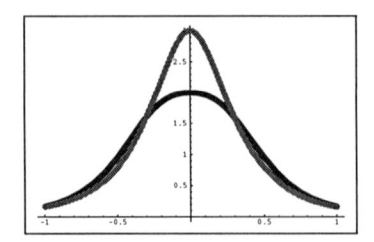

The curvature and torsion of $t \longmapsto (t, t^2, t^3)$

The black line represents the curvature and the gray line represents the torsion of **twicubic**.

For efficient computation of the curvature and torsion of an arbitrary-speed curve, we need formulas that bypass finding a unit-speed parametrization. Although we shall define the curvature and torsion of an arbitrary-speed curve in terms of the curvature and torsion of its unit-speed parametrization, ultimately we shall find formulas for these quantities that avoid finding a unit-speed parametrization explicitly.

Definition. *Let* $\alpha\colon (a,b) \longrightarrow \mathbb{R}^3$ *be a regular curve, and let* $\widetilde{\alpha}\colon (c,d) \longrightarrow \mathbb{R}^3$ *be a unit-speed reparametrization of* α. *Write* $\alpha(t) = \widetilde{\alpha}\bigl(s(t)\bigr)$ *(where $s(t)$ is just the arc*

8.3 Curvature and Torsion of Arbitrary-Speed Curves in $\mathbb{R}^3$

length function). Denote by $\widetilde{\kappa}$ and $\widetilde{\tau}$ the curvature and torsion of $\widetilde{\alpha}$, respectively. Also, let $\{\widetilde{\mathbf{T}}, \widetilde{\mathbf{N}}, \widetilde{\mathbf{B}}\}$ be the Frenet frame field of $\widetilde{\alpha}$. Then we define

$$\kappa(t) = \widetilde{\kappa}(s(t)), \qquad \tau(t) = \widetilde{\tau}(s(t)),$$

$$\mathbf{T}(t) = \widetilde{\mathbf{T}}(s(t)), \quad \mathbf{N}(t) = \widetilde{\mathbf{N}}(s(t)), \quad \mathbf{B}(t) = \widetilde{\mathbf{B}}(s(t)).$$

In other words, the curvature, torsion and Frenet frame field of an arbitrary-speed curve α are reparametrizations of those of a unit-speed parametrization of α.

Next, we generalize the Frenet formulas (8.10) to arbitrary-speed curves.

Theorem 8.6. Let $\alpha\colon (a,b) \longrightarrow \mathbb{R}^3$ be a regular curve with speed $v = \|\alpha'\| = s'$. Then the following generalizations of the Frenet formulas hold:

(8.15)
$$\begin{cases} \mathbf{T}' = & v\,\kappa\,\mathbf{N}, \\ \mathbf{N}' = -v\,\kappa\,\mathbf{T} & +v\,\tau\,\mathbf{B}, \\ \mathbf{B}' = & -v\,\tau\,\mathbf{N}. \end{cases}$$

Proof. By the chain rule we have

(8.16)
$$\begin{cases} \mathbf{T}'(t) = s'(t)\widetilde{\mathbf{T}}'(s(t)) = v(t)\widetilde{\mathbf{T}}'(s(t)), \\ \mathbf{N}'(t) = s'(t)\widetilde{\mathbf{N}}'(s(t)) = v(t)\widetilde{\mathbf{N}}'(s(t)), \\ \mathbf{B}'(t) = s'(t)\widetilde{\mathbf{B}}'(s(t)) = v(t)\widetilde{\mathbf{B}}'(s(t)). \end{cases}$$

Thus from the Frenet formulas for $\widetilde{\alpha}$, it follows that

$$\mathbf{T}'(t) = v(t)\widetilde{\kappa}(s(t))\widetilde{\mathbf{N}}(s(t)) = v(t)\kappa(t)\mathbf{N}(t).$$

The other two formulas of (8.16) are proved similarly. ∎

Lemma 8.7. *The velocity α' and acceleration α'' of a regular curve α are given by*

(8.17) $$\alpha' = v\mathbf{T},$$
(8.18) $$\alpha'' = \frac{dv}{dt}\mathbf{T} + v^2\kappa\,\mathbf{N},$$

where v denotes the speed of α.

Proof. Write $\alpha(t) = \tilde{\alpha}(s(t))$, where $\tilde{\alpha}$ is a unit-speed parametrization of α. By the chain rule we have

$$\alpha'(t) = \tilde{\alpha}'(s(t))s'(t) = v(t)\tilde{\mathbf{T}}(s(t)) = v(t)\mathbf{T}(t),$$

proving (8.17). Next, we take the derivative of (8.17) and use the first equation of (8.15) to get

$$\alpha'' = \frac{dv}{dt}\mathbf{T} + v\mathbf{T}' = \frac{dv}{dt}\mathbf{T} + v^2\kappa\mathbf{N}. \blacksquare$$

Now we can derive useful formulas for the curvature and torsion for an arbitrary-speed curve. These formulas avoid finding a unit-speed reparametrization.

Theorem 8.8. *Let $\alpha \colon (a, b) \longrightarrow \mathbb{R}^3$ be a regular curve with nonzero curvature. Then*

(8.19) $$\mathbf{T} = \frac{\alpha'}{\|\alpha'\|},$$

(8.20) $$\mathbf{N} = \mathbf{B} \times \mathbf{T},$$

(8.21) $$\mathbf{B} = \frac{\alpha' \times \alpha''}{\|\alpha' \times \alpha''\|},$$

(8.22) $$\kappa = \frac{\|\alpha' \times \alpha''\|}{\|\alpha'\|^3},$$

(8.23) $$\tau = \frac{\alpha' \times \alpha'' \cdot \alpha'''}{\|\alpha' \times \alpha''\|^2}.$$

Proof. Clearly, (8.19) is equivalent to (8.17), and (8.20) is an algebraic consequence of the definition of vector cross product. Furthermore, it follows from (8.17) and (8.18) that

(8.24) $$\alpha' \times \alpha'' = (v\,\mathbf{T}) \times \left(\frac{dv}{dt}\mathbf{T} + \kappa v^2\mathbf{N}\right)$$
$$= v\frac{dv}{dt}\mathbf{T} \times \mathbf{T} + \kappa v^3\mathbf{T} \times \mathbf{N} = \kappa v^3\mathbf{B}.$$

Taking norms in (8.24), we get

(8.25) $$\|\alpha' \times \alpha''\| = \|\kappa v^3\mathbf{B}\| = \kappa v^3.$$

8.3 Curvature and Torsion of Arbitrary-Speed Curves in $\mathbb{R}^3$

Then (8.25) implies (8.22). Furthermore, (8.24) and (8.25) imply that

$$\mathbf{B} = \frac{\boldsymbol{\alpha}' \times \boldsymbol{\alpha}''}{v^3 \kappa} = \frac{\boldsymbol{\alpha}' \times \boldsymbol{\alpha}''}{\|\boldsymbol{\alpha}' \times \boldsymbol{\alpha}''\|},$$

and so (8.21) is proved.

To prove (8.23) we need a formula for $\boldsymbol{\alpha}'''$ analogous to (8.17) and (8.18). Actually, all we need is the component of $\boldsymbol{\alpha}'''$ in the $\mathbf{B}$ direction, because we want to take the dot product of $\boldsymbol{\alpha}'''$ with $\boldsymbol{\alpha}' \times \boldsymbol{\alpha}''$. So we compute

$$(8.26) \qquad \boldsymbol{\alpha}''' = \left(\frac{dv}{dt}\mathbf{T} + \kappa v^2 \mathbf{N}\right)'$$
$$= \kappa v^2 \mathbf{N}' + \cdots$$
$$= \kappa \tau v^3 \mathbf{B} + \cdots.$$

(Here the dots represent irrelevant terms.) It follows from (8.24) and (8.26) that

$$(8.27) \qquad \boldsymbol{\alpha}' \times \boldsymbol{\alpha}'' \cdot \boldsymbol{\alpha}''' = \kappa^2 \tau v^6.$$

Now (8.23) follows from (8.27) and (8.25). ∎

Example

Let us compute the curvature and torsion of the curve **twicubic** defined on page 189. Instead of finding a unit-speed reparametrization of **twicubic**, we use Theorem 8.8. The computations are easy enough to do by hand:

$$\textbf{twicubic}'(t) = (1, 2t, 3t^2), \qquad \textbf{twicubic}''(t) = (0, 2, 6t),$$

$$\textbf{twicubic}'(t) \times \textbf{twicubic}''(t) = (6t^2, -6t, 2), \qquad \textbf{twicubic}'''(t) = (0, 0, 6),$$

so that

$$\|\textbf{twicubic}'(t)\| = (1 + 4t^2 + 9t^4)^{1/2},$$

$$\|\textbf{twicubic}'(t) \times \textbf{twicubic}''(t)\| = (4 + 36t^2 + 36t^4)^{1/2}.$$

Therefore, the curvature and torsion of **twicubic** are given by

$$\kappa(t) = \frac{(4 + 36t^2 + 36t^4)^{1/2}}{(1 + 4t^2 + 9t^4)^{3/2}} \quad \text{and} \quad \tau(t) = \frac{3}{1 + 9t^2 + 9t^4}.$$

8.4 Computing Curvature and Torsion with Mathematica

Computing curvature of mildly complicated curves by hand can be quite tedious. But before showing how to compute κ, τ, **T**, **N** and **B** using *Mathematica*, we must discuss how to define the vector cross product in *Mathematica*.

One definition of the cross product in *Mathematica* is essentially the one given in Section 8.1. The *Mathematica* command `IdentityMatrix[3]` yields

    {{1, 0, 0}, {0, 1, 0}, {0, 0, 1}}

This is exactly what we want as the first row in the definition of cross product. So let us define `cross1` by

```
cross1[p_,q_] := Det[{IdentityMatrix[3],p,q}]
```

where `p` and `q` are lists of length 3. This is the definition of vector cross product contained in the package `LinearAlgebra/Cross.m`, which is contained in Version 1.2 of *Mathematica*. Version 2.0 has other definitions of vector cross product in its packages `LinearAlgebra/Cross.m` and `Calculus/VectorAnalysis.m`. It turns out, however, that there is a faster alternative to both of these definitions, which will be our preferred *Mathematica* definition of the vector cross product. (See [De] for the following definition and a discussion of the relative merits of the various possibilities for vector cross products in *Mathematica*.) We put

```
cross[u_List,v_List] :=
    RotateLeft[u RotateLeft[v] -
    RotateLeft[u] v] /;
    Length[u] == Length[v] == 3
```

The details of this command are as follows. The command `RotateLeft` cyclically permutes a list. For example,

   `RotateLeft[{p1,p2,p3}]`

        {p2, p3, p1}

Note also that *Mathematica* permits multiplication of lists. (This multiplication is *not* the dot product.) For example,

8.4 Computing Curvature and Torsion with Mathematica

```
{p1,p2,p3}{q1,q2,q3}
```

yields

```
{p1 q1, p2 q2, p3 q3}
```

Finally, **Length[p]** is the number of elements in the list **p**. Using these definitions, it is easy to check that **cross** does indeed give the correct formula for the vector cross product.

Here are the formulas to compute the curvature and torsion of any curve **alpha**:

```
kappa[alpha_][t_]:=
    Simplify[Factor[cross[D[alpha[tt],tt],
    D[alpha[tt],{tt,2}]]].
    cross[D[alpha[tt],tt],D[alpha[tt],{tt,2}]]]]^(1/2)/
    Simplify[Factor[D[alpha[tt],tt].
    D[alpha[tt],tt]]]^(3/2)  /. tt->t
```

```
tau[alpha_][t_]:= Simplify[Det[
    {D[alpha[tt],tt],D[alpha[tt],{tt,2}],
    D[alpha[tt],{tt,3}]}]]/
    Simplify[Factor[cross[D[alpha[tt],tt],
    D[alpha[tt],{tt,2}]].cross[D[alpha[tt],tt],
    D[alpha[tt],{tt,2}]]]]  /. tt->t
```

and here are the formulas for **T**, **N** and **B**:

```
tangent[alpha_][t_]:= D[alpha[tt],tt]/
    Simplify[Factor[D[alpha[tt],tt].
      D[alpha[tt],tt]]]^(1/2)   /. tt->t
```

```
binormal[alpha_][t_]:= Simplify[cross[
    D[alpha[tt],tt],D[alpha[tt],{tt,2}]]]/
    Simplify[Factor[cross[D[alpha[tt],tt],
    D[alpha[tt],{tt,2}]].
    cross[D[alpha[tt],tt],D[alpha[tt],
    {tt,2}]]]]^(1/2)    /. tt->t
```

```
normal[alpha_][t_] := cross[binormal[alpha][t],
                            tangent[alpha][t]]
```

(We have changed the names of **T**, **N**, **B** to **tangent**, **normal**, **binormal** for use in *Mathematica* because frequently capital letters have been defined internally in *Mathematica*. For example, **N** means "compute the numerical value".)

The six commands **cross**, **kappa**, **tau**, **tangent**, **normal**, **binormal** are contained in the package **CSPROGS.m**. To use these commands to compute curvature and torsion, we first need to define the curve in *Mathematica*. For example, the curve γ given in usual mathematical notation as

$$\gamma(t) = (3t - t^3, 3t^2, 3t + t^3)$$

can be defined in *Mathematica* as

```
gamma[t_] := {3t - t^3, 3t^2, 3t + t^3}
```

As a trial, let us type **gamma'[t]**; *Mathematica* answers

$$\{3 - 3\,t^2,\ 6\,t,\ 3 + 3\,t^2\}$$

To compute the curvature we enter **kappa[gamma][t]**; *Mathematica* responds with

$$\frac{1}{3\,(1 + t^2)^2}$$

We need to use **Simplify**. in order to compute the torsion. *Mathematica's* response to
tau[gamma][t]//Simplify is

$$\frac{1}{3\,(1 + t^2)^2}$$

For **tangent[gamma]** we use the **tangent[gamma][t]//Simplify**, obtaining

$$\left\{\frac{1 - t^2}{\sqrt{2}\ \sqrt{(1 + t^2)^2}},\ \frac{\sqrt{2}\ t}{\sqrt{(1 + t^2)^2}},\ \frac{1 + t^2}{\sqrt{2}\ \sqrt{(1 + t^2)^2}}\right\}$$

8.4 Computing Curvature and Torsion with Mathematica

A glance at the square roots in this expression tells us that we can safely use **PowerExpand** to simplify the expression further:

tangent[gamma][t]//Simplify//PowerExpand

$$\left\{\frac{1-t^2}{\text{Sqrt}[2]\,(1+t^2)},\ \frac{\text{Sqrt}[2]\,t}{1+t^2},\ \frac{1}{\text{Sqrt}[2]}\right\}$$

Similarly, **normal[gamma][t]//Simplify** yields

$$\left\{\frac{-2t}{1+t^2},\ \frac{1-t^2}{1+t^2},\ 0\right\}$$

and **binormal[gamma][t]//Simplify//PowerExpand** results in

$$\left\{\frac{-1+t^2}{\text{Sqrt}[2]\,(1+t^2)},\ -\left(\frac{\text{Sqrt}[2]\,t}{1+t^2}\right),\ \frac{1}{\text{Sqrt}[2]}\right\}$$

As we have seen, it is often essential to use *Mathematica's* commands for simplifying algebraic expressions. These include **Simplify**, **Cancel**, **Factor**. **PowerExpand** can also be used with care.[8] Frequently, however, the situation is much worse; sometimes it is not clear how to simplify complicated expressions. In many cases a graph best explains curvature and torsion.

The *Mathematica* command[9] to plot both curves and surfaces in three dimensions is **ParametricPlot3D**. To plot the curve **gamma** we use

```
ParametricPlot3D[
Evaluate[gamma[t]],{t,-1,1},
PlotPoints->200,
PlotRange->{{-2,2},{0,3},{-4,4}}];
```

---

[8]Trigonometric simplifications in versions 2.0 and greater are either automatic or are accomplished with the option **Trig->True**. But in version 1.2 the package **Algebra/Trigonometry** must be read in. The most useful command in that package is **TrigCanonical** which usually convinces *Mathematica* that indeed $\sin^2 + \cos^2 = 1$.

[9]In version 1.2 in order to plot space curves the package **Graphics/ParametricPlot3D.m** must first be read in. The command **SpaceCurve** is then used to plot space curves. Furthermore, the option **PlotPoints** is unavailable for **SpaceCurve**, and **Release** must sometimes be used. Thus the command for plotting **gamma** in version 1.2 is: **SpaceCurve[Release[gamma[t]],{t,-1,1,0.01}]**.

In this particular case **kappa[gamma][t]** coincides with **tau[gamma][t]**; we use **Plot** to draw its graph, specifically,

```
Plot[Evaluate[kappa[gamma][t]],{t,-2,2},
Ticks->{Automatic,Range[0,0.35,0.1]}];
```

Here are the results:

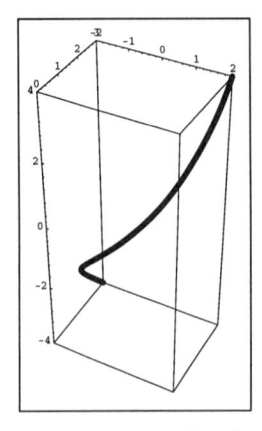

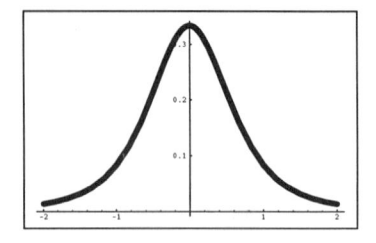

The curve $t \longmapsto (3t - t^3, 3t^2, 3t + t^3)$

The curvature and torsion of
$t \longmapsto (3t - t^3, 3t^2, 3t + t^3)$

An interesting collection of space curves is contained in the package **CURVES.m**

8.5 The Helix and its Generalizations

The (circular) **helix** is a curve in $\mathbb{R}^3$ that resembles a spring. Its natural parametrization is
(8.28) $$\mathbf{helix}[a, b](t) = (a \cos t, a \sin t, b t),$$
where a is the radius and b is the slope of the helix. This parametrization is simpler than the unit-speed parametrization of the helix that we discussed on page 188. Notice that the projection of $\mathbb{R}^3$ onto the xy-plane maps the helix onto a circle. In *Mathematica* the helix is defined by

```
helix[a_,b_][t_]:= {a Cos[t],a Sin[t],b t}
```

The command to draw the helix in version 2.0 of *Mathematica* (but not version 1.2) is, for example,

8.5 The Helix and its Generalizations

```
ParametricPlot3D[Evaluate[
    helix[1,0.2][t]],{t,0,4Pi},PlotPoints->200,
    PlotRange->{{-1,1},{-1,1},{0,0.8Pi}}];
```

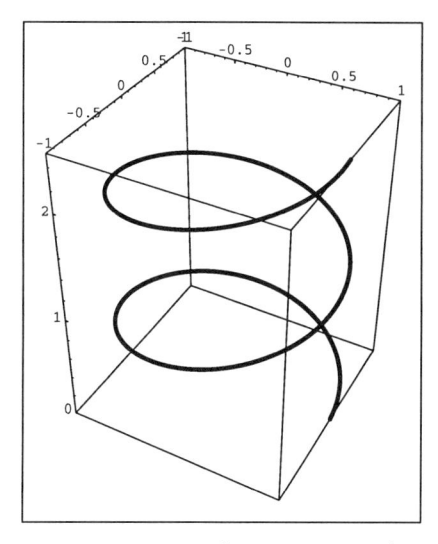

The helix $t \longmapsto (\cos t, \sin t, 0.2\,t)$

To compute the curvature and torsion of the helix, we use the commands `kappa[helix[a,b]][t]//PowerExpand` and `tau[helix[a,b]][t]`, obtaining

$$\frac{a}{a^2+b^2} \quad \text{and} \quad \frac{b}{a^2+b^2}$$

Since the curvature and torsion are both constant, there is no need to graph them.

Helical Curves over Plane Curves

Just as the helix sits over the circle, there are helical analogs of many other plane curves. *Mathematica* can be used to define the helix with incline **c** over a curve **alpha** via the command

```
helical[alpha_,c_][t_]:=
    {alpha[t][[1]],alpha[t][[2]],c t}
```

For example, to get the helical curve over the logarithmic spiral defined on page 40 we type into *Mathematica* the command

```
helical[logspiral[a,b],c][t]
```

and *Mathematica* answers

$$\{a\, E^{b\,t}\, \cos[t],\ a\, E^{b\,t}\, \sin[t],\ c\, t\}$$

Here is the *Mathematica*-generated picture of the helical curve over a logarithmic spiral, together with a simultaneous graph of the curvature and torsion.

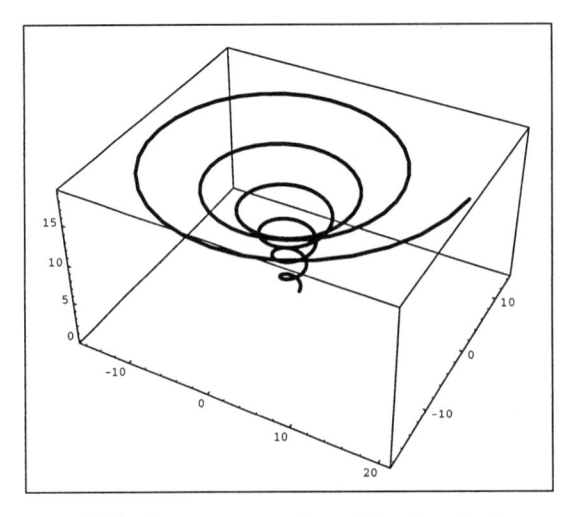

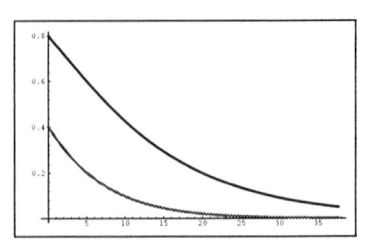

Helical curve over a logarithmic spiral

The curvature and torsion of a helical curve over a logarithmic spiral

The commands to plot the helical logarithmic spiral and graph its curvature and torsion are:

```
ParametricPlot3D[Evaluate[
helical[logspiral[1,0.08],0.5][t]],
{t,0,12Pi},PlotPoints->200];
```

and

```
Plot[Evaluate[
{kappa[helical[logspiral[1,0.08],0.5]][t],
 tau[helical[logspiral[1,0.08],0.5]][t]}],{t,0,12Pi},
PlotStyle->{{GrayLevel[0]},{GrayLevel[0.4]}}];
```

8.6 Viviani's Curve

Viviani's curve is the intersection of the cylinder $(x-a)^2 + y^2 = a^2$ and the sphere $x^2 + y^2 + z^2 = 4a^2$. It is parametrized by

$$\mathbf{viviani}[a](t) = a\bigl(1 + \cos t, \sin t, 2\sin(t/2)\bigr), \qquad (-2\pi \leq t \leq 2\pi),$$

or

```
viviani[a_][t_] := a{1 + Cos[t],Sin[t],2Sin[t/2]}
```

Viviani[10] studied the curve in 1692. See [Stru2, pages 10–11] and [Gomes, volume 2, pages 311–320]. Here is how to plot Viviani's curve in a such a way as to show how it lies on a sphere.

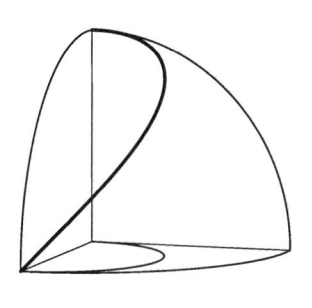

```
ParametricPlot3D[
{viviani[1][t],
{1+Cos[t],Sin[t],0},
{2Cos[t/2],0,2Sin[t/2]},
{2Cos[t/2],2Sin[t/2],0},
{0,2Cos[t/2],2Sin[t/2]}}
//Evaluate,{t,0,Pi},
Boxed->False,Ticks->None,
AxesEdge->
{{-1,-1},{-1,-1},{-1,-1}},
PlotRange->
{{0,2},{0,2},{0,2}},
ViewPoint->{10,3,1}];
```

Viviani's curve

The expressions for the curvature and torsion of Viviani's curve are easily computed with *Mathematica*:

```
PowerExpand[kappa[viviani[a]][t],a]

PowerExpand[tau[viviani[a]][t],a]
```

---

[10] Vincenzo Viviani (1622-1703). Student and biographer of Galileo. In 1660, together with Borelli, Viviani measured the velocity of sound by timing the difference between the flash and the sound of a cannon. In 1692 Viviani proposed the following problem which aroused much interest. How is it possible that a hemisphere has 4 equal windows of such a size that the remaining surface can be exactly squared? The answer involved the Viviani's curve.

It is best to make these two expressions into *Mathematica* functions so that the graphs can be drawn more quickly:

```
kappa[viviani[a_]][t_]:=
Sqrt[13 + 3Cos[t]]/(a(3 + Cos[t])^(3/2))

tau[viviani[a_]][t_]:= (6Cos[t/2])/((13 + 3Cos[t])a)
```

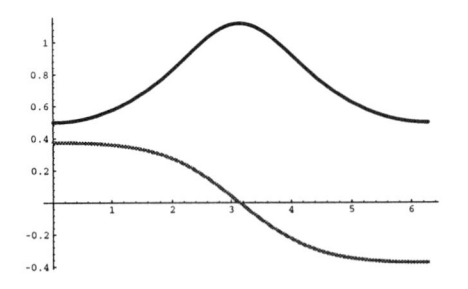

```
Plot[
{kappa[viviani[1]][t],
tau[viviani[1]][t]
}//Evaluate,
{t,0,2Pi},
PlotStyle->
{{GrayLevel[0]},
{GrayLevel[0.4]}}];
```

The curvature and torsion of Viviani's curve

8.7 Exercises

1. Establish the following identities:
    (a) $(\mathbf{a} \times \mathbf{b}) \times (\mathbf{c} \times \mathbf{d}) = (\mathbf{acd})\mathbf{b} - (\mathbf{bcd})\mathbf{a}$.
    (b) $\mathbf{d} \times (\mathbf{a} \times \mathbf{b}) \cdot (\mathbf{a} \times \mathbf{c}) = (\mathbf{abc})(\mathbf{a} \cdot \mathbf{d})$.
    (c) $\mathbf{a} \times (\mathbf{b} \times \mathbf{c}) + \mathbf{c} \times (\mathbf{a} \times \mathbf{b}) + \mathbf{b} \times (\mathbf{c} \times \mathbf{a}) = \mathbf{0}$.
    (d) $(\mathbf{a} \times \mathbf{b}) \cdot (\mathbf{b} \times \mathbf{c}) \times (\mathbf{c} \times \mathbf{a}) = (\mathbf{abc})^2$.

2. Plot the following space curves and graph their curvature and torsion:
    (a) $s \longmapsto \left(\dfrac{(1+s)^{3/2}}{3}, \dfrac{(1-s)^{3/2}}{3}, \dfrac{s}{\sqrt{2}} \right)$.
    (b) $t \longmapsto$
    $\left(\sqrt{1 - 0.09(\cos(10t))^2} \cos t, \sqrt{1 - 0.09(\cos(10t))^2} \sin t, 0.3\cos(10t) \right)$.
    (c) $t \longmapsto \left(0.5(1 + \cos(10t))\cos t, 0.5(1 + \cos(10t))\sin t, 0.5(1 + \cos(10t)) \right)$.
    (d) $t \longmapsto (a\cosh t, a\sinh t, bt)$.
    (e) $t \longmapsto \left(e^t t, e^{-t}, \sqrt{2} t \right)$.

8.7 Exercises

(f) $t \longmapsto a\bigl(t - \sin t, 1 - \cos t, 4\cos(t/2)\bigr)$.

3. Graph helical curves over a cycloid, a cardioid and a figure eight curve, and plot the curvature and torsion of each.

4. Let $\boldsymbol{\alpha}$ be an arbitrary-speed curve in $\mathbb{R}^n$. Define

$$\kappa n[\alpha] = \frac{\bigl\|\,\|\alpha'\|\alpha'' - \|\alpha'\|'\alpha'\,\bigr\|}{\|\alpha'\|^3}$$

or

```
kappan[alpha_][t_]:=
    Simplify[Sqrt[Factor[
    (Sqrt[Factor[D[alpha[tt],tt].
                D[alpha[tt],tt]]]D[alpha[tt],{tt,2}]-
    D[Sqrt[Factor[D[alpha[tt],tt].
                D[alpha[tt],tt]]],tt]D[alpha[tt],tt]).
    (Sqrt[Factor[D[alpha[tt],tt].
                D[alpha[tt],tt]]]D[alpha[tt],{tt,2}]-
    D[Sqrt[Factor[D[alpha[tt],tt].
                D[alpha[tt],tt]]],tt]*
    D[alpha[tt],tt])]]]/
    Simplify[Factor[D[alpha[tt],tt].
                D[alpha[tt],tt]]]^(3/2) /. tt->t
```

Show that $\kappa n[\alpha]$ is the curvature of $\boldsymbol{\alpha}$ for $n = 3$.

5. Let $\beta\colon (a,b) \longrightarrow \mathbb{R}^3$ be a straight line parametrized as $\beta(s) = s\,\mathbf{e}_1 + \mathbf{q}$, where $\mathbf{e}_1, \mathbf{q} \in \mathbb{R}^3$ with $\mathbf{e}_1$ a unit vector. Let $\mathbf{e}_2, \mathbf{e}_3 \in \mathbb{R}^3$ be such that $\{\mathbf{e}_1, \mathbf{e}_2, \mathbf{e}_3\}$ is an orthonormal basis of $\mathbb{R}^3$. Define vector fields $\{\mathbf{T}, \mathbf{N}, \mathbf{B}\}$ by

$$\mathbf{T}(s) = \mathbf{e}_1, \qquad \mathbf{N}(s) = \mathbf{e}_2, \qquad \mathbf{B}(s) = \mathbf{e}_3$$

for $a < s < b$. Show that the Frenet formulas (8.10) hold provided we take $\kappa(s) = \tau(s) = 0$ for $a < s < b$.

6. A 3-dimensional version of the astroid is defined by

```
ast3d[n_,a_,b_][t_]:= {a Cos[t]^n,b Sin[t]^n,Cos[2t]}
```

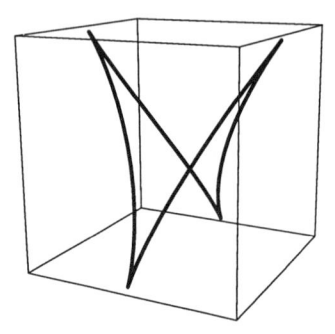

```
ParametricPlot3D[
Append[ast3d[3,1,1][t],
Thickness[0.01]]
//Evaluate,
{t,0,2Pi},
Axes->None,
ViewPoint->
{1.510,-2.926,0.779}];
```

Three dimensional astroid

Compute the curvature and torsion of **ast3d[n,1,1]** and plot **ast3d[n,1,1]** for $n = 1, 3$.

7. Show that each component of the intersection of a cylinder of radius a and a cylinder of radius b can be parametrized in *Mathematica* by

```
bicylinder[a_,b_,pm_][t_]:= {a Cos[t],a Sin[t],
                              pm Sqrt[b^2 - a^2 Sin[t]^2]}
```

where **pm** is ± 1.

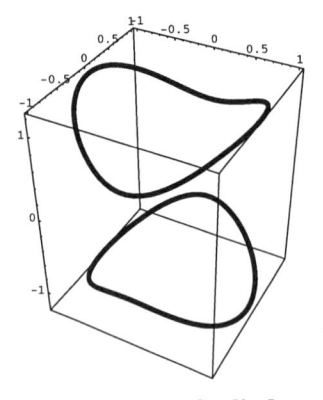

```
ParametricPlot3D[
{Append[
bicylinder[1,1.2,1][t],
Thickness[0.015]],
Append[
bicylinder[1,1.2,-1][t],
Thickness[0.015]]}
//Evaluate,
{t,0,2Pi}];
```

The intersection of cylinders of radii 1 and 1.2

8. An **elliptical helix** is parametrized in *Mathematica* by

```
helix[a_,b_,c_][t_]:= {a Cos[t],b Sin[t],c t}
```

Plot an elliptical helix, its curvature and torsion. Show that an elliptical helix has constant curvature if and only if $a^2 = b^2$. [Hint: Use *Mathematica* to compute the derivative of the curvature of an elliptical helix and evaluate it at $\pi/4$.]

9. The **Darboux**[11] **vector field** along a unit-speed curve $\beta\colon (a,b) \longrightarrow \mathbb{R}^3$ is defined by
$$\mathbf{D} = \tau\,\mathbf{T} + \kappa\,\mathbf{B}.$$

(a) Show that $\mathbf{D}$ satisfies the following equations:
$$\mathbf{T}' = \mathbf{D} \times \mathbf{T},$$
$$\mathbf{N}' = \mathbf{D} \times \mathbf{N},$$
$$\mathbf{B}' = \mathbf{D} \times \mathbf{B}.$$

(b) For an *arbitrary-speed curve* **gamma** show that the Darboux vector field can be defined in *Mathematica* by

```
darbouxvector[gamma_][t_]:=
    Sqrt[D[gamma[tt],tt].D[gamma[tt],tt]](
        tau[gamma][tt]tangent[gamma][tt] +
        kappa[gamma][tt]binormal[gamma][tt]) /. tt->t
```

(c) Plot the curves traced out by the Darboux vectors of an elliptical helix, Viviani's curve and a bicylinder.

---

[11] Jean Gaston Darboux (1842-1917). French mathematician, who is best known for his contributions to differential geometry. His four-volume work **Leçons sur la théorie general de surfaces** remains the bible of surface theory. In 1875 he provided new insight into the Riemann integral, first defining upper and lower sums and then defining a function to be integrable if the difference between the upper and lower sums tends to zero as the mesh size gets smaller.

10. A curve constructed from the Bessel functions J_a, J_b, J_c is defined by

```
besselcurve[a_,b_,c_][t_]:=
    {BesselJ[a,t],BesselJ[b,t],BesselJ[c,t]}
```

Plot **besselcurve[0,1,2]**.

11. A generalization **twicubic** is the curve **twistedn**$[n, a]: \mathbb{R} \longrightarrow \mathbb{R}^n$ given by

$$\text{twistedn}[n, a](t) = (t, t^2, \ldots, t^n),$$

or

```
twistedn[n_,a_][t_]:=Table[a t^k,{k,1,n}]
```

Compute the curvature of **twistedn**$[n, a]$ for $1 \leq n \leq 6$.

8.8 Animation of the Frenet Frame of Viviani's Curve

The following command creates an animation of a labeled Frenet frame moving around Viviani's curve.

```
Do[Show[ParametricPlot3D[
Append[viviani[1.2][t],
{RGBColor[1,0,0],AbsoluteThickness[2]}]//Evaluate,
{t,-2Pi,2Pi},
DisplayFunction->Identity],
Graphics3D[{AbsoluteThickness[3.5],
frenetframe[viviani[1.2],1.4,
"Helvetica-Bold",12][a]}],
ViewPoint->{4,-4,1},Axes->None,
PlotRange->{{-1,3.5},{-2.2,2.2},{-3.5,3.5}},
DisplayFunction->$DisplayFunction],
{a,-2Pi,2Pi,0.02Pi}]
```

9

TUBES AND KNOTS

Knots form one of the most interesting classes of space curves. In this chapter we study *torus knots*, that is, knots which lie on the surface of a torus.

Visualization of torus knots is considerably enhanced by making them thicker. This leads us to tubes. Directions for finding and plotting tubes about space curves are given in Section 9.1.

Torus knots are discussed in Section 9.2; we show how to plot them and, using *Mathematica*, how to plot small tubes surrounding them. It turns out that a tube about a space curve can twist on itself; we explain how the curvature and torsion of the torus knot influence this phenomenon.

9.1 Tubes about Curves

By definition the **tube** of radius r a set $\mathbf{B}$ is the set of points at a distance r from $\mathbf{B}$. In particular, let $\boldsymbol{\gamma}\colon (a,b) \longrightarrow \mathbb{R}^3$ be a regular curve whose curvature does not vanish. Since the normal $\mathbf{N}$ and binormal $\mathbf{B}$ are perpendicular to $\boldsymbol{\gamma}$, the circle $\theta \longmapsto \cos\theta\,\mathbf{N}(t) + \sin\theta\,\mathbf{B}(t)$ is perpendicular to $\boldsymbol{\gamma}$ at $\boldsymbol{\gamma}(t)$. As this circle moves along $\boldsymbol{\gamma}$ it traces out a surface about $\boldsymbol{\gamma}$, which will be the tube about $\boldsymbol{\gamma}$, provided r is not too large. We can parametrize this surface by

(9.1) $\quad \mathbf{tubecurve}[\boldsymbol{\gamma}][r](t,\theta) = \boldsymbol{\gamma}(t) + r\bigl(-\cos\theta\,\mathbf{N}(t) + \sin\theta\,\mathbf{B}(t)\bigr),$

(where $a \leq t \leq b$ and $0 \leq \theta \leq 2\pi$), or in *Mathematica*

```
tubecurve[gamma_][r_][t_,theta_]:= gamma[t] +
    r(-Cos[theta] normal[gamma][t] +
    Sin[theta] binormal[gamma][t])
```

The command **tubecurve** is contained in the package **CSPROGS.m**.

For example, let us find and draw a tube about the helix $t \longmapsto \mathbf{helix}[2, \mathbf{0.5}](t)$. To use *Mathematica* we first need either to type in the definition of the helix given on page 198 or to read in the file **CURVES.m** that contains the definition. To see the parametrization of a tube about a general helix, we type into *Mathematica* the command

tubecurve[helix[a,b]][r][t,theta]//PowerExpand//Simplify

This mildly complicated command requires *Mathematica* to work a little, but eventually it answers

$$\{a \cos[t] + r \left(\cos[t] \cos[\theta] + \frac{b \sin[t] \sin[\theta]}{\sqrt{a^2 + b^2}}\right),$$

$$a \sin[t] + r \left(\cos[\theta] \sin[t] - \frac{b \cos[t] \sin[\theta]}{\sqrt{a^2 + b^2}}\right),$$

$$b t + \frac{a r \sin[\theta]}{\sqrt{a^2 + b^2}}\}$$

The command to plot the tube is

```
ParametricPlot3D[Evaluate[
tubecurve[helix[2,0.5]][0.5][t,theta]],{t,0,4Pi},
{theta,0,2Pi},PlotPoints->{60,20},Axes->None];
```

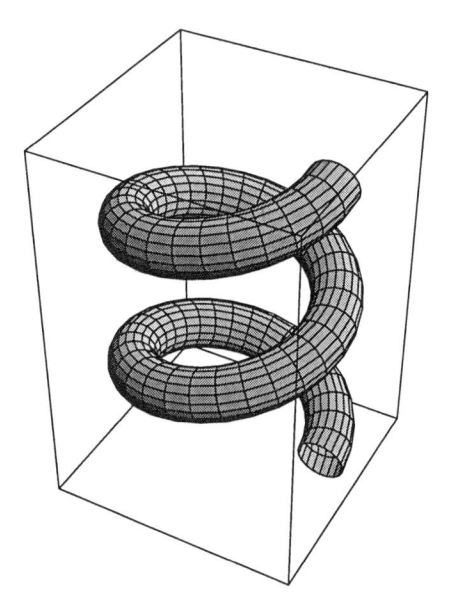

Tube of radius 0.5 about the helix
$$t \longmapsto \text{helix}[2, 0.5](t)$$

Tubes about curves are our first examples of surfaces in $\mathbb{R}^3$. The detailed study of surfaces begins with Chapter 12. A tube about a curve γ in $\mathbb{R}^3$ has the following interesting property: the volume depends only on the length of γ and radius of the tube. In particular, the volume of the tube does not depend on the curvature or torsion of γ. Thus, for example, tubes of the same radius about a circle and a helix of the same length will have the same volume. For the proofs of these facts and the study of tubes in higher dimensions, see [Gray1].

Canal surfaces are generalizations of tubes that we shall study in Chapter 35.

9.2 Torus Knots

Curves that wind around a torus are frequently knotted. Let us define a **torus knot** by

torusknot$[a, b, c][p, q](t) = \bigl((a + b\cos qt)\cos pt, (a + b\cos qt)\sin pt, c\sin qt\bigr),$

or in *Mathematica*

```
torusknot[a_,b_,c_][p_,q_][t_]:=
    {(a + b Cos[q t])Cos[p t],
     (a + b Cos[q t])Sin[p t],
     c Sin[q t]}
```

The surface formed by rotating an ellipse in a vertical plane about the vertical axis in $\mathbb{R}^3$ is called an **elliptical torus**; it will be discussed in more detail in Section 13.4.

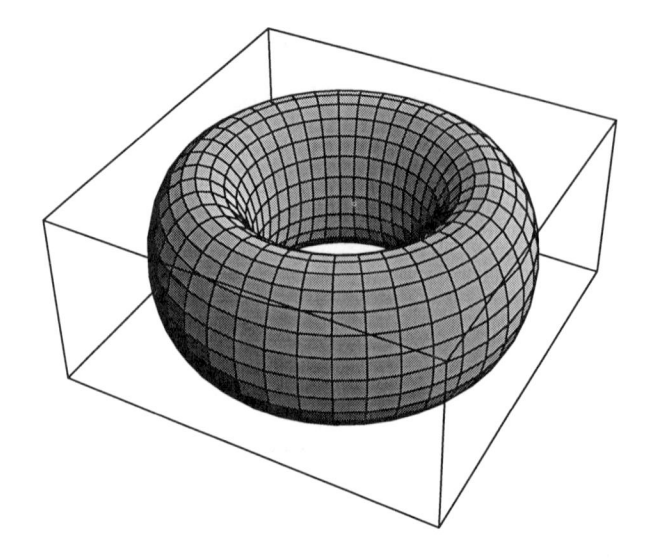

Elliptical torus

We shall see (equation (13.4), page 304) that $t \longmapsto$ **torusknot**$[a, b, c][p, q](t)$ lies on an elliptical torus. For this reason, we have called the curve a **torus knot**; it may or may not be knotted, depending on p and q. The torus knot $t \longmapsto$ **torusknot**$[8, 3, 5][2, 3](t)$ is usually called a **trefoil knot**.

In this section we only scratch the surface of knot theory. Of the many books about this interesting subject we mention [BuZi], [Kauf1], [Kauf2] and [Rolf].

We can plot $t \longmapsto$ **torusknot**$[8, 3, 5][2, 5](t)$ with the command

9.2 Torus Knots

```
ParametricPlot3D[
torusknot[8,3,5][2,5][t]//Evaluate,
{t,0,3Pi},PlotPoints->200,
ViewPoint->{20,30,70}];
```

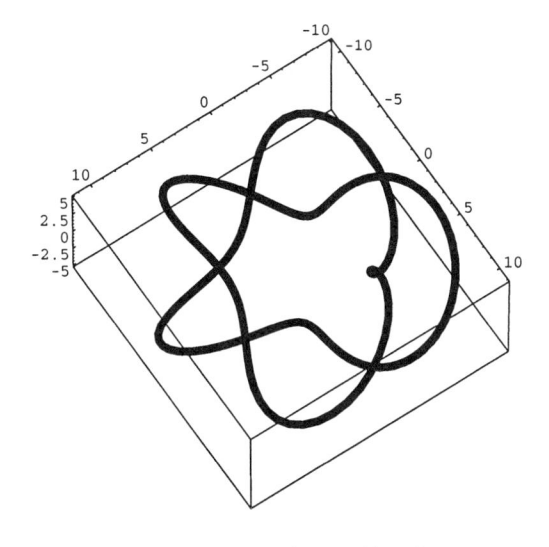

$t \longmapsto \mathbf{torusknot[8,3,5][2,5]}(t)$

This *Mathematica* picture is not as informative as we would like, because at an apparent crossing it is not clear which part of the curve is in front and which part of the curve is in back. Nevertheless, we see that the curve winds around the torus 5 times.

To see what really is going on let us draw a tube about $t \longmapsto \mathbf{torusknot[8,3,5][2,5]}(t)$ using the tube construction of Section 9.1. The plotting command we need is

```
ParametricPlot3D[Evaluate[
tubecurve[torusknot[8,3,5][2,5]][1.3][t,theta]],
{t,0,2Pi},{theta,0,2Pi},
PlotPoints->{160,10},Axes->None,
ViewPoint->{20,30,70}]//Timing
```

To see how long it takes to do the plot we have replaced the usual semicolon at the end of the command by **//Timing**. After the plot is finished *Mathematica* gives an answer such as

```
{75.2833 Second, -Graphics3D-}
```

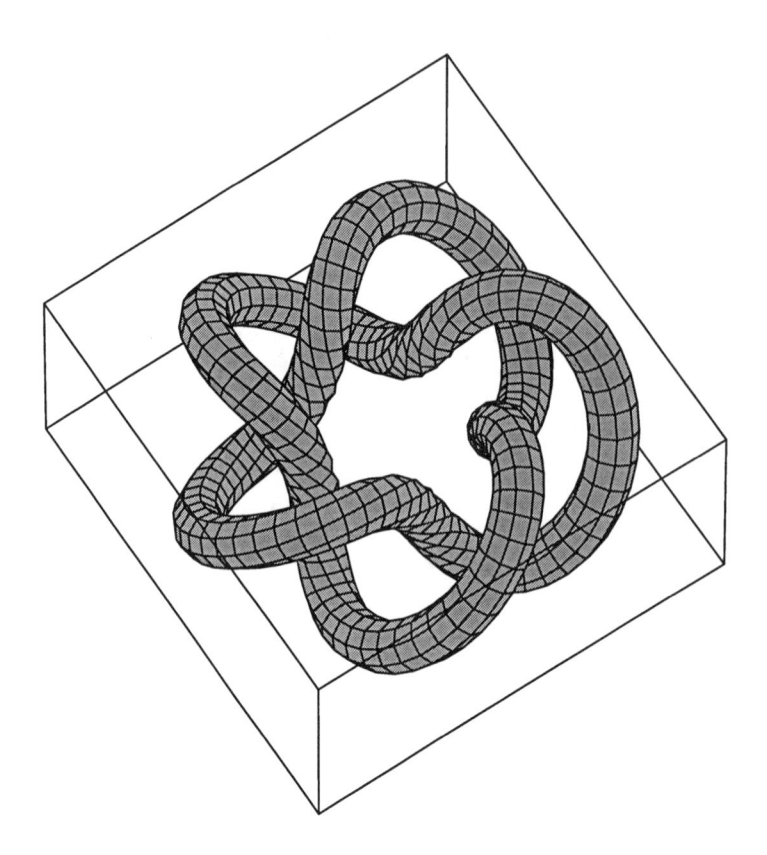

Tube of radius 1.3 about the torus knot
$$t \longmapsto \mathbf{torusknot}[8, 3, 5][2, 5](t)$$

The crossings are now clear; furthermore, we can describe the winding a little more precisely by saying that $t \longmapsto$ **torusknot**$[8, 3, 5][2, 5](t)$ spirals around the torus 5 times in the meridianal sense and 2 times in the longitudinal sense. More generally, $t \longmapsto$ **torusknot**$[a, b, c][p, q](t)$ will spiral around an elliptical torus p times in the longitudinal sense and q times in the meridianal sense.

However, a new issue has been raised. The tube itself seems to twist violently in several places. A careful look at the plot of the tube shows that this increased

9.2 Torus Knots

twisting occurs in five different places, in spite of the fact that other parts of the tube partially obscure the view. Moreover, nearby to where the violent twisting occurs the tube appears to be especially cylinder-like. This evidence leads us to suspect that there are five points on the original curve $t \longmapsto$ **torusknot**$[8, 3, 5][2, 5](t)$ where simultaneously the curvature κ is small and the absolute value of the torsion τ is large.

We can easily graph the curvature and torsion of $t \longmapsto$ **torusknot**$[8, 3, 5][2, 5](t)$ to test this empirical evidence. The plotting commands to do this are by now standard:

```
Plot[kappa[torusknot[8,3,5][2,5]][t]
//Evaluate,{t,0,2Pi}];

Plot[tau[torusknot[8,3,5][2,5]][t]
//Evaluate,{t,0,2Pi}];
```

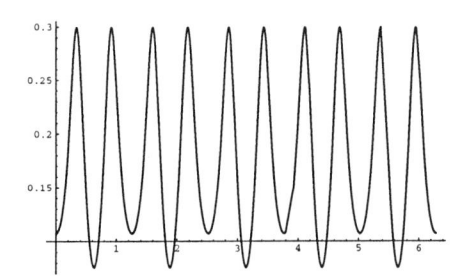

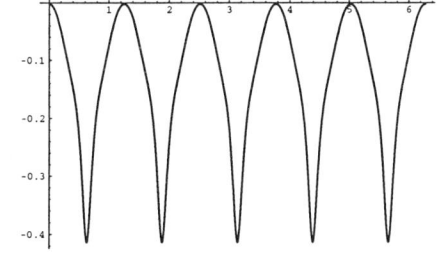

Curvature of the torus knot
$t \longmapsto$ torusknot$[8, 3, 5][2, 5](t)$

Torsion of the torus knot
$t \longmapsto$ torusknot$[8, 3, 5][2, 5](t)$

The graphs conclusively confirm that indeed there are five values of t at which the curvature of $t \longmapsto$ **torusknot**$[8, 3, 5][2, 5](t)$ is small when the absolute value of the torsion is large. Furthermore, the graph seems to indicate these points are $\pi/5$, $3\pi/5$, π, $7\pi/5$ and $9\pi/5$. In fact, this is the case; to find the numerical values of κ and τ at $\pi/5$ we enter the command

```
{kappa[torusknot[8,3,5][2,5]][Pi/5]//N,
tau[torusknot[8,3,5][2,5]][Pi/5]//N}
```

and *Mathematica* answers

```
{0.0758621, -0.413793}
```

In contrast, at $t = 2\pi/5$ the curvature is a little larger and the absolute value of the torsion is quite small:

```
kappa[torusknot[8,3,5][2,5]][2Pi/5]//N,
tau[torusknot[8,3,5][2,5]][2Pi/5]//N
```

{0.107304, -0.00227323}

The Frenet formulas (8.15), page 191, explain why small curvature and large absolute torsion produce so much twisting in the tube. The construction of the tube involves **N** and **B**, but not **T**. When the derivatives of **N** and **B** are large, the twisting in the tube will also be large. But a glance at the Frenet formulas shows that small curvature and large absolute torsion create exactly this effect.

Finally, we note that the curve $t \longmapsto$ **torusknot**[8, 3, 5][1, 10](t) is unknotted.

Tube of radius 1 about the torus knot
$t \longmapsto$ torusknot[8, 3, 5][1, 10](t)

This picture was drawn with the command

```
ParametricPlot3D[tubecurve[
torusknot[8,3,5][1,10]][1.0][t,theta]
//Evaluate, {t,0,2Pi},{theta,0,2Pi},
 PlotPoints->{180,12}, Axes->None,
 ViewPoint->{20,30,70}]//Timing
```

{321.633 Second, -Graphics3D-}

9.3 Exercises

1. Plot the torus knot with $p = 3$ and $q = 2$ and graph its curvature and torsion. Plot a tubular surface surrounding it.

2. A **figure eight knot** can be parametrized by

```
eightknot[t_]:=
    {10(Cos[t] + Cos[3t]) + Cos[2t] + Cos[4t],
     6Sin[t] + 10Sin[3t],
     4Sin[3t]Sin[5t/2] + 4Sin[4t] - 2Sin[6t]}
```

Plot the tube about the figure eight knot.

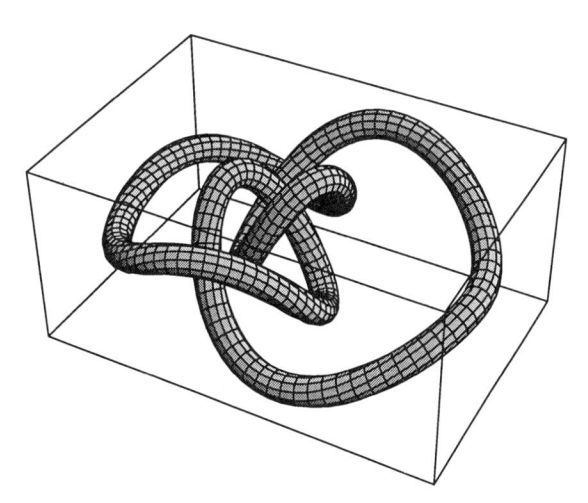

Tube of radius 1.5 about an eight knot

3. Find the formula for a tube about Viviani's curve (defined in Section 8.6) and plot it.

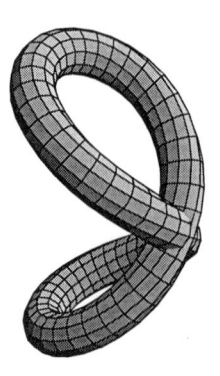

Tube around Viviani's curve

4. The definition of a spherical spiral is similar to that of a torus knot:

```
sphericalspiral[a_][m_,n_][t_]:=
    a{Cos[m t]Cos[n t],Sin[m t]Cos[n t],Sin[n t]}
```

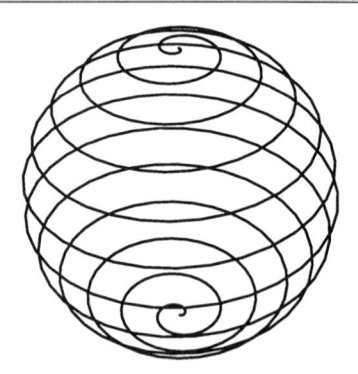

The spherical spiral
$$t \longmapsto (\cos t \cos 24t, \cos t \sin 24t, \sin t).$$

Use *Mathematica* to find the formulas for the curvature and torsion of a spherical spiral.

10

CONSTRUCTION OF SPACE CURVES

The analog of the Fundamental Theorem of Plane Curves (Theorem 6.8) is the *Fundamental Theorem of Space Curves*. This theorem, proved in Section 10.1, states that two curves with the same torsion and nonzero curvature differ only by a Euclidean motion of $\mathbb{R}^3$. Also, we show that space curves exist with prescribed torsion and curvature. In Section 10.2 we give a *Mathematica* miniprogram that plots a space curve with specified curvature and torsion.

The notion of contact between two curves or between a curve and a surface is defined in Section 10.3. Space curves that lie on spheres are characterized in Section 10.4. Curves of constant slope are generalizations of a helix. In Section 10.5 we discuss them and find interesting curves on a sphere with constant slope. Loxodromes on spheres are defined, discussed and plotted in Section 10.6. Animations of generalized helices spinning on spheres are given in Section 10.8.

10.1 The Fundamental Theorem of Space Curves

The curvature and torsion of a space curve determine a space curve in very much the same way as the curvature $\kappa 2$ determines a plane curve. First, we establish the invariance of curvature and torsion under Euclidean motions of $\mathbb{R}^3$. For that we need a fact about the vector cross product.

Lemma 10.1. *Let* $A \colon \mathbb{R}^3 \longrightarrow \mathbb{R}^3$ *be a linear map and* $\mathbf{a}, \mathbf{b}, \mathbf{c} \in \mathbb{R}^3$. *Then*
$$A\mathbf{a} \cdot A\mathbf{b} \times A\mathbf{c} = \det(A)(\mathbf{a} \cdot \mathbf{b} \times \mathbf{c}).$$

Proof. We regard $\mathbf{a}, \mathbf{b}, \mathbf{c}$ as row vectors. Then
$$A\mathbf{a} \cdot A\mathbf{b} \times A\mathbf{c} = \det \begin{pmatrix} A\mathbf{a} \\ A\mathbf{b} \\ A\mathbf{c} \end{pmatrix} = \det \left(A \begin{pmatrix} \mathbf{a} \\ \mathbf{b} \\ \mathbf{c} \end{pmatrix} \right)$$
$$= \det(A) \det \begin{pmatrix} \mathbf{a} \\ \mathbf{b} \\ \mathbf{c} \end{pmatrix} = \det(A)(\mathbf{a} \cdot \mathbf{b} \times \mathbf{c}). \blacksquare$$

Now we can determine the effect of a Euclidean motion on arc length, curvature and torsion.

Theorem 10.2. *The derivative of arc length, the curvature and the absolute value of the torsion are invariant under Euclidean motions of* $\mathbb{R}^3$. *The torsion changes sign under an orientation-reversing Euclidean motion of* $\mathbb{R}^3$.

Proof. Let $\alpha \colon (a,b) \longrightarrow \mathbb{R}^3$ be a curve, and let $F \colon \mathbb{R}^3 \longrightarrow \mathbb{R}^3$ be a Euclidean motion. Denote by A the linear part of F, so that for all $\mathbf{p} \in \mathbb{R}^3$ we have $F(\mathbf{p}) = A\mathbf{p} + F(0)$. We define a curve $\gamma \colon (a,b) \longrightarrow \mathbb{R}^3$ by $\gamma = F \circ \alpha$; then for $a < t < b$ we have
$$\gamma(t) = A\alpha(t) + F(0).$$

Hence $\gamma'(t) = A\alpha'(t)$, $\gamma''(t) = A\alpha''(t)$ and $\gamma'''(t) = A\alpha'''(t)$. Let s_α and s_γ denote the arc length functions with respect to α and γ. Since A is an orthogonal transformation, we have
$$s'_\gamma(t) = \|\gamma'(t)\| = \|A\alpha'(t)\| = s'_\alpha(t).$$

We compute the curvature of γ, making use of the Lagrange identity (8.2), page 183.
$$\kappa[\gamma](t) = \frac{\|\gamma'(t) \times \gamma''(t)\|}{\|\gamma'(t)\|^3} = \frac{\|A\alpha'(t) \times A\alpha''(t)\|}{\|A\alpha'(t)\|^3}$$
$$= \frac{\sqrt{\|A\alpha'(t)\|^2 \|A\alpha''(t)\|^2 - (A\alpha'(t) \cdot A\alpha''(t))^2}}{\|A\alpha'(t)\|^3}$$

10.1 The Fundamental Theorem of Space Curves

$$= \frac{\sqrt{\|\alpha'(t)\|^2 \|\alpha''(t)\|^2 - (\alpha'(t) \cdot \alpha''(t))^2}}{\|\alpha'(t)\|^3} = \frac{\|\alpha'(t) \times \alpha''(t)\|}{\|\alpha'(t)\|^3} = \kappa[\alpha](t).$$

Similarly, Lemma 10.1 and the fact that $\det(A) = \pm 1$ implies that

$$\tau[\gamma](t) = \frac{\gamma'(t) \cdot \gamma''(t) \times \gamma'''(t)}{\|\gamma'(t) \times \gamma''(t)\|^2} = \frac{A\alpha'(t) \cdot A\alpha''(t) \times A\alpha'''(t)}{\|A\alpha'(t) \times A\alpha''(t)\|^2}$$

$$= \frac{\det(A)\alpha'(t) \cdot \alpha''(t) \times \alpha'(t)}{\|\alpha'(t) \times \alpha''(t)\|^2} = \det(A)\tau[\alpha](t) = \pm \tau[\alpha](t). \blacksquare$$

In Theorem 6.8, page 136, we showed that a plane curve is determined up to a Euclidean motion of the plane by its signed curvature. This leads us to the conjecture that a space curve is determined up to a Euclidean motion of $\mathbb{R}^3$ by its curvature and torsion. We prove that this is true, with one additional (but essential) assumption.

Theorem 10.3. (Fundamental Theorem of Space Curves, Uniqueness) *Let α and γ be unit-speed curves in $\mathbb{R}^3$ defined on the same interval (a, b), and assume they have the same torsion and the same positive curvature. Then there is a Euclidean motion F of $\mathbb{R}^3$ that maps α onto γ.*

Proof. Since both α and γ have nonzero curvature, both of the Frenet frames $\{\mathbf{T}_\alpha, \mathbf{N}_\alpha, \mathbf{B}_\alpha\}$ and $\{\mathbf{T}_\gamma, \mathbf{N}_\gamma, \mathbf{B}_\gamma\}$ are well-defined. Fix s_0 with $a < s_0 < b$. There exists a translation of $\mathbb{R}^3$ taking $\alpha(s_0)$ into $\gamma(s_0)$. Moreover, we can find a 1-parameter family of rotations of $\mathbb{R}^3$ that map $\alpha'(s_0)$ onto $\gamma'(s_0)$. Among these is one rotation A that also maps $\mathbf{N}_\alpha(s_0)$ onto $\mathbf{N}_\gamma(s_0)$. Since A is a rotation, Lemma 10.1 implies that A maps $\mathbf{B}_\alpha(s_0)$ onto $\mathbf{B}_\gamma(s_0)$. Thus there exists a Euclidean motion F of $\mathbb{R}^3$ such that

$$F(\alpha(s_0)) = \gamma(s_0), \qquad A(\mathbf{T}_\alpha(s_0)) = \mathbf{T}_\gamma(s_0),$$
$$A(\mathbf{N}_\alpha(s_0)) = \mathbf{N}_\gamma(s_0), \qquad A(\mathbf{B}_\alpha(s_0)) = \mathbf{B}_\gamma(s_0),$$

To show that $F \circ \alpha$ coincides with γ, we define a real-valued function f by

$$f(s) = \|(A \circ \mathbf{T}_\alpha)(s) - \mathbf{T}_\gamma(s)\|^2 + \|(A \circ \mathbf{N}_\alpha)(s) - \mathbf{N}_\gamma(s)\|^2 + \|(A \circ \mathbf{B}_\alpha)(s) - \mathbf{B}_\gamma(s)\|^2$$

for $a < s < b$. A computation similar to that of (6.4), page 137, (using the assumptions that $\kappa[\alpha] = \kappa[\gamma]$ and $\tau[\alpha] = \tau[\gamma]$, see exercise 4) shows that the derivative of f is 0. Since $f(s_0) = 0$, we conclude that $f(s) = 0$ for all s. Hence $(F \circ \alpha)'(s) = (A \circ \mathbf{T}_\alpha)(s) = \gamma'(s)$ for all s, and so there exists $\mathbf{q} \in \mathbb{R}^3$ such that

$$(F \circ \alpha)(s) = \gamma(s) - \mathbf{q}$$

for all s. In fact $\mathbf{q} = 0$. Thus the Euclidean motion F maps α onto γ. ∎

Suppose that $\gamma\colon (a,b) \longrightarrow \mathbb{R}^3$ is a curve whose torsion vanishes everywhere except for a single t_0 with $a < t_0 < b$, and that the torsion of γ is undefined at t_0; then it is reasonable to say that γ has zero torsion. With this expanded definition of torsion, the assumption in Theorem 10.3 that α and γ have nonzero curvature is essential. There are indeed two space curves hat have the same curvature and torsion for which it is impossible to find a Euclidean motion mapping one onto the other. For example, consider curves **veryflat[1]** and **veryflat[2]** defined by

$$\mathbf{veryflat[1]}(t) = \begin{cases} 0 & \text{if } t = 0, \\ (t, 0, 5e^{-1/t^2}) & \text{if } t \neq 0, \end{cases}$$

and

$$\mathbf{veryflat[2]}(t) = \begin{cases} (t, 5e^{-1/t^2}, 0) & \text{if } t < 0, \\ 0 & \text{if } t = 0, \\ (t, 0, 5e^{-1/t^2}) & \text{if } t > 0. \end{cases}$$

The function $t \mapsto e^{-1/t^2}$ and all its derivatives vanish at 0, as is well-known from calculus. Hence the curvatures of both **veryflat[1]** and **veryflat[2]** vanish for $t = 0$. Furthermore, the torsion of **veryflat[1]** vanishes identically, since it is a plane curve. The torsion of **veryflat[2]** also vanishes identically for $t \neq 0$, so it is reasonable to define the torsion of **veryflat[2]** to be zero when $t = 0$. Any Euclidean motion mapping **veryflat[1]** onto **veryflat[2]** would have to be the identity on one portion of $\mathbb{R}^3$ and a rotation on another portion of $\mathbb{R}^3$, which is impossible. Thus **veryflat[1]** and **veryflat[2]** have the same curvature and torsion, but there is no

10.1 The Fundamental Theorem of Space Curves

Euclidean motion mapping one onto the other.

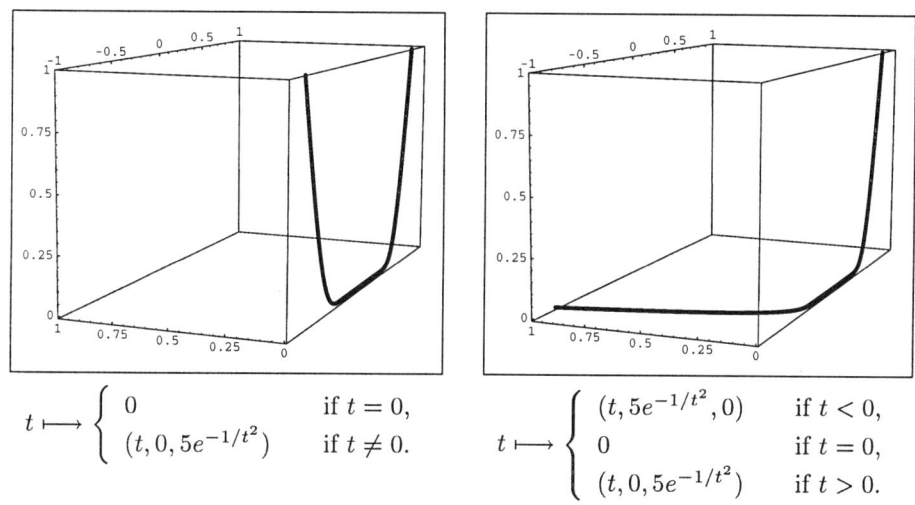

$$t \longmapsto \begin{cases} 0 & \text{if } t = 0, \\ (t, 0, 5e^{-1/t^2}) & \text{if } t \neq 0. \end{cases} \qquad t \longmapsto \begin{cases} (t, 5e^{-1/t^2}, 0) & \text{if } t < 0, \\ 0 & \text{if } t = 0, \\ (t, 0, 5e^{-1/t^2}) & \text{if } t > 0. \end{cases}$$

Now we turn to the question of the existence of curves with prescribed curvature and torsion.

Theorem 10.4. (Fundamental Theorem of Space Curves, Existence) *Let $\mathbf{k}: (a,b) \longrightarrow \mathbb{R}$ and $\mathbf{t}: (a,b) \longrightarrow \mathbb{R}$ be differentiable functions with $\mathbf{k} > 0$. Then there exists a unit-speed curve $\beta: (a,b) \longrightarrow \mathbb{R}^3$ whose curvature and torsion are $\mathbf{k}$ and $\mathbf{t}$. For $a < s_0 < b$ the value $\beta(s_0)$ can be prescribed arbitrarily. Also, the values of $\mathbf{T}(s_0)$ and $\mathbf{N}(s_0)$ can be prescribed subject to the conditions that $\mathbf{T}(s_0) = \mathbf{N}(s_0) = 1$ and $\mathbf{T}(s_0) \cdot \mathbf{N}(s_0) = 0$.*

Proof. Consider the following system of 12 differential equations

(10.1)
$$\begin{cases} x_1'(s) = t_1(s), & t_1'(s) = \mathbf{k}(s)n_1(s), \\ x_2'(s) = t_2(s), & t_2'(s) = \mathbf{k}(s)n_2(s), \\ x_3'(s) = t_3(s), & t_3'(s) = \mathbf{k}(s)n_3(s), \\ \\ n_1'(s) = -\mathbf{k}(s)t_1(s) + \mathbf{t}(s)b_1(s), & b_1'(s) = -\mathbf{t}(s)n_1(s), \\ n_2'(s) = -\mathbf{k}(s)t_2(s) + \mathbf{t}(s)b_2(s), & b_2'(s) = -\mathbf{t}(s)n_2(s), \\ n_3'(s) = -\mathbf{k}(s)t_3(s) + \mathbf{t}(s)b_3(s), & b_3'(s) = -\mathbf{t}(s)n_2(s), \end{cases}$$

together with the initial conditions

(10.2)
$$\begin{cases} x_1(s_0) = p_1, & t_1(s_0) = q_1, \\ x_2(s_0) = p_2, & t_2(s_0) = q_2, \\ x_3(s_0) = p_3, & t_3(s_0) = q_3, \\ \\ n_1(s_0) = r_1, & b_1(s_0) = q_2 r_3 - q_3 r_2, \\ n_2(s_0) = r_2, & b_2(s_0) = q_3 r_1 - q_1 r_3, \\ n_3(s_0) = r_3, & b_3(s_0) = q_1 r_2 - q_2 r_1. \end{cases}$$

In (10.2) we require that

$$(t_1(s_0), t_2(s_0), t_3(s_0)) \cdot (t_1(s_0), t_2(s_0), t_3(s_0))$$
$$= (n_1(s_0), n_2(s_0), n_3(s_0)) \cdot (n_1(s_0), n_2(s_0), n_3(s_0)) = 1$$

and

$$(t_1(s_0), t_2(s_0), t_3(s_0)) \cdot (n_1(s_0), n_2(s_0), n_3(s_0)) = 0.$$

From the theory of systems of differential equations, we know that this linear system has a unique solution. If we put $\beta = (x_1, x_2, x_3)$ and

$$\mathbf{T} = (t_1, t_2, t_3), \quad \mathbf{N} = (n_1, n_2, n_3), \quad \mathbf{B} = (b_1, b_2, b_3),$$

we see that (10.1) can be compressed into $\beta' = \mathbf{T}$ together with

(10.3)
$$\begin{cases} \mathbf{T}' = & k\mathbf{N}, \\ \mathbf{N}' = -t\mathbf{T} & +k\mathbf{B}, \\ \mathbf{B}' = & -t\mathbf{N}. \end{cases}$$

Comparison with the Frenet formulas (8.10) shows that **k** is the curvature of the curve β, and **t** is the torsion. ∎

10.2 Drawing Space Curves with Assigned Curvature

Mathematica's command **NDSolve** can be used to find space curves with assigned curvature and torsion. The miniprogram below is somewhat different from the corresponding program for plane curves. It constructs and plots a curve with specified

10.2 Drawing Space Curves with Assigned Curvature

curvature **kk** and torsion **tt**. The idea of the miniprogram **plotintrinsic3d** is to use **NDSolve** to solve the system (10.1) (which is equivalent to the Frenet formulas (8.10)) together with the initial conditions (10.2).

Actually the miniprogram solves (10.1) and (10.2) whether or not the function κ (in *Mathematica* **kk**) is assumed to be positive. But when we choose for **kk** a positive function and for **tt** an arbitrary function, the miniprogram plots a space curve with curvature **kk** and torsion **tt**.

```
plotintrinsic3d[{kk_,tt_},{a_:0,{p1_:0,p2_:0,p3_:0},
   {q1_:1,q2_:0,q3_:0},{r1_:0,r2_:1,r3_:0}},
      {smin_:-10,smax_:10},opts___]:=
ParametricPlot3D[Module[
  {x1,x2,x3,t1,t2,t3,n1,n2,n3,b1,b2,b3},
  {x1[s],x2[s],x3[s]} /.
  NDSolve[{x1'[ss]==t1[ss],
       x2'[ss]==t2[ss],
       x3'[ss]==t3[ss],
    t1'[ss]==kk[ss]n1[ss],
    t2'[ss]==kk[ss]n2[ss],
    t3'[ss]==kk[ss]n3[ss],
       n1'[ss]==-kk[ss]t1[ss] + tt[ss]b1[ss],
       n2'[ss]==-kk[ss]t2[ss] + tt[ss]b2[ss],
       n3'[ss]==-kk[ss]t3[ss] + tt[ss]b3[ss],
         b1'[ss]==-tt[ss]n1[ss],
         b2'[ss]==-tt[ss]n2[ss],
         b3'[ss]==-tt[ss]n2[ss],
  x1[a]==p1,x2[a]==p2,x3[a]==p3,
  t1[a]==q1,t2[a]==q2,t3[a]==q3,
  n1[a]==r1,n2[a]==r2,n3[a]==r3,
  b1[a]==q2 r3 - q3 r2,b2[a]==q3 r1 - q1 r3,
  b3[a]==q1 r2 - q2 r1},
 {x1,x2,x3,t1,t2,t3,n1,n2,n3,b1,b2,b3},
 {ss,smin,smax}]]//Evaluate,
 {s,smin,smax},opts];
```

In order that **plotintrinsic3d** give correct results, it is necessary that **{p1,p2,p3}** and **{q1,q2,q3}** be orthogonal unit vectors.

Here are some examples illustrating the use of **plotintrinsic3d**.

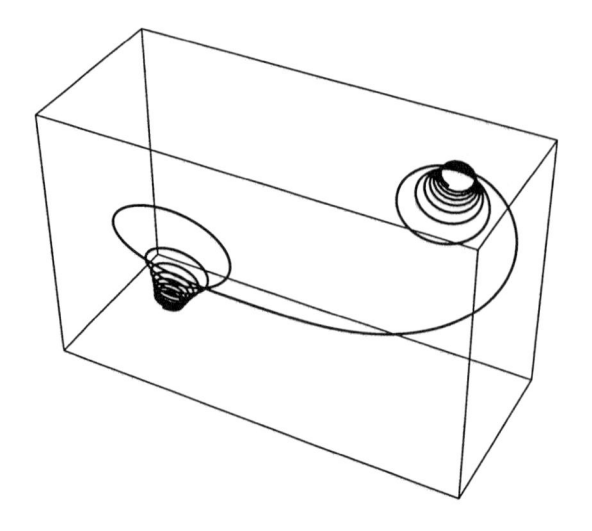

```
plotintrinsic3d[
 {Abs[#]&,0.3&},
 {0,{0,0,0},{1,0,0},
 {0,1,0}},{-10,10},
 Axes->None,
 PlotPoints->500];
```

Curve with $\kappa(s) = |s|$ **and** $\tau(s) = 0.3$

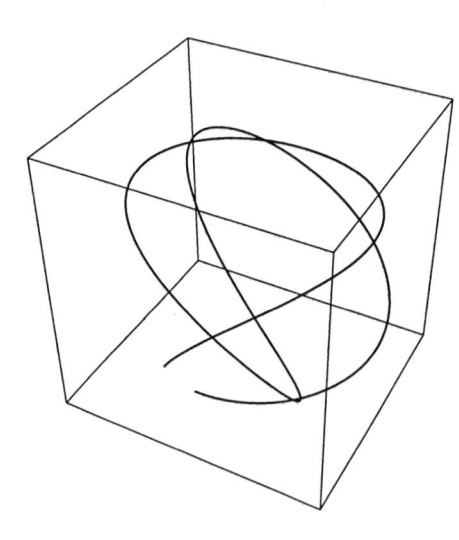

```
plotintrinsic3d[
 {1.3&,0.5Sin[#]&},
 {0,{0,0,0},
 {1,0,0},{0,1,0}},
 {0,4Pi},
 Axes->None,
 PlotPoints->200];
```

Curve with $\kappa(s) = 1.3$
and $\tau(s) = 0.5 \sin s$

10.3 Contact

In Section 5.1 we defined the evolute of a plane curve as the locus of its centers of curvature. In this section we show how to define the evolute of a space curve. This requires that we generalize the notion of center of curvature to space curves. For this, we need the notion of contact. Since this concept is also important for plane curves, we discuss it in a generality that will apply to both cases.

Definition. *An* **implicitly defined hypersurface** *in $\mathbb{R}^n$ is the set of zeros of a differentiable function $F \colon \mathbb{R}^n \longrightarrow \mathbb{R}$. We denote the set of zeros by $F^{-1}(0)$.*

Clearly, a hypersurface in $\mathbb{R}^2$ is an implicitly defined curve, and a hypersurface in $\mathbb{R}^3$ is an implicitly defined surface. These are the main cases that we need.

The simplest hypersurfaces are hyperplanes and hyperspheres. A **hyperplane** in $\mathbb{R}^n$ consists of the points containing some vector $\mathbf{u}$ that are perpendicular to another vector $\mathbf{v}$, that is, a subset of the form

$$\{\, \mathbf{p} \in \mathbb{R}^n \mid (\mathbf{p} - \mathbf{u}) \cdot \mathbf{v} = 0 \,\}.$$

Thus a hyperplane in $\mathbb{R}^2$ is a line and a hyperplane in $\mathbb{R}^3$ is a plane.

A **hypersphere** in $\mathbb{R}^n$ of radius r centered at $\mathbf{q} \in \mathbb{R}^n$ is the set of points at a distance r from $\mathbf{q}$; that is,

$$\{\, \mathbf{p} \in \mathbb{R}^n \mid \|\mathbf{p} - \mathbf{q}\| = r \,\}.$$

In particular, a hypersphere in $\mathbb{R}^2$ is a circle and a hypersphere in $\mathbb{R}^3$ is a sphere. We leave it to the reader to find differentiable functions whose zero sets are precisely $\{\, \mathbf{p} \in \mathbb{R}^n \mid (\mathbf{p} - \mathbf{u}) \cdot \mathbf{v} = 0 \,\}$ and $\{\, \mathbf{p} \in \mathbb{R}^n \mid \|\mathbf{p} - \mathbf{q}\| = r \,\}$.

Among all lines passing through a point $\mathbf{p}$ on a curve $\boldsymbol{\alpha}$, the tangent line affords the best approximation. Other curves through $\mathbf{p}$ may approximate the curve even more closely. The mathematical concept that makes precise this idea is the following:

Definition. *Let $\boldsymbol{\alpha} \colon (a,b) \longrightarrow \mathbb{R}^n$ be a regular curve, and let $F \colon \mathbb{R}^n \longrightarrow \mathbb{R}$ be a differentiable function. We say that a parametrically defined curve $\boldsymbol{\alpha}$ and an implicitly defined hypersurface $F^{-1}(0)$ have* **contact of order** k *at $\boldsymbol{\alpha}(t_0)$ provided we have*

(10.4) $$(F \circ \boldsymbol{\alpha})(t_0) = (F \circ \boldsymbol{\alpha})'(t_0) = \cdots = (F \circ \boldsymbol{\alpha})^{(k)}(t_0) = 0,$$

but

(10.5) $$(F \circ \boldsymbol{\alpha})^{(k+1)}(t_0) \neq 0.$$

To check that contact is a geometric concept, we prove:

Lemma 10.5. *The definition of contact between α and $F^{-1}(0)$ is independent of the parametrization of α.*

Proof. Let $\gamma\colon(c,d)\longrightarrow\mathbb{R}^n$ be a reparametrization of α; then there exists a differentiable function $t\colon(c,d)\longrightarrow(a,b)$ such that $\gamma(u)=\alpha(t(u))$ for $c<u<d$. Let u_0 be such that $t(u_0)=t_0$. Then

$$(F\circ\gamma)(u_0) = (F\circ\alpha)(t_0) = 0,$$

$$(F\circ\gamma)'(u_0) = (F\circ\alpha)'(t_0)u'(t_0) = 0,$$

$$(F\circ\gamma)''(u_0) = (F\circ\alpha)''(t_0)u'(t_0)^2 + (F\circ\alpha)'(t_0)u''(t_0) = 0,$$

$$(F\circ\gamma)'''(u_0) = (F\circ\alpha)'''(t_0)u'(t_0)^3 + 3(F\circ\alpha)'(t_0)u'(t_0)^2 u''(t_0)$$
$$+(F\circ\alpha)'(t_0)u'''(t_0) = 0,$$

and so forth. Thus we see that (10.4) implies that

$$(F\circ\gamma)(u_0) = (F\circ\gamma)'(u_0) = \cdots = (F\circ\gamma)^{(k)}(u_0) = 0,$$

and (10.5) implies that

$$(F\circ\gamma)^{(k+1)}(u_0) \neq 0. \blacksquare$$

For example, the horizontal line through $(0,0)$ consists of the set of zeros of the function $F\colon\mathbb{R}^2\longrightarrow\mathbb{R}$ given by $F(x,y)=y$. If n is a positive integer, the **generalized parabola** $t\longmapsto(t,t^n)$ has contact of order $n-1$ with this line at $(0,0)$.

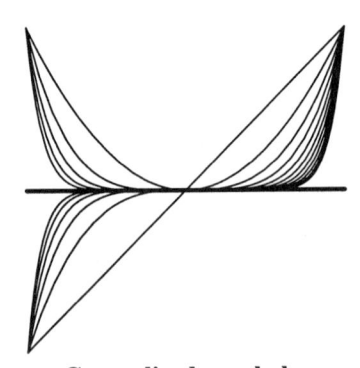

Generalized parabolas

We discuss next contact between hyperplanes and curves in $\mathbb{R}^n$.

10.3 Contact

Lemma 10.6. *Let $\beta\colon (a,b) \longrightarrow \mathbb{R}^n$ be a unit-speed curve, let $a < s_0 < b$ and let $\mathbf{v} \in \mathbb{R}^n$. Then the hyperplane $\{\,\mathbf{q} \mid (\mathbf{q}-\beta(s_0)) \cdot \mathbf{v} = 0\,\}$ has at least order 1 contact with β at $\beta(s_0)$ if and only if $\mathbf{v}$ is perpendicular to $\beta'(s_0)$.*

Proof. Since the hyperplane $\{\,\mathbf{q} \mid (\mathbf{q}-\beta(s_0)) \cdot \mathbf{v} = 0\,\}$ is the zero set of the function $F\colon \mathbb{R}^n \longrightarrow \mathbb{R}$ given by $F(\mathbf{q}) = (\mathbf{q} - \beta(s_0)) \cdot \mathbf{v}$, we define $f(s) = (\beta(s) - \beta(s_0)) \cdot \mathbf{v}$. Then
$$f'(s) = \beta'(s) \cdot \mathbf{v}.$$
Hence $f'(s_0) = 0$ if and only if $\beta'(s_0) \cdot \mathbf{v} = 0$. ∎

Thus at each point on a unit-speed curve, there are hyperplanes with at least order 1 contact with the curve. As we have just seen, higher order contact with a hyperplane is possible for special curves.

We turn to contact between hyperspheres and curves. Intuitively, it should be the case that a hypersphere can be chosen to have higher order contact with a curve at a given point than is possible with a hyperplane. We first consider plane curves.

Lemma 10.7. *Let $\beta\colon (a,b) \longrightarrow \mathbb{R}^2$ be a unit-speed curve. Choose s_0 such that $a < s_0 < b$ and let $\mathbf{v} \in \mathbb{R}^2$. The distance from $\mathbf{v}$ to $\beta(s_0)$ is $\|\mathbf{v} - \beta(s_0)\| = r$.*

(i) The circle $\{\,\mathbf{q} \mid \|\mathbf{q} - \mathbf{v}\| = r\,\}$ has at least order 1 contact with β at $\beta(s_0)$ if and only if the center $\mathbf{v}$ of the circle lies on the line perpendicular to β at $\beta(s_0)$.

(ii) Suppose that $\kappa 2(s_0) \neq 0$. Then the circle $\{\,\mathbf{q} \mid \|\mathbf{q} - \mathbf{v}\| = r\,\}$ has at least order 2 contact with β at $\beta(s_0)$ if and only if the center $\mathbf{v}$ of the circle satisfies
$$\mathbf{v} = \beta(s_0) + \frac{1}{\kappa 2(s_0)} J\mathbf{T}(s_0).$$
When this is the case, the radius r of the circle equals $1/|\kappa 2(s_0)|$.

Proof. Define $g(s) = \|\beta(s) - \mathbf{v}\|^2$. Then $g' = 2\beta' \cdot (\beta - \mathbf{v}) = 2\mathbf{T} \cdot (\beta - \mathbf{v})$, so that $g'(s_0) = 0$ if and only if $\beta'(s_0) \cdot (\beta(s_0) - \mathbf{v}) = 0$. This proves (i).

It follows from (i) that there is a number λ such that $\beta(s_0) - \mathbf{v} = \lambda J\mathbf{T}(s_0)$. Also, we have
$$g'' = 2(\mathbf{T} \cdot \mathbf{T} + \mathbf{T}' \cdot (\beta - \mathbf{v})) = 2(1 + \kappa 2\, J\mathbf{T} \cdot (\beta - \mathbf{v}));$$
hence
$$g''(s_0) = 2(1 + \kappa 2(s_0)\lambda).$$
Thus $g''(s_0) = 0$ if and only if $\lambda = -1/\kappa 2(s_0)$, proving (ii). ∎

Part (ii) of Lemma 10.7 yields a characterization of the osculating circle defined in Section 5.6:

Corollary 10.8. Let $\alpha\colon (a,b) \longrightarrow \mathbb{R}^2$ be a regular plane curve, and let $a < t_0 < b$ be such that $\kappa_2[\alpha](t_0) \neq 0$. Then the osculating circle of α at $\alpha(t_0)$ is the unique circle which has at least order 2 contact with α at $\alpha(t_0)$.

We consider next contact between space curves and spheres. In many cases contact of order 3 is possible.

Theorem 10.9. Let $\beta\colon (a,b) \longrightarrow \mathbb{R}^3$ be a unit-speed curve. Choose s_0 such that $a < s_0 < b$ and let $\mathbf{v} \in \mathbb{R}^3$. The distance from $\mathbf{v}$ to $\beta(s_0)$ is $\|\mathbf{v} - \beta(s_0)\| = r$. We abbreviate $\kappa[\beta]$ and $\tau[\beta]$ to κ and τ.

(i) The sphere $\{\mathbf{q} \mid \|\mathbf{q} - \mathbf{v}\| = r\}$ has at least order 1 contact with β at $\beta(s_0)$ if and only if the center $\mathbf{v}$ of the sphere lies on the line perpendicular to β at $\beta(s_0)$.

(ii) Suppose that $\kappa(s_0) \neq 0$. Then the sphere $\{\mathbf{q} \mid \|\mathbf{q} - \mathbf{v}\| = r\}$ has at least order 2 contact with β at $\beta(s_0)$ if and only if

$$\mathbf{v} - \beta(s_0) = \frac{1}{\kappa(s_0)} \mathbf{N}(s_0) - b\,\mathbf{B}(s_0)$$

for some b.

(iii) Suppose that $\kappa(s_0) \neq 0$ and $\tau(s_0) \neq 0$. Then the sphere $\{\mathbf{q} \mid \|\mathbf{q} - \mathbf{v}\| = r\}$ has at least order 3 contact with β at s_0 if and only if

$$\mathbf{v} - \beta(s_0) = \frac{1}{\kappa(s_0)} \mathbf{N}(s_0) - \frac{\kappa'(s_0)}{\kappa(s_0)^2 \tau(s_0)} \mathbf{B}(s_0).$$

Proof. Define $g(s) = \|\beta(s) - \mathbf{v}\|^2$. We calculate as in part (i) of Lemma 10.7:

$$g' = 2\beta' \cdot (\beta - \mathbf{v}) = 2\mathbf{T} \cdot (\beta - \mathbf{v}),$$

so that $g'(s_0) = 0$ if and only if $\beta'(s_0) \cdot (\beta(s_0) - \mathbf{v}) = 0$, proving (i). It follows from (i) that there are numbers a and b such that

$$\beta(s_0) - \mathbf{v} = a\,\mathbf{N}(s_0) + b\,\mathbf{B}(s_0).$$

Also, computing as in part (ii) of Lemma 10.7, we find that

(10.6) $$g'' = 2\big(\mathbf{T} \cdot \mathbf{T} + \mathbf{T}' \cdot (\beta - \mathbf{v})\big) = 2\big(1 + \kappa\,\mathbf{N} \cdot (\beta - \mathbf{v})\big).$$

Hence

$$g''(s_0) = 2\big(1 + \kappa(s_0)\mathbf{N}(s_0) \cdot (a\,\mathbf{N}(s_0) + b\,\mathbf{B}(s_0))\big) = 2\big(1 + \kappa(s_0)a\big).$$

10.3 Contact

Thus $g''(s_0) = 0$ if and only if $a = -1/\kappa(s_0)$, proving (ii).

For (iii) we use (10.6) to compute

$$g''' = 2\bigl((\kappa'\mathbf{N} + \kappa\mathbf{N}') \cdot (\boldsymbol{\beta} - \mathbf{v}) + \kappa\mathbf{N} \cdot \mathbf{T}\bigr) = \bigl(\kappa'\mathbf{N} - \kappa^2\mathbf{T} + \kappa\tau\mathbf{B}\bigr) \cdot (\boldsymbol{\beta} - \mathbf{v}).$$

In particular,

$$g'''(s_0) = 2\bigl((\kappa'(s_0)\mathbf{N}(s_0) - \kappa(s_0)^2\mathbf{T}(s_0) + \kappa(s_0)\tau(s_0)\mathbf{B}(s_0))\bigr) \cdot \left(-\frac{\mathbf{N}(s_0)}{\kappa(s_0)} + b\mathbf{B}(s_0)\right)$$

$$= -\frac{\kappa'(s_0)}{\kappa(s_0)} + \kappa(s_0)\tau(s_0)b.$$

Thus $g'''(s_0) = 0$ if and only if $b = \kappa'(s_0)/(\kappa(s_0)^2\tau(s_0))$. Hence we obtain (iii). ∎

The space curve analog of an osculating circle is an osculating sphere, defined precisely as follows:

Definition. Let $\alpha\colon (a,b) \longrightarrow \mathbb{R}^3$ be a regular space curve, and let $a < t_0 < b$ be such that $\kappa[\alpha](t_0) \neq 0$ and $\tau[\alpha](t_0) \neq 0$. Then the **osculating sphere** of α at $\alpha(t_0)$ is the unique sphere which has at least order 3 contact with α at $\alpha(t_0)$.

Corollary 10.10. *The osculating sphere at $\alpha(t_0)$ to a regular space curve $\alpha\colon (a,b) \longrightarrow \mathbb{R}^3$ whose curvature and torsion do not vanish at $\alpha(t_0)$ is a sphere of radius*

$$(10.7) \qquad \sqrt{\frac{1}{\kappa[\alpha](t_0)^2} + \frac{\kappa[\alpha]'(t_0)^2}{\|\alpha'(t_0)\|^2 \tau[\alpha](t_0)^2 \kappa[\alpha](t_0)^4}}$$

and center

$$(10.8) \qquad \mathbf{v} = \alpha(t_0) + \frac{1}{\kappa[\alpha](t_0)}\mathbf{N}(t_0) + \frac{\kappa[\alpha]'(t_0)}{\|\alpha'(t_0)\|\kappa[\alpha](t_0)^2\tau[\alpha](t_0)}\mathbf{B}(t_0).$$

Proof. Let β be a unit-speed reparametrization of α with $\alpha(t) = \beta(s(t))$; then by the chain rule

$$\kappa[\beta]'(s(t))s'(t) = \kappa[\alpha]'(t).$$

From this fact and part (iii) of Theorem 10.9 we get (10.8). Since the radius of the osculating sphere at $\alpha(t_0)$ is $\|\mathbf{v} - \alpha(t_0)\|$, equation (10.7) follows from (10.8). ∎

In Section 5.1 we defined the evolute of a plane curve to be the locus of the centers of the osculating circles to the curve. It is natural, therefore, to define the evolute of a space curve to be the locus of the centers of the osculating spheres.

Definition. *The* **evolute** *of a regular space curve* α *is the space curve given by*

$$\text{evolute3d}[\alpha](t) = \alpha(t) + \frac{1}{\kappa[\alpha](t)}\mathbf{N}(t) - \frac{\kappa[\alpha]'(t)}{\|\alpha'(t)\|\kappa[\alpha](t)^2\tau[\alpha](t)}\mathbf{B}(t).$$

The corresponding *Mathematica* definition is:

```
evolute3d[gamma_][t_]:= gamma[tt] +
    (1/kappa[gamma][tt])normal[gamma][tt] -
    (D[kappa[gamma][tt],tt]/(arclengthprime[gamma][tt]*
    kappa[gamma][tt]^2 tau[gamma][tt]))*
    binormal[gamma][tt]  /. tt->t
```

For example, let us find the evolute of **helix[a,b]**. We use

```
Simplify[PowerExpand[evolute3d[helix[a,b]][t]]]
```

obtaining

```
      2                2
     b  Cos[t]         b  Sin[t]
{-(-----------), -(-----------), b t}
        a                a
```

Thus the evolute of a helix is another helix.

```
ParametricPlot3D[
Evaluate[{helix[1,0.3][t],
evolute3d[helix[1,0.3]][t]}],
{t,0,4Pi},Axes->None];
```

The curve $t \longmapsto \text{helix}[1,0.3](t)$
and its evolute

10.4 Space Curves that Lie on a Sphere

If a space curve (such as Viviani's curve) lies on a sphere, then it has contact of order n with that sphere for any n. In this section we show that certain relations between the curvature and torsion of a space curve are necessary and sufficient for the space curve to lie on a sphere.

Theorem 10.11. *Let* $\beta\colon (a,b) \longrightarrow \mathbb{R}^3$ *be a unit-speed curve. Suppose that* β *lies on the sphere* $\{\mathbf{p} \mid \|\mathbf{p} - \mathbf{q}\| = c\}$ *of radius* $c > 0$ *centered at* $\mathbf{q} \in \mathbb{R}^3$. *Then:*

(i) $\kappa \geq 1/c$;

(ii) *the curvature* κ *and torsion* τ *of* β *are related by*

$$\tag{10.9} \tau^2 \left(c^2 - \frac{1}{\kappa^2} \right) = \left(\frac{\kappa'}{\kappa^2} \right)^2;$$

(iii) $\kappa'(t_0) = 0$ *implies* $\tau(t_0) = 0$ *or* $\kappa(t_0) = 1/c$;

(iv) $\tau \left(\dfrac{\kappa''}{\kappa^2} - \dfrac{2\kappa'^2}{\kappa^3} \right) = \dfrac{\tau'\kappa'}{\kappa^2} + \dfrac{\tau^3}{\kappa}$.

(v) $\kappa'(t_0) = \kappa''(t_0) = 0$ *implies* $\tau(t_0) = 0$;

(vi) $\kappa \equiv 1/a$ *if and only if* β *is a circle of radius* a, *necessarily contained in some plane.*

Proof. Differentiation of $\|\beta - \mathbf{q}\|^2 = c^2$ yields $(\beta - \mathbf{q}) \cdot \mathbf{T} = 0$. Another differentiation results in

$$1 + (\beta - \mathbf{q}) \cdot \mathbf{T}' = 0.$$

Hence $\mathbf{T}'$ is nonzero, which implies the Frenet frame $\{\mathbf{T}, \mathbf{N}, \mathbf{B}\}$ is well-defined on β. Thus

$$\tag{10.10} 1 + (\beta - \mathbf{q}) \cdot (\kappa \mathbf{N}) = 0.$$

The Cauchy-Schwarz inequality implies that

$$1 = |(\beta - \mathbf{q}) \cdot (\kappa \mathbf{N})| \leq \kappa \|\beta - \mathbf{q}\| \, \|\mathbf{N}\| = \kappa c,$$

proving (i). Differentiating (10.10), we obtain

$$\tag{10.11} \begin{aligned} 0 &= \mathbf{T} \cdot (\kappa \mathbf{N}) + (\beta - \mathbf{q}) \cdot \left(\kappa' \mathbf{N} + \kappa(-\kappa \mathbf{T} + \tau \mathbf{B}) \right) \\ &= \kappa'(\beta - \mathbf{q}) \cdot \mathbf{N} + \kappa \tau (\beta - \mathbf{q}) \cdot \mathbf{B} \\ &= -\frac{\kappa'}{\kappa} + \kappa \tau (\beta - \mathbf{q}) \cdot \mathbf{B}. \end{aligned}$$

It follows from (10.10) and (10.11) that

(10.12) $\tau(\beta - \mathbf{q}) = (\tau(\beta - \mathbf{q}) \cdot \mathbf{T})\mathbf{T} + (\tau(\beta - \mathbf{q}) \cdot \mathbf{N})\mathbf{N} + (\tau(\beta - \mathbf{q}) \cdot \mathbf{B})\mathbf{B}$

$$= -\frac{\tau}{\kappa}\mathbf{N} + \frac{\kappa'}{\kappa^2}\mathbf{B}.$$

Then (10.12) implies that

$$\tau^2 c^2 = \tau^2 \|\beta - \mathbf{q}\|^2 = \left(\frac{\tau}{\kappa}\right)^2 + \left(\frac{\kappa'}{\kappa^2}\right)^2,$$

from which we obtain (ii). Then (ii) implies (iii).

To prove (iv), we note that (10.12) implies that

(10.13) $\qquad \tau(\beta - \mathbf{q}) \cdot \mathbf{B} = \dfrac{\kappa'}{\kappa^2}.$

Differentiation of (10.13) together with (10.10) yields

(10.14) $\qquad \left(\dfrac{\kappa'}{\kappa^2}\right)' = \tau'(\beta - \mathbf{q}) \cdot \mathbf{B} + \tau(\mathbf{T} \cdot \mathbf{B} + (\beta - \mathbf{q}) \cdot (-\tau \mathbf{N}))$

$$= \tau'(\beta - \mathbf{q}) \cdot \mathbf{B} + \frac{\tau^2}{\kappa}.$$

Then (10.14) and (10.13) imply (iv). Furthermore, (iv) implies (v).

Finally, if $\kappa \equiv 1/a$, then (v) implies that $\tau = 0$. Hence by Lemma 8.5 we see that α is a plane curve of constant curvature. From Theorem 1.12, page 16 it follows that α is a circle. ∎

It is interesting to check that for Viviani's curve (which lies on a sphere of radius $2a$) that τ and κ' have the same zeros. The relevant graph is:

```
Plot[Evaluate[
  {kappa[viviani[1]]'[t],
   tau[viviani[1]][t]}],
  {t,-2Pi,2Pi},
  PlotStyle->{
    {GrayLevel[0]},
    {GrayLevel[0.5]}}]
```

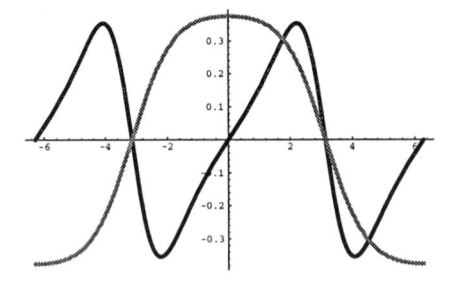

κ' and τ for Viviani's Curve

10.4 Space Curves that Lie on a Sphere

Notice that equation (10.9) also holds for a helix of slope b over a circle of radius $\sqrt{c^2 - b^2}$, but that a helix with nonzero torsion lies on no sphere. Therefore, we cannot expect to characterize spherical curves by (10.9) alone. Nevertheless, we have the following result, which uses the technique of Theorem 1.12.

Theorem 10.12. *Let $\beta\colon (a,b) \longrightarrow \mathbb{R}^3$ be a unit-speed curve whose curvature κ and torsion τ satisfy (10.9). Assume that κ, τ and κ' vanish only at isolated points. Then β lies on a sphere of radius c centered at some $\mathbf{q} \in \mathbb{R}^3$.*

Proof. Let
$$\gamma = \beta + \frac{1}{\kappa}\mathbf{N} - \left(\frac{\kappa'}{\tau\kappa^2}\right)\mathbf{B}.$$

Then (10.9) implies that

$$(10.15) \quad \gamma' = \mathbf{T} - \frac{\kappa'}{\kappa^2}\mathbf{N} + \frac{1}{\kappa}(-\kappa\mathbf{T} + \tau\mathbf{B}) - \left(\frac{\kappa'}{\tau\kappa^2}\right)'\mathbf{B} + \left(\frac{\kappa'}{\tau\kappa^2}\right)(\tau\mathbf{N})$$

$$= \left(\frac{\tau}{\kappa} - \left(\frac{\kappa'}{\tau\kappa^2}\right)'\right)\mathbf{B}.$$

Differentiation of (10.9) yields

$$(10.16) \quad \left(\frac{\kappa'}{\tau\kappa^2}\right)\left(\frac{\kappa'}{\tau\kappa^2}\right)' = -\frac{\kappa'}{\kappa^3}.$$

From (10.15) and (10.16) we get

$$\left(\frac{\kappa'}{\tau\kappa^2}\right)\gamma' = \left(\frac{\tau}{\kappa}\frac{\kappa'}{\tau\kappa^2} - \frac{\kappa'}{\kappa^3}\right)\mathbf{B} = 0$$

or $\kappa'\gamma' = 0$. Thus $\gamma'(t) = 0$ whenever t is such that $\kappa'(t) \neq 0$. Since γ' is continuous, $\gamma'(t) = 0$ for all t. Hence there exists $\mathbf{q} \in \mathbb{R}^3$ such that $\gamma(t) = \mathbf{q}$ for all t. In other words,

$$(10.17) \quad \beta(t) - \mathbf{q} = -\frac{1}{\kappa(t)}\mathbf{N}(t) + \left(\frac{\kappa'(t)}{\tau(t)\kappa(t)^2}\right)\mathbf{B}(t)$$

for all t. Then (10.9) implies that

$$\|\beta(t) - \mathbf{q}\|^2 = c^2.$$

Hence β lies on a sphere of radius c centered at $\mathbf{q}$. ∎

Corollary 10.13. *Let $\beta\colon (a,b) \longrightarrow \mathbb{R}^3$ be a unit-speed curve whose curvature κ and torsion τ satisfy (10.9). Then the osculating spheres of β coincide and β lies on the common sphere.*

10.5 Curves of Constant Slope

Since a helix has constant curvature κ and torsion τ, the ratio τ/κ is constant. Are there other curves with this property? In order to answer this question, we first show that the constancy of τ/κ for a curve is equivalent to the constancy of the angle between the curve and a fixed vector. More precisely:

Definition. *A curve $\gamma\colon(a,b) \longrightarrow \mathbb{R}^3$ is said to have* **constant slope** *with respect to a unit vector $\mathbf{u} \in \mathbb{R}^3$, provided the angle α between $\mathbf{u}$ and γ is constant. The analytical condition is*
$$\mathbf{T} \cdot \mathbf{u} = \cos\alpha;$$
we call $\cos\alpha$ the **slope** *of γ.*

Lemma 10.14. *If a curve $\gamma\colon(a,b) \longrightarrow \mathbb{R}^3$ has nonzero curvature κ and constant slope $\cos\alpha$ with respect to some unit vector $\mathbf{u}$, then the ratio τ/κ has the constant value $\pm\cot\alpha$.*

Proof. Without loss of generality, γ has unit-speed. We differentiate $\mathbf{T}\cdot\mathbf{u} = \cos\alpha$, obtaining $(\kappa\mathbf{N})\cdot\mathbf{u} = 0$. Therefore, $\mathbf{u}$ is perpendicular to $\mathbf{N}$, and so we can write
$$\mathbf{u} = \cos\alpha\,\mathbf{T} \pm \sin\alpha\,\mathbf{B}.$$
When we differentiate this equation, we obtain
$$0 = \bigl(\cos\alpha\,\kappa \mp \sin\alpha\,\tau\bigr)\mathbf{N}.$$
Hence the lemma follows. ∎

The converse of Lemma 10.14 is true.

Lemma 10.15. *Let $\gamma\colon(a,b) \longrightarrow \mathbb{R}^3$ be a curve for which the ratio τ/κ is constant. Write $\tau/\kappa = \cot\alpha$. Then γ has constant slope $\cos\alpha$ with respect to some unit vector $\mathbf{u} \in \mathbb{R}^3$.*

Proof. Without loss of generality, γ has unit-speed. We have
$$0 = (\kappa\cos\alpha - \tau\sin\alpha)\mathbf{N} = \frac{d}{ds}(\cos\alpha\,\mathbf{T} + \sin\alpha\,\mathbf{B}).$$
It follows that there exists a constant unit vector $\mathbf{u} \in \mathbb{R}^3$ such that
$$\mathbf{u} = \cos\alpha\,\mathbf{T} + \sin\alpha\,\mathbf{B}.$$

10.5 Curves of Constant Slope

In particular, $\mathbf{u} \cdot \mathbf{T}$ is constant. ∎

For example, a helix has constant slope with respect to $(0,0,1)$ (see exercise 6). Moreover, it is easy to prove the following fact (see exercise 7).

Lemma 10.16. *Any plane curve has constant zero slope with respect to $(0,0,1)$.*

Next, we show that any unit-speed plane curve gives rise to a constant slope space curve in the same way that a circle gives rise to a helix.

Lemma 10.17. *Let $\beta\colon (a,b) \longrightarrow \mathbb{R}^2$ be a unit-speed plane curve, and write $\beta(s) = \bigl(b_1(s), b_2(s)\bigr)$. Define a space curve β by*

$$\gamma(t) = \bigl(b_1(t), b_2(t), t\cos\alpha\bigr),$$

where α is a constant angle. Then γ has constant speed $\sqrt{1 + (\cos\alpha)^2}$ and constant slope $\cos\alpha / \sqrt{1 + (\cos\alpha)^2}$ with respect to $(0,0,1)$.

Proof. It is easy to check that γ has constant speed $\sqrt{1 + (\cos\alpha)^2}$. Using this fact, we compute the cosine of the angle between $\gamma'(t)$ and $(0,0,1)$ as

$$\frac{\gamma'(t) \cdot (0,0,1)}{\|\gamma'(t)\|} = \frac{\cos\alpha}{\sqrt{1 + (\cos\alpha)^2}}. \quad \blacksquare$$

The reverse construction also works: a space curve of constant slope projects to a plane curve in a natural way.

Lemma 10.18. *Let $\gamma\colon (a,b) \longrightarrow \mathbb{R}^3$ be a unit-speed curve that has constant slope $\cos\alpha$ with respect to a unit vector $\mathbf{u} \in \mathbb{R}^3$. Assume that $\cos\alpha \neq 1$. Let β be the projection of γ on a plane perpendicular to $\mathbf{u}$; that is,*

$$\beta(s) = \gamma(s) - \bigl(\gamma(s) \cdot \mathbf{u}\bigr)\mathbf{u}$$

for $a < s < b$. Then β has constant speed $|\sin\alpha|$; furthermore, the curvature of γ and the signed curvature of β are related by

(10.18) $$\kappa_2[\beta] = \frac{\pm\kappa[\gamma]}{(\sin\alpha)^2}.$$

Proof. Let v denote the speed of β, and let $\mathbf{T}_\beta$ and $\mathbf{T}_\gamma$ denote the unit tangent vectors of β and γ. We have

$$v\mathbf{T}_\beta = \beta'(s) = \mathbf{T}_\gamma - (\mathbf{T}_\gamma \cdot \mathbf{u})\mathbf{u} = \mathbf{T}_\gamma - (\cos\alpha)\mathbf{u}.$$

Thus
$$1 = \|\mathbf{T}_\gamma\|^2 = v^2 + (\cos\alpha)^2,$$
proving that $v = |\sin\alpha|$. Also, we have $J\mathbf{T}_\beta = \pm\mathbf{N}_\gamma$. Hence
$$\kappa[\gamma]\mathbf{N}_\gamma = \mathbf{T}'_\gamma = v\mathbf{T}'_\beta = v^2\kappa2[\beta]J\mathbf{T}_\beta = \pm(\sin\alpha)^2\kappa2[\beta]\mathbf{N}_\gamma.$$
Since $\sin\alpha \neq 0$, we get (10.18). ∎

To find interesting curves of constant slope other than a helix or a plane curve, we look for curves of constant slope that lie on some sphere. First, we determine what relations must exist between the curvature and torsion of such curves.

Lemma 10.19. *Let $\beta\colon (a,b) \longrightarrow \mathbb{R}^3$ be a unit-speed curve that has constant slope $\cos\alpha$ with respect to a unit vector $\mathbf{u} \in \mathbb{R}^3$, where $0 < \alpha < \pi/2$. Assume also that β lies on a sphere of radius $c > 0$.*

(i) *The curvature and torsion of β are given by*

(10.19) $$\left(\frac{1}{\kappa[\beta](s)}\right)^2 = c^2 - s^2(\cot\alpha)^2 \quad\text{and}\quad \left(\frac{1}{\tau[\beta](s)}\right)^2 = c^2(\tan\alpha)^2 - s^2.$$

(ii) *Let γ be the projection of β on a plane perpendicular to $\mathbf{u}$. Then the signed curvature of γ satisfies*

(10.20) $$\left(\frac{1}{\kappa2[\gamma](s_1)}\right)^2 = (\sin\alpha)^4 c^2 - (\cos\alpha)^2 s_1^2,$$

where $s_1 = (\sin\alpha)s$ is the arc length function of γ.

(iii) *Let $c = a+2b$ and $\cos\alpha = a/(a+2b)$. Then the natural equation (10.20) of γ is the same as that of the epicycloid $t \longmapsto$ **epicycloid**$[a,b](t)$ defined on page 151.*

Proof. Since $\tau = \kappa\cot\alpha$, from (10.9) we get
$$\left(\frac{\kappa'}{\kappa^2}\right)^2 = \tau^2\left(c^2 - \frac{1}{\kappa^2}\right) = (\kappa^2 c^2 - 1)(\cot\alpha)^2,$$

or

(10.21) $$\frac{\kappa'}{\kappa^2\sqrt{\kappa^2 c^2 - 1}} = \pm\cot\alpha.$$

Integrating (10.21), we obtain

(10.22) $$\frac{1}{\kappa}\sqrt{\kappa^2 c^2 - 1} = \pm s\cot\alpha.$$

10.5 Curves of Constant Slope

When we solve (10.22) for κ, we obtain (10.19), proving (i). Then (ii) follows from (10.18) and (i).

For (iii) we first compute

$$(\sin \alpha)^4 = \left(1 - \frac{a^2}{(a+2b)^2}\right)^2 = \frac{16b^2(a+b)^2}{(a+2b)^4}.$$

Hence (10.20) becomes

$$(10.23) \quad \left(\frac{1}{\kappa 2[\gamma](s_1)}\right)^2 = \frac{16b^2(a+b)^2}{(a+2b)^2} - \frac{a^2 s_1^2}{(a+2b)^2}.$$

Then (10.23) is the same as (6.13) with s replaced by s_1. ∎

Lemma 10.19 suggests the definition of a new kind of space curve.

Definition. *Write* **epicycloid**$[a, b](t) = (a_1(t), a_2(t))$. *Then the* **spherical helix** *of parameters a and b is defined by*

$$\textbf{sphericalhelix}[a, b](t) = \left(a_1(t), a_2(t), 2\sqrt{ab+b^2} \cos\left(\frac{at}{2b}\right)\right).$$

From Lemma 10.19 we have immediately

Lemma 10.20. *The space curve* **sphericalhelix**$[a, b]$ *has constant slope with respect to $(0, 0, 1)$ and lies on a sphere of radius $a + 2b$ centered at the origin.*

For example, on the unit sphere $S^2(1)$ the curves of constant slope that project to a cardioid and a nephroid are **sphericalhelix**$[1/3, 1/3]$ and **sphericalhelix**$[1/2, 1/4]$.

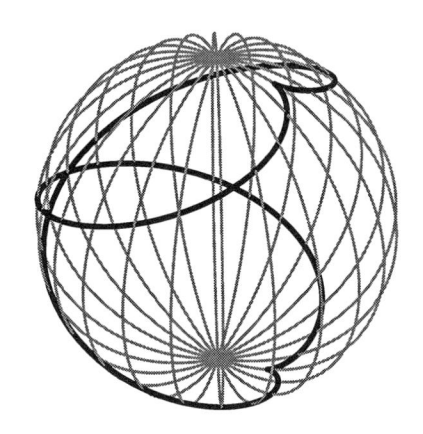

Spherical cardioid

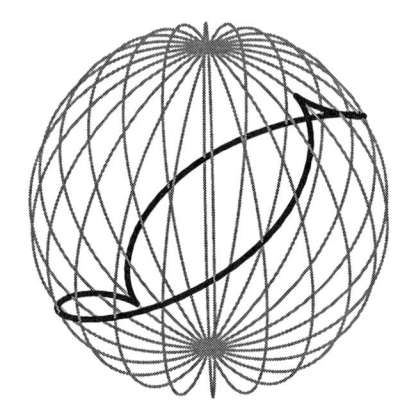

Spherical nephroid

The *Mathematica* definition of a spherical helix is

```
sphericalhelix[a_,b_][t_]:=
    {(a + b)Cos[t] - b Cos[(a + b)t/b],
     (a + b)Sin[t] - b Sin[(a + b)t/b],
     2Sqrt[a b + b^2]Cos[a t/(2b)]}
```

10.6 Loxodromes on Spheres

A **meridian** on a sphere is a great circle that passes through the north and south poles. A **parallel** on a sphere is a circle parallel to the equator. More generally:

Definition. *A* **spherical loxodrome** *or* **rhumb line** *is a curve on a sphere which meets each meridian of the sphere at the same angle, which we call the* **pitch** *of the loxodrome.*

A loxodrome has the north and south poles as asymptotic points. In the early years of terrestrial navigation, many sailors thought that a loxodrome was the same as a great circle, but Nuñez[1] distinguished the two.

A spherical loxodrome of pitch 0.2

To explain how to obtain the parametrization of a spherical loxodrome, we need an important map $\Upsilon \colon \mathbb{R}^2 \longrightarrow S^2(1)$ called the **stereographic map**. It is defined

---

[1] Pedro Nuñez Salaciense (1502-1578). The first great Portuguese mathematician. Professor at Coimbra.

by
$$\Upsilon(p_1, p_2) = \frac{(2p_1, 2p_2, p_1^2 + p_2^2 - 1)}{p_1^2 + p_2^2 + 1}.$$

It is easy to see that Υ is differentiable and that $\|\Upsilon(\mathbf{p})\| = 1$ for all $\mathbf{p} \in \mathbb{R}^2$.

Lemma 10.21. *The stereographic map Υ has the following properties*:

(i) Υ *maps circles in the plane onto circles on the sphere.*

(ii) Υ *maps straight lines in the plane onto circles on the sphere that pass through the north pole $(0, 0, 1)$.*

(iii) Υ *maps straight lines thorough the origin onto meridians.*

(iv) Υ *preserves angles.*

Proof. Any circle or line in the plane is given implicitly by an equation of the form

(10.24) $$a(x^2 + y^2) + b\,x + c\,y + d = 0.$$

Let $W = 1 + x^2 + y^2$, $X = 2x/W$, $Y = 2y/W$ and $Z = 1 - 2/W$. Under Υ the equation (10.24) is transformed into

(10.25) $$bX + cY + (a-d)Z + (d+a) = 0,$$

which is the equation of a plane in $\mathbb{R}^3$. This plane meets the sphere $S^2(1)$ in a great circle. This proves (i).

In the case of a straight line in the plane, $a = 0$ in (10.24). From (10.25), we see that $(0,0,1)$ is on the image curve, proving (ii). If in addition the straight line passes through the origin, then $a = d = 0$ in (10.24). Then from (10.25) we see that the plane containing the image curve also contains the Z-axis. Hence the curve is a meridian, proving (iii).

(iv) will be proved in Lemma 15.10, page 350. ∎

Lemma 10.22. *A spherical loxodrome is the image under a stereographic map of a logarithmic spiral.*

Proof. Lemma 2.2, page 41, implies that a logarithmic spiral meets every radial line from the origin at the same angle. The stereographic map Υ maps each of these radial lines into a meridian of the sphere. Since Υ preserves angles, it must map each logarithmic spiral onto a spherical loxodrome. ∎

The *Mathematica* version of the stereographic map Υ is

```
stereographicsphere[a_][u_,v_]:=
    {2a u/(u^2 + v^2 + 1),2a v/(u^2 + v^2 + 1),
    (a(u^2 + v^2 - 1))/(u^2 + v^2 + 1)}
```

The following command finds the composition of a surface and a curve.

```
curveonsurf[surf_,alpha_][t_]:=
    surf[alpha[t][[1]],alpha[t][[2]]]
```

To find a parametrization of a spherical loxodrome, we use

```
Simplify[curveonsurf[stereographicsphere[a],
                     logspiral[b,c]][t]]
```

obtaining

$$\left\{\frac{2\,a\,b\,E^{c\,t}\,Cos[t]}{1 + b^2\,E^{2\,c\,t}},\, \frac{2\,a\,b\,E^{c\,t}\,Sin[t]}{1 + b^2\,E^{2\,c\,t}},\, \frac{a\,(-1 + b^2\,E^{2\,c\,t})}{1 + b^2\,E^{2\,c\,t}}\right\}$$

This leads us to the *Mathematica* definition of a spherical loxodrome:

```
sphericalloxodrome[a_,b_,c_][t_]:=
    {2a b E^(c t)Cos[t]/(1 + b^2 E^(2c t)),
     2a b E^(c t)Sin[t]/(1 + b^2 E^(2c t)),
     a(-1 + b^2 E^(2c t))/(1 + b^2 E^(2c t))}
```

or

$$\text{sphericalloxodrome}[a,b,c](t) = \left(\frac{2a\,b\,e^{ct}\cos(t)}{1 + b^2 e^{2ct}}, \frac{2a\,b\,e^{ct}\sin(t)}{1 + b^2 e^{2ct}}, \frac{a(-1 + b^2 e^{2ct})}{1 + b^2 e^{2ct}}\right).$$

10.7 Exercises

1. The command

```
kappa23d[mag_,alpha_][t_]:=
    Append[alpha[t],mag kappa2[alpha][t]]
```

10.7 Exercises

constructs from a plane curve $\boldsymbol{\alpha}$ a space curve whose first two components are those of $\boldsymbol{\alpha}$ and whose third component is the signed curvature of $\boldsymbol{\alpha}$. For example:

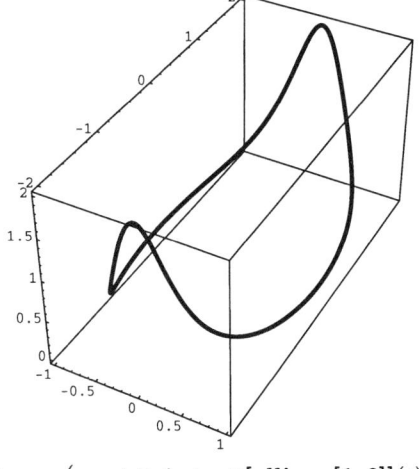

$$t \longmapsto \big(\cos t, 2\sin t, \kappa 2[\text{ellipse}[1,2]](t)\big)$$

Use **ParametricPlot3D** to plot **kappa23d** applied to the following curves: ellipse, figure eight, cardioid, astroid, cycloid.

2. Another twisted generalization of the plane curve $t \longmapsto \boldsymbol{\alpha}(t)$ is the space curve $t \longmapsto \mathbf{writhe}[\boldsymbol{\alpha}](t)$ given by

$$\mathbf{writhe}[\boldsymbol{\alpha}](t) = \big(a_1(t), a_2(t)\cos t, a_2(t)\sin t\big)$$

where $\boldsymbol{\alpha} = (a_1, a_2)$. The *Mathematica* version is

```
writhe[alpha_][t_]:= {alpha[t][[1]],
      Cos[t] alpha[t][[2]],Sin[t] alpha[t][[2]]}
```

Compute the curvature and torsion of **writhe[circle[a]]**. Use *Mathematica* to show that $t \longmapsto \mathbf{writhe}[\text{circle}[a]](t)$ and $t \longmapsto \mathbf{viviani}[a/2](2t+\pi)$ have the same curvature and torsion. Conclude that there is a Euclidean motion that takes one curve into the other. Then plot the curves.

3. Compute the curvature and torsion of $t \longmapsto \mathbf{writhe}[\text{lemniscate}[1]](t)$ and plot the curve.

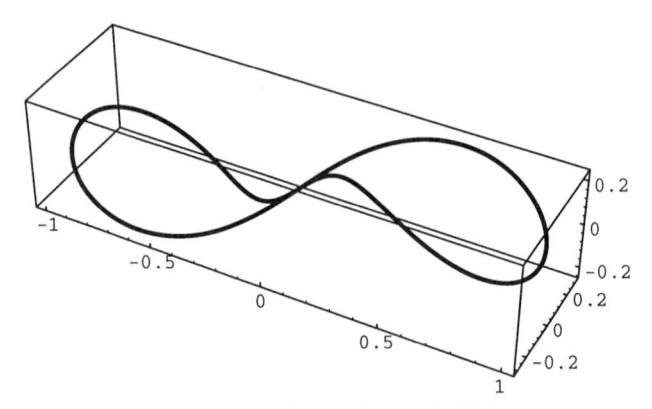

$t \longmapsto$ **writhe[lemniscate[1]]**(t)

4. Fill in the details of the proof of Theorem 10.3.

5. Find and plot the evolute of Viviani's curve.

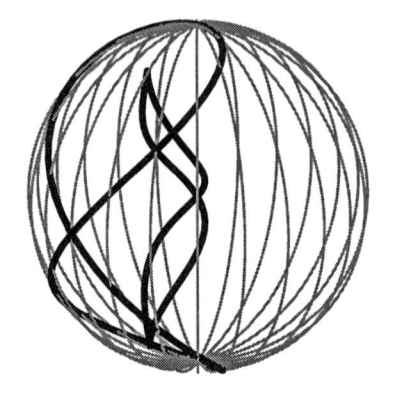

Viviani's curve and its evolute

6. Show that a helix has constant slope with respect to $(0,0,1)$.

7. Show that any plane curve has constant zero slope with respect to $(0,0,1)$.

8. Let $\theta\colon (a,b) \longrightarrow \mathbb{R}$ be an arbitrary differentiable function and define $\alpha\colon (a,b) \longrightarrow \mathbb{R}^3$ by

$$\alpha(t) = \left(\int \cos\theta(t)dt, \int \sin\theta(t)dt, ct \right),$$

10.8 Animation of Curves on a Sphere

where c is a constant. Compute the curvature and torsion of α and show that α has constant slope.

9. Use *Mathematica* to compute the curvature and torsion of **sphericalhelix[a, 2b]**. Check that the ratio of the curvature to torsion is constant and that **sphericalhelix[a, 2b]** is a curve on a sphere. Plot several spherical helices.

10. The plane curve **teardrop[a,b,c,d,n]** is defined in *Mathematica* by

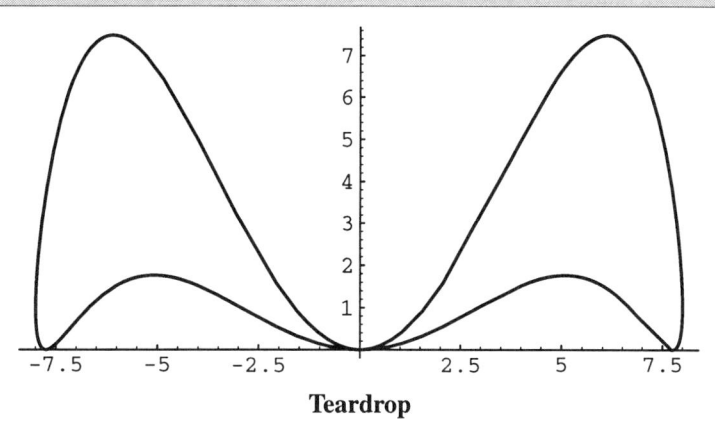

```
teardrop[a_,b_,c_,d_,n_][t_]:=
    {a Sin[t],d (b Sin[t] + c Sin[2t])^n}
```

Teardrop

Find the order of contact between **teardrop[8,1,2,1,n]** and the horizontal axis.

10.8 Animation of Curves on a Sphere

We need two packages; they can be read into a *Mathematica* session with the commands

```
Needs["CAS`"]
Needs["Graphics`Animation`"];
```

In order to spin a curve on a sphere we need to show the sphere in such a way that it does not overshadow the curve. For this purpose we define, but do not show, a few meridians with the command

```
meridians=ParametricPlot3D[Evaluate[Table[
Append[sphere[1][u,v],{AbsoluteThickness[1],
GrayLevel[0.5]}],
{u,0,2Pi,Pi/12}]],{v,-Pi/2,Pi/2},
DisplayFunction->Identity];
```

Next, we enter the command

```
loxredplot=ParametricPlot3D[Evaluate[
Append[sphericalloxodrome[1,1,0.2][t],
{RGBColor[1,0,0],AbsoluteThickness[2]}]],
{t,-20,20},PlotPoints->200,
DisplayFunction->Identity];
```

Then to generate a spinning red loxodrome, we use

```
loxshow=Show[{loxredplot,meridians},
Boxed->False,Axes->None,
DisplayFunction->Identity];
SpinShow[loxshow,Frames->30];
```

Similarly, a spinning spherical cardioid can be generated with the commands

```
sphericalcardioidplot=
ParametricPlot3D[
Append[sphericalhelix[1/3,1/3][t],
{RGBColor[1,0,0],AbsoluteThickness[2]}]//Evaluate,
{t,0,4Pi},PlotPoints->200,DisplayFunction->Identity];
```

and

```
sphericalcardioidshow=
Show[{sphericalcardioidplot,meridians},
Boxed->False,Axes->None,DisplayFunction->Identity];
SpinShow[sphericalcardioidshow];
```

11

CALCULUS ON EUCLIDEAN SPACE

More than the algebraic operations on $\mathbb{R}^n$ are needed for differential geometry. We need to know how to differentiate various geometric objects, and we need to know the relation between differentiation and the algebraic operations on $\mathbb{R}^n$. This chapter establishes the theoretical foundations of the theory of differentiation in $\mathbb{R}^n$ that we shall need for later chapters.

In Section 11.1 we define the notion of *tangent vector* to $\mathbb{R}^n$. The interpretation of a tangent vector to $\mathbb{R}^n$ as a directional derivative is described in Section 11.2. Tangent maps corresponding to differentiable maps between Euclidean spaces $\mathbb{R}^n$ and $\mathbb{R}^m$ are discussed in Section 11.3.

It is important to understand how tangent vectors vary from point to point, so we begin the study of *vector fields* on $\mathbb{R}^n$ in Section 11.4. Derivatives of vector fields on $\mathbb{R}^n$ with respect to tangent vectors are considered in Section 11.5. In Section 11.5 we also prove some elementary facts about the linear map on vector fields induced by a diffeomorphism. The classical notions of gradient, divergence and Laplacian, as well as their *Mathematica* implementations, are defined in Sections 11.4 and 11.5. Finally, in Section 11.6 we return to our study of curves, using tangent vectors to provide a different perspective.

11.1 Tangent Vectors to $\mathbb{R}^n$

In Section 1.1 we defined the notion of vector **v** in $\mathbb{R}^n$. So far we have considered vectors in $\mathbb{R}^n$ to be points in $\mathbb{R}^n$; however, a vector **v** can also be viewed as an arrow which starts at the origin and ends at the point **v**. From this point of view we should also study arrows that start at points other than the origin.

Definition. *A* **tangent vector** $\mathbf{v}_\mathbf{p}$ *to Euclidean space $\mathbb{R}^n$ consists of a pair of points $\mathbf{v}, \mathbf{p} \in \mathbb{R}^n$; **v** is called the* **vector part** *and **p** is called the* **point of application** *of $\mathbf{v}_\mathbf{p}$.*

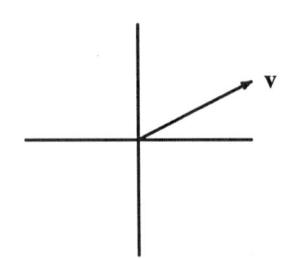

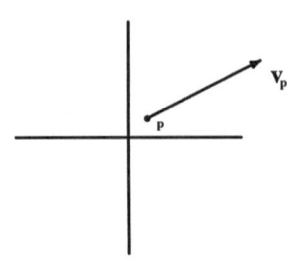

Notice that $\mathbf{v}_\mathbf{p}$ can be viewed as the arrow from **p** to $\mathbf{p} + \mathbf{v}$. Furthermore, two tangent vectors $\mathbf{v}_\mathbf{p}$ and $\mathbf{w}_\mathbf{q}$ are equal if and only if their vector parts and points of application coincide: $\mathbf{v} = \mathbf{w}$ and $\mathbf{p} = \mathbf{q}$.

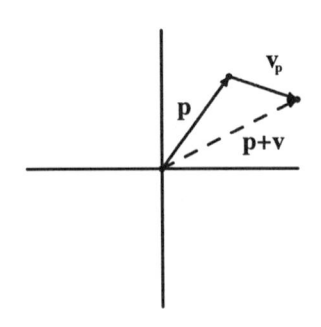

Definition. *Let $\mathbf{p} \in \mathbb{R}^n$. The* **tangent space** *of $\mathbb{R}^n$ at **p** is the set*

$$\mathbb{R}^n_\mathbf{p} = \{\, \mathbf{v}_\mathbf{p} \mid \mathbf{v} \in \mathbb{R}^n \,\}.$$

11.1 Tangent Vectors to $\mathbb{R}^n$

The tangent space $\mathbb{R}_p^n$ is a carbon copy of the vector space $\mathbb{R}^n$; in fact, there is a canonical isomorphism between $\mathbb{R}_p^n$ and $\mathbb{R}^n$ given by $\mathbf{v}_p \longmapsto \mathbf{v}$. This canonical isomorphism allows us to turn $\mathbb{R}_p^n$ into a vector space; that is, we can add elements in $\mathbb{R}_p^n$ and multiply by scalars:

$$\mathbf{v}_p + \mathbf{w}_p = (\mathbf{v} + \mathbf{w})_p,$$
$$\lambda \mathbf{v}_p = (\lambda \mathbf{v})_p.$$

Thus $\mathbb{R}^n$ has a tangent space $\mathbb{R}_p^n$ attached to each of its points, and the tangent space looks like $\mathbb{R}^n$ itself. For this reason, it is possible to transfer the dot product of $\mathbb{R}^n$ to each of its tangent spaces. We simply put

$$\mathbf{v}_p \cdot \mathbf{w}_p = \mathbf{v} \cdot \mathbf{w}$$

for $\mathbf{v}_p, \mathbf{w}_p \in \mathbb{R}_p^n$.

It is clear that there is a Cauchy-Schwarz inequality for tangent vectors, namely,

$$|\mathbf{v}_p \cdot \mathbf{w}_p| \leq \|\mathbf{v}_p\| \, \|\mathbf{w}_p\|$$

for $\mathbf{v}_p, \mathbf{w}_p \in \mathbb{R}_p^n$. Angles between tangent vectors are defined in exactly the same way as angles between ordinary vectors (see page 3).

Finally, let us consider tangent vectors to $\mathbb{R}^2$ and $\mathbb{R}^3$. The complex structure of $\mathbb{R}^2$ that we defined on page 4 can be naturally extended to tangent vectors by putting

$$J\mathbf{v}_p = (J\mathbf{v})_p$$

for a tangent vector $\mathbf{v}_p$ to $\mathbb{R}^2$. It is easy to see that $J\mathbf{v}_p$ is also a tangent vector to $\mathbb{R}^2$ and that $J^2 = -I$ on tangent vectors. Furthermore, $J\mathbf{v}_p \cdot J\mathbf{w}_p = \mathbf{v}_p \cdot \mathbf{w}_p$ for $\mathbf{v}_p, \mathbf{w}_p \in \mathbb{R}_p^2$.

Similarly, on $\mathbb{R}^3$ we extend the vector cross product to tangent vectors via

$$\mathbf{v}_p \times \mathbf{w}_p = (\mathbf{v} \times \mathbf{w})_p,$$

for $\mathbf{v}_p, \mathbf{w}_p \in \mathbb{R}_p^3$. The usual identities hold for vector cross products of tangent vectors. For example,

$$\|\mathbf{v}_p \times \mathbf{w}_p\|^2 = \|\mathbf{v}_p\|^2 \|\mathbf{w}_p\|^2 - (\mathbf{v}_p \cdot \mathbf{w}_p)^2.$$

11.2 Tangent Vectors as Directional Derivatives

Throughout this section, and from now on, the words "f is a differentiable function" will mean that f has derivatives of all orders. (For details on why this is a reasonable definition see Chapter 23.) Frequently, it is important to know how a real-valued function $f\colon \mathbb{R}^n \longrightarrow \mathbb{R}$ varies in different directions. For example, the partial derivative $\partial f/\partial u_i$ measures how much f varies in the u_i direction, for $i = 1, \ldots, n$. However, it is possible to measure the variation of f in other directions as well. To measure the variation of f at $\mathbf{p} \in \mathbb{R}^n$ along the straight line $t \longmapsto \mathbf{p}+t\mathbf{v}$, we need the tangent vector $\mathbf{v}_\mathbf{p}$. Therefore, we shall let $\mathbf{v}_\mathbf{p}$ operate on functions.

Definition. *Let $f\colon \mathbb{R}^n \longrightarrow \mathbb{R}$ be a differentiable function, and let $\mathbf{v}_\mathbf{p}$ be a tangent vector to $\mathbb{R}^n$ at $\mathbf{p} \in \mathbb{R}^n$. We put*

$$(11.1) \qquad \mathbf{v}_\mathbf{p}[f] = \frac{d}{dt}\bigl(f(\mathbf{p}+t\mathbf{v})\bigr)\bigg|_{t=0}.$$

In elementary calculus the definition (11.1) goes under the name of **directional derivative in the $\mathbf{v}_\mathbf{p}$ direction**, but we prefer to consider (11.1) as defining how a tangent vector can operate on functions. A more explicit way to write (11.1) is

$$\mathbf{v}_\mathbf{p}[f] = \lim_{t \to 0} \frac{f(\mathbf{p}+t\mathbf{v}) - f(\mathbf{p})}{t}.$$

For example, let $f(x,y) = \sin(xy)$, $\mathbf{p} = (1,2)$ and $\mathbf{v} = (2,3)$. Then

$$f(\mathbf{p}+t\mathbf{v}) = f\bigl((1,2)+t(2,3)\bigr) = f(2t+1, 3t+2) = \sin\bigl((2t+1)(3t+2)\bigr) = \sin\bigl(6t^2+7t+2\bigr),$$

so that

$$\frac{d}{dt}\bigl(f(\mathbf{p}+t\mathbf{v})\bigr) = (12t+7)\cos\bigl(6t^2+7t+2\bigr).$$

Thus

$$\mathbf{v}_\mathbf{p}[f] = (12t+7)\cos\bigl(6t^2+7t+2\bigr)\bigg|_{t=0} = 7\cos(2).$$

Although it is possible to compute $\mathbf{v}_\mathbf{p}[f]$ directly from the definition, it is usually easier to use a general formula, which we now derive.

11.2 Tangent Vectors as Directional Derivatives

Lemma 11.1. *Let $\mathbf{v_p} = (v_1,\ldots,v_n)_\mathbf{p}$ be a tangent vector to $\mathbb{R}^n$, and let $f\colon \mathbb{R}^n \longrightarrow \mathbb{R}$ be a differentiable function. Then*

$$(11.2) \qquad \mathbf{v_p}[f] = \sum_{j=1}^{n} v_j \frac{\partial f}{\partial u_j}(\mathbf{p}).$$

Proof. Let $\mathbf{p} = (p_1,\ldots,p_n)$, so that $\mathbf{p} + t\mathbf{v} = (p_1 + tv_1,\ldots,p_n + tv_n)$. The chain rule implies that

$$\begin{aligned}
\frac{d}{dt}\bigl(f(\mathbf{p}+t\mathbf{v})\bigr) &= \frac{d}{dt}\bigl(f(p_1+tv_1,\ldots,p_n+tv_n)\bigr) \\
&= \sum_{j=1}^{n} \frac{\partial f}{\partial u_j}(p_1+tv_1,\ldots,p_n+tv_n)\frac{d(p_j+tv_j)}{dt} \\
&= \sum_{j=1}^{n} \frac{\partial f}{\partial u_j}(p_1+tv_1,\ldots,p_n+tv_n) v_j.
\end{aligned}$$

When we put $t = 0$ in this equation, we get (11.2). ∎

Next, we list some algebraic properties of the directional derivative.

Lemma 11.2. *Let $f, g\colon \mathbb{R}^n \longrightarrow \mathbb{R}$ be differentiable functions, $\mathbf{v_p}, \mathbf{w_p}$ tangent vectors to $\mathbb{R}^n$ at $\mathbf{p} \in \mathbb{R}^n$ and a, b real numbers. Then*

$$(11.3) \qquad (a\mathbf{v_p} + b\mathbf{w_p})[f] = a\mathbf{v_p}[f] + b\mathbf{w_p}[f],$$

$$(11.4) \qquad \mathbf{v_p}[af + bg] = a\mathbf{v_p}[f] + b\mathbf{v_p}[g],$$

$$(11.5) \qquad \mathbf{v_p}[f\,g] = f(\mathbf{p})\mathbf{v_p}[g] + g(\mathbf{p})\mathbf{v_p}[f].$$

Proof. For example, we prove (11.5):

$$\begin{aligned}
\mathbf{v_p}[f\,g] &= \frac{d}{dt}\bigl((f\,g)(\mathbf{p}+t\mathbf{v})\bigr)\Big|_{t=0} \\
&= \left\{ f(\mathbf{p}+t\mathbf{v})\frac{d}{dt}\bigl(g(\mathbf{p}+t\mathbf{v})\bigr) + g(\mathbf{p}+t\mathbf{v})\frac{d}{dt}\bigl(f(\mathbf{p}+t\mathbf{v})\bigr) \right\}\Big|_{t=0} \\
&= f(\mathbf{p})\mathbf{v_p}[g] + g(\mathbf{p})\mathbf{v_p}[f]. \quad\blacksquare
\end{aligned}$$

There is a chain rule for tangent vectors:

Lemma 11.3. (The Chain Rule) *Let* $g_1, \ldots, g_k \colon \mathbb{R}^n \longrightarrow \mathbb{R}$ *and* $F \colon \mathbb{R}^k \longrightarrow \mathbb{R}$ *be differentiable functions and let* $\mathbf{v}_\mathbf{p} \in \mathbb{R}^n_\mathbf{p}$ *where* $\mathbf{p} \in \mathbb{R}^n$. *Write* $f = F(g_1, \ldots, g_k)$. *Then*

$$\mathbf{v}_\mathbf{p}[f] = \sum_{j=1}^{k} \frac{\partial F}{\partial u_j}(g_1(\mathbf{p}), \ldots, g_k(\mathbf{p})) \mathbf{v}_\mathbf{p}[g_j].$$

Proof. Let $\mathbf{v}_\mathbf{p} = (v_1, \ldots, v_n)_\mathbf{p}$. We use the ordinary chain rule to compute

$$\mathbf{v}_\mathbf{p}[f] = \sum_{i=1}^{n} v_i \frac{\partial f}{\partial u_i}(\mathbf{p}) = \sum_{i=1}^{n} \sum_{j=1}^{k} v_i \frac{\partial F}{\partial u_j}(g_1(\mathbf{p}), \ldots, g_k(\mathbf{p})) \frac{\partial g_j}{\partial u_i}(\mathbf{p})$$

$$= \sum_{j=1}^{k} \frac{\partial F}{\partial u_j}(g_1(\mathbf{p}), \ldots, g_k(\mathbf{p})) \mathbf{v}_\mathbf{p}[g_j]. \blacksquare$$

11.3 Tangent Maps

We shall need to deal with functions $F \colon \mathbb{R}^n \longrightarrow \mathbb{R}^m$ other than linear maps. Note that for any function $F \colon \mathbb{R}^n \longrightarrow \mathbb{R}^m$ we can write

(11.6) $$F(\mathbf{p}) = (f_1(\mathbf{p}), \ldots, f_m(\mathbf{p})),$$

where $f_j \colon \mathbb{R}^n \longrightarrow \mathbb{R}$ for $j = 1, \ldots, m$. Recall that a subset $\mathcal{U}$ of $\mathbb{R}^n$ is said to be **open**, provided that for all $\mathbf{p} \in \mathcal{U}$ there is an $\varepsilon > 0$ such that the ball

$$\{\mathbf{q} \in \mathbb{R}^m \mid \|\mathbf{q} - \mathbf{p}\| < \varepsilon\}$$

is entirely contained in $\mathcal{U}$. Clearly, (11.6) also holds for a function $F \colon \mathcal{U} \longrightarrow \mathbb{R}^m$. It will be convenient to use the notation

$$F = (f_1, \ldots, f_m)$$

as a shorthand for (11.6).

Definition. *Let* $\mathcal{U} \subset \mathbb{R}^n$ *be open. A function* $F \colon \mathcal{U} \longrightarrow \mathbb{R}^m$ *is said to be* **differentiable** *provided that each* f_j *is differentiable (that is, the partial derivatives of all orders of each* f_j *exist). More generally, if* A *is any subset of* $\mathbb{R}^n$, *we say that* $F \colon A \longrightarrow \mathbb{R}^m$ *is differentiable provided we can find an open set* $\mathcal{U}$ *such that* $A \subset \mathcal{U}$ *and a differentiable map* $\widetilde{F} \colon \mathcal{U} \longrightarrow \mathbb{R}^m$ *such that the restriction of* $\widetilde{F}$ *is* F. *A* **diffeomorphism** F *between open subsets* $\mathcal{U}$ *and* $\mathcal{V}$ *of* $\mathbb{R}^n$ *is a differentiable map* $F \colon \mathcal{U} \longrightarrow \mathcal{V}$ *which has a differentiable inverse* $F^{-1} \colon \mathcal{V} \longrightarrow \mathcal{U}$.

11.3 Tangent Maps

It is easy to prove that the composition of differentiable functions is differentiable, that is:

Lemma 11.4. *Suppose*

$$\mathcal{U} \subset \mathbb{R}^\ell \xrightarrow{F} \mathcal{V} \subset \mathbb{R}^m \xrightarrow{G} \mathbb{R}^n$$

are differentiable maps. Then the composition $G \circ F : \mathcal{U} \longrightarrow \mathbb{R}^n$ *is differentiable.*

Any differentiable map $F: \mathcal{U} \longrightarrow \mathbb{R}^m$ gives rise in a natural way to a linear map between tangent spaces. To establish this fact, first note that if $F: \mathcal{U} \longrightarrow \mathbb{R}^m$ is differentiable, then so is $t \longmapsto F(\mathbf{p} + t\mathbf{v})$ for $\mathbf{p} + t\mathbf{v} \in \mathcal{U}$, because it is the composition of differentiable functions. Therefore, the following definition makes sense:

Definition. Let $F: \mathcal{U} \longrightarrow \mathbb{R}^m$ be a differentiable map, where $\mathcal{U}$ is an open subset of $\mathbb{R}^n$, and let $\mathbf{p} \in \mathcal{U}$. For each tangent vector $\mathbf{v}_\mathbf{p}$ to $\mathbb{R}^n$ let $F_{*\mathbf{p}}(\mathbf{v}_\mathbf{p})$ be the initial velocity of the curve $t \longmapsto F(\mathbf{p} + t\mathbf{v})$, that is,

$$F_{*\mathbf{p}}(\mathbf{v}_\mathbf{p}) = F(\mathbf{p} + t\mathbf{v})'(0).$$

Then $F_{*\mathbf{p}}: \mathbb{R}^n_\mathbf{p} \longrightarrow \mathbb{R}^m_{F(\mathbf{p})}$ is called the **tangent map** of F at $\mathbf{p}$. Sometimes we abbreviate $F_{*\mathbf{p}}$ to F_*.

For example, let $\mathbf{p} = (1, 2, 3)$, $\mathbf{v} = (1, 0, -1)$ and $F(x, y, z) = (y\,z, z\,x, x\,y)$. Then

$$F(\mathbf{p} + t\mathbf{v}) = \big(6 - 2t, 3 + 2t - t^2, 2(1 + t)\big),$$

so that

$$F_{*\mathbf{p}}(\mathbf{v}_\mathbf{p}) = \left(\frac{d}{dt}\big(6 - 2t, 3 + 2t - t^2, 2(1 + t)\big)\bigg|_{t=0}\right)_{F(\mathbf{p})} = (-2, 2, 2)_{(6,3,2)}.$$

The next lemma provides a useful way of computing tangent maps.

Lemma 11.5. *Let* $F: \mathcal{U} \longrightarrow \mathbb{R}^m$ *be a differentiable map, where* $\mathcal{U}$ *is an open subset of* $\mathbb{R}^n$, *and write* $F = (f_1, \ldots, f_m)$. *If* $\mathbf{p} \in \mathcal{U}$ *and* $\mathbf{v}_\mathbf{p}$ *is a tangent vector to* $\mathbb{R}^n$ *at* $\mathbf{p}$, *then*

(11.7) $$F_{*\mathbf{p}}(\mathbf{v}_\mathbf{p}) = \big(\mathbf{v}_\mathbf{p}[f_1], \ldots, \mathbf{v}_\mathbf{p}[f_m]\big)_{F(\mathbf{p})}.$$

Proof. We replace $\mathbf{p}$ by $\mathbf{p} + t\mathbf{v}$ in (11.6) and differentiate with respect to t:

(11.8) $$F(\mathbf{p} + t\mathbf{v})' = \big(f_1(\mathbf{p} + t\mathbf{v})', \ldots, f_m(\mathbf{p} + t\mathbf{v})'\big).$$

When we evaluate both sides of (11.8) at $t = 0$ and use (11.1), we obtain (11.7). ∎

Corollary 11.6. *If $F: \mathcal{U} \longrightarrow \mathbb{R}^m$ is a differentiable map, where $\mathcal{U}$ is an open subset of $\mathbb{R}^n$, then at each point $\mathbf{p} \in \mathcal{U}$ the tangent map*

$$F_{*\mathbf{p}}: \mathbb{R}^n_{\mathbf{p}} \longrightarrow \mathbb{R}^m_{F(\mathbf{p})}$$

is a linear transformation.

Proof. That $F_{*\mathbf{p}}$ is linear is an easy consequence of (11.7) and (11.3). ∎

The tangent map of a differentiable map F can be thought of as the *best linear approximation* to F at a given point $\mathbf{p}$. We have already seen examples of tangent maps; for example, the velocity of a curve is a tangent map in disguised form. To explain this, we let $1_{\mathbf{p}}$ be the tangent vector to $\mathbb{R}$ whose vector part is 1 and whose point of application is $\mathbf{p}$. Note that from the definitions it follows that

$$(11.9) \qquad 1_{\mathbf{p}}[f] = \frac{d}{dt}\left(f(\mathbf{p} + t1)\right)\bigg|_{t=0} = f'(\mathbf{p})$$

for any differentiable function $f: \mathbb{R} \longrightarrow \mathbb{R}$.

Lemma 11.7. *Let $\alpha: (a,b) \longrightarrow \mathbb{R}^n$ be a curve. Then for $\mathbf{p} \in (a,b)$ we have*

$$\alpha_*(1_{\mathbf{p}}) = \alpha'(\mathbf{p}).$$

Proof. If we write $\alpha = (a_1, \ldots, a_n)$, then from (11.9) and (11.7) we get

$$\alpha_*(1_{\mathbf{p}}) = \left(1_{\mathbf{p}}[a_1], \ldots, 1_{\mathbf{p}}[a_n]\right) = \left(a'_1(\mathbf{p}), \ldots, a'_n(\mathbf{p})\right) = \alpha'(\mathbf{p}). \quad \blacksquare$$

If $\alpha: (a,b) \longrightarrow \mathcal{U} \subseteq \mathbb{R}^n$ is a differentiable curve, and if $F: \mathcal{U} \longrightarrow \mathbb{R}^m$ is a differentiable map, then $F \circ \alpha: (a,b) \longrightarrow \mathbb{R}^m$ is a differentiable curve by Lemma 11.4. It is reasonable to suspect that F_* maps the velocity of α into the velocity of $F \circ \alpha$. We show that this is indeed the case.

Lemma 11.8. *Let $F: \mathcal{U} \longrightarrow \mathbb{R}^m$ be a differentiable map, where $\mathcal{U}$ is an open subset of $\mathbb{R}^n$, and let $\alpha: (a,b) \longrightarrow \mathcal{U}$ be a curve. Then*

$$(F \circ \alpha)'(t) = F_*\left(\alpha'(t)\right)$$

for $a < t < b$.

11.3 Tangent Maps

Proof. Write $F = (f_1, \ldots, f_m)$ and $\alpha = (a_1, \ldots, a_n)$. Lemmas 11.1, 11.5 and the chain rule imply that

$$F_*(\alpha'(t)) = (\alpha'(t)[f_1], \ldots, \alpha'(t)[f_m])$$

$$= \left(\sum_{j=1}^n a_j'(t) \frac{\partial f_1}{\partial u_j}, \ldots, \sum_{j=1}^n a_j'(t) \frac{\partial f_m}{\partial u_j} \right)_{\alpha(t)}$$

$$= ((f_1 \circ \alpha)'(t), \ldots, (f_m \circ \alpha)'(t))$$

$$= (F \circ \alpha)'(t). \blacksquare$$

Corollary 11.9. *Let $F: \mathcal{U} \longrightarrow \mathbb{R}^m$ be a differentiable map, where $\mathcal{U}$ is an open subset of $\mathbb{R}^n$, and let $\mathbf{p} \in \mathcal{U}$. Let $\mathbf{v}_\mathbf{p} \in \mathbb{R}^n_\mathbf{p}$ and let $\alpha: (a,b) \longrightarrow \mathcal{U}$ be a curve such that $\alpha'(0) = \mathbf{v}_\mathbf{p}$. Then*

$$F_{*\mathbf{p}}(\mathbf{v}_\mathbf{p}) = (F \circ \alpha)'(0).$$

We originally defined $F_{*\mathbf{p}}(\mathbf{v}_\mathbf{p})$ to be the velocity vector at 0 of the image under F of the straight line $t \longrightarrow \mathbf{p} + t\mathbf{v}$. Corollary 11.9 says that $F_{*\mathbf{p}}(\mathbf{v}_\mathbf{p})$ equals the velocity vector at 0 of the image under F of any curve α such that $\alpha(0) = \mathbf{p}$ and $\alpha'(0) = \mathbf{v}_\mathbf{p}$.

Corollary 11.10. *Let*

$$\mathcal{U} \subset \mathbb{R}^\ell \xrightarrow{F} \mathcal{V} \subset \mathbb{R}^m \xrightarrow{G} \mathbb{R}^n$$

be differentiable. Then

$$(G \circ F)_* = G_* \circ F_*.$$

Proof. Let $\mathbf{v}_\mathbf{p}$ be a tangent vector to $\mathbb{R}^\ell$ at $\mathbf{p} \in \mathbb{R}^\ell$. Then $t \longmapsto F(\mathbf{p} + t\mathbf{v})$ is a curve in $\mathbb{R}^m$ such that $F_{*\mathbf{p}}(\mathbf{v}_\mathbf{p}) = F(\mathbf{p} + t\mathbf{v})'(0)$. Hence

$$(G \circ F)_*(\mathbf{v}_\mathbf{p}) = ((G \circ F)(\mathbf{p} + t\mathbf{v}))'(0) = G(F(\mathbf{p} + t\mathbf{v}))'(0)$$

$$= G_*(F(\mathbf{p} + t\mathbf{v})'(0)) = G_*(F_{*\mathbf{p}}(\mathbf{v}_\mathbf{p})). \blacksquare$$

Recall that if $L: \mathbb{R}^n \longrightarrow \mathbb{R}^m$ is a linear map, then the matrix associated with L is the $m \times n$ matrix $A = (a_{ij})$ such that

$$L\mathbf{v} = A\mathbf{v}$$

for all $\mathbf{v} \in \mathbb{R}^n$. Let $\{\mathbf{e}_1, \ldots, \mathbf{e}_n\}$ be the standard basis of $\mathbb{R}^n$, that is, $\mathbf{e}_i$ is the n-tuple with 1 in the ith spot and zeros elsewhere; similarly, let $\{\mathbf{e}_1, \ldots, \mathbf{e}_m\}$ be the standard basis of $\mathbb{R}^m$. Then the formula that relates A to L is

$$L\mathbf{e}_i = \sum_{j=1}^{m} a_{ji} \mathbf{e}_j$$

for $i = 1, \ldots, n$.

If $F: \mathcal{U} \longrightarrow \mathbb{R}^m$ is a differentiable map, where $\mathcal{U}$ is an open subset of $\mathbb{R}^n$, then for each $\mathbf{p} \in \mathcal{U}$ the tangent map $F_{*\mathbf{p}}$ is a linear map between the vector spaces $\mathbb{R}^n_{\mathbf{p}}$ and $\mathbb{R}^m_{F(\mathbf{p})}$. Since $\mathbb{R}^n_{\mathbf{p}}$ is canonically isomorphic to $\mathbb{R}^n$ and $\mathbb{R}^m_{F(\mathbf{p})}$ is canonically isomorphic to $\mathbb{R}^m$, we can associate an $m \times n$ matrix B with $F_{*\mathbf{p}}$. This matrix (called the Jacobian matrix of F at $\mathbf{p}$) has the property that

$$F_{*\mathbf{p}} \mathbf{v}_{\mathbf{p}} = B \mathbf{v}_{\mathbf{p}}$$

for all $\mathbf{v}_{\mathbf{p}} \in \mathbb{R}^n_{\mathbf{p}}$. The precise definition is as follows.

Definition. *Let $F: \mathcal{U} \longrightarrow \mathbb{R}^m$ be a differentiable map, where $\mathcal{U}$ is an open subset of $\mathbb{R}^n$, and write $F = (f_1, \ldots, f_m)$. The **Jacobian**[1] matrix is the matrix-valued function $\mathcal{J}(F)$ of F given by*

$$\mathcal{J}(F)(\mathbf{p}) = \begin{pmatrix} \dfrac{\partial f_1}{\partial u_1}(\mathbf{p}) & \cdots & \dfrac{\partial f_1}{\partial u_n}(\mathbf{p}) \\ \vdots & \ddots & \vdots \\ \dfrac{\partial f_m}{\partial u_1}(\mathbf{p}) & \cdots & \dfrac{\partial f_m}{\partial u_n}(\mathbf{p}) \end{pmatrix}$$

for $\mathbf{p} \in \mathcal{U}$.

Recall that the **rank** of an $m \times n$ matrix A is the dimension of the space spanned by the columns of A. If this dimension equals n, then A is said to have **maximum rank**.

Lemma 11.11. *Let $F: \mathcal{U} \longrightarrow \mathbb{R}^m$ be a differentiable map, where $\mathcal{U}$ is an open subset of $\mathbb{R}^n$. Then each tangent map $F_{*\mathbf{p}}: \mathbb{R}^n_{\mathbf{p}} \longrightarrow \mathbb{R}^m_{F(\mathbf{p})}$ is injective if and only if $\mathcal{J}(F)(\mathbf{p})$ has maximum rank.*

---

[1] Carl Gustav Jacob Jacobi (1804-1851). Professor of Mathematics at Königsberg and Berlin. He is remembered for his work on dynamics and elliptic functions. His **Fundamenta Nova Theoria Functionum Ellipticarum**, in which elliptic function theory is based on four theta functions, was published in 1829.

11.4 Vector Fields on $\mathbb{R}^n$

Proof. Let $\mathbf{v}_\mathbf{p} \in \mathbb{R}^n_\mathbf{p}$ and write $\mathbf{v} = (v_1, \ldots, v_n)$; also put $F = (f_1, \ldots, f_m)$. Suppose that $F_{*\mathbf{p}}(\mathbf{v}_\mathbf{p}) = 0$ and $\mathbf{v} \neq 0$. Then Lemma 11.5 implies that

$$\mathbf{v}_\mathbf{p}[f_1] = \cdots = \mathbf{v}_\mathbf{p}[f_m] = 0.$$

Hence by Lemma 11.1 we have

$$\sum_{i=1}^n v_i \frac{\partial f_j}{\partial u_i}(\mathbf{p}) = 0$$

for $j = 1, \ldots, m$. It follows that

$$\sum_{i=1}^n v_i \left(\frac{\partial f_1}{\partial u_i}(\mathbf{p}), \ldots, \frac{\partial f_m}{\partial u_i}(\mathbf{p}) \right) = 0.$$

Therefore, the columns of the Jacobian matrix $\mathcal{J}(F)(\mathbf{p})$ are linearly dependent, and consequently $\mathcal{J}(F)(\mathbf{p})$ does not have maximum rank.

Conversely, if $\mathcal{J}(F)(\mathbf{p})$ does not have maximum rank, the steps in the above proof can be reversed to conclude that $F_{*\mathbf{p}}(\mathbf{v}_\mathbf{p}) = 0$ for some nonzero $\mathbf{v}_\mathbf{p} \in \mathbb{R}^n_\mathbf{p}$. ∎

11.4 Vector Fields on $\mathbb{R}^n$

It will be necessary to consider tangent vectors as they vary from point to point.

Definition. *A* **vector field** *V on an open subset $\mathcal{U}$ of $\mathbb{R}^n$ is a function that assigns to each $\mathbf{p} \in \mathcal{U}$ a tangent vector $V(\mathbf{p}) \in \mathbb{R}^n_\mathbf{p}$. If $f: \mathcal{U} \longrightarrow \mathbb{R}$ is differentiable, we let V act on f via*

$$V[f](\mathbf{p}) = V(\mathbf{p})[f].$$

The vector field V is said to be **differentiable**, *provided that $V[f]: \mathcal{U} \longrightarrow \mathbb{R}$ is differentiable whenever f is.*

It is important to distinguish between the ith coordinate of a point $\mathbf{p} \in \mathbb{R}^n$ and the function that assigns to $\mathbf{p}$ its ith coordinate.

Definition. *The* **natural coordinate functions** *of $\mathbb{R}^n$ are the functions $\{u_1, \ldots, u_n\}$ defined by*

$$u_i(\mathbf{p}) = p_i$$

for $\mathbf{p} = (p_1, \ldots, p_n)$.

In the special cases of $\mathbb{R}$, $\mathbb{R}^2$ and $\mathbb{R}^3$ we shall usually denote the natural coordinate functions by t, $\{u, v\}$ and $\{x, y, z\}$, respectively.

We shall need some standard vector fields on $\mathbb{R}^n$.

Definition. *The vector field U_j on $\mathbb{R}^n$ is defined by*

$$U_j(\mathbf{p}) = (0, \ldots, 0, 1, 0, \ldots, 0)_\mathbf{p},$$

*where 1 occurs in the jth position. We call $\{U_1, \ldots, U_n\}$ the **natural frame field** of $\mathbb{R}^n$.*

It is obvious from Lemma 11.1 that $U_j[f] = \partial f/\partial u_j$ for any differentiable function f. Therefore, we sometimes write $\partial/\partial u_j$ for U_j. Note that the restrictions of $U_1, \ldots, U_n$ to any open subset $\mathcal{U} \subset \mathbb{R}^n$ are vector fields on $\mathcal{U}$.

We note some algebraic properties of vector fields; the proofs are easy.

Lemma 11.12. *Let X and Y be vector fields on an open subset $\mathcal{U} \subset \mathbb{R}^n$, let $a \in \mathbb{R}$ and let $g: \mathbb{R}^n \longrightarrow \mathbb{R}$. Define $X + Y$, aX, gX and $X \cdot Y$ by*

$$(X+Y)[f] = X[f] + Y[f], \qquad (aX)[f] = aX[f],$$

$$(gX)[f] = gX[f], \qquad (X \cdot Y)(\mathbf{p}) = X(\mathbf{p}) \cdot Y(\mathbf{p}),$$

where $f: \mathcal{U} \longrightarrow \mathbb{R}$ is differentiable and $\mathbf{p} \in \mathcal{U}$. Then $X + Y$, aX, gX are vector fields on $\mathcal{U}$ and $X \cdot Y: \mathcal{U} \longrightarrow \mathbb{R}$ is a function. If X, Y and g are differentiable, so are the vector fields $X + Y$, aX, gX, and the function $X \cdot Y: \mathcal{U} \longrightarrow \mathbb{R}$.

It is easy to see that any vector field on an open subset of $\mathbb{R}^n$ can be expressed in terms of the U_j's.

Lemma 11.13. *If V is any vector field on an open subset $\mathcal{U} \subset \mathbb{R}^n$, there exist functions $v_j: \mathcal{U} \longrightarrow \mathbb{R}$ for $j = 1, \ldots, n$ such that*

(11.10) $$V = \sum_{j=1}^{n} v_j U_j.$$

Proof. Since $V(\mathbf{p}) \in \mathbb{R}^n_\mathbf{p}$, we can write $V(\mathbf{p}) = \bigl(v_1(\mathbf{p}), \ldots, v_n(\mathbf{p})\bigr)_\mathbf{p}$. In this way the functions $v_j: \mathcal{U} \longrightarrow \mathbb{R}$ are defined. On the other hand,

$$\begin{aligned}
\bigl(v_1(\mathbf{p}), \ldots, v_n(\mathbf{p})\bigr)_\mathbf{p} &= v_1(\mathbf{p})(1, 0, \ldots, 0)_\mathbf{p} + \cdots + v_n(\mathbf{p})(0, \ldots, 0, 1)_\mathbf{p} \\
&= \sum_{j=1}^{n} v_j(\mathbf{p}) U_j(\mathbf{p}).
\end{aligned}$$

11.4 Vector Fields on $\mathbb{R}^n$

Since this equation holds for arbitrary $\mathbf{p} \in \mathcal{U}$, we get (11.10). ∎

Equation (11.10) yields a useful criterion for the differentiability of a vector field.

Corollary 11.14. *A vector field V on an open subset $\mathcal{U} \subset \mathbb{R}^n$ is differentiable if and only if the functions $v_1, \ldots, v_n$ of V given in (11.10) are differentiable.*

There is an important way to construct vector fields from functions on $\mathbb{R}^n$.

Definition. *Let $g: \mathbb{R}^n \longrightarrow \mathbb{R}$ be a differentiable function. The **gradient** of g is the vector field $\operatorname{\mathbf{grad}} g$ given by*

$$\operatorname{\mathbf{grad}} g = \sum_{j=1}^{n} \frac{\partial g}{\partial u_j} U_j.$$

The operator **grad** is characterized by a useful formula.

Lemma 11.15. *The gradient of a function g is the unique differentiable vector field $\operatorname{\mathbf{grad}} g$ on $\mathbb{R}^n$ for which*

$$\operatorname{\mathbf{grad}} g \cdot V = V[g]$$

for any differentiable vector field V on $\mathbb{R}^n$.

Proof. Write $V = \sum_{j=1}^{n} v_j U_j$. Then

$$\operatorname{\mathbf{grad}} g \cdot V = \left(\sum_{j=1}^{n} U_j \frac{\partial g}{\partial u_j} \right) \cdot V = \sum_{j=1}^{n} v_j \frac{\partial g}{\partial u_j}$$

$$= \sum_{j=1}^{n} v_j U_j[g] = V[g].$$

Conversely, let Z be a vector field for which $Z \cdot V = V[g]$ for any vector field V on $\mathbb{R}^n$. We can write

$$Z = \sum_{i=1}^{n} f_i U_i,$$

where $f_i: \mathbb{R}^n \longrightarrow \mathbb{R}$ is differentiable for $i = 1, \ldots, n$. Then for $j = 1, \ldots, n$ we have

$$f_j = Z \cdot U_j = U_j[g] = \frac{\partial g}{\partial u_j},$$

so that $Z = \mathbf{grad}\, g$. ∎

The following **Mathematica** miniprogram defines the gradient of a function.

```
grad[expr_,coor_]:=
    Table[D[expr,coor[[i]]],{i,Length[coor]}]
```

For example, let us compute the gradient of $(x,y,z) \longmapsto (x\,y\,z)^3$:

```
grad[(x y z)^3,{x,y,z}]
         2 3 3     3 2 3     3 3 2
      {3 x y z , 3 x y z , 3 x y z }
```

Finally, we note some special operations on vector fields on $\mathbb{R}^2$ and $\mathbb{R}^3$.

Lemma 11.16. *For a vector field Y on an open subset $\mathcal{U} \subset \mathbb{R}^2$ we define JY by*

$$(JY)(\mathbf{p}) = J(Y(\mathbf{p}))$$

for $\mathbf{p} \in \mathcal{U}$. Then JY is a vector field on $\mathcal{U}$ which is differentiable when Y is differentiable.

Similarly, for vector fields Y, Z on an open subset $\mathcal{V} \subset \mathbb{R}^3$ we define $Y \times Z$ in a pointwise fashion, namely

$$(Y \times Z)(\mathbf{p}) = Y(\mathbf{p}) \times Z(\mathbf{p})$$

for $\mathbf{p} \in \mathcal{V}$. Then $Y \times Z$ is a vector field on $\mathcal{V}$ which is differentiable when Y and Z are differentiable.

11.5 Derivatives of Vector Fields on $\mathbb{R}^n$

In Section 11.2 we learned how to differentiate functions by means of tangent vectors. A key concept in differential geometry is the notion of how to differentiate a vector field. For vector fields on open subsets of $\mathbb{R}^n$, the definition is natural.

If W is a vector field on an open subset $\mathcal{U} \subset \mathbb{R}^n$, let $\widetilde{W}(\mathbf{q})$ be the element of $\mathbb{R}^n$ such that

$$\widetilde{W}(\mathbf{q})_\mathbf{q} = W(\mathbf{q}).$$

We restrict W to that portion of the straight line $t \longmapsto \mathbf{p} + t\mathbf{v}$ contained in $\mathcal{U}$. Then $t \longmapsto \widetilde{W}(\mathbf{p} + t\mathbf{v})$ is the curve in $\mathbb{R}^n$ such that

$$\widetilde{W}(\mathbf{p} + t\mathbf{v})_{\mathbf{p}+t\mathbf{v}} = W(\mathbf{p} + t\mathbf{v}).$$

11.5 Derivatives of Vector Fields on $\mathbb{R}^n$

Definition. *Let W be a differentiable vector field on an open subset $\mathcal{U} \subset \mathbb{R}^n$, and let $\mathbf{v}_\mathbf{p}$ be a tangent vector to $\mathbb{R}^n$ at $\mathbf{p} \in \mathcal{U}$. Then the **derivative** of W with respect to $\mathbf{v}_\mathbf{p}$ is the tangent vector $D_\mathbf{v} W \in \mathbb{R}^n_\mathbf{p}$ given by*

$$D_\mathbf{v} W = \widetilde{W}(\mathbf{p}+t\mathbf{v})'(0)_\mathbf{p} = \lim_{t \to 0} \left. \frac{\widetilde{W}(\mathbf{p}+t\mathbf{v}) - \widetilde{W}(\mathbf{p})}{t} \right|_\mathbf{p}.$$

Although it is possible to compute $D_\mathbf{v} W$ directly from the definition, there is an alternate formula (similar to equation (11.7)), which is usually more convenient.

Lemma 11.17. *Let W be a differentiable vector field on an open subset $\mathcal{U} \subset \mathbb{R}^n$ and write*

$$W = \sum_{i=1}^n w_i U_i.$$

Then for $\mathbf{p} \in \mathcal{U}$ and $\mathbf{v}_\mathbf{p} \in \mathbb{R}^n_\mathbf{p}$ we have

(11.11) $$D_\mathbf{v} W = \sum_{i=1}^n \mathbf{v}_\mathbf{p}[w_i] U_i(\mathbf{p}) = \left(\mathbf{v}_\mathbf{p}[w_1], \ldots, \mathbf{v}_\mathbf{p}[w_n] \right)_\mathbf{p}.$$

Proof. We have

$$W(\mathbf{p}+t\mathbf{v}) = \sum_{i=1}^n w_i(\mathbf{p}+t\mathbf{v}) U_i(\mathbf{p}+t\mathbf{v})$$
$$= \left(w_1(\mathbf{p}+t\mathbf{v}), \ldots, w_n(\mathbf{p}+t\mathbf{v}) \right)_{\mathbf{p}+t\mathbf{v}}.$$

Hence $\widetilde{W}(\mathbf{p}+t\mathbf{v}) = \left(w_1(\mathbf{p}+t\mathbf{v}), \ldots, w_n(\mathbf{p}+t\mathbf{v}) \right)$, so that

$$\widetilde{W}(\mathbf{p}+t\mathbf{v})' = \left(w_1(\mathbf{p}+t\mathbf{v})', \ldots, w_n(\mathbf{p}+t\mathbf{v})' \right).$$

But by the definition of tangent vector, we have

$$w_j(\mathbf{p}+t\mathbf{v})'(0) = \mathbf{v}_\mathbf{p}[w_j]$$

for $j = 1, \ldots, n$. Hence

(11.12) $$\widetilde{W}(\mathbf{p}+t\mathbf{v})'(0) = \left(\mathbf{v}_\mathbf{p}[w_1], \ldots, \mathbf{v}_\mathbf{p}[w_n] \right).$$

Then (11.12) implies (11.11). ∎

It is easy to compute the derivatives of the vector fields $U_1, \ldots, U_n$.

Corollary 11.18. *For any tangent vector $\mathbf{v}_p$ to $\mathbb{R}^n$ we have*

(11.13) $$D_v U_j = 0$$

for $j = 1, \ldots, n$.

Proof. If we use (11.10) to express U_j in terms of $U_1, \ldots, U_n$, then the functions $v_1, \ldots, v_n$ in (11.10) are all 0 except for one. In particular, they are all constants. Hence (11.11) implies (11.13). ∎

Next, we show that the operator D behaves naturally with respect to the operations on vector fields given in Lemma 11.12.

Lemma 11.19. *Let Y and Z be differentiable vector fields on an open subset $\mathcal{U} \subset \mathbb{R}^n$, and let $\mathbf{v}_p, \mathbf{w}_p$ be tangent vectors to $\mathbb{R}^n$ at $\mathbf{p} \in \mathcal{U}$. Then*

(11.14) $$D_{a\mathbf{v}+b\mathbf{w}} Y = a D_v Y + b D_w Y,$$

(11.15) $$D_v(aY + bZ) = a D_v Y + b D_v Z,$$

(11.16) $$D_v(fY) = \mathbf{v}_p[f] Y(\mathbf{p}) + f(\mathbf{p}) D_v Y,$$

(11.17) $$\mathbf{v}_p[Y \cdot Z] = D_v Y \cdot Z(\mathbf{p}) + Y(\mathbf{p}) \cdot D_v Z,$$

where a, b are real numbers and $f: \mathcal{U} \longrightarrow \mathbb{R}$ is a differentiable function.

Proof. We prove (11.17); the proofs of the other formulas are similar. Write

$$Y = \sum_{i=1}^n y_i U_i \quad \text{and} \quad Z = \sum_{i=1}^n z_i U_i.$$

From (11.4) and (11.5) it follows that

$$\mathbf{v}_p[Y \cdot Z] = \sum_{i=1}^n \mathbf{v}_p[y_i z_i]$$
$$= \sum_{i=1}^n \left\{ \mathbf{v}_p[y_i] z_i(\mathbf{p}) + y_i(\mathbf{p}) \mathbf{v}_p[z_i] \right\}$$
$$= \left\{ \sum_{i=1}^n \mathbf{v}_p[y_i] U_i(\mathbf{p}) \right\} \cdot \left\{ \sum_{i=1}^n z_i(\mathbf{p}) U_i(\mathbf{p}) \right\}$$
$$+ \left\{ \sum_{i=1}^n y_i(\mathbf{p}) U_i(\mathbf{p}) \right\} \cdot \left\{ \sum_{i=1}^n \mathbf{v}_p[z_i] U_i(\mathbf{p}) \right\}$$
$$= D_v Y \cdot Z(\mathbf{p}) + Y(\mathbf{p}) \cdot D_v Z. \quad \blacksquare$$

11.5 Derivatives of Vector Fields on $\mathbb{R}^n$

The operator D also behaves naturally with respect to the complex structure J of $\mathbb{R}^2$ and the vector cross product $\times$ of $\mathbb{R}^3$.

Lemma 11.20. *Let Y and Z be differentiable vector fields on an open subset $\mathcal{U} \subset \mathbb{R}^n$, where $n = 2$ or 3, and let $\mathbf{v}_p, \mathbf{w}_p$ be tangent vectors to $\mathbb{R}^n$. Then*

(11.18) $\qquad D_v(JY) = J\bigl(D_v(Y)\bigr) \qquad$ (for $n = 2$),

(11.19) $\qquad D_v(Y \times Z) = D_vY \times Z(\mathbf{p}) + Y(\mathbf{p}) \times D_vZ \qquad$ (for $n = 3$).

Proof. If Y is a vector field on $\mathbb{R}^2$, we can write $Y = y_1 U_1 + y_2 U_2$; then $JY = -y_2 U_1 + y_1 U_2$. It follows from (11.11) that

$$D_v(JY) = \bigl(-\mathbf{v}_p[y_2], \mathbf{v}_p[y_1]\bigr)_p = J\bigl(\mathbf{v}_p[y_1], \mathbf{v}_p[y_2]\bigr)_p = J(D_vY),$$

proving (11.18). The proof of (11.19) is similar. ∎

It is also possible to define the derivative of one vector field by another.

Lemma 11.21. *Let V and W be vector fields defined on an open subset $\mathcal{U} \subset \mathbb{R}^n$, and assume that W is differentiable. Define $D_V W$ by*

$$(D_V W)(\mathbf{p}) = D_{V(\mathbf{p})} W$$

for $\mathbf{p} \in \mathcal{U}$. Then $D_V W$ is a vector field on $\mathcal{U}$ which is differentiable if V and W are differentiable. Moreover, if

$$W = \sum_{i=1}^{n} w_i U_i,$$

then

(11.20) $$D_V W = \sum_{i=1}^{n} V[w_i] U_i.$$

Proof. Equation (11.20) is an immediate consequence of (11.11). ∎

Let $F: \mathcal{U} \longrightarrow \mathcal{V}$ be a diffeomorphism between open subsets $\mathcal{U}$ and $\mathcal{V}$ of $\mathbb{R}^n$, and let V be a differentiable vector field on $\mathcal{U}$. We want to define the image $F_*(V)$ of V under F; we must have consistency with (11.7). The following definition does the job:

Definition. Let $F: \mathcal{U} \longrightarrow \mathcal{V}$ be a diffeomorphism between open subsets $\mathcal{U}$ and $\mathcal{V}$ of $\mathbb{R}^n$ and let V be a differentiable vector field on $\mathcal{U}$. Write $F = (f_1, \ldots, f_n)$. Then $F_*(V)$ is the vector field on $\mathcal{V}$ given by

$$(11.21) \qquad F_*(V) = \sum_{i=1}^n \left(V[f_i] \circ F^{-1}\right) U_i.$$

It is easy to prove

Lemma 11.22. Suppose that $F: \mathcal{U} \longrightarrow \mathcal{V}$ is a diffeomorphism between open subsets $\mathcal{U}$ and $\mathcal{V}$ of $\mathbb{R}^n$. Let Y and Z be differentiable vector fields on an open subset $\mathcal{U}$. Then

$$(11.22) \qquad F_*(aY + bZ) = a F_*(Y) + b F_*(Z),$$

$$(11.23) \qquad F_*(fY) = (f \circ F^{-1}) F_*(Y),$$

$$(11.24) \qquad F_*(Y)[f \circ F^{-1}] = Y[f] \circ F^{-1},$$

where a, b are real numbers and $f: \mathcal{U} \longrightarrow \mathbb{R}$ is a differentiable function.

Next, we show that certain maps of $\mathbb{R}^n$ into itself map the derivatives of vector fields in a nice way.

Lemma 11.23. Suppose that $F: \mathbb{R}^n \longrightarrow \mathbb{R}^n$ is an affine transformation. Then

$$(11.25) \qquad \boldsymbol{D}_{F_*(V)}\bigl(F_*(W)\bigr) = F_*(\boldsymbol{D}_V W).$$

Proof. By assumption
$$F(\mathbf{p}) = A\mathbf{p} + \mathbf{q},$$

where $A: \mathbb{R}^n \longrightarrow \mathbb{R}^n$ is a linear transformation. Write $A = (a_{ij})$; then for $i = 1, \ldots, n$ we can write

$$(11.26) \qquad F_*(U_i) = \sum_{j=1}^n a_{ij} U_j,$$

where each a_{ij} is constant. From (11.22) and (11.23) we get

$$(11.27) \qquad \boldsymbol{D}_{F_*(V)} F_*(W) = \boldsymbol{D}_{F_*(V)} \left(F_*\left(\sum_{i=1}^n w_i U_i \right) \right)$$

$$= \boldsymbol{D}_{F_*(V)} \left(\sum_{i=1}^n (w_i \circ F^{-1}) F_*(U_i) \right)$$

11.5 Derivatives of Vector Fields on $\mathbb{R}^n$

$$= \sum_{i=1}^{n} \boldsymbol{D}_{F_*(V)} \left((w_i \circ F^{-1}) F_*(U_i) \right).$$

Then (11.26) implies that

(11.28) $\quad \boldsymbol{D}_{F_*(V)} \left((w_i \circ F^{-1}) F_*(U_i) \right) = \boldsymbol{D}_{F_*(V)} \left((w_i \circ F^{-1}) \sum_{j=1}^{n} a_{ij} U_j \right)$

$$= \sum_{j=1}^{n} F_*(V)[(w_i \circ F^{-1}) a_{ij}] U_j.$$

Since each a_{ij} is constant,

(11.29) $\quad F_*(V)[(w_i \circ F^{-1}) a_{ij}] = F_*(V)[w_i \circ F^{-1}] a_{ij}.$

From (11.27)—(11.29) we get

(11.30) $\quad \boldsymbol{D}_{F_*(V)} F_*(W) = \sum_{ij=1}^{n} F_*(V)[w_i \circ F^{-1}] a_{ij} U_j$

$$= \sum_{i=1}^{n} (V[w_i] \circ F^{-1}) F_*(U_i)$$

$$= \sum_{i=1}^{n} F_*(V[w_i] U_i).$$

Now (11.25) follows from (11.30) and (11.20). ∎

We conclude this section by defining some standard vector analysis functions.

Definition. *Let V be a differentiable vector field defined on an open subset $\mathcal{U} \subset \mathbb{R}^n$. The **divergence** of V is the function*

$$\operatorname{div} V = \sum_{i=1}^{n} (\boldsymbol{D}_{U_i} V \cdot U_i).$$

*If $f: \mathcal{U} \longrightarrow \mathbb{R}$ is differentiable, then the **Laplacian** of f is the function*

$$\Delta f = \operatorname{div}(\operatorname{grad} f).$$

Lemma 11.24. *If* $f: \mathcal{U} \longrightarrow \mathbb{R}$ *is a differentiable function, where* $\mathcal{U} \subset \mathbb{R}^n$ *is open, then*

$$\Delta f = \sum_{i=1}^{n} U_i^2 f = \sum_{i=1}^{n} \frac{\partial^2 f}{\partial u_i^2}.$$

Proof. We have

$$\Delta f = \operatorname{div}(\operatorname{grad} f) = \sum_{i=1}^{n} D_{U_i}(\operatorname{grad} f) \cdot U_i$$

$$= \sum_{i=1}^{n} U_i[\operatorname{grad} f \cdot U_i]$$

$$= \sum_{i=1}^{n} U_i[U_i[f]]. \blacksquare$$

Here are the *Mathematica* definitions of the divergence of a differentiable vector field and the Laplacian of a differentiable function.

```
divergence[ww_,coor_]:= Sum[D[ww,coor[[i]]],
                              {i,Length[coor]}]
```

```
laplacian[expr_,coor_]:=
    Sum[D[expr,{coor[[i]],2}],{i,Length[coor]}]
```

For example,

```
divergence[{x^a,y^b,z^c},{x,y,z}]
```

yields

$$a\, x^{-1+a} + b\, y^{-1+b} + c\, z^{-1+c}$$

and

```
laplacian[(x y z)^3,{x,y,z}]
```

gives

$$6 x\, y^3\, z^3 + 6 x^3\, y\, z^3 + 6 x^3\, y^3\, z$$

11.6 Curves Revisited

In the study of curves and surfaces it is frequently useful to make no distinction between a tangent vector v_p and its vector part. For example, on page 6 we defined the velocity $\alpha'(t)$ of a curve α to be a vector in $\mathbb{R}^n$. More properly, it should be a tangent vector at the point $\alpha(t)$. (In this section we assume for simplicity that all curves are differentiable.)

Definition. (*Revised*) The **velocity** *of a curve* α *is the function*

$$t \longmapsto \alpha'(t)_{\alpha(t)},$$

and the **acceleration** *of* α *is the function*

$$t \longmapsto \alpha''(t)_{\alpha(t)}.$$

Since $\alpha'(t)_{\alpha(t)}$ is a tangent vector, we can apply it to a differentiable function $f: \mathbb{R}^n \longrightarrow \mathbb{R}$.

Lemma 11.25. *Let* $\alpha: (a,b) \longrightarrow \mathbb{R}^n$ *be a curve and* $f: \mathbb{R}^n \longrightarrow \mathbb{R}$ *a differentiable function. Then*

$$\alpha'(t)_{\alpha(t)}[f] = (f \circ \alpha)'(t).$$

Proof. We write $\alpha'(t) = \bigl(a_1'(t), \ldots, a_n'(t)\bigr)$. It follows from Lemma 11.1 that

(11.31) $$\alpha'(t)_{\alpha(t)}[f] = \sum_{j=1}^{n} a_j'(t) \frac{\partial f}{\partial u_j}(\alpha(t)).$$

But the chain rule says that the right-hand side of (11.31) is $(f \circ \alpha)'(t)$. ∎

In Section 1.4 we defined the notion of vector field along a curve α in $\mathbb{R}^n$. A vector field W on $\mathbb{R}^n$ gives rise to a vector field on α; it is simply $W \circ \alpha$. Note that for each t, $(W \circ \alpha)(t)$ is a vector in $\mathbb{R}^n$. A closely related concept is the restriction of W to α, which is the function

$$t \longmapsto W(\alpha(t))_{\alpha(t)}.$$

Frequently, it is useful to make no distinction between the vector $(W \circ \alpha)(t) \in \mathbb{R}^n$ and the tangent vector $W(\alpha(t)) = W(\alpha(t))_{\alpha(t)} \in \mathbb{R}^n_{\alpha(t)}$, mainly because it is too much trouble to write the subscript. Similarly, unless complete clarity is required, we shall not distinguish between $\alpha'(t)$ and $\alpha'(t)_{\alpha(t)}$.

In Section 1.4 we also defined the derivative of a vector field along a curve. In particular, we know how to compute $(W \circ \alpha)'$. Let us make precise the relation between $(W \circ \alpha)'$ and the derivative of W with respect to $\alpha'(t)$.

Lemma 11.26. *If W is a differentiable vector field on $\mathbb{R}^n$ and $\alpha: (a,b) \longrightarrow \mathbb{R}^n$ is a curve, then*
$$D_{\alpha'(t)} W = (W \circ \alpha)'(t)_{\alpha(t)}.$$

Proof. By Lemma 11.17 we have
$$D_{\alpha'(t)} W = \sum_{i=1}^{n} \alpha'(t)[w_i] U_i.$$

On the other hand, Lemma 11.25 implies that $\alpha'(t)[w_i] = (w_i \circ \alpha)'(t)$. Thus we have
$$\begin{aligned} D_{\alpha'(t)} W &= \sum_{i=1}^{n} \alpha'(t)[w_i] U_i \\ &= \sum_{i=1}^{n} (w_i \circ \alpha)'(t) U_i \\ &= (W \circ \alpha)'(t)_{\alpha(t)}. \end{aligned}$$ ∎

11.7 Exercises

1. Let $\mathbf{v} = (v_1, v_2, v_3)$ and $\mathbf{p} = (2, -1, 4)$. Define $f: \mathbb{R}^3 \longrightarrow \mathbb{R}$ by
$$f(x, y, z) = x y^2 z^4.$$
Compute $\mathbf{v}_\mathbf{p}[f]$.

2. Fill in the details of the proofs of Lemma 11.2, Lemma 11.4, Corollary 11.6 and Corollary 11.9.

3. Define $F: \mathbb{R}^3 \longrightarrow \mathbb{R}^3$ by $F(x, y, z) = (x y, y z, z x)$. Determine the following sets:
$$A = \{ \mathbf{p} \in \mathbb{R}^3 \mid \|\mathbf{p}\| = 1, \ F(\mathbf{p}) = \mathbf{0} \},$$
$$B = \{ \mathbf{p} \in \mathbb{R}^3 \mid \|\mathbf{p}\| = 1, \ F_*(\mathbf{v}_\mathbf{p}) = \mathbf{0} \text{ for some } \mathbf{v}_\mathbf{p} \in \mathbb{R}^3_\mathbf{p} \}.$$

11.7 Exercises

4. Compute by hand the gradients of the following functions:
   (a) $(x, y, z) \longmapsto e^{az} \cos(ax) - \cos(ay)$.
   (b) $(x, y, z) \longmapsto \sin(z) - \sinh(x) \sinh(y)$.
   (c) $(x, y, z) \longmapsto \dfrac{x^2}{a^2} - \dfrac{y^2}{b^2} - cz$.
   (d) $(x, y, z) \longmapsto x^m + y^n + z^p$.

5. Use *Mathematica* to compute the gradient of the following functions:
   (a) `g1[x_,y_,z_]:= x^2 + 2y^2 + 3z^2 - x z + y z - x y`
   (b) `g2[a_][x_,y_,z_]:= x^3 + y^3 + z^3 - 3a x y z`
   (c) `g3[x_,y_,z_]:= x y z Exp[x + y + z]`
   (d) `g4[x_,y_,z_]:=`
   `ArcTan[x + y + z - x y z/(1 - x y - y z - x z)]`

6. Compute by hand the divergence of each of the following functions:
   (a) $(x, y, z) \longmapsto \bigl(a\,y\,z\cos(x\,y), b\,y\,z\sin(x\,y), c\,x\,y\bigr)$.
   (b) $(x, y, z) \longmapsto \bigl(\sin(xy), \cos(yz), -3\cos^2(yz)\sin(xy) + \sin^3(xy)\bigr)$.
   (c) $(x, y, z) \longmapsto \bigl((x\,y\,z)^a, (x\,y\,z)^b, (x\,y\,z)^c\bigr)$.
   (d) $(x, y, z) \longmapsto \bigl(\log(x + y - z), e^{x-y+z}, (-x + y + z)^5\bigr)$.

7. Use *Mathematica* to compute the divergence of each of the following vector fields:
   (a) `ww1[x_,y_,z_] := {x y z, 2x + 3y + z, x^2 + z^2}`
   (b) `ww2[x_,y_,z_]:=`
   `{6x^2y^2 - z^3 + y z - 5, 4x^3 y + x z + 2, x y - 3 x z^2 -3}`

8. Compute the Laplacian of the functions in exercises 4 and 5.

9. Let $f: \mathbb{R}^n \longrightarrow \mathbb{R}$ be a differentiable function. A point $\mathbf{p} \in \mathbb{R}^n$ is said to be a **critical point** of f provided the tangent map $f_*: \mathbb{R}^n_{\mathbf{p}} \longrightarrow \mathbb{R}_{F(\mathbf{p})}$ maps some nonzero tangent vector to zero. If $\mathbf{p}$ is a critical point of f, we define the **Hessian** of f at $\mathbf{p}$ to be

$$\text{Hessian}[f](\mathbf{v}_\mathbf{p}, \mathbf{w}_\mathbf{p}) = \mathbf{v}_\mathbf{p}(Wf)(\mathbf{p})$$

for $\mathbf{v}_\mathbf{p}, \mathbf{w}_\mathbf{p} \in \mathbb{R}^n_\mathbf{p}$. Here W is any vector field on $\mathbb{R}^n$ such that $W(\mathbf{p}) = \mathbf{w}_\mathbf{p}$.

(a) Show that
$$\text{Hessian}[f](\mathbf{v}_p, \mathbf{w}_p) = \sum_{i=1}^{n} v_i w_j \frac{\partial^2 f}{\partial u_i \partial u_j}.$$

(b) Conclude that the definition of the Hessian is independent of the choice of the vector field W.

10. Fill in the details of the proofs of Lemmas 11.19, 11.20 and 11.22.

12

SURFACES IN EUCLIDEAN SPACE

The 2-dimensional analog of a curve is a surface. However, surfaces in general are much more complicated than curves. In this introductory chapter we give basic definitions that will be used throughout the rest of the book.

The most straightforward 2-dimensional generalization of a curve in $\mathbb{R}^n$ is a *patch* or local surface, which we define in Section 12.1. Patches in $\mathbb{R}^3$ are discussed in Section 12.2. In Section 12.3 we describe a **Mathematica** program to compute the unit normal to a patch in $\mathbb{R}^3$.

A second generalization of a curve in $\mathbb{R}^n$ is a *regular surface* in $\mathbb{R}^n$; this notion is defined in Section 12.4. In Sections 12.5 and 12.6 we define tangent vector to a regular surface and surface mapping; these are the surface analogs of a tangent vector to $\mathbb{R}^n$ and a mapping from $\mathbb{R}^n$ to $\mathbb{R}^n$.

12.1 Patches in $\mathbb{R}^n$

Since a curve in $\mathbb{R}^n$ is a vector-valued function of 1 variable, it is reasonable to consider vector-valued functions of 2 variables. Such an object is called a patch. First, we give the precise definition.

Definition. *A* **patch** *or* **local surface** *is a differentiable mapping*

$$\mathbf{x}: \mathcal{U} \longrightarrow \mathbb{R}^n,$$

where $\mathcal{U}$ is an open subset of $\mathbb{R}^2$. More generally, if A is any subset of $\mathbb{R}^2$, we say that a map $\mathbf{x}\colon A \longrightarrow \mathbb{R}^n$ is a patch provided that $\mathbf{x}$ can be extended to a differentiable mapping from $\mathcal{U}$ into $\mathbb{R}^n$, where $\mathcal{U}$ is an open set containing A. We call $\mathbf{x}(\mathcal{U})$ (or more generally $\mathbf{x}(A)$) the **trace** of $\mathbf{x}$.

Although the domain of definition of a patch can in theory be any set, more often than not it is an open or closed rectangle.

The standard local parametrization of the **sphere** $S^2(a)$ of radius a is the map **sphere[a]**: $[0, 2\pi] \times [-\pi/2, \pi/2] \longrightarrow \mathbb{R}^3$ given by

$$\mathbf{sphere}[a](u, v) = (a\cos v \cos u, a\cos v \sin u, a\sin v).$$

This parametrization is chosen so that u measures longitude and v measures latitude. The corresponding *Mathematica* definition is as follows:

```
sphere[a_][u_,v_]:=
    a{Cos[v]Cos[u],Cos[v]Sin[u],Sin[v]}
```

To find out what it looks like using *Mathematica*, it is useful to plot only a portion of the sphere, so that we can see inside.

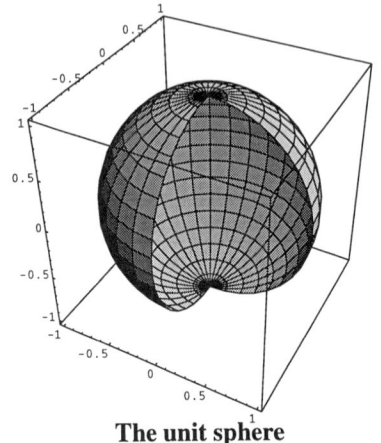

```
ParametricPlot3D[
sphere[1][
u,v]//Evaluate,
{u,0,3Pi/2},
{v,-Pi/2,Pi/2},
PlotPoints->{24,20}];
```

The unit sphere

$(u, v) \longmapsto (\cos v \cos u, \cos v \sin u, \sin v)$

We have used **ParametricPlot3D** to plot the portion of a surface, in this case a sphere. On page 197 it was explained how to use **ParametricPlot3D** to plot a space curve. The only difference between the two modes is that surface plotting requires two parameters, while space curve plotting requires only one.

12.1 Patches in $\mathbb{R}^n$

Although **sphere**$[a]$ is defined on the closed rectangle $[0, 2\pi] \times [-\pi/2, \pi/2]$, it has a differentiable extension to an open set containing $[0, 2\pi] \times [-\pi/2, \pi/2]$. However, any such extension will multiply cover substantial parts of $S^2(a)$, while **sphere**$[a]$ multiply covers only a semicircle from the north pole $(0, 0, a)$ to the south pole $(0, 0, -a)$.

Since a patch can be written as an n-tuple of functions

(12.1) $$\mathbf{x}(u, v) = \big(x_1(u, v), \ldots, x_n(u, v)\big),$$

we can define the partial derivative $\mathbf{x}_u$ of $\mathbf{x}$ with respect to u:

(12.2) $$\mathbf{x}_u(u, v) = \left(\frac{\partial x_1}{\partial u}(u, v), \ldots, \frac{\partial x_n}{\partial u}(u, v) \right).$$

The other partial derivatives of $\mathbf{x}$, for example, $\mathbf{x}_v$, $\mathbf{x}_{uu}$, $\mathbf{x}_{uv}$, are defined similarly. Frequently, we abbreviate (12.1) and (12.2) to

$$\mathbf{x} = (x_1, \ldots, x_n) \qquad \text{and} \qquad \mathbf{x}_u = \left(\frac{\partial x_1}{\partial u}, \ldots, \frac{\partial x_n}{\partial u} \right).$$

The partial derivatives $\mathbf{x}_u$ and $\mathbf{x}_v$ can be expressed in terms of the tangent map of the patch $\mathbf{x}$.

Lemma 12.1. *Let* $\mathbf{x}: \mathcal{U} \longrightarrow \mathbb{R}^n$ *be a patch and let* $\mathbf{q} \in \mathcal{U}$. *Then*

$$\mathbf{x}_*\big(U_1(\mathbf{q})\big) = \mathbf{x}_u(\mathbf{q}) \qquad \text{and} \qquad \mathbf{x}_*\big(U_2(\mathbf{q})\big) = \mathbf{x}_v(\mathbf{q}),$$

where $\mathbf{x}_*$ *denotes the tangent map of* $\mathbf{x}$, *and* $\{U_1, U_2\}$ *denotes the natural frame field of* $\mathbb{R}^2$.

Proof. By Lemma 11.5 we have

$$\mathbf{x}_*\big(U_1(\mathbf{q})\big) = \big(U_1(\mathbf{q})[x_1], \ldots, U_1(\mathbf{q})[x_n]\big)_{\mathbf{x}(\mathbf{q})} = \left(\frac{\partial x_1}{\partial u}(\mathbf{q}), \ldots, \frac{\partial x_n}{\partial u}(\mathbf{q}) \right)_{\mathbf{x}(\mathbf{q})} = \mathbf{x}_u(\mathbf{q}),$$

and similarly for $\mathbf{x}_*\big(U_2(\mathbf{q})\big)$. ∎

On page 254 we defined the Jacobian matrix of a differentiable map. For the case of a patch we modify this definition slightly.

Definition. *The* **Jacobian matrix** *of a patch* $\mathbf{x}: \mathcal{U} \longrightarrow \mathbb{R}^n$ *is the matrix-valued function* $\mathcal{J}(\mathbf{x})$ *given by*

(12.3) $$\mathcal{J}(\mathbf{x})(u, v) = \begin{pmatrix} \frac{\partial x_1}{\partial u}(u, v) & \cdots & \frac{\partial x_n}{\partial u}(u, v) \\ \frac{\partial x_1}{\partial v}(u, v) & \cdots & \frac{\partial x_n}{\partial v}(u, v) \end{pmatrix} = \begin{pmatrix} \mathbf{x}_u(u, v) \\ \mathbf{x}_v(u, v) \end{pmatrix}.$$

The matrix in (12.3) is the transpose of the Jacobian matrix defined on page 254; otherwise, we would have to define $\mathbf{x}_u$ and $\mathbf{x}_v$ as column vectors, which is typographically inconvenient.

Recall that the rank of a matrix a is the largest integer m such that a has an $m \times m$ submatrix whose determinant is nonzero. The following lemma is a consequence of well-known facts from linear algebra.

Lemma 12.2. *Let* $\mathbf{p}, \mathbf{q} \in \mathbb{R}^n$. *The following conditions are equivalent*:

(i) $\mathbf{p}$ *and* $\mathbf{q}$ *are linearly dependent*;

(ii) $\det \begin{pmatrix} \mathbf{p} \cdot \mathbf{p} & \mathbf{p} \cdot \mathbf{q} \\ \mathbf{q} \cdot \mathbf{p} & \mathbf{q} \cdot \mathbf{q} \end{pmatrix} = 0$;

(iii) *the* $2 \times n$ *matrix* $\begin{pmatrix} \mathbf{p} \\ \mathbf{q} \end{pmatrix}$ *has rank less than* 2.

Proof. If $\mathbf{p}$ and $\mathbf{q}$ are linearly dependent, then either $\mathbf{p}$ is a multiple of $\mathbf{q}$, or vice versa. For example, if $\mathbf{p} = \lambda \mathbf{q}$, then

$$\det \begin{pmatrix} \mathbf{p} \cdot \mathbf{p} & \mathbf{p} \cdot \mathbf{q} \\ \mathbf{q} \cdot \mathbf{p} & \mathbf{q} \cdot \mathbf{q} \end{pmatrix} = \det \begin{pmatrix} \lambda \mathbf{q} \cdot \lambda \mathbf{q} & \lambda \mathbf{q} \cdot \mathbf{q} \\ \lambda \mathbf{q} \cdot \mathbf{q} & \mathbf{q} \cdot \mathbf{q} \end{pmatrix} = \lambda^2 \det \begin{pmatrix} \mathbf{q} \cdot \mathbf{q} & \mathbf{q} \cdot \mathbf{q} \\ \mathbf{q} \cdot \mathbf{q} & \mathbf{q} \cdot \mathbf{q} \end{pmatrix} = 0.$$

Thus (i) implies (ii).

Next, suppose

$$\det \begin{pmatrix} \mathbf{p} \cdot \mathbf{p} & \mathbf{p} \cdot \mathbf{q} \\ \mathbf{q} \cdot \mathbf{p} & \mathbf{q} \cdot \mathbf{q} \end{pmatrix} = 0.$$

Write $\mathbf{p} = (p_1, \ldots, p_n)$ and $\mathbf{q} = (q_1, \ldots, q_n)$. Then

$$0 = \|\mathbf{p}\|^2 \|\mathbf{q}\|^2 - (\mathbf{p} \cdot \mathbf{q})^2 = \left(\sum_{i=1}^n p_i^2 \right) \left(\sum_{j=1}^n q_j^2 \right) - \left(\sum_{i=1}^n p_i q_i \right)^2$$

$$= \sum_{1 \leq i < j \leq n} (p_i q_j - p_j q_i)^2.$$

It follows that $p_i q_j = p_j q_i$ for all i and j. Hence (ii) implies (iii).

Finally, suppose (iii) holds. Then $p_i q_j = p_j q_i$ for all i and j. Without loss of generality, we can suppose that $q_i \neq 0$ for some i. Then

$$p_j = \frac{p_i}{q_i} q_j$$

for all j. Hence
$$\mathbf{p} = (p_1, \ldots, p_n) = \frac{p_i}{q_i}(q_1, \ldots, q_n) = \frac{p_i}{q_i}\mathbf{q}.$$

It follows that $\mathbf{p}$ and $\mathbf{q}$ are linearly dependent. ∎

Corollary 12.3. *Let* $\mathbf{x} : \mathcal{U} \longrightarrow \mathbb{R}^n$ *be a patch. Then the following conditions are equivalent*:

(i) $\mathbf{x}_u(u_0, v_0)$ *and* $\mathbf{x}_v(u_0, v_0)$ *are linearly dependent*;

(ii) $\det \begin{pmatrix} \mathbf{x}_u \cdot \mathbf{x}_u & \mathbf{x}_u \cdot \mathbf{x}_v \\ \mathbf{x}_v \cdot \mathbf{x}_u & \mathbf{x}_v \cdot \mathbf{x}_v \end{pmatrix}$ *vanishes at* (u_0, v_0);

(iii) *the Jacobian matrix* $\mathcal{J}(\mathbf{x})$ *has rank less than* 2 *at* (u_0, v_0).

We shall usually want a stronger notion of patch $\mathbf{x}$ in order that the trace $\mathbf{x}(\mathcal{U})$ resemble the open set $\mathcal{U}$ more closely.

Definition. *A* **regular patch** *is a patch* $\mathbf{x} : \mathcal{U} \longrightarrow \mathbb{R}^n$ *for which the Jacobian matrix* $\mathcal{J}(\mathbf{x})(u, v)$ *has rank* 2 *for all* $(u, v) \in \mathcal{U}$. *An* **injective patch** *is a patch such that* $\mathbf{x}(u_1, v_1) = \mathbf{x}(u_2, v_2)$ *implies that* $u_1 = u_2$ *and* $v_1 = v_2$ *for all* (u_1, v_1) *and* (u_2, v_2).

There are regular patches which are not injective (for example, the circular cylinder defined by $\mathbf{x}(u, v) = (\cos u, \sin u, v)$, where $-\infty < u < \infty$, $-2 < v < 2$). Also, there are injective patches which are not regular (for example, the patch $\mathbf{y}$ defined by $\mathbf{y}(u, v) = (u^3, v^3, uv)$, where $-1 < u, v < 1$).

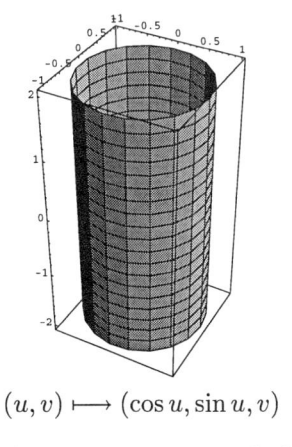

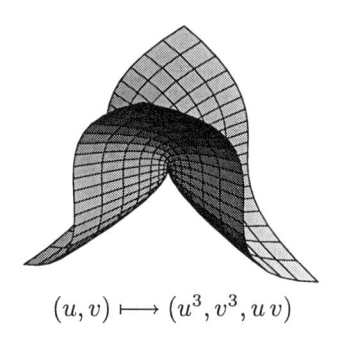

$(u, v) \longmapsto (\cos u, \sin u, v)$

$(u, v) \longmapsto (u^3, v^3, uv)$

Usually, it is necessary to deal with patches that are not regular, but fail to be regular at just a few points. To handle such cases, we make the following definition:

Definition. *A patch* $\mathbf{x}\colon \mathcal{U} \longrightarrow \mathbb{R}^n$ *is* **regular** *at a point* $(u_0, v_0) \in \mathcal{U}$ (*or sometimes we say at* $\mathbf{x}(u_0, v_0)$) *provided its Jacobian matrix has rank 2 at* (u_0, v_0).

As a consequence of Lemma 12.1 and Corollary 12.3 we have

Lemma 12.4. *A patch* $\mathbf{x}\colon \mathcal{U} \longrightarrow \mathbb{R}^n$ *is regular at* $\mathbf{q} \in \mathcal{U}$ *if and only if its tangent map* $\mathbf{x}_*\colon \mathbb{R}^2_\mathbf{q} \longrightarrow \mathbb{R}^n_{\mathbf{x}(\mathbf{q})}$ *is injective.*

The following lemma is a consequence of the inverse function theorem (as given, for example, in [Buck], page 276).

Lemma 12.5. *Let* $\mathbf{x}\colon \mathcal{U} \longrightarrow \mathbb{R}^n$ *be a regular patch and let* $\mathbf{q} \in \mathcal{U}$. *Then there exists a neighborhood* $\mathcal{U}_\mathbf{q}$ *of* $\mathbf{q}$ *such that* $\mathbf{x}$ *maps* $\mathcal{U}_\mathbf{q}$ *diffeomorphically onto* $\mathbf{x}(\mathcal{U}_\mathbf{q})$.

Proof. Write $\mathbf{x} = (x_1, \ldots, x_n)$. Since $\mathbf{x}$ is regular, its Jacobian matrix has a 2×2 submatrix with nonzero determinant. By renaming $x_1, \ldots, x_n$ if necessary, we can suppose that

$$\det \begin{pmatrix} \dfrac{\partial x_1}{\partial u}(u,v) & \dfrac{\partial x_2}{\partial u}(u,v) \\ \dfrac{\partial x_1}{\partial v}(u,v) & \dfrac{\partial x_2}{\partial v}(u,v) \end{pmatrix} \neq 0$$

for $(u,v) \in \mathcal{U}$. We extend $\mathbf{x}\colon \mathcal{U} \longrightarrow \mathbb{R}^n$ to a map $\widetilde{\mathbf{x}}\colon \mathcal{U} \times \mathbb{R}^{n-2} \longrightarrow \mathbb{R}^n$ by defining

$$\widetilde{\mathbf{x}}(u, v, t_3, \ldots, t_n) = \big(x_1(u,v), x_2(u,v), x_3(u,v) + t_3, \ldots, x_n(u,v) + t_n\big).$$

It is clear that $\widetilde{\mathbf{x}}$ is differentiable; moreover, the determinant of the Jacobian matrix of $\widetilde{\mathbf{x}}$ is

$$\det\big(\mathcal{J}(\widetilde{\mathbf{x}})(u,v)\big) = \det \begin{pmatrix} \dfrac{\partial x_1}{\partial u}(u,v) & \dfrac{\partial x_2}{\partial u}(u,v) & \dfrac{\partial x_3}{\partial u}(u,v) & \cdots & \dfrac{\partial x_n}{\partial u}(u,v) \\ \dfrac{\partial x_1}{\partial v}(u,v) & \dfrac{\partial x_2}{\partial v}(u,v) & \dfrac{\partial x_3}{\partial v}(u,v) & \cdots & \dfrac{\partial x_n}{\partial v}(u,v) \\ 0 & 0 & 1 & \cdots & 0 \\ \vdots & \vdots & \vdots & \ddots & \vdots \\ 0 & 0 & 0 & \cdots & 1 \end{pmatrix}$$

$$= \det \begin{pmatrix} \dfrac{\partial x_1}{\partial u}(u,v) & \dfrac{\partial x_2}{\partial u}(u,v) \\ \dfrac{\partial x_1}{\partial v}(u,v) & \dfrac{\partial x_2}{\partial v}(u,v) \end{pmatrix} \neq 0.$$

12.1 Patches in $\mathbb{R}^n$

Because $\det(\mathcal{J}(\tilde{\mathbf{x}}))(\mathbf{q}) \neq 0$, the inverse function theorem says that there is a neighborhood $\tilde{\mathcal{U}}_\mathbf{q}$ of $(\mathbf{q},\mathbf{0})$ upon which $\tilde{\mathbf{x}}$ has a differentiable inverse. Let $\mathcal{U}_\mathbf{q} = \tilde{\mathcal{U}}_\mathbf{q} \cap \mathcal{U}$. Since $\tilde{\mathbf{x}}\colon \tilde{\mathcal{U}}_\mathbf{q} \longrightarrow \tilde{\mathbf{x}}(\tilde{\mathcal{U}}_\mathbf{q})$ is a diffeomorphism, so is $\mathbf{x}\colon \mathcal{U}_\mathbf{q} \longrightarrow \mathbf{x}(\mathcal{U}_\mathbf{q})$. ∎

Corollary 12.6. *Let* $\mathbf{x}\colon \mathcal{U} \longrightarrow \mathbb{R}^n$ *be an injective regular patch. Then* $\mathbf{x}$ *maps* $\mathcal{U}$ *diffeomorphically onto* $\mathbf{x}(\mathcal{U})$.

Let us define a *Mathematica* version of the Jacobian matrix $\mathcal{J}(\mathbf{x})$ of a patch $\mathbf{x}$.

```
surfacejacobianmatrix[x_][u_,v_]:=
    {D[x[uu,vv],uu],D[x[uu,vv],vv]}   /. {uu->u,vv->v}
```

Similarly, the *Mathematica* version of

$$x \longmapsto \det\begin{pmatrix} \mathbf{x}_u \cdot \mathbf{x}_u & \mathbf{x}_u \cdot \mathbf{x}_v \\ \mathbf{x}_v \cdot \mathbf{x}_u & \mathbf{x}_v \cdot \mathbf{x}_v \end{pmatrix}$$

is given by

```
surfacejacobian[x_][u_,v_]:= Det[
    surfacejacobianmatrix[x][u,v].
    Transpose[surfacejacobianmatrix[x][u,v]]]
```

Here **Transpose** is *Mathematica*'s command to find the transpose of a matrix. For example, let us find when the standard parametrization of a sphere fails to be regular. The command

```
surfacejacobian[sphere[a]][u,v]//Simplify
```

elicits the response

$$a^4 \, \text{Cos}[v]^2$$

We conclude that this parametrization of the sphere fails to be regular when $v = \pm\pi/2$, that is, at the north and south poles.

Associated with any patch are some naturally defined curves.

Definition. *Let* $\mathbf{x}\colon \mathcal{U} \longrightarrow \mathbb{R}^n$ *be a patch, and fix* $(u_0, v_0) \in \mathcal{U}$. *The curves*

$$u \longmapsto \mathbf{x}(u, v_0) \quad \text{and} \quad v \longmapsto \mathbf{x}(u_0, v)$$

are called u-**parameter** *and* v-**parameter** *curves of* $\mathbf{x}$.

These are the curves that *Mathematica* usually displays when it draws a patch.

There is also a convenient way to represent a general curve whose trace is contained in the trace of a patch.

Lemma 12.7. *Let $\alpha\colon (a,b) \longrightarrow \mathbb{R}^n$ be a curve whose trace lies on the trace $\mathbf{x}(\mathcal{U})$ of a single regular patch $\mathbf{x}\colon \mathcal{U} \longrightarrow \mathbb{R}^n$ such that $\mathbf{x}\colon \mathcal{U} \longrightarrow \mathbf{x}(\mathcal{U})$ is a homeomorphism. Then there exist unique differentiable functions $u,v\colon (a,b) \longrightarrow \mathbb{R}$ such that*

$$(12.4) \qquad \alpha(t) = \mathbf{x}\big(u(t), v(t)\big)$$

for $a < t < b$.

Proof. For $a < t < b$ we can write

$$(\mathbf{x}^{-1} \circ \alpha)(t) = \big(u(t), v(t)\big);$$

this equation is equivalent to (12.4). It is clear that u and v are unique; by Lemma 12.5 they are also differentiable. ∎

Now we can define the notion of tangent vector to a patch.

Definition. Let $\mathbf{x}\colon \mathcal{U} \longrightarrow \mathbb{R}^n$ be an injective patch, and let $\mathbf{p} \in \mathbf{x}(\mathcal{U})$. A **tangent vector** *to $\mathbf{x}$ at $\mathbf{p}$ is a tangent vector $\mathbf{v}_\mathbf{p} \in \mathbb{R}^n_\mathbf{p}$ for which there exists a curve $\alpha\colon (a,b) \longrightarrow \mathbb{R}^n$ that can be written as*

$$(12.5) \qquad \alpha(t) = \mathbf{x}\big(u(t), v(t)\big) \qquad (a < t < b),$$

such that $\alpha(0) = \mathbf{p}$ and $\alpha'(0) = \mathbf{v}_\mathbf{p}$. We denote the set of tangent vectors to $\mathbf{x}$ at $\mathbf{p}$ by $\mathbf{x}(\mathcal{U})_\mathbf{p}$.

Lemma 12.8. *The set $\mathbf{x}(\mathcal{U})_\mathbf{p}$ of all tangent vectors to a patch $\mathbf{x}\colon \mathcal{U} \longrightarrow \mathbb{R}^n$ at a point $\mathbf{p} = \mathbf{x}(u_0, v_0) \in \mathbf{x}(\mathcal{U})$ forms a vector space that is spanned by $\mathbf{x}_u(u_0, v_0)$ and $\mathbf{x}_v(u_0, v_0)$.*

Proof. By definition of tangent vector $\mathbf{v}_\mathbf{p}$ to $\mathbf{x}$, there exists a curve α of the form (12.5) such that $\alpha(0) = \mathbf{p}$ and $\alpha'(0) = \mathbf{v}_\mathbf{p}$. The chain rule for curves (Lemma 1.3, page 7) implies that

$$\alpha'(t) = u'(t)\mathbf{x}_u\big(u(t), v(t)\big) + v'(t)\mathbf{x}_v\big(u(t), v(t)\big).$$

In particular,

$$\begin{aligned}\mathbf{v}_\mathbf{p} &= \alpha'(0) = u'(0)\mathbf{x}_u\big(u(0), v(0)\big) + v'(0)\mathbf{x}_v\big(u(0), v(0)\big) \\ &= u'(0)\mathbf{x}_u(u_0, v_0) + v'(0)\mathbf{x}_v(u_0, v_0).\end{aligned}$$

Conversely, if $\mathbf{v_p} = c_1\mathbf{x}_u(u_0,v_0) + c_2\mathbf{x}_v(u_0,v_0)$, then an easy calculation shows that $\mathbf{v_p}$ is the velocity vector at $\mathbf{x}(u_0,v_0) = \mathbf{p}$ of the curve $t \longmapsto \mathbf{x}(u_0 + tc_1, v_0 + tc_2)$. ∎

The proof of Lemma 12.8 yields the following result:

Corollary 12.9. *Let $\boldsymbol{\alpha}: (a,b) \longrightarrow \mathbb{R}^n$ be a curve whose trace lies on the trace $\mathbf{x}(\mathcal{U})$ of a single regular patch $\mathbf{x}:\mathcal{U} \longrightarrow \mathbb{R}^n$ such that $\mathbf{x}:\mathcal{U} \longrightarrow \mathbf{x}(\mathcal{U})$ is a homeomorphism. Then there exist unique differentiable functions $u, v: (a,b) \longrightarrow \mathbb{R}$ such that*

(12.6) $$\boldsymbol{\alpha}' = u'\mathbf{x}_u + v'\mathbf{x}_v.$$

Definition. Let $\mathbf{x}:\mathcal{U} \longrightarrow \mathbb{R}^n$ be an injective patch, and let $\mathbf{z_p} \in \mathbb{R}^n_\mathbf{p}$ with $\mathbf{p} \in \mathbf{x}(\mathcal{U})$. We say that $\mathbf{z_p}$ is **normal** or **perpendicular** to $\mathbf{x}$ at $\mathbf{p}$, provided $\mathbf{z_p} \cdot \mathbf{v_p} = 0$ for all vectors $\mathbf{v_p}$ tangent to $\mathbf{x}$ at $\mathbf{p}$.

It is easy to see that if $\mathbf{x}(\mathcal{U})_\mathbf{p}^\perp$ denotes the space of vectors in $\mathbb{R}^n_\mathbf{p}$ perpendicular to $\mathbf{x}(\mathcal{U})_\mathbf{p}$, then

$$\mathbb{R}^n_\mathbf{p} = \mathbf{x}(\mathcal{U})_\mathbf{p} \oplus \mathbf{x}(\mathcal{U})_\mathbf{p}^\perp.$$

We shall also need vector fields on patches.

Definition. *A **vector field** V on a patch $\mathbf{x}:\mathcal{U} \longrightarrow \mathbb{R}^n$ is a function that assigns to each $\mathbf{q} \in \mathcal{U}$ a tangent vector $V(\mathbf{q}) \in \mathbb{R}^n_\mathbf{p}$, where $\mathbf{p} = \mathbf{x}(\mathbf{q})$. We say that V is **tangent** to $\mathbf{x}$ if $V(\mathbf{q}) \in \mathbf{x}(\mathcal{U})_{\mathbf{x}(\mathbf{q})}$ for all $\mathbf{q} \in \mathcal{U}$. Similarly, a vector field W on $\mathbf{x}$ is **normal** or **perpendicular** to $\mathbf{x}$ if $W(\mathbf{q}) \cdot \mathbf{v_p} = 0$ for all $\mathbf{v_p} \in \mathbf{x}(\mathcal{U})_{\mathbf{x}(\mathbf{q})}$ for all $\mathbf{q} \in \mathcal{U}$.*

12.2 Patches in $\mathbb{R}^3$

We now restrict our attention to patches in $\mathbb{R}^3$ because they are the ones that are easiest to visualize. Furthermore, because there is a vector cross product on $\mathbb{R}^3$, computations become simpler. Here is a useful criterion for the regularity for a patch in $\mathbb{R}^3$:

Lemma 12.10. *A patch $\mathbf{x}:\mathcal{U} \longrightarrow \mathbb{R}^3$ is regular at $(u_0,v_0) \in \mathcal{U}$ if and only if $\mathbf{x}_u \times \mathbf{x}_v$ is nonzero at (u_0,v_0).*

Proof. The 2×2 submatrices of $\mathcal{J}(\mathbf{x})$ are

$$\begin{pmatrix} \dfrac{\partial x_1}{\partial u} & \dfrac{\partial x_2}{\partial u} \\ \dfrac{\partial x_1}{\partial v} & \dfrac{\partial x_2}{\partial v} \end{pmatrix}, \quad \begin{pmatrix} \dfrac{\partial x_1}{\partial u} & \dfrac{\partial x_3}{\partial u} \\ \dfrac{\partial x_1}{\partial v} & \dfrac{\partial x_3}{\partial v} \end{pmatrix}, \quad \begin{pmatrix} \dfrac{\partial x_2}{\partial u} & \dfrac{\partial x_3}{\partial u} \\ \dfrac{\partial x_2}{\partial v} & \dfrac{\partial x_3}{\partial v} \end{pmatrix}.$$

On the other hand,

$$\mathbf{x}_u \times \mathbf{x}_v = \det \begin{pmatrix} \mathbf{i} & \mathbf{j} & \mathbf{k} \\ \dfrac{\partial x_1}{\partial u} & \dfrac{\partial x_2}{\partial u} & \dfrac{\partial x_3}{\partial u} \\ \dfrac{\partial x_1}{\partial v} & \dfrac{\partial x_2}{\partial v} & \dfrac{\partial x_3}{\partial v} \end{pmatrix}$$

$$= \det \begin{pmatrix} \dfrac{\partial x_2}{\partial u} & \dfrac{\partial x_3}{\partial u} \\ \dfrac{\partial x_2}{\partial v} & \dfrac{\partial x_3}{\partial v} \end{pmatrix} \mathbf{i} - \det \begin{pmatrix} \dfrac{\partial x_1}{\partial u} & \dfrac{\partial x_3}{\partial u} \\ \dfrac{\partial x_1}{\partial v} & \dfrac{\partial x_3}{\partial v} \end{pmatrix} \mathbf{j} + \det \begin{pmatrix} \dfrac{\partial x_1}{\partial u} & \dfrac{\partial x_2}{\partial u} \\ \dfrac{\partial x_1}{\partial v} & \dfrac{\partial x_2}{\partial v} \end{pmatrix} \mathbf{k}.$$

Thus $(\mathbf{x}_u \times \mathbf{x}_v)(u_0, v_0) = \mathbf{0}$ if and only if the rank of $\mathcal{J}(\mathbf{x})(u_0, v_0)$ is less than 2. Hence $\mathbf{x}$ is regular at (u_0, v_0) if and only if $(\mathbf{x}_u \times \mathbf{x}_v)(u_0, v_0) \neq \mathbf{0}$. ∎

The notion of a vector perpendicular to a surface in $\mathbb{R}^3$ is intuitively clear. The vector cross product gives rise to a convenient way to find such a vector.

Lemma 12.11. *Let* $\mathbf{x}: \mathcal{U} \longrightarrow \mathbb{R}^3$ *be an injective patch. Then the vector field* $(u,v) \longmapsto \mathbf{x}_u \times \mathbf{x}_v$ *is everywhere perpendicular to* $\mathbf{x}(\mathcal{U})$.

Proof. Property (8.4) implies that $\mathbf{x}_u \times \mathbf{x}_v$ is perpendicular to both $\mathbf{x}_u$ and $\mathbf{x}_v$. Since any tangent vector $\mathbf{v}_p$ to $\mathbf{x}$ is a linear combination of $\mathbf{x}_u$ and $\mathbf{x}_v$ at $\mathbf{p}$, it follows from Lemma 12.8 that $(u,v) \longmapsto \mathbf{x}_u \times \mathbf{x}_v$ is perpendicular to $\mathbf{x}(\mathcal{U})$. ∎

Now we can define a vector field of unit length that is perpendicular to a patch.

Definition. *For an injective patch* $\mathbf{x}: \mathcal{U} \longrightarrow \mathbb{R}^3$ *the* **unit normal vector field** *or* **surface normal U** *is defined by*

$$(12.7) \qquad \mathbf{U}(u,v) = \frac{\mathbf{x}_u \times \mathbf{x}_v}{\|\mathbf{x}_u \times \mathbf{x}_v\|}(u,v)$$

at those points $(u,v) \in \mathcal{U}$ *at which* $\mathbf{x}_u \times \mathbf{x}_v$ *does not vanish.*

The notion of regularity of a patch has the following geometric interpretation.

Corollary 12.12. *Let* $\mathbf{x}: \mathcal{U} \longrightarrow \mathbb{R}^3$ *be an injective patch. Then* $\mathbf{x}$ *is regular if and only if the unit normal vector field* $\mathbf{U}$ *is everywhere well-defined.*

12.3 The Local Gauss Map

One of the key elements necessary for the study of surfaces is the map that assigns to each point **p** on a surface $\mathcal{M} \subset \mathbb{R}^3$ the point on the unit sphere $S^2(1) \subset \mathbb{R}^3$ that is parallel to the unit normal **U(p)**. It is called the Gauss map. This map may not be defined for a general surface for reasons that we take up in Chapter 14. However, the Gauss[1] map is defined for any patch.

Definition. *Let* $\mathbf{x}: \mathcal{U} \longrightarrow \mathbb{R}^3$ *be a patch. Then the unit normal* **U** *to* **x** *viewed as a mapping from* $\mathcal{U}$ *to the unit sphere* $S^2(1) \subset \mathbb{R}^3$ *is called the* **Gauss map** *of* **x**.

Note that **U** is undefined at nonregular points of **x**. Although geometrically $\mathbf{U}(u, v)$ represents the unit normal to **x** at $\mathbf{x}(u, v)$, the map $(u, v) \longmapsto \mathbf{U}(u, v)$ can also be considered to be a patch whose image is contained in $S^2(1)$.

To calculate the unit normal to a patch, we need the *Mathematica* function **cross** described on page 194. If a patch **x** has been defined, then we can use *Mathematica* to compute $\mathbf{U}(u, v)$ (which we call **unitnormal** in *Mathematica*) with the miniprogram

---

[1] Carl Friedrich Gauss (1777-1855). Gauss's **Disquisitiones Generales Circa Superficies Curvas** (published in 1828) revolutionized differential geometry as his **Disquisitiones Arithmeticae** had revolutionized number theory. In particular, his approach depended only on the intrinsic properties of surfaces, not on their embedding in Euclidean 3-space.

Gauss' genius extended into nearly every branch of mathematics. While a student at Göttingen he made his first major original discovery–the constructibility of the 17-sided regular polygon, closely followed by the first of his four proofs of the fundamental theorem of algebra (his doctoral dissertation) and his calculation of the orbit of the newly discovered asteroid Ceres. Gauss's interest in geodesy led to the invention of the heliotrope and the development of least squares approximation, a technique he also used in his investigation of the distribution of prime numbers.

Gauss' work on terrestrial magnetism led to his construction, in collaboration with Wilhelm Weber, of the first operating electric telegraph. Other applied discoveries of significance were those in optics, potential theory, and astronomy. Gauss spent his entire career in Brunswick, under the patronage of the local royalty; for many years he directed the observatory there. Perhaps due to his deeply ingrained conservatism or to his distaste for controversy, Gauss did not publish his development of the theory of non-Euclidean geometry nor his work anticipating Hamilton's investigation of quaternions. Although he worked for the most part in mathematical isolation, his results laid the basis for new departures in number theory and statistics, as well as in differential geometry.

```
unitnormal[x_][u_,v_]:= Module[{xu,xv,N1},
    xu = D[x[uu,vv],uu];
    xv = D[x[uu,vv],vv];
    N1 = cross[xu,xv];
    Simplify[N1.N1]^(-1/2)N1] /. {uu->u,vv->v}
```

This command is contained in the file **CSPROGS.m**, or in Appendix A. The *Mathematica* command[2] **Module** allows one to define functions via a miniprogram that involves several steps and local variables. In the case of **unitnormal** the local variables are **xu**, **xv**, and **N1**. For example, let us find the unit normal of the hyperbolic paraboloid, defined by

```
hyperbolicparaboloid[u_,v_] := {u,v,u v}
```

When we enter

```
unitnormal[hyperbolicparaboloid][u,v]
```

Mathematica answers

$$\left\{\frac{-v}{\sqrt{1+u^2+v^2}}, \frac{-u}{\sqrt{1+u^2+v^2}}, \frac{1}{\sqrt{1+u^2+v^2}}\right\}$$

Since *Mathematica* sometimes has trouble simplifying under square roots, it is useful to test the regularity of a patch **x** by computing the square of the determinant of the Jacobian matrix of **x**. A modification, let us call it **jacobian2**, of the **unitnormal** program does the trick:

```
jacobian2[x_][u_,v_]:= Module[{xu,xv,N1},
    xu = D[x[uu,vv],uu];
    xv = D[x[uu,vv],vv];
    N1 = cross[xu,xv];
    Simplify[N1.N1]] /. {uu->u,vv->v}
```

For patches that map into $\mathbb{R}^3$, **jacobian2** performs the same task as the command **surfacejacobian**, but is faster.

---

[2]In version 1.2 of *Mathematica* the command **Block** should be used instead of **Module**.

12.4 The Definition of a Regular Surface in $\mathbb{R}^n$

There are some subsets of $\mathbb{R}^n$ that we would like to call surfaces, but they do not fit into the framework of Section 12.1. Such a subset cannot be expressed as the image of a single regular one-to-one patch defined on an open set; spheres and tori are examples. In this section we define the notion of regular surface; roughly speaking, what we shall do is to combine several patches. The resulting definition is more complicated, but it is what is needed in many situations.

Definition. *A subset $M \subset \mathbb{R}^n$ is a* **regular surface**, *provided that for each point $\mathbf{p} \in M$ there exist a neighborhood $\mathcal{V}$ of $\mathbf{p}$ in $\mathbb{R}^n$ and a map $\mathbf{x}:\mathcal{U} \longrightarrow \mathbb{R}^n$ of an open set $\mathcal{U} \subset \mathbb{R}^2$ onto $\mathcal{V} \cap M$ such that:*

(i) *$\mathbf{x}$ is differentiable;*

(ii) *$\mathbf{x}:\mathcal{U} \longrightarrow \mathcal{V} \cap M$ is a homeomorphism. This means that $\mathbf{x}$ has a continuous inverse $\mathbf{x}^{-1}:\mathcal{V} \cap M \longrightarrow \mathcal{U}$ such that $\mathbf{x}^{-1}$ is the restriction to $\mathcal{V} \cap M$ of a continuous map $F:\mathcal{W} \longrightarrow \mathbb{R}^2$, where $\mathcal{W}$ is an open subset of $\mathbb{R}^n$ that contains $\mathcal{V} \cap M$;*

(iii) *each map $\mathbf{x}:\mathcal{U} \longrightarrow M$ is a regular patch.*

Each map $\mathbf{x}:\mathcal{U} \longrightarrow M$ is called a **local chart** *or* **system of local coordinates** *in a neighborhood of $\mathbf{p} \in M$.*

We shall need the following fact.

Lemma 12.13. *Let $\mathcal{W}$ be an open subset of $\mathbb{R}^n$, and suppose that $G:\mathcal{W} \longrightarrow \mathbb{R}^m$ is a map such that $G:\mathcal{W} \longrightarrow G(\mathcal{W})$ is a diffeomorphism. If $M \subseteq \mathcal{W}$ is a regular surface, then $G(M)$ is also a regular surface.*

Proof. If $\mathbf{x}$ is a patch on M satisfying (i), (ii) and (iii) in the definition of regular surface, then $G \circ \mathbf{x}:\mathcal{W} \longrightarrow \mathbb{R}^m$ also satisfies (i), (ii) and (iii). ∎

There is an easy way to find new regular surfaces inside a given regular surface M. A subset $\mathcal{V}$ of a regular surface $M \subset \mathbb{R}^n$ is said to be **open** provided it is the intersection of an open subset of $\mathbb{R}^n$ with M.

Lemma 12.14. *An open subset $\mathcal{W}$ of a regular surface $\mathcal{M}$ is also a regular surface.*

Proof. Let $\mathbf{x}$ be a patch on $\mathcal{M}$ satisfying (i), (ii) and (iii) in the definition of regular surface; then $\mathbf{x}$ is continuous. Since the inverse image of an open set is open, $\mathcal{U} = \mathbf{x}^{-1}(\mathcal{W})$ is open. If $\mathcal{U}$ is nonempty, the restriction $\mathbf{x}|\mathcal{U}$ becomes a patch on $\mathcal{W}$. It is clear that $\mathbf{x}|\mathcal{U}$ satisfies (i), (ii) and (iii) in the definition of regular surface. Thus $\mathcal{W}$ becomes a regular surface. ∎

Let us clarify the relation between patches and regular surfaces. Certainly an arbitrary patch will fail to be a regular surface if its trace is 0- or 1-dimensional. Even a regular patch can fail to be a regular surface if it has self-intersections. (Condition (ii) in the definition of regular surface precludes a regular surface from having self-intersections.) Nevertheless, let us prove that when we restrict the domain of definition of a regular patch we obtain a regular surface.

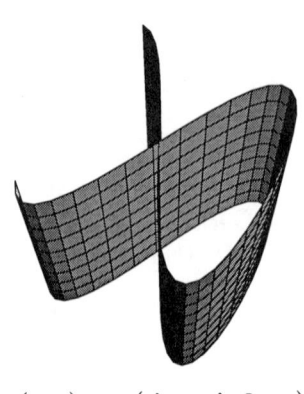

$(u, v) \longmapsto (\sin u, \sin 2u, v)$

Lemma 12.15. *Let $\mathbf{x}: \mathcal{U} \longrightarrow \mathbb{R}^n$ be a regular patch. For any $\mathbf{q} \in \mathcal{U}$ there exists a neighborhood $\mathcal{U}_\mathbf{q}$ of $\mathbf{q}$ such that $\mathbf{x}(\mathcal{U}_\mathbf{q})$ is a regular surface.*

Proof. Lemma 12.5 states that $\mathbf{q}$ has a neighborhood $\mathcal{U}_\mathbf{q}$ such that $\mathbf{x}$ is a diffeomorphism between $\mathcal{U}_\mathbf{q}$ and $\mathbf{x}(\mathcal{U}_\mathbf{q})$. The neighborhood $\mathcal{U}_\mathbf{q}$ is an open set in the plane, and hence a regular surface by Lemma 12.14. Then Lemma 12.13 implies that $\mathbf{x}(\mathcal{U}_\mathbf{q})$ is a regular surface. ∎

Corollary 12.16. *Let $\mathbf{x}: \mathcal{U} \longrightarrow \mathbb{R}^n$ be an regular injective patch. Then $\mathbf{x}(\mathcal{U}_\mathbf{q})$ is a regular surface.*

On the other hand, a sphere is an example of a regular surface which needs at least two patches to cover it. Here are two patches which together cover the sphere.

12.4 The Definition of a Regular Surface in $\mathbb{R}^n$

Both are defined on $\{(u,v) \mid 0 < u < 7\pi/4,\ -\pi/2 < v < \pi/2\}$.

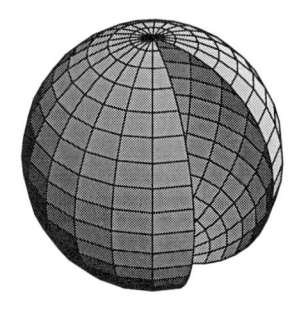

$(u,v) \longmapsto (\cos u \cos v, \sin u \cos v, \sin v)$

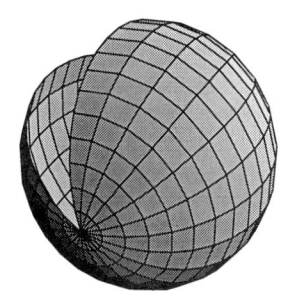

$(u,v) \longmapsto \big(\cos(0.6\pi - u)\cos v, \sin v,$
$\sin(0.6\pi - u)\cos v\big)$

Definition. *Let $\mathcal{M}$ be a regular surface in $\mathbb{R}^n$. If there exists a single regular injective patch $\mathbf{x}:\mathcal{U} \longrightarrow \mathbb{R}^n$ such that $\mathbf{x}(\mathcal{U}) = \mathcal{M}$, we say that $\mathcal{M}$ is **parametrized by $\mathbf{x}$**.*

Next, we establish a fact about differentiable maps into a regular surface.

Theorem 12.17. *Let $\mathcal{M} \subset \mathbb{R}^n$ be a regular surface, and let $\mathcal{V} \subset \mathbb{R}^m$ be an open subset. Suppose that $F:\mathcal{V} \longrightarrow \mathbb{R}^n$ is a differentiable mapping such that $F(\mathcal{V}) \subseteq \mathcal{M}$, and that $\mathbf{x}:\mathcal{U} \longrightarrow \mathcal{M}$ is a regular patch on $\mathcal{M}$ such that $F(\mathcal{V}) \subseteq \mathbf{x}(\mathcal{U})$. Then $\mathbf{x}^{-1} \circ F:\mathcal{V} \longrightarrow \mathcal{U}$ is differentiable.*

Proof. It is tempting to try to prove this theorem by saying that since $\mathbf{x}^{-1}$ and F are differentiable, so is $\mathbf{x}^{-1} \circ F$. Unfortunately, this will not work, because we do not know at this point what it means for a function defined on a regular surface to be differentiable. Therefore, we are forced to prove the theorem in a more roundabout fashion. We modify the proof of Lemma 12.5. Write $\mathbf{x} = (x_1, \ldots, x_n)$. Since $\mathbf{x}$ is regular, its Jacobian matrix has a 2×2 submatrix with nonzero determinant. By renaming $x_1, \ldots, x_n$ if necessary, we can suppose that

$$\det \begin{pmatrix} \dfrac{\partial x_1}{\partial u}(u,v) & \dfrac{\partial x_2}{\partial u}(u,v) \\ \dfrac{\partial x_1}{\partial v}(u,v) & \dfrac{\partial x_2}{\partial v}(u,v) \end{pmatrix} \neq 0$$

for $(u,v) \in \mathcal{U}$. We extend $\mathbf{x}: \mathcal{U} \longrightarrow \mathbb{R}^n$ to a map $\widetilde{\mathbf{x}}: \mathcal{U} \times \mathbb{R}^{n-2} \longrightarrow \mathbb{R}^n$ by defining

$$\widetilde{\mathbf{x}}(u,v,t_3,\ldots,t_n) = \big(x_1(u,v), x_2(u,v), x_3(u,v) + t_3, \ldots, x_n(u,v) + t_n\big).$$

It is clear that $\widetilde{\mathbf{x}}$ is differentiable; moreover, the determinant of the Jacobian matrix of $\widetilde{\mathbf{x}}$ is

$$\det\big(\mathcal{J}(\widetilde{\mathbf{x}})(u,v)\big) = \det \begin{pmatrix} \dfrac{\partial x_1}{\partial u}(u,v) & \dfrac{\partial x_2}{\partial u}(u,v) & \dfrac{\partial x_3}{\partial u}(u,v) & \cdots & \dfrac{\partial x_n}{\partial u}(u,v) \\ \dfrac{\partial x_1}{\partial v}(u,v) & \dfrac{\partial x_2}{\partial v}(u,v) & \dfrac{\partial x_3}{\partial v}(u,v) & \cdots & \dfrac{\partial x_n}{\partial v}(u,v) \\ 0 & 0 & 1 & \cdots & 0 \\ \vdots & \vdots & \vdots & \ddots & \vdots \\ 0 & 0 & 0 & \cdots & 1 \end{pmatrix}$$

$$= \det \begin{pmatrix} \dfrac{\partial x_1}{\partial u}(u,v) & \dfrac{\partial x_2}{\partial u}(u,v) \\ \dfrac{\partial x_1}{\partial v}(u,v) & \dfrac{\partial x_2}{\partial v}(u,v) \end{pmatrix} \neq 0.$$

Let $\mathbf{p} \in \mathcal{V}$; then $F(\mathbf{p}) \in \mathbf{x}(\mathcal{U}) \subseteq \mathcal{M}$. Because $\det\big(\mathcal{J}(\widetilde{\mathbf{x}})\big) \neq 0$ on $\mathcal{U}$, the inverse function theorem says that $\widetilde{\mathbf{x}}$ has an inverse $\widetilde{\mathbf{x}}^{-1}$ on a neighborhood $\widetilde{\mathcal{U}} \subseteq \mathbb{R}^n$ of $F(\mathbf{p})$, and that $\widetilde{\mathbf{x}}^{-1}$ is differentiable on $\widetilde{\mathcal{U}}$. Since F is continuous, there exists a neighborhood $\widetilde{\mathcal{V}}$ of $\mathbf{p}$ such that $F(\widetilde{\mathcal{V}}) \subset \widetilde{\mathcal{U}}$. Moreover,

$$\widetilde{\mathbf{x}}^{-1} \circ F|\widetilde{\mathcal{V}} = \mathbf{x}^{-1} \circ F|\widetilde{\mathcal{V}}.$$

Since $\widetilde{\mathbf{x}}^{-1} \circ F|\widetilde{\mathcal{V}}$ is differentiable at $\mathbf{p}$, so is $\mathbf{x}^{-1} \circ F|\widetilde{\mathcal{V}}$. Since $\mathbf{p} \in \mathcal{V}$ is arbitrary, it follows that $\mathbf{x}^{-1} \circ F$ is differentiable on all of $\mathcal{V}$. ∎

An important special case of Theorem 12.17 is:

Corollary 12.18. *Let $\mathcal{M}$ be a regular surface, and suppose we are given regular patches $\mathbf{x}: \mathcal{U} \longrightarrow \mathcal{M}$ and $\mathbf{y}: \mathcal{V} \longrightarrow \mathcal{M}$ such that $\mathbf{x}(\mathcal{U}) \cap \mathbf{y}(\mathcal{V}) = \mathcal{W}$ is nonempty. Then the* **change of coordinates**

(12.8) $$(\mathbf{x}^{-1} \circ \mathbf{y}): \mathbf{y}^{-1}(\mathcal{W}) \longrightarrow \mathbf{x}^{-1}(\mathcal{W})$$

is a diffeomorphism between open subsets of $\mathbb{R}^2$.

12.4 The Definition of a Regular Surface in $\mathbb{R}^n$

Proof. That $\mathbf{x}^{-1}\circ\mathbf{y}$ and $\mathbf{y}^{-1}\circ\mathbf{x}$ are differentiable is a consequence of Theorem 12.17. Since both $\mathbf{x}$ and $\mathbf{y}$ are homeomorphisms, so are $\mathbf{x}^{-1}\circ\mathbf{y}$ and $\mathbf{y}^{-1}\circ\mathbf{x}$. Moreover, these maps are inverses of each other. Hence $\mathbf{x}^{-1}\circ\mathbf{y}$ and $\mathbf{y}^{-1}\circ\mathbf{x}$ are diffeomorphisms. ∎

We can write (12.8) more explicitly as follows: there exist differentiable functions $\bar{u},\bar{v}$ such that
$$\mathbf{y}(u,v) = \mathbf{x}\bigl(\bar{u}(u,v),\bar{v}(u,v)\bigr).$$

Note that $(\bar{u},\bar{v}) = \mathbf{x}^{-1}\circ\mathbf{y}$.

Lemma 12.19. *Let* $\mathbf{x}:\mathcal{U} \longrightarrow \mathcal{M}$ *and* $\mathbf{y}:\mathcal{V} \longrightarrow \mathcal{M}$ *be patches on a regular surface* $\mathcal{M}$ *with* $\mathbf{x}(\mathcal{U})\cap\mathbf{y}(\mathcal{V})$ *nonempty. Let* $\mathbf{x}^{-1}\circ\mathbf{y} = (\bar{u},\bar{v}):\mathcal{U}\cap\mathcal{V} \longrightarrow \mathcal{U}\cap\mathcal{V}$ *be the associated change of coordinates, so that*

(12.9) $$\mathbf{y}(u,v) = \mathbf{x}\bigl(\bar{u}(u,v),\bar{v}(u,v)\bigr).$$

Then
(12.10) $$\mathbf{y}_u = \frac{\partial \bar{u}}{\partial u}\mathbf{x}_{\bar{u}} + \frac{\partial \bar{v}}{\partial u}\mathbf{x}_{\bar{v}} \quad and \quad \mathbf{y}_v = \frac{\partial \bar{u}}{\partial v}\mathbf{x}_{\bar{u}} + \frac{\partial \bar{v}}{\partial v}\mathbf{x}_{\bar{v}}.$$

Proof. (12.10) is an immediate consequence of (12.9) and the chain rule. ∎

We need to extend the calculus that we developed in Chapter 11 for $\mathbb{R}^n$ to regular surfaces. Our first task is to define what it means for a real-valued function on a regular surface to be differentiable.

Definition. *Let* $f:\mathcal{W} \subset \mathcal{M} \longrightarrow \mathbb{R}$ *be a function defined on an open subset* $\mathcal{W}$ *of a regular surface* $\mathcal{M}$. *We say that* f *is* **differentiable** *at* $\mathbf{p} \in \mathcal{W}$ *provided that for some patch* $\mathbf{x}:\mathcal{U} \subset \mathbb{R}^2 \longrightarrow \mathcal{M}$ *with* $\mathbf{p} \in \mathbf{x}(\mathcal{U}) \subset \mathcal{W}$, *the composition*

$$f\circ\mathbf{x}:\mathcal{U} \subset \mathbb{R}^2 \longrightarrow \mathbb{R}$$

is differentiable at $\mathbf{x}^{-1}(\mathbf{p})$. *If* f *is differentiable at all points of* $\mathcal{W}$, *we say that* f *is* **differentiable** *on* $\mathcal{W}$.

Lemma 12.20. *The definition of differentiability of a real-valued function on a regular surface does not depend on the choice of patch.*

Proof. If $\mathbf{x}$ and $\mathbf{y}$ are patches on a regular surface $\mathcal{M}$, then $\mathbf{y}^{-1}\circ\mathbf{x}$ is differentiable by Corollary 12.18. Since the composition of differentiable functions is differentiable, the result follows. ∎

Lemma 12.21. *Let M be a regular surface in $\mathbb{R}^n$. Then the restriction $f|M$ of any differentiable function $f\colon \mathbb{R}^n \longrightarrow \mathbb{R}$ to M is differentiable.*

Proof. If $\mathbf{x}$ is any patch on M, then $(f|M)\circ\mathbf{x} = f\circ\mathbf{x}$ is differentiable, since it is the composition of differentiable functions. Thus by definition $f|M$ is differentiable. ∎

A simple consequence of Lemma 12.21 is:

Corollary 12.22. *The restrictions to any regular surface $M \subset \mathbb{R}^n$ of the natural coordinate functions $u_1, \ldots, u_n$ of $\mathbb{R}^n$ are differentiable.*

Here are two important examples of differentiable functions on a regular surface.

Definition. *Let M be a regular surface and let $\mathbf{v} \in \mathbb{R}^n$. The **height function** and the **square of the distance function** of M relative to $\mathbf{v}$ are the functions $h, f\colon M \longrightarrow \mathbb{R}$ defined by*

$$h(\mathbf{p}) = \mathbf{p} \cdot \mathbf{v} \quad \text{and} \quad f(\mathbf{p}) = \|\mathbf{p} - \mathbf{v}\|^2.$$

Lemma 12.23. *For any regular surface $M \subset \mathbb{R}^n$, the height function and the square of distance function are differentiable.*

Proof. Both the height functions and the square of distance functions are algebraic combinations of the natural coordinate functions of $\mathbb{R}^n$. By Lemma 12.21 they are differentiable. ∎

We shall also need the notion of a curve on a regular surface.

Definition. *A **curve on a regular surface** $M \subset \mathbb{R}^n$ is a curve $\boldsymbol{\alpha}\colon (a,b) \longrightarrow \mathbb{R}^n$ such that $\boldsymbol{\alpha}(t) \in M$ for $a < t < b$. The curve $\boldsymbol{\alpha}$ is said to be **differentiable** provided that $(\mathbf{x}^{-1} \circ \boldsymbol{\alpha})\colon (a,b) \longrightarrow \mathcal{U}$ is differentiable for any patch $\mathbf{x}\colon \mathcal{U} \longrightarrow M$, where $\mathcal{U} \subseteq \mathbb{R}^2$.*

12.5 Tangent Vectors to Regular Surfaces in $\mathbb{R}^n$

We are now ready to discuss the important notion of tangent vector to a regular surface. This is the next step in the extension to regular surfaces of the calculus of $\mathbb{R}^n$ that we developed in Chapter 11.

12.5 Tangent Vectors to Regular Surfaces in $\mathbb{R}^n$

Definition. Let $\mathcal{M}$ be a regular surface in $\mathbb{R}^n$ and let $\mathbf{p} \in \mathcal{M}$. We say that $\mathbf{v}_\mathbf{p} \in \mathbb{R}^n_\mathbf{p}$ is **tangent** to $\mathcal{M}$ at $\mathbf{p}$ provided there exists a curve $\boldsymbol{\alpha}\colon (a,b) \longrightarrow \mathbb{R}^n$ such that $\boldsymbol{\alpha}(0) = \mathbf{p}$, $\boldsymbol{\alpha}'(0) = \mathbf{v}_\mathbf{p}$ and $\boldsymbol{\alpha}(t) \in \mathcal{M}$ for $a < t < b$. The **tangent space** to $\mathcal{M}$ at $\mathbf{p}$ is the set

$$\mathcal{M}_\mathbf{p} = \{\, \mathbf{v}_\mathbf{p} \in \mathbb{R}^n_\mathbf{p} \mid \mathbf{v}_\mathbf{p} \text{ is tangent to } \mathcal{M} \text{ at } \mathbf{p}\,\}.$$

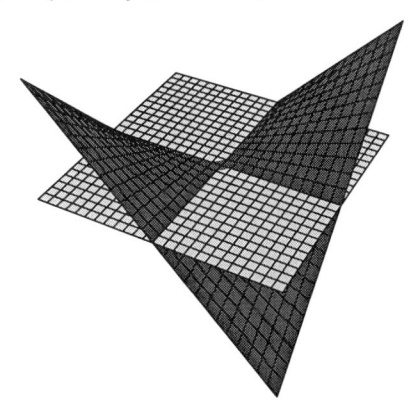

**A hyperbolic paraboloid
and one of its tangent planes**

On page 271 we defined $\mathbf{x}_u(u,v)$ and $\mathbf{x}_v(u,v)$ to be vectors in $\mathbb{R}^n$. Sometimes it is useful to modify this definition so that $\mathbf{x}_u(u,v)$ and $\mathbf{x}_v(u,v)$ are tangent vectors at $\mathbf{x}(u,v)$. Thus if $\mathbf{x}(u,v) = \bigl(x_1(u,v), \ldots, x_n(u,v)\bigr)$ we can redefine

$$\mathbf{x}_u(u,v) = \left(\frac{\partial x_1}{\partial u}(u,v), \ldots, \frac{\partial x_n}{\partial u}(u,v)\right)_{\mathbf{x}(u,v)}$$

and

$$\mathbf{x}_v(u,v) = \left(\frac{\partial x_1}{\partial v}(u,v), \ldots, \frac{\partial x_n}{\partial v}(u,v)\right)_{\mathbf{x}(u,v)}.$$

Similar remarks hold for $\mathbf{x}_{uu}$ and the other higher partial derivatives of $\mathbf{x}$. However, more often than not, such fine distinctions are unnecessary. For example, when we need to calculate the partial derivatives of a patch explicitly (either by hand or by *Mathematica*) we consider them to be vectors in $\mathbb{R}^n$. In summary, we regard $\mathbf{x}_u$ and the other partial derivatives of $\mathbf{x}$ to be vectors in $\mathbb{R}^n$ unless for some theoretical reason we require them to be tangent vectors at the point of application.

Lemma 12.24. *Let $\mathbf{p}$ be a point on a regular surface $\mathcal{M} \subset \mathbb{R}^n$. Then the tangent space $\mathcal{M}_\mathbf{p}$ to $\mathcal{M}$ at $\mathbf{p}$ is a 2-dimensional vector subspace of $\mathbb{R}^n_\mathbf{p}$. If $\mathbf{x}\colon \mathcal{U} \longrightarrow \mathcal{M}$ is* any regular patch on $\mathcal{M}$ with $\mathbf{p} = \mathbf{x}(\mathbf{q})$, then

$$\mathbf{x}_*(\mathbb{R}_\mathbf{q}^2) = \mathcal{M}_\mathbf{p}.$$

Proof. It follows from Lemma 12.1 that $\mathbf{x}_*(\mathbb{R}_\mathbf{q}^2)$ is spanned by $\mathbf{x}_u(\mathbf{q})$ and $\mathbf{x}_v(\mathbf{q})$. The regularity of $\mathbf{x}$ at $\mathbf{q}$ implies that $\mathbf{x}_*: \mathbb{R}_\mathbf{q}^2 \longrightarrow \mathbb{R}_\mathbf{p}^n$ is injective; consequently, $\mathbf{x}_u(\mathbf{q})$ and $\mathbf{x}_v(\mathbf{q})$ are linearly independent and $\dim \mathbf{x}_*(\mathbb{R}_\mathbf{q}^2) = 2$ by Lemma 12.4.

On the other hand, Lemma 12.8 implies that any tangent vector to $\mathcal{M}$ at $\mathbf{q}$ is a linear combination of $\mathbf{x}_u(\mathbf{q})$ and $\mathbf{x}_v(\mathbf{q})$. Hence the lemma follows. ∎

Frequently, we shall need the notion of tangent vector perpendicular to a surface.

Definition. *Let $\mathcal{M}$ be a regular surface in $\mathbb{R}^n$ and let $\mathbf{z}_\mathbf{p} \in \mathbb{R}_\mathbf{p}^n$ with $\mathbf{p} \in \mathcal{M}$. We say that $\mathbf{z}_\mathbf{p}$ is* **normal** *or* **perpendicular** *to $\mathcal{M}$ at $\mathbf{p}$, provided $\mathbf{z}_\mathbf{p} \cdot \mathbf{v}_\mathbf{p} = 0$ for all tangent vectors $\mathbf{v}_\mathbf{p} \in \mathcal{M}_\mathbf{p}$. We denote the set of normal vectors to $\mathcal{M}$ at $\mathbf{p}$ by $\mathcal{M}_\mathbf{p}^\perp$; then $\mathbb{R}_\mathbf{p}^n = \mathcal{M}_\mathbf{p} \oplus \mathcal{M}_\mathbf{p}^\perp$.*

The notions of tangent and normal vector fields also make sense.

Definition. *A* **vector field** *V on a regular surface $\mathcal{M}$ is a function which assigns to each point $\mathbf{p} \in \mathcal{M}$ a tangent vector $V(\mathbf{p}) \in \mathbb{R}_\mathbf{p}^n$. We say that V is* **tangent** *to $\mathcal{M}$ if $V(\mathbf{p}) \in \mathcal{M}_\mathbf{p}$ for all $\mathbf{p} \in \mathcal{M}$ and that V is* **normal** *or* **perpendicular** *to $\mathcal{M}$ if $V(\mathbf{p}) \in \mathcal{M}_\mathbf{p}^\perp$ for all $\mathbf{p} \in \mathcal{M}$.*

12.6 Surface Mappings

The definition of differentiability of a mapping between regular surfaces is similar to that of a real-valued function on a regular surface.

Definition. *A function $F: \mathcal{M} \longrightarrow \mathcal{N}$ from one regular surface to another is* **differentiable**, *provided that for any two regular injective patches $\mathbf{x}$ of $\mathcal{M}$ and $\mathbf{y}$ of $\mathcal{N}$ the composition $\mathbf{y}^{-1} \circ F \circ \mathbf{x}$ is differentiable. When this is the case, we call F a* **surface mapping**.

The simplest example of a surface mapping is the identity map $\mathbf{1}_\mathcal{M}: \mathcal{M} \longrightarrow \mathcal{M}$ (defined by $\mathbf{1}_\mathcal{M}(\mathbf{p}) = \mathbf{p}$ for $\mathbf{p} \in \mathcal{M}$).

12.6 Surface Mappings

Definition. A **diffeomorphism** between regular surfaces $\mathcal{M}$ and $\mathcal{N}$ is a differentiable map $F:\mathcal{M} \longrightarrow \mathcal{N}$ which has a differentiable inverse, that is, a surface mapping $G:\mathcal{N} \longrightarrow \mathcal{M}$ such that

$$G \circ F = \mathbf{1}_{\mathcal{M}} \quad \text{and} \quad F \circ G = \mathbf{1}_{\mathcal{N}},$$

where $\mathbf{1}_{\mathcal{M}}$ and $\mathbf{1}_{\mathcal{N}}$ denote the identity maps of $\mathcal{M}$ and $\mathcal{N}$.

Definition. Let $\mathcal{U}$ be an open subset of $\mathcal{M}$. We shall say that a mapping

$$\varphi : \mathcal{U} \subset \mathcal{M} \longrightarrow \mathcal{N}$$

is a **local diffeomorphism** at $\mathbf{p} \in \mathcal{U}$, provided there exists a neighborhood $\mathcal{V} \subset \mathcal{U}$ of $\mathbf{p}$ such that the restriction of φ to $\mathcal{V}$ is a diffeomorphism of $\mathcal{V}$ onto an open subset $\varphi(\mathcal{V}) \subseteq \mathcal{N}$.

Just as a regular surface has a tangent space, a surface mapping has a tangent map.

Definition. Let $\mathcal{M}$ and $\mathcal{N}$ be regular surfaces in $\mathbb{R}^n$, and let $\mathbf{p} \in \mathcal{M}$. Let $F:\mathcal{M} \longrightarrow \mathcal{N}$ be a surface mapping. Then the **tangent map** of F at $\mathbf{p}$ is the map

$$F_* : \mathcal{M}_{\mathbf{p}} \longrightarrow \mathcal{N}_{F(\mathbf{p})}$$

given as follows. For $\mathbf{v}_{\mathbf{p}} \in \mathcal{M}_{\mathbf{p}}$, choose a curve $\boldsymbol{\alpha}: (a,b) \longrightarrow \mathcal{M}$ such that $\boldsymbol{\alpha}'(0) = \mathbf{v}_{\mathbf{p}}$. Then we define $F_*(\mathbf{v}_{\mathbf{p}})$ to be the initial velocity of the image curve $F \circ \boldsymbol{\alpha}: \longrightarrow \mathcal{N}$; that is,

$$F_*(\mathbf{v}_{\mathbf{p}}) = (F \circ \boldsymbol{\alpha})'(0).$$

Lemma 12.25. *Let $F:\mathcal{M} \longrightarrow \mathcal{N}$ be a surface mapping, and let $\mathbf{v}_{\mathbf{p}} \in \mathcal{M}_{\mathbf{p}}$. The definition of $F_*(\mathbf{v}_{\mathbf{p}})$ is independent of the choice of curve $\boldsymbol{\alpha}$ with $\boldsymbol{\alpha}'(0) = \mathbf{v}_{\mathbf{p}}$. Furthermore, $F_*:\mathcal{M}_{\mathbf{p}} \longrightarrow \mathcal{N}_{F(\mathbf{p})}$ is a linear map.*

Proof. Let $\boldsymbol{\alpha}$ and $\widetilde{\boldsymbol{\alpha}}$ be curves in $\mathcal{M}$ with $\boldsymbol{\alpha}(0) = \widetilde{\boldsymbol{\alpha}}(0) = \mathbf{p}$ and $\boldsymbol{\alpha}'(0) = \widetilde{\boldsymbol{\alpha}}'(0) = \mathbf{v}_{\mathbf{p}}$. Let $\mathbf{x}:\mathcal{U} \longrightarrow \mathbb{R}^n$ be a regular patch on $\mathcal{M}$ such that $\mathbf{x}(u_0, v_0) = \mathbf{p}$. Write

$$\boldsymbol{\alpha}(t) = \mathbf{x}\big(u(t), v(t)\big) \quad \text{and} \quad \widetilde{\boldsymbol{\alpha}}(t) = \mathbf{x}\big(\widetilde{u}(t), \widetilde{v}(t)\big).$$

Just as in the proof of Lemma 12.8, the chain rule for curves implies that

$$(F \circ \boldsymbol{\alpha})'(t) = u'(t)(F \circ \mathbf{x})_u(u(t), v(t)) + v'(t)(F \circ \mathbf{x})_v(u(t), v(t))$$

and
$$(F \circ \widetilde{\alpha})'(t) = \widetilde{u}'(t)(F \circ \mathbf{x})_u\bigl(\widetilde{u}(t), \widetilde{v}(t)\bigr) + \widetilde{v}'(t)(F \circ \mathbf{x})_v\bigl(\widetilde{u}(t), \widetilde{v}(t)\bigr).$$

Since $u(0) = \widetilde{u}(0) = u_0$, $v(0) = \widetilde{v}(0) = v_0$, $u'(0) = \widetilde{u}'(0)$ and $v'(0) = \widetilde{v}'(0)$, it follows that
$$F_*(\mathbf{v_p}) = (F \circ \alpha)'(0) = (F \circ \widetilde{\alpha})'(0).$$

This proves that $F_*(\mathbf{v_p})$ does not depend on the choice of α. To prove the linearity of F_*, we note that
$$F_*(\mathbf{v_p}) = u'(0)(F \circ \mathbf{x})_u\bigl(u(0), v(0)\bigr) + v'(0)(F \circ \mathbf{x})_v\bigl(u(0), v(0)\bigr).$$

Since $u'(0)$ and $v'(0)$ depend linearly on $\mathbf{v_p}$, it follows that F_* must be linear. ∎

We have the following consequence of the inverse function theorem for $\mathbb{R}^2$.

Theorem 12.26. *If $\mathcal{M}$ and $\mathcal{N}$ are regular surfaces and $\varphi: \mathcal{W} \longrightarrow \mathcal{N}$ is a differentiable mapping of an open subset $\mathcal{W} \subseteq \mathcal{M}$ such that the tangent map φ_* of φ at $\mathbf{p} \in \mathcal{W}$ is an isomorphism, then φ is a local diffeomorphism at $\mathbf{p}$.*

Proof. Let $\mathbf{x}: \mathcal{U} \longrightarrow \mathcal{M}$ and $\mathbf{y}: \mathcal{V} \longrightarrow \mathcal{N}$ be injective regular patches on $\mathcal{M}$ and $\mathcal{N}$ such that $(\varphi \circ \mathbf{x})(\mathcal{U}) \cap \mathbf{y}(\mathcal{V})$ is nonempty. By restricting the domains of definition of $\mathbf{x}$ and $\mathbf{y}$ if necessary, we can assume that $(\varphi \circ \mathbf{x})(\mathcal{U}) = \mathbf{y}(\mathcal{V})$. Then the tangent map of $\mathbf{y}^{-1} \circ \varphi \circ \mathbf{x}$ is an isomorphism, and the inverse function theorem for $\mathbb{R}^2$ implies that the map $\mathbf{y}^{-1} \circ \varphi \circ \mathbf{x}$ possesses a local inverse. Hence φ also has a local inverse. ∎

Examples of Surface Mappings

1. Let $\mathbf{y}: \mathcal{U} \longrightarrow \mathcal{M}$ be a regular patch on a regular surface $\mathcal{M}$. Corollary 12.18 implies that $\mathbf{y}$ is a diffeomorphism between the regular surfaces $\mathcal{U}$ and $\mathbf{y}(\mathcal{U})$.

2. Let $S^2(a) = \{\mathbf{p} \mid \|\mathbf{p}\| = a\}$ be the sphere of radius a in $\mathbb{R}^3$. Then the **antipodal map** is the function **antipodal**: $S^2(a) \longrightarrow S^2(a)$ is defined by **antipodal**$(\mathbf{p}) = -\mathbf{p}$. It is a diffeomorphism.

3. Let $F: \mathbb{R}^n \longrightarrow \mathbb{R}^n$ be a diffeomorphism, and let $\mathcal{M} \subset \mathbb{R}^n$ be a regular surface. Then the restriction $F|\mathcal{M}: \mathcal{M} \longrightarrow F(\mathcal{M})$ is a diffeomorphism of regular surfaces. In particular, any Euclidean motion of $\mathbb{R}^n$ gives rise to a surface mapping.

12.7 Level Surfaces in $\mathbb{R}^3$

So far we have been considering regular surfaces defined by patches. Such a description is called a **parametric representation**. Another way to describe a regular surface $\mathcal{M}$ is by means of a **nonparametric representation**. For a regular surface in $\mathbb{R}^3$ this means that $\mathcal{M}$ is the set of points mapped by a differentiable function $g\colon \mathbb{R}^3 \longrightarrow \mathbb{R}$ into the same real number.

Definition. *Let $g\colon \mathbb{R}^3 \longrightarrow \mathbb{R}$ be a differentiable function and c a real number. Then the set*

$$\mathcal{M}(c) = \{\, \mathbf{p} \in \mathbb{R}^3 \mid g(\mathbf{p}) = c \,\}$$

*is called the **level surface** of g corresponding to c.*

Theorem 12.27. *Let $g\colon \mathbb{R}^3 \longrightarrow \mathbb{R}$ be a differentiable function and c a real number. Then the level surface $\mathcal{M}(c)$ of g is a regular surface if it is nonempty and the gradient $\mathbf{grad}\, g$ is nonzero at all points of $\mathcal{M}(c)$. When these conditions are satisfied, $\mathbf{grad}\, g$ is everywhere perpendicular to $\mathcal{M}(c)$.*

Proof. For each $\mathbf{p} \in \mathcal{M}(c)$ we must find a regular patch on a neighborhood of $\mathbf{p}$. The hypothesis that $(\mathbf{grad}\, g)(\mathbf{p}) \neq 0$ is equivalent to saying that at least one of the partial derivatives

$$\left\{ \frac{\partial g}{\partial x}, \frac{\partial g}{\partial y}, \frac{\partial g}{\partial z} \right\}$$

does not vanish at $\mathbf{p}$. Let us suppose that $(\partial g/\partial z)(\mathbf{p}) \neq 0$. The implicit function theorem states that the equation $g(x, y, z) = c$ can be solved for z. More precisely, there exists a function h such that

$$g\bigl(x, y, h(x, y)\bigr) = c.$$

Then the required patch is defined by

$$\mathbf{x}(u, v) = \bigl(u, v, h(u, v)\bigr).$$

Furthermore, let $\mathbf{v}_\mathbf{p} = (v_1, v_2, v_3)_\mathbf{p} \in \mathcal{M}(c)_\mathbf{p}$. Then there exists a curve $\boldsymbol{\alpha}$ on $\mathcal{M}(c)$ with $\boldsymbol{\alpha}(0) = \mathbf{p}$ and $\boldsymbol{\alpha}'(0) = \mathbf{v}_\mathbf{p}$. Write $\boldsymbol{\alpha}(t) = \bigl(a_1(t), a_2(t), a_3(t)\bigr)$. Since $\boldsymbol{\alpha}$ lies on $\mathcal{M}(c)$, we have $g\bigl(a_1(t), a_2(t), a_3(t)\bigr) = c$ for all t. The chain rule implies that

$$\sum_{i=1}^{3} \left(\frac{\partial g}{\partial u_i} \circ \boldsymbol{\alpha} \right) \frac{da_i}{dt} = 0.$$

In particular,

$$0 = \sum_{i=1}^{3} \frac{\partial g}{\partial u_i}(\boldsymbol{\alpha}(0))\frac{da_i}{dt}(0) = \sum_{i=1}^{3} \frac{\partial g}{\partial u_i}(\mathbf{p})v_i = (\mathbf{grad}\, g)(\mathbf{p}) \cdot \mathbf{v}_\mathbf{p}.$$

Hence $(\mathbf{grad}\, g)(\mathbf{p})$ is perpendicular to $\mathcal{M}(c)_\mathbf{p}$ for each $\mathbf{p} \in \mathcal{M}(c)$. ∎

For example, consider the function $g[n]\colon \mathbb{R}^3 \longrightarrow \mathbb{R}$ defined by

$$g[n](x,y,z) = x^n + y^n + z^n,$$

where n is an even integer greater than or equal to 2. The *Mathematica* version of this function is

```
g[n_][x_,y_,z_] := x^n + y^n + z^n
```

The gradient of $g[n]$ is easily computed either by hand or with *Mathematica*:

```
grad[g[n]][x,y,z]
                -1 + n      -1 + n      -1 + n
           {n x      , n y      , n z        }
```

Hence $\mathbf{grad}\,(g[n])(x,y,z) = 0$ if and only if $x = y = z = 0$. Thus we see that the set $\{\mathbf{p} \in \mathbb{R}^3 \mid g[n](\mathbf{p}) = c\}$ is a regular surface for any $c > 0$. Obviously, $\{\mathbf{p} \in \mathbb{R}^3 \mid g[2](\mathbf{p}) = c\}$ is a sphere. As n becomes larger and larger, $\{\mathbf{p} \in \mathbb{R}^3 \mid g[n](\mathbf{p}) = c\}$ becomes more and more cube-like.

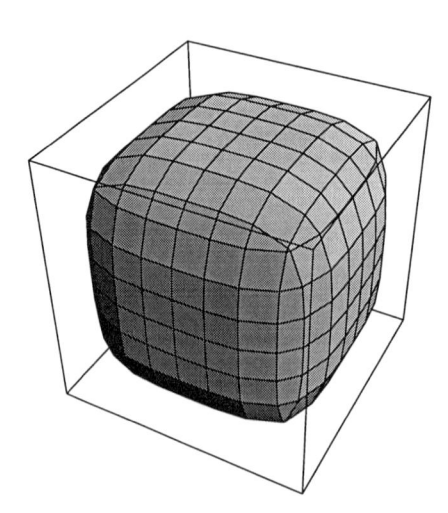

The regular surface
$x^4 + y^4 + z^4 = 1$

12.8 Exercises

Note that there are functions g for which $\{\mathbf{p} \in \mathbb{R}^3 \mid g(\mathbf{p}) = c\}$ consists of several components. For example, the regular surface defined by

$$(x^2 + y^2 + z^2 - 1)((x-2)^2 + y^2 + z^2 - 1) = 0$$

consists of two disjoint spheres.

Because the gradient of a function $g \colon \mathbb{R}^3 \longrightarrow \mathbb{R}$ is always perpendicular to each of its level surfaces, there is an easy way to obtain the Gauss map for such surfaces.

Lemma 12.28. *Let* $g \colon \mathbb{R}^3 \longrightarrow \mathbb{R}$ *be a differentiable function and c a number such that* **grad** g *is nonzero on all points of the level surface* $\mathcal{M}(c) = \{\mathbf{p} \in \mathbb{R}^3 \mid g(\mathbf{p}) = c\}$. *Then the vector field* $\mathbf{U}$ *defined by*

$$\mathbf{U}(\mathbf{p}) = \frac{\operatorname{grad} g(\mathbf{p})}{\|\operatorname{grad} g(\mathbf{p})\|}$$

is a globally defined unit normal on $\mathcal{M}(c)$. *Hence we can define the Gauss map of* $\mathcal{M}(c)$ *to be* $\mathbf{U}$ *considered as a mapping* $\mathbf{U} \colon \mathcal{M}(c) \longrightarrow S^2(1)$.

12.8 Exercises

1. Denote by A^T the transpose of a matrix A. Show that the Jacobian matrix of a patch $\mathbf{x} \colon \mathcal{U} \longrightarrow \mathbb{R}^n$ is related to $\mathbf{x}_u$ and $\mathbf{x}_v$ by the formula

$$\mathcal{J}(\mathbf{x})\mathcal{J}(\mathbf{x})^T = \begin{pmatrix} \mathbf{x}_u \cdot \mathbf{x}_u & \mathbf{x}_u \cdot \mathbf{x}_v \\ \mathbf{x}_v \cdot \mathbf{x}_u & \mathbf{x}_v \cdot \mathbf{x}_v \end{pmatrix}.$$

   When $n = 3$ show that

$$\det\left(\mathcal{J}(\mathbf{x})\mathcal{J}(\mathbf{x})^T\right) = \|\mathbf{x}_u \times \mathbf{x}_v\|^2.$$

2. The following miniprograms draw a patch and the tangent space to the patch at a specified point:

```
tangentplane[x_][u0_,v0_][u_,v_] := x[u0,v0] +
    D[x[uu0,vv0],uu0]u + D[x[uu0,vv0],vv0]v /.
    {uu0->u0,vv0->v0}
```

```
plottp[surf_,{u0_,v0_},{umin_,umax_},
    {vmin_,vmax_},opts___]:=
  ParametricPlot3D[{surf[u0 + u,v0 + v],
      tangentplane[surf][u0,v0][u,v]}//Evaluate,
    {u,umin,umax},{v,vmin,vmax},
    Boxed->False,Axes->None,opts];
```

Use them to draw a portion of a sphere together with its tangent plane at some point.

3. Determine where the patch
$$(u,v) \longmapsto (\cos u \cos v \sin v, \sin u \cos v \sin v, \sin v)$$
is regular.

4. A torus can be defined as a level surface $\mathcal{M}(b^2)$ of the function $f\colon \mathbb{R}^3 \longrightarrow \mathbb{R}$ defined by
$$f(x,y,z) = z^2 + \left(\sqrt{x^2+y^2} - a\right)^2,$$
where $a > b > 0$. Show that such a torus is a regular surface and compute its unit normal.

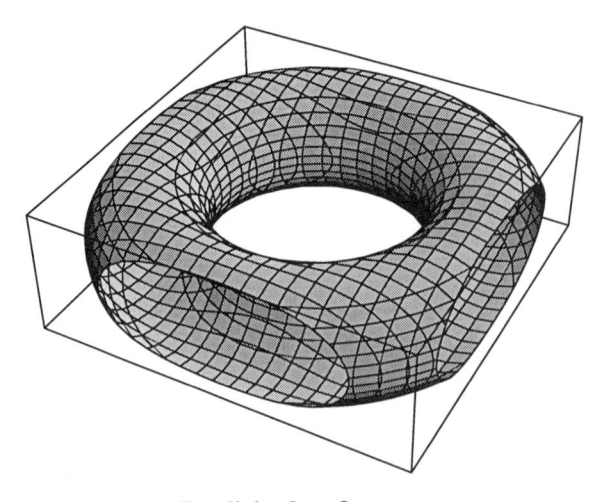

Implicit plot of a torus

13
EXAMPLES OF SURFACES

The intuitive idea of a surface is a 2-dimensional set of points. Globally the surface may be rather complicated and not look at all like a plane, but any sufficiently small piece of the surface should look like a warped portion of a plane. Here are two candidates for a surface.

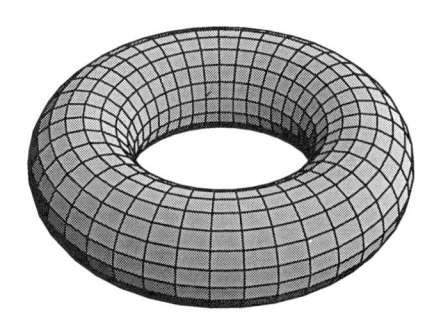

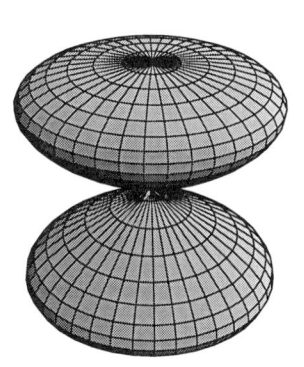

The first is a torus; it is a regular surface as defined in Section 12.4. The second candidate (which consists of a figure eight revolved about the z-axis) is not a regular surface. However, it is the trace of a patch, namely, the patch is defined by

$$\mathbf{eightsurface}(u,v) = (\cos u \cos v \sin v, \sin u \cos v \sin v, \sin v).$$

295

The eight surface becomes a regular surface only when the center point is excluded. We shall have little difficulty working with it and other surfaces with singularities using *Mathematica*.

It is quite easy to draw surfaces using *Mathematica*. The command **Plot3D** is mainly useful to graph functions of two variables. More important for us is **ParametricPlot3D**. Although this command is slower than **Plot3D**, it is much more versatile. Any patch, regular or not, can be plotted with the command **ParametricPlot3D** provided the three functions that define the surface are known to *Mathematica*.

Visualization of a surface in $\mathbb{R}^3$ is usually enhanced if the surface is drawn with a grid. As we know from Section 12.4, the net of u- and v-parameter curves of a patch **x** constitute such a grid. It is fortuitous that **ParametricPlot3D** reproduces this net. Let us emphasize this fact.

> **ParametricPlot3D** *draws the u- and v-parameter curves.*

Consequently, *Mathematica* is a very useful tool in visualizing analytical properties of the u- and v-parameter curves of a patch.

In this chapter we show how to use *Mathematica* to draw some well-known surfaces, and also a few interesting but not so well-known surfaces. In later chapters we shall show how to compute for each of these surfaces various geometric quantities such as curvature.

Mathematica only draws a polyhedral surface that is a linear approximation to an actual surface. This approximation can be as fine as we require. However, there is a trade off, because very detailed approximations can sometimes take a very long time to draw. On the first attempt to draw a surface it is usually best to use *Mathematica's* defaults. After any syntax errors are corrected, the command to draw the surface can be modified to give a finer, hence more realistic, approximation.

13.1 The Graph of a Function of Two Variables

Let $h: \mathcal{U} \longrightarrow \mathbb{R}$ be a real-valued differentiable function of two variables, where $\mathcal{U}$ is an open subset of $\mathbb{R}^2$. Then $\mathbf{x}: \mathcal{U} \longrightarrow \mathbb{R}^3$ defined by

$$\mathbf{x}(u,v) = \bigl(u, v, h(u,v)\bigr)$$

13.1 The Graph of a Function of Two Variables

is a surface, which in fact is the graph of h. It is a regular patch, because the Jacobian matrix

$$\mathcal{J}(\mathbf{x}) = \begin{pmatrix} 1 & 0 & \dfrac{\partial h}{\partial u} \\ 0 & 1 & \dfrac{\partial h}{\partial v} \end{pmatrix}$$

obviously has rank 2. The nonparametric form of the graph of a function is simply $z = h(x, y)$.

The Hyperbolic Paraboloid

The simplest parametrization of the **hyperbolic paraboloid** is

(13.1) $$\text{hyperbolicparaboloid}(u, v) = (u, v, u\,v),$$

or

```
hyperbolicparaboloid[u_,v_] := {u,v,u v}
```

It may be drawn in *Mathematica*, for example, with the command

```
ParametricPlot3D[Evaluate[
hyperbolicparaboloid[u,v]],
{u,-1.5,1.5},{v,-1.5,1.5},
ViewPoint->{1,3,1},
PlotPoints->{20,20},
Axes->None,Boxed->False];
```

(The option **Boxed->False** draws the surface without the box containing it, and **Axes->None** eliminates the coordinate axes.)

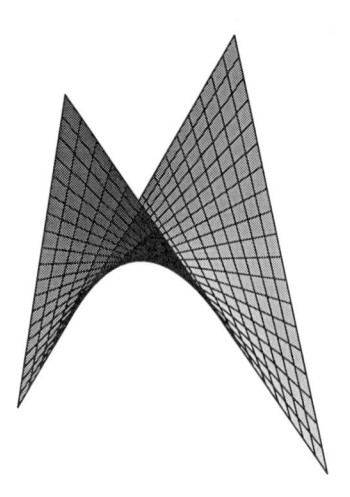

The hyperbolic paraboloid $(u,v) \longmapsto (u,v,uv)$
drawn with `ParametricPlot3D`

For graphs of functions, the graphics command `Plot3D` is available; this command is much faster than the command `ParametricPlot3D`. For example, we can use the command `Plot3D[u v,{u,-1.5,1.5},{v,-1.5,1.5}]` to draw the hyperbolic paraboloid:

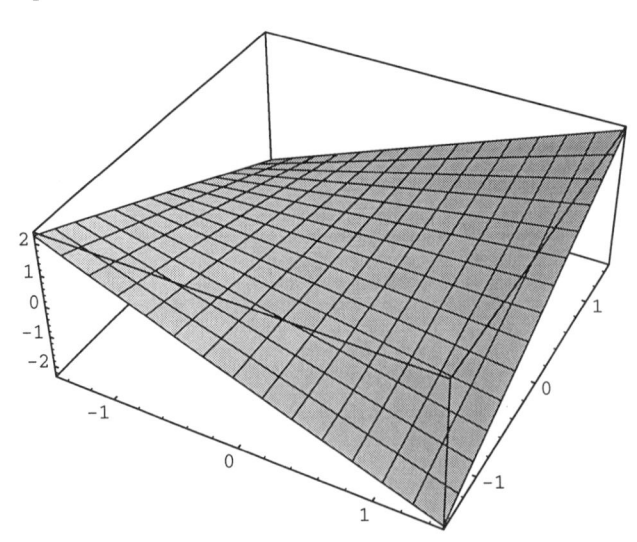

The hyperbolic paraboloid $(u,v) \longmapsto (u,v,uv)$
drawn with `Plot3D`

The Monkey Saddle

Because of its shape the hyperbolic paraboloid is often called a **saddle surface**. A person could sit comfortably on a hyperbolic paraboloid because there are indentations for legs. However, a monkey would have difficulty because there is no space for a tail. The **monkey saddle**, on the other hand, would be just right for a monkey; it is the patch given by

(13.2) $$\mathbf{monkeysaddle}(u,v) = (u, v, u^3 - 3uv^2),$$

or

```
monkeysaddle[u_,v_]:= {u,v,u^3 - 3u v^2}
```

It can be drawn in *Mathematica* with the command

```
ParametricPlot3D[Evaluate[
monkeysaddle[u,v]],{u,-1.5,1.5},{v,-1.5,1.5},
ViewPoint->{1,3,1},PlotPoints->{20,20},
Axes->None,Boxed->False,AspectRatio->1.2];
```

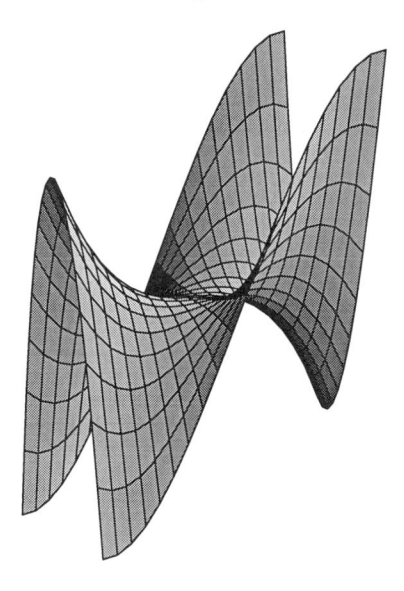

The monkey saddle $(u,v) \longmapsto (u, v, u^3 - 3uv^2)$

The hyperbolic paraboloid and the monkey saddle are the first two members of a series of surfaces. These are the graphs of the real (or imaginary) part of the function $z \longmapsto z^n$, where z is a complex number, or more explicitly,

$$\mathbf{monkey}[n](u,v) = \big(u, v, \Re((u+iv)^n)\big).$$

The monkey saddle for a monkey with two legs and three tails can be drawn nicely with **Plot3D**, specifically the command

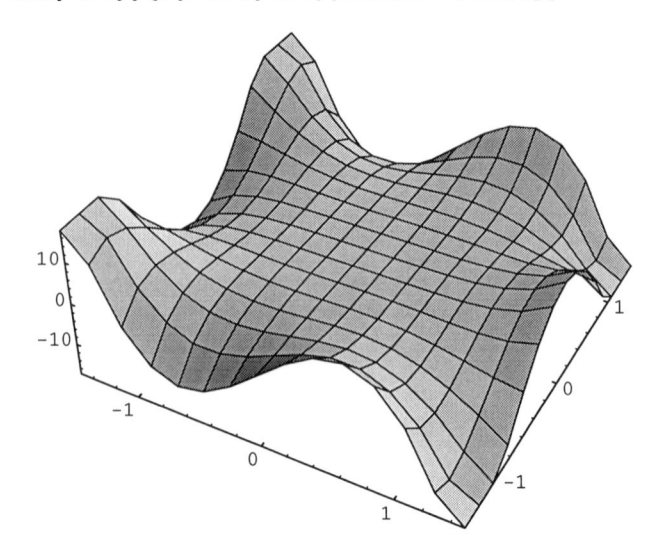

```
Plot3D[Re[(u + I v)^5],
{u,-1.5,1.5},{v,-1.5,1.5},Boxed->False];
```

The generalized monkey saddle $(u,v) \longmapsto \Re\big((u+iv)^5\big)$
drawn with **Plot3D**

The command **ContourPlot** is useful for understanding the generalized monkey saddle. The syntax of **ContourPlot** is similar to that of **Plot3D**. For example, the command to draw the **ContourPlot** version of the saddle for the 3-tailed monkey is

```
ContourPlot[Re[(u + I v)^5],
{u,-1.5,1.5},{v,-1.5,1.5},PlotPoints->40];
```

But instead of drawing a 3-dimensional object, it draws on a surface those lines that have the same height above the xy-plane.

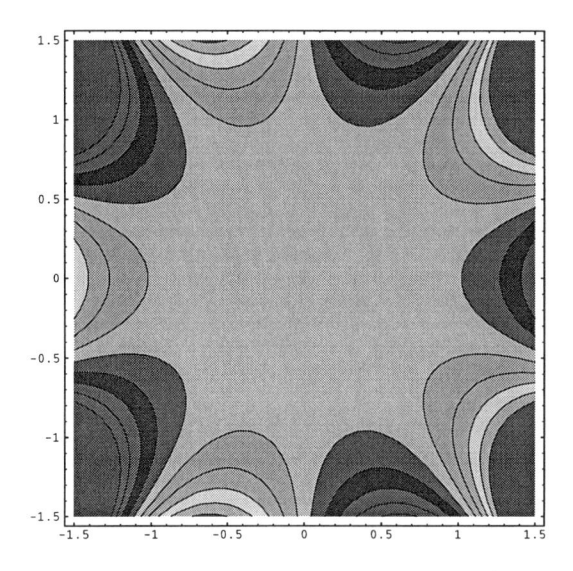

The generalized monkey saddle $(u,v) \longmapsto \mathfrak{Re}\bigl((u+iv)^5\bigr)$
drawn with `ContourPlot`

13.2 The Ellipsoid

The nonparametric equation that defines **the ellipsoid** is

(13.3) $$\frac{x^2}{a^2} + \frac{y^2}{b^2} + \frac{z^2}{c^2} = 1.$$

Here, a, b and c are the lengths of the axes of the ellipsoid. Because they are surfaces of revolution, ellipsoids, with only two of a, b and c distinct, are considerably simpler than the general ellipsoids. Such an ellipsoid is called an **ellipsoid of revolution**; it is the surface obtained by rotating an ellipse about a line, which can be taken to be the z-axis. (Surfaces of revolution will be studied in detail in Chapter 20.) Here are pictures that illustrate the difference between an ellipsoid of revolution and a general ellipsoid.

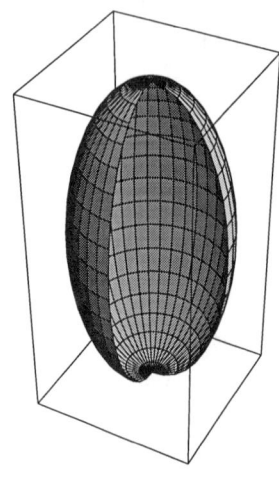

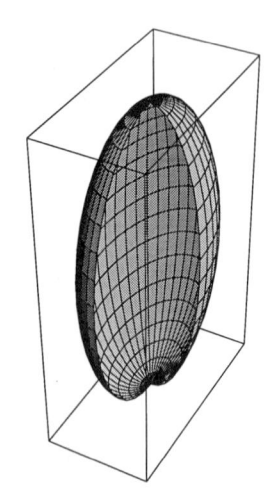

$$x^2 + y^2 + \frac{z^2}{4} = 1 \qquad\qquad x^2 + \frac{y^2}{4} + \frac{z^2}{9} = 1$$

The parametric form of (13.3) is

$$\mathbf{ellipsoid}[a, b, c](u, v) = (a \cos v \cos u, b \cos v \sin u, c \sin v),$$

or

```
ellipsoid[a_,b_,c_][u_,v_]:=
    {a Cos[v]Cos[u],b Cos[v]Sin[u],c Sin[v]}
```

The **ellipsoid**$[a, b, c]$: $[0, 2\pi] \times [-\pi/2, \pi/2] \longrightarrow \mathbb{R}^3$ is regular except at the north and south poles (which are $(0, 0, \pm c)$).

13.3 The Stereographic Ellipsoid

In this section we define and plot a different parametrization of an ellipsoid, the **stereographic ellipsoid**. It is a generalization of the stereographic projection of the sphere that can be found in many books on complex variables. For viewing convenience, it is rotated so that the "north pole" appears at the front of the picture.

13.3 The Stereographic Ellipsoid

First, the definition:

$$\text{stereographicellipsoid}[a, b, c](u, v)$$
$$= \left(\frac{2a\,u}{1+u^2+v^2}, \frac{2b\,v}{1+u^2+v^2}, \frac{c(u^2+v^2-1)}{1+u^2+v^2} \right),$$

or

```
stereographicellipsoid[a_,b_,c_][u_,v_]:=
    {2a u/(u^2 + v^2 + 1),2b v/(u^2 + v^2 + 1),
    c(u^2 + v^2 - 1)/(u^2 + v^2 + 1)}
```

Here is the plot:

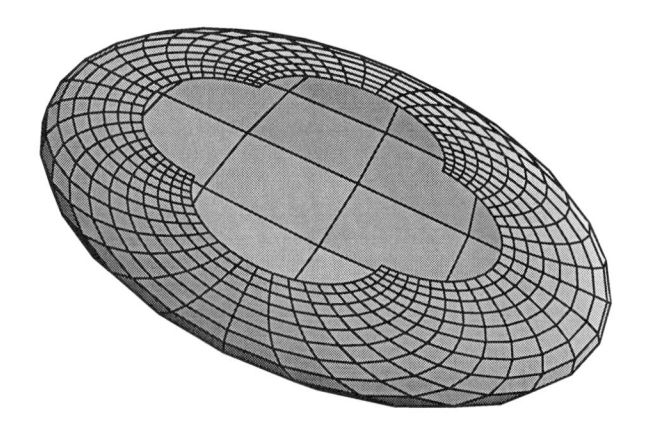

$(u, v) \longmapsto \text{stereographicellipsoid}[5, 3, 1](u, v)$

```
ParametricPlot3D[Evaluate[
stereographicellipsoid[5,3,1][u,v]],
{u,-2.8,2.8},{v,-2.8,2.8},
ViewPoint->{1,-1.7,2.7},
Boxed->False,Axes->None,PlotPoints->25];
```

13.4 Tori

The patch **torus**$[a, b, c]$: $[0, 2\pi) \times [0, 2\pi) \longrightarrow \mathbb{R}^3$ defined by

(13.4) **torus**$[a, b, c](u, v) = \big((a + b\cos v)\cos u, (a + b\cos v)\sin u, c\sin v\big)$

is a parametrization of an **elliptical torus** in $\mathbb{R}^3$. The *Mathematica* version is

```
torus[a_,b_,c_][u_,v_]:=
    {(a + b Cos[v])Cos[u],(a + b Cos[v])Sin[u],c Sin[v]}
```

When $b = c < a$, we get a standard round torus, where a is the **wheel radius** and b the **tube radius** of the torus. A torus with $a > b$ and $a > c$ is called a **ring torus**. Here is an elliptical ring torus of wheel radius 8, specifically **torus[8,3,7]**.

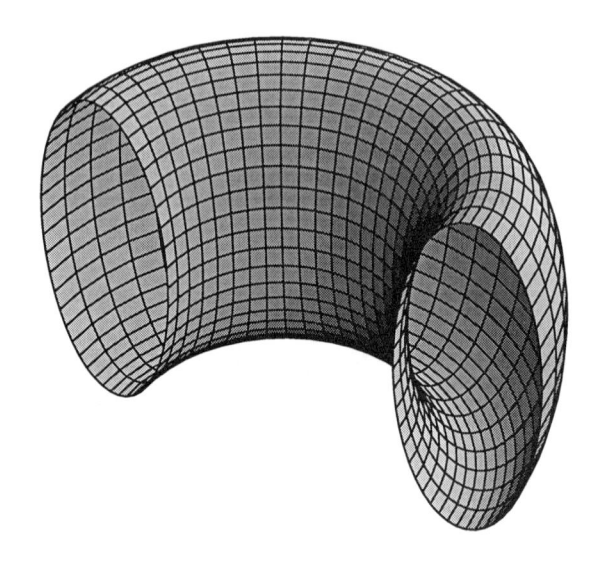

The elliptical ring torus
$(u, v) \longmapsto \big((8 + 3\cos v)\cos u, (8 + 3\cos v)\sin u, 7\sin v\big)$

13.4 Tori

It was drawn with the command

```
ParametricPlot3D[Evaluate[
torus[8,3,7][u,v]],
{u,0,Pi},{v,0,2Pi},
PlotPoints->{24,48},
Axes->None,Boxed->False];
```

An elliptical ring torus is a regular surface in the sense of Section 12.4. Here are four patches which cover it.

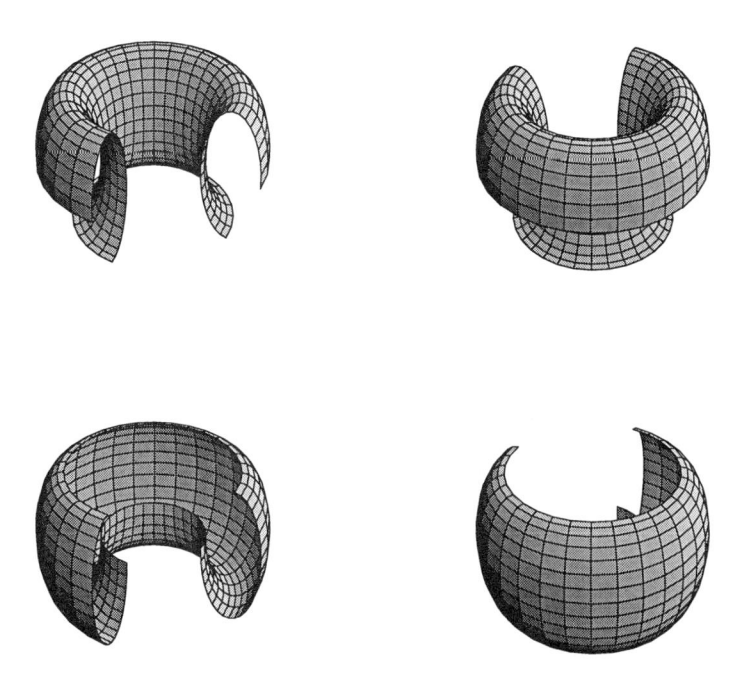

**Four patches that cover an
elliptical ring torus**

If the wheel radius is less than the tube radius, the resulting torus becomes a self-intersecting surface. The following torus has wheel radius 3 and tube radius 8:

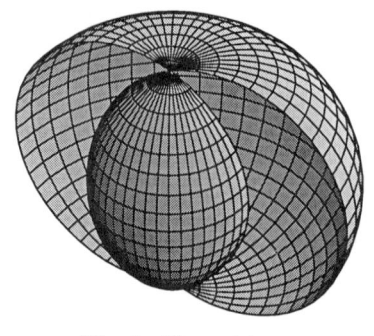

```
ParametricPlot3D[
torus[3,8,8][
u,v]//Evaluate,
{u,0,Pi},{v,0,2Pi},
PlotPoints->{24,48},
Axes->None,
Boxed->False];
```

The inside-out torus
$$(u,v) \longmapsto$$
$$\bigl(\cos u(3+8\cos v), \sin u(3+8\cos v), 8\sin v\bigr)$$

There are three kinds of tori. The torus **torus[8, 3, 7]** is an example of a **ring torus** and an inside-out torus is sometimes called a **horn torus**. The intermediate case when the wheel radius equals the tube radius is called a **spindle torus**. Here is an example:

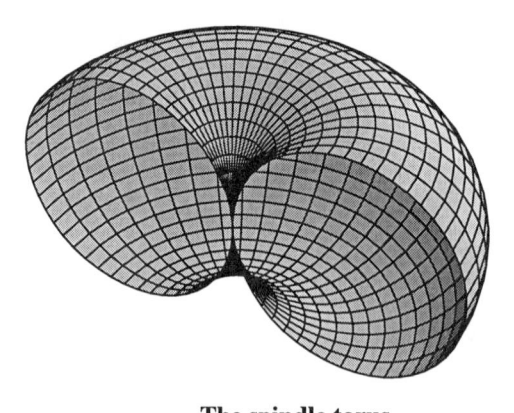

```
ParametricPlot3D[
torus[4,4,4][
u,v]//Evaluate,
{u,0,Pi},{v,0,2Pi},
PlotPoints->{24,48},
Axes->None,
Boxed->False];
```

The spindle torus
$$(u,v) \longmapsto$$
$$\bigl(\cos u(4+4\cos v), \sin u(4+4\cos v), 4\sin v\bigr)$$

13.5 The Paraboloid

The nonparametric form of the **paraboloid** is

$$z = a(x^2 + y^2),$$

and the corresponding parametrization is

$$\mathbf{paraboloid}[a](u,v) = (u, v, a(u^2 + v^2)),$$

or

```
paraboloid[a_][u_,v_] := {u,v,a(u^2 + v^2)}
```

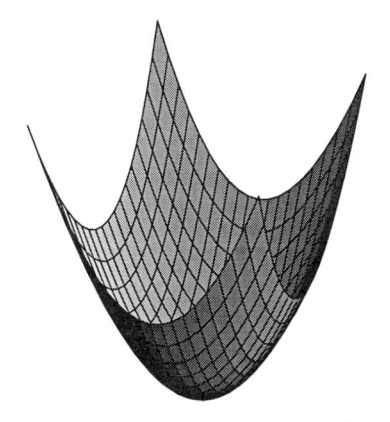

The paraboloid $(u, v) \longmapsto (u, v, u^2 + v^2)$

```
ParametricPlot3D[Evaluate[
paraboloid[1][u,v]],
{u,-1.5,1.5},{v,-1.5,1.5},
Axes->None,Boxed->False];
```

Perhaps a clearer representation is obtained if we use polar coordinates (r, θ) instead of the rectangular coordinates (u, v). The relation between the two systems is

$$u = r\cos\theta, \qquad v = r\sin\theta.$$

So the polar parametric representation of the paraboloid is

$$\mathbf{paraboloid}(r\cos\theta, r\sin\theta) = (r\cos\theta, r\sin\theta, r^2).$$

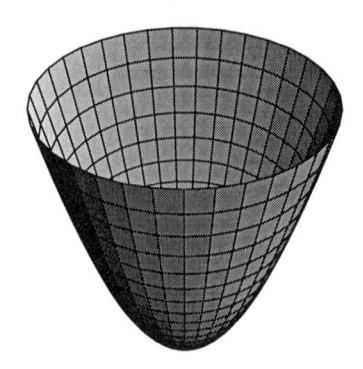

The paraboloid $(r,\theta) \longmapsto (r\cos\theta, r\sin\theta, r^2)$

One way to plot the paraboloid in polar coordinates is

```
ParametricPlot3D[Evaluate[
paraboloid[1][r Cos[theta],r Sin[theta]]],
{r,0,2},{theta,0,2Pi},
PlotPoints->{20,36},
Axes->None,Boxed->False];
```

Since we shall need to convert other surfaces from rectangular to polar coordinates, let us define a simple *Mathematica* function to do the conversion.

```
polarize[func_][r_,theta_]:=
    func[r Cos[theta],r Sin[theta]]
```

Note that **polarize** will convert either a function or an *n*-tuple of functions from rectangular to polar coordinates.

polarize[paraboloid[1]][r,theta]//Simplify

{r Cos[theta], r Sin[theta], r^2}

13.6 Sea Shells

In Section 9.1 we showed how to construct tubes about curves in $\mathbb{R}^3$. The construction can be easily modified to allow the radius of the tube to change from point to point. For example, given a curve $\gamma\colon (a,b) \longrightarrow \mathbb{R}^3$, let us define a **sea shell**

$$\mathbf{seashell}[\gamma][r](t,\theta) = \gamma(t) + r\,t\,\bigl(-\cos\theta\,\mathbf{N}(t) + \sin\theta\,\mathbf{B}(t)\bigr),$$

where **N** and **B** are the normal and binormal to γ. The corresponding *Mathematica* definition is

```
seashell[gamma_][r_][t_,theta_]:= gamma[t] +
    r t(-Cos[theta]normal[gamma][t] +
        Sin[theta]binormal[gamma][t])
```

If we take γ to be a helix, then this construction yields a sea shell:

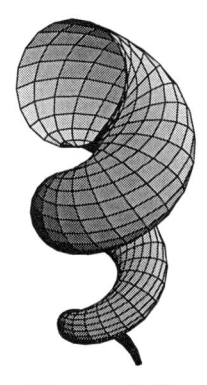

```
ParametricPlot3D[
seashell[
helix[1,0.6]][0.1][t,
theta]//Evaluate,
{t,Pi/4,5Pi},
{theta,0,2Pi},
PlotPoints->{48,20},
Boxed->False,
Axes->None];
```

The sea shell
$(t,\theta) \longmapsto \mathbf{seashell}[\mathbf{helix}[1,0.6]][0.1](t,\theta)$

13.7 Patches with Singularities

Patches can have singular points where they are 1-dimensional or even 0-dimensional. Self-intersections are also possible. If the singularities are mild, *Mathematica* has no difficulty plotting the surface.

The Eight Surface

A good example of a surface with a self-intersection is the **eight surface**. It is defined by

(13.5) $\quad$ **eightsurface**$(u,v) = (\cos u \cos v \sin v, \sin u \cos v \sin v, \sin v)$,

or

```
eightsurface[u_,v_]:= {Cos[u]Cos[v]Sin[v],
    Sin[u]Cos[v]Sin[v],Sin[v]}
```

In spite of the fact that this surface has a singular point, *Mathematica* has no difficulty plotting it.

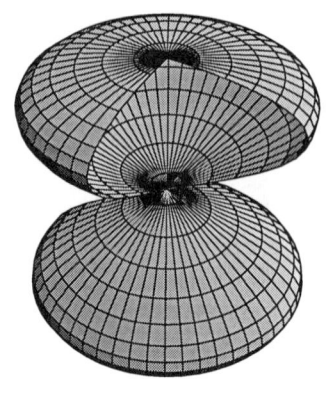

```
ParametricPlot3D[
eightsurface[u,v]
//Evaluate,
{u,0,3Pi/2},
{v,-Pi/2,Pi/2},
PlotPoints->{40,30},
Axes->None,
Boxed->False];
```

The eight surface

$(u,v) \longmapsto$

$(\cos u \cos v \sin v, \sin u \cos v \sin v, \sin v)$

The Whitney Umbrella

An example of a surface with a **pinch point** is the **Whitney umbrella**, defined by

$$\text{whitneyumbrella}(u,v) = (uv, u, v^2),$$

or

13.8 Implicit Plots of Surfaces

```
whitneyumbrella[u_,v_]:= {u v,u,v^2}
```

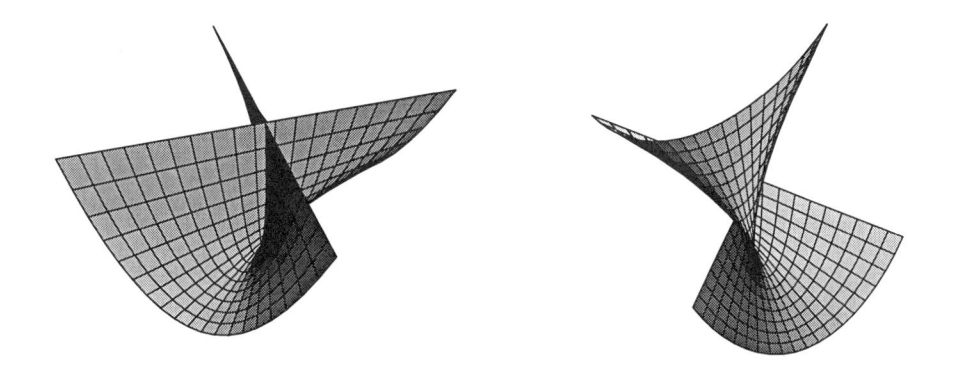

Two views of the Whitney umbrella $(u,v) \longmapsto (uv, u, v^2)$

For a discussion of this surface see [Francis, pages 8–9].

13.8 Implicit Plots of Surfaces

Steven Wilkinson has written a package **ImplicitPlot3D.m** that can be used to draw surfaces that are defined nonparametrically. Here is an example:

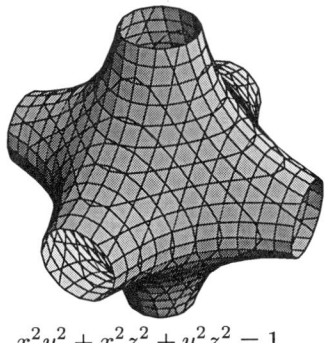

```
ImplicitPlot3D[
x^2y^2 + y^2z^2 + z^2x^2==1,
{x,-2,2},{y,-2,2},{z,-2,2},
PlotPoints->{{5,5,5},
{2,2,2},{2,2,2}},
Boxed->False]
```

$x^2y^2 + x^2z^2 + y^2z^2 = 1$

Alternatively, the package **ContourPlot3D** can be used. It can be read into a *Mathematica* session with the command

```
<<Graphics/ContourPlot3D.m
```

Then the command

```
ContourPlot3D[x^2y^2  + y^2z^2 + z^2x^2,
{x,-1,2},{y,-2,2},{z,-2,2},
Contours->{1.},PlotPoints->5,
Boxed->False];
```

produces a picture similar to the one above.

13.9 Exercises

1. Graph the hyperbolic paraboloid and monkey saddle using polar coordinates.

2. Use *Mathematica* to find unit normals to the monkey saddle, the sphere, the torus and the eight surface.

3. Define the **Mercator**[1] **parametrization of an ellipsoid** by

    mercatorellipsoid$[a, b, c](u, v) = (a \operatorname{sech} v \, \cos u, b \operatorname{sech} v \, \sin u, c \tanh v)$

    or

    ```
    mercatorellipsoid[a_,b_,c_][u_,v_]:=
        {a Sech[v]Cos[u],b Sech[v]Sin[u],c Tanh[v]}
    ```

    Plot the surface.

---

[1] Gerardus Mercator (Latinized name of Gerhard Kremer) (1512-1594). Flemish cartographer. In 1569 he first used the map projection which bears his name.

13.9 Exercises

4. If **surface[u,v]** is a surface defined in *Mathematica*, then **surface[u,v]^n** is the surface whose x-, y- and z-entries are the nth powers of those of **surface[u,v]**. Plot the 3rd power of **torus[8,3,8]**.

5. Plot the following surfaces described in Appendix C, or in **SURFS.m**: handkerchief, menn, perturbedms, shoe.

6. Use **ImplicitPlot3D** to plot **Kummer's**[2] surface, which is defined by

```
kummerimplicit[x_,y_,z_]:= x^4 + y^4 + z^4 -
    (y^2 z^2 + z^2 x^2 + x^2 y^2) - (x^2 + y^2 + z^2) + 1
```

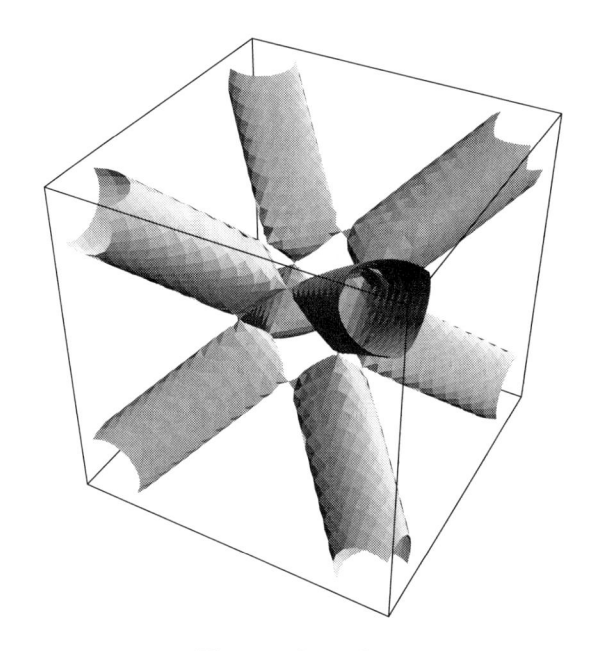

Kummer's surface

---

[2] Ernst Eduard Kummer (1810-1893). Kummer is remembered for his work in hypergeometric functions, number theory and algebraic geometry.

7. Use **ImplicitPlot3D** to plot **Goursat's**[3] **surface**, which is defined by

```
goursatimplicit[a_,b_,c_][x_,y_,z_]:= x^4 + y^4 + z^4 +
    a(x^2 + y^2 + z^2)^2 + b(x^2 + y^2 + z^2) + c
```

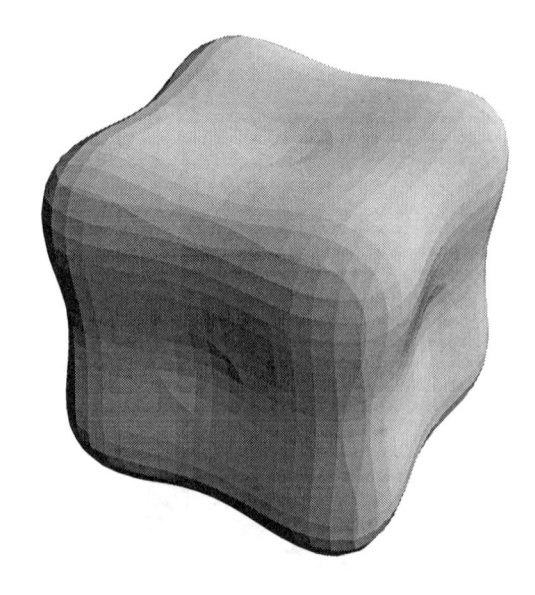

Goursat's surface

8. Use **ImplicitPlot3D** to plot a surface with two cusps defined by

```
twocuspsimplicit[x_,y_,z_]:= (z - 1)^2(x^2 - z^2) -
    (x^2 - z)^2 - y^4 - y^2(2x^2 + z^2 + 2z - 1)
```

---

[3] Édouard Jean Babtiste Goursat (1858-1936). Goursat was a leading analyst of his day. His **Cours d'analyse mathématique** has long been a classic text in France.

13.9 Exercises

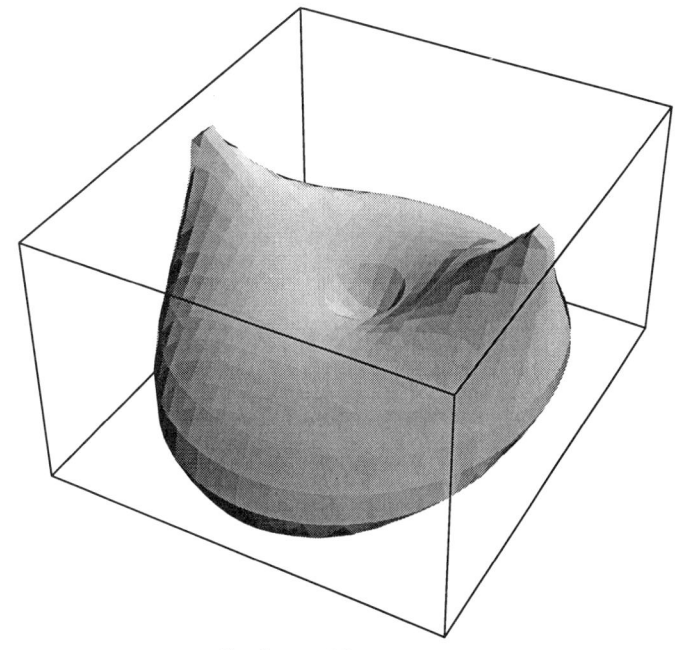

Surface with two cusps

9. Plot the **sine surface** defined by

```
sinsurface[a_][u_,v_] := {a*Sin[u],a*Sin[v],a*Sin[u + v]}
```

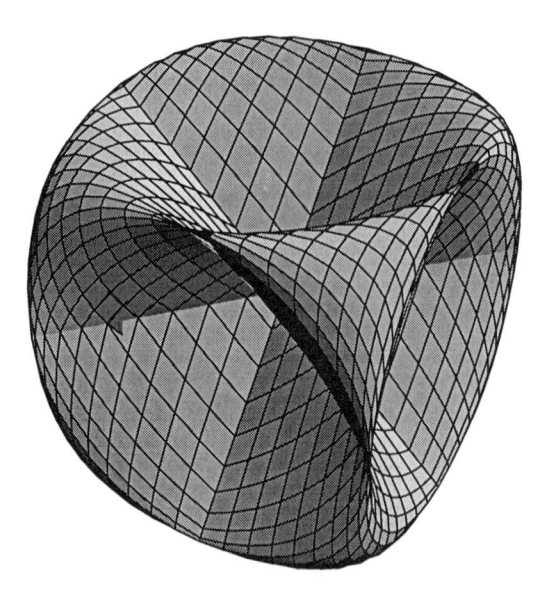

Sin surface

Then find the nonparametric form of the surface and plot it using `ImplicitPlot3D`.

14

NONORIENTABLE SURFACES

An important discovery of the nineteenth century was that nonorientable surfaces exist. The goal of the present chapter is to gain an understanding of such surfaces. In Section 14.1 we define precisely what it means for a regular surface to be orientable. Examples of nonorientable surfaces are first described in Section 14.2 by means of certain identifications of the edges of a square. These are topological descriptions of some of the simpler nonorientable surfaces. The rest of the chapter is devoted to realizing nonorientable surfaces as surfaces in $\mathbb{R}^3$. In Sections 14.3, 14.4, 14.5, and 14.6 we show how to plot the Möbius strip, the Klein bottle, and the real projective plane.

14.1 *Orientability of Surfaces*

If V is a 2-dimensional vector space, we call a linear map $J\colon V \longrightarrow V$ such that $J^2 = -I$ a **complex structure** on V. When $V = \mathbb{R}^2$ this is the same complex structure that we defined on page 4. We also observed on page 247 that each tangent space to $\mathbb{R}^2$ has a complex structure. More generally, because all 2-dimensional vector spaces are isomorphic, each tangent space $\mathcal{M}_{\mathbf{p}}$ to a regular surface $\mathcal{M}$ has a complex structure $J_{\mathbf{p}}\colon \mathcal{M}_{\mathbf{p}} \longrightarrow \mathcal{M}_{\mathbf{p}}$. However, it may or may not happen that $\mathbf{p} \longmapsto J_{\mathbf{p}}$ is a continuous function of $\mathbf{p}$. (Continuity of $\mathbf{p} \longmapsto J_{\mathbf{p}}$ means that $\mathbf{p} \longmapsto J_{\mathbf{p}}X$ is continuous for each continuous vector field X tangent to $\mathcal{M}$.) This leads to the following definition.

Definition. *A regular surface $M \subset \mathbb{R}^n$ is called* **orientable** *provided each tangent space M_p has a complex structure $J_p \colon M_p \longrightarrow M_p$ such that $\mathbf{p} \longmapsto J_p$ is a continuous function. An* **oriented** *regular surface $M \subset \mathbb{R}^3$ is an orientable regular surface together with a choice of the complex structure $\mathbf{p} \longmapsto J_p$.*

We shall see in Chapter 25 that this definition works as well for surfaces not contained in $\mathbb{R}^n$. For regular surfaces contained in $\mathbb{R}^3$ there is a more intuitive way to describe orientability using the vector cross product.

Theorem 14.1. *A regular surface $M \subset \mathbb{R}^3$ is orientable if and only if there is a continuous map $\mathbf{p} \longmapsto \mathbf{U}(\mathbf{p})$ that assigns to each $\mathbf{p} \in M$ a unit normal vector $\mathbf{U}(\mathbf{p}) \in M_p^\perp$.*

Proof. Suppose we are given $\mathbf{p} \longmapsto \mathbf{U}(\mathbf{p})$. Then for each $\mathbf{p} \in M$ we define $J_p \colon M_p \longrightarrow M_p$ by

$$J_p \mathbf{v}_p = \mathbf{U}(\mathbf{p}) \times \mathbf{v}_p,$$

where $\times$ denotes the vector cross product of $\mathbb{R}^3$. It is easy to check that J_p maps M_p into M_p and not merely into $\mathbb{R}_p^3$, and that $\mathbf{p} \longmapsto J_p$ is continuous. From (8.3), page 183, it follows that

$$\begin{aligned} J_p^2 \mathbf{v}_p &= \mathbf{U}(\mathbf{p}) \times \big(\mathbf{U}(\mathbf{p}) \times \mathbf{v}_p\big) \\ &= \big(\mathbf{U}(\mathbf{p}) \cdot \mathbf{v}_p\big)\mathbf{U}(\mathbf{p}) - \big(\mathbf{U}(\mathbf{p}) \cdot \mathbf{U}(\mathbf{p})\big)\mathbf{v}_p = -\mathbf{v}_p. \end{aligned}$$

Conversely, if we are given a regular surface $M \subset \mathbb{R}^3$ with a globally defined continuous complex structure $\mathbf{p} \longmapsto J_p$, we define $\mathbf{U}(\mathbf{p}) \in \mathbb{R}_p^3$ by

$$(14.1) \qquad \mathbf{U}(\mathbf{p}) = \frac{\mathbf{v}_p \times J_p \mathbf{v}_p}{\|\mathbf{v}_p \times J_p \mathbf{v}_p\|}$$

for any nonzero $\mathbf{v}_p \in M_p$. Then $\mathbf{U}(\mathbf{p})$ is perpendicular to both $\mathbf{v}_p$ and $J_p \mathbf{v}_p$. Since $\mathbf{v}_p$ and $J_p \mathbf{v}_p$ form a basis for M_p, it follows that $\mathbf{U}(\mathbf{p})$ is perpendicular to M_p. To check that the $\mathbf{U}(\mathbf{p})$ defined by (14.1) is independent of the choice of $\mathbf{v}_p$, let $\mathbf{w}_p$ be another nonzero tangent vector in M_p. Then $\mathbf{w}_p$ is a linear combination of $\mathbf{v}_p$ and $J_p \mathbf{v}_p$:

$$\mathbf{w}_p = a\mathbf{v}_p + bJ_p\mathbf{v}_p.$$

Hence

$$\mathbf{w}_p \times J_p \mathbf{w}_p = (a\mathbf{v}_p + bJ_p\mathbf{v}_p) \times (-b\mathbf{v}_p + aJ_p\mathbf{v}_p) = (a^2 + b^2)(\mathbf{v}_p \times J_p\mathbf{v}_p),$$

14.1 Orientability of Surfaces

and so
$$\frac{\mathbf{w_p} \times J_p\mathbf{w_p}}{\|\mathbf{w_p} \times J_p\mathbf{w_p}\|} = \frac{(a^2+b^2)(\mathbf{v_p} \times J_p\mathbf{v_p})}{\|(a^2+b^2)(\mathbf{v_p} \times J_p\mathbf{v_p})\|} = \frac{\mathbf{v_p} \times J_p\mathbf{v_p}}{\|\mathbf{v_p} \times J_p\mathbf{v_p}\|}.$$

Thus $\mathbf{U}(\mathbf{p})$ in (14.1) is unambiguously defined. Since $\mathbf{p} \longmapsto J_p$ is continuous, so is $\mathbf{p} \longmapsto \mathbf{U}(\mathbf{p})$. Hence $\mathbf{p} \longmapsto \mathbf{U}(\mathbf{p})$ is a continuous globally defined surface normal to $\mathcal{M}$. ∎

Theorem 14.1 permits us to define the Gauss map of an arbitrary orientable regular surface in $\mathbb{R}^3$.

Definition. *Let $\mathcal{M}$ be an oriented regular surface in $\mathbb{R}^3$, and let $\mathbf{U}$ be a globally defined unit normal vector field on $\mathcal{M}$ that defines the orientation of $\mathcal{M}$. Let $S^2(1)$ denote the unit sphere in $\mathbb{R}^3$. Then $\mathbf{U}$, viewed as a map*
$$\mathbf{U}\colon \mathcal{M} \longrightarrow S^2(1),$$
is called the **Gauss map** *of $\mathcal{M}$.*

It is easy to find examples of orientable regular surfaces in $\mathbb{R}^3$.

Lemma 14.2. *The graph $\mathcal{M}_h$ of a function $h\colon \mathcal{U} \subseteq \mathbb{R}^2 \longrightarrow \mathbb{R}$ is an orientable regular surface.*

Proof. Just as in Section 13.1 we define a patch $\mathbf{x}\colon \mathcal{U} \longrightarrow \mathcal{M}_h$ by
$$\mathbf{x}(u,v) = \bigl(u,v,h(u,v)\bigr).$$

Then $\mathbf{x}$ covers all of $\mathcal{M}_h$; that is, $\mathbf{x}(\mathcal{U}) = \mathcal{M}_h$. Furthermore, $\mathbf{x}$ is regular and injective. The surface normal $\mathbf{U}$ to $\mathcal{M}_h$ is given by
$$\mathbf{U} \circ \mathbf{x} = \frac{\mathbf{x}_u \times \mathbf{x}_v}{\|\mathbf{x}_u \times \mathbf{x}_v\|} = \frac{(-h_u, -h_v, 1)}{\sqrt{1 + h_u^2 + h_v^2}}.$$

Since $\mathbf{U}$ is everywhere nonzero, it follows from Theorem 14.1 that $\mathcal{M}_h$ is orientable. ∎

More generally, the method of proof of Lemma 14.2 yields:

Lemma 14.3. *Any surface $\mathcal{M} \subset \mathbb{R}^3$ which is the trace of a single injective regular patch $\mathbf{x}$ is orientable.*

Here is another generalization of Lemma 14.2.

Lemma 14.4. *Let $g: \mathbb{R}^3 \longrightarrow \mathbb{R}$ be a differentiable function and c a number such that $\mathbf{grad}\ g$ is nonzero on all points of the level surface $\mathcal{M}(c) = \{\, \mathbf{p} \in \mathbb{R}^3 \mid g(\mathbf{p}) = c \,\}$. Then $\mathcal{M}(c)$ as well as every component of $\mathcal{M}(c)$ is orientable.*

Proof. Theorem 12.27, page 291, implies that $\mathcal{M}(c)$ is a regular surface and that $\mathbf{grad}\ g$ is everywhere perpendicular to $\mathcal{M}(c)$. We are assuming that $\mathbf{grad}\ g$ never vanishes on $\mathcal{M}(c)$. Putting these facts together we see that

$$\frac{\mathbf{grad}\ g}{\|\mathbf{grad}\ g\|}$$

is a unit vector field that is well-defined at all points of $\mathcal{M}(c)$ and everywhere perpendicular to $\mathcal{M}(c)$. Then Theorem 14.1 says that $\mathcal{M}(c)$ and each of its components are orientable. ∎

Recall from general topology that a topological space X is **connected** provided any decomposition $X = X_1 \cup X_2$ where $X_1 \cap X_2$ is empty must be trivial; that is, either $X = X_1$ or $X = X_2$.

Lemma 14.5. *Let $\mathcal{M}$ be a connected orientable regular surface in $\mathbb{R}^3$. Then $\mathcal{M}$ has exactly two globally defined unit normal vector fields.*

Proof. Since $\mathcal{M}$ is orientable, it has at least one globally defined unit normal vector field $\mathbf{p} \longmapsto \mathbf{U}(\mathbf{p})$. Let $\mathbf{p} \longmapsto \mathbf{V}(\mathbf{p})$ be any other globally defined unit normal vector field on $\mathcal{M}$. The sets $\mathcal{W}_\pm = \{\, \mathbf{p} \in \mathcal{M} \mid \mathbf{U}(\mathbf{p}) = \pm\mathbf{V}(\mathbf{p}) \,\}$ are closed, because $\mathbf{U}$ and $\mathbf{V}$ are continuous. Also $\mathcal{M} = \mathcal{W}_+ \cup \mathcal{W}_-$. The connectedness of $\mathcal{M}$ implies that $\mathcal{M}$ coincides with either $\mathcal{W}_+$ or $\mathcal{W}_-$. Consequently, the globally defined unit vector fields on $\mathcal{M}$ are $\mathbf{p} \longmapsto \mathbf{U}(\mathbf{p})$ and $\mathbf{p} \longmapsto -\mathbf{U}(\mathbf{p})$. ∎

Definition. *Let $\mathcal{M}$ be an orientable regular surface in $\mathbb{R}^3$. An **orientation** of $\mathcal{M}$ is a choice of the globally defined unit normal vector field on $\mathcal{M}$.*

In general, it may not be possible to cover a regular surface with a single patch. If we have a family of patches we must know how their orientations are related in order to define the orientation of a regular surface.

Lemma 14.6. *Let $\mathcal{M} \subset \mathbb{R}^3$ be a regular surface. If $\mathbf{x}: \mathcal{U} \longrightarrow \mathcal{M}$ and $\mathbf{y}: \mathcal{V} \longrightarrow \mathcal{M}$ are patches such that $\mathbf{x}(\mathcal{U}) \cap \mathbf{y}(\mathcal{V})$ is nonempty, then*

$$\mathbf{y}_u \times \mathbf{y}_v = \det\left(\mathcal{J}(\mathbf{x}^{-1} \circ \mathbf{y})\right) \mathbf{x}_{\bar{u}} \times \mathbf{x}_{\bar{v}},$$

where $\mathcal{J}(\mathbf{x}^{-1} \circ \mathbf{y})$ denotes the Jacobian matrix of $\mathbf{x}^{-1} \circ \mathbf{y}$.

14.1 Orientability of Surfaces

Proof. It follows from Lemma 12.19, page 285, that

$$\mathbf{y}_u \times \mathbf{y}_v = \left(\frac{\partial \bar{u}}{\partial u}\mathbf{x}_{\bar{u}} + \frac{\partial \bar{v}}{\partial u}\mathbf{x}_{\bar{v}}\right) \times \left(\frac{\partial \bar{u}}{\partial v}\mathbf{x}_{\bar{u}} + \frac{\partial \bar{v}}{\partial v}\mathbf{x}_{\bar{v}}\right)$$

$$= \det \begin{pmatrix} \dfrac{\partial \bar{u}}{\partial u} & \dfrac{\partial \bar{u}}{\partial v} \\ \dfrac{\partial \bar{v}}{\partial u} & \dfrac{\partial \bar{v}}{\partial v} \end{pmatrix} \mathbf{x}_{\bar{u}} \times \mathbf{x}_{\bar{v}}$$

$$= \det \left(\mathcal{J}(\mathbf{x}^{-1} \circ \mathbf{y})\right) \mathbf{x}_{\bar{u}} \times \mathbf{x}_{\bar{v}}. \blacksquare$$

Definition. *We say that patches* $\mathbf{x}: \mathcal{U} \longrightarrow \mathcal{M}$ *and* $\mathbf{y}: \mathcal{V} \longrightarrow \mathcal{M}$ *on a regular surface* $\mathcal{M}$ *with* $\mathbf{x}(\mathcal{U}) \cap \mathbf{y}(\mathcal{V})$ *nonempty are* **coherently oriented**, *provided the determinant of the Jacobian matrix* $\mathcal{J}(\mathbf{x}^{-1} \circ \mathbf{y})$ *is positive on* $\mathbf{x}(\mathcal{U}) \cap \mathbf{y}(\mathcal{V})$.

Theorem 14.7. *A regular surface* $\mathcal{M} \subset \mathbb{R}^3$ *is orientable if and only if it is possible to cover* $\mathcal{M}$ *with a family* $\mathfrak{B}$ *of regular injective patches such that any two patches* $(\mathbf{x}, \mathcal{U})$, $(\mathbf{y}, \mathcal{V})$ *with* $\mathbf{x}(\mathcal{U}) \cap \mathbf{y}(\mathcal{V})$ *nonempty are coherently oriented.*

Proof. Suppose that $\mathcal{M}$ is orientable. Theorem 14.1 implies that $\mathcal{M}$ has a globally defined surface normal $\mathbf{p} \longmapsto \mathbf{U}(\mathbf{p})$. Let $\mathfrak{A}$ be a family of regular injective patches whose union covers $\mathcal{M}$. Without loss of generality, we may suppose that the domain of definition of each patch in $\mathfrak{A}$ is connected. We must construct from $\mathfrak{A}$ a family $\mathfrak{B}$ of coherently oriented patches whose union covers $\mathcal{M}$.

To this end, we first note that if $\mathbf{x}$ is any regular injective patch on $\mathcal{M}$, then $\widetilde{\mathbf{x}}$ defined by

$$\widetilde{\mathbf{x}}(u, v) = \mathbf{x}(v, u)$$

is also a regular injective patch on $\mathcal{M}$. Moreover, $\widetilde{\mathbf{x}}$ and $\mathbf{x}$ are oppositely oriented, because $\widetilde{\mathbf{x}}_u \times \widetilde{\mathbf{x}}_v = \mathbf{x}_v \times \mathbf{x}_u$. Now it is clear how to choose the patches that are to be members of the family $\mathfrak{B}$. If $\mathbf{x}$ is a patch in $\mathfrak{A}$ and

$$\frac{\mathbf{x}_u \times \mathbf{x}_v}{\|\mathbf{x}_u \times \mathbf{x}_v\|} = \mathbf{U}$$

we put $\mathbf{x}$ into $\mathfrak{B}$, but if

$$\frac{\mathbf{x}_u \times \mathbf{x}_v}{\|\mathbf{x}_u \times \mathbf{x}_v\|} = -\mathbf{U}$$

we put $\widetilde{\mathbf{x}}$ into $\mathfrak{B}$. Then $\mathfrak{B}$ is a family of coherently oriented patches that covers $\mathcal{M}$.

Conversely, suppose that $\mathcal{M}$ is covered by a family $\mathfrak{B}$ of coherently oriented patches. If $\mathbf{x}$ is a patch in $\mathfrak{B}$ defined on $\mathcal{U} \subset \mathbb{R}^2$, we define $\mathbf{U}_\mathbf{x}$ on $\mathbf{x}(\mathcal{U})$ by

$$\mathbf{U}_\mathbf{x}\bigl(\mathbf{x}(u, v)\bigr) = \frac{\mathbf{x}_u \times \mathbf{x}_v}{\|\mathbf{x}_u \times \mathbf{x}_v\|}(u, v).$$

Let $\mathbf{y}\colon\mathcal{V}\longrightarrow\mathcal{M}$ be another patch in $\mathfrak{B}$ with $\mathbf{x}(\mathcal{U})\cap\mathbf{y}(\mathcal{V})$ nonempty. Lemma 14.6 implies that on $\mathbf{x}(\mathcal{U})\cap\mathbf{y}(\mathcal{V})$ we have

$$\mathbf{U}_\mathbf{y}\bigl(\mathbf{y}(u,v)\bigr) = \frac{\mathbf{y}_u\times\mathbf{y}_v}{\|\mathbf{y}_u\times\mathbf{y}_v\|}(u,v) = \left(\frac{\mathcal{J}(\mathbf{x}^{-1}\circ\mathbf{y})}{|\mathcal{J}(\mathbf{x}^{-1}\circ\mathbf{y})|}\frac{\mathbf{x}_u\times\mathbf{x}_v}{\|\mathbf{x}_u\times\mathbf{x}_v\|}\right)(u,v)$$

$$= \frac{\mathbf{x}_u\times\mathbf{x}_v}{\|\mathbf{x}_u\times\mathbf{x}_v\|}(u,v) = \mathbf{U}_\mathbf{x}\bigl(\mathbf{x}(u,v)\bigr).$$

Thus we get a well-defined surface normal $\mathbf{U}$ on $\mathcal{M}$ by putting $\mathbf{U}=\mathbf{U}_\mathbf{x}$ on $\mathbf{x}(\mathcal{U})$ for any patch $\mathbf{x}$ in $\mathfrak{B}$. By Theorem 14.1 $\mathcal{M}$ is orientable. ∎

The proof of Theorem 14.7 shows:

Corollary 14.8. *A family of coherently oriented regular injective patches on a regular surface in $\mathbb{R}^3$ defines globally a unit normal vector field on a surface $\mathcal{M}$ in $\mathbb{R}^3$, that is, an orientation of $\mathcal{M}$.*

14.2 Nonorientable Surfaces Described by Identifications

There are useful topological descriptions of some elementary surfaces that are obtained from identifying edges in a square. For example, if we identify the top and bottom edges of a square we obtain a cylinder. We describe this identification by means of a square with an arrow along the top edge and an arrow pointing in the same direction along the bottom edge.

Now consider a square with the top and bottom edges identified, but in reverse order. We indicate this identification by means of an arrow along the top edge of the square pointing in one direction and an arrow along the bottom edge pointing in the opposite direction. The resulting surface is called a **Möbius**[1] **strip**.

---

[1] August Ferdinand Möbius (1790-1868). Professor at the University of Leipzig. Möbius' name is attached to several important mathematical objects such as the Möbius function, the Möbius inversion formula and Möbius transformations. He discovered the Möbius strip at age 71.

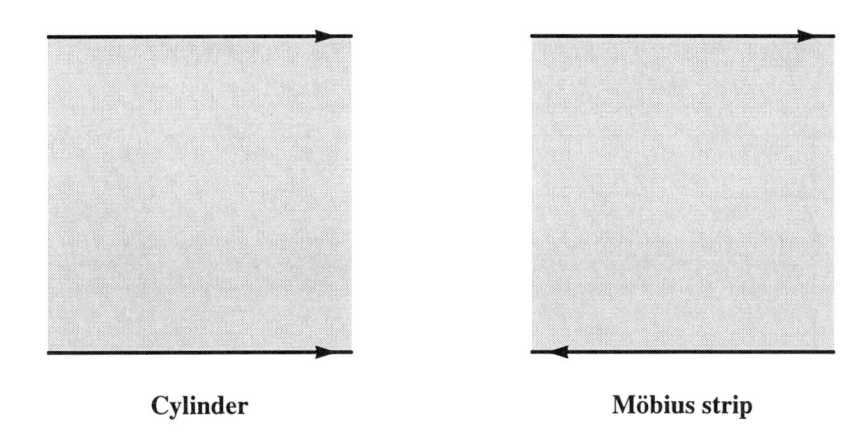

14.2 Nonorientable Surfaces Described by Identifications

Cylinder   **Möbius strip**

It is easy to put a cylinder into Euclidean space $\mathbb{R}^3$, but to find the actual parametrization of a Möbius strip in $\mathbb{R}^3$ is more difficult. We shall return to this problem shortly.

Now let us see what happens when we identify the vertical as well as the horizontal edges of a square. There are three possibilities. If the vertical arrows point in the same direction, and if the direction of the horizontal arrows is also the same, we obtain a torus. We have already seen on page 304 how to parametrize a torus in $\mathbb{R}^3$ as **torus**$[a, b, c]$. For different values of a, b, c, we get tori which are different in shape, but are topologically all the same.

The **Klein**[2] **bottle** is the surface that results when the edges of the square are identified with the vertical arrows pointing in the same direction, but the horizontal arrows pointing in opposite directions. Clearly, the same surface results if we interchange horizontal with vertical.

Finally, the surface that results when we identify the edges of the square with both the two vertical arrows pointing in different directions and the two horizontal

---

[2] Christian Felix Klein (1849-1925). Klein made fruitful contributions to many branches of mathematics, including applied mathematics and mathematical physics. His Erlanger Programm (1872) instituted research directions in geometry for a half century; his subsequent work on Riemann surfaces established their essential role in function theory. In his writings Klein concerned himself with what he saw as a developing gap between the increasing abstraction of mathematics and applied fields whose practitioners did not appreciate the fundamental rôle of mathematics, as well as with mathematics instruction at the secondary level.

Klein discussed nonorientable surfaces in 1874 in [Klein].

arrows pointing in different directions is called the **real projective plane**. This surface can also be thought of as a sphere with antipodal points identified.

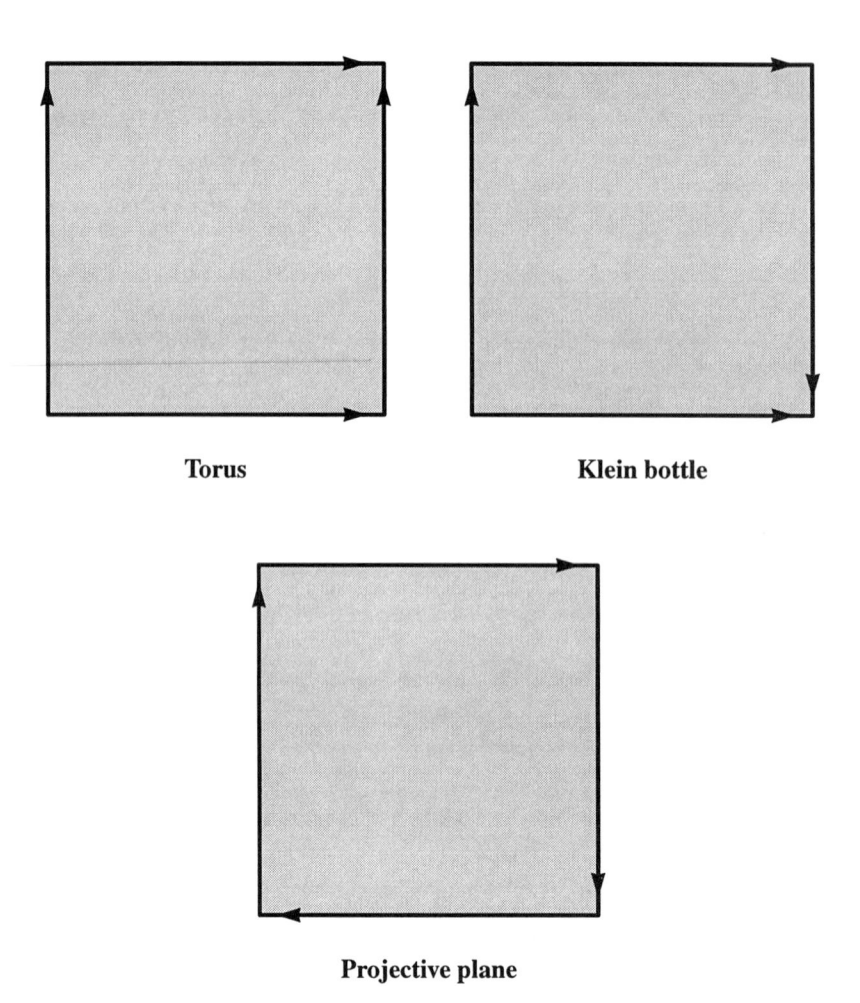

Torus

Klein bottle

Projective plane

We have described the Möbius strip, the Klein bottle and the real projective plane topologically. All these surfaces turn out to be nonorientable. It is quite another matter, however, to find parametrizations of these surfaces in $\mathbb{R}^3$. That is the subject of the rest of this chapter.

14.3 The Möbius Strip

Another useful topological description of the Möbius strip is as the surface resulting from revolving a line segment around an axis, but putting one twist in the line segment as it goes around the axis.

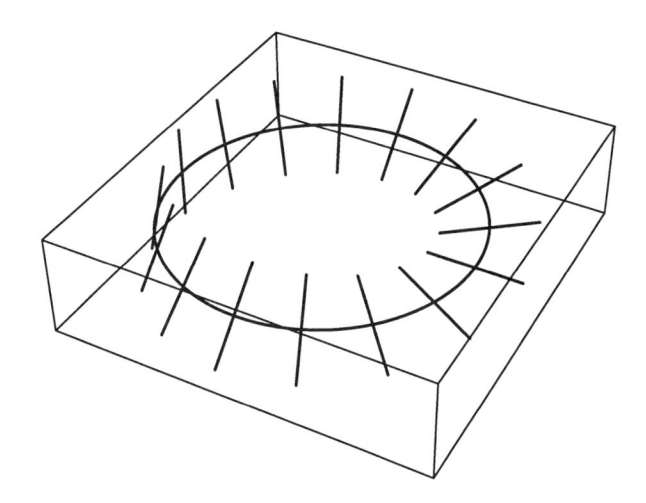

Lines twisting to form a Möbius strip

We can use this model to get a parametrization of the Möbius strip:

$$(14.2) \quad \mathbf{moebiusstrip}[a](u,v) = a\bigg(\cos u + v \cos\left(\frac{u}{2}\right)\cos u,$$

$$\sin u + v\cos\left(\frac{u}{2}\right)\sin u, v\sin\left(\frac{u}{2}\right)\bigg),$$

or in *Mathematica*,

```
moebiusstrip[a_][u_,v_]:= a{Cos[u] + v Cos[u/2]Cos[u],
      Sin[u] + v Cos[u/2]Sin[u],v Sin[u/2]}
```

It is easy to see that $u \longmapsto \mathbf{moebiusstrip}[a](u,0)$ is the central circle of the Möbius strip, and that each $v \longmapsto \mathbf{moebiusstrip}[a](u_0,v)$ is a line segment meeting the central circle. As u_0 goes from 0 to 2π the angle between the z-axis the line

$v \longmapsto \text{moebiusstrip}[a](u_0, v)$ changes from 0 to π.

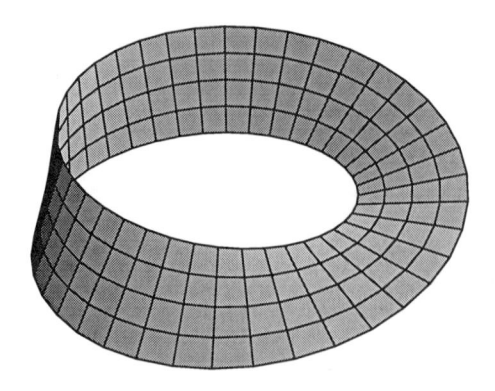

```
ParametricPlot3D[
moebiusstrip[1][u,v]
//Evaluate,
{u,0,2Pi},
{v,-0.3,0.3},
PlotPoints->{40,5},
Axes->None,
Boxed->False];
```

Möbius strip

Any attempt to define a unit normal vector on all of the Möbius strip is doomed to failure. Of course, it is possible to define a unit normal locally. But if one tries to extend the definition of the unit normal so that it is defined on all of the Möbius strip by going around the center circle, then the unit normal comes back on the other side of the surface.

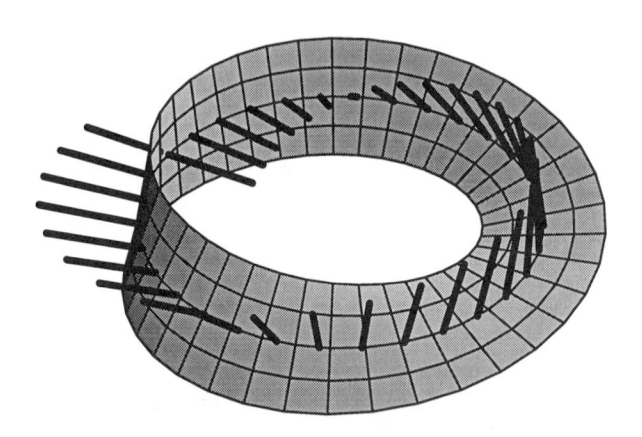

Normals to a Möbius strip

14.4 The Klein Bottle

The Klein bottle is also a nonorientable surface, but in contrast to the Möbius strip it is compact. One way to define the Klein bottle is as follows. It is the surface that results from rotating a figure of eight about an axis, but putting a twist in it. Thus the rotation and twisting are the same as a Möbius strip, but instead of a line segment one uses a figure of eight.

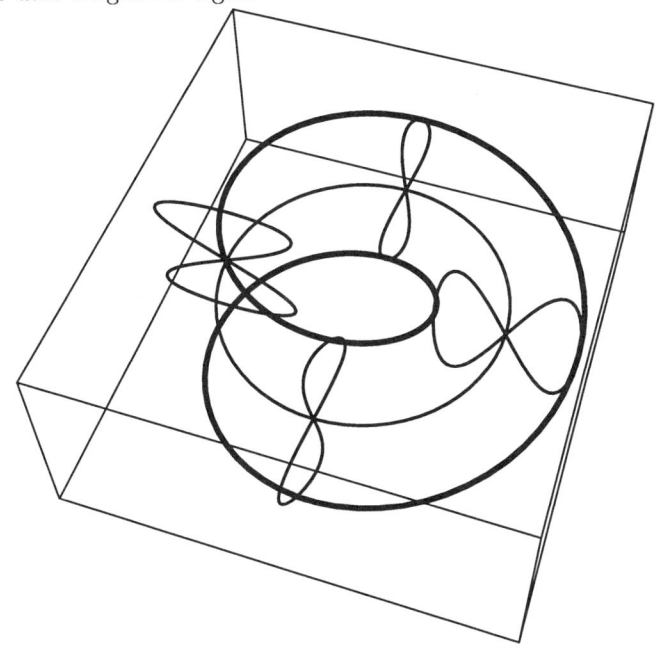

Figure eights twisting to form a Klein bottle

Here is a parametrization of the Klein bottle that uses this construction.

$$\mathbf{kleinbottle}[a](u,v) = \left(\left(a + \cos\left(\frac{u}{2}\right)\sin v - \sin\left(\frac{u}{2}\right)\sin 2v\right)\cos u, \right.$$
$$\left(a + \cos\left(\frac{u}{2}\right)\sin v - \sin\left(\frac{u}{2}\right)\sin 2v\right)\sin u,$$
$$\left. \sin\left(\frac{u}{2}\right)\sin v + \cos\left(\frac{u}{2}\right)\sin 2v \right),$$

or in *Mathematica*,

```
kleinbottle[a_][u_,v_]:= {(a + Cos[u/2]Sin[v]
    - Sin[u/2]Sin[2 v])Cos[u],
    (a + Cos[u/2]Sin[v]
    - Sin[u/2]Sin[2 v])Sin[u],
    Sin[u/2]Sin[v] + Cos[u/2]Sin[2 v]}
```

The central circle of the Klein bottle is $u \longmapsto$ **kleinbottle**$[a](u, 0)$, and each of the curves $v \longmapsto$ **kleinbottle**$[a](u_0, v)$ is a figure of eight. As u varies from 0 to 2π, the figures of eight twist from 0 to π; this is the same twisting that we encountered in the parametrization (14.2) of the Möbius strip.

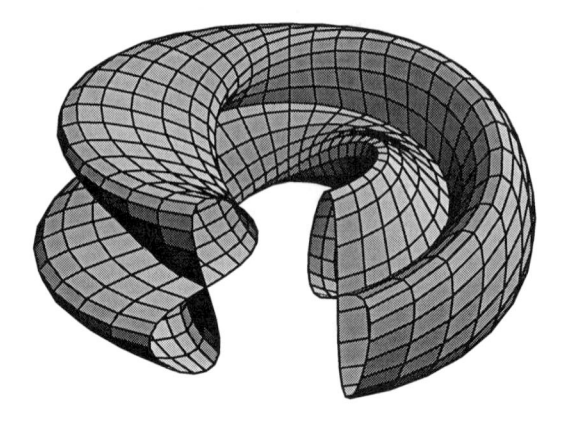

The Klein bottle

```
ParametricPlot3D[kleinbottle[2][u,v]//Evaluate,
{v,0,2Pi},{u,-Pi/4,3Pi/2},
PlotPoints->{32,32},
Axes->None,Boxed->False];
```

The Klein bottle is not a regular surface in $\mathbb{R}^3$ because it has self-intersections. However, the Klein bottle can be shown to be an abstract surface. This concept will be defined in Chapter 25.

14.5 A Different Klein Bottle

There are two kinds of Klein bottles in $\mathbb{R}^3$. The first one, $\mathcal{K}_1$ described in Section 14.4, has the feature that a neighborhood $\mathcal{V}_1$ of the self-intersection curve is nonorientable. In fact, it is clear that $\mathcal{V}_1$ is a nonorientable "X" that rotates and twists about an axis in the same way that the figures of eight move when they form the surface.

The original description of Klein (see [HC-V, pages 308-311]) of his surface was much different, and is described as follows. Consider a tube $\mathcal{T}$ of variable radius about a line. Topologically, a torus is formed from $\mathcal{T}$ by bending the tube until the ends meet and then gluing the boundary circles together. Another way to glue the ends is as follows. Let one end of $\mathcal{T}$ be a little smaller than the other. We bend the smaller end, then push it through the surface of the tube, and move it so that it is a concentric circle with the larger end, lying in the same plane. We complete the surface by adjoining a torus on the other side of the plane. The result is shown in the following picture.

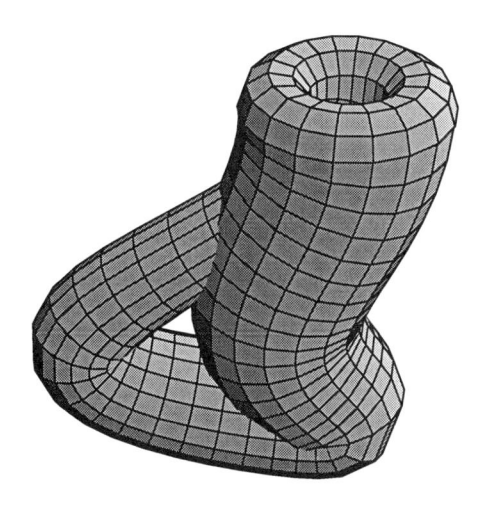

```
ParametricPlot3D[Evaluate[
kleinbottlebis[1,2][u,v]],
{u,-Pi,29.9517},
{v,-Pi,Pi},
ViewPoint->{-2,0,-3},
PlotPoints->{49,19},
Axes->None,Boxed->False]
```

Klein bottle with an orientable neighborhood of the self-intersection curve

To see that this new Klein bottle $\mathcal{K}_2$ is (as a surface in $\mathbb{R}^3$) distinct from the Klein bottle $\mathcal{K}_1$ formed by twisting figures of eight, consider a neighborhood $\mathcal{V}_2$ of the self-intersection curve of $\mathcal{K}_2$ This neighborhood is an orientable "X". Thus the difference between $\mathcal{K}_1$ and $\mathcal{K}_2$ is that $\mathcal{V}_1$ is nonorientable, but $\mathcal{V}_2$ is orientable.

formed by identifying sides of squares, as described in Section 14.2.

We call the new version of the Klein bottle **kleinbottlebis**; here is the parametrization in *Mathematica*:

```
kleinbottlebis[e_,f_][u_,v_]:=If[u<0,
    {f Sin[u]/2,(-e - f/2 + f Cos[u]/2)Cos[v],
    (e + f/2 - f Cos[u]/2)Sin[v]},
    {10Sin[0.105u] + 2(e + 0.033f u)
    (Cos[0.105u] + 4Cos[0.21u])Cos[v]
    (Sin[0.105u] + 2Sin[0.21u])/
    Sqrt[100Cos[0.105u]^2 + 4(Cos[0.105u] +
      4Cos[0.21u])^2(Sin[0.105u] + 2Sin[0.21u])^2],
    (Sin[0.105u] + 2Sin[0.21u])^2 -
    (10(e + 0.033f u)Cos[0.105u]Cos[v])/
    Sqrt[100Cos[0.105u]^2 + 4(Cos[0.105u] +
      4Cos[0.21u])^2(Sin[0.105u] + 2Sin[0.21u])^2],
    (e + 0.033f u)Sin[v]}]
```

14.6 Realizations of the Real Projective Plane

Let $S^2(a) = \{\mathbf{p} \mid \|\mathbf{p}\| = a\}$ be the sphere of radius a in $\mathbb{R}^3$. Recall that the **antipodal map** of the sphere $S^2(a) \subset \mathbb{R}^3$ is the diffeomorphism **antipodal**: $S^2(a) \longrightarrow S^2(a)$ defined by **antipodal**$(\mathbf{p}) = -\mathbf{p}$. The real projective plane $\mathbb{R}P^2(a)$ can be defined as the set that results when antipodal points of $S^2(a)$ are identified; thus

$$\mathbb{R}P^2(a) = \{\, \{\mathbf{p}, -\mathbf{p}\} \mid \|\mathbf{p}\| = a \,\}.$$

To realize the real projective plane as a surface in $\mathbb{R}^3$ let us look for a map of $\mathbb{R}^3$ into itself with a special property.

Definition. *A map* $F: \mathbb{R}^3 \longrightarrow \mathbb{R}^3$ *such that*

(14.3) $$F(-\mathbf{p}) = F(\mathbf{p})$$

is said to have the **antipodal property**.

14.6 Realizations of the Real Projective Plane

That we can use a map with the antipodal property to realize the real projective plane is a consequence of the following easily-proven lemma:

Lemma 14.9. *A map* $F \colon \mathbb{R}^3 \longrightarrow \mathbb{R}^3$ *with the antipodal property gives rise to a map* $\widetilde{F} \colon \mathbb{R}P^2(a) \longrightarrow F(S^2(a)) \subset \mathbb{R}^3$ *defined by*

$$\widetilde{F}(\{\mathbf{p}, -\mathbf{p}\}) = F(\mathbf{p}).$$

Thus we can realize $\mathbb{R}P^2(a)$ as the image of $S^2(a)$ under a map F which has the antipodal property. (Of course, we should choose F so that its Jacobian matrix has determinant zero at only a few points. This can be accomplished by choosing the components of F to be certain quadratic polynomials.) Moreover, any patch $\mathbf{x} \colon \mathcal{U} \longrightarrow S^2(a)$ will give rise to a patch $\widetilde{\mathbf{x}} \colon \mathcal{U} \longrightarrow F(S^2(a))$ defined by

$$\widetilde{\mathbf{x}}(u, v) = F(\mathbf{x}(u, v)).$$

Let us give two examples of maps with the antipodal property and use them to obtain two realizations of $\mathbb{R}P^2(a)$.

Steiner's Roman Surface

When Jakob Steiner visited Rome in 1844 he developed the concept of a surface that is now called Steiner's Roman surface. (See [Apéry, page 37].) It is a realization of the real projective plane. To describe it, we first define a map **romanmap**: $\mathbb{R}^3 \longmapsto \mathbb{R}^3$ by

$$\mathbf{romanmap}(x, y, z) = (x\,y, y\,z, z\,x)$$

or

```
romanmap[{x_,y_,z_}] := {x y, y z, z x}
```

It is obvious that **romanmap** has the antipodal property; hence **romanmap** induces a map of $\mathbb{R}P^2(a)$ onto $\mathbf{romanmap}(S^2(a))$. We call this image **Steiner's**[3] **Roman surface** of radius a. Moreover, we can plot a portion of $\mathbf{romanmap}(S^2(a))$

---

[3] Jakob Steiner (1796-1863). Swiss mathematician who was professor at the University of Berlin. Steiner did not learn to read and write until he was 14 and only went to school at the age of 18, against the wishes of his parents. Synthetic geometry was revolutionized by Steiner. He hated analysis as thoroughly as Lagrange hated geometry, according to [Cajori]. He believed that calculation replaces thinking while geometry stimulates thinking.

by composing **romanmap** with any patch on $S^2(a)$, for example, the standard parametrization **sphere**[a] $(u,v) \longmapsto (a\cos v \cos u, a\cos v \sin u, a\sin v)$. Then the composition **romanmap** ∘ **sphere**[a] parametrizes all of Steiner's Roman surface of radius a.

To carry out this symbolic computation, we need the *Mathematica* function **Composition**, which composes functions. For example,

```
Composition[Sin,Cos][x]
```

yields

```
Sin[Cos[x]]
```

We can get a parametrization of Steiner's Roman surface with the command:

```
Composition[romanmap,sphere[a]][u,v]//Simplify
```

$$\left\{ \frac{a^2 \text{Cos}[v]^2 \text{Sin}[2\,u]}{2}, \frac{a^2 \text{Sin}[u] \text{Sin}[2\,v]}{2}, \frac{a^2 \text{Cos}[u] \text{Sin}[2\,v]}{2} \right\}$$

Therefore, we define

$$\mathbf{roman}[a](u,v) = \left(\frac{a^2}{2}\right)\left(\sin 2u (\cos v)^2, \sin u \sin 2v, \cos u \sin 2v\right)$$

or

```
roman[a_][u_,v_] := (a^2/2){Sin[2u]Cos[v]^2,
                            Sin[u]Sin[2v],Cos[u]Sin[2v]}
```

14.6   Realizations of the Real Projective Plane

Here are two views of Steiner's Roman surface.

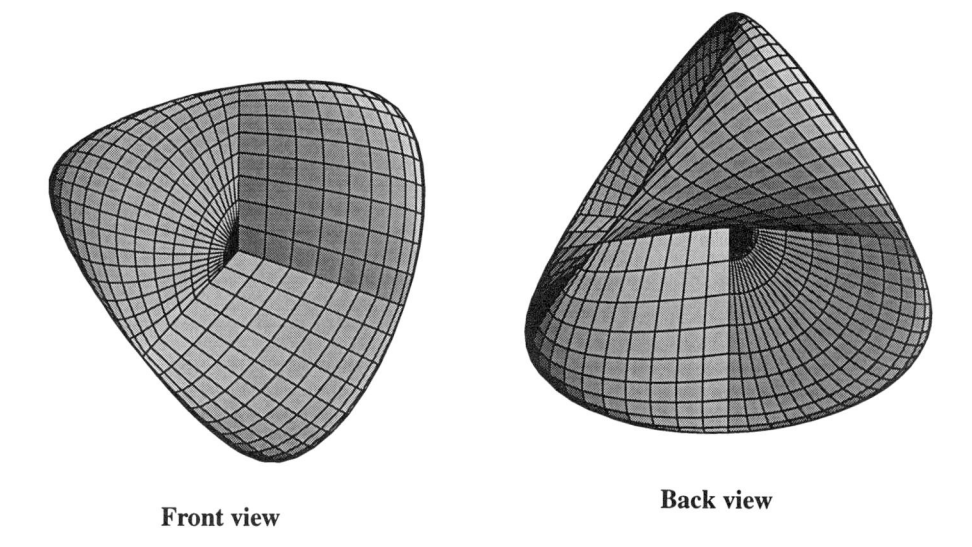

Front view   **Back view**

Steiner's Roman surface

The origin is a triple point of Steiner's Roman surface, because

$$\mathbf{romanmap}(\pm 1, 0, 0) = \mathbf{romanmap}(0, \pm 1, 0) = \mathbf{romanmap}(0, 0, \pm 1) = \mathbf{0}.$$

The Cross Cap

Another map with the antipodal property is given by

$$\mathbf{crosscapmap}(x, y, z) = \left(y\,z, 2\,x\,y, x^2 - y^2\right)$$

or

```
crosscapmap[{x_,y_,z_}]:= {y z,2 x y, x^2 - y^2}
```

We get a parametrization of the cross cap just as we did for Steiner's Roman surface:

```
Composition[crosscapmap,sphere[a]][u,v]//Simplify
```

yields

$$\left\{ \frac{a^2 \mathrm{Sin}[u]\, \mathrm{Sin}[2\,v]}{2},\ a^2\, \mathrm{Cos}[v]^2\, \mathrm{Sin}[2\,u],\ a^2\, \mathrm{Cos}[2\,u]\, \mathrm{Cos}[v]^2 \right\}$$

Hence we define

$$\mathbf{crosscap}[a](u,v) = a^2 \left(\frac{\sin u \sin 2v}{2}, \sin 2u \cos^2 v, \cos 2u \cos^2 v \right)$$

or

```
crosscap[a_][u_,v_]:= a^2{(Sin[u]Sin[2v])/2,
              Sin[2u]Cos[v]^2,Cos[2u]Cos[v]^2}
```

```
ParametricPlot3D[
crosscap[1][u,v]
//Evaluate,
{u,0,Pi},
{v,-Pi/2,Pi/2},
PlotPoints->{40,40},
Boxed->False,
Axes->None,
ViewPoint->
{2.73,-1.47,1.36}];
```

A cross cap

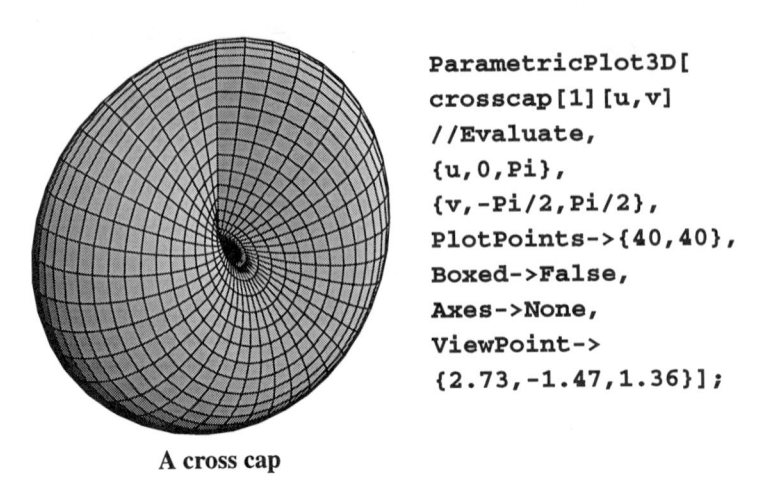

Effective use of the option **PlotRange** can be used to split open a cross cap.

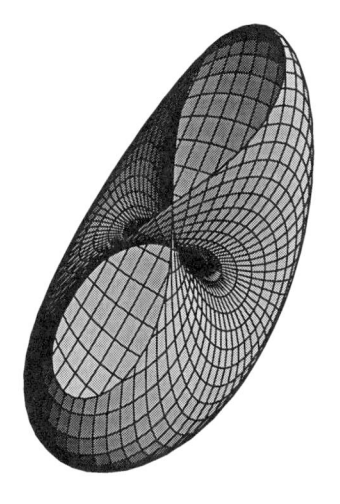

```
ParametricPlot3D[
crosscap[1][u,v]
//Evaluate,
{u,0,Pi},
{v,-Pi/2,Pi/2},
PlotPoints->{40,40},
Boxed->False,
Axes->None,
PlotRange->
{Automatic,Automatic,
{-1,0.5}},
ViewPoint->
{0.51,-1.0,3.2}];
```

A cross cap with its top cut off

14.7 Coloring Surfaces with Mathematica

There is an extended version of *Mathematica's* command **ParametricPlot3D** that permits coloring and shading. This extended version plots a 4-tuple, the first three entries of which are coordinates of a surface. The last entry of the 4-tuple specifies the coloring or shading, for example, using *Mathematica's* command **RGBColor**. The file **Colors.m** (which can be read into a *Mathematica* session with the command **Needs["Graphics`Colors`"]**) contains the definitions of many common colors, so that, for example, one can type in **Purple** instead of its hard-to-remember equivalent **RGBColor[0.31,0.18,0.31]**.

In order to color a surface we need a way to include the color information with the parametrization. This can be accomplished by using *Mathematica's* command **Append**, which appends an element to a list. *Mathematica* has a graphics primitive **FaceForm**, which can be used to color the sides of a surface different colors. Here is how to combine **Append** with **FaceForm** to color a Klein bottle green and yellow.

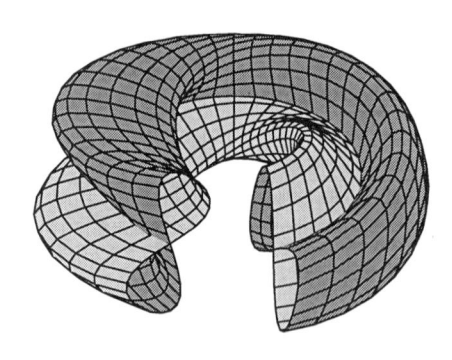

336　　　　　　　　　　　　　　　　　　　　　　　*Chapter 14　　Nonorientable Surfaces*

```
ParametricPlot3D[
Append[
kleinbottle[2][u,v],
FaceForm[Green,Yellow]]
//Evaluate,
{v,0,2Pi},
{u,-Pi/4,3Pi/2},
PlotPoints->{32,32},
Lighting->False,
Axes->None,
Boxed->False];
```

Colored Klein bottle

As another example we color the inside and outside of Steiner's Roman surface different colors and cut it open.

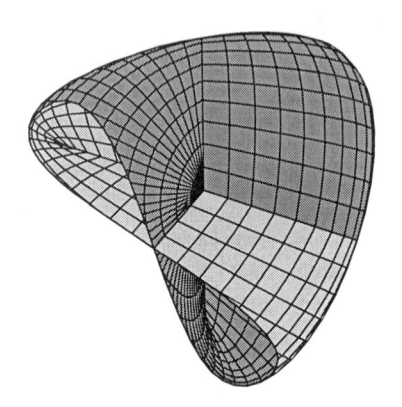

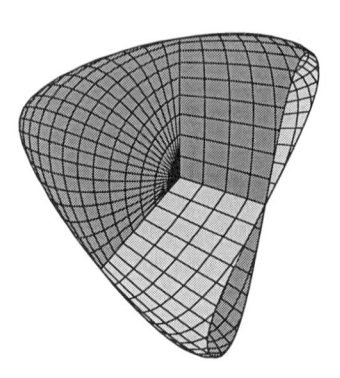

Cut views of Steiner's Roman surface

A typical plotting command is

```
ParametricPlot3D[Evaluate[Append[roman[1][u,v],
FaceForm[Green,Yellow]]],{u,0,Pi},{v,0,Pi},
Lighting->False,PlotPoints->{39,39},
PlotRange->{Automatic,{-0.3,0.3},Automatic},
Boxed->False,Axes->None,
ViewPoint->{1.300,-2.400,2.000}];
```

14.8 Exercises

1. Prove Corollary 14.8.

2. Let us put a family of figure of eight curves into $\mathbb{R}^3$ in such a way that the first and last figures of eight reduce to points. This defines a surface, which we call a **pseudocrosscap**, that can be parametrized as follows:

$$\textbf{pseudocrosscap}(u,v) = \big((1-u^2)\sin v, (1-u^2)\sin 2v, u\big);$$

The *Mathematica* version of the pseudocrosscap is:

```
pseudocrosscap[u_,v_]:= {(1 - u^2)Sin[v],
                        (1 - u^2)Sin[2v],u}
```

Pseudocrosscap

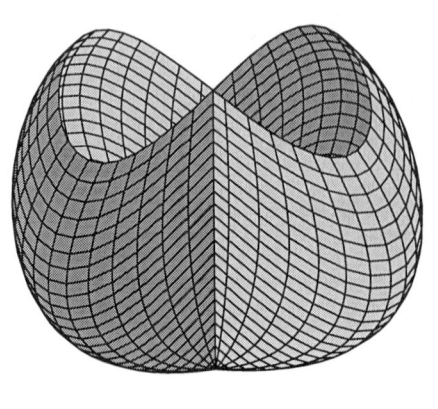

```
ParametricPlot3D[
Append[
pseudocrosscap[u,v],
FaceForm[
Green,Yellow]]
//Evaluate,
{u,-1,0.5},{v,0,2Pi},
Boxed->False,
Axes->None,
PlotPoints->{24,48},
ViewPoint->
{0.2,5,2.6},
Lighting->False];
```

Show that **pseudocrosscap** is not regular at the points $(0, 0, \pm 1)$. Why should the pseudocrosscap be considered orientable, even though it is not a regular surface?

3. Use *Mathematica* to draw the picture on page 325 of the lines twisting to form a Möbius strip.

4. Use *Mathematica* to draw the picture on page 326 of a Möbius strip with normal lines.

5. Use *Mathematica* to draw the picture on page 327 of the figures of eight twisting to form a Klein bottle.

14.8 Exercises

6. Show that the image under **crosscapmap** of a torus centered at the origin is topologically a Klein bottle.

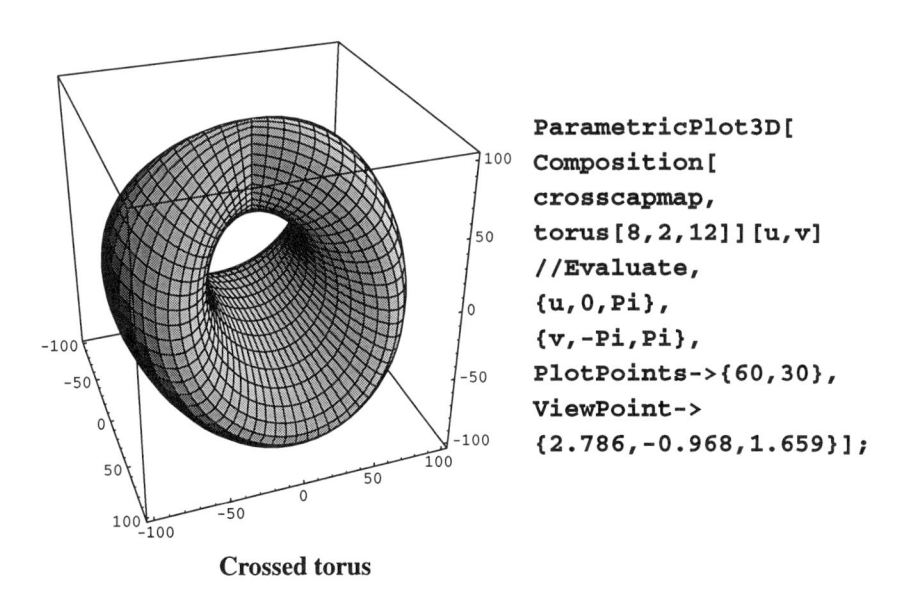

```
ParametricPlot3D[
Composition[
crosscapmap,
torus[8,2,12]][u,v]
//Evaluate,
{u,0,Pi},
{v,-Pi,Pi},
PlotPoints->{60,30},
ViewPoint->
{2.786,-0.968,1.659}];
```

Crossed torus

7. A cross cap can be defined implicitly by

```
crosscapimplicit[a_,b_][x_,y_]:=
    (a x^2 + b y^2)(x^2 + y^2 + z^2) -2 z (x^2 + y^2)
```

Use **ImplicitPlot3D** to plot **crosscapimplicit**.

14.9 Animation of Steiner's Roman Surface

We need two packages; they can be read into a *Mathematica* session with the commands

```
Needs["CAS`"]
Needs["Graphics`Animation`"];
```

The following command begins with Steiner's Roman surface, then gradually cuts it open:

```
Do[ParametricPlot3D[Evaluate[
Append[roman[1][u,v],FaceForm[Green,Yellow]]],
{u,0,Pi},{v,0,Pi},Lighting->False,PlotPoints->{39,39},
PlotRange->{{-0.51,0.51},{-0.51,0.51},{-0.51,0.51}},
Boxed->False,Axes->None],{a,0.5,-0.5,-0.01}]
```

15
METRICS ON SURFACES

In this chapter we begin the study of the geometry of surfaces from the point of view of distance and area. When mathematicians began to study surfaces at the end of the eighteenth century, they used the notions of infinitesimal distance and area. In Sections 15.1 and 15.3 we explain these notions intuitively, but from a modern standpoint. In Section 15.2 we define and discuss surface isometries. Section 15.4 contains *Mathematica* programs for computing metrics and areas of surfaces. Some simple examples of how the *Mathematica* formulas of Section 15.4 can be used to compute metrics and areas are given in Section 15.5.

15.1 The Intuitive Idea of Distance on a Surface

So far we have not discussed how to measure distances on a surface. One of the key facts about distance in Euclidean space $\mathbb{R}^n$ is that the Pythagorean Theorem holds. This means that if $\mathbf{p} = (p_1, \ldots, p_n)$ and $\mathbf{q} = (q_1, \ldots, q_n)$ are points in $\mathbb{R}^n$, then the distance s from $\mathbf{p}$ to $\mathbf{q}$ is given by

$$(15.1) \qquad s^2 = (p_1 - q_1)^2 + \cdots + (p_n - q_n)^2.$$

How is this notion different for a surface? Because a general surface is curved, distance on it is not the same as in Euclidean space; in particular, (15.1) is in general false. To describe how to measure distance on a surface, we need the mathematically imprecise notion of infinitesimal. The infinitesimal version of (15.1) for $n = 2$ is

(15.2) $$ds^2 = dx^2 + dy^2.$$

We can think of dx and dy as small quantities in the x and y directions. Formula (15.2) is valid for $\mathbb{R}^2$. For a surface (or more precisely a patch) the corresponding equation is

(15.3) $$ds^2 = E\,du^2 + 2F\,du\,dv + G\,dv^2.$$

This is the classical notation[1] for a metric on a surface. We may consider (15.3) as an infinitesimal warped version of the Pythagorean Theorem.

Among the many ways to define metrics on surfaces, one is especially simple and important. Let $\mathcal{M}$ be a regular surface in $\mathbb{R}^n$. The scalar product $\cdot$ of $\mathbb{R}^n$ gives rise to a scalar product on $\mathcal{M}$ by restriction. If $\mathbf{v}_p$ and $\mathbf{w}_p$ are tangent vectors to $\mathcal{M}$ at $\mathbf{p} \in \mathcal{M}$, we can take the scalar product $\mathbf{v}_p \cdot \mathbf{w}_p$ because $\mathbf{v}_p$ and $\mathbf{w}_p$ are also tangent vectors to $\mathbb{R}^n$ at $\mathbf{p}$.

In this chapter we deal only with local properties of distance, so without loss of generality, we can assume that $\mathcal{M}$ is the image of an injective regular patch. However, the following definition makes sense for any patch.

Definition. *Let* $\mathbf{x}: \mathcal{U} \longrightarrow \mathbb{R}^n$ *be a patch. We define functions* $E, F, G: \mathcal{U} \longrightarrow \mathbb{R}$ *by*

$$E = \|\mathbf{x}_u\|^2, \qquad F = \mathbf{x}_u \cdot \mathbf{x}_v, \qquad G = \|\mathbf{x}_v\|^2.$$

Then $ds^2 = E\,du^2 + 2F\,du\,dv + G\,dv^2$ *is the* **Riemannian metric** *or* **first fundamental form** *of the patch* $\mathbf{x}$. *Furthermore,* E, F *and* G *are called the* **coefficients of the first fundamental form** *of* $\mathbf{x}$.

The equivalence of the formal definition of E, F and G with the infinitesimal version (15.3) can be better understood if we consider curves on a patch.

---

[1] This notation (including the choice of letters E, F and G) was already in use in the early part of the 19th century; it can be found, for example, in the works of Gauss (see [Gauss] and [Dom]). Gauss had the idea to study properties of a surface that are independent of the way the surface sits in space, that is, properties of a surface that are expressible in terms of E, F and G alone. One such property is Gauss's *Theorema Egregium*, which we shall prove in Section 22.2.

15.1 The Intuitive Idea of Distance on a Surface

Lemma 15.1. *Let $\alpha\colon (a,b) \longrightarrow \mathbb{R}^n$ be a curve that lies on a regular injective patch $\mathbf{x}\colon \mathcal{U} \longrightarrow \mathbb{R}^n$. Then the arc length function s of α starting at $\alpha(c)$ is given by*

$$(15.4) \qquad s(t) = \int_c^t \sqrt{E\left(\frac{du}{dt}\right)^2 + 2F\frac{du}{dt}\frac{dv}{dt} + G\left(\frac{dv}{dt}\right)^2}\, dt.$$

Proof. Write $\alpha(t) = \mathbf{x}(u(t), v(t))$ for $a < t < b$. By the definition of the arc length function and Corollary 12.9, page 277, we have

$$\begin{aligned}s(t) &= \int_c^t \|\alpha'(t)\| dt = \int_c^t \|u'(t)\mathbf{x}_u(u(t),v(t)) + v'(t)\mathbf{x}_v(u(t),v(t))\| dt \\ &= \int_c^t \sqrt{u'(t)^2\|\mathbf{x}_u\|^2 + 2u'(t)v'(t)\mathbf{x}_u\cdot\mathbf{x}_v + v'(t)^2\|\mathbf{x}_v\|^2}\, dt \\ &= \int_c^t \sqrt{E\left(\frac{du}{dt}\right)^2 + 2F\frac{du}{dt}\frac{dv}{dt} + G\left(\frac{dv}{dt}\right)^2}\, dt.\end{aligned}$$ ∎

It follows from (15.4) that

$$(15.5) \qquad \frac{ds}{dt} = \sqrt{E\left(\frac{du}{dt}\right)^2 + 2F\frac{du}{dt}\frac{dv}{dt} + G\left(\frac{dv}{dt}\right)^2}.$$

Now we can make sense of (15.3). We square both sides of (15.5) and multiply through by dt^2. Although strictly speaking multiplication by the infinitesimal dt^2 is not permitted, at least formally we obtain (15.3).

Notice that the right-hand side of (15.3) does not involve the parameter t, except insofar as u and v depend on t. We may think of ds as the **infinitesimal arc length**, because it gives the arc length function when integrated over any curve. Geometrically, ds can be interpreted as the **infinitesimal distance** from a point $\mathbf{x}(u,v)$ to a point $\mathbf{x}(u+du, v+dv)$ measured along the surface.

Notice also that

$$\alpha(t+dt) \approx \alpha(t) + \alpha'(t)dt = \alpha(t) + \mathbf{x}_u u'(t)dt + \mathbf{x}_v v'(t)dt,$$

so that

$$\begin{aligned}\|\alpha(t+dt) - \alpha(t)\| &\approx \|\mathbf{x}_u u'(t) + \mathbf{x}_v v'(t)\| dt \\ &= \sqrt{E\, u'^2 + 2F\, u'v' + G\, v'^2}\, dt = ds.\end{aligned}$$

A standard notion from calculus of several variables is that of the **differential** of a function $f:\mathbb{R}^2 \longrightarrow \mathbb{R}$; it is given by

$$df = \frac{\partial f}{\partial u}du + \frac{\partial f}{\partial v}dv.$$

More generally, if $\mathbf{x}:\mathcal{U} \longrightarrow \mathcal{M}$ is a regular injective patch and $f:\mathcal{M} \longrightarrow \mathbb{R}$ is a differentiable function, we put

$$df = \frac{\partial(f \circ \mathbf{x})}{\partial u}du + \frac{\partial(f \circ \mathbf{x})}{\partial v}dv.$$

We call df the **differential** of f. The differentials of the functions $\mathbf{x}(u,v) \longmapsto u$ and $\mathbf{x}(u,v) \longmapsto v$ are denoted by du and dv. In spite of its appearance, ds will hardly ever be the differential of a function on a surface. But formally we have $d\mathbf{x} = \mathbf{x}_u du + \mathbf{x}_v dv$, so that

$$d\mathbf{x} \cdot d\mathbf{x} = (\mathbf{x}_u du + \mathbf{x}_v dv) \cdot (\mathbf{x}_u du + \mathbf{x}_v dv) = \|\mathbf{x}_u\|^2 du^2 + \mathbf{x}_u \cdot \mathbf{x}_v \, du \, dv + \|\mathbf{x}_v\|^2 dv^2$$
$$= E\,du^2 + 2F\,du\,dv + G\,dv^2 = ds^2.$$

Equation (15.2) represents ds^2 of $\mathbb{R}^2$ written in terms of the Cartesian coordinates x and y. There is a different expression for ds^2 in polar coordinates.

Lemma 15.2. *The metric ds^2 on $\mathbb{R}^2$, given in Cartesian coordinates as*

$$ds^2 = dx^2 + dy^2,$$

in polar coordinates becomes

$$ds^2 = dr^2 + r^2 d\theta^2.$$

Proof. We have the standard change of variable formulas from rectangular to polar coordinates:

$$\begin{cases} x = r\cos\theta, \\ y = r\sin\theta. \end{cases}$$

15.1 The Intuitive Idea of Distance on a Surface

Therefore,

(15.6)
$$\begin{cases} dx = -r\sin\theta\, d\theta + \cos\theta\, dr, \\ dy = r\cos\theta\, d\theta + \sin\theta\, dr. \end{cases}$$

Hence

$$\begin{aligned} dx^2 + dy^2 &= (-r\sin\theta\, d\theta + \cos\theta\, dr)^2 + (r\cos\theta\, d\theta + \sin\theta\, dr)^2 \\ &= dr^2 + r^2 d\theta^2. \end{aligned}$$ ∎

We need to know how the expression for a metric changes under a change of coordinates.

Lemma 15.3. *Let* $\mathbf{x}\colon \mathcal{U} \longrightarrow \mathcal{M}$ *and* $\mathbf{y}\colon \mathcal{V} \longrightarrow \mathcal{M}$ *be patches on a regular surface* $\mathcal{M}$ *with* $\mathbf{x}(\mathcal{U}) \cap \mathbf{y}(\mathcal{V})$ *nonempty. Let* $\mathbf{x}^{-1} \circ \mathbf{y} = (\bar{u}, \bar{v})\colon \mathcal{U} \cap \mathcal{V} \longrightarrow \mathcal{U} \cap \mathcal{V}$ *be the associated change of coordinates, so that*

$$\mathbf{y}(u, v) = \mathbf{x}\big(\bar{u}(u, v), \bar{v}(u, v)\big).$$

Suppose that $\mathcal{M}$ *has a metric and denote the induced metrics on* $\mathbf{x}$ *and* $\mathbf{y}$ *by*

$$ds_{\mathbf{x}}^2 = E_{\mathbf{x}} d\bar{u}^2 + 2F_{\mathbf{x}} d\bar{u}\, d\bar{v} + G_{\mathbf{x}} d\bar{v}^2 \qquad \text{and} \qquad ds_{\mathbf{y}}^2 = E_{\mathbf{y}} du^2 + 2F_{\mathbf{y}} du\, dv + G_{\mathbf{y}} dv^2.$$

Then

(15.7)
$$\begin{cases} E_{\mathbf{y}} = E_{\mathbf{x}} \left(\dfrac{\partial \bar{u}}{\partial u}\right)^2 + 2F_{\mathbf{x}} \dfrac{\partial \bar{u}}{\partial u}\dfrac{\partial \bar{v}}{\partial u} + G_{\mathbf{x}} \left(\dfrac{\partial \bar{v}}{\partial u}\right)^2, \\[6pt] F_{\mathbf{y}} = E_{\mathbf{x}} \dfrac{\partial \bar{u}}{\partial u}\dfrac{\partial \bar{u}}{\partial v} + F_{\mathbf{x}} \left(\dfrac{\partial \bar{u}}{\partial u}\dfrac{\partial \bar{v}}{\partial v} + \dfrac{\partial \bar{u}}{\partial v}\dfrac{\partial \bar{v}}{\partial u}\right) + G_{\mathbf{x}} \dfrac{\partial \bar{v}}{\partial u}\dfrac{\partial \bar{v}}{\partial v}, \\[6pt] G_{\mathbf{y}} = E_{\mathbf{x}} \left(\dfrac{\partial \bar{u}}{\partial v}\right)^2 + 2F_{\mathbf{x}} \dfrac{\partial \bar{u}}{\partial v}\dfrac{\partial \bar{v}}{\partial v} + G_{\mathbf{x}} \left(\dfrac{\partial \bar{v}}{\partial v}\right)^2. \end{cases}$$

Proof. We use Lemma 12.19, page 285, to compute

$$\begin{aligned} E_{\mathbf{y}} = \mathbf{y}_u \cdot \mathbf{y}_u &= \left(\frac{\partial \bar{u}}{\partial u}\mathbf{x}_{\bar{u}} + \frac{\partial \bar{v}}{\partial u}\mathbf{x}_{\bar{v}}\right) \cdot \left(\frac{\partial \bar{u}}{\partial u}\mathbf{x}_{\bar{u}} + \frac{\partial \bar{v}}{\partial u}\mathbf{x}_{\bar{v}}\right) \\ &= \left(\frac{\partial \bar{u}}{\partial u}\right)^2 \mathbf{x}_{\bar{u}}\cdot\mathbf{x}_{\bar{u}} + 2\left(\frac{\partial \bar{u}}{\partial u}\right)\left(\frac{\partial \bar{v}}{\partial v}\right)\mathbf{x}_{\bar{u}}\cdot\mathbf{x}_{\bar{v}} + \left(\frac{\partial \bar{v}}{\partial v}\right)^2 \mathbf{x}_{\bar{v}}\cdot\mathbf{x}_{\bar{v}} \\ &= \left(\frac{\partial \bar{u}}{\partial u}\right)^2 E_{\mathbf{x}} + 2\left(\frac{\partial \bar{u}}{\partial u}\right)\left(\frac{\partial \bar{v}}{\partial u}\right) F_{\mathbf{x}} + \left(\frac{\partial \bar{v}}{\partial u}\right)^2 G_{\mathbf{x}}. \end{aligned}$$

The other equations are proved similarly. ∎

15.2 Isometries and Conformal Maps of Surfaces

In Section 12.6 we defined the notion of a mapping Φ between surfaces in $\mathbb{R}^n$. Even if Φ is a local diffeomorphism, $\Phi(\mathcal{M})$ can be quite different from $\mathcal{M}$. (Good examples are the maps **romanmap** and **crosscapmap** of Section 14.6 that map a sphere onto Steiner's Roman surface and a cross cap.) Differentiable maps that preserve infinitesimal distances, on the other hand, distort much less. The search for such maps leads to the following definition.

Definition. *Let $\mathcal{M}_1$ and $\mathcal{M}_2$ be regular surfaces in $\mathbb{R}^n$. A map $\Phi \colon \mathcal{M}_1 \longrightarrow \mathcal{M}_2$ is called a* **local isometry** *provided its tangent map satisfies*

$$(15.8) \qquad \|\Phi_*(\mathbf{v_p})\| = \|\mathbf{v_p}\|$$

for all tangent vectors $\mathbf{v_p}$ to $\mathcal{M}_1$. An **isometry** *is a surface mapping which is simultaneously a local isometry and a diffeomorphism.*

Lemma 15.4. *A local isometry is a local diffeomorphism.*

Proof. It is easy to see that (15.8) implies that each tangent map of a local isometry Φ is injective. Then the inverse function theorem implies that Φ is a local diffeomorphism. ∎

Every isometry of $\mathbb{R}^n$ which maps a regular surface $\mathcal{M}_1$ onto a regular surface $\mathcal{M}_2$ obviously restricts to an isometry between $\mathcal{M}_1$ and $\mathcal{M}_2$. When we study minimal surfaces in Chapter 30, we shall see that there are isometries between surfaces that do not arise in this fashion.

Here are some elementary reformulations of the definition of isometry.

Lemma 15.5. *Condition (15.8) is equivalent to each of the following conditions.*

(i) *For all tangent vectors $\mathbf{v_p}$ and $\mathbf{w_p}$ to $\mathcal{M}_1$ we have*

$$(15.9) \qquad \Phi_*(\mathbf{v_p}) \cdot \Phi_*(\mathbf{w_p}) = \mathbf{v_p} \cdot \mathbf{w_p}.$$

(ii) *For each $\mathbf{p} \in \mathcal{M}_1$ there are linearly independent tangent vectors $\mathbf{v_p}, \mathbf{w_p} \in \mathcal{M}_{1\mathbf{p}}$ such that*

15.2 Isometries and Conformal Maps of Surfaces

$$(15.10) \quad \begin{cases} \Phi_*(\mathbf{v_p}) \cdot \Phi_*(\mathbf{v_p}) = \mathbf{v_p} \cdot \mathbf{v_p}, \\ \Phi_*(\mathbf{w_p}) \cdot \Phi_*(\mathbf{w_p}) = \mathbf{w_p} \cdot \mathbf{w_p}, \\ \Phi_*(\mathbf{v_p}) \cdot \Phi_*(\mathbf{w_p}) = \mathbf{v_p} \cdot \mathbf{w_p}. \end{cases}$$

Proof. It is clear that (15.9) implies (15.8). Conversely, suppose (15.8) holds. Since Φ_* is linear, we have

$$\Phi_*(\mathbf{v_p}) \cdot \Phi_*(\mathbf{w_p}) = \frac{1}{2}\left(\|\Phi_*(\mathbf{v_p} + \mathbf{w_p})\|^2 - \|\Phi_*(\mathbf{v_p})\|^2 - \|\Phi_*(\mathbf{w_p})\|^2\right)$$

$$= \frac{1}{2}\left(\|(\mathbf{v_p} + \mathbf{w_p})\|^2 - \|\mathbf{v_p}\|^2 - \|\mathbf{w_p}\|^2\right) = \mathbf{v_p} \cdot \mathbf{w_p}.$$

Also, it is clear that (15.9) implies (15.10). Finally, suppose (15.10) holds. Then for all $a, b \in \mathbb{R}$ we have

$$\|\Phi_*(a\,\mathbf{v_p} + b\,\mathbf{w_p})\|^2 = \|a\,\Phi_*(\mathbf{v_p}) + b\,\Phi_*(\mathbf{w_p})\|^2$$

$$= a^2\|\Phi_*(\mathbf{v_p})\|^2 + b^2\|\Phi_*(\mathbf{w_p})\|^2 + 2ab\,\Phi_*(\mathbf{v_p}) \cdot \Phi_*(\mathbf{w_p})$$

$$= a^2\|\mathbf{v_p}\|^2 + b^2\|\mathbf{w_p}\|^2 + 2ab\,\mathbf{v_p} \cdot \mathbf{w_p} = \|a\,\mathbf{v_p} + b\,\mathbf{w_p}\|^2.$$

Hence we get (15.8). ∎

Next, we show that a surface mapping is an isometry if and only if it preserves Riemannian metrics.

Lemma 15.6. *Let* $\mathbf{x}: \mathcal{U} \longrightarrow \mathbb{R}^n$ *be a regular injective patch and let* $\mathbf{y}: \mathcal{U} \longrightarrow \mathbb{R}^n$ *be any patch. Let*

$$ds_\mathbf{x}^2 = E_x du^2 + 2F_x du\,dv + G_x dv^2 \quad \text{and} \quad ds_\mathbf{y}^2 = E_y du^2 + 2F_y du\,dv + G_y dv^2$$

denote the induced Riemannian metrics on $\mathbf{x}$ *and* $\mathbf{y}$. *Then the map*

$$\Phi = \mathbf{y} \circ \mathbf{x}^{-1}: \mathbf{x}(\mathcal{U}) \longrightarrow \mathbf{y}(\mathcal{U})$$

is a local isometry if and only if

$$(15.11) \quad ds_\mathbf{x}^2 = ds_\mathbf{y}^2.$$

Proof. First, note that

$$\Phi_*(\mathbf{x}_u) = ((\mathbf{y}\circ\mathbf{x}^{-1})_* \circ \mathbf{x}_*)\left(\frac{\partial}{\partial u}\right) = \mathbf{y}_*\left(\frac{\partial}{\partial u}\right) = \mathbf{y}_u;$$

similarly, $\Phi_*(\mathbf{x}_v) = \mathbf{y}_v$. If Φ is a local isometry, then

$$E_\mathbf{y} = \|\mathbf{y}_u\|^2 = \|\Phi_*(\mathbf{x}_u)\|^2 = \|\mathbf{x}_u\|^2 = E_\mathbf{x}.$$

In the same way, $F_\mathbf{y} = F_\mathbf{x}$ and $G_\mathbf{y} = G_\mathbf{x}$; thus (15.11) holds.

To prove the converse, consider a curve α of the form

$$\alpha(t) = \mathbf{x}(u(t), v(t));$$

then $(\Phi \circ \alpha)(t) = \mathbf{y}(u(t), v(t))$. From Corollary 12.9 we know that

$$\alpha' = u'\mathbf{x}_u + v'\mathbf{x}_v \quad \text{and} \quad (\Phi \circ \alpha)' = u'\mathbf{y}_u + v'\mathbf{y}_v.$$

Hence

(15.12)
$$\begin{cases} \|\alpha'\|^2 = E_\mathbf{x} u'^2 + 2F_\mathbf{x} u'v' + G_\mathbf{x} v'^2, \\ \|(\Phi \circ \alpha)'\|^2 = E_\mathbf{y} u'^2 + 2F_\mathbf{y} u'v' + G_\mathbf{y} v'^2. \end{cases}$$

Then (15.11) and (15.12) imply that

(15.13)
$$\|\alpha'(t)\| = \|(\Phi \circ \alpha)'(t)\|.$$

Since every tangent vector to $\mathbf{x}(\mathcal{U})$ can be represented as $\alpha'(0)$ for some curve α, it follows that Φ is an isometry. ∎

Corollary 15.7. *Let $\Phi\colon \mathcal{M}_1 \longrightarrow \mathcal{M}_2$ be a surface mapping, and let $\mathbf{x}\colon \mathcal{U} \longrightarrow \mathcal{M}_1$ be a patch on $\mathcal{M}_1$. Let $\mathbf{y} = \Phi \circ \mathbf{x}$. Then Φ is a local isometry if and only if for each regular injective patch on $\mathcal{M}_1$ we have $ds_\mathbf{x}^2 = ds_\mathbf{y}^2$.*

A Riemannian metric determines a function which measures distances on a regular surface.

Definition. *Let $\mathcal{M} \subset \mathbb{R}^n$ be a regular surface, and let $\mathbf{p}, \mathbf{q} \in \mathcal{M}$. Then the **intrinsic distance** $\rho(\mathbf{p}, \mathbf{q})$ is the greatest lower bound of the lengths of all piecewise-differentiable curves α that lie entirely on $\mathcal{M}$ and connect $\mathbf{p}$ to $\mathbf{q}$. We call ρ the **distance function** of $\mathcal{M}$.*

15.2 Isometries and Conformal Maps of Surfaces

In general the intrinsic distance $\rho(\mathbf{p}, \mathbf{q})$ will be greater than the Euclidean distance $\|\mathbf{p} - \mathbf{q}\|$.

An isometry also preserves the intrinsic distance.

Lemma 15.8. *Let $\Phi: \mathcal{M}_1 \longrightarrow \mathcal{M}_2$ be an isometry. Then Φ preserves the intrinsic distances ρ_1, ρ_2 of $\mathcal{M}_1$ and $\mathcal{M}_2$ in the following sense:*

$$(15.14) \qquad \rho_2\big(\Phi(\mathbf{p}), \Phi(\mathbf{q})\big) = \rho_1(\mathbf{p}, \mathbf{q}),$$

for $\mathbf{p}, \mathbf{q} \in \mathcal{M}_1$.

Proof. Let $\alpha: (c, d) \longrightarrow \mathcal{M}$ be a piecewise-differentiable curve with $\alpha(a) = \mathbf{p}$ and $\alpha(b) = \mathbf{q}$, where $c < a < b < d$. From the definition of local isometry it follows that

$$\mathbf{Length}[\alpha] = \int_a^b \|\alpha'(t)\| dt = \int_a^b \|(\Phi \circ \alpha)'(t)\| dt = \mathbf{Length}[\Phi \circ \alpha].$$

Since Φ is a diffeomorphism, there is a one-to-one correspondence between the piecewise-differentiable curves on $\mathcal{M}_1$ connecting $\mathbf{p}$ to $\mathbf{q}$ and the piecewise-differentiable curves on $\mathcal{M}_2$ connecting $\Phi(\mathbf{p})$ to $\Phi(\mathbf{q})$. Since corresponding curves have equal lengths, we obtain (15.14). ∎

Next, let us consider a generalization of the notion of local isometry.

Definition. *Let $\mathcal{M}_1$ and $\mathcal{M}_2$ be regular surfaces in $\mathbb{R}^n$. A map $\Phi: \mathcal{M}_1 \longrightarrow \mathcal{M}_2$ is called a* **conformal map** *provided there is a differentiable everywhere positive function $\lambda: \mathcal{M}_1 \longrightarrow \mathbb{R}$ such that*

$$(15.15) \qquad \|\Phi_*(\mathbf{v}_\mathbf{p})\| = \lambda(\mathbf{p}) \|\mathbf{v}_\mathbf{p}\|$$

for all $\mathbf{p} \in \mathcal{M}_1$ and all tangent vectors $\mathbf{v}_\mathbf{p}$ to $\mathcal{M}_1$ at $\mathbf{p}$. We call λ the **scale factor**. *A* **conformal diffeomorphism** *is a surface mapping which is simultaneously a conformal map and a diffeomorphism.*

Since the tangent map Φ_* of an isometry preserves inner products, it also preserves angles. The tangent map of a conformal diffeomorphism in general will change the lengths of tangent vectors, but will still preserve angles. More precisely:

Lemma 15.9. *A conformal map $\Phi: \mathcal{M}_1 \longrightarrow \mathcal{M}_2$ preserves the angles between nonzero tangent vectors.*

Proof. The proof that (15.9) is equivalent to (15.8) can be easily modified to prove that (15.15) is equivalent to

(15.16) $$\Phi_*(\mathbf{v_p}) \cdot \Phi_*(\mathbf{w_p}) = \lambda(\mathbf{p})^2 \mathbf{v_p} \cdot \mathbf{w_p}$$

for all nonzero tangent vectors $\mathbf{v_p}$ and $\mathbf{w_p}$ to $\mathcal{M}_1$. Then (15.16) implies that

$$\frac{\Phi_*(\mathbf{v_p}) \cdot \Phi_*(\mathbf{w_p})}{\|\Phi_*(\mathbf{v_p})\| \|\Phi_*(\mathbf{w_p})\|} = \frac{\mathbf{v_p} \cdot \mathbf{w_p}}{\|\mathbf{v_p}\| \|\mathbf{w_p}\|}.$$

From the definition of angle on page 3 it follows that Φ_* preserves angles between tangent vectors. ∎

Here is an important example of a conformal map.

Definition. Let $S^2(1)$ denote the unit sphere in $\mathbb{R}^3$. The **stereographic map** $\Upsilon: \mathbb{R}^2 \longrightarrow S^2(1)$ is defined by

$$\Upsilon(p_1, p_2) = \frac{(2p_1, 2p_2, p_1^2 + p_2^2 - 1)}{p_1^2 + p_2^2 + 1}.$$

We abbreviate the definition of the stereographic map to

(15.17) $$\Upsilon(\mathbf{p}) = \frac{(2\mathbf{p}, \|\mathbf{p}\|^2 - 1)}{1 + \|\mathbf{p}\|^2},$$

for $\mathbf{p} \in S^2(1)$. It is easy to see that Υ is differentiable, and that $\|\Upsilon(\mathbf{p})\| = 1$. Moreover,

Lemma 15.10. Υ *preserves angles between vectors.*

Proof. Let $\alpha: (a, b) \longrightarrow \mathbb{R}^2$ be a curve. It follows from (15.17) that the image of α by Υ is the curve

$$\Upsilon \circ \alpha = \frac{(2\alpha, \|\alpha\|^2 - 1)}{1 + \|\alpha\|^2};$$

this equation can be rewritten as

(15.18) $$(1 + \|\alpha\|^2)(\Upsilon \circ \alpha) = (2\alpha, \|\alpha\|^2 - 1).$$

Differentiating (15.18), we obtain

(15.19) $$2(\alpha \cdot \alpha')(\Upsilon \circ \alpha) + (1 + \|\alpha\|^2)(\Upsilon \circ \alpha)' = 2(\alpha', \alpha \cdot \alpha').$$

Since $\Upsilon(\mathbb{R}^2) \subset S^2(1)$, we have

(15.20) $\qquad \|\Upsilon \circ \alpha\|^2 = 1 \quad$ and $\quad (\Upsilon \circ \alpha) \cdot (\Upsilon \circ \alpha)' = 0.$

From (15.19) and (15.20) we get

$$\begin{aligned} 4\|\alpha'\|^2 + 4(\alpha \cdot \alpha')^2 &= 4\|(\alpha', \alpha \cdot \alpha')\|^2 \\ &= \left\|2(\alpha \cdot \alpha')(\Upsilon \circ \alpha) + (1 + \|\alpha\|^2)(\Upsilon \circ \alpha)'\right\|^2 \\ &= 4(\alpha \cdot \alpha')^2\|(\Upsilon \circ \alpha)\|^2 + (1 + \|\alpha\|^2)^2\|(\Upsilon \circ \alpha)'\|^2 \\ &= 4(\alpha \cdot \alpha')^2 + (1 + \|\alpha\|^2)^2\|(\Upsilon \circ \alpha)'\|^2. \end{aligned}$$

Therefore,

(15.21) $\qquad \|(\Upsilon \circ \alpha)'\| = \dfrac{2\|\alpha'\|}{1 + \|\alpha\|^2}.$

From (15.21) we conclude that

(15.22) $\qquad \|\Upsilon_*(\mathbf{v}_\mathbf{p})\| = \dfrac{2\|\mathbf{v}_\mathbf{p}\|}{1 + \|\mathbf{p}\|^2}.$

for all $\mathbf{p}$ and all tangent vectors $\mathbf{v}_\mathbf{p}$ to $\mathbb{R}^2$ at $\mathbf{p}$. Hence Υ is conformal with scale factor $\lambda(\mathbf{p}) = 2/(1 + \|\mathbf{p}\|^2)$. ∎

15.3 The Intuitive Idea of Area on a Surface

In $\mathbb{R}^n$ the infinitesimal hypercube bounded by $dx_1, \ldots, dx_n$ has as its volume the product[2]

$$dV = dx_1 \wedge \ldots \wedge dx_n;$$

we call dV the **infinitesimal volume element** of $\mathbb{R}^n$. The corresponding concept for a surface with metric $ds^2 = E\,du^2 + 2F\,du\,dv + G\,dv^2$ is the **element of area**

---

[2] There are two kinds of products between infinitesimals du and dv: the **symmetric product** $du\,dv$ that we used in the definition of the metric ds and the **wedge product** $du \wedge dv$. The main difference is that $du\,dv = dv\,du$, but $du \wedge dv = -dv \wedge du$.

dA; it is given by

(15.23) $$dA = \sqrt{EG - F^2}\, du \wedge dv.$$

For a patch $\mathbf{x}: \mathcal{U} \longrightarrow \mathbb{R}^3$ it is easy to see why this is the appropriate definition of element of area. By (8.6), page 183, we have

$$\sqrt{EG - F^2} = \|\mathbf{x}_u \times \mathbf{x}_v\| = \|\mathbf{x}_u\|\,\|\mathbf{x}_v\| \sin \theta,$$

where θ is the positive angle between the vectors $\mathbf{x}_u$ and $\mathbf{x}_v$. On the other hand, the quantity $\|\mathbf{x}_u\|\,\|\mathbf{x}_v\| \sin \theta$ represents the area of the parallelogram with sides $\mathbf{x}_u$ and $\mathbf{x}_v$ and angle θ between the sides.

Let us compute the element of area for the usual metric on the plane, but in polar coordinates. We use the method of Lemma 15.2.

Lemma 15.11. *The metric ds^2 on $\mathbb{R}^2$ given by (15.2) has as its element of area*

(15.24) $$dA = dx \wedge dy = r\, dr \wedge d\theta.$$

Proof. If follows from (15.6), using the fact that the wedge product is antisymmetric, that

$$\begin{aligned} dA &= dx \wedge dy \\ &= (-r \sin \theta\, d\theta + \cos \theta\, dr) \wedge (r \cos \theta\, d\theta + \sin \theta\, dr) \\ &= -r^2 \sin \theta \cos \theta\, d\theta \wedge d\theta + \cos \theta \sin \theta\, dr \wedge dr \\ &\quad - r \sin^2 \theta\, d\theta \wedge dr + r \cos^2 \theta\, dr \wedge d\theta \\ &= r\, dr \wedge d\theta. \quad\blacksquare \end{aligned}$$

Equation (15.23) motivates the following definition of the area of a closed subset of the trace of a patch. Recall that a subset S of $\mathbb{R}^n$ is **bounded**, provided there exists a number M such that $\|\mathbf{p}\| \leq M$ for all $\mathbf{p} \in S$. A **compact** subset of $\mathbb{R}^n$ is a subset which is closed and bounded.

Definition. *Let* $\mathbf{x}: \mathcal{U} \longrightarrow \mathbb{R}^n$ *be an injective regular patch, and let S be a compact subset of $\mathbf{x}(\mathcal{U})$. Then the **area** of S is*

$$\text{area}(S) = \iint_{\mathbf{x}^{-1}(S)} \sqrt{EG - F^2}\, du \wedge dv. \tag{15.25}$$

That this definition is geometric is a consequence of the following lemma:

Lemma 15.12. *The definition of area is independent of the choice of patch.*

Proof. Let $\mathbf{x}: \mathcal{V} \longrightarrow \mathbb{R}^n$ and $\mathbf{y}: \mathcal{W} \longrightarrow \mathbb{R}^n$ be injective regular patches, and assume that $S \subseteq \mathbf{x}(\mathcal{W}) \cap \mathbf{y}(\mathcal{V})$. A long but straightforward computation using Lemma 15.3 shows that

$$\sqrt{E_\mathbf{y} G_\mathbf{y} - F_\mathbf{y}^2} = \sqrt{E_\mathbf{x} G_\mathbf{x} - F_\mathbf{x}^2} \left| \frac{\partial \bar{u}}{\partial u} \frac{\partial \bar{v}}{\partial v} - \frac{\partial \bar{u}}{\partial v} \frac{\partial \bar{v}}{\partial v} \right|. \tag{15.26}$$

But

$$\left| \frac{\partial \bar{u}}{\partial u} \frac{\partial \bar{v}}{\partial v} - \frac{\partial \bar{u}}{\partial v} \frac{\partial \bar{v}}{\partial v} \right|$$

is the determinant of the Jacobian matrix of $\mathbf{x}^{-1} \circ \mathbf{y}$. Hence by the change of variables formulas for multiple integrals, we have

$$\iint_{\mathbf{y}^{-1}(S)} \sqrt{E_\mathbf{y} G_\mathbf{y} - F_\mathbf{y}^2}\, d\bar{u} \wedge d\bar{v} = \iint_{\mathbf{y}^{-1}(S)} \sqrt{E_\mathbf{x} G_\mathbf{x} - F_\mathbf{x}^2} \left| \frac{\partial \bar{u}}{\partial u} \frac{\partial \bar{v}}{\partial v} - \frac{\partial \bar{u}}{\partial v} \frac{\partial \bar{v}}{\partial v} \right| d\bar{u} \wedge d\bar{v}$$

$$= \iint_{\mathbf{x}^{-1}(S)} \sqrt{E_\mathbf{x} G_\mathbf{x} - F_\mathbf{x}^2}\, du \wedge dv. \blacksquare$$

Surface area cannot, in general, be computed by taking the limit of the area of approximating polyhedra. For example, see [Krey1, pages 115–117]. Thus the analog for surfaces of Theorem 1.5, page 9, is false.

15.4 Programs for Computing Metrics and Areas on a Surface

We shall use **ee**, **ff**, **gg** in *Mathematica* for E, F and G. (This is to avoid collision with *Mathematica*'s use of **E** as $2.718\ldots$) If a patch $\mathbf{x}: \mathcal{U} \longrightarrow \mathcal{M} \subset \mathbb{R}^n$ has been defined in *Mathematica*, we can find **ee**, **ff** and **gg** by means of the formulas

```
ee[x_][u_,v_]:= Simplify[D[x[uu,vv],uu].
               D[x[uu,vv],uu]] /. {uu->u,vv->v}

ff[x_][u_,v_]:= Simplify[D[x[uu,vv],uu].
               D[x[uu,vv],vv]] /. {uu->u,vv->v}

gg[x_][u_,v_]:= Simplify[D[x[uu,vv],vv].
               D[x[uu,vv],vv]] /. {uu->u,vv->v}
```

The infinitesimal area is defined similarly in *Mathematica*:

```
infarea[x_][u_,v_]:= Sqrt[Simplify[ee[x][u,v]gg[x][u,v]
                    - ff[x][u,v]^2]]
```

The area of a patch can be computed by integrating **infarea** over the domain of definition of the patch.

15.5 Examples of Metrics

The Sphere

The components of the metric and the infinitesimal area of a sphere $S^2(a)$ of radius a are easily computed with *Mathematica*. The *Mathematica* definition of **sphere[a]** is given on page 270. We defined the functions **ee**, **ff**, **gg** and **infarea** in Section 15.4. These functions are also contained in Appendix A, or in the package **CSPROGS.m**, so if that package is accessible, it is easier to read in that package than to retype the formulas. The metric of the sphere $S^2(a)$ is then given by the following sequence of questions (commands) and answers:

```
ee[sphere[a]][u,v]
       2    2
      a  Cos[v]

ff[sphere[a]][u,v]
       0
```

15.5 Examples of Metrics

```
gg[sphere[a]][u,v]
```

$$a^2$$

Hence the Riemannian metric of the standard parametrization of $S^2(a)$ is given by

(15.27) $$ds^2 = a^2\left((\cos v)^2 du^2 + dv^2\right).$$

Similarly,

```
infarea[sphere[a]][u,v]//PowerExpand
```

gives

$$a^2 \, \text{Cos}[v]$$

so that the element of area is

$$dA = a^2 \cos v \, du \wedge dv.$$

If we note that, except for one meridian, `sphere[a]` covers the sphere exactly once when $0 < u < 2\pi$ and $-\pi/2 < v < \pi/2$, we can compute the total area of a sphere by using the *Mathematica* command `Integrate` on `infarea[sphere[a]][u,v]`:

```
Integrate[Evaluate[PowerExpand[
    infarea[sphere[a]][u,v]]],
    {u,0,2Pi},{v,-Pi/2,Pi/2}]
```

As expected, the result is

$$4 \, a^2 \, \text{Pi}$$

The Monkey Saddle

We can also use *Mathematica* to find the metric and infinitesimal area of the monkey saddle defined on page 299:

```
ee[monkeysaddle][u,v]
```

$$1 + (3 u^2 - 3 v^2)^2$$

```
ff[monkeysaddle][u,v]
```

$$-18\,u\,v\,(u^2 - v^2)$$

gg[monkeysaddle][u,v]

$$1 + 36\,u^2\,v^2$$

Hence the Riemannian metric of the monkey saddle is given by

$$ds^2 = \left(1 + (3u^2 - 3v^2)^2\right) du^2 - 36u\,v(u^2 - v^2)du\,dv + \left(1 + 36u^2 v^2\right)dv^2.$$

Furthermore,

infarea[monkeysaddle][u,v]

yields

$$\text{Sqrt}[1 + 9\,u^4 + 18\,u^2\,v^2 + 9\,v^4]$$

so that

$$dA = \sqrt{1 + 9u^4 + 18u^2 v^2 + 9v^4}\,du \wedge dv.$$

We can use the numerical integration command **NIntegrate** to find the area of the square-like portion { **monkeysaddle**(p,q) | $-1 < p, q < 1$ } of the monkey saddle. Thus

NIntegrate[Evaluate[
   infarea[monkeysaddle][u,v]],
   {u,0,1},{v,0,1}]

yields

2.33003

15.6 Exercises

1. Find the metric, infinitesimal area, and total area of a circular torus defined by

```
torus[a_,b_][u_,v_]:= {(a + b Cos[v])Cos[u],
                      (a + b Cos[v])Sin[u],b Sin[v]}
```

15.6 Exercises

2. Fill in the details of the proof of Lemma 15.9.

3. The geometric definition of the **helicoid** is the surface generated by a line **L** attached orthogonally to an axis **A** such that **L** moves along **A** and also rotates, both at constant speed.

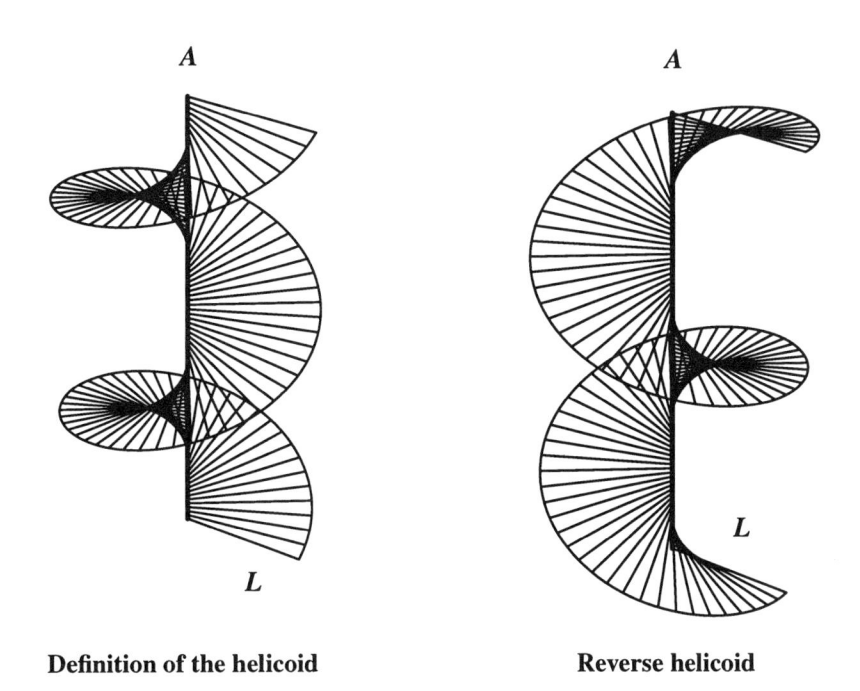

Definition of the helicoid

Reverse helicoid

Find the metric and infinitesimal area of the helicoid parametrized by

(15.28) $\quad$ **helicoid**$[a, b](u, v) = (a\,v \cos u, a\,v \sin u, b\,u)$,

or

```
helicoid[a_,b_][u_,v_]:= {a v Cos[u],a v Sin[u],b u}
```

Find the area of the set { **helicoid**$[a, b](p, q) \mid 0 < p < 2\pi, c < q < d$ }. Also, plot the helicoid with $a = b = 1$.

4. Find the metric and infinitesimal area of **Enneper's surface** [Enn1] defined by

$$\mathbf{enneper}(u,v) = \left(u - \frac{u^3}{3} + uv^2, -v + \frac{v^3}{3} - vu^2, u^2 - v^2\right),$$

or

```
enneper[u_,v_]:= {u - u^3/3 + u v^2,
                  -v + v^3/3 - v u^2, u^2 - v^2}
```

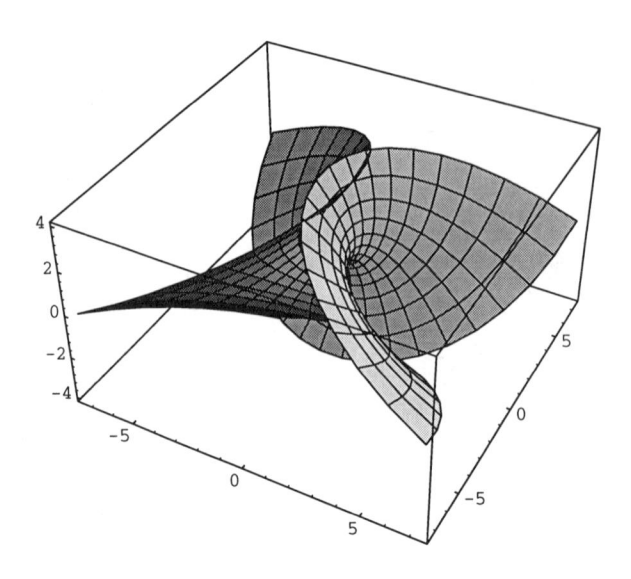

Enneper's surface

Also, plot Enneper's surface with various viewpoints. Find at least one view that shows that Enneper's surface has a hole in its middle. Finally, compute the area of the image under **enneper** of the set $[-1,1] \times [-1,1]$.

5. Find the formulas for the components of the metrics of the following surfaces: hyperbolic paraboloid, Whitney umbrella, eight surface, elliptic paraboloid, ellipsoid, Möbius strip, pseudocrosscap, Klein bottle.

16

SURFACES IN 3-DIMENSIONAL SPACE

In this chapter we study the relationship between the geometry of a regular surface $\mathcal{M}$ in 3-dimensional space $\mathbb{R}^3$ and the geometry of $\mathbb{R}^3$. The basic tool is the *shape operator* S defined in Section 16.1. The shape operator at a point $\mathbf{p} \in \mathcal{M}$ is a linear transformation $S \colon \mathcal{M}_\mathbf{p} \longrightarrow \mathcal{M}_\mathbf{p}$ that measures how $\mathcal{M}$ bends in different directions. We also prove (Lemma 16.4) that the shape operator can be considered to be the negative of the tangent map of the *Gauss map* of $\mathcal{M}$. In Section 16.2 we define a variant of the shape operator called the *normal curvature*; for $\mathbf{v_p} \in \mathcal{M}_\mathbf{p}$ the normal curvature $\mathbf{k}(\mathbf{v_p})$ is a real number that measures how a surface bends in the direction $\mathbf{v_p}$. The normal curvature $\mathbf{k}(\mathbf{v_p})$ is easy to understand geometrically because up to sign it is the signed curvature of the plane curve formed by the intersection of $\mathcal{M}$ with the plane passing through $\mathbf{v_p}$ meeting $\mathcal{M}$ perpendicularly. (See the pictures on page 366.) Techniques for computing the shape operator and normal curvature are given in Section 16.3. The eigenvalues of the shape operator at $\mathbf{p} \in \mathcal{M}$ turn out to be the maximum and minimum of the normal curvature at $\mathbf{p}$; they are studied in Section 16.4. The most important curvatures of a surface in $\mathbb{R}^3$ are the *Gaussian* and *mean curvatures*; they are defined in Section 16.5. The definitions of the *first*, *second* and *third fundamental forms* are given in Section 16.6. Techniques for computing curvature by hand from the parametric representations of surfaces are described in Section 16.7. (*Mathematica* techniques are postponed until Chapter 17.) Finally, Section 16.8 is devoted to a global curvature theorem.

16.1 The Shape Operator

We want to measure how a regular surface $\mathcal{M}$ bends in $\mathbb{R}^3$. A good way to do this is to estimate how the surface normal **U** changes from point to point. We use a linear operator called the shape operator to calculate the bending of $\mathcal{M}$. The shape operator came into wide use after its introduction in O'Neill's book [ON1]; however, it occurs much earlier, for example, implicitly in Blashke's[1] classical book [Blas2]. and explicitly in [BuBu].[2]

The shape operator applied to a tangent vector $\mathbf{v_p}$ is the negative of the derivative of **U** in the direction $\mathbf{v_p}$.

Definition. Let $\mathcal{M} \subset \mathbb{R}^3$ be a regular surface, and let **U** be a surface normal to $\mathcal{M}$ defined in a neighborhood of a point $\mathbf{p} \in \mathcal{M}$. For a tangent vector $\mathbf{v_p}$ to $\mathcal{M}$ at $\mathbf{p}$ we put

$$S(\mathbf{v_p}) = -\boldsymbol{D_v}\mathbf{U}.$$

Then S is called the **shape operator**.

(For the definition of $\boldsymbol{D_v}\mathbf{U}$ see Section 11.5.) It is easy to see that the shape operator of a plane is identically zero at all points of the plane. For a nonplanar surface the surface normal **U** will twist and turn from point to point, and S will be nonzero.

At any point of an orientable regular surface there are two choices for the unit normal: **U** and $-\mathbf{U}$. The shape operator corresponding to $-\mathbf{U}$ is the negative of the shape operator corresponding to **U**.

If $\mathcal{M}$ is nonorientable, a surface normal **U** cannot be defined continuously on all of $\mathcal{M}$. This does not matter in the present chapter, because all calculations involving **U** are local. It suffices to perform the local calculations on an open subset $\mathcal{U}$ of $\mathcal{M}$ where **U** is defined continuously.

---

[1] Wilhelm Johann Eugen Blaschke (1885-1962). Austrian-German mathematician. In 1919 he was appointed to a chair in Hamburg, where he built an important school of differential geometry.

[2] Around 1900 the Gibbs-Heaviside vector analysis notation (which one can read about in the interesting book [Crowe]) spread to differential geometry, although its use was controversial for another 30 years. Blashke's [Blas2] was one of the first differential geometry books to use vector analysis. The book of Burali-Forti and Burgatti [BuBu] contains an amusing attack on those resisting the new vector notation. In the multivolume works of Darboux [Darb1] and Bianchi [Bian] most formulas are written component by component. Compact vector notation, of course, is indispensable nowadays both for humans and for computers.

16.1 The Shape Operator

Recall (Lemma 12.15, page 282) that any point $\mathbf{q}$ in the domain of definition of a regular patch $\mathbf{x}: \mathcal{U} \longrightarrow \mathbb{R}^3$ has a neighborhood $\mathcal{U}_\mathbf{q}$ of $\mathbf{q}$ such that $\mathbf{x}(\mathcal{U}_\mathbf{q})$ is a regular surface. Therefore, the shape operator of a regular patch is also defined. Conversely, we can use patches on a regular surface $\mathcal{M} \subset \mathbb{R}^3$ to calculate the shape operator of $\mathcal{M}$.

In the next several lemmas we establish some elementary properties of the shape operator.

Lemma 16.1. *Let* $\mathbf{x}: \mathcal{U} \longrightarrow \mathbb{R}^3$ *be a regular patch. Then*

$$S(\mathbf{x}_u) = -\mathbf{U}_u \quad \text{and} \quad S(\mathbf{x}_v) = -\mathbf{U}_v.$$

Proof. Fix v and define a curve $\boldsymbol{\alpha}$ by $\boldsymbol{\alpha}(u) = \mathbf{x}(u, v)$. Then by Lemma 11.26, page 266, we have

$$S(\mathbf{x}_u(u,v)) = S(\boldsymbol{\alpha}'(u)) = -D_{\boldsymbol{\alpha}'(u)}\mathbf{U} = -(\mathbf{U} \circ \boldsymbol{\alpha})'(u).$$

But $(\mathbf{U} \circ \boldsymbol{\alpha})'$ is just $\mathbf{U}_u$. Similarly, $S(\mathbf{x}_v) = -\mathbf{U}_v$. ∎

Lemma 16.2. *At each point* $\mathbf{p}$ *of a regular surface* $\mathcal{M} \subset \mathbb{R}^3$*, the shape operator is a linear map*

$$S: \mathcal{M}_\mathbf{p} \longrightarrow \mathcal{M}_\mathbf{p}.$$

Proof. That S is linear follows from the fact that $D_{a\mathbf{v}+b\mathbf{w}} = aD_\mathbf{v} + bD_\mathbf{w}$. To prove that S maps $\mathcal{M}_\mathbf{p}$ into $\mathcal{M}_\mathbf{p}$ (instead of merely into $\mathbb{R}^3_\mathbf{p}$), we differentiate the equation $\mathbf{U} \cdot \mathbf{U} = 1$ and use (11.17), page 260:

$$0 = \mathbf{v}_\mathbf{p}[\mathbf{U} \cdot \mathbf{U}] = 2D_\mathbf{v}\mathbf{U} \cdot \mathbf{U}(\mathbf{p}) = -2S(\mathbf{v}_\mathbf{p}) \cdot \mathbf{U}(\mathbf{p}),$$

for any tangent vector $\mathbf{v}_\mathbf{p}$. Since $S(\mathbf{v}_\mathbf{p})$ is perpendicular to $\mathbf{U}(\mathbf{p})$, it must be tangent to $\mathcal{M}$; that is, $S(\mathbf{v}_\mathbf{p}) \in \mathcal{M}_\mathbf{p}$. ∎

Next, we find an important relation between the shape operator of a surface and the acceleration of a curve on the surface.

Lemma 16.3. *If* $\boldsymbol{\alpha}$ *is a curve on a regular surface* $\mathcal{M} \subset \mathbb{R}^3$*, then*

$$\boldsymbol{\alpha}'' \cdot \mathbf{U} = S(\boldsymbol{\alpha}') \cdot \boldsymbol{\alpha}'.$$

Proof. We restrict the vector field $\mathbf{U}$ to the curve $\boldsymbol{\alpha}$ and use Lemma 11.26. Since $\boldsymbol{\alpha}(t) \in \mathcal{M}$ for all t, the velocity $\boldsymbol{\alpha}'$ is always tangent to $\mathcal{M}$; therefore,

$$\boldsymbol{\alpha}' \cdot \mathbf{U} = 0.$$

When we differentiate this equation and use Lemmas 16.1 and 16.2, we obtain

$$\boldsymbol{\alpha}'' \cdot \mathbf{U} = -\mathbf{U}' \cdot \boldsymbol{\alpha}' = S(\boldsymbol{\alpha}') \cdot \boldsymbol{\alpha}'. \blacksquare$$

Note that $\boldsymbol{\alpha}'' \cdot \mathbf{U}$ can be interpreted geometrically as *the component of the acceleration of $\boldsymbol{\alpha}$ that is perpendicular to $\mathcal{M}$*.

It turns out that the shape operator is essentially the tangent map of the Gauss map.

Lemma 16.4. *Let $\mathcal{M}$ be a regular surface in $\mathbb{R}^3$ oriented by a unit normal vector field $\mathbf{U}$. View $\mathbf{U}$ as the Gauss map $\mathbf{U}\colon \mathcal{M} \longrightarrow S^2(1)$, where $S^2(1)$ denotes the unit sphere in $\mathbb{R}^3$. If $\mathbf{v}_\mathbf{p}$ is a tangent vector to $\mathcal{M}$ at $\mathbf{p} \in \mathcal{M}$, then $\mathbf{U}_*(\mathbf{v}_\mathbf{p})$ is parallel to $-S(\mathbf{v}_\mathbf{p}) \in \mathcal{M}_\mathbf{p}$.*

Proof. By Lemma 11.5, page 251, we have

$$\mathbf{U}_*(\mathbf{v}_\mathbf{p}) = \big(\mathbf{v}_\mathbf{p}[u_1], \mathbf{v}_\mathbf{p}[u_2], \mathbf{v}_\mathbf{p}[u_3]\big)_{\mathbf{U}(\mathbf{p})}.$$

On the other hand, Lemma 11.17, page 259, implies that

$$-S(\mathbf{v}_\mathbf{p}) = \boldsymbol{D}_\mathbf{v}(\mathbf{U}) = \big(\mathbf{v}_\mathbf{p}[u_1], \mathbf{v}_\mathbf{p}[u_2], \mathbf{v}_\mathbf{p}[u_3]\big)_\mathbf{p}.$$

Since the vectors $\big(\mathbf{v}_\mathbf{p}[u_1], \mathbf{v}_\mathbf{p}[u_2], \mathbf{v}_\mathbf{p}[u_3]\big)_{\mathbf{U}(\mathbf{p})}$ and $\big(\mathbf{v}_\mathbf{p}[u_1], \mathbf{v}_\mathbf{p}[u_2], \mathbf{v}_\mathbf{p}[u_3]\big)_\mathbf{p}$ are parallel, the lemma follows. $\blacksquare$

We conclude this introductory section by noting a fundamental relationship between shape operators of surfaces and Euclidean motions of $\mathbb{R}^3$.

Theorem 16.5. *Let $F\colon \mathbb{R}^3 \longrightarrow \mathbb{R}^3$ be an orientation preserving Euclidean motion, and let $\mathcal{M}_1$ and $\mathcal{M}_2$ be oriented regular surfaces such that $F(\mathcal{M}_1) = \mathcal{M}_2$. Then*

(i) *the unit normals $\mathbf{U}_1$ and $\mathbf{U}_2$ of $\mathcal{M}_1$ and $\mathcal{M}_2$ can be chosen so that*

$$F_*(\mathbf{U}_1) = \mathbf{U}_2;$$

(ii) *the shape operators S_1 and S_2 of the two surfaces (with the choice of unit normals given by (i)) are related by*

$$S_2 \circ F_* = F_* \circ S_1.$$

Proof. Since F_* preserves lengths and inner products, it follows that $F_*(\mathbf{U}_1)$ is perpendicular to $\mathcal{M}_2$ and has unit length. Hence $F_*(\mathbf{U}_1) = \pm \mathbf{U}_2$; we choose the plus sign at all points of $\mathcal{M}_2$. Let $\mathbf{p} \in \mathcal{M}_1$ and $\mathbf{v_p} \in \mathcal{M}_{1\mathbf{p}}$, and let $\mathbf{w}_{F(\mathbf{p})} = F_*(\mathbf{v_p})$. A Euclidean motion is an affine transformation, so by Lemma 11.23, page 262, we have

$$(S_2 \circ F_*)(\mathbf{v_p}) = S_2(\mathbf{w}_{F(\mathbf{p})}) = -D_\mathbf{w} \mathbf{U}_2 = -D_\mathbf{w} F_*(\mathbf{U}_1)$$
$$= -F_*(D_\mathbf{v} \mathbf{U}_1) = -(F_* \circ S_1)(\mathbf{v_p}).$$

Because $\mathbf{v_p}$ is arbitrary, we have $S_2 \circ F_* = F_* \circ S_1$. ∎

16.2 Normal Curvature

Although the shape operator does the job of measuring the bending of a surface in different directions, frequently it is useful to have a real-valued function, called the normal curvature, that does the same thing. Although we define it in terms of the shape operator, the normal curvature is a much older concept (see [Euler3], [Meu] and Corollary 16.14). First, we need to make precise the notion of direction on a surface.

Definition. *A **direction** L on a regular surface $\mathcal{M}$ is a 1-dimensional subspace (that is, a line) of a tangent space to $\mathcal{M}$.*

Since a nonzero vector $\mathbf{v_p}$ in a tangent space $\mathcal{M}_\mathbf{p}$ determines a unique 1-dimensional subspace L, we sometimes use the terminology "the direction $\mathbf{v_p}$" to mean L.

Definition. *Let $\mathbf{u_p}$ be a tangent vector to a regular surface $\mathcal{M} \subset \mathbb{R}^3$ with $\|\mathbf{u_p}\| = 1$. Then the **normal curvature of $\mathcal{M}$ in the direction $\mathbf{u_p}$** is*

$$\mathbf{k}(\mathbf{u_p}) = S(\mathbf{u_p}) \cdot \mathbf{u_p}.$$

More generally, if $\mathbf{v_p}$ is any nonzero tangent vector to $\mathcal{M}$ at $\mathbf{p}$, we put

(16.1) $$\mathbf{k}(\mathbf{v_p}) = \frac{S(\mathbf{v_p}) \cdot \mathbf{v_p}}{\|\mathbf{v_p}\|^2}.$$

It is easy to prove (see exercise 3):

Lemma 16.6. *Let L be a direction in a tangent space $\mathcal{M}_p$ to a regular surface $\mathcal{M} \subset \mathbb{R}^3$. Then the normal curvature $\mathbf{k}(\mathbf{v}_p)$ is the same for all nonzero tangent vectors $\mathbf{v}_p$ in L. Therefore, we call the common value of the normal curvature the* **normal curvature of the direction L**.

Let us single out two kinds of directions.

Definition. *Let L be a direction in a tangent space $\mathcal{M}_p$, where $\mathcal{M} \subset \mathbb{R}^3$ is a regular surface. If the normal curvature of L is zero, we say that L is an* **asymptotic direction**. *Similarly, if the normal curvature vanishes on a tangent vector $\mathbf{v}_p$ to $\mathcal{M}$, we say that $\mathbf{v}_p$ is an* **asymptotic vector**.

Definition. *Let $\mathcal{M}$ be a regular surface in $\mathbb{R}^3$ and let $\mathbf{p} \in \mathcal{M}$. The maximum and minimum values of the normal curvature $\mathbf{k}$ of $\mathcal{M}$ at $\mathbf{p}$ are called the* **principal curvatures** *of $\mathcal{M}$ at $\mathbf{p}$ and are denoted by $\mathbf{k}_1$ and $\mathbf{k}_2$. Unit vectors $\mathbf{e}_1, \mathbf{e}_2 \in \mathcal{M}_p$ at which these extreme values occur are called* **principal vectors**. *The corresponding directions are called* **principal directions**.

The principal curvatures measure the maximum and minimum bending of a regular surface $\mathcal{M}$ at each point $\mathbf{p} \in \mathcal{M}$.

There is an important interpretation of normal curvature of a regular surface as the curvature of a space curve.

Lemma 16.7. *Let $\mathbf{u}_p$ be a unit tangent vector to $\mathcal{M}$ at $\mathbf{p}$, and let β be a unit-speed curve in $\mathcal{M}$ with $\beta(0) = \mathbf{p}$ and $\beta'(0) = \mathbf{u}_p$. Then*

$$(16.2) \qquad \mathbf{k}(\mathbf{u}_p) = \kappa[\beta](0) \cos\theta,$$

where $\kappa[\beta](0)$ is the curvature at 0 of β and $\cos\theta$ is the cosine of the angle between the normal $\mathbf{N}(0)$ of β and the surface normal $\mathbf{U}(\mathbf{p})$.

Proof. Suppose that $\kappa[\beta](0) \neq 0$. By Lemma 16.3 and Theorem 8.4, page 185, we have

$$(16.3) \qquad \begin{aligned} \mathbf{k}(\mathbf{u}_p) &= S(\mathbf{u}_p) \cdot \mathbf{u}_p = \beta''(0) \cdot \mathbf{U}(\mathbf{p}) \\ &= \kappa[\beta](0)\mathbf{N}(0) \cdot \mathbf{U}(\mathbf{p}) = \kappa[\beta](0) \cos\theta. \end{aligned}$$

In the exceptional case that $\kappa[\beta](0) = 0$, the normal $\mathbf{N}(0)$ is not defined, but we still have $\mathbf{k}(\mathbf{u}_p) = 0$. ∎

To understand the meaning of normal curvature geometrically, we need to find curves on a surface to which we can apply Lemma 16.7.

16.2 Normal Curvature

Definition. Let $\mathcal{M} \subset \mathbb{R}^3$ be a regular surface and $\mathbf{u_p}$ a unit tangent vector to $\mathcal{M}$. Denote by $\Pi(\mathbf{u_p}, \mathbf{U}(\mathbf{p}))$ the plane determined by $\mathbf{u_p}$ and the surface normal $\mathbf{U}(\mathbf{p})$. The **normal section** of $\mathcal{M}$ in the $\mathbf{u_p}$ direction is the intersection of $\Pi(\mathbf{u_p}, \mathbf{U}(\mathbf{p}))$ and $\mathcal{M}$.

Corollary 16.8. *Let β be a unit-speed curve which lies in the intersection of a regular surface $\mathcal{M} \subset \mathbb{R}^3$ and one of its normal sections Π through $\mathbf{p} \in \mathcal{M}$. Assume that $\beta(0) = \mathbf{p}$ and put $\mathbf{u_p} = \beta'(0)$. Then the normal curvature $\mathbf{k}(\mathbf{u_p})$ of $\mathcal{M}$ and the curvature of β are related by*

$$(16.4) \qquad \mathbf{k}(\mathbf{u_p}) = \pm \kappa[\beta](0).$$

Proof. If $\kappa[\beta](0) = 0$, then (16.4) is an obvious consequence of (16.2), so assume that $\kappa[\beta](0) \neq 0$. Since β has unit-speed, $\kappa[\beta](0)\mathbf{N}(0) = \beta''(0)$ is perpendicular to $\beta'(0)$. On the other hand, both $\mathbf{U}(\mathbf{p})$ and $\mathbf{N}(0)$ lie in Π, so the only possibility is $\mathbf{N}(0) = \pm\mathbf{U}(\mathbf{p})$. Hence $\cos\theta = \pm 1$ in (16.2), and so we obtain (16.4). ∎

Corollary 16.8 gives an excellent method for estimating normal curvature visually. For a regular surface $\mathcal{M} \subset \mathbb{R}^3$, suppose we want to know the normal curvature in various directions at $\mathbf{p} \in \mathcal{M}$. Each unit vector $\mathbf{u_p} \in \mathcal{M_p}$, together with the surface normal $\mathbf{U}(\mathbf{p})$, determines a plane $\Pi(\mathbf{u_p}, \mathbf{U}(\mathbf{p}))$. Then the normal section in the direction $\mathbf{u_p}$ is the intersection of $\Pi(\mathbf{u_p}, \mathbf{U}(\mathbf{p}))$ and $\mathcal{M}$. This is a plane curve in $\Pi(\mathbf{u_p}, \mathbf{U}(\mathbf{p}))$ whose curvature is given by (16.4). There are three cases:

- If $\mathbf{k}(\mathbf{u_p}) > 0$, then the normal section is bending in the same direction as $\mathbf{U}(\mathbf{p})$. Hence in the $\mathbf{u_p}$ direction $\mathcal{M}$ is bending toward $\mathbf{U}(\mathbf{p})$.

- If $\mathbf{k}(\mathbf{u_p}) < 0$, then the normal section is bending in the opposite direction from $\mathbf{U}(\mathbf{p})$. Hence in the $\mathbf{u_p}$ direction $\mathcal{M}$ is bending away from $\mathbf{U}(\mathbf{p})$.

- If $\mathbf{k}(\mathbf{u_p}) = 0$, then the curvature of the normal section vanishes at $\mathbf{p}$ so that the normal to a curve in the normal section is undefined. It is impossible to conclude that there is no bending of $\mathcal{M}$ in the $\mathbf{u_p}$ direction since $\kappa[\beta]$ might vanish only at $\mathbf{p}$. But in some sense the bending is small.

As the unit tangent vector $\mathbf{u_p}$ turns, the surface may bend in quite different ways. A good example of this occurs at the center point of a hyperbolic paraboloid.

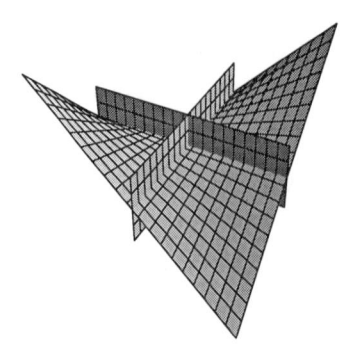

Normal sections to a hyperbolic paraboloid passing through asymptotic curves

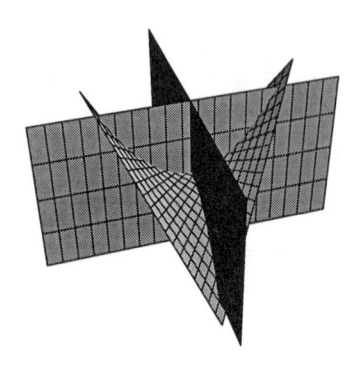

Normal sections to a hyperbolic paraboloid passing through principal curves

In the first picture both normal sections intersect the hyperbolic paraboloid in straight lines; thus the normal curvature determined by each of these sections vanishes. In contrast, in the second picture, the normal curvature determined by one normal section is positive, and that determined by the other normal section is negative.

The normal sections at the center of a monkey saddle are similar to those of the hyperbolic paraboloid, but more complicated. In this case, there are three asymptotic directions passing through the center point of the monkey saddle.

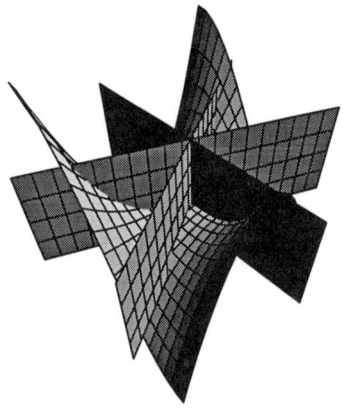

Normal sections to a monkey saddle

These pictures of surfaces and their normal sections are good examples of how *Mathematica* can be used to display intersecting surfaces. For example, the plot of the monkey saddle together with its normal sections was drawn with the command

```
Show[ParametricPlot3D[{u,0,v},{u,-3,3},{v,-3,9},
PlotPoints->{Automatic,5},
DisplayFunction->Identity],
ParametricPlot3D[{Cos[2Pi/3]u,Sin[2Pi/3]u,v},{u,-3,3},
{v,-3,9},PlotPoints->{Automatic,5},
DisplayFunction->Identity],
ParametricPlot3D[{Cos[4Pi/3]u,Sin[4Pi/3]u,v},{u,-3,3},
{v,-3,9},PlotPoints->{Automatic,5},
DisplayFunction->Identity],
ParametricPlot3D[Evaluate[
monkeysaddle[u,v]],{u,-2,2},{v,-2,2},
DisplayFunction->Identity],DisplayFunction->$DisplayFunction,
ViewPoint->{Cos[-Pi/4],Sin[-Pi/4],1},
BoxRatios->{1,1,1},Axes->None,Boxed->False];
```

16.3 Calculation of the Shape Operator

Symmetric linear operators on a vector space are much easier to work with than general linear operators. It is fortunate that the shape operator is symmetric. We now prove this fact.

Lemma 16.9. *The shape operator of a regular surface $\mathcal{M}$ is symmetric; that is,*

$$S(\mathbf{v_p}) \cdot \mathbf{w_p} = \mathbf{v_p} \cdot S(\mathbf{w_p})$$

for all tangent vectors $\mathbf{v_p}, \mathbf{w_p}$ to $\mathcal{M}$.

Proof. Let $\mathbf{x}$ be an injective regular patch on $\mathcal{M}$. We differentiate the formula $\mathbf{U} \cdot \mathbf{x}_u = 0$ with respect to v and obtain

(16.5) $$0 = \frac{\partial}{\partial v}(\mathbf{U} \cdot \mathbf{x}_u) = \mathbf{U}_v \cdot \mathbf{x}_u + \mathbf{U} \cdot \mathbf{x}_{uv},$$

where $\mathbf{U}_v$ is the derivative of the vector field $v \longmapsto \mathbf{U}(u,v)$ along any v-parameter curve. Since $\mathbf{U}_v = -S(\mathbf{x}_v)$, equation (16.5) becomes

(16.6) $$S(\mathbf{x}_v) \cdot \mathbf{x}_u = \mathbf{U} \cdot \mathbf{x}_{uv}.$$

Similarly,
(16.7) $$S(\mathbf{x}_u) \cdot \mathbf{x}_v = \mathbf{U} \cdot \mathbf{x}_{vu}.$$

From (16.6), (16.7) and the fact that $\mathbf{x}_{uv} = \mathbf{x}_{vu}$ we get

(16.8) $$S(\mathbf{x}_u) \cdot \mathbf{x}_v = \mathbf{U} \cdot \mathbf{x}_{vu} = \mathbf{U} \cdot \mathbf{x}_{uv} = S(\mathbf{x}_v) \cdot \mathbf{x}_u.$$

Next, let $\mathbf{p} \in \mathcal{M}$ and $\mathbf{v_p}, \mathbf{w_p} \in \mathcal{M_p}$. Now we choose the patch $\mathbf{x}$ so that $\mathbf{x}(u_0, v_0) = \mathbf{p}$. Then we can write

(16.9)
$$\begin{cases} \mathbf{v_p} = a_{11}\mathbf{x}_u(u_0, v_0) + a_{12}\mathbf{x}_v(u_0, v_0), \\ \mathbf{w_p} = a_{21}\mathbf{x}_u(u_0, v_0) + a_{22}\mathbf{x}_v(u_0, v_0). \end{cases}$$

It follows from (16.7) (omitting the evaluation at (u_0, v_0) from the notation) that

$$\begin{aligned} S(\mathbf{v_p}) \cdot \mathbf{w_p} &= a_{11}a_{21}S(\mathbf{x}_u) \cdot \mathbf{x}_u + a_{11}a_{22}S(\mathbf{x}_u) \cdot \mathbf{x}_v \\ &\quad + a_{12}a_{21}S(\mathbf{x}_v) \cdot \mathbf{x}_u + a_{12}a_{22}S(\mathbf{x}_v) \cdot \mathbf{x}_v \\ &= a_{11}a_{21}S(\mathbf{x}_u) \cdot \mathbf{x}_u + (a_{11}a_{22} + a_{12}a_{21})S(\mathbf{x}_u) \cdot \mathbf{x}_v + a_{21}a_{22}S(\mathbf{x}_v) \cdot \mathbf{x}_v \\ &= \cdots = S(\mathbf{w_p}) \cdot \mathbf{v_p}, \end{aligned}$$

and so we obtain (16.5). ∎

Definition. *Let* $\mathbf{x} : \mathcal{U} \longrightarrow \mathbb{R}^3$ *be a regular patch. Then*

(16.10)
$$\begin{cases} e = -\mathbf{U}_u \cdot \mathbf{x}_u = \mathbf{U} \cdot \mathbf{x}_{uu}, \\ f = -\mathbf{U}_v \cdot \mathbf{x}_u = \mathbf{U} \cdot \mathbf{x}_{uv} = \mathbf{U} \cdot \mathbf{x}_{vu} = -\mathbf{U}_u \cdot \mathbf{x}_v, \\ g = -\mathbf{U}_v \cdot \mathbf{x}_v = \mathbf{U} \cdot \mathbf{x}_{vv}. \end{cases}$$

Classically, e, f and g are called the **coefficients of the second fundamental form** of $\mathbf{x}^3$; similarly, E, F and G (defined in Section 15.1) are called the **coefficients of the first fundamental form** of $\mathbf{x}$. We shall also use this terminology. More details are given in Section 16.6. In Section 15.1 we wrote

$$ds^2 = E\,du^2 + 2F\,du\,dv + G\,dv^2$$

for the first fundamental form. Analogously, we can write

$$sff = e\,du^2 + 2f\,du\,dv + g\,dv^2;$$

for the second fundamental form. This is the notation used in most classical differential geometry books.

Next, we show how to express the shape operator in terms of e, f, g, E, F and G.

---

[3]Some books use L, M, N for e, f and g. Gauss used the notation D, D', D'' for the quantities $e\sqrt{EG - F^2}$, $f\sqrt{EG - F^2}$ and $g\sqrt{EG - F^2}$.

16.3 Calculation of the Shape Operator

Theorem 16.10. (The Weingarten[4] equations) *Let* $\mathbf{x}: \mathcal{U} \longrightarrow \mathbb{R}^3$ *be a regular patch. Then the shape operator* S *of* $\mathbf{x}$ *is given in terms of the basis* $\{\mathbf{x}_u, \mathbf{x}_v\}$ *by*

(16.11)
$$\begin{cases} -S(\mathbf{x}_u) = \mathbf{U}_u = \dfrac{fF - eG}{EG - F^2}\mathbf{x}_u + \dfrac{eF - fE}{EG - F^2}\mathbf{x}_v, \\ -S(\mathbf{x}_v) = \mathbf{U}_v = \dfrac{gF - fG}{EG - F^2}\mathbf{x}_u + \dfrac{fF - gE}{EG - F^2}\mathbf{x}_v. \end{cases}$$

Proof. Since $\mathbf{x}$ is regular, and $\mathbf{x}_u$ and $\mathbf{x}_v$ are linearly independent, we can write

(16.12)
$$\begin{cases} -S(\mathbf{x}_u) = \mathbf{U}_u = a_{11}\mathbf{x}_u + a_{12}\mathbf{x}_v, \\ -S(\mathbf{x}_v) = \mathbf{U}_v = a_{21}\mathbf{x}_u + a_{22}\mathbf{x}_v, \end{cases}$$

for some functions $a_{11}, a_{12}, a_{21}, a_{22}$. To prove (16.11), we need to compute the coefficients $a_{11}, a_{12}, a_{21}, a_{22}$ in (16.12). We take the scalar product of each of the equations in (16.12) with $\mathbf{x}_u$ and $\mathbf{x}_v$ and obtain

(16.13)
$$\begin{cases} -e = a_{11}E + a_{12}F, \\ -f = a_{11}F + a_{12}G, \\ -f = a_{21}E + a_{22}F, \\ -g = a_{21}F + a_{22}G. \end{cases}$$

Equations (16.13) can be written more concisely in terms of matrices:

$$-\begin{pmatrix} e & f \\ f & g \end{pmatrix} = \begin{pmatrix} a_{11} & a_{12} \\ a_{21} & a_{22} \end{pmatrix} \begin{pmatrix} E & F \\ F & G \end{pmatrix};$$

hence

$$\begin{pmatrix} a_{11} & a_{12} \\ a_{21} & a_{22} \end{pmatrix} = -\begin{pmatrix} e & f \\ f & g \end{pmatrix} \begin{pmatrix} E & F \\ F & G \end{pmatrix}^{-1}.$$

---

[4] Julius Weingarten (1836-1910). Professor at the Technische Universität in Berlin. A surface for which there is a definite functional relation between the principal curvatures is called a **Weingarten surface**.

It can be directly verified that

$$(16.14) \quad \begin{pmatrix} E & F \\ F & G \end{pmatrix}^{-1} = \frac{1}{EG-F^2} \begin{pmatrix} G & -F \\ -F & E \end{pmatrix},$$

so that

$$\begin{pmatrix} a_{11} & a_{12} \\ a_{21} & a_{22} \end{pmatrix} = \frac{-1}{EG-F^2} \begin{pmatrix} e & f \\ f & g \end{pmatrix} \begin{pmatrix} G & -F \\ -F & E \end{pmatrix}$$

$$= \frac{-1}{EG-F^2} \begin{pmatrix} eG - fF & -eF + fE \\ fG - gF & -fF + gE \end{pmatrix}.$$

Thus we obtain

$$(16.15) \quad \begin{cases} a_{11} = \dfrac{fF - eG}{EG - F^2}, & a_{12} = \dfrac{eF - fE}{EG - F^2}, \\[6pt] a_{21} = \dfrac{gF - fG}{EG - F^2}, & a_{22} = \dfrac{fF - gE}{EG - F^2}. \end{cases}$$

From (16.15) and (16.12) we get (16.11). ∎

Note that although S is a symmetric linear operator, its matrix relative to $\mathbf{x}_u$ and $\mathbf{x}_v$ need not be symmetric, because $\mathbf{x}_u$ and $\mathbf{x}_v$ in general are not perpendicular to one another.

There is also a way to express the normal curvature in terms of E, F, G, e, f and g.

Lemma 16.11. *Let $M \subset \mathbb{R}^3$ be a regular surface and let $\mathbf{p} \in M$. Let $\mathbf{x}$ be an injective regular patch on M with $\mathbf{p} = \mathbf{x}(u_0, v_0)$. Let $\mathbf{v}_\mathbf{p} \in M_\mathbf{p}$ and write*

$$\mathbf{v}_\mathbf{p} = a\,\mathbf{x}_u(u_0, v_0) + b\,\mathbf{x}_v(u_0, v_0).$$

Then the normal curvature of M in the direction $\mathbf{v}_\mathbf{p}$ is

$$k(\mathbf{v}_\mathbf{p}) = \frac{e\,a^2 + 2f\,a\,b + g\,b^2}{E\,a^2 + 2F\,a\,b + G\,b^2}.$$

Proof. We have

$$\|\mathbf{v}_\mathbf{p}\|^2 = \|a\,\mathbf{x}_u + b\,\mathbf{x}_v\|^2 = a^2 E + 2ab\,F + b^2 G,$$

and
$$S(\mathbf{v_p}) \cdot \mathbf{v_p} = \bigl(a\,S(\mathbf{x}_u) + b\,S(\mathbf{x}_v)\bigr) \cdot (a\,\mathbf{x}_u + b\,\mathbf{x}_v) = a^2 e + 2ab f + b^2 g.$$
Hence by (16.1),
$$\mathbf{k}(\mathbf{v_p}) = \frac{S(\mathbf{v_p}) \cdot \mathbf{v_p}}{\|\mathbf{v_p}\|^2} = \frac{e\,a^2 + 2f\,a\,b + g\,b^2}{E\,a^2 + 2F\,a\,b + G\,b^2}. \quad \blacksquare$$

16.4 The Eigenvalues of the Shape Operator

We recall an important fact from linear algebra.

Lemma 16.12. *Let V be an n-dimensional vector space with an inner product and let $A\colon V \longrightarrow V$ be a linear transformation that is symmetric with respect to the inner product. Then the eigenvalues of A are real and A is diagonalizable; that is, there is an orthonormal basis $\{\mathbf{e}_1, \ldots, \mathbf{e}_n\}$ of V such that*
$$A\mathbf{e}_j = \lambda_j \mathbf{e}_j \qquad \text{for } j = 1, \ldots, n,$$
where $\{\lambda_1, \ldots, \lambda_n\}$ are the eigenvalues of A.

We know by Lemma 16.9 that the shape operator S is a symmetric linear operator on each tangent space to a regular surface in $\mathbb{R}^3$. Therefore, Lemma 16.12 tells us that the eigenvalues of S are real. These eigenvalues are important geometric quantities associated with each regular surface in $\mathbb{R}^3$.

Instead of proving Lemma 16.12 in its full generality, we prove it for the special case of the shape operator.

Lemma 16.13. *The eigenvalues of the shape operator S of a regular surface $\mathcal{M} \subset \mathbb{R}^3$ at $\mathbf{p} \in \mathcal{M}$ are precisely the principal curvatures $\mathbf{k}_1$ and $\mathbf{k}_2$ of $\mathcal{M}$ at $\mathbf{p}$. The corresponding unit eigenvectors are unit principal vectors, and vice versa. If $\mathbf{k}_1 = \mathbf{k}_2$, then S is scalar multiplication by the common value of $\mathbf{k}_1$ and $\mathbf{k}_2$. Otherwise, the eigenvectors $\mathbf{e}_1$ and $\mathbf{e}_2$ of S are orthogonal, and S is given by*

(16.16) $$S\mathbf{e}_1 = \mathbf{k}_1\mathbf{e}_1, \qquad S\mathbf{e}_2 = \mathbf{k}_2\mathbf{e}_2.$$

Proof. Consider the normal curvature as a function $\mathbf{k}\colon S^1_{\mathbf{p}} \longrightarrow \mathbb{R}$, where $S^1_{\mathbf{p}}$ is the set of unit tangent vectors in the tangent space $\mathcal{M}_{\mathbf{p}}$. Since $S^1_{\mathbf{p}}$ is a circle, it is compact, and so **k** achieves its maximum at some unit vector, call it $\mathbf{e}_1 \in S_\mathbf{p}^1$. Choose $\mathbf{e}_2$ to be any vector in $S_\mathbf{p}^1$ perpendicular to $\mathbf{e}_1$. Since $\{\mathbf{e}_1, \mathbf{e}_2\}$ is an orthonormal basis of $\mathcal{M}_\mathbf{p}$, we have

(16.17) $$\begin{cases} S(\mathbf{e}_1) = \big(S(\mathbf{e}_1) \cdot \mathbf{e}_1\big)\mathbf{e}_1 + \big(S(\mathbf{e}_1) \cdot \mathbf{e}_2\big)\mathbf{e}_2, \\ S(\mathbf{e}_2) = \big(S(\mathbf{e}_2) \cdot \mathbf{e}_1\big)\mathbf{e}_1 + \big(S(\mathbf{e}_2) \cdot \mathbf{e}_2\big)\mathbf{e}_2. \end{cases}$$

We show that $\mathbf{e}_1$ and $\mathbf{e}_2$ are eigenvectors of S.

Define a function $\mathbf{u} = \mathbf{u}(\theta)$ by $\mathbf{u}(\theta) = (\cos\theta)\mathbf{e}_1 + (\sin\theta)\mathbf{e}_2$, and write $\mathbf{k}(\theta) = \mathbf{k}(\mathbf{u}(\theta))$. Then

(16.18) $$\mathbf{k}(\theta) = \big(S(\mathbf{e}_1) \cdot \mathbf{e}_1\big)\cos^2\theta + 2\big(S(\mathbf{e}_1) \cdot \mathbf{e}_2\big)\sin\theta\cos\theta \\ + \big(S(\mathbf{e}_2) \cdot \mathbf{e}_2\big)\sin^2\theta,$$

so that

$$\frac{d}{d\theta}\mathbf{k}(\theta) = 2\big(S(\mathbf{e}_2) \cdot \mathbf{e}_2 - S(\mathbf{e}_1) \cdot \mathbf{e}_1\big)\sin\theta\cos\theta + 2S(\mathbf{e}_1) \cdot \mathbf{e}_2(\cos^2\theta - \sin^2\theta).$$

In particular,

(16.19) $$0 = \frac{d\mathbf{k}}{d\theta}(0) = 2S(\mathbf{e}_1) \cdot \mathbf{e}_2,$$

because $\mathbf{k}(\theta)$ has a maximum at $\theta = 0$. Then (16.17) and (16.19) imply (16.16). From (16.16) it follows that both $\mathbf{e}_1$ and $\mathbf{e}_2$ are eigenvectors of S, and from (16.18) it follows that the principal curvatures of $\mathcal{M}$ at $\mathbf{p}$ are the eigenvalues of S. (See also (16.20).) Hence the lemma follows. ∎

The principal curvatures determine the normal curvature completely.

Corollary 16.14. (Euler) *Let $\mathbf{k}_1(\mathbf{p})$ and $\mathbf{k}_2(\mathbf{p})$ be the principal curvatures of a regular surface $\mathcal{M} \subset \mathbb{R}^3$ at $\mathbf{p} \in \mathcal{M}$, and let $\mathbf{e}_1, \mathbf{e}_2$ be the corresponding unit principal vectors. Let θ denote the oriented angle from $\mathbf{e}_1$ to $\mathbf{u}_\mathbf{p}$, so that $\mathbf{u}_\mathbf{p} = (\cos\theta)\mathbf{e}_1 + (\sin\theta)\mathbf{e}_2$. Then the normal curvature $\mathbf{k}(\mathbf{u}_\mathbf{p})$ is given by*

(16.20) $$\mathbf{k}(\mathbf{u}_\mathbf{p}) = \mathbf{k}_1(\mathbf{p})\cos^2\theta + \mathbf{k}_2(\mathbf{p})\sin^2\theta.$$

Proof. Since $S(\mathbf{e}_1) \cdot \mathbf{e}_2 = 0$, (16.18) reduces to (16.20). ∎

16.5 The Gaussian and Mean Curvatures

The notion of the curvature of a surface is a great deal more complicated than the notion of curvature of a curve. Let α be a curve in $\mathbb{R}^3$, and let $\mathbf{p}$ be a point on the trace of α. The curvature of α at $\mathbf{p}$ measures the rate at which α leaves the tangent line to α at $\mathbf{p}$. By analogy, the curvature of a surface $\mathcal{M} \subset \mathbb{R}^3$ at $\mathbf{p} \in \mathcal{M}$ should measure the rate at which $\mathcal{M}$ leaves the tangent plane to $\mathcal{M}$ at $\mathbf{p}$. But a difficulty arises for surfaces that was not present for curves: although a curve can separate from one of its tangent lines in only two directions, a surface separates from one of its tangent planes in infinitely many directions. In general, the rate of departure of a surface from one of its tangent planes depends on the direction.

In fact, there are several competing notions for the curvature of a surface in $\mathbb{R}^3$; these include:

- the normal curvature $\mathbf{k}$;
- the principal curvatures $\mathbf{k}_1$ and $\mathbf{k}_2$;
- the mean curvature H;
- the Gaussian curvature K.

We have already defined the normal curvature and the principal curvatures of a surface $\mathcal{M} \subset \mathbb{R}^3$ in Section 16.2. In the present section we give the definitions of the Gaussian and mean curvatures; these are the most important curvatures in surface theory.

First, we recall some useful facts from linear algebra about the determinant and trace of a matrix.

Lemma 16.15. *If $A\colon V \longrightarrow V$ is a linear transformation on a vector space V, put*

$$\det A = \det\begin{pmatrix} a_{11} & \cdots & a_{1n} \\ \vdots & \ddots & \vdots \\ a_{n1} & \cdots & a_{nn} \end{pmatrix} \quad \text{and} \quad \operatorname{tr} A = \operatorname{tr}\begin{pmatrix} a_{11} & \cdots & a_{1n} \\ \vdots & \ddots & \vdots \\ a_{n1} & \cdots & a_{nn} \end{pmatrix},$$

where

$$a = \begin{pmatrix} a_{11} & \cdots & a_{1n} \\ \vdots & \ddots & \vdots \\ a_{n1} & \cdots & a_{nn} \end{pmatrix}$$

is any matrix representing A. Then $\det A$ and $\operatorname{tr} A$ are independent of the choice of matrix representing A.

Proof. We need two facts from matrix algebra. If a and b are $n \times n$ matrices, then

$$(16.21) \qquad \det(a\,b) = \det(a)\det(b),$$

and

$$(16.22) \qquad \operatorname{tr}(a\,b) = \operatorname{tr}(b\,a).$$

If a is a matrix representing the linear transformation A, then any other matrix which represents A is of the form $c^{-1}a\,c$, where c is a nonsingular matrix. But if a and c are $n \times n$ matrices with c nonsingular, then it follows from (16.21) that

$$\det(c^{-1}a\,c) = \det(c)^{-1}\det(a)\det(c) = \det(a),$$

and from (16.22) it follows that

$$\operatorname{tr}(c^{-1}a\,c) = \operatorname{tr}(a\,c\,c^{-1}) = \operatorname{tr}(a). \quad \blacksquare$$

Definition. *Let $\mathcal{M}$ be a regular surface in $\mathbb{R}^3$. The **Gaussian curvature** K and **mean curvature** H of $\mathcal{M}$ are the functions $K, H \colon \mathcal{M} \longrightarrow \mathbb{R}$ defined by*

$$(16.23) \qquad K(\mathbf{p}) = \det\bigl(S(\mathbf{p})\bigr) \qquad \text{and} \qquad H(\mathbf{p}) = \frac{1}{2}\operatorname{tr}\bigl(S(\mathbf{p})\bigr).$$

Lemma 16.15 implies that the definitions of K and H do not depend on which matrix we choose to represent the shape operator. Note also that although the shape operator S and the mean curvature H depend on the choice of unit normal $\mathbf{U}$, the Gaussian curvature K is independent of that choice. The name "mean curvature" is due to Germain.[5]

Definition. *A **minimal surface** in $\mathbb{R}^3$ is a regular surface for which the mean curvature vanishes identically. A regular surface is **flat** if and only if its Gaussian curvature vanishes identically.*

---

[5] Sophie Germain (1776-1831). French mathematician, best known for her work on elasticity and Fermat's last theorem. Germain (under the pseudonym "M. Blanc") corresponded with Gauss regarding her results in geometry and number theory.

16.5 The Gaussian and Mean Curvatures

We shall see in Chapter 30 that surfaces of minimal area are minimal in the sense that the mean curvature vanishes.

The Gaussian curvature permits us to distinguish four kinds of points on a surface.

Definition. *Let $\mathbf{p}$ be a point on a regular surface $\mathcal{M} \subset \mathbb{R}^3$. We say that*

- $\mathbf{p}$ *is* **elliptic** *if $K(\mathbf{p}) > 0$ (equivalently $\mathbf{k}_1$ and $\mathbf{k}_2$ have the same sign);*

- $\mathbf{p}$ *is* **hyperbolic** *if $K(\mathbf{p}) < 0$ (equivalently $\mathbf{k}_1$ and $\mathbf{k}_2$ have opposite signs);*

- $\mathbf{p}$ *is* **parabolic** *if $K(\mathbf{p}) = 0$, but $S(\mathbf{p}) \neq 0$ (equivalently exactly one of $\mathbf{k}_1$ and $\mathbf{k}_2$ is zero);*

- $\mathbf{p}$ *is* **planar** *if $K(\mathbf{p}) = 0$ and $S(\mathbf{p}) = 0$ (equivalently $\mathbf{k}_1 = \mathbf{k}_2 = 0$).*

It is usually possible to glance at almost any surface and recognize which points are elliptic, hyperbolic, parabolic or planar. On page 385 is a plot of a torus indicating which points are elliptic, hyperbolic or parabolic. Here are examples of surfaces made up entirely of each of the four kinds of points.

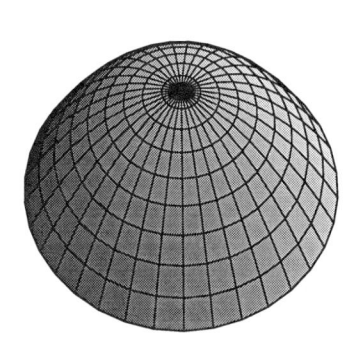

A surface consisting of elliptic points

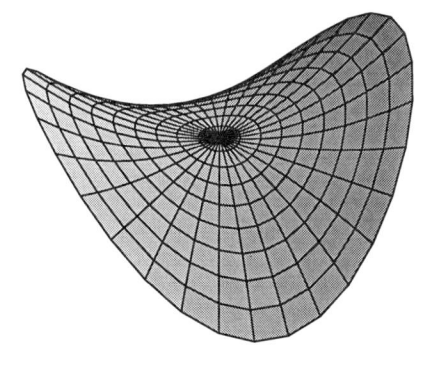

A surface consisting of hyperbolic points

A surface consisting of parabolic points

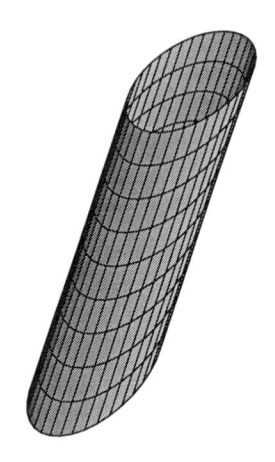

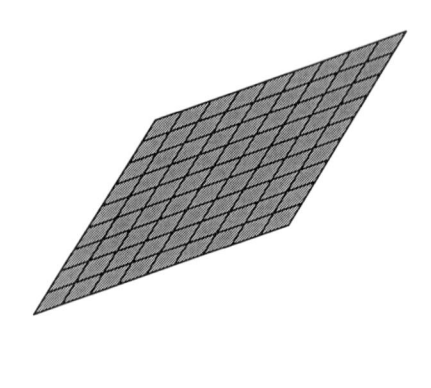

A surface consisting of planar points

There are two especially useful ways of choosing a basis of a tangent space to a surface in $\mathbb{R}^3$. Each gives rise to important formulas for the Gaussian and mean curvatures, which we give in the next two theorems.

Theorem 16.16. *The Gaussian curvature and mean curvature of a regular surface $\mathcal{M} \subset \mathbb{R}^3$ are related to the principal curvatures by*

$$K = \mathbf{k}_1 \mathbf{k}_2 \quad \text{and} \quad H = \frac{1}{2}(\mathbf{k}_1 + \mathbf{k}_2).$$

Proof. If we choose an orthonormal basis of eigenvectors of S for $\mathcal{M}_\mathbf{p}$, the matrix of S with respect to this basis is

$$\begin{pmatrix} \mathbf{k}_1 & 0 \\ 0 & \mathbf{k}_2 \end{pmatrix},$$

so that

$$K = \det \begin{pmatrix} \mathbf{k}_1 & 0 \\ 0 & \mathbf{k}_2 \end{pmatrix} = \mathbf{k}_1 \mathbf{k}_2 \quad \text{and} \quad H = \frac{1}{2}\operatorname{tr}\begin{pmatrix} \mathbf{k}_1 & 0 \\ 0 & \mathbf{k}_2 \end{pmatrix} = \frac{1}{2}(\mathbf{k}_1 + \mathbf{k}_2). \blacksquare$$

Let $\mathcal{M} \subset \mathbb{R}^3$ be a regular surface. The Gaussian and mean curvatures are functions $K, H: \mathcal{M} \longrightarrow \mathbb{R}$; we write $K(\mathbf{p})$ and $H(\mathbf{p})$ for the value of K and H at $\mathbf{p} \in \mathcal{M}$.

16.5 The Gaussian and Mean Curvatures

We need a slightly different notation for a regular patch $\mathbf{x}:\mathcal{U} \longrightarrow \mathbb{R}^3$. Strictly speaking, $K, H:\mathbf{x}(\mathcal{U}) \longrightarrow \mathbb{R}$. However, in this case we shall follow conventional notation and abbreviate $K \circ \mathbf{x}$ to K and $H \circ \mathbf{x}$ to H. Thus $K(u,v)$ and $H(u,v)$ will denote the values of the Gaussian and mean curvatures at $\mathbf{x}(u,v)$.

Theorem 16.17. *Let* $\mathbf{x}:\mathcal{U} \longrightarrow \mathbb{R}^3$ *be a regular patch. Then the Gaussian curvature and mean curvature of* $\mathbf{x}$ *are given by the formulas*

$$(16.24) \qquad K = \frac{eg - f^2}{EG - F^2},$$

$$(16.25) \qquad H = \frac{eG - 2fF + gE}{2(EG - F^2)},$$

where e, f *and* g *are the coefficients of the second fundamental form relative to* $\mathbf{x}$ *and* E, F, G *are the coefficients of the first fundamental form.*

Proof. This time we compute K and H using the basis $\{\mathbf{x}_u, \mathbf{x}_v\}$. From the Weingarten equations (Theorem 16.10) we get

$$K = \det \begin{pmatrix} -\dfrac{fF - eG}{EG - F^2} & -\dfrac{eF - fE}{EG - F^2} \\ -\dfrac{gF - fG}{EG - F^2} & -\dfrac{fF - gE}{EG - F^2} \end{pmatrix}$$

$$= \frac{(fF - eG)(fF - gE) - (eF - fE)(gF - fG)}{(EG - F^2)^2} = \frac{eg - f^2}{EG - F^2}.$$

Similarly,

$$H = \frac{1}{2} \operatorname{tr} \begin{pmatrix} -\dfrac{fF - eG}{EG - F^2} & -\dfrac{eF - fE}{EG - F^2} \\ -\dfrac{gF - fG}{EG - F^2} & -\dfrac{fF - gE}{EG - F^2} \end{pmatrix}$$

$$= -\frac{(fF - eG) + (fF - gE)}{2(EG - F^2)} = \frac{eG - 2fF + gE}{2(EG - F^2)}. \blacksquare$$

The importance of Theorem 16.16 is theoretical, while that of Theorem 16.17 is computational. Usually one first uses Theorem 16.17 to compute K and H. Afterwards Theorem 16.16 can be used to find the principal curvatures. More explicitly,

Corollary 16.18. *The principal curvatures* $\mathbf{k}_1$ *and* $\mathbf{k}_2$ *are the roots of the quadratic equation*

$$\mathbf{k}^2 - 2H\mathbf{k} + K = 0.$$

Thus we can choose $\mathbf{k}_1$ *and* $\mathbf{k}_2$ *so that*

(16.26) $\qquad \mathbf{k}_1 = H + \sqrt{H^2 - K} \qquad$ and $\qquad \mathbf{k}_2 = H - \sqrt{H^2 - K}.$

Corollary 16.19. *Suppose that* $\mathcal{M} \subset \mathbb{R}^3$ *has negative Gaussian curvature* K *at* $\mathbf{p}$. *Then:*

(i) *there are exactly two asymptotic directions at* $\mathbf{p}$, *and they are bisected by the principal directions;*

(ii) *the two asymptotic directions at* $\mathbf{p}$ *are perpendicular if and only if the mean curvature* H *of* $\mathcal{M}$ *vanishes at* $\mathbf{p}$.

Proof. Let $\mathbf{e}_1$ and $\mathbf{e}_2$ be unit principal vectors corresponding to $\mathbf{k}_1(\mathbf{p})$ and $\mathbf{k}_2(\mathbf{p})$. Then $K(\mathbf{p}) = \mathbf{k}_1(\mathbf{p})\mathbf{k}_2(\mathbf{p}) < 0$ implies that $\mathbf{k}_1(\mathbf{p})$ and $\mathbf{k}_1(\mathbf{p})$ have opposite signs. Corollary 16.14 implies that there exists θ with $-\pi/2 < \theta < \pi/2$ such that

$$\tan^2 \theta = -\frac{\mathbf{k}_1(\mathbf{p})}{\mathbf{k}_2(\mathbf{p})}.$$

Put $\mathbf{u}_\mathbf{p}(\theta) = (\cos\theta)\mathbf{e}_1 + (\sin\theta)\mathbf{e}_2$. Then (16.20) implies that $\mathbf{u}_\mathbf{p}(\theta)$ and $\mathbf{u}_\mathbf{p}(-\theta)$ are linearly independent asymptotic vectors at $\mathbf{p}$. The angle between $\mathbf{u}_\mathbf{p}(\theta)$ and $\mathbf{u}_\mathbf{p}(-\theta)$ is 2θ, and it is clear that $\mathbf{e}_1$ bisects the angle between $\mathbf{u}_\mathbf{p}(\theta)$ and $\mathbf{u}_\mathbf{p}(-\theta)$. This proves (i). For (ii) we observe that $H(\mathbf{p}) = 0$ if and only if $\theta = \pm\pi/4$. ∎

It is neither enlightening nor useful to write out the formulas for the principal curvatures in terms of E, F, G, e, f, g in the general case. However, there is one special case where such formulas for $\mathbf{k}_1$ and $\mathbf{k}_2$ are worth noting.

Corollary 16.20. *Let* $\mathbf{x}: \mathcal{U} \longrightarrow \mathbb{R}^3$ *be a regular patch for which* $f = F = 0$. *Then with respect to this patch the principal curvatures are* e/E *and* g/G.

Proof. When $F = f = 0$, the Weingarten equations (16.11) reduce to

$$S(\mathbf{x}_u) = \frac{e}{E}\mathbf{x}_u \qquad \text{and} \qquad S(\mathbf{x}_v) = \frac{g}{G}\mathbf{x}_v.$$

Thus by definition e/E and g/G are the eigenvalues of the shape operator S. ∎

Classically, the standard formulas for computing K and H for a patch $\mathbf{x}$ are (16.24) and (16.25). It is usually too tedious to compute K and H by hand in one

16.5 The Gaussian and Mean Curvatures

step. Therefore, the functions E, F, G and e, f, g need to be computed before any of the curvature functions are calculated.

The computation of E, F, G is straightforward: first one computes the first derivatives $\mathbf{x}_u$ and $\mathbf{x}_v$ and then the dot products $E = \mathbf{x}_u \cdot \mathbf{x}_u$, $F = \mathbf{x}_u \cdot \mathbf{x}_v$ and $G = \mathbf{x}_v \cdot \mathbf{x}_v$.

There are two methods for computing e, f and g. The direct approach using the definitions necessitates computing the surface normal via formula (12.7), page 278, and then using the definition (16.10). The other method, explained in the next lemma, avoids the computation of the surface normal; instead it uses the vector triple product, which can be computed as a determinant.

Lemma 16.21. *Let* $\mathbf{x}:\mathcal{U} \longrightarrow \mathbb{R}^3$ *be a regular patch. Then*

(16.27)
$$\begin{cases} e = \dfrac{(\mathbf{x}_{uu}\mathbf{x}_u\mathbf{x}_v)}{\sqrt{EG-F^2}} = \dfrac{\det\begin{pmatrix} \mathbf{x}_{uu} \\ \mathbf{x}_u \\ \mathbf{x}_v \end{pmatrix}}{\sqrt{EG-F^2}}, \\[2em] f = \dfrac{(\mathbf{x}_{uv}\mathbf{x}_u\mathbf{x}_v)}{\sqrt{EG-F^2}} = \dfrac{\det\begin{pmatrix} \mathbf{x}_{uv} \\ \mathbf{x}_u \\ \mathbf{x}_v \end{pmatrix}}{\sqrt{EG-F^2}}, \\[2em] g = \dfrac{(\mathbf{x}_{vv}\mathbf{x}_u\mathbf{x}_v)}{\sqrt{EG-F^2}} = \dfrac{\det\begin{pmatrix} \mathbf{x}_{vv} \\ \mathbf{x}_u \\ \mathbf{x}_v \end{pmatrix}}{\sqrt{EG-F^2}}. \end{cases}$$

Proof. From (8.5) and (16.10) it follows that

$$e = \mathbf{x}_{uu} \cdot \mathbf{U} = \mathbf{x}_{uu} \cdot \frac{\mathbf{x}_u \times \mathbf{x}_v}{\|\mathbf{x}_u \times \mathbf{x}_v\|} = \frac{(\mathbf{x}_{uu}\mathbf{x}_u\mathbf{x}_v)}{\sqrt{EG-F^2}}.$$

The other formulas are proved similarly. ∎

On the other hand, at least theoretically, we can compute K and H for a regular patch $\mathbf{x}$ directly in terms of the first and second derivatives of $\mathbf{x}$. Here are the relevant formulas.

Corollary 16.22. *Let* $\mathbf{x}:\mathcal{U} \longrightarrow \mathbb{R}^3$ *be a regular patch. Then the Gaussian curvature and mean curvature of* $\mathcal{M}$ *are given by the formulas*

$$(16.28) \quad K = \frac{(\mathbf{x}_{uu}\mathbf{x}_u\mathbf{x}_v)(\mathbf{x}_{vv}\mathbf{x}_u\mathbf{x}_v) - (\mathbf{x}_{uv}\mathbf{x}_u\mathbf{x}_v)^2}{\left(\|\mathbf{x}_u\|^2\|\mathbf{x}_v\|^2 - (\mathbf{x}_u\cdot\mathbf{x}_v)^2\right)^2},$$

$$(16.29) \quad H = \frac{(\mathbf{x}_{uu}\mathbf{x}_u\mathbf{x}_v)\|\mathbf{x}_v\|^2 - 2(\mathbf{x}_{uv}\mathbf{x}_u\mathbf{x}_v)(\mathbf{x}_u\cdot\mathbf{x}_v) + (\mathbf{x}_{vv}\mathbf{x}_u\mathbf{x}_v)\|\mathbf{x}_u\|^2}{2\left(\|\mathbf{x}_u\|^2\|\mathbf{x}_v\|^2 - (\mathbf{x}_u\cdot\mathbf{x}_v)^2\right)^{\frac{3}{2}}}.$$

Proof. Equations (16.28) and (16.29) follow from (16.24), (16.24) and Lemma 16.21 when we write out e, f, g, E, F and G explicitly in terms of dot products. ∎

Formulas (16.28) and (16.29) are usually too complicated for hand calculation. However, if we use **Mathematica** to compute K and H (which we shall do in Section 17.1), (16.28) and (16.29) will be quite effective.

16.6 The Three Fundamental Forms

In classical differential geometry, for example in Eisenhart's[6] influential textbook [Eisen1], there are frequent references to the "second fundamental form" of a surface in $\mathbb{R}^3$. This notion is essentially equivalent to the shape operator S.

Definition. *Let* $\mathcal{M}$ *be a regular surface in* $\mathbb{R}^3$. *Then the* **second fundamental form** *is the symmetric bilinear form* $\mathbf{II}$ *on a tangent space* $\mathcal{M}_\mathbf{p}$ *given by*

$$\mathbf{II}(\mathbf{v}_\mathbf{p}, \mathbf{w}_\mathbf{p}) = S(\mathbf{v}_\mathbf{p}) \cdot \mathbf{w}_\mathbf{p}$$

for $\mathbf{v}_\mathbf{p}, \mathbf{w}_\mathbf{p} \in \mathcal{M}_\mathbf{p}$.

Since there is a second fundamental form, there must also be a **first fundamental form**. It is nothing but the inner product between tangent vectors:

$$\mathbf{I}(\mathbf{v}_\mathbf{p}, \mathbf{w}_\mathbf{p}) = \mathbf{v}_\mathbf{p} \cdot \mathbf{w}_\mathbf{p}.$$

[6] Luther Pfahler Eisenhart (1876-1965). American differential geometer and dean at Princeton University.

16.6 The Three Fundamental Forms

Note that the first fundamental form **I** is defined whether or not the surface is in $\mathbb{R}^3$. (See Chapters 15 and 25.)

The following lemma is an immediate consequence of the definitions.

Lemma 16.23. *Let* $\mathbf{x}\colon \mathcal{U} \longrightarrow \mathbb{R}^3$ *be a regular patch. Then*

(16.30) $$\mathbf{I}(a\,\mathbf{x}_u + b\,\mathbf{x}_v, a\,\mathbf{x}_u + b\,\mathbf{x}_v) = E\,a^2 + 2F\,a\,b + G\,b^2,$$

and

(16.31) $$\mathbf{II}(a\,\mathbf{x}_u + b\,\mathbf{x}_v, a\,\mathbf{x}_u + b\,\mathbf{x}_v) = e\,a^2 + 2f\,a\,b + g\,b^2.$$

Hence

$$\mathbf{k}(\mathbf{v}_\mathbf{p}) = \frac{\mathbf{II}(\mathbf{v}_\mathbf{p}, \mathbf{v}_\mathbf{p})}{\mathbf{I}(\mathbf{v}_\mathbf{p}, \mathbf{v}_\mathbf{p})}$$

for any nonzero tangent vector $\mathbf{v}_\mathbf{p}$.

Equations (16.30) and (16.31) explain why we call E, F, G the coefficients of the first fundamental form, and e, f, g the coefficients of the second fundamental form.

Finally, there is also a **third fundamental form III** for a surface in $\mathbb{R}^3$ given by

$$\mathbf{III}(\mathbf{v}_\mathbf{p}, \mathbf{w}_\mathbf{p}) = S(\mathbf{v}_\mathbf{p}) \cdot S(\mathbf{w}_\mathbf{p})$$

for $\mathbf{v}_\mathbf{p}, \mathbf{w}_\mathbf{p} \in \mathcal{M}_\mathbf{p}$. Note that **III**, in contrast to **II**, does not depend on the choice of surface normal **U**. The third fundamental form **III** contains no new information, since it is expressible in terms of **I** and **II**, as we now prove.

Lemma 16.24. *Let* $\mathcal{M} \subset \mathbb{R}^3$ *be a regular surface. Then the following relation holds between the first, second and third fundamental forms of* $\mathcal{M}$:

(16.32) $$\mathbf{III} - 2H\mathbf{II} + K\mathbf{I} = 0,$$

where H *and* K *denote the mean curvature and Gaussian curvature of* $\mathcal{M}$.

Proof. Although (16.32) follows from Corollary 16.18 and the Cayley-Hamilton[7] Theorem (which states that a matrix is a root of its own characteristic polynomial),

---

[7] Sir William Rowan Hamilton (1805-1865). Irish mathematician, best known for having been struck with the concept of quaternions as he crossed Brougham Bridge in Dublin, and for his work in dynamics.

we prefer to give a direct proof. First, note that $H\mathbf{II}$ is independent of the choice of surface normal $\mathbf{U}$. Hence (16.32) makes sense whether or not $\mathcal{M}$ is orientable.

To prove (16.32), we observe that since $\mathbf{III} - 2H\mathbf{II} + K\mathbf{I}$ is a symmetric bilinear form, it suffices to show that for each $\mathbf{p} \in \mathcal{M}$ and some basis $\{\mathbf{e}_1, \mathbf{e}_2\}$ of $\mathcal{M}_\mathbf{p}$ we have

(16.33) $$(\mathbf{III} - 2H\mathbf{II} + K\mathbf{I})(\mathbf{e}_i, \mathbf{e}_j) = 0,$$

for $i, j = 1, 2$. We choose $\mathbf{e}_1$ and $\mathbf{e}_2$ to be linearly independent principal vectors at $\mathbf{p}$. Then
$$(\mathbf{III} - 2H\mathbf{II} + K\mathbf{I})(\mathbf{e}_1, \mathbf{e}_2) = 0$$

because each term vanishes separately. Furthermore,

$$(\mathbf{III} - 2H\mathbf{II} + K\mathbf{I})(\mathbf{e}_1, \mathbf{e}_1) = \mathbf{k}_1^2 - (\mathbf{k}_1 + \mathbf{k}_2)\mathbf{k}_1 + \mathbf{k}_1\mathbf{k}_2 = 0,$$

and similarly, $(\mathbf{III} - 2H\mathbf{II} + K\mathbf{I})(\mathbf{e}_2, \mathbf{e}_2) = 0$. Hence (16.32) holds in general. ∎

16.7 Examples of Curvature Calculations by Hand

In this section we show how to compute by hand the Gaussian curvature K and the mean curvature H for a monkey saddle and a torus. Along the way we compute the coefficients of the first and second fundamental forms.

The monkey saddle

For the monkey saddle parametrized by

$$\mathbf{monkeysaddle}(u, v) = (u, v, u^3 - 3u\,v^2)$$

we easily compute

$$\begin{aligned}
\mathbf{monkeysaddle}_u(u, v) &= (1, 0, 3u^2 - 3v^2), \\
\mathbf{monkeysaddle}_v(u, v) &= (0, 1, -6u\,v), \\
\mathbf{monkeysaddle}_{uu}(u, v) &= (0, 0, 6u), \\
\mathbf{monkeysaddle}_{uv}(u, v) &= (0, 0, -6v), \\
\mathbf{monkeysaddle}_{vv}(u, v) &= (0, 0, -6u).
\end{aligned}$$

16.7 Examples of Curvature Calculations by Hand

Therefore,
$$E = 1 + (3u^2 - 3v^2)^2, \qquad F = -6uv(3u^2 - 3v^2), \qquad G = 1 + 36u^2v^2.$$

Furthermore, by inspection a surface normal is
$$\mathbf{U} = \frac{(-3u^2 + 3v^2, 6uv, 1)}{\sqrt{1 + 9u^4 + 18u^2v^2 + 9v^4}},$$

so that
$$e = \mathbf{U} \cdot \mathbf{monkeysaddle}_{uu} = \frac{6u}{\sqrt{1 + 9u^4 + 18u^2v^2 + 9v^4}},$$
$$f = \mathbf{U} \cdot \mathbf{monkeysaddle}_{uv} = \frac{-6v}{\sqrt{1 + 9u^4 + 18u^2v^2 + 9v^4}},$$
$$g = \mathbf{U} \cdot \mathbf{monkeysaddle}_{vv} = \frac{-6u}{\sqrt{1 + 9u^4 + 18u^2v^2 + 9v^4}}.$$

Therefore,
$$K = \frac{-36(u^2 + v^2)}{(1 + 9u^4 + 18u^2v^2 + 9v^4)^2}, \qquad H = \frac{-27u^5 + 54u^3v^2 + 81uv^4}{(1 + 9u^4 + 18u^2v^2 + 9v^4)^{\frac{3}{2}}}.$$

A glance at the expression for K shows that $(0, 0, 0)$ is a planar point of the monkey saddle and that every other point is hyperbolic. Furthermore, the Gaussian curvature of the monkey saddle is invariant under all rotations about the z-axis, even though the monkey saddle itself does not have this property.

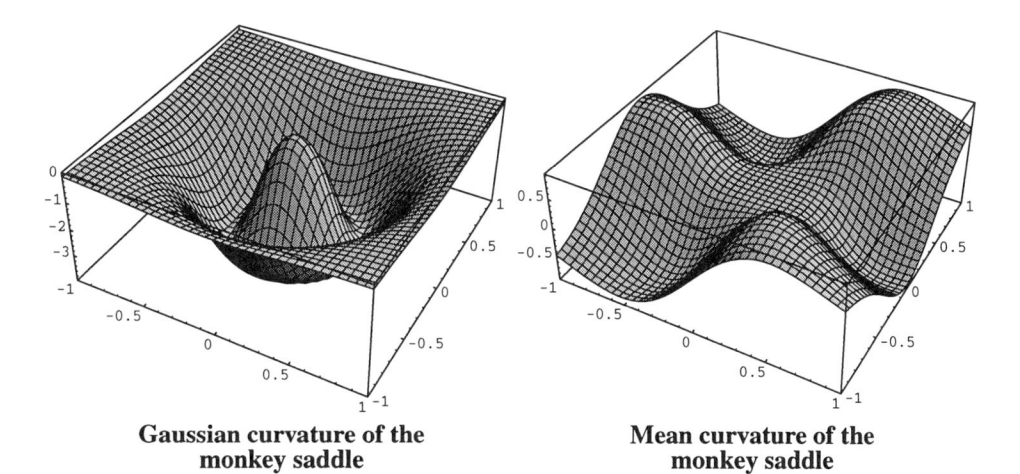

Gaussian curvature of the monkey saddle

Mean curvature of the monkey saddle

The torus

We compute the Gaussian and mean curvatures of the torus parametrized by

$$\mathbf{torus}(u,v) = \big((a + b\cos v)\cos u, (a + b\cos v)\sin u, b\sin v\big).$$

Here $\mathbf{torus}: [0, 2\pi) \times [0, 2\pi) \longrightarrow \mathbb{R}^3$. We find that

$$\mathbf{torus}_u(u,v) = \big(-(a + b\cos v)\sin u, (a + b\cos v)\cos u, 0\big),$$
$$\mathbf{torus}_v(u,v) = \big(-b\sin v\cos u, -b\sin v\sin u, b\cos v\big),$$

and so

$$E = (a + b\cos v)^2, \qquad F = 0, \qquad G = b^2.$$

Furthermore,

$$\mathbf{torus}_{uu}(u,v) = \big(-(a + b\cos v)\cos u, -(a + b\cos v)\sin u, 0\big),$$
$$\mathbf{torus}_{uv}(u,v) = \big(b\sin v\sin u, -b\sin v\cos u, 0\big),$$
$$\mathbf{torus}_{vv}(u,v) = \big(-b\cos v\cos u, -b\cos v\sin u, -b\sin v\big),$$

so that

$$e = \frac{\det\begin{pmatrix} -(a+b\cos v)\cos u & -(a+b\cos v)\sin u & 0 \\ -(a+b\cos v)\sin u & (a+b\cos v)\cos u & 0 \\ -b\sin v\cos u & -b\sin v\sin u & b\cos v \end{pmatrix}}{b(a+b\cos v)}$$

$$= -\cos v(a + b\cos v),$$

$$f = \frac{\det\begin{pmatrix} b\sin v\sin u & -b\sin v\cos u & 0 \\ -(a+b\cos v)\sin u & (a+b\cos v)\cos u & 0 \\ -b\sin v\cos u & -b\sin v\sin u & b\cos v \end{pmatrix}}{b(a+b\cos v)} = 0,$$

$$g = \frac{\det\begin{pmatrix} -b\cos v\cos u & -b\cos v\sin u & -b\sin v \\ -(a+b\cos v)\sin u & (a+b\cos v)\cos u & 0 \\ -b\sin v\cos u & -b\sin v\sin u & b\cos v \end{pmatrix}}{b(a+b\cos v)} = -b.$$

Therefore,
$$K = \frac{\cos v}{b(a + b\cos v)}, \qquad H = -\frac{a + 2b\cos v}{2b(a + b\cos v)},$$
$$\mathbf{k}_1 = -\frac{\cos v}{a + b\cos v}, \qquad \mathbf{k}_2 = -\frac{1}{b}.$$

Suppose that $a > b$; we see that the Gaussian curvature K of the **torus** vanishes along the curves given by $v = \pm\pi/2$. These are the parabolic points of the torus. The set of hyperbolic points is

$$\left\{ \mathbf{torus}(u,v) \;\middle|\; \frac{\pi}{2} < v < \frac{3\pi}{2} \right\},$$

and the set of elliptic points is

$$\left\{ \mathbf{torus}(u,v) \;\middle|\; -\frac{\pi}{2} < v < \frac{\pi}{2} \right\}.$$

The elliptic and hyperbolic points can be seen on the torus by using an extended version of **ParametricPlot3D**, which allows a fourth variable to specify a coloring or shading function. Let us see how this works for the surface **torus[8,3]**. To distinguish the hyperbolic points and elliptic points we use **GrayLevel[N[If[v<=Pi/2,0.8,0.2]]]**. In the following picture the dark points are hyperbolic, the light points are elliptic and the parabolic points have an intermediate color. A more interesting picture can be obtained by plotting the surface in color, as explained in Section 14.7.

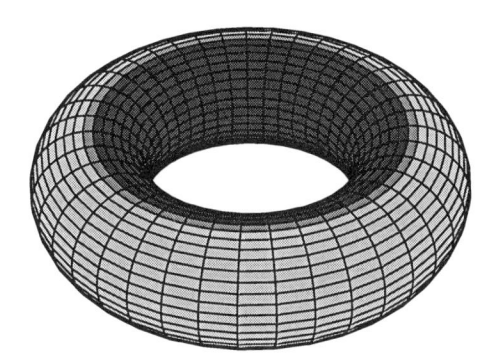

Elliptic, hyperbolic and parabolic points on a torus

The *Mathematica* command to draw this picture is:

```
ParametricPlot3D[
{Cos[u](8 + 3Cos[v]),(8 + 3Cos[v])Sin[u],3Sin[v],
FaceForm[GrayLevel[N[If[v<=Pi/2,0.8,0.2]]]]},
{u,0,2Pi},{v,-Pi/2,3Pi/2},
PlotPoints->{40,40},
Lighting->False,
Axes->None,Boxed->False];
```

16.8 A Global Curvature Theorem

We recall the following fundamental fact about compact subsets of $\mathbb{R}^n$ (see page 353):

Lemma 16.25. *Let S be a compact subset of $\mathbb{R}^n$, and let $f: S \longrightarrow \mathbb{R}$ be a continuous function. Then f assumes its maximum value at some point $\mathbf{p} \in S$.*

For a proof of this fundamental lemma, see [Buck, page 74].

Intuitively, it is reasonable that for each compact surface $\mathcal{M} \subset \mathbb{R}^3$, there is a point $\mathbf{p}_0 \in \mathcal{M}$ that is furthest from the origin, and at $\mathbf{p}_0$ the surface bends towards the origin. Thus it appears that the Gaussian curvature K of $\mathcal{M}$ is positive at $\mathbf{p}_0$. We now prove that this is indeed the case. The proof uses standard facts from calculus concerning a maximum of a differentiable function of one variable.

Theorem 16.26. *If $\mathcal{M}$ is a compact regular surface in $\mathbb{R}^3$, there is a point $\mathbf{p} \in \mathbb{R}^3$ at which the Gaussian curvature K is strictly positive.*

Proof. Let $f: \mathbb{R}^3 \longrightarrow \mathbb{R}$ be defined by $f(\mathbf{p}) = \|\mathbf{p}\|^2$. Then f is continuous (in fact, differentiable), since it can be expressed in terms of the natural coordinate functions of $\mathbb{R}^3$ as $f = u_1^2 + u_2^2 + u_3^2$. By Lemma 16.25, f assumes its maximum value at some point $\mathbf{p}_0 \in \mathcal{M}$. Let $\mathbf{v} \in \mathcal{M}_\mathbf{p}$ be a unit tangent vector, and choose a unit-speed curve $\boldsymbol{\alpha}: (a, b) \longrightarrow \mathcal{M}$ such that $a < 0 < b$, $\boldsymbol{\alpha}(0) = \mathbf{p}_0$ and $\boldsymbol{\alpha}'(0) = \mathbf{v}$. Since the function $g: (a, b) \longrightarrow \mathbb{R}$ defined by $g = f \circ \boldsymbol{\alpha}$ has a maximum at 0, it follows that

$$g'(0) = 0 \quad \text{and} \quad g''(0) \leq 0.$$

But $g(t) = \boldsymbol{\alpha}(t) \cdot \boldsymbol{\alpha}(t)$, so that

(16.34) $$0 = g'(0) = 2\boldsymbol{\alpha}'(0) \cdot \boldsymbol{\alpha}(0) = 2\mathbf{v} \cdot \mathbf{p}_0.$$

In (16.34), $\mathbf{v}$ can be an arbitrary unit tangent vector, and so $\mathbf{p}$ must be normal to $\mathcal{M}$ at $\mathbf{p}_0$. Clearly, $\mathbf{p}_0 \neq 0$, so that (16.34) implies that $\mathbf{p}_0/\|\mathbf{p}_0\|$ is a unit normal vector to $\mathcal{M}$ at $\mathbf{p}_0$. Furthermore,

$$0 \geq g''(0) = 2\boldsymbol{\alpha}''(0) \cdot \boldsymbol{\alpha}(0) + 2\boldsymbol{\alpha}'(0) \cdot \boldsymbol{\alpha}'(0) = 2\big(\boldsymbol{\alpha}''(0) \cdot \mathbf{p}_0 + 1\big),$$

so that $\alpha''(0) \cdot \mathbf{p}_0 \leq -1$, or

(16.35) $$\mathbf{k}(\mathbf{v}) = \alpha''(0) \cdot \frac{\mathbf{p}_0}{\|\mathbf{p}_0\|} \leq -\frac{1}{\|\mathbf{p}_0\|},$$

where $\mathbf{k}(\mathbf{v})$ is the normal curvature determined by the tangent vector $\mathbf{v}$ and the unit normal vector $\mathbf{p}_0/\|\mathbf{p}_0\|$. In particular, the principal curvatures of $\mathcal{M}$ at $\mathbf{p}_0$ (with respect to $\mathbf{p}_0/\|\mathbf{p}_0\|$) satisfy

$$\mathbf{k}_1(\mathbf{p}_0), \mathbf{k}_2(\mathbf{p}_0) \leq -1/\|\mathbf{p}_0\|.$$

This implies that the Gaussian curvature of $\mathcal{M}$ at $\mathbf{p}_0$ satisfies

$$K(\mathbf{p}_0) = \mathbf{k}_1(\mathbf{p}_0)\mathbf{k}_2(\mathbf{p}_0) \geq \frac{1}{\|\mathbf{p}_0\|^2} > 0. \blacksquare$$

On the other hand, noncompact surfaces of positive Gaussian curvature exist (see exercise 14). As for surfaces of negative curvature, we have the following result (see exercise 15):

Corollary 16.27. *Any surface in $\mathbb{R}^3$ whose Gaussian curvature is everywhere negative must be noncompact.*

16.9 Exercises

1. For the following surfaces describe the sets of elliptic, hyperbolic, parabolic and planar points. No calculation is necessary.

    (a) A sphere.
    (b) An ellipsoid.
    (c) A paraboloid.
    (d) A hyperbolic paraboloid.
    (e) A hyperboloid of one sheet.
    (f) A hyperboloid of two sheets.
    (g) A cylinder over an ellipse.
    (h) A cylinder over a parabola.
    (i) A cylinder over a hyperbola.
    (j) A cylinder over $y = x^3$.
    (k) A cone over a circle.

2. Describe the general shape of the image of the Gauss map of the surfaces listed in problem 1. No calculation is necessary.

3. Prove Lemma 16.6.

4. Show that the first fundamental form of the Gauss map of a patch **x** is the third fundamental form of **x**.

5. Show that orientation preserving Euclidean motion $F\colon \mathbb{R}^3 \longrightarrow \mathbb{R}^3$ preserves both principal curvatures and principal vectors.

6. Show that the mean curvature $H(\mathbf{p})$ of a surface $\mathcal{M} \subset \mathbb{R}^3$ at $\mathbf{p} \in \mathcal{M}$ is given by

$$H(\mathbf{p}) = \frac{1}{\pi}\int_0^\pi \mathbf{k}(\theta)d\theta,$$

where $\mathbf{k}(\theta)$ is the notation for the normal curvature used in Lemma 16.13.

7. (Continuation) Let n be an integer larger than 2 and for $1 \le i \le n$ put $\theta_i = \psi + 2\pi i/n$, where ψ is some angle. Show that

$$H(\mathbf{p}) = \frac{1}{n}\sum_{i=1}^n \mathbf{k}(\theta_i).$$

8. Compute by hand the coefficients of the first fundamental form, the coefficients of the second fundamental form, the unit normal, the mean curvature and the principal curvatures of the following surfaces:

   a. the torus defined on page 356. Plot the Gaussian and mean curvatures.

   b. the helicoid defined on page 357.

   c. Enneper's minimal surface defined on page 358.

9. Compute by hand the first fundamental form, the second fundamental form, the unit normal, the Gaussian curvature, the mean curvature and the principal curvatures of the patch defined by $\mathbf{x}_n(u,v) = (u^n, v^n, uv)$. (Clearly, $\mathbf{x}_1$ is a hyperbolic paraboloid. For $n=3$ see the picture on page 273.)

10. Compute the first fundamental form, the second fundamental form, the unit normal, the Gaussian curvature, the mean curvature and the principal curvatures of the patch defined by $\mathbf{y}(u,v) = (u^2+v, v^2+u, uv)$.

11. The **translation surface** determined by curves $\boldsymbol{\alpha}, \boldsymbol{\gamma}\colon (a,b) \longrightarrow \mathbb{R}^3$ is the patch
$$(u,v) \longmapsto \boldsymbol{\alpha}(u) + \boldsymbol{\gamma}(v).$$

16.9 Exercises

It is the surface formed by moving α parallel to itself in such a way that a point of the curve moves along γ. Show that $f = 0$ for a translation surface.

12. Show that the **Bohemian dome** defined by

$$\mathbf{bohdom}[a, b, c](u, v) = (a\cos u, a\sin u + b\cos v + c\sin v)$$

or

```
bohdom[a_,b_,c_][u_,v_]:= {a Cos[u],
                 a Sin[u] + b Cos[v],c Sin[v]}
```

is the translation surface (see exercise 11) of two ellipses. Compute by hand the Gaussian curvature, the mean curvature and the principal curvatures of $(u, v) \longmapsto \mathbf{bohdom}[a, b, b](u, v)$.

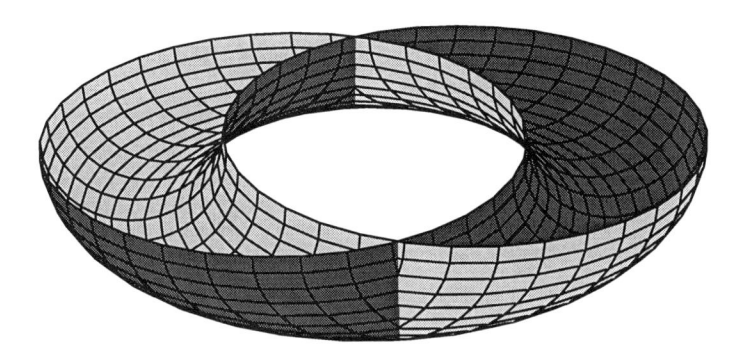

A Bohemian dome
$(u, v) \longmapsto \mathbf{bohdom[8,3,3][u,v]}$

13. Prove the following lemma.

Lemma 16.28. *Let* $\mathcal{M} \subset \mathbb{R}^3$ *be a surface, and let* $\mathbf{x}: \mathcal{U} \longrightarrow \mathcal{M}$ *and* $\mathbf{y}: \mathcal{V} \longrightarrow \mathcal{M}$ *be coherently oriented patches on* $\mathcal{M}$ *with* $\mathbf{x}(\mathcal{U}) \cap \mathbf{y}(\mathcal{V})$ *nonempty. Let* $\mathbf{x}^{-1} \circ \mathbf{y} = (\bar{u}, \bar{v}): \mathcal{U} \cap \mathcal{V} \longrightarrow \mathcal{U} \cap \mathcal{V}$ *be the associated change of coordinates, so that*

$$\mathbf{y}(u, v) = \mathbf{x}\bigl(\bar{u}(u, v), \bar{v}(u, v)\bigr).$$

Let e_x, f_x and g_x denote the coefficients of the second fundamental form of **x**, and let e_y, f_y and g_y denote the coefficients of the second fundamental form of **y**. Then

(16.36)
$$\begin{cases} e_y = e_x \left(\dfrac{\partial \bar{u}}{\partial u}\right)^2 + 2f_x \dfrac{\partial \bar{u}}{\partial u}\dfrac{\partial \bar{v}}{\partial u} + g_x \left(\dfrac{\partial \bar{v}}{\partial u}\right)^2, \\ f_y = e_x \dfrac{\partial \bar{u}}{\partial u}\dfrac{\partial \bar{u}}{\partial v} + f_x \left(\dfrac{\partial \bar{u}}{\partial u}\dfrac{\partial \bar{v}}{\partial v} + \dfrac{\partial \bar{u}}{\partial v}\dfrac{\partial \bar{v}}{\partial u}\right) + g_x \dfrac{\partial \bar{v}}{\partial u}\dfrac{\partial \bar{v}}{\partial v}, \\ g_y = e_x \left(\dfrac{\partial \bar{u}}{\partial v}\right)^2 + 2f_x \dfrac{\partial \bar{u}}{\partial v}\dfrac{\partial \bar{v}}{\partial v} + g_x \left(\dfrac{\partial \bar{v}}{\partial v}\right)^2. \end{cases}$$

14. Give examples of the following:

    a. A noncompact surface whose Gaussian curvature is negative;

    b. A noncompact surface whose Gaussian curvature is identically zero;

    c. A noncompact surface whose Gaussian curvature is positive;

    d. A noncompact surface containing elliptic, hyperbolic, parabolic and planar points.

15. Prove Corollary 16.27.

16. Show that there are no compact minimal surfaces in $\mathbb{R}^3$.

17

SURFACES IN 3-DIMENSIONAL SPACE VIA MATHEMATICA

Mathematica programs for computing Gaussian and mean curvatures are given in Section 17.1. For most surfaces in $\mathbb{R}^3$, calculation of the shape operator and curvatures is more efficiently and accurately done with *Mathematica* than by hand. Another advantage of computing curvatures with *Mathematica* is that the curvatures can be easily plotted. Often, visualization of curvatures provides more insight than do formulas; good examples are the plots of the Gauss and mean curvatures of the monkey saddle on page 383. Computation and graphing of the Gaussian and mean curvatures of several surfaces are discussed in Section 17.2 and in the exercises. *Mathematica* formulas for principal curvatures are given in Section 17.3, and we use them to illustrate the nondifferentiability of the principal curvatures at the planar point of a monkey saddle. In Section 17.4 we show how to use *Mathematica* to draw the image of a surface under the Gauss map. Finally, in Section 17.5 we discuss how to compute the Gaussian and mean curvatures of nonparametrically defined surfaces.

17.1 Programs for Computing the Shape Operator and Curvature

There is already in *Mathematica* a convenient formula for computing the vector triple product of three vectors **v1**, **v2**, **v3** in $\mathbb{R}^3$, namely **Det[v1,v2,v3]**. In order to avoid collision with *Mathematica's* internal symbols, we use **eee**, **fff** and **ggg** for e, f and g. Here are the *Mathematica* formulas to compute these functions:

```
eee[x_][u_,v_] := Simplify[Det[{
    D[x[uu,vv],uu,uu],D[x[uu,vv],uu],D[x[uu,vv],vv]}]/
    Sqrt[Simplify[D[x[uu,vv],uu].D[x[uu,vv],uu]*
        D[x[uu,vv],vv].D[x[uu,vv],vv] -
        D[x[uu,vv],uu].
        D[x[uu,vv],vv]^2]] /. {uu->u,vv->v}
```

```
fff[x_][u_,v_] := Simplify[Det[{
    D[x[uu,vv],uu,vv],D[x[uu,vv],uu],D[x[uu,vv],vv]}]/
    Sqrt[Simplify[D[x[uu,vv],uu].D[x[uu,vv],uu]*
        D[x[uu,vv],vv].D[x[uu,vv],vv] -
        D[x[uu,vv],uu].
        D[x[uu,vv],vv]^2]] /. {uu->u,vv->v}
```

```
ggg[x_][u_,v_] := Simplify[Det[{
    D[x[uu,vv],vv,vv],D[x[uu,vv],uu],D[x[uu,vv],vv]}]/
    Sqrt[Simplify[D[x[uu,vv],uu].D[x[uu,vv],uu]*
        D[x[uu,vv],vv].D[x[uu,vv],vv] -
        D[x[uu,vv],uu].
        D[x[uu,vv],vv]^2]] /. {uu->u,vv->v}
```

Appendix A (or the package **CSPROGS.m**) contains all the commands mentioned in this section. The *Mathematica* command to find the right-hand sides of the Weingarten equations (16.11) (as a 2 × 2 matrix) is

17.1 Programs for Computing the Shape Operator and Curvature

```
weingarten[x_][u_,v_]:=
    {{eee[x][u,v]gg[x][u,v] - fff[x][u,v]ff[x][u,v],
      fff[x][u,v]ee[x][u,v] - eee[x][u,v]ff[x][u,v]},
     {fff[x][u,v]gg[x][u,v] - ggg[x][u,v]ff[x][u,v],
      ggg[x][u,v]ee[x][u,v] - fff[x][u,v]ff[x][u,v]}}/
    (ee[x][u,v]gg[x][u,v] - ff[x][u,v]^2)
```

weingarten can be used with *Mathematica's* internal command **MatrixForm** to display the matrix of the shape operator S relative to $\mathbf{x}_u$ and $\mathbf{x}_v$. For example,

**MatrixForm[Simplify[
    weingarten[hyperbolicparaboloid][u,v]]]**

yields

$$\begin{pmatrix} -\dfrac{u\,v}{(1+u^2+v^2)^{3/2}} & \dfrac{1+v^2}{(1+u^2+v^2)^{3/2}} \\[1em] \dfrac{1+u^2}{(1+u^2+v^2)^{3/2}} & -\dfrac{u\,v}{(1+u^2+v^2)^{3/2}} \end{pmatrix}$$

We describe two ways that *Mathematica* can be used to compute K and H. The first approach imitates the traditional hand calculation method described in Section 16.7. It uses (16.24) and (16.25); these equations define K and H in terms of e, f, g, E, F, G. The commands **gaussiancurvature** and **meancurvature** are defined in terms of **ee**, **ff**, **gg** (on page 354), and **eee**, **fff**, **ggg** as follows:

```
gaussiancurvature[x_][u_,v_]:= Simplify[
    (eee[x][u,v]ggg[x][u,v] - fff[x][u,v]^2)/
    (ee[x][u,v]gg[x][u,v] - ff[x][u,v]^2)]
```

```
meancurvature[x_][u_,v_]:= Simplify[
    (eee[x][u,v]gg[x][u,v] - 2fff[x][u,v]ff[x][u,v] +
     ggg[x][u,v]ee[x][u,v])/
    (2(ee[x][u,v]gg[x][u,v] - ff[x][u,v]^2))]
```

Another way to compute K and H using *Mathematica* is to mimic (16.28) and (16.29) via the commands **gcurvature** and **mcurvature**, whose definitions are as follows.

```
gcurvature[x_][u_,v_]:= Simplify[
   (Det[{D[x[uu,vv],uu,uu],D[x[uu,vv],uu],
        D[x[uu,vv],vv]}]*
    Det[{D[x[uu,vv],vv,vv],D[x[uu,vv],uu],
        D[x[uu,vv],vv]}] -
    Det[{D[x[uu,vv],uu,vv],D[x[uu,vv],uu],
        D[x[uu,vv],vv]}]^2)/
   (D[x[uu,vv],uu].D[x[uu,vv],uu]*
        D[x[uu,vv],vv].D[x[uu,vv],vv] -
    D[x[uu,vv],uu].
        D[x[uu,vv],vv]^2)^2] /. {uu->u,vv->v}
```

```
mcurvature[x_][u_,v_]:= Simplify[
   (Det[{D[x[uu,vv],uu,uu],D[x[uu,vv],uu],
        D[x[uu,vv],vv]}]*
    D[x[uu,vv],vv].D[x[uu,vv],vv] -
    2Det[{D[x[uu,vv],uu,vv],D[x[uu,vv],uu],
        D[x[uu,vv],vv]}]*
    D[x[uu,vv],uu].D[x[uu,vv],vv] +
    Det[{D[x[uu,vv],vv,vv],D[x[uu,vv],uu],
        D[x[uu,vv],vv]}]*
    D[x[uu,vv],uu].D[x[uu,vv],uu])/
   (2(D[x[uu,vv],uu].D[x[uu,vv],uu]*
        D[x[uu,vv],vv].D[x[uu,vv],vv] -
    D[x[uu,vv],uu].
        D[x[uu,vv],vv]^2)^(3/2))] /. {uu->u,vv->v}
```

The commands **gcurvature** and **mcurvature** are usually more than twice as fast as **gaussiancurvature** and **meancurvature**, so for most purposes the commands **gcurvature** and **mcurvature** are the preferred ones. The commands **gaussiancurvature** and **meancurvature** might be used, for example, if the coefficients of the first and second fundamental forms have already been computed and saved to a file. The precalculated values in that file could then be read in and used by **gaussiancurvature** and **meancurvature**.

17.2 Examples of Curvature Calculations with Mathematica

The Sphere

The calculation of e, f and g for the sphere $S^2(a)$ is similar to the calculation of E, F and G given in Section 15.5. Before starting, we need all the *Mathematica* definitions given in Sections 15.4 and 17.1. For convenience, these functions are contained in Appendix A (or in the package **CSPROGS.m**). We also need the definition of **sphere[a]** given on page 270 (or in Appendix C); it can be obtained by reading in the file **SURFS.m**.

Here is the *Mathematica* dialog that computes the coefficients of the second fundamental form of $S^2(a)$:

eee[sphere[a]][u,v]//PowerExpand

$$-(a\,\text{Cos}[v]^2)$$

fff[sphere[a]][u,v]

$$0$$

ggg[sphere[a]][u,v]//PowerExpand

$$-a$$

The matrix of the shape operator is displayed with

weingarten[sphere[a]][u,v]//PowerExpand//MatrixForm

$$\begin{pmatrix} -\left(\dfrac{1}{a}\right) & 0 \\ 0 & -\left(\dfrac{1}{a}\right) \end{pmatrix}$$

Also, for comparison we time the computations of the Gaussian and mean curvatures:

```
gaussiancurvature[sphere[a]][u,v]//PowerExpand//Timing
```
$$\{12.6667 \text{ Second}, a^{-2}\}$$

```
gcurvature[sphere[a]][u,v]//Timing
```
$$\{4.95 \text{ Second}, a^{-2}\}$$

```
meancurvature[sphere[a]][u,v]//PowerExpand//Timing
```
$$\{14.8667 \text{ Second}, -(\frac{1}{a})\}$$

```
mcurvature[sphere[a]][u,v]//PowerExpand//Timing
```
$$\{5.88333 \text{ Second}, -(\frac{1}{a})\}$$

The Astroidial Ellipsoid

If we modify the standard parametrization of the ellipsoid given on page 302 by replacing each coordinate by its cube, we obtain the **astroidial ellipsoid**. Explicitly, it is given by

$$\text{astell}[a, b, c](u, v) = \left((a \cos u \cos v)^3, (b \sin u \cos v)^3, (c \sin v)^3\right),$$

or

```
astell[a_,b_,c_][u_,v_] := {a Cos[u]Cos[v],
                            b Sin[u]Cos[v],c Sin[v]}^3
```

Of course, the nonparametric equation of the astroidal ellipsoid is

$$\frac{x^{2/3}}{a^2} + \frac{y^{2/3}}{b^2} + \frac{z^{2/3}}{c^2} = 1.$$

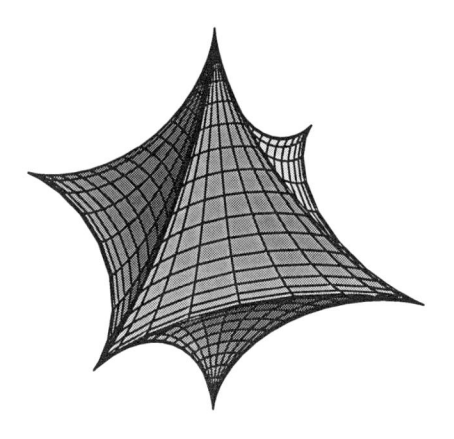

17.2 Examples of Curvature Calculations with Mathematica

```
ParametricPlot3D[
astell[1,1,1][u,v]
//Evaluate,
{u,0,2Pi},
{v,-Pi/2,Pi/2},
PlotPoints->{40,40},
Boxed->False,
Axes->None,
PlotRange->
{{-1,1},
{-1,1},
{-1,1}}];
```

The astroidial sphere

$$(u,v) \longmapsto \left((\cos u \cos v)^3, (\sin u \cos v)^3, (\sin v)^3\right)$$

Since **astell**[a, a, a] has the nonparametric equation

$$x^{2/3} + y^{2/3} + x^{2/3} = a^2,$$

it corresponds to a sphere, so we call it an **astroidial sphere**. Let us compute and graph its Gaussian curvature.

gcurvature[astell[a,a,a]][u,v]

$$\frac{1024 \operatorname{Sec}[v]^4}{9 a^6 (-18 + 2 \operatorname{Cos}[4u] + \operatorname{Cos}[4u - 2v] + 14 \operatorname{Cos}[2v] + \operatorname{Cos}[4u + 2v])^2}$$

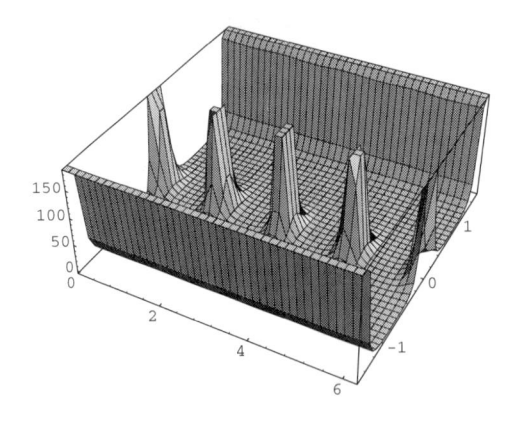

```
Plot3D[gcurvature[
astell[1,1,1]][u,v]
//Evaluate,
{u,0,2Pi},
{v,-Pi/2,Pi/2},
PlotPoints->{40,40}];
```

The Gaussian curvature of
$$(u,v) \longmapsto$$
$$\bigl((\cos u\cos v)^3, (\sin u\cos v)^3, (\sin v)^3\bigr)$$

The picture shows a surprise: except at the vertices, the Gaussian curvature of the astroidial sphere is continuous on the edges.

A Monge Patch

An important class of surfaces in $\mathbb{R}^3$ consists of those that are graphs of a real-valued function of two variables.

Definition. *A* Monge[1] *patch is a patch* $\mathbf{x}: \mathcal{U} \longrightarrow \mathbb{R}^3$ *of the form*

$$\mathbf{x}(u,v) = \bigl(u, v, h(u,v)\bigr),$$

[1] Gaspard Monge (1746-1818). French mathematician, erstwhile secretary of the navy, founding director of the École Polytechnique, which played a leading role in the development and organization of scientific research and education in France. Monge's work on fortifications led him to descriptive geometry, and from there he went on to a broad exposition of the differential geometry of space curves. His major contribution to the development of differential geometry was the integration of geometrical facts and intuition with the use of partial differential equations. A fervent supporter first of the Revolution and then of Napoleon, Monge was expelled from the Institut de France following the Battle of Waterloo. Many consider Monge to be the father of French differential geometry.

17.2 Examples of Curvature Calculations with Mathematica

where $\mathcal{U}$ is an open set in $\mathbb{R}^2$ and $h: \mathcal{U} \longrightarrow \mathbb{R}$ is a differentiable function.

It is not hard to compute the Gaussian and mean curvatures by hand (see, for example, [dC1, pages 162–163]). However, it is informative to see how *Mathematica* does the computations. All we need is an operator that assigns to each real-valued function h of two variables its Monge patch. We put

```
monge[h_][u_,v_] := {u,v,h[u,v]}
```

Then we can compute the functions **ee**, ... ending with the computations of

        gcurvature   and   **mcurvature**.

For example,

```
gcurvature[monge[h]][u,v]
```

$$\frac{-h^{(1,1)}[u,v]^2 + h^{(0,2)}[u,v]\, h^{(2,0)}[u,v]}{(1 + h^{(0,1)}[u,v]^2 + h^{(1,0)}[u,v]^2)^2}$$

The results are summarized in the following lemma:

Lemma 17.1. *For a Monge patch* $(u, v) \longmapsto (u, v, h(u, v))$ *we have*

$$E = 1 + h_u^2, \qquad F = h_u h_v, \qquad G = 1 + h_v^2,$$

$$e = \frac{h_{uu}}{(1 + h_u^2 + h_v^2)^{1/2}}, \quad f = \frac{h_{uv}}{(1 + h_u^2 + h_v^2)^{1/2}}, \quad g = \frac{h_{vv}}{(1 + h_u^2 + h_v^2)^{1/2}},$$

$$K = \frac{h_{uu} h_{vv} - h_{uv}^2}{(1 + h_u^2 + h_v^2)^2}, \qquad H = \frac{(1 + h_v^2)h_{uu} - 2h_u h_v h_{uv} + (1 + h_u^2)h_{vv}}{2(1 + h_u^2 + h_v^2)^{3/2}}.$$

For the details of the proof of Lemma 17.1 using *Mathematica*, see exercise 1. Once these general calculations have been done with *Mathematica*, they can be applied to specific functions. For example, suppose we want to know the formula for the curvature of the graph of the function $u^m v^n$. We define a function **p** in *Mathematica* by

```
p[m_,n_][u_,v_] := u^m v^n
```

Then the response to

gcurvature[monge[p[m,n]]][u,v]//Timing

is

$$\left\{3.45 \text{ Second}, \frac{m(1-m-n)n\, u^{-2+2m} v^{-2+2n}}{(1+m^2 u^{-2+2m} v^{2n} + n^2 u^{2m} v^{-2+2n})^2}\right\}$$

We see from this formula that the graph of the function $u^m v^n$ has nonpositive Gaussian curvature at all points, provided $n + m > 1$. Let us plot $u^2 v^4$ and its Gaussian curvature.

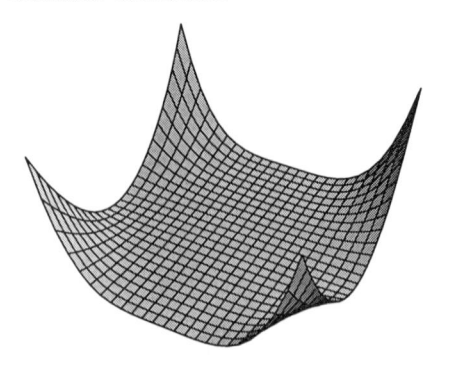

The surface $(u,v) \longmapsto (u,v,u^2v^4)$

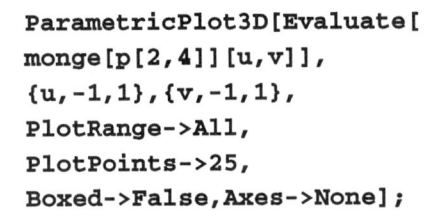

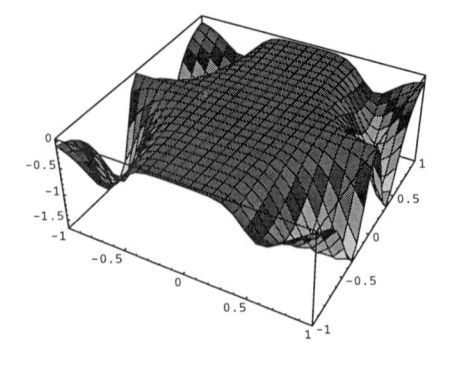

The Gaussian curvature of $(u,v) \longmapsto (u,v,u^2v^4)$

```
ParametricPlot3D[Evaluate[
monge[p[2,4]][u,v]],
{u,-1,1},{v,-1,1},
PlotRange->All,
PlotPoints->25,
Boxed->False,Axes->None];
```

```
Plot3D[Evaluate[
gcurvature[
monge[p[2,4]]][u,v]],
{u,-1,1},{v,-1,1},
PlotPoints->25,
ColorFunction->Hue];
```

Indeed the graph shows that the Gaussian curvature of u^2v^4 is nonpositive; it is also more complicated than we might have expected before seeing the right-hand graph.[2] However, the left-hand graph does not show very clearly that all points are either hyperbolic or planar. To illustrate this fact we redo the plot of u^2v^4 in polar coordinates using the command

---

[2] The useful option **ColorFunction->Hue** is valid for **Plot3D** and **ContourPlot**, but not **ParametricPlot3D**. Height is used for the coloration.

17.2 Examples of Curvature Calculations with Mathematica

```
ParametricPlot3D[Evaluate[
monge[p[2,4]][r Cos[theta],r Sin[theta]]],
{r,0,Sqrt[2]},{theta,0,2Pi},
PlotRange->All,PlotPoints->{20,60},
Boxed->False,Axes->None];
```

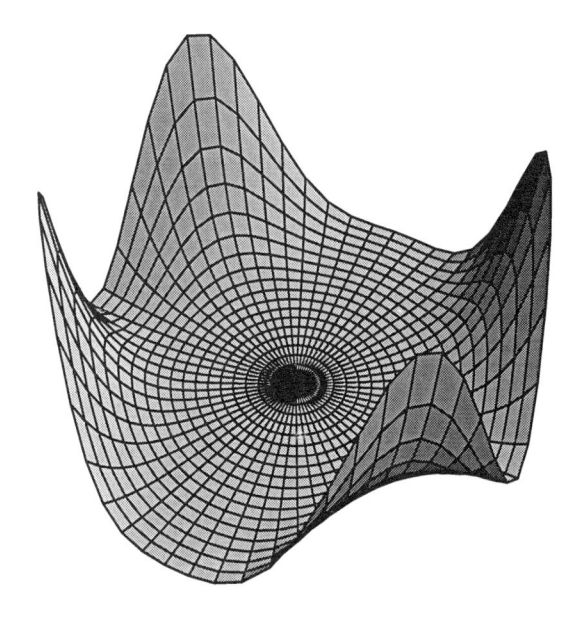

The surface $(u,v) \longmapsto (u,v,u^2v^4)$
in polar coordinates

The Whitney Umbrella

We compute and plot the Gaussian curvature and the mean curvature of the Whitney umbrella, whose picture is on page 310.

```
gcurvature[whitneyumbrella][u,v]
```

$$\frac{-4v^2}{(u^2 + 4v^2 + 4v^4)^2}$$

mcurvature[whitneyumbrella][u,v]

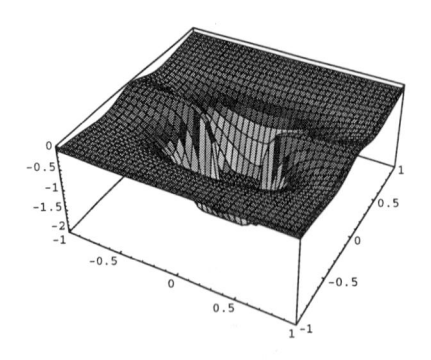

$$-\left(\frac{u\,(1+3\,v^2)}{(u^2+4\,v^2+4\,v^4)^{3/2}}\right)$$

The Gaussian curvature of the Whitney umbrella

$(u,v) \longmapsto (u\,v, u, v^2)$

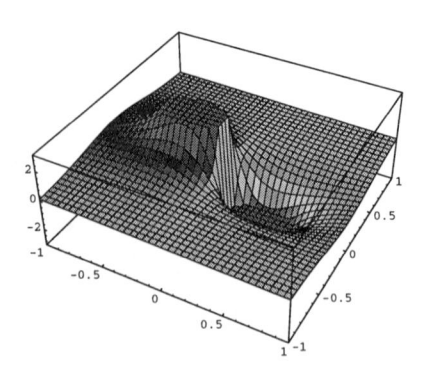

The mean curvature of the Whitney umbrella

$(u,v) \longmapsto (u\,v, u, v^2)$

For example, the first plot can be made with the command

```
Plot3D[gcurvature[whitneyumbrella][u,v]//Evaluate,
{u,-1,1},{v,-1,1},PlotPoints->{40,40},
ColorFunction->Hue];
```

17.3 Principal Curvatures via Mathematica

The *Mathematica* version of formulas (16.26), page 378, for the principal curvatures k_1 and k_2 is:

```
k1[x_][u_,v_]:= mcurvature[x][u,v] +
    Sqrt[Simplify[mcurvature[x][u,v]^2 -
    gcurvature[x][u,v]]]
```

```
k2[x_][u_,v_] := mcurvature[x][u,v] -
    Sqrt[Simplify[mcurvature[x][u,v]^2 -
        gcurvature[x][u,v]]]
```

Because the definitions of k_1 and k_2 (formula (16.26)) involve square roots, the functions k_1 and k_2 need not be differentiable. For example, let us compute and plot the principal curvatures of a monkey saddle. The formulas for k_1 and k_2 of $(u,v) \longmapsto \text{monkeysaddle}(u,v)$ can be computed with the commands

    k1[monkeysaddle][u,v]   and   **k2[monkeysaddle][u,v]**.

Mathematica finds the formulas easily enough, but we suppress the mildly complicated output. Instead we use *Mathematica* to plot the principal curvatures in a neighborhood of the origin, which is a planar point.

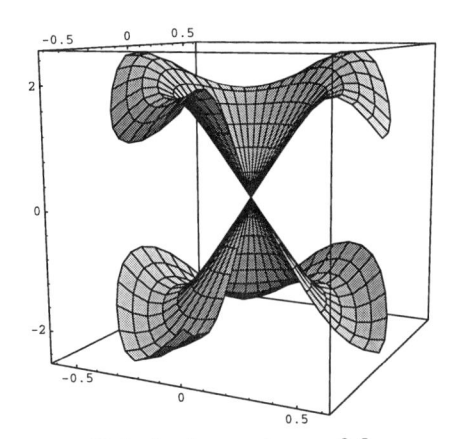

```
ParametricPlot3D[Evaluate[{
    {r Cos[theta],r Sin[theta],
    k1[monkeysaddle][
    r Cos[theta],r Sin[theta]]},
    {r Cos[theta],r Sin[theta],
    k2[monkeysaddle][
    r Cos[theta],r Sin[theta]]}}],
    {r,0,0.6},{theta,0,3Pi/2},
    PlotPoints->{10,30},
    BoxRatios->{1,1,1},
    ViewPoint->{1.5,-3.0, 0.6}]
```

Principal curvatures of the monkey saddle near its planar point

17.4 *The Gauss Map via* Mathematica

Frequently, the image of the Gauss map of a surface has many self-intersections. For this reason, a variant of **ParametricPlot3D** which draws only the edges of the approximating polygons and not their interiors is frequently useful:

```
wireframe[surface_,{u_,u0_,u1_},{v_,v0_,v1_},opts___]:=
    Module[{plottmp,grtmp},
        plottmp = ParametricPlot3D[
        surface//Evaluate,{u,u0,u1},{v,v0,v1},
        DisplayFunction->Identity,opts];
        grtmp = plottmp /.(Polygon[pts_]:>
        Line[Append[pts,First[pts]]]);
    Show[grtmp,DisplayFunction->$DisplayFunction]]
```

The command **wireframe** uses `ParametricPlot3D` to compute the necessary information for the plot, but the option `DisplayFunction->Identity` suppresses the output. Next,

   `Polygon[pts_]:> Line[Append[pts,First[pts]]]`

deletes the interior of each face. Then **Show** with the option `DisplayFunction->$DisplayFunction` outputs the display.

We describe the Gauss map of several surfaces, using **wireframe** or `ParametricPlot3D`, whichever is more appropriate.

The Gauss Map of a Hyperboloid of One Sheet

The hyperboloid of one sheet is defined by

$$\text{hyperboloid}[a, b, c](u, v) = (a \cosh v \cos u, b \cosh v \sin u, c \sinh v).$$

or

```
hyperboloid[a_,b_,c_][u_,v_]:=
    {a Cosh[v]Cos[u],b Cosh[v]Sin[u],c Sinh[v]}
```

The nonparametric equation that defines the hyperboloid of one sheet is

$$\frac{x^2}{a^2} + \frac{y^2}{b^2} - \frac{z^2}{c^2} = 1.$$

To plot the image under the Gauss map of the equatorial region

$$\left\{ \text{hyperboloid}[1, 1, 1](u, v) \mid 0 \leq u \leq 2\pi, -1 \leq v \leq 1 \right\}$$

of **hyperboloid**[1, 1, 1], we use

17.4 The Gauss Map via Mathematica

```
wireframe[unitnormal[hyperboloid[1,1,1]][u,v],
{u,0,2Pi},{v,-1,1},Axes->None,Boxed->False,
ViewPoint->{1.0,-1.8,2.7}];
```

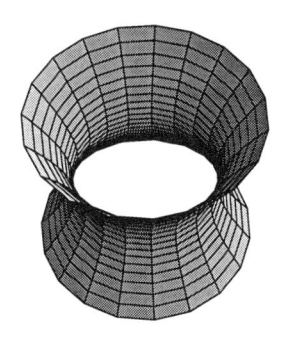

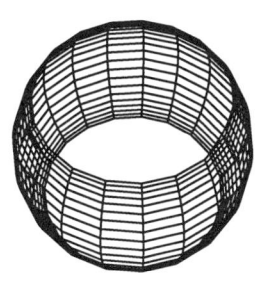

An equatorial region of the hyperboloid of one sheet
$$x^2 + y^2 - z^2 = 1$$
and its image under the Gauss map

The image of the whole hyperboloid of one sheet is also an equatorial region, because the normals to the surface approach those of an asymptotic cone, as the following picture shows:

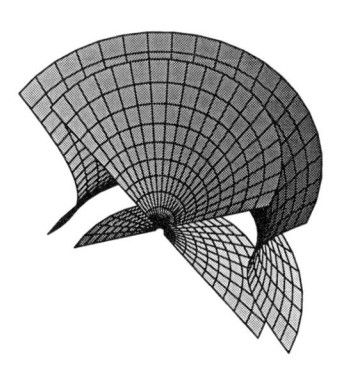

The hyperboloid of one sheet
$$x^2 + y^2 - z^2 = 1$$
and its asymptotic cone

The Gauss Map of a Hyperboloid of Two Sheets

The hyperboloid of two sheets is defined by

$$\mathbf{hy2sheet}[a, b, c](u, v) = (a \cosh v \cosh u, b \cosh v \sinh u, c \sinh v),$$

or

```
hy2sheet[a_,b_,c_][u_,v_]:=
    {a Cosh[u]Cosh[v],b Sinh[u]Cosh[v],c Sinh[v]}
```

The nonparametric equation that defines the hyperboloid of two sheets is

$$\frac{x^2}{a^2} - \frac{y^2}{b^2} - \frac{z^2}{c^2} = 1.$$

To plot the image under the Gauss map of the region

$$\left\{ \mathbf{hy2sheet}[1, 1, 1](u, v) \mid -1 \leq u, v \leq 1 \right\}$$

of hy2sheet[1, 1, 1], we use

```
wireframe[
{{-0.4,0,0} + unitnormal[hy2sheet[1,1,1]][u,v],
{0.4,0,0}-unitnormal[hy2sheet[-1,1,1]][u,v]}//Evaluate,
{u,-1,1},{v,-1,1},Axes->None,Boxed->False,
ViewPoint->{0.4,-3.0,1.2}];
```

to get

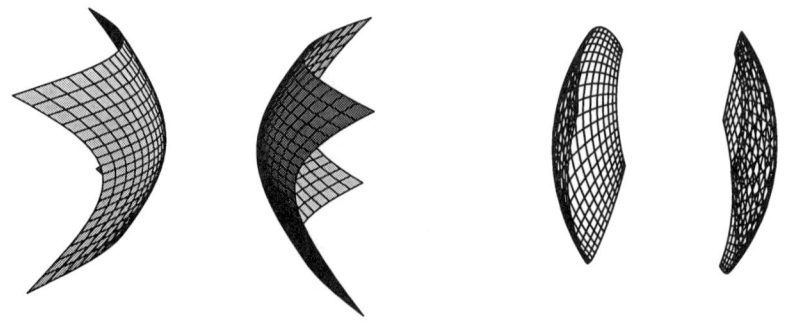

A hyperboloid of two sheets $x^2 - y^2 - z^2 = 1$
and its image under the Gauss map

The image of the whole hyperboloid of two sheets under the Gauss map consists of two antipodal disks.

The Gauss Map of an Ellipsoid

There must be some distortion when the Gauss map maps an ellipsoid onto a sphere because the two surfaces are not isometric.

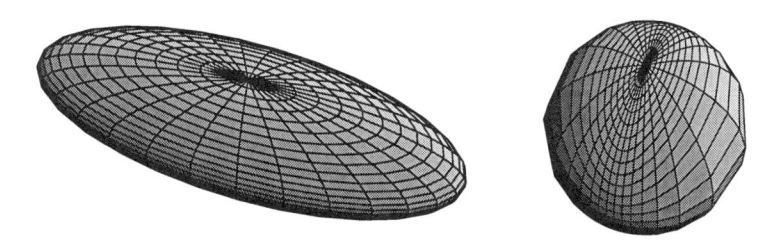

The ellipsoid $\quad \dfrac{x^2}{9} + \dfrac{y^2}{4} + z^2 = 1 \quad$ **and its image under the Gauss map**

The Gauss Map of a Möbius strip

The nonorientability of the Möbius strip means that the Gauss map is not well-defined on the whole surface. However, the Gauss map is defined on any orientable portion, for example, on a Möbius strip minus a line orthogonal to the middle circle. So we get a formula for the Gauss map, namely

```
unitnormal[moebiusstrip[1]][u,v]//Simplify
```

$$\left\{ \frac{(-(v \cos[\frac{u}{2}]) + 2 \cos[u] + v \cos[\frac{3u}{2}]) \sin[\frac{u}{2}]}{\sqrt{4 + 3 v^2 + 8 v \cos[\frac{u}{2}] + 2 v^2 \cos[u]}}, \right.$$

$$\frac{v + 2 \cos[\frac{u}{2}] + 2 v \cos[u] - 2 \cos[\frac{3u}{2}] - v \cos[2 u]}{2 \sqrt{4 + 3 v^2 + 8 v \cos[\frac{u}{2}] + 2 v^2 \cos[u]}},$$

$$\left. -\left(\frac{\cos[\frac{u}{2}] (1 + v \cos[\frac{u}{2}])}{\sqrt{1 + \frac{3 v^2}{4} + 2 v \cos[\frac{u}{2}] + \frac{v^2 \cos[u]}{2}}} \right) \right\}$$

We can use this formula to see what happens when we go around the Möbius strip twice:

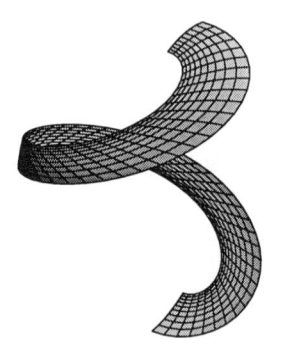

The image under the Gauss map of a Möbius strip traversed once

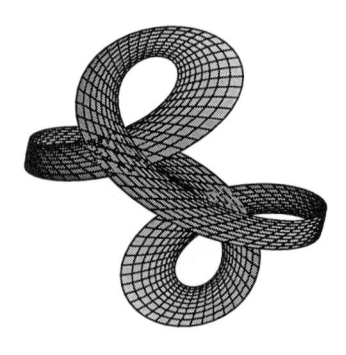

The image under the Gauss map of a Möbius strip traversed twice

The Gauss Map of a Monkey Saddle

We plot a monkey saddle and its image under the Gauss map. The image of the monkey saddle's planar point under the Gauss map is easy to find.

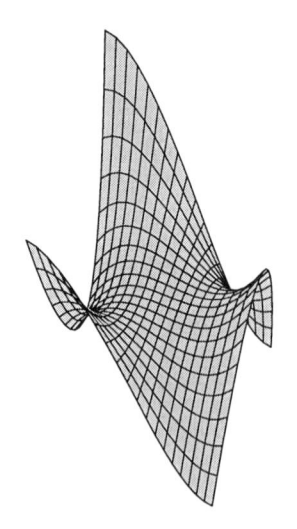

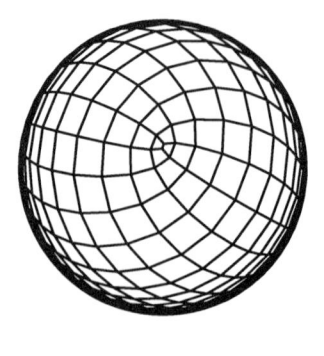

The monkey saddle $(u,v) \longmapsto (u, v, \mathfrak{Re}((u+iv)^3))$
and its image under the Gauss map

17.5 The Curvature of Nonparametrically Defined Surfaces

So far we have discussed computing the curvature of a surface from its parametric representation. In this section we show how in some cases the curvature can be computed from the nonparametric form of a surface.

Lemma 17.2. *Let $\mathbf{p}$ be a point on a regular surface $\mathcal{M} \subset \mathbb{R}^3$, and let $\mathbf{v}_\mathbf{p}$ and $\mathbf{w}_\mathbf{p}$ be tangent vectors to $\mathcal{M}$ at $\mathbf{p}$. Then the Gaussian and mean curvatures of $\mathcal{M}$ at $\mathbf{p}$ are related to the shape operator by the formulas*

(17.1) $$S(\mathbf{v}_\mathbf{p}) \times S(\mathbf{w}_\mathbf{p}) = K(\mathbf{p})\mathbf{v}_\mathbf{p} \times \mathbf{w}_\mathbf{p},$$

(17.2) $$S(\mathbf{v}_\mathbf{p}) \times \mathbf{w}_\mathbf{p} + \mathbf{v}_\mathbf{p} \times S(\mathbf{w}_\mathbf{p}) = 2H(\mathbf{p})\mathbf{v}_\mathbf{p} \times \mathbf{w}_\mathbf{p}.$$

Proof. First, assume that $\mathbf{v}_\mathbf{p}$ and $\mathbf{w}_\mathbf{p}$ are linearly independent. Then we can write

$$S(\mathbf{v}_\mathbf{p}) = a\mathbf{v}_\mathbf{p} + b\mathbf{w}_\mathbf{p} \quad \text{and} \quad S(\mathbf{w}_\mathbf{p}) = c\mathbf{v}_\mathbf{p} + d\mathbf{w}_\mathbf{p},$$

so that

$$\begin{pmatrix} a & b \\ c & d \end{pmatrix}$$

is the matrix of S with respect to $\mathbf{v}_\mathbf{p}$ and $\mathbf{w}_\mathbf{p}$. It follows from (16.23), page 374, that

$$\begin{aligned} S(\mathbf{v}_\mathbf{p}) \times \mathbf{w}_\mathbf{p} + \mathbf{v}_\mathbf{p} \times S(\mathbf{w}_\mathbf{p}) &= (a\mathbf{v}_\mathbf{p} + b\mathbf{w}_\mathbf{p}) \times \mathbf{w}_\mathbf{p} + \mathbf{v}_\mathbf{p} \times (c\mathbf{v}_\mathbf{p} + d\mathbf{w}_\mathbf{p}) \\ &= (a+d)\mathbf{v}_\mathbf{p} \times \mathbf{w}_\mathbf{p} \\ &= \operatorname{tr} S(\mathbf{p})\mathbf{v}_\mathbf{p} \times \mathbf{w}_\mathbf{p} = 2H(\mathbf{p})\mathbf{v}_\mathbf{p} \times \mathbf{w}_\mathbf{p}, \end{aligned}$$

proving (17.2) in the case that $\mathbf{v}_\mathbf{p}$ and $\mathbf{w}_\mathbf{p}$ are linearly independent.

If $\mathbf{v}_\mathbf{p}$ and $\mathbf{w}_\mathbf{p}$ are linearly dependent, they are still the limits of linearly independent tangent vectors. Since both sides of (17.2) are continuous in $\mathbf{v}_\mathbf{p}$ and $\mathbf{w}_\mathbf{p}$, we get (17.2) in the general case.

Equation (17.1) is proved by the same method (see exercise 8). ∎

Theorem 17.3. *Let $\mathbf{Z}$ be a nonvanishing vector field on a regular surface $\mathcal{M} \subset \mathbb{R}^3$ which is everywhere perpendicular to $\mathcal{M}$. Let V and W be vector fields tangent to $\mathcal{M}$ such that $V \times W = \mathbf{Z}$. Then*

$$(17.3) \qquad K = \frac{\mathbf{Z} \cdot (D_V \mathbf{Z} \times D_W \mathbf{Z})}{\|\mathbf{Z}\|^4},$$

$$(17.4) \qquad H = \frac{-\mathbf{Z} \cdot (D_V \mathbf{Z} \times W + V \times D_W \mathbf{Z})}{2\|\mathbf{Z}\|^3}.$$

Proof. Let $\mathbf{U} = \mathbf{Z}/\|\mathbf{Z}\|$; then (11.5), page 249, implies that

$$D_V \mathbf{U} = \frac{D_V \mathbf{Z}}{\|\mathbf{Z}\|} + V\left[\frac{1}{\|\mathbf{Z}\|}\right] \mathbf{Z}.$$

Therefore,

$$S(V) = -D_V \mathbf{U} = \frac{-D_V \mathbf{Z}}{\|\mathbf{Z}\|} + N_V,$$

where N_V is a vector field normal to $\mathcal{M}$. Hence by Lemma 17.2 we have

$$(17.5) \qquad KV \times W = S(V) \times S(W)$$

$$= \left(\frac{-D_V \mathbf{Z}}{\|\mathbf{Z}\|} + N_V\right) \times \left(\frac{-D_W \mathbf{Z}}{\|\mathbf{Z}\|} + N_W\right).$$

Since N_V and N_W are linearly dependent, it follows from (17.5) that

$$(17.6) \qquad K\mathbf{Z} = \frac{D_V \mathbf{Z} \times D_W \mathbf{Z}}{\|\mathbf{Z}\|^2} + \text{some vector field tangent to } \mathcal{M}.$$

We take the scalar product of both sides of (17.6) with $\mathbf{Z}$ and obtain (17.3). Equation (17.4) is proved in a similar fashion. ∎

In order to make use of Theorem 17.3, we need an important function that measures the distance from the origin of each tangent plane to a surface.

Definition. *Let $\mathcal{M}$ be an oriented regular surface in $\mathbb{R}^3$ with surface normal $\mathbf{U}$. Then the* **support function** *of $\mathcal{M}$ is the function $\mathbf{h}: \mathcal{M} \longrightarrow \mathbb{R}$ given by*

$$\mathbf{h}(\mathbf{p}) = \mathbf{p} \cdot \mathbf{U}(\mathbf{p}).$$

Geometrically, $\mathbf{h}(\mathbf{p})$ is the distance from the origin to the tangent space $\mathcal{M}_\mathbf{p}$.

17.5 The Curvature of Nonparametrically Defined Surfaces

Corollary 17.4. Let $\mathcal{M}$ be the surface

$$\{(p_1, p_2, p_3) \in \mathbb{R}^3 \mid f_1 p_1^k + f_2 p_2^k + f_3 p_3^k = 1\},$$

where f_1, f_2, f_3 are constants, not all zero, and k is a nonzero real number. Then the support function, Gaussian curvature and mean curvature of $\mathcal{M}$ are given by

$$(17.7) \quad \mathbf{h} = \frac{1}{(f_1^2 u_1^{2k-2} + f_2^2 u_2^{2k-2} + f_3^2 u_3^{2k-2})^{1/2}},$$

$$(17.8) \quad K = \frac{(k-1)^2 f_1 f_2 f_3 (u_1 u_2 u_3)^{k-2}}{\left(\sum_{i=1}^{3} f_i^2 u_i^{2k-2}\right)^2} = \mathbf{h}^4 (k-1)^2 f_1 f_2 f_3 (u_1 u_2 u_3)^{k-2},$$

$$(17.9) \quad H = \frac{k-1}{2\left(\sum_{i=1}^{3} f_i^2 u_i^{2k-2}\right)^{\frac{3}{2}}} \Big(f_1 f_2 (u_1 u_2)^{k-2}(f_1 u_1^k + f_2 u_2^k)$$

$$+ f_2 f_3 (u_2 u_3)^{k-2}(f_2 u_2^k + f_3 u_3^k) + f_3 f_1 (u_3 u_1)^{k-2}(f_3 u_3^k + f_1 u_1^k)\Big).$$

Proof. Let $g(p_1, p_2, p_3) = f_1 p_1^k + f_2 p_2^k + f_3 p_3^k - 1$, so that

$$\mathcal{M} = \{\mathbf{p} \in \mathbb{R}^3 \mid g(\mathbf{p}) = 0\}.$$

Then $\mathbf{Z} = \mathbf{grad}\, g$ is a nonvanishing vector field which is everywhere perpendicular to $\mathcal{M}$; explicitly, $\mathbf{Z}$ is given by

$$\mathbf{Z} = k \sum_{i=1}^{3} f_i u_i^{k-1} U_i,$$

where u_1, u_2, u_3 are the natural coordinate functions of $\mathbb{R}^3$. Let $\mathbf{X}$ be the vector field defined by

$$\mathbf{X} = \sum_{i=1}^{3} u_i U_i.$$

Then

$$(17.10) \quad \mathbf{X} \cdot \mathbf{Z} = k \sum_{i=1}^{3} f_i u_i^k,$$

so that the support function **h** of $\mathcal{M}$ is given by

$$\mathbf{h} = \mathbf{X} \cdot \frac{\mathbf{Z}}{\|\mathbf{Z}\|} = \frac{\sum_{i=1}^{3} f_i u_i^k}{\sqrt{\sum_{i=1}^{3} f_i^2 u_i^{2k-2}}}.$$

Since $\sum f_i u_i^k$ equals 1 on $\mathcal{M}$, we get (17.7).

Next, let

$$V = \sum_{i=1}^{3} v_i U_i \quad \text{and} \quad W = \sum_{i=1}^{3} w_i U_i$$

be vector fields on $\mathbb{R}^3$. Since f_1, f_2, f_3 are constants, we have

$$\mathbf{D}_V \mathbf{Z} = k \sum_{i=1}^{3} V[f_i u_i^{k-1}] U_i = k(k-1) \sum_{i=1}^{3} f_i v_i u_i^{k-2} U_i.$$

and similarly for W. Therefore,

(17.11) $\quad \mathbf{Z} \cdot \mathbf{D}_V \mathbf{Z} \times \mathbf{D}_W \mathbf{Z}$

$$= \det \begin{pmatrix} k f_1 u_1^{k-1} & k f_2 u_2^{k-1} & k f_3 u_3^{k-1} \\ k(k-1) f_1 v_1 u_1^{k-2} & k(k-1) f_2 v_2 u_2^{k-2} & k(k-1) f_3 v_3 u_3^{k-2} \\ k(k-1) f_1 w_1 u_1^{k-2} & k(k-1) f_2 w_2 u_2^{k-2} & k(k-1) f_3 w_3 u_3^{k-2} \end{pmatrix}$$

$$= k^2 (k-1)^2 f_1 f_2 f_3 u_1^{k-2} u_2^{k-2} u_3^{k-2} \mathbf{X} \cdot V \times W.$$

Now we choose V and W so that they are tangent to $\mathcal{M}$ and $V \times W = \mathbf{Z}$. Then (17.11) and (17.3) imply

$$K = \frac{k^2(k-1)^2 f_1 f_2 f_3 u_1^{k-2} u_2^{k-2} u_3^{k-2} \mathbf{X} \cdot \mathbf{Z}}{\|\mathbf{Z}\|^4} = \frac{(k-1)^2 f_1 f_2 f_3 u_1^{k-2} u_2^{k-2} u_3^{k-2}}{\left(\sum_{i=1}^{3} f_i^2 u_i^{2k-2}\right)^2}.$$

This proves (17.8). The proof of (17.9) is similar. ∎

Let us translate the formulas just obtained into **Mathematica**. We use (17.8) and (17.9) to define new **Mathematica** versions of the Gaussian and mean curvature functions.

17.5 The Curvature of Nonparametrically Defined Surfaces

```
gcursq[k_][f1_,f2_,f3_][u1_,u2_,u3_]:=
    (k - 1)^2 f1 f2 f3 (u1 u2 u3)^(k - 2)/
    (f1^2 u1^(2k - 2) +  f2^2 u2^(2k - 2)  +
                    f3^2 u3^(2k - 2))^2
```

```
hcursq[k_][f1_,f2_,f3_][u1_,u2_,u3_]:=
    (k-1)( f1 f2 (u1 u2)^(k - 2)(f1 u1^k + f2 u2^k) +
           f2 f3 (u2 u3)^(k - 2)(f2 u2^k + f3 u3^k) +
           f3 f1 (u3 u1)^(k - 2)(f3 u3^k + f1 u1^k))/
    (2(f1^2 u1^(2k - 2) +  f2^2 u2^(2k-2)  +
                    f3^2 u3^(2k - 2))^(3/2))
```

For example, **gcursq[2][1/a^2,1/b^2,-1/c^2][u1,u2,u3]** gives

$$-\left(\frac{1}{a^2 b^2 c^2 \left(\frac{u1^2}{a^4} + \frac{u2^2}{b^4} + \frac{u3^2}{c^4}\right)^2}\right)$$

and

hcursq[2][1/a^2,1/b^2,-1/c^2][u1,u2,u3]//Simplify//PowerExpand

gives

$$\frac{\dfrac{u1^2}{a^4 b^2} - \dfrac{u1^2}{a^4 c^2} + \dfrac{u2^2}{a^2 b^4} - \dfrac{u2^2}{b^4 c^2} + \dfrac{u3^2}{a^2 c^4} + \dfrac{u3^2}{b^2 c^4}}{2 \left(\dfrac{u1^2}{a^4} + \dfrac{u2^2}{b^4} + \dfrac{u3^2}{c^4}\right)^{3/2}}$$

Similar calculations with *Mathematica* yield three important special cases of Corollary 17.4.

Corollary 17.5. *The support function and Gaussian curvature of the ellipsoid*

$$\frac{x^2}{a^2} + \frac{y^2}{b^2} + \frac{z^2}{c^2} = 1$$

are given by

$$\begin{cases} \mathbf{h} = \dfrac{1}{\left(\dfrac{x^2}{a^4} + \dfrac{y^2}{b^4} + \dfrac{z^2}{c^4}\right)^{1/2}}, \\ K = \dfrac{\mathbf{h}^4}{a^2 b^2 c^2} > 0. \end{cases}$$

Corollary 17.6. *The support function and Gaussian curvature of the hyperboloid of one sheet*

$$\frac{x^2}{a^2} + \frac{y^2}{b^2} - \frac{z^2}{c^2} = 1$$

are given by

$$\begin{cases} \mathbf{h} = \dfrac{1}{\left(\dfrac{x^2}{a^4} + \dfrac{y^2}{b^4} + \dfrac{z^2}{c^4}\right)^{1/2}}, \\ K = -\dfrac{\mathbf{h}^4}{a^2 b^2 c^2} < 0. \end{cases}$$

Corollary 17.7. *The support function and Gaussian curvature of the hyperboloid of two sheets*

$$\frac{x^2}{a^2} - \frac{y^2}{b^2} - \frac{z^2}{c^2} = 1$$

are given by

$$\begin{cases} \mathbf{h} = \dfrac{1}{\left(\dfrac{x^2}{a^4} + \dfrac{y^2}{b^4} + \dfrac{z^2}{c^4}\right)^{1/2}}, \\ K = \dfrac{\mathbf{h}^4}{a^2 b^2 c^2} > 0. \end{cases}$$

We shall return to these quadratic surfaces in Section 29.1, where we describe geometrically useful parametrizations. We conclude the present section by computing the Gaussian curvature of a **super quadric**, that is a surface of the form

$$f_1 x^k + f_2 y^k + f_3 z^k = 1.$$

We do the special case $k = 2/3$, which is an **astroidial ellipsoid**. (See page 396.)

```
gcursq[2/3][f1,f2,f3][u1,u2,u3]//Together//PowerExpand
```

$$\frac{f1\ f2\ f3}{9\ (f3^2\ u1^{2/3}\ u2^{2/3} + f2^2\ u1^{2/3}\ u3^{2/3} + f1^2\ u2^{2/3}\ u3^{2/3})^2}$$

Corollary 17.8. *The Gaussian curvature of the super quadric*

$$f_1 x^{2/3} + f_2 y^{2/3} + f_3 z^{2/3} = 1$$

is given by

$$K = \frac{f_1 f_2 f_3}{9\left(f_1^2(xy)^{2/3} + f_2^2(yz)^{2/3} + f_2^2(zx)^{2/3}\right)^2}.$$

17.6 Exercises

Use *Mathematica* in the following exercises.

1. Use *Mathematica* to find the coefficients the first fundamental form, to find the coefficients the second fundamental form, the unit normal, the mean curvature and the principal curvatures of the following surfaces:

    a. the torus defined on page 356. Plot the Gaussian and mean curvatures.

    b. the helicoid defined on page 357. Plot the Gaussian curvature.

    c. Enneper's minimal surface defined on page 358. Plot the Gaussian curvature using **Plot3D** and **ContourPlot**

    d. a Monge patch defined on page 398.

2. Plot the graph of $u^2 v^3$ and its Gaussian curvature.

3. Find the formulas for the coefficients of the second fundamental forms of the following surfaces: hyperbolic paraboloid, eight surface, elliptic paraboloid, ellipsoid, Möbius strip, cross cap, Klein bottle.

4. Find the formulas for the Gaussian and mean curvatures of the following surfaces: hyperbolic paraboloid, eight surface, elliptic paraboloid, ellipsoid, Möbius strip, pseudocrosscap, Klein bottle.

5. Draw the picture on page 405 of the hyperboloid of one sheet with its asymptotic cone.

6. Plot the principal curvatures of the Whitney umbrella.

7. The *Mathematica* command to create a translation surface (see exercise 10 of Chapter 16) from curves $\alpha, \gamma \colon (a, b) \longrightarrow \mathbb{R}^n$ is

```
transsurf[alpha_,gamma_][u_,v_] := alpha[u] + gamma[v]
```

Plot the translation surface formed by a circle and a lemniscate lying in perpendicular planes.

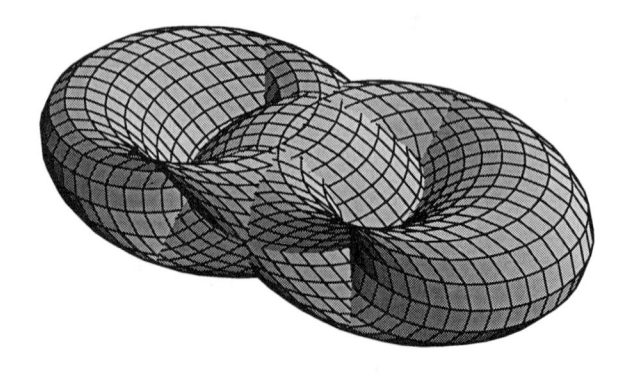

Translation surface formed by moving a lemniscate around a circle

8. Prove equation (17.1).

18
ASYMPTOTIC CURVES ON SURFACES

In this chapter we begin our study of special curves on surfaces in $\mathbb{R}^3$. An *asymptotic curve* on a surface $\mathcal{M} \subset \mathbb{R}^3$ is a curve whose velocity vector always points in a direction in which the normal curvature of $\mathcal{M}$ vanishes. In some sense $\mathcal{M}$ bends less along an asymptotic curve than it does along a general curve. For example, the straight lines $v \longmapsto (\cos u, \sin u, v)$ are asymptotic curves on the cylinder $(u,v) \longmapsto (\cos u, \sin u, v)$. It is true (Corollary 18.6) that any straight line contained in a surface is an asymptotic curve. However, curves other than straight lines can also be asymptotic curves. In Section 18.1 we derive the differential equation that must be satisfied in order that a curve be asymptotic. We also prove Theorem 18.7, which relates the Gaussian curvature of $\mathcal{M}$ to the torsion of any asymptotic curve in $\mathcal{M}$.

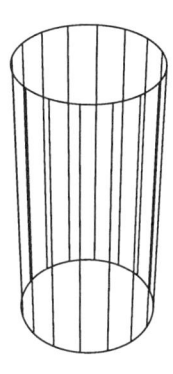

In Section 18.2 we give examples of surfaces with interesting asymptotic curves. A *Mathematica* program for finding asymptotic curves on surfaces is described in Section 18.3.

Since we shall be doing mostly local calculations in this chapter, we shall assume that all surfaces are orientable, unless explicitly stated otherwise. This means that we can choose once and for all a globally defined unit normal **U** for any surface $\mathcal{M}$.

18.1 Asymptotic Curves

Let $\mathcal{M}$ be a regular surface in $\mathbb{R}^3$. On page 364 we defined an asymptotic direction at a point $\mathbf{p} \in \mathcal{M}$ to be a direction in which the normal curvature of $\mathcal{M}$ vanishes. The following lemma is obvious from the definitions.

Lemma 18.1. *Let $\mathcal{M} \subset \mathbb{R}^3$ be a regular surface.*

(i) *At an elliptic point of $\mathcal{M}$ there are no asymptotic directions.*

(ii) *At a hyperbolic point of $\mathcal{M}$ there are exactly two asymptotic directions.*

(iii) *At a parabolic point of $\mathcal{M}$ there is exactly one asymptotic direction.*

(iv) *At a planar point of $\mathcal{M}$ every direction is asymptotic.*

Hyperbolic points are considered in more detail in Theorem 18.4 below.

Definition. *An **asymptotic curve** is a curve $\boldsymbol{\alpha}$ in $\mathcal{M}$ for which the normal curvature vanishes in the direction $\boldsymbol{\alpha}'$; that is,*

$$\mathbf{k}(\boldsymbol{\alpha}'(t)) = 0$$

for all t in the domain of definition of $\boldsymbol{\alpha}$.

Here is an alternate description of an asymptotic curve.

Lemma 18.2. *A curve $\boldsymbol{\alpha}$ in a regular surface $\mathcal{M} \subset \mathbb{R}^3$ is asymptotic if and only if its acceleration is always tangent to $\mathcal{M}$.*

Proof. Without loss of generality, we can assume that $\boldsymbol{\alpha}$ is a unit-speed curve. If **U** denotes the surface normal to $\mathcal{M}$, then differentiation of $\boldsymbol{\alpha}' \cdot \mathbf{U} = 0$ yields

$$0 = \boldsymbol{\alpha}' \cdot \mathbf{U}' + \boldsymbol{\alpha}'' \cdot \mathbf{U} = -\mathbf{k}(\boldsymbol{\alpha}') + \boldsymbol{\alpha}'' \cdot \mathbf{U}.$$

Hence $\mathbf{k}(\boldsymbol{\alpha}')$ vanishes if and only if $\boldsymbol{\alpha}''$ is perpendicular to **U**. ∎

Next, we derive the differential equation for the asymptotic curves.

18.1 Asymptotic Curves

Lemma 18.3. *Let α be a curve that lies in the image of a patch $\mathbf{x}$. Write $\alpha(t) = \mathbf{x}(u(t), v(t))$. Then α is an asymptotic curve if and only if one of the following equivalent conditions is satisfied for all t:*

(i)
$$e(\alpha(t))u'(t)^2 + 2f(\alpha(t))u'(t)v'(t) + g(\alpha(t))v'(t)^2 = 0; \tag{18.1}$$

(ii) $\mathbf{II}(\alpha'(t), \alpha'(t)) = 0$, *where $\mathbf{II}$ is defined on page* 380.

Proof. Corollary 12.9, page 277, implies that $\alpha' = \mathbf{x}_u u' + \mathbf{x}_v v'$. Hence by Lemma 16.11, page 370, we have

$$\mathbf{k}(\alpha'(t)) = \frac{e(\alpha(t))u'(t)^2 + 2f(\alpha(t))u'(t)v'(t) + g(\alpha(t))v'(t)^2}{E(\alpha(t))u'(t)^2 + 2F(\alpha(t))u'(t)v'(t) + G(\alpha(t))v'(t)^2}. \tag{18.2}$$

Then (18.1) is clear from (18.2); furthermore, (ii) is a restatement of (i). ∎

Equation (18.1) can be written more succinctly as

$$e\,u'^2 + 2f\,u'v' + g\,v'^2 = 0. \tag{18.3}$$

We call (18.3) the **differential equation for the asymptotic curves** of a surface. Since e, f and g have the same denominator, (18.3) is equivalent to

$$\widetilde{e}u'^2 + 2\widetilde{f}u'v' + \widetilde{g}v'^2 = 0, \tag{18.4}$$

where

$$\widetilde{e} = \det\begin{pmatrix} \mathbf{x}_{uu} \\ \mathbf{x}_u \\ \mathbf{x}_v \end{pmatrix}, \qquad \widetilde{f} = \det\begin{pmatrix} \mathbf{x}_{uv} \\ \mathbf{x}_u \\ \mathbf{x}_v \end{pmatrix}, \qquad \widetilde{g} = \det\begin{pmatrix} \mathbf{x}_{vv} \\ \mathbf{x}_u \\ \mathbf{x}_v \end{pmatrix}.$$

Theorem 18.4. *In a neighborhood of a hyperbolic point $\mathbf{p}$ of a regular surface $\mathcal{M} \subset \mathbb{R}^3$ there exist two distinct families of asymptotic curves.*

Proof. In a neighborhood of a hyperbolic point $\mathbf{p}$ the equation

$$\widetilde{e}a^2 + 2\widetilde{f}ab + \widetilde{g}b^2 = 0$$

has real roots. Hence we have a factorization

$$\widetilde{e}a^2 + 2\widetilde{f}ab + \widetilde{g}b^2 = (A\,a + B\,b)(C\,a + D\,b),$$

where A, B, C, D are real. Thus the differential equation for the asymptotic curves also factors:

(18.5) $$(A\,u' + B\,v')(C\,u' + D\,v') = 0,$$

where now A, B, C, D are real functions. One family consists of the solution curves to $A\,u' + B\,v' = 0$, and the other family consists of the solution curves to $C\,u' + D\,v' = 0$. ∎

Next, we give a characterization of an asymptotic curve in terms of its curvature as a curve in $\mathbb{R}^3$.

Theorem 18.5. *Let α be a regular curve on a regular surface $\mathcal{M} \subset \mathbb{R}^3$, and denote by $\{\mathbf{T}, \mathbf{B}, \mathbf{N}\}$ the Frenet frame and by $\kappa[\alpha]$ the curvature of α. Also, let $\mathbf{U}$ be the surface normal of $\mathcal{M}$. Then α is an asymptotic curve if and only if at every point $\alpha(t)$, either $\kappa[\alpha](t) = 0$ or $\mathbf{N}(t) \cdot \mathbf{U} = 0$.*

Proof. Without loss of generality, we can assume that α has unit-speed. The first Frenet formula (see page 186) in the case of nonzero curvature $\kappa[\alpha]$ is

(18.6) $$\mathbf{T}' = \kappa[\alpha]\mathbf{N}.$$

Notice, however, that (18.6) also makes sense when $\kappa[\alpha]$ vanishes, because then $\mathbf{N}$ can be chosen to be an arbitrary vector field along α. Then by (16.3) we have

$$\mathbf{k}(\alpha') = \alpha'' \cdot \mathbf{U} = \mathbf{T}' \cdot \mathbf{U} = \kappa[\alpha]\mathbf{N} \cdot \mathbf{U}.$$

Hence the conclusion follows. ∎

Corollary 18.6. *A straight line that is contained in a regular surface is always asymptotic.*

Proof. The curvature of a straight line vanishes identically. ∎

Now we can establish an important relation between the torsion of an asymptotic curve and the Gaussian curvature of the surface containing the curve.

Theorem 18.7. (Beltrami[1]-Enneper[2]) *Let α be an asymptotic curve on a reg-*

---

[1] Eugenio Beltrami (1835-1900). Professor at the Universities of Bologna, Pisa, Padua and Rome. He found the first concrete model of non-Euclidean geometry, published in [Beltr]. In this paper he showed how possible contradictions in non-Euclidean geometry would reveal themselves in the Euclidean geometry of surfaces. Beltrami also made known the work of the Jesuit mathematician Saccheri, whose work **Euclides ab Omni Naevo Vindicatus** foreshadowed non-Euclidean geometry.

[2] Alfred Enneper (1830-1885). Professor at the University of Göttingen. Enneper also studied

18.1 Asymptotic Curves

ular surface $\mathcal{M} \subset \mathbb{R}^3$, and assume the curvature $\kappa[\alpha]$ of α does not vanish. Then the torsion $\tau[\alpha]$ of α and the Gaussian curvature K of $\mathcal{M}$ are related (along α) by

(18.7) $$K \circ \alpha = -\tau[\alpha]^2.$$

Proof. Without loss of generality, we can assume that α is a unit-speed curve. Since α is an asymptotic curve with nonvanishing curvature, it follows from Theorem 18.5 that
$$\mathbf{T} \cdot \mathbf{U} = \mathbf{N} \cdot \mathbf{U} = 0.$$
Therefore, $\mathbf{U} = \pm \mathbf{B}$ along α. Then the third Frenet formula of (8.10) implies that

(18.8) $$\tau[\alpha]^2 = \mathbf{B}' \cdot \mathbf{B}' = \frac{d\mathbf{U}}{ds} \cdot \frac{d\mathbf{U}}{ds}.$$

On the other hand, since the second fundamental form $\mathbf{II}$ of $\mathcal{M}$ vanishes along α, equation (16.32) for the third fundamental form reduces to

(18.9) $$0 = (\mathbf{III} + K\,\mathbf{I})(\alpha', \alpha') = \frac{d\mathbf{U}}{ds} \cdot \frac{d\mathbf{U}}{ds} + K.$$

Now (18.7) follows from (18.8) and (18.9). ∎

Corollary 18.8. *An asymptotic curve on a regular surface $\mathcal{M} \subset \mathbb{R}^3$ with constant negative curvature has constant torsion.*

Examples of surfaces in $\mathbb{R}^3$ with constant negative curvature will be given in Sections 20.8 and 21.3.

Next, we consider the problem of building patches from asymptotic curves.

Definition. *An **asymptotic patch** on a regular surface $\mathcal{M} \subset \mathbb{R}^3$ is a patch for which the u- and v-parameter curves are asymptotic curves.*

Theorem 18.9. *Let $\mathbf{x}$ be a patch for which f never vanishes. Then $\mathbf{x}$ is an asymptotic patch if and only if e and g vanish identically.*

Proof. If the curve $u \longmapsto \mathbf{x}(u,v)$ is asymptotic, it follows from Lemma 18.2 that $e = \mathbf{x}_{uu} \cdot \mathbf{U} = 0$. Similarly, if $v \longmapsto \mathbf{x}(u,v)$ is asymptotic, then $g = 0$. Conversely, if e and g vanish identically, the differential equation for the asymptotic curves becomes
$$f\, u'v' = 0.$$
Clearly, $u = u_0$ and $v = v_0$ are solutions of this differential equation. Otherwise said, the u- and v-parameter curves are asymptotic. ∎

---
surfaces of constant negative curvature.

18.2 Examples of Asymptotic Curves

The Hyperbolic Paraboloid

We parametrize the hyperbolic paraboloid $z = xy$ by (13.1); it is easy to compute $e = g = 0$ and $f = 1/\sqrt{1+u^2+v^2}$. Hence by Theorem 18.9 each of the curves $u \longmapsto (u,v,uv)$ and $v \longmapsto (u,v,uv)$ is asymptotic. In fact, these curves are straight lines. Thus $(u,v) \longmapsto (u,v,uv)$ is an asymptotic patch. The polygonal paths that are made by **ParametricPlot3D** of this parametrization are in fact asymptotic curves. This is clearly visible in the picture on page 366.

The Elliptical Helicoid

The helicoid defined on page 357 is a special case of the **elliptical helicoid** defined by
$$\text{helicoid}[a, b, c](u, v) = (a\,v\cos u, b\,v\sin u, c\,u),$$
or

```
helicoid[a_,b_,c_][u_,v_] := {a v Cos[u],b v Sin[u],c u}
```

Like the hyperbolic paraboloid, the elliptical helicoid is an asymptotic patch, but unlike the hyperbolic paraboloid, only the v-parameter curves are straight lines. The u-parameter curves are, of course, elliptical helices. Theorem 18.7 tells us that the Gaussian curvature of the elliptical helicoid, when restricted to one of these elliptical helices, equals the negative of the square of the torsion of the elliptical helix. Let us use *Mathematica* to check this fact.

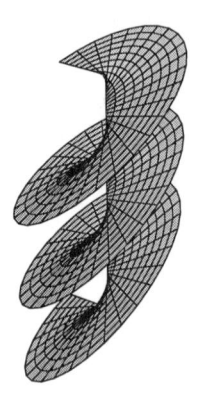

The elliptical helicoid
$(u,v) \longmapsto$
$(v\cos u, 3v\sin u, 0.3u)$

18.2 Examples of Asymptotic Curves

First, we compute the Gaussian curvature of the helicoid:

gcurvature[helicoid[a,b,c]][u,v]

$$\frac{-4 a^2 b^2 c^2}{(a^2 c^2 + b^2 c^2 + 2 a^2 b^2 v^2 + a^2 c^2 \cos[2u] - b^2 c^2 \cos[2u])^2}$$

Next, we compute the negative of the square of the torsion of the elliptical helix $u \longmapsto (av\cos u, bv\sin v, cu)$:

-tau[helix[a v,b v,c]][u]^2

$$\frac{-4 a^2 b^2 c^2}{(a^2 c^2 + b^2 c^2 + 2 a^2 b^2 v^2 + a^2 c^2 \cos[2u] - b^2 c^2 \cos[2u])^2}$$

The two quantities are the same.

The Monkey Saddle

We parametrize the monkey saddle $z = \mathfrak{Re}(x+iy)^3$ by (13.2), page 299. The point $(0,0,0)$ is planar, so that every direction at $(0,0,0)$ is asymptotic. This does not mean, however, that there are asymptotic curves in every direction. But it is easy to find three straight lines passing through $(0,0,0)$ that lie entirely in the surface:

$$v \longmapsto (0, v, 0),$$
$$v \longmapsto (v\sqrt{3}, v, 0),$$
$$v \longmapsto (-v\sqrt{3}, v, 0).$$

Independently, it can be checked that each of these curves satisfies the differential equation for the asymptotic curves, which is

$$u\,u'^2 - 2v\,u'v' - u\,v'^2 = 0.$$

The Funnel Surface

Consider the regular surface which is the image of the patch $\mathbf{x}: (0,\infty) \times [0, 2\pi] \longrightarrow \mathbb{R}^3$ defined by

$$\mathbf{x}(r, \theta) = (r\cos\theta, r\sin\theta, \log r).$$

This is a parametrization of the surface defined by the equation

$$z = \frac{1}{2}\log(x^2 + y^2).$$

We compute

$$e = \frac{-1}{r\sqrt{1+r^2}}, \qquad f = 0, \qquad g = \frac{r}{\sqrt{1+r^2}}.$$

Thus the differential equation for the asymptotic curves becomes

(18.10) $$-\frac{r'^2}{r} + r\theta'^2 = 0, \qquad \text{or} \qquad \theta' = \pm\frac{r'}{r}.$$

The two equations $\theta' = \pm r'/r$ have the solutions

$$\theta + 2u = \log r, \qquad \theta + 2v = -\log r,$$

where u and v are constants of integration. Solving for r and θ in terms of u and v gives

$$\theta = -u - v \qquad \text{and} \qquad r = e^{u-v}.$$

Thus **y** defined by

$$\mathbf{y}(u,v) = \mathbf{x}(r,\theta) = \left(e^{u-v}\cos(u+v), -e^{u-v}\sin(u+v), u-v\right)$$

is an asymptotic patch.

Let us plot the funnel with the two parametrizations that we have found. In *Mathematica* we define

```
funnel[a_,b_,c_][u_,v_] := {a v Cos[u],
                            b v Sin[u], c Log[v]}
```

and

```
funnelasym[c_][p_,q_] := {c E^(p - q)Cos[p + q],
                          -c E^(p - q)Sin[p + q], c(p - q)}
```

so $\mathbf{x}(r,\theta)$ corresponds to **funnel[1,1,1][theta,r]** and $\mathbf{y}(u,v)$ corresponds to **funnelasym[1][u,v]**. Here is the plot of $\mathbf{x}(r,\theta)$.

18.2 Examples of Asymptotic Curves

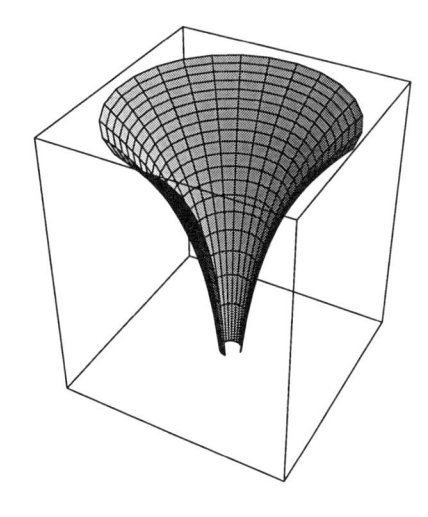

```
ParametricPlot3D[
{r Cos[t],r Sin[t],
Log[r]},
{r,0.1,1},{t,0,3Pi/2},
Axes->None];
```

Plot of
$(r, \theta) \longmapsto (r\cos\theta, r\sin\theta, \log r)$

When we use **ParametricPlot3D** to plot the parametrization of the funnel by asymptotic curves (using **funnelasym**), the result is ugly, because some of the approximating polygons have vertices whose x or y coordinates are large. *Mathematica's* command **Select** can be used to discard unwanted polygons. Here is a miniprogram (contained in **PLTPROGS.m**) that selects from a list of polygons those whose vertices have their coordinates bounded by some number **bound**:

```
selectgraphics3d[graphics3dobj_,bound_,opts___]:=
    Show[Graphics3D[Select[graphics3dobj,
    (Abs[#[[1,1,1]]]<bound && Abs[#[[1,1,2]]]<bound &&
    Abs[#[[1,1,3]]]<bound && Abs[#[[1,2,1]]]<bound &&
    Abs[#[[1,2,2]]]<bound && Abs[#[[1,2,3]]]<bound &&
    Abs[#[[1,3,1]]]<bound && Abs[#[[1,3,2]]]<bound &&
    Abs[#[[1,3,3]]]<bound && Abs[#[[1,4,1]]]<bound &&
    Abs[#[[1,4,2]]]<bound && Abs[#[[1,4,2]]]<bound
    )&]],opts]
```

To use this miniprogram, we first use **ParametricPlot3D** to create the polygons for a plot.

```
fp=ParametricPlot3D[Evaluate[funnelasym[1][u,v]],
    {u,0,Pi},{v,0,Pi}]
```

With this command *Mathematica* made a list **fp** consisting of two elements. The second element of this list is a sublist of graphics options. The first element of **fp** is the sublist of polygons created by **ParametricPlot3D**. To throw away all polygons with a vertex (x_1, x_2, x_3) for which $|x_j| \geq 1.8$ for some j, we use

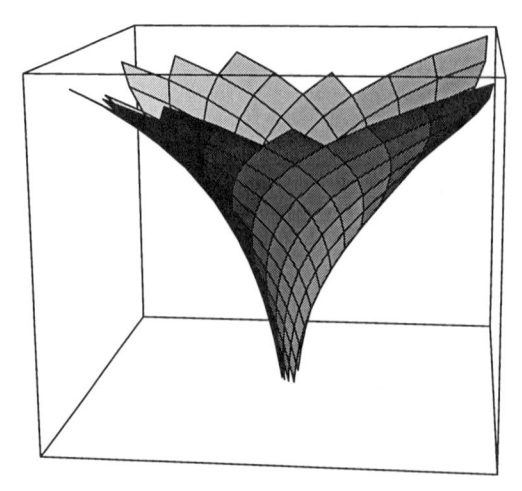

```
selectgraphics3d[
fp[[1]],1.8,
PlotRange->All,
ViewPoint->
{0.384,-3.285,0.713}];
```

A funnel parametrized by asymptotic curves

18.3 Using Mathematica *to Find Asymptotic Curves*

We first define $\widetilde{e}$, $\widetilde{f}$ and $\widetilde{g}$ in *Mathematica* by

```
eeee[x_][u_,v_]:= Simplify[Det[{
    D[x[uu,vv],uu,uu],
    D[x[uu,vv],uu],
    D[x[uu,vv],vv]}]] /. {uu->u,vv->v}
```

18.3 Using Mathematica to Find Asymptotic Curves

```
ffff[x_][u_,v_]:= Simplify[Det[{
    D[x[uu,vv],uu,vv],
    D[x[uu,vv],uu],
    D[x[uu,vv],vv]}]] /. {uu->u,vv->v}
```

```
gggg[x_][u_,v_]:= Simplify[Det[{
    D[x[uu,vv],vv,vv],
    D[x[uu,vv],uu],
    D[x[uu,vv],vv]}]] /. {uu->u,vv->v}
```

The *Mathematica* analog of (18.4) is either of the functions

```
asymeq1[x_][u_,v_]:=
    Module[{uuu,a,xxx,eqa1,eqa2},
    xxx=Solve[eeee[x][uuu,v]a^2 +
    2ffff[x][uuu,v]a +
    gggg[x][uuu,v ] == 0,a];
    eqa1=u'[v] == a /. xxx[[1]];
    eqa2=u'[v] == a /. xxx[[2]];
    {eqa1,eqa2} /. {a->u'[v],uuu->u[v]}]
```

```
asymeq2[x_][u_,v_]:=
    Module[{vvv,b,xxx,eqb1,eqb2},
    xxx=Solve[eeee[x][u,vvv] +
    2ffff[x][u,vvv]b +
    gggg[x][u,vvv]b^2 == 0,b];
    eqb1=v'[u] == b /. xxx[[1]];
    eqb2=v'[u] == b /. xxx[[2]];
    {eqb1,eqb2} /. {b->v'[u],vvv->v[u]}]
```

For example, let us consider the surface defined by

```
exptwist[a_,c_][u_,v_]:= {a v Cos[u],
                         a v Sin[u],a E^(c u)}
```

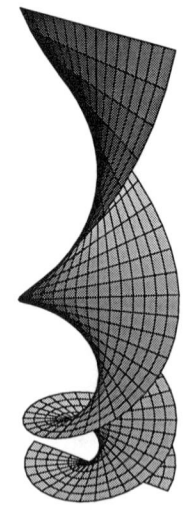

```
ParametricPlot3D[
exptwist[1,0.3][u,v]
//Evaluate,
{u,-Pi,2Pi},{v,-Pi,2Pi},
PlotPoints->{48,15},
Boxed->False,Axes->None];
```

Plot of
$(u,v) \longmapsto (v\cos u, v\sin u, e^{0.3u})$

Thus we see that **exptwist[a,c][u,v]** is a helicoidial-like surface whose twisting varies exponentially. To find an asymptotic patch on this surface, we first find the differential equations of the asymptotic curves:

deqs=asymeq1[exptwist[a,c]][u,v]

$$\{u'[v] == 0, \; u'[v] == \frac{2}{c\,v}\}$$

Next, we solve separately each of these differential equations:

dsol1=DSolve[deqs[[1]],u[v],v] /. C[1] -> p

$$\{\{u[v] \to p\}\}$$

dsol2=DSolve[deqs[[2]],u[v],v] /. C[1] -> q

$$\{\{u[v] \to q + \frac{2\,\text{Log}[v]}{c}\}\}$$

From this output we see how to define the asymptotic patch:

```
exptwistasym[a_,c_][p_,q_]:=
    {a E^((c (p - q))/2) Cos[p],
     a E^((c (p - q))/2) Sin[p],a E^(c p)}
```

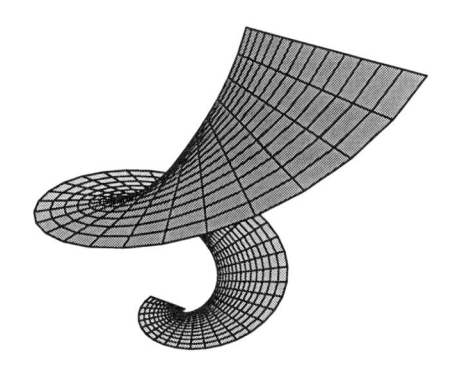

```
ParametricPlot3D[
exptwistasym[1,0.3][p,q]
//Evaluate,
{p,-Pi,2Pi},{q,-Pi,2Pi},
PlotPoints->{48,15},
Boxed->False,Axes->None];
```

Plot by asymptotic curves of
$(u,v) \longmapsto (v\cos u, v\sin u, e^{0.3t})$

18.4 Exercises

1. Show that the differential equation for the asymptotic curves on a Monge patch $\mathbf{x}(u,v) = (u,v,h(u,v))$ is

$$h_{uu}u'^2 + 2h_{uv}u'v' + h_{vv}v'^2 = 0.$$

2. Show that the differential equation for the asymptotic curves on a polar patch of the form $\mathbf{x}(r,\theta) = (r\cos\theta, r\sin\theta, h(r))$ is

$$h''(r)r'^2 + h'(r)r\,\theta'^2 = 0.$$

3. Find the differential equation for the asymptotic curves for the torus
$(u, v) \longmapsto ((a + b\cos v)\cos u, (a + b\cos v)\sin u, b\sin v)$.

4. Find an asymptotic patch for the surface $z = (x^2 + y^2)^\alpha$, where $0 \neq \alpha < 1/2$.

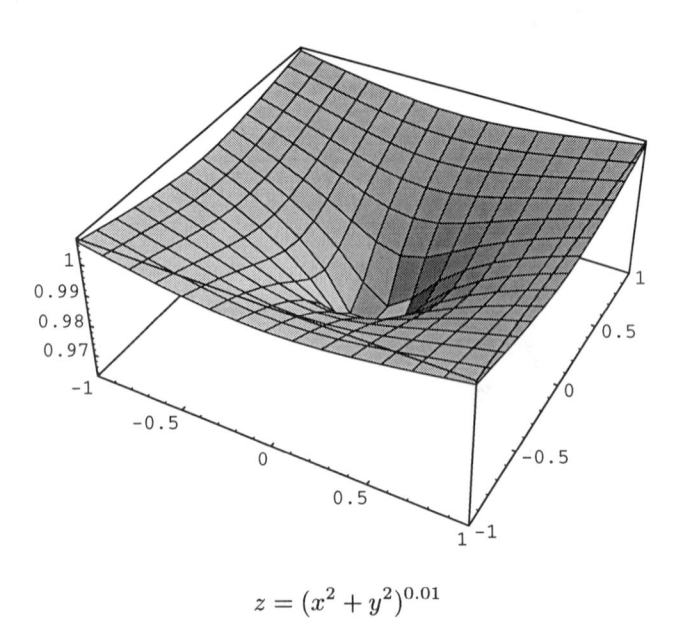

$$z = (x^2 + y^2)^{0.01}$$

5. Find and display the asymptotic curves of the generalized hyperbolic paraboloid $z = x^{2n} - y^{2n}$.

6. Find and display the asymptotic curves of the surface defined by $z = x^m y^n$.

7. Show that the Gaussian and mean curvatures of the (circular) helicoid defined on page 357 are given by

$$K = -\frac{c^2}{(c^2 + a^2 v^2)^2}, \qquad H = 0.$$

Conclude that the restriction of the Gaussian curvature to each u-parameter curve is constant.

19
RULED SURFACES

We describe in this chapter an important class of surfaces that contain straight lines. Such straight lines are automatically asymptotic curves by Corollary 18.6. A *ruled surface* is a surface generated by a straight line moving along a curve.

Definition. *A* **ruled surface** *M in $\mathbb{R}^3$ is a surface which contains at least one 1-parameter family of straight lines. Thus a ruled surface has a parametrization $\mathbf{x}: \mathcal{U} \longrightarrow M$ of the form*
$$\mathbf{x}(u,v) = \boldsymbol{\alpha}(u) + v\,\boldsymbol{\gamma}(u),$$
where $\boldsymbol{\alpha}$ and $\boldsymbol{\gamma}$ are curves in $\mathbb{R}^3$ with $\boldsymbol{\alpha}'$ never 0. We call $\mathbf{x}$ a **ruled patch**. *The curve $\boldsymbol{\alpha}$ is called the* **directrix** *or* **base curve** *of the ruled surface, and $\boldsymbol{\gamma}$ is called the* **director curve**. *The* **rulings** *of the ruled surface are the straight lines $v \longmapsto \boldsymbol{\alpha}(u) + v\,\boldsymbol{\gamma}(u)$.*

Definition. *Sometimes a ruled surface M has two distinct ruled patches on it. In this case, we say that $\mathbf{x}$ is* **doubly ruled**.

As a special case of Corollary 18.6, page 420, we have

Lemma 19.1. *The rulings of a ruled surface $M \subset \mathbb{R}^3$ are asymptotic curves.*

Furthermore,

Lemma 19.2. *The Gaussian curvature of a ruled surface $M \subset \mathbb{R}^3$ is everywhere nonpositive.*

Proof. If **x** is a ruled patch on $\mathcal{M}$, then $\mathbf{x}_{vv} = 0$; consequently $g = 0$. Hence it follows from Theorem 16.17, page 377, that

$$(19.1) \qquad K = \frac{-f^2}{EG - F^2} \leq 0. \quad \blacksquare$$

We shall study flat ruled surfaces (that is, ruled surfaces with zero Gaussian curvature) in Section 19.2 and nonflat ruled surfaces in Section 19.4. Tangent developables are discussed in Section 19.3. First, some examples.

19.1 Examples of Ruled Surfaces

The Elliptical Hyperboloid of One Sheet

The **elliptical hyperboloid of one sheet** is defined nonparametrically by

$$(19.2) \qquad \frac{x^2}{a^2} + \frac{y^2}{b^2} - \frac{z^2}{c^2} = 1.$$

Planes perpendicular to the z-axis intersect the surface in ellipses, while planes parallel to the z-axis intersect it in hyperbolas. On page 404 we gave the standard parametrization of the hyperboloid of one sheet; that parametrization does not show rulings.

Let us show that the elliptical hyperboloid of one sheet is a doubly ruled surface by finding two ruled patches on it. This can be done by defining

$$\mathbf{ellipticalhyperboloid}^{\pm}[a, b, c](u, v)$$
$$= \mathbf{ellipse}[a, b](u) \pm v\bigl(\mathbf{ellipse}[a, b]'(u) + (0, 0, c)\bigr),$$

where **ellipse**[a, b] is the standard parametrization of the ellipse

$$\frac{x^2}{a^2} + \frac{y^2}{b^2} = 1$$

in the xy-plane given by

$$(19.3) \qquad \mathbf{ellipse}[a, b](u) = (a \cos u, b \sin u, 0).$$

Then it can be checked (for example, using *Mathematica*) that

$(19.4) \quad \mathbf{ellipticalhyperboloid}^{\pm}[a, b, c]$
$$= \bigl(a(\cos u \mp v \sin u), b(\sin u \pm v \cos u), \pm c\,v\bigr).$$

19.1 Examples of Ruled Surfaces

An easy calculation using (19.4) shows that both **ellipticalhyperboloid**$^+[a, b, c]$ and **ellipticalhyperboloid**$^-[a, b, c]$ are parametrizations of the elliptical hyperboloid of one sheet defined nonparametrically by (19.2). Hence the elliptic hyperboloid is doubly ruled. The base curve in both cases is an ellipse given by (19.3).

The *Mathematica* definitions of **ellipticalhyperboloid**$^+$ and **ellipticalhyperboloid**$^-$ are translations of formula (19.4):

```
ellipticalhyperboloidplus[a_,b_,c_][u_,v_]:=
    {a Cos[u] - a v Sin[u],
     b v Cos[u] + b Sin[u], c v}

ellipticalhyperboloidminus[a_,b_,c_][u_,v_]:=
    {a Cos[u] + a v Sin[u],
     -b v Cos[u] + b Sin[u], c v}
```

We use these definitions to compute the Gaussian and mean curvatures of the important special cases

ellipticalhyperboloid$^+[a, a, c]$ and **ellipticalhyperboloid**$^-[a, a, c]$.

Thus

```
gcurvature[
ellipticalhyperboloidplus[a,a,c]][u,v]
```

gives

$$-\left(\frac{c^2}{(c^2 + a^2 v^2 + c^2 v^2)^2}\right)$$

and

```
PowerExpand[mcurvature[
ellipticalhyperboloidplus[a,a,c]][u,v],a]
```

gives

$$\frac{c^2(a^2 - c^2 - a^2 v^2 - c^2 v^2)}{2 a (c^2 + a^2 v^2 + c^2 v^2)^{3/2}}$$

(Note the new variation of **PowerExpand**.) The Gaussian and mean curvatures of **ellipticalhyperboloidminus[a,a,c]** turn out to be the same as those of **ellipticalhyperboloidplus[a,a,c]**.

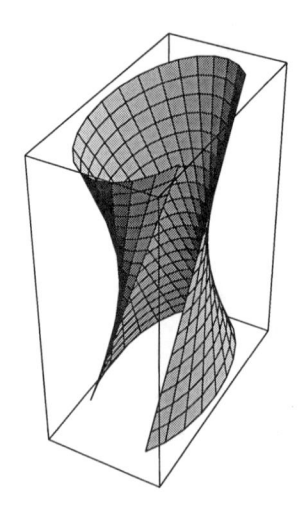

ellipticalhyperboloid$^+$ 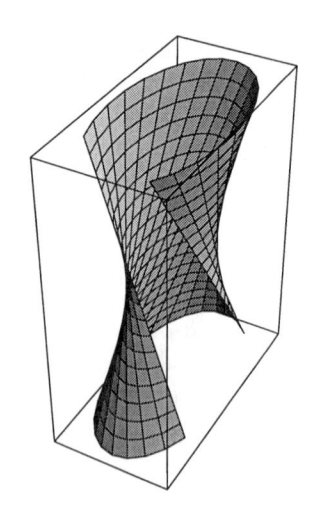   ellipticalhyperboloid$^-$

The *Mathematica* commands for these graphs are:

```
ParametricPlot3D[
ellipticalhyperboloidplus[
1,2,3][u,v]//Evaluate,
{u,0,3Pi/2},{v,-1.5,1.5},
Axes->None];
```

```
ParametricPlot3D[
ellipticalhyperboloidminus[
1,2,3][u,v]//Evaluate,
{u,0,3Pi/2},{v,-1.5,1.5},
Axes->None];
```

The General Hyperbolic Paraboloid

The general hyperbolic paraboloid is defined nonparametrically by

$$z = \frac{x^2}{a^2} - \frac{y^2}{b^2}.$$

It is doubly ruled since it can be parametrized in two ways:

$$\begin{aligned}\mathbf{genhypparab}[a,b]^{\pm}(u,v) &= (a\,u, 0, u^2) + v(a, \pm b, 2u) \\ &= (a(u+v), \pm b\,v, u^2 + 2u\,v).\end{aligned}$$

19.1 Examples of Ruled Surfaces

Both parametrizations can be obtained by the same formula by allowing b to be positive or negative; thus in *Mathematica* we define

```
genhypparab[a_,b_][u_,v_] := {a*(u + v),b*v,u^2 + 2*u*v}
```

See exercise 1.

Plücker's Conoid

The surface defined nonparametrically by $z = 2xy/(x^2 + y^2)$ is called **Plücker's**[1] **conoid**. The Monge parametrization of this surface is

$$\mathbf{plucker}(u,v) = \left(u, v, \frac{2uv}{u^2 + v^2}\right),$$

or

```
plucker[u_,v_] := {u,v,2 u v/(u^2 + v^2)}
```

With this parametrization *Mathematica* does not display the rulings.

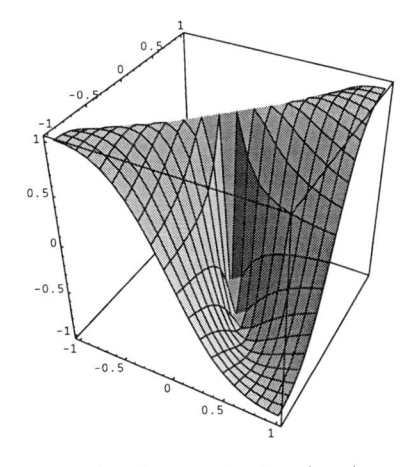

```
ParametricPlot3D[
plucker[u,v]//Evaluate,
{u,-1,1},{v,-1,1}];
```

$(u,v) \longmapsto \mathbf{plucker}(u,v)$

---

[1] Julius Plücker (1801-1868). German mathematician. Until 1846 Plücker's original research was in analytic geometry, but starting in 1846 as professor of physics in Bonn, he devoted his energies to experimental physics for nearly twenty years. At the end of his life he returned to mathematics, inventing line geometry.

To see the rulings, we need to convert (u, v) to polar coordinates, as follows. Let us write:

$$\mathbf{plucker}(r \cos \theta, r \sin \theta) = (r \cos \theta, r \sin \theta, 2 \cos \theta \sin \theta)$$

$$= (0, 0, 2 \cos \theta \sin \theta) + r(\cos \theta, \sin \theta, 0);$$

this shows that the z-axis is the base curve and the circle $\theta \longmapsto (\cos \theta, \sin \theta)$ is the director curve of $(r, \theta) \longmapsto \mathbf{plucker}(r \cos \theta, r \sin \theta)$.

The *Mathematica* plot of this parametrization of the Plücker conoid displays the rulings; all of them pass through the z-axis.

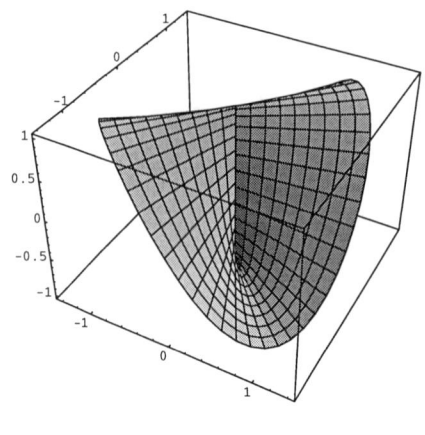

```
ParametricPlot3D[
plucker[r Cos[theta],
r Sin[theta]]
//Simplify//Evaluate,
{r,0,Sqrt[2]},
{theta,0,2Pi},
PlotPoints->{10,60}];
```

$(r, \theta) \longmapsto (r \cos \theta, r \sin \theta, 2 \cos \theta \sin \theta)$

It is easy to define a generalization of Plücker's conoid that has n folds instead of 2:

$$\mathbf{pluckerpolar}[n](r, \theta) = (r \cos \theta, r \sin \theta, \sin n\theta),$$

or

```
pluckerpolar[n_][r_,theta_]:=
    {r Cos[theta],r Sin[theta],Sin[n theta]}
```

Each surface **pluckerpolar**[n] is a ruled surface with the rulings passing through the z-axis.

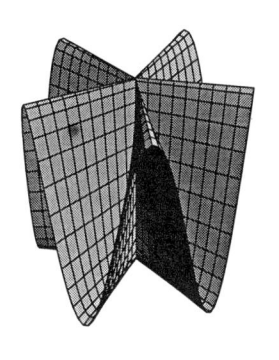

19.1 Examples of Ruled Surfaces

$(r, \theta) \longmapsto (r\cos\theta, r\sin\theta, \sin 5\theta)$ 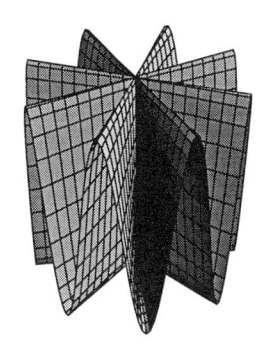 $\qquad$ $(r, \theta) \longmapsto (r\cos\theta, r\sin\theta, \sin 8\theta)$

For a generalization of **pluckerpolar** that includes a polar parametrization of a monkey saddle, see exercise 3.

The Möbius Strip as a Ruled Surface

We can rewrite the definition (14.2), page 325, of the Möbius strip as

(19.5) $\qquad$ **moebiusstrip**$[a](u, v) = a(\cos u, \sin u, 0)$

$$+ a\, v \left(\cos\left(\frac{u}{2}\right) \cos u, \cos\left(\frac{u}{2}\right) \sin u, \sin\left(\frac{u}{2}\right) \right).$$

We see from (19.5) that **moebiusstrip**$[a]$ is a ruled surface with base curve

$$u \longmapsto (\cos u, \sin u, 0)$$

and director curve

$$u \longmapsto a \left(\cos\left(\frac{u}{2}\right) \cos u, \cos\left(\frac{u}{2}\right) \sin u, \sin\left(\frac{u}{2}\right) \right),$$

which is a curve on a sphere of radius a. *Mathematica* can be used to compute the coefficients of the first and second fundamental forms, the Gaussian curvature and the mean curvature. We use the *Mathematica* definition **moebiusstrip[a][u,v]** given on page 325. For example, the *Mathematica*-computed values for F and G are 0 and 1; as for E:

ee[moebiusstrip[a]][u,v]

$$\frac{a^2 (4 + 3v^2 + 8v\cos[\frac{u}{2}] + 2v^2 \cos[u])}{4}$$

Furthermore, we can compute the Gauss and mean curvatures of the Möbius strip with *Mathematica*:

gcurvature[moebiusstrip[a]][u,v]

$$\frac{-4}{a^2 (4 + 3v^2 + 8v\cos[\frac{u}{2}] + 2v^2 \cos[u])^2}$$

mcurvature[moebiusstrip[a]][u,v]

$$\frac{-2a^5 (2 + 2v^2 + 4v\cos[\frac{u}{2}] + v^2 \cos[u]) \sin[\frac{u}{2}]}{(a^4 (4 + 3v^2 + 8v\cos[\frac{u}{2}] + 2v^2 \cos[u]))^{3/2}}$$

Here is a plot of the Gaussian curvature of the Möbius strip.

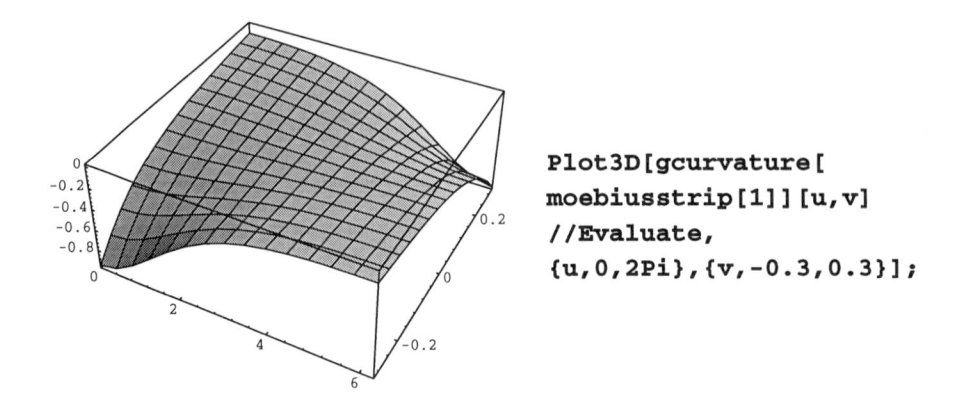

```
Plot3D[gcurvature[
moebiusstrip[1]][u,v]
//Evaluate,
{u,0,2Pi},{v,-0.3,0.3}];
```

Curvature of a Möbius strip

Thus the Gaussian curvature of this particular parametrization never vanishes, which means that this parametrization is not locally isometric to the flat Möbius strip that one makes out of a piece of paper.

19.2 Flat Ruled Surfaces

Recall that a flat surface is a surface whose Gaussian curvature vanishes everywhere. Such a surface is classically called a **developable surface**. Let us give a criterion for flatness of a ruled surface.

Lemma 19.3. *Let $\mathcal{M} \subset \mathbb{R}^3$ be a ruled surface. Then the component g of the second fundamental form of $\mathcal{M}$ vanishes. Consequently, $\mathcal{M}$ is flat if and only if $f = 0$.*

Proof. This is an easy consequence of (19.1). ∎

Next, we consider three classical flat ruled surfaces.

Definition. *Let $\mathcal{M} \subset \mathbb{R}^3$ be a surface. Then:*

(i) $\mathcal{M}$ *is the* **tangent developable** *of a curve $\boldsymbol{\alpha}\colon (a,b) \longrightarrow \mathbb{R}^3$, provided $\mathcal{M}$ can be parametrized as*

$$\mathbf{x}(u,v) = \boldsymbol{\alpha}(u) + v\,\boldsymbol{\alpha}'(u);$$

(ii) $\mathcal{M}$ *is a* **generalized cylinder**, *over a curve $\boldsymbol{\alpha}\colon (a,b) \longrightarrow \mathbb{R}^3$, provided $\mathcal{M}$ can be parametrized as*

$$\mathbf{y}(u,v) = \boldsymbol{\alpha}(u) + v\,\mathbf{q},$$

where $\mathbf{q} \in \mathbb{R}^3$ is a fixed point;

(iii) $\mathcal{M}$ *is a* **generalized cone**, *over a curve $\boldsymbol{\alpha}\colon (a,b) \longrightarrow \mathbb{R}^3$, provided $\mathcal{M}$ can be parametrized as*

$$\mathbf{z}(u,v) = \mathbf{p} + v\,\boldsymbol{\alpha}(u),$$

where $\mathbf{p} \in \mathbb{R}^3$ is a fixed point. (Hence $\mathbf{p}$ can be interpreted as the **vertex** *of the cone.)*

Lemma 19.4. *The criteria for the regularity of tangent developables, generalized cylinders and generalized cones are as follows.*

(i) *Let $\boldsymbol{\alpha}\colon (a,b) \longrightarrow \mathbb{R}^3$ be a regular curve whose curvature $\kappa[\boldsymbol{\alpha}]$ is everywhere nonzero. Then the tangent developable $\mathbf{x}$ of $\boldsymbol{\alpha}$ is regular everywhere except along $\boldsymbol{\alpha}$.*

(ii) *A generalized cylinder $\mathbf{y}(u,v) = \boldsymbol{\alpha}(u) + v\,\mathbf{q}$ is regular wherever $\boldsymbol{\alpha}' \times \mathbf{q}$ does not vanish.*

(iii) *A generalized cone $\mathbf{z}(u,v) = \mathbf{p} + v\,\boldsymbol{\alpha}(u)$ is regular wherever $v\,\boldsymbol{\alpha} \times \boldsymbol{\alpha}'$ is nonzero. A generalized cone is never regular at its vertex.*

Proof. For a tangent developable **x** we have

(19.6) $$(\mathbf{x}_u \times \mathbf{x}_v)(u,v) = (\boldsymbol{\alpha}' + v\boldsymbol{\alpha}'') \times \boldsymbol{\alpha}' = v\boldsymbol{\alpha}'' \times \boldsymbol{\alpha}'.$$

If $\kappa[\boldsymbol{\alpha}] \neq 0$, then $\boldsymbol{\alpha}'' \times \boldsymbol{\alpha}'$ is everywhere nonzero by (8.22). Thus (19.6) implies that **x** is regular whenever $v \neq 0$. The other statements have similar proofs. ∎

Theorem 19.5. *If $\mathcal{M}$ is a tangent developable, a generalized cylinder or a generalized cone, then $\mathcal{M}$ is flat.*

Proof. By Lemma 19.3 it suffices to show that $f = 0$ in each of the three cases. For a tangent developable **x** we have $\mathbf{x}_u = \boldsymbol{\alpha}' + v\boldsymbol{\alpha}''$, $\mathbf{x}_v = \boldsymbol{\alpha}'$, and $\mathbf{x}_{uv} = \boldsymbol{\alpha}''$; hence

$$f = \frac{\det\begin{pmatrix} \boldsymbol{\alpha}'' \\ \boldsymbol{\alpha}' + v\boldsymbol{\alpha}'' \\ \boldsymbol{\alpha}' \end{pmatrix}}{\|\mathbf{x}_u \times \mathbf{x}_v\|} = 0.$$

It is obvious that $f = 0$ for a generalized cylinder **y**, since $\mathbf{y}_{uv} = 0$. Finally, for a generalized cone, we compute

$$f = \frac{\det\begin{pmatrix} \boldsymbol{\gamma}' \\ v\boldsymbol{\gamma}' \\ \boldsymbol{\gamma} \end{pmatrix}}{\|\mathbf{x}_u \times \mathbf{x}_v\|} = 0. \qquad \blacksquare$$

In Theorem 19.12 we show that in some sense a developable surface is the union of tangent developables, generalized cylinders and generalized cones.

Next, we describe *Mathematica* miniprograms to draw cylinders and cones over curves in $\mathbb{R}^3$. Since we shall also want to consider cylinders and cones over plane curves, we first define a *Mathematica* function that converts plane curves into space curves. The function **inscurve** will insert the value **a** as the nth coordinate into the parametrization of a curve:

```
inscurve[n_,a_,alpha_][t_] := Insert[alpha[t],a,n]
```

The *Mathematica* programs to construct cylinders and cones over curves mimic the mathematical definitions very closely:

```
cylinder[{q1_,q2_,q3_},beta_][u_,v_] :=
                    beta[u] + v{q1,q2,q3}
```

19.3 Tangent Developables

```
cone[{p1_,p2_,p3_},gamma_][u_,v_]:=
                    {p1,p2,p3} + v gamma[u]
```

Here are typical plots of a cone and cylinder over a figure eight.

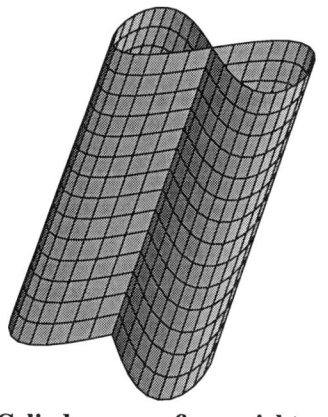

```
ParametricPlot3D[
cylinder[{0,1,1},
inscurve[3,0,eight]][u,v]
//Evaluate,
{u,0,2Pi},{v,0,2},
Axes->None,
Boxed->False,
PlotPoints->{40,15}];
```

Cylinder over a figure eight

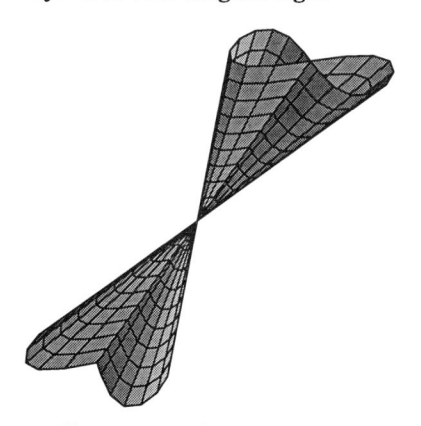

```
ParametricPlot3D[
cone[{0,0,0},
inscurve[3,2,
(eight[#]+{1,1})&]][u,v]
//Evaluate,
{u,0,2Pi},{v,-2,2},
Axes->None,
Boxed->False];
```

Cone over a figure eight

19.3 Tangent Developables

We know from Lemma 19.4 that the surface **tandev**[α] is singular along the curve α. We prove next that **tandev**[α] is made up of two sheets which meet along the curve α in a sharp edge, called the **edge of regression**.

Theorem 19.6. *Let $\alpha\colon (a,b) \longrightarrow \mathbb{R}^3$ be an analytic unit-speed curve with $a<0<b$, and let $\mathbf{x}$ be the tangent developable of α. Then the intersection of the trace of $\mathbf{x}$ with the plane perpendicular to α at $\alpha(0)$ is a semicubical parabola with a cusp at $\alpha(0)$.*

Proof. We compute the first four derivatives of α using the Frenet formulas (Theorem 8.4, page 185):

$$(19.7) \quad \begin{cases} \alpha'(s) = \mathbf{T}(s), \\ \alpha''(s) = \kappa(s)\mathbf{N}(s), \\ \alpha'''(s) = -\kappa(s)^2\mathbf{T}(s) + \kappa'(s)\mathbf{N}(s) + \kappa(s)\tau(s)\mathbf{B}(s), \\ \alpha''''(s) = -3\kappa(s)\kappa'(s)\mathbf{T}(s) + \bigl(-\kappa(s)^3 + \kappa''(s) - \kappa(s)\tau(s)^2\bigr)\mathbf{N}(s) \\ \qquad\qquad + \bigl(2\kappa'(s)\tau(s) + \kappa(s)\tau'(s)\bigr)\mathbf{B}(s). \end{cases}$$

When we expand α in a power series about $\alpha(0)$ and use (19.7), we get

$$(19.8) \quad \alpha(s) = \alpha(0) + s\alpha'(0) + \frac{s^2}{2}\alpha''(0) + \frac{s^3}{6}\alpha'''(0) + \frac{s^4}{24}\alpha''''(0) + O(s^5)$$

$$= \alpha(0) + \mathbf{T}(0)\left(s - \frac{s^3}{6}\kappa(0)^2 - \frac{s^4}{8}\kappa(0)\kappa'(0)\right)$$

$$+ \mathbf{N}(0)\left(\frac{s^2}{2}\kappa(0) + \frac{s^3}{6}\kappa'(0) + \frac{s^4}{24}\bigl(-\kappa(0)^3 + \kappa''(0) - \kappa(0)\tau(0)^2\bigr)\right)$$

$$+ \mathbf{B}(0)\left(\frac{s^3}{6}\kappa(0)\tau(0) + \frac{s^4}{24}\bigl(2\kappa'(0)\tau(0) + \kappa(0)\tau'(0)\bigr)\right) + O(s^5).$$

Taking the derivative of both sides of (19.8), we obtain

$$(19.9) \quad \alpha'(s) = \mathbf{T}(0)\left(1 - \frac{s^2}{2}\kappa(0)^2 - \frac{s^3}{2}\kappa(0)\kappa'(0)\right)$$

$$+ \mathbf{N}(0)\left(s\kappa(0) + \frac{s^2}{2}\kappa'(0) + \frac{s^3}{6}\bigl(-\kappa(0)^3 + \kappa''(0) - \kappa(0)\tau(0)^2\bigr)\right)$$

$$+ \mathbf{B}(0)\left(\frac{s^2}{2}\kappa(0)\tau(0) + \frac{s^3}{6}\bigl(2\kappa'(0)\tau(0) + \kappa(0)\tau'(0)\bigr)\right) + O(s^4).$$

19.3 Tangent Developables

Next, let $\mathbf{x}$ be the tangent developable of $\boldsymbol{\alpha}$. From (19.8) and (19.9) we obtain

(19.10) $\mathbf{x}(u,v) = \boldsymbol{\alpha}(u) + v\,\boldsymbol{\alpha}'(u)$

$$= \boldsymbol{\alpha}(0) + \mathbf{T}(0)\left(u - \frac{u^3}{6}\kappa(0)^2 + v - \frac{u^2 v}{2}\kappa(0)^2 - \frac{u^3 v}{2}\kappa(0)\kappa'(0)\right)$$

$$+ \mathbf{N}(0)\left(\frac{u^2}{2}\kappa(0) + \frac{u^3}{6}\kappa'(0) + uv\,\kappa(0) + \frac{u^2 v}{2}\kappa'(0)\right.$$

$$\left. + \frac{u^3 v}{6}\bigl(-\kappa(0)^3 + \kappa''(0) - \kappa(0)\tau(0)^2\bigr)\right)$$

$$+ \mathbf{B}(0)\left(\left(\frac{u^3}{6} + \frac{u^2 v}{2}\right)\kappa(0)\tau(0)\right.$$

$$\left. + \frac{u^3 v}{6}\bigl(2\kappa'(0)\tau(0) + \kappa(0)\tau'(0)\bigr)\right) + O(u^4).$$

We want to determine the intersection of $\mathbf{x}$ with the plane perpendicular to $\boldsymbol{\alpha}$ at $\boldsymbol{\alpha}(0)$. Therefore, we set the coefficient of $\mathbf{T}(0)$ in (19.10) equal to zero and solve for v. The result is

$$v = -\frac{u - \dfrac{u^3}{6}\kappa(0)^2 + O(u^4)}{1 - \dfrac{u^2}{2}\kappa(0)^2 - \dfrac{u^3}{2}\kappa(0)\kappa'(0) + O(u^4)} = -u - \kappa(0)^2\frac{u^3}{3} + O(u^4).$$

We insert this value of v into (19.10) and get power series expansions for the coefficients of $\mathbf{N}(0)$ and $\mathbf{B}(0)$:

(19.11) $\begin{cases} \text{coefficient of } \mathbf{N}(0) = -\dfrac{u^2}{2}\kappa(0) - \dfrac{u^3}{3}\kappa'(0) + \cdots, \\ \text{coefficient of } \mathbf{B}(0) = -\dfrac{u^3}{3}\kappa(0)^2\tau(0) + \cdots. \end{cases}$

If we ignore the higher order terms, we see that the plane curve described by (19.11) is given implicitly by

$$\frac{x^2}{2}\kappa(0) + \frac{x^3}{3}\kappa'(0) = \frac{y^3}{3}\kappa(0)^2\tau(0),$$

which is a semicubical parabola. ∎

The following lemma is easy to prove.

Lemma 19.7. *Let $\beta\colon (c,d) \longrightarrow \mathbb{R}^3$ be a unit-speed curve. Then the coefficients of the metric of the tangent developable* **tandev**$[\beta]$ *depend only on the curvature of β. Explicitly:*

$$E = 1 + v^2 \kappa[\beta]^2, \qquad F = 1, \qquad G = 1.$$

Since a helix and a circle of the same radius both have the same constant curvature, Lemma 19.7 implies that the tangent developable to a helix can be constructed from a piece of paper by cutting out a disk and twisting the remaining portion around a cylinder.

Let us examine tangent developables more closely with *Mathematica*. First, we need a *Mathematica* program to construct from a curve $\alpha\colon (a,b) \longrightarrow \mathbb{R}^3$ its tangent developable:

```
tandev[alpha_][u_,v_]:= alpha[uu] + v D[alpha[uu],uu]/
    Sqrt[D[alpha[uu],uu].D[alpha[uu],uu]] /. uu->u
```

Here is how to draw the tangent developable to a circular helix:

```
ParametricPlot3D[
tandev[helix[1,0.4]][u,v]
//Evaluate,
{u,0,4Pi},{v,-2,2},
PlotPoints->{40,10},
Axes->None,
Boxed->False];
```

Tangent developable to a circular helix

19.4 Noncylindrical Ruled Surfaces

At the other extreme from flat ruled surfaces are the noncylindrical ruled surfaces.

Definition. *A ruled surface parametrized by* $\mathbf{x}(u,v) = \boldsymbol{\beta}(u) + v\boldsymbol{\gamma}(u)$ *is said to be* **noncylindrical** *provided* $\boldsymbol{\gamma} \times \boldsymbol{\gamma}'$ *never vanishes.*

Thus the rulings are always changing directions on a noncylindrical ruled surface. We show how to find a useful reference curve on a noncylindrical ruled surface. This curve, called a striction curve, is a generalization of the edge of regression of a tangent developable.

Lemma 19.8. *Let* $\tilde{\mathbf{x}}$ *be a parametrization of a noncylindrical ruled surface of the form* $\tilde{\mathbf{x}}(u,v) = \boldsymbol{\beta}(u) + v\boldsymbol{\gamma}(u)$. *Then* $\tilde{\mathbf{x}}$ *has a reparametrization of the form*

$$(19.12) \qquad \mathbf{x}(u,v) = \boldsymbol{\sigma}(u) + v\boldsymbol{\delta}(u),$$

where $\|\boldsymbol{\delta}\| = 1$ *and* $\boldsymbol{\sigma}' \cdot \boldsymbol{\delta}' = 0$. *The curve* $\boldsymbol{\sigma}$ *is called the* **striction curve** *of* $\tilde{\mathbf{x}}$.

Proof. Since $\boldsymbol{\gamma} \times \boldsymbol{\gamma}'$ is never zero, $\boldsymbol{\gamma}$ is never zero. We define a reparametrization $\tilde{\tilde{\mathbf{x}}}$ of $\tilde{\mathbf{x}}$ by

$$\tilde{\tilde{\mathbf{x}}}(u,v) = \tilde{\mathbf{x}}\left(u, \frac{v}{\|\boldsymbol{\gamma}(u)\|}\right) = \boldsymbol{\beta}(u) + \frac{v\boldsymbol{\gamma}(u)}{\|\boldsymbol{\gamma}(u)\|}.$$

Clearly, $\tilde{\tilde{\mathbf{x}}}$ has the same trace as $\tilde{\mathbf{x}}$. Furthermore, if we put $\boldsymbol{\delta}(u) = \boldsymbol{\gamma}(u)/\|\boldsymbol{\gamma}(u)\|$ then

$$\tilde{\tilde{\mathbf{x}}}(u,v) = \boldsymbol{\beta}(u) + v\boldsymbol{\delta}(u),$$

where $\|\boldsymbol{\delta}(u)\| = 1$. Then also $\boldsymbol{\delta}(u) \cdot \boldsymbol{\delta}'(u) = 0$.

Next, we need to find a curve $\boldsymbol{\sigma}$ such that $\boldsymbol{\sigma}'(u) \cdot \boldsymbol{\delta}'(u) = 0$. To this end, we write

$$(19.13) \qquad \boldsymbol{\sigma}(u) = \boldsymbol{\beta}(u) + t(u)\boldsymbol{\delta}(u)$$

for some function $t = t(u)$ and determine t. We differentiate (19.13), obtaining

$$(19.14) \qquad \boldsymbol{\sigma}'(u) = \boldsymbol{\beta}'(u) + t'(u)\boldsymbol{\delta}(u) + t(u)\boldsymbol{\delta}'(u).$$

Since $\boldsymbol{\delta}(u) \cdot \boldsymbol{\delta}'(u) = 0$, it follows from (19.14) that

$$\boldsymbol{\sigma}'(u) \cdot \boldsymbol{\delta}'(u) = \boldsymbol{\beta}'(u) \cdot \boldsymbol{\delta}'(u) + t(u)\boldsymbol{\delta}'(u) \cdot \boldsymbol{\delta}'(u).$$

Since $\gamma \times \gamma'$ never vanishes, γ and γ' are always linearly independent, and consequently δ' never vanishes. Thus if we define $t(u)$ by

$$t(u) = -\frac{\beta'(u) \cdot \delta'(u)}{\|\delta'(u)\|^2},$$

we get $\sigma'(u) \cdot \delta'(u) = 0$. Now define

$$\mathbf{x}(u,v) = \widetilde{\mathbf{x}}(u, t(u) + v).$$

Then $\mathbf{x}(u,v) = \beta(u) + (t(u) + v)\delta(u) = \sigma(u) + v\,\delta(u)$, so that $\mathbf{x}$, $\widetilde{\mathbf{x}}$ and $\widetilde{\widetilde{\mathbf{x}}}$ all have the same trace, and $\mathbf{x}$ satisfies (19.12). ∎

Lemma 19.9. *The striction curve of a noncylindrical ruled surface $\mathbf{x}$ does not depend on the choice of base curve.*

Proof. Let β and $\widetilde{\beta}$ be two base curves for $\mathbf{x}$. Then we can write

(19.15) $$\mathbf{x}(u,v) = \beta(u) + v\,\delta(u) = \widetilde{\beta}(u) + w(v)\delta(u)$$

for some function $w = w(v)$. Let σ and $\widetilde{\sigma}$ be the corresponding striction curves. Then

$$\sigma(u) = \beta(u) - \frac{\beta'(u) \cdot \delta'(u)}{\|\delta'(u)\|^2}\delta(u)$$

and

$$\widetilde{\sigma}(u) = \widetilde{\beta}(u) - \frac{\widetilde{\beta}'(u) \cdot \delta'(u)}{\|\delta'(u)\|^2}\delta(u),$$

so that

(19.16) $$\sigma - \widetilde{\sigma} = \beta - \widetilde{\beta} - \frac{(\beta' - \widetilde{\beta}') \cdot \delta'}{\|\delta'\|^2}\delta.$$

On the other hand, it follows from (19.15) that

(19.17) $$\beta - \widetilde{\beta} = (w(v) - v)\delta.$$

From (19.16), (19.17) and the fact that $\delta \cdot \delta' = 0$, we get

$$\sigma - \widetilde{\sigma} = \left(w(v) - v - \frac{(w(v) - v)\delta' \cdot \delta'}{\|\delta'\|^2}\right)\delta = 0. \quad \blacksquare$$

There is a nice geometric interpretation of the striction curve σ. Let $\varepsilon > 0$ be small. Since nearby rulings are not parallel to each other, there is a unique point $P(\varepsilon)$ on $v \longmapsto v\,\delta(u)$ that is closest to $v \longmapsto v\,\delta(u+\varepsilon)$. As $\varepsilon \longrightarrow 0$, $P(\varepsilon) \longrightarrow \sigma(u)$.

19.4   Noncylindrical Ruled Surfaces

Definition. *Let* **x** *be a noncylindrical ruled surface given by* (19.12). *Then the* **distribution parameter** *of* **x** *is the function p defined by*

$$(19.18) \qquad p = \frac{(\sigma' \delta \delta')}{\delta' \cdot \delta'}.$$

Lemma 19.10. *Let* $\mathcal{M}$ *be a noncylindrical ruled surface, so that it can be parametrized by a patch* **x** *of the form* (19.12). *Then* **x** *is regular whenever* $v \neq 0$, *or when* $v = 0$ *and* $p(u) \neq 0$. *Furthermore, the Gaussian curvature of a ruled surface* **x** *is given in terms of its distribution parameter by*

$$(19.19) \qquad K = \frac{-p(u)^2}{\big(p(u)^2 + v^2\big)^2}.$$

Also,

$$E = \|\sigma'\|^2 + v^2\|\delta'\|^2, \quad F = \sigma' \cdot \delta, \quad G = 1, \quad EG - F^2 = (p^2 + v^2)\|\delta'\|^2,$$

and

$$g = 0, \qquad f = \frac{p\|\delta'\|}{\sqrt{p^2 + v^2}}.$$

Proof. First, we observe that both $\sigma' \times \delta$ and δ' are perpendicular to both δ and σ'. Therefore, $\sigma' \times \delta$ must be a multiple of δ':

$$\sigma' \times \delta = \frac{(\sigma' \delta \delta')}{\delta' \cdot \delta'} \delta' = p\, \delta'.$$

Since $\mathbf{x}_u = \sigma' + v\, \delta'$ and $\mathbf{x}_v = \delta$, we have

$$\mathbf{x}_u \times \mathbf{x}_v = p\, \delta' + v\, \delta' \times \delta,$$

so that

$$(19.20) \qquad \|\mathbf{x}_u \times \mathbf{x}_v\|^2 = \|p\, \delta'\|^2 + \|v\, \delta' \times \delta\|^2 = (p^2 + v^2)\|\delta'\|^2.$$

From (19.20) it is clear that regularity of **x** is as stated.

Next, $\mathbf{x}_{uv} = \delta'$ and $\mathbf{x}_{vv} = 0$, so that $g = 0$ and

$$f = \frac{(\mathbf{x}_{uv}\mathbf{x}_u\mathbf{x}_v)}{\|\mathbf{x}_u \times \mathbf{x}_v\|} = \frac{\delta' \cdot (p\, \delta' + v\, \delta' \times \delta)}{\sqrt{(p^2 + v^2)}\|\delta'\|} = \frac{p\|\delta'\|}{\sqrt{p^2 + v^2}}.$$

Therefore,

$$K = \frac{-f^2}{\|\mathbf{x}_u \times \mathbf{x}_v\|^2} = \frac{-\left(\dfrac{p\|\delta'\|}{\sqrt{p^2+v^2}}\right)^2}{(p^2 + v^2)\|\delta'\|^2} = \frac{-p^2}{(p^2 + v^2)^2}. \quad \blacksquare$$

Notice that from (19.19) it follows that the Gaussian curvature of a noncylindrical ruled surface is nonpositive. But this observation is also a consequence of the fact that a ruled surface has asymptotic curves, namely, the rulings. However, more can be said about the Gaussian curvature of a noncylindrical ruled surface.

Corollary 19.11. *Let M be a noncylindrical ruled surface given by (19.12) with distribution parameter p.*

(i) *Along a ruling $v \longmapsto v\,\delta(u)$ the Gaussian curvature $K(u,v)$ tends to 0 as $v \longrightarrow \infty$.*

(ii) *$K(u,v) = 0$ if and only if $p(u) = 0$.*

(iii) *If the distribution parameter p never vanishes, then $K(u,v)$ is continuous and $|K(u,v)|$ assumes its maximum at $v = 0$.*

Proof. All of these statements follow from (19.19). ∎

Next, we prove a partial converse of Theorem 19.5.

Theorem 19.12. *Let $\mathbf{x}(u,v) = \beta(u) + v\,\delta(u)$ with $\|\delta(u)\| = 1$ parametrize a flat ruled surface M.*

(i) *If $\beta'(u) \equiv 0$, then M is a cone.*

(ii) *If $\delta'(u) \equiv 0$, then M is a cylinder.*

(iii) *If both β' and δ' never vanish, then M is the tangent developable of its striction curve.*

Proof. Parts (i) and (ii) are immediate from the definitions, so it suffices to prove (iii). We can assume that β is a unit-speed striction curve, so that

$$(19.21) \qquad \beta' \cdot \delta' \equiv 0.$$

Since $K \equiv 0$, it follows from (19.19) and (19.18) that

$$(19.22) \qquad (\beta'\delta\delta') \equiv 0.$$

Then (19.21) and (19.22) imply that β' and δ are collinear. ∎

Of course, cases (i), (ii) and (iii) of Theorem 19.12 do not exhaust all of the possibilities. If there is a clustering of the zeros β or δ, the surface can be complicated. In any case, away from the cluster points a developable surface is the union of pieces of cylinders, cones and tangent developables.

19.5 Examples of Striction Curves of Noncylindrical Ruled Surfaces

The Helicoid

The parametrization (15.28) of the circular helicoid can be rewritten as

$$\mathbf{helicoid}[a,b](u,v) = (0,0,bu) + av(\cos u, \sin u, 0),$$

which shows that it is a ruled surface. The striction curve σ and director curve δ are given by

$$\sigma(u) = (0,0,bu) \quad \text{and} \quad \delta(u) = (\cos u, \sin u, 0).$$

Then

$$\mathbf{helicoid}[a,b](u,v/a) = \sigma(u) + v\,\delta(u) = (\cos u, \sin u, 0).$$

An easy computation shows that $p(u) \equiv b$.

The Hyperbolic Paraboloid

The hyperbolic paraboloid (13.1), page 297, when parametrized as

$$\mathbf{hyperbolicparaboloid}(u,v) = (u,0,0) + v(0,1,u),$$

has $\sigma(u) = (u,0,0)$ as its striction curve and

$$\delta(u) = \frac{(0,1,u)}{\sqrt{1+u^2}}$$

as its director curve. Thus

$$\mathbf{hyperbolicparaboloid}\left(u, v\sqrt{1+u^2}\right) = \sigma(u) + v\,\delta(u),$$

and the distribution parameter is given by $p(u) = 1 + u^2$.

The computation of $p(u)$ can be carried out with *Mathematica*. First, we define `delta[u_]:={0,1,u}/Sqrt[1 + u^2]` and `sigma[u_]:={u,0,0}`. Then

```
Simplify[Det[{sigma'[u],delta[u],delta'[u]}]/
        (delta'[u].delta'[u])]
```

yields

```
        2
  1 + u
```

19.6 A Program for Ruled Surfaces

We define a *Mathematica* command **ruled** that constructs a ruled surface **ruled[alpha,gamma]** from space curves **alpha** and **gamma**:

```
ruled[alpha_,gamma_][u_,v_] := alpha[u] + v gamma[u]
```

This definition is sufficient for computing the components of the first fundamental form, using the commands in Appendix A or **CSPROGS.m**. For example,

> **ee[ruled[alpha,gamma]][u,v]**

elicits the response

> (alpha'[u] + v gamma'[u]).(alpha'[u] + v gamma'[u])

However, the components of the second fundamental form, as well as the Gaussian curvature and mean curvature, involve the computation of the determinant of a 3 × 3 matrix. To compute these components, the curves **alpha** and **gamma** must be specified explicitly as 3-tuples:

> **alpha[u_] := {a1[u],a2[u],a3[u]}**
>
> **gamma[u_] := {c1[u],c2[u],c3[u]}**

After making these definitions, we use

> **ggg[ruled[alpha,gamma]][u,v]**

to obtain 0. The expressions for the other components of the second fundamental form, the Gaussian curvature and the mean curvature are too complicated to be useful in the general case. We show how to make use of these formulas in an explicit example.

More general than Plücker's conoid or its generalization given on page 436 is a **right conoid**, which is a ruled surface with rulings parallel to a plane and passing through a line that is perpendicular to the plane. For example, if we take the plane to be the xy-plane and the line to be the z-axis, a right conoid will have the form

$$\mathbf{rightconoid}[\vartheta, h](u,v) = \big(v \cos \vartheta(u), v \sin \vartheta(u), h(u)\big).$$

In *Mathematica*, we put

19.6 A Program for Ruled Surfaces

```
betaconoid[u_]:= {0,0,h[u]}

gammaconoid[theta_][u_]:=
         {Cos[theta[u]],Sin[theta[u]],0}
```

We have specified neither of the functions **h** or **theta**, so they will appear in the *Mathematica*-generated formulas for the Gaussian and mean curvatures of a right conoid:

ruled[betaconoid,gammaconoid[theta]][u,v]

{v Cos[theta[u]], v Sin[theta[u]], h[u]}

**gcurvature[
ruled[betaconoid,gammaconoid[theta]]][u,v]**

$$-\left(\frac{h'[u]^2 \, theta'[u]^2}{(h'[u]^2 + v^2 \, theta'[u]^2)^2}\right)$$

**mcurvature[
ruled[betaconoid,gammaconoid[theta]]][u,v]**

$$\frac{v \, (-(theta'[u] \, h''[u]) + h'[u] \, theta''[u])}{2 \, (h'[u]^2 + v^2 \, theta'[u]^2)^{3/2}}$$

The choice $h(u) = 2u$, $\vartheta(u) = u$ yields a helicoid. Here is the right conoid generated with the choice $h(u) = 2\sin u$, $\vartheta(u) = u$:

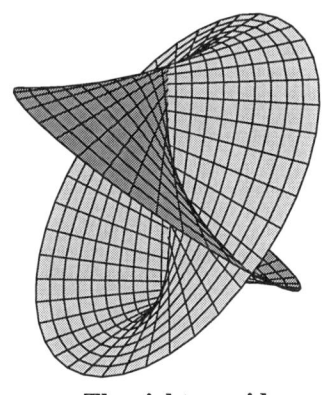

```
ParametricPlot3D[
{v Cos[u],v Sin[u],2 Sin[u],
FaceForm[Green,Yellow]},
{u,-Pi,Pi},{v,-2,2},
PlotPoints->{40,20},
PlotRange->All,
Lighting->False,
Axes->None,
Boxed->False,
ViewPoint->
{2.257,-1.461,2.054}];
```

The right conoid

$(u,v) \longmapsto (v\cos u, v\sin u, 2\sin u)$

The right conoid $(u, v) \longmapsto (v \cos u, v \sin u, 2 \sin u)$ is a reparametrization of

$$(u, v) \longrightarrow 2\mathbf{pluckerpolar}[1](u, v).$$

19.7 Other Examples of Ruled Surfaces

The ruled surfaces generated by the normal or binormal lines to a space curve **alpha** have definitions that are similar to that of tangent developable. The *Mathematica* formulas are

```
normalsurf[alpha_][u_,v_]:=
    alpha[u] + v normal[alpha][u]
```

and

```
binormalsurf[alpha_][u_,v_]:=
    alpha[u] + v binormal[alpha][u]
```

Note that in contrast to a tangent developable, the normal and binormal surfaces may not be flat. Let us compute the normal and binormal surfaces to Viviani's curve (defined in Section 8.6):

normalsurf[viviani[a]][u,v]//PowerExpand//Simplify

$$\left\{ 2a\cos\left[\frac{u}{2}\right] - \frac{v(3 + 12\cos[u] + \cos[2u])}{2\sqrt{3 + \cos[u]}\sqrt{13 + 3\cos[u]}}, \right.$$

$$a\sin[u] - \frac{v(12\sin[u] + \sin[2u])}{2\sqrt{3 + \cos[u]}\sqrt{13 + 3\cos[u]}},$$

$$\left. 2a\sin\left[\frac{u}{2}\right] - \frac{2v\sin\left[\frac{u}{2}\right]}{\sqrt{3 + \cos[u]}\sqrt{13 + 3\cos[u]}} \right\}$$

binormalsurf[viviani[a]][u,v]//PowerExpand//Simplify

19.7 Other Examples of Ruled Surfaces

$$\{2a\cos[\frac{u}{2}] + \frac{v\,(3\sin[\frac{u}{2}] + \sin[\frac{3u}{2}])}{\sqrt{2}\,\sqrt{13 + 3\cos[u]}},$$

$$\frac{-2\sqrt{2}\,v\cos[\frac{u}{2}]^3}{\sqrt{13 + 3\cos[u]}} + a\sin[u],\ \frac{2\sqrt{2}\,v}{\sqrt{13 + 3\cos[u]}} + 2a\sin[\frac{u}{2}]\}$$

As usual, the plots of these formulas clarify their meaning:

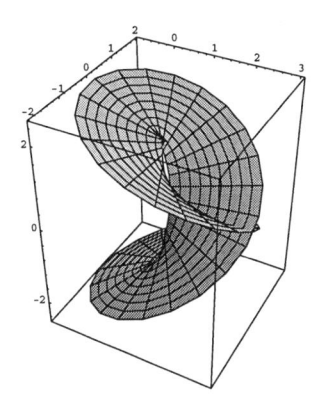

```
ParametricPlot3D[
normalsurf[viviani[
1]][u,v]//Evaluate,
{u,-2Pi,2Pi},{v,-1,1},
PlotPoints->{40,10}];
```

The normal surface to Viviani's curve

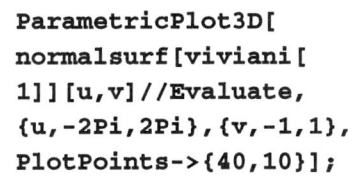

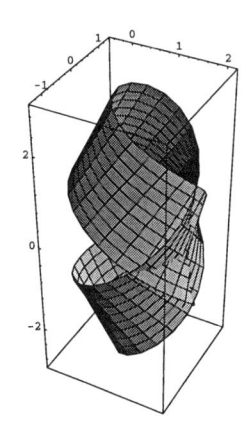

```
ParametricPlot3D[
binormalsurf[viviani[
1]][u,v]//Evaluate,
{u,-2Pi,2Pi},{v,-1,1},
PlotPoints->{40,10}];
```

The binormal surface to Viviani's curve

19.8 Exercises

1. Compute the Gaussian and mean curvatures of the generalized hyperbolic paraboloid defined on page 434. Plot the surface two ways showing the rulings.

2. Show that the patch **exptwist** defined on page 427 is a ruled patch and find the rulings.

3. Show that the following generalization of **pluckerpolar** (on page 436) also generalizes the monkey saddle (on page 299).

```
pluckerpolar[m_,n_,a_][r_,theta_]
    {a r Cos[theta],a r Sin[theta],a r^m Sin[n theta]}
```

Compute the Gaussian and mean curvature of the patch.

4. The **conical edge of Wallis**[2] is defined by

$$\text{wallis}[a, b, c](u, v) = \left(v \cos u, v \sin u, c\sqrt{a^2 - b^2 \cos^2 u} \right),$$

or

```
wallis[a_,b_,c_][u_,v_]:=
    {v Cos[u],v Sin[u],c Sqrt[a^2 - b^2 Cos[u]^2]}
```

[2] John Wallis (1616-1703). English mathematician. Although ordained as a minister, Wallis was appointed Savilian professor of geometry at Oxford in 1649. He was one of the first to use Cartesian methods to study conic sections instead of employing the traditional synthetic approach. The sign ∞ for infinity (probably adapted from the late Roman symbol for 1000) was first introduced by Wallis.

19.8 Exercises

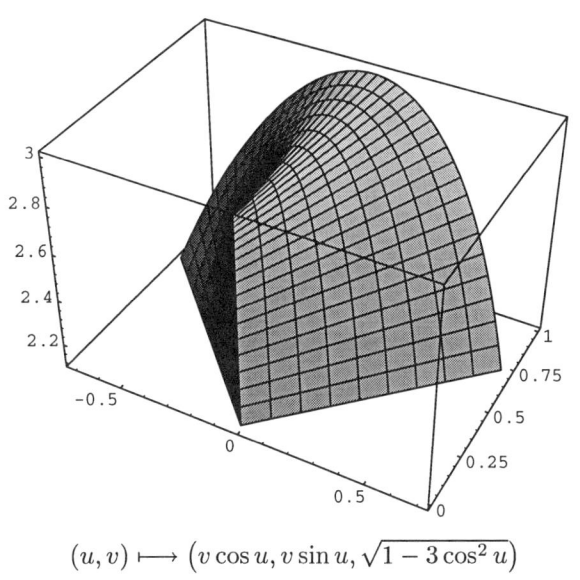

$$(u, v) \longmapsto \left(v \cos u, v \sin u, \sqrt{1 - 3\cos^2 u}\right)$$

Show that **wallis[a, b, c]** is a right conoid. Compute and plot its Gaussian and mean curvatures.

5. Complete the proof of Lemma 19.4 and prove Lemma 19.7.

6. Plot the tangent developable to the twisted cubic defined on page 189.

Tangent developable to the twisted cubic

456   Chapter 19   Ruled Surfaces

7. Plot the tangent developable to the Viviani curve defined on page 201.

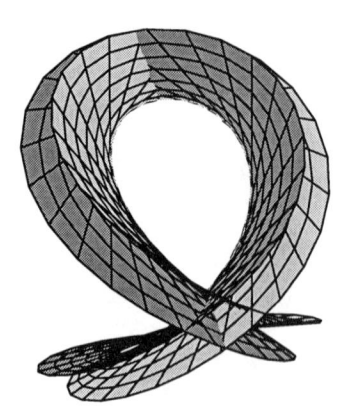

Tangent developable to Viviani's curve

8. Plot the tangent developable to the bicylinder defined on page 204.

9. Carry out the calculations of Theorem 19.6 using *Mathematica*.

10. Compute the Gaussian and mean curvatures of the right conoid whose picture is on page 451. Find the asymptotic curves.

11. Show that the normal surface to a circular helix is a helicoid. Draw the binormal surface to a circular helix and compute its Gaussian curvature.

12. For any space curve **alpha** there are other surfaces which lie between the normal surface and the binormal surface. In *Mathematica* we can define

```
perpsurf[phi_,alpha_][u_, v_]:=
    alpha[u] + v(Cos[phi]normal[alpha][u] +
        Sin[phi]binormal[alpha][u])
```

Clearly, **perpsurf[0,alpha]** is the normal surface of **alpha**, and **perpsurf[Pi/2,alpha]** is the binormal surface of **alpha**. Use **perpsurf** to construct several of these intermediate surfaces for a helix.

20

SURFACES OF REVOLUTION

Surfaces of revolution form one of the simplest nontrivial classes of surfaces. The sphere, torus and paraboloid (defined in Section 13.5) are all surfaces of revolution. An ellipsoid is a surface of revolution provided that two of its axes are equal. Many objects from everyday life such as cans, table glasses, and furniture legs are surfaces of revolution. A surface of revolution is formed by revolving a plane curve about a line in $\mathbb{R}^3$. More precisely:

Definition. *Let Π be a plane in $\mathbb{R}^3$, let A be a line in Π, and let C be a point set in Π. When C is rotated in $\mathbb{R}^3$ about A, the resulting point set M is called the* **surface of revolution** *generated by C, and C is called the* **profile curve** *of M. The line A is called the* **axis of revolution** *of M.*

For convenience, we choose Π to be the xz-plane and A to be the z-axis. We shall assume that the point set C has a parametrization $\alpha\colon (a,b) \longrightarrow C$ that is differentiable. Write $\alpha = (\varphi, \psi)$.

Definition. *The patch* **surfrev**$[\alpha]\colon [0, 2\pi] \times (a,b) \longrightarrow \mathbb{R}^3$ *defined by*

(20.1) $\qquad \text{surfrev}[\alpha](u,v) = \bigl(\varphi(v)\cos u,\, \varphi(v)\sin u,\, \psi(v)\bigr).$

is called the **standard parametrization** *of the surface of revolution M.*

One usually assumes $\varphi(v) \geq 0$ in (20.1) to ensure that the profile curve does not cross the axis of revolution; however, we do not make this assumption initially. In particular, the curvature formulas that we derive hold in the general case.

A meridian on the earth is a great circle that passes through the north and south poles. A parallel is a circle on the earth parallel to the equator. These notions generalize to an arbitrary surface of revolution.

Definition. *Let C be a point set in a plane $\Pi \subset \mathbb{R}^3$, and let $\mathcal{M}[C]$ be the surface of revolution in $\mathbb{R}^3$ generated by revolving C about a line $\mathbf{A} \subset \Pi$. A* **meridian** *on $\mathcal{M}[C]$ is the intersection of $\mathcal{M}[C]$ with a plane containing the axis of the surface of revolution $\mathbf{A}$. A* **parallel** *on $\mathcal{M}[C]$ is the intersection of $\mathcal{M}[C]$ with a plane orthogonal to the axis of the surface of revolution.*

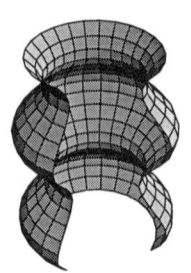

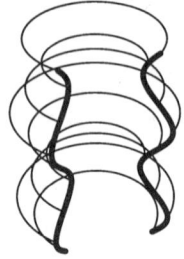

The surface of revolution generated by the curve
$$t \longmapsto (2 + (1/2)\sin 2t, t)$$
together with its meridians and parallels

The quantities $|\varphi(v)|$ and $\psi(v)$ in (20.1) have geometric interpretations: $|\varphi(v)|$ represents the radius of the parallel $u \longmapsto \bigl(\varphi(v)\cos u, \varphi(v)\sin u, \psi(v)\bigr)$, and $\psi(v)$ can be interpreted as the distance (measured positively or negatively) of the center of $u \longmapsto \bigl(\varphi(v)\cos u, \varphi(v)\sin u, \psi(v)\bigr)$ from the origin.

We prove in Theorem 20.3 that all meridians and parallels to a surface of revolution are principal curves. (See the definition of principal curve on page 459.) Furthermore, for a surface of revolution parametrized by (20.1), the meridians and parallels are coordinate curves. Since **ParametricPlot3D** always plots coordinate curves, we see that *Mathematica* naturally displays polygonal approximations of the meridians and parallels on any surface of revolution.

In Section 20.1 we establish some general facts about principal curves. The Gaussian and mean curvatures of a surface of revolution are computed in Section 20.2. In Section 20.3 we show how to generate a surface of revolution using *Mathematica*. We prove in Section 20.4 that any surface of revolution which is also a minimal surface is contained in a catenoid or a plane. The catenoid is compared with a hyperboloid of revolution in Section 20.5. We show in Section 20.6 how to find and plot the surface of revolution of a plane curve that has been constructed from its signed curvature.

Some of the most interesting surfaces of revolution are those of constant curvature. We consider them in Chapter 21.

Darboux ([Darb2, Volume 1, page 128]) put it very well when he observed that a surface of revolution has an important kinematic property: when it rotates about its axis, it glides over itself. In more modern language, we would say that a surface of revolution is the orbit of a curve in a plane Π by a 1-parameter group of rotations of $\mathbb{R}^3$ about a line in Π. A generalization of a rotation in $\mathbb{R}^3$ is a *screw motion*, that is, a Euclidean motion that is the composition of a rotation about an axis and a translation along the axis. In Section 20.8 we discuss *generalized helicoids*. These surfaces also enjoy a kinematic property: when a generalized helicoid is subjected to a screw motion, it glides over itself. In modern language, a generalized helicoid is the orbit of a curve in a plane Π by a 1-parameter group of screw motions of $\mathbb{R}^3$ about a line in Π.

20.1 Principal Curves

Principal curves are of equal importance to the asymptotic curves that we studied in Chapter 18. Just as in Chapter 18, we shall deal only with orientable surfaces. So for any surface $\mathcal{M}$ we choose a globally defined surface unit normal **U**.

Definition. *A curve $\boldsymbol{\alpha}$ on a regular surface $\mathcal{M} \subset \mathbb{R}^3$ is called a* **principal curve** *if and only if the velocity $\boldsymbol{\alpha}'$ always points in a principal direction; that is,*

$$S(\boldsymbol{\alpha}') = \mathbf{k}_i \boldsymbol{\alpha}',$$

where $\mathbf{k}_i$ is a principal curvature of $\mathcal{M}$. Here S denotes the shape operator of $\mathcal{M}$ with respect to **U**.

A useful characterization of a principal vector is provided by the following lemma.

Lemma 20.1. *A nonzero tangent vector $\mathbf{v}_p$ to a regular surface $\mathcal{M} \subset \mathbb{R}^3$ is principal if and only if*

$$S(\mathbf{v}_p) \times \mathbf{v}_p = \mathbf{0}.$$

Hence a curve $\boldsymbol{\alpha}$ on $\mathcal{M}$ is a principal curve if and only if $S(\boldsymbol{\alpha}') \times \boldsymbol{\alpha}' = 0$.

Proof. If $S(\mathbf{v}_p) = \mathbf{k}_i \mathbf{v}_p$, then $S(\mathbf{v}_p) \times \mathbf{v}_p = \mathbf{k}_i \mathbf{v}_p \times \mathbf{v}_p = 0$. Conversely, if $S(\mathbf{v}_p) \times \mathbf{v}_p = 0$, then $S(\mathbf{v}_p)$ and $\mathbf{v}_p$ are linearly dependent. ∎

More often than not, principal curves can be found by geometrical considerations. For example, in Chapter 29 we shall use triply orthogonal families of surfaces to construct principal curves. Another example is the next theorem, due to Joachimsthal[1] (see [Joach1]); it provides a simple but extremely useful way of finding principal curves. It gives a simple criterion for the intersection of two surfaces to be a principal curve on both.

Theorem 20.2. (Joachimsthal) *Let $\boldsymbol{\alpha}$ be a curve which lies on the intersection of regular surfaces $\mathcal{M}_1, \mathcal{M}_2 \subset \mathbb{R}^3$. Denote by $\mathbf{U}_i$ the unit surface normal to $\mathcal{M}_i$, $i = 1, 2$. Suppose that along $\boldsymbol{\alpha}$ the surfaces $\mathcal{M}_1$, $\mathcal{M}_2$ meet at a constant angle; that is, $\mathbf{U}_1 \cdot \mathbf{U}_2$ is constant along $\boldsymbol{\alpha}$. Then $\boldsymbol{\alpha}$ is a principal curve in $\mathcal{M}_1$ if and only if it is a principal curve in $\mathcal{M}_2$.*

Proof. Along the curve of intersection we have

$$(20.2) \qquad 0 = \frac{d}{dt}(\mathbf{U}_1 \cdot \mathbf{U}_2) = \left(\frac{d}{dt}\mathbf{U}_1\right) \cdot \mathbf{U}_2 + \mathbf{U}_1 \cdot \left(\frac{d}{dt}\mathbf{U}_2\right).$$

Suppose that the curve of intersection $\boldsymbol{\alpha}$ is a principal curve in $\mathcal{M}_1$. Then

$$(20.3) \qquad \frac{d}{dt}\mathbf{U}_1 = -\mathbf{k}_1 \boldsymbol{\alpha}',$$

where $\mathbf{k}_1$ is a principal curvature on $\mathcal{M}_1$. But $\boldsymbol{\alpha}'$ is also orthogonal to $\mathbf{U}_2$, so we conclude from (20.2) and (20.3) that

$$(20.4) \qquad \mathbf{U}_1 \cdot \left(\frac{d}{dt}\mathbf{U}_2\right) = 0.$$

Since $d\mathbf{U}_2/dt$ is also perpendicular to $\mathbf{U}_2$, it follows from (20.4) that

$$\frac{d}{dt}\mathbf{U}_2 = -\mathbf{k}_2 \boldsymbol{\alpha}'$$

for some $\mathbf{k}_2$. In other words, $\boldsymbol{\alpha}$ is a principal curve in $\mathcal{M}_2$. ∎

As an important application of Theorem 20.2 we find principal curves on a surface of revolution.

---

[1] Ferdinand Joachimsthal (1818-1861). German mathematician, student of Kummer and professor at Halle and Breslau. Theorem 20.2 is Joachimsthal's principal contribution to differential geometry. Joachimsthal was a great teacher. His book [Joach2] was one of the first to explain the results of the Monge school and Gauss.

Theorem 20.3. *Suppose the surface of revolution $\mathcal{M}[\alpha]$ generated by a plane curve α is a regular surface. Then the meridians and parallels on $\mathcal{M}[\alpha]$ are principal curves.*

Proof. Each meridian is sliced from $\mathcal{M}[\alpha]$ by a plane Π_{meridian} containing the axis of rotation of $\mathcal{M}[\alpha]$. For $\mathbf{p} \in \mathcal{M}[\alpha] \cap \Pi_{\text{meridian}}$ it is clear that the surface normal $\mathbf{U}(\mathbf{p})$ of $\mathcal{M}[\alpha]$ lies in Π_{meridian}. Hence $\mathbf{U}(\mathbf{p})$ and the unit surface normal of Π_{meridian} are orthogonal. Therefore, Theorem 20.2 implies that the meridians are principal curves of $\mathcal{M}[\alpha]$.

Next, let Π_{parallel} be a plane orthogonal to the axis of $\mathcal{M}[\alpha]$. By rotational symmetry, the unit surface normal $\mathbf{U}$ of $\mathcal{M}[\alpha]$ makes a constant angle with the unit surface normal of Π_{parallel}. Again, Theorem 20.2 implies that the parallels are principal curves. ∎

Principal curves give rise to an important class of patches for which curvature computations are especially simple.

Definition. *A* **principal patch** *is a patch $\mathbf{x} : \mathcal{U} \longrightarrow \mathbb{R}^3$ for which the curves*

$$u \longmapsto \mathbf{x}(u,v) \quad \text{and} \quad v \longmapsto \mathbf{x}(u,v)$$

are principal curves.

Lemma 20.4. *A patch $\mathbf{x} : \mathcal{U} \longrightarrow \mathbb{R}^3$ is principal if and only if*

(20.5) $$F = f = 0.$$

Proof. If $\mathbf{x}$ is principal, then both $\mathbf{x}_u$ and $\mathbf{x}_v$ are eigenvectors of the shape operator. Then Lemma 16.13 (page 371) implies that (20.5) holds. Conversely, (20.5) and the Weingarten equations (Theorem 16.10, page 369) imply that $\mathbf{x}_u$ and $\mathbf{x}_v$ are eigenvectors of the shape operator. ∎

We shall see in Section 20.2 that the standard parametrization (20.1) of a surface of revolution is a principal patch.

20.2 The Curvature of a Surface of Revolution

First, we compute the coefficients of the first and second fundamental forms, and also the unit surface normal for a general surface of revolution.

Lemma 20.5. *Let M be a surface of revolution with profile curve $\alpha = (\varphi, \psi)$. Let $\mathbf{x}: \mathcal{U} \longrightarrow \mathbb{R}^3$ be the standard parametrization (20.1) of M. Then*

(20.6) $$E = \varphi^2, \qquad F = 0, \qquad G = \varphi'^2 + \psi'^2.$$

Thus $\mathbf{x}$ is regular wherever φ and $\varphi'^2 + \psi'^2$ are nonzero. When this is the case,

(20.7) $$e = \frac{-|\varphi|\psi'}{\sqrt{\varphi'^2 + \psi'^2}}, \qquad f = 0, \qquad g = \frac{\operatorname{sign}(\varphi)(\varphi''\psi' - \varphi'\psi'')}{\sqrt{\varphi'^2 + \psi'^2}},$$

and the unit surface normal is given by

(20.8) $$\mathbf{U}(u,v) = \operatorname{sign}(\varphi) \frac{(\psi' \cos u, \psi' \sin u, \varphi')}{\sqrt{\varphi'^2 + \psi'^2}}.$$

Proof. From (20.1) it follows that the first partial derivatives of $\mathbf{x}$ are given by

(20.9) $$\begin{cases} \mathbf{x}_u = \bigl(-\varphi(v) \sin u, \varphi(v) \cos u, 0\bigr), \\ \mathbf{x}_v = \bigl(\varphi'(v) \cos u, \varphi'(v) \sin u, \psi'(v)\bigr). \end{cases}$$

Then (20.6) is immediate from (20.9) and the definitions of E, F and G. Similarly, the second partial derivatives of $\mathbf{x}$ are

(20.10) $$\begin{cases} \mathbf{x}_{uu} = \bigl(-\varphi(v) \cos u, -\varphi(v) \sin u, 0\bigr), \\ \mathbf{x}_{uv} = \bigl(-\varphi'(v) \sin u, \varphi'(v) \cos u, 0\bigr), \\ \mathbf{x}_{vv} = \bigl(\varphi''(v) \cos u, \varphi''(v) \sin u, \psi''(v)\bigr). \end{cases}$$

Next, we compute e, f and g using Lemma 16.21, page 379:

$$e = \frac{\det\begin{pmatrix} \mathbf{x}_{uu} \\ \mathbf{x}_u \\ \mathbf{x}_v \end{pmatrix}}{\sqrt{EG - F^2}} = \frac{\det\begin{pmatrix} -\varphi \cos u & -\varphi \sin u & 0 \\ -\varphi \sin u & \varphi \cos u & 0 \\ \varphi' \cos u & \varphi' \sin u & \psi' \end{pmatrix}}{\sqrt{\varphi^2(\varphi'^2 + \psi'^2)}}$$

$$= \frac{-\varphi^2 \psi'}{|\varphi|\sqrt{\varphi'^2 + \psi'^2}} = \frac{-|\varphi|\psi'}{\sqrt{\varphi'^2 + \psi'^2}},$$

20.2 The Curvature of a Surface of Revolution

$$f = \frac{\det\begin{pmatrix} \mathbf{x}_{uv} \\ \mathbf{x}_u \\ \mathbf{x}_v \end{pmatrix}}{\sqrt{EG-F^2}} = \frac{\det\begin{pmatrix} -\varphi'\sin u & \varphi'\cos u & 0 \\ -\varphi\sin u & \varphi\cos u & 0 \\ \varphi'\cos u & \varphi'\sin u & \psi' \end{pmatrix}}{\sqrt{\varphi^2(\varphi'^2+\psi'^2)}} = 0,$$

$$g = \frac{\det\begin{pmatrix} \mathbf{x}_{vv} \\ \mathbf{x}_u \\ \mathbf{x}_v \end{pmatrix}}{\sqrt{EG-F^2}} = \frac{\det\begin{pmatrix} \varphi''\cos u & \varphi''\sin u & \psi'' \\ -\varphi\sin u & \varphi\cos u & 0 \\ \varphi'\cos u & \varphi'\sin u & \psi' \end{pmatrix}}{\sqrt{\varphi^2(\varphi'^2+\psi'^2)}}$$

$$= \frac{\varphi(\varphi''\psi' - \varphi'\psi'')}{|\varphi|\sqrt{\varphi'^2+\psi'^2}} = \frac{\operatorname{sign}(\varphi)(\varphi''\psi' - \varphi'\psi'')}{\sqrt{\varphi'^2+\psi'^2}}.$$

Finally, for (20.8) we use (20.9) and compute:

$$\mathbf{U} = \frac{\det\begin{pmatrix} \mathbf{i} & \mathbf{j} & \mathbf{k} \\ \mathbf{x}_u \\ \mathbf{x}_v \end{pmatrix}}{\sqrt{EG-F^2}} = \frac{\det\begin{pmatrix} \mathbf{i} & \mathbf{j} & \mathbf{k} \\ -\varphi\sin u & \varphi\cos u & 0 \\ \varphi'\cos u & \varphi'\sin u & \psi' \end{pmatrix}}{\sqrt{\varphi^2(\varphi'^2+\psi'^2)}}$$

$$= \frac{(\varphi\psi'\cos u, \varphi\psi'\sin u, \varphi\varphi')}{|\varphi|\sqrt{\varphi'^2+\psi'^2}} = \operatorname{sign}(\varphi)\frac{(\psi'\cos u, \psi'\sin u, \varphi')}{\sqrt{\varphi'^2+\psi'^2}}. \blacksquare$$

Corollary 20.6. *The standard parametrization (20.1) of a surface of revolution is a principal patch.*

Proof. Since $F = f = 0$ for $\mathbf{x}$, the corollary is an immediate consequence of Lemma 20.4. ∎

At this point we could compute the Gaussian curvature K using Theorem 16.17; however, we prefer to compute the principal curvatures first. Recall that $u \longmapsto \mathbf{x}(u,v)$ is a parallel and $v \longmapsto \mathbf{x}(u,v)$ is a meridian. We know from Theorem 20.3 that these coordinate curves are principal curves. Hence the principal curvatures for a surface of revolution have special meaning, so we denote them by $\mathbf{k}_\pi$ and $\mathbf{k}_\mu$ instead of the usual $\mathbf{k}_1$ and $\mathbf{k}_2$.

Here $\mathbf{k}_\pi$ is the curvature of the parallel $u \longmapsto \mathbf{x}(u,v)$ and $\mathbf{k}_\mu$ is the curvature of the meridian $v \longmapsto \mathbf{x}(u,v)$.

Theorem 20.7. *The principal curvatures of a surface of revolution parametrized by the standard parametrization* (20.1) *are given by*

(20.11)
$$\begin{cases} \mathbf{k}_\mu = \dfrac{g}{G} = \dfrac{\operatorname{sign}(\varphi)(\varphi''\psi' - \varphi'\psi'')}{(\varphi'^2 + \psi'^2)^{3/2}}, \\ \mathbf{k}_\pi = \dfrac{e}{E} = \dfrac{-\psi'}{|\varphi|\sqrt{\varphi'^2 + \psi'^2}}. \end{cases}$$

The Gaussian curvature is given by

(20.12)
$$K = \frac{-\psi'^2\varphi'' + \varphi'\psi'\psi''}{\varphi(\varphi'^2 + \psi'^2)^2},$$

and the mean curvature is given by

(20.13)
$$H = \frac{\varphi(\varphi''\psi' - \varphi'\psi'') - \psi'(\varphi'^2 + \psi'^2)}{2|\varphi|(\varphi'^2 + \psi'^2)^{3/2}}.$$

Proof. Since $F = f = 0$, it follows that

$$\left\{ \frac{\mathbf{x}_u}{\|\mathbf{x}_u\|}, \frac{\mathbf{x}_v}{\|\mathbf{x}_v\|} \right\}$$

forms an orthonormal basis which diagonalizes the shape operator S wherever $\mathbf{x}$ is regular. Hence by Theorem 16.10, page 369, we have

(20.14)
$$S(\mathbf{x}_u) = \frac{e}{E}\mathbf{x}_u \quad \text{and} \quad S(\mathbf{x}_v) = \frac{g}{G}\mathbf{x}_v.$$

Thus by definition $\mathbf{k}_\pi = e/E$ and $\mathbf{k}_\mu = g/G$. Now (20.11) follows from (20.6) and (20.7).

Finally, (20.12) and (20.13) follow from the facts that $K = \mathbf{k}_\pi \mathbf{k}_\mu$ and $H = (\mathbf{k}_\pi + \mathbf{k}_\mu)/2$. ∎

Corollary 20.8. *For a surface of revolution the curvatures K, H, $\mathbf{k}_\mu$, $\mathbf{k}_\pi$ as well as the functions E, F, G, e, f, g are constant along parallels.*

Proof. All of these functions are expressible in terms of φ and ψ and their derivatives. But φ and ψ do not depend on u. ∎

We know from the geometric description of the normal curvature given in Section 16.2 that $\mathbf{k}_\mu$ at $\mathbf{p} \in \mathcal{M}$ coincides up to sign with the curvature κ of the meridian through $\mathbf{p}$. The first equation of (20.11) confirms this fact.

Any profile curve may be assumed to have unit-speed. When this is the case, the formulas we have derived so far for a surface of revolution simplify considerably.

Corollary 20.9. *Let* **x** *be the standard parametrization* (20.1) *of a surface of revolution in* $\mathbb{R}^3$ *whose profile curve* $\alpha = (\varphi, \psi)$ *has unit-speed. Then*

$$E = \varphi^2, \qquad F = 0, \qquad G = 1,$$

$$e = -|\varphi|\psi', \qquad f = 0, \qquad g = \operatorname{sign}(\varphi)(\varphi''\psi' - \varphi'\psi''),$$

$$\mathbf{k}_\mu = \operatorname{sign}(\varphi)(\varphi''\psi' - \varphi'\psi''), \qquad\qquad \mathbf{k}_\pi = \frac{-\psi'}{|\varphi|},$$

$$H = \frac{1}{2}\left(\operatorname{sign}(\varphi)(\varphi''\psi' - \varphi'\psi'') - \frac{\psi'}{|\varphi|}\right), \qquad K = \frac{-\varphi''}{\varphi}.$$

Proof. Since $1 = \|\alpha'\| = \sqrt{\varphi'^2 + \psi'^2}$, all formulas, except for the last, are immediate from (20.6), (20.7) and (20.11). The formula for K follows from (20.12) and the fact that $\varphi'\varphi'' + \psi'\psi'' = 0$. ∎

20.3 Generating a Surface of Revolution with Mathematica

Suppose that a plane curve **alpha** has been defined in *Mathematica*. The following command, which is the translation into *Mathematica* of (20.1), generates a surface of revolution from **alpha**:

```
surfrev[alpha_][u_,v_]:= {Cos[u] alpha[v][[1]],
    Sin[u] alpha[v][[1]],alpha[v][[2]]}
```

In general, if **a** is a list, **a[[k]]** extracts the kth element of the list. So the two components of the plane curve **alpha** are **alpha[v][[1]]** and **alpha[v][[2]]**.

For example, a circle with center $(a, 0)$ and radius b is parametrized by

$$t \longmapsto \mathbf{circle}[b](t) + (a, 0).$$

To obtain the surface of revolution generated by this circle, we apply **surfrev** to **circle[b]** in the following way:

surfrev[(circle[b][#] + {a,0})&][u,v]

Mathematica answers

{Cos[u] (a + b Cos[v]), (a + b Cos[v]) Sin[u], b Sin[v]}

This is the parametrization of a torus.

We can use *Mathematica* to compute **ee** and the other coefficients of the metric and second fundamental form of any surface of revolution, as well as **gcurvature** and **mcurvature**. As an example, we show how to compute these quantities for a general surface of revolution.

It is essential that *Mathematica* knows that **alpha** is a plane curve before we apply **surfrev** to **alpha**; this is best done by defining

alpha[t_] := {phi[t],psi[t]}

The functions **phi** and **psi** do not need to be defined in advance. The surface of revolution generated by **alpha** can be obtained by applying the *Mathematica* function **surfrev** (defined as above or in Appendix A or **CSPROGS.m**) to **alpha**:

surfrev[alpha][u,v]

To this command we get the response

{phi[v] Cos[u], phi[v] Sin[u], psi[v]}

Then, for example, we can compute the third component of the first fundamental form of a general surface of revolution with

gg[surfrev[alpha]][u,v]

$$phi'[v]^2 + psi'[v]^2$$

Similarly,

eee[surfrev[alpha]][u,v]

yields

$$-\left(\frac{phi[v]^2 \; psi'[v]}{\text{Sqrt}[phi[v]^2 \; (phi'[v]^2 + psi'[v]^2)]} \right)$$

We could use **PowerExpand** on `eee[surfrev[alpha]][u,v]`, but this would be incorrect when $\varphi < 0$. The reduction to (20.7) must be done by hand. For the computation via *Mathematica* of the Gaussian and mean curvatures of a general surface of revolution, see exercise 2.

The usual way to compute principal curvatures of a surface is first to compute K and H (for example, by Theorem 16.17) and then to find the principal curvatures using Corollary 16.18. However, there is a more direct way to find the principal curvatures for those surfaces for which $f = F = 0$, namely formula (20.11). We define the *Mathematica* functions

```
kmu[x_][u_,v_] := ggg[x][u,v]/gg[x][u,v]

kpi[x_][u_,v_] := eee[x][u,v]/ee[x][u,v]
```

In particular, these formulas are valid for surfaces of revolution. When they can be used, the commands **kmu** and **kpi** are much faster than the commands **k1** and **k2** (given in Section 17.3) for computing principal curvatures.

20.4 The Catenoid

The **catenoid** is the surface of revolution generated by a catenary. (See page 56.) Since `surfrev[catenary[c]][u,v]` elicits the response

$$\{c\, \text{Cos}[u]\, \text{Cosh}[\tfrac{v}{c}],\ c\, \text{Cosh}[\tfrac{v}{c}]\, \text{Sin}[u],\ v\}$$

we define

$$\mathbf{catenoid}[c](u,v) = \left(c \cos u \cosh\left(\frac{v}{c}\right), c \sin u \cosh\left(\frac{v}{c}\right), v\right),$$

or

```
catenoid[c_][u_,v_] :=
    {c Cos[u]Cosh[v/c],c Sin[u]Cosh[v/c],v}
```

It is easy to compute the principal curvatures of the catenoid by hand or by using the *Mathematica* formulas given in Section 20.3 for **kpi** and **kmu**. For example,

```
PowerExpand[kmu[catenoid[c]][u,v]]
```

yields

```
     v 2
Sech[-]
     c
———————
   c
```

Similarly, `PowerExpand[kpi[catenoid[c]][u,v]]` gives the negative of `PowerExpand[kmu[catenoid[c]][u,v]]`. Hence

$$\mathbf{k}_\mu = -\mathbf{k}_\pi = \frac{1}{c\cosh\left(\frac{v}{c}\right)^2}.$$

Thus $H = 0$ and

$$K = \frac{-1}{c^2 \cosh\left(\frac{v}{c}\right)^4}.$$

It follows that the catenoid is a minimal surface, as defined on page 374. A plane is also a minimal surface, since its shape operator vanishes, and hence both principal curvatures vanish. We now prove that these are essentially the only surfaces of revolution that are simultaneously minimal surfaces.

Theorem 20.10. *A surface of revolution $\mathcal{M}$ which is a minimal surface is contained in either a plane or a catenoid.*

Proof. Let $\mathbf{x}$ be a patch whose trace is contained in $\mathcal{M}$, and let $\boldsymbol{\alpha} = (\varphi, \psi)$ be the profile curve. Then $\mathbf{x}$ is given by (20.1). There are three cases.

Case 1. ψ' is identically 0. Then ψ is constant, so that $\boldsymbol{\alpha}$ is a horizontal line and $\mathcal{M}$ is part of a plane perpendicular to the axis of revolution.

Case 2. ψ' is never 0. Then by the inverse function theorem, ψ has an inverse ψ^{-1}. Define

$$\widetilde{\boldsymbol{\alpha}}(t) = \boldsymbol{\alpha}(\psi^{-1}(t)) = (h(t), t),$$

where $h = \varphi \circ \psi^{-1}$, and a new patch $\mathbf{y}$ by

$$\mathbf{y}(u,v) = (h(v)\cos u, h(v)\sin u, v).$$

Since $\widetilde{\boldsymbol{\alpha}}$ is a reparametrization of $\boldsymbol{\alpha}$, it follows that $\mathbf{x}$ and $\mathbf{y}$ have the same trace. Thus it suffices to show that the surface of revolution $\mathbf{y}$ is part of a catenoid.

20.4 The Catenoid

Equations (20.11) reduce to

(20.15)
$$\begin{cases} \mathbf{k}_\mu = \dfrac{g}{G} = -\dfrac{h''}{(h'^2+1)^{3/2}}, \\ \mathbf{k}_\pi = \dfrac{e}{E} = \dfrac{1}{h\sqrt{h'^2+1}}. \end{cases}$$

It follows from the assumption that $H=0$ and (20.15) that h must satisfy the differential equation

(20.16) $$h''h = 1 + h'^2.$$

To solve (20.16) we first rewrite it as

(20.17) $$\frac{2h'h''}{1+h'^2} = \frac{2h'}{h}.$$

We can integrate both sides of (20.17); the result is

$$\log(1+h'^2) = \log(h^2) - \log(c^2)$$

for some constant $c \neq 0$. Exponentiating this equation, we obtain

(20.18) $$1 + h'^2 = \left(\frac{h}{c}\right)^2.$$

In turn, this differential equation can be written as

(20.19) $$\frac{h'/c}{\sqrt{(h/c)^2 - 1}} = \frac{1}{c}.$$

Both sides of (20.19) can be integrated to yield

$$\cosh^{-1}\left(\frac{h}{c}\right) = \frac{v}{c} + b.$$

Thus the solution of (20.16) is

$$h(v) = c \cosh\left(\frac{v}{c} + b\right),$$

and so $\mathcal{M}$ is part of a catenoid.

Case 3. ψ' is zero at some points, but nonzero at others. In fact, this case cannot occur. Suppose, for example, that $\psi'(v_0) = 0$, but $\psi'(v) > 0$ for $v < v_0$. By **Case 2** the profile curve is a catenary for $v < v_0$, whose slope is given by φ'/ψ'. Then $\psi'(v_0) = 0$ implies that the slope becomes infinite at v_0. But this is impossible, since the profile curve is the graph of the function $\cosh$. ∎

For more information on the catenoid, in particular its relation to the helicoid, see Section 30.2.

20.5 The Hyperboloid of Revolution

The **hyperboloid of revolution** (generated as a surface of revolution) is a special case of the elliptical hyperboloid considered in Chapter 17. The shape of a small portion of the hyperboloid of revolution around the equator is similar to that of the catenoid, but large portions are quite different, as are the Gauss maps. The image of the Gauss map of the whole hyperboloid of revolution omits disks around the north and south poles, whereas the image of the Gauss map of the whole catenoid omits only the north and south poles.

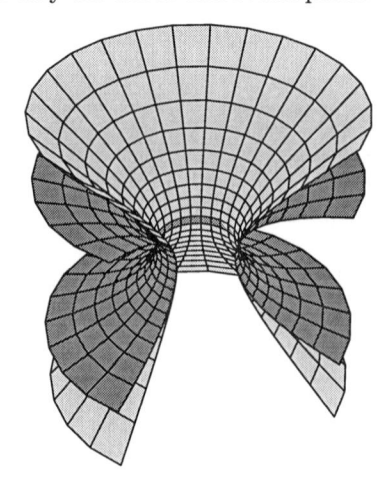

```
ParametricPlot3D[
{Append[
hyperboloid[1,1,1][u,v],
FaceForm[Yellow]],
Append[
catenoid[1][u,v],
FaceForm[LimeGreen]]}
//Evaluate,
{u,0,3Pi/2},
{v,-2,2},
Lighting->False,
Boxed->False,
Axes->None];
```

Hyperboloid of revolution and catenoid

Mathematica tells us that the Gaussian curvature of the hyperboloid of revolution is given by

```
gcurvature[hyperboloid[a,b]][u,v]
```

$$\frac{-4b^2}{(-a^2 + b^2 + a^2 \operatorname{Cosh}[2v] + b^2 \operatorname{Cosh}[2v])^2}$$

20.6 Surfaces of Revolution of Curves with Specified Curvature

In Section 6.4 we gave a *Mathematica* miniprogram **intrinsic** that constructed a plane curve from its signed curvature. The output of **intrinsic** is called an interpolating function (**InterpolatingFunction**) in *Mathematica*. An interpolating function is generated by numerical data, but can be used for input to symbolic calculations. In particular, we can use a curve constructed using **intrinsic** as input to the operator **surfrev**. Here are several examples.

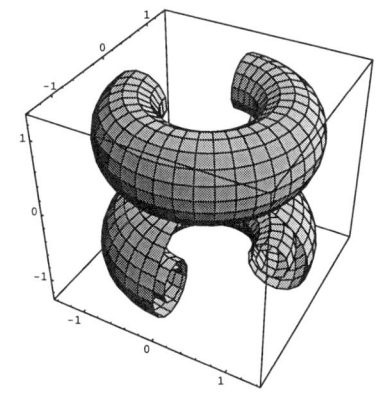

```
surfrevclothoid=
ParametricPlot3D[
surfrev[intrinsic[
-#&,0,{0,0},
{-5,5}]][u,v]
//Evaluate,
{u,0,3Pi/2},{v,-5,5},
PlotPoints->{30,60}]
```

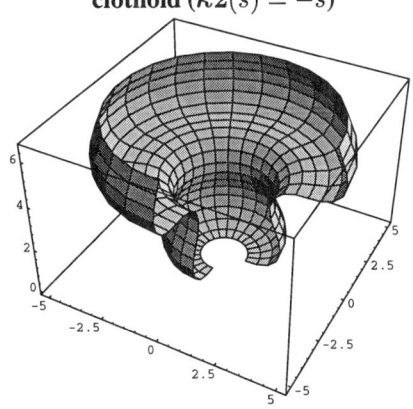

Surface of revolution of a clothoid ($\kappa 2(s) = -s$)

```
ParametricPlot3D[surfrev[
intrinsic[Sin[#]&,0,{1,0},
{0,10}]][u,v]//Evaluate,
{u,0,3Pi/2},{v,0,10}];
```

Surface of revolution of a curve whose curvature is

$$\kappa 2(s) = \sin s$$

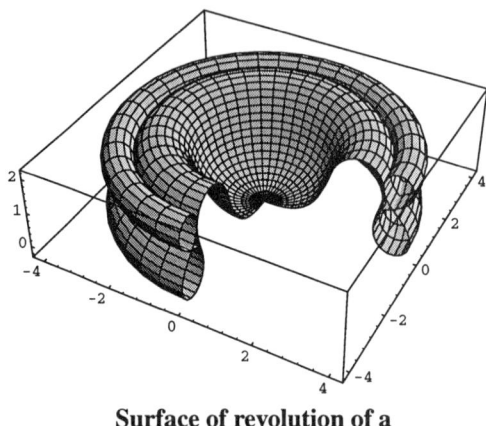

Surface of revolution of a
curve whose curvature is

$$\kappa 2(s) = sJ_0(s)$$

```
ParametricPlot3D[
surfrev[
intrinsic[
#BesselJ[0,#]&,
0,{0,0},
{0,12}]][u,v]
//Evaluate,
{u,0,3Pi/2},
{v,0,12},
PlotPoints->
{30,60}];
```

20.7 Surfaces of Revolution Generated by Data

In this section we show how to construct a surface of revolution from a 2-dimensional graphics object consisting of lines. The process is analogous to lathing a piece of wood from a profile. First, we need a miniprogram **linetolines** to change one *Mathematica* representation of a line into another.

Mathematica's command **Line** constructs the broken line segment connecting a sequence of points. Its form is

    Line[{point1,point2,...}]

The result is a **Graphics** object if the points are ordered pairs, or a **Graphics3D** object if the points are ordered triples. A completely equivalent graphics object is a list of lines connecting pairs of points:

    {Line[{point1,point2}],Line[{point2,point3}],...}

The following *Mathematica* program converts **Line[{point1,point2,...}]** into **{Line[{point1,point2}],Line[{point2,point3}],...}**.

20.7 Surfaces of Revolution Generated by Data

```
linetolines[Line[pts_]] := Table[Line[{pts[[i]],
    pts[[i + 1]]}],{i,Length[pts] - 1}]
```

Next, we use **linetolines** to construct a surface of revolution from a **Graphics** list of lines. The following command **graphics2dtosurfrev** applied to a line generates **Graphics3D** polygons that go around the surface of revolution.

```
graphics2dtosurfrev[r_,n_][Line[{{a_,b_},{c_,d_}}]] :=
    Table[Polygon[{
    {r a Cos[2Pi(k - 1)/n],r a Sin[2Pi(k - 1)/n],b},
    {r a Cos[2Pi k/n],r a Sin[2Pi k/n],b},
    {r c Cos[2Pi k/n],r c Sin[2Pi k/n],d},
    {r c Cos[2Pi (k - 1)/n],
        r c Sin[2Pi(k - 1)/n],d}}],{k,n}]
```

We have included a parameter **r** to allow changing the radius of the surface of revolution and a parameter **n** to specify the number of lines that approximate each circle of latitude of the surface of revolution.

Then we can use **Map** to apply **graphics2dtosurfrev** to all the line segments in order to generate the surface of revolution.

For example, suppose we are given a broken line segment

```
line1 = Line[{
{0.832049,0.773834}, {0.893102,0.676868},
{0.842823,0.612224}, {0.911058,0.518849},
{0.957746,0.400335}, {0.929015,0.353648},
{0.889510,0.310551}, {0.896693,0.249499},
{0.903876,0.192037}, {0.860780,0.138167}}]
```

Then a surface of revolution lathed from **line1** is given by

```
Show[Graphics3D[Map[graphics2dtosurfrev[0.5,15],
    linetolines[line1]]],Axes->True]
```

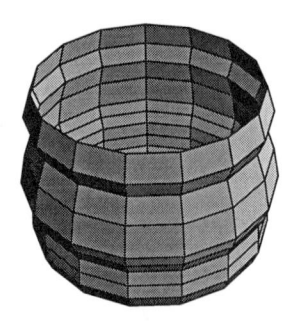

A lathed surface

The data points in **line1** might be generated by an external program. Window-based front ends to *Mathematica* can also generate lists of points by holding down the command key while moving the mouse in a graphics cell. For more details, see the booklet that accompanies the particular window-based version of *Mathematica* that is being used.

To get a surface of revolution with no edge lines we need to manipulate the graphics output. First, we define

```
dip[ins_][g_]:= $DisplayFunction[Insert[g,ins,{1,1}]]
```

Then

```
Show[Graphics3D[Map[graphics2dtosurfrev[0.5,60],
linetolines[line1]]],Boxed->False,
DisplayFunction->dip[EdgeForm[]]]
```

plots the surface of revolution with no edge lines.

A smooth lathed surface

20.8 Generalized Helicoids

Both helicoids and surfaces of revolution are examples of the class of surfaces known as generalized helicoids. Generalized helicoids were first studied by Minding[2] in 1839 (see [Mind]).

Definition. *Let Π be a plane in $\mathbb{R}^3$, let A be a line in Π, and let C be a point set in Π. Suppose that C is rotated in $\mathbb{R}^3$ about A and simultaneously displaced parallel to A so that the speed of displacement is proportional to the speed of rotation. Then the resulting point set $\mathcal{M}$ is called the **generalized helicoid** generated by C, and C is called the **profile curve** of $\mathcal{M}$. The line A is called the **axis** of $\mathcal{M}$. The ratio of the speed of displacement to the speed of rotation is called the **slant** of the generalized helicoid $\mathcal{M}$.*

The Euclidean motion (consisting of a simultaneous translation and rotation) used in this definition is called a **screw motion**. Clearly, a surface of revolution is a generalized helicoid of slant 0, and a generalized helicoid reduces to an ordinary helicoid when the profile curve is $t \longmapsto (a\,t, 0)$.

Just as we did for surfaces of revolution on page 457, we choose Π to be the xz-plane and A to be the z-axis. We shall assume that the point set C has a parametrization $\boldsymbol{\alpha}\colon (a,b) \longrightarrow C$ which is differentiable. Write $\boldsymbol{\alpha} = (\varphi, \psi)$.

Definition. *The patch $\mathbf{genhel}[\boldsymbol{\alpha}]\colon [e, f] \times (a, b) \longrightarrow \mathbb{R}^3$ defined by*

$$(20.20) \qquad \mathbf{genhel}[c, \boldsymbol{\alpha}](u, v) = \bigl(\varphi(v) \cos u,\, \varphi(v) \sin u,\, c\,u + \psi(v)\bigr)$$

*is called the **standard parametrization** of the generalized helicoid $\mathcal{M}$.*

It is also possible to define the notion of meridian for a generalized helicoid.

---

[2] Ernst Ferdinand Adolf Minding (1806-1885). German professor, later dean of the faculty, at the University of Dorpat (now Tartu) in Estonia. Minding was a self-taught mathematician. While a school teacher, he studied for his doctorate, which was awarded by Halle for a thesis on approximating the values of double integrals. In 1864 Minding became a Russian citizen and in the same year was elected to the St. Petersburg Academy. In the 1830s Minding was one of the first mathematicians to use Gauss's approach to the differential geometry of surfaces.

Definition. Let C be a point set in a plane $\Pi \subset \mathbb{R}^3$, and let $\mathcal{M}[C, c]$ be the generalized helicoid in $\mathbb{R}^3$ of slant c generated by revolving C about a line $A \subset \Pi$ and simultaneously displacing it parallel to A. A **meridian** on $\mathcal{M}[C, c]$ is the intersection of $\mathcal{M}[C, c]$ with a plane containing the axis A.

The generalization to generalized helicoids of the notion of parallel given on page 458 is more complicated. For a surface of revolution a parallel is a circle; for a generalized helicoid the corresponding curve is a helical curve, which we call a **parallel helical curve**. The v-parameter curves are the meridians and the u-parameter curves are the parallel helical curves. In contrast to the situation with a surface of revolution, the meridians and parallel helical curves will not be perpendicular to one another for a generalized helicoid of nonzero slope. In Theorem 21.6, page 495, we shall show that a generalized helicoid with the property that its meridians are principal curves must either be a surface of revolution or Dini's surface.

Generalized helicoids can be considered to be twisted surfaces of revolution. For example, here is a twisted version of the picture on page 458.

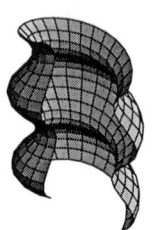

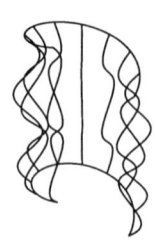

The generalized helicoid of slant 0.5 generated by the curve
$$t \longmapsto (2 + (1/2) \sin 2t, t)$$
together with its meridians and parallel helical curves

The *Mathematica* version of the generalized helicoid is

```
genhel[alpha_][sl_][u_,v_] := {Cos[u]alpha[v][[1]],
                               Sin[u]alpha[v][[1]],
                               sl u + alpha[v][[2]]}
```

The proof of the following lemma is an easy extension of the proof of Lemma 20.5, page 462.

Lemma 20.11. Let $\mathbf{x}$ be a generalized helicoid in $\mathbb{R}^3$ whose profile curve is $\boldsymbol{\alpha} = (\varphi, \psi)$. Then
$$E = \varphi^2 + c^2, \qquad F = c\psi', \qquad G = \varphi'^2 + \psi'^2,$$

20.8 Generalized Helicoids

and
$$EG - F^2 = \varphi^2(\varphi'^2 + \psi'^2) + c^2\varphi'^2.$$

Thus **x** *is regular wherever* $\varphi^2(\varphi'^2 + \psi'^2) + c^2\varphi'^2$ *is nonzero. When this is the case,*

$$e = \frac{-\varphi^2 \psi'}{\sqrt{\varphi^2(\varphi'^2 + \psi'^2) + c^2\varphi'^2}},$$

$$f = \frac{c\varphi'^2}{\sqrt{\varphi^2(\varphi'^2 + \psi'^2) + c^2\varphi'^2}},$$

$$g = \frac{\varphi(\varphi''\psi' - \varphi'\psi'')}{\sqrt{\varphi^2(\varphi'^2 + \psi'^2) + c^2\varphi'^2}}.$$

Furthermore,

(20.21)
$$K = \frac{-c^2\varphi'^4 + \varphi^3\psi'(-\psi'\varphi'' + \varphi'\psi'')}{(\varphi^2(\varphi'^2 + \psi'^2) + c^2\varphi'^2)^2},$$

and
$$H = \frac{-2c^2\varphi'^2\psi' - \varphi^2\psi'(\varphi'^2 + \psi'^2) + \varphi(c^2 + \varphi^2)(\psi'\varphi'' - \varphi'\psi'')}{2(\varphi^2(\varphi'^2 + \psi'^2) + c^2\varphi'^2)^{3/2}}.$$

The Corkscrew Surface

We call the twisted version of the sphere (obtained by using **genhel** on a circle) the **corkscrew surface** or **twisted sphere**. Explicitly, it is given by

$$\textbf{twisphere}[a, b](u, v) = (a\cos u \cos v, a \sin u \cos v, a \sin v + bu),$$

or

```
twisphere[a_,b_][u_,v_]:=
    {a Cos[u]Cos[v],a Sin[u]Cos[v],a Sin[v] + b u}
```

To compute the curvature with version 2.2 of *Mathematica*, we need to read in a package with

Needs["Algebra`Trigonometry`"]

(This is unnecessary in version 3.0). Now the command **TrigReduce** is available. The Gaussian curvature of the twisted sphere is computed with

```
Simplify[
gcurvature[twisphere[a,b]][u,v]//Cancel//TrigReduce,
Trig->False] /. b^2 - b^2 Cos[v]^2 -> b^2 Sin[v]^2
```

$$\frac{a^2 \operatorname{Cos}[v]^4 - b^4 \operatorname{Sin}[v]^2}{(a^2 \operatorname{Cos}[v]^2 + b^2 \operatorname{Sin}[v]^2)^2}$$

The same command works in version 3.0, but the output is more complicated. Here is the plot of the corkscrew surface.

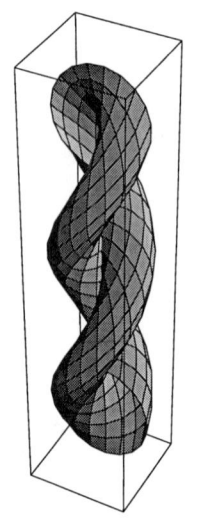

```
ParametricPlot3D[
twisphere[1,1][u,v]
//Evaluate,
{u,0,2Pi},{v,-Pi,Pi}];
```

Corkscrew surface

20.9 Exercises

1. Show that the surface of revolution generated by the graph of a function $h: \mathbb{R} \longrightarrow \mathbb{R}$, which we can parametrize as

$$\mathbf{graph}[h](t) = \bigl(t, h(t)\bigr),$$

or

```
graph[h_][t_] := {t,h[t]}
```

20.9 Exercises

is given by

$$\textbf{surfrev}[\textbf{graph}[h]](u,v) = \bigl(v\cos u, v\sin u, h(v)\bigr).$$

Find E, F, G, e, f, g, $\mathbf{k}_\mu$, $\mathbf{k}_\pi$ and K.

2. Use *Mathematica* to find the formulas for the Gaussian, mean and principal curvatures of a general surface of revolution whose profile curve is given in *Mathematica* by

```
alpha[t_]:= {phi[t],psi[t]}
```

3. Plot the surfaces of revolution corresponding to each of the following curves: cycloid, cissoid, logarithmic spiral, lemniscate, cardioid, astroid, deltoid, nephroid, triangle.

4. Use *Mathematica* to draw the three plots on page 458.

5. Prove the following converse to Joachimsthal's Theorem (Theorem 20.2).

Theorem 20.12. *Let α be a curve which lies on the intersection of regular surfaces $\mathcal{M}_1, \mathcal{M}_2 \subset \mathbb{R}^3$. Suppose that α is a principal curve in $\mathcal{M}_1$ and also in $\mathcal{M}_2$. Then the normals to $\mathcal{M}_1$ and $\mathcal{M}_2$ meet at a constant angle along α.*

6. Let α be an asymptotic curve defined by $\alpha(t) = \textbf{surfrev}[\alpha]\bigl(u(t), v(t)\bigr)$, where $\textbf{surfrev}[\alpha]$ is the standard parametrization of a surface of revolution given by (20.1). Show that u and v satisfy the differential equation

$$u'(t)^2 = \left(\frac{\varphi''(t)\psi'(t) - \varphi'(t)\psi''(t)}{\varphi(t)\psi'(t)}\right)v'(t)^2.$$

7. Compute the Gaussian curvature and mean curvature of the generalized helicoid of a catenary. Plot the generalized helicoid of `catenary[1]` with slant `0.5`.

8. A surface of revolution is formed by moving a curve γ in a plane Π_1 about a circle C in a plane Π_2, where Π_1 and Π_2 are perpendicular. There is a more general construction in which C is replaced by an arbitrary curve α in Π_2. The *Mathematica* definition is

```
gensurfrev[alpha_,gamma_][u_,v_]:=
    {alpha[u][[1]]gamma[v][[1]],
     alpha[u][[2]]gamma[v][[1]],gamma[v][[2]]}
```

Here is an example

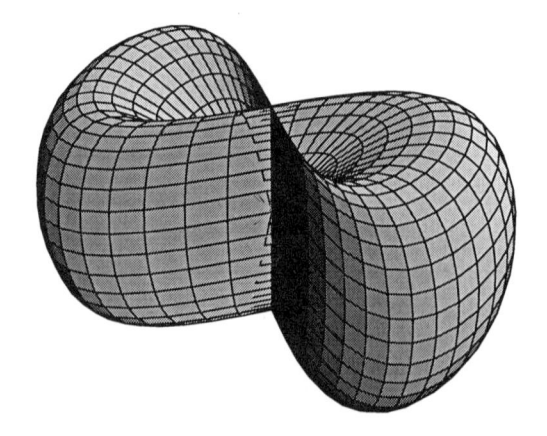

**Surface generated by moving an
eight curve along an eight curve**

Determine when a generalized surface of revolution is a principal patch. Compute the Gaussian and mean curvatures of the generalized surface of revolution formed by moving an eight curve along an eight curve.

21
SURFACES OF CONSTANT GAUSSIAN CURVATURE

The sphere, plane, cylinder and cone are the best known examples of surfaces of constant Gaussian curvature, but there are many others. In Sections 21.2 and 21.3 we discuss the surfaces of revolution that have constant Gaussian curvature. For that we need to know a little about the *elliptic integrals* of Legendre;[1] this we tackle in Section 21.1. Surfaces of revolution of constant positive curvature are discussed in Section 21.2. There are two other types besides spheres; one is shaped like an American football, the other is barrel-shaped. The corresponding result for surfaces of revolution of constant negative curvature is proved in Section 21.3. Again, there are three types. The most famous of these is the pseudosphere, which is the surface of revolution generated by a tractrix.

There are many exotic surfaces with constant Gaussian curvature. For example, we define and discuss a flat generalized helicoid in Section 21.4, Dini's surface in Section 21.5 and Kuen's surface in Section 21.6.

---
[1] Adrien Marie Legendre (1752-1833). French mathematician, who made numerous contributions to number theory and the theory of elliptic functions. In 1782 Legendre determined the attractive force for certain solids of revolution by introducing an infinite series of polynomials that are now called Legendre polynomials. In his three-volume work **Traité des fonctions elliptiques** (1825,1826,1830) Legendre founded the theory of elliptic integrals.

21.1 The Elliptic Integral of the Second Kind

We shall make a slight detour into the complicated subject of elliptic functions and integrals, which we shall need when we study surfaces of revolution of constant curvature.

Definition. *The* elliptic integral of the second kind *is defined by*

$$E(\phi|m) = \int_0^\phi (1 - m\sin^2\theta)^{\frac{1}{2}} d\theta.$$

The complete elliptic integral of the second kind *is*

$$E\left(\tfrac{\pi}{2}|m\right) = \int_0^{\frac{\pi}{2}} (1 - m\sin^2\theta)^{\frac{1}{2}} d\theta.$$

In *Mathematica* $E(\phi|m)$ is denoted by `EllipticE[phi,m]` and $E(\tfrac{\pi}{2}|m)$ is denoted by `EllipticE[m]`.

Notice that for $-\pi/2 \leq \phi \leq \pi/2$ we have $E(\phi|1) = \sin\phi$; thus $E(\phi|m)$ can be considered to be a generalization of the function sin. The corresponding generalization of sinh turns out to be $-i\,E(i\phi|-m)$ because

$$-i\,E(i\phi|-m) = \int_0^\phi (1 - m\sinh^2\theta)^{\frac{1}{2}} d\theta,$$

as can be checked by changing variables in the integral and using the identity $\sinh i x = i \sin x$. We shall need $\phi \longmapsto E(\phi|m)$ in our study of surfaces of revolution of constant positive curvature and $\phi \longmapsto -i\,E(i\phi|-m)$ in our study of surfaces of revolution of constant negative curvature.

21.2 Surfaces of Revolution of Constant Positive Curvature

We know that the sphere $S^2(a)$ is a surface of revolution and can be parametrized by

$$\mathbf{sphere}[a](u,v) = (a\cos v\cos u, a\cos v\sin u, a\sin v);$$

21.2 Surfaces of Revolution of Constant Positive Curvature

furthermore, the sphere has constant positive Gaussian curvature $K = 1/a^2$. The sphere is not the only surface in $\mathbb{R}^3$ that has constant curvature. We can take a spherical cap made up of some thin inelastic material, for example, half of a ping-pong ball. Then the spherical cap can be bent without stretching into many different shapes. Since there is no stretching involved and the Gaussian curvature is an isometric invariant (see Chapter 22), a bent cap also has constant positive curvature. On the other hand, it is clear intuitively that a whole ping-pong ball cannot be bent. Liebmann's Theorem (Theorem 28.12, page 653) states that indeed a whole sphere $S^2(a)$ is rigid.

It is remarkable that there exist other surfaces of revolution in $\mathbb{R}^3$ that have constant positive curvature. To find them we proceed backwards: we assume that we are given a surface of revolution $\mathcal{M}$ with constant positive curvature and then find restrictions on a parametrization $\mathbf{x}$ of $\mathcal{M}$.

First, we determine the profile curves of a surface of constant positive Gaussian curvature.

Theorem 21.1. *Let $\mathcal{M}$ be a surface of revolution whose Gaussian curvature is a positive constant $1/a^2$, where $a > 0$. Then $\mathcal{M}$ is part of a surface parametrized by a patch $\mathbf{x}$ of the form*

$$\mathbf{x}(u,v) = \big(\varphi(v)\cos u, \varphi(v)\sin u, \psi(v)\big),$$

where

(21.1)
$$\begin{cases} \varphi(v) = b\cos\left(\dfrac{v}{a}\right), \\ \psi(v) = \displaystyle\int_0^v \sqrt{1 - \dfrac{b^2}{a^2}\left(\sin\left(\dfrac{t}{a}\right)\right)^2}\, dt = a\, E\left(\dfrac{v}{a}\,\Big|\,\dfrac{b^2}{a^2}\right), \end{cases}$$

for some constant $b > 0$. The parameter v has one of the following ranges:

$$\text{if } b = a, \quad \text{then} \quad -\frac{\pi}{2} \le v \le \frac{\pi}{2};$$

$$\text{if } b < a, \quad \text{then} \quad -\infty \le v \le \infty;$$

$$\text{if } b > a, \quad \text{then} \quad -a\arcsin\left(\frac{a}{b}\right) \le v \le a\arcsin\left(\frac{a}{b}\right).$$

The patch $\mathbf{x}$ is regular at (u,v) if and only if $\varphi(v) \ne 0$; that is, $v \ne (n + 1/2)\pi a$.

Proof. We can assume without loss of generality that $\mathbf{x}$ is given by (20.1), page 457, and that the profile curve $\boldsymbol{\alpha} = (\varphi, \psi)$ has unit-speed; then $\varphi'^2 + \psi'^2 = 1$. If

$\mathcal{M}$ has constant positive curvature $1/a^2$, then Corollary 20.9, page 465, implies that φ satisfies the differential equation $\varphi'' + \varphi/a^2 = 0$, whose general solution is $\varphi(v) = b\cos(v/a + c)$. Without loss of generality, we can assume that $c = 0$; this amounts to translating the profile curve along the axis of revolution so that the profile curve is farthest from the axis of revolution when $v = 0$. Also, by taking the mirror image of the surface if necessary, we may assume that $b > 0$. Thus we get the first equation of (21.1).

We can assume that $\psi'(v) \geq 0$ for all v; otherwise, we replace v by $-v$. Then $\varphi'^2 + \psi'^2 = 1$ implies that $\psi'(v) = \sqrt{1 - (b^2/a^2)(\sin(v/a))^2}$, and when we integrate this equation from 0 to v we get the second equation of (21.1).

In order that $\psi(v)$ be well-defined, it is necessary that $1 - (b^2/a^2)(\sin(v/a))^2$ be positive. For $b < a$ this quantity is always positive, so $\psi(v)$ is defined for all v. When $b = a$ the profile curve is a part of a circle; in that case the requirement that $-\pi/2 \leq v \leq \pi/2$ ensures that the profile curve does not overlap itself and is thus a semicircle. If $b > a$ then $1 - (b^2/a^2)(\sin(v/a))^2 \geq 0$ if and only if $-a\arcsin(a/b) \leq v \leq a\arcsin(a/b)$. ∎

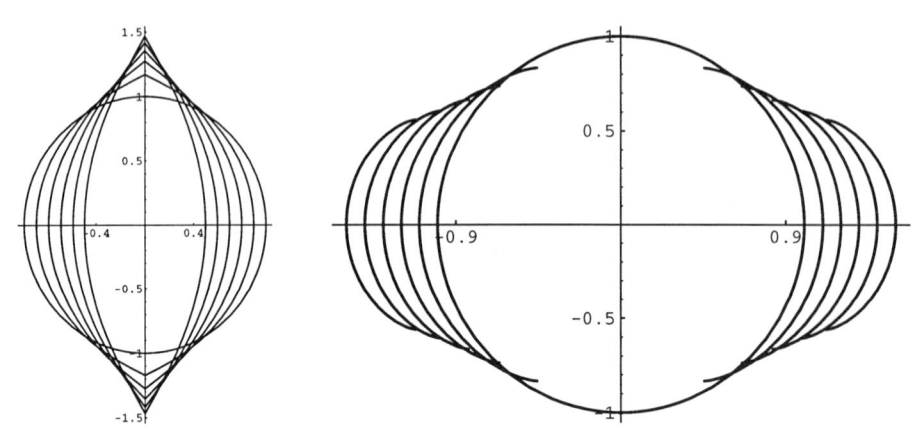

Profile curves with $b \leq a$ **Profile curves with $b \geq a$**

Theorem 21.2. *Let $S(a,b)$ be the surface of revolution whose profile curve is $\alpha = (\varphi, \psi)$ with φ, ψ given by (21.1).*

(i) *$S(a,a)$ is an ordinary sphere of radius a.*

(ii) *(Spindle or football type) If $0 < b < a$, then $S(a,b)$ is a surface of revolution that resembles an infinite string of beads, each of which is shaped like a football with its vertices on the axis of revolution.*

21.2 Surfaces of Revolution of Constant Positive Curvature

(iii) (*Bulge or barrel type*) If $0 < a < b$, then $S(a,b)$ is barrel-shaped and does not meet the axis of revolution.

Proof. When $b = a$, the profile curve is

$$v \longmapsto \left(a\cos\left(\frac{v}{a}\right), a\sin\left(\frac{v}{a}\right)\right)$$

for $-\pi/2 \leq v/a \leq \pi/2$. Thus in case (i) the profile curve is a semicircle of length πa, which when revolved about the z-axis yields a sphere $S^2(a)$ of radius a.

In case (ii) the expression inside the square root in the definition of ψ in (21.1) is always positive, so that φ is defined for $-\pi/2 \leq v/a \leq \pi/2$, just as in case (i). The profile curve is

$$v \longmapsto \left(a\cos\left(\frac{v}{a}\right), \int_0^v \sqrt{1 - \frac{b^2}{a^2}\left(\sin\left(\frac{t}{a}\right)\right)^2}\,dt\right) = \left(a\cos\left(\frac{v}{a}\right), aE\left(\frac{v}{a}\bigg|\frac{b^2}{a^2}\right)\right).$$

Thus the profile curve makes a shallower arc and meets the z-axis at

$$\pm \int_0^{\frac{\pi a}{2}} \sqrt{1 - \frac{b^2}{a^2}\left(\sin\left(\frac{t}{a}\right)\right)^2}\,dt = \pm a\, E\left(\frac{\pi}{2}\bigg|\frac{b^2}{a^2}\right).$$

The profile curve crosses the z-axis at these points, and its length between these points is πa, which is the same as the length of the semicircle in case (i). The main difference between case (ii) and case (i) is that the profile curve winds its way up and down the z-axis instead of forming a circle.

Finally, when $b > a$ the profile curve is only defined on the interval

$$-a\arcsin\left(\frac{a}{b}\right) \leq t \leq a\arcsin\left(\frac{a}{b}\right),$$

because outside that interval the expression under the square root becomes negative. The resulting surface of revolution resembles a barrel when b is moderately larger than a. ∎

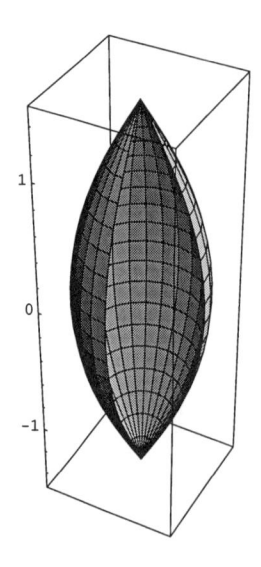

Surface $\mathcal{S}(a,b)$ **of constant curvature** $1/a^2$ **with** $b < a$

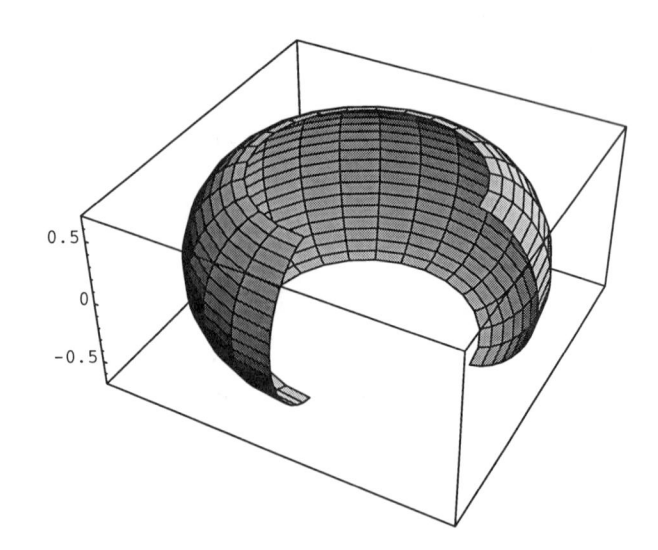

Surface $\mathcal{S}(a,b)$ **of constant curvature** $1/a^2$ **with** $b > a$

21.3 Surfaces of Revolution of Constant Negative Curvature

The determination of the surfaces of revolution of constant negative curvature proceeds along the same lines as what we did in Section 21.2. However, the resulting surfaces are quite different.

Theorem 21.3. *Let* $\mathcal{M}$ *be a surface of revolution whose Gaussian curvature is a negative constant* $-1/a^2$. *Then* $\mathcal{M}$ *is part of a surface parametrized by a patch* **x** *such that*

$$\mathbf{x}(u,v) = \big(\varphi(v)\cos u, \varphi(v)\sin u, \psi(v)\big),$$

where the profile curve $\boldsymbol{\alpha} = (\varphi, \psi)$ *is one of the following types:*

21.3 Surfaces of Revolution of Constant Negative Curvature

(i) (*Pseudosphere*)

$$(21.2) \quad \alpha(v) = \begin{cases} \left(a e^{-v/a}, \int_0^v \sqrt{1 - e^{-2t/a}}\, dt \right) & \text{for } 0 \leq v < \infty, \\ \left(a e^{v/a}, \int_0^v \sqrt{1 - e^{2t/a}}\, dt \right) & \text{for } -\infty < v \leq 0. \end{cases}$$

(ii) (*Hyperboloid type*)

$$(21.3) \quad \begin{cases} \varphi(v) = b \cosh\left(\dfrac{v}{a}\right), \\ \psi(v) = \displaystyle\int_0^v \sqrt{1 - \dfrac{b^2}{a^2}\left(\sinh\left(\dfrac{t}{a}\right)\right)^2}\, dt = -i\, a\, E\left(\dfrac{i v}{a}\,\bigg|\,\dfrac{-b^2}{a^2}\right), \end{cases}$$

for some constant $b > 0$, and v is limited by

$$-a \sinh^{-1}\left(\dfrac{a}{b}\right) \leq v \leq a \sinh^{-1}\left(\dfrac{a}{b}\right).$$

(iii) (*Conic type*)

$$(21.4) \quad \begin{cases} \varphi(v) = b \sinh\left(\dfrac{v}{a}\right), \\ \psi(v) = \displaystyle\int_0^v \sqrt{1 - \dfrac{b^2}{a^2}\left(\cosh\left(\dfrac{t}{a}\right)\right)^2}\, dt \\ \qquad\quad = -i\sqrt{a^2 - b^2}\, E\left(\dfrac{i v}{a}\,\bigg|\,\dfrac{-b^2}{a^2 - b^2}\right), \end{cases}$$

for some constant b with $0 < b \leq a$, and v is limited by

$$-a \sinh^{-1}\left(\dfrac{\sqrt{a^2 - b^2}}{b}\right) \leq v \leq a \sinh^{-1}\left(\dfrac{\sqrt{a^2 - b^2}}{b}\right).$$

Proof. Without loss of generality, $a > 0$. The general solution of $\varphi'' - \varphi/a^2 = 0$ is

$$(21.5) \quad \varphi(v) = A\, e^{v/a} + B\, e^{-v/a}.$$

Case 1. First, suppose that A is zero in (21.5). We can assume that $B > 0$ by changing v to $-v$ if necessary. Moreover, making use of the change of variables $v \longmapsto v + a \log B - a \log a$, we can assume that $B = a$; then $\varphi(v) = ae^{-v/a}$. Since

$$0 \leq \psi'(v)^2 = 1 - \varphi'(v)^2 = 1 - e^{-2v/a},$$

we must have $v \geq 0$. Thus we get the first alternative in (21.2). Similarly, $B = 0$ leads to the second alternative.

Next, suppose that A and B are both different from zero in (21.5). Using the change of variables

$$v \longmapsto v + \frac{a}{2} \log \left| \frac{B}{A} \right|$$

if necessary, we may assume that $|A| = |B|$.

Case 2. When $A = B$, we can (using the mirror image of the profile curve if necessary) assume that $A > 0$. Then

$$\varphi(v) = A\left(e^{v/a} + e^{-v/a}\right) = 2A \cosh\left(\frac{v}{a}\right).$$

We put $b = A/2$ and obtain (21.3).

Case 3. If $A = -B$, we can assume that $A > 0$. (If necessary, we can change v to $-v$.) Thus (21.5) becomes

$$\varphi(v) = A\left(e^{v/a} - e^{-v/a}\right) = 2A \sinh\left(\frac{v}{a}\right).$$

We put $b = A/2$ and obtain (21.4). ∎

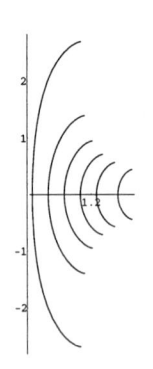

Profile curves for
constant negative curvature
surfaces of hyperboloid type

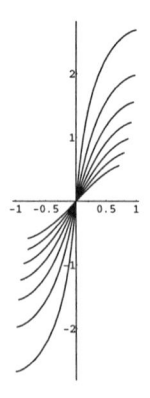

Profile curves for
constant negative curvature
surfaces of conic type

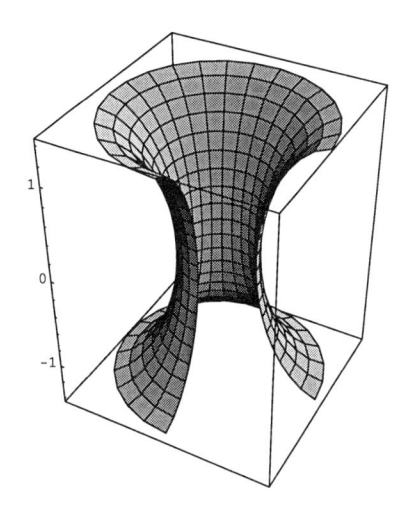

21.3 Surfaces of Revolution of Constant Negative Curvature

Constant negative curvature surface of hyperboloid type

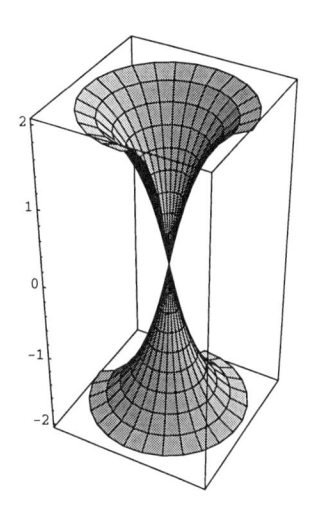

Constant negative curvature surface of conic type

Corollary 21.4. *The surface of revolution whose profile curve is given by* (21.2) *is a* **pseudosphere** *or* **tractoid**; *that is, the surface of revolution of a tractrix.*

Proof. This is an immediate consequence of Lemma 3.1, page 63, and part (i) of Theorem 21.3. ∎

According to Lemma 3.1, we can define the pseudosphere by

(21.6) $\mathbf{pseudosphere}[a](u,v) = a\bigl(\cos u \sin v, \sin u \sin v, \cos v + \log(\tan(v/2))\bigr),$

or in *Mathematica* by

```
pseudosphere[a_][u_,v_] := a{Cos[u]Sin[v],
        Sin[u]Sin[v],Cos[v] + Log[Tan[v/2]]}
```

Indeed, it is easy to check with *Mathematica* that **pseudosphere**[a] has constant curvature $-1/a^2$.

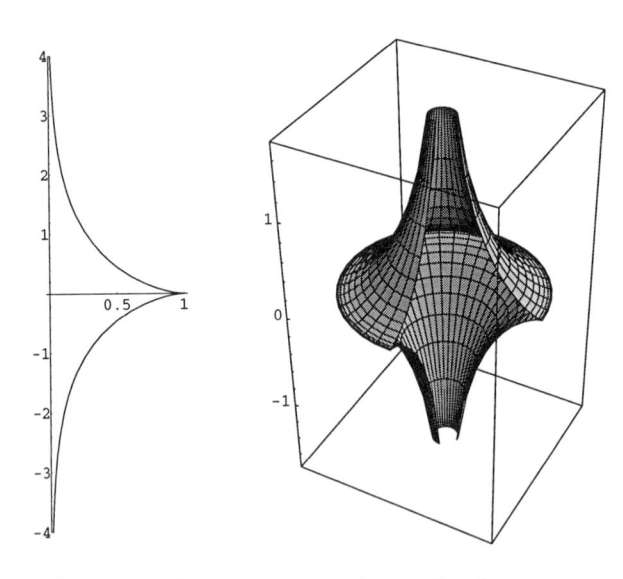

Profile curve of a pseudosphere

A pseudosphere

21.4 Flat Generalized Helicoids

In this section we determine the flat generalized helicoids.

Theorem 21.5. *A generalized helicoid is flat if and only if its profile curve α can be parametrized as*

$$(21.7) \qquad \alpha(t) = \left(t, \pm\left(t\sqrt{a^2 - \frac{c^2}{t^2}} + c\arcsin\left(\frac{c}{at}\right)\right)\right).$$

Proof. Equation (20.21), page 477, implies that a generalized helicoid is flat if and only if its profile curve $\alpha = (\varphi, \psi)$ satisfies the differential equation

$$(21.8) \qquad \varphi^3(\varphi'\psi'\psi'' - \psi'^2\varphi'') - c^2\varphi'^4 = 0.$$

The profile curve can be parametrized so that $\varphi(t) = t$; then (21.8) reduces to

$$(21.9) \qquad t^3\psi'\psi'' - c^2 = 0.$$

21.4 Flat Generalized Helicoids

We can integrate (21.9) once by inspection:

(21.10) $$\psi'(t) = \pm\sqrt{a^2 - \frac{c^2}{t^2}},$$

where a is a constant of integration, which can be assumed to be positive. Then *Mathematica* can be used to integrate the right-hand side of (21.10):

```
Integrate[Sqrt[a^2 - c^2/t^2],t]
```

```
            2    2  2
          -c  + a  t                       c
  t Sqrt[-----------] + c ArcSin[-----------]
              2                       2
              t                   Sqrt[a ] t
```

Thus we find that up to a constant of integration

(21.11) $$\psi(t) = \pm\left(t\sqrt{a^2 - \frac{c^2}{t^2}} + c \arcsin\left(\frac{c}{a\,t}\right)\right). \blacksquare$$

Theorem 21.5 motivates the following definition in *Mathematica*:

```
zcprofile[pm_,a_,c_][t_]:=
    {t,pm(t Sqrt[a^2 - c^2/t^2] + c ArcSin[c/(a t)])}
```

```
                              Show[
                               {ParametricPlot[
                                Evaluate[
                                zcprofile[1,1,1][t]],
                                {t,-3,-1},
                                AspectRatio->
                                Automatic,
                                DisplayFunction->
                                Identity],
                                ParametricPlot[
                                Evaluate[
                                zcprofile[1,1,1][t]],
                                {t,1,3},
                                AspectRatio->
                                Automatic,
                                DisplayFunction->
                                Identity]},
                                DisplayFunction->
                                $DisplayFunction];
```

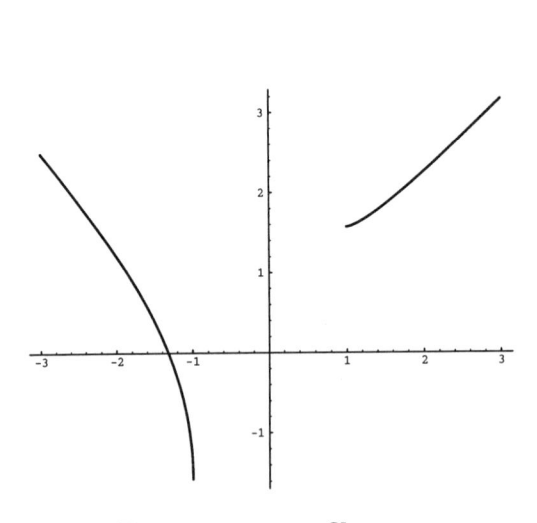

Zero curvature profile curve

Let us use *Mathematica* to verify that the generalized helicoid generated by $t \longmapsto (t, \psi(t))$ does indeed have zero curvature. First, we find the parametrization of the generalized helicoid whose profile curve is $t \longmapsto (t, \psi(t))$ with the command

```
genhel[zcprofile[pm,a,c]][c][u,v]
```

$$\left\{v \cos[u],\ v \sin[u],\ c\,u + pm\left(\text{Sqrt}\left[a^2 - \frac{c^2}{v^2}\right] v + c \arcsin\left[\frac{c}{a\,v}\right]\right)\right\}$$

Next, we compute the curvature of the generalized helicoid generated by **zcprofile**.

```
gcurvature[genhel[zcprofile[
    pm,a,c]][c]][u,v]//Together//PowerExpand//Simplify
```

yields

21.5 Dini's Surface

$$\frac{c^2(-1+\text{pm}^2)}{(c^2-c^2\,\text{pm}^2+v^2+a^2\,\text{pm}^2\,v^2)}$$

Since **pm** is ± 1, we find that indeed the Gaussian curvature of **zcprofile** vanishes.

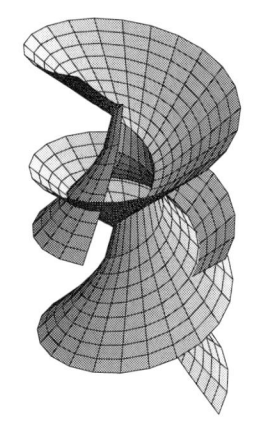

```
ParametricPlot3D[
{genhel[zcprofile[1,
1,1]][1][u,v],
genhel[zcprofile[-1,
1,1]][1][u,v]}//Evaluate,
{u,0,7Pi/2},{v,1,5},
PlotPoints->{40,10},
Boxed->False,Axes->None];
```

**Zero curvature
generalized helicoid**

21.5 Dini's Surface

The twisted pseudosphere is called **Dini's**[2] **surface**; the explicit parametrization (obtained using **genhel** on a tractrix) is:

$$\mathbf{dini}[a,b](u,v) = \bigl(a\cos u\sin v,\, a\sin u\sin v,\, a\sin v,\, a\cos u + \log(\tan(v/2)) + b\,u\bigr),$$

or

[2] Ulisse Dini (1845-1918). Italian mathematician who worked mainly in Pisa. He made fundamental contributions to surface theory and real analysis. His statue can be found a short distance from the Leaning Tower.

```
dini[a_,b_][u_,v_] := {a Cos[u] Sin[v],
                      a Sin[u] Sin[v],
                      a(Cos[v] + Log[Tan[v/2]]) + b u}
```

Let us compute the curvature of Dini's surface using *Mathematica*:

Simplify[gcurvature[dini[a,b]][u,v]]

$$-\left(\frac{1}{a^2 + b^2}\right)$$

Thus in contrast to the twisted sphere, Dini's surface has constant curvature. Here is a plot of Dini's surface, showing clearly that it is a deformation by twisting of the pseudosphere:

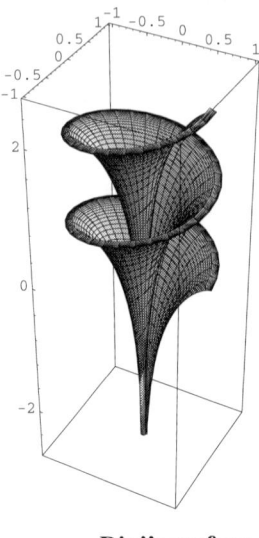

```
ParametricPlot3D[
dini[1,0.2][u,v]
//Evaluate,
{u,0,4Pi},{v,0.001,2},
PlotPoints->{60,40}];
```

Dini's surface

The following theorem is taken from volume 1 page 353 of Luigi Bianchi's[3] classic

---

[3] Luigi Bianchi (1856-1928). Italian mathematician who worked mainly in Pisa. Although he is most remembered for the "Bianchi identities", he also made fundamental contributions to surface theory.

text, **Lezioni di Geometria Differenziale** ([Bian]).

Theorem 21.6. *Let $\mathcal{M}$ be a generalized helicoid with the property that the meridians are principal curves. Then $\mathcal{M}$ is part of Dini's surface.*

Proof. Without loss of generality, we can suppose that the profile curve of the generalized helicoid is of the form

$$\alpha(t) = (t, \psi(t)),$$

so that

$$\mathbf{genhel}[c, \alpha](u, v) = (v \cos u, v \sin u, c u + \psi(v)).$$

It can easily be checked with *Mathematica* that the formula for the unit normal $\mathbf{U}$ to $\mathbf{genhel}[c, \alpha]$ is given by

$$\mathbf{U}(u, v) = \frac{(-c \sin u + v \psi'(v) \cos u, c \cos u + v \psi'(v) \sin u, -v)}{\sqrt{c^2 + v^2 + v^2 \psi'(v)^2}}.$$

The meridian

$$v \longmapsto \mathbf{genhel}[c, \alpha](u, v)$$

of the generalized helicoid is the intersection of the generalized helicoid with a plane Π through its axis. The unit normal to this plane is given by

$$\mathbf{V}(u, v) = (-\sin u, \cos u, 0).$$

Now suppose that α is a principal curve on the generalized helicoid. Since α is automatically a principal curve on Π, Theorem 20.12, page 479, (the converse to Joachimsthal's Theorem) implies that the vector fields $\mathbf{U}$ and $\mathbf{V}$ meet at a constant angle σ along α. Thus

(21.12) $$\cos \sigma = \mathbf{U} \cdot \mathbf{V} = \frac{c}{\sqrt{c^2 + v^2 + v^2 \psi'(v)^2}}.$$

Then (21.12) implies that ψ satisfies the differential equation

(21.13) $$v^2 (1 + \psi'(v)^2) = c^2 \tan^2 \sigma.$$

But (21.13) implies that α is a tractrix, because the differential equation given on page 62 is satisfied. Hence the generalized helicoid is a surface of Dini. ∎

21.6 Kuen's Surface

A more complicated surface of constant negative curvature is **Kuen's**[4] **surface** ([Kuen]). It can be parametrized by

$$\mathbf{kuen}[a](u,v) = \left(\frac{2a(\cos u + u \sin u)\sin v}{1 + u^2 \sin^2 v}, \right.$$
$$\frac{2a(\sin u - u \cos u)\sin v}{1 + u^2 \sin^2 v},$$
$$\left. a\left(\log(\tan(v/2)) + \frac{2\cos v}{1 + u^2 \sin^2 v}\right)\right)$$

or

```
kuen[a_][u_,v_]:=
    {2a(Cos[u] + u Sin[u])Sin[v]/(1 + u^2 Sin[v]^2),
     2a(Sin[u] - u Cos[u])Sin[v]/(1 + u^2 Sin[v]^2),
     a(Log[Tan[v/2]] + 2Cos[v]/(1 + u^2 Sin[v]^2))}
```

An extraordinary plaster model was made of this surface; plate 86 of [Fischer] is a photo of this plaster model. Reckziegel has given an excellent description of Kuen's surface. (See pages 30–41 of the Commentaries to [Fischer].)

Mathematica can be used to compute the various curvatures of Kuen's surface and to draw it. It turns out that $F = f = 0$ so that the u- and v-parameter curves of the parametrization **kuen** are principal; these are the curves visible in the picture below. Furthermore, *Mathematica* checks for us that indeed the Gaussian curvature of **kuen** is given by $K = -a^{-2}$. Here are the principal curvatures as computed with *Mathematica*:

```
kmu[kuen[a]][u,v]//PowerExpand

        4 u Sin[v]
---------------------
      2    2
a (2 - u + u Cos[2 v])

kpi[kuen[a]][u,v]//PowerExpand

        2    2
-((2 - u + u Cos[2 v]) Csc[v])
-------------------------------
            4 a u
```

[4]Th. Kuen used Bianchi's parametrization to make a plaster model of the surface (see [Kuen]).

Here is one view of Kuen's surface using *Mathematica*:

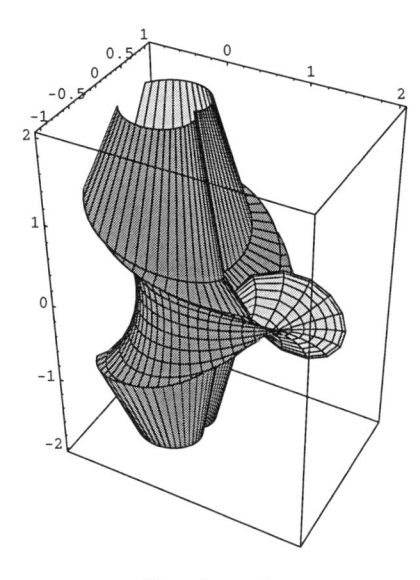

Kuen's surface

It was drawn with the command

```
ParametricPlot3D[
Append[kuen[1][u,v],
FaceForm[Green,Yellow]]//Evaluate,
{u,-4,4},{v,0.01,Pi - 0.01},Lighting->False,
PlotPoints->{60,15}];
```

21.7 Exercises

1. **Flat surfaces of revolution.** Show that a surface of revolution whose Gaussian curvature is 0 is a part of a plane, circular cone or circular cylinder.

2. Plot $\phi \mapsto E(\phi|m)$ and $\phi \mapsto -i\,E(i\phi|-m)$ for $m = 0, 0.2, 0.4, 0.6, 0.8, 1.0$. Use **Table**.

3. The *Mathematica* definition of the profile curve of a surface of revolution of constant positive curvature is

```
cpcprofile[pm_,a_,b_][t_]:=
    {pm b Cos[t/a],a EllipticE[t/a,b^2/a^2]}
```

where **pm** is ±1. Use it to draw the profile curves on page 484.

4. Show that a surface of revolution of constant positive curvature can be defined in *Mathematica* by

```
cpcsurfrev[a_,b_][u_,v_]:=
    {b Cos[u] Cos[v/a],b Sin[u]Cos[v/a],
                a EllipticE[v/a,b^2/a^2]}
```

Use it to draw the surfaces of revolution of constant positive curvature of football and barrel types on page 486.

5. Find the principal curvatures and the Gauss map of a pseudosphere.

6. The *Mathematica* definitions of the profile curves of surfaces of revolution of constant negative curvature of hyperboloid and conic types are

```
cnchyprofile[pm_,a_,b_][t_]:=
    {pm b Cosh[t/a],-I a EllipticE[I t/a,-b^2/a^2]}
```

and

```
cnccoprofile[pm_,a_,b_][t_]:=
    {pm b Sinh[t/a],-I Sqrt[a^2 - b^2]EllipticE[I t/a,
                -b^2/(a^2 - b^2)]}
```

where **pm** is ±1. Use these formulas to draw the profile curves on page 488.

7. Show that surfaces of revolution of constant negative curvature of hyperboloid and conic types can be defined in *Mathematica* by

```
cnchysurfrev[a_,b_][u_,v_]:= {b Cos[u]Cosh[v/a],
                              b Sin[u]Cosh[v/a],
                              - I a EllipticE[I v/a,-b^2/a^2]}
```

and

```
cnccosurfrev[a_,b_][u_,v_]:= {b Cos[u]Sinh[v/a],
                              b Sin[u]Sinh[v/a],
                              - I Sqrt[a^2 - b^2]EllipticE[I v/a,
                              -b^2/(a^2 - b^2)]}
```

Use these formulas to draw the surfaces of revolution of constant negative curvature on page 489. Also, compute the mean curvature of each of these two surfaces.

8. Find the formula in terms of the elliptic integral E for the length of the ellipse $\dfrac{x^2}{a^2} + \dfrac{y^2}{b^2} = 1$.

9. Plot Sievert's surface (a surface of constant positive curvature, see [Siev] and [Fischer, Commentary, page 38]) defined by

```
sievert[a_][u_,v_]:=
    {(2/(a + 1 - a Sin[v]^2Cos[u]^2))*
     (Sqrt[(a + 1)(1 + a Sin[u]^2)]Sin[v]/Sqrt[a])*
     Cos[-u/Sqrt[a + 1] + ArcTan[Sqrt[a + 1]Tan[u]]],
     (2/(a + 1 - a Sin[v]^2Cos[u]^2))*
     (Sqrt[(a + 1)(1 + a Sin[u]^2)]Sin[v]/Sqrt[a])*
     Sin[-u/Sqrt[a + 1] + ArcTan[Sqrt[a + 1]Tan[u]]],
     Log[Tan[v/2]]/Sqrt[a] + 2(a + 1)Cos[v]/
                ((a + 1 - a Sin[v]^2Cos[u]^2)Sqrt[a])}
```

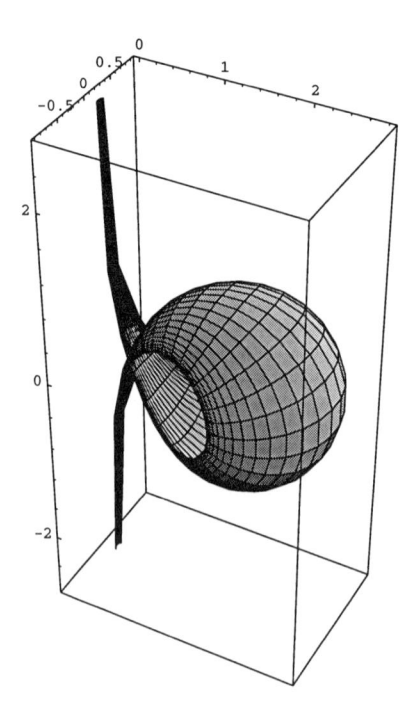

Sievert's surface

10. Fill in the details of the proof of Theorem 21.6.

22

INTRINSIC SURFACE GEOMETRY

The second fundamental form of a regular surface $\mathcal{M} \subset \mathbb{R}^3$ describes how $\mathcal{M}$ sits in $\mathbb{R}^3$. The first fundamental form of $\mathcal{M}$, on the other hand, can be used to measure distance on $\mathcal{M}$, which is intrinsic to $\mathcal{M}$. In other words, distance on $\mathcal{M}$ does not depend on the second fundamental form of $\mathcal{M}$.

It is a remarkable fact that the Gaussian curvature K of a regular surface $\mathcal{M} \subset \mathbb{R}^3$ is also intrinsic to $\mathcal{M}$, in spite of the fact that K is the product of two principal curvatures, each of which is nonintrinsic. In order to prove this fact, we establish Brioschi's formula in Section 22.1. It is an explicit (but complicated) formula for K in terms of E, F, G and their first and second partial derivatives. Brioschi's formula is used in Section 22.2 to prove Gauss's *Theorema Egregium*, which states that a surface isometry preserves the Gaussian curvature.

The *Theorema Egregium* has enormous practical importance as well as theoretical importance. It implies, for example, that no portion of the earth, which is curved, can be mapped isometrically onto a flat surface. That is why all flat maps of the earth distort distances in some way or another.

Christoffel symbols of surfaces are defined and discussed in Section 22.3. In Section 22.4 we define and study the *geodesic curvature* κ_g and *geodesic torsion* τ_g of a curve on a surface. The geodesic curvature is the generalization to surfaces of the signed curvature $\kappa 2$ that we used in Chapters 1–6 to study plane curves. Furthermore, an analog of the Frenet formulas (8.10), called the *Darboux formulas*, is established for a curve on a surface.

22.1 Intrinsic Formulas for the Gaussian Curvature

We begin by giving a formula for the Gaussian curvature of a patch **x** in terms of the dot products of the first and second partial derivatives of **x**.

Theorem 22.1. *Let* $\mathbf{x}:\mathcal{U} \longrightarrow \mathbb{R}^3$ *be a patch. Then the Gaussian curvature of* **x** *is given by*

$$(22.1) \quad K = \frac{1}{(EG - F^2)^2} \left\{ \det \begin{pmatrix} \mathbf{x}_{uu} \cdot \mathbf{x}_{vv} & \mathbf{x}_{uu} \cdot \mathbf{x}_u & \mathbf{x}_{uu} \cdot \mathbf{x}_v \\ \mathbf{x}_u \cdot \mathbf{x}_{vv} & \mathbf{x}_u \cdot \mathbf{x}_u & \mathbf{x}_u \cdot \mathbf{x}_v \\ \mathbf{x}_v \cdot \mathbf{x}_{vv} & \mathbf{x}_v \cdot \mathbf{x}_u & \mathbf{x}_v \cdot \mathbf{x}_v \end{pmatrix} \right.$$

$$\left. - \det \begin{pmatrix} \mathbf{x}_{uv} \cdot \mathbf{x}_{uv} & \mathbf{x}_{uv} \cdot \mathbf{x}_u & \mathbf{x}_{uv} \cdot \mathbf{x}_v \\ \mathbf{x}_u \cdot \mathbf{x}_{uv} & \mathbf{x}_u \cdot \mathbf{x}_u & \mathbf{x}_u \cdot \mathbf{x}_v \\ \mathbf{x}_v \cdot \mathbf{x}_{uv} & \mathbf{x}_v \cdot \mathbf{x}_u & \mathbf{x}_v \cdot \mathbf{x}_v \end{pmatrix} \right\}.$$

Proof. It follows from Lemma 16.21, page 379, and Theorem 16.17, page 377, that

$$(22.2) \quad K = \frac{eg - f^2}{EG - F^2} = \frac{(\mathbf{x}_{uu}\mathbf{x}_u\mathbf{x}_v)(\mathbf{x}_{vv}\mathbf{x}_u\mathbf{x}_v) - (\mathbf{x}_{uv}\mathbf{x}_u\mathbf{x}_v)^2}{(EG - F^2)^2}.$$

Also, if we write $\mathbf{x} = (x_1, x_2, x_3)$, then

$$(22.3) \quad (\mathbf{x}_{uu}\mathbf{x}_u\mathbf{x}_v)(\mathbf{x}_{vv}\mathbf{x}_u\mathbf{x}_v)$$

$$= \det \begin{pmatrix} x_{1uu} & x_{2uu} & x_{3uu} \\ x_{1u} & x_{2u} & x_{3u} \\ x_{1v} & x_{2v} & x_{3v} \end{pmatrix} \det \begin{pmatrix} x_{1vv} & x_{2vv} & x_{3vv} \\ x_{1u} & x_{2u} & x_{3u} \\ x_{1v} & x_{2v} & x_{3v} \end{pmatrix}$$

$$= \det \left(\begin{pmatrix} x_{1uu} & x_{2uu} & x_{3uu} \\ x_{1u} & x_{2u} & x_{3u} \\ x_{1v} & x_{2v} & x_{3v} \end{pmatrix} \begin{pmatrix} x_{1vv} & x_{1u} & x_{1v} \\ x_{2vv} & x_{2u} & x_{2v} \\ x_{3vv} & x_{3u} & x_{3v} \end{pmatrix} \right)$$

22.1 Intrinsic Formulas for the Gaussian Curvature

$$= \det \begin{pmatrix} \mathbf{x}_{uu} \cdot \mathbf{x}_{vv} & \mathbf{x}_{uu} \cdot \mathbf{x}_u & \mathbf{x}_{uu} \cdot \mathbf{x}_v \\ \mathbf{x}_u \cdot \mathbf{x}_{vv} & \mathbf{x}_u \cdot \mathbf{x}_u & \mathbf{x}_u \cdot \mathbf{x}_v \\ \mathbf{x}_v \cdot \mathbf{x}_{vv} & \mathbf{x}_v \cdot \mathbf{x}_u & \mathbf{x}_v \cdot \mathbf{x}_v \end{pmatrix}.$$

Similarly,

$$(22.4) \qquad (\mathbf{x}_{uv}\mathbf{x}_u\mathbf{x}_v)^2 = \det \begin{pmatrix} \mathbf{x}_{uv} \cdot \mathbf{x}_{uv} & \mathbf{x}_{uv} \cdot \mathbf{x}_u & \mathbf{x}_{uv} \cdot \mathbf{x}_v \\ \mathbf{x}_u \cdot \mathbf{x}_{uv} & \mathbf{x}_u \cdot \mathbf{x}_u & \mathbf{x}_u \cdot \mathbf{x}_v \\ \mathbf{x}_v \cdot \mathbf{x}_{uv} & \mathbf{x}_v \cdot \mathbf{x}_u & \mathbf{x}_v \cdot \mathbf{x}_v \end{pmatrix}.$$

Now (22.1) follows from (22.2), (22.3) and (22.4). ∎

Except for $\mathbf{x}_{uu} \cdot \mathbf{x}_{vv}$ and $\mathbf{x}_{uv} \cdot \mathbf{x}_{uv}$, we know how to express each of the dot products in (22.1) in terms of E, F, G and their derivatives. Next, we show that at least we can write the difference of the remaining two dot products in terms of the second derivatives of E, F and G.

Lemma 22.2. *Let* $\mathbf{x}: \mathcal{U} \longrightarrow \mathbb{R}^3$ *be a patch. Then*

$$(22.5) \qquad \mathbf{x}_{uu} \cdot \mathbf{x}_{vv} - \mathbf{x}_{uv} \cdot \mathbf{x}_{uv} = -\frac{1}{2}E_{vv} + F_{uv} - \frac{1}{2}G_{uu}.$$

Proof. We have

$$(22.6) \qquad \mathbf{x}_{uu} \cdot \mathbf{x}_{vv} - \mathbf{x}_{uv} \cdot \mathbf{x}_{uv} = (\mathbf{x}_u \cdot \mathbf{x}_{vv})_u - \mathbf{x}_u \cdot \mathbf{x}_{vvu} \\ -(\mathbf{x}_u \cdot \mathbf{x}_{uv})_v + \mathbf{x}_u \cdot \mathbf{x}_{uvv}.$$

Since $\mathbf{x}_{vvu} = \mathbf{x}_{uvv}$, equation (22.6) reduces to

$$\begin{aligned} \mathbf{x}_{uu} \cdot \mathbf{x}_{vv} - \mathbf{x}_{uv} \cdot \mathbf{x}_{uv} &= (\mathbf{x}_u \cdot \mathbf{x}_{vv})_u - (\mathbf{x}_u \cdot \mathbf{x}_{uv})_v \\ &= ((\mathbf{x}_u \cdot \mathbf{x}_v)_v - \mathbf{x}_{uv} \cdot \mathbf{x}_v)_u - \frac{1}{2}(\mathbf{x}_u \cdot \mathbf{x}_u)_{vv} \\ &= (\mathbf{x}_u \cdot \mathbf{x}_v)_{vu} - \frac{1}{2}(\mathbf{x}_v \cdot \mathbf{x}_v)_{uu} - \frac{1}{2}(\mathbf{x}_u \cdot \mathbf{x}_u)_{vv} \\ &= -\frac{1}{2}E_{vv} + F_{uv} - \frac{1}{2}G_{uu}. \quad \blacksquare \end{aligned}$$

Now we are ready to derive Brioschi's[1] formula (see [Brio]). It is a great deal more complicated than (22.2), but it has the advantage that it expresses the Gaussian curvature of a patch $\mathbf{x}: \mathcal{U} \longrightarrow \mathbb{R}^3$ entirely in terms of the first fundamental form.

Theorem 22.3. (Brioschi's Formula) *Let* $\mathbf{x}: \mathcal{U} \longrightarrow \mathbb{R}^3$ *be a patch. Then the Gaussian curvature of* $\mathbf{x}$ *is given by*

$$(22.7) \quad K = \frac{1}{(EG - F^2)^2} \left\{ \det \begin{pmatrix} -\frac{1}{2}E_{vv} + F_{uv} - \frac{1}{2}G_{uu} & \frac{1}{2}E_u & F_u - \frac{1}{2}E_v \\ F_v - \frac{1}{2}G_u & E & F \\ \frac{1}{2}G_v & F & G \end{pmatrix} - \det \begin{pmatrix} 0 & \frac{1}{2}E_v & \frac{1}{2}G_u \\ \frac{1}{2}E_v & E & F \\ \frac{1}{2}G_u & F & G \end{pmatrix} \right\}.$$

Proof. We use (22.5) to rewrite (22.1) as

$$(22.8) \quad K = \frac{1}{(EG - F^2)^2} \left\{ \det \begin{pmatrix} \mathbf{x}_{uu} \cdot \mathbf{x}_{vv} & \frac{1}{2}E_u & F_u - \frac{1}{2}E_v \\ F_v - \frac{1}{2}G_u & E & F \\ \frac{1}{2}G_v & F & G \end{pmatrix} - \det \begin{pmatrix} \mathbf{x}_{uv} \cdot \mathbf{x}_{uv} & \frac{1}{2}E_v & \frac{1}{2}G_u \\ \frac{1}{2}E_v & E & F \\ \frac{1}{2}G_u & F & G \end{pmatrix} \right\}.$$

---

[1] Francesco Brioschi (1824-1897). Italian mathematician. Professor first in Padua, then in Milan. Brioschi was one of the founders of modern mathematics in Italy. In addition to his work in differential geometry, Brioschi used elliptical modular functions to solve equations of the fifth and sixth degrees.

22.1 Intrinsic Formulas for the Gaussian Curvature

It is possible to expand each 3×3 determinant in (22.8) by means of 2×2 determinants. In the resulting expansion each of the expressions $\mathbf{x}_{uu} \cdot \mathbf{x}_{vv}$ and $\mathbf{x}_{uv} \cdot \mathbf{x}_{uv}$ occurs with the same factor $EG - F^2$. Thus we can manipulate the determinants in (22.8) using (22.5) so that (22.8) reduces to (22.7). ∎

Brioschi's formula in its full generality is tedious to use in hand calculations. However, in a frequently occurring case Brioschi's formula simplifies considerably.

Corollary 22.4. *Let* $\mathbf{x}: \mathcal{U} \longrightarrow \mathbb{R}^3$ *be a patch for which* $F = 0$. *Then the Gaussian curvature of* $\mathbf{x}$ *is given by*

$$(22.9) \qquad K = \frac{-1}{\sqrt{EG}} \left\{ \frac{\partial}{\partial u}\left(\frac{1}{\sqrt{E}} \frac{\partial \sqrt{G}}{\partial u} \right) + \frac{\partial}{\partial v}\left(\frac{1}{\sqrt{G}} \frac{\partial \sqrt{E}}{\partial v} \right) \right\}$$

$$= \frac{-1}{2\sqrt{EG}} \left\{ \frac{\partial}{\partial u}\left(\frac{G_u}{\sqrt{EG}} \right) + \frac{\partial}{\partial v}\left(\frac{E_v}{\sqrt{EG}} \right) \right\}.$$

Proof. When $F = 0$, equation (22.7) reduces to

$$(22.10) \qquad K = \frac{1}{(EG)^2}\left\{ \det\begin{pmatrix} -\frac{1}{2}E_{vv} - \frac{1}{2}G_{uu} & \frac{1}{2}E_u & -\frac{1}{2}E_v \\ -\frac{1}{2}G_u & E & 0 \\ \frac{1}{2}G_v & 0 & G \end{pmatrix} \right.$$

$$\left. - \det\begin{pmatrix} 0 & \frac{1}{2}E_v & \frac{1}{2}G_u \\ \frac{1}{2}E_v & E & 0 \\ \frac{1}{2}G_u & 0 & G \end{pmatrix} \right\}$$

$$= \frac{1}{(EG)^2}\left\{ \left(-\frac{1}{2}E_{vv} - \frac{1}{2}G_{uu}\right) EG \right.$$

$$\left. + \frac{1}{4}E_u G_u G + \frac{1}{4}E_v G_v E + \frac{1}{4}E_v^2 G + \frac{1}{4}G_u^2 E \right\}$$

$$= -\frac{1}{2}\frac{E_{vv}}{EG} + \frac{1}{4}\frac{E_v^2}{E^2 G} + \frac{1}{4}\frac{E_v G_v}{EG^2} - \frac{1}{2}\frac{G_{uu}}{EG} + \frac{1}{4}\frac{G_u^2}{EG^2} + \frac{1}{4}\frac{E_u G_u}{E^2 G}.$$

On the other hand,

$$(22.11) \qquad \frac{1}{\sqrt{EG}} \frac{\partial}{\partial v}\left(\frac{1}{\sqrt{G}} \frac{\partial \sqrt{E}}{\partial v} \right) = \frac{1}{\sqrt{EG}}\left\{ -\frac{G_v}{2G^{\frac{3}{2}}} \frac{\partial \sqrt{E}}{\partial v} + \frac{1}{\sqrt{G}} \frac{\partial^2 \sqrt{E}}{\partial v^2} \right\}$$

$$= \frac{1}{\sqrt{EG}} \left\{ -\frac{G_v E_v}{4G^{\frac{3}{2}}\sqrt{E}} + \frac{1}{2\sqrt{G}} \frac{\partial}{\partial v} \left(E^{-\frac{1}{2}} E_v \right) \right\}$$

$$= \frac{1}{\sqrt{EG}} \left\{ -\frac{G_v E_v}{4G^{\frac{3}{2}}\sqrt{E}} + \frac{1}{2\sqrt{G}} \left(-\frac{E_v^2}{2E^{\frac{3}{2}}} + \frac{E_{vv}}{\sqrt{E}} \right) \right\}$$

$$= -\frac{G_v E_v}{4G^2 E} - \frac{E_v^2}{4E^2 G} + \frac{E_{vv}}{2EG}.$$

Similarly,

(22.12) $$\frac{1}{\sqrt{EG}} \frac{\partial}{\partial u} \left(\frac{1}{\sqrt{E}} \frac{\partial \sqrt{G}}{\partial u} \right) = -\frac{E_u G_u}{4E^2 G} - \frac{G_u^2}{4EG^2} + \frac{G_{uu}}{2EG}.$$

Then (22.9) follows from (22.10)–(22.12). ∎

The *Mathematica* version of Brioschi's formula is

```
brioschicurvature[x_][u_,v_]:=
   (Det[{
      {-D[ee[x][uu,vv],{vv,2}]/2 +
        D[ff[x][uu,vv],uu,vv] -
        D[gg[x][uu,vv],{uu,2}]/2,
       D[ee[x][uu,vv],uu]/2,
       D[ff[x][uu,vv],uu] -
        D[ee[x][uu,vv],vv]/2},
      {D[ff[x][uu,vv],vv] -
        D[gg[x][uu,vv],uu]/2,
       ee[x][uu,vv],ff[x][uu,vv]},
      {D[gg[x][uu,vv],vv]/2,
       ff[x][uu,vv],gg[x][uu,vv]}}]
   -Det[{
      {0,D[ee[x][uu,vv],vv]/2,
        D[gg[x][uu,vv],uu]/2},
      {D[ee[x][uu,vv],vv]/2,
        ee[x][uu,vv],ff[x][uu,vv]},
      {D[gg[x][uu,vv],uu]/2,
        ff[x][uu,vv],gg[x][uu,vv]}}])/
   (ee[x][uu,vv]gg[x][uu,vv]-ff[x][uu,vv]^2)^2 /.
    {uu->u,vv->v}
```

Although the command **brioschicurvature[x][u,v]** will compute the curvature of a patch **x** in $\mathbb{R}^3$, for most purposes it is much faster to use **gcurvature**. For example, let us compare how long it takes each function to compute the Gaussian curvature of a monkey saddle:

brioschicurvature[monkeysaddle][u,v]//Simplify//Timing

$$\left\{3.3\ \text{Second},\ \frac{-36(u^2+v^2)}{(1+9u^4+18u^2v^2+9v^4)^2}\right\}$$

gcurvature[monkeysaddle][u,v]//Simplify//Timing

$$\left\{0.933333\ \text{Second},\ \frac{-36(u^2+v^2)}{(1+9u^4+18u^2v^2+9v^4)^2}\right\}$$

22.2 Gauss's Theorema Egregium

We can now prove one of the most celebrated theorems of the 19th century. Mathematicians at the end of the 18th century such as Euler and Monge had used the Gaussian curvature, but only defined as the product of the principal curvatures. Since each principal curvature of a surface depends on the particular way the surface is embedded in $\mathbb{R}^3$, there is no obvious reason to suppose that the product of the principal curvatures is intrinsic to $\mathcal{M}$. Gauss's discovery ([Gauss], published in 1828) that the product of the principal curvatures depends only on the intrinsic geometry of the surface revolutionized differential geometry. This is the *Theorema Egregium*; Gauss's proof of it is not simple. The difficulty of the proof can be compressed into Brioschi's formula (22.7). (A formula equivalent to (22.7) already occurs in [Gauss].) As a matter of fact, Brioschi derived his formula in [Brio] in order to provide a new proof of the *Theorema Egregium*. (Another proof of the *Theorema Egregium* using geodesic balls is given in [BDP].)

Theorem 22.5. (Gauss's Theorema Egregium) *Let* $\Phi\colon \mathcal{M}_1 \longrightarrow \mathcal{M}_2$ *be a local isometry between regular surfaces* $\mathcal{M}_1, \mathcal{M}_2 \subset \mathbb{R}^3$. *Denote the Gaussian curvatures of* $\mathcal{M}_1$ *and* $\mathcal{M}_2$ *by* K_1 *and* K_2. *Then*

$$K_1 = K_2 \circ \Phi.$$

Proof. Let $\mathbf{x}: \mathcal{U} \longrightarrow \mathbb{R}^3$ be a regular patch on $\mathcal{M}_1$, and put $\mathbf{y} = \Phi \circ \mathbf{x}$. Then the restriction $\Phi|\mathbf{x}(\mathcal{U}): \mathbf{x}(\mathcal{U}) \longrightarrow \mathbf{y}(\mathcal{U})$ is a local isometry. Lemma 15.6, page 347, tells us that
$$E_x = E_y, \qquad F_x = F_y, \qquad G_x = G_y.$$
But then $(E_x)_u = (E_y)_u$, $(F_x)_u = (F_y)_u$, $(G_x)_u = (G_y)_u$, and similarly for the other partial derivatives. In particular, Theorem 22.3 implies that $K_x = K_y$, where K_x and K_y denote the Gaussian curvatures of $\mathbf{x}$ and $\mathbf{y}$. Hence
$$K_2 \circ \Phi = K_y \circ \mathbf{y}^{-1} \circ \Phi = K_x \circ \mathbf{x}^{-1} = K_1. \quad \blacksquare$$

For another proof of Theorem 22.5, see page 651.

Examples Illustrating Gauss's Theorema Egregium

(i) The patch $\mathbf{x}: \mathbb{R}^2 \longrightarrow \mathbb{R}^3$ defined by
$$\mathbf{x}(u, v) = \left(\cos\left(\frac{u}{a}\right), \sin\left(\frac{u}{a}\right), av \right)$$

can be considered to be a surface mapping between the plane and a circular cylinder. It is easy to check that both $E = G = 1$ and $F = 0$ for both the plane and $\mathbf{x}$; hence $\mathbf{x}$ is a local isometry. On the other hand, $\mathbf{x}$ preserves neither the second fundamental form nor the mean curvature.

(ii) We shall show in Section 30.2 that the helicoid is locally isometric to the catenoid.

(iii) A sphere and plane are not locally isometric because the Gaussian curvature of a sphere is nonzero, but the Gaussian curvature of a plane is zero. The earth is not exactly a sphere, but still has nonzero Gaussian curvature. This is the precise reason why any flat map of a portion of the earth must distort distances.

(iv) Theorem 22.5 states that a local isometry preserves curvature. There are, however, curvature preserving diffeomorphisms that are not local isometries. We describe a classical example of such a map Φ from the funnel surface (page 423)
$$(u, v) \longmapsto (a v \cos u, a v \sin u, b \log v)$$
to the helicoid (page 357)
$$(u, v) \longmapsto (a v \cos u, a v \sin u, b u).$$

It is given by

$$\Phi(a v \cos u, a v \sin u, b \log v) = (a v \cos u, a v \sin u, b u).$$

When the curvatures of these two surfaces are computed, they turn out to be the same:

$$\frac{-b^2}{b^2 + a^2 v^2}.$$

Hence Φ is a diffeomorphism which preserves curvature. The metric of the funnel surface is given by

$$ds^2 = a^2 v^2 du^2 + \left(a^2 + \frac{b^2}{v^2}\right) dv^2,$$

while the metric of the helicoid is given by

$$ds^2 = (b^2 + a^2 v^2) du^2 + a^2 dv^2.$$

There is no local isometry between the two surfaces, since the v-parameters would have to correspond to preserve K, but then E would not be preserved. Thus the converse to Theorem 22.5 is false.

(v) As we noted on page 383, any rotation about the z-axis preserves the Gaussian curvature of the monkey saddle. It can be checked (see exercise 10) that only the rotations by an angle which is an integer multiple of $2\pi/3$ are isometries. This example also shows that the converse to Theorem 22.5 is false.

22.3 Christoffel Symbols

Let $\mathbf{x}: \mathcal{U} \longrightarrow \mathbb{R}^3$ be a regular patch. According to the definitions (16.10), page 368, of e, f and g, the normal components of $\mathbf{x}_{uu}$, $\mathbf{x}_{uv}$ and $\mathbf{x}_{vv}$ are given by

$$\mathbf{x}_{uu}^\perp = e\,\mathbf{U}, \qquad \mathbf{x}_{uv}^\perp = f\,\mathbf{U}, \qquad \mathbf{x}_{vv}^\perp = g\,\mathbf{U},$$

where

$$\mathbf{U} = \frac{\mathbf{x}_u \times \mathbf{x}_v}{\|\mathbf{x}_u \times \mathbf{x}_v\|}$$

is the unit surface normal of $\mathbf{x}$. For many applications the tangential components $\mathbf{x}_{uu}$, $\mathbf{x}_{uv}$ and $\mathbf{x}_{vv}$ are also needed. It turns out that these tangential components can be expressed in terms of E, F and G alone. Since the expressions are, however, a little complicated, we first define eight (really six) functions of E, F and G, called the Christoffel[2] symbols, originally defined in [Chris]. The Christoffel symbols will also be needed when we study geodesics in Chapter 26, and for geodesic curvature in Sections 22.4 and 25.5.

Definition. Let $\mathbf{x}: \mathcal{U} \longrightarrow \mathbb{R}^n$ be a regular patch. Then the **Christoffel symbols** Γ^i_{jk} $(i, j, k = 1, 2)$ corresponding to $\mathbf{x}$ are defined by

(22.13)
$$\begin{cases} \Gamma^1_{11} = \dfrac{G\,E_u - 2F\,F_u + F\,E_v}{2(E\,G - F^2)}, & \Gamma^2_{11} = \dfrac{2E\,F_u - E\,E_v - F\,E_u}{2(E\,G - F^2)}, \\[2mm] \Gamma^1_{12} = \dfrac{G\,E_v - F\,G_u}{2(E\,G - F^2)}, & \Gamma^2_{12} = \dfrac{E\,G_u - F\,E_v}{2(E\,G - F^2)}, \\[2mm] \Gamma^1_{22} = \dfrac{2G\,F_v - G\,G_u - F\,G_v}{2(E\,G - F^2)}, & \Gamma^2_{22} = \dfrac{E\,G_v - 2FF_v + F\,G_u}{2(E\,G - F^2)}, \end{cases}$$

and $\Gamma^1_{21} = \Gamma^1_{12}$, $\Gamma^2_{21} = \Gamma^2_{12}$.

In one important case the relations (22.13) simplify greatly:

Lemma 22.6. *The formulas for the Christoffel symbols for a patch $\mathbf{x}$ for which $F = 0$ simplify to*

$$\Gamma^1_{11} = \frac{E_u}{2E}, \qquad \Gamma^2_{11} = \frac{-E_v}{2G},$$

$$\Gamma^1_{12} = \frac{E_v}{2E}, \qquad \Gamma^2_{12} = \frac{G_u}{2G},$$

$$\Gamma^1_{22} = \frac{-G_u}{2E}, \qquad \Gamma^2_{22} = \frac{G_v}{2G}.$$

In Theorem 16.10, page 369, we computed the normal components of $\mathbf{x}_{uu}$, $\mathbf{x}_{uv}$, $\mathbf{x}_{vv}$ in terms of E, F, G, e, f, g. Now we show that the tangential components are expressible in terms of the Christoffel symbols. Thus ultimately the tangential components of $\mathbf{x}_{uu}$, $\mathbf{x}_{uv}$, $\mathbf{x}_{vv}$ are expressible in terms of E, F, G and their first and second partial derivatives.

---

[2] Elwin Bruno Christoffel (1829-1900). German mathematician, professor in Strasbourg. He introduced the three-index symbols that we now call Christoffel symbols for use in invariant theory.

22.3 Christoffel Symbols

Theorem 22.7. (**Gauss Equations**) *Let* $\mathbf{x}$ *be a regular patch on a regular surface in* $\mathbb{R}^3$ *with surface normal* $\mathbf{U}$. *Then*

(22.14)
$$\begin{cases} \mathbf{x}_{uu} = \Gamma_{11}^1 \mathbf{x}_u + \Gamma_{11}^2 \mathbf{x}_v + e\,\mathbf{U}, \\ \mathbf{x}_{uv} = \Gamma_{12}^1 \mathbf{x}_u + \Gamma_{12}^2 \mathbf{x}_v + f\,\mathbf{U}, \\ \mathbf{x}_{vv} = \Gamma_{22}^1 \mathbf{x}_u + \Gamma_{22}^2 \mathbf{x}_v + g\,\mathbf{U}. \end{cases}$$

Proof. Since $\mathbf{x}$ is regular, $\mathbf{x}_u, \mathbf{x}_v$ and $\mathbf{U}$ form a basis for $\mathbb{R}_\mathbf{p}^3$ at each point $\mathbf{p}$ in the trace of $\mathbf{x}$. Therefore, we can write

(22.15)
$$\begin{cases} \mathbf{x}_{uu} = \alpha_1 \mathbf{x}_u + \alpha_2 \mathbf{x}_v + \alpha_3 \mathbf{U}, \\ \mathbf{x}_{uv} = \beta_1 \mathbf{x}_u + \beta_2 \mathbf{x}_v + \beta_3 \mathbf{U}, \\ \mathbf{x}_{vv} = \gamma_1 \mathbf{x}_u + \gamma_2 \mathbf{x}_v + \gamma_3 \mathbf{U}. \end{cases}$$

We must compute the coefficients $\alpha_1, \alpha_2, \alpha_3, \beta_1, \beta_2, \beta_3, \gamma_1, \gamma_2, \gamma_3$ in (22.15). Note that

(22.16) $$\mathbf{U} \cdot \mathbf{U} = 1, \qquad \mathbf{U} \cdot \mathbf{x}_u = \mathbf{U} \cdot \mathbf{x}_v = 0.$$

It is clear from the definitions (16.10) of e, f, g and (22.16) that

(22.17) $$e = \alpha_3, \qquad f = \beta_3, \qquad g = \gamma_3.$$

To determine the other coefficients in (22.15), we take the scalar product of each of the equations in (22.15) with $\mathbf{x}_u$ and $\mathbf{x}_v$. We get:

(22.18)
$$\begin{cases} \alpha_1 E + \alpha_2 F = \mathbf{x}_{uu} \cdot \mathbf{x}_u = \dfrac{1}{2}\dfrac{\partial}{\partial u}\|\mathbf{x}_u\|^2 = \dfrac{E_u}{2}, \\ \alpha_1 F + \alpha_2 G = \mathbf{x}_{uu} \cdot \mathbf{x}_v = \dfrac{\partial}{\partial u}(\mathbf{x}_u \cdot \mathbf{x}_v) - \mathbf{x}_u \cdot \mathbf{x}_{uv} = F_u - \dfrac{E_v}{2}, \\ \beta_1 E + \beta_2 F = \mathbf{x}_{uv} \cdot \mathbf{x}_u = \dfrac{1}{2}\dfrac{\partial}{\partial v}\|\mathbf{x}_u\|^2 = \dfrac{E_v}{2}, \\ \beta_1 F + \beta_2 G = \mathbf{x}_{uv} \cdot \mathbf{x}_v = \dfrac{1}{2}\dfrac{\partial}{\partial u}\|\mathbf{x}_v\|^2 = \dfrac{G_u}{2}, \\ \gamma_1 E + \gamma_2 F = \mathbf{x}_{vv} \cdot \mathbf{x}_u = \dfrac{\partial}{\partial v}(\mathbf{x}_u \cdot \mathbf{x}_v) - \mathbf{x}_v \cdot \mathbf{x}_{uv} = F_v - \dfrac{G_u}{2}, \\ \gamma_1 F + \gamma_2 G = \mathbf{x}_{vv} \cdot \mathbf{x}_v = \dfrac{1}{2}\dfrac{\partial}{\partial v}\|\mathbf{x}_v\|^2 = \dfrac{G_v}{2}. \end{cases}$$

The first two equations of (22.18) can be solved for α_1 and α_2, yielding $\alpha_1 = \Gamma^1_{11}$ and $\alpha_2 = \Gamma^2_{11}$. The other equations have similar solutions, and so the proof is complete. ∎

In the next lemma, we obtain some additional useful formulas involving the Christoffel symbols and E, F and G.

Lemma 22.8. *Let* **x** *be a regular patch. Then the following formulas give the Christoffel symbols in terms of E, F, G:*

$$(22.19) \quad \begin{cases} \Gamma^1_{11} E + \Gamma^2_{11} F = \dfrac{E_u}{2}, & \Gamma^1_{11} F + \Gamma^2_{11} G = F_u - \dfrac{E_v}{2}, \\[4pt] \Gamma^1_{12} E + \Gamma^2_{12} F = \dfrac{E_v}{2}, & \Gamma^1_{12} F + \Gamma^2_{12} G = \dfrac{G_u}{2}, \\[4pt] \Gamma^1_{22} E + \Gamma^2_{22} F = F_v - \dfrac{G_u}{2}, & \Gamma^1_{22} F + \Gamma^2_{22} G = \dfrac{G_v}{2}, \end{cases}$$

and

$$(22.20) \quad \begin{cases} \Gamma^1_{11} + \Gamma^2_{12} = \left(\log \sqrt{EG - F^2}\right)_u, \\[4pt] \Gamma^1_{12} + \Gamma^2_{22} = \left(\log \sqrt{EG - F^2}\right)_v. \end{cases}$$

Proof. Equations (22.19) follow from the definitions of the Christoffel symbols. We prove the first equation of (22.20). From (22.19) we compute

$$2(EG - F^2)(\Gamma^1_{11} + \Gamma^2_{12}) = G E_u - 2F F_u + E G_u = (EG - F^2)_u.$$

Hence

$$\Gamma^1_{11} + \Gamma^2_{12} = \frac{(EG - F^2)_u}{2(EG - F^2)} = \left(\log \sqrt{EG - F^2}\right)_u. \quad \blacksquare$$

The Christoffel symbol Γ^1_{11} of a regular patch **x** can be computed in *Mathematica* with the commands:

```
christoffel[1,1,1][x_][u_,v_]:=
    Simplify[(D[ee[x][uu,vv],vv]*ff[x][uu,vv] -
    2ff[x][uu,vv]*D[ff[x][uu,vv],uu] +
    D[ee[x][uu,vv],uu]*gg[x][uu,vv])/
    (2*(ee[x][uu,vv]gg[x][uu,vv] -
    ff[x][uu,vv]^2))] /. {uu->u,vv->v}
```

For example,

christoffel[1,1,1][monkeysaddle][u,v]

$$\frac{18\,u\,(u^2 - v^2)}{1 + 9\,u^4 + 18\,u^2\,v^2 + 9\,v^4}$$

Similar commands for the other Christoffel symbols are given in Appendix A or in **CSPROGS.m**.

22.4 Geodesic Curvature and Torsion

Let $\mathcal{M} \subset \mathbb{R}^n$ be a regular surface and $\beta\colon (c,d) \longrightarrow \mathcal{M}$ a unit-speed curve. We know that the acceleration of β in $\mathbb{R}^n$ has a component perpendicular to $\mathcal{M}$ and a component tangent to $\mathcal{M}$. Therefore, we can write

$$(22.21) \qquad \beta''(s) = \beta''(s)^\top + \beta''(s)^\perp$$

for $c < s < d$, where $\beta''(s)^\top$ is tangent to $\mathcal{M}$ and $\beta''(s)^\perp$ is perpendicular to $\mathcal{M}$.

Definition. *The* **unsigned geodesic curvature** $\widetilde{\kappa}_g[\beta]$ *and the* **normal curvature** $\kappa_n[\beta]$ *of a unit-speed curve* $\beta\colon (c,d) \longrightarrow \mathcal{M} \subset \mathbb{R}^n$ *are given by*

$$(22.22) \qquad \widetilde{\kappa}_g[\beta](s) = \|\beta''(s)^\top\| \quad \text{and} \quad \kappa_n[\beta](s) = \|\beta''(s)^\perp\|.$$

The normal and unsigned geodesic curvatures of an arbitrary-speed curve $\alpha\colon (a,b) \longrightarrow \mathcal{M} \subset \mathbb{R}^n$ *are defined to be the normal and geodesic curvatures of any unit-speed reparametrization of* α.

There is a simple relation between $\widetilde{\kappa}_g[\beta]$, $\kappa_n[\beta]$ and the curvature $\kappa[\beta]$ that we defined in Section 8.2.

Lemma 22.9. *Let* $\mathcal{M} \subset \mathbb{R}^n$ *be a regular surface and* $\beta\colon (c,d) \longrightarrow \mathcal{M}$ *a unit-speed curve. Then the unsigned geodesic and normal curvatures of* β *are related to the curvature* $\kappa[\beta]$ *of* β *as a curve in* $\mathbb{R}^n$ *by the formula*

$$\kappa[\beta]^2 = \widetilde{\kappa}_g[\beta]^2 + \kappa_n[\beta]^2.$$

Proof. We have

$$\kappa[\beta](s)^2 = \|\beta''(s)\|^2 = \|\beta''(s)^\top\|^2 + \|\beta''(s)^\perp\|^2 = \widetilde{\kappa}_g[\beta](s)^2 + \kappa_n[\beta](s)^2. \blacksquare$$

Next, we specialize to an oriented regular surface $\mathcal{M} \subset \mathbb{R}^3$ with unit normal $\mathbf{U}$, so that we can define a refined version of $\widetilde{\kappa}_g[\beta]$. Let $\beta\colon (c,d) \longrightarrow \mathcal{M}$ be a unit-speed curve. Then $\beta''(s)^\top$ is a multiple of $J\beta'(s) = \mathbf{U} \times \beta'(s)$ for $c < s < d$.

Definition. Let $\mathcal{M}$ be an oriented regular surface in $\mathbb{R}^3$. The **geodesic curvature** $\kappa_g[\beta]$ of a curve $\beta\colon (c,d) \longrightarrow \mathcal{M}$ is given (for $c < s < d$) by

$$(22.23) \qquad \beta''(s)^\top = \kappa_g[\beta](s) \cdot J\beta'(s) = \kappa_g[\beta](s) \mathbf{U} \times \beta'(s).$$

Here is the refined version of Lemma 22.9.

Corollary 22.10. *Let $\mathcal{M}$ be an oriented regular surface in $\mathbb{R}^3$ and let $\beta\colon (c,d) \longrightarrow \mathcal{M}$ be a unit-speed curve. Then*

$$|\kappa_g[\beta]| = \widetilde{\kappa}_g[\beta] \qquad \text{and} \qquad \kappa[\beta]^2 = \kappa_g[\beta]^2 + \kappa_n[\beta]^2.$$

Proof. We have

$$\widetilde{\kappa}_g[\beta](s) = \|\beta''(s)^\top\| = \|\kappa_g[\beta](s) J\beta'(s)\| = |\kappa_g[\beta](s)| \, \|J\beta'(s)\| = |\kappa_g[\beta](s)|. \quad \blacksquare$$

We collect some elementary facts about the curvatures κ_g and κ_n for the case of an oriented regular surface $\mathcal{M} \subset \mathbb{R}^3$. First, we show that the normal curvature of a unit-speed curve contains the same geometric information as the normal curvature that we defined in Section 16.2.

Lemma 22.11. *Let $\mathcal{M} \subset \mathbb{R}^3$ be an oriented regular surface, and choose a unit surface normal $\mathbf{U}$ to $\mathcal{M}$. Let $\mathbf{u_p}$ be a unit tangent vector to $\mathcal{M}$ at $\mathbf{p}$. If β is any unit-speed curve in $\mathcal{M}$ with $\beta(s) = \mathbf{p}$ and $\beta'(s) = \mathbf{u_p}$, then*

$$\kappa_n[\beta](s) = \mathbf{k}(\mathbf{u_p}),$$

where $\mathbf{k}(\mathbf{u_p})$ denotes the normal curvature of $\mathcal{M}$ in the direction $\mathbf{u_p}$.

Proof. By definition (see page 363)

$$\mathbf{k}(\mathbf{u_p}) = S(\mathbf{u_p}) \cdot \mathbf{u_p} = S(\beta'(s)) \cdot \beta'(s).$$

But according to Lemma 16.3, page 361, we have $S(\beta'(s)) \cdot \beta'(s) = \beta''(s) \cdot \mathbf{U}$. Because $(\beta''(s) \cdot \mathbf{U})\mathbf{U}$ is the component of $\beta''(s)$ perpendicular to $\mathcal{M}$, the lemma follows from the definition of $\kappa_n[\beta]$. $\blacksquare$

Next, we find simple formulas for the geodesic and normal curvatures of an arbitrary-speed curve on an oriented regular surface in $\mathbb{R}^3$.

22.4 Geodesic Curvature and Torsion

Lemma 22.12. Let $\alpha\colon (a,b) \longrightarrow \mathcal{M} \subset \mathbb{R}^3$ be an arbitrary-speed curve in an oriented regular surface $\mathcal{M}$ in $\mathbb{R}^3$. Let $\mathbf{U}$ be a surface normal to $\mathcal{M}$. Then the normal and geodesic curvatures of α are given (for $a < t < b$) by

$$\kappa_n[\alpha](t) = \frac{\alpha''(t) \cdot \mathbf{U}}{\|\alpha'(t)\|^2} \tag{22.24}$$

and

$$\kappa_g[\alpha](t) = \frac{\big(\alpha''(t)\mathbf{U}\alpha'(t)\big)}{\|\alpha'(t)\|^3} = \frac{\alpha''(t) \cdot J\alpha'(t)}{\|\alpha'(t)\|^3}. \tag{22.25}$$

Proof. Let β be a unit-speed reparametrization of α and write

$$\alpha(t) = \beta\big(s(t)\big)$$

for $a < t < b$. Then

$$\alpha'(t) = s'(t)\beta'\big(s(t)\big) \quad \text{and} \quad \alpha''(t) = s''(t)\beta'\big(s(t)\big) + s'(t)^2\beta''\big(s(t)\big),$$

so that

$$\kappa_n[\alpha](t) = \kappa_n[\beta]\big(s(t)\big) = \beta''\big(s(t)\big) \cdot \mathbf{U} = \frac{\alpha''(t) \cdot \mathbf{U}}{s'(t)^2}.$$

Since $s'(t) = \|\alpha'(t)\|$, we get (22.24).

To prove (22.25) we first note that

$$\alpha''(t) \times \alpha'(t) = \big(s''(t)\beta'\big(s(t)\big) + s'(t)^2\beta''\big(s(t)\big)\big) \times s'(t)\beta'\big(s(t)\big)$$
$$= s'(t)^3 \beta''\big(s(t)\big) \times \beta'\big(s(t)\big).$$

Therefore,

$$\big(\alpha''(t)\mathbf{U}\alpha'(t)\big) = -\mathbf{U} \cdot \big(\alpha''(t) \times \alpha'(t)\big) = -s'(t)^3 \mathbf{U} \cdot \big(\beta''\big(s(t)\big) \times \beta'\big(s(t)\big)\big).$$

Furthermore,

$$\kappa_g[\alpha](t) = \kappa_g[\beta]\big(s(t)\big) = \beta''\big(s(t)\big) \cdot J\beta'\big(s(t)\big)$$
$$= \beta''\big(s(t)\big) \cdot \big(\mathbf{U} \times \beta'\big(s(t)\big)\big) = \frac{\big(\alpha''(t)\mathbf{U}\alpha'(t)\big)}{s'(t)^3},$$

and so we get (22.25). ∎

Corollary 22.13. *In the case of a plane curve the geodesic curvature κ_g coincides with the signed curvature $\kappa 2$.*

Proof. Formulas (1.9) and (22.25) are the same. ∎

Thus the geodesic curvature is the generalization to an arbitrary surface of the signed curvature of a plane curve.

Next, we show that the geodesic curvature κ_g is intrinsic to $\mathcal{M} \subset \mathbb{R}^3$ by expressing κ_g in terms of the coefficients E, F, G of the first fundamental form with respect to a patch. Since the computation is local, there is no loss of generality by assuming that $\mathcal{M}$ is oriented.

Theorem 22.14. *Let $\boldsymbol{\alpha}: (a,b) \longrightarrow \mathcal{M} \subset \mathbb{R}^3$ be a curve whose trace is contained in $\mathbf{x}(\mathcal{U})$, where $\mathbf{x}: \mathcal{U} \longrightarrow \mathcal{M}$ is a patch. Write $\boldsymbol{\alpha}(t) = \mathbf{x}(u(t), v(t))$ for $a < t < b$. Then*

$$(22.26) \quad \kappa_g[\boldsymbol{\alpha}] \|\boldsymbol{\alpha}'\|^3 = \sqrt{EG - F^2}\Big(-\Gamma_{11}^2 u'^3 + \Gamma_{22}^1 v'^3$$

$$- (2\Gamma_{12}^2 - \Gamma_{11}^1)u'^2 v' + (2\Gamma_{12}^1 - \Gamma_{22}^2)u' v'^2 + u''v' - v''u' \Big).$$

Proof. We have $\boldsymbol{\alpha}' = \mathbf{x}_u u' + \mathbf{x}_v v'$, and also

$$\boldsymbol{\alpha}'' = \mathbf{x}_{uu} u'^2 + 2\mathbf{x}_{uv} u' v' + \mathbf{x}_{vv} v'^2 + \mathbf{x}_u u'' + \mathbf{x}_v v'',$$

so that

$$\boldsymbol{\alpha}'' \times \boldsymbol{\alpha}' = (\mathbf{x}_{uu} \times \mathbf{x}_u) u'^3 + (\mathbf{x}_{vv} \times \mathbf{x}_v) v'^3 + (\mathbf{x}_{uu} \times \mathbf{x}_v + 2\mathbf{x}_{uv} \times \mathbf{x}_u) u'^2 v'$$

$$+ (\mathbf{x}_{vv} \times \mathbf{x}_u + 2\mathbf{x}_{uv} \times \mathbf{x}_v) u' v'^2 + (u''v' - v''u') \mathbf{x}_u \times \mathbf{x}_v.$$

Therefore,

$$(22.27) \quad (\boldsymbol{\alpha}'' \times \boldsymbol{\alpha}') \cdot (\mathbf{x}_u \times \mathbf{x}_v) = (\mathbf{x}_{uu} \times \mathbf{x}_u) \cdot (\mathbf{x}_u \times \mathbf{x}_v) u'^3$$

$$+ (\mathbf{x}_{vv} \times \mathbf{x}_v) \cdot (\mathbf{x}_u \times \mathbf{x}_v) v'^3 + (\mathbf{x}_{uu} \times \mathbf{x}_v + 2\mathbf{x}_{uv} \times \mathbf{x}_u) \cdot (\mathbf{x}_u \times \mathbf{x}_v) u'^2 v'$$

$$+ (\mathbf{x}_{vv} \times \mathbf{x}_u + 2\mathbf{x}_{uv} \times \mathbf{x}_v) \cdot (\mathbf{x}_u \times \mathbf{x}_v) u' v'^2 + (u''v' - v''u') \|\mathbf{x}_u \times \mathbf{x}_v\|^2.$$

We next show how to express the coefficients of u'^3, v'^3, $u'^2 v'$, $u' v'^2$ on the right-hand side of (22.27) in terms of E, F, G and the Christoffel symbols. First, we use the Lagrange identity (8.1), page 182, and (22.13) to compute

$$(22.28) \quad (\mathbf{x}_{uu} \times \mathbf{x}_u) \cdot (\mathbf{x}_u \times \mathbf{x}_v) = (\mathbf{x}_{uu} \cdot \mathbf{x}_u)(\mathbf{x}_u \cdot \mathbf{x}_v) - (\mathbf{x}_{uu} \cdot \mathbf{x}_v) \|\mathbf{x}_u\|^2$$

$$= \frac{1}{2} E_u F - \left(F_u - \frac{1}{2} E_v \right) E = -(EG - F^2) \Gamma_{11}^2;$$

22.4 Geodesic Curvature and Torsion

similarly,
$$(22.29) \qquad (\mathbf{x}_{vv} \times \mathbf{x}_v) \cdot (\mathbf{x}_u \times \mathbf{x}_v) = (EG - F^2)\Gamma_{22}^1.$$

Also,

$$(22.30) \quad (\mathbf{x}_{uu} \times \mathbf{x}_v + 2\mathbf{x}_{uv} \times \mathbf{x}_u) \cdot (\mathbf{x}_u \times \mathbf{x}_v)$$

$$= (\mathbf{x}_{uu} \cdot \mathbf{x}_u)\|\mathbf{x}_v\|^2 - (\mathbf{x}_{uu} \cdot \mathbf{x}_v)(\mathbf{x}_u \cdot \mathbf{x}_v) + 2(\mathbf{x}_{uv} \cdot \mathbf{x}_u)(\mathbf{x}_u \cdot \mathbf{x}_v) - 2(\mathbf{x}_{uv} \cdot \mathbf{x}_v)\|\mathbf{x}_u\|^2$$

$$= \frac{1}{2}E_u G - \left(F_u - \frac{1}{2}E_v\right)E + E_v F - G_u E = (EG - F^2)(\Gamma_{11}^1 - 2\Gamma_{12}^2)$$

and

$$(22.31) \qquad (\mathbf{x}_{vv} \times \mathbf{x}_u + 2\mathbf{x}_{uv} \times \mathbf{x}_v) \cdot (\mathbf{x}_u \times \mathbf{x}_v) = -(EG - F^2)(\Gamma_{22}^2 - 2\Gamma_{12}^1).$$

When (22.28)–(22.31) are substituted into (22.27), we get (22.26). ∎

Just as the geodesic curvature $\kappa_g[\alpha]$ of a curve α is a variant of the ordinary curvature $\kappa[\alpha]$, there is a variant of the torsion of α.

Definition. *Let $\mathcal{M}$ be an oriented regular surface in $\mathbb{R}^n$. The **geodesic torsion** $\kappa_g[\alpha]$ of a curve $\alpha:(c,d) \longrightarrow \mathcal{M}$ is given by*

$$(22.32) \qquad \tau_g[\alpha](t) = \frac{S\alpha'(t) \cdot J\alpha'(t)}{\|\alpha'(t)\|^2}.$$

The geodesic torsion measures the failure of a curve to be a principal curve because of the following easily proved lemma (see exercise 3):

Lemma 22.15. *The definition of geodesic torsion does not depend on the parametrization. Furthermore, a curve on a surface has geodesic torsion zero if and only if it is a principal curve.*

There is also an analog of the Frenet formulas for a curve α on an oriented surface $\mathcal{M} \subset \mathbb{R}^3$. Just as we did in Theorem 8.4, we put $\mathbf{T} = \alpha'$. However, since we want to study the relation between the geometry of α and $\mathcal{M}$, we use the vector fields $J\mathbf{T}$ and $\mathbf{U}$ instead of $\mathbf{N}$ and $\mathbf{B}$. (Note that $J\mathbf{T}$ is always tangent to $\mathcal{M}$, and $\mathbf{U}$ is always perpendicular to $\mathcal{M}$; in contrast, $\mathbf{N}$ and $\mathbf{B}$ in general will be neither tangent nor perpendicular to $\mathcal{M}$, and therefore are not useful to describe the geometry of $\mathcal{M}$.)

Definition. Let $\mathcal{M}$ be an oriented regular surface in $\mathbb{R}^n$ with unit normal $\mathbf{U}$, and let $\boldsymbol{\alpha}\colon(c,d) \longrightarrow \mathcal{M}$ be a curve on $\mathcal{M}$. Denote by J the complex structure of $\mathcal{M}$ determined by $\mathbf{U}$, and write $\mathbf{T} = \boldsymbol{\alpha}'/\|\boldsymbol{\alpha}'\|$. The **Darboux frame field** of $\boldsymbol{\alpha}$ with respect to $\mathcal{M}$ consists of the triple of vector fields $\{\mathbf{T}, J\mathbf{T}, \mathbf{U}\}$.

Corresponding to the Frenet formulas are the **Darboux formulas**; we first describe them for a unit-speed curve $\boldsymbol{\beta}$.

Theorem 22.16. *Let $\boldsymbol{\beta}\colon(c,d) \longrightarrow \mathcal{M}$ be a unit-speed curve, and let $\{\mathbf{T}, J\mathbf{T}, \mathbf{U}\}$ be the Darboux frame field of $\boldsymbol{\beta}$ with respect to $\mathcal{M}$. Then*

$$(22.33) \quad \begin{cases} \mathbf{T}' = \kappa_g\, J\mathbf{T} + \kappa_n\, \mathbf{U}, \\ (J\mathbf{T})' = -\kappa_g\, \mathbf{T} + \tau_g\, \mathbf{U}, \\ \mathbf{U}' = -\kappa_n\, \mathbf{T} - \tau_g\, J\mathbf{T}. \end{cases}$$

Proof. We have

$$\mathbf{T}' = (\mathbf{T}'\cdot J\mathbf{T})J\mathbf{T} + (\mathbf{T}'\cdot\mathbf{U})\mathbf{U}$$
$$= (\boldsymbol{\beta}''\cdot J\mathbf{T})J\mathbf{T} + (\boldsymbol{\beta}''\cdot\mathbf{U})\mathbf{U} = \kappa_g\, J\mathbf{T} + \kappa_n\, \mathbf{U}.$$

The other formulas of (22.33) are proved in a similar fashion. ∎

The Darboux formulas for an arbitrary-speed curve can be derived in the same way that we derived the Frenet formulas for an arbitrary-speed curve on page 191. Here is the result.

Corollary 22.17. *Let $\boldsymbol{\alpha}\colon(c,d) \longrightarrow \mathcal{M}$ be a curve with speed $v = \|\boldsymbol{\alpha}'\|$, and let $\{\mathbf{T}, J\mathbf{T}, \mathbf{U}\}$ be the Darboux frame field of $\boldsymbol{\alpha}$ with respect to $\mathcal{M}$. Then*

$$(22.34) \quad \begin{cases} \mathbf{T}' = v\,\kappa_g\, J\mathbf{T} + v\,\kappa_n\, \mathbf{U}, \\ (J\mathbf{T})' = -v\,\kappa_g\, \mathbf{T} + v\,\tau_g\, \mathbf{U}, \\ \mathbf{U}' = -v\,\kappa_n\, \mathbf{T} - v\,\tau_g\, J\mathbf{T}. \end{cases}$$

Frame fields are a very useful tool for studying curves and surfaces and differential geometry in general. See É. Cartan's[3] paper [Cartan] for a general discussion of this subject.

---

[3] Élie Cartan (1869-1951). French mathematician. One of the preeminent mathematicians of the 20th century. His important contributions include the classifications of compact Lie groups and symmetric spaces and the study of various geometries using moving frames and the structure equations.

22.5 Exercises

1. Compute the Gaussian curvature with both **brioschicurvature** and **gcurvature** of the following surfaces. Use **Timing** to compare the speeds of computation.

    (a) a sphere.
    (b) a torus.
    (c) a catenoid.
    (d) Enneper's minimal surface.

2. Compute the Christoffel symbols for each of the surfaces in exercise 1.

3. Prove Lemma 22.15 and Corollary 22.17. Complete the proof of Theorem 22.16.

4. The following command **normalcurvature** calculates the normal curvature of the curve $t \longmapsto$ **surf[alpha[t]]** contained in the parametrized surface **surf**: $\mathcal{U} \longrightarrow \mathbb{R}^3$, where **alpha** is a plane curve whose image is contained in the subset $\mathcal{U}$ of $\mathbb{R}^2$.

```
normalcurvature[surf_,alpha_][t_]:=
Module[{unc,tmp,crv1,crv2},
    unc=unitnormal[surf][alpha[tt][[1]],alpha[tt][[2]]];
    tmp=surf[alpha[tt][[1]],alpha[tt][[2]]];
    crv1=D[tmp,tt];crv2=D[tmp,{tt,2}];
    Simplify[PowerExpand[
            (crv2.unc)/(crv1.crv1)  /. tt->t]]]
```

Use the miniprogram **normalcurvature** to compute the normal curvature of a spherical spiral (see exercise 4 of Chapter 9).

5. The following command **geodesiccurvature** calculates the geodesic curvature of the curve $t \longmapsto$ **surf[alpha[t]]** contained in the parametrized surface **surf**: $\mathcal{U} \longrightarrow \mathbb{R}^3$, where **alpha** is a plane curve whose image is contained in the subset $\mathcal{U}$ of $\mathbb{R}^2$.

```
geodesiccurvature[surf_,alpha_][t_]:=
    Module[{unc,tmp,crv1,crv2},
    unc=unitnormal[surf][alpha[tt][[1]],alpha[tt][[2]]];
    tmp=surf[alpha[tt][[1]],alpha[tt][[2]]];
    crv1=D[tmp,tt];crv2=D[tmp,{tt,2}];
    Simplify[PowerExpand[Det[
    {crv2,unc,crv1}]/(crv1.crv1)^(3/2) /. tt->t]]]
```

Use the miniprogram **geodesiccurvature** to compute the geodesic curvature of a spherical spiral.

6. Show that the geodesic torsion of any curve on a sphere vanishes.

7. Compute the geodesic curvature and torsion of the curve $t \longmapsto (t, at, at^2)$ on the hyperbolic paraboloid $(u, v) \longmapsto (u, v, uv)$.

8. Use the following miniprogram to compute the geodesic torsion of a winding line on a torus.

```
geodesictorsion[surf_,alpha_][t_]:=
    Module[{unc,tmp,crv1,unc1},
    unc=unitnormal[surf][alpha[tt][[1]],alpha[tt][[2]]];
    tmp=surf[alpha[tt][[1]],alpha[tt][[2]]];
    crv1=D[tmp,tt];unc1=D[unc,tt];
    Simplify[PowerExpand[Det[
    {unc,unc1,crv1}]/(crv1.crv1)^(3/2) /. tt->t]]]
```

9. Let $\mathbf{p} \in \mathcal{M}$ where $\mathcal{M}$ is a surface in $\mathbb{R}^3$, and let $\alpha \colon (a, b) \longrightarrow \mathcal{M}$ be a curve with $\alpha(0) = \mathbf{p}$. Show that the geodesic torsion of α at $\mathbf{p}$ is given by

(22.35) $$\tau_g[\alpha](0) = \frac{1}{2}(\mathbf{k}_2(\mathbf{p}) - \mathbf{k}_1(\mathbf{p}))\sin(2\theta),$$

where θ denotes the oriented angle from $\mathbf{e}_1$ to $\mathbf{u}_\mathbf{p}$. Compare formulas (22.35) and (16.20).

10. Show that any rotation about the x-axis preserves the curvature of a monkey saddle, but only rotations of angles $\pm 2\pi/3$ are isometries.

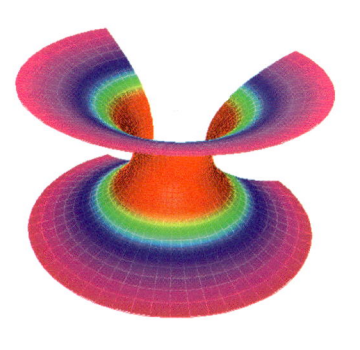

Catenoid colored by Gaussian curvature

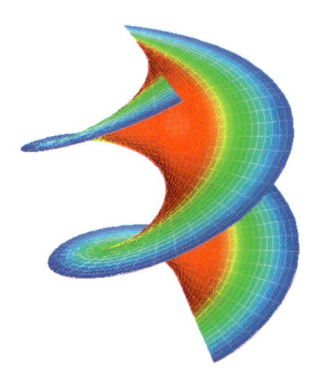

Helicoid colored by Gaussian curvature

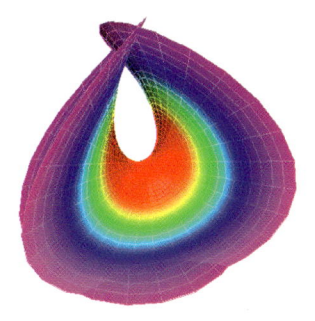

First Enneper surface colored by Gaussian curvature

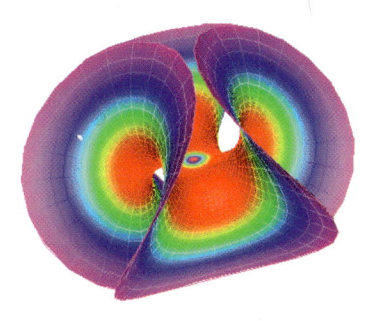

Second Enneper surface colored by Gaussian curvature

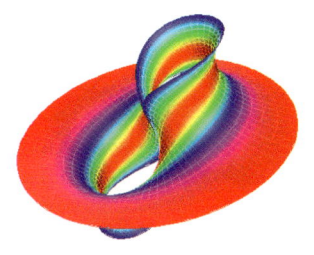

First Richmond surface colored by Gaussian curvature

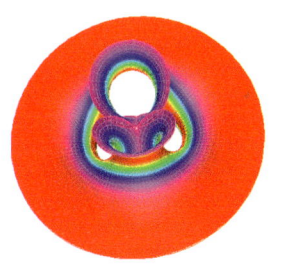

Second Richmond surface colored by Gaussian curvature

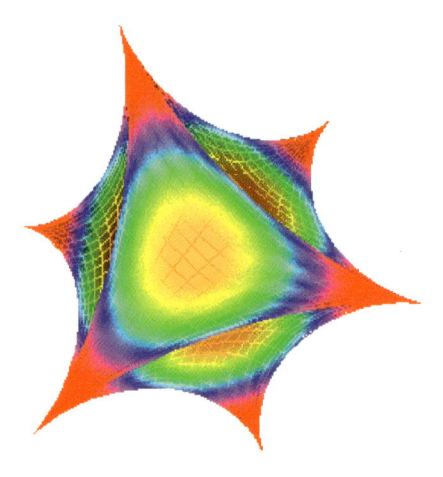

Astroidial sphere colored by Gaussian curvature

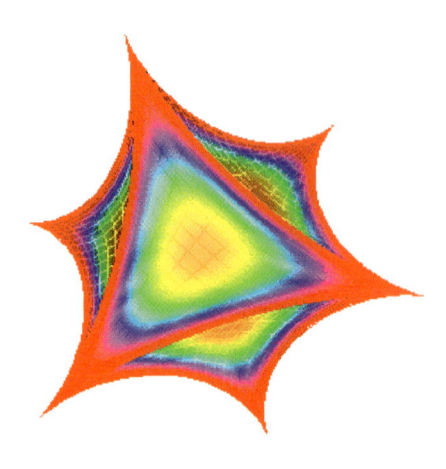

Astroidial sphere colored by Gaussian curvature

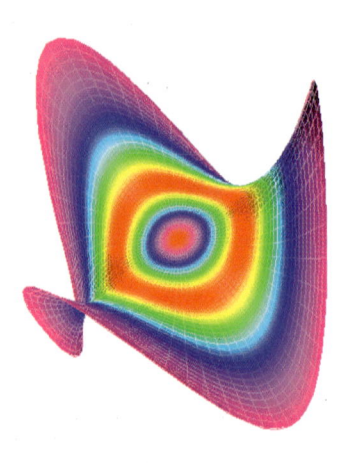

Monkey saddle colored by Gaussian curvature

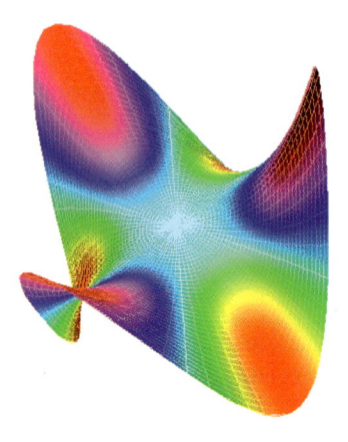

Monkey saddle colored by mean curvature

Torus knot colored by curvature

Torus knot colored by torsion

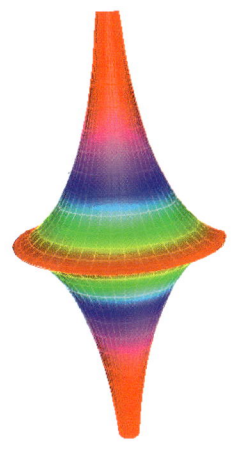

Pseudosphere colored by mean curvature, principal parametrization

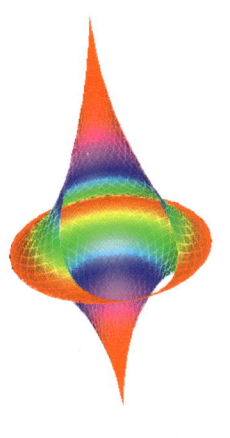

Pseudosphere colored by mean curvature, asymptotic parametrization

Dini's surface colored by mean curvature, principal parametrization

Dini's surface colored by mean curvature, asymptotic parametrization

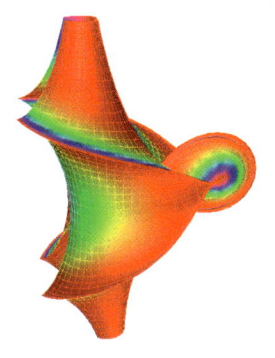

Kuen's surface colored by mean curvature, principal parametrization

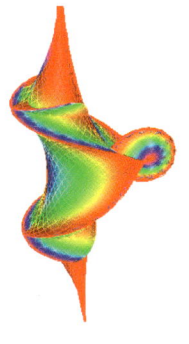

Kuen's surface colored by mean curvature, asymptotic parametrization

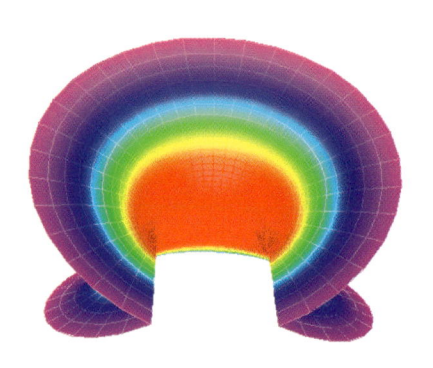

Catalan's minimal surface colored by Gaussian curvature

Minimal surface containing a clothoid as a geodesic

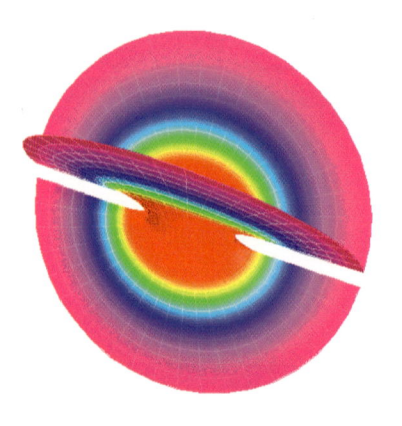

Minimal surface containing an ellipse as a geodesic

Enneper-catenoid colored by Gaussian curvature

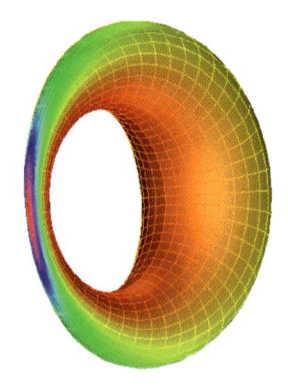

Ring cyclide colored by Gaussian curvature

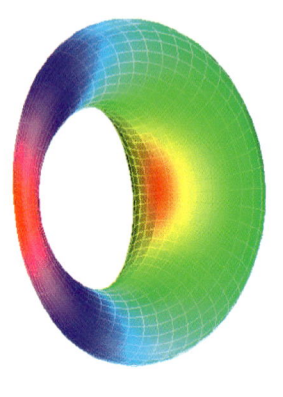

Ring cyclide colored by mean curvature

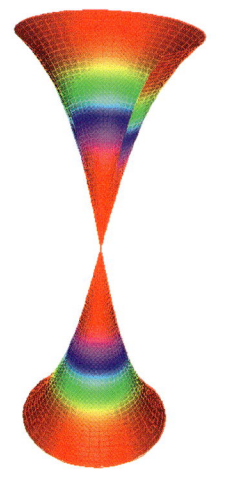

Constant curvature surface of revolution of conic type

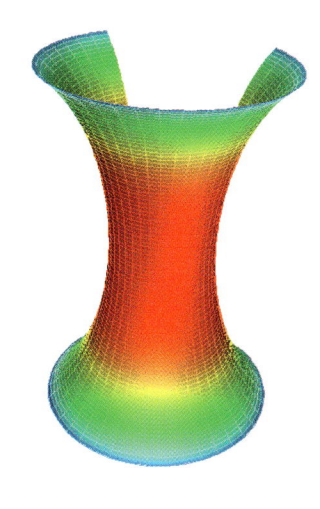

Constant curvature surface of revolution of hyperboloid type

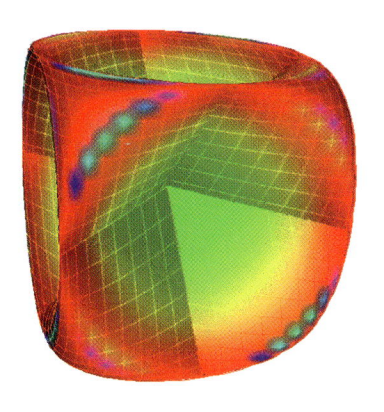

Sin surface

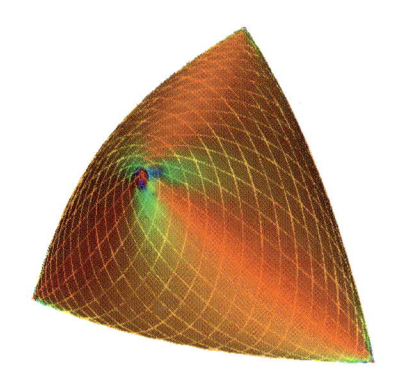

Cos surface

Twisted sphere colored by Gaussian curvature

Twisted sphere colored by mean curvature

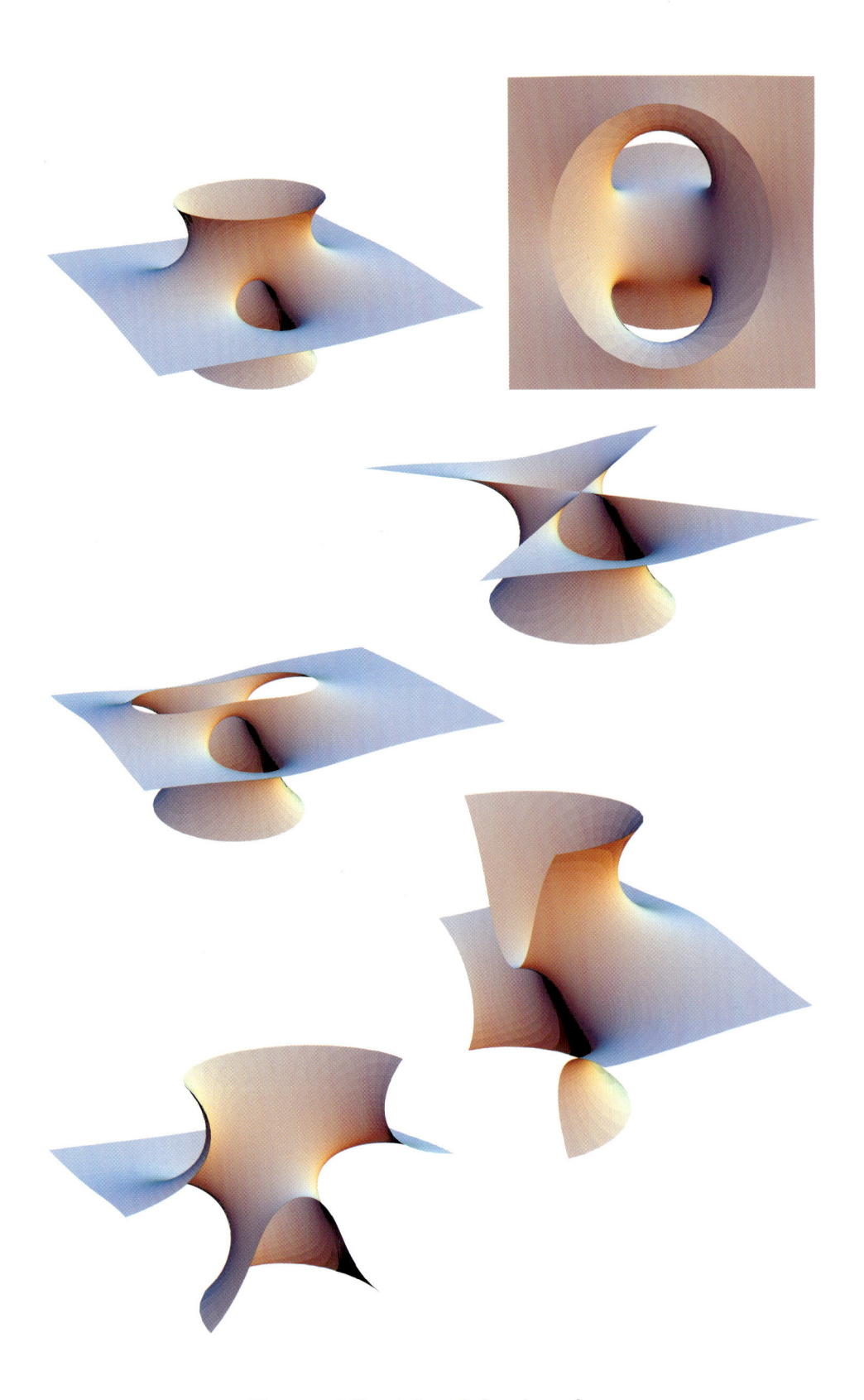

Views of Costa's minimal surface

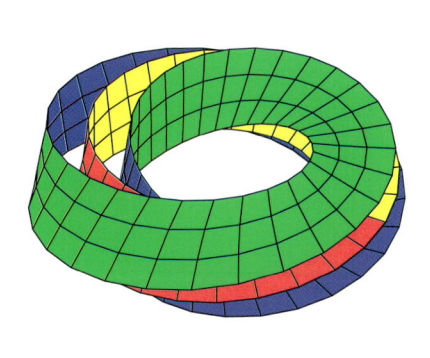

Mobius strip and parallel surface

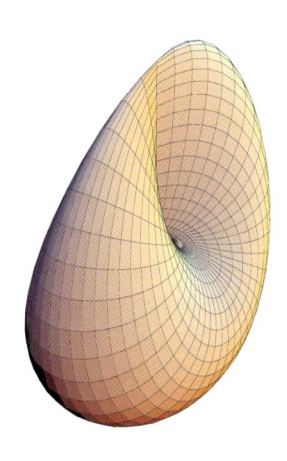

Cross cap

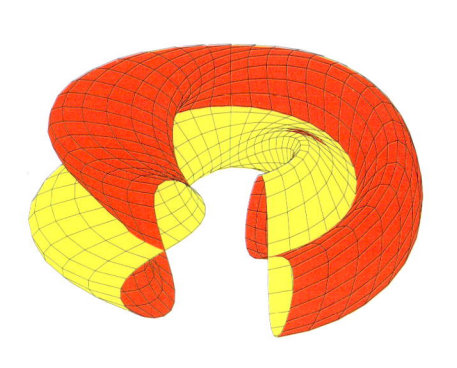

Klein bottle

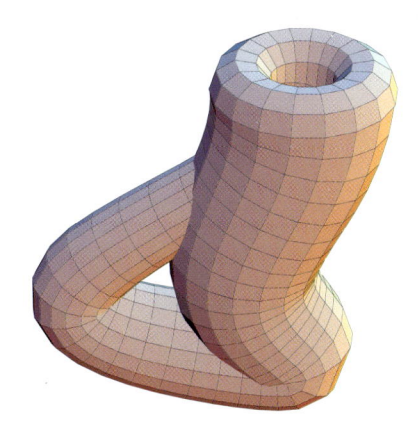

Klein bottle

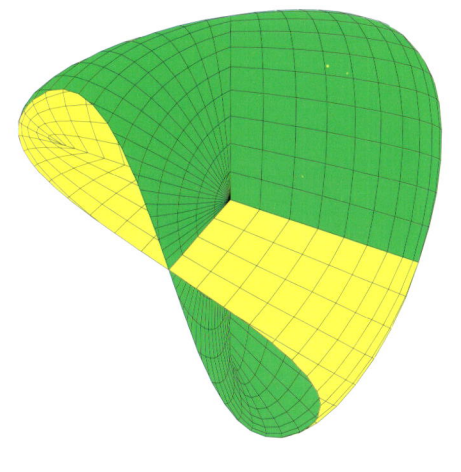

Steiner's roman surface

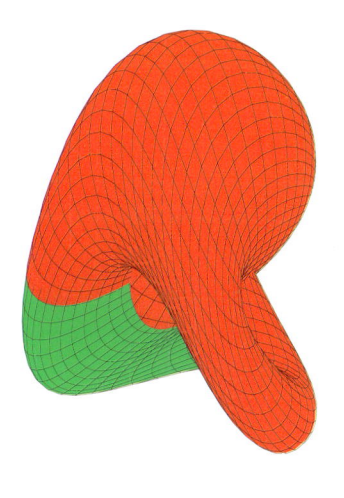

Boy's surface

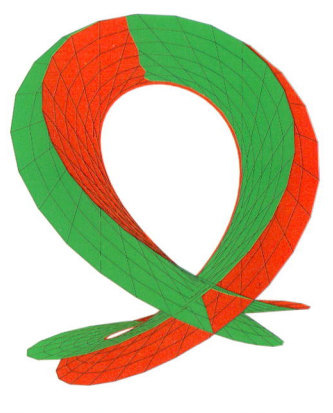

Tangent developable to Viviani's curve

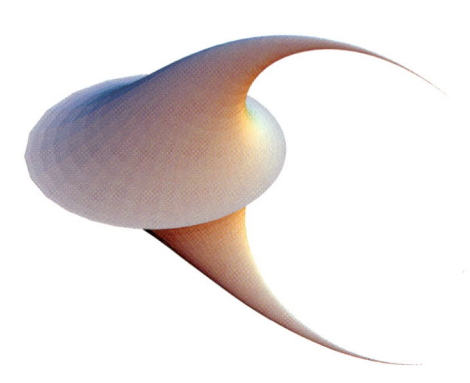

Image of a pseudosphere under an Inversion

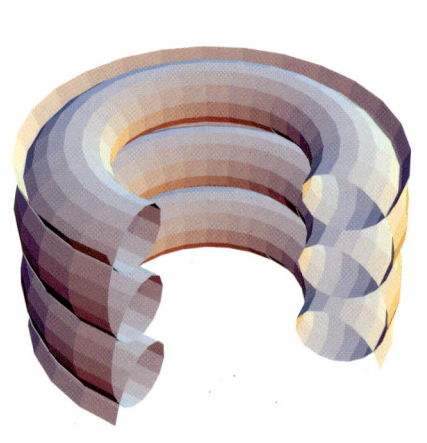

Nodoid

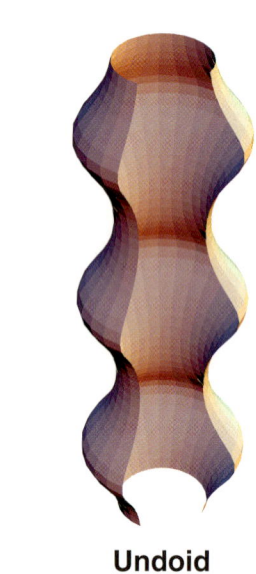

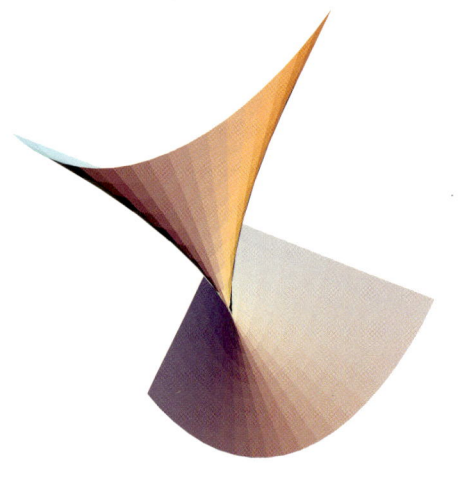

Undoid

Whitney umbrella

Singular monge surface

23

DIFFERENTIABLE MANIFOLDS

A **manifold**, generally speaking, is a topological space which resembles Euclidean space locally. A **differentiable manifold** is a manifold $\mathcal{M}$ for which this resemblance is sharp enough to allow partial differentiation and consequently all the features of differential calculus on $\mathcal{M}$. Thus the study of differentiable manifolds involves topology, since differentiability implies continuity. However, the metric properties of Euclidean space are not *a priori* included. In this chapter we define the notion of differentiable manifold and some of the standard apparatus associated with it.

We know that the set $\mathbb{R}^n$ of n-tuples of real numbers is a vector space; it is also a topological space, and the vector operations are continuous with respect to the topology. In addition, there is the notion of differentiability of real-valued functions on $\mathbb{R}^n$: we say that $f:\mathbb{R}^n \longrightarrow \mathbb{R}$ (where $\mathbb{R} = \mathbb{R}^1$) is **differentiable** provided that the mixed partial derivatives

$$\frac{\partial^{i_1+\cdots+i_r} f}{\partial u_1^{i_1} \ldots \partial u_r^{i_r}}$$

of all orders exist. Such functions are called C^∞-**functions**. They contrast, on the one hand, with the C^k-**functions** (the functions which have continuous partial derivatives up to order k), and, on the other hand, with the **analytic functions** (functions which have convergent power series). If we denote the respective classes by $C^\infty(\mathbb{R}^n)$, $C^k(\mathbb{R}^n)$ and $C^\omega(\mathbb{R}^n)$, then we have the strict inclusions

$$C^\omega(\mathbb{R}^n) \subset C^\infty(\mathbb{R}^n) \subset C^k(\mathbb{R}^n).$$

Almost always we choose to work with C^∞-functions for technical reasons: the C^k-functions are not smooth enough, and there are not enough C^ω-functions available.

We recall the definition of **natural coordinate functions** of $\mathbb{R}^n$ given on page 255. These are the functions $u_i: \mathbb{R}^n \longrightarrow \mathbb{R}$ defined by

$$u_i\big((p_1,\ldots,p_n)\big) = p_i$$

for $i = 1,\ldots,n$. We shall resolutely distinguish the functions u_i and the numbers p_i. We say that a function $\Psi: \mathbb{R}^m \longrightarrow \mathbb{R}^n$ is **differentiable, continuous** or **linear** if and only if each $u_i \circ \Psi$ is differentiable, continuous or linear, respectively. We shall assume that the reader is familiar with all the basic information about calculus of several variables, including such things as the implicit function theorem and the existence theorems for ordinary differential equations.

In Section 23.1 we give the definition of differentiable manifold and describe some simple examples. The algebra $\mathfrak{F}(\mathcal{M})$ of C^∞ functions on a differentiable manifold is described in Section 23.2. The tangent space to a differentiable manifold is defined in Section 23.3, and maps between differentiable manifolds are studied in Section 23.4. Vector fields and tensor fields are defined in Sections 23.5 and 23.6.

23.1 The Definition of Differentiable Manifold

We now make precise the "local resemblance" referred to above.

Definition. *A* **patch** *on a topological space $\mathcal{M}$ is a pair $(\mathbf{x}, \mathcal{U})$, where $\mathcal{U}$ is an open subset of $\mathbb{R}^n$ and*

$$\mathbf{x}: \mathcal{U} \longrightarrow \mathbf{x}(\mathcal{U}) \subset \mathcal{M}$$

is a homeomorphism of $\mathcal{U}$ onto an open set $\mathbf{x}(\mathcal{U})$ of $\mathcal{M}$. Here $\mathbf{x}$ is called the **local homeomorphism** *of the patch, and $\mathbf{x}(\mathcal{U})$ the* **coordinate neighborhood**. *Frequently, we refer to "the patch $\mathbf{x}$" when the domain $\mathcal{U}$ is understood. Let*

$$x_i = u_i \circ \mathbf{x}^{-1} : \mathbf{x}(\mathcal{U}) \longrightarrow \mathbb{R}$$

*for $i = 1,\ldots,n$. Then x_i is called the ith-***coordinate function**, *and $(x_1,\ldots,x_n)$ is called a* **system of local coordinates** *for $\mathcal{M}$.*

23.1 The Definition of Differentiable Manifold

The coordinate functions $x_1, \ldots, x_n$ contain the same information as the local homeomorphism $\mathbf{x}$. Often, we write

$$\mathbf{x}^{-1} = (x_1, \ldots, x_n)$$

with the meaning that

$$\mathbf{x}^{-1}(\mathbf{p}) = (x_1(\mathbf{p}), \ldots, x_n(\mathbf{p}))$$

for all $\mathbf{p} \in \mathbf{x}(\mathcal{U})$.

We are now ready to define the notion of differentiable manifold. Intuitively, the idea is this: in studying the geography of the earth, it is a great convenience to use geographical maps or **patches** instead of examining the earth directly. A collection of geographical maps that covers the earth is called an **atlas**, and it gives a complete picture of the earth. Roughly, we shall follow the same procedure with differentiable manifolds.

Definition. *An* **atlas** $\mathfrak{A}$ *on a topological space* $\mathcal{M}$ *is a collection of patches on* $\mathcal{M}$ *such that all the patches map from open subsets of the same Euclidean space* $\mathbb{R}^n$ *into* $\mathcal{M}$, *and* $\mathcal{M}$ *is the union of all the* $\mathbf{x}(\mathcal{U})$*'s such that* $(\mathbf{x}, \mathcal{U}) \in \mathfrak{A}$. *A topological space* $\mathcal{M}$ *equipped with an atlas is called a* **topological manifold**.

Let $\mathfrak{A}$ be an atlas on a topological space $\mathcal{M}$. Notice that if $(\mathbf{x}, \mathcal{U})$ and $(\mathbf{y}, \mathcal{V})$ are two patches in $\mathfrak{A}$ such that $\mathbf{x}(\mathcal{U}) \cap \mathbf{y}(\mathcal{V}) = \mathcal{W} \subset \mathcal{M}$ is nonempty, then the map

$$(23.1) \qquad (\mathbf{x}^{-1} \circ \mathbf{y}) \colon \mathbf{y}^{-1}(\mathcal{W}) \longrightarrow \mathbf{x}^{-1}(\mathcal{W})$$

is a homeomorphism between open subsets of $\mathbb{R}^n$. We call $\mathbf{x}^{-1} \circ \mathbf{y}$ a **change of coordinates**. From calculus we know what it means for such a map to be differentiable. This fact allows us to make the following definition.

Definition. *A* **differentiable manifold** *is a paracompact Hausdorff topological space* $\mathcal{M}$ *equipped with an atlas* $\mathfrak{A}$ *such that for any two patches* $(\mathbf{x}, \mathcal{U}), (\mathbf{y}, \mathcal{V}) \in \mathfrak{A}$ *with* $\mathbf{x}(\mathcal{U}) \cap \mathbf{y}(\mathcal{V}) = \mathcal{W}$ *nonempty, the change of coordinates* (23.1) *is differentiable (that is, of class* C^∞*) in the ordinary Euclidean sense. The* **dimension** *of the manifold* $\mathcal{M}$ *(denoted by* $\dim \mathcal{M}$*) is the dimension of* $\mathcal{U}$ *for any* $(\mathbf{x}, \mathcal{U}) \in \mathfrak{A}$. *Unless otherwise stated, we shall assume that all differentiable manifolds are connected.*

Remarks

(1) See any book on general topology for the definitions of Hausdorff and paracompact. The exact meanings of these words is not important for us. They serve to exclude certain pathological examples.

(2) Variants of this notion of differentiable manifold (that is, C^∞-manifold) are the notions of C^k-**manifold** and **analytic manifold**. These are obtained by replacing "C^∞-differentiability" by that of "C^k-differentiability" and "C^ω-differentiability", respectively, in the definition of differentiable manifold. We shall not consider C^k-manifolds, and analytic manifolds will appear only in passing. At any rate, analytic manifolds are C^∞-manifolds, which in turn are C^k-manifolds for any k.

(3) Recall that a function of n complex variables is called **holomorphic** provided it is expandable in a complex power series in a neighborhood of each point where it is defined. The definition of **complex manifold** is the same as that of a differentiable manifold, except that the local homeomorphisms are required to map from open subsets of the space $\mathbb{C}^n$ of n-tuples of complex numbers, and the changes of coordinates (23.1) are required to be holomorphic instead of C^∞-differentiable. Since $\mathbb{C}^n = \mathbb{R}^{2n}$, we may regard a complex manifold as a real differentiable manifold whose real dimension is twice the complex dimension, provided we use C^∞-patches instead of C^ω-patches.

(4) Topological terms applied to a manifold of course apply to its underlying topological space. Manifolds share many of the local properties of Euclidean space $\mathbb{R}^n$, for example, local connectedness and local compactness. Also, note that since we are assuming our manifolds are paracompact, they are normal.

(5) The notion of distance is *not* included in the definition of differentiable manifold.

(6) Implicitly associated with $\mathfrak{A}$ is a possibly larger atlas.

Definition. *The* **completion** $\widetilde{\mathfrak{A}}$ *of an atlas* $\mathfrak{A}$ *is the collection*

$$\widetilde{\mathfrak{A}} = \{\,(\mathbf{x}, \mathcal{U}) \mid \mathbf{x}^{-1} \circ \mathbf{y} \text{ and } \mathbf{y}^{-1} \circ \mathbf{x} \text{ are differentiable}$$
$$\text{for all } (\mathbf{y}, \mathcal{V}) \in \mathfrak{A}\,\}.$$

We say that an atlas $\mathfrak{A}$ *is* **complete** *if it coincides with its completion.*

We do not distinguish between differentiable manifolds $(\mathcal{M}, \mathfrak{A}_1)$ and $(\mathcal{M}, \mathfrak{A}_2)$ when the atlases $\mathfrak{A}_1$ and $\mathfrak{A}_2$ have the same completion. When we speak of a "differentiable manifold $\mathcal{M}$", we shall mean a differentiable manifold $\mathcal{M}$ equipped with a specific complete atlas.

There are two simple but important ways to construct new manifolds from old.

23.1 The Definition of Differentiable Manifold

Definition. Let $\mathcal{M}$ be a differentiable manifold defined by a complete atlas $\mathfrak{A}$, and let $\mathcal{V}$ be an open subset of $\mathcal{M}$. Define

$$\mathfrak{A}|\mathcal{V} = \{\, (\mathbf{x},\mathcal{U}) \in \mathfrak{A} \mid \mathbf{x}(\mathcal{U}) \subseteq \mathcal{V} \,\}.$$

Evidently, $\mathfrak{A}|\mathcal{V}$ is an atlas of $\mathcal{V}$. We call $\mathcal{V}$ equipped with the atlas $\mathfrak{A}(V)$ an **open submanifold** of $\mathcal{M}$.

Definition. Let $\mathcal{M}_1$ and $\mathcal{M}_2$ be differentiable manifolds of dimensions n_1 and n_2 defined by atlases $\mathfrak{A}_1$ and $\mathfrak{A}_2$. If $(\mathbf{x},\mathcal{U})$ is in $\mathfrak{A}_1$ and $(\mathbf{y},\mathcal{V})$ is in $\mathfrak{A}_2$, we define $\mathbf{x} \times \mathbf{y}\colon \mathcal{U} \times \mathcal{V} \longrightarrow \mathcal{M}_1 \times \mathcal{M}_2$ by

$$(\mathbf{x} \times \mathbf{y})(\mathbf{p},\mathbf{q}) = \bigl(\mathbf{x}(\mathbf{p}),\mathbf{y}(\mathbf{q})\bigr).$$

Then $\mathcal{M}_1 \times \mathcal{M}_2$ equipped with the completion of the atlas

$$\{\, (\mathbf{x} \times \mathbf{y},\mathcal{U} \times \mathcal{V}) \mid (\mathbf{x},\mathcal{U}) \in \mathfrak{A}_1,\ (\mathbf{y},\mathcal{V}) \in \mathfrak{A}_2 \,\}$$

is a differentiable manifold of dimension n_1+n_2, called the **product** of $\mathcal{M}_1$ and $\mathcal{M}_2$.

For the proof that $\mathcal{M}_1 \times \mathcal{M}_2$ is actually a differentiable manifold, see exercise 1.

Examples of Differentiable Manifolds
Euclidean Space $\mathbb{R}^n$

Let $\mathbf{1}\colon \mathbb{R}^n \longrightarrow \mathbb{R}^n$ be the identity map

$$\mathbf{1} = (u_1,,\ldots,u_n)\colon \mathbb{R}^n \longrightarrow \mathbb{R}^n.$$

Then $(\mathbf{1},\mathbb{R}^n)$ constitutes an atlas for $\mathbb{R}^n$ all by itself. (This formalizes the fact that the notion of manifold is a generalization of Euclidean space.)

The Sphere $S^n(a)$

Let $a > 0$ and put

$$S^n(a) = \left\{\, (t_0,\ldots,t_n) \in \mathbb{R}^{n+1} \,\middle|\, \sum_{j=0}^n t_j^2 = a^2 \,\right\}.$$

We shall make $S^n(a)$ into a differentiable manifold by defining an atlas

$$\mathfrak{A} = \{\, (\mathbf{north},\mathbb{R}^n),(\mathbf{south},\mathbb{R}^n) \,\},$$

where **north** and **south** are the "stereographic injections". Analytically these are the patches

$$\textbf{north}: \mathbb{R}^n \longrightarrow S^n(a) \setminus \{\textbf{sp}\} \quad \text{and} \quad \textbf{south}: \mathbb{R}^n \longrightarrow S^n(a) \setminus \{\textbf{np}\},$$

where $\textbf{np} = (a, 0, \ldots, 0)$ is the "north pole", and $\textbf{sp} = (-a, 0, \ldots, 0)$ is the "south pole". It is easily verified that **north** and **south** are injective because they are given by the formulas

$$\textbf{north} = (\Phi_0, \ldots, \Phi_n), \qquad \textbf{south} = (\Psi_0, \ldots, \Psi_n),$$

where

$$\Phi_0 = -\Psi_0 = a \left(\frac{\sum_{j=1}^{n} u_j^2 - a^2}{\sum_{j=1}^{n} u_j^2 + a^2} \right), \quad \text{and} \quad \Phi_k = \Psi_k = \frac{2a^2 u_k}{\sum_{j=1}^{n} u_j^2 + a^2}$$

for $k = 1, \ldots, n$. Thus $\textbf{north} \circ \textbf{south}^{-1}$ and $\textbf{south} \circ \textbf{north}^{-1}$ are differentiable.

A Surface in $\mathbb{R}^n$

Corollary 12.18, page 284, implies that a regular surface in $\mathbb{R}^n$, as defined on page 281, is a 2-dimensional differentiable manifold.

23.2 Differentiable Functions on Differentiable Manifolds

Differentiable manifolds are locally like Euclidean space. For their study it will be important to transfer to manifolds as much of the differential calculus of Euclidean space as we can. First on the agenda is the notion of differentiability for real-valued functions on a differentiable manifold. The definition of differentiability of a real-valued function on a differentiable manifold is almost the same as the corresponding definition that we gave on page 285.

Definition. *Let $f: \mathcal{W} \subset \mathcal{M} \longrightarrow \mathbb{R}$ be a function defined on an open subset $\mathcal{W}$ of a differentiable manifold $\mathcal{M}$. We say that f is **differentiable** at $\textbf{p} \in \mathcal{W}$, provided*

23.2 Differentiable Functions on Differentiable Manifolds

that for some patch $\mathbf{x}: \mathcal{U} \subset \mathbb{R}^n \longrightarrow \mathcal{M}$ with $\mathbf{p} \in \mathbf{x}(\mathcal{U}) \subset \mathcal{W}$, the composition $f \circ \mathbf{x}: \mathcal{U} \subset \mathbb{R}^n \longrightarrow \mathbb{R}$ is differentiable (in the ordinary Euclidean sense) at $\mathbf{x}^{-1}(\mathbf{p})$. If f is differentiable at all points of $\mathcal{W}$, we say that f is **differentiable** on $\mathcal{W}$.

This definition illustrates some of the force behind the definition of differentiable manifold. One might think that in the above definition it should be necessary to require that $f \circ \mathbf{x}$ be differentiable for every patch $(\mathbf{x}, \mathcal{U})$ in the atlas of $\mathcal{M}$. However, this fact is a consequence of the definition.

Lemma 23.1. *The definition of differentiability of a real-valued function on a differentiable manifold does not depend on the choice of patch.*

Proof. If $(\mathbf{x}, \mathcal{U})$ and $(\mathbf{y}, \mathcal{V})$ are patches on a manifold $\mathcal{M}$, then the change of coordinates $\mathbf{x}^{-1} \circ \mathbf{y}$ is differentiable by the definition of differentiable manifold. We can write $f \circ \mathbf{y}$ as

$$f \circ \mathbf{y} = (f \circ \mathbf{x}) \circ (\mathbf{x}^{-1} \circ \mathbf{y}).$$

Since the composition of the Euclidean-differentiable functions is differentiable, the differentiability of $f \circ \mathbf{x}$ implies the differentiability of $f \circ \mathbf{y}$, and conversely. ∎

Next, we consider the totality of differentiable functions on a differentiable manifold and define some algebraic structure on it.

Definition. *Let $\mathcal{M}$ be a differentiable manifold. We put*

$$\mathfrak{F}(\mathcal{M}) = \{\, f: \mathcal{M} \longrightarrow \mathbb{R} \mid f \text{ is differentiable}\,\}.$$

*We call $\mathfrak{F}(\mathcal{M})$ the **algebra of real-valued differentiable functions** $\mathcal{M}$. For $a, b \in \mathbb{R}$ and $f, g \in \mathfrak{F}(\mathcal{M})$ the functions $a f + b g$ and $f g$ are defined by*

$$(a f + b g)(\mathbf{m}) = a f(\mathbf{m}) + b g(\mathbf{m}) \qquad \text{and} \qquad (f g)(\mathbf{m}) = f(\mathbf{m}) g(\mathbf{m})$$

for $\mathbf{m} \in \mathcal{M}$. Also, we identify any $a \in \mathbb{R}$ with the constant function a given by $a(\mathbf{m}) = a$ for $\mathbf{m} \in \mathcal{M}$.

Let us note some of the algebraic properties of $\mathfrak{F}(\mathcal{M})$.

Lemma 23.2. *Let $\mathcal{M}$ be a differentiable manifold. Then $\mathfrak{F}(\mathcal{M})$ is a commutative ring with identity and an (infinite-dimensional) algebra over the real numbers $\mathbb{R}$.*

Proof. Let $f, g \in \mathfrak{F}(\mathcal{M})$ and $a, b \in \mathbb{R}$. If $(\mathbf{x}, \mathcal{U})$ is a patch on $\mathcal{M}$, then $f \circ \mathbf{x}$ and $g \circ \mathbf{x}$ are differentiable in the ordinary Euclidean sense; hence $a(f \circ \mathbf{x}) + b(g \circ \mathbf{x})$ and $(f \circ \mathbf{x})(g \circ \mathbf{x})$ are Euclidean differentiable. It follows easily that both $af + bg$ and fg are differentiable. Also, constant functions are obviously differentiable. The identity of the ring $\mathfrak{F}(\mathcal{M})$ is $1 \in \mathbb{R}$. Associativity, commutativity and distributivity are easy to prove. ∎

Note also that if $f \in \mathfrak{F}(\mathcal{M})$ is never zero, then $1/f \in \mathfrak{F}(\mathcal{M})$.

It will be important to know when we can extend a real-valued function on an open set $\mathcal{V} \subset \mathcal{M}$ to a function that is differentiable on all of $\mathcal{M}$.

Lemma 23.3. *Let $\mathcal{M}$ be a differentiable manifold. If $\mathcal{U}$ is an open neighborhood of $\mathbf{p} \in \mathcal{M}$, then there exist a function $k \in \mathfrak{F}(\mathcal{M})$ and neighborhoods $\mathcal{V}$ of $\mathbf{p}$ and $\mathcal{W}$ of $\mathcal{M} \backslash \mathcal{U}$ such that $\mathbf{p} \in \mathcal{V} \subset \mathcal{M} \backslash \mathcal{W} \subset \mathcal{U}$ and*

(i) $0 \leq k(\mathbf{m}) \leq 1$ *for* $\mathbf{m} \in \mathcal{M}$;

(ii) $k(\mathbf{m}) = 1$ *for* $\mathbf{m} \in \mathcal{V}$;

(iii) $k(\mathbf{m}) = 0$ *for* $\mathbf{m} \in \mathcal{W}$.

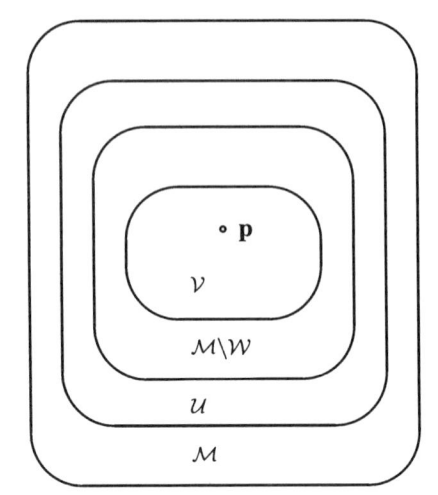

Proof. For $c > 0$ let

$$\mathcal{V}_c = \left\{ (p_1, \ldots, p_n) \in \mathbb{R}^n \;\middle|\; \sum_{j=1}^n p_j^2 < c \right\}.$$

There exist real numbers a and b with $0 < a < b$ and patches $(\mathbf{x}, \mathcal{V}_a)$, $(\mathbf{x}, \mathcal{V}_b)$ in the atlas of $\mathcal{M}$ such that

$$\mathbf{p} \in \mathbf{x}(\mathcal{V}_a) \subset \mathbf{x}(\mathcal{V}_b) \subset \overline{\mathbf{x}(\mathcal{V}_b)} \subset \mathcal{U},$$

23.2 Differentiable Functions on Differentiable Manifolds

where $\overline{\mathbf{x}(\mathcal{V}_b)}$ denotes the closure of $\mathbf{x}(\mathcal{V}_b)$. We choose $\mathcal{V} = \mathbf{x}(\mathcal{V}_a)$ and $\mathcal{W} = \mathcal{M}\setminus\overline{\mathbf{x}(\mathcal{V}_b)}$. Define $g:\mathbb{R} \longrightarrow \mathbb{R}$ by

$$g(x) = g[a,b](x) = \begin{cases} \exp\left(\dfrac{1}{x-b} - \dfrac{1}{x-a}\right) & \text{if } a < x < b, \\ 0 & \text{otherwise.} \end{cases}$$

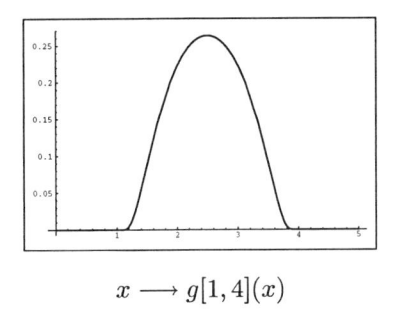

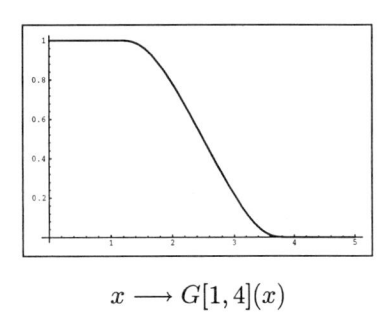

$x \longrightarrow g[1,4](x)$ \qquad\qquad $x \longrightarrow G[1,4](x)$

Then g is a C^∞-function (but not analytic). This fact is obvious at all points except a and b. It is clear at least that g is continuous at a and b. Moreover, it follows from L'Hôpital's rule that

$$\lim_{x \to a^+} \frac{g(x) - g(a)}{x - a} = \lim_{x \to a^+} \frac{\exp\left(\dfrac{1}{x-b} - \dfrac{1}{x-a}\right)}{x - a}$$

$$= \exp\left(\frac{1}{a-b}\right) \lim_{x \to a^+} \frac{\exp\left(-\dfrac{1}{x-a}\right)}{x-a} = \exp\left(\frac{1}{a-b}\right) \lim_{y \to 0^+} \frac{\dfrac{1}{y}}{\exp\left(\dfrac{1}{y}\right)}$$

$$= \exp\left(\frac{1}{a-b}\right) \lim_{y \to 0^+} \frac{-\left(\dfrac{1}{y}\right)^2}{\exp\left(\dfrac{1}{y}\right)\left(-\left(\dfrac{1}{y}\right)^2\right)} = \exp\left(\frac{1}{a-b}\right) \lim_{y \to 0^+} \exp\left(-\frac{1}{y}\right) = 0.$$

In a similar fashion, all other derivatives of g exist and vanish at both a and b. The functions $G:\mathbb{R} \longrightarrow \mathbb{R}$ and $\psi:\mathbb{R}^n \longrightarrow \mathbb{R}$ defined by

$$G(x) = G[a,b](x) = \frac{\int_x^b g(t)\,dt}{\int_a^b g(t)\,dt}, \qquad \psi(p_1,\ldots,p_n) = G(p_1^2 + \cdots + p_n^2)$$

are also differentiable. We define $k\colon \mathcal{M} \longrightarrow \mathbb{R}$ by

$$k(\mathbf{q}) = \begin{cases} \psi(\mathbf{x}^{-1}(\mathbf{q})), & \text{for } \mathbf{q} \in \mathbf{x}(\mathcal{V}_b), \\ 0, & \text{otherwise.} \end{cases}$$

Then k has all the required properties. ∎

If f is a real-valued differentiable function on an open subset $\mathcal{U}$ of a differentiable manifold $\mathcal{M}$, it might not be possible to extend f to a differentiable function defined on all of $\mathcal{M}$. (There might be some bizarre behavior of f at the boundary of $\mathcal{U}$.) However, we now show that each point of $\mathcal{U}$ has a neighborhood $\mathcal{V}$ such that we can extend the restriction $f|\mathcal{V}$ to all of $\mathcal{M}$.

Lemma 23.4. *Let $\mathcal{U} \subset \mathcal{M}$ be an open neighborhood of $\mathbf{p} \in \mathcal{M}$, and suppose $f \in \mathfrak{F}(\mathcal{U})$. Then there exist $\widetilde{f} \in \mathfrak{F}(\mathcal{M})$ and a neighborhood $\mathcal{V}$ of $\mathbf{p}$ with $\mathbf{p} \in \mathcal{V} \subseteq \mathcal{U}$ such that $\widetilde{f}|\mathcal{V} = f|\mathcal{V}$. We call $\widetilde{f}$ a **globalization** of f.*

Proof. By Lemma 23.3 there are neighborhoods $\mathcal{V}$ of $\mathbf{p}$ and $\mathcal{W}$ of $\mathcal{M}\backslash\mathcal{U}$ with the following properties:

(i) $\mathcal{V} \subseteq \mathcal{U}$;

(ii) there is a function $k \in \mathfrak{F}(\mathcal{M})$ such that $k(\mathbf{m}) = 1$ for $\mathbf{m} \in \mathcal{V}$ and $k(\mathbf{m}) = 0$ for $\mathbf{m} \in \mathcal{W}$.

Define $\widetilde{f}\colon \mathcal{M} \longrightarrow \mathbb{R}$ by

$$\widetilde{f}(\mathbf{q}) = \begin{cases} (kf)(\mathbf{q}) & \text{for } \mathbf{q} \in \mathcal{U}, \\ 0 & \text{for } \mathbf{q} \in \mathcal{M}\backslash\mathcal{U}. \end{cases}$$

Then $\widetilde{f} \in \mathfrak{F}(\mathcal{M})$, and on $\mathcal{V}$ we have $\widetilde{f} = kf = f$. ∎

Notice that Lemmas 23.3 and 23.4 are false for analytic functions on analytic manifolds.

Having defined the notion of real-valued differentiable function on a differentiable manifold, we are ready to define what it means for a map between manifolds to be differentiable.

Definition. *Let $\mathcal{M}$ and $\mathcal{N}$ be differentiable manifolds, and let $\Psi\colon \mathcal{M} \longrightarrow \mathcal{N}$ be a map. We say that Ψ is **differentiable**, provided that $\mathbf{y}^{-1} \circ \Psi \circ \mathbf{x}$ is differentiable for*

23.2 Differentiable Functions on Differentiable Manifolds

every patch $(\mathbf{x}, \mathcal{U})$ in the atlas of $\mathcal{M}$ and every patch $(\mathbf{y}, \mathcal{V})$ in the atlas of $\mathcal{N}$, where the compositions are defined. A **diffeomorphism** between manifolds $\mathcal{M}$ and $\mathcal{N}$ is a differentiable map $\Phi \colon \mathcal{M} \longrightarrow \mathcal{N}$ which has a differentiable inverse $\Phi^{-1} \colon \mathcal{N} \longrightarrow \mathcal{M}$. If such a map Φ exists, $\mathcal{M}$ and $\mathcal{N}$ are said to be **diffeomorphic**. A map $\Psi \colon \mathcal{M} \longrightarrow \mathcal{N}$ is called a **local diffeomorphism** provided each $\mathbf{p} \in \mathcal{M}$ has a neighborhood $\mathcal{U}$ such that $\Psi|\mathcal{U} \colon \mathcal{U} \longrightarrow \Psi(\mathcal{U})$ is a diffeomorphism.

The following lemma is an easy consequence of the definitions and the fact that the corresponding lemma for $\mathbb{R}^n$ is known.

Lemma 23.5. *Suppose $\mathcal{M} \xrightarrow{\Phi} \mathcal{N} \xrightarrow{\Psi} \mathcal{P}$ are differentiable maps between differentiable manifolds. Then the composition $\Psi \circ \Phi \colon \mathcal{M} \longrightarrow \mathcal{P}$ is differentiable. If Φ and Ψ are diffeomorphisms, then so is $\Psi \circ \Phi$ and*

$$(\Psi \circ \Phi)^{-1} = \Phi^{-1} \circ \Psi^{-1}.$$

The following lemma is to be expected.

Lemma 23.6. *Let $\mathcal{M}$ be an n-dimensional differentiable manifold and $\mathbf{x} \colon \mathcal{U} \longrightarrow \mathcal{M}$ be a patch. Write $\mathbf{x}^{-1} = (x_1, \ldots, x_n)$. Then*

(i) $\mathbf{x}$ is a differentiable mapping between the manifolds $\mathcal{U}$ and $\mathcal{M}$;

(ii) $\mathbf{x}^{-1} \colon \mathbf{x}(\mathcal{U}) \longrightarrow \mathbb{R}^n$ is differentiable;

(iii) for $i = 1, \ldots, n$ the coordinate function $x_i \colon \mathbf{x}(\mathcal{U}) \longrightarrow \mathbb{R}$ is differentiable.

Proof. Let $\mathbf{1} \colon \mathbb{R}^n \longrightarrow \mathbb{R}^n$ be the identity map. For each patch $(\mathbf{y}, \mathcal{V})$ in $\mathfrak{A}$ the maps $\mathbf{y}^{-1} \circ \mathbf{x} \circ \mathbf{1}$ and $\mathbf{1} \circ \mathbf{x}^{-1} \circ \mathbf{y}$ are $\mathbb{R}^n$-differentiable. Thus by definition $\mathbf{x}$ and $\mathbf{x}^{-1}$, considered as maps between the manifolds $\mathcal{U}$ and $\mathcal{M}$, are differentiable. That the coordinate function x_i is differentiable is part of the definition of differentiable manifold. ∎

A differentiable map between manifolds induces a correspondence between the algebras of differentiable functions on each manifold.

Lemma 23.7. *Let $\Phi \colon \mathcal{M} \longrightarrow \mathcal{N}$ be a differentiable mapping between manifolds. Then $f \in \mathfrak{F}(\mathcal{N})$ implies $f \circ \Phi \in \mathfrak{F}(\mathcal{M})$.*

Proof. Let $(\mathbf{x}, \mathcal{U})$ be a patch on $\mathcal{M}$ and $(\mathbf{y}, \mathcal{V})$ a patch on $\mathcal{N}$. By hypothesis $f \circ \mathbf{y} \colon \mathcal{V} \longrightarrow \mathbb{R}$ and $\mathbf{y}^{-1} \circ \Phi \circ \mathbf{x}$ is differentiable. Hence the composition

$$f \circ \Phi \circ \mathbf{x} = (f \circ \mathbf{y}) \circ (\mathbf{y}^{-1} \circ \Phi \circ \mathbf{x})$$

is differentiable. Since this is true for every path $(\mathbf{x}, \mathcal{U})$ in the atlas of $\mathcal{M}$, it follows that $f \circ \Phi$ is differentiable. ∎

23.3 Tangent Vectors on Differentiable Manifolds

By their very nature differentiable manifolds are "curved" spaces, and they can be very complicated objects to study. In comparison a vector space is much simpler. Furthermore, a vector space has a great deal of structure that facilitates its study.

In this section we define the notion of tangent space to a differentiable manifold $\mathcal{M}$ at a point $\mathbf{p} \in \mathcal{M}$. It can be thought of as the best linear approximation to $\mathcal{M}$ at $\mathbf{p}$.

The notion of a vector tangent to a curve or surface in $\mathbb{R}^n$ is intuitively clear, as we have seen in Chapters 1, 8 and 12; we want to define a similar concept for an arbitrary differentiable manifold. However, if we try to generalize directly the notion of tangent vector, we face a difficulty: the elementary definition of tangent vector that we gave in Section 12.5 makes a tangent vector to a surface a tangent vector to $\mathbb{R}^n$. But a surface, more generally an arbitrary manifold, is not *a priori* contained in any Euclidean space, so we need a definition of tangent vector that does not depend on any such assumption.

Working backwards, let us suppose we have a suitable definition of a vector $\mathbf{v}_\mathbf{p}$ tangent to a manifold $\mathcal{M}$ at a point $\mathbf{p} \in \mathcal{M}$. Then, just as in elementary calculus, we can speak of the derivative $\mathbf{v}_\mathbf{p}[f]$ of $f \in \mathfrak{F}(\mathcal{M})$ in the direction $\mathbf{v}_\mathbf{p}$. Roughly speaking, the number $\mathbf{v}_\mathbf{p}[f]$ is the ordinary derivative of f at $\mathbf{p}$ along a curve leaving $\mathbf{p}$ in the direction $\mathbf{v}_\mathbf{p}$. Now for a fixed $\mathbf{v}_\mathbf{p}$, the function $\mathfrak{F}(\mathcal{M}) \longrightarrow \mathbb{R}$ that maps f into $\mathbf{v}_\mathbf{p}[f]$ has the essential properties always possessed by differentiation: linearity and the Leibniz[1] product rule. Thus, in standard mathematical fashion, we shall define a tangent vector to be a function that has these properties. Although it is not clear from the outset that this definition will yield a mathematical object that has all the intuitive properties of a tangent vector, in fact, it does.

---

[1] Baron Gottfried Wilhelm Leibniz (1646-1716). German mathematician. One of the cofounders of calculus. Although Leibniz discovered calculus a few years later than Newton, it is the notation of Leibniz (such as dt and $\int$) that has gained the widest acceptance.

23.3 Tangent Vectors on Differentiable Manifolds

Definition. *Let* $\mathbf{p}$ *be a point of a manifold* $\mathcal{M}$. *A* **tangent vector** $\mathbf{v}_\mathbf{p}$ *to* $\mathcal{M}$ *at* $\mathbf{p}$ *is a real-valued function* $\mathbf{v}_\mathbf{p}: \mathfrak{F}(\mathcal{M}) \longrightarrow \mathbb{R}$ *such that*

$$\mathbf{v}_\mathbf{p}[a f + b g] = a\, \mathbf{v}_\mathbf{p}[f] + b\, \mathbf{v}_\mathbf{p}[g] \qquad \text{(the \textbf{linearity} property)},$$

$$\mathbf{v}_\mathbf{p}[f g] = f(\mathbf{p})\mathbf{v}_\mathbf{p}[g] + g(\mathbf{p})\mathbf{v}_\mathbf{p}[f] \quad \text{(the \textbf{Leibnizian} property)},$$

for all $a, b \in \mathbb{R}$ *and* $f, g \in \mathfrak{F}(\mathcal{M})$.

Clearly, a tangent vector to a surface in $\mathbb{R}^3$ as defined in Section 12.5 is a tangent vector in the sense of this definition.

We can easily give some nontrivial examples of tangent vectors. First, we need some new notation.

Definition. *Let* $\mathbf{x}: \mathcal{U} \longrightarrow \mathcal{M}$ *be a patch on a differentiable manifold* $\mathcal{M}$ *and write* $\mathbf{x}^{-1} = (x_1, \ldots, x_n)$. *For* $f \in \mathfrak{F}(\mathcal{M})$ *and* $\mathbf{p} \in \mathbf{x}(\mathcal{U})$, *write* $\mathbf{p} = \mathbf{x}(\mathbf{q})$, *and define*

$$(23.2) \qquad \frac{\partial f}{\partial x_i}(\mathbf{p}) = \left.\frac{\partial f}{\partial x_i}\right|_{\mathbf{x}(\mathbf{q})} = \left.\frac{\partial (f \circ \mathbf{x})}{\partial u_i}\right|_{\mathbf{q}}$$

for $i = 1, \ldots, n$. *Here, as usual, the* u_i's *are the natural coordinate functions of* $\mathbb{R}^n$, *and the ordinary Euclidean partial derivative appears on the right-hand side of* (23.2).

Note that for $\mathcal{M} = \mathbb{R}^n$ and $\mathbf{x}$ the identity map, the right-hand side of (23.2) reduces to the ordinary partial derivative. In general, we always have

$$\frac{\partial f}{\partial x_i} = \frac{\partial (f \circ \mathbf{x})}{\partial u_i} \circ \mathbf{x}^{-1}.$$

Lemma 23.8. *Let* $\mathbf{x}: \mathcal{U} \longrightarrow \mathcal{M}$ *be a patch on a differentiable manifold* $\mathcal{M}$ *and write* $\mathbf{x}^{-1} = (x_1, \ldots, x_n)$. *For* $\mathbf{p} \in \mathbf{x}(\mathcal{U})$ *the function*

$$\left.\frac{\partial}{\partial x_i}\right|_\mathbf{p} : \mathfrak{F}(\mathcal{M}) \longrightarrow \mathbb{R}$$

defined by

$$\left.\frac{\partial}{\partial x_i}\right|_\mathbf{p}[f] = \left.\frac{\partial f}{\partial x_i}\right|_\mathbf{p}$$

is a tangent vector to $\mathcal{M}$ *at* $\mathbf{p}$ *for* $i = 1, \ldots, n$.

Proof. Put $\mathbf{p} = \mathbf{x}(\mathbf{q})$, where $\mathbf{q} \in \mathcal{U}$. Let $a, b \in \mathbb{R}$ and $f, g \in \mathfrak{F}(\mathcal{M})$. We have

$$\left.\frac{\partial(a f + b g)}{\partial x_i}\right|_{\mathbf{x}(\mathbf{q})} = \left.\frac{\partial}{\partial u_i}\left((a f + b g) \circ \mathbf{x}\right)\right|_{\mathbf{q}} = \left.\frac{\partial}{\partial u_i}\left(a(f \circ \mathbf{x}) + b(g \circ \mathbf{x})\right)\right|_{\mathbf{q}}$$

$$= \left. a \frac{\partial(f \circ \mathbf{x})}{\partial u_i}\right|_{\mathbf{q}} + \left. b \frac{\partial(g \circ \mathbf{x})}{\partial u_i}\right|_{\mathbf{q}} = \left. a \frac{\partial f}{\partial x_i}\right|_{\mathbf{x}(\mathbf{q})} + \left. b \frac{\partial g}{\partial x_i}\right|_{\mathbf{x}(\mathbf{q})}.$$

Similarly,

$$\left.\frac{\partial(f g)}{\partial x_i}\right|_{\mathbf{x}(\mathbf{q})} = \left.\frac{\partial}{\partial u_i}\left((f g) \circ \mathbf{x}\right)\right|_{\mathbf{q}} = \left.\frac{\partial}{\partial u_i}\left((f \circ \mathbf{x})(g \circ \mathbf{x})\right)\right|_{\mathbf{q}}$$

$$= (g \circ \mathbf{x})(\mathbf{q}) \left.\frac{\partial(f \circ \mathbf{x})}{\partial u_i}\right|_{\mathbf{q}} + (f \circ \mathbf{x})(\mathbf{q}) \left.\frac{\partial(g \circ \mathbf{x})}{\partial u_i}\right|_{\mathbf{q}}$$

$$= g(\mathbf{x}(\mathbf{q})) \left.\frac{\partial f}{\partial x_i}\right|_{\mathbf{x}(\mathbf{q})} + f(\mathbf{x}(\mathbf{q})) \left.\frac{\partial g}{\partial x_i}\right|_{\mathbf{x}(\mathbf{q})}. \blacksquare$$

Now that we know that there are some tangent vectors at every point $\mathbf{p}$, let us consider the set of all tangent vectors at $\mathbf{p}$.

Definition. Let $\mathcal{M}$ be a differentiable manifold, and let $\mathbf{p}$ be a point in $\mathcal{M}$. The **tangent space** to $\mathcal{M}$ at $\mathbf{p}$ is the set of all tangent vectors to $\mathcal{M}$ at $\mathbf{p}$. Thus

$$\mathcal{M}_\mathbf{p} = \{\, \mathbf{v}_\mathbf{p} : \mathfrak{F}(\mathcal{M}) \longrightarrow \mathbb{R} \mid \mathbf{v}_\mathbf{p}[a f + b g] = a\, \mathbf{v}_\mathbf{p}[f] + b\, \mathbf{v}_\mathbf{p}[g]$$

$$\text{and } \mathbf{v}_\mathbf{p}[f g] = f(\mathbf{p}) \mathbf{v}_\mathbf{p}[g] + g(\mathbf{p}) \mathbf{v}_\mathbf{p}[f]$$

$$\text{for } a, b \in \mathbb{R} \text{ and } f, g \in \mathfrak{F}(\mathcal{M}) \,\}.$$

Lemma 23.9. *If $\mathcal{M}$ is a differentiable manifold and $\mathbf{p}$ is a point in $\mathcal{M}$, then the tangent space $\mathcal{M}_\mathbf{p}$ is naturally a vector space.*

Proof. We make $\mathcal{M}_\mathbf{p}$ into a vector space over $\mathbb{R}$ as follows. Let $\mathbf{v}_\mathbf{p}, \mathbf{w}_\mathbf{p} \in \mathcal{M}_\mathbf{p}$, $a \in \mathbb{R}$ and $f \in \mathfrak{F}(\mathcal{M})$. Then we define $\mathbf{v}_\mathbf{p} + \mathbf{w}_\mathbf{p}$ and $a\, \mathbf{v}_\mathbf{p}$ by

$$\begin{cases} (\mathbf{v}_\mathbf{p} + \mathbf{w}_\mathbf{p})[f] = \mathbf{v}_\mathbf{p}[f] + \mathbf{w}_\mathbf{p}[f], \\ (a\, \mathbf{v}_\mathbf{p})[f] = a\, \mathbf{v}_\mathbf{p}[f]. \end{cases}$$

It is easily verified by the same sort of proof as that of Lemma 23.8 that in fact $\mathbf{v}_\mathbf{p} + \mathbf{w}_\mathbf{p}, a\, \mathbf{v}_\mathbf{p} \in \mathcal{M}_\mathbf{p}$. Furthermore, we can also check that this addition and scalar multiplication makes $\mathcal{M}_\mathbf{p}$ into a vector space. $\blacksquare$

23.3 Tangent Vectors on Differentiable Manifolds

We would now like to prove that $\mathcal{M}_\mathbf{p}$ is a finite-dimensional vector space whose dimension is the same as that of $\mathcal{M}$. Furthermore, we have

$$\left\{ \left.\frac{\partial}{\partial x_1}\right|_\mathbf{p}, \ldots, \left.\frac{\partial}{\partial x_n}\right|_\mathbf{p} \right\}$$

as a natural candidate for a basis of $\mathcal{M}_\mathbf{p}$. However, there are some technical difficulties to overcome. First, we need:

Lemma 23.10. *Let $\mathcal{M}$ be a differentiable manifold and $\mathbf{p} \in \mathcal{M}$.*

(i) If $f \in \mathfrak{F}(\mathcal{M})$ can be expressed as the product of two functions $g, h \in \mathfrak{F}(\mathcal{M})$ both of which vanish at $\mathbf{p}$, then $\mathbf{v}_\mathbf{p}[f] = 0$ for all $\mathbf{v}_\mathbf{p} \in \mathcal{M}_\mathbf{p}$.

(ii) If c is a constant function on $\mathcal{M}$, then $\mathbf{v}_\mathbf{p}[c] = 0$ for all $\mathbf{v}_\mathbf{p} \in \mathcal{M}_\mathbf{p}$.

(iii) Tangent vectors are local; that is, if $\mathbf{v}_\mathbf{p} \in \mathcal{M}_\mathbf{p}$, then the number $\mathbf{v}_\mathbf{p}[f]$ for $f \in \mathfrak{F}(\mathcal{M})$ depends only on the values of f in a neighborhood of $\mathbf{p}$.

(iv) Let h be a differentiable function defined only on a neighborhood $\mathcal{U}$ of $\mathbf{p}$. If $\mathbf{v}_\mathbf{p} \in \mathcal{M}_\mathbf{p}$, then for any globalization $\widetilde{h}$ of h the value $\mathbf{v}_\mathbf{p}[\widetilde{h}]$ is the same; we call this value $\mathbf{v}_\mathbf{p}[h]$.

Proof. For (i) suppose that $f = gh$ where $g, h \in \mathfrak{F}(\mathcal{M})$ and $g(\mathbf{p}) = h(\mathbf{p}) = 0$. Then for $\mathbf{v}_\mathbf{p} \in \mathcal{M}_\mathbf{p}$ we have by the Leibnizian property that

$$\mathbf{v}_\mathbf{p}[f] = \mathbf{v}_\mathbf{p}[g\,h] = g(\mathbf{p})\mathbf{v}_\mathbf{p}[h] + h(\mathbf{p})\mathbf{v}_\mathbf{p}[g] = 0.$$

Next, for (ii) we first observe that

$$\mathbf{v}_\mathbf{p}[1] = \mathbf{v}_\mathbf{p}[1^2] = 1 \cdot \mathbf{v}_\mathbf{p}[1] + 1 \cdot \mathbf{v}_\mathbf{p}[1] = 2\mathbf{v}_\mathbf{p}[1],$$

so that $\mathbf{v}_\mathbf{p}[1] = 0$. For an arbitrary constant c we then have by the linearity property that

$$\mathbf{v}_\mathbf{p}[c] = \mathbf{v}_\mathbf{p}[c \cdot 1] = c\mathbf{v}_\mathbf{p}[1] = 0.$$

Since parts (iii) and (iv) are rephrasings of the same statement, we prove (iv). Let $\widetilde{h}$ and $\widehat{h}$ be globalizations of h which agree with h on a neighborhood $\mathcal{V}$ of $\mathbf{p}$, and let $g = \widetilde{h} - \widehat{h}$. Then g vanishes on $\mathcal{V}$. By Lemma 23.3 there exist a neighborhood $\mathcal{W} \subset \mathcal{V}$ of $\mathbf{p}$ and a function $k \in \mathfrak{F}(\mathcal{M})$ such that $k(\mathbf{m}) = 0$ for $\mathbf{m} \in \mathcal{W}$ and $k(\mathbf{m}) = 1$ for $\mathcal{M}\backslash\mathcal{V}$. Then $g = gk$ and $g(\mathbf{p}) = k(\mathbf{p}) = 0$. Hence by part (i) we have $0 = \mathbf{v}_\mathbf{p}[g] = \mathbf{v}_\mathbf{p}[\widetilde{h}] - \mathbf{v}_\mathbf{p}[\widehat{h}]$, so that $\mathbf{v}_\mathbf{p}[\widetilde{h}] = \mathbf{v}_\mathbf{p}[\widehat{h}]$. ∎

Next, we need a fact about $\mathfrak{F}(\mathbb{R}^n)$.

Lemma 23.11. Let $g: \mathbb{R}^n \longrightarrow \mathbb{R}$ be a differentiable function, and let $\mathbf{a} = (a_1, \ldots, a_n) \in \mathbb{R}^n$. Then there exist differentiable functions $g_i: \mathbb{R}^n \longrightarrow \mathbb{R}$ for $i = 1, \ldots, n$ such that

(23.3) $$g = \sum_{i=1}^n (u_i - a_i) g_i + g(\mathbf{a}).$$

Proof. Fix $\mathbf{t} = (t_1, \ldots, t_n) \in \mathbb{R}^n$ and consider the function $f: \mathbb{R} \longrightarrow \mathbb{R}$ defined by

$$f(s) = g\Big(s t_1 + (1-s)a_1, \ldots, s t_n + (1-s)a_n\Big) = g(s\mathbf{t} + (1-s)\mathbf{a}).$$

Since g is differentiable, so is f, and

$$f'(s) = \sum_{i=1}^n (t_i - a_i) \frac{\partial g}{\partial u_i}\Big(s t_1 + (1-s)a_1, \ldots, s t_n + (1-s)a_n\Big).$$

Now

$$\int_0^1 f'(s) ds = f(1) - f(0) = g(\mathbf{t}) - g(\mathbf{a}),$$

and so the fundamental theorem of calculus implies that

$$\begin{aligned} g(\mathbf{t}) - g(\mathbf{a}) &= \int_0^1 f'(s) ds \\ &= \sum_{i=1}^n (t_i - a_i) \int_0^1 \frac{\partial g}{\partial u_i}\Big(s t_1 + (1-s)a_1, \ldots, s t_n + (1-s)a_n\Big) ds \\ &= \sum_{i=1}^n \big(u_i(\mathbf{t}) - a_i\big) g_i(\mathbf{t}), \end{aligned}$$

where

$$g_i(\mathbf{t}) = \int_0^1 \frac{\partial g}{\partial u_i}\Big(s t_1 + (1-s)a_1, \ldots, s t_n + (1-s)a_n\Big) ds.$$

Clearly, the g_i's are differentiable. Thus we get (23.3). ∎

Note that the proof of this lemma fails for C^k-functions. If g is a C^k-function, we can only say that each g_i is a C^{k-1}-function.

We are now able to prove:

Theorem 23.12. (Basis Theorem) Let $\mathbf{x}: \mathcal{U} \longrightarrow \mathcal{M}$ be a patch on a differentiable manifold $\mathcal{M}$ with $\mathbf{p} \in \mathbf{x}(\mathcal{U})$ and write $\mathbf{x}^{-1} = (x_1, \ldots, x_n)$. If $\mathbf{v}_\mathbf{p} \in \mathcal{M}_\mathbf{p}$, then

$$\mathbf{v}_\mathbf{p} = \sum_{i=1}^n \mathbf{v}_\mathbf{p}[x_i] \frac{\partial}{\partial x_i}\bigg|_\mathbf{p}.$$

23.3 Tangent Vectors on Differentiable Manifolds

Proof. It suffices to show that

$$\mathbf{v_p}[f] = \sum_{i=1}^{n} \mathbf{v_p}[x_i] \frac{\partial f}{\partial x_i}\bigg|_{\mathbf{p}} \tag{23.4}$$

for all $f \in \mathfrak{F}(\mathcal{M})$. To this end, let $g \colon \mathbb{R}^n \longrightarrow \mathbb{R}$ be a globalization of $f \circ \mathbf{x}$ near $\mathbf{x}^{-1}(\mathbf{p})$. Then by Lemma 23.11 there exist differentiable functions $g_i \colon \mathbb{R}^n \longrightarrow \mathbb{R}$ for $i = 1, \ldots, n$ such that

$$g = \sum_{i=1}^{n} \bigl(u_i - u_i(\mathbf{x}^{-1}(\mathbf{p}))\bigr) g_i + g(\mathbf{x}^{-1}(\mathbf{p})).$$

Then near $\mathbf{p}$ we have

$$f = \sum_{i=1}^{n} (x_i - x_i(\mathbf{p})) f_i + f(\mathbf{p}), \tag{23.5}$$

where $f_i = g_i \circ \mathbf{x}^{-1}$. Lemma 23.10 (iii) allows us to use (23.5) to compute $\mathbf{v_p}[f]$. From (23.5) it follows that

$$\begin{aligned}
\mathbf{v_p}[f] &= \sum_{i=1}^{n} \Bigl\{ f_i(\mathbf{p}) \mathbf{v_p}[x_i - x_i(\mathbf{p})] + (x_i(\mathbf{p}) - x_i(\mathbf{p})) \mathbf{v_p}[f_i] \Bigr\} + \mathbf{v_p}[f(\mathbf{p})] \\
&= \sum_{i=1}^{n} f_i(\mathbf{p}) \mathbf{v_p}[x_i],
\end{aligned}$$

because $x_i(\mathbf{p})$ and $f(\mathbf{p})$ are constants. For the same reasons, from (23.5) it follows that

$$\frac{\partial f}{\partial x_j}\bigg|_{\mathbf{p}} = \sum_{i=1}^{n} f_i(\mathbf{p}) \frac{\partial x_i}{\partial x_j}(\mathbf{p}) = f_j(\mathbf{p}),$$

because[2]

$$\frac{\partial x_i}{\partial x_j}(\mathbf{p}) = \frac{\partial (x_i \circ \mathbf{x})}{\partial u_j}(\mathbf{x}^{-1}(\mathbf{p})) = \frac{\partial u_i}{\partial u_j}(\mathbf{x}^{-1}(\mathbf{p})) = \delta_{ij}.$$

Hence

$$\left(\sum_{j=1}^{n} \mathbf{v_p}[x_j] \frac{\partial}{\partial x_j} \right)\bigg|_{\mathbf{p}} [f] = \sum_{j=1}^{n} \mathbf{v_p}[x_j] \frac{\partial f}{\partial x_j}\bigg|_{\mathbf{p}} = \sum_{j=1}^{n} \mathbf{v_p}[x_j] f_j(\mathbf{p}) = \mathbf{v_p}[f].$$

---

[2] Here δ_{ij} denotes the Kronecker delta function. It is defined by

$$\delta_{ij} = \begin{cases} 1 & \text{for } i = j, \\ 0 & \text{for } i \neq j. \end{cases}$$

Thus we get (23.4). ∎

We have seen in Lemmas 23.3 and 23.4 that to show the existence of nonconstant differentiable functions on a differentiable manifold $\mathcal{M}$, it was necessary to use C^∞-functions and not C^ω-functions. Furthermore, in the proofs of Lemma 23.11 and Theorem 23.12 it is necessary to use C^∞-functions and not C^k-functions.

Corollary 23.13. *Let* $\mathbf{x}\colon \mathcal{U} \longrightarrow \mathcal{M}$ *be a patch on a differentiable manifold* $\mathcal{M}$. *Let* $\mathbf{p} \in \mathbf{x}(\mathcal{U})$ *and write* $\mathbf{x}^{-1} = (x_1, \ldots, x_n)$. *Then the vectors*

$$\left.\frac{\partial}{\partial x_1}\right|_{\mathbf{p}}, \ldots, \left.\frac{\partial}{\partial x_n}\right|_{\mathbf{p}}$$

form a basis for the tangent space $\mathcal{M}_{\mathbf{p}}$. *Hence the dimension of each tangent space* $\mathcal{M}_{\mathbf{p}}$ *as a vector space is the same as the dimension of* $\mathcal{M}$ *as a manifold.*

Proof. By Theorem 23.12 the vectors

$$\left.\frac{\partial}{\partial x_1}\right|_{\mathbf{p}}, \ldots, \left.\frac{\partial}{\partial x_n}\right|_{\mathbf{p}}$$

span the tangent space $\mathcal{M}_{\mathbf{p}}$. To prove they are linearly independent, suppose that $a_1, \ldots, a_n \in \mathbb{R}$ are such that

$$\sum_{i=1}^n a_i \left.\frac{\partial}{\partial xi}\right|_{\mathbf{p}} = 0.$$

Then

$$0 = \sum_{i=1}^n a_i \left.\frac{\partial x_j}{\partial x_i}\right|_{\mathbf{p}} = \sum_{i=1}^n a_i \frac{\partial(x_j \circ \mathbf{x})}{\partial u_i}(\mathbf{x}^{-1}(\mathbf{p})) = \sum_{i=1}^n a_i \delta_{ij} = a_j$$

for $j = 1, \ldots, n$. ∎

In order to differentiate functions on manifolds (that is, apply tangent vectors to them) as easily as we would differentiate functions on $\mathbb{R}^n$, we shall need:

Lemma 23.14. (**The Chain Rule**) *Let* $\mathcal{M}$ *be a differentiable manifold. Suppose* $g_1, \ldots, g_k \in \mathfrak{F}(\mathcal{M})$ *and* $F \in \mathfrak{F}(\mathbb{R}^k)$. *Let*

$$f = F(g_1, \ldots, g_k)$$

(that is, $f(\mathbf{p}) = F\bigl(g_1(\mathbf{p}), \ldots, g_k(\mathbf{p})\bigr)$ *for* $\mathbf{p} \in \mathcal{M}$*). Then for* $\mathbf{v}_{\mathbf{p}} \in \mathcal{M}_{\mathbf{p}}$ *we have*

(23.6) $$\mathbf{v}_{\mathbf{p}}[f] = \sum_{i=1}^k \frac{\partial F}{\partial u_i}\bigl(g_1(\mathbf{p}), \ldots, g_k(\mathbf{p})\bigr)\mathbf{v}_{\mathbf{p}}[g_i].$$

23.3 Tangent Vectors on Differentiable Manifolds

Proof. Let $g = (g_1, \ldots, g_k)$. Define $\mathbf{w}\colon \mathfrak{F}(\mathbb{R}^k) \longrightarrow \mathbb{R}$ by

$$\mathbf{w}[H] = \mathbf{v}_\mathbf{p}[H \circ g]$$

for $H \in \mathfrak{F}(\mathbb{R}^k)$. Then it is easy to check from the definition that $\mathbf{w}$ is a tangent vector in $\mathbb{R}^k_{g(\mathbf{p})}$, because it is linear and Leibnizian. Therefore, using Theorem 23.12 with the patch $(u_1, \ldots, u_k)$ at $g(\mathbf{p})$ we get

$$(23.7) \qquad \mathbf{w}[F] = \sum_{i=1}^k \mathbf{w}[u_i] \frac{\partial F}{\partial u_i}(g(\mathbf{p})).$$

Now $\mathbf{w}[F] = \mathbf{v}_\mathbf{p}[F \circ g] = \mathbf{v}_\mathbf{p}[f]$ and $\mathbf{w}[u_i] = \mathbf{v}_\mathbf{p}[u_i \circ g] = \mathbf{v}_\mathbf{p}[g_i]$, so that (23.7) becomes

$$(23.8) \qquad \mathbf{v}_\mathbf{p}[f] = \sum_{i=1}^k \frac{\partial F}{\partial u_i}(g(\mathbf{p})) \mathbf{v}_\mathbf{p}[g_i].$$

But (23.8) is another way of writing (23.6). ∎

Note that (23.6) allows us to transfer standard differentiation formulas from $\mathbb{R}^n$ to manifolds. For example, if $\mathbf{v}_\mathbf{p}$ is a tangent vector to a manifold $\mathcal{M}$ at $\mathbf{p}$ and $f \in \mathfrak{F}(\mathcal{M})$, then

$$\mathbf{v}_\mathbf{p}[\sin f] = \cos(f(\mathbf{p})) \mathbf{v}_\mathbf{p}[f].$$

For abstract surface theory it will be useful to have an alternate notation for tangent vectors.

Definition. Let $\mathbf{x}\colon \mathcal{U} \longrightarrow \mathcal{M}$ be a patch on a differentiable manifold $\mathcal{M}$, and let $\mathbf{q} \in \mathcal{U}$. Then

$$\mathbf{x}_{u_i}(\mathbf{q})\colon \mathfrak{F}(\mathcal{M}) \longrightarrow \mathbb{R}$$

is the operator given by

$$(23.9) \qquad \mathbf{x}_{u_i}(\mathbf{q})[f] = \left.\frac{\partial(f \circ \mathbf{x})}{\partial u_i}\right|_\mathbf{q}$$

for $\mathbf{q} \in \mathcal{U}$ and $f \in \mathfrak{F}(\mathcal{M})$.

Then (23.9) is just a generalization of $\mathbf{x}_u$ and $\mathbf{x}_v$ defined by (12.2) on page 271.

23.4 Induced Maps

In the previous section we showed that to each point $\mathbf{p}$ of a differentiable manifold $\mathcal{M}$ there is associated a vector space called the tangent space $\mathcal{M}_\mathbf{p}$. In the present section we show how a differentiable map $\Psi \colon \mathcal{M} \longrightarrow \mathcal{N}$ between differentiable manifolds $\mathcal{M}$ and $\mathcal{N}$ gives rise to a linear map between tangent spaces. Just as the tangent space is the best linear approximation of a differentiable manifold, the tangent map is the best linear approximation to a differentiable map between manifolds.

Definition. Let $\Psi \colon \mathcal{M} \longrightarrow \mathcal{N}$ be a differentiable map between differentiable manifolds $\mathcal{M}$ and $\mathcal{N}$, and let $\mathbf{p} \in \mathcal{M}$. Then the **tangent map** of Ψ at $\mathbf{p}$ is the map

$$\Psi_{*\mathbf{p}} \colon \mathcal{M}_\mathbf{p} \longrightarrow \mathcal{N}_{\Psi(\mathbf{p})}$$

given by

(23.10) $$\Psi_{*\mathbf{p}}(\mathbf{v}_\mathbf{p})[f] = \mathbf{v}_\mathbf{p}[f \circ \Psi]$$

for each $f \in \mathfrak{F}(\mathcal{N})$ and $\mathbf{v}_\mathbf{p} \in \mathcal{M}_\mathbf{p}$.

In order for this definition to make sense, we must be sure that the image of $\Psi_{*\mathbf{p}}$ is actually contained in the tangent space $\mathcal{N}_{\Psi(\mathbf{p})}$.

Lemma 23.15. *Let $\Psi \colon \mathcal{M} \longrightarrow \mathcal{N}$ be a differentiable map, and let $\mathbf{p} \in \mathcal{M}$, $\mathbf{v}_\mathbf{p} \in \mathcal{M}_\mathbf{p}$. Define*

$$\Psi_{*\mathbf{p}}(\mathbf{v}_\mathbf{p}) \colon \mathfrak{F}(\mathcal{N}) \longrightarrow \mathbb{R}$$

by (23.10). Then $\Psi_{*\mathbf{p}}(\mathbf{v}_\mathbf{p}) \in \mathcal{N}_{\Psi(\mathbf{p})}$.

Proof. For example, we show that $\Psi_{*\mathbf{p}}(\mathbf{v}_\mathbf{p})$ is Leibnizian. Using the fact that $\mathbf{v}_\mathbf{p}$ is Leibnizian, we have

$$\begin{aligned}\Psi_{*\mathbf{p}}(\mathbf{v}_\mathbf{p})[fg] &= \mathbf{v}_\mathbf{p}[(fg) \circ \Psi] = \mathbf{v}_\mathbf{p}\big[(f \circ \Psi)(g \circ \Psi)\big] \\ &= (f \circ \Psi)(\mathbf{p})\mathbf{v}_\mathbf{p}[g \circ \Psi] + (g \circ \Psi)(\mathbf{p})\mathbf{v}_\mathbf{p}[f \circ \Psi] \\ &= f\big(\Psi(\mathbf{p})\big)\Psi_{*\mathbf{p}}(\mathbf{v}_\mathbf{p})[g] + g\big(\Psi(\mathbf{p})\big)\Psi_{*\mathbf{p}}(\mathbf{v}_\mathbf{p})[f]. \blacksquare\end{aligned}$$

Lemma 23.16. *Let $\mathcal{M}$ be an n-dimensional differentiable manifold and let $\mathbf{x} \colon \mathcal{U} \longrightarrow \mathcal{M}$ be a patch, where $\mathcal{U}$ is an open subset of $\mathbb{R}^n$. Write $\mathbf{x}^{-1} = (x_1, \ldots, x_n)$. Then*

(23.11) $$\mathbf{x}_{*\mathbf{q}}\left(\left.\frac{\partial}{\partial u_i}\right|_\mathbf{q}\right) = \mathbf{x}_{u_i}(\mathbf{q}) = \left.\frac{\partial}{\partial x_i}\right|_{\mathbf{x}(\mathbf{q})}$$

for $\mathbf{q} \in \mathcal{U}$ and $i = 1, \ldots, n$.

23.4 Induced Maps

Proof. For $f \in \mathfrak{F}(\mathcal{M})$ we have by (23.2) that

$$\mathbf{x}_{*q}\left(\left.\frac{\partial}{\partial u_i}\right|_q\right)[f] = \left.\frac{\partial}{\partial u_i}\right|_q [f \circ \mathbf{x}] = \left.\frac{\partial f}{\partial x_i}\right|_{\mathbf{x}(q)} = \left.\frac{\partial}{\partial x_i}\right|_{\mathbf{x}(q)}[f]. \blacksquare$$

Thus the tangent map $\Psi_{*\mathbf{p}}$ can be considered as a kind of dual of the homomorphism $\mathfrak{F}(\mathcal{N}) \longrightarrow \mathfrak{F}(\mathcal{M}): g \longmapsto g \circ \Psi$, which in turn can be considered as a dual of the original function $\Psi: \mathcal{M} \longrightarrow \mathcal{N}$.

Lemma 23.17. Let $\Psi: \mathcal{M} \longrightarrow \mathcal{N}$ be a differentiable map and $\mathbf{p} \in \mathcal{M}$.

(i) The map $\Psi_{*\mathbf{p}}: \mathcal{M}_\mathbf{p} \longrightarrow \mathcal{N}_{\Psi(\mathbf{p})}$ is a linear map between vector spaces.

(ii) If $\mathcal{M} \xrightarrow{\Psi} \mathcal{N} \xrightarrow{\Phi} \mathcal{P}$ are differentiable maps, then

$$(\Phi \circ \Psi)_{*\mathbf{p}} = \Phi_{*\Psi(\mathbf{p})} \circ \Psi_{*\mathbf{p}}.$$

(iii) Let $\mathbf{x}: \mathcal{U} \longrightarrow \mathcal{M}$ be a patch on $\mathcal{M}$ at $\mathbf{p}$ and $\mathbf{y}: \mathcal{V} \longrightarrow \mathcal{N}$ a patch on $\mathcal{N}$ at $\Psi(\mathbf{p})$. Write $\mathbf{x}^{-1} = (x_1, \ldots, x_m)$ and $\mathbf{y}^{-1} = (y_1, \ldots, y_n)$, where $m = \dim \mathcal{M}$ and $n = \dim \mathcal{N}$. Then

$$(23.12) \qquad \Psi_{*\mathbf{p}}\left(\left.\frac{\partial}{\partial x_i}\right|_\mathbf{p}\right) = \sum_{j=1}^n \frac{\partial(y_j \circ \Psi)}{\partial x_i}(\mathbf{p}) \left.\frac{\partial}{\partial y_j}\right|_{\Psi(\mathbf{p})}$$

for $i = 1, \ldots, m$.

Proof. The proofs of (i) and (ii) are straightforward. To prove (iii), we use the basis theorem (Theorem 23.12). Since

$$\Psi_{*\mathbf{p}}\left(\left.\frac{\partial}{\partial x_i}\right|_\mathbf{p}\right) \in \mathcal{N}_{\Psi(\mathbf{p})},$$

and

$$\left\{\left.\frac{\partial}{\partial y_1}\right|_{\Psi(\mathbf{p})}, \ldots, \left.\frac{\partial}{\partial y_n}\right|_{\Psi(\mathbf{p})}\right\}$$

is a basis for $\mathcal{N}_{\Psi(\mathbf{p})}$, we can write

$$(23.13) \qquad \Psi_{*\mathbf{p}}\left(\left.\frac{\partial}{\partial x_i}\right|_\mathbf{p}\right) = \sum_{j=1}^n a_{ij} \left.\frac{\partial}{\partial y_j}\right|_{\Psi(\mathbf{p})}.$$

To compute the a_{ij}'s in (23.13), we apply both sides of (23.13) to y_k. We have

$$\left(\sum_{j=1}^n a_{ij} \frac{\partial}{\partial y_j}\bigg|_{\Psi(\mathbf{p})}\right)[y_k] = a_{ik}$$

while

$$\Psi_{*\mathbf{p}}\left(\frac{\partial}{\partial x_i}\bigg|_{\mathbf{p}}\right)[y_k] = \left(\frac{\partial}{\partial x_i}\bigg|_{\mathbf{p}}\right)[y_k \circ \Psi] = \frac{\partial(y_k \circ \Psi)}{\partial x_i}(\mathbf{p}).$$

Hence

$$a_{ik} = \frac{\partial(y_k \circ \Psi)}{\partial x_i}(\mathbf{p}),$$

and we get (23.12). ∎

Definition. Let $\Psi \colon \mathcal{M} \longrightarrow \mathcal{N}$ be a differentiable map, where $\mathcal{M}$ is an m-dimensional differentiable manifold and $\mathcal{N}$ is an n-dimensional differentiable manifold. Let $\mathbf{p} \in \mathcal{M}$, $(\mathbf{x}, \mathcal{U})$ a patch on $\mathcal{M}$ at $\mathbf{p}$ and $(\mathbf{y}, \mathcal{V})$ a patch on $\mathcal{N}$ at $\Psi(\mathbf{p})$. The **Jacobian matrix** $\mathcal{J}(\Psi)(\mathbf{p})$ of Ψ at $\mathbf{p}$ relative to $\mathbf{x}$ and $\mathbf{y}$ is the matrix of $\Psi_{*\mathbf{p}}$ relative to the bases

$$\left\{\frac{\partial}{\partial x_1}\bigg|_{\mathbf{p}}, \ldots, \frac{\partial}{\partial x_m}\bigg|_{\mathbf{p}}\right\} \quad \text{and} \quad \left\{\frac{\partial}{\partial y_1}\bigg|_{\Psi(\mathbf{p})}, \ldots, \frac{\partial}{\partial y_n}\bigg|_{\Psi(\mathbf{p})}\right\}.$$

Explicitly, $\mathcal{J}(\Psi)(\mathbf{p})$ is the matrix

$$\left(\frac{\partial(y_j \circ \Psi)}{\partial x_i}(\mathbf{p})\right).$$

We can now get the manifold formulation of the inverse function theorem from multivariable calculus.

Theorem 23.18. (Inverse Function Theorem) *Let $\Psi \colon \mathcal{M} \longrightarrow \mathcal{N}$ be a differentiable map, and let $\mathbf{p} \in \mathcal{M}$. Suppose $\dim \mathcal{M} = \dim \mathcal{N} = n$. Then the following conditions are equivalent:*

(i) *$\Psi_{*\mathbf{p}}$ is a linear isomorphism;*

(ii) *the Jacobian matrix $\mathcal{J}(\Psi)(\mathbf{p})$ is invertible;*

(iii) *there exist neighborhoods $\mathcal{W}$ of $\mathbf{p}$ and $\mathcal{V}$ of $\Psi(\mathbf{p})$ such that the restriction $\Psi|\mathcal{W} \colon \mathcal{W} \longrightarrow \mathcal{V}$ has a differentiable inverse $(\Psi|\mathcal{W})^{-1} \colon \mathcal{V} \longrightarrow \mathcal{W}$.*

23.4 Induced Maps

Proof. That (i) implies (ii) is a standard fact of linear algebra. Also, it is easy to see from Lemma 23.17 (ii) that (iii) implies (i).

To show that (ii) implies (iii), consider $\mathbf{y}^{-1} \circ \Psi \circ \mathbf{x}$, where $(\mathbf{x}, \mathcal{U})$ is a patch on $\mathcal{M}$ with $\mathbf{p} \in \mathcal{U}$ and $(\mathbf{y}, \mathcal{V})$ is a patch on $\mathcal{N}$ with $\Psi(\mathbf{q}) \in \mathcal{V}$. Without loss of generality, we can assume that
$$(\mathbf{y}^{-1} \circ \Psi \circ \mathbf{x})(0) = 0.$$
Write $\mathbf{x}^{-1} = (x_1, \ldots, x_n)$ and $\mathbf{y}^{-1} = (y_1, \ldots, y_n)$. By assumption
$$\det\left(\left(\frac{\partial(u_j \circ \mathbf{y}^{-1} \circ \Psi \circ \mathbf{x})}{\partial u_i}(0)\right)\right) = \det\left(\left(\frac{\partial(y_j \circ \Psi)}{\partial x_i}(\mathbf{p})\right)\right) = \mathcal{J}(\Psi)(\mathbf{p}) \neq 0.$$
The inverse function theorem for $\mathbb{R}^n$ implies that $\mathbf{y}^{-1} \circ \Psi \circ \mathbf{x}$ has a local inverse. Hence so does Ψ. ∎

Corollary 23.19. *Let $(x_1, \ldots, x_n)$ be a coordinate system for $\mathcal{M}$ at $\mathbf{p}$, where each x_i is defined on a neighborhood $\mathcal{U}$ of $\mathbf{p}$. Then functions $f_1, \ldots, f_n \in \mathfrak{F}(\mathcal{U})$ form a coordinate system for $\mathcal{M}$ at $\mathbf{p}$ if and only if*
$$\det\left(\frac{\partial f_j}{\partial x_i}\right) \neq 0$$
on $\mathcal{U}$.

Definition. *A differentiable map $\Psi \colon \mathcal{M} \longrightarrow \mathcal{N}$ is called* **regular**, *provided $\Psi_{*\mathbf{p}}$ is a nonsingular linear transformation for each $\mathbf{p} \in \mathcal{M}$.*

Corollary 23.20. *A map $\Psi \colon \mathcal{M} \longrightarrow \mathcal{N}$ is a local diffeomorphism if and only if each $\Psi_{*\mathbf{p}}$ is a linear isomorphism.*

Next, we discuss the notion of curve in a differentiable manifold using the concepts we have developed. In what follows I denotes an open interval of the real line $\mathbb{R}$ and d/du the natural coordinate vector field on it. For each $t \in I$ we have a canonical tangent vector in $\mathbb{R}_t$, namely
$$\left.\frac{d}{du}\right|_t.$$

Definition. *A* **curve** *on a differentiable manifold $\mathcal{M}$ is a differentiable function $\alpha \colon I \longrightarrow \mathcal{M}$. If $t \in I$ the* **velocity vector** *of α at t is the tangent vector*
$$\alpha'(t) = \alpha_{*t}\left(\left.\frac{d}{du}\right|_t\right) \in \mathcal{M}_{\alpha(t)}.$$

We do not exclude the possibility that the interval I is half-infinite or all of $\mathbb{R}$. One can usually arrange to have zero in the domain I of a curve α; it is a convenient reference point. A curve on a surface in $\mathbb{R}^3$ is also a curve in the sense of this definition and the two notions of the velocity vectors are equivalent. We have not defined the acceleration vector of a curve in an abstract manifold because that requires a connection on the manifold. (See Section 24.1.)

Definition. *We say that a curve α starts at $\mathbf{p} \in \mathcal{M}$ provided that $\alpha(0) = \mathbf{p}$, and that α has initial velocity $\mathbf{v}_\mathbf{p} \in \mathcal{M}_\mathbf{p}$, provided $\alpha'(0) = \mathbf{v}_\mathbf{p}$.*

There is one subtle point in the notation. Suppose that $h: I \longrightarrow \mathbb{R}$ is a differentiable function. Then
$$\frac{dh}{du}(t)$$
is a real number for $t \in I$, whereas $h'(t)$ is a tangent vector in the tangent space $I_{h(t)} = \mathbb{R}_{h(t)}$. On the other hand, there is a canonical isomorphism between $\mathbb{R}$ and $\mathbb{R}_{h(t)}$ which identifies
$$\frac{dh}{du}(t)$$
with $h'(t)$.

Lemma 23.21. *Let $\alpha: I \longrightarrow \mathcal{M}$ be a curve. Then the value of the velocity vector $\alpha'(t)$ on a function $f \in \mathfrak{F}(\mathcal{M})$ is given by*
$$\alpha'(t)[f] = \frac{d(f \circ \alpha)}{du}(t).$$

Proof. We have
$$\alpha'(t)[f] = \alpha_{*t}\left(\left.\frac{d}{du}\right|_t\right)[f] = \frac{d(f \circ \alpha)}{du}(t). \quad\blacksquare$$

This makes rigorous the earlier suggestion that for a tangent vector $\mathbf{v}_\mathbf{p}$ the number $\mathbf{v}_\mathbf{p}[f]$ is the derivative of f in the direction $\mathbf{v}_\mathbf{p}$. This is because for any curve α with initial velocity $\mathbf{v}_\mathbf{p}$ we have
$$\mathbf{v}_\mathbf{p}[f] = \frac{d(f \circ \alpha)}{du}(0).$$

Lemma 23.22. *Let α be a curve in $\mathcal{M}$, and let $(x_1, \ldots, x_n)$ be a coordinate system at $\alpha(t) \in \mathcal{M}$. Then*
$$\alpha'(t) = \sum_{i=1}^n \frac{d(x_i \circ \alpha)}{du}(t) \left.\frac{\partial}{\partial x_i}\right|_{\alpha(t)}.$$

Next, we prove the generalization to differentiable manifolds of Lemma 1.3, page 7. Although the following lemma can be proved by showing that it is a special case of Lemma 23.14, it is easier to prove it directly.

Lemma 23.23. (**The Chain Rule for Curves on a Manifold**) *Let $\alpha\colon (a,b) \longrightarrow \mathcal{M}$ and $h\colon (c,d) \longrightarrow (a,b)$ be differentiable. Put $\beta = \alpha \circ h$. Then*

$$\beta' = h'(\alpha' \circ h).$$

Proof. We have

$$\beta'(t) = \beta_{*t}\left(\frac{d}{du}\bigg|_t\right) = \alpha_{*h(t)} \circ h_{*t}\left(\frac{d}{du}\bigg|_t\right)$$

$$= \alpha_{*h(t)}\left(h'(t)\frac{d}{du}\bigg|_{h(t)}\right) = h'(t)\alpha'\bigl(h(t)\bigr). \blacksquare$$

23.5 Vector Fields on Differentiable Manifolds

The notion of derivation of an algebra comes up frequently in differential geometry. Let us give the general definition and then specialize it to show that a vector field is a derivation.

Definition. *Let A be an algebra (not necessarily associative nor commutative) over a field $\mathbb{F}$. A* **derivation** *of A is a mapping $D\colon A \longrightarrow A$ such that*

$$\begin{cases} D(af + bg) = a\,Df + b\,Dg, \\ D(fg) = f(Dg) + (Df)g \end{cases}$$

for all $a, b \in \mathbb{F}$ and $f, g \in A$.

Shortly, we shall define a vector field as a derivation of a special sort. But first let us study derivations in general.

Definition. If D_1 and D_2 are derivations of an algebra, then the **bracket** $[D_1, D_2]$ of D_1 and D_2 is defined by

$$[D_1, D_2] = D_1 D_2 - D_2 D_1.$$

The proofs of the next two lemmas are easy.

Lemma 23.24. *If D_1 and D_2 are derivations of an algebra, so are $[D_1, D_2]$, $D_1 + D_2$ and $a D_1$ for $a \in \mathbb{F}$.*

Lemma 23.25. *Let D_1, D_2 and D_3 be derivations of an algebra. Then:*

(i) $[D_1, D_2] = -[D_2, D_1]$.

(ii) *The* **Jacobi identity** *holds*:

$$[[D_1, D_2], D_3] + [[D_3, D_1], D_2] + [[D_2, D_3], D_1] = 0.$$

We sometimes abbreviate the Jacobi identity to

$$\mathop{\mathfrak{S}}_{123} [[D_1, D_2], D_3] = 0,$$

where $\mathfrak{S}$ denotes the cyclic sum. It is useful to introduce another abstract definition.

Definition. *Let $\mathbb{F}$ be a field. A* **Lie algebra** *over $\mathbb{F}$ is a vector space V over $\mathbb{F}$ with a bracket $[\ ,\ ]\colon V \times V \longrightarrow V$ such that*

(23.14) $$[X, X] = 0,$$

(23.15) $$[a X + b Y, Z] = a[X, Z] + b[Y, Z],$$

(23.16) $$[Z, a X + b Y] = a[Z, X] + b[Z, Y],$$

(23.17) $$\mathop{\mathfrak{S}}_{XYZ} [X, [Y, Z]] = 0$$

for each $X, Y, Z \in V$ and $a, b \in \mathbb{F}$.

We have a ready-made example of a Lie algebra.

Lemma 23.26. *Let A be a commutative algebra over a field $\mathbb{F}$ and denote by $\mathfrak{D}(A)$ the derivations of A. Then $\mathfrak{D}(A)$ is a module over A and a Lie algebra over $\mathbb{F}$.*

23.5 Vector Fields on Differentiable Manifolds

Although derivations arise in many different contexts in differential geometry, the one of immediate interest is the following:

Definition. *A* **vector field** *X on a differentiable manifold $\mathcal{M}$ is a derivation of the algebra $\mathfrak{F}(\mathcal{M})$ of real-valued differentiable functions on $\mathcal{M}$.*

Thus a vector field X on $\mathcal{M}$ is a mapping

$$X \colon \mathfrak{F}(\mathcal{M}) \longrightarrow \mathfrak{F}(\mathcal{M})$$

satisfying

$$X[af + bg] = aX[f] + bX[g] \qquad \text{(the \textbf{linearity} property)},$$
$$X[fg] = fX[g] + gX[f] \qquad \text{(the \textbf{Leibnizian} property)},$$

for $a, b \in \mathbb{R}$ and $f, g \in \mathfrak{F}(\mathcal{M})$. Also, the **bracket** $[X, Y]$ of vector fields X and Y is given by

$$[X, Y] = XY - YX.$$

Note that XY is a well-defined operator on $\mathfrak{F}(\mathcal{M})$, but not a vector field because the Leibnizian property is not satisfied. We put

$$\mathfrak{X}(\mathcal{M}) = \{\, X \mid X \text{ is a vector field on } \mathcal{M}\,\}.$$

We know from Lemma 23.26 that $\mathfrak{X}(\mathcal{M})$ is a module over $\mathfrak{F}(\mathcal{M})$ and a Lie algebra over $\mathbb{R}$. In other words, we may multiply vector fields by either real numbers or real-valued functions. Furthermore, we have a bracket "multiplication" $[\,,\,]$ between vector fields that yields another vector field in $\mathfrak{X}(\mathcal{M})$. The bracket is *not* multilinear with respect to functions; instead, there is a more complicated rule:

Lemma 23.27. *Let $X, Y \in \mathfrak{X}(\mathcal{M})$ and $f, g \in \mathfrak{F}(\mathcal{M})$. Then*

$$(23.18) \qquad [fX, gY] = fg[X,Y] + fX[g]Y - gY[f]X.$$

Of course, the definition of vector field that we have given is very abstract. Probably a more intuitive notion of vector field is that of a function that assigns to each point $\mathbf{p}$ of a manifold $\mathcal{M}$ a tangent vector $X_\mathbf{p}$. We now show that this intuitive notion is equivalent to the definition that we have given. First, we introduce the important notion of tangent bundle.

Definition. *The* **tangent bundle** *of a differentiable manifold $\mathcal{M}$ is the set*

$$T(\mathcal{M}) = \{\, (\mathbf{p}, \mathbf{v}_\mathbf{p}) \mid \mathbf{v}_\mathbf{p} \in \mathcal{M}_\mathbf{p},\ \mathbf{p} \in \mathcal{M}\,\}.$$

Theorem 23.28. *The tangent bundle $T(\mathcal{M})$ of a differentiable manifold $\mathcal{M}$ is naturally a differentiable manifold whose dimension is twice the dimension of $\mathcal{M}$. Furthermore, the* **projection map** *$\pi \colon T(\mathcal{M}) \longrightarrow \mathcal{M}$ defined by $\pi\big((\mathbf{p},\mathbf{v_p})\big) = \mathbf{p}$ is a differentiable mapping.*

Lemma 23.29. *Associated with each vector field $X \in \mathfrak{X}(\mathcal{M})$ is a* **section** *of $T(\mathcal{M})$, that is, a differentiable map $X \colon \mathcal{M} \longrightarrow T(\mathcal{M})$ such that $\pi \circ X = \mathbf{1}_{\mathcal{M}}$, where $\mathbf{1}_{\mathcal{M}}$ denotes the identity map on $\mathcal{M}$. In particular, X is a map that associates a tangent vector $X_\mathbf{p} \in \mathcal{M}_\mathbf{p}$ with each point $\mathbf{p} \in \mathcal{M}$.*

Conversely, each section of the tangent bundle gives rise in a natural way to an element of $\mathfrak{X}(\mathcal{M})$.

From now on we consider vector fields as derivations of $\mathfrak{F}(\mathcal{M})$, or as sections of $T(\mathcal{M})$, whichever seems more convenient. If $X \in \mathfrak{X}(\mathcal{M})$, we write $X_\mathbf{p}$ for the tangent vector in $\mathcal{M}_\mathbf{p}$ determined by X.

Definition. Let $(x_1, \ldots, x_n)$ be a system of local coordinates for $\mathcal{M}$ at $\mathbf{p}$ defined on an open set $\mathcal{U} \subset \mathcal{M}$. Then

$$\frac{\partial}{\partial x_i} \colon \mathfrak{F}(\mathcal{U}) \longrightarrow \mathfrak{F}(\mathcal{U})$$

is the operator that assigns to each function $f \in \mathfrak{F}(\mathcal{U})$ its partial derivative $\partial f / \partial x_i$.

The next lemma is an obvious consequence of Lemma 23.8.

Lemma 23.30. *For $i = 1, \ldots, n$ we have*

$$\frac{\partial}{\partial x_i} \in \mathfrak{X}(\mathcal{U}) \quad \text{and} \quad \left.\frac{\partial}{\partial x_i}\right|_\mathbf{p} \in \mathcal{M}_\mathbf{p}$$

for each $\mathbf{p} \in \mathcal{U}$.

Next, we give a version of the basis theorem (Theorem 23.12) for vector fields.

Corollary 23.31. *Let $X \in \mathfrak{X}(\mathcal{M})$ and let $\mathbf{x}^{-1} = (x_1, \ldots, x_n)$ be a system of local coordinates defined on an open set $\mathcal{U}$. Then X can be written as*

$$(23.19) \qquad X = \sum_{i=1}^n X[x_i] \frac{\partial}{\partial x_i}$$

on $\mathcal{U}$.

23.5 Vector Fields on Differentiable Manifolds

Proof. It follows from Theorem 23.12 that

$$(23.20) \qquad X_{\mathbf{p}} = \sum_{i=1}^{n} X_{\mathbf{p}}[x_i] \frac{\partial}{\partial x_i}\bigg|_{\mathbf{p}}.$$

Since (23.20) holds for all $\mathbf{p} \in \mathcal{U}$, we get (23.19). ∎

We now consider the effect of a differentiable map $\Psi: \mathcal{M} \longrightarrow \mathcal{N}$ on vector fields. Unfortunately, it is not always possible to transfer vector fields on $\mathcal{M}$ to vector fields on $\mathcal{N}$. The problem is that if $X \in \mathfrak{X}(\mathcal{M})$ and $\mathbf{p}, \mathbf{q} \in \mathcal{M}$ are points such that $\Psi(\mathbf{p}) = \Psi(\mathbf{q})$, we may have that

$$\Psi_{*\mathbf{p}}(X_{\mathbf{p}}) \neq \Psi_{*\mathbf{q}}(X_{\mathbf{q}}).$$

There is, however, a favorable case.

Definition. Let $\Psi: \mathcal{M} \longrightarrow \mathcal{N}$ be a differentiable map, and let $X \in \mathfrak{X}(\mathcal{M})$, $Y \in \mathfrak{X}(\mathcal{N})$. We say that X and Y are Ψ-*related* provided that

$$\Psi_{*\mathbf{p}}(X_{\mathbf{p}}) = Y_{\Psi(\mathbf{p})}$$

for all $\mathbf{p} \in \mathcal{M}$. We use the notation $X^{\Psi} = Y$ or $\Psi_*(X) = Y$.

Lemma 23.32. *Let $\Psi: \mathcal{M} \longrightarrow \mathcal{N}$ be a differentiable map.*

(i) *A vector field $X \in \mathfrak{X}(\mathcal{M})$ is Ψ-related to a vector field $Y \in \mathfrak{X}(\mathcal{N})$ if and only if*

$$Y[f] \circ \Psi = X[f \circ \Psi]$$

for all $f \in \mathfrak{F}(\mathcal{N})$.

(ii) *Let $W, X \in \mathfrak{X}(\mathcal{M})$ and $Y, Z \in \mathfrak{X}(\mathcal{N})$ with $W^{\Psi} = Y$ and $X^{\Psi} = Z$. Then $[W, X]$ is Ψ-related to $[Y, Z]$ and*

$$[W^{\Psi}, X^{\Psi}] = [Y, Z] = [W, X]^{\Psi}.$$

(iii) *If $\Psi: \mathcal{M} \longrightarrow \mathcal{N}$ is a diffeomorphism, then for any $X \in \mathfrak{X}(\mathcal{M})$ there exists a unique $Y \in \mathfrak{X}(\mathcal{N})$ such that $X^{\Psi} = Y$.*

Corollary 23.33. *Let $\mathcal{M}$ be an n-dimensional differentiable manifold and let $\mathbf{x}: \mathcal{U} \longrightarrow \mathcal{M}$ be a patch, where $\mathcal{U}$ is an open subset of $\mathbb{R}^n$. Write $\mathbf{x}^{-1} = (x_1, \ldots, x_n)$. Then*

(23.21) $$\begin{cases} \mathbf{x}_*\left(\dfrac{\partial}{\partial u_i}\right) = \dfrac{\partial}{\partial x_i} = \mathbf{x}_{u_i} \circ \mathbf{x}^{-1}, \\ \left[\dfrac{\partial}{\partial x_i}, \dfrac{\partial}{\partial x_j}\right] = \left[\mathbf{x}_{u_i}, \mathbf{x}_{u_j}\right] = 0 \end{cases}$$

for $1 \leq i, j \leq n$.

Proof. The first equation (23.21) is the vector field version of Lemma 23.16. Then

$$\left[\frac{\partial}{\partial x_i}, \frac{\partial}{\partial x_j}\right] = \left[\mathbf{x}_*\left(\frac{\partial}{\partial u_i}\right), \mathbf{x}_*\left(\frac{\partial}{\partial u_j}\right)\right] = \mathbf{x}_*\left(\left[\frac{\partial}{\partial u_i}, \frac{\partial}{\partial u_j}\right]\right) = 0,$$

proving the second equation of (23.21). ∎

23.6 Tensor Fields on Differentiable Manifolds

It will also be necessary to consider **tensor fields**. Tensor analysis was developed by Ricci-Curbastro[3] (see [Ricci]). The modern invariant notation described below was introduced by Koszul (see [Kosz]).

Instead of developing the most general situation, it will be sufficient for our purposes to treat two special cases.

Definition. *A* **covariant tensor field** *of degree r on a differentiable manifold $\mathcal{M}$ is a mapping*

$$\alpha \colon \mathfrak{X}(\mathcal{M}) \times \cdots \times \mathfrak{X}(\mathcal{M}) \longrightarrow \mathfrak{F}(\mathcal{M})$$

that satisfies

(23.22) $\alpha(X_1, \ldots, f_i Y_i + g_i Z_i, \ldots, X_r) = f_i \alpha(X_1, \ldots, Y_i, \ldots, X_r)$
$\qquad\qquad\qquad\qquad\qquad\qquad + g_i \alpha(X_1, \ldots, Z_i, \ldots, X_r)$

for all $f_i, g_i \in \mathfrak{F}(\mathcal{M})$, $X_1, \ldots, X_r, Y_i, Z_i \in \mathfrak{X}(\mathcal{M})$.

---
[3] Gregorio Ricci-Curbastro (1853-1925). Italian mathematician, professor at the University of Padua. He invented the absolute differential calculus between 1884 and 1894. It became the foundation of the tensor analysis that was used by Einstein in his theory of general relativity.

23.6 Tensor Fields on Differentiable Manifolds

A **vectorvariant tensor field** of **degree** r is a mapping

$$\Phi: \mathfrak{X}(\mathcal{M}) \times \cdots \times \mathfrak{X}(\mathcal{M}) \longrightarrow \mathfrak{X}(\mathcal{M})$$

that satisfies (23.22).

Remarks

(1) Condition (23.22) means that α and Φ are *multilinear with respect to functions*. On the other hand, the Lie bracket [,] is not a tensor field because it is not multilinear with respect to functions; instead [,] satisfies (23.18) which is more complicated.

(2) Sometimes a covariant tensor field of degree r is said to be a tensor field of type $(r, 0)$, and a vectorvariant tensor field of degree r is said to be a tensor field of type $(r, 1)$. It is also possible to define tensor fields of type (r, s). But tensor fields of type (r, s) with $s \geq 2$ occur so rarely that we shall not bother to define them.

(3) There is a definition in linear algebra (that has nothing to do with differentiable manifolds) that is very similar to the definition of tensor field. This is the notion of **tensor**. Since we shall need this definition in our study of a differentiable manifold $\mathcal{M}$, we give it at each point $\mathbf{m} \in \mathcal{M}$.

Definition. *A* **covariant tensor** *at a point* $\mathbf{m}$ *in a differentiable manifold* $\mathcal{M}$ *is a multilinear map*

$$\alpha: \mathcal{M}_\mathbf{m} \times \cdots \times \mathcal{M}_\mathbf{m} \longrightarrow \mathbb{R}$$

that satisfies

(23.23) $\quad \alpha(\mathbf{x}_1, \ldots, a_i \mathbf{y}_i + b_i \mathbf{z}_i, \ldots, \mathbf{x}_r) = a_i \alpha(\mathbf{x}_1, \ldots, \mathbf{y}_i, \ldots, \mathbf{x}_r)$
$\qquad\qquad\qquad\qquad\qquad\qquad\qquad + b_i \alpha(\mathbf{x}_1, \ldots, \mathbf{z}_i, \ldots, \mathbf{x}_r)$

for all $a_i, b_i \in \mathbb{R}$, *and all* $\mathbf{x}_1, \ldots, \mathbf{x}_r, \mathbf{y}_i, \mathbf{z}_i \in \mathcal{M}_\mathbf{m}$. *Similarly, a* **vectorvariant tensor** *at a point* $\mathbf{m}$ *in a differentiable manifold* $\mathcal{M}$ *is a multilinear map*

$$\Phi: \mathcal{M}_\mathbf{m} \times \cdots \times \mathcal{M}_\mathbf{m} \longrightarrow \mathcal{M}_\mathbf{m}$$

that satisfies

(23.24) $\quad \Phi(\mathbf{x}_1, \ldots, a_i \mathbf{y}_i + b_i \mathbf{z}_i, \ldots, \mathbf{x}_r) = a_i \Phi(\mathbf{x}_1, \ldots, \mathbf{y}_i, \ldots, \mathbf{x}_r)$
$\qquad\qquad\qquad\qquad\qquad\qquad\qquad + b_i \Phi(\mathbf{x}_1, \ldots, \mathbf{z}_i, \ldots, \mathbf{x}_r)$

for all $a_i, b_i \in \mathbb{R}$, *and all* $\mathbf{x}_1, \ldots, \mathbf{x}_r, \mathbf{y}_i, \mathbf{z}_i \in \mathcal{M}_\mathbf{m}$.

On page 549 we noted that some differentiable mappings between manifolds $\mathcal{M}$ and $\mathcal{N}$ do not always induce a map between the algebras of vector fields of $\mathcal{M}$ and $\mathcal{N}$. The situation is much better with covariant tensor fields.

Definition. *Let $\Psi: \mathcal{M} \longrightarrow \mathcal{N}$ be a differentiable mapping, and let α be a covariant tensor field of degree r on $\mathcal{N}$. Then the **pull back** of α is the covariant tensor field $\Psi^*(\alpha)$ on $\mathcal{M}$ given by*

$$\Psi^*(\alpha)(\mathbf{v}_1, \ldots, \mathbf{v}_r) = \alpha\big(\Psi_{*p}(\mathbf{v}_1), \ldots, \Psi_{*p}(\mathbf{v}_r)\big),$$

where $\mathbf{v}_1, \ldots, \mathbf{v}_r$ are arbitrary tangent vectors at an arbitrary point $\mathbf{p} \in \mathcal{M}$.

The following lemma is easy to prove.

Lemma 23.34. *Let $\Psi: \mathcal{M} \longrightarrow \mathcal{N}$ be a differentiable mapping, and let α and β be covariant tensor fields of degree r on $\mathcal{N}$. Also, let $f, g \in \mathfrak{F}(\mathcal{N})$. Then*

$$\Psi^*(f\,\alpha + g\,\beta) = (f \circ \Psi)\Psi^*(\alpha) + (g \circ \Psi)\Psi^*(\beta).$$

Let us single out one of the simplest types of covariant tensor fields.

Definition. *A **differential 1-form** on a manifold $\mathcal{M}$ is an $\mathfrak{F}(\mathcal{M})$-linear map*

$$\omega: \mathfrak{X}(\mathcal{M}) \longrightarrow \mathfrak{F}(\mathcal{M}).$$

In other words, a 1-form ω has the property that

$$\omega(f\,X + g\,Y) = f\,\omega(X) + g\,\omega(Y)$$

for $f, g \in \mathfrak{F}(\mathcal{M})$ and $X, Y \in \mathfrak{X}(\mathcal{M})$. We denote the collection of 1-forms on $\mathcal{M}$ by $\mathfrak{X}(\mathcal{M})^*$, and we make it into a module over $\mathfrak{F}(\mathcal{M})$ by defining

$$(f\,\omega + g\,\theta)(X) = f\,\omega(X) + g\,\theta(X)$$

for $f, g \in \mathfrak{F}(\mathcal{M})$, $\omega, \theta \in \mathfrak{X}(\mathcal{M})^*$ and $X \in \mathfrak{X}(\mathcal{M})$.

Clearly, $\mathfrak{X}(\mathcal{M})^*$ is the dual module of $\mathfrak{X}(\mathcal{M})$.

Lemma 23.35. *Suppose $X \in \mathfrak{X}(\mathcal{M})$ and $\mathbf{p} \in \mathcal{M}$ are such that $X_\mathbf{p} = 0$. Then*

$$\omega(X)(\mathbf{p}) = 0$$

for all $\omega \in \mathfrak{X}(\mathcal{M})^*$.

23.6 Tensor Fields on Differentiable Manifolds

Proof. Let $(\mathbf{x}, \mathcal{U})$ be a patch on $\mathcal{M}$ at $\mathbf{p}$ with $\mathbf{x}^{-1} = (x_1, \ldots, x_n)$. By the basis theorem (Corollary 23.31) we have

$$(23.25) \qquad X = \sum_{i=1}^{n} X[x_i] \frac{\partial}{\partial x_i}$$

near $\mathbf{p}$. Also, $X_\mathbf{p} = 0$ implies that

$$(23.26) \qquad X[x_i](\mathbf{p}) = 0$$

for $i = 1, \ldots, n$. Then from (23.25) and (23.26) we get

$$\omega(X)(\mathbf{p}) = \omega\left(\sum_{i=1}^{n} X[x_i]\left(\frac{\partial}{\partial x_i}\right)\right)(\mathbf{p}) = \sum_{i=1}^{n} X[x_i](\mathbf{p})\, \omega\left(\frac{\partial}{\partial x_i}\right)(\mathbf{p}) = 0. \ \blacksquare$$

We can evaluate 1-forms at a point in the same way that we evaluate vector fields at a point.

Definition. Let $\omega \in \mathfrak{X}(\mathcal{M})^*$ and $\mathbf{p} \in \mathcal{M}$. Then $\omega_\mathbf{p} \colon \mathcal{M}_\mathbf{p} \longrightarrow \mathbb{R}$ is defined by

$$\omega_\mathbf{p}(\mathbf{x}) = \omega(X)(\mathbf{p}),$$

where $X \in \mathfrak{X}(\mathcal{M})$ is any vector field for which $X_\mathbf{p} = \mathbf{x}$.

Lemma 23.36. *The definition of $\omega_\mathbf{p}$ does not depend on the choice of the vector field X.*

Proof. Let X and $\widetilde{X}$ be vector fields such that $X_\mathbf{p} = \widetilde{X}_\mathbf{p} = \mathbf{x}$. Then $(X - \widetilde{X})_\mathbf{p} = 0$, and so by Lemma 23.35 it follows that

$$\omega(X)(\mathbf{p}) - \omega(\widetilde{X})(\mathbf{p}) = \omega(X - \widetilde{X})(\mathbf{p}) = 0. \ \blacksquare$$

There is a particularly important kind of 1-form.

Definition. For $f \in \mathfrak{F}(\mathcal{M})$ the **differential** df of f is the 1-form defined by

$$(23.27) \qquad df(X) = X[f]$$

for $X \in \mathfrak{X}(\mathcal{M})$.

If V is any vector space over any field $\mathbb{F}$, we denote its dual space by V^*. Thus

$$V^* = \{\, \alpha \colon V \longrightarrow \mathbb{F} \mid \alpha \text{ is linear}\,\}.$$

Lemma 23.37. *Let $(x_1, \ldots, x_n)$ be a system of coordinates defined on an open set $\mathcal{U} \subset \mathcal{M}$. Then*

(i) *For each $\mathbf{p} \in \mathcal{U}$ the 1-forms $dx_1(\mathbf{p}), \ldots, dx_n(\mathbf{p})$ form a basis of $\mathcal{M}_\mathbf{p}^*$. Hence*

$$\{\omega_\mathbf{p} \mid \omega \in \mathfrak{X}(\mathcal{M})^*\} = \mathcal{M}_\mathbf{p}^*.$$

(ii) *The set $\{dx_1, \ldots, dx_n\}$ is a basis for $\mathfrak{X}(\mathcal{U})^*$. Thus for any $\omega \in \mathfrak{X}(\mathcal{U})^*$ we can write*

$$\omega = \sum_{i=1}^n \omega\left(\frac{\partial}{\partial x_i}\right) dx_i.$$

23.7 Exercises

1. If $\mathcal{M}_1$ and $\mathcal{M}_2$ are differentiable manifolds, show that $\mathcal{M}_1 \times \mathcal{M}_2$ is a differentiable manifold.

2. Show that each tangent space $\mathcal{M}_\mathbf{p}$ to a differentiable manifold $\mathcal{M}$ is itself a differentiable manifold.

3. Let $\Phi: \mathbb{R} \longrightarrow \mathbb{R}$ be defined by $\Phi(t) = t^3$. Show that the patch $(\Phi, \mathbb{R})$ determines an atlas $\mathfrak{A}_2$ on $\mathbb{R}$ that is different from the usual atlas $\mathfrak{A}_1$, but that $(\mathbb{R}, \mathfrak{A}_1)$ and $(\mathbb{R}, \mathfrak{A}_2)$ are diffeomorphic.

4. Define an equivalence relation $\sim$ on $S^n(1) \subset \mathbb{R}^{n+1}$ by writing $\mathbf{a} \sim \mathbf{b}$ if and only if $\mathbf{a} = \pm \mathbf{b}$. The resulting quotient space is called **real projective space** of dimension n and denoted by $\mathbb{R}P^n$. Define $p: S^n(1) \longmapsto \mathbb{R}P^n$ by $p(\mathbf{a}) = \{\pm \mathbf{a}\}$. Let

$$\mathcal{U}_j = \{p(a_1, \ldots, a_n) \mid a_j \neq 0\}.$$

   Show that:

   (a) Each $\mathcal{U}_j$ is open in $\mathbb{R}P^n$.

   (b) $\mathbb{R}P^n$ is the union of all the $\mathcal{U}_j$'s.

   (c) For each j there is a homeomorphism $\psi_j: \mathcal{U}_j \longrightarrow \mathbb{R}^n$.

   (d) $\mathbb{R}P^n$ is a compact differentiable manifold of dimension n.

23.7 Exercises

5. On $\mathbb{R}^3$ consider the vector fields
$$X = x^2 \frac{\partial}{\partial x} + y \frac{\partial}{\partial z} \quad \text{and} \quad Y = y^3 \frac{\partial}{\partial y}.$$
and the function $f: \mathbb{R}^3 \longrightarrow \mathbb{R}$ defined by $f(p_1, p_2, p_3) = p_1^2 p_2 p_3$. Compute

   (a) $[X, Y]_{(1,0,1)}$
   (c) $(Yf)(1, 0, 1)$
   (b) $(fY)_{(1,0,1)}$
   (d) $f_{*(1,0,1)} Y_{(1,0,1)}$.

   Give examples of vector fields X and Y on $\mathbb{R}^2$ for which
$$X_{(0,0)} = Y_{(0,0)} \quad \text{but} \quad [X, Y]_{(0,0)} \neq \mathbf{0}.$$

6. Let $\Phi: \mathbb{R}^3 \longrightarrow \mathbb{R}^2$ be defined by $\Phi((p_1, p_2, p_3)) = (p_1, p_2)$, and let Z be the vector field on $\mathbb{R}^3$ given by
$$Z = \cos z + \sin z \frac{\partial}{\partial y} + \frac{\partial}{\partial z},$$
where $\{x, y, z\}$ are the natural coordinate functions of $\mathbb{R}^3$. Show that there is no vector field Y on $\mathbb{R}^2$ for which
$$\Phi_{*\mathbf{p}}(Z) = Y_{\Phi(\mathbf{p})}$$
for all $\mathbf{p} \in \mathbb{R}^3$.

7. Let $f: \mathbb{R}^2 \longrightarrow \mathbb{R}^2$ and $g: \mathbb{R}^2 \longrightarrow \mathbb{R}^3$ be defined by
$$f(u, v) = (u^2 \sin v, e^{uv}) \quad \text{and} \quad g(u, v) = (uv^2, \log(uv), v \sin u).$$
Compute the matrices of f_* and g_* with respect to the standard bases of vector fields on $\mathbb{R}^2$ and $\mathbb{R}^3$.

8. Prove Lemma 23.34.

9. For each patch $(\mathbf{x}, \mathcal{U})$ on an n-dimensional differentiable manifold $\mathcal{M}$, define $\widetilde{\mathbf{x}}: \mathcal{U} \times \mathbb{R}^n \longrightarrow T(\mathcal{M})$ by
$$\widetilde{\mathbf{x}}(\mathbf{p}, \mathbf{q}) = \left(\mathbf{x}(\mathbf{p}), \mathbf{x}_{*\mathbf{p}} \left(\sum_{i=1}^{n} q_i \frac{\partial}{\partial u_i} \right) \right),$$

where $\mathbf{p} \in \mathcal{U}$ and $\mathbf{q} = (q_1, \ldots, q_n)$. Show that the tangent bundle $T(\mathcal{M})$ becomes a differentiable manifold with atlas

$$\{ (\tilde{\mathbf{x}}, \mathcal{U} \times \mathbb{R}^n) \mid (\mathbf{x}, \mathcal{U}) \text{ is a patch on } \mathcal{M} \}.$$

10. Use *Mathematica* to draw the pictures on page 529.

11. Define in *Mathematica*

```
stereoinj[a_][p_]:= Append[2a^2 p,
                           a(p.p - a^2)]/(p.p + a^2)

stereoproj[a_][q_]:= a Drop[q,-1]/(a - Last[q])
```

Work out a few examples in different dimensions to show that **stereoinj[a]** and **stereoinj[-a]** are the stereographic injections **north** and **south** defined on page 526. Also, show (again by doing a few examples) that the composition of **stereoinj[a]** and **stereoproj[a]** is the identity map.

12. On $\mathbb{R}^3$ let

$$\omega = y\,dx + z^2\,dy + x^3\,dz \quad \text{and} \quad X = z\frac{\partial}{\partial x} + x^2\frac{\partial}{\partial x} + y^3\frac{\partial}{\partial x}.$$

Compute $\omega(X)$.

24

RIEMANNIAN MANIFOLDS

Our ultimate goal is the study of the geometry of *abstract surfaces*, that is, 2-dimensional manifolds that have a notion of metric compatible with the manifold structure. Regular surfaces in $\mathbb{R}^3$ are abstract surfaces, but there are other abstract surfaces that cannot be put into $\mathbb{R}^3$.

In this chapter we approach abstract surfaces from the point of view of more general concepts, namely, those of differentiable manifold, covariant derivative and Riemannian manifold. Although we are primarily interested in dimension 2, it is just as easy to define these concepts for arbitrary dimensions. Specialization to dimension 2 will take place in Chapter 25.

The theory of differentiable manifolds developed in Chapter 23 is insufficient to measure lengths and areas. For that we need to impose some additional structure so that we end up with a generalization of a Riemannian metric as defined in Section 15.1. Covariant derivatives are defined and their basic properties are developed in Section 24.1; covariant derivatives will be needed when we study geodesics in Chapter 26. Pseudo-Riemannian manifolds are defined in Section 24.2; every pseudo-Riemannian manifold has a covariant derivative. A Riemannian manifold is a pseudo-Riemannian manifold whose metric is positive definite. Section 24.3 is devoted to bridging the gap between the modern techniques of Sections 24.1–24.2 and the classical treatment of metrics.

24.1 Covariant Derivatives

Since it is important to study vector fields on a differentiable manifold $\mathcal{M}$, it is natural to try to define how to differentiate vector fields on $\mathcal{M}$. For $\mathbb{R}^n$ there is a natural way to do this. Let $Y \in \mathfrak{X}(\mathbb{R}^n)$. In terms of the vector fields

$$\frac{\partial}{\partial u_1}, \ldots, \frac{\partial}{\partial u_n}$$

determined by the natural coordinate functions $u_1, \ldots, u_n$, we can write

$$Y = \sum_{i=1}^{n} f_i \frac{\partial}{\partial u_i}$$

for some functions $f_1, \ldots, f_n \in \mathfrak{F}(\mathbb{R}^n)$. As a matter of fact, in calculus the vector field Y is frequently identified with the n-tuple $(f_1, \ldots, f_n)$. Then for $Y \in \mathfrak{X}(\mathbb{R}^n)$ there is a natural candidate for the derivative $D_X Y$ of Y in the direction X, namely,

(24.1) $$D_X Y = (X[f_1], \ldots, X[f_n]).$$

In fact, in Lemma 11.21, page 261, we wrote $\boldsymbol{D}_X Y$ for $D_X Y$.

Unfortunately, for a general differentiable manifold, there is no natural notion of differentiation of vector fields. Instead we must add a new element of structure:

Definition. *A* **connection** *or* **covariant derivative** *on a differentiable manifold $\mathcal{M}$ is a map*

$$D: \mathfrak{X}(\mathcal{M}) \times \mathfrak{X}(\mathcal{M}) \longrightarrow \mathfrak{X}(\mathcal{M}),$$

which for $X \in \mathfrak{X}(\mathcal{M})$ we regard as a map

$$D_X: \mathfrak{X}(\mathcal{M}) \longrightarrow \mathfrak{X}(\mathcal{M}).$$

It is required to have the following properties:

(24.2) $$D_{fX+gY} = fD_X + gD_Y,$$

(24.3) $$D_X(Y+Z) = D_X Y + D_X Z,$$

(24.4) $$D_X(fY) = X[f]Y + fD_X Y,$$

for $X, Y, Z \in \mathfrak{X}(\mathcal{M})$ and $f, g \in \mathfrak{F}(\mathcal{M})$.

Remarks

(1) A connection D is not a tensor field in the sense of Section 23.6, because it is not linear with respect to functions in its second argument.

(2) Equations (24.3) and (24.4) imply that

(24.5) $$D_X(aY + bZ) = a\,D_X Y + b\,D_X Z,$$

for $a, b \in \mathbb{R}$ and $X, Y, Z \in \mathfrak{X}(\mathcal{M})$.

(3) Usually there is no canonical choice of connection on a differentiable manifold. However, we shall see that every Riemannian metric has a special connection associated with it.

Levi-Civita[1] introduced the covariant derivative in his book **The Absolute Differential Calculus** [LeviC] to explain parallelism in Riemannian manifolds.

Example: the natural connection on $\mathbb{R}^n$

Define
$$D: \mathfrak{X}(\mathbb{R}^n) \times \mathfrak{X}(\mathbb{R}^n) \longrightarrow \mathfrak{X}(\mathbb{R}^n)$$
by first putting
$$D_{\frac{\partial}{\partial u_i}} \frac{\partial}{\partial u_j} = 0$$
for $1 \leq i, j \leq n$, where $u_1, \ldots, u_n$ are the natural coordinate functions of $\mathbb{R}^n$. More generally, for
$$Y = \sum_{i=1}^n f_i \frac{\partial}{\partial u_i}$$
we are forced by (24.4) and (24.5) to require that

(24.6) $$D_X Y = \sum_{i=1}^n X[f_i] \frac{\partial}{\partial u_i},$$

for $1 \leq i \leq n$. In fact, (24.6) defines a natural connection on $\mathbb{R}^n$; moreover, (24.6) is equivalent to (24.1).

---

[1] Tullio Levi-Civita (1873-1941). Italian mathematician. Professor at Padua and Rome. He also wrote on hydrodynamics and rational mechanics. Levi-Civita was an outspoken opponent of fascism, and Italy's fascist government forced him to give up teaching in 1938.

There is a variant of a connection analogous to the definition of $D_\mathbf{v}$ that we gave in Section 11.5. Let $\mathcal{M}$ be a differentiable manifold and let $\mathbf{p} \in \mathcal{M}$. A connection D on $\mathcal{M}$ gives rise to a map

$$D: \mathcal{M}_\mathbf{p} \times \mathfrak{X}(\mathcal{M}) \longrightarrow \mathcal{M}_\mathbf{p}.$$

Definition. Let $\mathbf{v}_\mathbf{p}$ be a tangent vector to $\mathcal{M}$ at $\mathbf{p} \in \mathcal{M}$. Choose $V \in \mathfrak{X}(\mathcal{M})$ such that $V_\mathbf{p} = \mathbf{v}_\mathbf{p}$. For $W \in \mathfrak{X}(\mathcal{M})$ we define $D_\mathbf{v}W$ to be the tangent vector in $\mathcal{M}_\mathbf{p}$ given by

$$D_\mathbf{v}W = (D_V W)_\mathbf{p}.$$

Lemma 24.1. *Fix $W \in \mathfrak{X}(\mathcal{M})$. The definition of $D_\mathbf{v}W$ is independent of the choice of $V \in \mathfrak{X}(\mathcal{M})$ such that $V_\mathbf{p} = \mathbf{v}_\mathbf{p}$.*

Proof. Let $V, Y \in \mathfrak{X}(\mathcal{M})$ be such that $V_\mathbf{p} = Y_\mathbf{p}$, and let $(x_1, \ldots, x_n)$ be a coordinate system defined in an open set containing $\mathbf{p}$. By Lemma 23.31, page 548, we can write

$$V = \sum_{i=1}^n V[x_i] \frac{\partial}{\partial x_i}, \quad \text{and} \quad Y = \sum_{i=1}^n Y[x_i] \frac{\partial}{\partial x_i}.$$

Then (24.2) and (24.4) imply that

$$(D_V W)_\mathbf{p} = \sum_{i=1}^n \left(V[x_i] D_{\frac{\partial}{\partial x_i}} W\right)_\mathbf{p} = \sum_{i=1}^n V[x_i](\mathbf{p}) \left(D_{\frac{\partial}{\partial x_i}} W\right)_\mathbf{p}$$

$$= \sum_{i=1}^n Y[x_i](\mathbf{p}) \left(D_{\frac{\partial}{\partial x_i}} W\right)_\mathbf{p} = (D_Y W)_\mathbf{p}. \blacksquare$$

In contrast, it may happen that $Y_\mathbf{p} = Z_\mathbf{p}$ but $(D_X Y)_\mathbf{p} \neq (D_X Z)_\mathbf{p}$. In spite of this difficulty we prove:

Lemma 24.2. *Let $X \in \mathfrak{X}(\mathcal{M})$ and let $\alpha: (a, b) \longrightarrow \mathcal{M}$ be a curve such that*

$$\alpha'(t) = X_{\alpha(t)}$$

*(that is, α is an **integral curve** of X). Let $Y, Z \in \mathfrak{X}(\mathcal{M})$ be such that*

$$Y_{\alpha(t)} = Z_{\alpha(t)}$$

for $a < t < b$. Then

$$(D_X Y)_{\alpha(t)} = (D_X Z)_{\alpha(t)}$$

for $a < t < b$.

24.1 Covariant Derivatives

Proof. We can assume without loss of generality that there is a system of coordinates $(x_1, \ldots, x_n)$ defined on an open set $\mathcal{U} \subset \mathcal{M}$ such that the trace of $\boldsymbol{\alpha}$ is contained in $\mathcal{U}$.

It suffices to show that for $Y \in \mathfrak{X}(\mathcal{M})$ we have

$$\begin{cases} t \longmapsto Y_{\alpha(t)} \text{ vanishes identically} \\ \\ \text{implies} \\ \\ t \longmapsto (D_X Y)_{\alpha(t)} \text{ vanishes identically.} \end{cases}$$

By Lemma 23.31 we can write

$$(24.7) \qquad Y = \sum_{i=1}^{n} Y[x_i] \frac{\partial}{\partial x_i}$$

on $\mathcal{U}$. Suppose that $t \longmapsto Y_{\alpha(t)}$ vanishes identically. Then (24.7) implies that

$$(24.8) \qquad Y[x_i]\bigl(\boldsymbol{\alpha}(t)\bigr) = 0$$

for $a < t < b$ and $i = 1, \ldots, n$. When we differentiate (24.7) we obtain

$$(24.9) \qquad D_X Y = \sum_{i=1}^{n} \left(XY[x_i] \frac{\partial}{\partial x_i} + Y[x_i] D_X \frac{\partial}{\partial x_i} \right).$$

Next, we restrict (24.9) to $\boldsymbol{\alpha}$ and use (24.8) to get

$$(24.10) \qquad (D_X Y)_{\alpha(t)} = \sum_{i=1}^{n} \left(XY[x_i] \frac{\partial}{\partial x_i} \right) \bigg|_{\alpha(t)}.$$

But by (24.8) and Lemma 23.21 we have

$$XY[x_i]\bigl(\boldsymbol{\alpha}(t)\bigr) = X_{\alpha(t)}[Y[x_i]] = \boldsymbol{\alpha}'(t)[Y[x_i]] = \frac{d(Y[x_i] \circ \boldsymbol{\alpha})}{du}(t) = 0.$$

Thus the coefficients in (24.10) all vanish, and so $(D_X Y)_{\alpha(t)} = 0$ for $a < t < b$. ∎

In Section 1.4 we considered vector fields along a curve in $\mathbb{R}^n$. This notion generalizes to manifolds.

Definition. *Let $\boldsymbol{\alpha}: (a, b) \longrightarrow \mathcal{M}$ be a curve in a differentiable manifold $\mathcal{M}$. A **vector field along** $\boldsymbol{\alpha}$ is a function $\mathbf{Y}$ that assigns to each t with $a < t < b$ a tangent vector $\mathbf{Y}(t) \in \mathcal{M}_{\alpha(t)}$.*

Lemma 24.2 implies that we can define unambiguously the covariant derivative of a vector field along a curve.

Definition. Let $\alpha\colon (a,b) \longrightarrow \mathcal{M}$ be an injective regular curve in a differentiable manifold $\mathcal{M}$, and let $\mathbf{Y}$ be a vector field along α. Then

$$\mathbf{Y}'(t) = D_{\alpha'(t)}\widetilde{Y}$$

where $\widetilde{Y} \in \mathfrak{X}(\mathcal{M})$ is any vector field such that $\widetilde{Y}_{\alpha(t)} = \mathbf{Y}(t)$ for $a < t < b$.

Vector fields along a curve in a manifold $\mathcal{M}$ have most of the properties enjoyed by vector fields along a curve in $\mathbb{R}^n$. For example, the following lemma generalizes Lemma 1.8, page 13.

Lemma 24.3. Let $\mathbf{Y}$ and $\mathbf{Z}$ be vector fields along a curve $\alpha\colon (a,b) \longrightarrow \mathcal{M}$ in a differentiable manifold $\mathcal{M}$, and let $f\colon (a,b) \longrightarrow \mathbb{R}$ be differentiable. Then

(24.11) $$(\mathbf{Y} + \mathbf{Z})' = \mathbf{Y}' + \mathbf{Z}',$$

(24.12) $$(f\mathbf{Y})' = f'\mathbf{Y} + f\mathbf{Y}'.$$

Proof. We prove (24.12). Let $\widetilde{f} \in \mathfrak{F}(\mathcal{M})$ and $\widetilde{Y} \in \mathfrak{X}(\mathcal{M})$ be such that $\widetilde{f}(\alpha(t)) = f(t)$ and $\widetilde{Y}(\alpha(t)) = \mathbf{Y}(t)$ for $a < t < b$. Then

$$(f\mathbf{Y})'(t) = D_{\alpha'(t)}(\widetilde{f}\widetilde{Y})$$

$$= \alpha'(t)[\widetilde{f}]\widetilde{Y}_{\alpha(t)} + \widetilde{f}(\alpha(t))D_{\alpha'(t)}\widetilde{Y}$$

$$= (\widetilde{f} \circ \alpha)'(t)\mathbf{Y}(t) + f(t)\mathbf{Y}'(t) = f'(t)\mathbf{Y}(t) + f(t)\mathbf{Y}'(t). \quad \blacksquare$$

There is also a chain rule for vector fields along a curve.

Lemma 24.4. Let $\mathbf{Y}$ be a vector field along a curve $\alpha\colon (a,b) \longrightarrow \mathcal{M}$ in a differentiable manifold $\mathcal{M}$, and let $h\colon (c,d) \longrightarrow (a,b)$ be differentiable. Then

(24.13) $$(\mathbf{Y} \circ h)'(t) = h'(t)\mathbf{Y}'(h(t))$$

for $c < t < d$.

Proof. If $\widetilde{Y} \in \mathfrak{X}(\mathcal{M})$ is such that $\widetilde{Y}(\alpha(h(t))) = \mathbf{Y}(h(t))$ for $c < t < d$, then by the chain rule for curves (Lemma 23.23, page 545) we have

$$(\mathbf{Y} \circ h)'(t) = D_{(\alpha \circ h)'(t)}\widetilde{Y} = D_{h'(t)(\alpha' \circ h)(t)}\widetilde{Y}$$

$$= h'(t)D_{(\alpha' \circ h)(t)}\widetilde{Y} = h'(t)\mathbf{Y}'(h(t)). \quad \blacksquare$$

24.1 Covariant Derivatives

Now we are ready to define the notion of acceleration of a curve in a manifold. It depends on the choice of covariant derivative.

Definition. *Let $\alpha\colon (a,b) \longrightarrow \mathcal{M}$ be a curve in a differentiable manifold $\mathcal{M}$, and let D be a connection on $\mathcal{M}$. The* **acceleration** *of α with respect to D is the vector field*
$$t \longmapsto D_{\alpha'(t)}\alpha'(t),$$
which we write more simply as $D_{\alpha'}\alpha'$. We say that a curve $\alpha\colon(a,b) \longrightarrow \mathcal{M}$ is a **geodesic** *with respect to D provided*
$$D_{\alpha'(t)}\alpha'(t) = 0$$
for $a < t < b$.

(Geodesics on abstract surfaces will be studied in detail in Chapter 26.) We have denoted the acceleration of a curve α in a manifold $\mathcal{M}$ by $D_{\alpha'}\alpha'$ to avoid confusion with the acceleration of a curve in $\mathbb{R}^n$. For example, if α is a curve in a surface $\mathcal{M} \subset \mathbb{R}^n$, the two accelerations $D_{\alpha'}\alpha'$ and α'' are usually different.

Next, we prove the manifold version of equation (1.13), page 16, which tells us the relation between the acceleration of a curve α and the acceleration of a reparametrization of α.

Lemma 24.5. *Let D be a connection on a differentiable manifold $\mathcal{M}$, and let $\alpha\colon(a,b) \longrightarrow \mathcal{M}$ be a curve on $\mathcal{M}$. Suppose that $\beta = \alpha \circ h$ is a reparametrization of α, where $h\colon(c,d) \longrightarrow (a,b)$ is differentiable. Then*

$$(24.14) \qquad D_{\beta'}\beta' = h'^2 \left(D_{\alpha'}\alpha'\right) \circ h + h''(\alpha' \circ h).$$

Proof. From Lemma 23.23 we know that

$$(24.15) \qquad \beta' = h'(\alpha' \circ h).$$

We differentiate (24.15) and use (24.13) and (24.12). For $c < t < d$ we have

$$(24.16) \quad \begin{aligned} D_{\beta'(t)}\beta'(t) &= D_{\beta'(t)}(\alpha \circ h)'(t) = D_{\beta'(t)}\bigl(h'(t)\alpha'\bigl(h(t)\bigr)\bigr) \\ &= h''(t)\alpha'\bigl(h(t)\bigr) + h'(t)D_{\beta'(t)}\alpha'\bigl(h(t)\bigr) \\ &= h''(t)\alpha'\bigl(h(t)\bigr) + h'(t)D_{h'(t)\alpha'\bigl(h(t)\bigr)}\alpha'\bigl(h(t)\bigr) \\ &= h''(t)\alpha'\bigl(h(t)\bigr) + h'(t)^2 D_{\alpha'\bigl(h(t)\bigr)}\alpha'\bigl(h(t)\bigr). \end{aligned}$$

Since (24.16) holds for all t we get (24.14). ∎

24.2 Indefinite Riemannian Metrics

After studying 1-forms on a manifold in Section 23.6, it is natural to investigate covariant tensor fields of degree 2. In this generality not much can be said. To make progress one must make a restrictive assumption on the covariant tensor field.

Definition. *Let α be a covariant tensor field of degree 2 on a differentiable manifold $\mathcal{M}$. We say that α is a* **symmetric bilinear form** *provided that*

(24.17) $$\alpha(X, Y) = \alpha(Y, X)$$

for all $X, Y \in \mathfrak{X}(\mathcal{M})$. Similarly, α is called an **antisymmetric bilinear form** *if*

(24.18) $$\alpha(X, Y) = -\alpha(Y, X)$$

for all $X, Y \in \mathfrak{X}(\mathcal{M})$.

We shall concentrate on symmetric bilinear forms that satisfy a nondegeneracy condition that we now explain. Notice that any covariant tensor field α on a manifold $\mathcal{M}$ (whether or not it is symmetric) gives rise to a tensor $\alpha_\mathbf{p}$ on each tangent space $\mathcal{M}_\mathbf{p}$. Explicitly, $\alpha_\mathbf{p}$ is given as follows. Let $\mathbf{x}, \mathbf{y} \in \mathcal{M}_\mathbf{p}$; then

$$\alpha_\mathbf{p}(\mathbf{x}, \mathbf{y}) = \alpha(X, Y)(\mathbf{p}),$$

where $X, Y \in \mathfrak{X}(\mathcal{M})$ are such that $X_\mathbf{p} = \mathbf{x}$ and $Y_\mathbf{p} = \mathbf{y}$. Since α is a tensor field, it is linear in both its arguments with respect to functions. This fact implies that the definition of $\alpha_\mathbf{p}$ does not depend on the choice of X and Y such that $X_\mathbf{p} = \mathbf{x}$ and $Y_\mathbf{p} = \mathbf{y}$.

Definition. *A symmetric bilinear form α on a differentiable manifold $\mathcal{M}$ is called* **nondegenerate**, *provided that $\alpha_\mathbf{p}$ is nondegenerate on each tangent space $\mathcal{M}_\mathbf{p}$ for each point $\mathbf{p} \in \mathcal{M}$; that is,*

$$\alpha_\mathbf{p}(\mathbf{x}, \mathbf{y}) = 0 \text{ for all } \mathbf{y} \in \mathcal{M}_\mathbf{p} \quad \text{implies} \quad \mathbf{x} = 0.$$

A nondegenerate symmetric bilinear form is called a **pseudo-Riemannian metric**.

A theorem from linear algebra states that a symmetric bilinear form α on an n-dimensional vector space V over $\mathbb{R}$ can be diagonalized. This means that there exists a basis $\{f_1, \ldots, f_n\}$ of V for which

$$\alpha(f_i, f_j) = 0$$

24.2 Indefinite Riemannian Metrics

for $i \neq j$. Moreover, in the case that α is nondegenerate the basis $\{f_1, \ldots, f_n\}$ can be chosen so that

$$\alpha(f_i, f_i) = \pm 1.$$

The number p of $+$ signs and the number q of $-$ signs does not depend on the choice of the basis $\{f_1, \ldots, f_n\}$. We call (p, q) the **signature** of the symmetric bilinear form.

A particularly important special case occurs when the signature of the symmetric bilinear form is $(n, 0)$.

Definition. *A symmetric bilinear form α on a differentiable manifold $\mathcal{M}$ is called* **positive definite***, provided that α_p is positive definite on $\mathcal{M}_p$ for each point $\mathbf{p} \in \mathcal{M}$; that is,*

$$\alpha_p(\mathbf{x}, \mathbf{x}) > 0 \text{ for all nonzero } \mathbf{x} \in \mathcal{M}_p.$$

In this case, we say that α is a **Riemannian metric***.*

We shall use the word **metric** to mean either a pseudo-Riemannian metric or a Riemannian metric. Usually we denote a metric by $\langle \ , \ \rangle$. When $\langle \ , \ \rangle$ is positive definite, we write

$$\|\mathbf{x}\| = \sqrt{\langle \mathbf{x}, \mathbf{x} \rangle}$$

for any tangent vector $\mathbf{x}$ to $\mathcal{M}$. Even when $\langle \ , \ \rangle$ is not positive definite we shall write $\|\mathbf{x}\|^2$ for $\langle \mathbf{x}, \mathbf{x} \rangle$.

Definition. *A* **pseudo-Riemannian manifold** *is a differentiable manifold $\mathcal{M}$ equipped with a metric $\langle \ , \ \rangle$. If $\mathbf{x}, \mathbf{y}$ are tangent vectors to $\mathcal{M}$, we call $\langle \mathbf{x}, \mathbf{y} \rangle$ the* **inner product** *of $\mathbf{x}$ and $\mathbf{y}$. Similarly, we can speak about the inner product of vector fields. If $X, Y \in \mathfrak{X}(\mathcal{M})$, then $\langle X, Y \rangle$ is the function in $\mathfrak{F}(\mathcal{M})$ defined by*

$$\langle X, Y \rangle(\mathbf{p}) = \langle X_p, Y_p \rangle.$$

A **Riemannian manifold** *is a differentiable manifold $\mathcal{M}$ equipped with a positive definite metric $\langle \ , \ \rangle$.*

Example: metrics on $\mathbb{R}^n$

There is a natural positive definite metric on $\mathbb{R}^n$ given by

(24.19) $$\left\langle \sum_{i=1}^{n} f_i \frac{\partial}{\partial u_i}, \sum_{j=1}^{n} g_j \frac{\partial}{\partial u_j} \right\rangle = \sum_{i=1}^{n} f_i g_i.$$

This metric coincides with the standard dot product · on $\mathbb{R}^n$ that we introduced in Section 1.1. A more general metric on $\mathbb{R}^n$ is given by

$$(24.20) \qquad \left\langle \sum_{i=1}^n f_i \frac{\partial}{\partial u_i}, \sum_{j=1}^n g_j \frac{\partial}{\partial u_j} \right\rangle = \sum_{i=1}^n a_i f_i g_i,$$

where $a_1, \ldots, a_n$ are nonzero constants. The metric (24.20) is positive definite if and only if all the a_i's are positive. An even more general metric is given by

$$(24.21) \qquad \left\langle \sum_{i=1}^n f_i \frac{\partial}{\partial u_i}, \sum_{j=1}^n g_j \frac{\partial}{\partial u_j} \right\rangle = \sum_{ij=1}^n a_{ij} f_i g_j,$$

where (a_{ij}) is a matrix of constants whose determinant is nonzero.

The notion of gradient that we defined on page 257 has a generalization to pseudo-Riemannian manifolds.

Definition. *Let θ be a 1-form on a pseudo-Riemannian manifold $\mathcal{M}$. The **dual** of θ is the vector field $\theta^* \in \mathfrak{X}(\mathcal{M})$ such that*

$$\langle \theta^*, Y \rangle = \theta(Y)$$

*for all $Y \in \mathfrak{X}(\mathcal{M})$. In the special case when $\theta = df$ for some $f \in \mathfrak{F}(\mathcal{M})$ we call df^* the **gradient** of f and denote it by $\operatorname{grad} f$; it is given by*

$$\langle \operatorname{grad} f, Y \rangle = Y[f].$$

That θ^* is well-defined is a consequence of an elementary linear algebra fact:

Lemma 24.6. *Let V be a vector space with an inner product $\langle\ ,\ \rangle$. Each $\mathbf{x} \in V$ is completely determined by all the "inner products" $\langle \mathbf{x}, \mathbf{y} \rangle$ for all $\mathbf{y} \in V$.*

Proof. If $\langle \mathbf{x}, \mathbf{y} \rangle = \langle \mathbf{x}', \mathbf{y} \rangle$ for all $\mathbf{y} \in V$, then $\langle \mathbf{x} - \mathbf{x}', \mathbf{y} \rangle = 0$ for all $\mathbf{y} \in V$. It follows from the nondegeneracy of $\langle\ ,\ \rangle$ that $\mathbf{x} = \mathbf{x}'$. ∎

It is fortunate for the study of pseudo-Riemannian manifolds that there is a canonical choice for the covariant derivative.

Definition. *Let $\mathcal{M}$ be a pseudo-Riemannian manifold. Then the **Riemannian connection** $\nabla \colon \mathfrak{X}(\mathcal{M}) \times \mathfrak{X}(\mathcal{M}) \longrightarrow \mathfrak{X}(\mathcal{M})$ of $\mathcal{M}$ is defined by*

$$(24.22) \qquad 2\langle \nabla_X Y, Z \rangle = X\langle Y, Z \rangle + Y\langle X, Z \rangle - Z\langle X, Y \rangle$$
$$- \langle X, [Y, Z] \rangle - \langle Y, [X, Z] \rangle + \langle Z, [X, Y] \rangle$$

for $X, Y, Z \in \mathfrak{X}(\mathcal{M})$.

The proof of the following lemma is straightforward:

Lemma 24.7. *A Riemannian connection ∇ has the following properties*:

(24.23) $$\nabla_{fX+gY} = f\nabla_X + g\nabla_Y,$$

(24.24) $$\nabla_X(aY+bZ) = a\nabla_X Y + b\nabla_X Z,$$

(24.25) $$\nabla_X fY = X[f]Y + f\nabla_X Y,$$

(24.26) $$\nabla_X Y - \nabla_Y X = [X,Y],$$

(24.27) $$X\langle Y,Z\rangle = \langle \nabla_X Y, Z\rangle + \langle Y, \nabla_X Z\rangle,$$

for $X, Y, Z \in \mathfrak{X}(\mathcal{M})$, $f, g \in \mathfrak{F}(\mathcal{M})$ and $a, b \in \mathbb{R}$. Conversely, if

$$\nabla \colon \mathfrak{X}(\mathcal{M}) \times \mathfrak{X}(\mathcal{M}) \longrightarrow \mathfrak{X}(\mathcal{M})$$

satisfies (24.23)—(24.27), *then it is given by* (24.22).

Remarks

(1) Lemma 24.6 implies that $\nabla_X Y$ is well-defined by (24.22).

(2) A general connection D on a differentiable manifold may or may not satisfy condition (24.26). A connection that does satisfy (24.26) is said **to have torsion zero**.

It is possible to characterize a Riemannian connection.

Lemma 24.8. *Let $\langle\ ,\ \rangle$ be a pseudo-Riemannian metric on a manifold $\mathcal{M}$. Then there is a unique connection ∇ on $\mathcal{M}$ that satisfies* (24.26) *and* (24.27).

24.3 The Classical Treatment of Metrics

In the literature one frequently sees the notation ds^2 for a metric, together with the equation

(24.28) $$ds^2 = \sum_{ij=1}^{n} g_{ij} dx_i dx_j.$$

We gave an intuitive interpretation of (24.28) in Chapter 15 in terms of infinitesimals. To make sense of (24.28) in modern terminology, let $\mathcal{M}$ be a Riemannian manifold with metric $\langle\ ,\ \rangle$. We write

$$\langle\ ,\ \rangle = ds^2$$

and choose a coordinate system $(x_1, \ldots, x_n)$ on $\mathcal{M}$. This coordinate system gives rise to 1-forms $dx_1, \ldots, dx_n$. The **symmetric product** of dx_i and dx_j is the covariant tensor field of degree 2 given by

$$(dx_i dx_j)(X, Y) = \frac{1}{2}\big(dx_i(X)dx_j(Y) + dx_i(Y)dx_j(X)\big)$$

$$= \frac{1}{2}\big(X[x_i]Y[x_j] + Y[x_i]X[x_j]\big)$$

for $X, Y \in \mathfrak{X}(\mathcal{M})$. Finally, let

$$g_{ij} = \left\langle \frac{\partial}{\partial x_i}, \frac{\partial}{\partial x_j} \right\rangle$$

for $1 \leq i, j \leq n$. The g_{ij}'s are called the **components of the metric** $\langle\ ,\ \rangle$.

Lemma 24.9. *A metric $ds^2 = \langle\ ,\ \rangle$ is given in terms of the g_{ij}'s by the formula*

(24.29) $$ds^2 = \sum_{ij=1}^{n} g_{ij} dx_i dx_j.$$

Proof. By definition

$$ds^2\left(\frac{\partial}{\partial x_k}, \frac{\partial}{\partial x_l}\right) = \left\langle \frac{\partial}{\partial x_k}, \frac{\partial}{\partial x_l} \right\rangle = g_{kl}.$$

On the other hand, we have

$$(g_{ij} dx_i dx_j)\left(\frac{\partial}{\partial x_k}, \frac{\partial}{\partial x_l}\right) = \frac{g_{ij}}{2}\left\{\frac{\partial x_i}{\partial x_k}\frac{\partial x_j}{\partial x_l} + \frac{\partial x_i}{\partial x_l}\frac{\partial x_j}{\partial x_k}\right\} = \frac{g_{ij}}{2}(\delta_{ik}\delta_{jl} + \delta_{il}\delta_{jk}),$$

so that

(24.30) $$\left(\sum_{ij=1}^{n} g_{ij} dx_i dx_j\right)\left(\frac{\partial}{\partial x_k}, \frac{\partial}{\partial x_l}\right) = \sum_{ij=1}^{n} \frac{g_{ij}}{2}(\delta_{ik}\delta_{jl} + \delta_{il}\delta_{jk}) = g_{kl}.$$

24.3 The Classical Treatment of Metrics

Thus it follows that

$$ds^2\left(\frac{\partial}{\partial x_k}, \frac{\partial}{\partial x_l}\right) = g_{kl} = \left(\sum_{ij=1}^n g_{ij} dx_i dx_j\right)\left(\frac{\partial}{\partial x_k}, \frac{\partial}{\partial x_l}\right).$$

for $1 \leq k, l \leq n$. For general $X, Y \in \mathfrak{X}(\mathcal{M})$ we write

$$X = \sum_{k=1}^n a_k \frac{\partial}{\partial x_k} \quad \text{and} \quad Y = \sum_{l=1}^n b_l \frac{\partial}{\partial x_l}.$$

Then (24.30) implies that

$$(24.31) \qquad ds^2(X, Y) = \sum_{kl=1}^n a_k b_l \, ds^2\left(\frac{\partial}{\partial x_k}, \frac{\partial}{\partial x_l}\right)$$

$$= \sum_{kl=1}^n a_k b_l \left(\sum_{ij=1}^n g_{ij} dx_i dx_j\right)\left(\frac{\partial}{\partial x_k}, \frac{\partial}{\partial x_l}\right)$$

$$= \left(\sum_{ij=1}^n g_{ij} dx_i dx_j\right)(X, Y).$$

Since (24.31) holds for all $X, Y \in \mathfrak{X}(\mathcal{M})$, we get (24.29). ∎

Here is some additional classical terminology.

Definition. *Let g_{ij} ($1 \leq i, j \leq n$) be the components of a metric with respect to a coordinate system $(x_1, \ldots, x_n)$; then (g^{ij}) is the inverse matrix of (g_{ij}). Frequently, the g^{ij}'s are called the **upper** g-ij's, while the g_{ij}'s are called the **lower** g-ij's. Finally, we put*

$$g = \det(g_{ij}).$$

Lemma 24.10. *For any coordinate system $(x_1, \ldots, x_n)$ on a Riemannian manifold $\mathcal{M}$, each 1-form dx_i is dual to*

$$\sum_{h=1}^n g^{ih} \frac{\partial}{\partial x_h}$$

for $i = 1, \ldots, n$; that is,

$$dx_i^* = \sum_{h=1}^n g^{ih} \frac{\partial}{\partial x_h}.$$

Proof. By definition of inverse matrix, we have

$$\delta_{ij} = \sum_{h=1}^{n} g^{ih} g_{hj} = \sum_{h=1}^{n} g^{ih} \left\langle \frac{\partial}{\partial x_h}, \frac{\partial}{\partial x_j} \right\rangle = \left\langle \sum_{h=1}^{n} g^{ih} \frac{\partial}{\partial x_h}, \frac{\partial}{\partial x_j} \right\rangle.$$

On the other hand,

$$\left\langle dx_i^*, \frac{\partial}{\partial x_j} \right\rangle = dx^i \left(\frac{\partial}{\partial x_j} \right) = \delta_{ij}.$$

Then Lemma 24.6 implies that dx_i^* and

$$\sum_{h=1}^{n} g^{ih} \frac{\partial}{\partial x_h}$$

coincide. ∎

Sometimes we need to express the Riemannian connection in terms of local coordinates, so we make the following definition.

Definition. Let $\langle\ ,\ \rangle$ be a metric on an n-dimensional manifold M endowed with a metric $ds^2 = \langle\ ,\ \rangle$, and let ∇ be the Riemannian connection of ds^2. If $(x_1, \ldots, x_n)$ is a system of local coordinates on M, then the **Christoffel symbols** Γ_{ij}^k of ∇ relative to $(x_1, \ldots, x_n)$ are given by

$$(24.32) \qquad \Gamma_{ij}^k = dx_k \left(\nabla_{\frac{\partial}{\partial x_i}} \frac{\partial}{\partial x_j} \right)$$

for $1 \leq i, j, k \leq n$.

The next lemma is an easy consequence of the definition of the Christoffel symbols and the basis theorem (Theorem 23.12, page 536).

Lemma 24.11. Let $(x_1, \ldots, x_n)$ be a coordinate system on a manifold M endowed with a metric $ds^2 = \langle\ ,\ \rangle$, and let ∇ be the Riemannian connection of ds^2. Then

$$(24.33) \qquad \nabla_{\frac{\partial}{\partial x_i}} \frac{\partial}{\partial x_j} = \sum_{k=1}^{n} \Gamma_{ij}^k \frac{\partial}{\partial x_k},$$

for $1 \leq i, j \leq n$.

It is possible, of course, to define the Christoffel symbols for a general connection on a manifold. Here is a symmetry property of the Christoffel symbols of a Riemannian connection that may not hold for a general connection.

24.3 The Classical Treatment of Metrics

Lemma 24.12. *Christoffel symbols of a Riemannian connection satisfy the symmetry relation*

(24.34) $$\Gamma_{ij}^k = \Gamma_{ji}^k$$

for $i, j, k = 1, \ldots, n$.

Proof. Equation (24.34) is an immediate consequence of two facts: (1) the Riemannian connection ∇ has zero torsion (that is, equation (24.26) holds), and (2) the brackets of coordinate vector fields vanish. ∎

Next, we derive the classical expression for the Γ_{jk}^i's in terms of the g_{ij}'s and the g^{ij}'s.

Lemma 24.13. *The Γ_{jk}^i's of a Riemannian connection are given by*

(24.35) $$\Gamma_{jk}^i = \sum_{h=1}^n \frac{g^{ih}}{2}\left(\frac{\partial g_{hj}}{\partial x_k} + \frac{\partial g_{hk}}{\partial x_j} - \frac{\partial g_{jk}}{\partial x_h}\right).$$

Proof. Since brackets of coordinate vector fields are zero, (24.22) reduces to

(24.36) $$\left\langle \nabla_{\frac{\partial}{\partial x_j}} \frac{\partial}{\partial x_k}, \frac{\partial}{\partial x_h} \right\rangle = \frac{1}{2}\left(\frac{\partial}{\partial x_j}\left\langle \frac{\partial}{\partial x_k}, \frac{\partial}{\partial x_h}\right\rangle \right.$$
$$+ \frac{\partial}{\partial x_k}\left\langle \frac{\partial}{\partial x_j}, \frac{\partial}{\partial x_h}\right\rangle - \left.\frac{\partial}{\partial x_h}\left\langle \frac{\partial}{\partial x_j}, \frac{\partial}{\partial x_k}\right\rangle\right)$$
$$= \frac{1}{2}\left(\frac{\partial g_{kh}}{\partial x_j} + \frac{\partial g_{jh}}{\partial x_k} - \frac{\partial g_{jk}}{\partial x_h}\right).$$

It follows from the definition of Γ_{jk}^i and Lemma 24.10 that

(24.37) $$\Gamma_{jk}^i = dx_i\left(\nabla_{\frac{\partial}{\partial x_j}} \frac{\partial}{\partial x_k}\right) = \left\langle \nabla_{\frac{\partial}{\partial x_j}} \frac{\partial}{\partial x_k}, dx_i^*\right\rangle = \sum_{h=1}^n g^{ih}\left\langle \nabla_{\frac{\partial}{\partial x_j}} \frac{\partial}{\partial x_k}, \frac{\partial}{\partial x_h}\right\rangle.$$

Now (24.35) follows from (24.36) and (24.37). ∎

Finally, we generalize (24.33).

Lemma 24.14. *Let $X, Y \in \mathfrak{X}(\mathcal{M})$, and let $(x_1, \ldots, x_n)$ be a coordinate system on a Riemannian manifold $\mathcal{M}$. Then*

(24.38) $$\nabla_X Y = \sum_{k=1}^n \left(XY[x_k] + \sum_{ij=1}^n X[x_i]Y[x_j]\Gamma_{ij}^k\right)\frac{\partial}{\partial x_k}.$$

Proof. We have

$$\nabla_X Y = \nabla_{\left(\sum X[x_i]\frac{\partial}{\partial x_i}\right)} \left(\sum Y[x_j]\frac{\partial}{\partial x_j}\right)$$

$$= \sum_{ij=1}^n X[x_i]\left(\frac{\partial}{\partial x_i} Y[x_j]\frac{\partial}{\partial x_j} + Y[x_j]\nabla_{\frac{\partial}{\partial x_i}}\frac{\partial}{\partial x_j}\right)$$

$$= \sum_{j=1}^n XY[x_j]\frac{\partial}{\partial x_j} + \sum_{ijk=1}^n \left(X[x_i]Y[x_j]\Gamma_{ij}^k\frac{\partial}{\partial x_k}\right)$$

$$= \sum_{k=1}^n \left\{XY[x_k] + \sum_{ij=1}^n X[x_i]Y[x_j]\Gamma_{ij}^k\right\}\frac{\partial}{\partial x_k}. \blacksquare$$

24.4 Exercises

1. Complete the proof of Lemma 24.3.

2. Give examples of vector fields X, Y, Z on $\mathbb{R}$ such that $Y_0 = Z_0$, but $(D_X Y)_0 \neq (D_X Z)_0$.

3. Prove Lemmas 24.7 and 24.8.

4. The **torsion** of a connection D on a differentiable manifold $\mathcal{M}$ is defined by

$$T(X, Y) = D_X Y - D_Y X - [X, Y].$$

Show that T is a vectorvariant tensor field on $\mathcal{M}$ and that $T(Y, X) = -T(X, Y)$.

25

ABSTRACT SURFACES

A revolution in surface theory was initiated by Gauss in the first half of the nineteenth century. Before Gauss, surface theory meant the study of surfaces in ordinary space $\mathbb{R}^3$. Gauss's *Theorema Egregium* (Theorem 22.5, page 507), stating that the curvature K of a surface in $\mathbb{R}^3$ does not depend on how the surface is embedded in $\mathbb{R}^3$, led mathematicians to question the necessity of a surface being embedded in $\mathbb{R}^3$ in the first place. Thus the idea of *abstract surface* arose. Riemann[1] in his thesis [Riem] (written in 1854 but published only in 1867) introduced the metric ds^2 in n dimensions and described how the curvature of ds^2 can be measured. (Such a metric ds^2 is an obvious generalization of the notion of metric that we introduced intuitively in Section 15.1.) The modern concept of a differentiable manifold, due

---

[1] Georg Friedrich Bernhard Riemann (1826-1866). German mathematician. Although he died young and published few papers, Riemann's ideas have had a profound effect on the development of mathematics and physics. In an early paper he clarified the notion of integral by defining what is now called the Riemann integral. Riemann's Habilitationsvortrag [Riem], **Über die Hypothesen welche der Geometrie zu Grunde liegen**, (impressing even Gauss) became a classic, and its results were incorporated into Einstein's general relativity. The Riemann hypothesis, which concerns the zeros of the Riemann zeta function

$$\zeta(z) = \sum_{n=1}^{\infty} \frac{1}{n^z},$$

has enormous implications for number theory but remains unsolved.

to Weyl[2] (see [Weyl]), did not appear until the early twentieth century.

First, let us note the following fact.

Lemma 25.1. *A regular surface $\mathcal{M}$ in $\mathbb{R}^n$ is a differentiable manifold.*

Proof. For the atlas of $\mathcal{M}$ we choose the regular injective patches on $\mathcal{M}$. Corollary 12.18, page 284, implies that each change of coordinates is differentiable. Hence $\mathcal{M}$ is a differentiable manifold. ∎

Since we have introduced differentiable manifolds in Chapter 23 and Riemannian metrics in Chapter 24, we are ready to consider the geometry of surfaces freed from the requirement that the surfaces be embedded in $\mathbb{R}^3$ or even $\mathbb{R}^n$.

Definition. *A 2-dimensional differentiable manifold equipped with a Riemannian or pseudo-Riemannian metric $\langle \ , \ \rangle$ is called an* **abstract surface**.

The metric on an abstract surface is a generalization of the induced metrics that we studied in Chapter 15. Lemma 25.1 implies:

Lemma 25.2. *A regular surface $\mathcal{M}$ equipped with the metric induced from $\mathbb{R}^n$ is an abstract surface.*

The pre-Gauss notion of Gaussian curvature K was introduced in Chapter 16. Brioschi's formula (22.7), page 504, shows that K depends only on the metric $\langle \ , \ \rangle$. Therefore, we make the following definition.

Definition. *The Gaussian curvature K of an abstract surface equipped with a Riemannian metric ds^2 is given by Brioschi's formula (22.7).*

We need to specialize to the 2-dimensional case the general theory of pseudo-Riemannian manifolds that we developed in Chapter 24. Section 25.1 is devoted to pinning down the exact relation between the notation of Section 24.3 for the g_{ij}'s and Christoffel symbols, and that of Chapters 16 and 22. Examples of abstract surfaces are given in Section 25.2. Then in Section 25.3 we use a *Mathematica*

---

[2] Hermann Klaus Hugo Weyl (1885-1955). German mathematician, who together with É. Cartan created the basis for the modern theory of Lie groups and their representations. With his application of group theory to quantum mechanics, he set up the modern theory of the subject. He was professor in Zürich and Göttingen. From 1933 until he retired in 1952 he worked at the Institute of Advanced Study at Princeton.

miniprogram implementing Brioschi's formula to compute the curvature of some of the abstract surfaces described in Section 25.2. *Orientability* for abstract surfaces is defined in Section 25.4, and *geodesic curvature* is discussed in Section 25.5. The *mean curvature vector field* of a surface in $\mathbb{R}^n$ is studied in Section 25.6 and is used to extend the definition of minimal surfaces to surfaces in $\mathbb{R}^n$.

25.1 Metrics and Christoffel Symbols for Abstract Surfaces

The definitions of E, F and G given on page 342 make sense for a patch on an abstract surface. All that needs to be done is to use the abstract surface metric in place of the induced metric. We begin by making precise the relation between the g_{ij}'s introduced in Section 24.3 and the coefficients of the first fundamental form E, F, G. We denote the natural coordinate functions of $\mathbb{R}^2$ either by $\{u,v\}$ or by $\{u_1, u_2\}$.

Lemma 25.3. *Let* $\mathbf{x}: \mathcal{U} \longrightarrow \mathcal{M}$ *be a patch on an abstract surface* $\mathcal{M}$ *endowed with a metric* ds^2. *Then the four lower* g_{ij}'s *are related to* E, F *and* G *by the formulas*

(25.1)
$$\begin{cases} g_{11} \circ \mathbf{x} = E, & g_{12} \circ \mathbf{x} = F, \\ g_{21} \circ \mathbf{x} = F, & g_{22} \circ \mathbf{x} = G. \end{cases}$$

Hence the pull back (see page 552) of ds^2 *by* $\mathbf{x}$ *is given by*

$$\mathbf{x}^*(ds^2) = \mathbf{x}^*\left(g_{11}dx_1^2 + 2g_{12}dx_1dx_2 + g_{22}dx_2^2\right) = E\,du^2 + 2F\,du\,dv + G\,dv^2.$$

Proof. It follows from (23.21), page 550, that

$$g_{ij} \circ \mathbf{x} = \left\langle \frac{\partial}{\partial x_i}, \frac{\partial}{\partial x_j} \right\rangle \circ \mathbf{x} = \left\langle \mathbf{x}_*\left(\frac{\partial}{\partial u_i}\right), \mathbf{x}_*\left(\frac{\partial}{\partial u_j}\right) \right\rangle \circ \mathbf{x} = \langle \mathbf{x}_{u_i}, \mathbf{x}_{u_j} \rangle.$$

In particular, $g_{11} \circ \mathbf{x} = \langle \mathbf{x}_{u_1}, \mathbf{x}_{u_1} \rangle = E$. The other formulas of (25.1) are proved similarly. Also, we have

$$\mathbf{x}^*\left(g_{11}dx_1^2 + 2g_{12}dx_1dx_2 + g_{22}dx_2^2\right)$$
$$= (g_{11} \circ \mathbf{x})\mathbf{x}^*(dx_1)^2 + 2(g_{12} \circ \mathbf{x})\mathbf{x}^*(dx_1)\mathbf{x}^*(dx_2) + (g_{22} \circ \mathbf{x})\mathbf{x}^*(dx_2)^2$$

$$= (g_{11} \circ \mathbf{x})\big(d(x_1 \circ \mathbf{x})\big)^2 + 2(g_{12} \circ \mathbf{x})d(x_1 \circ \mathbf{x})d(x_2 \circ \mathbf{x}) + (g_{22} \circ \mathbf{x})\big(d(x_2 \circ \mathbf{x})\big)^2$$

$$= (g_{11} \circ \mathbf{x})du^2 + 2(g_{12} \circ \mathbf{x})dudv + (g_{22} \circ \mathbf{x})dv^2$$

$$= E\,du^2 + 2F\,dudv + G\,dv^2. \blacksquare$$

Next, we find the relation between the upper g^{ij}'s and the metric coefficients E, F and G.

Lemma 25.4. *The upper g^{ij}'s of a Riemannian metric ds^2 on a patch $\mathbf{x}$ on an abstract surface $\mathcal{M}$ are given by*

(25.2)
$$\begin{cases} g^{11} \circ \mathbf{x} = \dfrac{G}{E\,G - F^2}, & g^{12} \circ \mathbf{x} = \dfrac{-F}{E\,G - F^2}, \\ g^{21} \circ \mathbf{x} = \dfrac{-F}{E\,G - F^2}, & g^{22} \circ \mathbf{x} = \dfrac{E}{E\,G - F^2}. \end{cases}$$

Proof. It can be verified by hand (see (16.14), page 370) or with *Mathematica* that

(25.3)
$$\begin{pmatrix} E & F \\ F & G \end{pmatrix}^{-1} = \frac{1}{E\,G - F^2} \begin{pmatrix} G & -F \\ -F & E \end{pmatrix}.$$

Then (25.2) is clear from (25.1) and (25.3). $\blacksquare$

We have already defined the Christoffel symbols for a surface in $\mathbb{R}^3$ by (22.13), page 510. Since (22.13) defines the Γ^i_{jk}'s in terms of nothing more than E, F, G and their derivatives, we can just use the same formulas for a general abstract surface.

Definition. *Let $\mathcal{M}$ be an abstract surface with a metric given locally by*

$$\langle\,,\,\rangle = ds^2 = E\,du^2 + 2F\,dudv + G\,dv^2.$$

*Then the **Christoffel symbols** Γ^i_{jk} of $\langle\,,\,\rangle$ are given by*

(25.4)
$$\begin{cases} \Gamma^1_{11} = \dfrac{G\,E_u - 2F\,F_u + F\,E_v}{2(E\,G - F^2)}, & \Gamma^2_{11} = \dfrac{2E\,F_u - E\,E_v - F\,E_u}{2(E\,G - F^2)}, \\ \Gamma^1_{12} = \dfrac{G\,E_v - F\,G_u}{2(E\,G - F^2)}, & \Gamma^2_{12} = \dfrac{E\,G_u - F\,E_v}{2(E\,G - F^2)}, \\ \Gamma^1_{22} = \dfrac{2G\,F_v - G\,G_u - F\,G_v}{2(E\,G - F^2)}, & \Gamma^2_{22} = \dfrac{E\,G_v - 2F\,F_v + F\,G_u}{2(E\,G - F^2)}, \end{cases}$$

and $\Gamma^1_{21} = \Gamma^1_{12}$, $\Gamma^2_{21} = \Gamma^2_{12}$.

25.1 Metrics and Christoffel Symbols for Abstract Surfaces

Fortunately, there is a close relation between the Γ^i_{jk}'s of (25.4) and the $\boldsymbol{\Gamma}^i_{jk}$'s given by (24.35), page 571.

Lemma 25.5. *The Christoffel symbols $\boldsymbol{\Gamma}^i_{jk}$ of a metric ds^2 on a patch $\mathbf{x}$ of an abstract surface $\mathcal{M}$ given by (24.35) are related to the Christoffel symbols Γ^i_{jk} given by (25.4) by means of the formula*

(25.5) $$\boldsymbol{\Gamma}^i_{jk} \circ \mathbf{x} = \Gamma^i_{jk}$$

for $1 \le i, j, k \le 2$.

Proof. For example, we show that $\boldsymbol{\Gamma}^1_{11} \circ \mathbf{x}$ given by (24.35) coincides with Γ^1_{11} given by (25.4). First, we compute

(25.6) $$\boldsymbol{\Gamma}^1_{11} = \sum_{h=1}^{2} \frac{g^{1h}}{2}\left\{2\frac{\partial g_{h1}}{\partial x_1} - \frac{\partial g_{11}}{\partial x_h}\right\} = \frac{g^{11}}{2}\frac{\partial g_{11}}{\partial x_1} + g^{12}\frac{\partial g_{21}}{\partial x_1} - \frac{g^{12}}{2}\frac{\partial g_{11}}{\partial x_2}.$$

On the other hand, (25.1) implies that

$$\frac{\partial g_{11}}{\partial x_1} \circ \mathbf{x} = \mathbf{x}_*\left(\frac{\partial}{\partial u}\right)[g_{11}] = \frac{\partial(g_{11} \circ \mathbf{x})}{\partial u} = E_u.$$

Similarly,

$$\frac{\partial g_{21}}{\partial x_1} \circ \mathbf{x} = F_u \quad \text{and} \quad \frac{\partial g_{11}}{\partial x_2} \circ \mathbf{x} = E_v.$$

We substitute these expressions and (25.2) into (25.6) and get

$$\boldsymbol{\Gamma}^1_{11} \circ \mathbf{x} = \frac{G\,E_u - 2F\,F_u + F\,E_v}{2(E\,G - F^2)} = \Gamma^1_{11}.\quad\blacksquare$$

Isometries of Abstract Surfaces

Isometries and local isometries between abstract surfaces are defined exactly the same way as we defined isometries and local isometries between surfaces in $\mathbb{R}^n$ in Section 15.2, page 346. Lemmas 15.4, 15.5, 15.6, 15.8 and Corollary 15.7 hold for abstract surfaces. In particular, a local isometry preserves E, F and G; this means that a local isometry preserves any function on a surface that can be expressed in terms of E, F and G. Thus a local isometry preserves Christoffel symbols. Furthermore, Gauss's *Theorema Egregium* (Theorem 22.5) is valid for isometries between abstract surfaces.

25.2 Examples of Metrics on Abstract Surfaces

There are many interesting metrics that do not arise from the dot product on $\mathbb{R}^3$. Here are a few.

The Poincaré Metric on the Upper Half-Plane

Let (u,v) be the standard coordinates on the upper half-plane
$$\mathbb{R}^{2+} = \{\,(p,q) \in \mathbb{R}^2 \mid q > 0\,\}.$$
Let $\lambda > 0$. The Poincaré[3] metric on $\mathbb{R}^{2+}$ is defined by

(25.7) $$ds^2 = \frac{du^2 + dv^2}{\lambda^2 v^2}.$$

A Generalization of the Poincaré Metric on the Upper Half-Plane

The metric (25.7) can be generalized by putting

(25.8) $$ds^2 = \frac{du^2}{\lambda^2 v^m} + \frac{dv^2}{\lambda^2 v^n}.$$

The Poincaré Metric on a Disk

Let (u,v) be the standard coordinates on a disk $\mathcal{D}(\lambda)$ of radius λ in $\mathbb{R}^2$ centered at the origin:
$$\mathcal{D}(\lambda) = \{\,(p,q) \in \mathbb{R}^2 \mid p^2 + q^2 < \lambda^2\,\}.$$
Let $\lambda > 0$. The Poincaré metric on $\mathcal{D}(\lambda)$ is defined by
$$ds^2 = \frac{du^2 + dv^2}{\left(1 - \dfrac{\lambda^2(u^2+v^2)}{4}\right)^2}.$$

---

[3] Jules Henri Poincaré (1854-1912). French mathematician, widely regarded as the leading mathematician of his time. He developed the concept of automorphic functions and invented much of algebraic topology. He also worked in algebraic geometry, analytic functions of several complex variables and number theory.

Geodesic Polar Metric

Let $f: \mathbb{R}^2 \setminus \{(0,0)\} \longrightarrow \mathbb{R}$ be any differentiable function. The **geodesic polar metric** relative to f is defined by

$$ds^2 = du^2 + f(u,v)^2 dv^2.$$

For example, $du^2 + u^2 dv^2$ is the standard flat metric of $\mathbb{R}^2$ in polar coordinates.

A Flat Metric on a Torus

The torus is a rather complicated surface in $\mathbb{R}^3$. There is a much simpler version of the torus in $\mathbb{R}^4$, however. Let us define a patch **flattorus**$[a, b]: [0, 2\pi) \times [0, 2\pi) \longrightarrow \mathbb{R}^4$ by

$$\textbf{flattorus}[a, b](u, v) = (a \cos u, a \sin u, b \cos v, b \sin v).$$

Then $E = a^2$, $G = b^2$ and $F = 0$. Since E and G are constant, it is immediate from Corollary 22.4, page 505, that the Gaussian curvature of **flattorus**$[a, b]$ vanishes identically. For this reason, **flattorus**$[a, b]$ is called a **flat torus**.

Let us also remark that **flattorus**$[a, b]$ is naturally a submanifold of the 3-dimensional sphere of radius $\sqrt{a^2 + b^2}$ in $\mathbb{R}^4$ centered at the origin.

The Schwarzschild Surface

The **Schwarzschild**[4] **metric** used in general relativity is the 4-dimensional metric defined by

$$(25.9) \qquad ds^2 = \frac{dv^2}{\left(1 - \frac{2m}{v}\right)} + v^2 d\theta^2 + v^2 \sin^2 \theta^2 d\phi^2 - \left(1 - \frac{2m}{v}\right) du^2.$$

---

[4] Karl Schwarzschild (1873-1916). German mathematician. Professor at Göttingen and Potsdam. He died soon after returning from military service in Russia during World War I.

Taking θ and ϕ to be constants in (25.9), we get a 2-dimensional metric on either the upper half-plane or the lower half-plane:

$$(25.10) \qquad ds^2 = \frac{dv^2}{\left(1 - \dfrac{2m}{v}\right)} - \left(1 - \frac{2m}{v}\right) du^2.$$

We define the **Schwarzschild surface** to be the abstract surface consisting of either the upper half-plane or the lower half-plane together with the metric (25.10). See [ON2, page 152] for a discussion of the Schwarzschild surface.

25.3 Computing Curvature of Metrics on Abstract Surfaces

We can use Brioschi's formula (22.7) to compute the curvature of a metric $ds^2 = E\,du^2 + 2F\,du\,dv + G\,dv^2$ on an abstract surface. However, to do so, the coefficients E, F and G must be specified explicitly. The following examples illustrate how to use *Mathematica* for the computation.

The Curvature of the Poincaré Metric

For example, let us compute the Gaussian curvature of the Poincaré metric (25.7) on the upper half-plane. First, we need the *Mathematica* definitions of the components of the metric.

```
ee[poincare[lambda_]][u_,v_] := 1/(lambda v)^2

ff[poincare[lambda_]][u_,v_] := 0

gg[poincare[lambda_]][u_,v_] := 1/(lambda v)^2
```

Then the command

```
brioschicurvature[poincare[lambda]][u,v]
```

elicits the response

25.3 Computing Curvature of Metrics on Abstract Surfaces

$$-\text{lambda}^2$$

Thus the Poincaré metric on the upper half-plane has constant negative curvature $-\lambda^2$. We have already seen in Chapter 21 several examples of surfaces in $\mathbb{R}^3$ with constant negative curvature, including the pseudosphere and Kuen's surface. But all of these surfaces are much more complicated than the upper half-plane with its Poincaré metric, even though they are all locally isometric to a portion of it. Thus it is sometimes advantageous to study an abstract surface without the requirement that the surface be realized in $\mathbb{R}^3$. The Poincaré metric on the disk also has constant negative curvature $-\lambda^2$.

The Curvature of the Generalized Poincaré Metric

The generalized Poincaré metric (25.8) can be treated in the same way. First, we define

```
ee[poincare[m_,n_,lambda_]][u_,v_] := 1/(lambda v)^m

ff[poincare[m_,n_,lambda_]][u_,v_] := 0

gg[poincare[m_,n_,lambda_]][u_,v_] := 1/(lambda v)^n
```

This time

```
brioschicurvature[poincare[m,n,lambda]][u,v] //Simplify
```

elicits the response

$$\frac{m\,(-2 - m + n)\,(\text{lambda}\,v)^n}{4\,v^2}$$

The Curvature of the Schwarzschild Surface

The metric of the Schwarzschild surface is given by

```
ee[schwarzschildplane[m_]][u_,v_] := -(1 - 2m/v)

ff[schwarzschildplane[m_]][u_,v_] := 0

gg[schwarzschildplane[m_]][u_,v_] := (1 - 2m/v)^-1
```

Then

```
brioschicurvature[schwarzschildplane[m]][u,v]//Simplify
```

tells us that the curvature is given by

$$\frac{2m}{v^3}$$

25.4 Orientability of an Abstract Surface

An abstract surface may or may not be contained in Euclidean space $\mathbb{R}^3$. For surfaces in $\mathbb{R}^3$ the existence of a continuous unit normal vector field is a convenient intuitive criterion for orientability. This criterion is not available for abstract surfaces, but we can define orientability in terms of the existence of a complex structure. The definition for surfaces in $\mathbb{R}^3$ given on page 318 goes over almost *verbatim* to abstract surfaces:

Definition. *Let $\mathcal{M}$ be an abstract surface. Then $\mathcal{M}$ is* **orientable** *provided each tangent space $\mathcal{M}_\mathbf{p}$ has a complex structure $J_\mathbf{p}$ such that $\mathbf{p} \longmapsto J_\mathbf{p}$ is a continuous function. An* **oriented** *abstract surface $(\mathcal{M}, J)$ consists of an orientable abstract surface together with a choice of the complex structure $\mathbf{p} \longmapsto J_\mathbf{p}$.*

Note that we can regard J as an $\mathfrak{F}(\mathcal{M})$-linear map $J\colon \mathfrak{X}(\mathcal{M}) \longrightarrow \mathfrak{X}(\mathcal{M})$ such that $J^2 = -I$. We also need an explicit formula for J on an abstract surface.

Lemma 25.6. *Let $\mathbf{x}\colon \mathcal{U} \longrightarrow \mathcal{M}$ be a patch on an oriented surface $(\mathcal{M}, J)$. Assume that $\mathcal{M}$ has a Riemannian metric, which when written in terms of $\mathbf{x}$ is $ds^2 = E\,du^2 + 2F\,du\,dv + G\,dv^2$. Then*

$$(25.11) \qquad J\mathbf{x}_u = \pm \frac{-F\mathbf{x}_u + E\mathbf{x}_v}{\sqrt{EG - F^2}} \qquad \text{and} \qquad J\mathbf{x}_v = \pm \frac{-G\mathbf{x}_u + F\mathbf{x}_v}{\sqrt{EG - F^2}}.$$

25.5 Geodesic Curvature for Abstract Surfaces

Proof. Since $\mathbf{x}$ is regular, $\mathbf{x}_u$ and $\mathbf{x}_v$ are linearly independent, and we can write

$$J\mathbf{x}_u = b_{21}\mathbf{x}_u + b_{22}\mathbf{x}_v,$$

for some functions b_{21} and b_{22}, where $b_{22} \neq 0$. Then

$$E = \|\mathbf{x}_u\|^2 = \|J\mathbf{x}_u\|^2 = b_{21}^2 E + 2b_{21}b_{22}F + b_{22}^2 G$$

and

$$0 = \langle J\mathbf{x}_u, \mathbf{x}_u \rangle = b_{21}E + b_{22}F.$$

Hence $b_{21} = -b_{22}F/E$ and

$$E = b_{22}^2 \left(\frac{F^2}{E} - 2\frac{F^2}{E} + G \right) = \frac{b_{22}^2(EG - F^2)}{E}.$$

Thus

$$b_{22} = \frac{\pm E}{\sqrt{EG - F^2}} \quad \text{and} \quad b_{21} = \frac{\mp F}{\sqrt{EG - F^2}},$$

and we obtain the first equation of (25.11); the second equation is proved similarly. ∎

25.5 Geodesic Curvature for Abstract Surfaces

In this section we generalize the curvature of a curve in the plane, namely $\kappa\mathbf{2}$, to curves in an arbitrary orientable surface. The appropriate substitute for the acceleration $\boldsymbol{\alpha}''$ is $\nabla_{\boldsymbol{\alpha}'}\boldsymbol{\alpha}'$.

Definition. *Let $(\mathcal{M}, J)$ be an oriented abstract surface and $\boldsymbol{\alpha}: (a, b) \longrightarrow \mathcal{M}$ a curve. Then for those values of t for which $\boldsymbol{\alpha}'(t) \neq 0$ the* **geodesic curvature** *$\kappa_g[\boldsymbol{\alpha}]$ of $\boldsymbol{\alpha}$ is given by*

$$(25.12) \qquad \kappa_g[\boldsymbol{\alpha}](t) = \frac{\langle \nabla_{\boldsymbol{\alpha}'(t)} \boldsymbol{\alpha}'(t), J\boldsymbol{\alpha}'(t) \rangle}{\|\boldsymbol{\alpha}'(t)\|^3}.$$

The motivation for (25.12) is of course (22.25), page 515; in fact (25.12) reduces to (22.25) for surfaces in $\mathbb{R}^3$.

Next, we show that the geodesic curvature κ_g behaves in exactly the same way as $\kappa\mathbf{2}$ under reparametrizations.

Theorem 25.7. *Let $\mathcal{M}$ be an oriented abstract surface and $\boldsymbol{\alpha}\colon (a,b) \longrightarrow \mathcal{M}$ a curve, and let $\boldsymbol{\gamma}\colon (c,d) \longrightarrow \mathcal{M}$ be a reparametrization of $\boldsymbol{\alpha}$. Write $\boldsymbol{\gamma} = \boldsymbol{\alpha} \circ h$, where $h\colon (c,d) \longrightarrow (a,b)$ is piecewise-differentiable. Then the geodesic curvatures of $\boldsymbol{\alpha}$ and $\boldsymbol{\gamma}$ are related by*

(25.13) $$\kappa_g[\boldsymbol{\gamma}](t) = \bigl(\operatorname{sign} h'(t)\bigr)\kappa_g[\boldsymbol{\alpha}]\bigl(h(t)\bigr)$$

wherever $\operatorname{sign} h'(t)$ is defined.

Proof. It follows from Lemma 23.23, page 545, and Lemma 24.5, page 563, that

$$\kappa_g[\boldsymbol{\gamma}] = \frac{\bigl\langle h'^2\,(\nabla_{\boldsymbol{\alpha}'}\boldsymbol{\alpha}') \circ h + h''\boldsymbol{\alpha}' \circ h,\, h'J(\boldsymbol{\alpha}' \circ h)\bigr\rangle}{\bigl\|(\boldsymbol{\alpha}' \circ h)h'\bigr\|^3}$$

$$= \left(\frac{h'^3}{|h'|^3}\right) \frac{\bigl\langle (\nabla_{\boldsymbol{\alpha}'}\boldsymbol{\alpha}') \circ h,\, J(\boldsymbol{\alpha}' \circ h)\bigr\rangle}{\bigl\|(\boldsymbol{\alpha}' \circ h)\bigr\|^3}$$

$$= (\operatorname{sign} h')\kappa_g[\boldsymbol{\alpha}] \circ h.$$

Hence we get (25.13). ∎

The geodesic curvature is especially simple for unit-speed curves.

Lemma 25.8. *Let $\boldsymbol{\beta}\colon (a,b) \longrightarrow \mathcal{M}$ be a unit-speed curve in an oriented abstract surface $\mathcal{M}$. Then*

(25.14) $$\nabla_{\boldsymbol{\beta}'}\boldsymbol{\beta}' = \kappa_g[\boldsymbol{\beta}]J\boldsymbol{\beta}'.$$

Proof. The proof is almost word-for-word that of Lemma 1.11, page 16. ∎

The Turning Angle of a Curve on an Abstract Surface

Next, we generalize the notion of turning angle given in Section 1.7 to abstract surfaces. Since in general there is no notion of "horizontal" for a general abstract surface, we must choose a reference vector to measure the turning.

Lemma 25.9. *Let $\mathcal{M}$ be an oriented surface, let $\boldsymbol{\alpha}\colon (a,b) \longrightarrow \mathcal{M}$ be a regular curve and let $\mathbf{X}$ be an everywhere nonzero vector field along $\boldsymbol{\alpha}$. Fix t_0 with $a < t_0 < b$. Let θ_0 be a number such that*

$$\frac{\boldsymbol{\alpha}'(t_0)}{\|\boldsymbol{\alpha}'(t_0)\|} = \cos\theta_0 \frac{\mathbf{X}(t_0)}{\|\mathbf{X}(t_0)\|} + \sin\theta_0 \frac{J\mathbf{X}(t_0)}{\|\mathbf{X}(t_0)\|}.$$

25.5 Geodesic Curvature for Abstract Surfaces

Then there exists a unique differentiable function $\theta[\alpha, \mathbf{X}]\colon (a, b) \longrightarrow \mathbb{R}$ *such that* $\theta[\alpha, \mathbf{X}](t_0) = \theta_0$ *and*

$$(25.15) \qquad \frac{\alpha'(t)}{\|\alpha'(t)\|} = \frac{1}{\|\mathbf{X}(t)\|} \Big(\cos\theta[\alpha, \mathbf{X}](t)\mathbf{X}(t) + \sin\theta[\alpha, \mathbf{X}](t)J\mathbf{X}(t) \Big).$$

for $a < t < b$. *We call* $\theta[\alpha, \mathbf{X}]$ *the* **turning angle** *of* α *with respect to* $\mathbf{X}$ *determined by* θ_0 *and* t_0.

Proof. Corollary 1.14, page 18, implies the existence of a unique function $\theta[\alpha]\colon (a, b) \longrightarrow \mathbb{R}$ such that $\theta[\alpha](t_0) = \theta_0$ and

$$(25.16) \qquad \begin{cases} \cos\theta[\alpha](t) = \dfrac{\alpha'(t) \cdot \mathbf{X}(t)}{\|\alpha'(t)\| \, \|\mathbf{X}(t)\|}, \\[1em] \sin\theta[\alpha](t) = \dfrac{\alpha'(t) \cdot J\mathbf{X}(t)}{\|\alpha'(t)\| \, \|\mathbf{X}(t)\|}. \end{cases}$$

Then (25.16) is equivalent to (25.15). ∎

We can now generalize Lemma 1.16, page 20, to abstract surfaces.

Lemma 25.10. *Let* $\mathcal{M}$ *be an oriented surface, let* $\alpha\colon (a, b) \longrightarrow \mathcal{M}$ *be a regular curve and let* $\mathbf{X}$ *be an everywhere nonzero vector field along* α. *Then the turning angle* $\theta[\alpha, \mathbf{X}]$ *is related to the geodesic curvature* κ_g *by the formula*

$$(25.17) \qquad \theta[\alpha, \mathbf{X}]'(t) = \kappa_g[\alpha](t) \, \|\alpha'(t)\| - \frac{\langle \nabla_{\alpha'}\mathbf{X}, J\mathbf{X}\rangle(t)}{\|\mathbf{X}(t)\|^2}.$$

Proof. Just as in the proof of Lemma 1.16, we have

$$\nabla_{\alpha'(t)}\!\left(\frac{\alpha'(t)}{\|\alpha'(t)\|}\right) = \frac{\alpha''(t)}{\|\alpha'(t)\|} + \alpha'(t)\frac{d}{dt}\!\left(\frac{1}{\|\alpha'(t)\|}\right);$$

hence

$$(25.18) \qquad \left\langle \nabla_{\alpha'(t)}\!\left(\frac{\alpha'(t)}{\|\alpha'(t)\|}\right), \frac{J\alpha'(t)}{\|\alpha(t)\|}\right\rangle = \frac{\langle \alpha''(t), J\alpha'(t)\rangle}{\|\alpha'(t)\|^2} = \kappa_g[\alpha](t)\|\alpha'(t)\|.$$

On the other hand, (writing $\alpha = \alpha(t)$, $\mathbf{X} = \mathbf{X}(t)$ and $\theta = \theta[\alpha, \mathbf{X}]$) we use (25.15) to compute

$$\nabla_{\alpha'}\left(\frac{\alpha'}{\|\alpha'\|}\right) = \nabla_{\alpha'}\left(\frac{\cos\theta \mathbf{X} + \sin\theta J\mathbf{X}}{\|\mathbf{X}\|}\right)$$

$$= -\frac{\|\mathbf{X}\|'}{\|\mathbf{X}\|^2}(\cos\theta \mathbf{X} + \sin\theta J\mathbf{X}) + \left(\frac{-\sin\theta \mathbf{X} + \cos\theta J\mathbf{X}}{\|\mathbf{X}\|}\right)\theta'$$

$$+ \frac{(\cos\theta \nabla_{\alpha'}\mathbf{X} + \sin\theta J\nabla_{\alpha'}\mathbf{X})}{\|\mathbf{X}\|}$$

$$= -\frac{\|\mathbf{X}\|'}{\|\mathbf{X}\|}\frac{\alpha'}{\|\alpha'\|} + \frac{J\alpha'}{\|\alpha'\|}\theta' + \frac{(\cos\theta \nabla_{\alpha'}\mathbf{X} + \sin\theta J\nabla_{\alpha'}\mathbf{X})}{\|\mathbf{X}\|},$$

so that

(25.19) $\qquad \left\langle \nabla_{\alpha'}\left(\dfrac{\alpha'}{\|\alpha'\|}\right), \dfrac{J\alpha'}{\|\alpha'\|}\right\rangle = \theta' + \dfrac{\langle \nabla_{\alpha'}\mathbf{X}, J\mathbf{X}\rangle}{\|\mathbf{X}\|^2}.$

Now (25.17) follows from (25.18) and (25.19). ∎

Liouville's Theorem

A natural choice for the vector field $\mathbf{X}$ of Lemma 25.10 is $\mathbf{x}_u$, where $\mathbf{x}$ is a patch on an abstract surface.

Lemma 25.11. *Let $\mathcal{M}$ be an oriented surface and let $\alpha\colon (a,b) \longrightarrow \mathcal{M}$ be a regular curve whose trace is contained in $\mathbf{x}(\mathcal{U})$, where $\mathbf{x}\colon \mathcal{U} \longrightarrow \mathcal{M}$ is a patch coherent with respect to the orientation of $\mathcal{M}$. Denote by $ds^2 = E\,du^2 + 2F\,du\,dv + G\,dv^2$ the metric of $\mathcal{M}$ and write $\alpha(t) = \mathbf{x}\bigl(u(t), v(t)\bigr)$ for $a < t < b$. Then*

(25.20) $\qquad \kappa_g[\alpha]\|\alpha'\| = \theta[\alpha, \mathbf{x}_u]' + \dfrac{-\dfrac{F}{E}(E_u u' + E_v v') - E_v u' + G_u v'}{2\sqrt{EG - F^2}}.$

Proof. We have $\alpha' = u'\mathbf{x}_u + v'\mathbf{x}_v$; thus from (25.11) it follows that

$$\frac{\langle \nabla_{\alpha'}\mathbf{x}_u, J\mathbf{x}_u\rangle}{\|\mathbf{x}_u\|^2} = \frac{1}{E}\left\langle u'\nabla_{\mathbf{x}_u}\mathbf{x}_u + v'\nabla_{\mathbf{x}_v}\mathbf{x}_u, \frac{-F\mathbf{x}_u + E\mathbf{x}_v}{\sqrt{EG-F^2}}\right\rangle$$

$$= \frac{1}{E\sqrt{EG-F^2}}\Bigl(-u'F\langle \nabla_{\mathbf{x}_u}\mathbf{x}_u, \mathbf{x}_u\rangle + u'E\langle \nabla_{\mathbf{x}_u}\mathbf{x}_u, \mathbf{x}_v\rangle$$

$$-v'F\langle \nabla_{\mathbf{x}_v}\mathbf{x}_u, \mathbf{x}_u\rangle + v'E\langle \nabla_{\mathbf{x}_v}\mathbf{x}_u, \mathbf{x}_v\rangle\Bigr).$$

25.5 Geodesic Curvature for Abstract Surfaces

Since $\langle \nabla_{\mathbf{x}_u}\mathbf{x}_u, \mathbf{x}_u\rangle = E_u/2$, $\langle \nabla_{\mathbf{x}_u}\mathbf{x}_u, \mathbf{x}_v\rangle = -\langle \nabla_{\mathbf{x}_v}\mathbf{x}_u, \mathbf{x}_u\rangle = -E_v/2$ and $\langle \nabla_{\mathbf{x}_v}\mathbf{x}_u, \mathbf{x}_v\rangle = G_u/2$, we obtain

$$\frac{\langle \nabla_{\alpha'}\mathbf{x}_u, J\mathbf{x}_u\rangle}{\|\mathbf{x}_u\|^2} = \frac{1}{E\sqrt{EG-F^2}}\left(-\frac{u'}{2}(FE_u + EE_v) + \frac{v'}{2}(-FE_v + EG_u)\right).$$

From this equation and (25.17) we get (25.20). ∎

Corollary 25.12. *Suppose the hypotheses of Lemma 25.11 are satisfied and that in addition $F = 0$. Then*

$$(25.21) \qquad \kappa_g[\alpha]\,\|\alpha'\| = \theta[\alpha, \mathbf{x}_u]' + \frac{1}{2\sqrt{EG}}(G_u v' - E_v u').$$

Next, we find a formula for the geodesic curvature of a curve on a patch $\mathbf{x}$ in terms of the geodesic curvatures of the curves $u \longmapsto \mathbf{x}(u,v)$ and $v \longmapsto \mathbf{x}(u,v)$.

Theorem 25.13. (**Liouville**[5]) *Let $\mathcal{M}$ be an oriented surface and let $\alpha\colon (a,b) \longrightarrow \mathcal{M}$ be a regular curve whose trace is contained in $\mathbf{x}(\mathcal{U})$, where $\mathbf{x}\colon \mathcal{U} \longrightarrow \mathcal{M}$ is a patch coherent with respect to the orientation of $\mathcal{M}$ for which $F = 0$. Let $(\kappa_g)_1$ denote the geodesic curvature of $u \longmapsto \mathbf{x}(u,v)$, and let $(\kappa_g)_2$ denote the geodesic curvature of $v \longmapsto \mathbf{x}(u,v)$. Then*

$$(25.22) \qquad \kappa_g[\alpha] = (\kappa_g)_1 \cos\theta + (\kappa_g)_2 \sin\theta + \frac{\theta'}{\|\alpha'\|},$$

where θ is the angle between α and $u \longmapsto \mathbf{x}(u,v)$.

Proof. Applying (25.21) to the curves $u \longmapsto \mathbf{x}(u,v)$ and $v \longmapsto \mathbf{x}(u,v)$, we obtain

$$(\kappa_g)_1 \sqrt{E} = -\frac{E_v}{2\sqrt{EG}} \qquad \text{and} \qquad (\kappa_g)_2 \sqrt{G} = \frac{G_u}{2\sqrt{EG}}.$$

For a general curve α we have

$$u' = \frac{\|\alpha'\|}{\sqrt{E}}\cos\theta \qquad \text{and} \qquad v' = \frac{\|\alpha'\|}{\sqrt{G}}\sin\theta,$$

---

[5] Joseph Liouville (1809-1882). French mathematician, who worked in number theory, differential equations, dynamics, differential geometry and complex variables. Liouville played a large role in introducing Gauss's ideas to France.

so that

$$(25.23) \quad -\frac{E_v u'}{2\sqrt{EG}} = \|\boldsymbol{\alpha}'\|(\boldsymbol{\kappa}_g)_1 \cos\theta \quad \text{and} \quad \frac{G_u v'}{2\sqrt{EG}} = \|\boldsymbol{\alpha}'\|(\boldsymbol{\kappa}_g)_2 \sin\theta.$$

Then (25.22) follows from (25.21) and (25.23). ∎

Notice the deceptive similarity between (25.22) and Euler's formula (16.20).

Total Geodesic Curvature

The total signed curvature of a plane curve (defined in Section 7.1) generalizes to abstract surfaces as the total geodesic curvature.

Definition. *Let $\mathcal{M}$ be an abstract surface. The* **total geodesic curvature** *of a curve $\boldsymbol{\alpha}:[a,b] \longrightarrow \mathcal{M}$ is*

$$\int_{\boldsymbol{\alpha}} \kappa_g ds = \mathbf{TGC}[\boldsymbol{\alpha}] = \int_a^b \kappa_g[\boldsymbol{\alpha}](t)\|\boldsymbol{\alpha}'(t)\|dt,$$

where $\kappa_g[\boldsymbol{\alpha}]$ denotes the geodesic curvature of $\boldsymbol{\alpha}$.

The generalization of Lemma 7.1, page 154, is

Lemma 25.14. *The total geodesic curvature of a curve on a surface remains unchanged under a positive reparametrization, but changes sign under a negative reparametrization.*

The proofs of Lemmas 25.14 and 7.1 are essentially the same (see exercise 8). We also have the following generalization of Lemma 7.2, which follows upon integration of (25.17):

Lemma 25.15. *Let $\mathcal{M}$ be an oriented surface, let $\boldsymbol{\alpha}:(a,b) \longrightarrow \mathcal{M}$ be a regular curve and let $\mathbf{X}$ be an everywhere nonzero vector field along $\boldsymbol{\alpha}$. The total geodesic curvature can be expressed in terms of a turning angle $\theta[\boldsymbol{\alpha}]$ of $\boldsymbol{\alpha}$ by*

$$(25.24) \quad \int_{\boldsymbol{\alpha}} \kappa_g ds = \mathbf{TGC}[\boldsymbol{\alpha}] = \theta[\boldsymbol{\alpha}, \mathbf{X}](b) - \theta[\boldsymbol{\alpha}, \mathbf{X}](a) + \int_a^b \frac{\langle \nabla_{\boldsymbol{\alpha}'}\mathbf{X}, J\mathbf{X} \rangle}{\|\mathbf{X}\|^2} dt.$$

25.6 The Mean Curvature Vector Field

In contrast to the Gaussian curvature, the mean curvature of a surface $\mathcal{M}$ in $\mathbb{R}^3$ is *not* intrinsic. In other words, the mean curvature of $\mathcal{M}$ depends on the particular way that $\mathcal{M}$ is embedded in $\mathbb{R}^3$. For example, a cylinder over a circle of radius a has mean curvature $-1/(2a)$ and a plane has mean curvature 0. (The calculations can be done easily with **Mathematica** or by hand.) But like a roll of tape, the cylinder can be unrolled isometrically onto the plane.

Therefore, it makes no sense to define the mean curvature of an abstract surface. However, it should be possible to define a notion of mean curvature of a surface in $\mathbb{R}^n$. To find the appropriate generalization of the mean curvature H of a surface in $\mathbb{R}^3$, let us examine the formula (16.25), page 377, for the mean curvature H of a patch $\mathbf{x}: \mathcal{U} \longrightarrow \mathbb{R}^3$. Using the definitions of e, f and g, we can rewrite (16.25) as

$$(25.25) \qquad H = \left(\frac{G\mathbf{x}_{uu} - 2F\mathbf{x}_{uv} + E\mathbf{x}_{vv}}{2(EG - F^2)} \right) \cdot \mathbf{U},$$

where, as usual, $\mathbf{U}$ denotes the unit normal to the patch. This formula suggests that we consider the vector field

$$\mathbf{P} = \frac{G\mathbf{x}_{uu} - 2F\mathbf{x}_{uv} + E\mathbf{x}_{vv}}{2(EG - F^2)}$$

instead of the scalar function H. We are interested only in the normal component of $\mathbf{P}$; it is found by subtracting the components of $\mathbf{P}$ that are tangent to the patch. This leads to the following definition.

Definition. *Let* $\mathbf{x}: \mathcal{U} \longrightarrow \mathbb{R}^n$ *be a patch. The* **mean curvature vector field H** *of* $\mathbf{x}$ *is defined to be the component of*

$$\frac{G\mathbf{x}_{uu} - 2F\mathbf{x}_{uv} + E\mathbf{x}_{vv}}{2(EG - F^2)}$$

that is perpendicular to $\mathbf{x}_u$ *and* $\mathbf{x}_v$.

The explicit formula for **H** is complicated.

Lemma 25.16. *The mean curvature vector field* **H** *of a regular patch* $\mathbf{x}:\mathcal{U} \longrightarrow \mathbb{R}^n$ *is given by*

$$(25.26) \quad \mathbf{H} = \frac{G\mathbf{x}_{uu} - 2F\mathbf{x}_{uv} + E\mathbf{x}_{vv}}{2(EG - F^2)}$$

$$+ \frac{1}{2(EG - F^2)^2} \left\{ \left(\frac{G^2 E_u}{2} - \frac{FGE_v}{2} + EGF_v \right. \right.$$

$$+ \left(F^2 - \frac{EG}{2} \right) G_u - FGF_u - \left. \frac{EFG_v}{2} \right) \mathbf{x}_u$$

$$+ \left(\frac{E^2 G_v}{2} - \frac{FEG_u}{2} + EGF_u \right.$$

$$+ \left(F^2 - \frac{EG}{2} \right) E_v - FEF_v - \left. \frac{GFE_u}{2} \right) \mathbf{x}_v \bigg\}.$$

Proof. Let $\mathbf{Q} = G\mathbf{x}_{uu} - 2F\mathbf{x}_{uv} + E\mathbf{x}_{vv}$; there exist unique functions a, b and c such that

$$\mathbf{Q} = a\mathbf{x}_u + 2b\mathbf{x}_v + c\mathbf{U}.$$

We compute a and b by finding the scalar product of $\mathbf{Q}$ with $\mathbf{x}_u$ and $\mathbf{x}_v$. Thus

$$(25.27) \quad aE + bF = \mathbf{Q} \cdot \mathbf{x}_u = G\mathbf{x}_{uu} \cdot \mathbf{x}_u - 2F\mathbf{x}_{uv} \cdot \mathbf{x}_u + E\mathbf{x}_{vv} \cdot \mathbf{x}_u$$

$$= \frac{GE_u}{2} - FE_v + E\left(F_v - \frac{G_u}{2} \right),$$

and similarly,

$$(25.28) \quad aF + bG = G\left(F_u - \frac{E_v}{2} \right) - FG_u + \frac{EG_v}{2}.$$

When we solve (25.27) and (25.28) for a and b, we find

$$(25.29) \quad a = \frac{1}{EG - F^2} \left\{ \frac{G^2 E_u}{2} - \frac{FGE_v}{2} + EGF_v \right.$$

$$+ \left(F^2 - \frac{EG}{2} \right) G_u - FGF_u - \left. \frac{EFG_v}{2} \right\}.$$

and

$$(25.30) \quad b = \frac{1}{EG - F^2} \left\{ \frac{E^2 G_v}{2} - \frac{FEG_u}{2} + EGF_u \right.$$

$$+ \left(F^2 - \frac{EG}{2} \right) E_v - FEF_v - \left. \frac{GFE_u}{2} \right\}.$$

Since
$$\mathbf{P} = \frac{\mathbf{Q}}{2(EG - F^2)},$$
it follows from (25.29) and (25.30) that the component of **P** tangent to **x** is

$$\frac{1}{2(EG-F^2)^2}\Bigg\{ \left(\frac{G^2 E_u}{2} - \frac{FGE_v}{2} + EGF_v \right)$$
$$+ \left(F^2 - \frac{EG}{2} \right) G_u - FGF_u - \frac{EFG_v}{2} \Bigg) \mathbf{x}_u$$
$$+ \left(\frac{E^2 G_v}{2} - \frac{FEG_u}{2} + EGF_u \right)$$
$$+ \left(F^2 - \frac{EG}{2} \right) E_v - FEF_v - \frac{GFE_u}{2} \Bigg) \mathbf{x}_u \Bigg\}.$$

Hence we obtain (25.26). ∎

25.7 Exercises

1. Use *Mathematica* to find the curvature of the following metrics:

   (a) $ds^2 = (u^m + v^n)du^2 + dv^2$;

   (b) $ds^2 = \dfrac{dv^2}{\left(1 - \dfrac{2m}{v} + f(v)\right)} - \left(1 - \dfrac{2m}{v} + f(v)\right)du^2.$

2. A generalization of the flat torus is defined in *Mathematica* by

   ```
   flattorus[a_,b_,c_,d_][u_,v_]:=
       {a Cos[u],b Sin[u],c Cos[v],d Sin[v]}
   ```

   Use *Mathematica* to show that the coefficients of the induced metric from $\mathbb{R}^4$ are nonconstant, but still the Gaussian curvature is zero.

3. A patch of the form
$$ds^2 = \big(f(u) + g(v)\big)(du^2 + dv^2)$$

is called a **Liouville patch**. Compute the Gaussian curvature and Christoffel symbols of a Liouville patch.

4. An embedding of a Klein bottle into $\mathbb{R}^4$ is given by

$$\mathbf{klein4}[a,b](u,v) = \big((a+b\cos v)\cos u, (a+b\cos v)\sin u,$$
$$b\sin(v)\cos(u/2), b\sin v \sin(u/2)\big),$$

or

```
klein4[a_,b_][u_,v_]:=
    {(a + b Cos[v])Cos[u],(a + b Cos[v])Sin[u],
     b Sin[v]Cos[u/2],b Sin[v]Sin[u/2]}
```

Show that **klein4**[a, b] has nonzero Gaussian curvature. Then use **ortho4d** (described in Appendix D) to display four projections of **klein4**[8, 3] into $\mathbb{R}^3$.

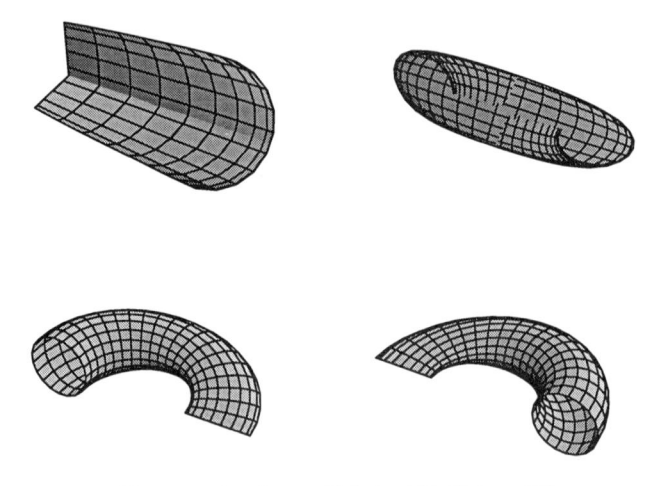

Four projections of klein4[8, 3] into $\mathbb{R}^3$

5. Show that the mean curvature vector field **H** can be defined in *Mathematica* by

25.7 Exercises

```
meancurvaturevf[x_][u_,v_]:=
    Module[{h0,ar,h1,h2,h3},
    h0=gg[x][uu,vv]D[x[uu,vv],uu,uu] -
        2ff[x][uu,vv]D[x[uu,vv],uu,vv] +
        ee[x][uu,vv]D[x[uu,vv],vv,vv];
    ar=Simplify[(ee[x][uu,vv]gg[x][uu,vv] -
            ff[x][uu,vv]^2)];
    h1 = (h0.D[x[uu,vv],uu])gg[x][uu,vv] -
        (h0.D[x[uu,vv],vv])ff[x][uu,vv];
    h2 = (h0.D[x[uu,vv],vv])ee[x][uu,vv] -
        (h0.D[x[uu,vv],uu])ff[x][uu,vv];
    Simplify[h0/(2ar) - h1 D[x[uu,vv],uu]/(2ar^2) -
        h2 D[x[uu,vv],vv]/(2ar^2) /. {uu->u,vv->v}]]
```

Compute the mean curvature vector fields of **flattorus** and **klein4**.

6. A **Tchebyshef**[6] net is a patch on an abstract surface $\mathcal{M}$ for which $E = G = 1$ and $F = \cos\theta$. Use **brioschicurvature** to show that the Gaussian curvature of $\mathcal{M}$ is given by

$$K = \frac{-\theta_{uv}}{\sin\theta}.$$

7. Prove the following result:

   Theorem 25.17. (Gauss Equations for abstract surfaces) *Let* **x** *be a regular patch on an abstract surface* $\mathcal{M}$. *Then*

   (25.31) $\begin{cases} \nabla_{\mathbf{x}_u}(\mathbf{x}_u) = \Gamma^1_{11}\mathbf{x}_u + \Gamma^2_{11}\mathbf{x}_v, \\ \nabla_{\mathbf{x}_u}(\mathbf{x}_v) = \nabla_{\mathbf{x}_v}(\mathbf{x}_u) = \Gamma^1_{12}\mathbf{x}_u + \Gamma^2_{12}\mathbf{x}_v, \\ \nabla_{\mathbf{x}_v}(\mathbf{x}_u) = \Gamma^1_{22}\mathbf{x}_u + \Gamma^2_{22}\mathbf{x}_v. \end{cases}$

8. Prove Lemma 25.14.

---

[6] Pafnuty Lvovich Tchebyshef (1821-1894). A leading Russian mathematician of his time. He worked in probability theory, number theory and differential geometry.

9. The **tangent developable** of a curve $\alpha\colon (a,b) \longrightarrow \mathbb{R}^n$ is a patch in $\mathbb{R}^n$. It is defined by the same formula as that of a curve in $\mathbb{R}^3$, namely

$$\mathbf{x}(u,v) = \alpha(u) + v\,\alpha'(u).$$

   (See Sections 19.2 and 19.3.) Use Brioschi's formula to show that the Gaussian curvature of the tangent developable of α vanishes, and compute the mean curvature vector.

10. A **Veronese**[7] **surface** in $\mathbb{R}^5$ is defined by

$$\mathbf{veronese}[a](u,v) = a(u, v, u^2, u\,v, v^2).$$

    Compute the metric, Gaussian curvature and mean curvature vector of **veronese**[a]. Plot the Gaussian curvature. What is the relation of a Veronese surface to a Whitney umbrella?

11. The following miniprogram can be used to display the metric of any patch in $\mathbb{R}^n$ or any abstract patch:

```
ds2[x_][u_,v_][du_,dv_]:= ee[x][u,v]**du^2 +
    ff[x][u,v]**(du*dv) + gg[x][u,v]**dv^2 /.
    {0**_->0,_**0->0}
```

    Use it to display the metric of each of the following patches: **sphere[a]**, **hyperbolicparaboloid**, **monkeysaddle**, **enneper**, **catenoid[a]**, **flattorus[a,b,c,d]**, **klein4[a,b]**, **veronese[a]**, **schwarzschildplane[m,f]**, **liouville[f,g]**, and **poincare[m,n,lambda]**.

12. Christoffel symbols can be computed with **christoffel** in the same way as **brioschicurvature** was used to compute Gaussian curvature. The command **Table** can be combined with **christoffel** to compute all the Christoffel symbols of a given metric or patch in $\mathbb{R}^n$. Compute the Christoffel symbols of the metrics and patches of exercise 11.

---

[7]Giuseppe Veronese (1854-1917). Famous Italian algebraic geometer, professor at the University of Padua.

26

GEODESICS ON SURFACES

Roughly speaking, a *geodesic* on a surface $\mathcal{M}$ is the shortest curve among all piecewise-differentiable curves on $\mathcal{M}$ connecting two points. This notion makes sense not only for surfaces in $\mathbb{R}^3$, but also for abstract surfaces, and more generally for Riemannian manifolds. Geodesics were first studied by Johann Bernoulli[1] in 1697; the name geodesic is due to Liouville in 1844.

In Section 26.1 we determine the differential equations for a geodesic. Clairaut patches, for which the geodesic equations are partially solvable, are discussed in Section 26.2, and several examples are given. In Section 26.3 we show how to solve geodesic equations numerically, and we also discuss plotting of geodesics. The *exponential map*, which naturally maps a tangent space to a surface onto the surface is defined in Section 26.4; then it is used to prove Lemma 26.26, known as the *Gauss Lemma*. Length minimizing properties of geodesics are discussed in Section 26.5, and it is shown that an abstract surface is a *metric space* in Section 26.6.

---

[1] Johann Bernoulli (1667-1748) occupied the chair of mathematics at Groningen from 1695 to 1705, and at Basel, where he succeeded his elder brother Jakob, from 1705 to 1748. Although Johann was tutored by Jakob, he became involved in controversies with him, and also with his own son Daniel (1700-1782). Discoveries of Johann Bernoulli include the exponential calculus, the treatment of trigonometry as a branch of analysis, the determination of orthogonal trajectories and the solution of the brachistochrone problem. Other mathematician sons of Johann were Nicolaus (1695-1726) and Johann (II) (1710-1790).

26.1 The Geodesic Equations

To motivate the definition of geodesic, let us first consider the case of a surface $M \subset \mathbb{R}^n$. Since a curve α in M is also a curve in $\mathbb{R}^n$, there are two competing accelerations of α, namely, the second derivative α'' and the covariant derivative $\nabla_{\alpha'}\alpha'$, where ∇ denotes the Riemannian connection of M defined in Section 24.1. The two accelerations are related as follows.

Lemma 26.1. *Let M be a surface in $\mathbb{R}^n$ and let $\alpha\colon (a,b) \longrightarrow M$ be a curve. Then the tangential component of the $\mathbb{R}^n$-acceleration α'' of α is the abstract surface acceleration $\nabla_{\alpha'}\alpha'$ of α.*

Proof. Let X and Y be vector fields tangent to M. It follows from (24.22), page 566, that $\nabla_X Y$ is the tangential component of $D_X Y$, where D denotes the natural connection of $\mathbb{R}^n$. In particular, $\nabla_{\alpha'}\alpha'$ is the tangential component of $\alpha'' = D_{\alpha'}\alpha'$. ∎

More can be said in the case $n = 3$.

Lemma 26.2. *Let M be an orientable surface in $\mathbb{R}^3$ and choose a unit normal vector field $\mathbf{U}$ to M. If $\alpha\colon (a,b) \longrightarrow M$ is a curve, then*

$$(26.1) \qquad \alpha'' = \nabla_{\alpha'}\alpha' + \big(S(\alpha') \cdot \alpha'\big)\mathbf{U},$$

where S denotes the shape operator of M defined by $\mathbf{U}$.

Proof. The decomposition of α'' into its tangential and normal components is

$$\alpha'' = \nabla_{\alpha'}\alpha' + (\alpha'' \cdot \mathbf{U})\mathbf{U}.$$

But from Lemma 26.1 and Lemma 16.3, page 361, we know that $\alpha'' \cdot \mathbf{U} = S(\alpha') \cdot \alpha'$, and so we get (26.1). ∎

Before giving the definition of geodesic on an abstract surface, we give the definition of geodesic on a surface in $\mathbb{R}^3$.

Definition. Let $M \subset \mathbb{R}^3$ be a surface and $\alpha\colon (a,b) \longrightarrow M$ a curve. We say that α is a **geodesic** on M provided that the tangential component $\nabla_{\alpha'}\alpha'$ of the acceleration α'' vanishes.

26.1 The Geodesic Equations

Thus the definition of geodesic is in some sense complementary to that of asymptotic curve. Recall (see page 364) that $\alpha \colon (a,b) \longrightarrow \mathcal{M} \subset \mathbb{R}^3$ is an asymptotic curve provided the normal component of α'' vanishes. Of course, a curve in $\mathcal{M}$ is simultaneously a geodesic and an asymptotic curve if and only if it is a straight line in $\mathbb{R}^3$. Moreover:

Lemma 26.3. *Any straight line in $\mathbb{R}^3$ that is contained in a surface $\mathcal{M}$ is automatically a geodesic.*

Next, we determine the differential equations that any geodesic on a surface in $\mathbb{R}^3$ must satisfy.

Lemma 26.4. *Let $\mathcal{M} \subset \mathbb{R}^3$ be a surface parametrized by a regular patch $\mathbf{x} \colon \mathcal{U} \longrightarrow \mathbb{R}^3$, where $\mathcal{U} \subset \mathbb{R}^2$. Then the geodesics on $\mathcal{M}$ are determined by two second-order differential equations:*

$$(26.2) \quad \begin{cases} u'' + \Gamma^1_{11} u'^2 + 2\Gamma^1_{12} u'v' + \Gamma^1_{22} v'^2 = 0, \\ v'' + \Gamma^2_{11} u'^2 + 2\Gamma^2_{12} u'v' + \Gamma^2_{22} v'^2 = 0, \end{cases}$$

where the Γ^i_{jk} are Christoffel symbols of $\mathbf{x}$.

Proof. Let $\alpha \colon (a,b) \longrightarrow \mathcal{M}$ be a curve. We can write

$$(26.3) \qquad \alpha(t) = \mathbf{x}\bigl(u(t), v(t)\bigr),$$

where $u, v \colon (a,b) \longrightarrow \mathbb{R}$ are differentiable functions. When we differentiate (26.3) twice using the chain rule, we get

$$(26.4) \qquad \alpha'' = \mathbf{x}_{uu} u'^2 + \mathbf{x}_u u'' + 2\mathbf{x}_{uv} u'v' + \mathbf{x}_{vv} v'^2 + \mathbf{x}_v v''.$$

Then (22.14), page 511, and (26.4) imply that

$$(26.5) \qquad \begin{aligned} \alpha''(t) = {} & \bigl(u'' + \Gamma^1_{11} u'^2 + 2\Gamma^1_{12} u'v' + \Gamma^1_{22} v'^2\bigr) \mathbf{x}_u \\ & + \bigl(v'' + \Gamma^2_{11} u'^2 + 2\Gamma^2_{12} u'v' + \Gamma^2_{22} v'^2\bigr) \mathbf{x}_v \\ & + \bigl(e\, u'^2 + 2f\, u'v' + g\, v'^2\bigr) \mathbf{U}, \end{aligned}$$

where

$$\mathbf{U} = \frac{\mathbf{x}_u \times \mathbf{x}_v}{\|\mathbf{x}_u \times \mathbf{x}_v\|}$$

is the unit normal vector field to $\mathcal{M}$. In order that α be a geodesic, the coefficients of $\mathbf{x}_u$ and $\mathbf{x}_v$ in (26.5) must each vanish. Hence we get (26.2). ∎

To define the notion of geodesic for an abstract surface, we recall that the definition of Riemannian connection (page 566), and also the definitions (25.4), page 576, of the Christoffel symbols, make sense for an abstract surface M. This is because each Christoffel symbol is defined in terms of the coefficients of the first fundamental form of M, and the Riemannian connection can be expressed in terms of the Christoffel symbols (see Section 24.3).

Definition. *Let M be an abstract surface and $\alpha\colon (a,b) \longrightarrow M$ a curve. Then α is a **geodesic** provided $\nabla_{\alpha'}\alpha' = 0$, where ∇ denotes the Riemannian connection of M.*

Here is a simple but much-used property of a geodesic.

Lemma 26.5. *A geodesic α on an abstract surface M has constant speed.*

Proof. We compute
$$\frac{d}{dt}\|\alpha'\|^2 = 2\langle \nabla_{\alpha'}\alpha', \alpha'\rangle = 0. \quad \blacksquare$$

We show next that the differential equations for a geodesic on an abstract surface are exactly the same as those for a surface in $\mathbb{R}^3$.

Lemma 26.6. *Let M be an abstract surface, and let $\mathbf{x}\colon \mathcal{U} \longrightarrow M$ be a regular patch that parametrizes M. Then the geodesics on M are determined by two second-order differential equations (26.2), where the Γ^i_{jk} are the Christoffel symbols of $\mathbf{x}$.*

Proof. Let $\alpha\colon (a,b) \longrightarrow M$ be a curve whose trace is contained in $\mathbf{x}(\mathcal{U})$, and let $X \in \mathfrak{X}(M)$ be a vector field such that $X_{\alpha(t)} = \alpha'(t)$ for $a < t < b$. Write $\mathbf{x}^{-1} = (x_1, x_2)$; then $x_1 = u \circ \mathbf{x}^{-1}$ and $x_2 = v \circ \mathbf{x}^{-1}$, where u and v are the natural coordinate functions of $\mathbb{R}^2$. By Lemma 23.21, page 544, we have $x_1 \circ \alpha = u$ and $x_2 \circ \alpha = v$, so that

(26.6) $\qquad X[x_1]\bigl(\alpha(t)\bigr) = X_{\alpha(t)}[x_1] = \alpha'(t)[x_1] = (x_1 \circ \alpha)'(t) = u'(t).$

Moreover,

(26.7) $\qquad X^2[x_1]\bigl(\alpha(t)\bigr) = X_{\alpha(t)}[X[x_1]] = \alpha'(t)[X[x_1]]$

$\qquad\qquad\qquad\quad = \bigl(X[x_1] \circ \alpha\bigr)'(t) = (x_1 \circ \alpha)''(t) = u''(t).$

Similar equations hold for $X[x_2]\bigl(\alpha(t)\bigr)$ and $X^2[x_1]\bigl(\alpha(t)\bigr)$. Lemma 23.16, page 540, implies that

(26.8) $\qquad \dfrac{\partial}{\partial x_1} = \mathbf{x}_u \circ \mathbf{x}^{-1} \qquad \text{and} \qquad \dfrac{\partial}{\partial x_2} = \mathbf{x}_v \circ \mathbf{x}^{-1}.$

26.1 The Geodesic Equations

As a special case of (24.38), page 571, we have

$$(26.9) \qquad \nabla_X X = \sum_{k=1}^{2} \left\{ X^2[x_k] + \sum_{ij=1}^{2} X[x_i] X[x_j] \Gamma_{ij}^k \right\} \frac{\partial}{\partial x_k}.$$

From (26.6)—(26.9) we obtain

$$(26.10) \quad (\nabla_{\alpha'} \alpha')(t) = (\nabla_X X)(\alpha(t))$$

$$= \left(u'' + \Gamma_{11}^1 u'^2 + 2\Gamma_{12}^1 u'v' + \Gamma_{22}^1 v'^2 \right) \mathbf{x}_u + \left(v'' + \Gamma_{11}^2 u'^2 + 2\Gamma_{12}^2 u'v' + \Gamma_{22}^2 v'^2 \right) \mathbf{x}_v.$$

Hence the vanishing of $\nabla_{\alpha'}\alpha'$ is equivalent to (26.2). ∎

Now we can obtain another important property of geodesics.

Corollary 26.7. *Isometries and local isometries preserve geodesics.*

Proof. Since the Christoffel symbols are expressed in terms of E, F and G, and since Corollary 15.7 holds for abstract surfaces (see page 577), local isometries preserve the geodesic equations (26.2). ∎

Next, we prove that from each point on an abstract surface there is a geodesic in every direction. More precisely:

Theorem 26.8. *Let $\mathbf{p}$ be a point on an abstract surface $\mathcal{M}$ and $\mathbf{v}_\mathbf{p}$ a tangent vector to $\mathcal{M}$ at $\mathbf{p}$. Then there exists a geodesic γ parametrized on some interval containing 0 such that $\gamma(0) = \mathbf{p}$ and $\gamma'(0) = \mathbf{v}_\mathbf{p}$. Furthermore, any two geodesics with these initial conditions coincide on any interval containing 0 on which they are defined.*

Proof. Let $\mathbf{x}: \mathcal{U} \longrightarrow \mathcal{M}$ be a regular patch with $\mathbf{x}(0,0) = \mathbf{p}$. Then (26.2) is the system of differential equations determining the geodesics on trace of $\mathbf{x}$. Write $\gamma(t) = \mathbf{x}(u(t), v(t))$ and $\mathbf{v}_\mathbf{p} = (v_1, v_2)$. The initial conditions for γ translate into

$$(26.11) \qquad \begin{cases} u(0) = 0, \\ v(0) = 0, \end{cases} \quad \text{and} \quad \begin{cases} u'(0) = v_1, \\ v'(0) = v_2. \end{cases}$$

From the theory of ordinary differential equations, we know that the system (26.2) together with the four initial conditions (26.11) has a unique solution on some interval containing 0. Hence the theorem follows. ∎

Theorem 26.8 gives no information concerning the size of the interval on which the geodesic satisfying given initial conditions is defined. Let γ_1 and γ_2 be geodesics satisfying the same initial conditions. Since the two geodesics must coincide on some interval, it is clear that there is a geodesic γ_3 whose domain of definition contains those of both γ_1 and γ_2 and coincides wherever possible with the two geodesics. Applying this consistency result, we obtain a single **maximal geodesic**. The domain of a maximal geodesic may or may not be the whole real line, but in any case we can make the following definition.

Definition. *An abstract surface M is **geodesically complete** provided that every maximal geodesic is defined on all of $\mathbb{R}$.*

Trivially, a constant curve is a maximal geodesic. Excluding this case, we see that geodesic completeness means that every maximal geodesic is infinitely long in both directions. It is easy to check that spheres and planes are geodesically complete. More generally, any compact surface is geodesically complete. A famous theorem of Hopf and Rinow, which we shall discuss in Section 26.6, states that any Riemannian manifold is geodesically complete if and only if it is complete as a metric space.

It is easy to find examples of geodesically incomplete surfaces. For example, $\mathbb{R}^2 \setminus (0,0)$ cannot be geodesically complete, because the geodesic $t \longmapsto (t,0)$ cannot be extended to a map of the whole real line into $\mathbb{R}^2 \setminus (0,0)$. A more complicated example of a geodesically incomplete surface is the pseudosphere.

It is important to distinguish between a geodesic and the trace of a geodesic, because not every reparametrization of a geodesic has constant speed. Therefore, we introduce the following notion:

Definition. *A curve α in a surface M is called a **pregeodesic** provided there is a reparametrization $\alpha \circ h$ of α such that $\alpha \circ h$ is a geodesic.*

Here are several characterizations of pregeodesics. See also exercise 1.

Lemma 26.9. *Let M be an abstract surface and let $\alpha \colon (a,b) \longrightarrow M$ be a regular curve. Then the following conditions are equivalent:*

(i) *α is a pregeodesic;*

(ii) *there exists a function $f \colon (a,b) \longrightarrow \mathbb{R}$ such that*

$$(\nabla_{\alpha'}\alpha')(t) = f(t)\alpha'(t)$$

for $a < t < b$;

(iii) *the geodesic curvature of α vanishes.*

26.1 The Geodesic Equations

Proof. Suppose $\beta = \alpha \circ h$ is a geodesic. Without loss of generality, β' never vanishes. Then $\beta' = (\alpha' \circ h)h'$, so that h' never vanishes. Furthermore, it follows from (24.14) that

$$0 = \nabla_{\beta'}\beta' = h'^2(\nabla_{\alpha'}\alpha') \circ h + h''\alpha' \circ h,$$

so that

$$\nabla_{\alpha'}\alpha' = -\left(\frac{h''}{h'^2} \circ h^{-1}\right)\alpha';$$

thus we can take $f = -(h''/h'^2) \circ h^{-1}$.

Conversely, assume that $\nabla_{\alpha'}\alpha' = f\alpha'$ and let h be a nonzero solution of the differential equation $h'' + h'^2(f \circ h) = 0$. Put $\beta = \alpha \circ h$; then

$$\nabla_{\beta'}\beta' = h'^2(\nabla_{\alpha'}\alpha') \circ h + h''\alpha' \circ h = \big((f \circ h)h'^2 + h''\big)\alpha' \circ h = 0.$$

This shows the equivalence of (i) and (ii). The equivalence of (i) and (iii) is obvious from the definition of geodesic curvature (see equation (25.12), page 583). ∎

Before proceeding further with the general theory, we find geodesics in three important cases.

Meridians on a Surface of Revolution

Finding *all* of the geodesics on a general surface of revolution is complicated. However, we can find some of them with no computation.

Theorem 26.10. *Any meridian of a surface of revolution $\mathcal{M} \subset \mathbb{R}^3$ can be parametrized as a geodesic.*

Proof. Let β be a unit-speed parametrization of a meridian of $\mathcal{M}$, and let Π be a plane passing through β and the axis of revolution. Since β is a plane curve in Π, the unit normal $\mathbf{N} = J\beta'$ also lies in Π. Theorem 20.12, page 479, implies that the plane Π meets $\mathcal{M}$ perpendicularly, and so $\mathbf{N}$ is also perpendicular to $\mathcal{M}$. But then $\beta'' = \kappa_2\mathbf{N}$ is perpendicular to $\mathcal{M}$. Hence by definition β is a geodesic. ∎

For a generalization of Theorem 26.10, see exercise 2.

Geodesics on a Sphere

First, we prove a general result relating the shape operator of a surface in $\mathbb{R}^3$ and the Frenet frame of a curve on the surface.

Lemma 26.11. *Let $\mathcal{M}$ be an orientable surface in $\mathbb{R}^3$, and let $\alpha\colon(a,b) \longrightarrow \mathcal{M}$ be a unit-speed geodesic on $\mathcal{M}$ with nonzero curvature. Denote by $\{\mathbf{T},\mathbf{N},\mathbf{B}\}$ the Frenet frame field of α and by κ and τ the curvature and torsion of α. Then it is possible to choose a unit normal vector field $\mathbf{U}$ to $\mathcal{M}$ such that*

(26.12) $$S(\mathbf{T}) = \kappa\,\mathbf{T} - \tau\,\mathbf{B},$$

where S denotes the shape operator of $\mathcal{M}$ with respect to $\mathbf{U}$.

Proof. Since α has nonzero curvature, the vector field $\mathbf{N}$ is defined on all of (a,b). The assumption that α is a geodesic implies that $\mathbf{N} = \alpha''/\kappa 2$ is everywhere perpendicular to $\mathcal{M}$. Thus it is possible to choose $\mathbf{U}$ so that

$$\mathbf{N}(t) = \mathbf{U}(\alpha(t))$$

for $a < t < b$. By the definition of the shape operator and by the Frenet formulas (8.10), page 186, we have

$$S(\mathbf{T}) = -\frac{d}{dt}\mathbf{U}(\alpha(t)) = -\mathbf{N}' = \kappa\mathbf{T} - \tau\mathbf{B}. \blacksquare$$

Now we use Lemma 26.11 to determine the geodesics of a sphere.

Lemma 26.12. *The geodesics on a sphere of radius $c > 0$ are parts of great circles.*

Proof. Let β be a geodesic on a sphere of radius c in $\mathbb{R}^3$. By part (i) of Theorem 10.11, page 231, the curvature of β is nonzero, so that the Frenet frame $\{\mathbf{T},\mathbf{N},\mathbf{B}\}$ of β is well-defined. Since every point of a sphere is an umbilic point, the shape operator of the sphere satisfies (with a proper choice of the unit normal)

(26.13) $$S(\mathbf{T}) = \frac{1}{c}\mathbf{T}.$$

It follows from (26.12) and (26.13) that the curvature of β has the constant value $1/c$, and that the torsion vanishes. Moreover, part (vi) of Theorem 10.11 implies that β is part of a circle. Since the radius of the circle is the maximum possible, β must be part of a great circle. $\blacksquare$

Geodesics on a Generalized Cylinder

Geodesics on a generalized cylinder (see page 439) are characterized by the following theorem.

26.1 The Geodesic Equations

Theorem 26.13. *Let $\mathcal{M}$ be the generalized cylinder whose base curve $\boldsymbol{\alpha}$ is a plane curve perpendicular to the rulings of $\mathcal{M}$, and denote by $\mathbf{u}$ a unit vector tangent to $\mathcal{M}$ and perpendicular to $\boldsymbol{\alpha}$, so that $\mathcal{M}$ is parametrized as*

$$\mathbf{x}(u, v) = \boldsymbol{\alpha}(u) + v\,\mathbf{u}.$$

Then the geodesics of $\mathcal{M}$ are precisely those constant speed curves on $\mathcal{M}$ that have constant slope with respect to $\mathbf{u}$.

Proof. Let $\boldsymbol{\gamma}: (a, b) \longrightarrow \mathcal{M}$ be a curve on the generalized cylinder; we can write

(26.14) $$\boldsymbol{\gamma}(t) = \boldsymbol{\alpha}\bigl(u(t)\bigr) + v(t)\mathbf{u}.$$

for some functions $t \longmapsto u(t)$ and $t \longmapsto v(t)$. Then

(26.15) $$\boldsymbol{\gamma}' = \boldsymbol{\alpha}'u' + v'\mathbf{u} \quad \text{and} \quad \boldsymbol{\gamma}'' = \boldsymbol{\alpha}''u'^2 + \boldsymbol{\alpha}'u'' + v''\mathbf{u}.$$

Now assume that $\boldsymbol{\gamma}$ is a geodesic. Since any normal vector to $\mathcal{M}$ is always perpendicular to $\mathbf{u}$, it follows from the second equation of (26.15) that $v'' = 0$, so that v' is a constant. Thus $\boldsymbol{\gamma}$ makes a constant angle with $\mathbf{u}$, and so $\boldsymbol{\gamma}$ must have constant slope (see Section 10.5).

Conversely, let $\boldsymbol{\gamma}: (a, b) \longrightarrow \mathcal{M}$ be a constant speed curve that has constant slope with respect to $\mathbf{u}$. Then (26.14) holds, so that $\boldsymbol{\alpha} \circ u$ is the projection of $\boldsymbol{\gamma}$ onto the plane perpendicular to $\mathbf{u}$. Then v' is constant, so that $\boldsymbol{\gamma}''$ is perpendicular to $\mathbf{u}$. Moreover, it follows from (26.14) that $\boldsymbol{\gamma}'' = (\boldsymbol{\alpha} \circ u)''$. Also, Lemma 10.18, page 235, implies that $\boldsymbol{\alpha} \circ u$ has constant speed. Hence $\boldsymbol{\gamma}''$ is perpendicular to both $\mathbf{x}_u = \boldsymbol{\alpha}'$ and $\mathbf{u} = \mathbf{x}_v$, proving that $\boldsymbol{\gamma}$ is a geodesic. ∎

For example, the geodesics on a circular or elliptical cylinder are circles lines or helices. Here is a more general example.

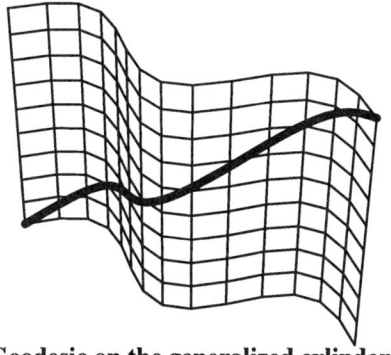

Geodesic on the generalized cylinder
$$(u, v) \longmapsto (u, \sin u, v)$$

26.2 Clairaut Patches

In general, the geodesic equations (26.2) are difficult to solve explicitly. However, there are two important cases where their solution can be reduced to computing integrals.

Definition. Let $\mathcal{M}$ be an abstract surface with metric
$$ds^2 = E\,du^2 + 2F\,du\,dv + G\,dv^2.$$

(i) A u-**Clairaut patch** on $\mathcal{M}$ is a patch $\mathbf{x}\colon\mathcal{U} \longrightarrow \mathcal{M}$ for which
$$E_v = G_v = F = 0.$$

(ii) A v-**Clairaut patch** on $\mathcal{M}$ is a patch $\mathbf{x}\colon\mathcal{U} \longrightarrow \mathcal{M}$ for which
$$E_u = G_u = F = 0.$$

We work mainly with v-Clairaut patches, leaving the corresponding results for u-Clairaut patches to the reader. It is easy to see that the identity map on the upper half-plane equipped with the Poincaré metric (25.7) is an example of a v-Clairaut patch; we shall see other examples shortly. The Christoffel symbols for a v-Clairaut patch are considerably simpler than those of a general patch. The following lemma is an immediate consequence of the definition (25.4) of Christoffel symbols for an abstract surface.

Lemma 26.14. *For a v-Clairaut patch with $ds^2 = E\,du^2 + G\,dv^2$ we have*

(26.16)
$$\begin{cases} \Gamma^1_{11} = 0, & \Gamma^2_{11} = \dfrac{-E_v}{2G}, \\[4pt] \Gamma^1_{12} = \dfrac{E_v}{2E}, & \Gamma^2_{12} = 0, \\[4pt] \Gamma^1_{22} = 0, & \Gamma^2_{22} = \dfrac{G_v}{2G}, \end{cases}$$

so that the geodesic equations (26.2) reduce to

(26.17)
$$\begin{cases} u'' + \dfrac{E_v}{E} u'v' = 0, \\[4pt] v'' - \dfrac{E_v}{2G} u'^2 + \dfrac{G_v}{2G} v'^2 = 0. \end{cases}$$

26.2 Clairaut Patches

Some geodesics on a Clairaut patch are easy to determine.

Lemma 26.15. *Let* $\mathbf{x}:\mathcal{U} \longrightarrow \mathcal{M}$ *be a v-Clairaut patch. Then*:

(i) *any v-parameter curve is a pregeodesic*;

(ii) *a u-parameter curve* $u \longmapsto \mathbf{x}(u, v_0)$ *is a geodesic if and only if* $E_v(u, v_0)$ *vanishes along* $u \longmapsto \mathbf{x}(u, v_0)$.

Proof. To prove (i), it suffices by Lemma 26.9 to show that $\nabla_{\mathbf{x}_v}\mathbf{x}_v$ is a scalar multiple of $\mathbf{x}_v$ along $v \longmapsto \mathbf{x}(u, v)$. Since $\mathbf{x}_u$ and $\mathbf{x}_v$ are orthogonal, $\nabla_{\mathbf{x}_v}\mathbf{x}_v$ will be a multiple of $\mathbf{x}_v$ provided $\langle \nabla_{\mathbf{x}_v}\mathbf{x}_v, \mathbf{x}_u \rangle = 0$. But in general,

$$\langle \nabla_{\mathbf{x}_v}\mathbf{x}_v, \mathbf{x}_u \rangle = \langle \mathbf{x}_u, \mathbf{x}_v \rangle_v - \langle \nabla_{\mathbf{x}_v}\mathbf{x}_u, \mathbf{x}_v \rangle = F_v - \frac{1}{2}G_u = 0.$$

In particular, $\langle \nabla_{\mathbf{x}_v}\mathbf{x}_v, \mathbf{x}_u \rangle$ vanishes along $v \longmapsto \mathbf{x}(u, v)$.

To prove (ii), we first observe that if $u \longmapsto \mathbf{x}(u, v_0)$ is a geodesic, then

$$E_v(u, v_0) = 2\langle \nabla_{\mathbf{x}_u}\mathbf{x}_v, \mathbf{x}_u \rangle(u, v_0) = 2F_u(u, v_0) - 2\langle \nabla_{\mathbf{x}_u}\mathbf{x}_u, \mathbf{x}_v \rangle(u, v_0) = 0.$$

Conversely, if $E_v(u, v_0) = 0$, then

$$\langle \nabla_{\mathbf{x}_u}\mathbf{x}_u, \mathbf{x}_v \rangle = \langle \mathbf{x}_u, \mathbf{x}_v \rangle_u - \langle \mathbf{x}_u, \nabla_{\mathbf{x}_u}\mathbf{x}_v \rangle = F_u - \frac{1}{2}E_v = -\frac{1}{2}E_v,$$

so that

$$\langle \nabla_{\mathbf{x}_u}\mathbf{x}_u, \mathbf{x}_v \rangle(u, v_0) = -\frac{1}{2}E_v(u, v_0) = 0.$$

But also,

$$\langle \nabla_{\mathbf{x}_u}\mathbf{x}_u, \mathbf{x}_u \rangle = \frac{1}{2}E_u = 0,$$

so that $u \longmapsto \mathbf{x}(u, v_0)$ is a geodesic. ∎

We turn now to the problem of finding other geodesics on a Clairaut patch. The key result we need is:

Theorem 26.16. (*Clairaut's relation*) *Let* $\mathbf{x}:\mathcal{U} \longrightarrow \mathcal{M}$ *be a v-Clairaut patch, and let* β *be a geodesic whose trace is contained in* $\mathbf{x}(\mathcal{U})$. *Let* θ *be the angle between* β *and* $\mathbf{x}_u$. *Then*

(26.18) $$\sqrt{E}\,\|\beta'\|\cos\theta = E\,u' \quad \text{is constant along } \beta.$$

Proof. We write $\beta(t) = \mathbf{x}(u(t), v(t))$. From the assumption that $E_u = 0$ and the first equation of (26.17), it follows that

$$(E u')' = E' u' + E u'' = E_v v' u' + E u'' = 0.$$

Hence there is a constant c such that

(26.19) $$E u' = c.$$

Furthermore,

(26.20) $$\|\beta'\| \cos \theta = \frac{\langle \mathbf{x}_u, \beta' \rangle}{\|\mathbf{x}_u\|} = \frac{\langle \mathbf{x}_u, \mathbf{x}_u u' + \mathbf{x}_v v' \rangle}{\|\mathbf{x}_u\|} = \|\mathbf{x}_u\| u' = \sqrt{E}\, u'.$$

Then (26.18) follows from (26.19) and (26.20). ∎

Definition. *We call the constant c given by (26.19) the **slant** of the geodesic β in the v-Clairaut patch $\mathbf{x}$. Then the angle between β and $\mathbf{x}_u$ is given by*

$$\cos \theta = \frac{c}{\|\beta'\|\sqrt{E}}.$$

We next obtain information about geodesics on a Clairaut patch other than those covered by Lemma 26.15.

Lemma 26.17. *Let $\mathbf{x}: \mathcal{U} \longrightarrow \mathcal{M}$ be a v-Clairaut patch, and let $\beta: (a,b) \longrightarrow \mathbb{R}$ be a unit-speed curve whose trace is contained in $\mathbf{x}(\mathcal{U})$. Write $\beta(t) = \mathbf{x}(u(t), v(t))$. If β is a geodesic, then there is a constant c such that*

(26.21) $$\begin{cases} u' = \dfrac{c}{E}, \\ v' = \pm \dfrac{\sqrt{E - c^2}}{\sqrt{E G}}. \end{cases}$$

Conversely, if (26.21) holds, and if $v'(t) \neq 0$ for $a < t < b$, or if $v'(t) = 0$ for $a < t < b$, then β is a geodesic.

Proof. Suppose that β is a unit-speed geodesic. The first equation of (26.21) follows from (26.19). Furthermore,

$$1 = \langle \beta', \beta' \rangle = E u'^2 + G v'^2 = \frac{c^2}{E} + G v'^2, \quad \text{so that} \quad v'^2 = \frac{1}{G}\left(1 - \frac{c^2}{E}\right) = \frac{E - c^2}{E G}.$$

26.2 Clairaut Patches

Thus the second equation of (26.21) also holds.

Conversely, suppose that (26.21) holds. The first equation of (26.21) implies that
$$u'' = \left(\frac{c}{E}\right)' = -\frac{c\,E_v v'}{E^2} = -\frac{E_v u' v'}{E}.$$
Thus we see that the first equation of (26.17) is satisfied. Furthermore, β has unit-speed because
$$\langle \beta', \beta' \rangle = \langle \mathbf{x}_u u' + \mathbf{x}_v v', \mathbf{x}_u u' + \mathbf{x}_v v' \rangle$$
$$= E u'^2 + G v'^2 = E\left(\frac{c}{E}\right)^2 + G\left(\frac{E-c^2}{EG}\right) = 1.$$

Next, when we differentiate $G v'^2 = 1 - E u'^2$, we obtain

(26.22) $\quad G_v v'^3 + 2G v'' v' = (G v'^2)' = -E_v v' u'^2 - 2E u' u''$
$$= -E_v v' u'^2 + 2E u'\left(\frac{E_v}{E} u' v'\right) = E_v v' u'^2.$$

Then (26.22) implies that the second equation of (26.17) is satisfied on any interval where v' is different from zero. On the other hand, on an interval where v' vanishes identically, the second equation of (26.21) implies that E is a constant on that interval. Again, the second equation of (26.17) is satisfied. ∎

Corollary 26.18. *Let* $\mathbf{x}: \mathcal{U} \longrightarrow \mathcal{M}$ *be a v-Clairaut patch. Then a curve* $\alpha: (a, b) \longrightarrow \mathcal{M}$ *of the form*
$$\alpha(u) = \mathbf{x}(u(v), u)$$
is a pregeodesic if and only if there is a constant c such that

(26.23) $\qquad\qquad \dfrac{du}{dv} = \pm c \sqrt{\dfrac{G}{E(E - c^2)}}.$

Then c is the slant of α with respect to $\mathbf{x}$.

Proof. There exists a unit-speed geodesic β reparametrizing α such that
$$\beta(s) = \beta(u(s)) = \mathbf{x}(u(v(s)), v(s)).$$
Then Lemma 26.17 implies that
$$\frac{du}{ds} = u' = \frac{c}{E} \quad \text{and} \quad \frac{dv}{ds} = \pm\frac{\sqrt{E - c^2}}{\sqrt{EG}}.$$

Hence
$$\frac{du}{dv} = \frac{\frac{du}{ds}}{\frac{dv}{ds}} = \frac{\frac{c}{E}}{\pm\frac{\sqrt{E-c^2}}{\sqrt{EG}}} = \pm c\sqrt{\frac{G}{E(E-c^2)}}.$$

Conversely, if (26.23) holds, we define u' and v' by

$$\frac{u'}{v'} = \pm c\sqrt{\frac{G}{E(E-c^2)}} \quad \text{and} \quad E u'^2 + G v'^2 = 1.$$

An easy calculation then shows that (26.21) holds, and so we get a pregeodesic. ∎

Let us consider several examples of Clairaut patches and find their geodesics.

The Euclidean Plane in Polar Coordinates

The polar coordinate parametrization of the plane is

$$\mathbf{polarplane}(u,v) = (v\cos u, v\sin u, 0).$$

It is easy to compute that

$$E = v^2, \quad F = 0, \quad G = 1,$$

from which it is clear that **polarplane** is a v-Clairaut patch. Equation (26.23) becomes

$$\frac{du}{dv} = \pm\frac{c}{\sqrt{v^2(v^2-c^2)}}, \quad \text{or equivalently,} \quad \frac{c\,dv}{\sqrt{v^2(v^2-c^2)}} = \pm du.$$

To integrate this equation with *Mathematica*, we use

```
Integrate[c/Sqrt[v^2(v^2 - c^2)],v]//PowerExpand
```

to which *Mathematica* answers

```
             c
  -ArcTan[ ─────────────── ]
                  2    2
           Sqrt[-c  + v ]
```

Thus

$$\arctan\left(\frac{c}{\sqrt{v^2-c^2}}\right) = \pm(u-u_0), \quad \text{or} \quad c = \pm v\sin(u-u_0)$$

for some constant u_0. This is the equation of a general straight line in polar coordinates.

Finding the geodesics on the plane using Cartesian coordinates is easier; see exercise 3.

The Poincarè Metric on the Upper Half-Plane

For the upper half-plane with the Poincaré metric we have

$$E = G = \frac{1}{\lambda^2 v^2} \quad \text{and} \quad F = 0.$$

Thus the identity map is a v-Clairaut patch, and equation (26.23) becomes

$$\frac{du}{dv} = \frac{c}{\pm\sqrt{\frac{1}{\lambda^2 v^2} - c^2}}, \quad \text{or equivalently,} \quad \frac{c^2 dv}{\sqrt{\frac{1}{\lambda^2 v^2} - c^2}} = \pm c \, du.$$

To integrate this equation, we use

```
Integrate[c^2/Sqrt[1/(lambda^2 v^2) - c^2],v]//Simplify
```

resulting in

```
            2       1
-(Sqrt[-c  + ------------] v)
                  2   2
              lambda  v
```

Therefore,

$$-v\sqrt{\frac{1}{\lambda^2 v^2} - c^2} = \pm c(u - u_0),$$

or equivalently,

$$v^2 + (u - u_0)^2 = \frac{1}{c^2 \lambda^2}.$$

for some constant u_0. Thus the geodesics of the Poincaré upper half-plane are Euclidean lines or circles that are perpendicular to the horizontal axis.

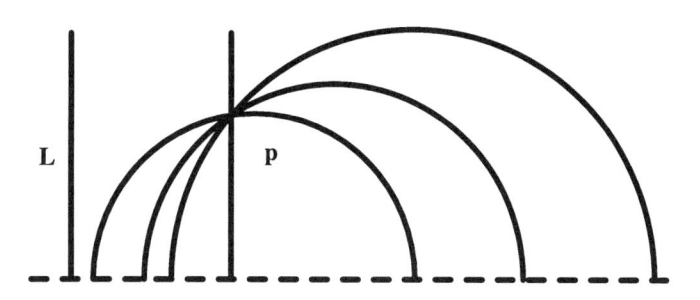

Geodesics in the Poincaré upper half plane

Geodesics are the analogs for a general surface of straight lines in the plane. The famous parallel postulate of Euclid is equivalent to

Playfair's[2] **axiom**: Given a line in the plane and a point not on the line, a unique coplanar line can be drawn through the point parallel to the given line.

This axiom clearly does not hold for the Poincaré upper half plane; in the picture above infinitely many geodesics through the point **p** do not intersect the line **L**. Many mathematicians had doubts about the parallel postulate, but it was Gauss, Lobachevsky[3] and Bolyai[4] who first observed that the parallel postulate was independent of the other postulates of Euclid, and new geometries result when the parallel postulate is replaced by either the

Hyperbolic axiom: Given a line and a point not on the line, more than one line can be drawn through the point parallel to the given line.

or the

Elliptic axiom: Given a line in the plane and a point not on the line, there are no lines parallel to the given line.

Almost 40 years elapsed until explicit models of these non-Euclidean geometries were given by Beltrami, Klein and Poincaré. In [Beltr] Beltrami showed by models that an elliptic or hyperbolic geometry is consistent, provided Euclidean geometry is consistent. For more details see [McCl].

A Surface of Revolution

The standard parametrization of a surface of revolution in $\mathbb{R}^3$ is

$$\mathbf{x}(u,v) = \big(\varphi(v)\cos u, \varphi(v)\sin u, \psi(v)\big).$$

---

[2] John Playfair (1748-1819). Scottish mathematician.

[3] Nikolai Ivanovich Lobachevsky (1792-1856). Russian mathematician, rector of the University of Kazan. His version of non-Euclidean geometry, published in 1829 was the first account of the subject to appear in print.

[4] János Bolyai (1802-1860). Hungarian mathematician. Son of Farkas Wolfgang Bolyai. János Bolyai published his version of non-Euclidean geometry as an appendix to his father's book **Tentamen**.

26.2 Clairaut Patches

Let us assume that $\varphi(v) > 0$; then $\varphi(v) > 0$ can be interpreted as the radius of the parallel $u \longmapsto \bigl(\varphi(v)\cos u, \varphi(v)\sin u, \psi(v)\bigr)$. Since

$$E = \varphi(v)^2, \qquad F = 0 \qquad \text{and} \qquad G = \varphi'(v)^2 + \psi'(v)^2,$$

we see that $E_u = G_u = F = 0$; thus **x** is a v-Clairaut patch, and the equation for the geodesics is

$$\frac{du}{dv} = \pm \frac{\sqrt{\varphi'^2 + \psi'^2}}{\varphi\sqrt{\varphi^2 - c^2}}.$$

Furthermore, since the v-parameter curves are meridians and the u-parameter curves are parallels, we obtain the following theorem as a special case of Lemma 26.15:

Theorem 26.19. *Let $\mathcal{M}$ be a surface of revolution in $\mathbb{R}^3$. Then:*

(i) *each meridian of $\mathcal{M}$ is a geodesic;*

(ii) *a parallel is a geodesic if an only if $\varphi'(v_0) = 0$.*

Geodesics on a Pseudosphere

First we note:

Theorem 26.20. *A pseudosphere in $\mathbb{R}^3$ is not geodesically complete.*

Proof. By Corollary 21.4, page 489, each meridian of a pseudosphere is a tractrix. The curve $\boldsymbol{\alpha}$ given by (3.13), page 63, is a unit-speed parametrization of a tractrix. Hence it gives rise to a geodesic whose trace is a meridian. From (3.13) we see that $\boldsymbol{\alpha}$ is not differentiable at 0, so that $\boldsymbol{\alpha}$ cannot be a geodesic defined on all of $\mathbb{R}$. More generally, no constant speed parametrization of a tractrix can be defined on the whole real line $\mathbb{R}$. Hence a pseudosphere must be incomplete. ∎

To find the geodesics which are not meridians with **Mathematica**, we introduce the following miniprogram:

```
clairautv[pm_,c_,u0_][x_][u_,t_]:=
    PowerExpand[u0 + Integrate[pm Sqrt[c^2gg[x][u,tt]/
    (ee[x][u,tt](ee[x][u,tt] - c^2))],tt]] /. tt->t
```

which solves the differential equation (26.23) for a given patch. In particular,

```
FullSimplify[clairautv[pm,c,u0][pseudosphere[a]][u,t]]
```

yields

$$u0 + \frac{pm\, \text{Sqrt}[a^2 - 2c^2 - a^2 \text{Cos}[2t]]\, \text{Csc}[t]}{\text{Sqrt}[2]\, c}$$

and so we define

```
pseudospheregeodesic[a_,c_,u0_,pm_][t_]:=
    {a Sin[t]Cos[u0 + pm Sqrt[a^2 - 2c^2 - a^2Cos[2t]]/
    (Sqrt[2]c Sin[t])],
    a Sin[t]Sin[u0 + pm Sqrt[a^2 - 2c^2 - a^2Cos[2t]]/
    (Sqrt[2]c Sin[t])],a(Cos[t] + Log[Tan[t/2]])}
```

Here is a typical plot of a geodesic on a pseudosphere:

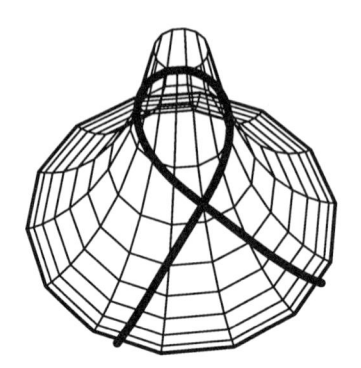

Geodesic on a pseudosphere

The *Mathematica* commands to make this plot are:

```
bb=wireframe[pseudosphere[1][u,v],{u,0,2Pi},
    {v,0.5Pi,0.95Pi},PlotPoints->{15,10}];
a25=ParametricPlot3D[Evaluate[{
Append[pseudospheregeodesic[1,0.25,Pi,1][t],
        AbsoluteThickness[3]],
Append[pseudospheregeodesic[1,0.25,Pi,-1][t],
        AbsoluteThickness[3]]}],
{t,0.5Pi,0.919569Pi},PlotRange->All,PlotPoints->1000,
DisplayFunction->Identity];
Show[bb,a25,DisplayFunction->$DisplayFunction,
PlotRange->All,Axes->None,Boxed->False,ViewPoint->{3,-1,4}];
```

26.3 Finding Geodesics Numerically with Mathematica

We describe a method for using *Mathematica* to generate geodesics on surfaces by solving the system (26.2). First, we need to rewrite (26.2) as an equivalent system of four first-order equations:

(26.24)
$$\begin{cases} u' = p, \\ v' = q, \\ p' = -\Gamma^1_{11} p^2 - 2\Gamma^1_{12} p\, q - \Gamma^1_{22} q^2, \\ q' = -\Gamma^2_{11} p^2 - 2\Gamma^2_{12} p\, q - \Gamma^2_{22} q^2. \end{cases}$$

We need the *Mathematica* function that describes the right-hand sides of the equations in (26.2):

```
geo[x_][p_,q_,u_,v_]:=
    {p,q,-christoffel[1,1,1][x][u,v]p^2-
    2christoffel[1,1,2][x][u,v]p q -
    christoffel[1,2,2][x][u,v]q^2,
    -christoffel[2,1,1][x][u,v]p^2
    -2christoffel[2,1,2][x][u,v]p q
    -christoffel[2,2,2][x][u,v]q^2}
```

To solve the geodesic equations, we use

```
solvegeoeqs[x_,a_:0,{u0_:0,v0_:0},tfin_,div_,
optsnd___]:=
    Flatten[Map[NDSolve[#,{u,v},{ss,0,tfin}]&,
    Table[Join[MapThread[Equal,
    {{u''[ss],v''[ss]},
    Take[geo[x][u'[ss],v'[ss],u[ss],v[ss]],{3,4}]}],
    {u[a]==u0,v[a]==v0,
    u'[a]==Cos[theta],v'[a]==Sin[theta]}],
    {theta,2Pi/div,2Pi,2Pi/div}]],1]
```

To make use of **geo** and **solvegeoeqs** to find the geodesics of a surface, a good strategy is first to use *Mathematica* to compute **geo** of the surface and to save the result as a function in a file. (The file **GEO.m** contains the geodesic equations for may common surfaces.) Then we can use the saved function in numerical calculations; in this way *Mathematica* does not need to redo the symbolic calculations each time it does a numerical calculation. For example, to compute the geodesics of the monkey saddle, we first read the files **CSPROGS.m** and **SURFS.m** into a *Mathematica* session with the command **<<CAS.m**. Then we give the command

```
geo[monkeysaddle][p,q,u,v]>>MONKEYSA.geo
```

This writes *Mathematica's* version of the geodesic equations to the file **MONKEYSA.geo**. The file **MONKEYSA.geo** then needs to be edited so that it contains the definition of a new function **geo[monkeysaddle]**, which is

```
geo[monkeysaddle][p_,q_,u_,v_]:=
    {p,q,(-18*p^2*u*(u^2 - v^2))/
    (1 + 9*u^4 + 18*u^2*v^2 + 9*v^4) -
    (18*q^2*u*(-u^2 + v^2))/
    (1 + 9*u^4 + 18*u^2*v^2 + 9*v^4) -
    36*p*q*v*(-u^2 + v^2)/
    (1 + 9*u^4 + 18*u^2*v^2 + 9*v^4),
    36*p^2*u^2*v/(1 + 9*u^4 + 18*u^2*v^2 + 9*v^4) -
    36*q^2*u^2*v/(1 + 9*u^4 + 18*u^2*v^2 + 9*v^4) -
    72*p*q*u*v^2/(1 + 9*u^4 + 18*u^2*v^2 + 9*v^4)}
```

Next, we use **solvegeoeqs**. The variables **tfin** and **div** specify the length and number of geodesics issuing from the initial point. This miniprogram may require a considerable amount of time. The output of **solvegeoeqs**, which consists of interpolating functions, can be used as the input of various graphing functions.

Let us illustrate the procedure for a monkey saddle. First, we use **solvegeoeqs** to create 30 geodesics of length 2 starting from the center of the monkey saddle.

```
sd=solvegeoeqs[monkeysaddle,0,{0,0},2,30];
```

The output of this command is not interesting to see, so we suppress it. However, we name the result **sd** so that we can use it. The 2-dimensional picture of the geodesics is

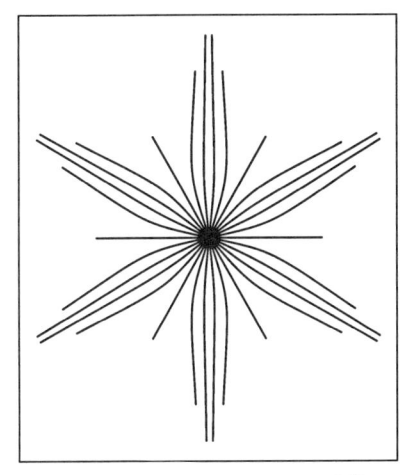

26.3 Finding Geodesics Numerically with Mathematica

Geodesics on a monkey saddle

```
ParametricPlot[
Evaluate[
{u[t],v[t]} /. sd],
{t,0,2},
AspectRatio->Automatic,
Axes->None];
```

and the 3-dimensional picture is

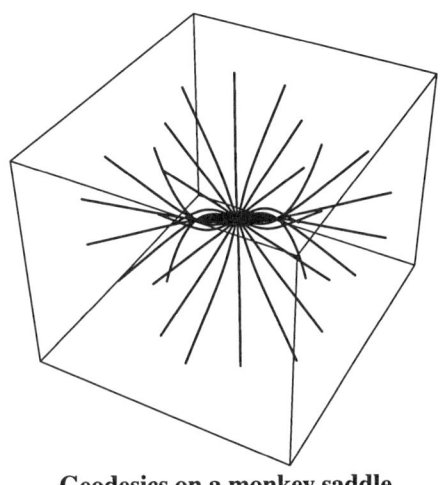

Geodesics on a monkey saddle

```
ParametricPlot3D[
Evaluate[
monkeysaddle@@Sequence[
{u[t],v[t]}] /. sd],
{t,0,2},Axes->None];
```

We can also plot circles at fixed distances from the center of the monkey saddle.

Geodesic circles on a monkey saddle

```
Show[Graphics[
Table[Line[
Append[
{u[t],v[t]} /. sd,
First[
{u[t],v[t]} /. sd]]],
{t,0,2,0.25}]],
AspectRatio->Automatic];
```

Here is the 3-dimensional version:

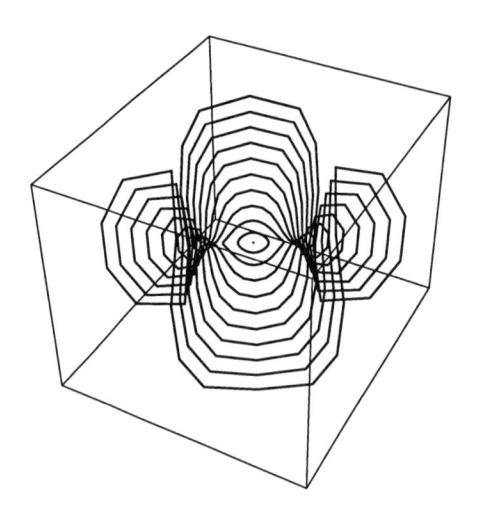

Geodesic circles on a monkey saddle

```
Show[Graphics3D[
Table[Line[
Append[
monkeysaddle@@Sequence[
{u[t],v[t]}] /. sd,
First[
monkeysaddle@@Sequence[
{u[t],v[t]}] /. sd]]],
{t,0,2,0.2}]]];
```

26.4 The Exponential Map and the Gauss Lemma

The exponential map[5] provides a natural way using geodesics to map each tangent space to an abstract surface onto the surface. Let $\mathcal{M}$ be an abstract surface with $\mathbf{p} \in \mathcal{M}$, and let $\mathbf{v} \in \mathcal{M}_\mathbf{p}$. By Theorem 26.8 we know that for small $b > 0$ there exists a unique geodesic $\boldsymbol{\alpha}\colon [0,b) \longrightarrow \mathcal{M}$ such that $\boldsymbol{\alpha}(0) = \mathbf{p}$ and $\boldsymbol{\alpha}'(0) = \mathbf{v}$.

Definition. *Let $\mathcal{M}$ be an abstract surface and let $\mathbf{p} \in \mathcal{M}$. The **exponential map** of $\mathcal{M}$ at $\mathbf{p}$ is the map $\exp_\mathbf{p}$ defined on a neighborhood $\mathcal{U}$ of $\mathbf{p}$ by*

$$\exp_\mathbf{p}(\mathbf{v}) = \boldsymbol{\alpha}_\mathbf{v}(1),$$

where $\boldsymbol{\alpha}_\mathbf{v}$ is the geodesic in $\mathcal{M}$ with $\boldsymbol{\alpha}_\mathbf{v}(0) = \mathbf{p}$ and $\boldsymbol{\alpha}'_\mathbf{v}(0) = \mathbf{v}$.

We first note:

Lemma 26.21. *For any real number a, the exponential map $\exp_\mathbf{p}$ satisfies*

$$\exp_\mathbf{p}(a\,\mathbf{v}) = \boldsymbol{\alpha}_\mathbf{v}(a\,t),$$

where $\boldsymbol{\alpha}_\mathbf{v}$ is the geodesic in $\mathcal{M}$ with $\boldsymbol{\alpha}_\mathbf{v}(0) = \mathbf{p}$ and $\boldsymbol{\alpha}'_\mathbf{v}(0) = \mathbf{v}_\mathbf{p}$.

Proof. Both of the curves $\boldsymbol{\alpha}_{a\mathbf{v}}$ and $t \longmapsto \boldsymbol{\alpha}_\mathbf{v}(a\,t)$ are geodesics with the same initial point and the same initial velocity, namely, $a\,\mathbf{v}_\mathbf{p}$. Hence by Theorem 26.8 they coincide on some interval containing 0. ∎

The next order of business is to show that $\exp_\mathbf{p}$ is a diffeomorphism near the origin $\mathbf{0} \in \mathcal{M}_\mathbf{p}$. For this, we need the elementary observation that each tangent space to an abstract surface is itself an abstract surface.

---

[5] The reason for the name "exponential map" is briefly as follows. The ordinary exponential function from real and complex variables, that is,

$$(26.25) \qquad z \longmapsto e^z = \sum_{j=0}^{\infty} \frac{z^j}{j!},$$

can be extended to square matrices, because the right-hand side of (26.25) makes sense when z is replaced by a square matrix A. This new map $A \longmapsto e^A$ provides a method of mapping the vector space of all $n \times n$ matrices onto the group of all nonsingular $n \times n$ matrices. In particular, it maps the vector space of all antisymmetric $n \times n$ matrices onto the group $\mathbf{SO}(n)$ of all orthogonal $n \times n$ matrices of determinant 1. It turns out that there is a natural Riemannian metric on $\mathbf{SO}(n)$ for which $t \longmapsto e^{tA}$ is a geodesic in $\mathbf{SO}(n)$ for any $n \times n$ antisymmetric matrix A. This fact can be generalized to any Riemannian manifold; the result is exp.

Lemma 26.22. *Let $\mathcal{M}$ be an abstract surface. Then for each $\mathbf{p} \in \mathcal{M}$ there is a neighborhood $\mathcal{U}_\mathbf{p}$ of $\mathbf{0} \in \mathcal{M}_\mathbf{p}$ such that $\exp_\mathbf{p}$ maps $\mathcal{U}_\mathbf{p}$ diffeomorphically onto an open subset of $\mathcal{M}$.*

Proof. The differentiability of the exponential map follows from the fact from the theory of differential equations that the solutions of the system (26.2) depend differentiably on the initial conditions (26.11).

To show that $\exp_\mathbf{p}$ is a diffeomorphism near the origin, we need radial curves in the tangent space $\mathcal{M}_\mathbf{p}$ emanating from the origin. For $\mathbf{v} \in \mathcal{M}_\mathbf{p}$, let $\rho_\mathbf{v}: \mathbb{R} \longrightarrow \mathcal{M}_\mathbf{p}$ be the curve defined by $\rho_\mathbf{v}(t) = t\mathbf{v}$. Then $\rho_\mathbf{v}$ is the unique geodesic in the tangent space $\mathcal{M}_\mathbf{p}$ (considered as an abstract surface) starting at $\mathbf{p}$ with initial velocity $\mathbf{v}$. Similarly, Lemma 26.21 implies that the curve $t \longmapsto \xi_\mathbf{v}(t)$ in $\mathcal{M}$ defined by $\xi_\mathbf{v} = \exp_\mathbf{p} \circ \rho_\mathbf{v}$ is the unique geodesic in $\mathcal{M}$ starting at $\mathbf{p}$ with initial velocity $\mathbf{v}_\mathbf{p}$. The manifold $\mathcal{M}_\mathbf{p}$ is canonically identified with its tangent space $(\mathcal{M}_\mathbf{p})_0$ at the origin as follows. For $\mathbf{v} \in \mathcal{M}_\mathbf{p}$, let $\mathbf{v}_0 \in (\mathcal{M}_\mathbf{p})_0$ be the tangent vector defined by $(\mathbf{v}_\mathbf{p})_0 = \rho'_\mathbf{v}(0)$. In fact, the tangent map of the exponential map $((\exp_\mathbf{p})_*)_0 : (\mathcal{M}_\mathbf{p})_0 \longrightarrow \mathcal{M}_\mathbf{p}$ is this canonical identification because

$$((\exp_\mathbf{p})_*)_0\big((\mathbf{v}_\mathbf{p})_0\big) = (\exp_\mathbf{p} \circ \rho_\mathbf{p})'(0) = \xi'(0) = \mathbf{v}_\mathbf{p}.$$

The inverse function theorem implies that $\exp_\mathbf{p}$ maps a neighborhood of $0 \in \mathcal{M}_\mathbf{p}$ diffeomorphically onto an open subset of $\mathcal{M}$. ∎

Definition. *Let $\mathcal{M}$ be an abstract surface and let $\mathbf{p} \in \mathcal{M}$. We say that an open subset $\mathcal{N}_\mathbf{p}$ of $\mathcal{M}$ containing $\mathbf{p}$ is a* **normal neighborhood** *of $\mathbf{p}$, provided there is an open subset $\mathcal{O}_\mathbf{p}$ of the tangent space $\mathcal{M}_\mathbf{p}$ such that $\exp_\mathbf{p}$ maps $\mathcal{O}_\mathbf{p}$ diffeomorphically onto $\mathcal{N}_\mathbf{p}$.*

In the special case of the surface $\mathbb{R}^2$, a geodesic starting at $\mathbf{p} \in \mathbb{R}^2$ with initial velocity $\mathbf{v}_\mathbf{p}$ is of the form $t \longmapsto \mathbf{p} + t\mathbf{v}$. Hence the exponential map at $\mathbf{p} \in \mathbb{R}^2$ is given simply by

$$\exp_\mathbf{p}(\mathbf{v}_\mathbf{p}) = \mathbf{p} + \mathbf{v}.$$

Thus any open subset of $\mathbb{R}^2$ containing $\mathbf{p}$ is a normal neighborhood of $\mathbf{p}$.

Next, we use the exponential map to define natural generalizations of rectangular and polar coordinates.

Definition. *Let $\mathcal{M}$ be an abstract surface and let $\mathbf{p} \in \mathcal{M}$. Choose an orthonormal basis $\{\mathbf{e}_1, \mathbf{e}_2\}$ of the tangent space $\mathcal{M}_\mathbf{p}$, and let $\mathcal{N}_\mathbf{p} = \exp_\mathbf{p}(\mathcal{O}_\mathbf{p})$ be a normal neighborhood of $\mathbf{p}$.*

26.4 The Exponential Map and the Gauss Lemma

(i) A **normal patch** of $\mathcal{M}$ at $\mathbf{p}$ (relative to $\{\mathbf{e}_1, \mathbf{e}_2\}$) is the map

$$\mathbf{x} \colon (a,b) \times (c,d) \longrightarrow \mathcal{M} \quad \text{given by} \quad \mathbf{x}(u,v) = \exp_{\mathbf{p}}(u\,\mathbf{e}_1 + v\,\mathbf{e}_2),$$

where the rectangle $\{\, u\,\mathbf{e}_1 + v\,\mathbf{e}_2 \mid a < u < b, c < v < d\,\}$ is contained in $\mathcal{O}_{\mathbf{p}}$.

(ii) A **geodesic polar patch** of $\mathcal{M}$ at $\mathbf{p}$ (relative to $\{\mathbf{e}_1, \mathbf{e}_2\}$) is the map

$$\mathbf{y} \colon (0,c) \times [0, 2\pi] \longrightarrow \mathcal{M} \quad \text{given by} \quad \mathbf{y}(r,\theta) = \exp_{\mathbf{p}}\bigl(r\cos(\theta)\mathbf{e}_1 + r\sin(\theta)\mathbf{e}_2\bigr),$$

where the punctured disk

$$\{\, r\cos(\theta)\mathbf{e}_1 + r\sin(\theta)\mathbf{e}_2 \mid 0 < r < c, 0 \leq \theta \leq 2\pi \,\}$$

is contained in $\mathcal{O}_{\mathbf{p}}$.

Since a normal patch is the inverse of the exponential map, we have the following fact (see exercise 9):

Lemma 26.23. *A normal patch is regular near the origin; a geodesic polar patch is normal near the origin except at the origin itself.*

Although regularity of a geodesic polar patch $\mathbf{y}$ fails at the origin, still $\mathbf{y}$ can be extended to a differentiable map

(26.26) $$\mathbf{y} \colon [0,c) \times (-\infty, \infty) \longrightarrow \mathcal{M}.$$

Lemma 26.24. *Let $\mathcal{N}_{\mathbf{p}}$ be a normal neighborhood on an abstract surface $\mathcal{M}$ and let $\mathbf{q} \in \mathcal{N}_{\mathbf{p}}$. Then there is one and only one geodesic from $\mathbf{q}$ to $\mathbf{p}$ that lies entirely in $\mathcal{N}_{\mathbf{p}}$.*

Proof. Any unit-speed geodesic starting at $\mathbf{p}$ is a r-parameter curve of the polar parametrization. Since $\exp_{\mathbf{p}} \colon \mathcal{N}_{\mathbf{p}} \longrightarrow \mathcal{N}_{\mathbf{p}}$ is a diffeomorphism, one such geodesic reaches $\mathbf{q}$ and is unique. ∎

Next, let us compute the first fundamental form of a geodesic polar patch.

Lemma 26.25. *Let $\mathbf{y}$ be a geodesic polar parametrization at a point $\mathbf{p}$ on an abstract surface $\mathcal{M}$. Then the coefficients of the first fundamental form of $\mathbf{y}$ have the property that*

$$E = 1 \quad \text{and} \quad F = 0,$$

so that the metric of $\mathbf{y}$ is given by

$$ds^2 = dr^2 + G(r,\theta)d\theta^2.$$

Proof. Since the r-parameter curves of $\mathbf{y}$ are unit-speed geodesics, we have
$$E = \langle \mathbf{y}_r, \mathbf{y}_r \rangle = 1 \quad \text{and} \quad \mathbf{y}_{rr} = 0.$$
Moreover,
$$F_r = \frac{\partial \langle \mathbf{y}_r, \mathbf{y}_\theta \rangle}{\partial r} = \langle \mathbf{y}_r, \mathbf{y}_{\theta r} \rangle = \langle \mathbf{y}_r, \mathbf{y}_{r\theta} \rangle = \frac{E_\theta}{2} = 0,$$
so that F is constant on each r-parameter curve. The r-parameter curve $\theta \longmapsto \mathbf{y}(0, \theta)$ is just the constant curve at $\mathbf{p}$; hence $\mathbf{y}_\theta(0, \theta) = 0$. This implies that $F(0, \theta) = 0$. (F is defined at $(0, \theta)$ because of (26.26).) Since F is constant on each r-parameter curve, we conclude that F is identically 0. ∎

Let $\mathcal{M}$ be an abstract surface and let $\mathbf{p} \in \mathcal{M}$. It turns out that although the exponential map $\exp_\mathbf{p}$ is usually *not* an isometry, it is a partial isometry, as we now explain. Consider the tangent space $\mathcal{M}_\mathbf{p}$ to be an abstract surface isometric to $\mathbb{R}^2$. We call the geodesics in $\mathcal{M}_\mathbf{p}$ emanating from $0 \in \mathcal{M}_\mathbf{p}$ **radial**, and we call a tangent vector to $\mathcal{M}_\mathbf{p}$ at $\mathbf{v} \in \mathcal{M}_\mathbf{p}$ **radial**, provided $\mathbf{v}$ is tangent to a radial geodesic. Similarly, we call a tangent vector to $\mathcal{M}_\mathbf{p}$ at $\mathbf{v}$ **coradial**, if it is perpendicular to the radial tangent vectors at $\mathbf{v}$. The radial geodesics in $\mathcal{M}_\mathbf{p}$ are mapped by $\exp_\mathbf{p}$ into the geodesics in $\mathcal{M}$ starting at $\mathbf{p}$. Now observe that the notions of "radial" and "coradial" also make sense for tangent vectors to $\mathcal{M}$.

In paragraph 15 of his paper [Gauss], Gauss states:

> If on a curved surface an infinite number of shortest lines [that is, geodesics] be drawn from the same initial point, the line joining their extremities will be normal to each of the lines.

This is the first enunciation of what is now known as the **Gauss Lemma**. In terms of the exponential map, this important lemma is expressed as follows.

Lemma 26.26. (**Gauss Lemma**) *Let $\mathcal{M}$ be an abstract surface and let $\mathbf{p} \in \mathcal{M}$. Then:*

(i) $\exp_\mathbf{p}$ *preserves the lengths of radial tangent vectors;*

(ii) $\exp_\mathbf{p}$ *preserves the orthogonality between coradial and radial tangent vectors.*

Proof. Let $\mathbf{y}$ be a geodesic polar patch on $\mathcal{M}$ at $\mathbf{p}$; then the composition $\mathbf{y} \circ \exp_\mathbf{p}$ gives the ordinary polar coordinates at $\mathbf{p}$. Let E, F and G denote the coefficients of the first fundamental form of $\mathbf{y}$. Part (i) is a reformulation of the fact from Lemma 26.25 that $E = 1$, and part (ii) is a reformulation of the fact that $F = 0$. ∎

On the other hand, most of the time $\exp_\mathbf{p}$ does change the lengths and inner products of the coradial vectors; this accounts for the fact that it is almost never an isometry.

26.5 Length Minimizing Properties of Geodesics

The definition of distance function ρ given on page 348 for a surface in $\mathbb{R}^n$ also works for an abstract surface $\mathcal{M}$. Thus for $\mathbf{p}, \mathbf{q} \in \mathcal{M}$,

$$\rho(\mathbf{p}, \mathbf{q}) = \text{the greatest lower bound of the lengths of all piecewise-differentiable curves } \alpha \text{ connecting } \mathbf{p} \text{ to } \mathbf{q}.$$

Definition. *Let $\mathcal{M}$ be an abstract surface, and let α be a piecewise-differentiable curve connecting $\mathbf{p}, \mathbf{q} \in \mathcal{M}$. We say that α **minimizes arc length from $\mathbf{p}$ to $\mathbf{q}$** provided*

(26.27) $$\rho(\mathbf{p}, \mathbf{q}) = \text{Length}(\alpha),$$

and any other piecewise-differentiable curve connecting $\mathbf{p}$ and $\mathbf{q}$ is a reparametrization of α.

Next, if $\mathbf{p}$ and $\mathbf{q}$ are not too far apart, we find that a curve whose length equals $\rho(\mathbf{p}, \mathbf{q})$ is a geodesic.

Theorem 26.27. *For each point $\mathbf{q}$ in a normal neighborhood $\mathcal{N}_\mathbf{p}$ of an abstract surface $\mathcal{M}$, the radial geodesic from $\mathbf{p}$ to $\mathbf{q}$ uniquely minimizes arc length.*

Proof. Let $\mathbf{y}$ be a geodesic polar patch that parametrizes $\mathcal{N}_\mathbf{p}$. We can write $\mathbf{q} = \mathbf{y}(u_0, v_0)$; then the unique geodesic connecting $\mathbf{p}$ and $\mathbf{q}$ is the radial geodesic $\gamma \colon [0, u_0] \longrightarrow \mathcal{N}_\mathbf{p}$ defined by $\gamma(u) = \mathbf{y}(u, v_0)$. Any other curve α from $\mathbf{p}$ to $\mathbf{q}$ can be assumed to be parametrized on the same interval $[0, u_0]$.

We want to show that

(26.28) $$\text{Length}(\gamma) \leq \text{Length}(\alpha).$$

Discarding any loop that α makes at $\mathbf{p}$ only shortens α; therefore, we can assume that α never returns to $\mathbf{p}$.

First, consider the case when α never leaves $\mathcal{N}_\mathbf{p}$. Since α never returns to $\mathbf{p}$, there exist differentiable functions $a_1, a_2 \colon [0, u_0] \longrightarrow \mathbb{R}$ such that

$$\alpha(t) = \mathbf{x}(a_1(t), a_2(t)).$$

Since $\alpha(0) = \mathbf{p}$ and $\alpha(u_0) = \mathbf{q}$, we have

(26.29) $\quad a_1(0) = 0, \quad a_1(u_0) = u_0, \quad a_2(0) = v_0 \quad$ and $\quad a_2(u_0) = v_0 + 2n\pi,$

for some integer n. Since $\mathbf{y}$ is a geodesic polar patch, we have $E = 1$ and $F = 0$, and so the speed of $\boldsymbol{\alpha}$ satisfies

$$\|\boldsymbol{\alpha}'\| = \sqrt{a_1'^2 + G\, a_2'^2} \geq |a_1'| \geq a_1'.$$

Hence (26.29) implies that

(26.30) $\quad \mathbf{Length}(\boldsymbol{\alpha}) = \int_0^{u_0} \sqrt{a_1'^2 + G\, a_2'^2}\, dt \geq \int_0^{u_0} a_1'\, dt = a_1(u_0) - a_1(0) = u_0.$

The radial geodesic $\boldsymbol{\gamma}$ has unit-speed; that is,

(26.31) $\quad\quad\quad\quad\quad\quad \mathbf{Length}(\boldsymbol{\gamma}) = \int_0^{u_0} dt = u_0.$

Then (26.28) follows from (26.30) and (26.31).

On the other hand, if $\boldsymbol{\alpha}$ does not stay in $\mathcal{N}_\mathbf{p}$, then it must cross the circle $v \longmapsto \mathbf{y}(u_0, v)$; by the proof above, the portion of $\boldsymbol{\alpha}$ inside this circle must have length at least $\mathbf{Length}(\boldsymbol{\gamma})$. It follows that in this case $\mathbf{Length}(\boldsymbol{\gamma}) < \mathbf{Length}(\boldsymbol{\alpha})$.

Finally, suppose that $\mathbf{Length}(\boldsymbol{\gamma}) = \mathbf{Length}(\boldsymbol{\alpha})$. Then $\boldsymbol{\alpha}$ cannot cross the circle $v \longmapsto \mathbf{y}(u_0, v)$ and (26.30) becomes an equality. Hence

(26.32) $\quad\quad\quad\quad\quad\quad \sqrt{a_1'^2 + G\, a_2'^2} = a_1'.$

Since $G > 0$, (26.32) implies that

$$a_2' = 0.$$

Thus a_2 has the constant value v_0, and so

$$\boldsymbol{\alpha}(t) = \mathbf{y}\big(a_1(t), v_0\big),$$

showing that $\boldsymbol{\alpha}$ is indeed a reparametrization of $\boldsymbol{\gamma}$. ∎

Lemma 26.28. *Let $\mathcal{M}$ be an abstract surface, and let $\mathbf{p}, \mathbf{q} \in \mathcal{M}$. Suppose that $\boldsymbol{\alpha} \colon [a, b] \longrightarrow \mathcal{M}$ is a curve on $\mathcal{M}$ such that the length of $\boldsymbol{\alpha}$ is shortest among all curves connecting $\mathbf{p} = \boldsymbol{\alpha}(a)$ to $\mathbf{q} = \boldsymbol{\alpha}(b)$. Then $\boldsymbol{\alpha}$ is a geodesic.*

Proof. Under the assumption that $\boldsymbol{\alpha}$ is not a geodesic, we show that

$$\mathbf{Length}(\boldsymbol{\alpha}) > \rho(\mathbf{p}, \mathbf{q}).$$

If $\boldsymbol{\alpha}$ is not a geodesic, then there exists t_0 with $\big(\nabla_{\boldsymbol{\alpha}'} \boldsymbol{\alpha}\big)'(t_0) \neq 0$. Since $\nabla_{\boldsymbol{\alpha}'} \boldsymbol{\alpha}'$ is continuous, it is nonzero near t_0; thus without loss of generality, $a < t_0 < b$. There exists $\epsilon > 0$ such that the restriction of α to the interval $[t_0 - \epsilon, t_0 + \epsilon]$ is not a geodesic but lies in a normal neighborhood of $\alpha(t_0)$. Then Theorem 26.27 implies that the length $\mathbf{Length}(\alpha|[t_0, t_0+\epsilon])$ of the restriction of α to the interval $[t_0, t_0+\epsilon]$ is strictly greater than $\rho(\alpha(t_0), \alpha(t_0 + \epsilon))$. But then, using the triangle inequality, we have

$$\mathbf{Length}(\alpha) = \mathbf{Length}(\alpha|[a, t_0]) + \mathbf{Length}(\alpha|[t_0, t_0 + \epsilon]) + \mathbf{Length}(\alpha|[t_0 + \epsilon, b])$$

$$> \rho(\mathbf{p}, \alpha(t_0)) + \rho(\alpha(t_0), \alpha(t_0 + \epsilon)) + \rho(\alpha(t_0 + \epsilon), \mathbf{q}) \geq \rho(\mathbf{p}, \mathbf{q}). \blacksquare$$

26.6 An Abstract Surface as a Metric Space

Let us recall an important definition from topology.

Definition. *A set S is called a **metric space** provided there exists a function $d: S \times S \longrightarrow S$ (called the **distance function** of S) with the following properties*:

(i) $d(\mathbf{p}, \mathbf{q}) = d(\mathbf{q}, \mathbf{p})$;

(ii) $d(\mathbf{p}, \mathbf{r}) \leq d(\mathbf{p}, \mathbf{q}) + d(\mathbf{q}, \mathbf{r})$;

(iii) $d(\mathbf{p}, \mathbf{q}) \geq 0$;

(iv) $d(\mathbf{p}, \mathbf{q}) = 0$ *if and only if* $\mathbf{p} = \mathbf{q}$

for $\mathbf{p}, \mathbf{q} \in S$.

The simplest nontrivial example of a metric space is $\mathbb{R}^n$ with the metric the standard distance function, defined by $\mathbf{distance}(\mathbf{p}, \mathbf{q}) = \|\mathbf{p} - \mathbf{q}\|$ for $\mathbf{p}, \mathbf{q} \in \mathbb{R}^n$. However, the notion of metric space is much more general. It is important for the theory of abstract surfaces, because of the following result:

Theorem 26.29. *The distance function ρ of an abstract surface $\mathcal{M}$ gives $\mathcal{M}$ the structure of a metric space.*

Proof. That ρ satisfies (i) in the definition of metric space is clear from the fact that each parametrized curve gives rise to a parametrized curve in the opposite direction, and (iii) is obvious from the fact that the length of any curve is nonnegative. To prove that (ii) is satisfied, let $\epsilon > 0$ and let α and β be piecewise-differentiable curves on $\mathcal{M}$ connecting **p** to **q** and **q** to **r**, respectively, such that

$$\mathbf{Length}(\alpha) < \rho(\mathbf{p},\mathbf{q}) + \frac{\epsilon}{2} \quad \text{and} \quad \mathbf{Length}(\beta) < \rho(\mathbf{q},\mathbf{r}) + \frac{\epsilon}{2}.$$

We can parametrize the union of the traces of α and β as a new piecewise-differentiable curve γ; then

$$\rho(\mathbf{p},\mathbf{r}) \leq \mathbf{Length}(\gamma) = \mathbf{Length}(\alpha) + \mathbf{Length}(\beta) < \rho(\mathbf{p},\mathbf{q}) + \rho(\mathbf{q},\mathbf{r}) + \epsilon.$$

Since ϵ is arbitrary, we get (ii).

Finally, to prove that (iv) is satisfied, we proceed as follows. Suppose that $\rho(\mathbf{p},\mathbf{q}) = 0$, but $\mathbf{p} \neq \mathbf{q}$. Let $\mathcal{N}_\mathbf{p}$ be a normal neighborhood of **p**; there exists $\epsilon > 0$ such that

$$\{\mathbf{r} \in \mathcal{N}_\mathbf{p} \mid \rho(\mathbf{p},\mathbf{r}) \leq \epsilon\} \subseteq \mathcal{N}_\mathbf{p}.$$

By definition of ρ, there exists a piecewise-differentiable curve $\alpha \colon [a,b] \longrightarrow \mathcal{M}$ with $\alpha(a) = \mathbf{p}$ and $\alpha(b) = \mathbf{q}$ such that $\mathbf{Length}(\alpha) < \epsilon$. Since $\alpha([a,b])$ is connected, there exists t_0 with $a < t_0 < b$ such that $\rho(\mathbf{p}, \alpha(t_0)) = \epsilon$. But then Theorem 26.27 implies that $\mathbf{Length}(\alpha) \geq \epsilon$. This contradiction shows that $\mathbf{p} = \mathbf{q}$. ∎

There is a notion of completeness for metric spaces.

Definition. *Let S be a metric space with distance function d. A sequence of points $\{\mathbf{p}_n\}$ in S is called a **Cauchy sequence**, provided that for each $\epsilon > 0$ there is an index $N = N(\epsilon)$ such that $d(\mathbf{p}_i, \mathbf{p}_j) < \epsilon$ whenever $i, j \geq N$. We say that S is a **complete metric space**, provided every Cauchy sequence converges to some point.*

For example, $\mathbb{R}^2 \setminus (0,0)$ with the standard metric is not complete because the sequence $\{(1/n, 1/n)\}$ converges to no point in $\mathbb{R}^2 \setminus (0,0)$.

Thus for an abstract surface we have two competing notions of completeness. In fact, they coincide.

Lemma 26.30. *Let $\mathcal{M}$ be a connected abstract surface. If $(\mathcal{M}, \rho)$ is a complete metric space, then $\mathcal{M}$ is geodesically complete.*

Proof. Let $\beta \colon [0,b) \longmapsto \mathcal{M}$ be a unit-speed geodesic. If $\{t_j\}$ is a sequence in $[0,b)$ with $t_j \longrightarrow b$, then $\{\beta(t_j)\}$ is a Cauchy sequence in $\mathcal{M}$, since

$$\rho(\beta(t_j), \beta(t_k)) \leq \mathbf{Length}(\beta|[t_j, t_k]) = |t_j - t_k|.$$

Hence $\{\beta(t_j)\}$ converges to some $\mathbf{q} \in \mathcal{M}$. Let $\mathcal{N}_\mathbf{q}$ be a normal neighborhood of **q**; since β is a geodesic in $\mathcal{N}_\mathbf{q}$ that reaches **q**, the exponential map at **q** can be used to extend β as a geodesic past $\mathbf{q}$. In this way β can be extended to a geodesic defined on all of $[0, \infty)$. Similarly, β can be extended to a geodesic defined on all of $(-\infty, 0]$, and thus on the whole real line. ∎

Lemma 26.30 has the following generalization:

Theorem 26.31. (Hopf-Rinow) *Let $\mathcal{M}$ be a connected abstract surface. Then the following conditions are equivalent*:

(i) $(\mathcal{M}, \rho)$ *is a complete metric space*;

(ii) $\mathcal{M}$ *is geodesically complete*;

(iii) *for each $\mathbf{p} \in \mathcal{M}$, the exponential map $\exp_{\mathbf{p}}$ is defined on all of the tangent space $\mathcal{M}_{\mathbf{p}}$.*

Furthermore, each of (i), (ii) *and* (iii) *implies that for any $\mathbf{p}, \mathbf{q} \in \mathcal{M}$, there is a geodesic $\beta \colon [a, b] \longrightarrow \mathcal{M}$ such that $\beta(a) = \mathbf{p}$, $\beta(b) = \mathbf{q}$ and $\rho(\mathbf{p}, \mathbf{q}) = \mathrm{Length}(\beta)$.*

For a proof [Hicks] or [ON2].

26.7 Exercises

1. Show that a pregeodesic is a geodesic if and only if it has constant speed.

2. Show that a meridian on a generalized helicoid can be parametrized as a geodesic. [Hint: Generalize Theorem 26.10.]

3. Use Lemma 26.17 to find the geodesics on the plane parametrized by Cartesian coordinates.

4. Plot simultaneously several geodesics on a pseudosphere[6] using different colors. Prove that each geodesic which is not a tractrix has exactly one point of self-intersection. Find it.

5. Find a parametrization of a general geodesic on a catenoid and draw several of them.

---

[6]For more information on geodesics on the pseudosphere and interesting pictures of them, see F. Schilling's **Die Pseudosphäre und die nichteuklidische Geometrie** [Schill].

6. Let β be a geodesic on a catenoid other than the center circle, and suppose the initial velocity of β is perpendicular to the axis of revolution. Use Clairaut's relation (Theorem 26.16) to show that β approaches the center curve asymptotically, but never reaches it.

7. Draw geodesics and geodesic circles on a sphere. using **solvegeoeqs**.

8. Draw geodesics and geodesic circles on ellipsoids with two distinct axes and on ellipsoids with three distinct axes using **solvegeoeqs**.

9. Prove Lemma 26.14, and give the details of the proof of Lemma 26.23.

10. Let $\mathbf{p}, \mathbf{v} \in S^2(1) \subset \mathbb{R}^3$ with $\mathbf{p} \cdot \mathbf{v} = 0$, so that $\mathbf{v}_\mathbf{p} \in S^2(1)_\mathbf{p}$. Find the formula for $\exp_\mathbf{p}(t\mathbf{v}_\mathbf{p})$ for $t \in \mathbb{R}$.

11. Let $\{\mathbf{p}_j\}$ and $\{\mathbf{q}_j\}$ be sequences of points on an abstract surface $\mathcal{M}$ such that in the topology of the surface

$$\lim_{j \to \infty} \mathbf{p}_j = \mathbf{p} \quad \text{and} \quad \lim_{j \to \infty} \mathbf{q}_j = \mathbf{q}.$$

Show that

$$\lim_{j \to \infty} \rho(\mathbf{p}_j, \mathbf{q}_j) = \rho(\mathbf{p}, \mathbf{q}).$$

27

THE GAUSS-BONNET THEOREM

That the sum of the interior angles of a triangle in the plane equals π was one of the first mathematical facts established by the Greeks. In 1603 Harriot[1] showed that on a sphere of radius 1 the area of a spherical triangle with angles α, β and γ is given by
$$\alpha + \beta + \gamma - \pi.$$
Both of these theorems are special cases of the Gauss-Bonnet[2] Theorem, which will be proved in this chapter.

The Gauss-Bonnet Theorem consists of a formula for the integral of the Gaussian curvature over all or part of an abstract surface. For example, we know that the curvature and area of a sphere $S^2(a)$ are $1/a^2$ and $4\pi a^2$. Thus
$$\int_{S^2(a)} K\, dA = \frac{1}{a^2} \int_{S^2(a)} dA = \frac{1}{a^2} 4\pi a^2 = 4\pi,$$

---

[1] Thomas Harriot (1560-1621). English mathematician, adviser to Sir Walter Raleigh. He made a voyage to Virginia in 1585-86 and reported on the Native American languages and customs. His telescopic observations were the earliest in England; in particular, he discovered sunspots in 1610 and used them to deduce the period of the Sun's rotation.

[2] Pierre Ossian Bonnet (1819-1892). French mathematician, who made many important contributions to surface theory, including the Gauss-Bonnet Theorem. Bonnet was director of studies at the École Polytechnique, professor of astronomy in the faculty of sciences at the University of Paris, and member of the board of longitudes.

showing that the integral of the Gaussian curvature over a sphere is independent of the radius $1/a$ of the sphere. It is a remarkable fact that for any ellipsoid, or more generally for any convex surface $\mathcal{M}$ in $\mathbb{R}^3$, we have

$$\int_{\mathcal{M}} K \, dA = 4\pi.$$

Still more generally, for any compact surface $\mathcal{M}$ the Gauss-Bonnet Theorem expresses the integral of the Gaussian curvature in terms of a topological invariant, the Euler characteristic $\chi(\mathcal{M})$. This important number is always an integer, and is expressible in terms of the number of vertices, faces and edges of any triangulation of $\mathcal{M}$.

In Section 27.1 we prove a local version of the Gauss-Bonnet Theorem. A summary of topological facts about surfaces is given in Section 27.2 in order to establish global versions of the Gauss-Bonnet Theorem in Section 27.3. Finally, Section 27.4 is devoted to applications of the Gauss-Bonnet Theorem.

27.1 The Local Gauss-Bonnet Theorem

First, we need some facts about piecewise-differentiable curves on surfaces. Let $\mathcal{M}$ be an abstract surface, and let $\boldsymbol{\alpha}:[a,b] \longrightarrow \mathcal{M}$ be a piecewise-differentiable curve. Associated with $\boldsymbol{\alpha}$ is a subdivision

$$a = t_0 < t_1 < \cdots < t_k = b$$

of $[a,b]$ for which $\boldsymbol{\alpha}$ is differentiable and regular on $[t_j, t_{j+1}]$ for $j = 0, \ldots, k-1$. (At the end points of each subinterval the appropriate left and right derivatives of $\boldsymbol{\alpha}$ are required to exist.) The points $\boldsymbol{\alpha}(t_0), \ldots, \boldsymbol{\alpha}(t_k)$ are called the **vertices** of $\boldsymbol{\alpha}$.

The following theorem is an immediate consequence of Lemma 25.15, page 588.

Theorem 27.1. *Let $\mathcal{M}$ be an abstract surface, and let $\boldsymbol{\alpha}:[a,b] \longrightarrow \mathcal{M}$ be a piecewise-differentiable curve whose trace is contained in $\mathbf{x}(\mathcal{U})$, where $\mathbf{x}:\mathcal{U} \longrightarrow \mathcal{M}$ is a patch on $\mathcal{M}$ for which $F = 0$. Denote by $\boldsymbol{\alpha}(t_0), \ldots, \boldsymbol{\alpha}(t_k)$ the vertices of $\boldsymbol{\alpha}$. Then the total geodesic curvature of $\boldsymbol{\alpha}$ is given by*

$$(27.1) \quad \mathrm{TGC}[\boldsymbol{\alpha}] = \int_{\boldsymbol{\alpha}} \kappa_g \, ds = \sum_{j=0}^{k-1} \int_{t_j}^{t_{j+1}} \kappa_g[\boldsymbol{\alpha}](t) \|\boldsymbol{\alpha}'(t)\| \, dt$$

$$= \sum_{j=0}^{k-1} \int_{t_j}^{t_{j+1}} \left(\frac{G_u}{2\sqrt{EG}} \frac{dv}{dt} - \frac{E_v}{2\sqrt{EG}} \frac{du}{dt} \right) dt + \sum_{j=0}^{k-1} (\theta_{j+1}^- - \theta_j^+),$$

27.1 The Local Gauss-Bonnet Theorem

where

$$\theta_j^+ = \lim_{\substack{t \to t_j \\ t < t_j}} \theta[\alpha, \mathbf{x}_u](\alpha(t)) \quad \text{and} \quad \theta_j^- = \lim_{\substack{t \to t_j \\ t > t_j}} \theta[\alpha, \mathbf{x}_u](\alpha(t)).$$

We call $\varepsilon_j = \theta_j^+ - \theta_j^-$ the **exterior angle** of α at $\alpha(t_j)$ and $\iota_j = \pi - \varepsilon_j$ the **interior angle** at $\alpha(t_j)$.

In Section 15.3 we defined the notion of area of a closed subset of a surface; that notion admits the following generalization:

Definition. Let $\mathcal{M}$ be an abstract surface and let $\mathbf{x}: \mathcal{U} \longrightarrow \mathcal{M}$ be an injective regular patch, and let $S \subseteq \mathbf{x}(\mathcal{U})$ be a closed subset, and let $f: S \longrightarrow \mathbb{R}$ be a continuous function. Then the integral of f over S is given by

$$\iint_S f\, dA = \iint_{\mathbf{x}^{-1}(S)} (f \circ \mathbf{x}) \sqrt{EG - F^2}\, du \wedge dv.$$

Just as in Section 15.3, we need to show that this definition is geometric. The proof of the following lemma is similar to that of Lemma 15.12, page 353 (see exercise 1).

Lemma 27.2. *The definition of the integral of f is independent of the choice of patch.*

We recall the definitions of boundary and region from general topology. The **boundary** of a subset S of an abstract surface $\mathcal{M}$ is the set

$\{\mathbf{p} \in \mathcal{M} \mid \text{each neighborhood of } \mathbf{p} \text{ contains a point in } S \text{ and a point not in } S\}.$

The boundary of S is denoted by ∂S. A **region** of $\mathcal{M}$ is the union of an open connected subset with its boundary.

A key tool for proving the Gauss-Bonnet theorem is Green's Theorem. (See, for example, [Buck, page 406].)

Theorem 27.3. (Green[3]) *Let $\mathcal{R} \subset \mathbb{R}^2$ be a simply connected region, and let $P, Q: \mathcal{U} \longrightarrow \mathbb{R}$ be differentiable functions, where $\mathcal{U}$ is an open set containing $\mathcal{R}$. Then*

$$\iint_{\mathcal{R}} \left(\frac{\partial Q}{\partial u} - \frac{\partial P}{\partial v} \right) du\, dv = \int_{\partial \mathcal{R}} P\, du + Q\, dv.$$

---

[3]George Green (1793-1841). English mathematician, owner of a windmill in Nottingham. Although he published only 10 papers, he made profound contributions to mathematical physics. He invented Green's functions and coined the term "potential function". Where Green acquired his mathematical skills remains a mystery.

We are now ready to prove the first version of the Gauss-Bonnet Theorem.

Theorem 27.4. (Gauss-Bonnet, local version) *Let $\mathcal{M}$ be an oriented abstract surface and let $\mathbf{x}:\mathcal{U} \longrightarrow \mathcal{M}$ be a regular patch coherent with the orientation of $\mathcal{M}$. Let $\mathcal{R} = \mathbf{x}([a,b] \times [c,d])$, where $[a,b] \times [c,d] \subseteq \mathcal{U}$. Then*

$$(27.2) \qquad \iint_{\mathcal{R}} K\, dA + \int_{\partial \mathcal{R}} \kappa_g\, ds + \sum_{j=0}^{3} \varepsilon_j = 2\pi,$$

where $\varepsilon_0, \varepsilon_1, \varepsilon_2, \varepsilon_3$ are the exterior angles at the vertices $\mathbf{p}_0, \mathbf{p}_1, \mathbf{p}_2, \mathbf{p}_3$ of $\mathcal{R}$.

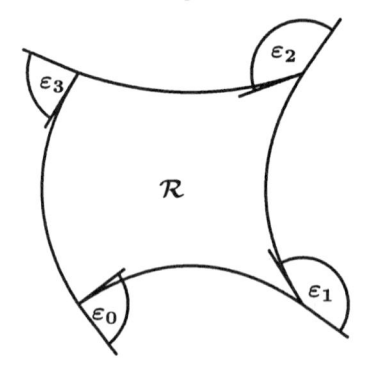

Proof. First, suppose that $\mathbf{x}$ is a patch with $F = 0$. The portion of $\partial \mathcal{R}$ between $\mathbf{p}_0$ and $\mathbf{p}_1$ is parametrized by $u \longmapsto \mathbf{x}(u,c)$, while the portion of $\partial \mathcal{R}$ between $\mathbf{p}_2$ and $\mathbf{p}_3$ is parametrized by $u \longmapsto \mathbf{x}(u,d)$; it follows that

$$\theta_0^+ = \theta_1^- = 0 \quad \text{and} \quad \theta_2^+ = \theta_3^- = \pi.$$

Thus the exterior angles are

$$\begin{cases} \varepsilon_0 = \theta_0^+ - \theta_0^- = -\theta_0^-, \\ \varepsilon_1 = \theta_1^+ - \theta_1^- = \theta_1^+, \\ \varepsilon_2 = \theta_2^+ - \theta_2^- = -\theta_2^- + \pi, \\ \varepsilon_3 = \theta_3^+ - \theta_3^- = \theta_3^+ - \pi, \end{cases}$$

so that

$$(27.3) \qquad \sum_{j=0}^{3} \left(\theta_{j+1}^- - \theta_j^+\right) = \left(\theta_2^- - \theta_1^+\right) + \left(\theta_0^- - \theta_3^+\right)$$

$$= -\varepsilon_2 + \pi - \varepsilon_1 - \varepsilon_0 - \varepsilon_3 + \pi = 2\pi - \sum_{j=0}^{3} \varepsilon_j.$$

27.1 The Local Gauss-Bonnet Theorem

By Green's Theorem (Theorem 27.3) and formula (22.9), page 505, we have

$$(27.4) \quad \int_{\partial \mathcal{R}} \left(\frac{G_u}{2\sqrt{EG}} \frac{dv}{dt} - \frac{E_v}{2\sqrt{EG}} \frac{du}{dt} \right) dt$$

$$= \int_a^b \int_c^d \left(\left(\frac{E_v}{2\sqrt{EG}} \right)_v + \left(\frac{G_u}{2\sqrt{EG}} \right)_u \right) du\, dv$$

$$= -\int_a^b \int_c^d K \sqrt{EG}\, du\, dv = -\iint_{\mathcal{R}} K\, dA.$$

From (27.3), (27.4) and (27.1) it follows that the total geodesic curvature of $\partial \mathcal{R}$ is given by

$$(27.5) \quad \int_{\partial \mathcal{R}} \kappa_g ds = \sum_{j=0}^{3} \int_{t_j}^{t_{j+1}} \kappa_g[\alpha](t) \|\alpha'(t)\| dt = 2\pi - \sum_{j=0}^{3} \varepsilon_j - \iint_{\mathcal{R}} K\, dA.$$

Then (27.5) is equivalent to (27.2).

Since all terms in (27.2) are invariantly defined, (27.2) holds for all patches. ∎

It will be convenient to rewrite (27.2) using interior angles instead of exterior angles.

Corollary 27.5. *Under the hypotheses of Theorem 27.4 we have*

$$(27.6) \quad \iint_{\mathcal{R}} K\, dA + \int_{\partial \mathcal{R}} \kappa_g ds = -2\pi + \sum_{j=0}^{3} \iota_j,$$

where $\iota_0, \iota_1, \iota_2, \iota_3$ are the interior angles at the vertices of $\mathcal{R}$.

In order to generalize Theorem 27.4 to more complicated regions, we introduce the following notion.

Definition. *Let $\mathcal{M}$ be an abstract surface. A **regular region** is a compact region whose boundary is the union of a finite number of piecewise-regular simple closed curves that do not intersect each other.*

Note that a compact regular surface can be considered as a regular region with empty boundary.

Each of the piecewise-regular simple closed curves that constitute the boundary $\partial \mathcal{Q}$ of a regular region $\mathcal{Q}$ has vertices and exterior angles; therefore, we can speak of the vertices $\mathbf{p}_1, \ldots, \mathbf{p}_m$ and the external angles $\varepsilon_1, \ldots, \varepsilon_m$ of the regular region $\mathcal{Q}$. Let us also call the portion of $\partial \mathcal{Q}$ between consecutive vertices an **edge** of $\mathcal{Q}$.

Definition. A **polygonal region** *is a regular region with a finite number of vertices, each of which has a nonzero external angle. A* **triangular region** *is a polygonal region with exactly 3 vertices.*

It will be necessary to decompose regular regions into smaller regions.

Definition. *A* **polygonal decomposition** *of a regular region Q of an abstract surface is a finite collection $\mathfrak{R}$ of polygonal regions called* **faces** *such that*

(i) *each $\mathcal{R} \in \mathfrak{R}$ is homeomorphic to a disk;*

(ii) *the union of all regions in $\mathfrak{R}$ is Q;*

(iii) *if $\mathcal{R}_1, \mathcal{R}_2 \in \mathfrak{R}$, then the intersection $\mathcal{R}_1 \cap \mathcal{R}_2$ is either a common edge or a common vertex of $\mathcal{R}_1$ and $\mathcal{R}_2$.*

An **oriented polygonal decomposition** *of Q is a polygonal decomposition in which the faces and edges are assigned orientations as follows:*

(iv) *each face of $\mathfrak{R}$ is assigned the orientation of $\mathcal{M}$;*

(v) *each edge of a face $\mathcal{R} \in \mathfrak{R}$, when parametrized as a curve $\boldsymbol{\alpha}$, is such that $J\boldsymbol{\alpha}'$ points toward the interior of $\mathcal{R}$.*

If all the faces of $\mathfrak{R}$ are triangular regions, then $\mathfrak{R}$ is called a **triangulation** *of Q.*

Notice that if γ is an edge of polygonal regions $\mathcal{R}_1$ and $\mathcal{R}_2$, then the orientation of γ with respect to $\mathcal{R}_1$ is the negative of the orientation of γ with respect to $\mathcal{R}_2$.

As a first step to generalizing Theorem 27.4, we derive the formula for the integral of the Gaussian curvature over a triangular region.

Corollary 27.6. *Let $\mathcal{M}$ be an oriented abstract surface, and let $\mathcal{T}$ be a triangular region of $\mathcal{M}$, such that $\mathcal{T} \subset \mathbf{x}(\mathcal{U})$, where $\mathbf{x}:\mathcal{U} \longrightarrow \mathcal{M}$ is a regular patch coherent with the orientation of $\mathcal{M}$. Then*

$$(27.7) \qquad \iint_{\mathcal{T}} K \, dA + \int_{\partial \mathcal{T}} \kappa_g \, ds = -\pi + \iota_1 + \iota_2 + \iota_3,$$

where $\iota_1, \iota_2, \iota_3$ are the interior angles at the vertices of $\mathcal{T}$.

27.1 The Local Gauss-Bonnet Theorem

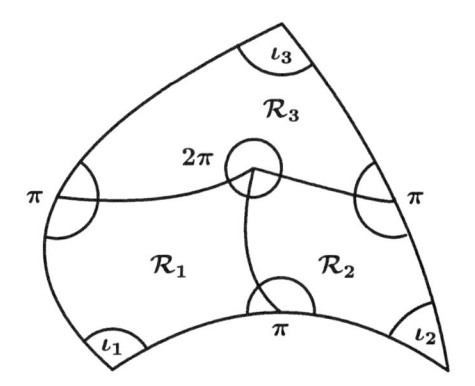

Proof. We construct an oriented polygonal decomposition of $\mathcal{T}$ by quadrilaterals as follows. Choose a point in the interior of the triangular region and points on each of the edges between the vertices. Join the three new points on the edges to the point in the center. Three quadrilaterals $\mathcal{R}_1$, $\mathcal{R}_2$, and $\mathcal{R}_3$ are formed; we choose them to have their orientations coherent with that of $\mathcal{T}$. Since $\mathcal{T}$ is contained in the trace of the patch $\mathbf{x}$, the three quadrilaterals can be chosen as images under $\mathbf{x}$ of ordinary quadrilaterals in $\mathbb{R}^2$. Therefore, because of additivity of integration over closed subsets of $\mathbb{R}^2$ and (27.6), we have

$$(27.8) \quad \iint_{\mathcal{T}} K\,dA = \sum_{k=1}^{3} \iint_{\mathcal{R}_k} K\,dA = -6\pi + I - \sum_{k=1}^{3} \int_{\partial \mathcal{R}_k} \kappa_g\,ds,$$

where I is the sum of all interior angles.

The boundaries $\partial \mathcal{R}_1$, $\partial \mathcal{R}_2$, and $\partial \mathcal{R}_3$ contribute six interior edges and six exterior edges. The integrals

$$\int \kappa_g\,ds$$

cancel in pairs over the interior edges and combine in pairs over the exterior edges; hence

$$(27.9) \quad \sum_{k=1}^{3} \int_{\partial \mathcal{R}_k} \kappa_g\,ds = \int_{\partial \mathcal{T}} \kappa_g\,ds.$$

Included in the sum I are the interior angles ι_1, ι_2, ι_3, together with the interior angles at the newly created vertices. It is easily seen that the interior angles at the newly created vertices add up to 5π, and so

$$(27.10) \quad I = 5\pi + \iota_1 + \iota_2 + \iota_3.$$

Now (27.7) follows from (27.8)–(27.10). ∎

Now we can state the original version of the theorem that Gauss proved in Section 20 of [Gauss].

Corollary 27.7. *Suppose that the sides of the triangular region T in Corollary 27.6 are geodesics. Then*

$$\iint_T K \, dA = -\pi + \iota_1 + \iota_2 + \iota_3, \tag{27.11}$$

where ι_1, ι_2, ι_3 are the interior angles at the vertices of T.

When $K = 0$, we obtain the familiar theorem that the sum of the angles of a triangle in the plane equals π. More generally:

Corollary 27.8. *Suppose that the sides of the triangular region T in Corollary 27.6 are geodesics, and that K has the constant value $a \neq 0$. Then the area of T is given by*

$$\mathbf{area}(T) = \frac{-\pi + \iota_1 + \iota_2 + \iota_3}{a}, \tag{27.12}$$

where ι_1, ι_2, ι_3 are the interior angles at the vertices of T.

When $a = 1$, equation (27.12) reduces to Harriot's formula for the area of a spherical triangle. Thus on a unit sphere, the sum of the interior angles of any geodesic triangle is greater than π, and the excess over π equals the area of the triangle.

27.2 Topology of Surfaces

In this chapter we summarize information about the topology of surfaces. For more information see [AhSa, Chapter 1], [Arm, Chapter 7], [Massey, Chapter 1] and [Springer, Chapter 5]. The most important theorem we need is:

Theorem 27.9. *Let $\mathcal{M}_1$ and $\mathcal{M}_2$ be compact abstract surfaces. The following conditions are equivalent:*

(i) *$\mathcal{M}_1$ and $\mathcal{M}_2$ are homeomorphic;*

(ii) *$\mathcal{M}_1$ and $\mathcal{M}_2$ are diffeomorphic;*

(iii) *$\mathcal{M}_1$ and $\mathcal{M}_2$ are both orientable or both nonorientable and have the same Euler characteristic.*

(In (i) and (ii) the homeomorphism and diffeomorphism are possibly different maps.) Here is the definition of Euler characteristic:

27.2 Topology of Surfaces

Definition. Let $\mathfrak{R}$ be a polygonal decomposition of a regular region $\mathcal{Q}$ of an abstract surface. We put

$$\begin{cases} V = \text{the number of vertices of } \mathfrak{R}, \\ E = \text{the number of edges of } \mathfrak{R}, \\ F = \text{the number of faces of } \mathfrak{R}. \end{cases}$$

Then $\chi(\mathcal{Q}) = F - E + V$ is called the **Euler characteristic** of $\mathcal{Q}$.

The Euler characteristic can also be described in terms of a sphere S_g with g handles. Here g is called the **geometric genus**. It is not difficult to show that $\chi(S_g) = 2(1-g)$.

For the proofs of the following propositions see [AhSa, Chapter 1], [Arm, Chapter 7], [Massey, Chapter 1] and [Springer, Chapter 5].

Proposition 27.10. *The Euler characteristic of a regular region of an abstract surface is invariant under the following processes:*

(i) *subdivision of an edge by adding a new vertex at the interior point of an edge;*

(ii) *subdivision of a polygonal region by connecting two vertices by a new edge;*

(iii) *introducing a new vertex into a polygonal region and connecting it by an edge to one of the vertices of the polygonal region.*

The proof of (27.10) is instructive and not difficult (see exercise 2). Clearly, it can be used to prove that any two polygonal decompositions of a regular region of an abstract surface are the same, provided the two polygonal decompositions have only a finite number of points in common. However, it can happen that two polygonal decompositions intersect each other in bizarre ways. Nevertheless:

Proposition 27.11. *The Euler characteristic of a regular region of an abstract surface is independent of the polygonal decomposition used to define it.*

Proposition 27.12. *A regular region of an abstract surface admits a triangulation.*

Proposition 27.13. *Let $\mathcal{M}$ be an oriented abstract surface, and suppose that $\mathfrak{A} = \{(\mathbf{x}_\alpha, \mathcal{U}_\alpha)\}$ is a family of parametrizations covering $\mathcal{M}$ compatible with the orientation of $\mathcal{M}$. Then there is an oriented triangulation $\mathfrak{T}$ of $\mathcal{M}$ such that every triangular region $\mathcal{T} \in \mathfrak{T}$ is contained in some coordinate neighborhood $\mathcal{U}_\mathcal{T}$ for some $(\mathbf{x}_\mathcal{T}, \mathcal{U}_\mathcal{T}) \in \mathfrak{A}$. Furthermore, if the boundary of every triangular region in $\mathfrak{T}$ is positively oriented, then adjacent triangles determine opposite orientations in common edges.*

Proposition 27.14. *Assume that M, $\mathfrak{A}$ and $\mathfrak{T}$ are as in Proposition 27.13, and let $ds_\alpha^2 = E_\alpha du_\alpha^2 + 2F_\alpha du_\alpha dv_\alpha + G_\alpha dv_\alpha^2$ be the expression of the metric of M in terms of the coordinates given by $\mathbf{x}_\alpha$. Let Q be a regular region, and let $f\colon Q \longrightarrow \mathbb{R}$ be a continuous function. Then*

$$\sum_{T \in \mathfrak{T}} \iint_{\mathbf{x}_\alpha^{-1}(T)} (f \circ \mathbf{x}_\alpha) \sqrt{E_\alpha G_\alpha - F_\alpha^2} \, du_\alpha \wedge dv_\alpha$$

is the same for any triangulation $\mathfrak{T}$ of Q, such that every triangular region $T \in \mathfrak{T}$ is contained in some coordinate neighborhood $\mathcal{U}_\alpha$ for some $(\mathbf{x}_\alpha, \mathcal{U}_\alpha) \in \mathfrak{A}$.

27.3 The Global Gauss-Bonnet Theorem

The Gauss-Bonnet Theorem is concerned with subsets of abstract surfaces whose boundaries are piecewise-regular curves. More precisely:

Theorem 27.15. *(Gauss-Bonnet, global version) Let Q be a regular region of an oriented abstract surface M, and let $C_1, \ldots, C_n$ be piecewise-regular curves which form the boundary of Q. Suppose that each C_i is positively oriented. Let $\varepsilon_1, \ldots, \varepsilon_p$ be the external angles of $C_1, \ldots, C_n$. Then*

$$(27.13) \qquad \iint_Q K \, dA + \int_{\partial Q} \kappa_g ds + \sum_{j=1}^p \varepsilon_j = 2\pi \chi(Q).$$

Proof. Let $\mathfrak{T}$ be a triangulation of Q such that every triangular region in $\mathfrak{T}$ is contained in a coordinate neighborhood of a family of orthogonal patches. We give the positive orientation to each triangular region in $T \in \mathfrak{T}$. In this way, adjacent triangular regions give opposite orientations to their common edge.

We apply Corollary 27.6 to every triangular region in $\mathfrak{T}$ and add up the results. The integrals of the geodesic curvature along the interior edges (that is, those edges that are not contained in one of the boundary curves) cancel one another. We obtain

$$(27.14) \qquad \iint_\mathcal{R} K \, dA + \sum_{i=1}^n \int_{C_i} \kappa_g ds = -\pi F + \sum \iota_{ij},$$

where F is the number of triangular regions in $\mathfrak{T}$ and $\sum \iota_{ij}$ is the sum of all interior angles. Let

$$V_e = \text{number of external vertices of } \mathfrak{T},$$

27.3 The Global Gauss-Bonnet Theorem

$$V_i = \text{number of internal vertices of } \mathfrak{T},$$
$$E_e = \text{number of external edges of } \mathfrak{T},$$
$$E_i = \text{number of internal edges of } \mathfrak{T}.$$

The angles around each internal vertex add up to 2π; hence the sum of all internal interior angles is $2\pi V_i$. A similar calculation computes the sum of the external interior angles, and so

$$\sum \iota_{ij} = 2\pi V_i + \pi V_e - \sum_{j=1}^{p} \varepsilon_j.$$

Thus (27.14) can be rewritten as

(27.15) $$\iint_{\mathcal{R}} K\,dA + \sum_{i=1}^{n} \int_{C_i} \kappa_g\,ds + \sum_{j=1}^{p} \varepsilon_j = \pi(-F + 2V_i + V_e).$$

We have the relations

(27.16) $$V_e = E_e, \quad V = V_e + V_i, \quad E = E_e + E_i.$$

Furthermore, $3F = 2E_i + E_e$, which can be written as

(27.17) $$-F = 2F - 2E_i - E_e.$$

From (27.15)—(27.17) we compute

$$\iint_{\mathcal{R}} K\,dA + \sum_{i=1}^{n} \int_{C_i} \kappa_g\,ds + \sum_{j=1}^{p} \varepsilon_j = \pi(2F - 2E_i - E_e + 2V_i + V_e)$$
$$= \pi(2F - 2E_i - E_e + 2V - 2V_e + V_e)$$
$$= \pi(2F - 2E + 2E_e - E_e + 2V - 2V_e + V_e)$$
$$= \pi(2F - 2E + 2V) = 2\pi\chi(\mathcal{Q}),$$

proving (27.13). ∎

In the special case of a compact surface we obtain:

Corollary 27.16. *Let $\mathcal{M}$ be a compact orientable abstract surface. Then*

(27.18) $$\iint_{\mathcal{M}} K\,dA = 2\pi\chi(\mathcal{M}),$$

where K is the Gaussian curvature with respect to any metric on $\mathcal{M}$.

Proof. A compact surface has no boundary. Hence (27.13) reduces to (27.18). ∎

In 1828 Gauss proved Theorem 27.15 for the special case of polygonal regions whose sides are geodesics, and Bonnet generalized it to the form above in 1848 (see [Gauss] and [Bonn1]). The use of Green's Theorem in proving Theorem 27.4 was clarified by Darboux ([Darb2, volume 3, pages 122–128]).

27.4 Applications of the Gauss-Bonnet Theorem

A simple but important application of Corollary 27.16 is the determination of the diffeomorphism type of compact orientable surfaces with positive Gaussian curvature.

Theorem 27.17. *A compact orientable surface $\mathcal{M}$ with positive Gaussian curvature is diffeomorphic to a sphere.*

Proof. If $\mathcal{M}$ has positive Gaussian curvature, Corollary 27.16 implies that $\mathcal{M}$ has positive Euler characteristic. Since $\mathcal{M}$ is orientable, it follows from Theorem 27.9 that $\mathcal{M}$ is diffeomorphic to a sphere. ∎

Next, we prove a theorem of Jacobi (see [Jacobi]).

Theorem 27.18. (Jacobi, 1843) *Let $\alpha\colon [a,b] \longrightarrow \mathbb{R}^3$ be a closed regular curve with nonzero curvature. Assume that the curve described by the normal $\mathbf{N}$ (considered as a curve in the unit sphere $S^2(1) \subset \mathbb{R}^3$) is a simple closed curve. Then $\mathbf{N}([a,b])$ divides $S^2(1)$ into two regions of equal areas.*

Proof. Without loss of generality, we can assume that α has unit-speed. Let s denote the arc length function of α, and let $\tilde{s}$ denote the arc length function of $\mathbf{N}$, considered as a curve on the unit sphere. The second Frenet formula of (8.10), page 186, implies that

$$\frac{d\mathbf{N}}{ds} = -\kappa\,\mathbf{T} + \tau\,\mathbf{B},$$

and

$$\frac{d^2\mathbf{N}}{ds^2} = -\frac{d\kappa}{ds}\mathbf{T} + \frac{d\tau}{ds}\mathbf{B} - (\kappa^2 + \tau^2)\mathbf{N}.$$

27.4 Applications of the Gauss-Bonnet Theorem

From these two equations we get

(27.19) $$\mathbf{N} \times \frac{d\mathbf{N}}{ds} \cdot \frac{d^2\mathbf{N}}{ds^2} = \kappa \frac{d\tau}{ds} - \tau \frac{d\kappa}{ds}.$$

and

(27.20) $$\frac{d\mathbf{N}}{ds} \cdot \frac{d\mathbf{N}}{ds} = \kappa^2 + \tau^2.$$

Equation (27.20) implies that

(27.21) $$1 = \frac{d\mathbf{N}}{d\tilde{s}} \cdot \frac{d\mathbf{N}}{d\tilde{s}} = \left(\frac{ds}{d\tilde{s}}\right)^2 (\kappa^2 + \tau^2).$$

From (27.19)—(27.21) we obtain

(27.22) $$\kappa_g(\mathbf{N}) = \frac{\mathbf{N} \times \frac{d\mathbf{N}}{ds} \cdot \frac{d^2\mathbf{N}}{ds^2}}{\left\|\frac{d\mathbf{N}}{ds}\right\|^3} = \frac{\kappa \frac{d\tau}{ds} - \tau \frac{d\kappa}{ds}}{(\kappa^2 + \tau^2)^{3/2}} = \frac{\kappa \frac{d\tau}{ds} - \tau \frac{d\kappa}{ds}}{\kappa^2 + \tau^2} \frac{ds}{d\tilde{s}}$$

$$= \frac{d}{ds} \arctan\left(\frac{\tau}{\kappa}\right) \frac{ds}{d\tilde{s}} = \frac{d}{d\tilde{s}} \arctan\left(\frac{\tau}{\kappa}\right).$$

Since $\mathbf{N}([a,b])$ is a closed curve, equation (27.22) implies that

(27.23) $$\int_{\mathbf{N}([a,b])} \kappa_g(\mathbf{N}) d\tilde{s} = \int_{\mathbf{N}([a,b])} \frac{d}{d\tilde{s}} \arctan\left(\frac{\tau}{\kappa}\right) d\tilde{s}$$

$$= \int_{\mathbf{N}([a,b])} d \arctan\left(\frac{\tau}{\kappa}\right) = 0.$$

Let $\mathcal{R}$ be the region enclosed by $\mathbf{N}([a,b])$; then $\chi(\mathcal{R}) = 1$ because $\mathbf{N}([a,b])$ is a simple curve. Therefore, from (27.13) and (27.23) we obtain

$$2\pi = 2\pi\chi(\mathcal{R}) = \int_{\mathcal{R}} K\, dA = \mathbf{area}(\mathcal{R}).$$

Since the total area of the sphere is 4π, the theorem follows. ∎

Finally, we state without proof an important generalization of Corollary 27.16 due to Cohn-Vossen [CV].

Theorem 27.19. *Let $\mathcal{M}$ be a complete orientable abstract surface. Then*

(27.24) $$\iint_{\mathcal{M}} K\, dA \leq 2\pi\chi(\mathcal{M}),$$

provided both sides of (27.24) are finite.

27.5 Exercises

1. Prove Lemma 27.2 and Corollary 27.5.

2. Prove Proposition 27.10.

3. Let $M \subset \mathbb{R}^3$ be a compact orientable surface not homeomorphic to a sphere. Show that there are points where the Gaussian curvature is positive, negative and zero.

4. Let
$$T(a,b) = \left\{ (x,y,z) \in \mathbb{R}^3 \ \big| \ z^2 + \left(\sqrt{x^2+y^2} - a\right)^2 = b^2 \right\}$$

   be a torus of wheel radius a and tube radius b (see Section 13.4). Use the parametrization of $T(a,b)$ on page 384 to compute
$$\int_{T(a,b)} K \, dA \quad \text{and} \quad \int_{T(a,b)} H^2 \, dA.$$

   Do the calculations both by hand and with **Mathematica**. What does the Gauss-Bonnet Theorem tell us about the integral of the Gaussian curvature over $T(a,b)$? Is the integral over a surface of the square of the mean curvature a topological invariant?

5. Show that
$$\int_{S^2(a)} H^{2n} \, dA.$$

   is independent of the radius of the sphere $S^2(a)$ if and only if $n = 2$.

6. Let M be an abstract surface homeomorphic to a cylinder with Gaussian curvature $K < 0$. Show that M has at most one simple closed geodesic.

28
PRINCIPAL CURVES AND UMBILIC POINTS

In this chapter we continue our study of special curves on surfaces. A *principal curve* on a surface is a curve whose velocity vector always points in a *principal direction*, that is, a direction in which the normal curvature is a maximum or a minimum. In Section 28.1 we derive the differential equation for the principal curves on a patch in $\mathbb{R}^3$ and give examples of its solution. An *umbilic point* on a surface is a point at which the principal curvatures are equal. Every point of a sphere is an umbilic point. We discuss umbilic points in Section 28.2; we show, for example, that if a regular surface $\mathcal{M} \subset \mathbb{R}^3$ consists entirely of umbilic points, then it is part of a plane or sphere. For surfaces such as an ellipsoid

$$\frac{x^2}{a^2} + \frac{y^2}{b^2} + \frac{z^2}{c^2} = 1$$

with a, b and c distinct, the umbilic points are isolated and can be considered to be degenerate principal curves. In fact, each looks very much like a navel, hence the name "umbilic". (The 4 umbilic points on an ellipsoid (with a, b and c distinct) are easy to locate visually, provided one draws the ellipsoid so that the principal curves are visible. See the picture on page 673.)

The coefficients of the first and second fundamental forms of a regular surface $\mathcal{M}$ in $\mathbb{R}^3$ are not independent of one another. The Peterson-Mainardi-Codazzi equations, derived in Section 28.3, consist of two equations that relate the derivatives of the coefficients of the second fundamental form in terms of the coefficients of the first and second fundamental forms themselves. In Section 28.4 we use the Peterson-

Mainardi-Codazzi equations to prove Hilbert's Lemma (Lemma 28.11), which in turn is used to prove a global result, Liebmann's Theorem (Theorem 28.12).

Finally, in Section 28.5 we prove the *Fundamental Theorem of Surfaces*; this important theorem is the surface analog of the *Fundamental Theorem of Plane Curves*, proved in Section 6.7 and the *Fundamental Theorem of Surface Curves*, proved in Section 10.1.

28.1 The Differential Equation for the Principal Curves of a Surface

In this section we first determine when a tangent vector to a patch is a principal vector. Then we derive the differential equation for the principal curves.

Lemma 28.1. *Let* $\mathbf{x}: \mathcal{U} \longrightarrow \mathbb{R}^3$ *be a regular patch. A tangent vector* $\mathbf{v_p} = v_1 \mathbf{x}_u + v_2 \mathbf{x}_v$ *is a principal vector if and only if*

$$(28.1) \qquad \det \begin{pmatrix} v_2^2 & -v_1 v_2 & v_1^2 \\ E & F & G \\ e & f & g \end{pmatrix} = 0.$$

Proof. Using the notation of Theorem 16.10, page 369, we compute $S(\mathbf{v_p}) \times \mathbf{v_p}$ and apply Lemma 20.1, page 459:

$$(28.2) \; S(\mathbf{v_p}) \times \mathbf{v_p} = S(v_1 \mathbf{x}_u + v_2 \mathbf{x}_v) \times (v_1 \mathbf{x}_u + v_2 \mathbf{x}_v)$$

$$= \left(-a_{12} v_1^2 + (a_{11} - a_{22}) v_1 v_2 + a_{21} v_2^2 \right) \mathbf{x}_u \times \mathbf{x}_v$$

$$= \left(\left(-\frac{eF - fE}{EG - F^2} \right) v_1^2 + \left(\frac{fF - eG}{EG - F^2} - \frac{fF - gE}{EG - F^2} \right) v_1 v_2 \right.$$

$$\left. + \left(\frac{gF - fG}{EG - F^2} \right) v_2^2 \right) \mathbf{x}_u \times \mathbf{x}_v$$

$$= \left((fE - eF) v_1^2 + (gE - eG) v_1 v_2 + (gF - fG) v_2^2 \right) \frac{\mathbf{x}_u \times \mathbf{x}_v}{EG - F^2}.$$

From Lemma 20.1, page 459, we know that $\mathbf{v_p}$ is principal if and only if $S(\mathbf{v_p}) \times \mathbf{v_p} = 0$. Hence from (28.2) it follows that $\mathbf{v_p}$ is principal if and only if

$$(28.3) \qquad (fE - eF) v_1^2 + (gE - eG) v_1 v_2 + (gF - fG) v_2^2 = 0.$$

28.1 The Differential Equation for the Principal Curves of a Surface

Then (28.3) is another way of writing (28.1). ∎

Here is the differential equation for the principal curves analogous to the differential equation (18.1), page 419, for the asymptotic curves.

Corollary 28.2. *Let α be a curve that lies on the trace of a patch* **x**. *Write* $\alpha(t) = \mathbf{x}(u(t), v(t))$. *Then α is a principal curve if and only if*

$$(28.4) \quad (fE - eF)(\alpha(t))u'(t)^2 + (gE - eG)(\alpha(t))u'(t)v'(t) \\ + (gF - fG)(\alpha(t))v'(t)^2 = 0$$

for all t.

Proof. Equation (28.4) is an immediate consequence of (28.3). ∎

Let us use (28.4) to find the principal curves on the helicoid given by

$$\mathbf{helicoid}[a, b](u, v) = (a\, v \cos u,\, a\, v \sin u,\, b\, u).$$

Even better, we find a principal patch that reparametrizes **helicoid**$[a, b]$. The coefficients of the first and second fundamental forms of **helicoid**$[a, b]$ are easily computed to be

$$F = e = g = 0, \quad E = b^2 + a^2 v^2, \quad G = a^2, \quad \text{and} \quad f = \frac{ab}{b^2 + a^2 v^2}.$$

The differential equation for the principal curves reduces to $fEu'^2 - fGv'^2 = 0$, or

$$(28.5) \quad (b^2 + a^2 v^2)u'^2 = a^2 v'^2.$$

We integrate (28.5), obtaining two solutions:

$$u - 2p = \sinh^{-1}\left(\frac{a v}{b}\right) \quad \text{and} \quad u - 2q = -\sinh^{-1}\left(\frac{a v}{b}\right).$$

Hence

$$u = p + q \quad \text{and} \quad v = \frac{b}{a}\sinh(-p + q).$$

Then to reparametrize **helicoid**$[a, b]$, we use

```
helicoid[a,b][u,v] /. {u->p + q,v->(b/a)Sinh[-p + q]}
```

to obtain

```
{-(b Cos[p + q] Sinh[p - q]), -(b Sin[p + q] Sinh[p - q]), b (p + q)}
```

644   Chapter 28   Principal Curves and Umbilic Points

Thus we get a principal patch reparametrizing the **helicoid**$[a, b]$ by defining

helicoidprin$[b](p, q) = \bigl(b\sinh(q - p)\cos(p + q), b\sinh(q - p)\sin(p + q), b(p + q)\bigr).$

or

```
helicoidprin[b_][p_,q_]:=  {b Cos[p + q]Sinh[q - p],
     b Sin[p + q]Sinh[q - p],b(p + q)}
```

To draw the principal curves on a helicoid, we first use **ParametricPlot3D** to create the polygons for the plot of the helicoid without displaying them:

```
hp=ParametricPlot3D[Evaluate[helicoidprin[1][p,q]],
     {p,-2Pi,2Pi},{q,-2Pi,2Pi},PlotPoints->{80,80},
     PlotRange->All,DisplayFunction->Identity];
```

With this command *Mathematica* has made a list **hp** consisting of two elements. The second element is a sublist of graphics options, and the first element is the sublist of polygons created by **ParametricPlot3D**. To throw away all polygons with a vertex (x_1, x_2, x_3) for which $|x_j| \geq 5$ for some j, we use the miniprogram **selectgraphics3d** contained in **PLTPROGS.m** and described on page 425:

```
selectgraphics3d[hp[[1]],5]
```

The result is

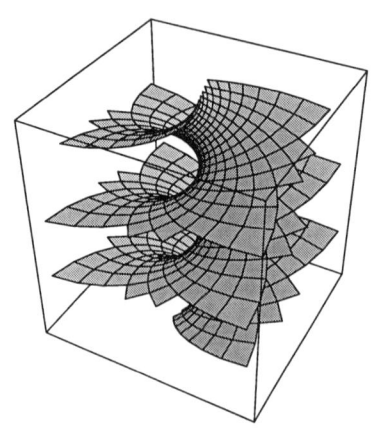

Principal curves on a helicoid

It is interesting to compare this plot of a helicoid by principal curves with the plot by asymptotic curves on page 686.

28.2 Umbilic Points

In this section we consider those points on a surface in $\mathbb{R}^3$ for which all the normal curvatures are equal.

Definition. *Let $\mathcal{M}$ be a regular surface in $\mathbb{R}^3$. A point $\mathbf{p} \in \mathcal{M}$ is called an **umbilic point** provided the principal curvatures at $\mathbf{p}$ are equal: $\mathbf{k}_1(\mathbf{p}) = \mathbf{k}_2(\mathbf{p})$.*

It is clear that all points of a plane or a sphere are umbilic points. We shall see shortly that the vertex $(0,0,0)$ of the paraboloid $z = x^2 + y^2$ is an isolated umbilic point.

Lemma 28.3. *A point $\mathbf{p}$ on a regular surface $\mathcal{M} \subset \mathbb{R}^3$ is an umbilic point if and only if the shape operator of $\mathcal{M}$ at $\mathbf{p}$ is a multiple of the identity.*

Proof. Let $S_\mathbf{p}$ denote the shape operator of $\mathcal{M}$ at $\mathbf{p}$. Then $S_\mathbf{p}$ is a multiple of the identity if and only if all of the normal curvatures of $\mathcal{M}$ at $\mathbf{p}$ coincide. This is true if and only if the maximum and minimum of the normal curvatures, namely, $\mathbf{k}_1(\mathbf{p})$ and $\mathbf{k}_2(\mathbf{p})$, coincide. Write $\mathbf{k}(\mathbf{p}) = \mathbf{k}_1(\mathbf{p}) = \mathbf{k}_2(\mathbf{p})$; then $S_\mathbf{p} = \mathbf{k}(\mathbf{p}) I_\mathbf{p}$, where $I_\mathbf{p}$ denotes the identity map on $\mathcal{M}_\mathbf{p}$. ∎

First, let us determine the regular surfaces that consist entirely of umbilic points.

Lemma 28.4. *If all of the points of a connected regular surface $\mathcal{M} \subset \mathbb{R}^3$ are umbilic, then $\mathcal{M}$ has constant Gaussian curvature $K \geq 0$.*

Proof. Let $\mathbf{k}: \mathcal{M} \longrightarrow \mathbb{R}$ be the function such that $\mathbf{k}(\mathbf{q})$ is the common value of the principal curvatures at $\mathbf{q} \in \mathcal{M}$. Let $\mathbf{p} \in \mathcal{M}$, and let $\mathbf{x}: \mathcal{U} \longrightarrow \mathcal{M}$ be a regular patch such that $\mathbf{p} \in \mathbf{x}(\mathcal{U})$. We can assume that $\mathcal{U}$ is connected. Then we have

$$(28.6) \qquad \mathbf{U}_u = -S(\mathbf{x}_u) = -\mathbf{k}\,\mathbf{x}_u,$$

$$(28.7) \qquad \mathbf{U}_v = -S(\mathbf{x}_v) = -\mathbf{k}\,\mathbf{x}_v.$$

We differentiate (28.6) with respect to v and (28.7) with respect to u, then subtract the results. Since $\mathbf{U}_{uv} = \mathbf{U}_{vu}$ and $\mathbf{x}_{uv} = \mathbf{x}_{vu}$, we obtain

$$(28.8) \qquad \mathbf{k}_v \mathbf{x}_u - \mathbf{k}_u \mathbf{x}_v = 0.$$

The linear independence of $\mathbf{x}_u$ and $\mathbf{x}_v$ together with (28.8) implies that $\mathbf{k}$ is constant on the neighborhood $\mathcal{U}$ of $\mathbf{p}$. Moreover, the Gaussian curvature K is also constant on $\mathcal{U}$, because it is given by $K = \mathbf{k}^2 \geq 0$.

Let $\mathcal{V} = \{\mathbf{q} \in \mathcal{M} \mid K(\mathbf{q}) = K(\mathbf{p})\}$. Then $\mathcal{V}$ is obviously closed, and by the previous argument it is also open. Since $\mathcal{M}$ is connected, $\mathcal{V} = \mathcal{M}$. Hence K is a constant nonnegative function on $\mathcal{M}$. ∎

To determine the connected regular surface consisting entirely of umbilic points, we imitate the proof of Theorem 1.12, page 16.

Theorem 28.5. *Let $\mathcal{M}$ be a connected regular surface in $\mathbb{R}^3$ consisting entirely of umbilic points. Then $\mathcal{M}$ is part of a plane or sphere.*

Proof. We know from Lemma 28.4 that $\mathcal{M}$ has constant curvature $K \geq 0$, and that $K = \mathbf{k}^2$, where $\mathbf{k}$ is the common value of the principal curvatures. If $K = 0$, then both $\mathbf{k}$ and S vanish identically. Hence $D_\mathbf{v} \mathbf{U} = 0$ for any tangent vector $\mathbf{v}$ to $\mathcal{M}$. Therefore, for each $\mathbf{p} \in \mathcal{M}$ we have $\mathbf{U}(\mathbf{p}) = (n_1, n_2, n_3)_\mathbf{p}$, where the n_i are constants. Then $\mathcal{M}$ is contained in the plane perpendicular to $(n_1, n_2, n_3)_\mathbf{p}$.

Next, suppose $K > 0$. Choose a point $\mathbf{p} \in \mathcal{M}$ and a unit normal vector $\mathbf{U}(\mathbf{p})$ to $\mathcal{M}$ at $\mathbf{p}$. We show that the point $\mathbf{c}$ defined by

$$\mathbf{c} = \mathbf{p} + \frac{1}{\mathbf{k}(\mathbf{p})}\mathbf{U}(\mathbf{p})$$

is equidistant from all points of $\mathcal{M}$. To this end, let $\mathbf{q}$ be any point of $\mathcal{M}$, and let $\alpha\colon (a,b) \longrightarrow \mathcal{M}$ be a curve with $a < 0 < 1 < b$ such that $\alpha(0) = \mathbf{p}$ and $\alpha(1) = \mathbf{q}$. We extend $\mathbf{U}(\mathbf{p})$ to a unit normal vector field $U \circ \alpha$ along α.

Define a new curve $\gamma\colon (a,b) \longrightarrow \mathbb{R}^3$ by

$$\gamma(t) = \alpha(t) + \frac{1}{\mathbf{k}(\alpha(t))}\mathbf{U}(\alpha(t)).$$

Since $t \longmapsto \mathbf{k}(\alpha(t))$ is constant, we have

(28.9) $$\gamma'(t) = \alpha'(t) + \frac{1}{\mathbf{k}(\alpha(t))}(U \circ \alpha)'(t).$$

But

$$(U \circ \alpha)'(t) = -S(\alpha'(t)) = -\mathbf{k}(\alpha(t))\alpha'(t),$$

so that (28.9) reduces to

$$\gamma' = 0.$$

Hence γ must be a constant, that is, a single point in $\mathbb{R}^3$, in fact, the point $\mathbf{c}$. Thus

(28.10) $$\mathbf{c} = \gamma(0) = \gamma(1) = \mathbf{q} + \frac{1}{\mathbf{k}(\mathbf{q})}\mathbf{U}(\mathbf{q}).$$

28.2 Umbilic Points

From (28.10) it is immediate that

$$\|\mathbf{c} - \mathbf{q}\| = \frac{1}{|\mathbf{k}(\mathbf{q})|}.$$

Thus $\mathcal{M}$ is part of a sphere of radius $1/|\mathbf{k}(\mathbf{q})|$ with center $\mathbf{c}$. ∎

Next, we give a useful criterion for finding umbilic points.

Lemma 28.6. *A point $\mathbf{p}$ on a patch $\mathbf{x}: \mathcal{U} \longrightarrow \mathbb{R}^3$ is an umbilic point if and only if there is a number k such that at $\mathbf{p}$ we have*

(28.11) $$e = kE, \qquad f = kF, \qquad g = kG.$$

Proof. $\mathbf{p}$ is umbilic if and only if equation (28.3) holds for all v_1 and v_2. It is easy to prove that this condition is equivalent to (28.11). ∎

Let us use *Mathematica* as an aid to finding the umbilics on the elliptic paraboloid

(28.12) $$z = \frac{x^2}{a^2} + \frac{y^2}{b^2}.$$

The *Mathematica* definition of the obvious parametrization of this surface is

```
ellipticparaboloid[a_,b_][u_,v_]:=
              {u,v,u^2/a^2 + v^2/b^2}
```

To find the umbilic points, it is probably best to combine *Mathematica* with ordinary mathematical reasoning. Although *Mathematica* usually is faster and more accurate than hand calculation for symbolic manipulation, frequently it is possible to use a mathematical argument that would be too tedious to program into *Mathematica*.

First, let us define a useful *Mathematica* function.

```
umeqs[k_][x_][u_,v_]:=
    {ee[x][u,v],ff[x][u,v],gg[x][u,v]} -
    k{eee[x][u,v],fff[x][u,v],ggg[x][u,v]}
```

The vanishing of **umeqs** is equivalent to (28.11). Then the command **umeqs[k][ellipticparaboloid[a,b]][u,v]** elicits the response

$$\{1 + \frac{4u^2}{a^4} - \frac{2k}{a^2 \, \text{Sqrt}[1 + \frac{4u^2}{a^4} + \frac{4v^2}{b^4}]}, \frac{4uv}{a^2 b^2},$$

$$1 + \frac{4v^2}{b^4} - \frac{2k}{b^2 \, \text{Sqrt}[1 + \frac{4u^2}{a^4} + \frac{4v^2}{b^4}]}\}$$

A glance at the middle coordinate tells us that at an umbilic point either $u = 0$ or $v = 0$. We try letting $u = 0$ and solving the resulting equations with *Mathematica*:

```
Solve[umeqs[k][ellipticparaboloid[a,b]][0,v][[1]] == 0,k]
```

to which *Mathematica* answers

$$\{\{k \to \frac{a^2 \, \text{Sqrt}[\frac{b^4 + 4v^2}{b^4}]}{2}\}\}$$

and

```
Solve[umeqs[k][ellipticparaboloid[a,b]][0,v][[3]] == 0,k]
```

to which *Mathematica* answers

$$\{\{k \to \frac{(b^4 + 4v^2) \, \text{Sqrt}[\frac{b^4 + 4v^2}{b^4}]}{2b^2}\}\}$$

This output shows that to determine **v**, we need the command

```
Solve[a^2 == (b^4 + 4 v^2)/b^2,v]
```

Mathematica answers

$$\{\{v \to \frac{b\sqrt{a^2-b^2}}{2}\}, \{v \to \frac{-(b\sqrt{a^2-b^2})}{2}\}\}$$

Thus if $a \geq b > 0$, we find the umbilic points to be $\left(0, \pm(b/2)\sqrt{a^2-b^2}\right)$. Hence the elliptic paraboloid given by (28.12) has one umbilic point if $a = b$, otherwise it has two umbilic points.

28.3 The Peterson-Mainardi-Codazzi Equations

The nine coefficients in the Gauss equations (22.14), page 511, are not independent. It turns out that there is a relation among e, f and g, their derivatives and the Christoffel symbols. These equations were proved by Gauss ([Gauss, paragraph 11]) using obscure notation, then reproved successively by Peterson[1] in his thesis [Pet], Mainardi[2] [Main] and Codazzi[3] [Codaz]. See [Reich, page 303-305] and [Cool2] for details on the history of these equations.

Theorem 28.7. (The Peterson-Mainardi-Codazzi Equations) Let $\mathbf{x}: \mathcal{U} \longrightarrow \mathbb{R}^3$ be a regular patch. Then

$$(28.13) \quad \begin{cases} \dfrac{\partial e}{\partial v} - \dfrac{\partial f}{\partial u} = e\,\Gamma_{12}^1 + f\left(\Gamma_{12}^2 - \Gamma_{11}^1\right) - g\,\Gamma_{11}^2, \\ \dfrac{\partial f}{\partial v} - \dfrac{\partial g}{\partial u} = e\,\Gamma_{22}^1 + f\left(\Gamma_{22}^2 - \Gamma_{12}^1\right) - g\,\Gamma_{12}^2. \end{cases}$$

Proof. The idea of the proof is to differentiate the equations of (22.14) and use the fact that $\mathbf{x}_{uuv} = \mathbf{x}_{uvu}$ and $\mathbf{x}_{uvv} = \mathbf{x}_{vvu}$. Thus from the first two equations of

---

[1] Karl Mikhailovich Peterson (1828-1881). A student of Minding and Senff in Dorpat (Tartu). Peterson was one of the founders of the Moscow Mathematical Society. Peterson's derivation on the equations of Theorem 28.7 was not generally known in his lifetime.

[2] Gaspare Mainardi (1800-1879). Professor at the University of Padua.

[3] Delfino Codazzi (1824-1873). Professor at the University of Padua. The equations (28.13) were proved by Mainardi [Main] in 1856. However, Codazzi's formulation was simpler because he was careful that his expressions had geometric meaning, and his applications were wider. Since Codazzi submitted an early version of his work to the French Academy of Sciences for a prize competition in 1859, his contribution was known long before those of Peterson and Mainardi. Codazzi also published papers on geodesic triangles, equiareal mapping and the stability of floating bodies.

(22.14) it follows that

(28.14) $\quad 0 = \mathbf{x}_{uuv} - \mathbf{x}_{uvu}$

$$= \frac{\partial}{\partial v}\left(\Gamma^1_{11}\mathbf{x}_u + \Gamma^2_{11}\mathbf{x}_v + e\,\mathbf{U}\right) - \frac{\partial}{\partial u}\left(\Gamma^1_{12}\mathbf{x}_u + \Gamma^2_{12}\mathbf{x}_v + f\,\mathbf{U}\right)$$

$$= \left(\frac{\partial \Gamma^1_{11}}{\partial v} - \frac{\partial \Gamma^1_{12}}{\partial u}\right)\mathbf{x}_u + \left(\frac{\partial \Gamma^2_{11}}{\partial v} - \frac{\partial \Gamma^2_{12}}{\partial u}\right)\mathbf{x}_v + (e_v - f_u)\mathbf{U}$$
$$- \Gamma^1_{12}\mathbf{x}_{uu} + \left(\Gamma^1_{11} - \Gamma^2_{12}\right)\mathbf{x}_{uv} + \Gamma^2_{11}\mathbf{x}_{vv} + e\,\mathbf{U}_v - f\,\mathbf{U}_u.$$

We expand (28.14) and use (22.14) again to expand $\mathbf{x}_{uu}$, $\mathbf{x}_{uv}$ and $\mathbf{x}_{vv}$. We also need the Weingarten equations (16.11), page 369. After collecting terms we obtain:

(28.15) $\quad 0 = \left(\dfrac{\partial \Gamma^1_{11}}{\partial v} - \dfrac{\partial \Gamma^1_{12}}{\partial u} - \Gamma^2_{12}\Gamma^1_{12} + \Gamma^2_{11}\Gamma^1_{22} + e\dfrac{gF - fG}{EG - F^2} - f\dfrac{fF - eG}{EG - F^2}\right)\mathbf{x}_u$

$\quad + \left(\dfrac{\partial \Gamma^2_{11}}{\partial v} - \dfrac{\partial \Gamma^2_{12}}{\partial u} + \Gamma^1_{11}\Gamma^2_{12} + \Gamma^2_{11}\Gamma^2_{22} - \Gamma^1_{12}\Gamma^2_{11} - \left(\Gamma^2_{12}\right)^2\right.$

$\quad\quad \left. + e\dfrac{fF - gE}{EG - F^2} - f\dfrac{eF - fE}{EG - F^2}\right)\mathbf{x}_v$

$\quad + \left(e_v - f_u - e\Gamma^1_{12} + f\left(\Gamma^1_{11} - \Gamma^2_{12}\right) + g\Gamma^2_{11}\right)\mathbf{U}$

$\quad = \left(\dfrac{\partial \Gamma^1_{11}}{\partial v} - \dfrac{\partial \Gamma^1_{12}}{\partial u} - \Gamma^2_{12}\Gamma^1_{12} + \Gamma^2_{11}\Gamma^1_{22} + F\dfrac{eg - f^2}{EG - F^2}\right)\mathbf{x}_u$

$\quad + \left(\dfrac{\partial \Gamma^2_{11}}{\partial v} - \dfrac{\partial \Gamma^2_{12}}{\partial u} + \Gamma^1_{11}\Gamma^2_{12} + \Gamma^2_{11}\Gamma^2_{22} - \Gamma^1_{12}\Gamma^2_{11} - \left(\Gamma^2_{12}\right)^2 - E\dfrac{eg - f^2}{EG - F^2}\right)\mathbf{x}_v$

$\quad + \left(e_v - f_u - e\Gamma^1_{12} + f\left(\Gamma^1_{11} - \Gamma^2_{12}\right) + g\Gamma^2_{11}\right)\mathbf{U}.$

When we take the normal components of (28.15), we get the first equation of (28.13).

A similar argument exploiting $\mathbf{x}_{uvv} = \mathbf{x}_{uvu}$ and the second and third equations of (22.14) yields the second equation of (28.13). ∎

More generally, let us consider the consequences of the equations

(28.16) $\quad \begin{cases} \mathbf{x}_{uuv} - \mathbf{x}_{uvu} = 0, \\ \mathbf{x}_{uvv} - \mathbf{x}_{vvu} = 0, \\ \mathbf{U}_{uv} - \mathbf{U}_{vu} = 0. \end{cases}$

28.3 The Peterson-Mainardi-Codazzi Equations

for a regular patch $\mathbf{x}:\mathcal{U} \longrightarrow \mathbb{R}^3$. When the Gauss equations (22.14) are substituted into (28.16), the result is three equations of the form

(28.17)
$$\begin{cases} A_1\mathbf{x}_u + B_1\mathbf{x}_v + C_1\mathbf{U} = 0, \\ A_2\mathbf{x}_u + B_2\mathbf{x}_v + C_2\mathbf{U} = 0, \\ A_3\mathbf{x}_u + B_3\mathbf{x}_v + C_3\mathbf{U} = 0. \end{cases}$$

Since $\mathbf{x}_u, \mathbf{x}_v, \mathbf{U}$ are linearly independent, we must have

$$A_j = B_j = C_j = 0$$

for $j = 1, 2, 3$. In the proof of Theorem 28.7, we obtained the Peterson-Mainardi-Codazzi equations (28.13) from $C_1 = C_2 = 0$. In fact, $A_3 = B_3 = 0$ also yield (28.13). Furthermore, $C_3 = 0$ because when it is computed, all terms cancel. From $A_1 = B_1 = A_2 = B_2 = 0$ we get the four equations

(28.18)
$$\begin{cases} EK = \dfrac{\partial \Gamma_{11}^2}{\partial v} - \dfrac{\partial \Gamma_{12}^2}{\partial u} + \Gamma_{11}^1\Gamma_{12}^2 + \Gamma_{11}^2\Gamma_{22}^2 - \Gamma_{12}^1\Gamma_{11}^2 - \left(\Gamma_{12}^2\right)^2, \\ -FK = \dfrac{\partial \Gamma_{11}^1}{\partial v} - \dfrac{\partial \Gamma_{12}^1}{\partial u} - \Gamma_{12}^2\Gamma_{12}^1 + \Gamma_{11}^2\Gamma_{22}^1, \\ GK = \dfrac{\partial \Gamma_{22}^1}{\partial u} - \dfrac{\partial \Gamma_{12}^1}{\partial v} + \Gamma_{22}^1\Gamma_{11}^1 + \Gamma_{22}^2\Gamma_{12}^1 - \left(\Gamma_{12}^1\right)^2 + \Gamma_{12}^2\Gamma_{22}^1, \\ -FK = \dfrac{\partial \Gamma_{22}^2}{\partial u} - \dfrac{\partial \Gamma_{12}^2}{\partial v} + \Gamma_{22}^1\Gamma_{11}^2 - \Gamma_{12}^1\Gamma_{12}^2. \end{cases}$$

These four equations turn out to be equivalent to one another. Since each of them (when the Christoffel symbols are expanded) expresses the Gaussian curvature in terms of E, F and G, any one of them provides a new proof of Gauss's *Theorema Egregium* (Theorem 22.5, page 507).

The Peterson-Mainardi-Codazzi equations simplify greatly for principal patches, see page 461.

Corollary 28.8. *Let* $\mathbf{x}:\mathcal{U} \longrightarrow \mathbb{R}^3$ *be a principal patch. Then*

(28.19)
$$\begin{cases} \dfrac{\partial e}{\partial v} = \dfrac{E_v}{2}\left(\dfrac{e}{E} + \dfrac{g}{G}\right), \\ \dfrac{\partial g}{\partial u} = \dfrac{G_u}{2}\left(\dfrac{e}{E} + \dfrac{g}{G}\right). \end{cases}$$

Proof. Equation (28.19) follows from Lemma 22.6, page 510, and Theorem 28.7 ∎

Now we can derive a formula linking principal curvatures and their derivatives. For a proof of the following corollary, see exercise 2.

Corollary 28.9. Let $\mathbf{x}: \mathcal{U} \to \mathbb{R}^3$ be a regular principal patch with principal curvatures $\mathbf{k}_1$ and $\mathbf{k}_2$. Then

(28.20)
$$\begin{cases} \mathbf{k}_{1v} = \dfrac{E_v}{2E}(\mathbf{k}_2 - \mathbf{k}_1), \\ \mathbf{k}_{2u} = \dfrac{G_u}{2G}(\mathbf{k}_1 - \mathbf{k}_2). \end{cases}$$

28.4 Hilbert's Lemma and Liebmann's Theorem

We need the following existence theorem.

Lemma 28.10. *Let $\mathbf{p}$ be a nonumbilical point of a regular surface $\mathcal{M}$ in $\mathbb{R}^3$. Then there exists a principal patch that parametrizes a neighborhood of $\mathbf{p}$.*

For a proof see [dC1, page 185].

Lemma 28.11. (Hilbert[4]) *Let $\mathcal{M}$ be a regular surface in $\mathbb{R}^3$, and let $\mathbf{p} \in \mathcal{M}$ be a point such that*

(i) $\mathbf{k}_1$ *has a local maximum at $\mathbf{p}$;*

(ii) $\mathbf{k}_2$ *has a local minimum at $\mathbf{p}$;*

(iii) $\mathbf{k}_1(\mathbf{p}) > \mathbf{k}_2(\mathbf{p})$.

Then $K(\mathbf{p}) \leq 0$.

Proof. At $\mathbf{p}$ we have

$$\mathbf{k}_{1v} = \mathbf{k}_{2u} = 0, \qquad \mathbf{k}_{1vv} \leq 0 \quad \text{and} \quad \mathbf{k}_{2uu} \geq 0$$

because of (i) and (ii). Condition (iii) and Lemma 28.10 imply that there exists a principal patch $\mathbf{x}$ parametrizing a neighborhood of $\mathbf{p}$. Let E, F and G denote the

---

[4] David Hilbert (1862-1943). Professor at Göttingen, the leading German mathematician of his time. Hilbert contributed to many branches of mathematics, including invariants, algebraic number fields, functional analysis, integral equations, mathematical physics, and the calculus of variations. Hilbert's most famous contribution to differential geometry is [Hil], in which he proved that no regular surface of constant negative curvature in $\mathbb{R}^3$ can be complete.

coefficients of the first fundamental form of **x**. From the Peterson-Codazzi-Mainardi equations for principal patches, that is, (28.20), it follows that

(28.21) $$E_v(\mathbf{p}) = G_u(\mathbf{p}) = 0.$$

Furthermore, if we differentiate (28.20) we obtain

(28.22) $$\begin{cases} \mathbf{k}_{1vv} = \left(\dfrac{E\,E_{vv} - E_v^2}{2E^2}\right)(\mathbf{k}_2 - \mathbf{k}_1) + \dfrac{E_v}{2E}(\mathbf{k}_2 - \mathbf{k}_1)_v, \\ \mathbf{k}_{2uu} = \left(\dfrac{G\,G_{uu} - G_u^2}{2G^2}\right)(\mathbf{k}_1 - \mathbf{k}_2) + \dfrac{G_u}{2G}(\mathbf{k}_1 - \mathbf{k}_2)_u. \end{cases}$$

When we evaluate (28.22) at **p**, we obtain

(28.23) $\quad 0 \geq \mathbf{k}_{1vv}(\mathbf{p}) = \left.\dfrac{E_{vv}}{2E}(\mathbf{k}_2 - \mathbf{k}_1)\right|_{\mathbf{p}} \quad\text{and}\quad 0 \leq \mathbf{k}_{2uu}(\mathbf{p}) = \left.\dfrac{G_{uu}}{2G}(\mathbf{k}_1 - \mathbf{k}_2)\right|_{\mathbf{p}}.$

It follows from (28.23) and (iii) that

$$E_{vv}(\mathbf{p}) \geq 0 \quad\text{and}\quad G_{uu}(\mathbf{p}) \geq 0.$$

Thus from (28.21) and (22.9), page 505, we get

$$K(\mathbf{p}) = -\left.\dfrac{E_{vv} + G_{uu}}{2EG}\right|_{\mathbf{p}} \leq 0. \quad\blacksquare$$

Now, we can prove the main result of this section.

Theorem 28.12. (Liebmann[5]) *Let $\mathcal{M}$ be a compact surface in $\mathbb{R}^3$ with constant Gaussian curvature K. Then $\mathcal{M}$ is a sphere of radius $1/\sqrt{K}$.*

Proof. We have

(28.24) $$H^2 - K = \left(\dfrac{\mathbf{k}_1 - \mathbf{k}_2}{2}\right)^2.$$

Since $\mathcal{M}$ is compact, the function $H^2 - K$ assumes its maximum value (which must be nonnegative) at some point $\mathbf{p} \in \mathcal{M}$. Assume that this maximum is positive; we shall obtain a contradiction.

Choose an oriented neighborhood $\mathcal{U}$ containing **p** for which $H^2 - K$ is positive on $\mathcal{U}$. The choice of orientation gives rise to a unit normal vector field **U** defined on all of $\mathcal{U}$. Let $\mathbf{k}_1$ and $\mathbf{k}_2$ be the principal curvatures defined with respect to **U**;

---

[5] Heinrich Liebmann (1874-1939). German mathematician, professor at Munich and Heidelberg.

since $K > 0$, we can arrange that $\mathbf{k}_1 \geq \mathbf{k}_2 > 0$ on $\mathcal{U}$. In fact, because of the identity (28.24), we have $\mathbf{k}_1 > \mathbf{k}_2 > 0$ on $\mathcal{U}$. Since $K = \mathbf{k}_1 \mathbf{k}_2$ is constant, it follows that $\mathbf{k}_1$ has a maximum at $\mathbf{p}$ and $\mathbf{k}_2$ has a minimum at $\mathbf{p}$. Hence Lemma 28.11 implies that $K(\mathbf{p}) \leq 0$, contradicting the assumption that $K(\mathbf{p}) > 0$.

Therefore, $H^2 - K$ is zero at $\mathbf{p}$. Equation (28.24) then implies that $H^2 - K$ vanishes identically on $\mathcal{M}$, so that $\mathcal{M}$ is all umbilic. Now Theorem 28.5 implies that $\mathcal{M}$ is part of a sphere of radius $1/\sqrt{K}$; since $\mathcal{M}$ is compact, $\mathcal{M}$ must be an entire sphere. ∎

28.5 The Fundamental Theorem of Surfaces

In Section 28.3 we found a dependence between the first and second fundamental forms of a surface, namely, equations (28.13) and (28.18). In the present section we deal with two questions:

- To what extent do the first and second fundamental forms determine a surface?

- Given two forms satisfying the compatibility conditions (28.13) and (28.18), does there exist a surface with the two forms as its first and second fundamental forms?

These questions are answered by Theorems 28.13 and 28.14, proved by Bonnet in 1865 (see [Bonn1]). These theorems, presented below, are the surface analogs of the *Fundamental Theorems of Plane and Surface Curves*, given in Sections 6.7 and 10.1. The proofs, however, are more difficult since they involve partial differential equations instead of ordinary differential equations. Together, these theorems constitute the **Fundamental Theorem of Surfaces**.

Theorem 28.13. (Fundamental Theorem of Surfaces, Uniqueness) *Let* $\mathbf{x}$ *and* $\widetilde{\mathbf{x}}$ *be regular patches in* $\mathbb{R}^3$ *defined on the same open subset* $\mathcal{U} \subseteq \mathbb{R}^2$, *and assume they have the same first and second fundamental forms. Then there is a Euclidean motion of* $\mathbb{R}^3$ *that maps* $\mathbf{x}$ *onto* $\widetilde{\mathbf{x}}$.

Proof. Let E, F, G, e, f and g denote the coefficients of the first and second fundamental forms of $\mathbf{x}$ and $\widetilde{\mathbf{x}}$. We can apply a Euclidean motion to $\widetilde{\mathbf{x}}$, so that

$$\widetilde{\mathbf{x}}(u_0, v_0) = \mathbf{x}(u_0, v_0), \qquad \widetilde{\mathbf{x}}_u(u_0, v_0) = \mathbf{x}_u(u_0, v_0), \qquad \widetilde{\mathbf{x}}_v(u_0, v_0) = \mathbf{x}_v(u_0, v_0).$$

28.5 The Fundamental Theorem of Surfaces

Let $t \longmapsto (u(t), v(t))$ be a curve in $\mathcal{U}$ with $u(t_0) = u_0$ and $v(t_0) = v_0$, and define curves $\boldsymbol{\alpha}, \boldsymbol{\beta}, \boldsymbol{\gamma}, \widetilde{\boldsymbol{\alpha}}, \widetilde{\boldsymbol{\beta}}, \widetilde{\boldsymbol{\gamma}}$ by

$$\begin{cases} \boldsymbol{\alpha}(t) = \mathbf{x}\bigl(u(t),v(t)\bigr), & \boldsymbol{\beta}(t) = \mathbf{x}_u\bigl(u(t),v(t)\bigr), & \boldsymbol{\gamma}(t) = \mathbf{x}_v\bigl(u(t),v(t)\bigr), \\ \widetilde{\boldsymbol{\alpha}}(t) = \widetilde{\mathbf{x}}\bigl(u(t),v(t)\bigr), & \widetilde{\boldsymbol{\beta}}(t) = \widetilde{\mathbf{x}}_u\bigl(u(t),v(t)\bigr), & \widetilde{\boldsymbol{\gamma}}(t) = \widetilde{\mathbf{x}}_v\bigl(u(t),v(t)\bigr). \end{cases}$$

We have
$$\boldsymbol{\alpha}'(t) = u'(t)\mathbf{x}_u + v'(t)\mathbf{x}_v = u'(t)\boldsymbol{\beta}(t) + v'(t)\boldsymbol{\gamma}(t),$$

and

$$\begin{aligned}\boldsymbol{\beta}'(t) &= u'(t)\mathbf{x}_{uu} + v'(t)\mathbf{x}_{uv} \\ &= \bigl(\Gamma_{11}^1 u'(t) + \Gamma_{12}^1 v'(t)\bigr)\mathbf{x}_u + \bigl(\Gamma_{11}^2 u'(t) + \Gamma_{12}^2 v'(t)\bigr)\mathbf{x}_v \\ &\quad + \frac{\bigl(e\,u'(t) + f\,v'(t)\bigr)}{\sqrt{EG - F^2}}\mathbf{x}_u \times \mathbf{x}_v \\ &= \bigl(\Gamma_{11}^1 u'(t) + \Gamma_{12}^1 v'(t)\bigr)\boldsymbol{\beta}(t) + \bigl(\Gamma_{11}^2 u'(t) + \Gamma_{12}^2 v'(t)\bigr)\boldsymbol{\gamma}(t) \\ &\quad + \frac{\bigl(e\,u'(t) + f\,v'(t)\bigr)}{\sqrt{EG - F^2}}\boldsymbol{\beta}(t) \times \boldsymbol{\gamma}(t).\end{aligned}$$

Similarly,

$$\begin{aligned}\boldsymbol{\gamma}'(t) &= u'(t)\mathbf{x}_{vu} + v'(t)\mathbf{x}_{vv} \\ &= \bigl(\Gamma_{12}^1 u'(t) + \Gamma_{22}^1 v'(t)\bigr)\boldsymbol{\beta}(t) + \bigl(\Gamma_{12}^2 u'(t) + \Gamma_{22}^2 v'(t)\bigr)\boldsymbol{\gamma}(t) \\ &\quad + \frac{\bigl(f\,u'(t) + g\,v'(t)\bigr)}{\sqrt{EG - F^2}}\boldsymbol{\beta}(t) \times \boldsymbol{\gamma}(t).\end{aligned}$$

Thus $t \longmapsto \bigl(\boldsymbol{\alpha}(t), \boldsymbol{\beta}(t), \boldsymbol{\gamma}(t)\bigr)$ is a solution of the system

$$(28.25) \quad \begin{cases} \boldsymbol{\lambda}'(t) = u'(t)\boldsymbol{\mu}(t) + v'(t)\boldsymbol{\nu}(t), \\ \boldsymbol{\mu}'(t) = \bigl(\Gamma_{11}^1 u'(t) + \Gamma_{12}^1 v'(t)\bigr)\boldsymbol{\mu}(t) + \bigl(\Gamma_{11}^2 u'(t) + \Gamma_{12}^2 v'(t)\bigr)\boldsymbol{\nu}(t) \\ \qquad + \dfrac{\bigl(e\,u'(t) + f\,v'(t)\bigr)}{\sqrt{EG - F^2}}\boldsymbol{\mu}(t) \times \boldsymbol{\nu}(t), \\ \boldsymbol{\nu}'(t) = \bigl(\Gamma_{12}^1 u'(t) + \Gamma_{22}^1 v'(t)\bigr)\boldsymbol{\mu}(t) + \bigl(\Gamma_{12}^2 u'(t) + \Gamma_{22}^2 v'(t)\bigr)\boldsymbol{\nu}(t) \\ \qquad + \dfrac{\bigl(f\,u'(t) + g\,v'(t)\bigr)}{\sqrt{EG - F^2}}\boldsymbol{\mu}(t) \times \boldsymbol{\nu}(t). \end{cases}$$

But $t \longmapsto (\widetilde{\alpha}(t), \widetilde{\beta}(t), \widetilde{\gamma}(t))$ is also a solution of (28.25). Also, initially

$$\widetilde{\alpha}(t_0) = \alpha(t_0), \qquad \widetilde{\beta}(t_0) = \beta(t_0), \qquad \widetilde{\gamma}(t_0) = \gamma(t_0).$$

It follows from the uniqueness theorem for ordinary differential equations that $\widetilde{\alpha}$, $\widetilde{\beta}$ and $\widetilde{\gamma}$ coincide with α, β and γ. Hence the patches $\widetilde{\mathbf{x}}$ and $\mathbf{x}$ coincide. ∎

Next, we turn to the question of existence. We must deal with finding a solution to a system of partial differential equations. In contrast to the situation with curves, certain integrability conditions must be satisfied.

Theorem 28.14. (**Fundamental Theorem of Surfaces, Existence**) *Let $\mathcal{V}$ be an open subset of $\mathbb{R}^2$ with $(u_0, v_0) \in \mathcal{V}$, and let $E, F, G, e, f, g : \mathcal{V} \longrightarrow \mathbb{R}$ be six functions. Define Γ^i_{jk} $(i, j, k = 1, 2)$ by*

$$\begin{cases} \Gamma^1_{11} = \dfrac{G E_u - 2F F_u + F E_v}{2(EG - F^2)}, & \Gamma^2_{11} = \dfrac{2E F_u - E E_v - F E_u}{2(EG - F^2)}, \\[6pt] \Gamma^1_{12} = \dfrac{G E_v - F G_u}{2(EG - F^2)}, & \Gamma^2_{12} = \dfrac{E G_u - F E_v}{2(EG - F^2)}, \\[6pt] \Gamma^1_{22} = \dfrac{2G F_v - G G_u - F G_v}{2(EG - F^2)}, & \Gamma^2_{22} = \dfrac{E G_v - 2F F_v + F G_u}{2(EG - F^2)}. \end{cases}$$

Assume that:

(i) $E > 0$, $G > 0$, $EG - F^2 > 0$;

(ii) *the compatibility conditions*

(28.26) $\quad \begin{cases} \dfrac{\partial e}{\partial v} - \dfrac{\partial f}{\partial u} = e\Gamma^1_{12} + f(\Gamma^2_{12} - \Gamma^1_{11}) - g\Gamma^2_{11}, \\[6pt] \dfrac{\partial f}{\partial v} - \dfrac{\partial g}{\partial u} = e\Gamma^1_{22} + f(\Gamma^2_{22} - \Gamma^1_{12}) - g\Gamma^2_{12}, \\[6pt] -F\left(\dfrac{eg - f^2}{EG - F^2}\right) = \dfrac{\partial \Gamma^1_{11}}{\partial v} - \dfrac{\partial \Gamma^1_{12}}{\partial u} - \Gamma^2_{12}\Gamma^1_{12} + \Gamma^2_{11}\Gamma^1_{22} \end{cases}$

hold.

Then there exist an open set $\mathcal{U}$ with $(u_0, v_0) \in \mathcal{U} \subset \mathcal{V}$ and a regular patch $\mathbf{x} : \mathcal{U} \longrightarrow \mathbb{R}^3$, such that E, F, G are the coefficients of the first fundamental form of $\mathbf{x}$, and e, f, g are the coefficients of the second fundamental form of $\mathbf{x}$.

28.5 The Fundamental Theorem of Surfaces

Proof. We consider the system of partial differential equations

(28.27)
$$\begin{cases} \mathbf{X}_u = \Gamma_{11}^1 \mathbf{X} + \Gamma_{11}^2 \mathbf{Y} + e\,\mathbf{Z}, \\ \mathbf{X}_v = \Gamma_{12}^1 \mathbf{X} + \Gamma_{12}^2 \mathbf{Y} + f\,\mathbf{Z}, \\ \mathbf{Y}_u = \Gamma_{12}^1 \mathbf{X} + \Gamma_{12}^2 \mathbf{Y} + f\,\mathbf{Z}, \\ \mathbf{Y}_v = \Gamma_{22}^1 \mathbf{X} + \Gamma_{22}^2 \mathbf{Y} + g\,\mathbf{Z}, \\ \mathbf{Z}_u = \dfrac{fF - eG}{EG - F^2}\mathbf{X} + \dfrac{eF - fE}{EG - F^2}\mathbf{Y}, \\ \mathbf{Z}_v = \dfrac{gF - fG}{EG - F^2}\mathbf{X} + \dfrac{fF - gE}{EG - F^2}\mathbf{Y}, \end{cases}$$

together with the initial conditions

(28.28)
$$\begin{cases} \mathbf{X}(u_0,v_0) = \left(\sqrt{E(u_0,v_0)},0,0\right), \\ \mathbf{Y}(u_0,v_0) = \left(\dfrac{F(u_0,v_0)}{E(u_0,v_0)}, \sqrt{\dfrac{E(u_0,v_0)G(u_0,v_0) - F(u_0,v_0)^2}{E(u_0,v_0)}}, 0\right), \\ \mathbf{Z}(u_0,v_0) = (0,0,1). \end{cases}$$

for the unknown vector-valued functions $\mathbf{X},\mathbf{Y},\mathbf{Z}\colon \mathcal{U} \longrightarrow \mathbb{R}^3$. Written out in components, (28.27) is a system of 18 first-order linear homogeneous partial differential equations. By the same calculation as that in Theorem 28.7, we have

(28.29)
$$\begin{cases} \mathbf{X}_{uv} - \mathbf{X}_{vu} = A_1 \mathbf{x}_u + B_1 \mathbf{x}_v + C_1 \mathbf{U}, \\ \mathbf{Y}_{uv} - \mathbf{Y}_{vu} = A_2 \mathbf{x}_u + B_2 \mathbf{x}_v + C_2 \mathbf{U}, \\ \mathbf{Z}_{uv} - \mathbf{Z}_{vu} = A_3 \mathbf{x}_u + B_3 \mathbf{x}_v + C_3 \mathbf{U}, \end{cases}$$

where $A_1,\dots,C_3$ are the same as in (28.18). Clearly, in order that there actually exist vector-valued functions $\mathbf{X}$, $\mathbf{Y}$ and $\mathbf{Z}$ satisfying (28.27), we must have

(28.30) $\qquad \mathbf{X}_{uv} = \mathbf{X}_{vu}, \qquad \mathbf{Y}_{uv} = \mathbf{Y}_{vu} \quad \text{and} \quad \mathbf{Z}_{uv} = \mathbf{Z}_{vu}.$

In fact, $C_3 = 0$ automatically, and the equations (28.26) guarantee that $A_1,\dots,C_2$ vanish. For a proof that the compatibility conditions (28.30) are sufficient to guarantee the existence of $\mathbf{X}$, $\mathbf{Y}$ and $\mathbf{Z}$ satisfying (28.27) see [Stoker, Appendix B]. Thus $\mathbf{X}$, $\mathbf{Y}$ and $\mathbf{Z}$ exist.

Next, let
$$\begin{cases} a_{11} = \dfrac{fF - eG}{EG - F^2}, & a_{12} = \dfrac{eF - fE}{EG - F^2}, \\ a_{21} = \dfrac{gF - fG}{EG - F^2}, & a_{22} = \dfrac{fF - gE}{EG - F^2}, \end{cases}$$
and consider the system of partial differential equations

(28.31)
$$\begin{cases} \widetilde{E}_u = 2\Gamma_{11}^1 \widetilde{E} + 2\Gamma_{11}^2 \widetilde{F} + 2e\,\widetilde{A}, & \widetilde{E}_v = 2\Gamma_{12}^1 \widetilde{E} + 2\Gamma_{12}^2 \widetilde{F} + 2f\,\widetilde{A}, \\ \widetilde{F}_u = \Gamma_{12}^1 \widetilde{E} + (\Gamma_{11}^1 + \Gamma_{12}^2)\widetilde{F} + \Gamma_{11}^2 \widetilde{G} + f\,\widetilde{A} + e\,\widetilde{B}, \\ \widetilde{F}_v = \Gamma_{22}^1 \widetilde{E} + (\Gamma_{12}^1 + \Gamma_{22}^2)\widetilde{F} + \Gamma_{12}^2 \widetilde{G} + g\,\widetilde{A} + f\,\widetilde{B}, \\ \widetilde{G}_u = 2\Gamma_{12}^1 \widetilde{F} + 2\Gamma_{12}^2 \widetilde{G} + 2f\,\widetilde{B}, & \widetilde{G}_v = 2\Gamma_{22}^1 \widetilde{F} + 2\Gamma_{22}^2 \widetilde{G} + 2g\,\widetilde{B}, \\ \widetilde{A}_u = a_{11}\widetilde{E} + a_{12}\widetilde{F} + \Gamma_{11}^1 \widetilde{A} + \Gamma_{11}^2 \widetilde{B} + e\,\widetilde{C}, \\ \widetilde{A}_v = a_{21}\widetilde{E} + a_{22}\widetilde{F} + \Gamma_{12}^1 \widetilde{A} + \Gamma_{12}^2 \widetilde{B} + f\,\widetilde{C}, \\ \widetilde{B}_u = a_{11}\widetilde{E} + a_{12}\widetilde{F} + \Gamma_{12}^1 \widetilde{A} + \Gamma_{12}^2 \widetilde{B} + f\,\widetilde{C}, \\ \widetilde{B}_v = a_{21}\widetilde{E} + a_{22}\widetilde{G} + \Gamma_{22}^1 \widetilde{A} + \Gamma_{22}^2 \widetilde{B} + g\,\widetilde{C}, \\ \widetilde{C}_u = 2a_{11}\widetilde{E} + 2a_{12}\widetilde{F}, & \widetilde{C}_v = 2a_{21}\widetilde{E} + 2a_{22}\widetilde{G}, \end{cases}$$

with the initial conditions

(28.32)
$$\begin{cases} \widetilde{E}(u_0, v_0) = E(u_0, v_0), & \widetilde{A}(u_0, v_0) = 0, \\ \widetilde{F}(u_0, v_0) = F(u_0, v_0), & \widetilde{B}(u_0, v_0) = 0, \\ \widetilde{G}(u_0, v_0) = G(u_0, v_0), & \widetilde{C}(u_0, v_0) = 1. \end{cases}$$

We know two solutions to (28.31) with the initial conditions (28.32), namely,
$$\begin{cases} \widetilde{E} = E, & \widetilde{A} = 0, \\ \widetilde{F} = F, & \widetilde{B} = 0, \\ \widetilde{G} = G, & \widetilde{C} = 1, \end{cases} \quad \text{and} \quad \begin{cases} \widetilde{E} = \mathbf{X} \cdot \mathbf{X}, & \widetilde{A} = \mathbf{X} \cdot \mathbf{Z}, \\ \widetilde{F} = \mathbf{X} \cdot \mathbf{Y}, & \widetilde{B} = \mathbf{Y} \cdot \mathbf{Z}, \\ \widetilde{G} = \mathbf{Y} \cdot \mathbf{Y}, & \widetilde{C} = \mathbf{Z} \cdot \mathbf{Z}. \end{cases}$$

Since the solution of (28.32) and (28.31) is unique, we must have

(28.33)
$$\begin{cases} \mathbf{X} \cdot \mathbf{X} = E, & \mathbf{X} \cdot \mathbf{Z} = 0, \\ \mathbf{X} \cdot \mathbf{Y} = F, & \mathbf{X} \cdot \mathbf{Z} = 0, \\ \mathbf{Y} \cdot \mathbf{Y} = G, & \mathbf{Z} \cdot \mathbf{Z} = 1. \end{cases}$$

We can now define a candidate for the patch $\mathbf{x}$ by

$$(28.34) \qquad \mathbf{x}(u,v) = \int_{u_0}^{u} \mathbf{X}(p,v) dp + \int_{v_0}^{v} \mathbf{Y}(u,q) dq.$$

Clearly,

$$\mathbf{x}_u = \mathbf{X}, \quad \text{and} \quad \mathbf{x}_v = \mathbf{Y}.$$

Since $\mathbf{x}_u \times \mathbf{x}_v = \mathbf{X} \times \mathbf{Y}$ and

$$\begin{aligned}\mathbf{x}_u(u_0,v_0) \times \mathbf{x}_v(u_0,v_0) &= \mathbf{X}(u_0,v_0) \times \mathbf{Y}(u_0,v_0) \\ &= \left(0,0,\sqrt{E(u_0,v_0)G(u_0,v_0) - F(u_0,v_0)^2}\right) \neq \mathbf{0},\end{aligned}$$

it follows that $(\mathbf{x}_u \times \mathbf{x}_v)(u,v)$ is nonzero for (u,v) in a neighborhood $\mathcal{U}_0$ of (u_0,v_0); hence $\mathbf{x}$ is regular in $\mathcal{U}_0$. It is also clear that

$$\begin{cases} \mathbf{x}_u \cdot \mathbf{x}_u = \mathbf{X} \cdot \mathbf{X} = E, \\ \mathbf{x}_u \cdot \mathbf{x}_v = \mathbf{X} \cdot \mathbf{Y} = F, \\ \mathbf{x}_v \cdot \mathbf{x}_v = \mathbf{Y} \cdot \mathbf{Y} = G, \end{cases}$$

and that

$$\mathbf{U} = \frac{\mathbf{x}_v \times \mathbf{x}_v}{\|\mathbf{x}_v \times \mathbf{x}_v\|} = \mathbf{Z}.$$

Finally, since $\mathbf{X} \cdot \mathbf{Z} = \mathbf{Y} \cdot \mathbf{Z} = 0$ and $\mathbf{Z} \cdot \mathbf{Z} = 1$, it follows from (28.27) that

$$\begin{cases} \mathbf{x}_{uu} \cdot \mathbf{U} = \mathbf{X}_u \cdot \mathbf{Z} = e, \\ \mathbf{x}_{uv} \cdot \mathbf{U} = \mathbf{X}_v \cdot \mathbf{Z} = f, \\ \mathbf{x}_{vv} \cdot \mathbf{U} = \mathbf{Y}_v \cdot \mathbf{Z} = g. \end{cases}$$

Hence the coefficients of the second fundamental form of $\mathbf{x}$ are e, f and g, completing the proof. ∎

28.6 Exercises

1. Prove the second equation of (28.13).

2. Prove Corollary 28.9.

3. The Peterson-Mainardi-Codazzi equations also simplify for an asymptotic patch. $\mathbf{x}: \mathcal{U} \longrightarrow \mathbb{R}^3$.

   Show that

   (28.35)
   $$\begin{cases} \dfrac{\partial \log f}{\partial u} = \Gamma^1_{11} - \Gamma^2_{12}, \\ \dfrac{\partial \log f}{\partial v} = \Gamma^2_{22} - \Gamma^1_{12}, \end{cases}$$

   and

   (28.36)
   $$\begin{cases} \dfrac{\partial}{\partial u}\left(\dfrac{\log f}{\sqrt{EG - F^2}}\right) = -2\Gamma^2_{12}, \\ \dfrac{\partial}{\partial v}\left(\dfrac{\log f}{\sqrt{EG - F^2}}\right) = -2\Gamma^1_{12}. \end{cases}$$

4. Find a patch $\mathbf{x}: \mathcal{U} \longrightarrow \mathbb{R}^3$ for which the coefficients of the first and second fundamental forms are
   $$\begin{cases} E = a^2 \cos^2 v, & F = 0, & G = a^2, \\ e = -a \cos^2 v, & f = 0, & g = -a. \end{cases}$$

5. Find a patch $\mathbf{x}: \mathcal{U} \longrightarrow \mathbb{R}^3$ for which the coefficients of the first and second fundamental forms are
   $$\begin{cases} E = a^2, & F = 0, & G = 1, \\ e = -a, & f = 0, & g = 0. \end{cases}$$

6. Find the most general patch $\mathbf{x}: \mathcal{U} \longrightarrow \mathbb{R}^3$ for which $\mathbf{x}_{uv} = 0$.

7. Show that there is no patch having $du^2 + dv^2$ and $du^2 - dv^2$ as its first and second fundamental forms.

8. Show that there is no patch having $du^2 + (\cos^2 u) dv^2$ and $(\cos^2 u) du^2 + dv^2$ as its first and second fundamental forms.

29

TRIPLY ORTHOGONAL SYSTEMS OF SURFACES

For certain surfaces such as the ellipsoid there is a much more effective way to find principal curves than solving the differential equation discussed in Section 28.1. It is based on the following notion.

Definition. *A* **triply orthogonal system of surfaces** *on an open set* $\mathcal{U} \subseteq \mathbb{R}^3$ *consists of three families* $\mathfrak{A}$, $\mathfrak{B}$, $\mathfrak{C}$, *of surfaces such that*

(i) *each point of* $\mathcal{U}$ *lies on one and only one member of each family;*

(ii) *each surface in each family meets every member of the other two families orthogonally.*

Triply orthogonal systems were introduced by Lamé[1] to study equations of mathematical physics (see, for example, [MoSp]). The connection between principal curvatures and triply orthogonal systems will become clear after we prove Dupin's Theorem (Theorem 29.4), which states that surfaces from different families intersect in principal curves. Simple examples of triply orthogonal systems are given in Section 29.1. Curvilinear patches are defined in Section 29.2 and used to prove Dupin's

---

[1] Gabriel Lamé (1795-1870). French engineer, mathematician and physicist. He worked on a wide variety of topics, including number theory, differential geometry (including triply orthogonal systems), elasticity (where two elastic constants are named after him) and diffusion in crystalline material.

Theorem. In Section 29.3 we define elliptic coordinates and use them to find and draw the principal curves on ellipsoids, hyperboloids of one sheet and hyperboloids of two sheets. Similarly, in Section 29.4 we use parabolic coordinates to find the principal curves of elliptic and hyperbolic paraboloids.

29.1 Examples of Triply Orthogonal Systems

The simplest example of a triply orthogonal system of surfaces is afforded by the three families of planes parallel to the xy-, xz- and yz-planes.

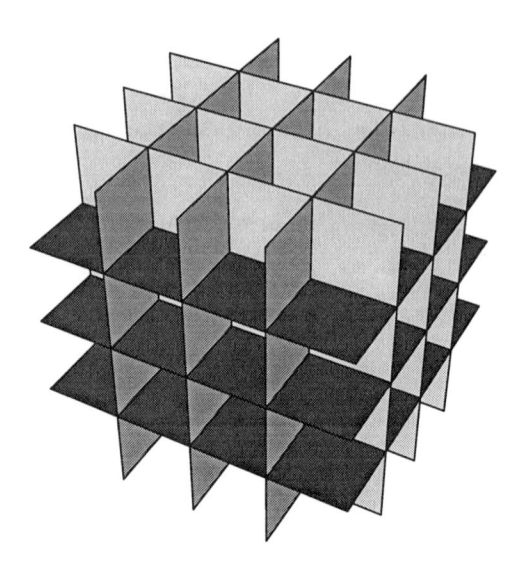

**A triply orthogonal system
consisting of three families of planes**

To give a second example of a triply orthogonal system of surfaces, let $\mathcal{U}$ be the complement of the vertical axis in $\mathbb{R}^3$; that is,

$$\mathcal{U} = \{\, (p_1, p_2, p_3) \in \mathbb{R}^3 \mid (p_1, p_2) \neq (0, 0) \,\}.$$

29.1 Examples of Triply Orthogonal Systems

Then the system consisting of

$\mathfrak{A}$ = planes through the z-axis,

$\mathfrak{B}$ = planes parallel to the xy-plane,

$\mathfrak{C}$ = circular cylinders around the z-axis

constitute a triply orthogonal system of surfaces on $\mathcal{U}$.

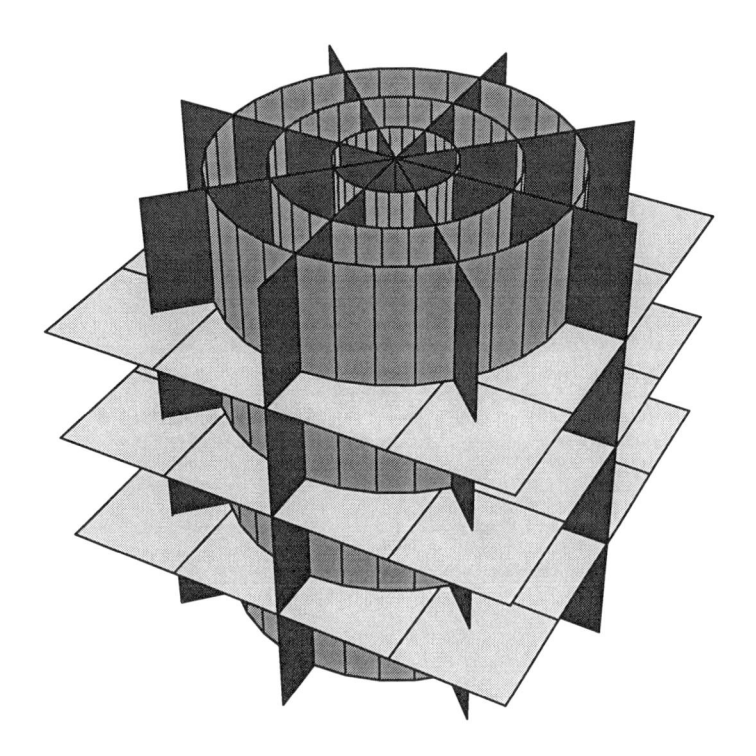

**A triply orthogonal system
consisting of two families of planes and
one family of cylinders**

For a third example, let $\mathcal{V}$ be all points of $\mathbb{R}^3$ except the origin; then

$\mathfrak{A}$ = planes parallel to the xy-plane,

$\mathfrak{B}$ = cones through the origin,

$\mathfrak{C}$ = spheres centered at the origin

constitute a triply orthogonal system of surfaces on $\mathcal{V}$.

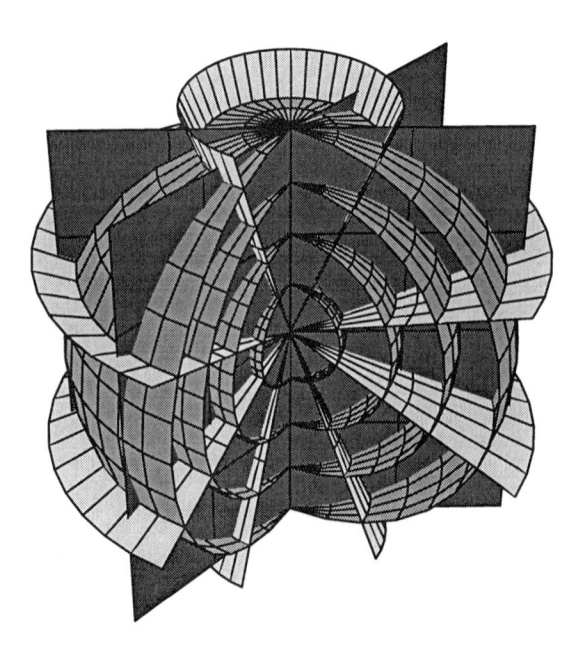

**A triply orthogonal system
consisting of one family of planes,
one family of cones and
one family of spheres**

Many classical differential geometry books have a section on triply orthogonal systems, see for example [Bian], [Eisen1] and [Wea]. An extensive list of triply orthogonal systems is given in [MoSp].

29.2 Curvilinear Patches and Dupin's Theorem

A curve in $\mathbb{R}^3$ has one parameter, a surface two. It is also useful to consider a function of three parameters in order to generate a triply orthogonal system.

Definition. A **curvilinear patch** *for* $\mathbb{R}^3$ *is a differentiable function* $\mathbf{x}:\mathcal{U} \longrightarrow \mathbb{R}^3$, *where* $\mathcal{U}$ *is some open set of* $\mathbb{R}^3$.

If $\mathbf{x}:\mathcal{U} \longrightarrow \mathbb{R}^3$ is a curvilinear patch, we can write

$$\mathbf{x}(u,v,w) = \bigl(x_1(u,v,w), x_2(u,v,w), x_3(u,v,w)\bigr).$$

Differentiation of a curvilinear patch is defined in the same way as differentiation of a patch in $\mathbb{R}^3$. Thus

$$\mathbf{x}_u(u,v,w) = \left(\frac{\partial x_1}{\partial u}(u,v,w), \frac{\partial x_2}{\partial u}(u,v,w), \frac{\partial x_3}{\partial u}(u,v,w)\right),$$

and so forth.

Definition. *A curvilinear patch* $\mathbf{x}:\mathcal{U} \longrightarrow \mathbb{R}^3$ *is* **regular** *provided the vector fields* $\mathbf{x}_u, \mathbf{x}_v, \mathbf{x}_w$ *are linearly independent throughout* $\mathcal{U}$, *or equivalently*

$$\det\begin{pmatrix} \mathbf{x}_u \\ \mathbf{x}_v \\ \mathbf{x}_w \end{pmatrix} = \mathbf{x}_u \times \mathbf{x}_v \cdot \mathbf{x}_w \neq 0.$$

We say that $\mathbf{x}$ *is* **orientation preserving** *provided*

$$\det\begin{pmatrix} \mathbf{x}_u \\ \mathbf{x}_v \\ \mathbf{x}_w \end{pmatrix} = \mathbf{x}_u \times \mathbf{x}_v \cdot \mathbf{x}_w > 0.$$

A triply orthogonal system of surfaces can be considered to be a special kind of curvilinear patch.

Lemma 29.1. *A regular curvilinear patch* $\mathbf{x}:\mathcal{U} \longrightarrow \mathbb{R}^3$ *satisfying*

(29.1) $$\mathbf{x}_u \cdot \mathbf{x}_v = \mathbf{x}_v \cdot \mathbf{x}_w = \mathbf{x}_w \cdot \mathbf{x}_u = 0$$

determines a triply orthogonal system of surfaces, each of which is regular.

Proof. For fixed w, the mapping
$$(u, v) \longmapsto \mathbf{x}(u, v, w)$$
is a 2-dimensional patch. It is regular because $\mathbf{x}_u \times \mathbf{x}_v$ is everywhere nonzero. Each of its tangent spaces is spanned by $\mathbf{x}_u$ and $\mathbf{x}_v$. Similarly, for fixed v the mapping
$$(u, w) \longmapsto \mathbf{x}(u, v, w)$$
is a regular 2-dimensional patch, each of whose tangent spaces is spanned by $\mathbf{x}_u$ and $\mathbf{x}_w$; and for fixed u the mapping
$$(v, w) \longmapsto \mathbf{x}(u, v, w)$$
is a regular 2-dimensional patch, each of whose tangent spaces is spanned by $\mathbf{x}_v$ and $\mathbf{x}_w$. Then (29.1) implies that the tangent spaces to $(u, v) \longmapsto \mathbf{x}(u, v, w)$, $(u, w) \longmapsto \mathbf{x}(u, v, w)$ and $(v, w) \longmapsto \mathbf{x}(u, v, w)$ are mutually perpendicular. Thus if we set
$$\mathfrak{A} = \{\, (u, v) \longmapsto \mathbf{x}(u, v, w) \,\},$$
$$\mathfrak{B} = \{\, (u, w) \longmapsto \mathbf{x}(u, v, w) \,\},$$
$$\mathfrak{C} = \{\, (v, w) \longmapsto \mathbf{x}(u, v, w) \,\},$$
then $\mathfrak{A}$, $\mathfrak{B}$, $\mathfrak{C}$ form a triply orthogonal system of surfaces.

That regularity of the curvilinear patch implies regularity of each of the surfaces in $\mathfrak{A}$, $\mathfrak{B}$, $\mathfrak{C}$ results from the following identities, which are consequences of (8.1):
$$(\mathbf{x}_u \times \mathbf{x}_v \cdot \mathbf{x}_w)^2 = \|\mathbf{x}_u\|^2 \|\mathbf{x}_v\|^2 \|\mathbf{x}_w\|^2,$$
and
$$\|\mathbf{x}_u \times \mathbf{x}_v\| = \|\mathbf{x}_u\| \|\mathbf{x}_v\|, \qquad \|\mathbf{x}_u \times \mathbf{x}_w\| = \|\mathbf{x}_u\| \|\mathbf{x}_w\|, \qquad \|\mathbf{x}_v \times \mathbf{x}_w\| = \|\mathbf{x}_v\| \|\mathbf{x}_w\|. \blacksquare$$

Since we shall deal only with local questions concerning triply orthogonal systems of surfaces, we need only consider those systems determined by a curvilinear patch $\mathbf{x} \colon \mathcal{U} \longrightarrow \mathbb{R}^3$ satisfying (29.1). If a curvilinear patch $\mathbf{x}$ is orientation reversing, then the curvilinear patch $\mathbf{y}$ defined by $\mathbf{y}(u, v, w) = \mathbf{x}(u, w, v)$ is orientation preserving.

Definition. *Let $\mathcal{U}$ be an open set in $\mathbb{R}^3$. A **triply orthogonal patch** is an orientation preserving regular curvilinear patch $\mathbf{x} \colon \mathcal{U} \longrightarrow \mathbb{R}^3$ such that (29.1) holds.*

29.2 Curvilinear Patches and Dupin's Theorem

Definition. Let $\mathbf{x}: \mathcal{U} \longrightarrow \mathbb{R}^3$ be a triply orthogonal patch. Then the **unit normals** determined by $\mathbf{x}$ are the vector fields $\mathbf{A}$, $\mathbf{B}$, $\mathbf{C}$ given by

$$\mathbf{A} = \frac{\mathbf{x}_u}{\|\mathbf{x}_u\|}, \qquad \mathbf{B} = \frac{\mathbf{x}_v}{\|\mathbf{x}_u\|}, \qquad \mathbf{C} = \frac{\mathbf{x}_w}{\|\mathbf{x}_w\|}.$$

Also, we put

$$p = \|\mathbf{x}_u\|, \qquad q = \|\mathbf{x}_v\|, \qquad r = \|\mathbf{x}_w\|.$$

Lemma 29.2. *A triply orthogonal patch $\mathbf{x}$ satisfies the following relations*:

(29.2) $\quad \mathbf{x}_v \times \mathbf{x}_w = \dfrac{qr}{p}\mathbf{x}_u, \qquad \mathbf{x}_w \times \mathbf{x}_u = \dfrac{rp}{q}\mathbf{x}_v, \qquad \mathbf{x}_u \times \mathbf{x}_v = \dfrac{pq}{r}\mathbf{x}_w,$

(29.3) $\quad \mathbf{x}_u \cdot \mathbf{x}_v \times \mathbf{x}_w = pqr,$

(29.4) $\quad \mathbf{x}_u \cdot \mathbf{x}_{vw} = 0, \qquad \mathbf{x}_v \cdot \mathbf{x}_{wu} = 0, \qquad \mathbf{x}_w \cdot \mathbf{x}_{uv} = 0.$

Proof. To prove (29.2), we compute

$$\mathbf{x}_v \times \mathbf{x}_w = q\mathbf{B} \times r\mathbf{C} = qr\mathbf{A} = \frac{qr}{p}\mathbf{x}_u,$$

and so forth. The proof of (29.3) is similar. To prove (29.4), we first differentiate the three equations of (29.1), obtaining:

(29.5) $\quad \begin{cases} \mathbf{x}_{wu} \cdot \mathbf{x}_v + \mathbf{x}_u \cdot \mathbf{x}_{wv} = 0, \\ \mathbf{x}_{uv} \cdot \mathbf{x}_w + \mathbf{x}_v \cdot \mathbf{x}_{uw} = 0, \\ \mathbf{x}_{vw} \cdot \mathbf{x}_u + \mathbf{x}_w \cdot \mathbf{x}_{vu} = 0. \end{cases}$

It is easy to see that (29.5) implies (29.4). ∎

Next, we show that the coefficients of the first and second fundamental forms of each of the surfaces determined by a triply orthogonal patch are especially simple.

Lemma 29.3. *Let $\mathbf{x}: \mathcal{U} \longrightarrow \mathbb{R}^3$ be a triply orthogonal patch. The coefficients E, F, G of the first fundamental form, the coefficients e, f, g of the second fundamental form, the principal curvatures $\mathbf{k}_1$, $\mathbf{k}_2$, the Gaussian curvature K and the mean curvature H of the patches*

(29.6) $\quad (v,w) \longmapsto \mathbf{x}(v,w), \qquad (w,u) \longmapsto \mathbf{x}(w,u), \qquad (u,v) \longmapsto \mathbf{x}(u,v)$

are given by the following table:

Patch	E F G	e	f	g	k_1	k_2	K	H
$(v,w) \longmapsto \mathbf{x}(v,w)$	$q^2 \ \ 0 \ \ r^2$	$-\dfrac{q\,q_u}{p}$	0	$-\dfrac{r\,r_u}{p}$	$-\dfrac{q_u}{pq}$	$-\dfrac{r_u}{pr}$	$\dfrac{q_u\,r_u}{p^2qr}$	$-\dfrac{(qr)_u}{2pqr}$
$(w,u) \longmapsto \mathbf{x}(w,u)$	$r^2 \ \ 0 \ \ p^2$	$-\dfrac{r\,r_v}{q}$	0	$-\dfrac{p\,p_v}{q}$	$-\dfrac{r_v}{qr}$	$-\dfrac{p_v}{pq}$	$\dfrac{p_v\,r_u}{pq^2r}$	$-\dfrac{(pr)_v}{2pqr}$
$(u,v) \longmapsto \mathbf{x}(u,v)$	$p^2 \ \ 0 \ \ q^2$	$-\dfrac{p\,p_w}{r}$	0	$-\dfrac{q\,q_w}{r}$	$-\dfrac{p_w}{pr}$	$-\dfrac{q_w}{qr}$	$\dfrac{p_w\,q_w}{pqr^2}$	$-\dfrac{(pq)_w}{2pqr}$

Proof. That $F = 0$ for each of the surfaces is a consequence of the assumption that the surfaces are pairwise orthogonal. Furthermore, the formulas for E and G are just the definitions of p, q and r.

As for the second fundamental forms, we first note that (29.5) implies that $f = 0$ for each of the surfaces in (29.6). Next, we compute e for the surface $(v, w) \longmapsto \mathbf{x}(v, w)$. Since the unit normal is $\mathbf{x}_u / \|\mathbf{x}_u\|$, we have

$$e = \mathbf{x}_{vv} \cdot \frac{\mathbf{x}_u}{\|\mathbf{x}_u\|} = \frac{-\|\mathbf{x}_v\|_u^2}{2\|\mathbf{x}_u\|} = \frac{-q\,q_u}{p}.$$

The computation of the other coefficients is similar. ∎

Now we can give a simple proof of the following important theorem:

Theorem 29.4. (Dupin[2]) *The curves of intersection of the surfaces (29.6) of a triply orthogonal patch are principal curves on each.*

Proof. Lemma 29.3 implies that $F = f = 0$; thus each of the patches given by (29.6) has the property that the parameter curves are principal curves. ∎

---

[2] Baron Pierre-Charles-Francois Dupin (1784-1873). French mathematician, a student of Monge. In addition to this theorem, Dupin's contributions to differential geometry include the indicatrix of Dupin and the theory of cyclides as the envelopes of spheres. He also invented the Dupin indicatrix which gives an indication of the local behavior of a surface in terms of its power series expansion. In 1830 Dupin was elected deputy for Tarn and continued in politics until 1870. Dupin also wrote extensively on economics and industry.

29.3 Elliptic Coordinates

In this section we show how to find triply orthogonal patches that give rise to principal patches on ellipsoids, hyperboloids of one sheet and hyperboloids of two sheets. Each of these surfaces can be rotated and translated so that it is described by a nonparametric equation of the form

(29.7) $$\frac{x^2}{a}+\frac{y^2}{b}+\frac{z^2}{c}=1$$

for appropriate a, b, c. Fix a, b and c and consider the surfaces defined by the equations

(29.8) $$\frac{x^2}{a-\lambda}+\frac{y^2}{b-\lambda}+\frac{z^2}{c-\lambda}=1$$

for different values of λ. Such surfaces are called the **confocals** of the surface defined by (29.7). The values of λ for the confocals through a given point $(x,y,z) \in \mathbb{R}^3$ are given by the cubic equation

(29.9) $$(a-\lambda)(b-\lambda)(c-\lambda)$$
$$-x^2(b-\lambda)(c-\lambda)-y^2(a-\lambda)(c-\lambda)-z^2(a-\lambda)(b-\lambda)=0.$$

Let u, v, w be the roots of this equation; thus

(29.10) $$(u-\lambda)(v-\lambda)(w-\lambda) = (a-\lambda)(b-\lambda)(c-\lambda)$$
$$-x^2(b-\lambda)(c-\lambda)-y^2(a-\lambda)(c-\lambda)-z^2(a-\lambda)(b-\lambda).$$

To find the values of x^2, y^2 and z^2 such that (29.10) holds for all λ, we give λ the values a, b, c in succession. The result is:

(29.11) $$\begin{cases} x^2 = \dfrac{(a-u)(a-v)(a-w)}{(b-a)(c-a)}, \\ y^2 = \dfrac{(b-u)(b-v)(b-w)}{(a-b)(c-b)}, \\ z^2 = \dfrac{(c-u)(c-v)(c-w)}{(a-c)(b-c)}. \end{cases}$$

The solutions to equations (29.11) form eight patches.

Definition. Let $a > b > c$. An **elliptic coordinate patch** *is one of the eight maps*

$$\text{ellipticcoor}[a, b, c]\colon \mathbb{R}^3 \longrightarrow \mathbb{R}^3$$

defined by

$$\text{ellipticcoor}[a, b, c](u, v, w) = \left(\pm\sqrt{\frac{(a-u)(a-v)(a-w)}{(b-a)(c-a)}}, \right.$$

$$\left. \pm\sqrt{\frac{(b-u)(b-v)(b-w)}{(a-b)(c-b)}}, \pm\sqrt{\frac{(c-u)(c-v)(c-w)}{(a-c)(b-c)}} \right).$$

Lemma 29.5. *For $a > b > c$ the curvilinear patch* $\text{ellipticcoor}[a, b, c]\colon \mathbb{R}^3 \longrightarrow \mathbb{R}^3$ *is triply orthogonal. It is regular on the set*

$$\{ (p_1, p_2, p_3) \in \mathbb{R}^3 \mid p_1, p_2, p_3 \text{ are distinct and } (p_1, p_2, p_3) \neq (a, b, c) \}.$$

Proof. Let $\text{ellipticcoor}[a, b, c] = \mathbf{x} = (x, y, z)$. It follows from (29.11) that

$$2x \frac{\partial x}{\partial u} = -\frac{(a-v)(a-w)}{(b-a)(c-a)} = \frac{x^2}{(u-a)}.$$

Similar formulas hold for $\partial x/\partial v$, $\partial y/\partial u$, $\partial y/\partial v$, $\partial z/\partial u$ and $\partial z/\partial v$. Therefore,

(29.12)
$$\begin{cases} \mathbf{x}_u = \left(\dfrac{x}{2(u-a)}, \dfrac{y}{2(u-b)}, \dfrac{z}{2(u-c)} \right), \\[6pt] \mathbf{x}_v = \left(\dfrac{x}{2(v-a)}, \dfrac{y}{2(v-b)}, \dfrac{z}{2(v-c)} \right), \\[6pt] \mathbf{x}_w = \left(\dfrac{x}{2(w-a)}, \dfrac{y}{2(w-b)}, \dfrac{z}{2(w-c)} \right). \end{cases}$$

Then (29.11) and (29.12) imply that

$$\mathbf{x}_u \cdot \mathbf{x}_v = \frac{1}{4} \left(\frac{x^2}{(u-a)(v-a)} + \frac{y^2}{(u-b)(v-b)} + \frac{z^2}{(u-c)(v-c)} \right)$$

$$= \frac{1}{4} \left(\frac{a-w}{(b-a)(c-a)} + \frac{b-w}{(a-b)(c-b)} + \frac{c-w}{(a-c)(b-c)} \right)$$

$$= \frac{-(a-w)(b-c) - (b-w)(c-a) - (c-w)(a-b)}{4(a-b)(b-c)(c-a)} = 0.$$

29.3 Elliptic Coordinates

Similarly, $\mathbf{x}_u \cdot \mathbf{x}_w = \mathbf{x}_v \cdot \mathbf{x}_w = 0$ and

$$p^2 = \|\mathbf{x}_u\|^2 = \frac{(u-v)(u-w)}{4(a-u)(b-u)(c-u)},$$

$$q^2 = \|\mathbf{x}_v\|^2 = \frac{(v-u)(v-w)}{4(a-v)(b-v)(c-v)},$$

$$r^2 = \|\mathbf{x}_w\|^2 = \frac{(w-u)(w-v)}{4(a-w)(b-w)(c-w)}.$$

Hence the Jacobian matrix of $\mathbf{x}$ has rank 3 on the set

$$\{\,(p_1, p_2, p_3) \in \mathbb{R}^3 \mid p_1, p_2, p_3 \text{ are distinct and } (p_1, p_2, p_3) \neq (a, b, c)\,\}. \blacksquare$$

Next, we determine the ranges of u and v in order to plot the surfaces determined by **ellipticcoor[a, b, c]**.

Lemma 29.6. *Let $a > b > c > 0$.*

(i) *Suppose $c > w$ so that*

$$\frac{x^2}{a-w} + \frac{y^2}{b-w} + \frac{z^2}{c-w} = 1$$

is an ellipsoid. In order that x, y, z in (29.11) be real, either

$$a \geq v \geq b \geq u \geq c \qquad \text{or} \qquad a \geq u \geq b \geq v \geq c.$$

(ii) *Suppose $b > w > c$, so that*

$$\frac{x^2}{a-w} + \frac{y^2}{b-w} + \frac{z^2}{c-w} = 1$$

is a hyperboloid of one sheet. In order that x, y, z in (29.11) be real, either

$$a \geq v \geq b \geq c \geq u \qquad \text{or} \qquad a \geq u \geq b > c \geq v.$$

(iii) *Suppose $a > w > b$, so that*

$$\frac{x^2}{a-w} + \frac{y^2}{b-w} + \frac{z^2}{c-w} = 1.$$

is a hyperboloid of two sheets. In order that x, y, z in (29.11) be real, either

$$a > b \geq v \geq c \geq u \qquad \text{or} \qquad a > b \geq u \geq c \geq v.$$

Proof. If $w < c < b < a$, it follows from (29.11) that

(29.13)
$$\begin{cases} (a-u)(a-v) > 0, \\ (b-u)(b-v) < 0, \\ (c-u)(c-v) > 0. \end{cases}$$

Without loss of generality, $v \geq u$. Then (29.13) implies (i). The proofs of (ii) and (iii) are similar. ∎

Next, we use elliptic coordinates to determine the umbilics of the surfaces determined by **ellipticcoor[a, b, c]**.

Theorem 29.7. *Suppose $a > b > c > 0$.*

(i) *If $c > w$, then the umbilic points of the ellipsoid*

$$\frac{x^2}{a-w} + \frac{y^2}{b-w} + \frac{z^2}{c-w} = 1$$

are the four points

$$\left(\pm\sqrt{\frac{(a-b)(a-w)}{(a-c)}}, 0, \pm\sqrt{\frac{(b-c)(c-w)}{(a-c)}} \right).$$

(ii) *If $b > w > c$, then the hyperboloid of one sheet*

$$\frac{x^2}{a-w} + \frac{y^2}{b-w} + \frac{z^2}{c-w} = 1$$

has no umbilic points.

(iii) *If $a > w > b$, then the umbilic points of the hyperboloid of two sheets*

$$\frac{x^2}{a-w} + \frac{y^2}{b-w} + \frac{z^2}{c-w} = 1$$

are the four points

$$\left(\pm\sqrt{\frac{(a-c)(a-w)}{(a-b)}}, \pm\sqrt{\frac{(b-c)(b-w)}{(a-b)}}, 0 \right).$$

29.3 Elliptic Coordinates

Proof. From Lemmas 29.3 and 29.5 it follows that the principal curvatures of $(u,v) \longmapsto \mathbf{ellipticcoor}[a,b,c](u,v)$ are given by

$$\mathbf{k}_1 = -\frac{1}{2r}\frac{\partial}{\partial w}\log(p^2) = \frac{1}{2(u-w)r} \quad \text{and} \quad \mathbf{k}_2 = -\frac{1}{2r}\frac{\partial}{\partial w}\log(q^2) = \frac{1}{2(v-w)r}.$$

Then $\mathbf{k}_1 = \mathbf{k}_2$ implies that $u = v$. In case (i) we have $u = v = b$, and in case (iii) we have $u = v = c$, but in case (ii) no umbilic points are possible. Hence the umbilic points are as stated. ∎

To make use of **ellipticcoor**$[a,b,c]$ in *Mathematica* we need a function that defines each of the three triply orthogonal surfaces, and also allows for the choice of signs. We put

```
ellipticcoor[d_,e_,f_][a_,b_,c_][w_][u_,v_]:=
    {d Sqrt[(a - u)(a - v)(a - w)/((a - b)(a - c))],
     e Sqrt[(b - u)(b - v)(b - w)/((b - a)(b - c))],
     f Sqrt[(c - u)(c - v)(c - w)/((c - a)(c - b))]}
```

The command to generate this ellipsoid using elliptical coordinates is complicated because each octant of the ellipsoid must be defined as a separate patch. We translate the lower four octants down by -2 so that we can peek inside the egg:

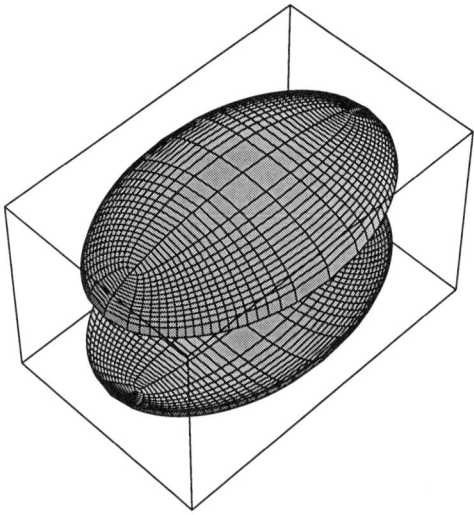

Principal curves on $\dfrac{x^2}{12} + \dfrac{y^2}{5} + z^2 = 1$

The command to generate this picture is:

```
ParametricPlot3D[{
ellipticcoor[1,1,1][12,5,1][0][u,v],
ellipticcoor[-1,1,1][12,5,1][0][u,v],
ellipticcoor[1,-1,1][12,5,1][0][u,v],
ellipticcoor[-1,-1,1][12,5,1][0][u,v],
ellipticcoor[1,1,-1][12,5,1][0][u,v] + {0,0,-2},
ellipticcoor[-1,1,-1][12,5,1][0][u,v] + {0,0,-2},
ellipticcoor[1,-1,-1][12,5,1][0][u,v] + {0,0,-2},
ellipticcoor[-1,-1,-1][12,5,1][0][u,v] + {0,0,-2}
}//Evaluate,{u,1,5},{v,5,12},
PlotPoints->{15,20},Axes->None,
ViewPoint->{2.7, 2.7, 4.2}];
```

Similar commands plot the hyperboloids of one and two sheets:

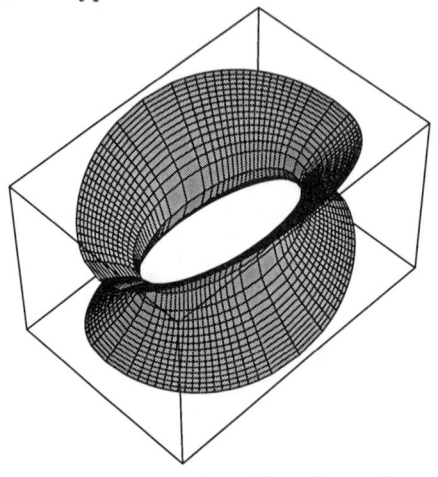

Principal curves on $\dfrac{x^2}{9} + \dfrac{y^2}{2} - \dfrac{z^2}{2} = 1$

29.3 Elliptic Coordinates

Principal curves on $\dfrac{x^2}{4} - \dfrac{y^2}{3} - \dfrac{z^2}{7} = 1$

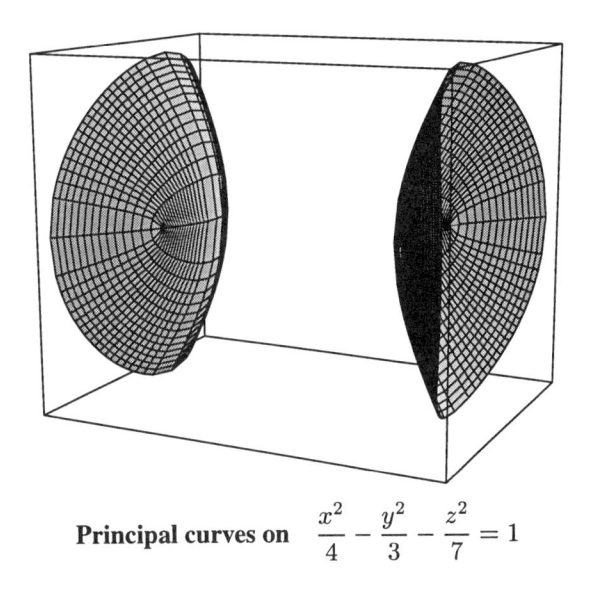

Finally, here is a picture illustrating how an ellipsoid, a hyperboloid of one sheet and a hyperboloid of two sheets intersect in principal curves.

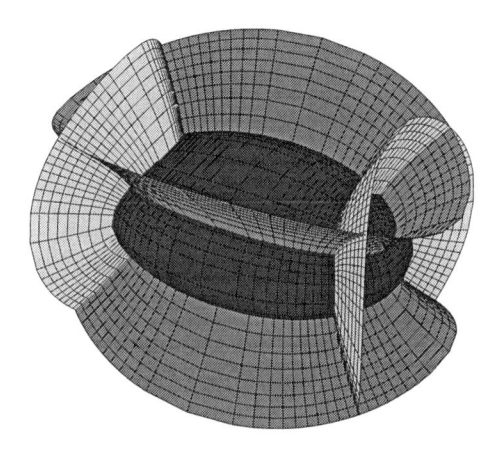

**Triply orthogonal system formed by a family of ellipsoids,
a family of hyperboloids of one sheet and
a family of hyperboloids of two sheets**

29.4 Parabolic Coordinates

The definition of parabolic coordinates is similar to that of elliptic coordinates. Parabolic coordinates give rise to principal patches on elliptic paraboloids and hyperbolic paraboloids. We proceed in the same way that we did in Section 29.3. Any surface defined by

$$(29.14) \qquad \frac{x^2}{a} + \frac{y^2}{b} = 2z - c.$$

is an elliptic or hyperbolic paraboloid, as are the confocals, which are defined by

$$(29.15) \qquad \frac{x^2}{a-\lambda} + \frac{y^2}{b-\lambda} = 2z - c - \lambda.$$

(Equation (29.15) is chosen so that the parabolas $y^2/(b-\lambda) = 2z-c-\lambda$ all have the same focus.) The values of λ for the confocals through a given point $(x, y, z) \in \mathbb{R}^3$ are given by the cubic equation

$$(a-\lambda)(b-\lambda)(2z-c-\lambda) - x^2(b-\lambda) - y^2(a-\lambda) = 0.$$

Let u, v, w be the roots of this equation; thus

$$(29.16) \quad (u-\lambda)(v-\lambda)(w-\lambda)$$
$$= (a-\lambda)(b-\lambda)(2z-c-\lambda) - x^2(b-\lambda) - y^2(a-\lambda).$$

In order that (29.16) hold for all λ, we must have

$$(29.17) \qquad \begin{cases} x^2 = \dfrac{(u-a)(v-a)(w-a)}{a-b}, \\[6pt] y^2 = \dfrac{(u-b)(v-b)(w-b)}{b-a}, \\[6pt] z = \dfrac{c-a-b+u+v+w}{2}. \end{cases}$$

Definition. *Suppose $a > b$. A **parabolic coordinate patch** is one of the four maps* $\mathbf{paraboliccoor}[a, b, c]\colon \mathbb{R}^3 \longrightarrow \mathbb{R}^3$ *defined by*

$$\mathbf{paraboliccoor}[a, b, c](u, v) = \left(\pm\sqrt{\frac{(a-u)(a-v)(a-w)}{b-a}}, \right.$$
$$\left. \pm\sqrt{\frac{(b-u)(b-v)(b-w)}{a-b}}, \frac{c-a-b+u+v+w}{2} \right).$$

29.4 Parabolic Coordinates

The *Mathematica* definition of parabolic coordinates is

```
paraboliccoor[d_,e_,f_][a_,b_,c_][w_][u_,v_]:=
    {d Sqrt[(a - u)(a - v)(a - w)/(b - a)],
     e Sqrt[(b - u)(b - v)(b - w)/(a - b)],
     f(c - a - b + u + v + w)/2}
```

Lemma 29.8. *For $a > b$ the curvilinear patch* **paraboliccoor**$[a, b, c]$ *is triply orthogonal and regular. It is regular on the set*

$$\{ (p_1, p_2, p_3) \in \mathbb{R}^3 \mid p_1, p_2, p_3 \text{ are distinct and } (p_1, p_2, p_3) \neq (a, b, c) \}.$$

Proof. Just as in the proof of Lemma 29.5, we write

$$\textbf{paraboliccoor}[a, b, c] = \mathbf{x} = (x, y, z).$$

By a calculation similar to that of (29.12) it follows from (29.17) that

$$(29.18) \quad \begin{cases} \mathbf{x}_u = \left(\dfrac{x}{2(u-a)}, \dfrac{y}{2(u-b)}, \dfrac{1}{2} \right), \\[1em] \mathbf{x}_v = \left(\dfrac{x}{2(v-a)}, \dfrac{y}{2(v-b)}, \dfrac{1}{2} \right), \\[1em] \mathbf{x}_w = \left(\dfrac{x}{2(w-a)}, \dfrac{y}{2(w-b)}, \dfrac{1}{2} \right). \end{cases}$$

Then (29.18) and (29.17) imply that $\mathbf{x}_u \cdot \mathbf{x}_v = \mathbf{x}_u \cdot \mathbf{x}_w = \mathbf{x}_v \cdot \mathbf{x}_w = 0$, and that

$$p^2 = \|\mathbf{x}_u\|^2 = \frac{(u-v)(u-w)}{4(a-u)(b-u)},$$

$$q^2 = \|\mathbf{x}_v\|^2 = \frac{(v-u)(v-w)}{4(a-v)(b-v)},$$

$$r^2 = \|\mathbf{x}_w\|^2 = \frac{(w-u)(w-v)}{4(a-w)(b-w)}.$$

Hence the Jacobian matrix of $\mathbf{x}$ has rank 3 on the set

$$\{ (p_1, p_2, p_3) \in \mathbb{R}^3 \mid p_1, p_2, p_3 \text{ are distinct and } (p_1, p_2, p_3) \neq (a, b, c) \}. \blacksquare$$

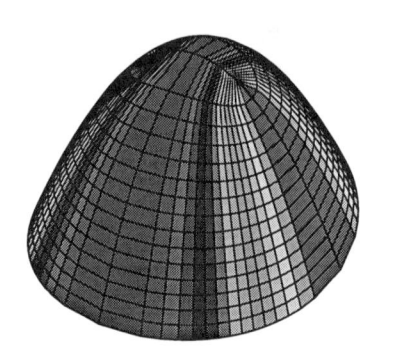

Principal curves on
$$\frac{x^2}{2} + \frac{y^2}{3} = -2z + 7$$

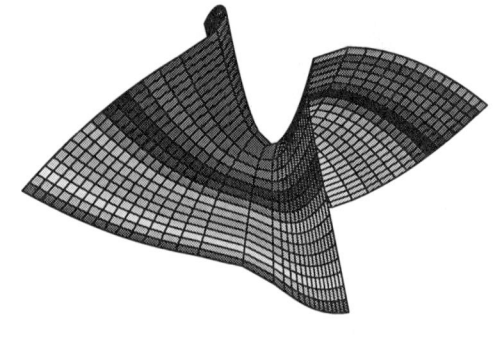

Principal curves on
$$x^2 - y^2 = 4z - 3$$

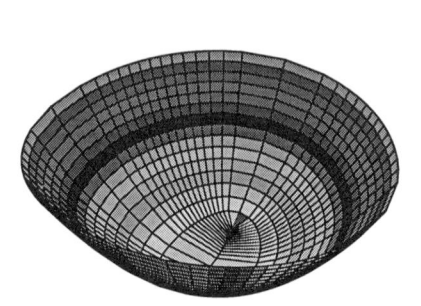

Principal curves on
$$\frac{x^2}{5} - \frac{y^2}{4} = 2z$$

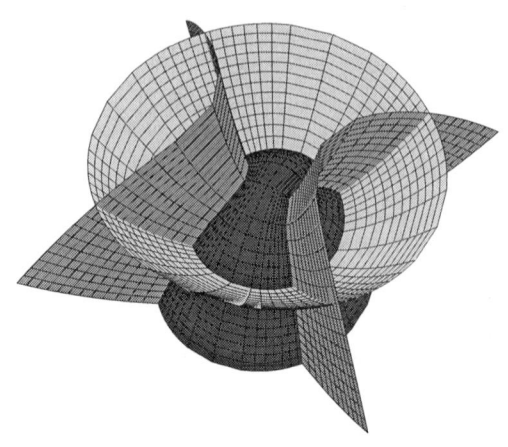

Triply orthogonal system formed by two families of elliptic paraboloids and one family of hyperbolic paraboloids

29.5 Exercises

1. Plot the systems of triply orthogonal surfaces on pages 662, 663 and 664.

29.5 Exercises

2. Use *Mathematica* to draw the principal curves on an ellipsoid, a hyperboloid of one sheet and a hyperboloid of two sheets.

The notion of a system of orthogonal curves in $\mathbb{R}^2$ corresponds to the notion of a system of triply orthogonal surfaces in $\mathbb{R}^3$:

Definition. *An* **orthogonal system of curves** *on an open set* $\mathcal{U} \subseteq \mathbb{R}^2$ *consists of two families* $\mathfrak{A}$, $\mathfrak{B}$, *of curves such that*

(i) *each point of* $\mathcal{U}$ *lies on one and only one member of each family;*

(ii) *each curve in each family meets every member of the other family orthogonally.*

3. Let $\mathcal{U} \subseteq \mathbb{R}^2$ be an open subset and $F: \mathcal{U} \longrightarrow \mathbb{C} = \mathbb{R}^2$ a holomorphic function. Let $\mathfrak{H}$ and $\mathfrak{V}$ be the families of horizontal and vertical lines in $\mathbb{R}^2$. Show that the images of $\mathfrak{H}$ and $\mathfrak{V}$ constitute a system of orthogonal curves in $\mathbb{R}^2$.

4. The following miniprogram **plotpatch2d**[3] plots the images of horizontal and vertical lines under the function $f: \mathcal{U} \longrightarrow \mathbb{R}^2$, where $\mathcal{U}$ is an open subset of $\mathbb{R}^2$.

```
plotpatch2d[expr_,{x_,xmin_,xmax_},{y_,ymin_,ymax_},
    PlotPoints->{xpoints_,ypoints_},opts___]:=
Show[ParametricPlot[Evaluate[
Table[ComplexExpand[{Re[expr],Im[expr]}],
    {y,ymin,ymax,(ymax - ymin)/(ypoints - 1)}]],
    {x,xmin,xmax},PlotPoints->xpoints,
DisplayFunction->Identity],
ParametricPlot[Evaluate[
Table[ComplexExpand[{Re[expr],Im[expr]}],
    {x,xmin,xmax,(xmax - xmin)/(xpoints -1)}]],
    {y,ymin,ymax},PlotPoints->ypoints,
DisplayFunction->Identity],
AspectRatio->Automatic,
DisplayFunction->$DisplayFunction,opts]
```

---

[3] The miniprogram **plotpatch2d** has the same functionality as Maeder's **CartesianMap** and **PolarMap**, but draws curved instead of straight lines.

Use it to draw the following picture:

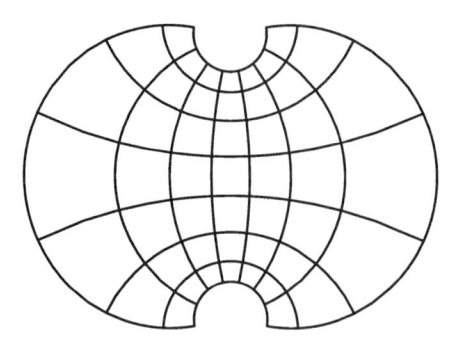

**Image of horizontal and vertical
lines by** $z \longmapsto \tan(z)$

5. Let $\mathfrak{A}$ and $\mathfrak{B}$ be a system of orthogonal curves in $\mathbb{R}^2$. Let $\widetilde{\mathfrak{A}}$ and $\widetilde{\mathfrak{B}}$ be the families of surfaces of revolution in $\mathbb{R}^3$ generated by $\mathfrak{A}$ and $\mathfrak{B}$. Show that $\widetilde{\mathfrak{A}}$ and $\widetilde{\mathfrak{B}}$ together with the family $\mathfrak{C}$ of planes through the axis of revolution constitute a triply orthogonal system of surfaces.

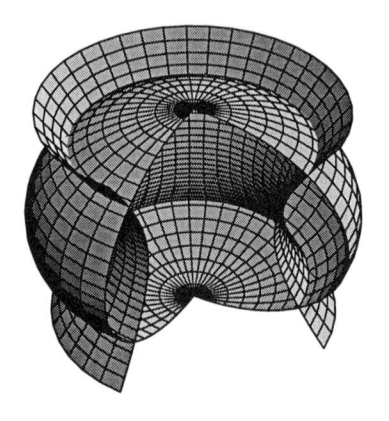

**Orthogonal ellipsoid and
hyperboloid of revolution**

30

MINIMAL SURFACES

We begin a more detailed study of minimal surfaces in this chapter. In Section 30.1 we show that in a suitable sense a minimal surface is a critical point of the area function. Several examples of minimal surfaces are described in Section 30.2. These include the deformation from the helicoid to the catenoid, and the surfaces of Enneper, Scherk, Henneberg and Catalan. In Section 30.3 we show that the Gauss map of a minimal surface is conformal.

In the theory of capillarity the importance of minimal surfaces as surfaces of least potential surface energy was illustrated by the experiments of Plateau,[1] who dipped wires in the form of space curves into a solution of soapy water and glycerin, thus realizing minimal surfaces. The problem of Plateau is the problem of determining the minimal surfaces through a given curve. It has been studied in great generality by many mathematicians.

30.1 Normal Variation

We defined a minimal surface on page 374 as a surface whose mean curvature vanishes. This is the definition of Lagrange, who first defined minimal surface in

---

[1] Joseph Antoine Ferdinand Plateau (1801-1883). Belgian physicist. His thesis concerned the impressions that light can have on the eye. Unfortunately, this led him to stare into the bright sun for approximately 25 seconds; after this experiment he gradually became blind. In 1840 he began a series of experiments with surfaces for which he became famous.

1760 [Lag]. A more intuitive meaning of minimal surface is the surface of least area among a family of surfaces having the same boundary. In order to show how these two definitions coincide, we define the normal variation of a surface $\mathcal{M}$ in $\mathbb{R}^3$ to be a family of surfaces $t \longmapsto \mathcal{M}(t)$ representing how $\mathcal{M}$ changes when pulled in a normal direction. Let $A(t)$ denote the area of $\mathcal{M}(t)$. We show that the mean curvature of $\mathcal{M}$ vanishes if and only if the first derivative of $t \longmapsto A(t)$ vanishes at $\mathcal{M}$.

First, we make precise the notion of normal variation. The definitions of region and boundary are given on page 629, and the general formula for the area of a bounded region of a surface in $\mathbb{R}^n$ is given by (15.25), page 353. In the case $n = 3$ there is a special formula.

Lemma 30.1. *Let $\mathcal{M}$ be a regular surface in $\mathbb{R}^3$ and let $\mathcal{Q}$ be a bounded region in $\mathcal{M}$ which is contained in the trace of a regular patch $\mathbf{x}:\mathcal{U} \longrightarrow \mathcal{M}$. Then the area of $\mathcal{Q}$ is given by*

$$\mathrm{Area}(\mathcal{Q}) = \iint_{\mathbf{x}^{-1}(\mathcal{Q})} \|\mathbf{x}_u \times \mathbf{x}_v\| \, du \, dv.$$

We wish to study how area changes with a small perturbation of a surface. The simplest perturbation is one that is normal to the surface.

Definition. *Let $\mathbf{x}:\mathcal{U} \longrightarrow \mathbb{R}^3$ be a regular patch, and choose a bounded region $\mathcal{Q} \subset \mathcal{U}$. Suppose that $h:\mathcal{Q} \longrightarrow \mathbb{R}$ is differentiable and $\varepsilon > 0$. Let $\mathbf{U}$ denote a unit vector field such that $\mathbf{U}(u,v)$ is perpendicular to $\mathbf{x}$ for all $(u,v) \in \mathcal{U}$; that is, $\mathbf{U} \cdot \mathbf{x}_u = \mathbf{U} \cdot \mathbf{x}_v = 0$. Then the **normal variation** of $\mathbf{x}$ and $\mathcal{Q}$, determined by h, is the map*

$$\mathbf{x}: (-\varepsilon, \varepsilon) \times \mathcal{Q} \longrightarrow \mathbb{R}^3$$

given by
(30.1) $$\mathbf{x}[t](u,v) = \mathbf{x}(u,v) + t\, h(u,v) \mathbf{U}(u,v)$$

for $(u,v) \in \mathcal{Q}$ and $-\varepsilon < t < \varepsilon$. Thus $\mathbf{x}[t]$ is a patch for each t with $-\varepsilon < t < \varepsilon$ for sufficiently small ε. Let

(30.2) $$\begin{cases} E(t) = \mathbf{x}[t]_u \cdot \mathbf{x}[t]_u, \\ F(t) = \mathbf{x}[t]_u \cdot \mathbf{x}[t]_v, \\ G(t) = \mathbf{x}[t]_v \cdot \mathbf{x}[t]_v. \end{cases}$$

Then $E = E(0)$, $F = F(0)$, $G = G(0)$, and the area of $\mathbf{x}[t](\mathcal{Q})$ is given by

(30.3) $$A(t) = \mathrm{Area}(\mathbf{x}[t](\mathcal{Q})) = \iint_{\mathcal{Q}} \sqrt{E(t)G(t) - F(t)^2} \, du \, dv.$$

30.1 Normal Variation

We compute the derivative at zero of the function $\mathcal{A}(t)$.

Lemma 30.2. *We have*

$$(30.4) \qquad A'(0) = -2 \iint_Q h\, H \sqrt{E\,G - F^2}\, du\, dv,$$

where H denotes the mean curvature of $\mathcal{M}$.

Proof. We differentiate (30.1) with respect to u and v, obtaining

$$(30.5) \qquad \begin{cases} \mathbf{x}[t]_u = \mathbf{x}_u + t\, h_u\, \mathbf{U} + t\, h\, \mathbf{U}_u, \\ \mathbf{x}[t]_v = \mathbf{x}_v + t\, h_v\, \mathbf{U} + t\, h\, \mathbf{U}_v. \end{cases}$$

From (16.10), (30.5) and the definitions (30.2) it follows that

$$(30.6) \qquad \begin{aligned} E(t) &= \mathbf{x}[t]_u \cdot \mathbf{x}[t]_u \\ &= (\mathbf{x}_u + t\, h_u\, \mathbf{U} + t\, h\, \mathbf{U}_u) \cdot (\mathbf{x}_u + t\, h_u\, \mathbf{U} + t\, h\, \mathbf{U}_u) \\ &= E + 2t\, h\, \mathbf{x}_u \cdot \mathbf{U}_u + O(t^2) \\ &= E - 2t\, h\, e + O(t^2). \end{aligned}$$

Similarly,

$$(30.7) \qquad F(t) = F - 2t\, h\, f + O(t^2),$$

$$(30.8) \qquad G(t) = G - 2t\, h\, g + O(t^2).$$

From (30.6)–(30.8) and Theorem 16.17, page 377, we get

$$\begin{aligned} E(t)G(t) - F(t)^2 &= \left(E - 2t\, h\, e + O(t^2)\right)\left(G - 2t\, h\, g + O(t^2)\right) \\ &\quad - \left(F - 2t\, h\, f + O(t^2)\right)^2 \\ &= E\,G - F^2 - 2t\, h(E\, g - 2F\, f + G\, e) + O(t^2) \\ &= (E\,G - F^2)(1 - 4t\, h\, H) + O(t^2), \end{aligned}$$

and hence

$$(30.9) \qquad \begin{aligned} \sqrt{E(t)G(t) - F(t)^2} &= \sqrt{(E\,G - F^2)(1 - 4t\, h\, H) + O(t^2)} \\ &= \sqrt{E\,G - F^2}\sqrt{(1 - 4t\, h\, H) + O(t^2)} \\ &= \sqrt{E\,G - F^2}\,(1 - 2t\, h\, H) + O(t^2). \end{aligned}$$

From (30.9) and (30.3) it follows that

$$(30.10) \quad A(t) = \text{Area}(\mathbf{x}[t](\mathcal{Q})) = \iint_{\mathcal{Q}} \left(\sqrt{EG - F^2} \, (1 - 2t\, h\, H) + O(t^2) \right) du\, dv$$

$$= \iint_{\mathcal{Q}} \sqrt{EG - F^2}\, du\, dv - 2t \iint_{\mathcal{Q}} h\, H \sqrt{EG - F^2}\, du\, dv + O(t^2).$$

When we differentiate (30.10) with respect to t and evaluate the resulting expression at 0, we obtain (30.4). ∎

Theorem 30.3. *Let* $\mathbf{x}: \mathcal{U} \longrightarrow \mathbb{R}^3$ *be a regular patch, and choose a bounded region* $\mathcal{Q} \subset \mathcal{U}$. *Then* $\mathbf{x}$ *is minimal on* $\mathcal{Q}$ *if and only if* $A'(0) = 0$ *for a normal variation of* $\mathbf{x}$ *and* $\mathcal{Q}$ *with respect to any* $h: \mathcal{Q} \longrightarrow \mathbb{R}$.

Proof. If H is identically zero for $\mathbf{x}$, then (30.4) implies that $A'(0) = 0$ for any h. Conversely, suppose that $A'(0) = 0$ for any h, but that there is $\mathbf{q} \in \mathcal{Q}$ for which $H(\mathbf{q}) \neq 0$. We choose $h: \mathcal{Q} \longrightarrow \mathbb{R}$ such that $h(\mathbf{q}) = H(\mathbf{q})$, and h is identically zero outside of a small neighborhood of $\mathbf{q}$. But then (30.4) implies that $A'(0) < 0$. This contradiction shows that $H(\mathbf{q}) = 0$. Since $\mathbf{q}$ is arbitrary, $\mathbf{x}$ is minimal. ∎

It should be pointed out that we have said nothing about the second derivative of A at 0, so that a minimal surface, although a critical point of A, may not actually be a minimum. More details, as well as extensions of the results of this section, can be found in [Nits, pages 90–116].

30.2 Examples of Minimal Surfaces

Enneper's Minimal Surface

One of the simplest minimal surfaces to define is **Enneper's minimal surface**, found by Enneper in 1864 (see [Enn1]). It is defined by

$$\mathbf{enneper}(u, v) = \left(u - \frac{u^3}{3} + u v^2,\, -v + \frac{v^3}{3} - v u^2,\, u^2 - v^2 \right).$$

or in *Mathematica*

```
enneper[u_,v_]:= {u - u^3/3 + u v^2,
                  -v + v^3/3 - v u^2, u^2 - v^2}
```

30.2 Examples of Minimal Surfaces

In spite of the simplicity of its definition, Enneper's surface is complicated because of self-intersections. The *Mathematica* package **Graphics3D.m** (which can be read in with the command **Needs["Graphics`Graphics3D`"]**) contains the useful command **Shadow**. Here is how to use **Shadow** to see the two holes in Enneper's surface:

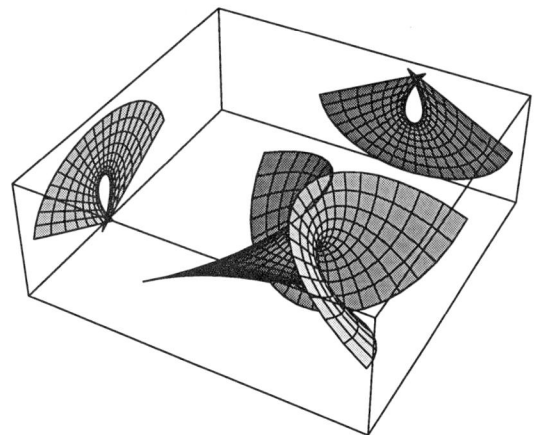

Enneper's surface with shadows

The command to draw this picture is

```
Shadow[ParametricPlot3D[enneper[u,v]//Evaluate,
{u,-2,2},{v,-2,2},PlotPoints->{20,20},
Axes->None,DisplayFunction->Identity],ZShadow->False,
DisplayFunction->$DisplayFunction];
```

The Deformation from the Helicoid to the Catenoid

We have already considered the helicoid on page 357 and the catenoid on page 467. In fact, these two surfaces are the beginning and end of a deformation through isometric minimal surfaces. More precisely, for each t with $0 \leq t \leq \pi/2$ we define

$$\mathbf{heltocat}[t](u,v) = \cos t(\sinh v \sin u, -\sinh v \cos u, u)$$
$$+ \sin t(\cosh v \cos u, \cosh v \sin u, v).$$

The *Mathematica* version of **heltocat**[t] is

```
heltocat[t_][u_,v_]:=
    Cos[t]{Sinh[v]Sin[u],-Sinh[v]Cos[u],u} +
    Sin[t]{Cosh[v]Cos[u],Cosh[v]Sin[u],v}
```

Mathematica can be used to determine the exact relation between the surfaces **heltocat**[*t*], **helicoid**[*a, b*] and **catenoid**[*c*]. We find that

$$\text{heltocat}[0](u,v) = \text{helicoid}[1,1](u - \pi/2, \sinh v)$$

and

$$\text{heltocat}[\pi/2](u,v) = \text{catenoid}[1](u,v).$$

Theorem 30.4. *The 1-parameter family of surfaces* $t \mapsto$ **heltocat**[*t*] *is a deformation from the helicoid to the catenoid such that* **heltocat**[0] *is (a reparametrization) of a helicoid and* **heltocat**[$\pi/2$] *is a catenoid. Furthermore, each* **heltocat**[*t*] *is a minimal surface which is locally isometric to* **heltocat**[0]. *In particular, the helicoid is locally isometric to the catenoid. Since the coefficients of the second fundamental form of* **heltocat**[*t*] *are nonconstant functions of t, the immersion of* **heltocat**[*t*] *depends on t.*

Proof. Let $E(t)$, $F(t)$, $G(t)$ denote the coefficients of the first fundamental form of **heltocat**[*t*]. An easy calculation (by hand or with *Mathematica*) shows that

$$E(t)(u,v) = G(t)(u,v) = \cosh^2 v, \qquad F(t)(u,v) = 0.$$

Hence $E(t)(u,v)$, $F(t)(u,v)$, $G(t)(u,v)$ are constant functions of t.

On the other hand, if $e(t)(u,v)$, $f(t)(u,v)$ and $g(t)(u,v)$ denote the coefficients of the second fundamental form of **heltocat**[*t*], then

$$e(t)(u,v) = -g(t)(u,v) = \sin t, \qquad f(t)(u,v) = \cos t. \blacksquare$$

Mathematica can be used to draw accurate pictures of the deformation $t \mapsto$ **heltocat**[*t*].

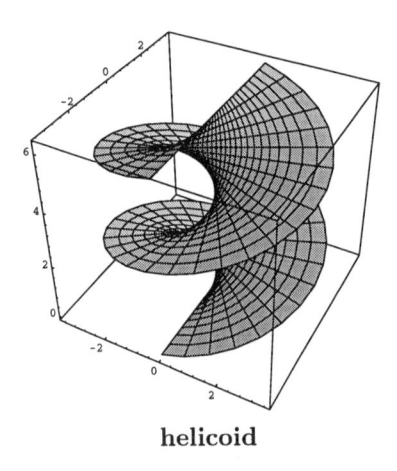

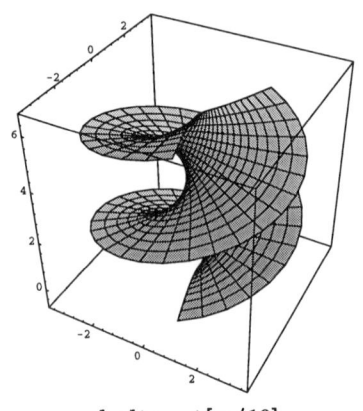

helicoid

heltocat[$\pi/10$]

30.2 Examples of Minimal Surfaces

heltocat[$\pi/5$]

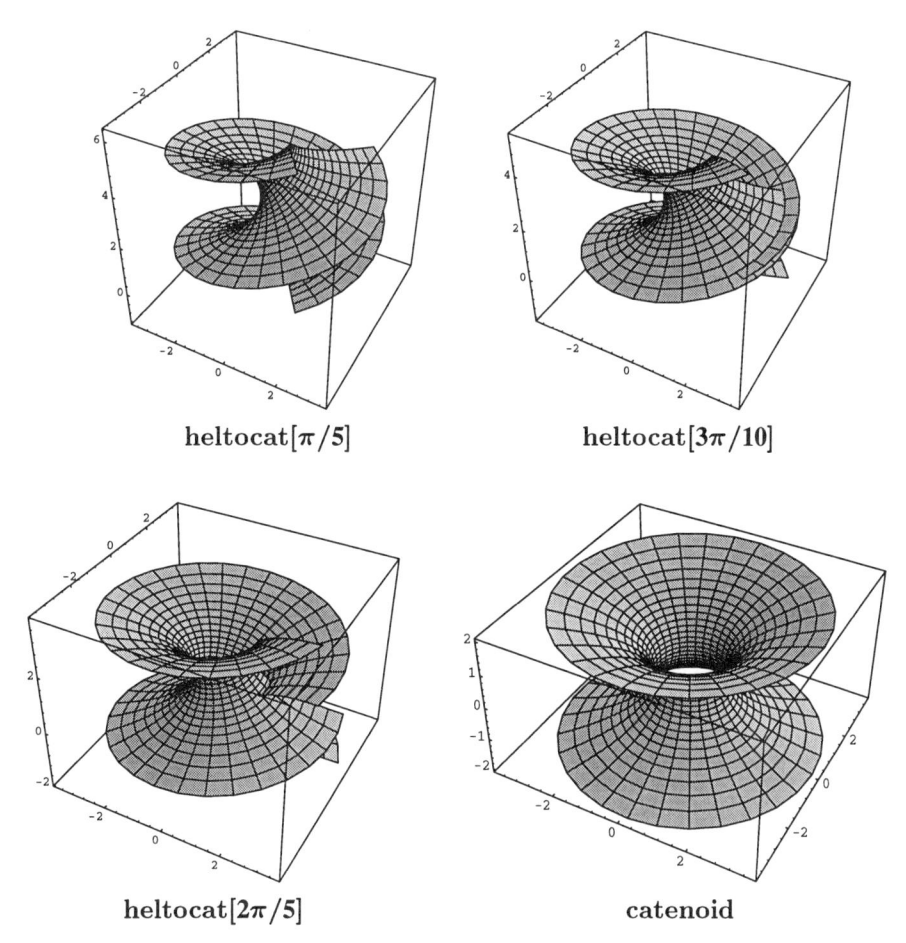

heltocat[$3\pi/10$]

heltocat[$2\pi/5$]

catenoid

Neither the catenoid nor the helicoid intersects itself; however, every intermediate surface **heltocat[t]** has self-intersections. The asymptotic curves on the helicoid are gradually transformed into principal curves on the catenoid.

For an animation of the deformation of a helicoid to a catenoid with coloring indicated by the Gaussian curvature, see Section 30.5.

Minimal Monge Patches and Scherk's Minimal Surface

We have the following immediate consequence of Lemma 17.1, page 399:

Lemma 30.5. *A Monge patch* $(u,v) \longmapsto \bigl(u,v,h(u,v)\bigr)$ *is a minimal surface if and only if*
$$(30.11) \qquad (1+h_v^2)h_{uu} - 2h_u h_v h_{uv} + (1+h_u^2)h_{vv} = 0.$$

One can hope to find interesting examples of minimal surfaces by assuming that h has a special form. In 1835 Scherk[2] (see [Scherk], [BaCo]) determined the minimal surfaces of the form $(u,v) \longmapsto \bigl(u,v,f(u)+g(v)\bigr)$.

Theorem 30.6. *If a Monge patch* $\mathbf{x}:\mathcal{U} \longrightarrow \mathcal{M}$ *with* $h(u,v) = f(u) + g(v)$ *is a minimal surface, then either* $\mathcal{M}$ *is part of a plane or there is a nonzero constant* a *such that*
$$(30.12) \qquad f(u) = -\frac{1}{a}\log\cos au, \quad \text{and} \quad g(v) = \frac{1}{a}\log\cos av,$$
or equivalently,
$$h(u,v) = \frac{1}{a}\log\left(\frac{\cos av}{\cos au}\right).$$

Proof. When $h(u,v) = f(u) + g(v)$, we have
$$h_{uu} = f''(u), \qquad h_{uv} = 0, \qquad h_{vv} = g''(v);$$
Hence equation (30.11) reduces to
$$(30.13) \qquad \frac{f''(u)}{1+f'(u)^2} = \frac{-g''(v)}{1+g'(v)^2}.$$

Here u and v are independent variables, so each side of (30.13) must equal a constant, call it a. If $a = 0$, then both f and g are linear, so that $\mathcal{M}$ is part of a plane. Otherwise, the two equations
$$(30.14) \qquad \frac{f''(u)}{1+f'(u)^2} = a = \frac{-g''(v)}{1+g'(v)^2}$$
are easily solved. The result is (30.12). ∎

It is instructive to use *Mathematica* to solve the differential equations (30.14); the relevant command is **DSolve**. *Mathematica* solves the first equation of (30.14) in two steps:

```
DSolve[f1'[t]/(1 + f1[t]^2) == 0,f1[t],t]
```

---
[2]Heinrich Ferdinand Scherk (1798-1885). German mathematician, who also worked in number theory. He studied with Bessel in Königsberg, and was elected rector of the University of Kiel three times, but was forced to leave in 1848 by the Danes.

30.2 Examples of Minimal Surfaces

```
{{f1[t] -> Tan[C[1]]}}

DSolve[f'[t] == Tan[C[1] t],f[t],t,DSolveConstants->C1]

                     Log[Cos[t C[1]]]
{{f[t] -> C1[1] -  ──────────────────}}
                          C[1]
```

As suggested by Theorem 30.6, we define **Scherk's minimal surface** by

$$\mathbf{scherk}[a](u,v) = \left(u, v, \frac{1}{a}\log\left(\frac{\cos av}{\cos au}\right)\right),$$

or in *Mathematica*,

```
scherk[a_][u_,v_] := {u,v,(1/a)Log[Cos[a v]/Cos[a u]]}
```

For simplicity we take $a = 1$. The surface **scherk[1]** is well-defined on the set

$$\mathcal{R} = \{(u,v) \mid \cos u \cos v > 0\}.$$

Then $\mathcal{R}$ can be imagined as the black squares in an infinite chess board with vertices

$$\left\{\left(m\pi + \frac{\pi}{2}, n\pi + \frac{\pi}{2}\right) \mid m \text{ and } n \text{ are integers}\right\}.$$

In fact, let us put

$$\mathcal{Q}(m,n) = \left\{(x,y) \mid m\pi - \frac{\pi}{2} < x < m\pi + \frac{\pi}{2} \text{ and } n\pi - \frac{\pi}{2} < y < n\pi + \frac{\pi}{2}\right\}.$$

We color the square $\mathcal{Q}(m,n)$ black if $m+n$ is even and white if $m+n$ is odd. Then

$$\mathcal{R} = \bigcup \{\mathcal{Q}(m,n) \mid m \text{ and } n \text{ are integers with } m+n \text{ even}\}$$

is made up of the black squares in an infinite chess board.

It is easy to see that

$$\text{scherk}[1](u + 2m\pi, v + 2n\pi) = \text{scherk}[1](u,v)$$

for all real u and v and all integers m and n. Hence the piece of Scherk's surface over a black square $\mathcal{Q}(m,n)$ is a translate of the portion over $\mathcal{Q}(0,0)$. Here is the graph of one such piece.

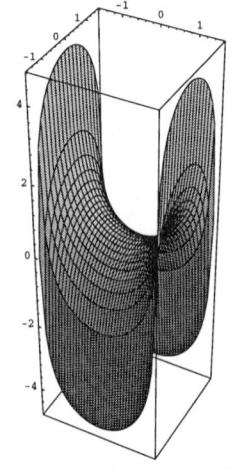

```
ParametricPlot3D[
Evaluate[
scherk[1][u,v]],
{u,-Pi/2 + 0.01,
Pi/2 - 0.01},
{v,-Pi/2 + 0.01,Pi/2 - 0.01},
PlotPoints->{40,20}];
```

Scherk's minimal surface

A better plot of Scherk's minimal surface is obtained using **ImplicitPlot3D**:

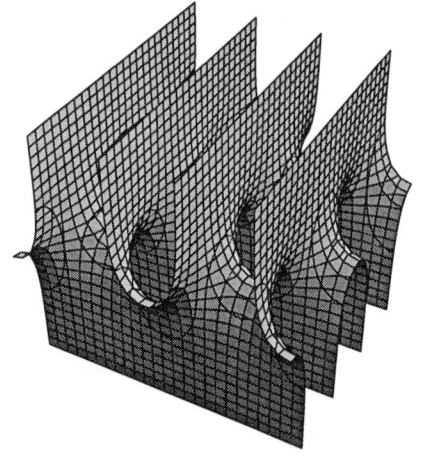

Implicit plot of Scherk's minimal surface

```
ImplicitPlot3D[scherkimplicit[1][x,y,z] == 0,
{x,-2Pi,2Pi},{y,-2Pi,2Pi},{z,-6,6},
PlotPoints->{{5},{5}},Boxed->False]
```

Henneberg's Minimal Surface

The following elementary parametrization of **Henneberg's**[3] **minimal surface** is given on page 144 of [Nits]:

$$\mathbf{henneberg}(u,v) = \left(2\sinh u \cos v - \frac{2}{3}\sinh 3u \cos 3v,\right.$$

$$\left. 2\sinh u \sin v + \frac{2}{3}\sinh 3u \sin 3v, \, 2\cosh 2u \cos 2v\right),$$

or

```
henneberg[u_,v_]:=
    {2Sinh[u]Cos[v] - (2/3)Sinh[3u]Cos[3v],
     2Sinh[u]Sin[v] + (2/3)Sinh[3u]Sin[3v],
     2Cosh[2u]Cos[2v]}
```

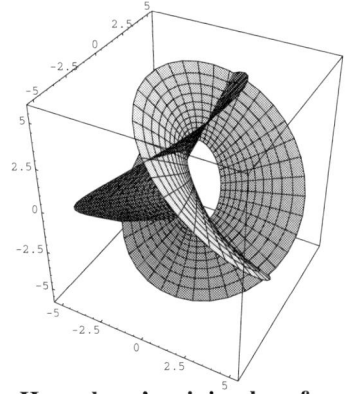

```
ParametricPlot3D[
henneberg[u,v]
//Evaluate,
{u,0.35,0.85},
{v,-Pi,Pi},
PlotPoints->{12,96},
PlotRange->All];
```

Henneberg's minimal surface

---

[3] Ernst Lebrecht Henneberg (1850-1922). German mathematician. Professor at the University of Darmstadt.

Let us determine the points at which the patch **henneberg** fails to be regular. We use

surfacejacobian[henneberg][u,v]//Simplify

$$64\, \text{Cosh}[u]^4\, (\text{Cos}[4\,v] - \text{Cosh}[4\,u])^2$$

Thus **henneberg** is regular except for the points $(0, n\pi/2)$, where n is an integer. Note that

(30.15) $\qquad$ **henneberg**$(u,v) = $ **henneberg**$(-u, v+\pi)$.

Hence for any region $\mathcal{U}$ in the right half-plane $\{\,(p,q) \mid p > 0\,\}$ there will be a region $\widetilde{\mathcal{U}}$ in the left half-plane $\{\,(p,q) \mid p < 0\,\}$ whose image under **henneberg** is the same subset of $\mathbb{R}^3$. But (30.15) implies that the unit normal **U** of **henneberg** satisfies

$$\mathbf{U}(u,v) = -\mathbf{U}(-u, v+\pi).$$

Hence **henneberg**$(\mathcal{U})$ and **henneberg**$(\widetilde{\mathcal{U}})$ will have opposite orientations, as illustrated in the following plots:

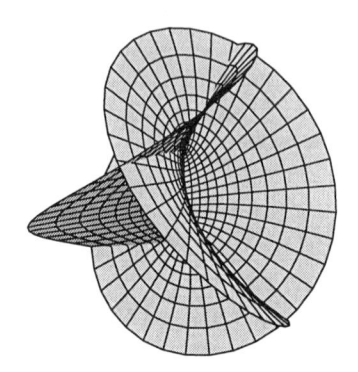

The image of
$\{\,(p,q) \mid p > 0,\ -\pi < q < \pi\,\}$
under henneberg

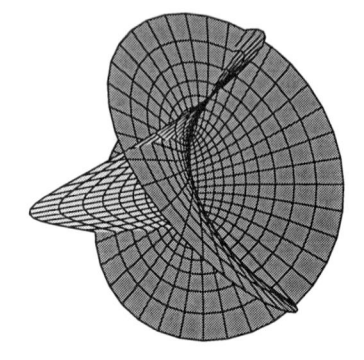

The image of
$\{\,(p,q) \mid p < 0,\ -\pi < q < \pi\,\}$
under henneberg

Catalan's Minimal Surface

An interesting problem in minimal surface theory is the determination of a minimal surface that contains a given curve as a geodesic (see Chapter 26 for the definition

30.2 Examples of Minimal Surfaces

of geodesic), an asymptotic curve or a principal curve. See [Nits, page 140]. For example, **Catalan's**[4] **minimal surface**, defined by

$$\mathbf{catalan}[a](u,v) = a\left(u - \sin u \cosh v, 1 - \cos u \cosh v, -4\sin\frac{u}{2}\sinh\frac{v}{2}\right)$$

contains a reparametrization of a cycloid as a geodesic. The *Mathematica* definition of Catalan's surface is:

```
catalan[a_][u_,v_] := {a(u - Sin[u]Cosh[v]),
                       a(1 - Cos[u]Cosh[v]),
                       -4a Sin[u/2]Sinh[v/2]}
```

In fact, $u \longmapsto \mathbf{catalan}[a](u,0)$ is easily seen to be a cycloid, and it can be proved that it is a pregeodesic, that is, a reparametrization of a geodesic. (See exercise 4.)

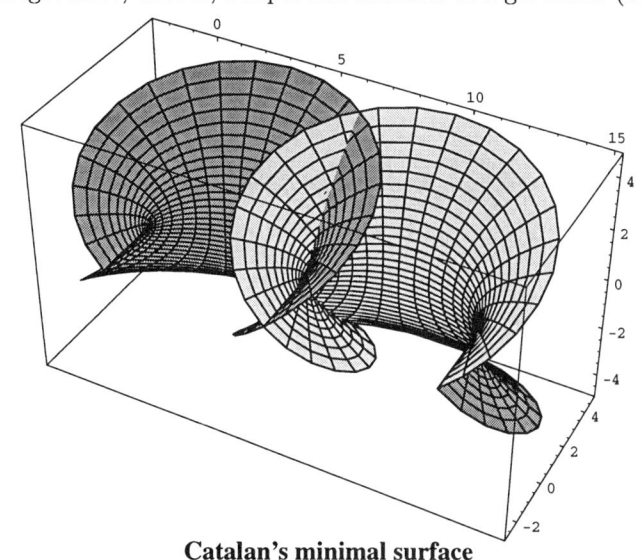

Catalan's minimal surface

```
ParametricPlot3D[Evaluate[Append[catalan[1][u,v],
FaceForm[Yellow,Green]]],{u,0,4Pi},{v,-2,2},
Lighting->False,PlotPoints->{45,30}];
```

---

[4] Eugène Charles Catalan (1814-1894). Belgian mathematician, who had difficulty obtaining a position in France because of his left-wing views. Catalan's constant is

$$\sum_{n=0}^{\infty}\frac{(-1)^n}{(2n+1)^2} \approx 0.915966.$$

30.3 The Gauss Map of a Minimal Surface

In this section we show that the Gauss map of a minimal surface has special properties that permit the use of complex analysis in the study of minimal surfaces. First, however, we need the following notion:

Definition. Let $\mathcal{M}_1$ and $\mathcal{M}_2$ be oriented regular surfaces in $\mathbb{R}^n$, and let J_1 and J_2 be the corresponding complex structures. A map $\Phi\colon \mathcal{M}_1 \longrightarrow \mathcal{M}_2$ is called a **complex map** provided

$$\Phi_* \circ J_1 = J_2 \circ \Phi_*. \tag{30.16}$$

Similarly, $\Phi\colon \mathcal{M}_1 \longrightarrow \mathcal{M}_2$ is called an **anticomplex map** provided

$$\Phi_* \circ J_1 = -J_2 \circ \Phi_*. \tag{30.17}$$

Next, we derive an elementary but important property of complex and anticomplex maps.

Lemma 30.7. *Let $\Phi\colon \mathcal{M}_1 \longrightarrow \mathcal{M}_2$ be a complex or anticomplex map. Then Φ is a conformal map.*

Proof. Let $\mathbf{p} \in \mathcal{M}_1$ and let $\mathbf{v_p}$ be a tangent vector to $\mathcal{M}_1$ at $\mathbf{p}$. Then there is a number $\lambda(\mathbf{p})$ such that

$$\|\Phi_*(\mathbf{v_p})\| = \lambda(\mathbf{p})\|\mathbf{v_p}\|. \tag{30.18}$$

Since Φ is complex or anticomplex, (30.18) implies that

$$\|\Phi_*(J_\mathbf{p}\mathbf{v_p})\| = \lambda(\mathbf{p})\|\mathbf{v_p}\| \quad \text{and} \quad \Phi_*(\mathbf{v_p}) \cdot \Phi_*(J_\mathbf{p}\mathbf{v_p}) = 0. \tag{30.19}$$

From (30.18) and (30.19) we see that

$$\|\Phi_*(a\mathbf{v_p} + bJ_\mathbf{p}\mathbf{v_p})\|^2 = \lambda(\mathbf{p})^2 \|a\mathbf{v_p} + bJ_\mathbf{p}\mathbf{v_p}\|^2$$

for all $a, b \in \mathbb{R}$. Hence Φ is conformal. ∎

We have a ready-made example of an anticomplex map.

Theorem 30.8. *The Gauss map of an oriented minimal surface $\mathcal{M} \subset \mathbb{R}^3$ is anticomplex. Furthermore, the Gauss map anticommutes with the shape operator:*

$$J_\mathbf{p} S_\mathbf{p} = -S_\mathbf{p} J_\mathbf{p} \tag{30.20}$$

for all $\mathbf{p} \in \mathcal{M}$.

30.3 The Gauss Map of a Minimal Surface

Proof. Let $\mathbf{p} \longmapsto \mathbf{U}(\mathbf{p})$ be the Gauss map of $\mathcal{M}$. Let $\mathbf{p} \in \mathcal{M}$, and let $\{\mathbf{e}_1, \mathbf{e}_2\}$ be an orthonormal basis of $\mathcal{M}_\mathbf{p}$ which diagonalizes the shape operator $S_\mathbf{p}$ of $\mathcal{M}$ at $\mathbf{p}$. Let $\mathbf{k}_1$ and $\mathbf{k}_2$ be the corresponding principal curvatures. Finally, let $J_\mathbf{p}$ denote the complex structure determined by $\mathbf{U}$; then $J_\mathbf{p}\mathbf{e}_1 = \pm\mathbf{e}_2$ and $J_\mathbf{p}\mathbf{e}_2 = \mp\mathbf{e}_1$. We use the fact that $\mathcal{M}$ is a minimal surface to compute

$$J_\mathbf{p} S_\mathbf{p} \mathbf{e}_1 = J_\mathbf{p} \mathbf{k}_1 \mathbf{e}_1 = \pm \mathbf{k}_1 \mathbf{e}_2 = \mp \mathbf{k}_2 \mathbf{e}_2 = \mp S_\mathbf{p} \mathbf{e}_2 = -S_\mathbf{p} J_\mathbf{p} \mathbf{e}_1.$$

Similarly, $J_\mathbf{p} S_\mathbf{p} \mathbf{e}_2 = -S_\mathbf{p} J_\mathbf{p} \mathbf{e}_2$. Thus we have established (30.20). But Lemma 16.4, page 362, tells us that $S_\mathbf{p}$ is the negative of the tangent map of $\mathbf{U}$ at $\mathbf{p}$. Hence (30.20) implies that $\mathbf{U}$ is anticomplex. ∎

From Lemma 30.7 and Theorem 30.8 we get

Corollary 30.9. *The Gauss map of a minimal surface is conformal.*

Corollary 30.9 implies that the Gauss map of a minimal surface preserves the proportions of infinitesimally small rectangles. When we plot a surface using **ParametricPlot3D** with a fine mesh, we get small rectangles. The ratios of the lengths of the sides of such small rectangles, and also the angles between the sides must be approximately preserved by the Gauss map. Let us examine this phenomenon by drawing Enneper's surface and its image under the Gauss map. The reader should compare the following pictures with the corresponding pictures of a nonminimal surface such as the monkey saddle and its image under the Gauss map on page 408.

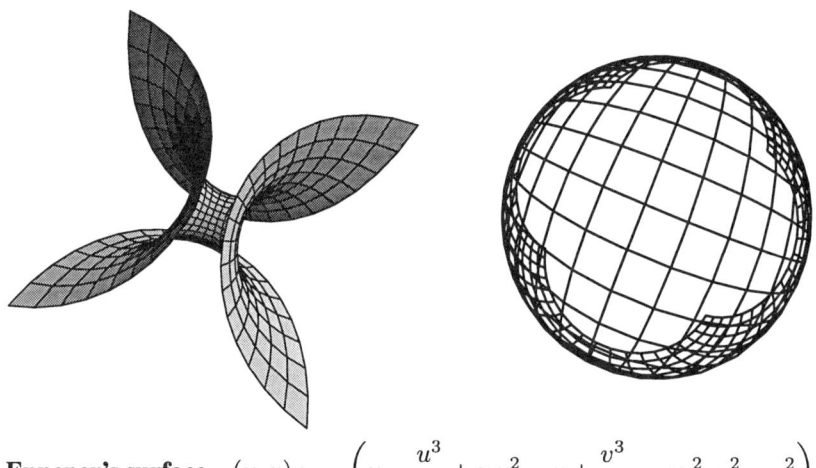

Enneper's surface $(u, v) \longmapsto \left(u - \dfrac{u^3}{3} + u v^2, -v + \dfrac{v^3}{3} - v u^2, u^2 - v^2\right)$
and its image under the Gauss map

30.4 Exercises

1. Show that the Gauss map of $(u,v) \longmapsto \mathbf{heltocat}[t](u,v)$ does not depend on t.

2. Use *Mathematica* to check that the surfaces of Henneberg and Catalan are indeed minimal surfaces and compute the Gaussian curvature of each.

3. Use the command **asymeq1** (see page 427) to find the differential equations for the asymptotic curves on a catenoid. Solve the differential equations. Define a new parametrization of the catenoid in which the u- and v-parameter curves are asymptotic curves. Plot the surface. Use **selectgraphics3d** to get rid of big polygons, and display the resulting surface.

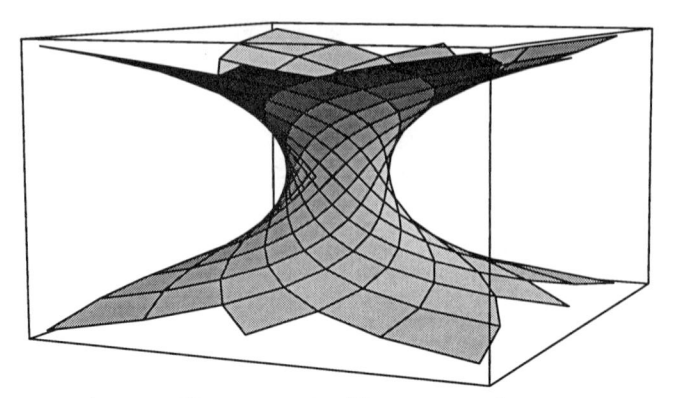

A catenoid parametrized by asymptotic curves

4. Show that $u \longmapsto \mathbf{catalan}[a](u,0)$ is a cycloid and a pregeodesic in Catalan's minimal surface.

5. A twisted generalization of Scherk's minimal surface is given by

$$\mathbf{scherk}[a, \theta] = \bigg(\sec(2\theta)(u\cos\theta + v\sin\theta), \sec(2\theta)(u\sin\theta + v\cos\theta),$$
$$\frac{1}{a}\log\bigg(\cos\bigg(\frac{\sec(2\theta)(u\sin\theta + v\cos\theta)}{\sec(2\theta)(u\cos\theta + v\sin\theta)}\bigg)\bigg)\bigg).$$

or

30.4 Exercises

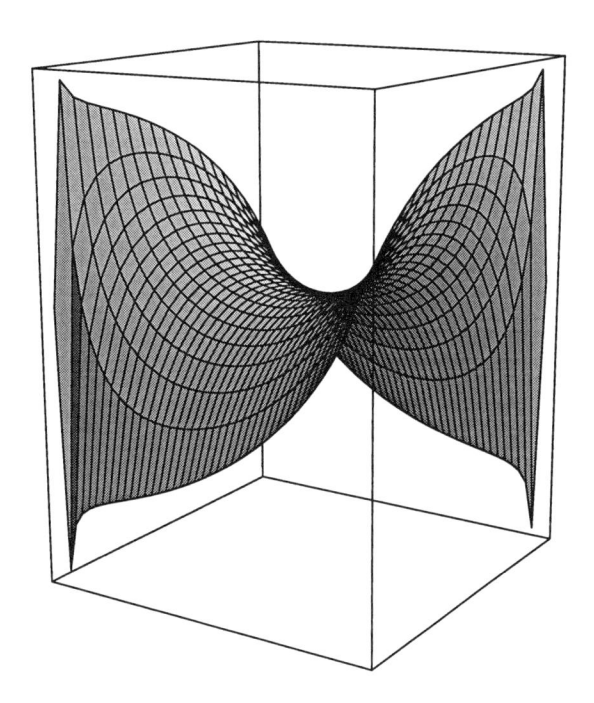

```
scherk[a_,theta_][u_,v_]:=
    {Sec[2theta](Cos[theta]u + Sin[theta]v),
     Sec[2theta](Sin[theta]u + Cos[theta]v),
     (1/a)Log[
     Cos[a Sec[2theta](Sin[theta]u + Cos[theta]v)]/
     Cos[a Sec[2theta](Cos[theta]u + Sin[theta]v)]]}
```

Scherk's minimal surface twisted by $\pi/16$

Show that **scherk[a,theta][u,v]** is a minimal surface and compute its Gaussian curvature.

6. **Scherk's fifth minimal surface** is defined implicitly by

$$\sin z = \sinh x \sinh y.$$

Show that it indeed is a minimal surface and plot it.

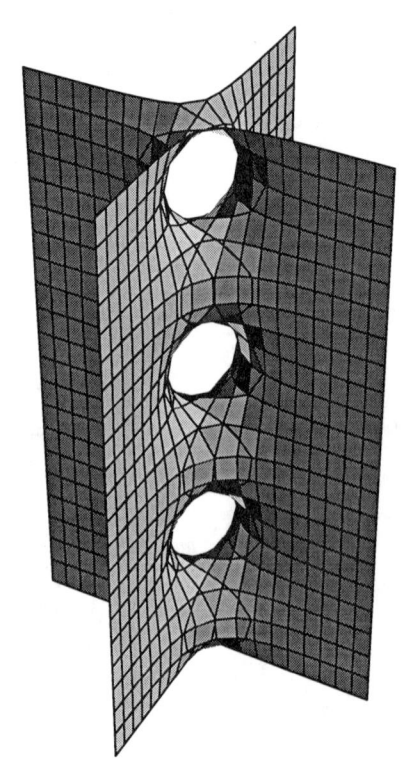

**Scherk's fifth minimal surface
plotted implicitly**

7. Portions of Scherk's fifth minimal surface are parametrized by

```
scherk5[a_,b_,c_][u_,v_]:=
     {a ArcSinh[u],b ArcSinh[v],c ArcSin[u v]}
```

where **a, b, c** are plus or minus 1. Show that **scherk5[a,b,c][u,v]** is a parametrized minimal surface and compute its Gaussian curvature.

30.5 Animations of Minimal Surfaces

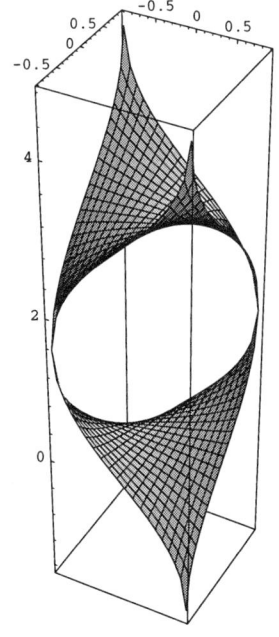

```
ParametricPlot3D[
{scherk5[1,1,1][u,v],
{0,0,Pi}+
scherk5[1,1,-1][u,v]}
//Evaluate,
{u,-1,1},{v,-1,1}];
```

Scherk's fifth minimal surface

30.5 Animations of Minimal Surfaces

To create an expanding Enneper's surface colored by the Gaussian curvature, we first define the Gaussian curvature with

```
gcurvature[enneper][u_,v_]:=-4/(1 + u^2 + v^2)^4
```

The animation is created with

```
Do[ParametricPlot3D[Evaluate[
Append[enneper[r Cos[theta],r Sin[theta]],
FaceForm[Hue[((gcurvature[
enneper][r Cos[theta],r Sin[theta]] + 4)/4)^40]]]],
{r,0,rr},{theta,0,2Pi},
Boxed->False,Axes->None,
Lighting->False,
PlotPoints->{Floor[10 rr],60},
BoxRatios->{1,1,1},
SphericalRegion->True,
PlotRange->{{-3.8,3.8},{-3.8,3.8},{-4,4}},
ViewPoint->{-0.121,-2.766,1.945}],
{rr,1,2,1/10}]
```

The following commands create an animation of a helicoid deforming into a catenoid. Each surface of the deformation is colored by its Gaussian curvature.

```
gcurvature[heltocat[c_]][u_,v_]:=-Sech[v]^4

Do[ParametricPlot3D[
Append[heltocat[t][u,v],
Hue[(-gcurvature[
heltocat[t]][u,v])]]//Evaluate,
{u,0,2Pi},{v,-2,2},
Boxed->False,Axes->None,
Lighting->False,
BoxRatios->{1,1,1}],
{t,0,Pi/2, Pi/40}]
```

31

MINIMAL SURFACES AND COMPLEX VARIABLES

In this chapter we use techniques from complex variables to study minimal surfaces. Many theorems from the theory of complex functions of one variable have minimal surface counterparts that will be powerful tools for us. *Isothermal coordinates* for surfaces are introduced in Section 31.1. In Section 31.2 we give a useful generalization to patches of the familiar *Cauchy-Riemann equations*; this allows us to define the notion of *minimal isothermal patch*, the preferred type of patch for studying minimal surfaces. The existence of *isometric deformations* is then proved. We introduce *complex derivatives* in Section 31.3. Section 31.4 contains preparatory material in which we point out the similarities and differences between complex vector algebra and real vector algebra. The important notion of *minimal curve* is introduced in Section 31.5. A simple algebraic method for constructing an isometric deformation from a minimal isothermal patch is given in Section 31.6. We also describe the **Mathematica** implementation of this method. This isometric deformation generalizes the deformation of the helicoid to the catenoid considered in Section 30.2. For example, we show how to draw deformations of Catalan's minimal surface with **Mathematica**. In Section 31.7 we describe a sequence of surfaces that generalize Enneper's minimal surface. An animation of Catalan's minimal surface is given in Section 31.9.

31.1 Isothermal Coordinates

We begin by defining an important type of patch that we shall need for our study of minimal surfaces.

Definition. Let $\mathcal{U} \subseteq \mathbb{R}^2$ be an open subset. A patch $\mathbf{x} : \mathcal{U} \longrightarrow \mathbb{R}^n$ is called **isothermal** provided there exists a differentiable function $\lambda : \mathcal{U} \longrightarrow \mathbb{R}$ such that

$$(31.1) \qquad \mathbf{x}_u \cdot \mathbf{x}_u = \mathbf{x}_v \cdot \mathbf{x}_v = \lambda^2 \quad \text{and} \quad \mathbf{x}_u \cdot \mathbf{x}_v = 0.$$

We call λ the **scaling function** of the isothermal patch.

The intuitive meaning of an isothermal patch $\mathbf{x}$ can be described as follows. Since $\mathbf{x}_u$ and $\mathbf{x}_v$ have the same length and are orthogonal, an isothermal patch maps an infinitesimal square in $\mathcal{U}$ into an infinitesimal square on its image. A more general patch would transform an infinitesimal square into an infinitesimal quadrilateral. The name isothermal is due to Lamé in 1833.

The following lemma is obvious from the definition of conformal map on page 349.

Lemma 31.1. *A patch* $\mathbf{x} : \mathcal{U} \longrightarrow \mathbb{R}^n$ *is isothermal if and only if it is conformal considered as a mapping* $\mathbf{x} : \mathcal{U} \longrightarrow \mathbf{x}(\mathcal{U})$.

For an isothemal patch on a sphere, see exercise 1. It is always possible to find an isothermal patch on a surface. More precisely:

Theorem 31.2. *Let* $\mathcal{M}$ *be a surface, and suppose* ds^2 *is a metric on* $\mathcal{M}$. *Let* $\mathbf{p} \in \mathcal{M}$. *Then there exists an open set* $\mathcal{U} \subset \mathbb{R}^2$ *and an isothermal patch* $\mathbf{x} : \mathcal{U} \longrightarrow \mathcal{M}$ *such that* $\mathbf{p} \in \mathbf{x}(\mathcal{U})$ *and*

$$ds^2 = \lambda^2 (du^2 + dv^2).$$

Proof. Let $\mathbf{y} : \mathcal{U} \longrightarrow \mathcal{M}$ be a patch; suppose that the metric expression for ds^2 in terms of the coordinates defined by $\mathbf{y}$ is $ds^2 = E\,dp^2 + 2F\,dp\,dq + G\,dq^2$. Then we can factor ds^2 as

$$(31.2) \quad ds^2 = \left(\sqrt{E}\,dp + \frac{F + i\sqrt{EG - F^2}}{\sqrt{E}}\,dq \right) \left(\sqrt{E}\,dp + \frac{F - i\sqrt{EG - F^2}}{\sqrt{E}}\,dq \right).$$

There exists an integrating factor μ such that μ times the first factor on the right-hand side of (31.2) is exact (see [AhSa, pages 125-126] for details); let us write

$$du + i\,dv = dz = \mu \left(\sqrt{E}\,dp + \frac{F + i\sqrt{EG - F^2}}{\sqrt{E}}\,dq \right),$$

31.1 Isothermal Coordinates

where u and v denote the real and imaginary parts of z. This differential equation defines p and q as functions of u and v. Then (31.2) becomes

$$(31.3) \qquad ds^2 = \frac{1}{|\mu|^2}(du + i\,dv)(du - i\,dv) = \frac{1}{|\mu|^2}(du^2 + dv^2).$$

We take $\lambda = |\mu|^{-1}$ in (31.3) and define a patch $\mathbf{x}$ by

$$\mathbf{x}(u,v) = \mathbf{y}\big(p(u,v), q(u,v)\big).$$

Then $\mathbf{x}$ is isothermal. ∎

We denote by Δ the **Laplacian** of $\mathbb{R}^2$; it is defined by

$$\Delta = \frac{\partial^2}{\partial u^2} + \frac{\partial^2}{\partial v^2}.$$

The formulas for the Gaussian curvature and Christoffel symbols of an isothermal patch are especially simple.

Theorem 31.3. *The Gaussian curvature K of an isothermal patch $\mathbf{x}: \mathcal{U} \longrightarrow \mathbb{R}^n$ is given in terms of the scaling function λ by*

$$(31.4) \qquad K = \frac{-\Delta \log \lambda}{\lambda^2}.$$

Furthermore, the Christoffel symbols of $\mathbf{x}$ are given by

$$(31.5) \qquad \Gamma^1_{11} = -\Gamma^1_{22} = \Gamma^2_{12} = \frac{\lambda_u}{\lambda} \quad \text{and} \quad \Gamma^1_{12} = -\Gamma^2_{11} = \Gamma^2_{22} = \frac{\lambda_v}{\lambda}.$$

Proof. Formula (31.4) is a special case of (22.9), page 505. In detail,

$$K = \frac{-1}{\lambda^2}\left\{\frac{\partial}{\partial u}\left(\frac{1}{\lambda}\frac{\partial \lambda}{\partial u}\right) + \frac{\partial}{\partial v}\left(\frac{1}{\lambda}\frac{\partial \lambda}{\partial v}\right)\right\}$$

$$= \frac{-1}{\lambda^2}\left\{\frac{\partial^2}{\partial u^2}\log\lambda + \frac{\partial^2}{\partial v^2}\log\lambda\right\}$$

$$= \frac{-\Delta \log \lambda}{\lambda^2}.$$

Similarly, (31.5) is a special case of Lemma 22.6, page 510. ∎

Next, we compute the mean curvature of an isothermal patch in $\mathbb{R}^3$.

Lemma 31.4. *Let* $\mathbf{x}: \mathcal{U} \longrightarrow \mathbb{R}^3$ *be a regular isothermal patch with scaling function* λ *and mean curvature* H. *Then the mean curvature of* $\mathbf{x}$ *is given by*

$$(31.6) \qquad H = \frac{e+g}{2\lambda^2}.$$

Furthermore,

$$(31.7) \qquad \mathbf{x}_{uu} + \mathbf{x}_{vv} = 2\lambda^2 H\, \mathbf{U},$$

where $\mathbf{U}$ *is the unit normal defined by*

$$\mathbf{U} = \frac{\mathbf{x}_u \times \mathbf{x}_v}{\|\mathbf{x}_u \times \mathbf{x}_v\|}.$$

Proof. Since $\mathbf{x}$ is isothermal, we can differentiate equations (31.1), obtaining

$$\mathbf{x}_{uu} \cdot \mathbf{x}_u = \mathbf{x}_{uv} \cdot \mathbf{x}_v \quad \text{and} \quad \mathbf{x}_{vv} \cdot \mathbf{x}_u = -\mathbf{x}_{vu} \cdot \mathbf{x}_v.$$

Therefore,

$$(\mathbf{x}_{uu} + \mathbf{x}_{vv}) \cdot \mathbf{x}_u = \mathbf{x}_{uv} \cdot \mathbf{x}_v - \mathbf{x}_{vu} \cdot \mathbf{x}_v = 0.$$

Similarly, $(\mathbf{x}_{uu} + \mathbf{x}_{vv}) \cdot \mathbf{x}_v = 0$. It follows that $\mathbf{x}_{uu} + \mathbf{x}_{vv}$ is normal to $\mathcal{M}$, and so $\mathbf{x}_{uu} + \mathbf{x}_{vv}$ is a multiple of $\mathbf{U}$. To find out which multiple, we use Theorem 16.17, page 377, and the assumption that $\mathbf{x}$ is isothermal to compute

$$H = \frac{eG - 2fF + gE}{2(EG - F^2)} = \frac{e+g}{2\lambda^2} = \frac{(\mathbf{x}_{uu} + \mathbf{x}_{vv}) \cdot \mathbf{U}}{2\lambda^2},$$

so that we get (31.6) and (31.7). ∎

Theorems 16.10, 16.17 and 22.7 simplify considerably for an isothermal patch $\mathbf{x}: \mathcal{U} \longrightarrow \mathbb{R}^3$.

Theorem 31.5. *Let* $\mathbf{x}: \mathcal{U} \longrightarrow \mathbb{R}^3$ *be an isothermal patch. Then the Gauss equations (22.14), page 511, for* $\mathbf{x}$ *reduce to*

$$(31.8) \qquad \begin{cases} \mathbf{x}_{uu} = \dfrac{\lambda_u}{\lambda}\mathbf{x}_u - \dfrac{\lambda_v}{\lambda}\mathbf{x}_v + e\,\mathbf{U}, \\[6pt] \mathbf{x}_{uv} = \dfrac{\lambda_v}{\lambda}\mathbf{x}_u + \dfrac{\lambda_u}{\lambda}\mathbf{x}_v + f\,\mathbf{U}, \\[6pt] \mathbf{x}_{vv} = -\dfrac{\lambda_u}{\lambda}\mathbf{x}_u + \dfrac{\lambda_v}{\lambda}\mathbf{x}_v + g\,\mathbf{U}. \end{cases}$$

31.2   Isometric Deformations of Minimal Surfaces   705

Furthermore, the Weingarten equations (16.11), *page* 369, *for* **x** *become*

(31.9)
$$\begin{cases} S(\mathbf{x}_u) = -\mathbf{U}_u = \dfrac{e}{\lambda^2}\mathbf{x}_u + \dfrac{f}{\lambda^2}\mathbf{x}_v, \\ S(\mathbf{x}_v) = -\mathbf{U}_v = \dfrac{f}{\lambda^2}\mathbf{x}_u + \dfrac{g}{\lambda^2}\mathbf{x}_v. \end{cases}$$

The Peterson-Mainardi-Codazzi equations (28.13), *page* 649, *also simplify for an isothermal patch*:

(31.10)
$$\begin{cases} \dfrac{\partial e}{\partial v} - \dfrac{\partial f}{\partial u} = 2\lambda\,\lambda_v, \\ \dfrac{\partial f}{\partial v} - \dfrac{\partial g}{\partial u} = 2\lambda\,\lambda_u. \end{cases}$$

31.2   Isometric Deformations of Minimal Surfaces

Minimal surfaces and complex analytic functions have many properties in common. Let us recall some relevant elementary properties of complex analytic functions of one variable. (For more details see, for example, [Ahlf] or [Hille].)

Let $\mathcal{U}$ be an open subset of $\mathbb{C}$. A function $f\colon \mathcal{U} \longrightarrow \mathbb{C}$ is called **analytic** provided f has a derivative in the complex sense at each point **p** of $\mathcal{U}$. This means that the limit
$$\lim_{\mathbf{q}\to\mathbf{p}} \frac{f(\mathbf{q}) - f(\mathbf{p})}{\mathbf{q} - \mathbf{p}}$$
exists no matter how **q** approaches **p**. A key theorem about such functions is that they have derivatives of all orders. Furthermore, if we write $g = \mathfrak{Re}\,f$ and $h = \mathfrak{Im}\,f$, then g and h are related by the Cauchy[1]-Riemann equations:

(31.11)
$$g_u = h_v, \qquad g_v = -h_u.$$

---

[1] Augustin Louis Cauchy (1789-1857). One of the leading French mathematicians of the first half of the 19th century. He introduced rigor into calculus, including precise definitions of continuity and integration. His best known results include the Cauchy integral formula and Cauchy-Riemann equations in complex analysis, and the Cauchy-Kovalevskaya existence theorem for the solution of partial differential equations. Cauchy's scientific output was enormous, second only to that of Euler. In politics and religion, Cauchy was an archconservative.

Moreover, taking second partial derivatives, we see that (31.11) implies that g and h are **harmonic**; that is, $g_{uu} + g_{vv} = h_{uu} + h_{vv} = 0$. When (31.11) holds, we say that g and h are **conjugate harmonic**.

The notions of harmonic and conjugate harmonic make sense for n-tuples of functions.

Definition. *Let $\mathcal{U}$ be an open subset of $\mathbb{R}^2$, and let $\mathbf{x}, \mathbf{y} : \mathcal{U} \longrightarrow \mathbb{R}^n$ be patches.*

(i) *We say that $\mathbf{x}$ is* **harmonic** *provided each of its coordinate functions is harmonic. Equivalently, $\mathbf{x}_{uu} + \mathbf{x}_{vv} = 0$.*

(ii) *$\mathbf{x}$ and $\mathbf{y}$ are said to* **satisfy the Cauchy-Riemann equations** *provided*

$$(31.12) \qquad \mathbf{x}_u = \mathbf{y}_v \quad \text{and} \quad \mathbf{x}_v = -\mathbf{y}_u.$$

Here the equations (31.12) generalize the Cauchy-Riemann equations (31.11).

Lemma 31.6. *If $\mathbf{x}, \mathbf{y} : \mathcal{U} \longrightarrow \mathbb{R}^n$ satisfy the Cauchy-Riemann equations, then they are harmonic. In this case, we say that $\mathbf{x}$ and $\mathbf{y}$ are* **conjugate harmonic**.

Proof. The proof is almost the same as the standard proof of (31.11) that one can find in any book on complex variables:

$$\mathbf{x}_{uu} + \mathbf{x}_{vv} = \mathbf{y}_{uv} - \mathbf{y}_{vu} = 0,$$

and similarly for $\mathbf{y}$. ∎

We shall discuss the problem of determining the conjugate of a harmonic patch in Section 31.6. As an immediate consequence of Lemma 31.4 we obtain:

Corollary 31.7. *A regular isothermal patch $\mathbf{x} : \mathcal{U} \longrightarrow \mathbb{R}^3$ is a minimal surface if and only if it is harmonic.*

So far we have only defined the notion of minimal surface in $\mathbb{R}^3$. Now Corollary 31.7 tells us how to define a minimal surface in $\mathbb{R}^n$ for any n. (See also Section 25.6.)

Definition. *A* **minimal isothermal patch** *$\mathbf{x} : \mathcal{U} \longrightarrow \mathbb{R}^n$ is a patch that is both isothermal and harmonic.*

Next, we describe a general method for obtaining a 1-parameter family of isometric minimal patches. This construction generalizes the deformation between the helicoid and the catenoid described on page 685.

31.2 Isometric Deformations of Minimal Surfaces

Definition. Let $\mathbf{x}, \mathbf{y}: \mathcal{U} \longrightarrow \mathbb{R}^n$ be conjugate harmonic minimal isothermal patches. The **associated family** of $\mathbf{x}$ and $\mathbf{y}$ is the 1-parameter family of patches $t \longmapsto \mathbf{z}[t]$, where $\mathbf{z}[t]: \mathcal{U} \longrightarrow \mathbb{R}^n$ is given by

$$(31.13) \qquad \mathbf{z}[t] = (\cos t)\mathbf{x} + (\sin t)\mathbf{y} = \mathfrak{Re}\left(e^{-it}(\mathbf{x} + i\,\mathbf{y})\right).$$

Next, we compute the derivatives of $\mathbf{z}[t]$ in terms of the derivatives of $\mathbf{x}$ and $\mathbf{y}$.

Lemma 31.8. Let $\mathbf{x}, \mathbf{y}: \mathcal{U} \longrightarrow \mathbb{R}^n$ be conjugate harmonic minimal isothermal patches. Then the associated family $t \longmapsto \mathbf{z}[t]$ of $\mathbf{x}$ and $\mathbf{y}$ satisfies

$$(31.14) \qquad \begin{cases} \mathbf{z}[t]_u = (\cos t)\mathbf{x}_u - (\sin t)\mathbf{x}_v, \\[4pt] \mathbf{z}[t]_v = (\sin t)\mathbf{x}_u + (\cos t)\mathbf{x}_v, \\[4pt] \mathbf{z}[t]_{uu} = -\mathbf{z}[t]_{vv} = (\cos t)\mathbf{x}_{uu} - (\sin t)\mathbf{x}_{uv}, \\[4pt] \mathbf{z}[t]_{uv} = (\sin t)\mathbf{x}_{uu} + (\cos t)\mathbf{x}_{uv}. \end{cases}$$

Furthermore, $\mathbf{z}[t]$ and $\mathbf{z}[t + \pi/2]$ are conjugate harmonic.

Proof. Equations (31.14) are a consequence of the Cauchy-Riemann equations and the assumption that $\mathbf{x}$ and $\mathbf{y}$ are isothermal. The last statement of the lemma can be proved as follows:

$$\begin{aligned} \mathbf{z}[t]_u &= \mathfrak{Re}\left(e^{-it}(\mathbf{x}_u + i\,\mathbf{y}_u)\right) = \mathfrak{Re}\left(e^{-it}(\mathbf{y}_v - i\,\mathbf{x}_v)\right) \\ &= \mathfrak{Re}\left(e^{-i(t+\pi/2)}(\mathbf{x}_v + i\,\mathbf{y}_v)\right) = \mathbf{z}[t + \pi/2]_v. \end{aligned}$$

Similarly, $\mathbf{z}[t]_v = -\mathbf{z}[t + \pi/2]_u$. ∎

Lemma 31.8 is a useful tool for obtaining information about the first and second fundamental forms of an associated family.

Theorem 31.9. Let $\mathbf{x}, \mathbf{y}: \mathcal{U} \longrightarrow \mathbb{R}^n$ be conjugate harmonic minimal isothermal patches, and let $t \longmapsto \mathbf{z}[t]$ be the associated family of $\mathbf{x}$ and $\mathbf{y}$. Then $\mathbf{z}[t]$ is also a minimal isothermal patch for each t. Furthermore, all of the patches in the associated family have the same first fundamental form. Consequently, $t \longmapsto \mathbf{z}[t]$ is an **isometric deformation**.

Proof. That $\mathbf{z}[t]$ is harmonic is a consequence of the third equation of (31.14). Let $E(t)$, $F(t)$ and $G(t)$ denote the coefficients of the first fundamental form of $\mathbf{z}[t]$. We use the first two equations of (31.14) to compute

(31.15) $$\begin{aligned} E(t) &= \mathbf{z}[t]_u \cdot \mathbf{z}[t]_u \\ &= \big((\cos t)\mathbf{x}_u - (\sin t)\mathbf{x}_v\big) \cdot \big((\cos t)\mathbf{x}_u - (\sin t)\mathbf{x}_v\big) \\ &= (\cos t)^2(\mathbf{x}_u \cdot \mathbf{x}_u) + (\sin t)^2(\mathbf{x}_v \cdot \mathbf{x}_v) - 2(\sin t \cos t)(\mathbf{x}_u \cdot \mathbf{x}_v). \end{aligned}$$

Since $\mathbf{x}_u \cdot \mathbf{x}_u = \mathbf{x}_v \cdot \mathbf{x}_v = E$ and $\mathbf{x}_u \cdot \mathbf{x}_v = 0$, equation (31.15) reduces to $E(t) = \mathbf{x}_u \cdot \mathbf{x}_u$. Similarly, $F(t) = 0$ and $G(t) = \mathbf{x}_v \cdot \mathbf{x}_v = E$ for all t. It follows that each $\mathbf{z}[t]$ is isothermal and has the same first fundamental form as $\mathbf{x}$. ∎

Next, we restrict our attention to minimal isothermal patches in $\mathbb{R}^3$. We show that up to parallel translation, all of the members of an associated family have the same unit normal, and consequently the same tangent space.

Lemma 31.10. Let $\mathbf{x}, \mathbf{y} \colon \mathcal{U} \longrightarrow \mathbb{R}^3$ be conjugate harmonic minimal isothermal patches.

(i) For all t,

$$\mathbf{z}[t]_u \times \mathbf{z}[t]_v = \mathbf{x}_u \times \mathbf{x}_v.$$

(ii) The unit normal $\mathbf{U}$ of the patch $\mathbf{z}[t]$ at $\mathbf{z}[t](u,v)$ is parallel to the unit normal $\mathbf{U}$ of the patch $\mathbf{x}$ at $\mathbf{x}(u,v)$.

(iii) The tangent space to the patch $\mathbf{z}[t]$ at $\mathbf{z}[t](u,v)$ is parallel to the tangent space to the patch $\mathbf{x}$ at $\mathbf{x}(u,v)$.

Proof. (i) is an easy calculation from the first two equations of (31.14); then (ii) follows from (i), and (iii) follows from (ii). ∎

In view of part (ii) of Lemma 31.10 we can identify the unit normal of $\mathbf{z}[t]$ with the unit normal $\mathbf{U}$ of $\mathbf{x}$. Thus

Corollary 31.11. All members of an associated family have the same Gauss map.

In spite of the fact that $\mathbf{z}[t]$ has the same first fundamental form and unit normal vector field as $\mathbf{x}$, the shape operators will not be the same in general. Let $e(t)$, $f(t)$ and $g(t)$ denote the coefficients of the second fundamental form of $\mathbf{z}[t]$, and let $S(t)$ denote the corresponding shape operator.

31.2 Isometric Deformations of Minimal Surfaces

Lemma 31.12. *The coefficients of the second fundamental form of* $\mathbf{z}[t]\colon \mathcal{U} \longrightarrow \mathbb{R}^3$ *are related to the coefficients of the second fundamental form of* $\mathbf{x}$ *by the formulas*

(31.16)
$$\begin{cases} e(t) = -g(t) = e\cos t - f\sin t, \\ f(t) = f\cos t + e\sin t. \end{cases}$$

Furthermore,

(31.17)
$$\begin{cases} S(t)\mathbf{z}[t]_u = S\mathbf{x}_u, \\ S(t)\mathbf{z}[t]_v = S\mathbf{x}_v. \end{cases}$$

Proof. We use the third equation of (31.14) to compute

$$-g(t) = e(t) = \mathbf{z}[t]_{uu} \cdot \mathbf{U}$$
$$= \big((\cos t)\mathbf{x}_{uu} - (\sin t)\mathbf{x}_{uu}\big) \cdot \mathbf{U}$$
$$= (\cos t)e - (\sin t)f.$$

Similarly, the fourth equation of (31.14) implies the second equation of (31.16).

Next, we use the Weingarten equations (31.9) for $\mathbf{z}[t]$ and Theorem 31.9 to write

(31.18)
$$S(t)\mathbf{z}[t]_u = \frac{1}{\lambda^2}\big(e(t)\mathbf{z}[t]_u + f(t)\mathbf{z}[t]_v\big).$$

We can calculate the right-hand side of (31.18) with the aid of (31.14) and (31.16):

$$S(t)\mathbf{z}[t]_u = \frac{1}{\lambda^2}\Big((e\cos t - f\sin t)\big((\cos t)\mathbf{x}_u - (\sin t)\mathbf{x}_v\big)$$
$$+ (f\cos t + e\sin t)\big((\sin t)\mathbf{x}_u + (\cos t)\mathbf{x}_v\big)\Big)$$
$$= \frac{1}{\lambda^2}\big(e\,\mathbf{x}_u + f\,\mathbf{x}_v\big) = S\mathbf{x}_u.$$

Similarly, $S(t)\mathbf{z}[t]_v = S\mathbf{x}_v$. ∎

Now we can get a simple formula for the shape operator $S(t)$ in terms of shape operator S and complex structure J of the minimal isothermal patch $\mathbf{x}$.

Lemma 31.13. *The shape operator* $S(t)$ *of* $\mathbf{z}[t]$ *is related to the shape operator* S *of* $\mathbf{x}$ *by the formula*

(31.19)
$$S(t) = (\cos t)S + (\sin t)SJ.$$

Proof. From (31.18) and (31.14) we get

$$(31.20) \quad \begin{cases} S\mathbf{x}_u = (\cos t)S(t)\mathbf{x}_u - (\sin t)S(t)\mathbf{x}_v, \\ S\mathbf{x}_v = (\sin t)S(t)\mathbf{x}_u + (\cos t)S(t)\mathbf{x}_v. \end{cases}$$

We solve the equations (31.20) for $S(t)\mathbf{x}_u$ and $S(t)\mathbf{x}_v$, obtaining

$$(31.21) \quad \begin{cases} S(t)\mathbf{x}_u = (\cos t)S\mathbf{x}_u + (\sin t)S\mathbf{x}_v, \\ S(t)\mathbf{x}_v = -(\sin t)S\mathbf{x}_u + (\cos t)S\mathbf{x}_v. \end{cases}$$

Since $\mathbf{x}_v = J\mathbf{x}_u$, we can rewrite (31.21) as

$$(31.22) \quad \begin{cases} S(t)\mathbf{x}_u = (\cos t)S\mathbf{x}_u + (\sin t)SJ\mathbf{x}_u, \\ S(t)\mathbf{x}_v = (\sin t)SJ\mathbf{x}_v + (\cos t)S\mathbf{x}_u. \end{cases}$$

Clearly, (31.22) is equivalent to (31.19). ∎

Corollary 31.14. *Let* $\mathbf{x}:\mathcal{U} \longrightarrow \mathbb{R}^3$ *be a harmonic minimal isothermal patch with shape operator* S. *Then the shape operator of the conjugate patch of* $\mathbf{x}$ *is* SJ, *where* J *is the complex structure of* $\mathbf{x}$.

Corollary 31.15. *Let* $\mathbf{x}, \mathbf{y}:\mathcal{U} \longrightarrow \mathbb{R}^3$ *be conjugate harmonic minimal isothermal patches, and let* $\mathbf{v}_p$ *be a tangent vector to* $\mathbf{x}$ *(which by Lemma 31.10 can also be considered to be a tangent vector to* $\mathbf{y}$). *Then*

(i) $\mathbf{v}_p$ *is an asymptotic vector of* $\mathbf{x}$ *if and only if it is a principal vector of* $\mathbf{y}$;

(ii) $\mathbf{v}_p$ *is a principal vector of* $\mathbf{x}$ *if and only if it is an asymptotic vector of* $\mathbf{y}$.

31.3 Complex Derivatives

Corollary 31.7 is the reason why harmonic functions are useful for the study of minimal surfaces. Since analytic function theory is so important for the study of harmonic functions of two variables, it is natural to introduce this subject into the study of minimal surfaces in $\mathbb{R}^n$.

31.3 Complex Derivatives

It is often advantageous to switch from the standard real coordinates u, v in $\mathbb{R}^2$ that we have been using to complex coordinates z and $\bar{z}$. The algebraic formulas relating the two kinds of coordinates are simple:

(31.23)
$$\begin{cases} z = u + iv, \\ \bar{z} = u - iv, \end{cases}$$

(31.24)
$$\begin{cases} u = \dfrac{z + \bar{z}}{2}, \\ v = \dfrac{z - \bar{z}}{2i}. \end{cases}$$

It is a little disconcerting at first to work with z and $\bar{z}$, because one tends to think that z determines $\bar{z}$. This is certainly the case when we start with u and v and define z and $\bar{z}$ by (31.23). But we can regard z and $\bar{z}$ as abstract coordinates for $\mathbb{R}^2 = \mathbb{C}$ (even though they are complex-valued), and then define u and v by (31.24). This point of view frequently leads to simpler formulas. Note that, except for the fact that they are complex-valued, $\{z, \bar{z}\}$ form a coordinate system for $\mathbb{R}^2$ that we can use in place of the standard coordinate system $\{u, v\}$.

In particular, it is convenient to introduce the operators

(31.25) $\quad \dfrac{\partial}{\partial z} = \dfrac{1}{2}\left(\dfrac{\partial}{\partial u} - i\dfrac{\partial}{\partial v}\right) \quad$ and $\quad \dfrac{\partial}{\partial \bar{z}} = \dfrac{1}{2}\left(\dfrac{\partial}{\partial u} + i\dfrac{\partial}{\partial v}\right).$

From (31.23) we have

(31.26) $\quad\quad\quad\quad dz = du + idv, \quad\quad d\bar{z} = du - idv.$

Note also the formulas
$$|dz|^2 = dz\, d\bar{z} = du^2 + dv^2$$

and
$$\Delta = \dfrac{\partial^2}{\partial u^2} + \dfrac{\partial^2}{\partial v^2} = 4\dfrac{\partial^2}{\partial \bar{z}\, \partial z}.$$

Even though a patch $\mathbf{x}$ in $\mathbb{R}^n$ consists of an n-tuple of real functions, it is very useful in minimal surface theory to apply $\partial/\partial z$ to $\mathbf{x}$.

Definition. *The* **complex derivative** *of a patch* $\mathbf{x}: \mathcal{U} \longrightarrow \mathbb{R}^n$ *is*

$$\dfrac{\partial \mathbf{x}}{\partial z}(z) = \dfrac{1}{2}(\mathbf{x}_u - i\mathbf{x}_v)(u, v), \quad\quad \text{where } z = u + iv.$$

Also, we write

$$\dfrac{\partial \mathbf{x}}{\partial z} = (\phi_1[\mathbf{x}], \ldots, \phi_n[\mathbf{x}]) = \dfrac{1}{2}\left(\dfrac{\partial x_1}{\partial u} - i\dfrac{\partial x_1}{\partial v}, \ldots, \dfrac{\partial x_n}{\partial u} - i\dfrac{\partial x_n}{\partial v}\right).$$

Lemma 31.16. *The complex derivative of a patch* $\mathbf{x}\colon \mathcal{U} \longrightarrow \mathbb{R}^n$ *satisfies the identities*

$$(31.27) \quad \sum_{k=1}^{n} \phi_k[\mathbf{x}]^2 = \frac{1}{4}(\mathbf{x}_u \cdot \mathbf{x}_u - \mathbf{x}_v \cdot \mathbf{x}_v - 2i\,\mathbf{x}_u \cdot \mathbf{x}_v) = \frac{1}{4}(E - G - 2i\,F),$$

and

$$(31.28) \quad \sum_{k=1}^{n} |\phi_k[\mathbf{x}]|^2 = \frac{1}{4}(\mathbf{x}_u \cdot \mathbf{x}_u + \mathbf{x}_v \cdot \mathbf{x}_v) = \frac{1}{4}(E + G).$$

Proof. We have

$$\sum_{k=1}^{n} \phi_k[\mathbf{x}]^2 = \frac{1}{4} \sum_{k=1}^{n} \left(\frac{\partial x_k}{\partial u} - i\frac{\partial x_k}{\partial v} \right)^2$$

$$= \frac{1}{4} \sum_{k=1}^{n} \left(\left(\frac{\partial x_k}{\partial u}\right)^2 - \left(\frac{\partial x_k}{\partial v}\right)^2 - 2i \left(\frac{\partial x_k}{\partial u}\right)\left(\frac{\partial x_k}{\partial v}\right) \right)$$

$$= \frac{1}{4}(\mathbf{x}_u \cdot \mathbf{x}_u - \mathbf{x}_v \cdot \mathbf{x}_v - 2i\,\mathbf{x}_u \cdot \mathbf{x}_v),$$

and

$$\sum_{k=1}^{n} |\phi_k[\mathbf{x}]|^2 = \frac{1}{4} \sum_{k=1}^{n} \left| \frac{\partial x_k}{\partial u} - i\frac{\partial x_k}{\partial v} \right|^2$$

$$= \frac{1}{4} \sum_{k=1}^{n} \left(\left|\frac{\partial x_k}{\partial u}\right|^2 + \left|\frac{\partial x_k}{\partial v}\right|^2 \right) = \frac{1}{4}(\mathbf{x}_u \cdot \mathbf{x}_u + \mathbf{x}_v \cdot \mathbf{x}_v). \blacksquare$$

The next theorem shows that an isothermal minimal patch gives rise to an n-tuple of analytic functions such that the sum of the squares of the components equals zero. As we shall see later, this alternate description of an isothermal patch yields much useful information, because it allows the use of powerful theorems from complex analysis.

Theorem 31.17. *Let* $\mathbf{x}\colon \mathcal{U} \longrightarrow \mathbb{R}^n$ *be a patch. Then:*

(i) $\mathbf{x}$ *is harmonic if and only if its complex derivative* $\partial \mathbf{x}/\partial z$ *is analytic;*

(ii) $\mathbf{x}$ *is isothermal if and only if*

$$(31.29) \quad \sum_{k=1}^{n} \phi_k[\mathbf{x}]^2 = 0;$$

31.3 Complex Derivatives

(iii) *if* **x** *is isothermal, then* **x** *is regular if and only if*

(31.30) $$\sum_{k=1}^{n} |\phi_k[\mathbf{x}]|^2 \neq 0.$$

Conversely, let $\mathcal{U}$ *be simply connected and let* $\phi_1, \ldots, \phi_n: \mathcal{U} \longrightarrow \mathbb{C}^n$ *be analytic functions satisfying*

(31.31) $$\sum_{k=1}^{n} \phi_k^2 = 0 \quad \text{and} \quad \sum_{k=1}^{n} |\phi_k|^2 \neq 0.$$

Then there exists a regular minimal isothermal patch $\mathbf{x}: \mathcal{U} \longrightarrow \mathbb{R}^n$ *such that* $\Phi = (\phi_1, \ldots, \phi_n)$ *is the complex derivative of* **x**.

Proof. (i) follows from the fact that the Cauchy-Riemann equations for $\partial \mathbf{x}/\partial z$ are simply $\mathbf{x}_{uu} + \mathbf{x}_{vv} = 0$ and $\mathbf{x}_{uv} - \mathbf{x}_{vu} = 0$. Then (ii) is a consequence of (31.27), and (iii) follows from (31.28). To prove the last statement, we suppose that (31.31) holds and put

$$\mathbf{x} = \mathfrak{Re}\left(\int (\phi_1(z), \ldots, \phi_n(z)) dz\right).$$

Then (31.27) implies that **x** is isothermal, and (31.28) implies that **x** is regular. Since **x** is the real part of an n-tuple of analytic functions, it is harmonic. Thus **x** is a regular minimal isothermal patch. ∎

An important fact in the theory of complex functions of one complex variable is that a pair of conjugate harmonic functions determines an analytic function. This suggests that we find an analog of a complex analytic function that we can associate with a pair of conjugate minimal surfaces.

Definition. *The* **complexification** *of a pair* $\mathbf{x}, \mathbf{y}: \mathcal{U} \longrightarrow \mathbb{R}^n$ *of conjugate harmonic minimal isothermal patches is the map* $\mathbf{x} + i\mathbf{y}: \mathcal{U} \longrightarrow \mathbb{C}^n$.

We already saw on page 707 that the associated family of conjugate harmonic minimal isothermal patches of **x** and **y** could be expressed in terms of the complexification of **x** and **y**.

Lemma 31.18. *The complexification* $\mathbf{x} + i\mathbf{y}: \mathcal{U} \longrightarrow \mathbb{C}^n$ *of a pair* $\mathbf{x}, \mathbf{y}: \mathcal{U} \longrightarrow \mathbb{R}^n$ *of conjugate minimal surfaces is analytic. Furthermore, the derivative of* $\mathbf{x} + i\mathbf{y}$ *is twice the complex derivative of* **x**, *and*

(31.32) $$2\frac{\partial \mathbf{x}}{\partial z} = \frac{d}{dz}(\mathbf{x} + i\mathbf{y}) = \mathbf{x}_u - i\mathbf{x}_v.$$

Proof. The first statement follows from the fact that $\mathbf{x}+i\,\mathbf{y}$ is an n-tuple of complex functions of one variable, each of which is analytic. For the second statement we compute

$$\frac{d}{dz}(\mathbf{x}+i\,\mathbf{y}) = \frac{1}{2}\left(\frac{\partial}{\partial u}-i\frac{\partial}{\partial v}\right)(\mathbf{x}+i\,\mathbf{y})$$

$$= \frac{1}{2}(\mathbf{x}_u - i\,\mathbf{x}_v + \mathbf{y}_v + i\mathbf{y}_u) = \mathbf{x}_u - i\,\mathbf{x}_v = 2\frac{\partial \mathbf{x}}{\partial z}. \blacksquare$$

31.4 Elementary Complex Vector Algebra

In preparation for our study of minimal curves in Section 31.5, we need to point out the differences between real vector algebra (studied in Sections 1.1 and 8.1) and complex vector algebra.

Definition. Complex Euclidean n-space $\mathbb{C}^n$ *consists of the set of all complex n-tuples*:

$$\mathbb{C}^n = \{\,(p_1,\ldots,p_n) \mid p_j \text{ is a complex number for } j=1,\ldots,n\,\}.$$

The elements of $\mathbb{C}^n$ are called **complex vectors**.

We make $\mathbb{C}^n$ into a complex vector space in the same way that we made $\mathbb{R}^n$ into a real vector space in Section 1.1. Thus if $\mathbf{p} = (p_1,\ldots,p_n)$ and $\mathbf{q} = (q_1,\ldots,q_n)$ are complex vectors, we define $\mathbf{p}+\mathbf{q}$ to be the element of $\mathbb{C}^n$ given by

$$\mathbf{p}+\mathbf{q} = (p_1+q_1,\ldots,p_n+q_n).$$

Similarly, for $\lambda \in \mathbb{C}$ the vector $\lambda\mathbf{p}$ is defined by

$$\lambda\mathbf{p} = (\lambda p_1,\ldots,\lambda p_n).$$

Furthermore, the **dot product** (or **scalar product**) of $\mathbb{C}^n$ is given by the same formula as its real counterpart:

$$\mathbf{p}\cdot\mathbf{q} = \sum_{j=1}^{n} p_j q_j.$$

31.4 Elementary Complex Vector Algebra

In addition to these operations, $\mathbb{C}^n$ also has a **conjugation** $\mathbf{p} \longmapsto \bar{\mathbf{p}}$, defined by

$$\bar{\mathbf{p}} = (\bar{p}_1, \ldots, \bar{p}_n).$$

It is easy to check that conjugation has the following properties:

$$\overline{\mathbf{p} + \mathbf{q}} = \bar{\mathbf{p}} + \bar{\mathbf{q}},$$

$$\overline{\lambda \mathbf{p}} = \bar{\lambda} \bar{\mathbf{p}},$$

$$\overline{\mathbf{p} \cdot \mathbf{q}} = \bar{\mathbf{p}} \cdot \bar{\mathbf{q}},$$

$$\mathbf{p} \cdot \bar{\mathbf{p}} \geq 0.$$

The **norm** of $\mathbb{C}^n$ has a slightly different definition from its real counterpart:

$$\|\mathbf{p}\| = \sqrt{\mathbf{p} \cdot \bar{\mathbf{p}}} = \sqrt{|p_1|^2 + \cdots + |p_n|^2}.$$

Thus $\mathbf{p} \cdot \mathbf{p}$ may be zero for nonzero $\mathbf{p}$, but $\|\mathbf{p}\| = 0$ implies $\mathbf{p} = 0$.

The definition of the vector cross product of complex 3-tuples is formally the same as that of real 3-tuples. Let $\mathbf{a} = (a_1, a_2, a_3)$ and $\mathbf{b} = (b_1, b_2, b_3)$ be vectors in $\mathbb{C}^3$; then the **vector cross product** of $\mathbf{a}$ and $\mathbf{b}$ is given by

$$\mathbf{a} \times \mathbf{b} = \det \begin{pmatrix} \mathbf{i} & \mathbf{j} & \mathbf{k} \\ a_1 & a_2 & a_3 \\ b_1 & b_2 & b_3 \end{pmatrix},$$

where

$$\mathbf{i} = (1, 0, 0), \quad \mathbf{j} = (0, 1, 0), \quad \mathbf{k} = (0, 0, 1).$$

It is easy to check that

$$\overline{\mathbf{a} \times \mathbf{b}} = \bar{\mathbf{a}} \times \bar{\mathbf{b}},$$

and that all of the identities for the real vector cross product given on page 182, with one exception, hold also for the complex vector cross product. The exception is the Lagrange identity (8.1), whose modification we include in the following lemma:

Lemma 31.19. *For* $\mathbf{a}, \mathbf{b} \in \mathbb{C}^3$ *we have*

(31.33) $\qquad \|\mathbf{a} \times \bar{\mathbf{b}}\|^2 = \|\mathbf{a}\|^2 \|\mathbf{b}\|^2 - |\mathbf{a} \cdot \bar{\mathbf{b}}|^2,$

(31.34) $\qquad \|\mathbf{a} \times \bar{\mathbf{a}}\| = \|\mathbf{a}\|^2,$

(31.35) $\qquad \mathbf{a} \times \bar{\mathbf{a}} = 2i\bigl(\mathfrak{Im}(a_2 \bar{a}_3), \mathfrak{Im}(a_3 \bar{a}_1), \mathfrak{Im}(a_1 \bar{a}_2)\bigr),$

where $\mathbf{a} = (a_1, a_2, a_3)$.

Proof. We prove (31.34):

$$\|\mathbf{a} \times \overline{\mathbf{a}}\|^2 = \mathbf{a} \times \overline{\mathbf{a}} \cdot \overline{\mathbf{a} \times \overline{\mathbf{a}}} = -(\mathbf{a} \times \overline{\mathbf{a}}) \cdot (\mathbf{a} \times \overline{\mathbf{a}}) = -(\mathbf{a} \cdot \mathbf{a})(\overline{\mathbf{a}} \cdot \overline{\mathbf{a}}) + (\mathbf{a} \cdot \overline{\mathbf{a}})^2 = \|\mathbf{a}\|^4. \blacksquare$$

31.5 Minimal Curves

In this section we look at functions that have some of the properties of the complex derivative of a minimal isothermal patch. First, we reformulate (31.31) as a definition.

Definition. *Let $\mathcal{U}$ be an open subset of $\mathbb{C}$. A **minimal curve** is an analytic function $\Psi: \mathcal{U} \longrightarrow \mathbb{C}^n$ such that*

(31.36) $$\Psi'(z) \cdot \Psi'(z) = 0$$

for $z \in \mathcal{U}$. If in addition

$$\Psi'(z) \cdot \overline{\Psi'(z)} \neq 0$$

*for $z \in \mathcal{U}$, we say that Ψ is a **regular minimal curve**.*

Minimal curves were first studied by Lie[2] in [Lie]. A minimal curve can be thought of as a generalization of a minimal isothermal patch. But in a different sense a minimal curve is a generalization of a real curve in $\mathbb{R}^n$. In the literature a parametrized curve satisfying (31.36) is called **isotropic**.

Lemma 31.20. *A minimal isothermal patch gives rise to a minimal curve, namely, its complexification. Conversely, given a minimal curve $\Psi: \mathcal{U} \longrightarrow \mathbb{C}^n$, the patches $\mathbf{x}, \mathbf{y}: \mathcal{U} \longrightarrow \mathbb{R}^n$ defined by*

(31.37) $$\mathbf{x}(u, v) = \mathfrak{Re}\big(\Psi(u + i\, v)\big) \quad \text{and} \quad \mathbf{y}(u, v) = \mathfrak{Im}\big(\Psi(u + i\, v)\big)$$

*are conjugate minimal isothermal patches. We call $\mathbf{x}$ and $\mathbf{y}$ the **conjugate minimal isothermal patches determined by** Ψ. Hence there is a one-to-one correspondence between minimal curves and pairs of conjugate minimal isothermal patches.*

---

[2] Marius Sophus Lie (1842-1899). Norwegian mathematician. Although Lie is best known for Lie groups and Lie algebras, he made many important contributions to differential equations, surface theory and the theory of contact transformations.

31.5 Minimal Curves

Proof. We have already seen in Theorem 31.17 that the complexification of a minimal isothermal patch is a minimal curve. Furthermore, it is easy to check that **x** and **y** defined by equation (31.37) are conjugate minimal isothermal patches. Furthermore, the regularity of Ψ implies the regularity of **x** and **y**. ∎

Thus the study of minimal isothermal patches is equivalent to the study of conjugate minimal isothermal patches. Hence the notion of associated family of a minimal curve makes sense.

Definition. *Let* $\Psi: \mathcal{U} \longrightarrow \mathbb{C}^n$ *be a minimal curve. The* **associated family** *of* Ψ *is the 1-parameter family of patches* $t \longmapsto \mathbf{z}[t]$, *where* $\mathbf{z}[t]: \mathcal{U} \longrightarrow \mathbb{R}^n$ *is given by*

$$\mathbf{z}[t](u,v) = \mathfrak{Re}\bigl(e^{-it}\Psi(u+iv)\bigr).$$

Thus the associated family of Ψ *is the same as the associated family of conjugate minimal isothermal patches determined by* Ψ.

We know from Theorem 31.9 that the associated family $t \longmapsto \mathbf{z}[t]$ of a minimal curve Ψ is an isometric deformation, and that $\mathbf{z}[t]$ is also a minimal isothermal patch for each t. Let us rewrite Lemma 31.8 in terms of minimal curves.

Corollary 31.21. *Let* $\Psi: \mathcal{U} \longrightarrow \mathbb{C}^n$ *be a minimal curve. Then the associated family* $t \longmapsto \mathbf{z}[t]$ *of* Ψ *satisfies*

$$\mathbf{z}[t]_u - i\mathbf{z}[t]_v = e^{-it}\Psi'. \tag{31.38}$$

Also,

$$(31.39) \quad \begin{cases} \mathbf{z}[t]_u(u,v) = \mathfrak{Re}\bigl(e^{-it}\Psi'(u+iv)\bigr), \\ \mathbf{z}[t]_v(u,v) = \mathfrak{Im}\bigl(e^{-it}\Psi'(u+iv)\bigr), \\ \mathbf{z}[t]_{uu}(u,v) = -\mathbf{z}[t]_{vv}(u,v) = \mathfrak{Re}\bigl(e^{-it}\Psi''(u+iv)\bigr), \\ \mathbf{z}[t]_{uv}(u,v) = \mathfrak{Im}\bigl(e^{-it}\Psi''(u+iv)\bigr). \end{cases}$$

Proof. From (31.32) we get (31.38); then (31.38) implies the first equation of (31.39). The other equations are proved similarly. ∎

It will be convenient to write

$$\frac{\Psi'}{2} = \mathbf{x}_z = \frac{\partial \mathbf{x}}{\partial z} = \frac{1}{2}(\mathbf{x}_u - i\,\mathbf{x}_v) = (\phi_1, \ldots, \phi_n)$$

for the complex derivative of a patch $\mathbf{x}: \mathcal{U} \longrightarrow \mathbb{R}^n$. Then

$$\frac{\overline{\Psi'}}{2} = \mathbf{x}_{\bar{z}} = \frac{\partial \mathbf{x}}{\partial \bar{z}} = \frac{1}{2}(\mathbf{x}_u + i\,\mathbf{x}_v) = (\overline{\phi}_1, \ldots, \overline{\phi}_n).$$

Next, we study the Gauss map of a minimal curve.

Definition. *The* **Gauss map of a minimal curve** Ψ *is the Gauss map of any member of an associated family of* Ψ. *As usual we identify the Gauss map with the unit normal* **U**.

The following lemma follows from (31.35):

Lemma 31.22. *Let* $\Psi: \mathcal{U} \longrightarrow \mathbb{C}^3$ *be a minimal curve. Then*

$$\frac{\Psi' \times \overline{\Psi'}}{4} = \mathbf{x}_z \times \mathbf{x}_{\bar{z}} = \frac{i}{2} \mathbf{x}_u \times \mathbf{x}_v$$

$$= 2i\big(\mathfrak{Im}(\phi_2 \overline{\phi}_3), \mathfrak{Im}(\phi_3 \overline{\phi}_1), \mathfrak{Im}(\phi_1 \overline{\phi}_2)\big),$$

and

(31.40) $$\mathbf{U} = \frac{\mathbf{x}_u \times \mathbf{x}_v}{\|\mathbf{x}_u \times \mathbf{x}_v\|} = \frac{\mathbf{x}_z \times \mathbf{x}_{\bar{z}}}{i\|\mathbf{x}_z\|^2} = \frac{\Psi' \times \overline{\Psi'}}{i\|\Psi'\|^2}$$

$$= \frac{2\big(\mathfrak{Im}(\phi_2 \overline{\phi}_3), \mathfrak{Im}(\phi_3 \overline{\phi}_1), \mathfrak{Im}(\phi_1 \overline{\phi}_2)\big)}{\|\Psi'\|^2}.$$

Proof. From (31.34) follows

$$\|\Psi'\|^2 = \|\Psi' \times \overline{\Psi'}\| = \frac{1}{2}\|\mathbf{x}_u \times \mathbf{x}_v\|.$$

Therefore, by (31.35) we have

$$\frac{\mathbf{x}_u \times \mathbf{x}_v}{\|\mathbf{x}_u \times \mathbf{x}_v\|} = \frac{(2/i)\Psi' \times \overline{\Psi'}}{2\|\Psi'\|^2} = \frac{\Psi' \times \overline{\Psi'}}{i\|\Psi'\|^2} = \frac{2\big(\mathfrak{Im}(\phi_2 \overline{\phi}_3), \mathfrak{Im}(\phi_3 \overline{\phi}_1), \mathfrak{Im}(\phi_1 \overline{\phi}_2)\big)}{\|\Psi'\|^2}. \blacksquare$$

We need an important map called the stereographic projection **st**, which can be used to transfer information between the complex plane and the sphere $S^2(1)$ of unit radius in $\mathbb{R}^3$ centered at the origin. We already discussed the inverse of **st** in Section 15.2, but now we use complex variables. Let $\mathbf{p} \in S^2(1)$ be any point other than the north pole $\mathbf{np} = (0, 0, 1)$. The line L through $\mathbf{p}$ and $\mathbf{np}$ meets the complex plane $\mathbb{C}$ at a point $\mathbf{q}$. Let us identify $\mathbb{C}$ and $\mathbb{R}^2$. Write $\mathbf{p} = (p_1, p_2, p_3)$ and $\mathbf{q} = (q_1, q_2)$. We can parametrize L as

$$t \longmapsto t\mathbf{q} + (1-t)\mathbf{np} = (t q_1, t q_2, 1-t).$$

Thus there is a t_0 such that

(31.41) $$(p_1, p_2, p_3) = \mathbf{p} = t_0 \mathbf{q} + (1 - t_0)\mathbf{np} = (t_0 q_1, t_0 q_2, 1 - t_0).$$

31.5 Minimal Curves

From (31.41) it follows that $t_0 = 1 - p_3$, so that

$$(31.42) \qquad q_1 = \frac{p_1}{t_0} = \frac{p_1}{1 - p_3} \quad \text{and} \quad q_2 = \frac{p_2}{t_0} = \frac{p_2}{1 - p_3}.$$

In view of (31.42) we make the following definition:

Definition. *The* **stereographic projection** *is the mapping* $\text{st}: S^2(1) \setminus \mathbf{np} \longrightarrow \mathbb{C}$ *given by*

$$\text{st}(x, y, z) = \frac{x + iy}{1 - z}.$$

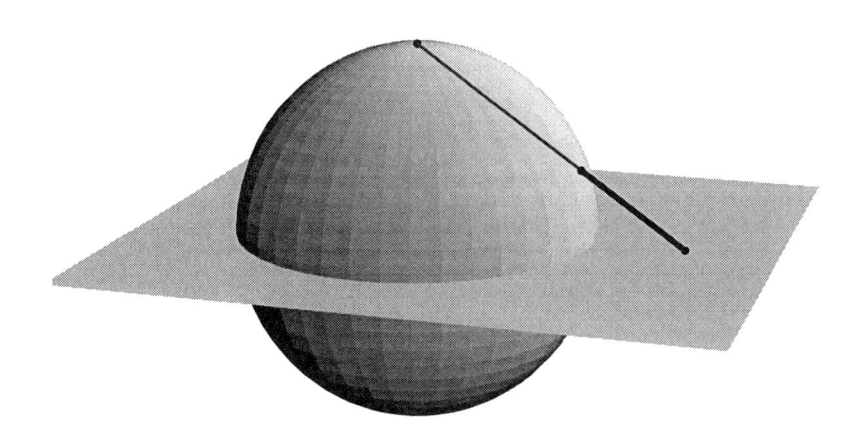

The stereographic projection

There is a simple formula for the composition of the stereographic projection and the Gauss map of a minimal curve.

Lemma 31.23. *Let* $\Psi: \mathcal{U} \longrightarrow \mathbb{C}^3$ *be a minimal curve, and write* $\Psi' = (\phi_1, \phi_2, \phi_3)$. *Let* $\mathbf{U}$ *be the unit normal vector field of the patch given by* (31.37). *Then*

$$(31.43) \qquad \text{st} \circ \mathbf{U} = \frac{\phi_3}{\phi_1 - i\phi_2}.$$

Proof. Starting with (31.40), we calculate

$$\text{st} \circ \mathbf{U} = \frac{\dfrac{2\bigl(\mathfrak{Im}(\phi_2\overline{\phi}_3) + i\,\mathfrak{Im}(\phi_3\overline{\phi}_1)\bigr)}{\|\Psi'\|^2}}{1 - \dfrac{2\mathfrak{Im}(\phi_1\overline{\phi}_2)}{\|\Psi'\|^2}} = \frac{2\bigl(\mathfrak{Im}(\phi_2\overline{\phi}_3) + i\,\mathfrak{Im}(\phi_3\overline{\phi}_1)\bigr)}{\|\Psi'\|^2 - 2\mathfrak{Im}(\phi_1\overline{\phi}_2)}$$

$$= \frac{\phi_2\overline{\phi}_3 - \overline{\phi}_2\phi_3 + i(\phi_3\overline{\phi}_1 - \overline{\phi}_3\phi_1)}{i(\|\Psi'\|^2 - 2\mathfrak{Im}(\phi_1\overline{\phi}_2))}$$

$$(31.44) \qquad = \frac{\phi_3(\overline{\phi}_1 + i\overline{\phi}_2) - \overline{\phi}_3(\phi_1 + i\phi_2)}{\|\Psi'\|^2 - 2\mathfrak{Im}(\phi_1\overline{\phi}_2)}.$$

Using the fact that
$$(\phi_1 - i\phi_2)(\phi_1 + i\phi_2) = -\phi_3^2,$$
we can rewrite the right-hand side of (31.44) as

$$\frac{\phi_3(\overline{\phi}_1 + i\overline{\phi}_2) - \overline{\phi}_3(\phi_1 + i\phi_2)}{\|\Psi'\|^2 - 2\mathfrak{Im}(\phi_1\overline{\phi}_2)} = \frac{\phi_3(\overline{\phi}_1 + i\overline{\phi}_2) + \overline{\phi}_3\left(\dfrac{\phi_3^2}{\phi_1 - i\phi_2}\right)}{\|\Psi'\|^2 - 2\mathfrak{Im}(\phi_1\overline{\phi}_2)}$$

$$= \frac{\phi_3\big((\overline{\phi}_1 + i\overline{\phi}_2)(\phi_1 - i\phi_2) + |\phi_3|^2\big)}{(\phi_1 - i\phi_2)(\|\Psi'\|^2 - 2\mathfrak{Im}(\phi_1\overline{\phi}_2))}$$

$$= \frac{\phi_3\big(\|\Psi'\|^2 + i\overline{\phi}_2\phi_1 - i\phi_2\overline{\phi}_1\big)}{(\phi_1 - i\phi_2)(\|\Psi'\|^2 - 2\mathfrak{Im}(\phi_1\overline{\phi}_2))} = \frac{\phi_3}{\phi_1 - i\phi_2}.$$

Hence we get (31.43). ∎

There are simple formulas for the coefficients of the second fundamental form of each member of the associated family of a minimal curve.

Lemma 31.24. *Let* $\Psi\colon\mathcal{U} \longrightarrow \mathbb{C}^3$ *be a minimal curve with unit normal vector field* **U** *and associated family* $t \longmapsto \mathbf{z}[t]$. *Then the coefficients of the second fundamental form of* $\mathbf{z}[t]$ *are given by*

$$(31.45) \qquad \begin{cases} e(t) = -g(t) = \mathfrak{Re}\big(e^{-it}\Psi''\big), \\ f(t) = -\mathfrak{Im}\big(e^{-it}\Psi''\big). \end{cases}$$

Proof. From (31.39) we get
$$\big(e^{-it}\Psi''\big)\cdot\mathbf{U} = \big(\mathbf{z}[t]_{uu} - i\mathbf{z}[t]_{uv}\big)\cdot\mathbf{U} = e(t) - i\,f(t).$$
and so (31.45) follows. ∎

Theorem 31.9 implies that all of the patches in the associated family of a minimal curve are isometric. Hence by Theorem 22.5 they all have the same Gaussian curvature. Therefore, we may speak of the **Gaussian curvature of a minimal curve**. We conclude this section by finding a formula for the Gaussian curvature of a minimal curve.

Lemma 31.25. *The Gaussian curvature of a minimal curve* $\Psi: \mathcal{U} \longmapsto \mathbb{C}^n$ *is given by*

$$K = \frac{-4\big(\|\Psi'\|^2\|\Psi''\|^2 - |\Psi'' \cdot \overline{\Psi'}|^2\big)}{\|\Psi'\|^6}.$$

Proof. Since $\|\Psi'\|^2 = 2\lambda^2$, we have

$$\Delta \log \lambda = \frac{1}{2}\Delta \log(\lambda^2) = 2\frac{\partial^2}{\partial \bar{z} \partial z}\log\big(\|\Psi'\|^2\big) = 2\frac{\partial^2}{\partial \bar{z} \partial z}\big(\log\big(\Psi' \cdot \overline{\Psi'}\big)\big)$$

$$= 2\frac{\partial}{\partial \bar{z}}\left(\frac{\Psi'' \cdot \overline{\Psi'}}{\|\Psi'\|^2}\right)$$

$$= 2\left(\frac{\|\Psi'\|^2\|\Psi''\|^2 - (\Psi'' \cdot \overline{\Psi'})(\Psi' \cdot \overline{\Psi''})}{\|\Psi'\|^4}\right)$$

$$= \frac{2\big(\|\Psi'\|^2\|\Psi''\|^2 - |\Psi'' \cdot \overline{\Psi'}|^2\big)}{\|\Psi'\|^4}.$$

Then (31.4) implies that

$$K = \frac{-\Delta \log \lambda}{\lambda^2} = -\left(\frac{2}{\|\Psi'\|^2}\right)\left(\frac{2\big(\|\Psi'\|^2\|\Psi''\|^2 - |\Psi'' \cdot \overline{\Psi'}|^2\big)}{\|\Psi'\|^4}\right)$$

$$= \frac{-4\big(\|\Psi'\|^2\|\Psi''\|^2 - |\Psi'' \cdot \overline{\Psi'}|^2\big)}{\|\Psi'\|^6}. \blacksquare$$

31.6 Finding Conjugate Minimal Surfaces

We know that a minimal isothermal patch $\mathbf{x}: \mathcal{U} \longrightarrow \mathbb{R}^n$ is the real part of an n-tuple $(\psi_1, \ldots, \psi_n)$ of analytic functions. How do we explicitly determine $(\psi_1, \ldots, \psi_n)$ from $\mathbf{x}$?

First, let us ask how we determine an analytic function $f: \mathcal{U} \longrightarrow \mathbb{C}$ such that the real part of f is a given harmonic function $h: \mathcal{U} \longrightarrow \mathbb{R}$. Of course, f is unique only up to an imaginary constant. The standard way (found in most text books on complex variables) is to use the Cauchy-Riemann equations; this method requires integration. There is, however, a more efficient method (see [Ahlf, pages 27–28], [Boas, pages 158–159] and [Lait]) that is completely algebraic, and hence involves no integration whatever. The idea is to substitute complex variables formally for the real variables x and y in the expression $h(x,y)$.

Lemma 31.26. *Let $h: \mathcal{U} \longrightarrow \mathbb{R}$ be a harmonic function where $\mathcal{U}$ is an open subset of $\mathbb{R}^2 = \mathbb{C}$, and let $z_0 = x_0 + i\,y_0 \in \mathcal{U}$.*

(i) Let $f: \mathcal{U} \longrightarrow \mathbb{C}$ be an analytic function such that $\mathfrak{Re}\, f(x+iy) = h(x,y)$ and $\mathfrak{Im}\, f(z_0) = 0$. Then

$$(31.46) \qquad f(z) = 2h\left(\frac{z+\overline{z_0}}{2}, \frac{z-\overline{z_0}}{2i}\right) - h(x_0, y_0).$$

(ii) Conversely, if f is defined by (31.46), then $f: \mathcal{U} \longrightarrow \mathbb{C}$ is an analytic function such that $f(z_0) = h(x_0, y_0)$.

Proof. For (i) we do the case $z_0 = 0$. Define g by $g(z) = \overline{f(\overline{z})}$. Then g is analytic, and we can express h in terms of f and g:

$$(31.47) \quad h(x,y) = \frac{1}{2}\left(f(x+iy) + \overline{f(x+iy)}\right) = \frac{1}{2}\left(f(x+iy) + g(x-iy)\right).$$

Since h is harmonic, it can be expanded in a double power series about any point $\mathbf{p} \in \mathcal{U}$. Hence, even though z is a complex variable, the expression

$$h\left(\frac{z}{2}, \frac{z}{2i}\right)$$

makes sense. Moreover, (31.47) implies that

$$h\left(\frac{z}{2}, \frac{z}{2i}\right) = \frac{1}{2}(f(z) + g(0)) = \frac{1}{2}(f(z) + h(0,0)).$$

For (ii) we compute

$$f_x = h_x - i\,h_y \qquad \text{and} \qquad f_y = i\,h_x + h_y.$$

Therefore,

$$\frac{\partial f}{\partial \overline{z}} = 0,$$

and so f is analytic. ∎

Corollary 31.27. *Let $\mathbf{x}: \mathcal{U} \longrightarrow \mathbb{R}^n$ be a minimal isothermal patch, where $\mathcal{U}$ is an open subset of $\mathbb{R}^2 = \mathbb{C}$. Let $\Psi = (\psi_1, \ldots, \psi_n)$ be the complexification of $\mathbf{x}$ such that $\mathfrak{Im}\, \Psi(0) = 0$. Then*

$$(31.48) \qquad \Psi(z) = 2\mathbf{x}\left(\frac{z}{2}, \frac{z}{2i}\right) - \mathbf{x}(0,0).$$

31.6 Finding Conjugate Minimal Surfaces

Furthermore, the conjugate minimal isothermal patch $\mathbf{y}$ *of* $\mathbf{x}$ *with* $\mathbf{y}(0,0) = (0,\ldots,0)$ *is given by*

(31.49) $$\mathbf{y}(u,v) = \mathfrak{Im}\left(2\mathbf{x}\left(\frac{u+iv}{2}, \frac{u+iv}{2i}\right) - \mathbf{x}(0,0)\right).$$

Proof. Write $\mathbf{x} = (x_1, \ldots, x_n)$, and for $j = 1, \ldots, n$ let ψ_j be the complexification of x_j such that $\mathfrak{Im}\psi_j(0) = 0$. Lemma 31.26 implies that

$$\psi_j(z) = 2x_j\left(\frac{z}{2}, \frac{z}{2i}\right) - x_j(0,0).$$

Hence we get (31.49). ∎

It is easy to adapt the construction of Corollary 31.27 to *Mathematica*. The general formula for the complexification of a patch **x** is:

```
harmonictoanalytic[x_][x0_,y0_][z_]:=
    Simplify[2x[(z + x0 - I y0)/2,
    (z - x0 + I y0)/(2 I)] - x[x0,y0]]
```

In the case that **x** is nonsingular at the origin we can use:

```
harmonictoanalytic[x_][z_]:=
    Simplify[2x[z/2,z/(2 I)] - x[0,0]]
```

Thus if **x** is a minimal isothermal patch, **harmonictoanalytic[x][x0,y0]** will be the n-tuple of analytic functions whose imaginary part vanishes at **{x0,y0}** and whose real part is **x**.

We also need a *Mathematica* operator that (using proper *Mathematica* syntax) constructs from a minimal curve **Psi** a minimal isothermal patch **x** such that the real part of **Psi** is **x**. Since we also want the harmonic conjugate of **x**, we might as well define a *Mathematica* operator that constructs the whole associated family; we call it **analytictominimal**.

```
analytictominimal[Psi_][t_][u_,v_]:=
    ComplexExpand[Re[E^(-I t)Psi[u + I v]]]
```

In particular, the conjugate of **x** is

```
analytictominimal[harmonictoanalytic][Pi/2]
```

Next, we give several examples of conjugate minimal surfaces.

The Associated Family of the Helicoid

Let us define in *Mathematica* a reparametrization of the general helicoid by

```
helicoid1[a_,b_,c_][u_,v_]:=
    {a Sinh[v] Cos[u],b Sinh[v] Sin[u],c u}
```

Mathematica can be used to check that **helicoid1[a,b,c]** is a minimal surface if and only if $a^2 = b^2$ or $c = 0$. Furthermore, **helicoid1[a,a,c]** is isothermal if and only if $a^2 = c^2$. We use *Mathematica* to find the associated family of **helicoid1[a,a,a]**:

```
analytictominimal[
harmonictoanalytic[helicoid1[a,a,a]]][t][u,v]//Simplify
        {-(a Cosh[v] Sin[t] Sin[u]) + a Cos[t] Cos[u] Sinh[v],
            a (-1 + Cos[u] Cosh[v]) Sin[t] +
            a Cos[t] Sin[u] Sinh[v], a (u Cos[t] + v Sin[t])}
```

In particular,

```
analytictominimal[
harmonictoanalytic[
    helicoid1[a,a,a]]][Pi/2][u,v]//Simplify
        {-(a Cosh[v] Sin[u]), a (-1 + Cos[u] Cosh[v]), a v}
```

Thus the conjugate of **helicoid1[a,a,a]** is a translate of a catenoid.

The Associated Family of Catalan's Surface

It can be checked that **catalan** (defined on page 693) is both minimal and isothermal. To find the associated family and harmonic conjugate of Catalan's minimal surface, we use

```
analytictominimal[
harmonictoanalytic[catalan[a]]][t][u,v]//Simplify
        {a Cos[t] (u - Cosh[v] Sin[u]) + a Sin[t] (v - Cos[u] Sinh[v]),
            a Cos[t] (1 - Cos[u] Cosh[v]) + a Sin[t] Sin[u] Sinh[v],
            4 a (1 - Cos[u/2] Cosh[v/2]) Sin[t] - 4 a Cos[t] Sin[u/2] Sinh[v/2]}
```

31.6 Finding Conjugate Minimal Surfaces

In particular, the conjugate of Catalan's minimal surface is parametrized by

```
analytictominimal[
  harmonictoanalytic[catalan[a]]][Pi/2][u,v]//Simplify
```

$$\{a\ (v\ -\ \text{Cos}[u]\ \text{Sinh}[v]),\ a\ \text{Sin}[u]\ \text{Sinh}[v],\ 4\ a\ (1\ -\ \text{Cos}[\tfrac{u}{2}]\ \text{Cosh}[\tfrac{v}{2}])\}$$

Here are some plots of stages in the deformation of Catalan's minimal surface.

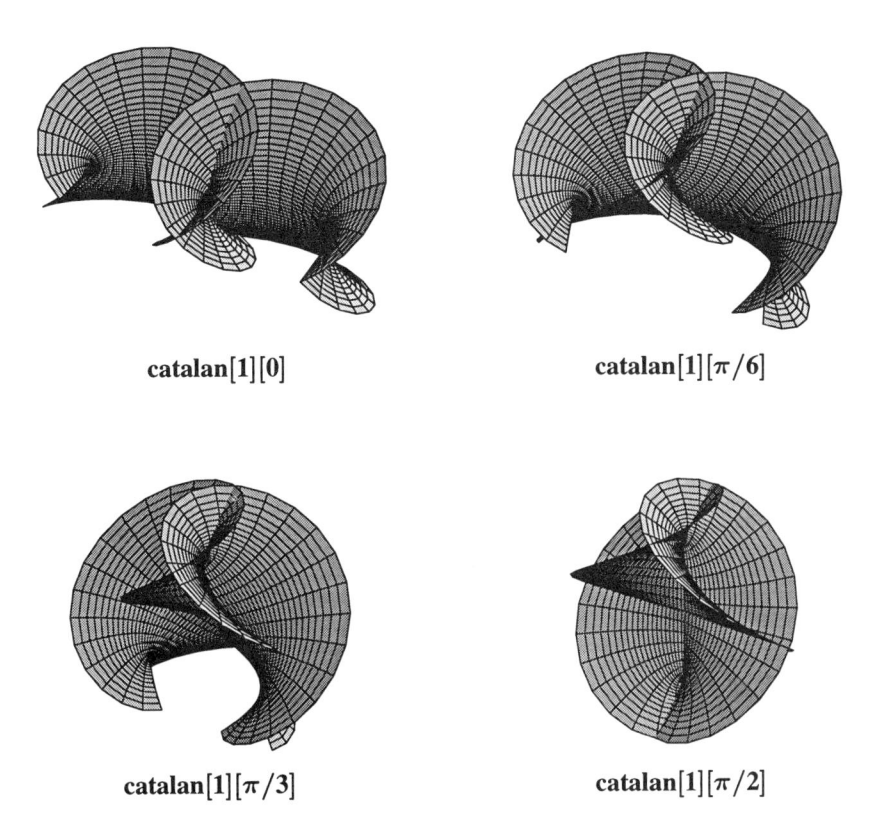

catalan[1][0]   catalan[1][$\pi/6$]

catalan[1][$\pi/3$]   catalan[1][$\pi/2$]

A typical command to draw one of these pictures is

```
ParametricPlot3D[Evaluate[
analytictominimal[
harmonictoanalytic[catalan[1]]][Pi/6][u,v]],
{u,0,4Pi},{v,-2,2},PlotPoints->{40,40}];
```

The Associated Family of Enneper's Surface

To find the complexification of **enneper** (defined on page 684), we query *Mathematica* with

```
harmonictoanalytic[enneper][z]
```

to which we get the response

$$\{z - \frac{z^3}{3}, \frac{I}{3} z (3 + z^2), z^2\}$$

To get the associated family of **enneper** we use

```
Apart[Together[
analytictominimal[harmonictoanalytic[
enneper]][t][u,v]]]
```

which elicits the response

$$\{\frac{u(3 - u^2 + 3v^2)\operatorname{Cos}[t]}{3} + \frac{v(3 - 3u^2 + v^2)\operatorname{Sin}[t]}{3},$$

$$\frac{v(-3 - 3u^2 + v^2)\operatorname{Cos}[t]}{3} + \frac{u(3 + u^2 - 3v^2)\operatorname{Sin}[t]}{3},$$

$$(u^2 - v^2)\operatorname{Cos}[t] + 2uv\operatorname{Sin}[t]\}$$

In particular, the minimal surface conjugate to **enneper** is given by

```
analytictominimal[
harmonictoanalytic[enneper]][Pi/2][u,v]//Simplify
```

$$\{v - u^2 v + \frac{v^3}{3}, u + \frac{u^3}{3} - uv^2, 2uv\}$$

The associated family of **enneper** yields no new surfaces because:

Lemma 31.28. *Let* $t \longmapsto$ **enneper**$[t]$ *denote the associated family of Enneper's surface. Then* **enneper**$[t]$ *coincides with a reparametrization of* **enneper** *rotated around the z-axis. Hence there is a Euclidean motion of* $\mathbb{R}^3$ *that maps* **enneper** *onto* **enneper**$[t]$.

31.6 Finding Conjugate Minimal Surfaces

Proof. It will be simpler to calculate using polar coordinates, so we first find the polar parametrization of each **enneper**[*t*].

```
analytictominimal[harmonictoanalytic[
enneper]][t][r Cos[theta],r Sin[theta]]//Simplify
```

$$\left\{ \frac{-(r^3 \operatorname{Cos}[t - 3\,\text{theta}])}{3} + r\,\operatorname{Cos}[t - \text{theta}], \right.$$

$$\frac{r^3\,\operatorname{Sin}[t - 3\,\text{theta}]}{3} + r\,\operatorname{Sin}[t - \text{theta}],$$

$$\left. r^2\,\operatorname{Cos}[t - 2\,\text{theta}] \right\}$$

In order to compare this parametrization with a rotated reparametrization of the patch **enneper**, it will be convenient to make use of the 3-dimensional rotation command in *Mathematica*'s standard rotation package. We read it in with

```
Needs["Geometry`Rotations`"]
```

The command **Rotate3D** to rotate a vector **{p1,p2,p3}** around the *z*-axis by an angle **psi** is now available. For example:

```
Rotate3D[{p1,p2,p3},psi,0,0]
```
      {p1 Cos[psi] + p2 Sin[psi], p2 Cos[psi] - p1 Sin[psi], p3}

We check that the output of

```
Rotate3D[
enneper[r Cos[theta-t/2],r Sin[theta-t/2]],
-t/2,0,0]//Simplify
```

is exactly the same output as

```
analytictominimal[harmonictoanalytic[
enneper]][t][r Cos[theta],r Sin[theta]]//Simplify
```

that we obtained above. Hence when rotated by an angle $-t/2$,

$$(r,\theta) \longmapsto \mathbf{enneper}\bigl(r\cos(\theta - t/2), r\sin(\theta - t/2)\bigr),$$

coincides with **enneper**[*t*]. ∎

31.7 Enneper's Surface of Degree n

The preceding discussion suggests how to generalize Enneper's surface. It is easier and more useful to define the minimal curve instead of the surface. We put

$$\mathbf{ennepermincurve}[n](z) = \left(z - \frac{z^{2n+1}}{2n+1}, i\left(z + \frac{z^{2n+1}}{2n+1}\right), \frac{2z^{n+1}}{n+1}\right).$$

or

```
ennepermincurve[n_][z_]:= {z - z^(2n + 1)/(2n + 1),
   I(z + z^(2n + 1)/(2n + 1)),2z^(n + 1)/(n + 1)}
```

If we write $\mathbf{ennepermincurve}[n]' = (\psi_1, \psi_2, \psi_3)$, it is easy to check that $\psi_1^2 + \psi_2^2 + \psi_3^2 = 0$, and that $|\psi_1|^2 + |\psi_2|^2 + |\psi_3|^2$ is never zero. Hence Theorem 31.17 says that indeed $(u, v) \longmapsto \Re(\mathbf{ennepermincurve}[n](u, v))$ is a minimal surface.

Like Enneper's surface, $(u, v) \longmapsto \Re(\mathbf{ennepermincurve}[n](u, v))$ is hard to visualize because of its many self-intersections. As an aid to understanding this surface, let us find the polar parametrization of each surface in the associated family of $z \longmapsto \mathbf{ennepermincurve}[n](z)$. We use

```
PowerExpand[Simplify[ComplexExpand[
analytictominimal[
ennepermincurve[n]][t][r Cos[theta],r Sin[theta]],
TargetFunctions->{Re,Im}]] /.
{ArcTan[r^2,0]->0,
ArcTan[r Cos[theta],r Sin[theta]]->theta}]
```

Mathematica's response is

```
                    1 + 2 n
                   r        Cos[t - (1 + 2 n) theta]
     {r Cos[t - theta] - ─────────────────────────────,
                              1 + 2 n

                         1 + 2 n
                        r        Sin[t - (1 + 2 n) theta]
          r Sin[t - theta] + ─────────────────────────────,
                                   1 + 2 n

           1 + n
        2 r      Cos[t - (1 + n) theta]
        ────────────────────────────────}
                    1 + n
```

31.7 Enneper's Surface of Degree n

Therefore, we define

$$\mathbf{enneperpolar}[n][t](r, \theta) = \left(r\cos(t - \theta) - \frac{r^{2n+1}\cos\left(t - (2n+1)\theta\right)}{2n + 1}, \right.$$

$$\left. r\sin(t - \theta) + \frac{r^{2n+1}\sin\left(t - (2n+1)\theta\right)}{2n + 1}, \frac{2r^{n+1}\cos(t - (n+1)\theta)}{n + 1} \right).$$

or

```
enneperpolar[n_][t_][r_,theta_]:=
   {r Cos[t - theta] - (r^(1 + 2n)
      Cos[t - (1 + 2n)theta])/(1 + 2n),
   r Sin[t - theta] + (r^(1 + 2n)
      Sin[t - (1 + 2n)theta])/(1 + 2n),
   (2r^(1 + n)Cos[t - (1 + n)theta])/(1 + n)}
```

The Gaussian curvature of $(r, \theta) \longmapsto \mathbf{enneperpolar}[n][t](r, \theta)$ is computed in *Mathematica* with

gcurvature[enneperpolar[n][t]][r,theta]

```
          2  -2 + 2 n
    -4 n  r
    ─────────────────
            2 n  4
      (1 + r   )
```

Thus the Gaussian curvature tends to 0 as $r \longrightarrow \infty$ and vanishes at the origin for $n \geq 2$. Also, we define

$\gamma[n, r](\theta) =$ the projection of $\mathbf{enneperpolar}[n][0](r, \theta)$ onto the xy-plane.

As r increases, $(r, \theta) \longmapsto \mathbf{enneperpolar}[n][0](r, \theta)$ becomes more and more complicated. The following pictures show the process in the case $n = 2$.

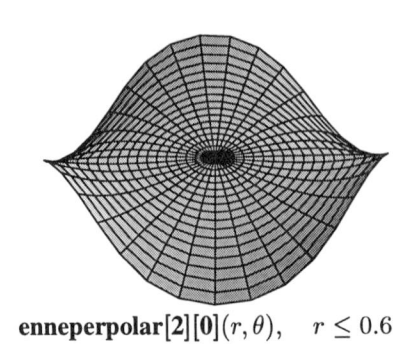

730　　　　　　　　　　　　　　　　Chapter 31　Minimal Surfaces and Complex Variables

enneperpolar[2][0](r,θ),　　$r \leq 0.6$

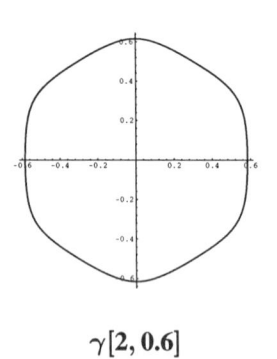

$\gamma[\mathbf{2}, \mathbf{0.6}]$

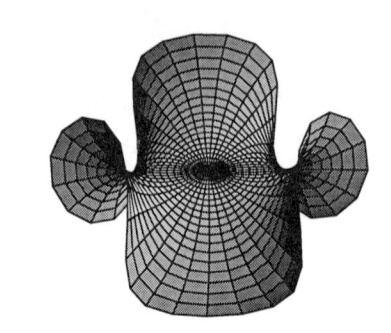

enneperpolar[2][0](r,θ),　　$r \leq 1.0$

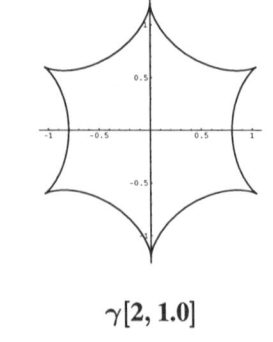

$\gamma[\mathbf{2}, \mathbf{1.0}]$

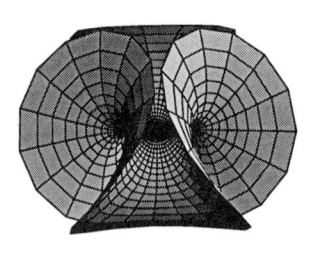

enneperpolar[2][0](r,θ),　　$r \leq 1.4$

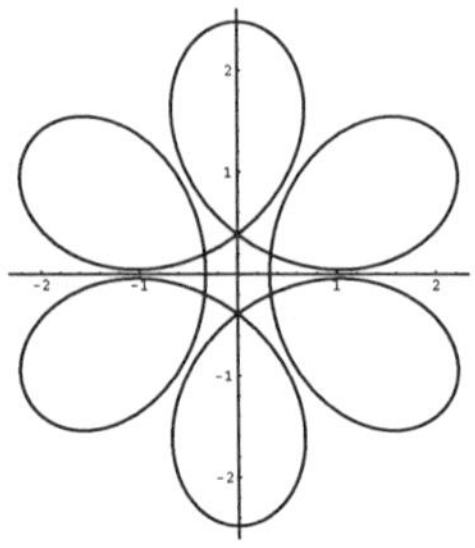

$\gamma[\mathbf{2}, \mathbf{1.4}]$

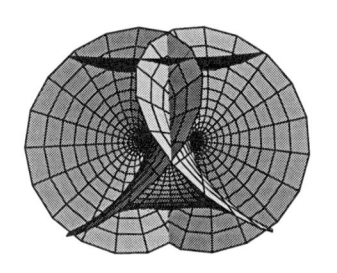

31.7 Enneper's Surface of Degree n 731

enneperpolar[2][0](r, θ), $r \leq 1.6$

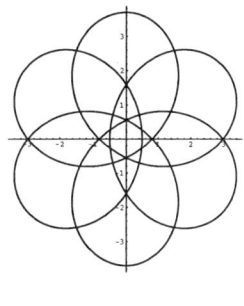

$\gamma[\mathbf{2}, \mathbf{1.6}]$

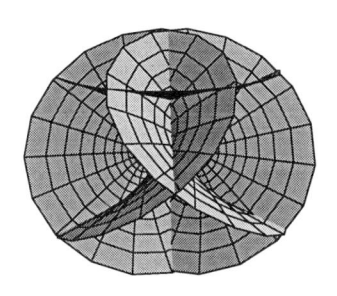

enneperpolar[2][0](r, θ), $r \leq 1.8$

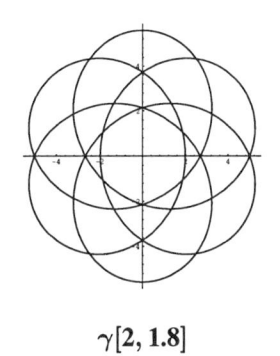

$\gamma[\mathbf{2}, \mathbf{1.8}]$

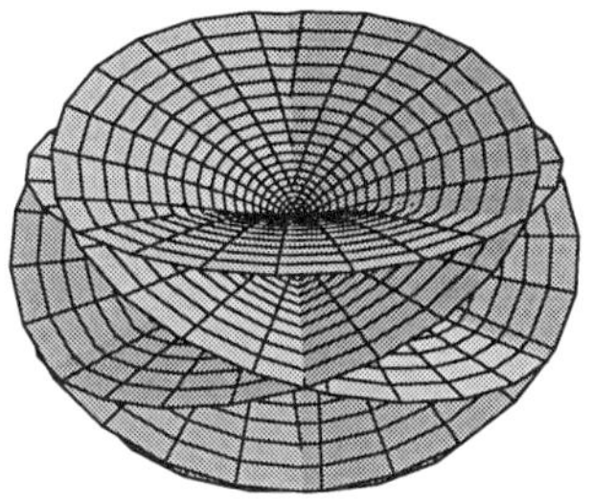

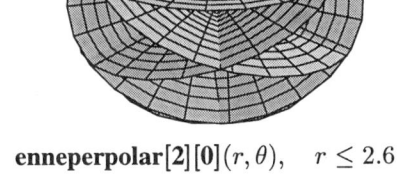

enneperpolar[2][0](r, θ), $r \leq 2.6$

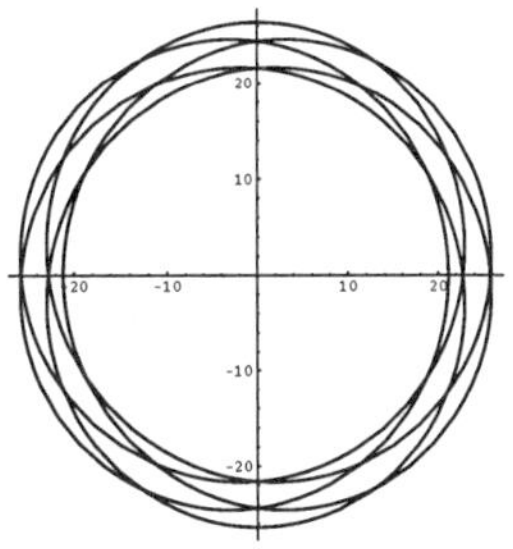

$\gamma[\mathbf{2}, \mathbf{2.6}]$

enneperpolar[2][0](r,θ), $\quad r \leq 3.4$

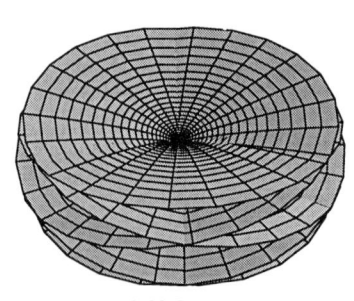

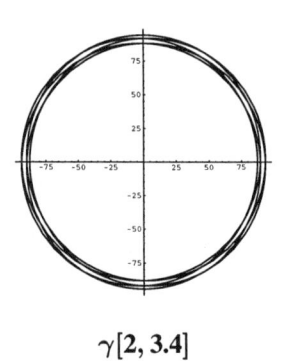

$\gamma[2, 3.4]$

Here are typical commands to draw these pictures:

```
ParametricPlot3D[
Evaluate[
enneperpolar[2][0]
[r,theta]],
{r,0,1.4},{theta,0,2Pi},
PlotPoints->{20,60},
PlotRange->All];
```

```
ParametricPlot[Evaluate[
Drop[enneperpolar[2][0]
[1.4,theta],-1]],
{theta,0,2Pi},
AspectRatio->Automatic];
```

31.8 Exercises

1. Show that an isothermal patch on a sphere $S^2(a)$ is given by

   $$\mathbf{y}(p,q) = a(\cos p \operatorname{sech} q, \sin p \operatorname{sech} q, \tanh q).$$

   (See exercise 3 of Chapter 13.)

2. Give the details of the proof of Lemma 31.8 and prove Corollary 31.15 and Lemma 31.19.

3. Define **Bour's**[3] **minimal curve bourmincurve[m]** by

   $$\mathbf{bourmincurve}[m](z) = \left(\frac{z^{m-1}}{m-1} - \frac{z^{m+1}}{m+1}, i\left(\frac{z^{m-1}}{m-1} + \frac{z^{m+1}}{m+1}\right), \frac{2z^m}{m}\right)$$

---

[3]Edmond Bour (1832-1866). French mathematician, who made significant contributions to analysis, algebra, geometry and applied mechanics despite his early death.

31.8 Exercises

or

```
bourmincurve[m_][z_]:=
    {z^(m - 1)/(m - 1) - z^(m + 1)/(m + 1),
    I(z^(m - 1)/(m - 1) + z^(m + 1)/(m + 1)),
    2z^m/m}
```

Show that **bourmincurve[2][0]** coincides with Enneper's minimal surface. Find the polar parametrization $(r, \theta) \longmapsto$ **bourpolar**$[m][t](r, \theta)$ of each member of the associated family of **bourmincurve**$[m]$ and compute its Gaussian curvature. Plot **bourpolar[3][0]**.

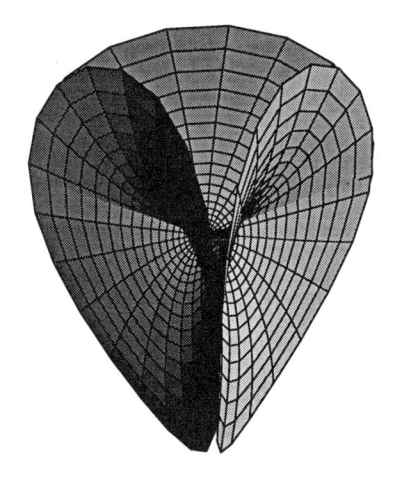

Bours's minimal surface of degree 3

The next three exercises are devoted to proving some elementary facts about minimal curves due to Study[4] (see Blaschke's classic book [Blas2, volume 1, pages 43–46 and 239–241]).

4. Show that if $\Psi: \mathcal{U} \longrightarrow \mathbb{C}^n$ is a minimal curve, then $\Psi' \cdot \Psi'' = 0$ and $\Psi' \cdot \Psi''' + \Psi'' \cdot \Psi'' = 0$.

---
[4]Eduard Study (1862-1930). German mathematician. He and Fubini gave the first description of the natural metric on complex projective space.

5. Let $\Psi:\mathcal{U} \longrightarrow \mathbb{C}^3$ be a minimal curve such that Ψ' and $\Psi'' \cdot \Psi''$ never vanish. Show that Ψ', Ψ'', Ψ''' are linearly independent at all points of $\mathcal{U}$.

6. Let $\Psi:\mathcal{U} \longrightarrow \mathbb{C}^3$ be a minimal curve such that Ψ' and $\Psi'' \cdot \Psi''$ never vanish. Show that
$$\Psi' \times \Psi'' = -\left(\frac{\Psi' \times \Psi'' \cdot \Psi'''}{\Psi'' \cdot \Psi''}\right)\Psi'.$$

31.9 Animations of Minimal Surfaces

The following command creates an animation of the associated family of Catalan's minimal surface.

```
Do[ParametricPlot3D[Evaluate[catalandef[t][u,v]],
{u,0,4Pi},{v,-2,2},Boxed->False,Axes->None,
PlotPoints->{40,20},BoxRatios->{1,1,2}],
{t,0,Pi/2,Pi/48}]
```

32

MINIMAL SURFACES VIA THE WEIERSTRASS REPRESENTATION

Minimal surface theory begins in 1744 with Euler's paper [Euler2], and in 1760 with Lagrange's paper [Lag]. Euler showed that the catenoid is a minimal surface, and Lagrange wrote down the partial differential equation that must be satisfied for a surface of the form $z = f(x, y)$ to be a minimal surface. In 1776 Meusnier[1] ([Meu]) rediscovered the catenoid and also showed that the helicoid is a minimal surface. The mathematical world had to wait over 50 years until other examples were found by Scherk. These include the surfaces now called "Scherk's minimal surface", "Scherk's fifth minimal surface", and a family of surfaces that includes both the catenoid and the helicoid.

In 1866, Weierstrass achieved a major breakthrough. He gave a formula that, on the one hand, permits the investigation of minimal surface problems in a general manner, and on the other hand, allows the easy generation of new minimal surfaces.

---

[1] Jean Babtiste Meusnier de la Place (1754-1793). French mathematician, a student of Monge. He was a general in the revolutionary army and died of battle wounds.

Weierstrass's[2] formula for generating minimal surfaces is described and used in Section 32.1. A *Mathematica* miniprogram to generate minimal curves by means of Weierstrass's formula is given in Section 32.2. We then use this miniprogram in Section 32.3 to generate the catenoid, the helicoid and Henneberg's minimal surface. A class of minimal surfaces with one planar end is studied in Section 32.4.

Few complete minimal surfaces are embedded; for example, Enneper's minimal surface has self-intersections in spite of its simple definition. Moreover, Weierstrass's formula usually produces self-intersecting minimal surfaces. The four complete minimal surfaces without self-intersections that we have seen are: the catenoid, the helicoid, Scherk's minimal surface and Scherk's fifth minimal surface. Another much more complicated complete minimal surface with no self-intersections is Costa's minimal surface; this spectacular minimal surface is discussed in Section 32.5 along with the theory of the Weierstrass $\wp$ function needed for the definition.

32.1 The Weierstrass Representation

Recall that a function $f\colon \mathbb{C} \longrightarrow \mathbb{C}$ is called **meromorphic** provided its only singularities are poles.

Definition. *Let $f(z)$ and $g(z)$ be meromorphic functions defined in a region $\mathcal{U}$ of the complex plane $\mathbb{C}$. Fix $z_0 \in \mathcal{U}$ and define*

(32.1)
$$\begin{cases} x_1(z) = \mathfrak{Re}\left(\int_{z_0}^{z} \frac{f(w)}{2}(1-g(w)^2)\,dw\right), \\ x_2(z) = \mathfrak{Re}\left(\int_{z_0}^{z} \frac{i\,f(w)}{2}(1+g(w)^2)\,dw\right), \\ x_3(z) = \mathfrak{Re}\left(\int_{z_0}^{z} f(w)g(w)\,dw\right), \end{cases}$$

---

[2] Karl Theodor Wilhelm Weierstrass (1815-1897). German mathematician. Weierstrass began his mathematical career as a high school teacher, but his work became recognized after he published a major paper on Abelian functions. This led to an appointment at the University of Berlin. His lectures in mathematics attracted students from all over the world. Topics of his lectures included mathematical physics, elliptic functions, and applications to problems in geometry and mechanics. Weierstrass's introduction of the Weierstrass $\wp$-function revolutionized the theory of elliptic functions.

32.1 The Weierstrass Representation

and

(32.2)
$$\begin{cases} y_1(z) = \mathfrak{Im}\left(\int_{z_0}^z \frac{f(w)}{2}(1-g(w)^2)\,dw\right), \\ y_2(z) = \mathfrak{Im}\left(\int_{z_0}^z \frac{i\,f(w)}{2}(1+g(w)^2)\,dw\right), \\ y_3(z) = \mathfrak{Im}\left(\int_{z_0}^z f(w)g(w)\,dw\right), \end{cases}$$

where $z = u + iv$. Put

$$\mathbf{x}(u,v) = (x_1(u,v), x_2(u,v), x_3(u,v))$$

and

$$\mathbf{y}(u,v) = (y_1(u,v), y_2(u,v), y_3(u,v)).$$

We call $\mathbf{x}$ the **Weierstrass patch** and $\mathbf{y}$ the **conjugate Weierstrass patch** determined by $f(z)$ and $g(z)$.

Theorem 32.1. *For any analytic function $f(z)$ and meromorphic function $g(z)$ both the Weierstrass patch and the conjugate Weierstrass patch are minimal isothermal patches. The metric of each is given by*

$$ds^2 = \frac{1}{4}|f(z)|^2\bigl(1+|g(z)|^2\bigr)^2|dz|^2.$$

In particular, $\mathbf{x}$ and $\mathbf{y}$ are isometric. They are regular except where f is zero or singular.

Proof. The complexification of $\mathbf{x}$ is given by

$$(\mathbf{x}+i\mathbf{y})(z) = \int_{z_0}^z \left(\frac{f(w)}{2}(1-g(w)^2), \frac{i\,f(w)}{2}(1+g(w)^2), f(w)g(w)\right)dw,$$

which shows that $(\mathbf{x}+i\mathbf{y})(z)$ is a triple of analytic functions. Since the real and imaginary parts of analytic functions are harmonic, it follows that

$$\mathbf{x}_{uu} + \mathbf{x}_{vv} = \mathbf{y}_{uu} + \mathbf{y}_{vv} = 0.$$

Also, the complex derivative of $\mathbf{x}$ (since it is $1/2$ the ordinary derivative) is given by

(32.3) $$\frac{1}{2}\left(\frac{f(z)}{2}(1-g(z)^2), i\frac{f(z)}{2}(1+g(z)^2), f(z)g(z)\right).$$

From (32.3) it is easy to see that (31.29), page 712, is satisfied, so that $(\mathbf{x}+i\mathbf{y})$ is a minimal curve and both $\mathbf{x}$ and $\mathbf{y}$ are isothermal. Finally, from (31.28), page 712, we have

$$E = G = 2\sum_{k=1}^{3} |\phi_k[\mathbf{x}]|^2$$

$$= \frac{1}{8}(|f|^2\,|1-g^2|^2 + |f|^2\,|1+g^2|^2 + 4|f\,g|^2) = \frac{1}{4}|f|^2(1+|g|^2)^2;$$

therefore, $\quad ds^2 = \frac{1}{4}|f|^2(1+|g|^2)^2(du^2+dv^2) = \frac{1}{4}|f|^2(1+|g|^2)^2|dz|^2.$ ∎

Corollary 32.2. *Let $f(z)$ and $g(z)$ be meromorphic functions defined in a region $\mathcal{U}$ of the complex plane $\mathbb{C}$, and let $\mathbf{x}$ be the Weierstrass patch and $\mathbf{y}$ the conjugate Weierstrass patch determined by $f(z)$ and $g(z)$. Then*

$$z \longmapsto (\mathbf{x}+i\mathbf{y})(z)$$

is a minimal curve.

We note that the converse of Corollary 32.2 also holds:

Lemma 32.3. *Let $\Psi\colon\mathcal{U} \longrightarrow \mathbb{C}^3$ be a minimal curve and write $\Psi' = (\varphi_1,\varphi_2,\varphi_3)$. Suppose that $\varphi_1 - i\varphi_2$ is not identically zero. Define*

(32.4) $$f = \varphi_1 - i\varphi_2, \quad \text{and} \quad g = \frac{\varphi_3}{\varphi_1 - i\varphi_2}.$$

Then f and g give rise to the Weierstrass representation of Ψ; that is,

$$\Psi' = \left(\frac{f}{2}(1-g^2), \frac{if}{2}(1+g^2), f\,g\right).$$

Proof. It is easy to check that if f and g are defined by (32.4), then

$$\varphi_1 = \frac{f}{2}(1-g^2), \quad \varphi_2 = \frac{if}{2}(1+g^2) \quad \text{and} \quad \varphi_3 = f\,g. \quad \blacksquare$$

Let us give a formula for the Weierstrass patch of the associated family of a Weierstrass patch.

Theorem 32.4. *For meromorphic functions $f(z)$ and $g(z)$, the associated family of the Weierstrass patch determined by f and g is given by*

$$t \longmapsto \mathbf{z}[t] = (z_1[t], z_2[t], z_3[t]),$$

32.1 The Weierstrass Representation

where

(32.5)
$$\begin{cases} z_1[t](z) = \mathfrak{Re}\left(e^{-it}\int_{z_0}^{z}\frac{f(w)}{2}(1-g(w)^2)dw\right), \\ z_2[t](z) = \mathfrak{Re}\left(e^{-it}\int_{z_0}^{z}\frac{if(w)}{2}(1+g(w)^2)dw\right), \\ z_3[t](z) = \mathfrak{Re}\left(e^{-it}\int_{z_0}^{z}f(w)g(w)dw\right). \end{cases}$$

Thus for each t, the patch $\mathbf{z}[t]$ is the Weierstrass patch determined by the meromorphic functions $e^{-it}f$ and g.

Proof. It is easy to see that $\mathbf{z}[t]$ given by (32.5) has the form (31.13). ∎

Since a Weierstrass patch and its conjugate are isometric, they have the same Gaussian curvature. Let us find the explicit formula.

Theorem 32.5. *The Gaussian curvature of a Weierstrass patch determined by meromorphic functions f and g is given by*

(32.6)
$$K = \frac{-16|g'|^2}{|f|^2(1+|g|^2)^4}.$$

Moreover, the same formula holds for the Gaussian curvature of each member of the associated family of the Weierstrass patch determined by f and g.

Proof. We have

(32.7) $\quad \Delta \log|f(z)| = 2\dfrac{\partial^2}{\partial\bar{z}\partial z}\log|f(z)|^2 = 2\dfrac{\partial^2}{\partial\bar{z}\partial z}\Big(\log f(z) + \log \overline{f(z)}\Big) = 0,$

and

(32.8) $\quad \Delta \log\big(1+|g(z)|^2\big) = 4\dfrac{\partial^2}{\partial\bar{z}\partial z}\log\big(1+g(z)\overline{g(z)}\big)$

$$= 4\frac{\partial}{\partial\bar{z}}\left(\frac{g'(z)\overline{g(z)}}{1+|g(z)|^2}\right) = \frac{4|g'(z)|^2}{(1+|g(z)|^2)^2}.$$

We know from Theorem 31.3, page 703, that the scaling factor is $\lambda = |f|(1+|g|^2)$. Therefore, from (32.7), (32.8) and (31.4) follows

$$K = \frac{-\Delta\log|f|(1+|g|^2)}{|f|^2(1+|g|^2)^2/4} = \frac{-\Delta\log(1+|g|^2)}{|f|^2(1+|g|^2)^2/4}$$

$$= \frac{-\dfrac{4|g'|^2}{(1+|g|^2)^2}}{|f|^2(1+|g|^2)^2/4} = \frac{-16|g'|^2}{|f|^2(1+|g|^2)^4}. \quad \blacksquare$$

Corollary 32.6. *The Gaussian curvature of a Weierstrass patch determined by meromorphic functions f and g vanishes precisely at the zeros of g'. Hence if g' is not identically zero, the zeros of the Gaussian curvature are isolated.*

Note that the zeros of the Gaussian curvature of any surface are both planar points and umbilic points. It is easy to see that the only umbilic points of a minimal surface are planar points.

There is a very simple formula for the Gauss map, or what amounts to the same thing, the unit normal, of a Weierstrass patch.

Theorem 32.7. *Let $\mathbf{U}$ be the unit normal of a Weierstrass patch determined by meromorphic functions f and g. Then $\mathrm{st} \circ \mathbf{U}$, considered as a function of a complex variable, is precisely the function g. Furthermore,*

$$(32.9) \qquad \mathbf{U}(z) = \left(\frac{2\mathfrak{Re}(g(z))}{|g(z)|^2 + 1}, \frac{2\mathfrak{Im}(g(z))}{|g(z)|^2 + 1}, \frac{|g(z)|^2 - 1}{|g(z)|^2 + 1} \right).$$

Proof. Lemma 31.23, page 719, implies that

$$(32.10) \qquad \mathrm{st} \circ \mathbf{U} = \frac{\varphi_3}{\varphi_1 - i\,\varphi_2} = \frac{f\,g}{f(1-g^2)/2 + f(1+g^2)/2} = g.$$

Furthermore, it is easily checked that the inverse of the function st defined on page 719 is the function $\mathrm{st}^{-1} \colon \mathbb{C} \longrightarrow S^2(1)$ given by

$$(32.11) \qquad \mathrm{st}^{-1}(w) = \left(\frac{2\mathfrak{Re}(w)}{|w|^2 + 1}, \frac{2\mathfrak{Im}(w)}{|w|^2 + 1}, \frac{|w|^2 - 1}{|w|^2 + 1} \right).$$

From (32.10) and (32.11) we get (32.9). ∎

We have already noted in Theorem 31.9 that the Gauss map is the same for all members of an associated family. So (32.9) is the formula for the unit normal of any member of the associated family of a Weierstrass patch.

32.2 Weierstrass Patches via *Mathematica*

Next, we describe how to use *Mathematica* to generate a Weierstrass patch and its associated family from meromorphic functions $f(z)$ and $g(z)$. For simplicity we take $z_0 = 0$. First, we define

32.3 Examples of Weierstrass Patches

```
            wei[f_,g_][z_]:= Module[{zz,tmp},
   tmp=Integrate[{(1 - g[zz]^2)f[zz]/2,
          I(1 + g[zz]^2)f[zz]/2,f[zz]g[zz]},zz];
        Simplify[tmp /. zz->z]]
```

(This definition is a little more complicated than usual in order to avoid technical problems when we use the operators **wei** and **analytictominimal** together.) The operator **wei** only generates a minimal curve. To get the parametrization of the minimal isothermal patch and its associated family, **wei** must be used together with the miniprogram **analytictominimal**.

The *Mathematica* version of formula (32.7) for the Gaussian curvature of a Weierstrass patch is:

```
weicurvature[f_,g_][z_]:= -16Abs[D[g[zz],zz]]^2/
       (Abs[f[zz]]^2(1 + Abs[g[zz]]^2)^4) /. zz->z
```

Let us use the program **wei** to generate several standard minimal surfaces. When making use of **wei** it is convenient to make use of the shorthand described on page 27 to represent functions.

32.3 Examples of Weierstrass Patches

The Catenoid and the Helicoid

The catenoid is obtained by means of the choice

$$f(z) = -e^{-z} \quad \text{and} \quad g(z) = -e^{z}.$$

To find the Weierstrass patch using *Mathematica*, we enter the command

```
wei[(-E^(-#))&,(-E^#)&][z]
```

The response is

Chapter 32 Minimal Surfaces via the Weierstrass Representation

$$\{\frac{1}{2E^z} + \frac{E^z}{2}, \frac{I}{E^z} - \frac{Iz}{2}E^z, z\}$$

The catenoid to helicoid deformation is obtained by using the command **analytictominimal** with **wei**. Thus

analytictominimal[wei[(-E^(-#))&,(-E^#)&]][t][u,v]

yields

$$\{Cos[t]\,(\frac{Cos[v]}{2E^u} + \frac{E^u\,Cos[v]}{2}) + Sin[t]\,(\frac{-Sin[v]}{2E^u} + \frac{E^u\,Sin[v]}{2}),$$

$$(\frac{Cos[v]}{2E^u} - \frac{E^u\,Cos[v]}{2})\,Sin[t] + Cos[t]\,(\frac{Sin[v]}{2E^u} + \frac{E^u\,Sin[v]}{2}),$$

$$u\,Cos[t] + v\,Sin[t]\}$$

In particular, the Weierstrass patch is given by

analytictominimal[wei[(-E^(-#))&,(-E^#)&]][0][u,v]

$$\{\frac{Cos[u]}{2E^v} + \frac{E^v\,Cos[u]}{2}, \frac{Sin[u]}{2E^v} + \frac{E^v\,Sin[u]}{2}, v\}$$

This tells us that the starting surface is exactly the catenoid **catenoid[1][u,v]** defined on page 467. Furthermore,

analytictominimal[wei[(-E^(-#))&,(-E^#)&]][Pi/2][v,u]

gives

$$\{\frac{-Sin[u]}{2E^v} + \frac{E^v\,Sin[u]}{2}, \frac{Cos[u]}{2E^v} - \frac{E^v\,Cos[u]}{2}, u\}$$

indicating that the conjugate Weierstrass patch is a helicoid. This parametrization is not quite the helicoid defined on page 357, but *Mathematica* will check that it is exactly the same as the variant

helicoid[1,1][u + Pi/2,-Sinh[v]] - {0,0,Pi/2}

32.3 Examples of Weierstrass Patches

Henneberg's Minimal Surface

Lemma 32.8. *The Weierstrass patch determined by the functions*

$$f(z) = \left(1 - \frac{1}{z^4}\right) \quad \text{and} \quad g(z) = z$$

is a reparametrization of the surface $(u, v) \longmapsto (1/2)\mathbf{henneberg}(u, -v)$.

Proof. The following *Mathematica* command composes the Weierstrass patch determined by f and g with $z \longmapsto e^z$:

```
Simplify[analytictominimal[
    Composition[wei[(1 - 1/#^4)&,#&],Exp]][0][u,v]]
```

$$\left\{\frac{-\cos[v]}{2\,E^u} + \frac{E^u\,\cos[v]}{2} + \frac{\cos[3\,v]}{6\,E^{3\,u}} - \frac{E^{3\,u}\,\cos[3\,v]}{6},\right.$$

$$\frac{\sin[v]}{2\,E^u} - \frac{E^u\,\sin[v]}{2} + \frac{\sin[3\,v]}{6\,E^{3\,u}} - \frac{E^{3\,u}\,\sin[3\,v]}{6},$$

$$\left.\frac{(1 + E^{4\,u})\,\cos[2\,v]}{2\,E^{2\,u}}\right\}$$

Moreover,

```
Simplify[analytictominimal[Composition[wei[(1 - 1/#^4)&,#&],
    Exp]][0][u,v] - henneberg[u,-v]/2]
```

yields

{0, 0, 0}

Hence `Composition[wei[(1 - 1/#&^4)&, #&],Exp]s[0][u,v]` and `henneberg[u,-v]/2` coincide. ∎

32.4 Minimal Surfaces with One Planar End

An interesting sequence of minimal surfaces more complicated than either the catenoid or Enneper's surface is generated by the Weierstrass representation when

$$f(z) = 1/z^2 \quad \text{and} \quad g(z) = z^{n+1}.$$

To find the minimal curve generated by f and g, we use

```
wei[1/#^2&,#^(n + 1)&][z]
```

to which *Mathematica* replies

$$\left\{\frac{-1}{2z} - \frac{z^{1+2n}}{2+4n}, \frac{-I}{2} + \frac{I z^{1+2n}}{2+4n}, \frac{z^n}{n}\right\}$$

The minimal surface generated by f and g when $n = 1$ is often called **Richmond's**[3] **minimal surface**; therefore, we define a minimal curve by

$$\text{richmondmincurve}[n](z) = \left(-\frac{1}{2z} - \frac{z^{2n+1}}{4n+2}, -\frac{i}{2z} + \frac{i z^{2n+1}}{4n+2}, \frac{z^n}{n}\right),$$

or

```
richmondmincurve[n_][z_]:=
    {-1/(2z) - z^(2n + 1)/(4n + 2)
    -I/(2z) + I z^(2n + 1)/(4n + 2),z^n/n}
```

The minimal curve $z \longmapsto \text{richmondmincurve}[n](z)$ (like $z \longmapsto \text{enneper}[n](z)$) can be studied effectively using polar coordinates. The following *Mathematica* command finds the general formula of the polar coordinate version of the associated family of $z \longmapsto \text{richmondmincurve}[n](z)$:

```
PowerExpand[Simplify[ComplexExpand[
analytictominimal[
richmondmincurve[n]][t][r Cos[theta],r Sin[theta]],
TargetFunctions->{Re,Im}]] /.
{ArcTan[r^2,0]->0,
ArcTan[r Cos[theta],r Sin[theta]]->theta}]
```

---
[3] Herbert William Richmond (1863-1949). English mathematician.

32.4 Minimal Surfaces with One Planar End

$$\left\{ \frac{-\cos[t+\theta]}{2r} - \frac{r^{1+2n}\cos[t-(1+2n)\theta]}{2+4n}, \right.$$

$$\frac{-\sin[t+\theta]}{2r} + \frac{r^{1+2n}\sin[t-(1+2n)\theta]}{2+4n},$$

$$\left. \frac{r^n \cos[t-n\theta]}{n} \right\}$$

Thus we define

$$\mathbf{richmondpolar}[n][t](r,\theta) = \left(\frac{-\cos(t+\theta)}{2r} - \frac{r^{2n+1}\cos\left(t-(2n+1)\theta\right)}{4n+2}, \right.$$

$$\left. \frac{-\sin(t+\theta)}{2r} + \frac{r^{2n+1}\sin\left(t-(2n+1)\theta\right)}{4n+2}, \frac{r^n \cos(t-n\theta)}{n} \right),$$

or

```
richmondpolar[n_][t_][r_,theta_]:=
   {-Cos[t + theta]/(2r) -
    (r^(1 + 2n)Cos[t - (1 + 2n)theta])/(2 + 4n),
    -Sin[t + theta]/(2*r) +
    (r^(1 + 2n)Sin[t - (1 + 2n)theta])/(2 + 4n),
    (r^n Cos[t - n theta])/n}
```

We can also use *Mathematica* to compute the Gaussian curvature of $(r, \theta) \mapsto$ **richmondpolar**$[n][t](r, \theta)$:

gcurvature[richmondpolar[n][t]][r,theta]

$$\frac{-16(1+n)^2 r^{4+2n}}{(1+r^{2+2n})^4}$$

To avoid frequent recomputation of the Gaussian curvature, we define

```
gcurvature[richmondpolar[n_][t_]][r_,theta_]:=
   -16(n + 1)^2 r^(2 n + 4)/(r^(2 n + 2) + 1)^4
```

Fortuitously, the Gaussian curvature K is a function of r alone. It obviously realizes its maximum value 0 when $r = 0$. Now let $n = 1$. To find the minimum value of the Gaussian curvature of $z \longmapsto$ **richmondpolar[1]**(z) we use

**Solve[D[
gcurvature[richmondpolar[1][t]][r,theta],r]==0,r]**

to which *Mathematica* replies

```
           3 1/4              3 1/4               3 1/4
{{r -> (-)     }, {r -> I (-)     }, {r -> -(-)     },
        5                  5                   5

         3 1/4
 {r -> -I (-)     }, {r -> 0}, {r -> 0}, {r -> 0},
          5

 {r -> 0}, {r -> 0}}
```

Thus $r = (3/5)^{1/4}$ is a critical point of the Gaussian curvature K. Moreover, it can be checked (for example, by plotting K as a function of r) that K has an absolute minimum at $r = (3/5)^{1/4}$. We can the find the value of K at $r = (3/5)^{1/4}$ with the command

gcurvature[richmondpolar[1][t]][(3/5)^(1/4),theta]

$$\frac{-\text{Sqrt}[84375]}{64}$$

We want to use *Mathematica's* command **Hue** to plot a portion of

$$(r, \theta) \longmapsto \textbf{richmondpolar}[1][0](r, \theta)$$

so that the color indicates the Gaussian curvature. But the arguments of **Hue** must lie between 0 and 1. Therefore, we require a normalization of

$$(r, \theta) \longmapsto \text{gcurvature}[\text{richmondpolar}[1][0]](r, \theta)$$

to a function whose range equals the interval $[0, 1]$. The obvious choice for such a function is

$$(r, \theta) \longmapsto \frac{\text{gcurvature}[\text{richmondpolar}[1][0]](r, \theta) + \sqrt{84375}/64}{\sqrt{84375}/64}.$$

Here then is the command to plot a portion of $(r, \theta) \longmapsto$ **richmondpolar**$[1][0](r, \theta)$ and color it by the Gaussian curvature:

```
ParametricPlot3D[
Append[
richmondpolar[1
][0][r,theta],
Hue[(gcurvature[
richmondpolar[1][0
]][r,theta]+
84375^(1/2)/64)/
(84375^(1/2)/64)]]
//Evaluate,
{r,0.3,1.3},
{theta,0,2Pi},
PlotPoints->{20,40},
Lighting->False,
Boxed->False,
Axes->None];
```

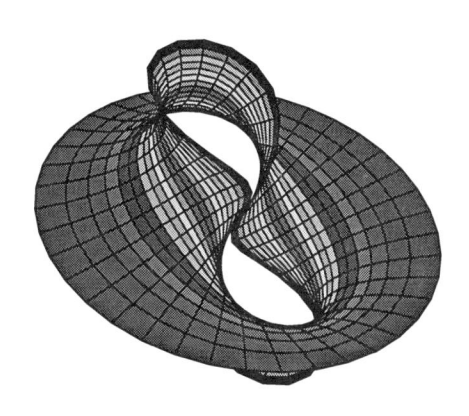

Minimal surface with one planar end

32.5 Costa's Minimal Surface

One of the most interesting minimal surfaces was defined by Costa in 1984 (see [Costa1]). Hoffman and Meeks (see [HoMe]) showed that Costa's minimal surface is a complete embedded surface. We shall adapt the treatment of Costa's minimal surface given in [BaCo] to find a parametrization[4] that can be used with *Mathematica*. For this, we need the Weierstrass Zeta function ζ. First, we explain the relevant theory of this function and also the Weierstrass function $\wp$.

The Weierstrass $\wp$ and ζ Functions

Let ω_1 and ω_2 be two complex numbers, both different from zero, and assume that $\Im(\omega_2/\omega_1)$ is positive. (Geometrically, this means that the angle between ω_1 and

---

[4] Helaman Ferguson has used this parametrization to make a sculpture of Costa's minimal surface. See [FeGrMa].

ω_2 is acute.) We define a lattice $\mathcal{L} \subset \mathbb{C}$ by

$$\mathcal{L} = \{\, m\omega_1 + n\omega_2 \mid m,n \in \mathbb{Z}^2 \,\}.$$

Put

(32.12) $$g_2 = 60 \sum_{\substack{\omega \in \mathcal{L} \\ \omega \neq 0}} \frac{1}{\omega^4} \quad \text{and} \quad g_3 = 140 \sum_{\substack{\omega \in \mathcal{L} \\ \omega \neq 0}} \frac{1}{\omega^6}.$$

The **Weierstrass $\wp$ function** associated with the lattice $\mathcal{L}$ is defined by

(32.13) $$\wp(z, \{g_2, g_3\}) = \frac{1}{z^2} + \sum_{\substack{\omega \in \mathcal{L} \\ \omega \neq 0}} \left(\frac{1}{(z-\omega)^2} - \frac{1}{\omega^2} \right).$$

Then $z \mapsto \wp(z, \{g_2, g_3\})$ is a doubly periodic function with periods ω_1 and ω_2. The set

$$\mathbf{FPP}(\{g_2, g_3\}) = \{\, z \in \mathbb{C} \mid z = \alpha\omega_1 + \beta\omega_2 \text{ with } 0 \leq \alpha, \beta < 1 \,\}$$

is called the **fundamental period parallelogram** of $\wp(z, \{g_2, g_3\})$; its importance lies in the fact that the image of $\mathbf{FPP}(g_2, g_3)$ under the Weierstrass $\wp$ function is the same as any translate $\mathbf{FPP}(\{g_2, g_3\}) + m\omega_1 + n\omega_2$ (where m and n are integers). The Weierstrass $\wp$ function satisfies an addition formula:

(32.14) $$\wp(z_1 + z_2) = \frac{1}{4}\left(\frac{\wp'(z_1) - \wp'(z_2)}{\wp(z_1) - \wp(z_2)} \right)^2 - \wp(z_1) - \wp(z_2).$$

We shall also need the **Weierstrass ζ function**, which is defined by

(32.15) $$\zeta(z, \{g_2, g_3\}) = \frac{1}{z} + \sum_{\substack{\omega \in \mathcal{L} \\ \omega \neq 0}} \left(\frac{1}{z-\omega} + \frac{1}{\omega} + \frac{z}{\omega^2} \right).$$

When g_2 and g_3 are understood, we write

$$\wp(z) = \wp(z, \{g_2, g_3\}) \quad \text{and} \quad \zeta(z) = \zeta(z, \{g_2, g_3\}).$$

It is well-known that $\wp$ satisfies the differential equation

(32.16) $$\wp'(z)^2 = 4\wp(z)^3 - g_2\wp(z) - g_3.$$

(See, for example, [Chand, page 29].) Furthermore, it is easy to see from (32.13) and (32.15) that ζ is related to $\wp$ by

(32.17) $$\zeta'(z) = -\wp(z).$$

32.5 Costa's Minimal Surface

Since $\wp'(z)$ is an odd function of z with period ω_1, we have

$$\wp'\left(\frac{\omega_1}{2}\right) = -\wp'\left(-\frac{\omega_1}{2}\right) = -\wp'\left(-\frac{\omega_1}{2}+\omega_1\right) = -\wp'\left(\frac{\omega_1}{2}\right),$$

showing that

$$\wp'\left(\frac{\omega_1}{2}\right) = 0.$$

Similarly,

$$\wp'\left(\frac{\omega_2}{2}\right) = \wp'\left(\frac{\omega_1+\omega_2}{2}\right) = 0.$$

Moreover, $\wp'$ has no other zeros in the fundamental period parallelogram besides $\omega_1/2$, $\omega_2/2$ and $(\omega_1+\omega_2)/2$. It follows from (32.16) that the three roots of the cubic polynomial
(32.18) $$4\wp(z)^3 - g_2\wp(z) - g_3$$
are precisely $\omega_1/2$, $\omega_2/2$ and $(\omega_1+\omega_2)/2$. If we put

$$e_1 = \wp\left(\frac{\omega_1}{2}\right), \qquad e_2 = \wp\left(\frac{\omega_2}{2}\right), \qquad e_3 = \wp\left(\frac{\omega_1+\omega_2}{2}\right),$$

then we can rewrite (32.16) in the form

(32.19) $$\wp'(z)^2 = 4\bigl(\wp(z)-e_1\bigr)\bigl(\wp(z)-e_2\bigr)\bigl(\wp(z)-e_3\bigr),$$

where

(32.20) $$\begin{cases} e_1 + e_2 + e_3 = 0, \\ e_1 e_2 + e_2 e_3 + e_3 e_1 = -\dfrac{g_2}{4}, \\ e_1 e_2 e_3 = \dfrac{g_3}{4}. \end{cases}$$

Moreover, e_1, e_2 and e_3 are all distinct; for details see [Chand, page 33].

The function ζ is not periodic, however,

(32.21) $$\zeta(z + m\omega_1 + n\omega_2) = \zeta(z) + 2m\,\zeta\left(\frac{\omega_1}{2}\right) + 2n\,\zeta\left(\frac{\omega_2}{2}\right).$$

Then (32.21) implies

(32.22) $$\zeta\left(\frac{\omega_1+\omega_2}{2}\right) = \zeta\left(\frac{\omega_1}{2}\right) + \zeta\left(\frac{\omega_2}{2}\right)$$

for integers m and n. We shall need Legendre's relation

(32.23) $$\zeta\left(\frac{\omega_1}{2}\right)\omega_2 - \zeta\left(\frac{\omega_2}{2}\right)\omega_1 = \pi.$$

For a proof of (32.23), see [Chand, page 50].

The Weierstrass $\wp$ and ζ Functions for a Square

The most symmetric fundamental period parallelogram is a square. Let us assume that the square is given by

(32.24) $\qquad \{\, z \in \mathbb{C} \mid 0 \leq \mathfrak{Re}(z) < 1 \text{ and } 0 \leq \mathfrak{Im}(z) < 1 \,\}.$

The Weierstrass $\wp$ function corresponding to this square has $\omega_1 = 1$ and $\omega_2 = i$, and is explicitly given by

(32.25) $\qquad \wp(z, \{g_2, g_3\}) = \dfrac{1}{z^2} + \displaystyle\sum_{\substack{(m,n) \in \mathbb{Z}^2 \\ (m,n) \neq (0,0)}} \left(\dfrac{1}{(z - m - ni)^2} - \dfrac{1}{(m + ni)^2} \right).$

Furthermore, the Weierstrass ζ function for the square (32.24) is given by

(32.26) $\qquad \zeta(z, \{g_2, g_3\}) = \dfrac{1}{z} + \displaystyle\sum_{\substack{(m,n) \in \mathbb{Z}^2 \\ (m,n) \neq (0,0)}} \left(\dfrac{1}{z - m - ni} + \dfrac{1}{m + ni} + \dfrac{z}{(m + ni)^2} \right).$

To determine g_2 and g_3 in (32.25) and (32.26), we use (32.25) to calculate

(32.27) $\qquad e_1 + e_2 = \wp\!\left(\dfrac{1}{2}, \{g_2, g_3\}\right) + \wp\!\left(\dfrac{i}{2}, \{g_2, g_3\}\right) = 0.$

Then it follows from (32.20) that

(32.28) $\qquad \begin{cases} e_3 = g_3 = 0, \\ g_2 = -4 e_1 e_2 = 4 e_1^2. \end{cases}$

In the case of the square given by (32.24), we use (32.28) to reduce the differential equation (32.16) to

(32.29) $\qquad \wp'(z)^2 = 4 \wp(z) \bigl(\wp(z)^2 - e_1^2 \bigr)$

In version 3.0 of *Mathematica*:

$\zeta(z, \{g_2, g_3\})$ is implemented by `WeierstrassZeta[z,{ g2,g3}]`,

$\wp(z, \{g_2, g_3\})$ is implemented by `WeierstrassP[z,{ g2,g3}]`,

$\wp'(z, \{g_2, g_3\})$ is implemented by `WeierstrassPPrime[z,{ g2,g3}]`.

Furthermore, the command `WeierstrassInvariants` can be used to compute g_2 and g_3 from ω_1 and ω_2. For example, the command

32.5 Costa's Minimal Surface

```
N[WeierstrassInvariants[{1/2,I/2}],10]
```

elicits the response

```
                        -13
{189.0727201, 2.071757156 10   }
```

telling us that the g_2 that corresponds to the periods 1 and i is approximately given by

(32.30) $$g_2 \approx 189.07272.$$

(We already know that $g_3 = 0$.) Therefore, the Weierstrass functions $\wp$ and ζ corresponding to the square (32.24) are $\wp(z, \{c, 0\})$ and $\zeta(z, \{c, 0\})$, where $c \approx 189.07272$. Furthermore,

```
Sqrt[N[WeierstrassInvariants[{1/2,I/2}],10][[1]]]/2
                  -7
{6.87519, 2.27583 10  }
```

tells us that $e_1 \approx 6.87519$.

We now introduce special notation for the Weierstrass $\wp$ and ζ functions corresponding to the square (32.24):

(32.31) $$\begin{cases} P(z) = \wp(z, \{c, 0\}), \\ Z(z) = \zeta(z, \{c, 0\}). \end{cases}$$

Lemma 32.9. *For all z we have*

(32.32) $$P\!\left(z - \frac{1}{2}\right) - P\!\left(z - \frac{i}{2}\right) - 2e_1 = \frac{16 e_1^3 P(z)}{P'(z)^2},$$

(32.33) $$iZ(iz) = Z(z),$$

(32.34) $$Z\!\left(\frac{1}{2}\right) = iZ\!\left(\frac{i}{2}\right) = \frac{\pi}{2},$$

(32.35) $$Z\!\left(\frac{1+i}{2}\right) = \frac{(1-i)\pi}{2}.$$

Proof. To prove (32.32), we use (32.14) to compute

(32.36) $$P\!\left(z - \frac{1}{2}\right) = \frac{P'(z)^2}{4\bigl(P(z) - e_1\bigr)^2} - P(z) - e_1.$$

Then (32.16) and (32.36) imply that

$$(32.37) \quad P\left(z - \frac{1}{2}\right) = \frac{P(z)\left(P(z)^2 - e_1^2\right)}{\left(P(z) - e_1\right)^2} - P(z) - e_1$$

$$= \frac{P(z)(P(z) + e_1)}{P(z) - e_1} - P(z) - e_1$$

$$= \frac{P(z)(P(z) + e_1) - P(z)^2 + e_1^2}{P(z) - e_1}$$

$$= \frac{e_1 P(z) + e_1^2}{P(z) - e_1}$$

$$= e_1 + \frac{2e_1^2}{P(z) - e_1}.$$

Similarly,

$$(32.38) \quad P\left(z - \frac{i}{2}\right) = -e_1 + \frac{2e_1^2}{P(z) + e_1}.$$

From (32.37) and (32.38) we get

$$(32.39) \quad P\left(z - \frac{1}{2}\right) - P\left(z - \frac{i}{2}\right) - 2e_1 = \frac{2e_1^2}{P(z) - e_1} - \frac{2e_1^2}{P(z) + e_1}$$

$$= \frac{4e_1^3}{P(z)^2 - e_1^2}.$$

Now (32.32) follows from (32.39) and (32.16).

Equation (32.33) is evident from (32.26). Then (32.34) follows from (32.33) and (32.23). Finally, (32.34) is a special case of (32.22). ∎

The Definition of Costa's Minimal Surface

A description of Costa's minimal surface using the Weierstrass function P has been given by Barbosa and Colares (see [BaCo, pages 84–90], and also [HoMe]). We adapt this description to *Mathematica*, but we use instead the Weierstrass Zeta function Z (given by (32.31)) to avoid one numerical integration. The use of the Weierstrass Zeta function Z instead of the Weierstrass P function speeds up the plotting of Costa's minimal surface by a factor of at least 50.

32.5 Costa's Minimal Surface

Costa's minimal surface can be defined as a Weierstrass patch using the functions f and g defined by

(32.40) $$f(z) = \boldsymbol{P}(z) \quad \text{and} \quad g(z) = \frac{A}{\boldsymbol{P}'(z)}.$$

In order that Costa's minimal surface have no self-intersections, we need to take

(32.41) $$A = \sqrt{2\pi\, g_2} = 2\sqrt{2\pi}\, e_1 \approx 34.46707.$$

(See [BaCo, page 89].)

Definition. *The* **Costa minimal curve** *is the meromorphic minimal curve* **costamincurve**$:\mathbb{C} \longrightarrow \mathbb{C}^3$ *defined as follows. First, we put*

(32.42) $$\textbf{costamincurve}'(z) = \frac{1}{2}\left(\frac{f(z)}{2}(1 - g(z)^2), i\frac{f(z)}{2}(1 + g(z)^2), f(z)g(z)\right),$$

where f and g are defined by (32.40). We take **costamincurve** *to be the antiderivative of* **costamincurve**$'$ *with the normalization*

$$\textbf{costamincurve}\left(\frac{1+i}{2}\right) = (0,0,0).$$

Next, we show how to use the Weierstrass function $\boldsymbol{Z}$ to express **costamincurve** without integrals.

Theorem 32.10. *The* **Costa minimal curve** *is given by*

$$\textbf{costamincurve}(z) = \bigl(\text{costamincurve}_1(z), \text{costamincurve}_2(z), \text{costamincurve}_3(z)\bigr),$$

where

$$\begin{cases}
\text{costamincurve}_1(z) = \dfrac{1}{2}\left(-\boldsymbol{Z}(z) + \pi z - i\pi + \dfrac{\pi^2(1+i)}{4e_1}\right.\\
\qquad\qquad\qquad\qquad \left. + \dfrac{\pi}{2e_1}\left(\boldsymbol{Z}\!\left(z - \dfrac{1}{2}\right) - \boldsymbol{Z}\!\left(z - \dfrac{i}{2}\right)\right)\right),\\[4pt]
\text{costamincurve}_2(z) = \dfrac{i}{2}\left(-\boldsymbol{Z}(z) - \pi z + \pi - \dfrac{\pi^2(1+i)}{4e_1}\right.\\
\qquad\qquad\qquad\qquad \left. - \dfrac{\pi}{2e_1}\left(\boldsymbol{Z}\!\left(z - \dfrac{1}{2}\right) - \boldsymbol{Z}\!\left(z - \dfrac{i}{2}\right)\right)\right),\\[4pt]
\text{costamincurve}_3(z) = \dfrac{\sqrt{2\pi}}{4}\left(\log\!\left(\dfrac{\boldsymbol{P}(z) - e_1}{\boldsymbol{P}(z) + e_1}\right) - \pi i\right).
\end{cases}$$

Proof. Since $\text{costamincurve}_3(z)$ is the simplest, we tackle it first. Using (32.16), we compute

$$(32.43) \qquad f(w)g(w) = \frac{AP(w)}{P'(w)} = \frac{AP'(w)P(w)}{P'(w)^2} = \frac{AP(w)}{4(P'(w)^2 - e_1^2)}.$$

We expand the right-hand side of (32.43) by partial fractions and use (32.41) to obtain

$$(32.44) \qquad f(w)g(w) = \frac{A}{8e_1} P'(w) \left(\frac{1}{P(w) - e_1} - \frac{1}{P(w) + e_1} \right)$$

$$= \frac{\sqrt{2\pi}}{4} \left(\frac{P'(w)}{P(w) - e_1} - \frac{P'(w)}{P(w) + e_1} \right).$$

We integrate both sides of (32.44), obtaining

$$\text{costamincurve}_3(z) = \int_{(1+i)/2}^{z} f(w)g(w)\,dw = \frac{\sqrt{2\pi}}{4} \log\left(\frac{P(w) - e_1}{P(w) + e_1} \right) \Bigg|_{(1+i)/2}^{z}$$

$$= \frac{\sqrt{2\pi}}{4} \left(\log\left(\frac{P(z) - e_1}{P(z) + e_1} \right) - \pi i \right).$$

Next,

$$(32.45) \qquad \frac{1}{2} f(w)(1 - g(w)^2) = \frac{1}{2} \left(P(w) - \frac{A^2 P(w)}{P'(w)^2} \right).$$

We use (32.32) and (32.41) to rewrite (32.45) as

$$(32.46) \qquad \frac{1}{2} f(w)(1 - g(w)^2) = \frac{1}{2} \left(P(w) - \frac{A^2}{16 e_1^3} \left(P\left(w - \frac{1}{2}\right) \right.\right.$$

$$\left.\left. - P\left(w - \frac{i}{2}\right) - 2 e_1 \right) \right)$$

$$= \frac{1}{2} \left(P(w) - \frac{\pi}{2e_1} \left(P\left(w - \frac{1}{2}\right) - P\left(w - \frac{i}{2}\right) - 2e_1 \right) \right)$$

$$= \frac{1}{2} \left(P(w) + \pi - \frac{\pi}{2e_1} \left(P\left(w - \frac{1}{2}\right) - P\left(w - \frac{i}{2}\right) \right) \right).$$

We integrate both sides of (32.46) from $(1+i)/2$ to z and use (32.34) and (32.35)

32.5 Costa's Minimal Surface

to get

$$(32.47) \quad \int_{(1+i)/2}^{z} \frac{1}{2} f(w)\bigl(1 - g(w)^2\bigr) dw$$

$$= \frac{1}{2}\left(-Z(w) + \pi w + \frac{\pi}{2e_1}\left(Z\left(w - \frac{1}{2}\right) - Z\left(w - \frac{i}{2}\right)\right)\right)\Big|_{(1+i)/2}^{z}$$

$$= \frac{1}{2}\bigg(-Z(z) + \pi z + \frac{\pi}{2e_1}\left(Z\left(z - \frac{1}{2}\right) - Z\left(z - \frac{i}{2}\right)\right)$$

$$+ Z\left(\frac{1+i}{2}\right) - \frac{\pi(1+i)}{2} - \frac{\pi}{2e_1}\left(Z\left(\frac{i}{2}\right) - Z\left(\frac{1}{2}\right)\right)\bigg)$$

$$= \frac{1}{2}\bigg(-Z(z) + \pi z + \frac{\pi}{2e_1}\left(Z\left(z - \frac{1}{2}\right) - Z\left(z - \frac{i}{2}\right)\right) - i\pi + \frac{\pi^2(1+i)}{4e_1}\bigg).$$

Similarly,

$$(32.48) \quad \int_{(1+i)/2}^{z} \frac{i}{2} f(w)\bigl(1 + g(w)^2\bigr) dw$$

$$= \frac{i}{2}\bigg(-Z(z) - \pi z - \frac{\pi}{2e_1}\left(Z\left(z - \frac{1}{2}\right) - Z\left(z - \frac{i}{2}\right)\right) + \pi - \frac{\pi^2(1+i)}{4e_1}\bigg). \quad \blacksquare$$

Corollary 32.11. Costa's minimal surface *is given by*

$$\mathbf{costa}(u, v) = \bigl(\mathrm{costa}_1(u, v), \mathrm{costa}_2(u, v), \mathrm{costa}_3(u, v)\bigr),$$

where

$$\begin{cases}
\mathrm{costa}_1(u, v) = \frac{1}{2}\mathfrak{Re}\bigg(-Z(u + iv) + \pi u + \frac{\pi^2}{4e_1} \\
\qquad\qquad\qquad + \frac{\pi}{2e_1}\left(Z\left(u + iv - \frac{1}{2}\right) - Z\left(u + iv - \frac{i}{2}\right)\right)\bigg), \\
\mathrm{costa}_2(u, v) = \frac{1}{2}\mathfrak{Re}\bigg(-iZ(u + iv) + \pi v + \frac{\pi^2}{4e_1} \\
\qquad\qquad\qquad - \frac{\pi i}{2e_1}\left(Z\left(u + iv - \frac{1}{2}\right) - Z\left(u + iv - \frac{i}{2}\right)\right)\bigg), \\
\mathrm{costa}_3(u, v) = \frac{\sqrt{2\pi}}{4}\log\left|\frac{P(u + iv) - e_1}{P(u + iv) + e_1}\right|.
\end{cases}$$

The Mathematica Implementation of Costa's Minimal Surface

We must use version 3.0 of *Mathematica* so that the Weierstrass Zeta function Z is available. First, we obtain numerical approximations to the constants c, e_1 and A. The command

```
c=N[WeierstrassInvariants[{1/2,I/2}],10][[1]]
```

tells us that $c \approx 189.07272$. Therefore, we enter the commands

```
e1=WeierstrassP[0.5,{189.07272,0}]

A=N[Sqrt[8Pi WeierstrassP[0.5,{189.07272,0}]^2],5]
```

telling us that $e_1 \approx 6.87519$ and $A \approx 34.467$.

Hence we define Costa's minimal surface in *Mathematica* by

```
costa[u_,v_]:=
   {(1/2)Re[-WeierstrassZeta[u + I v,{189.07272,0}] +
   3.14159 u + 0.35889 +
   0.228473(WeierstrassZeta[u + I v - 1/2,
                                    {189.07272,0}] -
       WeierstrassZeta[u + I v - I/2,{189.07272,0}])],
   (1/2)Re[-I WeierstrassZeta[u + I v,{189.07272,0}] +
   3.14159 v + 0.35889 -
   0.228473(I WeierstrassZeta[u + I v - 1/2,
                                    {189.07272,0}] -
       I WeierstrassZeta[u + I v - I/2,{189.07272,0}])],
   0.626657Log[Abs[(WeierstrassP[u + I v,{189.07272,0}] -
                                    6.87519)/
   (WeierstrassP[u + I v,{189.07272,0}] + 6.87519)]]}
```

Our first attempt to plot Costa's minimal surface with *Mathematica* is

```
costaplot80=ParametricPlot3D[costa[u,v],
    {u,0.001,1.001},{v,0.001,1.001},PlotPoints->80]
```

This plot is incorrect, so we do not display it. The difficulty is that *Mathematica* has included some polygons that have a vertex very far from the origin. To exclude

32.5 Costa's Minimal Surface

them we use the command **selectgraphics3d** described on page 425. We also need the command **dip** given on page 474 to eliminate the edges in the polygonal approximation of the surface. Here is the plotting command.

```
selectgraphics3d[costaplot80[[1]],8,
    Boxed->False,ViewPoint->{2.9,-1.4,1.2},
    PlotRange->{{-4,4},{-4,4},{-2, 2}},
    DisplayFunction->dip[EdgeForm[]]]
```

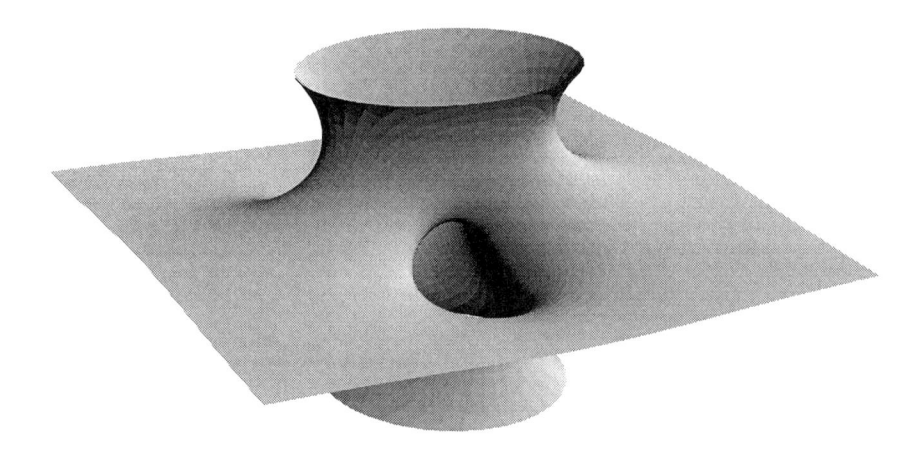

Costa's minimal surface

758   Chapter 32   Minimal Surfaces via the Weierstrass Representation

To get a top view of Costa's minimal surface, we use

```
selectgraphics3d[costaplot80[[1]],8,
    Boxed->False,ViewPoint->{0,0,4},
    PlotRange->{{-2.5,2.5},{-2.5,2.5},{-2,2}},
    DisplayFunction->dip[EdgeForm[]]]
```

Costa's minimal surface viewed from above

32.5 Costa's Minimal Surface

Costa's minimal surface contains two straight lines, as we can see from the following view:

```
selectgraphics3d[costaplot80[[1]],8,
    Boxed->False,ViewPoint->{2.9,-1.4,1.2},
    PlotRange->{{-2.5,2.5},{-2.5,2.5},{-2,2}},
    DisplayFunction->dip[EdgeForm[]]]
```

The lower half of Costa's minimal surface

32.6 Exercises

1. **The Gauss mapping of minimal surfaces.** Prove the following facts about the Gauss map:

    (a) The Gauss map of the catenoid is one-to-one; its image covers the sphere except for two points.

    (b) The image of the Gauss map of the helicoid omits exactly two points; otherwise, the Gauss map maps infinitely many points onto each point in its image.

    (c) The image of the Gauss map of Scherk's surface omits exactly four points; otherwise, the Gauss map maps infinitely many points onto each point in its image.

2. Show that a surface of revolution is isothermal if and only if $\psi'^2 = \varphi^2 - \varphi'^2$. Hence the surface **catenoid[a]** is isothermal if and only if $a = \pm 1$, even though it is a minimal surface for all a.

3. By computing the second fundamental form of the deformation

    $$t \longmapsto \mathtt{Composition[wei[(1 - 1/\#\&\hat{\ }4)\&,\#\&],Exp][t][u,v]}$$

    show that it is not the restriction of a Euclidean motion of $\mathbb{R}^3$. Plot several surfaces in the deformation.

4. The **trinoid** is the Weierstrass patch determined by $z \longmapsto 1/(z^3 - 1)^2$ and $z \longmapsto z^2$. Plot the trinoid using polar coordinates and compute its Gaussian curvature.

5. Plot the Weierstrass patch for which $f(z) = 1$ and $g(z) = z^{-1} + z^3$. The bottom of the resulting surface should resemble a catenoid and the top should resemble Enneper's minimal surface.

33

MINIMAL SURFACES VIA BJÖRLING'S FORMULA

In Section 33.1 we establish Björling's formula. Then in Section 33.2 we discuss an important special case of Björling's formula, which consists of a formula for constructing a minimal curve from a plane curve. This simple formula yields an effective way to construct from any plane curve γ a minimal surface that contains γ as a geodesic. Several examples of new minimal surfaces are given in Section 33.3. These include a generalization of the catenoid that has an ellipse instead of a circle as its center curve.

33.1 Björling's Formula

In this section we discuss an important way of constructing minimal curves from real curves; the minimal curves can then be used to construct minimal surfaces using the method of Section 31.3. The whole process can be effectively automated with *Mathematica*, as we shall see in Section 33.2.

In 1844 Björling[1] ([Bj]) asked if it is possible to find a minimal surface containing a given analytic strip. This question is frequently called Björling's problem. H.A.

---

[1]Emanuel Gabriel Björling (1808-1872). Swedish mathematician. Professor at the University of Upsala.

Schwarz[2] (see [Schwa, pages 179–189]) gave an elegant solution to it. For refinements of Schwarz's method see [Nits, pages 139–145] and [DHKW, pages 120–135].

To make precise the notion of minimal surface containing a given analytic strip, we make the following definition.

Definition. Let $\boldsymbol{\alpha}, \boldsymbol{\gamma}\colon (a,b) \longrightarrow \mathbb{R}^3$ be curves such that

(33.1) $$\|\boldsymbol{\gamma}\| = 1 \quad \text{and} \quad \boldsymbol{\alpha}' \cdot \boldsymbol{\gamma} = 0.$$

Suppose that $\boldsymbol{\alpha}$ and $\boldsymbol{\gamma}$ have holomorphic extensions $\boldsymbol{\alpha}, \boldsymbol{\gamma}\colon (a,b) \times (c,d) \longrightarrow \mathbb{C}^3$, such that (33.1) holds also for $z \in (a,b) \times (c,d)$. Fix $z_0 \in (a,b) \times (c,d)$. Then the **Björling curve** corresponding to $\boldsymbol{\alpha}$ and $\boldsymbol{\gamma}$ is given by

$$\mathbf{bjorling}[\boldsymbol{\alpha}, \boldsymbol{\gamma}](z) = \boldsymbol{\alpha}(z) - i \int_{z_0}^{z} \boldsymbol{\gamma}(z) \times \boldsymbol{\alpha}'(z) dz.$$

We now prove that a Björling curve is in fact a minimal curve, and we show how it is related to $\boldsymbol{\alpha}$ and $\boldsymbol{\gamma}$.

Theorem 33.1. Let $\boldsymbol{\alpha}, \boldsymbol{\gamma}\colon (a,b) \longrightarrow \mathbb{R}^3$ be curves satisfying (33.1). Suppose that $\boldsymbol{\alpha}$ and $\boldsymbol{\gamma}$ have holomorphic extensions $\boldsymbol{\alpha}, \boldsymbol{\gamma}\colon (a,b) \times (c,d) \longrightarrow \mathbb{C}^3$ such that (33.1) continues to hold for $z \in (a,b) \times (c,d)$.

(i) the Björling curve $z \longmapsto \mathbf{bjorling}[\boldsymbol{\alpha}, \boldsymbol{\gamma}](z)$ is a minimal curve;

(ii) for real u, the map $u \longmapsto \mathfrak{Re}\big(\mathbf{bjorling}[\boldsymbol{\alpha}, \boldsymbol{\gamma}](u)\big) - \boldsymbol{\alpha}(u)$ is constant;

(iii) for real u, $\mathbf{U}(u,0) = \boldsymbol{\gamma}(u)$, where $\mathbf{U}$ denotes the unit normal to the patch

$$(u,v) \longmapsto \mathfrak{Re}\big(\mathbf{bjorling}[\boldsymbol{\alpha}, \boldsymbol{\gamma}](u + i\,v)\big);$$

Proof. Write $\Psi = \mathbf{bjorling}[\boldsymbol{\alpha}, \boldsymbol{\gamma}]$; then $\Psi' = \boldsymbol{\alpha}' - i\boldsymbol{\gamma} \times \boldsymbol{\alpha}'$. Using (31.33), page 715, and (33.1), we find that

$$\begin{aligned}\Psi' \cdot \Psi' &= (\boldsymbol{\alpha}' - i\boldsymbol{\gamma} \times \boldsymbol{\alpha}') \cdot (\boldsymbol{\alpha}' - i\boldsymbol{\gamma} \times \boldsymbol{\alpha}') \\ &= \boldsymbol{\alpha}' \cdot \boldsymbol{\alpha}' - (\boldsymbol{\gamma} \times \boldsymbol{\alpha}') \cdot (\boldsymbol{\gamma} \times \boldsymbol{\alpha}') - 2i\,\boldsymbol{\alpha}' \cdot (\boldsymbol{\gamma} \times \boldsymbol{\alpha}') \\ &= \boldsymbol{\alpha}' \cdot \boldsymbol{\alpha}' - (\boldsymbol{\gamma} \cdot \boldsymbol{\gamma})(\boldsymbol{\alpha}' \cdot \boldsymbol{\alpha}') = 0.\end{aligned}$$

---

[2] Herman Amandus Schwarz (1843-1921). German mathematician, who made fundamental contributions to the theories of minimal surfaces and conformal mapping. His work on eigenvalues lead to what we now call the Cauchy-Schwarz inequality.

33.1 Björling's Formula

Hence Ψ is a minimal curve.

(ii) holds because $\alpha(u)$ and $\gamma(u)$ are real for real u. To prove (iii), we observe that

$$\Psi'(u) = \alpha'(u) - i\gamma(u) \times \alpha'(u) \quad \text{and} \quad \overline{\Psi'(u)} = \alpha'(u) + i\gamma(u) \times \alpha'(u)$$

for real u. Hence by (33.1)

$$\Psi'(u) \cdot \overline{\Psi'(u)} = 2\alpha'(u) \cdot \alpha'(u),$$

and by (8.3), page 183,

$$\Psi'(u) \times \overline{\Psi'(u)} = 2i\,\alpha'(u) \times \big(\gamma(u) \times \alpha'(u)\big) = 2i\big(\alpha'(u) \cdot \alpha'(u)\big)\gamma(u)$$

Using (31.40), page 718, we find that

$$\mathbf{U}(u,0) = \frac{\Psi'(u) \times \overline{\Psi'(u)}}{i\|\Psi'(u)\|^2} = \gamma(u). \quad \blacksquare$$

Corollary 33.2. *Suppose the hypotheses of Theorem* 33.1 *hold. Define a patch* $\mathbf{x}\colon (a,b) \times (c,d) \longmapsto \mathbb{R}^3$ *by*

$$\mathbf{x}(u,v) = \mathfrak{Re}\big(\text{bjorling}[\alpha,\gamma](u+iv)\big) = \alpha(u+iv) - i\int_{z_0}^{z} \gamma(z) \times \alpha'(z)dz\bigg|_{z \to u+iv}.$$

Then

(33.2) $$\mathbf{x}(u,-v) = \alpha(u+iv) + i\int_{z_0}^{z} \gamma(z) \times \alpha'(z)dz\bigg|_{z \to u+iv}.$$

Proof. Define $\mathbf{y}\colon (a,b) \times (c,d) \longmapsto \mathbb{R}^3$ by $\mathbf{y}(u,v) = \mathbf{x}(u,-v)$. It is easy to check that the unit normals $\mathbf{U}_x$ and $\mathbf{U}_y$ of the minimal patches $\mathbf{x}$ and $\mathbf{y}$ are related by

$$\mathbf{U}_y(u,v) = -\mathbf{U}_x(u,-v).$$

Hence up to a constant

$$\mathbf{x}(u,-v) = \mathbf{y}(u,v) = \mathfrak{Re}\big(\text{bjorling}[\alpha,-\gamma](u+iv)\big)$$

$$= \alpha(u+iv) + i\int_{z_0}^{z} \gamma(z) \times \alpha'(z)dz\bigg|_{z \to u+iv}. \quad \blacksquare$$

The *Mathematica* version of Björling's formula is

```
bjorling[alpha_,gamma_][z_]:= Module[{zz,tmp},
    tmp=I Integrate[cross[gamma[zz],alpha'[zz]],zz];
    Simplify[alpha[zz] + tmp /. zz->z]]
```

33.2 Minimal Surfaces from Plane Curves

We can use Theorem 33.1 to construct many interesting minimal surfaces. For example, let $\boldsymbol{\alpha}\colon (a,b) \longrightarrow \mathbb{R}^2$ be a regular plane curve which has a regular holomorphic extension $\boldsymbol{\alpha}\colon (a,b) \times (c,d) \longrightarrow \mathbb{C}^2$. We take $\boldsymbol{\gamma} = J\boldsymbol{\alpha}'/\|\boldsymbol{\alpha}'\|$. Then

$$\boldsymbol{\gamma} \times \boldsymbol{\alpha}' = -\|\boldsymbol{\alpha}'\|E_3,$$

where E_3 is the unit vector field perpendicular to $\mathbb{R}^2$. Thus

$$\mathbf{bjorling}[\boldsymbol{\alpha}, \boldsymbol{\gamma}](z) = \boldsymbol{\alpha}(z) + i\left(\int_{z_0}^{z} \sqrt{\boldsymbol{\alpha}'(z) \cdot \boldsymbol{\alpha}'(z)}\, dz\, E_3\right).$$

This leads to the following definition.

Definition. Let $\boldsymbol{\alpha}\colon (a,b) \longrightarrow \mathbb{R}^2$ be a plane curve, and suppose that $\boldsymbol{\alpha}$ has a holomorphic extension $\boldsymbol{\alpha}\colon (a,b) \times (c,d) \longrightarrow \mathbb{C}^2$. Write $\boldsymbol{\alpha} = (a_1, a_2)$. We define **geomincurve**$[\boldsymbol{\alpha}](z)$ by

$$\mathbf{geomincurve}[\boldsymbol{\alpha}](z) = \left(a_1(z), a_2(z), i\int_{z_0}^{z} \sqrt{a_1'(z)^2 + a_2'(z)^2}\, dz\right).$$

We need a result about the intersection of a plane and a surface in $\mathbb{R}^3$.

Theorem 33.3. *Let β be a unit-speed curve which lies on the intersection of a regular surface $\mathcal{M} \subset \mathbb{R}^3$ and a plane Π. Suppose that $\mathcal{M}$ meets Π perpendicularly along β. Then β is a geodesic in $\mathcal{M}$.*

Proof. Let $\mathbf{U}$ denote the unit normal vector field to $\mathcal{M}$. Then β', β'' and $\mathbf{U} \circ \beta$ all lie in Π, and β' is perpendicular to $\mathbf{U} \circ \beta$. Since β is a unit-speed curve, β'' is perpendicular to β'. Hence β'' is a multiple of $\mathbf{U} \circ \beta$, and so β is a geodesic in $\mathcal{M}$. ∎

Theorem 33.4. *Let $\beta\colon (a,b) \longrightarrow \mathbb{R}^2$ be a unit-speed plane curve which has a holomorphic extension $\beta\colon (a,b) \times (c,d) \longrightarrow \mathbb{C}^2$, where $c < 0 < d$. Then*

$$\mathfrak{Re}\bigl(\mathbf{geomincurve}[\beta]\bigr)\colon (a,b) \times (c,d) \longrightarrow \mathbb{R}^3$$

is a minimal surface which contains β as a geodesic.

33.2 Minimal Surfaces from Plane Curves

Proof. It is easy to check that

$$\textbf{geomincurve}'[\beta]\cdot\textbf{geomincurve}'[\beta]=0,$$

so that **geomincurve**[β] is a minimal curve. Furthermore, if we write $\beta = (b_1, b_2, b_3)$, then according to (31.40) the unit normal of **geomincurve**[β] is given by

$$\mathbf{U} = \frac{2\big(\Im\mathfrak{m}(-i\,\overline{b_2'\sqrt{b_1'^2+b_2'^2}}),\Im\mathfrak{m}(i\,\overline{b_1'}\sqrt{b_1'^2+b_2'^2}),\Im\mathfrak{m}(a_1'\overline{b_2'})\big)}{|b_1'|^2+|b_2'|^2+|\sqrt{b_1'^2+b_2'^2}|^2}.$$

Since $b_1'(z)$ and $b_2'(z)$ are real for real z, we get

$$\mathbf{U}(u,0) = \frac{2\big(-b_2'\sqrt{b_1'^2+b_2'^2},(b_1'\sqrt{b_1'^2+b_2'^2}),0\big)}{2(b_1'^2+b_2'^2)}\bigg|_{z=(u,0)} = \frac{2\big(-b_2',b_1',0\big)}{\sqrt{b_1'^2+b_2'^2}}\bigg|_{z=(u,0)}.$$

Therefore, **U** is perpendicular to E_3, which is just the unit normal of Π. Now Theorem 33.3 implies that α is a geodesic in $\Re\mathfrak{e}(\textbf{geomincurve}[\alpha])$. ∎

Corollary 33.5. *Let $\alpha\colon(a,b)\longrightarrow\mathbb{R}^2$ be a regular plane curve which has a holomorphic extension $\alpha\colon(a,b)\times(c,d)\longrightarrow\mathbb{C}^2$, where $c<0<d$. Then*

$$\Re\mathfrak{e}(\textbf{geomincurve}[\alpha])\colon(a,b)\times(c,d)\longrightarrow\mathbb{R}^3$$

is a minimal curve which contains α as a pregeodesic.

Proof. There exist an increasing analytic function $h\colon(e,f)\longrightarrow(c,d)$ such that $\beta=\alpha\circ h$ is a unit-speed plane curve with a holomorphic extension. Since

$$\textbf{geomincurve}[\beta](z)=\textbf{geomincurve}[\alpha]\big(h(z)\big)$$

and

$$\textbf{geomincurve}[\beta]'(z)=\textbf{geomincurve}[\alpha]\big(h(z)\big)h'(z),$$

it follows that **geomincurve**[β] is a minimal curve with the same range as α. Since β is a geodesic in $\Re\mathfrak{e}(\textbf{geomincurve}[\beta])=\Re\mathfrak{e}(\textbf{geomincurve}[\alpha])$ it follows that α is a pregeodesic in $\Re\mathfrak{e}(\textbf{geomincurve}[\alpha])$. ∎

The *Mathematica* version of **geomincurve**[α] is

```
geomincurve[alpha_][z_]:= Append[alpha[z],
    I Simplify[PowerExpand[arclength[alpha][z]]]]
```

Then **geomincurve** can be combined with **analytictominimal** to produce the parametrization of the minimal surface containing a given curve as a geodesic.

33.3 Examples of Minimal Surfaces Constructed from Plane Curves

A Minimal Surface Containing a Cycloid as a Geodesic

As a first example, let us find a minimal surface that contains a cycloid as a geodesic. We use

```
geomincurve[cycloid[a,a]][z]
```

to find the parametrization of the minimal curve:

```
{a z - a Sin[z], a - a Cos[z], -4 I a Cos[z/2]}
```

Also,

```
analytictominimal[geomincurve[cycloid[a,a]]][0][u,v]
```

gives us the parametrization of the corresponding minimal surface:

```
{u - Cosh[v] Sin[u], 1 - Cos[u] Cosh[v], -4 Sin[u/2] Sinh[v/2]}
```

Thus we see that the minimal surface which contains the cycloid

$$t \longmapsto (t - \sin t, 1 - \cos t)$$

as a pregeodesic is precisely Catalan's minimal surface, as defined on page 693.

A Minimal Surface Containing an Astroid as a Geodesic

Next, we find a minimal surface containing the astroid

$$x^{2/3} + y^{2/3} = \left(\frac{8}{3}\right)^{2/3}$$

as a geodesic. The minimal curve is given by

33.3 Examples of Minimal Surfaces Constructed from Plane Curves

`geomincurve[astroid[3,8/3,8/3]][z]`

$$\{\frac{8 \cos[z]^3}{3}, \frac{8 \sin[z]^3}{3}, -2 I \cos[2 z]\}$$

and the corresponding minimal surface is given by

`analytictominimal[geomincurve[`
`    astroid[3,8/3,8/3]]][0][u,v]//Simplify`

$$\{2 \cos[u] \cosh[v] + \frac{2 \cos[3 u] \cosh[3 v]}{3},$$
$$2 \cosh[v] \sin[u] - \frac{2 \cosh[3 v] \sin[3 u]}{3}, -2 \sin[2 u] \sinh[2 v]\}$$

This surface is a rotation of a conjugate of Henneberg's minimal surface.

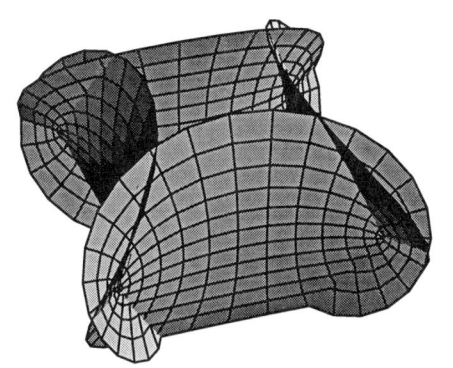

Minimal surface containing an astroid as a geodesic

A Minimal Surface Containing an Ellipse as a Geodesic

It is clear that a catenoid contains a circle as a geodesic, but what is the minimal surface that contains an ellipse as a geodesic? (See [Schwa, page 182].) Unfortunately, we cannot find a formula for the arc length of an ellipse in terms of elementary functions. However, versions of *Mathematica* 2.1 and greater have a package for the symbolic computation of certain integrals in terms of elliptic functions; it can be read in with the command

```
Needs["Calculus`EllipticIntegrate`"]
```

Then

```
arclength[ellipse[a,b]][t]//PowerExpand
```

tells us the formula for the arc length of an ellipse:

$$b \text{ EllipticE}[t, \frac{-a^2 + b^2}{b^2}]$$

```
ellipsemincurve[a_,b_][z_] := {a Cos[z],
                               b Sin[z],
                 I b EllipticE[z - a^2/b^2]}
```

Therefore, we define a 1-parameter family of minimal surfaces by

```
ellipsemin[a_,b_][t_][u_,v_] :=
   {a Cos[t]Cos[u]Cosh[v] - a Sin[t]Sin[u]Sinh[v],
    b Cos[t]Sin[u]Cosh[v] + b Sin[t]Cos[u]Sinh[v],
    -b(Cos[t]Im[EllipticE[u + I v,1 - a^2/b^2]] -
       Sin[t]Re[EllipticE[u + I v,1 - a^2/b^2]])}
```

33.3 Examples of Minimal Surfaces Constructed from Plane Curves

An elliptical catenoid is `ellipsemin[a,b][0]`.

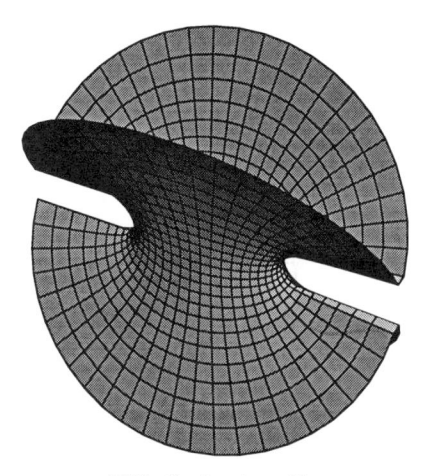

Elliptical catenoid

The conjugate minimal surface of the elliptical catenoid is an analog of the helicoid; in *Mathematica's* notation it is `ellipsemin[a,b][Pi/2]`.

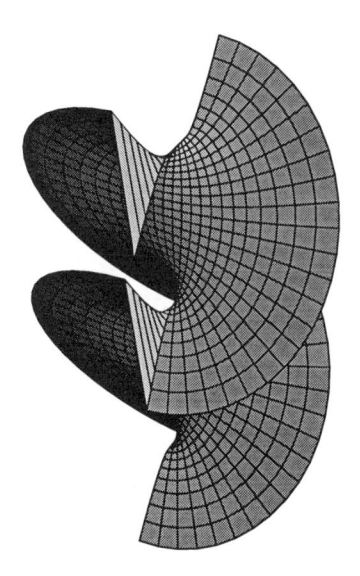

Elliptical helicoid

A Minimal Surface Containing a Lemniscate as a Geodesic

In a similar fashion, we can use the package **EllipticIntegrate** to find a minimal surface containing the lemniscate

$$t \longmapsto \left(\frac{a \cos t}{1 + \sin^2 t}, \frac{a \sin t \cos t}{1 + \sin^2 t} \right)$$

as a pregeodesic. The resulting minimal curve is given by

```
lemniscatemincurve[a_][z_]:=
    {(a Cos[z])/(1 + Sin[z]^2),
     (a Cos[z]Sin[z])/(1 + Sin[z]^2),
     I a EllipticF[z - 1]}
```

and the corresponding family of minimal surfaces is given by

```
lemniscatemin[a_][t_][u_,v_]:=
    {(-16 a Cos[u]Cosh[v] Sin[u]Sinh[v]
    (Cos[u]Cosh[v]Sin[t] + Cos[t]Sin[u]Sinh[v]))/
    (18 + Cos[4 u] - 12 Cos[2 u] Cosh[2 v] + Cosh[4v]) +
    (4a(3 - Cos[2u]Cosh[2v])
    (Cos[t]Cos[u]Cosh[v] - Sin[t]Sin[u]Sinh[v]))/
    (18 + Cos[4u] - 12Cos[2u]Cosh[2v] + Cosh[4v]),
    (a(-Sin[t - 4u] + 6Cosh[2v]Sin[t - 2u] -
    6Cosh[2v]Sin[t + 2u] + Sin[t + 4u] -
    6Sin[t - 2u]Sinh[2v] -
    6Sin[t + 2u]Sinh[2v] + 2Sin[t]Sinh[4v]))
    (2(-18 - Cos[4u] + 12Cos[2u]Cosh[2v] - Cosh[4v])),
    a(-Cos[t]Im[EllipticF[u + I v,-1]] +
    Sin[t]Re[EllipticF[u + I v,-1]])}
```

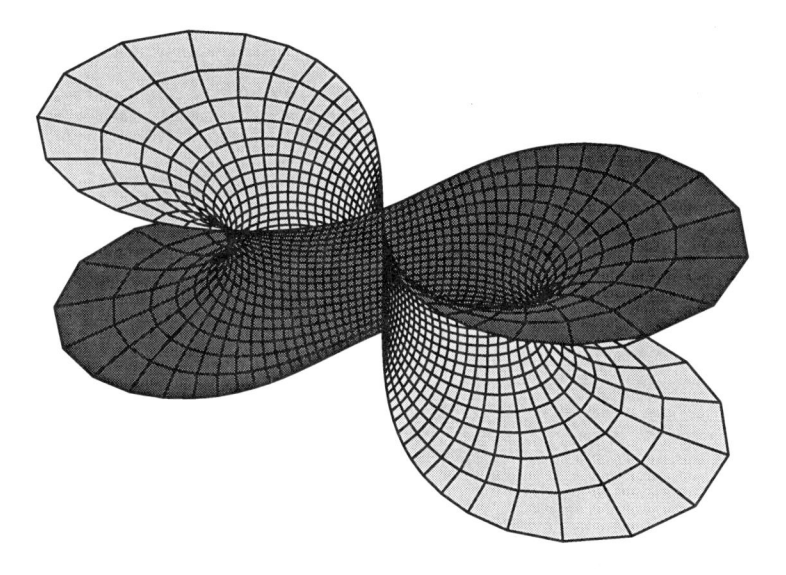

**Minimal surface containing a
lemniscate as a geodesic**

A Minimal Surface Containing a Clothoid as a Geodesic

Since the clothoid is one of the most elegant curves, it is no surprise that it gives rise to a beautiful minimal surface. The minimal curve constructed from the clothoid is given by

```
clothoidmincurve[a_][z_]:=
        {a Sqrt[Pi]FresnelS[z/Sqrt[Pi]],
         a Sqrt[Pi]FresnelC[z/Sqrt[Pi]],
         I a z}
```

and the family of minimal surfaces is given by

```
clothmin[a_][t_][u_,v_]:=
  {a Sqrt[Pi]Cos[t]Re[FresnelS[(u + I v)/Sqrt[Pi]]] +
   a Sqrt[Pi]Im[FresnelS[(u + I v)/Sqrt[Pi]]]Sin[t],
   a Sqrt[Pi]Cos[t]Re[FresnelC[(u + I v)/Sqrt[Pi]]] +
   a Sqrt[Pi]Im[FresnelC[(u + I v)/Sqrt[Pi]]]Sin[t],
   -(a v Cos[t]) + a u Sin[t]}
```

We draw $(u,v) \longmapsto$ **clothmin**[1][0](u,v), suppressing the polygonal edges using the command **dip** described on page 474.

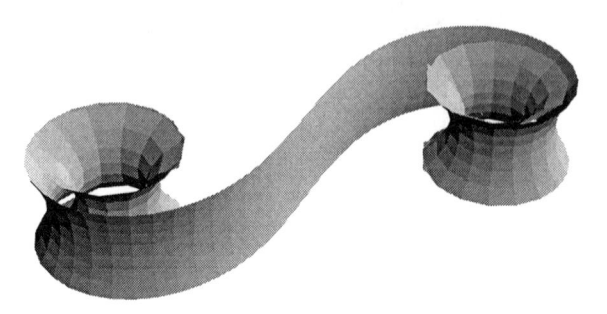

Minimal surface containing a clothoid as a geodesic

33.4 Exercises

1. Draw a minimal surface that contains a hyperbola as a geodesic.

2. Draw a minimal surface that contains a cardioid as a geodesic.

3. Draw a minimal surface that contains a nephroid as a geodesic.

34
CONSTRUCTION OF SURFACES

In this chapter we describe several methods of constructing new surfaces from old. In Section 34.1 we show how to construct the parallel surface $\mathcal{M}(t)$ to a given surface $\mathcal{M} \subseteq \mathbb{R}^3$. Parallel surfaces to a Möbius strip are orientable, as we see in Section 34.2. The shape operator $S(t)$ of $\mathcal{M}(t)$ is expressed in terms of the shape operator S of $\mathcal{M}$ in Section 34.3. In Section 34.4 we use the parallel surface construction to find a triply orthogonal system containing any given surface.

Pedal surfaces are discussed in Section 34.5; the parallel and pedal surfaces to a given surface are the surface analogs of parallel curve and pedal curve constructions that we gave in Sections 5.7 and 5.8. A twisting construction which generalizes both the Möbius strip and the Klein bottle is described in Section 34.6.

34.1 Parallel Surfaces

In Section 5.7 we constructed a parallel curve from a given plane curve. There is a similar definition of parallel surface.

Definition. *Let $\mathcal{M} \subset \mathbb{R}^3$ be a regular surface. The* **surface parallel to $\mathcal{M}$ at a distance** $t > 0$ *is the set*

$$\mathcal{M}(t) = \{\, \mathbf{q} \in \mathbb{R}^3 \,\big|\, \operatorname{distance}(\mathbf{q}, \mathcal{M}) = t \,\}.$$

If $\mathcal{M}$ is connected and orientable, then $\mathcal{M}(t)$ will consist of two components. (See, for example, [Gray1].) If t is large, $\mathcal{M}(t)$ may not be a regular surface, as the following picture shows.

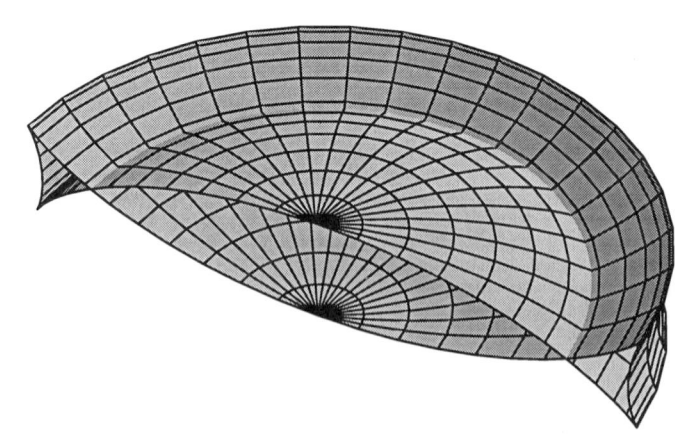

Parallel surface to the ellipsoid
$$\frac{4x^2}{9} + \frac{4y^2}{9} + z^2 = 1$$

However, for small t we prove that $\mathcal{M}(t)$ is a regular surface. First, let us define the notion of parallel patch. The idea is to move the patch a distance t along its normal $\mathbf{U}$. We want to be able to go in either direction along the normal, so we allow t to be either positive or negative.

Definition. *Let* $\mathbf{x}:\mathcal{U} \longrightarrow \mathbb{R}^3$ *be a regular patch. Then the* **patch parallel to x at distance** t *is the patch given by*

(34.1) $$\mathbf{x}[t](u,v) = \mathbf{x}(u,v) + t\,\mathbf{U}(u,v),$$

where
$$\mathbf{U} = \frac{\mathbf{x}_u \times \mathbf{x}_v}{\|\mathbf{x}_u \times \mathbf{x}_v\|}.$$

We give conditions for the patch parallel to $\mathbf{x}$ at distance t to be regular.

Lemma 34.1. *Let* $\mathcal{M} \subset \mathbb{R}^3$ *be a regular surface for which there exists a single regular patch* $\mathbf{x}:\mathcal{U} \longrightarrow \mathcal{M}$. *Also, assume that there is a number* $t_0 > 0$ *such that*

(34.2) $$\det(I - t\,S) > 0 \quad \text{for } |t| < t_0 \text{ on } \mathcal{M}.$$

34.1 Parallel Surfaces

Then for $|t| < t_0$, $\mathbf{x}[t]$ is a regular patch on $\mathcal{M}(t)$.

Proof. We have
$$\mathbf{x}[t]_u = \mathbf{x}_u + t\,\mathbf{U}_u = (I - t\,S)\mathbf{x}_u,$$

and similarly for $\mathbf{x}[t]_v$. Therefore,

(34.3) $\quad \mathbf{x}[t]_u \times \mathbf{x}[t]_v = (I - t\,S)\mathbf{x}_u \times (I - t\,S)\mathbf{x}_v = \det(I - t\,S)\mathbf{x}_u \times \mathbf{x}_v.$

Hence $\mathbf{x}[t]_u \times \mathbf{x}[t]_v$ is nonzero, and therefore $\mathbf{x}[t]$ is regular. Thus $\mathcal{M}(t)$ is a regular surface, since it is entirely covered by the regular patch $\mathbf{x}[t]$. ∎

Next, we consider parallel surfaces to arbitrary regular surfaces.

Lemma 34.2. *Let $\mathcal{M} \subset \mathbb{R}^3$ be a regular surface and let $\mathcal{M}(t)$ denote the surface parallel to $\mathcal{M}$ at a distance t. Assume that there is a number $t_0 > 0$ such that (34.2) holds. Then for $|t| < t_0$ the parallel surface $\mathcal{M}(t)$ is a regular surface.*

Proof. Let $\mathbf{x}: \mathcal{U} \longrightarrow \mathcal{M}$ be a regular injective patch on $\mathcal{M}$ and define a patch $\mathbf{x}[t]$ by (34.1), and define $\mathbf{y}: \mathcal{U} \times \{\,t \mid |t| < t_0\,\} \longrightarrow \mathbb{R}^3$ by

$$\mathbf{y}(u, v, t) = \mathbf{x}(u, v) + t\,\mathbf{U}(u, v).$$

The Jacobian matrix of $\mathbf{y}$ is

$$\mathcal{J}(\mathbf{y}) = \begin{pmatrix} \mathbf{U} \\ (I - t\,S)\mathbf{x}_u \\ (I - t\,S)\mathbf{x}_v \end{pmatrix}.$$

It is clear that $\det\bigl(\mathcal{J}(\mathbf{y})(t, u, v)\bigr) \neq 0$ for $(t, u, v) \in \mathcal{U} \times \{\,t \mid |t| < t_0\,\}$. Hence $\mathbf{y}$ is injective on a neighborhood of each point in $\mathcal{U} \times \{\,t \mid |t| < t_0\,\}$. In particular, $\mathbf{x}[t]$ is a regular injective patch on a neighborhood of each point of $\mathcal{U}$. Patches of the form $\mathbf{x}[t]$ cover $\mathcal{M}(t)$, so $\mathcal{M}(t)$ is a regular surface. ∎

The *Mathematica* definition of the parallel patch is

```
parsurf[x_][r_][u_,v_] := Module[{xu,xv,n1,n2,n3,n4},
    xu = D[x[uu,vv],uu];
    xv = D[x[uu,vv],vv];
    n1 = cross[xu, xv]//Simplify;
    n2 = Simplify[Factor[n1.n1]];
    n3 = PowerExpand[Sqrt[n2]]//Simplify;
    n4 = Simplify[n1/n3];
    Simplify[Together[x[uu,vv] +
        r n4]]] /. {uu->u,vv->v}
```

For example, let us read in **CAS.m** then query *Mathematica* with

```
parsurf[catenoid[a]][t][u,v]//PowerExpand
```

We get the response

$$\{a \cos[u] \cosh[\tfrac{v}{a}] + t \cos[u] \text{Sech}[\tfrac{v}{a}],$$
$$a \cosh[\tfrac{v}{a}] \sin[u] + t \text{Sech}[\tfrac{v}{a}] \sin[u], v - t \tanh[\tfrac{v}{a}]\}$$

In order to plot such a surface efficiently, we define a new function by

```
parsurf[catenoid[a_]][t_][u_,v_] :=
    {a Cos[u]Cosh[v/a] + t Cos[u]Sech[v/a],
     a Cosh[v/a]Sin[u] + t Sech[v/a]Sin[u],
     v - t Tanh[v/a]}
```

Here is the plot of a catenoid and two of its parallel surfaces.

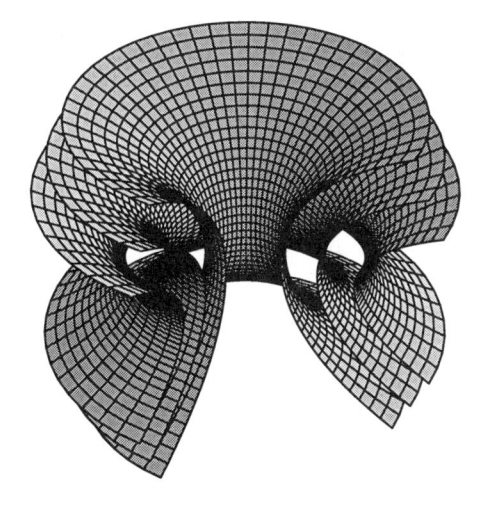

34.2 Parallel Surfaces to a Möbius Strip

```
ParametricPlot3D[
{parsurf[
catenoid[1]][-0.5][u,v],
catenoid[1][u,v],
parsurf[
catenoid[1]][0.5][u,v]}
//Evaluate,
{u,0,3Pi/2},
{v,-1.5,1.5},
PlotPoints->{40,40},
Axes->None,
Boxed->False];
```

**A catenoid and
two of its parallel surfaces**

34.2 Parallel Surfaces to a Möbius Strip

An informative picture of a Möbius strip can be obtained using the graphics command **FaceForm** to color different sides of the polygons that *Mathematica* uses to approximate a surface. Thus, for example,

```
ParametricPlot3D[Evaluate[Append[
moebiusstrip[1][u,v],FaceForm[Red,Yellow]]],
{u,0,2Pi},{v,-0.3,0.3},PlotPoints->{40,5},
Axes->None,Boxed->False,Lighting->False];
```

yields

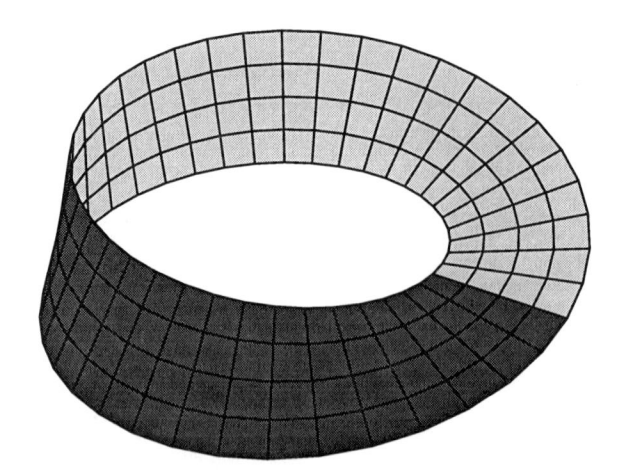

Colored Möbius strip

Since the Möbius strip is nonorientable, any attempt to cover all the backs and fronts of polygons consistently must fail; in other words, the two colors must meet.

On the other hand, since a parallel surface to a Möbius strip is orientable, it is possible to color its back and front with different colors in order to distinguish the two sides of the surface parallel to the Möbius strip. The four-color version of a Möbius strip and one of its parallel surfaces is generated by the command

```
Show[ParametricPlot3D[Evaluate[Append[
moebiusstrip[1][u,v],FaceForm[Red,Yellow]]],
{u,0,2Pi},{v,-0.3,0.3},PlotPoints->{40,5},
DisplayFunction->Identity],
ParametricPlot3D[Evaluate[
Append[parsurf[moebiusstrip[1]][0.3][u,v],
FaceForm[Green,Blue]]],
{u,0,4Pi},{v,-0.3,0.3},PlotPoints->{80,5},
DisplayFunction->Identity],Axes->None,Boxed->False,
Lighting->False,DisplayFunction->$DisplayFunction];
```

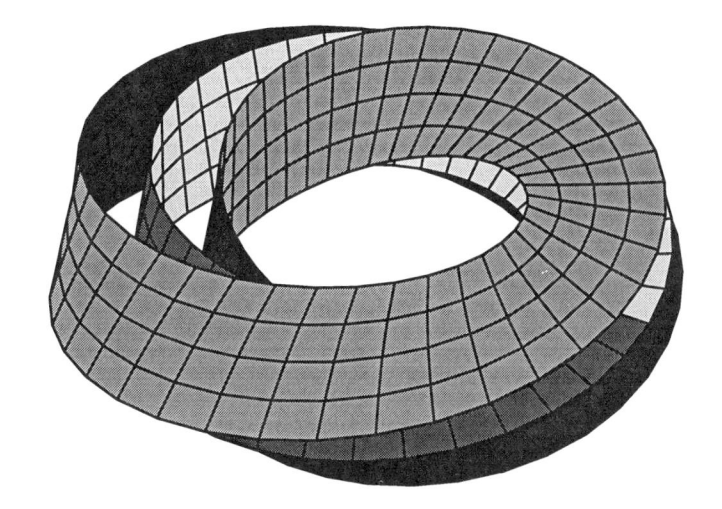

**Möbius strip and
parallel surface**

34.3 The Shape Operator of a Parallel Surface

First, we show that the shape operator of a regular surface $\mathcal{M} \subseteq \mathbb{R}^3$ determines the shape operator of each of its parallel surfaces.

Lemma 34.3. *Let $\mathcal{M} \subset \mathbb{R}^3$ be a regular surface and let $\mathcal{M}(t)$ denote the surface parallel to $\mathcal{M}$ at a distance t. Assume that condition (34.2) holds. Let S be the shape operator of $\mathcal{M}$ and $S(t)$ the shape operator of $\mathcal{M}(t)$. Then*

(i) *the matrix-valued function $t \longmapsto S(t)$ satisfies the differential equation*

$$S'(t) = S(t)^2;$$

(ii) $S(t) = S(I - tS)^{-1};$

(iii) *the principal curvatures of $\mathcal{M}(t)$ are given by*

(34.4) $$\mathbf{k}_i(t) = \frac{\mathbf{k}_i}{1 - t\,\mathbf{k}_i},$$

for $i = 1, 2$, where $\mathbf{k}_1$ and $\mathbf{k}_2$ are the principal curvatures of $\mathcal{M}$.

(iv) *The Gaussian curvature $K(t)$ and the mean curvature $H(t)$ of $\mathcal{M}(t)$ are given*

(34.5) $$K(t) = \frac{K}{1 - 2t\,H + t^2 K}$$

and

(34.6) $$H(t) = \frac{H - t\,K}{1 - 2t\,H + t^2 K},$$

where K and H denote the Gaussian and mean curvatures of $\mathcal{M}$.

Proof. Without loss of generality, $\mathcal{M}$ is parametrized by a single regular patch $\mathbf{x}\colon U \longrightarrow \mathcal{M}$. Then the patch $\mathbf{x}[t]$ on $\mathcal{M}(t)$ defined by (34.1) is regular. Let $\mathbf{U}[t]$ be defined by

$$\mathbf{U}[t] = \frac{\mathbf{x}[t]_u \times \mathbf{x}[t]_v}{\|\mathbf{x}[t]_u \times \mathbf{x}[t]_v\|}.$$

Since $\det(I - tS) > 0$, (34.3) implies that $\mathbf{U}[t]$ coincides with $\mathbf{U}$. Therefore,

(34.7) $$S(t)\mathbf{x}[t]_u = -\mathbf{U}[t]_u = -\mathbf{U}_u = S\mathbf{x}_u,$$

proving that the function $t \longmapsto S(t)\mathbf{x}[t]_u$ does not depend on t. Hence

(34.8) $$\bigl(S(t)\mathbf{x}[t]_u\bigr)' = 0,$$

where the prime denotes differentiation with respect to t. Also, from (34.1) we obtain

(34.9) $$\mathbf{x}[t]_u' = \mathbf{U}_u = -S\mathbf{x}_u.$$

Then (34.7)–(34.9) imply that

$$S'(t)\mathbf{x}[t]_u = -S(t)\mathbf{x}[t]_u' = S(t)S\mathbf{x}_u = S(t)^2\mathbf{x}[t]_u.$$

Similarly, $S'(t)\mathbf{x}[t]_v = S(t)^2\mathbf{x}[t]_v$. Equation (34.3) implies that $\mathbf{x}[t]_u$ and $\mathbf{x}[t]_v$ are linearly independent. Hence $S'(t) = S(t)^2$, proving (i).

To prove (ii), we define $\tilde{S}$ by

(34.10) $$\tilde{S}(t) = S(I - tS)^{-1} = \sum_{k=0}^{\infty} t^k S^{k+1}.$$

34.3 The Shape Operator of a Parallel Surface

It is clear from (34.10) that S commutes with $\tilde{S}(t)$ for all t, and so

$$\tilde{S}'(t) = \frac{S^2}{(I-tS)^2} = \tilde{S}(t)^2.$$

Since $\tilde{S}(0) = S$, we conclude that $\tilde{S}$ coincides with S.

For (iii) we compute the eigenvalues of $S(I-tS)^{-1}$ and use (ii). To prove (iv), we use (34.4) to compute

$$K(t) = \mathbf{k}_1(t)\mathbf{k}_2(t) = \frac{\mathbf{k}_1\mathbf{k}_2}{(1-t\mathbf{k}_1)(1-t\mathbf{k}_2)} = \frac{\mathbf{k}_1\mathbf{k}_2}{1-t(\mathbf{k}_1+\mathbf{k}_2)+t^2\mathbf{k}_1\mathbf{k}_2}$$

$$= \frac{K}{1-2tH+t^2K}.$$

and

$$2H(t) = \mathbf{k}_1(t) + \mathbf{k}_2(t) = \frac{\mathbf{k}_1}{1-t\mathbf{k}_1} + \frac{\mathbf{k}_2}{1-t\mathbf{k}_2} = \frac{\mathbf{k}_1(1-t\mathbf{k}_2) + \mathbf{k}_2(1-t\mathbf{k}_1)}{(1-t\mathbf{k}_1)(1-t\mathbf{k}_2)}$$

$$= \frac{\mathbf{k}_1+\mathbf{k}_2 - 2t\mathbf{k}_1\mathbf{k}_2}{(1-t\mathbf{k}_1)(1-t\mathbf{k}_2)} = \frac{2H-2tK}{1-2tH+t^2K}. \quad\blacksquare$$

Now we can prove an important result due to Bonnet. As Chern has remarked (see [Chern3]), Bonnet's Theorem implies that finding surfaces of constant mean curvature is equivalent to finding surfaces of constant positive curvature.

Theorem 34.4. (Bonnet) *Let $\mathcal{M} \subset \mathbb{R}^3$ be a regular surface.*

(i) *If $\mathcal{M}$ has constant mean curvature $H \equiv 1/(2c)$, then the parallel surface $\mathcal{M}(c)$ has constant Gaussian curvature $K(c) \equiv 1/c^2$, and the parallel surface $\mathcal{M}(2c)$ has constant mean curvature $H(2c) \equiv -1/(2c)$.*

(ii) *If $\mathcal{M}$ has constant Gaussian curvature $K \equiv c^2$, then the parallel surfaces $\mathcal{M}(\pm c)$ have constant mean curvature $H(\pm c) \equiv \mp 1/(2c)$.*

Proof. If $H \equiv 1/(2c)$, then it follows from (34.5), that

$$K(c) = \frac{K}{1-2/(2c)+Kc^2} = \frac{1}{c^2}.$$

The rest of (i) and (ii) follow from (34.5) in a similar fashion (see exercise 1). $\blacksquare$

Next, we prove a partial converse to Theorem 34.4.

Theorem 34.5. *Let $\mathcal{M} \subset \mathbb{R}^3$ be a regular surface with constant positive curvature a^{-2}, where $a > 0$. Let $\mathcal{M}(t)$ denote the surface parallel to $\mathcal{M}$ at a distance t. Suppose that the umbilic points of $\mathcal{M}$ are isolated. If $\mathcal{M}(t)$ has constant mean curvature, then $t = \pm a$.*

Proof. Fix t, and suppose that $H(t)$ is constant on $\mathcal{M}(t)$. Then (34.6) implies that

$$\mathbf{k}_1(1 - t\,\mathbf{k}_2) + \mathbf{k}_2(1 - t\,\mathbf{k}_1) = 2H(t)(1 - t\,\mathbf{k}_1)(1 - t\,\mathbf{k}_2),$$

or

$$\mathbf{k}_1 + \mathbf{k}_2 - 2t\,a^{-2} = 2H(t)\bigl(1 - t(\mathbf{k}_1 + \mathbf{k}_2) + t^2 a^{-2}\bigr).$$

Hence
(34.11) $\quad (\mathbf{k}_1 + \mathbf{k}_2)\bigl(1 + 2t\,H(t)\bigr) = 2\,H(t) + 2\,H(t)t^2 a^{-2} + 2t\,a^{-2}.$

By hypothesis, the right-hand of (34.11) is constant. But if the left-hand of (34.11) is constant, it must vanish at the nonumbilic points of $\mathcal{M}(t)$. Hence

(34.12) $\qquad\qquad 1 + 2t\,H(t) = 0$

at the nonumbilic points of $\mathcal{M}(t)$. Then (34.11) and (34.12) imply that

$$0 = 2\,H(t) + 2\,H(t)t^2 a^2 + 2t\,a^{-2} = 2\,H(t)(1 + t^2 a^{-2}) + 2t\,a^{-2}$$

$$= -\frac{1}{t}\left(1 + \frac{t^2}{a^2}\right) + \frac{2t}{a^2} = -\frac{1}{t} + \frac{t}{a^2}.$$

Therefore, $t = \pm a$. ∎

34.4 A General Construction of a Triply Orthogonal System

In this section we show that any principal patch in $\mathbb{R}^3$ is part of a triply orthogonal system of surfaces. We first prove

Theorem 34.6. *A necessary and sufficient condition that a curve on a surface be a principal curve is that the surface normals along the curve form a flat ruled surface.*

34.4 A General Construction of a Triply Orthogonal System

Proof. Let $\beta:(a,b) \longrightarrow \mathbb{R}^3$ be a unit-speed curve and a surface $\mathcal{M}$. Without loss of generality, we can assume that $\mathcal{M}$ is oriented with unit normal $\mathbf{U}$. Let $\mathbf{T}$ denote the unit-tangent vector of β, and let $\mathcal{N}$ be the surface formed by the normals along β. We parametrize $\mathcal{N}$ as

$$\mathbf{y}(s,v) = \mathbf{x}(s) + v\,\mathbf{U};$$

then we compute

(34.13) $\qquad \mathbf{y}_s = \mathbf{T} + \mathbf{U}_s, \qquad \mathbf{y}_v = \mathbf{U}, \qquad \mathbf{y}_{sv} = \mathbf{U}_s, \qquad \mathbf{y}_{vv} = 0.$

According to Corollary 16.22, page 379, and the last equation of (34.13), the Gaussian curvature of $\mathcal{N}$ is given by

$$K = \frac{(\mathbf{y}_{ss}\mathbf{y}_s\mathbf{y}_v)(\mathbf{y}_{vv}\mathbf{y}_s\mathbf{y}_v) - (\mathbf{y}_{sv}\mathbf{y}_s\mathbf{y}_v)^2}{\left(\|\mathbf{y}_s\|^2\|\mathbf{y}_v\|^2 - (\mathbf{y}_s\cdot\mathbf{y}_v)^2\right)^2} = \frac{-(\mathbf{y}_{sv}\mathbf{y}_s\mathbf{y}_v)^2}{\left(\|\mathbf{y}_s\|^2\|\mathbf{y}_v\|^2 - (\mathbf{y}_s\cdot\mathbf{y}_v)^2\right)^2}.$$

Thus $K = 0$ if and only if $(\mathbf{y}_{sv}\mathbf{y}_s\mathbf{y}_v) = 0$. From (34.13) we have

(34.14) $\qquad (\mathbf{y}_{sv}\mathbf{y}_s\mathbf{y}_v) = \mathbf{U}_s \times (\mathbf{T} + \mathbf{U}_s)\cdot\mathbf{U} = \mathbf{U}_s \times \mathbf{T}\cdot\mathbf{U}.$

Since both $\mathbf{T}$ and $\mathbf{U}_s$ are perpendicular to $\mathbf{U}$, equation (34.14) implies that $K = 0$ if and only if

(34.15) $\qquad 0 = \mathbf{U}_s \times \mathbf{T} = -S\mathbf{T} \times \mathbf{T},$

where S denotes the shape operator of $\mathcal{M}$. It follows from Lemma 20.1, page 459, that equation (34.15) is precisely the condition that β be a principal curve of $\mathcal{M}$. ∎

We are now able to construct the triply orthogonal system containing a given principal patch.

Theorem 34.7. *Let* $\mathbf{x}:\mathcal{U} \longrightarrow \mathbb{R}^3$ *be a principal patch. Define a curvilinear patch* $\mathbf{y}:\mathcal{U} \times \mathbf{R} \longrightarrow \mathbb{R}^3$ *by*

$$\mathbf{y}(u,v,w) = \mathbf{x}(u,v) + w\,\mathbf{U},$$

where $\mathbf{U}$ *is a unit normal to* $\mathbf{x}$. *Put*

$$\mathfrak{A} = \{\,(u,v) \longmapsto \mathbf{y}(u,v,w)\,\},$$

$$\mathfrak{B} = \{\,(u,w) \longmapsto \mathbf{y}(u,v,w)\,\},$$

$$\mathfrak{C} = \{\,(v,w) \longmapsto \mathbf{y}(u,v,w)\,\};$$

then $\mathfrak{A}$, $\mathfrak{B}$, $\mathfrak{C}$ *form a triply orthogonal system of surfaces. Each surface in* $\mathfrak{B}$ *and* $\mathfrak{C}$ *is a flat ruled surface and each surface in* $\mathfrak{A}$ *is a surface parallel to* $\mathbf{x}$.

Proof. Let $\mathbf{k}_1$ and $\mathbf{k}_2$ be the principal curvatures corresponding to $\mathbf{x}_u$ and $\mathbf{x}_v$. Then we have

$$\begin{cases} \mathbf{y}_u = \mathbf{x}_u + w\,\mathbf{U}_u = (1 - w\,\mathbf{k}_1)\mathbf{x}_u, \\ \mathbf{y}_v = \mathbf{x}_v + w\,\mathbf{U}_v = (1 - w\,\mathbf{k}_2)\mathbf{x}_v, \\ \mathbf{y}_w = \mathbf{U}. \end{cases}$$

Hence $\mathbf{y}_u, \mathbf{y}_v$, and $\mathbf{y}_w$ are mutually orthogonal, so that $\mathbf{y}$ is a triply orthogonal patch. Now $\mathbf{y}$ determines a triply orthogonal system of surfaces by Lemma 29.1, page 665. The last statement is a consequence of Theorem 34.6. ∎

The *Mathematica* command to create the triply orthogonal patch from a principal patch **x** by means of Theorem 34.7 is simply **parsurf[x][w][u,v]**. Here is an example.

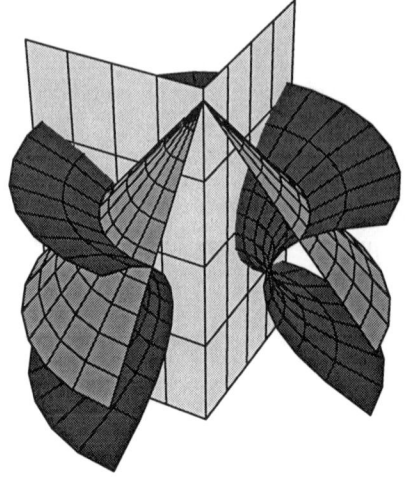

A triply orthogonal system consisting of surfaces parallel to a catenoid, cones and planes

34.5 Pedal Surfaces

There is a surface analog of pedal curve. It consists of the feet of the perpendiculars from a fixed point to the tangent spaces of a surface. More precisely:

34.5 Pedal Surfaces

Definition. *The* **pedal surface** *of a patch* $\mathbf{x}: U \longrightarrow \mathbb{R}^3$ *with respect to a point* $\mathbf{p} \in \mathbb{R}^3$ *is defined by*

$$\mathbf{pedsurf}[\mathbf{p}, \mathbf{x}](u, v) = \mathbf{p} + \left(\frac{(\mathbf{x}(u,v) - \mathbf{p}) \cdot \mathbf{x}_u \times \mathbf{x}_v}{\|\mathbf{x}_u \times \mathbf{x}_v\|^2} \right) \mathbf{x}_u \times \mathbf{x}_v.$$

The corresponding *Mathematica* definition is

```
pedsurf[{a_,b_,c_},x_][u_,v_]:=
Module[{xu,xv,n1,s2,s3,n4},
    xu = Derivative[1,0][x][u,v];
    xv = Derivative[0,1][x][u, v];
    n1 = cross[xu, xv]//Simplify;
    s2 = Simplify[Factor[n1.n1]];
    s3 = PowerExpand[Sqrt[s2]]//Simplify;
    n4 = Simplify[n1/s3];
    {a,b,c} + Simplify[Together[((x[u,v]
        - {a,b,c}).n4)n4]]]
```

Here is the pedal surface to the ellipsoid

$$\frac{x^2}{a^2} + \frac{y^2}{b^2} + \frac{z^2}{c^2} = 1$$

with respect to the origin:

```
ParametricPlot3D[
pedsurf[{0,0,0},
ellipsoid[1,2,3]][u,v]
//Evaluate,
{u,0,3Pi/2},
{v,-Pi/2,Pi/2},
Boxed->False,
Axes->None,
PlotPoints->{30,60}];
```

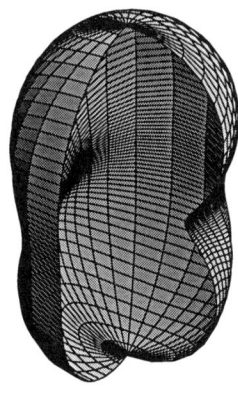

Pedal surface of an ellipsoid

34.6 Twisted Surfaces

In this section we define a class of twisted surfaces that generalize the Klein bottle and Möbius strip.

Definition. *Let α be a plane curve with the property that*

$$\alpha(-t) = -\alpha(t)$$

and write $\alpha(t) = \big(\varphi(t), \psi(t)\big)$. Then the **twisted surface with profile curve α** *is defined by*

$$\mathrm{twist}[a, b, \alpha](u, v) =$$

$$\big(a + \cos(b\,u)\varphi(v) - \sin(b\,u)\psi(v)\big)(\cos u, \sin u, 0)$$

$$+ \big(\sin(b\,u)\psi(v) + \cos(b\,u)\psi(v)\big)(0, 0, 1).$$

The *Mathematica* version is

```
twist[a_,b_,alpha_][u_,v_]:=
  (a + Cos[b u]alpha[v][[1]] - Sin[b u]alpha[v][[2]])*
  {Cos[u],Sin[u],0} +
  (Sin[b u]alpha[v][[1]] + Cos[b u]alpha[v][[2]])*
  {0,0,1}
```

We show that two of the surfaces that we considered in Chapter 14 can be constructed using `twist`.

Examples

The Möbius Strip

Define α by $\alpha(t) = (t, 0)$. Then

$$\mathrm{twist}[a, 1/2, \alpha](u, v) = a\bigg(\cos u + v \cos\left(\frac{u}{2}\right)\cos u,$$

$$\sin u + v\cos\left(\frac{u}{2}\right)\sin u, v \sin\left(\frac{u}{2}\right)\bigg)$$

$$= \mathbf{moebiusstrip}[a](u, v).$$

The Klein Bottle

Define γ by $\gamma(t) = (\sin t, \sin 2t)$. Then

$$\mathbf{twist}[a, 1/2, \gamma](u,v) = \left(\left(a + \cos\left(\frac{u}{2}\right)\sin v - \sin\left(\frac{u}{2}\right)\sin 2v\right)\cos u, \right.$$

$$\left. \left(a + \cos\left(\frac{u}{2}\right)\sin v - \sin\left(\frac{u}{2}\right)\sin 2v\right)\sin u, \sin\left(\frac{u}{2}\right)\sin v + \cos\left(\frac{u}{2}\right)\sin 2v \right)$$

$$= \mathbf{kleinbottle}[a](u,v).$$

The following lemma can be proved using *Mathematica*.

Lemma 34.8. *Let* **x** *be a twisted surface in* $\mathbb{R}^3$ *whose profile curve is* $\alpha = (\varphi, \psi)$. *Then*

$$E = a^2 + 2a\cos(bu)\varphi(v) + \varphi(v)^2\cos^2(bu) + \psi(v)^2\sin^2(bu)$$

$$+ b^2\left(\varphi(v)^2 + \psi(v)^2\right) + 2a\psi(v)\sin(bu) - \varphi(v)\psi(v)\sin(2bu),$$

$$F = -b\,\psi\varphi' + \varphi\psi',$$

$$G = \varphi'^2 + \psi'^2,$$

and $EG - F^2 = \varphi^2(\varphi'^2 + \psi'^2) + c^2\varphi'^2$. *Thus* **x** *is regular wherever* $\varphi^2(\varphi'^2 + \psi'^2) + c^2\varphi'^2$ *is nonzero.*

Twisted Surfaces of Lissajous Curves

Instead of twisting a figure eight around a circle we can twist a Lissajous curve:

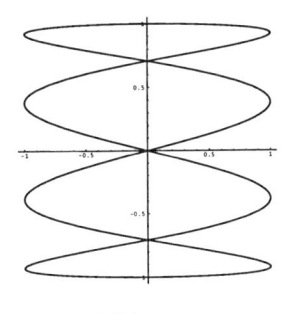

```
ParametricPlot[
lissajous[4,0,1,1][t]
//Evaluate,
{t,0,2Pi},
AspectRatio->Automatic];
```

A Lissajous curve

```
ParametricPlot3D[
twist[2,1/2,
lissajous[4,0,1,1]][u,v]
//Evaluate,
{u,-Pi/8,3Pi/2},
{v,0,2Pi},
Boxed->False,
Axes->None,
PlotPoints->{30,80}];
```

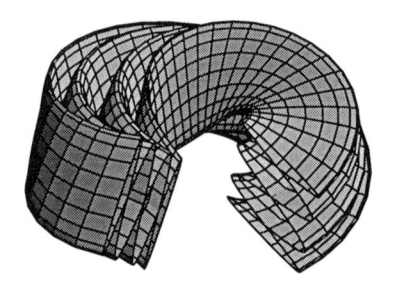

Twisted surface of a Lissajous curve

34.7 Exercises

1. Complete the proof of Theorem 34.4.

2. Plot the pedal surface with respect to the origin of `torus[8,3,3]`.

3. Use *Mathematica* to show that the surface of revolution generated by a parallel curve β to a curve α is the same as the parallel surface **y** to the surface of revolution **x** generated by α. To make this work, the signed distances between the parallel curves α and β should be the negative of the signed distances between the parallel surfaces **x** and **y**.

4. Use *Mathematica* to show that the surface of revolution generated by a pedal curve γ to a curve α is the same as the pedal surface to the surface of revolution generated by α.

5. Prove Lemma 34.8 using *Mathematica*.

35

CANAL SURFACES AND CYCLIDES OF DUPIN

Let $\mathcal{M}$ be a regular surface in $\mathbb{R}^3$ and let $\mathcal{W} \subseteq \mathcal{M}$ be an open set on which a unit normal $\mathbf{U}$ is defined and differentiable. Denote by $\mathbf{k}_1$ and $\mathbf{k}_2$ the principal curvatures with respect to $\mathbf{U}$. We order $\mathbf{k}_1$ and $\mathbf{k}_2$ so that $\mathbf{k}_1 \geq \mathbf{k}_2$ on $\mathcal{W}$. It is clear from (16.26), page 378, that each $\mathbf{k}_i$ is continuous on $\mathcal{W}$. On the other hand, in exceptional cases it can happen that $\mathbf{k}_1$ or $\mathbf{k}_2$ is not differentiable at an umbilic point. (This happens, for example, for the umbilic point of a monkey saddle. (See the picture on page 403).) But both $\mathbf{k}_1$ and $\mathbf{k}_2$ are differentiable, provided $\mathcal{W}$ contains no umbilic points (see [CeRy, page 134]). Also, for a sphere, for which every point is an umbilic point, clearly both $\mathbf{k}_1$ and $\mathbf{k}_2$ are differentiable because they are everywhere constant. In this chapter we assume that both $\mathbf{k}_1$ and $\mathbf{k}_2$ are differentiable on $\mathcal{W}$, except possibly for isolated points, and for convenience we take $\mathcal{W} = \mathcal{M}$.

Definition. *Let $\mathcal{M}$ be a regular surface in $\mathbb{R}^3$. We write*

$$\rho_1 = \frac{1}{\mathbf{k}_1} \quad \text{and} \quad \rho_2 = \frac{1}{\mathbf{k}_2}$$

for the reciprocals of the principal curvatures of a surface in $\mathbb{R}^3$. Each ρ_i is called a **principal radius of curvature** *of $\mathcal{M}$.*

For a point $\mathbf{q}$ in a regular surface $\mathcal{M}$ we denote by $L_\mathbf{q}$ the line normal to $\mathcal{M}$ at $\mathbf{q}$; then the surface normal $\mathbf{U}(\mathbf{q})$ is a vector in $L_\mathbf{q}$. Each normal section (see page 365) is a plane curve C in a plane Π containing L_q. Since $\mathbf{U}(\mathbf{q})$ is perpendicular to C at $\mathbf{q}$, the center of curvature (see page 98) of C lies on the line containing $\mathbf{U}(\mathbf{q})$. The centers of curvature at $\mathbf{q}$ fill out a connected subset F of L_q called the **focal interval** of $\mathcal{M}$ at $\mathbf{q}$; the focal interval is a point if $\mathbf{q}$ is an umbilic point; otherwise, it is a line segment. The extremities of F are called the **focal points** of $\mathcal{M}$ at $\mathbf{q}$. They coincide if and only if $\mathbf{q}$ is an umbilic point. If the Gaussian curvature of $\mathcal{M}$ vanishes at $\mathbf{q}$, then at least one focal point is at infinity.

Here are two examples showing typical behavior at a hyperbolic point and at an elliptic point.

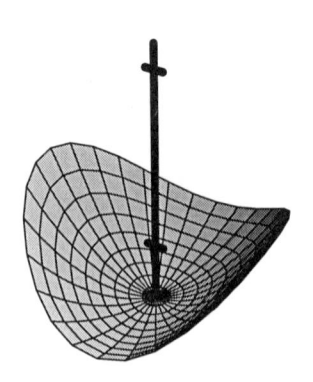

Focal interval at the origin of an elliptic paraboloid

Focal interval at the origin of an hyperbolic paraboloid

Definition. *The* **focal set** *of a regular surface* $\mathcal{M} \subset \mathbb{R}^3$ *is*

$$\mathbf{Focal}(\mathcal{M}) = \{\, \mathbf{p} \in \mathbb{R}^3 \mid \mathbf{p} \text{ is a focal point of some } \mathbf{q} \in \mathcal{M} \,\}.$$

Under favorable circumstances it is possible to make $\mathbf{Focal}(\mathcal{M})$ into a differentiable manifold (possibly with singularities). In the case that $\mathcal{M} \subset \mathbb{R}^3$ is connected and free of umbilic points the set $\mathbf{Focal}(\mathcal{M})$ has two components, one corresponding to each principal curvature. There are three possibilities:

Case 1. Each component of $\mathbf{Focal}(\mathcal{M})$ is 2-dimensional.

Case 2. One component of $\mathbf{Focal}(\mathcal{M})$ is 1-dimensional and the other 2-dimensional.

Case 3. Each component of $\mathbf{Focal}(\mathcal{M})$ is 1-dimensional.

Sections 35.1, 35.2 and 35.3 deal successively with each of these cases.

35.1 Surfaces Whose Focal Sets are 2-Dimensional

In this section we consider Case 1. The first order of business is to make each component of **Focal**($\mathcal{M}$) into a surface, possibly with singularities. This will be done by showing that each regular principal patch $\mathbf{x}:\mathcal{U} \longrightarrow \mathcal{M}$ gives rise to patches $\mathbf{z}_1:\mathcal{U} \longrightarrow$ **Focal**($\mathcal{M}$) and $\mathbf{z}_2:\mathcal{U} \longrightarrow$ **Focal**($\mathcal{M}$) parametrizing each of the components of **Focal**($\mathcal{M}$).

Definition. *Let* $\mathbf{x}:\mathcal{U} \longrightarrow \mathcal{M}$ *be a patch on a surface* $\mathcal{M} \subset \mathbb{R}^3$, *and denote by* ρ_1 *and* ρ_2 *the principal radii of curvature of* $\mathbf{x}$. *The* **focal patches** *corresponding to* $\mathbf{x}$ *are defined by*

(35.1) $\qquad \mathbf{z}_1(u,v) = \mathbf{x} + \rho_1 \mathbf{U} \quad and \quad \mathbf{z}_2(u,v) = \mathbf{x} + \rho_2 \mathbf{U}.$

By definition the image of each focal patch $\mathbf{z}_i$ is contained in **Focal**($\mathcal{M}$).

We want to compute the Gaussian curvatures of each component of the focal set of $\mathcal{M}$, and also the first and second fundamental forms. For this, we need to parametrize $\mathcal{M}$ by patches and then compute the first and second fundamental forms of each patch. The calculations are considerably easier when we use principal patches. It is easy to compute the derivatives and unit normals of the focal patches of a regular principal patch $\mathbf{x}$.

Theorem 35.1. *Let* $\mathbf{x}:\mathcal{U} \longrightarrow \mathbb{R}^3$ *be a regular principal patch and let* $\mathbf{z}_1$ *and* $\mathbf{z}_2$ *be the corresponding focal patches. Then*

(35.2) $\qquad \begin{cases} \mathbf{z}_{1u} = \rho_{1u}\mathbf{U}, \\[6pt] \mathbf{z}_{1v} = \left(1 - \dfrac{\rho_1}{\rho_2}\right)\mathbf{x}_v + \rho_{1v}\mathbf{U}, \\[6pt] \mathbf{U}_1 = -\mathrm{sign}\left(\rho_{1u}\left(1 - \dfrac{\rho_1}{\rho_2}\right)\right)\dfrac{\mathbf{x}_u}{\sqrt{E}}, \end{cases}$

and

(35.3) $\qquad \begin{cases} \mathbf{z}_{2u} = \left(1 - \dfrac{\rho_2}{\rho_1}\right)\mathbf{x}_u + \rho_{2u}\mathbf{U}, \\[6pt] \mathbf{z}_{2v} = \rho_{2v}\mathbf{U}, \\[6pt] \mathbf{U}_2 = -\mathrm{sign}\left(\rho_{2v}\left(1 - \dfrac{\rho_2}{\rho_1}\right)\right)\dfrac{\mathbf{x}_v}{\sqrt{G}}. \end{cases}$

Here $\mathbf{U}_1$ is the unit normal of $\mathbf{z}_1$ and $\mathbf{U}_2$ is the unit normal of $\mathbf{z}_2$. Hence $\mathbf{z}_1$ is regular on

$$\{(u,v) \in \mathcal{U} \mid \rho_1(u,v) \neq \rho_2(u,v) \text{ and } \rho_{1u}(u,v) \neq 0\},$$

and $\mathbf{z}_2$ is regular on

$$\{(u,v) \in \mathcal{U} \mid \rho_1(u,v) \neq \rho_2(u,v) \text{ and } \rho_{2v}(u,v) \neq 0\}.$$

Proof. We have

(35.4) $\quad -\mathbf{U}_u = S\mathbf{x}_u = \dfrac{1}{\rho_1}\mathbf{x}_u \quad$ and $\quad -\mathbf{U}_v = S\mathbf{x}_v = \dfrac{1}{\rho_2}\mathbf{x}_v.$

Differentiating (35.1) and using (35.4), we get

(35.5) $\quad \mathbf{z}_{1u} = \mathbf{x}_u + \rho_{1u}\mathbf{U} + \rho_1 \mathbf{U}_u = \mathbf{x}_u + \rho_{1u}\mathbf{U} + \rho_1\left(-\dfrac{1}{\rho_1}\mathbf{x}_u\right) = \rho_{1u}\mathbf{U}.$

Similarly,

(35.6) $\quad \mathbf{z}_{1v} = \mathbf{x}_v + \rho_{1v}\mathbf{U} + \rho_1 \mathbf{U}_v = \mathbf{x}_v + \rho_{1v}\mathbf{U} + \rho_1\left(-\dfrac{1}{\rho_2}\mathbf{x}_v\right)$

$$= \left(1 - \dfrac{\rho_1}{\rho_2}\right)\mathbf{x}_v + \rho_{1v}\mathbf{U}.$$

Using (35.5) and (35.6), we find that

$$\mathbf{z}_{1u} \times \mathbf{z}_{1v} = \rho_{1u}\mathbf{U} \times \left(\left(1 - \dfrac{\rho_1}{\rho_2}\right)\mathbf{x}_v + \rho_{1v}\mathbf{U}\right)$$

$$= \rho_{1u}\left(1 - \dfrac{\rho_1}{\rho_2}\right)\mathbf{U} \times \mathbf{x}_v.$$

Hence

$$\mathbf{U}_1 = \dfrac{\mathbf{z}_{1u} \times \mathbf{z}_{1u}}{\|\mathbf{z}_{1u} \times \mathbf{z}_{1u}\|} = \dfrac{\rho_{1u}\left(1 - \dfrac{\rho_1}{\rho_2}\right)\mathbf{U} \times \mathbf{x}_v}{\sqrt{G}\left|\rho_{1u}\left(1 - \dfrac{\rho_1}{\rho_2}\right)\right|} = -\text{sign}\left(\rho_{1u}\left(1 - \dfrac{\rho_1}{\rho_2}\right)\right)\dfrac{\mathbf{x}_u}{\sqrt{E}}.$$

This proves (35.2). Similar calculations yield (35.3). ∎

Next, we compute the coefficients of the first fundamental forms of the focal patches.

35.1 Surfaces Whose Focal Sets are 2-Dimensional

Lemma 35.2. *Let* $\mathbf{x}:\mathcal{U} \longrightarrow \mathbb{R}^3$ *be a regular principal patch. Then the coefficients of the first fundamental form of* $\mathbf{z}_1$ *are given by*

$$(35.7) \qquad E_1 = \rho_{1u}^2, \qquad F_1 = \rho_{1u}\rho_{1v}, \qquad G_1 = \left(1 - \frac{\rho_1}{\rho_2}\right)^2 G + \rho_{1v}^2,$$

and the coefficients of the first fundamental form of $\mathbf{z}_2$ *are given by*

$$(35.8) \qquad E_2 = \left(1 - \frac{\rho_2}{\rho_1}\right)^2 E + \rho_{2u}^2, \qquad F_2 = \rho_{2u}\rho_{2v}, \qquad G_2 = \rho_{2v}^2.$$

Thus the first fundamental forms of the focal patches are

$$ds_1^2 = \rho_{1u}^2 du^2 + \rho_{1u}\rho_{1v} du\, dv + \left(\left(1 - \frac{\rho_1}{\rho_2}\right)^2 G + \rho_{1v}^2\right) dv^2$$

and

$$ds_2^2 = \left(\left(1 - \frac{\rho_2}{\rho_1}\right)^2 E + +\rho_{2u}^2\right) du^2 + \rho_{2u}\rho_{2v} du\, dv + \rho_{2v}^2 dv^2.$$

Proof. From (35.2) we compute

$$E_1 = \mathbf{z}_{1u} \cdot \mathbf{z}_{1u} = \rho_{1u}^2, \qquad F_1 = \mathbf{z}_{1u} \cdot \mathbf{z}_{1v} = \rho_{1u}\rho_{1v}$$

and

$$G_1 = \mathbf{z}_{1v} \cdot \mathbf{z}_{1v} = \left(1 - \frac{\rho_1}{\rho_2}\right)^2 \mathbf{x}_v \cdot \mathbf{x}_v + \rho_{1v}^2 = \left(1 - \frac{\rho_1}{\rho_2}\right)^2 G + \rho_{1v}^2.$$

Similar calculations yield (35.8). ∎

Before computing the second fundamental forms of the focal patches, we note the following easy consequence of Lemma 28.9, page 652, (see exercise 5):

Lemma 35.3. *Let* $\mathbf{x}:\mathcal{U} \to \mathbb{R}^3$ *be a regular principal patch, and denote by* ρ_1 *and* ρ_2 *the principal radii of curvature of* $\mathbf{x}$. *Then*

$$(35.9) \qquad \begin{cases} \dfrac{\rho_{1v}}{\rho_1^2} = \dfrac{E_v}{2E}\left(\dfrac{1}{\rho_1} - \dfrac{1}{\rho_2}\right), \\[2mm] \dfrac{\rho_{2u}}{\rho_2^2} = \dfrac{G_u}{2G}\left(\dfrac{1}{\rho_2} - \dfrac{1}{\rho_1}\right). \end{cases}$$

Now we can compute the second fundamental forms of the focal patches.

Lemma 35.4. *Let* $\mathbf{x}: \mathcal{U} \longrightarrow \mathbf{R}^3$ *be a regular principal patch such that the two branches of the focal set of* $\mathbf{x}$ *are parametrized by* (35.1). *Then the coefficients of the second fundamental form of* $\mathbf{z}_1$ *are given by*

(35.10)
$$\begin{cases} e_1 = \operatorname{sign}\left(\rho_{1u}\left(1 - \dfrac{\rho_1}{\rho_2}\right)\right) \dfrac{\sqrt{E}\,\rho_{1u}}{\rho_1}, \\ f_1 = 0, \\ g_1 = \operatorname{sign}\left(\rho_{1u}\left(1 - \dfrac{\rho_1}{\rho_2}\right)\right)\left(1 - \dfrac{\rho_1}{\rho_2}\right)\dfrac{G_u}{2\sqrt{E}}, \end{cases}$$

and the coefficients of the second fundamental form of $\mathbf{z}_2$ *are given by*

(35.11)
$$\begin{cases} e_2 = \operatorname{sign}\left(\rho_{2v}\left(1 - \dfrac{\rho_2}{\rho_1}\right)\right)\left(1 - \dfrac{\rho_2}{\rho_1}\right)\dfrac{E_v}{2\sqrt{G}}, \\ f_2 = 0, \\ g_2 = \operatorname{sign}\left(\rho_{2v}\left(1 - \dfrac{\rho_2}{\rho_1}\right)\right)\dfrac{\sqrt{G}\,\rho_{2v}}{\rho_2}. \end{cases}$$

Furthermore, g_1 *and* e_2 *are given by the alternate formulas*

(35.12)
$$\begin{cases} g_1 = -\operatorname{sign}\left(\rho_{1u}\left(1 - \dfrac{\rho_1}{\rho_2}\right)\right)\dfrac{G\,\rho_1\rho_{2u}}{\rho_2^2\sqrt{E}}, \\ e_2 = -\operatorname{sign}\left(\rho_{2v}\left(1 - \dfrac{\rho_2}{\rho_1}\right)\right)\dfrac{E\,\rho_2\rho_{1v}}{\rho_1^2\sqrt{G}}. \end{cases}$$

Proof. From (35.2) we compute the second derivatives of $\mathbf{z}$:

(35.13)
$$\mathbf{z}_{1uu} = \rho_{1uu}\mathbf{U} + \rho_{1u}\mathbf{U}_u = \rho_{1uu}\mathbf{U} - \dfrac{\rho_{1u}}{\rho_1}\mathbf{x}_u,$$

(35.14)
$$\mathbf{z}_{1uv} = \rho_{1uv}\mathbf{U} + \rho_{1u}\mathbf{U}_v = \rho_{1uv}\mathbf{U} - \dfrac{\rho_{1u}}{\rho_2}\mathbf{x}_v.$$

Then (35.13) and (35.2) imply that

$$\begin{aligned} e_1 &= \mathbf{z}_{1uu} \cdot \mathbf{U}_1 = \left(\rho_{1uu}\mathbf{U} - \dfrac{\rho_{1u}}{\rho_1}\mathbf{x}_u\right) \cdot \left(-\operatorname{sign}\left(\rho_{1u}\left(1 - \dfrac{\rho_1}{\rho_2}\right)\right)\dfrac{\mathbf{x}_u}{\sqrt{E}}\right) \\ &= \operatorname{sign}\left(\rho_{1u}\left(1 - \dfrac{\rho_1}{\rho_2}\right)\right)\rho_{1u}\dfrac{\sqrt{E}}{\rho_1}. \end{aligned}$$

35.1 Surfaces Whose Focal Sets are 2-Dimensional

Similarly, (35.14) implies that

$$f_1 = \mathbf{z}_{1uv} \cdot \mathbf{U}_1 = \left(\rho_{1uv}\mathbf{U} - \frac{\rho_{1u}}{\rho_1}\mathbf{x}_v\right) \cdot \left(-\operatorname{sign}\left(\rho_{1u}\left(1 - \frac{\rho_1}{\rho_2}\right)\right)\frac{\mathbf{x}_u}{\sqrt{E}}\right) = 0,$$

because $F = 0$. To compute g_1, we use (35.2), (35.14) and the fact that $\mathbf{x}_u \cdot \mathbf{x}_v = \mathbf{x}_u \cdot \mathbf{U} = 0$; thus

(35.15) $\quad g_1 = \mathbf{z}_{1vv} \cdot \mathbf{U}_1$

$$= \left(-\left(\frac{\rho_1}{\rho_2}\right)_v \mathbf{x}_v + \left(1 - \frac{\rho_1}{\rho_2}\right)\mathbf{x}_{vv} + \rho_{1vv}\mathbf{U} + \rho_{1v}\mathbf{U}_v\right) \cdot$$

$$\left(-\operatorname{sign}\left(\rho_{1u}\left(1 - \frac{\rho_1}{\rho_2}\right)\right)\frac{\mathbf{x}_u}{\sqrt{E}}\right)$$

$$= -\operatorname{sign}\left(\rho_{1u}\left(1 - \frac{\rho_1}{\rho_2}\right)\right)\left(1 - \frac{\rho_1}{\rho_2}\right)\mathbf{x}_{vv} \cdot \frac{\mathbf{x}_u}{\sqrt{E}}.$$

Since $\mathbf{x}_{vv} \cdot \mathbf{x}_u = -G_u/2$, equation (35.15) can be written as

(35.16) $\quad g_1 = \operatorname{sign}\left(\rho_{1u}\left(1 - \frac{\rho_1}{\rho_2}\right)\right)\left(1 - \frac{\rho_1}{\rho_2}\right)\frac{G_u}{2\sqrt{E}}.$

Similar calculations yield (35.11). Finally, (35.12) is a consequence of (35.10), (35.11) and (35.9) (see exercise 5 for details). ∎

Finally, we find concise formulas for the Gaussian curvatures of the focal patches.

Corollary 35.5. *Let* $\mathbf{x}: \mathcal{U} \longrightarrow \mathbb{R}^3$ *be a regular principal patch such that the two branches of the focal set of* $\mathbf{x}$ *are parametrized by* (35.1). *Then the Gaussian curvatures of* $\mathbf{z}_1$ *and* $\mathbf{z}_2$ *are given by*

(35.17) $\quad \begin{cases} K_1 = -\dfrac{\rho_{2u}}{(\rho_1 - \rho_2)^2 \rho_{1u}}, \\ K_2 = -\dfrac{\rho_{1v}}{(\rho_1 - \rho_2)^2 \rho_{2v}}. \end{cases}$

Proof. The first equation (35.17) is an immediate consequence of formulas (35.7), (35.10), (35.12) and (16.24), page 377; the second equation (35.17) is found similarly (see exercise 5 for details). ∎

The *Mathematica* versions of the focal patches in the case of a general patch are:

```
focalk1[surf_][u_,v_]:= surf[uu,vv] +
    (1/k1[surf][uu,vv])*
    unitnormal[surf][uu,vv]  /. {uu->u,vv->v}
```

```
focalk2[surf_][u_,v_]:= surf[uu,vv] +
    (1/k2[surf][uu,vv])*
    unitnormal[surf][uu,vv]  /. {uu->u,vv->v}
```

For the special case of a principal patch we define

```
focalkmu[surf_][u_,v_]:= (surf[uu,vv] +
    (1/kmu[surf][uu,vv])*
    unitnormal[surf][uu,vv]  /. {uu->u,vv->v})
```

```
focalkpi[surf_][u_,v_]:= (surf[uu,vv] +
    (1/kpi[surf][uu,vv])*
    unitnormal[surf][uu,vv]  /. {uu->u,vv->v})
```

Focal Sets of a Hyperbolic Paraboloid

A hyperbolic paraboloid is parametrized by $\mathbf{x}(u,v) = (u,v,uv)$ (see page 297). We compute

$$\mathbf{x}_u(u,v) = (1,0,v), \qquad \mathbf{x}_v(u,v) = (0,1,v), \qquad \mathbf{x}_{uu}(u,v) = \mathbf{x}_{vv}(u,v) = (0,0,0),$$

and

$$\mathbf{x}_{uv}(u,v) = (0,0,1), \qquad \mathbf{U} = \frac{\mathbf{x}_u \times \mathbf{x}_v}{\|\mathbf{x}_u \times \mathbf{x}_v\|} = \frac{(-v,-u,1)}{\sqrt{1+u^2+v^2}}.$$

Furthermore, $E = 1+v^2$, $F = uv$, $G = 1+u^2$, $EG - F^2 = 1+u^2+v^2$, $e = g = 0$ and $f = 1/\sqrt{1+u^2+v^2}$, so that the Gaussian and mean curvatures are given by

$$K = \frac{eg - f^2}{EG - F^2} = -\frac{1}{(1+u^2+v^2)^2}$$

and

$$H = \frac{eG - 2fF + gG}{2(EG - F^2)^{3/2}} = -\frac{uv}{(1+u^2+v^2)^{3/2}}.$$

35.1 Surfaces Whose Focal Sets are 2-Dimensional

Thus
$$H^2 - K = \frac{(1+u^2)(1+v^2)}{(1+u^2+v^2)^3},$$

so that
$$\begin{cases} \mathbf{k}_1 = H + \sqrt{H^2 - K} = \dfrac{-u\,v + \sqrt{(1+u^2)(1+v^2)}}{(1+u^2+v^2)^{3/2}}, \\ \mathbf{k}_2 = H - \sqrt{H^2 - K} = \dfrac{-u\,v - \sqrt{(1+u^2)(1+v^2)}}{(1+u^2+v^2)^{3/2}}. \end{cases}$$

Parameterizations of the two focal patches can be found with *Mathematica*:

focalk1[hyperbolicparaboloid][u,v]//PowerExpand//Together

```
    2      3                  2            2
 v + 2 u  v + v  - u Sqrt[1 + u ] Sqrt[1 + v ]
{-----------------------------------------------,
              2                2
         u v - Sqrt[1 + u ] Sqrt[1 + v ]

       3        2              2              2
 u + u  + 2 u v  - Sqrt[1 + u ] v Sqrt[1 + v ]
 ----------------------------------------------,
              2                2
         u v - Sqrt[1 + u ] Sqrt[1 + v ]

        2    2    2  2           2              2
 -1 - u  - v  + u  v  - u Sqrt[1 + u ] v Sqrt[1 + v ]
 -----------------------------------------------------}
              2                2
         u v - Sqrt[1 + u ] Sqrt[1 + v ]
```

and

focalk2[hyperbolicparaboloid][u,v]//PowerExpand//Together

```
    2      3                  2            2
 v + 2 u  v + v  + u Sqrt[1 + u ] Sqrt[1 + v ]
{-----------------------------------------------,
              2                2
         u v + Sqrt[1 + u ] Sqrt[1 + v ]

       3        2              2              2
 u + u  + 2 u v  + Sqrt[1 + u ] v Sqrt[1 + v ]
 ----------------------------------------------,
              2                2
         u v + Sqrt[1 + u ] Sqrt[1 + v ]

        2    2    2  2           2              2
 -1 - u  - v  + u  v  + u Sqrt[1 + u ] v Sqrt[1 + v ]
 -----------------------------------------------------}
              2                2
         u v + Sqrt[1 + u ] Sqrt[1 + v ]
```

Here is a plot of portions of a hyperbolic paraboloid and the two components of its focal set:

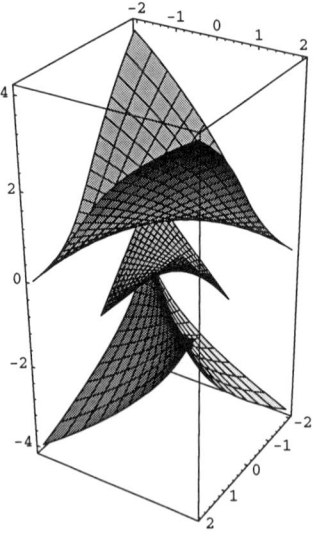

```
ParametricPlot3D[Evaluate[
{hyperbolicparaboloid[u,v],
focalk1[hyperbolicparaboloid][u,v],
focalk2[hyperbolicparaboloid][u,v]
}],
{u,-1,1}, {v,-1,1},
ViewPoint->{2.49,1.58,1.6}];
```

Hyperbolic paraboloid and its focal set

The principal curvatures of a hyperbolic paraboloid satisfy

$$|\mathbf{k}_1(u,v)|, |\mathbf{k}_2(u,v)| \geq \frac{1}{(1+u^2+v^2)^{3/2}}.$$

35.2 Canal Surfaces

An **envelope** of a 1-parameter family of surfaces is constructed in the same way that we constructed a 1-parameter family of curves in Section 7.6. The family is described by a differentiable function $F(x,y,z,\lambda) = 0$, where λ is a parameter. When λ can be eliminated from the equations

$$F(x,y,z,\lambda) = 0 \quad \text{and} \quad \frac{\partial F(x,y,z,\lambda)}{\partial \lambda} = 0,$$

we get the envelope, which is a surface described implicitly as $G(x,y,z) = 0$. For example, for a 1-parameter family of planes we get a developable surface. For an extensive discussion of envelopes of curves, see [BrGi, chapter 5] and [Bolt].

35.2 Canal Surfaces

Definition. *The envelope of a 1-parameter family $t \longmapsto S^2(t)$ of spheres in $\mathbb{R}^3$ is called a **canal surface**. The curve formed by the centers of the spheres is called the **center curve** of the canal surface. The **radius** of the canal surface is the function r such that $r(t)$ is the radius of the sphere $S^2(t)$.*

It is easy to find one principal curvature and its corresponding principal curve for a canal surface.

Lemma 35.6. *Let $t \longmapsto S^2(t)$ be the 1-parameter family of spheres that defines a canal surface $\mathcal{M}$. Then for each t the intersection $S^2(t) \cap \mathcal{M}$ is a circle and a principal curve on $\mathcal{M}$.*

Proof. Since $S^2(t)$ and $\mathcal{M}$ are tangent along $S^2(t) \cap \mathcal{M}$, the angle between their normals is zero. Furthermore, any curve on $S^2(t)$ is a principal curve, in particular, $S^2(t) \cap \mathcal{M}$. Joachimsthal's Theorem (Theorem 20.2, page 460) then implies that $S^2(t) \cap \mathcal{M}$ is a principal curve on $\mathcal{M}$. ∎

Now we can characterize canal surfaces.

Theorem 35.7. *Let $\mathcal{M} \subset \mathbb{R}^3$ be a regular surface whose focal set $\mathbf{Focal}(\mathcal{M})$ is a differentiable manifold. The following are equivalent:*

(i) *$\mathcal{M}$ is a canal surface;*

(ii) *one of the systems of principal curves of $\mathcal{M}$ consists of circles;*

(iii) *one of the components of $\mathbf{Focal}(\mathcal{M})$ is a curve.*

Proof. That (i) implies (ii) is a consequence of Lemma 35.6. The normal lines passing through points on any principal curve that is a circle meet at a point. As the circular principal curves move on $\mathcal{M}$, their centers generate a curve, which must be one of the components of $\mathbf{Focal}(\mathcal{M})$. Thus (ii) implies (iii). Finally, if (iii) holds, we can use the component of $\mathbf{Focal}(\mathcal{M})$ that is a curve to generate a canal surface using the distance between $\mathbf{Focal}(\mathcal{M})$ and $\mathcal{M}$ as the radius of the canal surface. Evidently, this canal surface coincides with $\mathcal{M}$. Hence (iii) implies (i). ∎

We can find a parametrization of a canal surface, provided we make mild assumptions about the center curve.

Theorem 35.8. *Suppose the center curve of a canal surface is a unit-speed curve $\gamma: (a,b) \longrightarrow \mathbb{R}^3$ with nonzero curvature. Then the canal surface can be parametrized* by the formula

(35.18) $\quad \mathbf{canalsurf}[\gamma][r](t,\theta) = \gamma(t) + r(t)\Big(-r'(t)\mathbf{T}$

$$\pm\sqrt{1-r'(t)^2}\big(-\cos\theta\,\mathbf{N} + \sin\theta\,\mathbf{B}\big)\Big),$$

where **T**, **N** and **B** denote the tangent, normal and binormal of γ.

Proof. Let **y** denote a patch that parametrizes the envelope of spheres defining the canal surface. Since the curvature of γ is nonzero, the Frenet frame $\{\mathbf{T},\mathbf{N},\mathbf{B}\}$ is well-defined, and we can write

(35.19) $\quad \mathbf{y}(t,\theta) - \gamma(t) = a(t,\theta)\mathbf{T}(t) + b(t,\theta)\mathbf{N}(t) + c(t,\theta)\mathbf{B}(t),$

where a, b and c are differentiable on the interval on which γ is defined. We must have

(35.20) $\quad \|\mathbf{y}(t,\theta) - \gamma(t)\|^2 = r(t)^2,$

Equation (35.20) expresses analytically the geometric fact that $\mathbf{y}(t,\theta)$ lies on a sphere $S^2(t)$ of radius $r(t)$ centered at $\gamma(t)$. Furthermore, $\mathbf{y}(t,\theta) - \gamma(t)$ is a normal vector to the canal surface; this fact implies that

(35.21) $\quad \big(\mathbf{y}(t,\theta) - \gamma(t)\big) \cdot \mathbf{y}_t = 0,$

(35.22) $\quad \big(\mathbf{y}(t,\theta) - \gamma(t)\big) \cdot \mathbf{y}_\theta = 0.$

Equations (35.21) and (35.22) say that the vectors $\mathbf{y}_t$ and $\mathbf{y}_\theta$ are tangent to $S^2(t)$.
From (35.19) and (35.20) we get

(35.23) $\quad \begin{cases} a^2 + b^2 + c^2 = r^2, \\ a\,a_t + b\,b_t + c\,c_t = r\,r'. \end{cases}$

When we differentiate (35.19) with respect to t and use the Frenet formulas (8.10), page 186, we obtain

(35.24) $\quad \mathbf{y}_t = (1 + a_t - b\kappa)\mathbf{T} + (a\kappa - c\tau + b_t)\mathbf{N} + (c_t + b\tau)\mathbf{B}.$

Then (35.23), (35.24), (35.19) and (35.21) imply that

(35.25) $\quad a + r\,r' = 0,$

35.2 Canal Surfaces

and from (35.23) and (35.25) we get

(35.26) $$b^2 + c^2 = r^2(1 - r'^2).$$

By Lemma 1.13, page 17, we can write $b = \mp r\sqrt{1 - r'^2}\cos\theta$ and $c = \pm r\sqrt{1 - r'^2}\sin\theta$. Thus (35.19) becomes

(35.27) $\mathbf{y}(t,\theta) - \gamma(t) = -r\,r'\mathbf{T} \mp r\sqrt{1 - r'^2}\cos\theta\,\mathbf{N} \pm r\sqrt{1 - r'^2}\sin\theta\,\mathbf{B}.$

Then (35.27) can be rewritten as (35.18). ∎

The notion of canal surface is a very general concept. In the next three subsections we consider three important subclasses of canal surfaces.

Tubes as Canal Surfaces

It is easy to see that when the radius function $r(t)$ is constant, the definition of canal surface reduces to the definition (9.1) of a tube given on page 207. In fact, we can characterize tubes among all canal surfaces.

Theorem 35.9. *The following conditions are equivalent for a canal surface* $\mathcal{M}$:

(i) $\mathcal{M}$ *is a tube parametrized by* (35.18);

(ii) *the radius of* $\mathcal{M}$ *is constant*;

(iii) *the radius vector of each sphere in the family that defines the canal surface* $\mathcal{M}$ *meets the center curve orthogonally*.

Proof. When we compare (35.18) and (9.1), we see that (i) is equivalent to (ii). That (ii) is equivalent to (iii) follows from (35.18). ∎

Let us compute the first and second fundamental forms of a tube parametrized by (35.18). We have
$$\mathbf{y}_\theta = r\bigl(-(\sin\theta)\mathbf{N} + (\cos\theta)\mathbf{B}\bigr),$$
and
$$\begin{aligned}\mathbf{y}_t &= s\mathbf{T} + r\cos\theta\bigl(-\kappa\mathbf{T} + \tau\mathbf{B}\bigr)s + r\sin\theta\bigl(-\tau\mathbf{N}\bigr)s \\ &= s(1 - r\kappa\cos\theta)\mathbf{T} + r\,s\,\tau\bigl(-(\sin\theta)\mathbf{N} + (\cos\theta)\mathbf{B}\bigr) \\ &= s(1 - r\kappa\cos\theta)\mathbf{T} + s\tau\,\mathbf{y}_\theta,\end{aligned}$$

using the Frenet formulas (8.10). Hence

$$\mathbf{y}_t \times \mathbf{y}_\theta = s(1 - r\kappa\cos\theta)\mathbf{T} \times r\bigl(-(\sin\theta)\mathbf{N} + (\cos\theta)\mathbf{B}\bigr)$$
$$= -rs(1 - r\kappa\cos\theta)\bigl((\cos\theta)\mathbf{N} + (\sin\theta)\mathbf{B}\bigr),$$

so that
$$\|\mathbf{y}_t \times \mathbf{y}_\theta\|^2 = r^2 s^2 (1 - r\kappa\cos\theta)^2$$

and
$$\mathbf{U} = \frac{\mathbf{y}_t \times \mathbf{y}_\theta}{\|\mathbf{y}_t \times \mathbf{y}_\theta\|} = -(\cos\theta)\mathbf{N} - (\sin\theta)\mathbf{B}.$$

Furthermore,
$$\begin{cases} E = \mathbf{y}_t \cdot \mathbf{y}_t = s^2(1 - r\kappa\cos\theta)^2 + (rs\tau)^2, \\ F = \mathbf{y}_t \cdot \mathbf{y}_\theta = r^2 s\tau, \\ G = \mathbf{y}_\theta \cdot \mathbf{y}_\theta = r^2. \end{cases}$$

Next, we compute

$$\mathbf{y}_{\theta\theta} = -r\bigl((\cos\theta)\mathbf{N} + (\sin\theta)\mathbf{B}\bigr),$$

$$\mathbf{y}_{\theta t} = rs\bigl(\kappa(\sin\theta)\mathbf{T} - \tau(\cos\theta)\mathbf{N} - \tau(\sin\theta)\mathbf{B}\bigr),$$

$$\mathbf{y}_{tt} = \bigl(s' - (sr\kappa)'\cos\theta\bigr)\mathbf{T} + s^2(1 - r\kappa\cos\theta)(\kappa\mathbf{N}) + (sr\tau)'\bigl(-\sin\theta\mathbf{N} + \cos\theta\mathbf{B}\bigr)$$
$$+ (s^2 r\tau)\bigl(-\sin\theta(-\kappa\mathbf{T} + \tau\mathbf{B}) + \cos\theta(-\tau)\mathbf{N}\bigr)$$

$$= \bigl(s' - (sr\kappa)'\cos\theta + s^2 r\kappa\tau\sin\theta\bigr)\mathbf{T}$$
$$+ \bigl(s^2(1 - r\kappa\cos\theta)\kappa - (sr\tau)'\sin\theta - s^2 r\tau^2\cos\theta\bigr)\mathbf{N}$$
$$+ \bigl((sr\tau)'\cos\theta - s^2 r\tau^2\sin\theta\bigr)\mathbf{B}$$

$$= \bigl(s' - (sr\kappa)'\cos\theta + s^2 r\kappa\tau\sin\theta\bigr)\mathbf{T}$$
$$+ \bigl(s^2\kappa - s^2 r(\kappa^2 + \tau^2)\cos\theta - (sr\tau)'\sin\theta\bigr)\mathbf{N}$$
$$+ \bigl((sr\tau)'\cos\theta - s^2 r\tau^2\sin\theta\bigr)\mathbf{B}.$$

Therefore,
$$\begin{cases} e = s^2\bigl(-\kappa\cos\theta + r\kappa^2\cos^2\theta + r\tau^2\bigr), \\ f = rs\tau, \\ g = r. \end{cases}$$

It follows that

$$K = \frac{eg - f^2}{EG - F^2} = \frac{rs^2(-\kappa\cos\theta + r\kappa^2\cos^2\theta + r\tau^2) - (rs\tau)^2}{r^2s^2(1 - r\kappa\cos\theta)^2} = \frac{-\kappa\cos\theta}{r(1 - r\kappa\cos\theta)}$$

and

$$H = \frac{eG - 2fF + gE}{2(EG - F^2)} = \frac{1}{2}\left(\frac{1}{r} + Kr\right).$$

Hence

$$H^2 - K = \frac{1}{4}\left(rK - \frac{1}{r}\right)^2,$$

so that

$$\mathbf{k}_1 = H + \sqrt{H^2 - K} = \frac{-\kappa\cos\theta}{1 - r\kappa\cos\theta}$$

and

$$\mathbf{k}_2 = H - \sqrt{H^2 - K} = \frac{1}{r}.$$

Surfaces of Revolution as Canal Surfaces

Almost any surface of revolution is a canal surface. The following theorem characterizes surfaces of revolution among all canal surfaces.

Theorem 35.10. *The center curve of a canal surface $\mathcal{M}$ is a straight line if and only if $\mathcal{M}$ is a surface of revolution for which no normal line to the surface is parallel to the axis of revolution. Furthermore, the parametrization (35.18) reduces to the standard parametrization (20.1) of a surface of revolution given on page 457.*

Proof. Let $\mathcal{M}$ be a surface of revolution and let $\mathbf{p} \in \mathcal{M}$. The normal line $L_\mathbf{p}$ perpendicular to $\mathcal{M}$ at $\mathbf{p}$ lies in the plane passing through the meridian of $\mathcal{M}$ passing through $\mathbf{p}$; this plane contains the axis of revolution A of $\mathcal{M}$. Hence either $L_\mathbf{p}$ intersects A or is parallel to it. By hypothesis, the latter possibility is excluded.

Let $L_\mathbf{p}$ intersect A at $\mathbf{q}$. The sphere $S^2(\mathbf{q})$ of radius $r(\mathbf{q}) = \|\mathbf{q} - \mathbf{p}\|$ and center $\mathbf{q}$ meets $L_\mathbf{p}$ perpendicularly at $\mathbf{p}$, and so it is tangent to $\mathcal{M}$ at $\mathbf{p}$. Hence $\mathcal{M}$ is the canal surface with center curve and variable radius $\mathbf{q} \longmapsto r(\mathbf{q})$.

Conversely, suppose that $\mathcal{M}$ is a canal surface whose center curve is a straight line. Instead of a Frenet Frame field along γ, we choose a parallel frame field $\{\mathbf{T}, \mathbf{N}, \mathbf{B}\}$. (See exercise 5 of Chapter 8.) We can proceed exactly as in the proof of Theorem 35.8; the only difference is that $\kappa = \tau = 0$ for the center curve. In particular, (35.18) holds.

To show that $\mathcal{M}$ is a surface of revolution, we use a Euclidean motion to move $\mathcal{M}$ so that the center curve is the z-axis and so that

$$\mathbf{T} = (0,0,1), \quad \mathbf{N} = (1,0,0), \quad \text{and} \quad \mathbf{B} = (0,1,0).$$

We may also assume that $\gamma(t) = (0,0,t)$. Then (35.18) becomes

(35.28) $\quad \mathbf{canalsurf}[\gamma][r](t,\theta) = \big(\mp r(t)\sqrt{1-r'(t)^2}\cos\theta,$
$$\pm r(t)\sqrt{1-r'(t)^2}\sin\theta, t - r(t)r'(t)\big).$$

We see that (35.28) coincides with (20.1) when we take

$$\theta = -u, \quad t = v, \quad \varphi(v) = \mp r(v)\sqrt{1-r'(v)^2} \quad \text{and} \quad \psi(v) = v - r(v)r'(v).$$

Hence $\mathcal{M}$ is a surface of revolution. ∎

There is a close connection between evolutes of plane curves and focal sets of surfaces of revolution:

Theorem 35.11. *One of the components of the focal set of a surface of revolution $\mathcal{M}$ generated by a plane curve C is the surface of revolution generated by the evolute of C.*

Proof. Let C be parametrized by a unit-speed curve $\boldsymbol{\alpha}$, and write $\boldsymbol{\alpha} = (\varphi, \psi)$. The standard parametrization of $\mathcal{M}$ is

$$\mathbf{x}(u,v) = \big(\varphi(v)\cos u, \varphi(v)\sin u, \psi(v)\big).$$

For simplicity we assume that $\varphi > 0$. According to Corollary 20.9, page 465, the principal curvatures of $\mathcal{M}$ are given by

$$\mathbf{k}_\mu = (\varphi''\psi' - \varphi'\psi'') \quad \text{and} \quad \mathbf{k}_\pi = \frac{-\psi'}{\varphi}.$$

Furthermore, by Lemma 20.5, page 462, the unit normal is given by

$$\mathbf{U} = (\psi'\cos u, \psi'\sin u, \varphi').$$

35.2 Canal Surfaces

Hence the formula for the nondegenerate focal patch of $\mathcal{M}$ is

(35.29) $\quad \mathbf{z}_1(u,v) = (\mathbf{x} + \rho_1 \mathbf{U})(u,v)$

$$= \big(\varphi(v)\cos u, \varphi(v)\sin u, \psi(v)\big) + \frac{(\psi'(v)\cos u, \psi'(v)\sin u, \varphi'(v))}{\varphi''(v)\psi'(v) - \varphi'(v)\psi''(v)}$$

$$= \bigg((\cos u)\bigg(\varphi(v) + \frac{\psi'(v)}{\varphi''(v)\psi'(v) - \varphi'(v)\psi''(v)}\bigg),$$

$$(\sin u)\bigg(\varphi(v) + \frac{\psi'(v)}{\varphi''(v)\psi'(v) - \varphi'(v)\psi''(v)}\bigg),$$

$$\psi(v) - \frac{\varphi'(v)}{\varphi''(v)\psi'(v) - \varphi'(v)\psi''(v)}\bigg).$$

On the other hand, the evolute of $\boldsymbol{\alpha}$ is given by

(35.30) $\quad \boldsymbol{\gamma}(t)$

$$= \bigg(\varphi(t) + \frac{\psi'(t)}{\varphi''(v)\psi'(v) - \varphi'(t)\psi''(t)}, \psi(t) - \frac{\varphi'(t)}{\varphi''(v)\psi'(v) - \varphi'(t)\psi''(t)}\bigg).$$

An easy calculation from (35.30) shows that the surface of revolution generated by $\boldsymbol{\gamma}$ is precisely the right-hand side of (35.29). ∎

For example, we know from Section 5.3 that the evolute of a tractrix is a catenoid. Hence the 2-dimensional component of the focal set of a pseudosphere is a catenary. The plot of a pseudosphere together with the 2-dimensional component of the focal set corresponding to the plot of a tractrix and its evolute on page 103 is

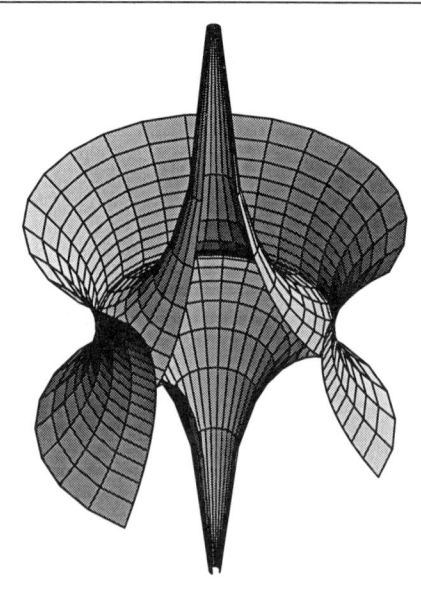

Pseudosphere and the catenoid component of its focal set

Canal Surfaces with Center Curve a Plane Curve

The parametrization (35.18) may or may not be principal. We have already seen (Lemma 35.6) that any canal surface has a system of circles as principal curves. Let us prove this fact by direct calculation.

Lemma 35.12. *Let $\mathcal{M}$ be a canal surface parametrized by (35.18). For fixed t the curve $\theta \longmapsto$ canalsurf$[\gamma][r](t,\theta)$ is a principal curve corresponding to the principal curvature $-1/r(t)$.*

Proof. Let $\mathbf{y}(t,\theta) =$ canalsurf$[\gamma][r](t,\theta)$. We can take for the unit normal of the canal surface the vector field

$$(35.31) \qquad \mathbf{U} = \frac{\mathbf{y}-\boldsymbol{\gamma}}{\|\mathbf{y}-\boldsymbol{\gamma}\|} = \frac{1}{r}(\mathbf{y}-\boldsymbol{\gamma}).$$

Thus
$$(35.32) \qquad \mathbf{y} = \boldsymbol{\gamma} + r\,\mathbf{U}.$$

When we differentiate (35.32) and use the fact that neither $\boldsymbol{\gamma}$ nor r depends on θ, we get

$$(35.33) \qquad \mathbf{y}_\theta = r\,\mathbf{U}_\theta = -r\,S\mathbf{y}_\theta,$$

where S is the shape operator of the canal surface. Hence the lemma follows. ∎

35.2 Canal Surfaces

Next, we characterize those canal surfaces whose center curve is a plane curve.

Theorem 35.13. *Let $\mathcal{M}$ be a canal surface parametrized by (35.18). The following conditions are equivalent.*

(i) *For each θ the curve $t \longmapsto$ **canalsurf**$[\gamma][r](t,\theta)$ is a principal curve.*

(ii) *Either the function r is constant and the torsion of the center curve γ vanishes, or both the curvature and torsion of γ vanish.*

(iii) *The center curve γ is a plane curve, and either the function r is constant or the curvature of γ vanishes.*

Proof. Assume that (i) holds. We first use (35.27) and (35.31) to write unit normal $\mathbf{U}$ to the canal surface as

$$(35.34) \qquad \mathbf{U} = -r'\mathbf{T} \pm \sqrt{1-r'^2}(\cos\theta\,\mathbf{N} + \sin\theta\,\mathbf{B}),$$

so that

$$(35.35) \qquad \mathbf{U}_\theta = \pm\sqrt{1-r'^2}\bigl(-\sin\theta\,\mathbf{N} + \cos\theta\,\mathbf{B}\bigr).$$

Also, (35.34) and the Frenet formulas (8.10) imply that

$$(35.36) \quad \begin{aligned}\mathbf{U}_t &= -r''\mathbf{T} - r'\kappa\,\mathbf{N} \pm \left(\sqrt{1-r'^2}\right)'\bigl(\cos\theta\,\mathbf{N} + \sin\theta\,\mathbf{B}\bigr) \\ &\quad \pm\sqrt{1-r'^2}\bigl(\cos\theta(-\kappa\,\mathbf{T} + \tau\,\mathbf{B}) + \sin\theta(-\tau\,\mathbf{N})\bigr) \\ &= -\left(r'' \pm \kappa\cos\theta\sqrt{1-r'^2}\right)\mathbf{T} - r'\kappa\,\mathbf{N} \\ &\quad \pm\left(\sqrt{1-r'^2}\right)'\bigl(\cos\theta\,\mathbf{N} + \sin\theta\,\mathbf{B}\bigr) \\ &\quad \pm\tau\sqrt{1-r'^2}\bigl(-\sin\theta\,\mathbf{N} + \cos\theta\,\mathbf{B}\bigr).\end{aligned}$$

From (35.35) and (35.36) we get

$$(35.37) \qquad \mathbf{U}_\theta \cdot \mathbf{U}_t = \pm r'\kappa\sin\theta\sqrt{1-r'^2} + \tau(1-r'^2).$$

On the other hand, when we differentiate (35.32) with respect to t, we find that

$$(35.38) \qquad \mathbf{y}_t = \mathbf{T} + r'\mathbf{U} + r\,\mathbf{U}_t.$$

Then (35.32), (35.33), (35.35), (35.37) and (35.38) imply that

$$(35.39) \quad \begin{aligned}\mathbf{y}_t \cdot \mathbf{y}_\theta &= \bigl(\mathbf{T} + r'\mathbf{U} + r\,\mathbf{U}_t\bigr)\cdot\bigl(r\,\mathbf{U}_\theta\bigr) = r^2\mathbf{U}_t\cdot\mathbf{U}_\theta \\ &= r^2\bigl(\pm r'\kappa\sin\theta\sqrt{1-r'^2} + \tau(1-r'^2)\bigr).\end{aligned}$$

If $\mathbf{y}_t$ is principal, then it must be perpendicular to $\mathbf{y}_\theta$. If this happens, (35.39) implies that both terms on the right-hand side of (35.39) vanish, since the first term depends on θ and the second does not. Thus $\tau = 0$ and either $r' = 0$ or $\kappa = 0$. Thus (i) implies (ii). Conversely, if (ii) holds, then (35.39) implies that $\mathbf{y}_t$ is principal. Finally, (ii) is clearly equivalent to (iii). ∎

Canal Surfaces via Mathematica

The *Mathematica* version of a canal surface is

```
canalsurf[gamma_][pm_,r_][t_,theta_]:=
    Simplify[gamma[tt] -
    r[tt]D[r[tt],tt]tangent[gamma][tt] +
    pm r[tt]Sqrt[1 - D[r[tt],tt]^2]
        (-Cos[theta]normal[gamma][tt] +
        Sin[theta]binormal[gamma][tt])] /. tt->t
```

where **pm** is $+1$ or -1.

Here is an example of a canal surface about a plane curve that is not a tube.

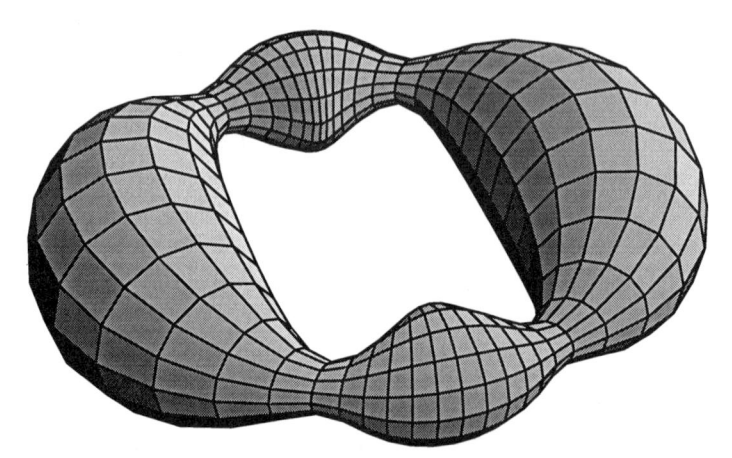

The canal surface generated by a sphere of radius
$1.8 + 0.48\sin(2t)$ **moving on a circle of radius** 5

35.3 Cyclides of Dupin via Focal Sets

In the early 1800s a class of surfaces all of whose lines of curvature are circles or lines was studied by Charles Dupin. Not only does this class include cones, cylinders and tori, but also distinct new surfaces. Dupin called such surfaces **cyclides** in his book [Dupin2]; the word "cyclide" seems to have been coined by Dupin. Since there is now a more general notion of cyclide (and to give homage to Dupin), the surfaces studied by Dupin are usually referred to as **cyclides of Dupin**.

Three methods are available for the study of cyclides of Dupin.

(i) Dupin's original method consisted of describing a cyclide as the envelope of all spheres tangent to three given spheres.

(ii) Liouville ([Liou]) showed that any cyclide of Dupin is the image under an inversion of a torus, circular cylinder or circular cone.

(iii) Maxwell[1] [Max] gave a direct construction of a cyclide of Dupin as the envelope of spheres whose centers move along ellipses, hyperbolas or parabolas; this construction directly generalizes the familiar construction of a torus as the envelope of spheres moving around a circle.

In this section we follow the approach of Maxwell, as modified by Darboux. See also [Darb4, pages 405-406] and [Bomp]. Inversions of tori, circular cylinders and cones are considered in Section 36.4.

First we define the cyclides of Dupin and give explicit parametrizations of them. From the construction it will be clear that the focal set of a cyclide of Dupin consists of two curves. In other words, a cyclide of Dupin can be considered to be a canal surface of either of the components of its focal set. Finally, we show that any surface for which the focal set consists of two curves must be a cyclide of Dupin. The following discussion is based on [Darb4, pages 262-263].

---

[1] James Clerk Maxwell (1831-1879). A leading scientist of the 19th century, celebrated for his **Treatise on Electricity and Magnetism**. The paper [Max] is informative and elegant. Before giving parametrizations of cyclides, Maxwell characterized them by a geometrical argument using almost no algebra. The paper also includes images viewable by a stereoscope together with Maxwell's recommendations for constructing the stereoscope.

Elliptic-Hyperbolic Cyclides

Let **ellipse**$_{xy}$ be the ellipse in the xy-plane given by

$$\frac{x_1^2}{a^2} + \frac{y_1^2}{a^2 - c^2} = 1,$$

and let **hyperbola**$_{xz}$ be the hyperbola in the xz-plane given by

$$\frac{x_2^2}{c^2} - \frac{z_2^2}{a^2 - c^2} = 1.$$

We assume that $a > c > 0$. Notice that one focus of the ellipse is $(c, 0, 0)$, which lies on the hyperbola, whereas one focus of the hyperbola is $(a, 0, 0)$, which lies on the ellipse.

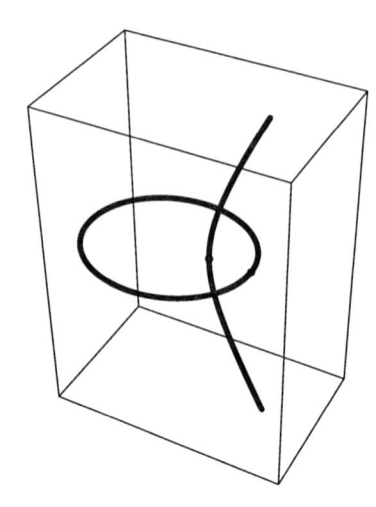

Ellipse and hyperbola in orthogonal planes

We first derive a simple formula for the distance between a point on the ellipse and a point on the hyperbola.

Lemma 35.14. *The distance between a point $(x_1, y_1, 0)$ on* **ellipse**$_{xy}$ *and a point $(x_2, 0, z_2)$ on the* **hyperbola**$_{xz}$ *is given by the formula*

(35.40) $$\mathbf{distance}((x_1, y_1, 0), (x_2, 0, z_2)) = \left| \frac{c}{a} x_1 - \frac{a}{c} x_2 \right|.$$

35.3 Cyclides of Dupin via Focal Sets

Proof. We compute as follows:

$$\Big(\text{distance}\big((x_1,y_1,0),(x_2,0,z_2)\big)\Big)^2 = (x_1 - x_2)^2 + y_1^2 + z_2^2$$

$$= (x_1 - x_2)^2 + (a^2 - c^2)\left(1 - \frac{x_1^2}{a^2} + \frac{x_2^2}{c^2} - 1\right)$$

$$= x_1^2 - 2x_1x_2 + x_2^2 - x_1^2 + \frac{c^2}{a^2}x_1^2 + \frac{a^2}{c^2}x_2^2 - x_2^2 = \left(\frac{c}{a}x_1 - \frac{a}{c}x_2\right)^2. \blacksquare$$

Next, we parametrize **ellipse**$_{xy}$ and **hyperbola**$_{xz}$ by

(35.41) $$\begin{cases} (x_1, y_1, 0) = \boldsymbol{\alpha}(u) = \big(a\cos u, \sqrt{a^2 - c^2}\sin u, 0\big), \\ (x_2, 0, z_2) = \boldsymbol{\beta}(v) = \big(c\sec v, 0, \sqrt{a^2 - c^2}\tan v\big). \end{cases}$$

Lemma 35.15. *Let* **sphere**$_1(u)$ *be a sphere with radius* $\rho_1(u)$ *centered at* $\boldsymbol{\alpha}(u)$ *and let* **sphere**$_2(v)$ *be a sphere with radius* $\rho_2(v)$ *centered at* $\boldsymbol{\beta}(v)$. *Suppose that the intersection of* **sphere**$_1(u)$ *and* **sphere**$_2(v)$ *consists of a single point* (x_3, y_3, z_3), *so that the two spheres are tangent at* (x_3, y_3, z_3). *Then there exists* k, *which is constant with respect to* u *and* v, *such that*

(35.42) $$\rho_1(u) = c\cos u - k \quad \text{and} \quad \rho_2(v) = -a\sec v + k.$$

Proof. We have

(35.43) $$\begin{cases} \text{distance}\big((x_1, y_1, 0), (x_3, y_3, z_3)\big) = \rho_1(u), \\ \text{distance}\big((x_2, 0, z_2), (x_3, y_3, z_3)\big) = \rho_2(v). \end{cases}$$

Since **sphere**$_1(u)$ and **sphere**$_2(v)$ are tangent at (x_3, y_3, z_3), Lemma 35.14 implies that

$$\rho_1(v) + \rho_2(u) = \text{distance}\big((x_1, y_1, 0), (x_3, y_3, z_3)\big) + \text{distance}\big((x_2, 0, z_2), (x_3, y_3, z_3)\big)$$

$$= \text{distance}\big((x_1, y_1, 0), (x_2, 0, z_2)\big) = \left|\frac{c}{a}x_1 - \frac{a}{c}x_2\right|$$

$$= |c\cos u - a\sec v| = a\sec v - c\cos u.$$

Thus

(35.44) $$\rho_1(u) - c\cos u = -\rho_2(v) + a\sec v.$$

Since the left-hand side of (35.44) depends only on u, while the right-hand side of (35.44) depends only on v, we get (35.42). $\blacksquare$

Let us now prove the converse of Lemma 35.15.

Lemma 35.16. *Let* $\mathrm{sphere}_1(u)$ *be a sphere with radius* $\rho_1(u)$ *centered at* $\alpha(u)$, *and let* $\mathrm{sphere}_2(v)$ *be a sphere with radius* $\rho_2(v)$ *centered at* $\beta(v)$. *Suppose that* $\rho_1(u)$ *and* $\rho_2(v)$ *are given by* (35.42) *for some* k. *Let* (x_3, y_3, z_3) *be the point on the line connecting the points* $\alpha(u)$ *and* $\beta(v)$ *at a distance* $\rho_1(u)$ *from* $\alpha(u)$. *Then the distance between the points* (x_3, y_3, z_3) *and* $\beta(v)$ *is* $\rho_2(v)$. *Furthermore,* $\mathrm{sphere}_1(u)$ *and* $\mathrm{sphere}_2(v)$ *are mutually tangent at* (x_3, y_3, z_3).

Proof. Lemma 35.14 implies that the distance from $\alpha(u)$ and $\beta(v)$ is

$$\mathrm{distance}\big((x_1, y_1, 0), (x_2, 0, z_2)\big) = \left| \frac{c}{a} x_1 - \frac{a}{c} x_2 \right| = |c \cos u - a \sec v| = a \sec v - c \cos u,$$

which precisely equals $\rho_1(u) + \rho_2(v)$. Therefore, (35.43) holds. This can only happen if $\mathrm{sphere}_1(u)$ and $\mathrm{sphere}_2(v)$ are mutually tangent at (x_3, y_3, z_3). ∎

As $\mathrm{sphere}_1(u)$ moves along the ellipse $\alpha(u)$ it traces out a surface, the envelope of the spheres. This envelope must coincide with the envelope formed by $\mathrm{sphere}_2(v)$ as it moves along the ellipse $\beta(v)$. The resulting surface is called a **cyclide of Dupin of elliptic-hyperbolic type**. It depends on three parameters a, c and k. In order to parametrize the cyclide, we define

$$\mathbf{x}(u, v) = \alpha(u) + \rho_1(u) V(u, v) = \beta(v) - \rho_2(v) V(u, v),$$

where $V(u, v)$ is a unit vector. We have

$$0 = \alpha(u) - \beta(v) + \big(\rho_1(u) + \rho_2(v)\big) V(u, v),$$

so that

$$V(u, v) = -\frac{\alpha(u) - \beta(v)}{\rho_1(u) + \rho_2(v)}.$$

Hence

$$\mathbf{x}(u, v) = \alpha(u) - \rho_1(u) \left(\frac{\alpha(u) - \beta(v)}{\rho_1(u) + \rho_2(v)} \right) = \frac{\rho_2(v) \alpha(u) + \rho_1(u) \beta(v)}{\rho_1(u) + \rho_2(v)}$$

$$= \frac{(-k + a \sec v) \alpha(u) + (k - c \cos u) \beta(v)}{a \sec v - c \cos u}$$

$$= \frac{(-k \cos v + a) \alpha(u) + (k - c \cos u)(\cos v) \beta(v)}{a - c \cos u \cos v}$$

$$= \frac{(-k \cos v + a)(a \cos u, \sqrt{a^2 - c^2} \sin u, 0)}{a - c \cos u \cos v}$$

$$+ \frac{(k - c \cos u) \cos v (c \sec v, 0, \sqrt{a^2 - c^2} \tan v)}{a - c \cos u \cos v}$$

35.3 Cyclides of Dupin via Focal Sets

$$= \left(\frac{c(k - c\cos u) + a\cos u(a - k\cos v)}{a - c\cos u \cos v}, \frac{\sqrt{a^2 - c^2}(a - k\cos v)\sin u}{a - c\cos u \cos v}, \right.$$

$$\left. \frac{\sqrt{a^2 - c^2}(k - c\cos u)\sin v}{a - c\cos u \cos v} \right).$$

The *Mathematica* version is

```
ellhypcyclide[a_,c_,k_][u_,v_]:=
    {(c(k - c Cos[u]) + a Cos[u](a - k Cos[v]))/
        (a - c Cos[u]Cos[v]),
    (Sqrt[a^2 - c^2](a - k Cos[v])Sin[u])/
        (a - c Cos[u]Cos[v]),
    (Sqrt[a^2 - c^2](k - c Cos[u])Sin[v])/
        (a - c Cos[u]Cos[v])}
```

We plot three examples of elliptic-hyperbolic cyclides. A typical **ring cyclide** can be plotted with command

```
ParametricPlot3D[Evaluate[
ellhypcyclide[12,3,4][u,v]],{u,0,2Pi},{v,0,2Pi},
PlotPoints->{45,25},PlotRange->All];
```

which yields the picture on the left below.

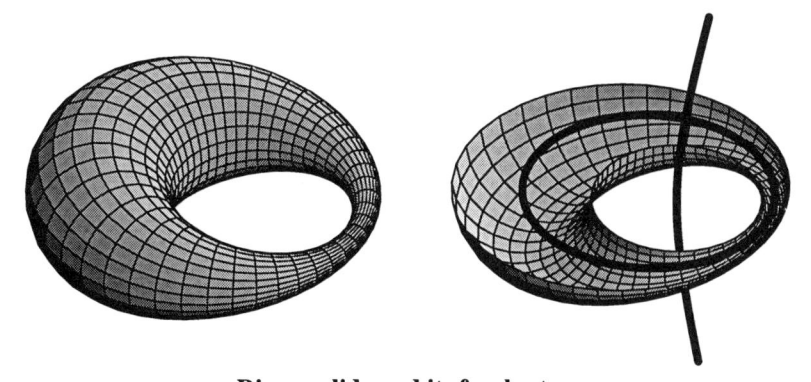

Ring cyclide and its focal set

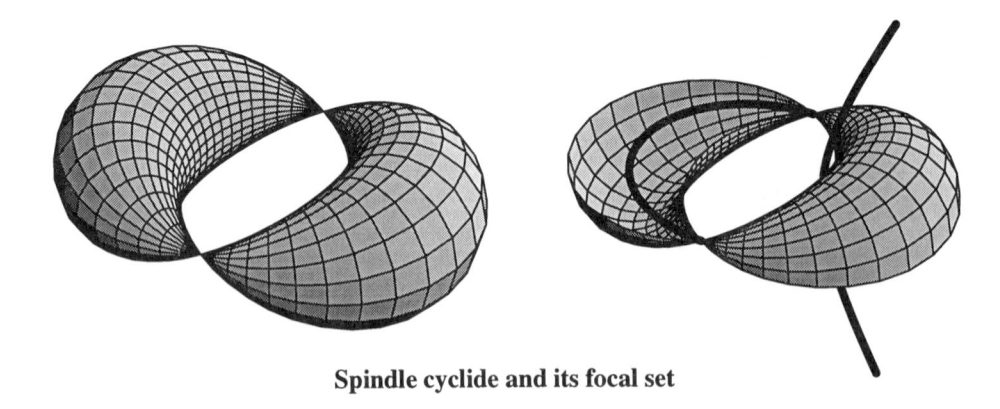

Spindle cyclide and its focal set

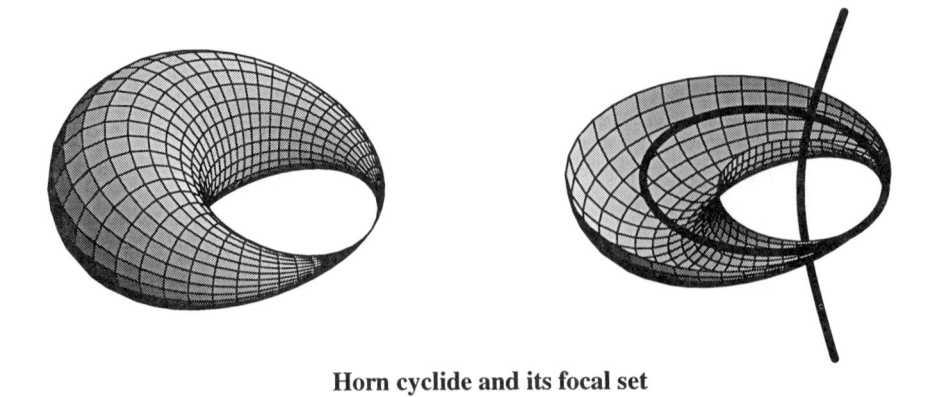

Horn cyclide and its focal set

Parabolic Cyclides

There is another type of cyclide whose focal set consist of two parabolas. Let **parabola**$_{xz}$ be the parabola in the xz-plane given by

$$z_1 = -\frac{x_1^2}{8a} + a,$$

and let **parabola**$_{yz}$ be the parabola in the yz-plane given by

$$z_2 = \frac{x_2^2}{8a} - a.$$

We parametrize **parabola**$_{xz}$ and **parabola**$_{yz}$ by

(35.45) $$\begin{cases} (x_1, 0, z_1) = \boldsymbol{\alpha}(u) = \left(u, 0, -\frac{u^2}{8a} + a\right), \\ (0, y_2, z_2) = \boldsymbol{\beta}(v) = \left(0, -v, \frac{v^2}{8a} - a\right). \end{cases}$$

Note that $\boldsymbol{\alpha}$ passes through the focus of $\boldsymbol{\beta}$ (namely $(0,0,a)$) and $\boldsymbol{\beta}$ passes through the focus of $\boldsymbol{\alpha}$ (namely $(0,0,-a)$).

Lemma 35.17. *Let* **sphere**$_1(u)$ *be a sphere with radius* $\rho_1(u)$ *centered at* $\boldsymbol{\alpha}(u)$, *and let* **sphere**$_2(v)$ *be a sphere with radius* $\rho_2(v)$ *centered at* $\boldsymbol{\beta}(v)$. *Suppose that the intersection of* **sphere**$_1(u)$ *and* **sphere**$_2(v)$ *consists of a point* (x_3, y_3, z_3), *and that the two spheres are tangent there. Then there exists k, which is constant with respect to u and v, such that*

(35.46) $$\rho_1(u) = \frac{u^2}{8a} - a + k \quad \text{and} \quad \rho_2(v) = \frac{v^2}{8a} + a - k.$$

Proof. We have

$$\|\boldsymbol{\alpha}(u) - \boldsymbol{\beta}(v)\|^2 = \left\|\left(u, v, -\frac{u^2}{8a} - \frac{v^2}{8a} - 2a\right)\right\|^2 = u^2 + v^2 + \left(2a - \frac{1}{8a}(u^2 + v^2)\right)^2$$

$$= \frac{1}{64a^2}\left((u^2 + v^2)^2 + 32a^2(u^2 + v^2) + 256a^4\right).$$

Hence

$$\|\boldsymbol{\alpha}(u) - \boldsymbol{\beta}(v)\| = \frac{u^2 + v^2 + 16a^2}{8a}.$$

Furthermore,

$$\text{distance}\big((x_1, 0, z_1), (x_3, y_3, z_3)\big) = \rho_1(u)$$

and
$$\text{distance}\big((0, y_2, z_2), (x_3, y_3, z_3)\big) = \rho_2(v).$$

Since $\mathbf{sphere}_1(u)$ and $\mathbf{sphere}_2(v)$ are tangent at (x_3, y_3, z_3), we must have

$$\begin{aligned}\rho_1(u) + \rho_2(v) &= \text{distance}\big((x_1, 0, z_1), (x_3, y_3, z_3)\big) + \text{distance}\big((0, y_2, z_2), (x_3, y_3, z_3)\big) \\ &= \text{distance}\big((x_1, 0, z_1), (0, y_2, z_2)\big) \\ &= \|\alpha(u) - \beta(v)\| = \frac{u^2 + v^2 + 16a^2}{8a},\end{aligned}$$

so that

(35.47) $$\rho_1(u) - \frac{u^2}{8a} + a = -\rho_2(v) - \frac{v^2}{8a} + a.$$

Just as in the proof of Lemma 35.15, the left-hand side of (35.47) depends only on u, while the right-hand side depends only on v. Thus there exists a constant k such that

$$\rho_1(u) - \frac{u^2}{8a} + a = \rho_2(v) + \frac{v^2}{8a} - a = k,$$

and we obtain (35.46). ∎

In *Mathematica* we define

```
paraboliccyclide[a_,k_][u_,v_]:=
    {(u(8a^2 + k + v^2))/(16a^2 + u^2 + v^2),
     ((8a^2 - k + u^2)v)/(16a^2 + u^2 + v^2),
     (16a^2k - 16a^2u^2 - k u^2 + 16a^2v^2 - k v^2)/
     (8a(16a^2 + u^2 + v^2))}
```

Finally, we determine all surfaces whose focal set consists of two curves.

Theorem 35.18. *Let $\mathcal{M}$ be a surface for which the focal set $\mathbf{Focal}(\mathcal{M})$ consists of two curves. Then each curve is a conic section (that is, an ellipse, hyperbola, parabola or straight line), and the planes of each component are perpendicular to one another.*

Proof. Write
$$\mathbf{Focal}(\mathcal{M}) = \mathbf{Focal}(\mathcal{M})_1 \cup \mathbf{Focal}(\mathcal{M})_2,$$
where the union is disjoint. Consider a point $\mathbf{p}$ in $\mathbf{Focal}(\mathcal{M})_1$. The set of normal lines to $\mathcal{M}$ that pass through $\mathbf{p}$ forms a cone $\mathbf{Cone}(\mathbf{p}, \mathcal{M})$ with $\mathbf{p}$ as its vertex.

By definition of focal set, each line in this cone also passes through a point of **Focal**($\mathcal{M}$)$_2$; furthermore, for each point $\mathbf{q} \in$ **Focal**($\mathcal{M}$)$_2$, the line containing $\mathbf{p}$ and $\mathbf{q}$ is in **Cone**($\mathbf{p}, \mathcal{M}$). It follows that the whole curve **Focal**($\mathcal{M}$)$_2$ must lie on **Cone**($\mathbf{p}, \mathcal{M}$).

This construction creates a cone for any $\mathbf{p} \in$ **Focal**($\mathcal{M}$)$_1$. In particular, it works for a point that is nearest to **Focal**($\mathcal{M}$)$_2$. For such a point, the corresponding cone is part of a plane. It follows that **Focal**($\mathcal{M}$)$_2$ is a plane curve.

Clearly, the same argument shows that **Focal**($\mathcal{M}$)$_1$ is also a plane curve lying on a cone. Hence both **Focal**($\mathcal{M}$)$_1$ and **Focal**($\mathcal{M}$)$_2$ are by definition conic sections. In fact **Cone**($\mathbf{p}, \mathcal{M}$) must be a right cone for $\mathbf{p} \in$ **Focal**($\mathcal{M}$)$_1$, since it contains a principal curve of $\mathcal{M}$, which is a circle. ∎

35.4 Exercises

1. Complete the proofs of Theorem 35.1, Lemma 35.2 and Lemma 35.4.

2. Plot an elliptic paraboloid together with its two focal hypersurfaces.

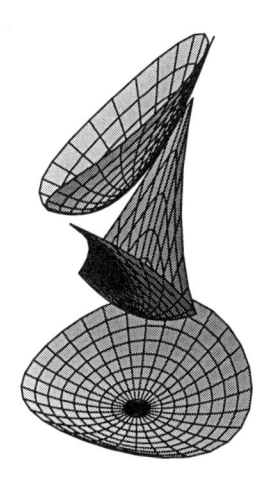

Elliptic paraboloid and its focal set

3. Let $\mathcal{M}$ be a surface of revolution generated by a curve C, not a part of a circle. Show that the surface of revolution generated by the evolute of C is one component of the focal set of $\mathcal{M}$.

4. Use *Mathematica* to find the two components of the focal set of a circular helicoid (see exercise 3 of Chapter 15).

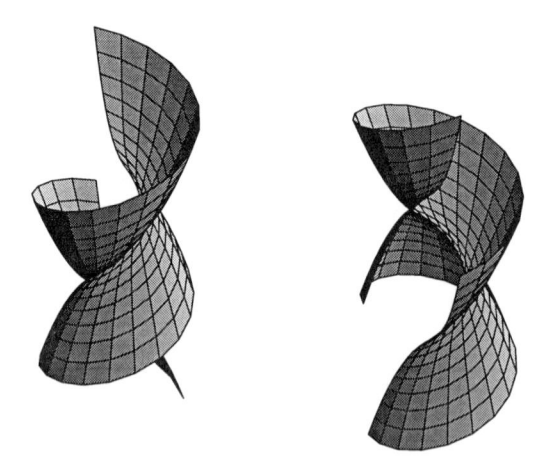

The two components of the focal set of a circular helicoid

5. Prove Lemma 35.3 and give the details of equations (35.9) and (35.17).

6. Let $\mathcal{M}$ be a surface with the property that $\rho_1 - \rho_2 = a$, where a is a constant. Assume that each component of the focal set of $\mathcal{M}$ is a regular surface. Show that each component of the focal set of $\mathcal{M}$ has constant negative curvature $-a^{-2}$.

7. Plot several parabolic cyclides.

36

INVERSIONS OF CURVES AND SURFACES

Perhaps the simplest type of mapping between open subsets of $\mathbb{R}^n$, other than an isometry or affine transformation, is an inversion. Although an inversion does not preserve distances, it does map the inside of a sphere into the outside in a natural way. Furthermore, as we shall show, it is a conformal map.

In Section 36.1 we give the precise definition of inversion and prove that any inversion preserves angles. Images of curves under inversion are studied in Section 36.2. The more complicated theory of inversion of surfaces is started in Section 36.3. Cyclides of Dupin are studied as inversions of tori in Section 36.4. Liouville's Theorem (which states that a conformal diffeomorphism between open subsets of $\mathbb{R}^3$ is a composition of inversions) is proved in Section 36.5.

36.1 The Definition of Inversion

We want to invert $\mathbb{R}^n$ with respect to a point $\mathbf{q} \in \mathbb{R}^n$; this means that points near to $\mathbf{q}$ are mapped far away in a reasonable way, and vice versa. We also want the composition of an inversion with itself to be the identity map. The precise definition is as follows:

Definition. *The* **inversion** *of $\mathbb{R}^n$ with respect to a point $\mathbf{q} \in \mathbb{R}^n$ with* **inversion radius** ρ *is the map* $\operatorname{inversion}[\mathbf{q}, \rho]: \mathbb{R}^n \setminus \mathbf{q} \longrightarrow \mathbb{R}^n \setminus \mathbf{q}$ *given by*

$$\operatorname{inversion}[\mathbf{q}, \rho](\mathbf{p}) = \mathbf{q} + \frac{\rho^2 (\mathbf{p} - \mathbf{q})}{\|\mathbf{p} - \mathbf{q}\|^2}. \tag{36.1}$$

We call $\mathbf{q}$ the **inversion center** *or* **pole** *of an inversion. In the special case $\mathbf{q} = 0$ we write*

$$\operatorname{inversion}[\rho](\mathbf{p}) = \frac{\rho^2 \mathbf{p}}{\|\mathbf{p}\|^2}. \tag{36.2}$$

It is easy to check that two applications of an inversion yield the identity map; that is, $\operatorname{inversion}[\mathbf{q}, \rho]^2$ is the identity map of $\mathbb{R}^n \setminus \mathbf{q}$.

An inversion is characterized by two properties.

Lemma 36.1. *Let $\mathbf{q} \in \mathbb{R}^n$. A map $\Phi: \mathbb{R}^n \setminus \mathbf{q} \longrightarrow \mathbb{R}^n \setminus \mathbf{q}$ is an inversion with inversion radius ρ if and only if there is a number $\lambda > 0$ such that*

$$\|\Phi(\mathbf{p}) - \mathbf{q}\| \, \|\mathbf{p} - \mathbf{q}\| = \rho^2 \quad \text{and} \quad \Phi(\mathbf{p}) - \mathbf{q} = \lambda (\mathbf{p} - \mathbf{q}) \tag{36.3}$$

for all $\mathbf{p}$.

Proof. It is easy to prove that an inversion satisfies (36.3). Conversely, suppose that Φ satisfies (36.3). Then $\lambda \|\mathbf{p} - \mathbf{q}\|^2 = \rho^2$, and so

$$\Phi(\mathbf{p}) - \mathbf{q} = \frac{\rho^2 (\mathbf{p} - \mathbf{q})}{\|\mathbf{p} - \mathbf{q}\|^2}. \quad \blacksquare$$

Because of Lemma 36.1, an inversion of $\mathbb{R}^n$ is sometimes called a **transformation by reciprocal radii** in the classical literature.

Next, we show that an inversion has the important property that its tangent map preserves angles between tangent vectors. For this, we make the following definition, which corresponds to the notion of conformality for surface mappings given on page 349.

Definition. *Let $\Phi: \mathcal{U} \longrightarrow \Phi(\mathcal{U}) \subset \mathbb{R}^n$ be a map, where $\mathcal{U}$ is an open subset of $\mathbb{R}^n$. We say that Φ is a* **conformal map** *of $\mathcal{U}$ onto $\Phi(\mathcal{U})$ provided there is a differentiable everywhere positive function $\lambda: \mathcal{U} \longrightarrow \mathbb{R}$ such that*

$$\|\Phi_*(\mathbf{v}_\mathbf{p})\| = \lambda(\mathbf{p}) \|\mathbf{v}_\mathbf{p}\| \tag{36.4}$$

for all $\mathbf{p} \in \mathcal{U}$. A **conformal diffeomorphism** *is a conformal map which is also a diffeomorphism.*

36.1 The Definition of Inversion

Lemma 36.2. *An inversion is a conformal diffeomorphism.*

Proof. Write $\Phi = \text{\bf inversion}[\mathbf{q}, \rho]$ and let $\boldsymbol{\alpha}\colon (a,b) \longrightarrow \mathbb{R}^n$ be a curve passing through $\mathbf{q}$. Then

$$(36.5) \qquad (\Phi \circ \boldsymbol{\alpha})(t) = \mathbf{q} + \frac{\rho^2 (\boldsymbol{\alpha}(t) - \mathbf{q})}{\|\boldsymbol{\alpha}(t) - \mathbf{q}\|^2}$$

for $a < t < b$. Differentiating (36.5), we obtain

$$(36.6) \qquad (\Phi \circ \boldsymbol{\alpha})'(t) = \frac{\rho^2 \boldsymbol{\alpha}'(t)}{\|\boldsymbol{\alpha}(t) - \mathbf{q}\|^2} - \frac{2\rho^2 ((\boldsymbol{\alpha}(t) - \mathbf{q}) \cdot \boldsymbol{\alpha}'(t))(\boldsymbol{\alpha}(t) - \mathbf{q})}{\|\boldsymbol{\alpha}(t) - \mathbf{q}\|^4}.$$

From (36.6) it follows that

$$
\begin{aligned}
(36.7) \quad \|(\Phi \circ \boldsymbol{\alpha})'(t)\|^2 &= \left\| \frac{\rho^2 \boldsymbol{\alpha}'(t)}{\|\boldsymbol{\alpha}(t) - \mathbf{q}\|^2} - \frac{2\rho^2 ((\boldsymbol{\alpha}(t) - \mathbf{q}) \cdot \boldsymbol{\alpha}'(t))(\boldsymbol{\alpha}(t) - \mathbf{q})}{\|\boldsymbol{\alpha}(t) - \mathbf{q}\|^4} \right\|^2 \\
&= \frac{\rho^4 \|\boldsymbol{\alpha}'(t)\|^2}{\|\boldsymbol{\alpha}(t) - \mathbf{q}\|^4} + \frac{4\rho^4 ((\boldsymbol{\alpha}(t) - \mathbf{q}) \cdot \boldsymbol{\alpha}'(t))^2 \|(\boldsymbol{\alpha}(t) - \mathbf{q})\|^2}{\|\boldsymbol{\alpha}(t) - \mathbf{q}\|^8} \\
&\quad - \frac{4\rho^4 ((\boldsymbol{\alpha}(t) - \mathbf{q}) \cdot \boldsymbol{\alpha}'(t))^2}{\|\boldsymbol{\alpha}(t) - \mathbf{q}\|^6} \\
&= \frac{\rho^4 \|\boldsymbol{\alpha}'(t)\|^2}{\|\boldsymbol{\alpha}(t) - \mathbf{q}\|^4}.
\end{aligned}
$$

Since any tangent vector can be realized as the velocity vector of a curve, (36.7) implies that

$$\|\Phi_*(\mathbf{v}_\mathbf{p})\| = \frac{\rho^2 \|\mathbf{v}_\mathbf{p}\|}{\|\mathbf{p} - \mathbf{q}\|^2}$$

for all tangent vectors $\mathbf{v}_\mathbf{p}$. Hence Φ distorts lengths of all tangent vectors at $\mathbf{p}$ by the same factor

$$\lambda(p) = \frac{\rho^2}{\|\mathbf{p} - \mathbf{q}\|^2},$$

and so Φ is a conformal map. ∎

Definition. *Let S be a subset of $\mathbb{R}^n$. The **inverse** of S with respect to $\mathbf{q} \in \mathbb{R}^n$ with radius of inversion ρ is the image of S under the inversion $\text{\bf inversion}[\mathbf{q}, \rho]$.*

Here are the *Mathematica* versions of inversions corresponding to (36.1) and (36.2).

```
inversion[q_,rho_][p_]:= q +
    (rho^2/((p - q).(p - q)))(p - q)

inversion[rho_][p_]:= (rho^2/p.p)p
```

36.2 Inversion of Curves

In this section we describe the image under inversion of several curves.

Definition. *The inversion of a curve* $\alpha: (a,b) \longrightarrow \mathbb{R}^n$ *is the curve* $t \longmapsto$ inversioncurve$[q, \rho, \alpha](t)$ *defined by*

(36.8) $\qquad$ inversioncurve$[q, \rho, \alpha](t) =$ inversion$[q, \rho]\big(\alpha(t)\big).$

The *Mathematica* definition of **inversioncurve** is as follows:

```
inversioncurve[q_List,rho_][curve_][t_]:=
                    inversion[q,rho][curve[t]]

inversioncurve[rho_][curve_][t_]:=
                    inversion[rho][curve[t]]
```

For example, the following command computes the inverse curve of a clothoid with respect to a generic point and a generic radius:

```
inversioncurve[{p1,p2},rho][clothoid[1,a]][t]
```

We plot the inverse curve of a clothoid with respect to the point $(2, 2)$ using inversion radius 1:

36.2 Inversion of Curves

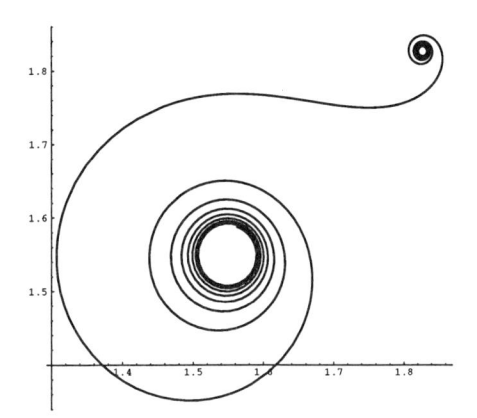

Inverse curve of a clothoid

Inversion provides a useful way to visualize what happens to a curve at infinity. For example, let us compare inverse curves of the catenary and the hyperbola. The inverse curve with inversion radius 1 with respect to the origin of a catenary is found with

inversioncurve[{0,0},1][catenary[1]][t]

resulting in

$$\{\frac{\operatorname{Cosh}[t]}{t^2 + \operatorname{Cosh}[t]^2}, \frac{t}{t^2 + \operatorname{Cosh}[t]^2}\}$$

Similarly, the inverse curve of a hyperbola with inversion radius 1 with respect to the origin is found with

inversioncurve[{0,0},1][hyperbola[1,1]][t]

$$\{\frac{\operatorname{Cosh}[t]}{\operatorname{Cosh}[t]^2 + \operatorname{Sinh}[t]^2}, \frac{\operatorname{Sinh}[t]}{\operatorname{Cosh}[t]^2 + \operatorname{Sinh}[t]^2}\}$$

The plots of these two curves with t varying from -10 to 10 are as follows.

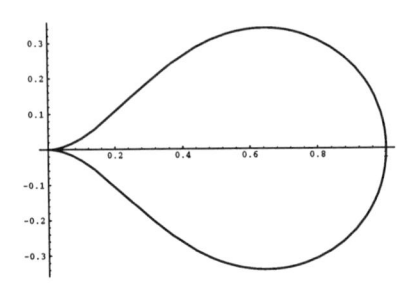

Inverse curve of a catenary 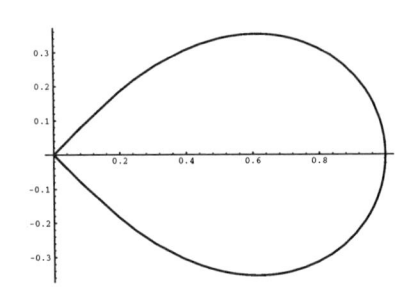   **Inverse curve of a hyperbola**

Although the shape of a hyperbola is not too different from that of a catenary, the inverse curves are quite different.

We can also find the images of space curves under inversion. The command

```
ParametricPlot3D[Evaluate[
    inversioncurve[{3,0,0},1][helix[1,0.1]][t]],
    {t,-30Pi,30Pi},PlotPoints->1000];
```

plots the inverse curve of a helix with respect to $(3,0,0)$ using inversion radius 1

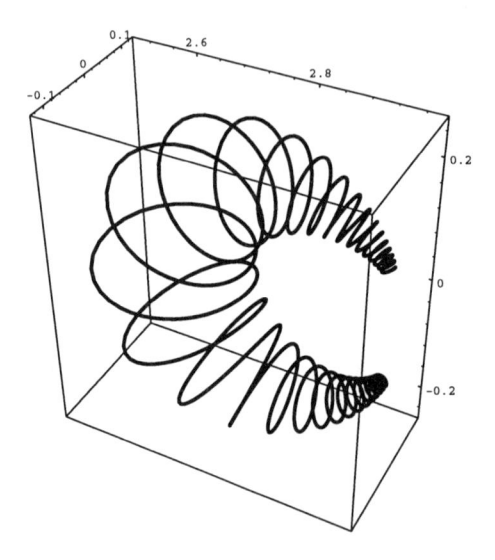

Inverse curve of a helix

36.3 Inversion of Surfaces

In this section we obtain formulas relating the first and second fundamental forms and the various curvatures of a patch and its composition with an inversion.

Definition. *Let* $\mathbf{x}: \mathcal{U} \longrightarrow \mathbb{R}^n$. *The* **inverse patch** *of* $\mathbf{x}$ *with respect to* $\mathbf{q} \in \mathbb{R}^n$ *with radius of inversion* ρ *is the patch given by*

$$\mathbf{y} = \text{inversion}[\mathbf{q}, \rho] \circ \mathbf{x}.$$

First, we find formulas for the tangent vectors $\mathbf{y}_u$ and $\mathbf{y}_v$ of an inverse patch $\mathbf{y}$ of a patch $\mathbf{x}$ in terms of $\mathbf{x}_u$ and $\mathbf{x}_v$.

Lemma 36.3. *Let* $\mathbf{x}$ *be a patch in* $\mathbb{R}^n$, *and let* $\mathbf{y}$ *be the inverse patch of* $\mathbf{x}$ *with respect to* $\mathbf{q} \in \mathbb{R}^n$ *with radius of inversion* ρ. *Then*

(36.9)
$$\begin{cases} \mathbf{y}_u = \dfrac{\rho^2 \mathbf{x}_u}{\|\mathbf{x} - \mathbf{q}\|^2} - \dfrac{2\rho^2 (\mathbf{x}_u \cdot (\mathbf{x} - \mathbf{q}))(\mathbf{x} - \mathbf{q})}{\|\mathbf{x} - \mathbf{q}\|^4}, \\ \mathbf{y}_v = \dfrac{\rho^2 \mathbf{x}_v}{\|\mathbf{x} - \mathbf{q}\|^2} - \dfrac{2\rho^2 (\mathbf{x}_v \cdot (\mathbf{x} - \mathbf{q}))(\mathbf{x} - \mathbf{q})}{\|\mathbf{x} - \mathbf{q}\|^4}. \end{cases}$$

Proof. The definition of inversion gives the following formula for $\mathbf{y}$:

(36.10)
$$\mathbf{y}(u, v) = \mathbf{q} + \frac{\rho^2 (\mathbf{x}(u, v) - \mathbf{q})}{\|\mathbf{x}(u, v) - \mathbf{q}\|^2}.$$

When we differentiate (36.10) with respect to u and v we get (36.9). ∎

Next, we express the coefficients of the first fundamental form of $\mathbf{y}$ in terms of the coefficients of the first fundamental form of $\mathbf{x}$.

Lemma 36.4. *Let* $\mathbf{x}$ *be a patch in* $\mathbb{R}^n$, *and let* $\mathbf{y}$ *be the inverse patch of* $\mathbf{x}$ *with respect to* $\mathbf{q} \in \mathbb{R}^n$ *with radius of inversion* ρ. *Then the coefficients of the first fundamental form of* $\mathbf{x}$ *are related to those of* $\mathbf{y}$ *by*

(36.11)
$$\begin{cases} E(\mathbf{y}) = \dfrac{\rho^4}{\|\mathbf{x} - \mathbf{q}\|^4} E(\mathbf{x}), \\ F(\mathbf{y}) = \dfrac{\rho^4}{\|\mathbf{x} - \mathbf{q}\|^4} F(\mathbf{x}), \\ G(\mathbf{y}) = \dfrac{\rho^4}{\|\mathbf{x} - \mathbf{q}\|^4} G(\mathbf{x}). \end{cases}$$

Proof. Equations (36.11) are consequences of (36.9) by a calculation similar to that of (36.7). ∎

Next, we specialize to patches in $\mathbb{R}^3$. To determine the formula relating the normals of a patch and its inverse, we first prove

Lemma 36.5. *Let $\mathbf{x}$ be a patch in $\mathbb{R}^3$, and let $\mathbf{y}$ be the inverse patch of $\mathbf{x}$ with respect to $\mathbf{q} \in \mathbb{R}^3$ with radius of inversion ρ. Then*

$$(36.12) \qquad \mathbf{y}_u \times \mathbf{y}_v = -\frac{\rho^4 \mathbf{x}_u \times \mathbf{x}_v}{\|\mathbf{x}-\mathbf{q}\|^4} + \frac{2\rho^4\big((\mathbf{x}-\mathbf{q}) \cdot (\mathbf{x}_u \times \mathbf{x}_v)\big)(\mathbf{x}-\mathbf{q})}{\|\mathbf{x}-\mathbf{q}\|^6}.$$

Proof. From (36.9) and the fact that $(\mathbf{x}-\mathbf{q}) \times (\mathbf{x}-\mathbf{q}) = 0$, we get

$$(36.13) \qquad \mathbf{y}_u \times \mathbf{y}_v = \frac{\rho^4 \mathbf{x}_u \times \mathbf{x}_v}{\|\mathbf{x}-\mathbf{q}\|^4} - \frac{2\rho^4 \big(\mathbf{x}_v \cdot (\mathbf{x}-\mathbf{q})\mathbf{x}_u - \mathbf{x}_u \cdot (\mathbf{x}-\mathbf{q})\mathbf{x}_v\big) \times (\mathbf{x}-\mathbf{q})}{\|\mathbf{x}-\mathbf{q}\|^6}.$$

We use (8.3), page 183, twice to obtain

$$(36.14) \qquad -\big(\mathbf{x}_v \cdot (\mathbf{x}-\mathbf{q})\mathbf{x}_u - \mathbf{x}_u \cdot (\mathbf{x}-\mathbf{q})\mathbf{x}_v\big) \times (\mathbf{x}-\mathbf{q})$$

$$= (\mathbf{x}-\mathbf{q}) \times \big((\mathbf{x}-\mathbf{q}) \times \mathbf{x}_u \times \mathbf{x}_v\big)$$

$$= \big((\mathbf{x}-\mathbf{q}) \cdot \mathbf{x}_u \times \mathbf{x}_v\big)(\mathbf{x}-\mathbf{q}) - \|\mathbf{x}-\mathbf{q}\|^2 \mathbf{x}_u \times \mathbf{x}_v.$$

Then (36.13) and (36.14) imply (36.12). ∎

Now we can find the formula for the unit normal $\mathbf{U}_y$ to an inverse patch $\mathbf{y}$ to a patch $\mathbf{x}$ in terms of the unit normal $\mathbf{U}_x$.

Corollary 36.6. *Let $\mathbf{x}$ be a patch in $\mathbb{R}^3$, and let $\mathbf{y}$ be the inverse patch of $\mathbf{x}$ with respect to $\mathbf{q} \in \mathbb{R}^3$ with radius of inversion ρ. Denote by $\mathbf{U}_x$ and $\mathbf{U}_y$ the respective normals. Then*

$$(36.15) \qquad \mathbf{U}_y = -\mathbf{U}_x + \frac{2\big(\mathbf{U}_x \cdot (\mathbf{x}-\mathbf{q})\big)(\mathbf{x}-\mathbf{q})}{\|\mathbf{x}-\mathbf{q}\|^2}.$$

Proof. First, we use (36.12) to compute

$$\|\mathbf{y}_u \times \mathbf{y}_v\|^2 = \left\|\frac{\rho^4 \mathbf{x}_u \times \mathbf{x}_v}{\|\mathbf{x}-\mathbf{q}\|^4} - \frac{2\rho^4\big((\mathbf{x}-\mathbf{q}) \cdot (\mathbf{x}_u \times \mathbf{x}_v)\big)(\mathbf{x}-\mathbf{q})}{\|\mathbf{x}-\mathbf{q}\|^6}\right\|^2$$

$$= \left\|\frac{\rho^4 \mathbf{x}_u \times \mathbf{x}_v}{\|\mathbf{x}-\mathbf{q}\|^4}\right\|^2 + \frac{4\rho^8 \big((\mathbf{x}-\mathbf{q}) \cdot (\mathbf{x}_u \times \mathbf{x}_v)\big)^2 \|\mathbf{x}-\mathbf{q}\|^2}{\|\mathbf{x}-\mathbf{q}\|^{12}} - \frac{4\rho^8 \big((\mathbf{x}-\mathbf{q}) \cdot (\mathbf{x}_u \times \mathbf{x}_v)\big)^2}{\|\mathbf{x}-\mathbf{q}\|^{10}}$$

$$= \frac{\rho^8 \|\mathbf{x}_u \times \mathbf{x}_v\|^2}{\|\mathbf{x}-\mathbf{q}\|^8}.$$

36.3 Inversion of Surfaces

Then (36.15) implies that

$$\mathbf{U}_\mathbf{y} = \frac{\mathbf{y}_u \times \mathbf{y}_v}{\|\mathbf{y}_u \times \mathbf{y}_v\|} = \frac{\|\mathbf{x}-\mathbf{q}\|^4}{\rho^4 \|\mathbf{x}_u \times \mathbf{x}_v\|} \left(-\frac{\rho^4 \mathbf{x}_u \times \mathbf{x}_v}{\|\mathbf{x}-\mathbf{q}\|^4} + \frac{2\rho^4\big((\mathbf{x}-\mathbf{q})\cdot(\mathbf{x}_u\times\mathbf{x}_v)\big)(\mathbf{x}-\mathbf{q})}{\|\mathbf{x}-\mathbf{q}\|^6}\right)$$

$$= -\mathbf{U}_\mathbf{x} + \frac{2\big(\mathbf{U}_\mathbf{x}\cdot(\mathbf{x}-\mathbf{q})\big)(\mathbf{x}-\mathbf{q})}{\|\mathbf{x}-\mathbf{q}\|^2}. \quad \blacksquare$$

Next, we determine the formulas relating the coefficients of the second fundamental forms of a patch and its inverse.

Lemma 36.7. *Let $\mathbf{x}$ be a patch in $\mathbb{R}^3$, and let $\mathbf{y}$ be the inverse patch of $\mathbf{x}$ with respect to $\mathbf{q} \in \mathbb{R}^3$ with radius of inversion ρ. Then the coefficients of the second fundamental form of $\mathbf{y}$ are related to those of $\mathbf{x}$ by*

(36.16)
$$\begin{cases} e(\mathbf{y}) = \dfrac{-\rho^2 e(\mathbf{x})}{\|\mathbf{x}-\mathbf{q}\|^2} - \dfrac{2\rho^2 E(\mathbf{x})\big(\mathbf{U}_\mathbf{x}\cdot(\mathbf{x}-\mathbf{q})\big)}{\|\mathbf{x}-\mathbf{q}\|^4}, \\[6pt] f(\mathbf{y}) = \dfrac{-\rho^2 f(\mathbf{x})}{\|\mathbf{x}-\mathbf{q}\|^2} - \dfrac{2\rho^2 F(\mathbf{x})\big(\mathbf{U}_\mathbf{x}\cdot(\mathbf{x}-\mathbf{q})\big)}{\|\mathbf{x}-\mathbf{q}\|^4}, \\[6pt] g(\mathbf{y}) = \dfrac{-\rho^2 g(\mathbf{x})}{\|\mathbf{x}-\mathbf{q}\|^2} - \dfrac{2\rho^2 G(\mathbf{x})\big(\mathbf{U}_\mathbf{x}\cdot(\mathbf{x}-\mathbf{q})\big)}{\|\mathbf{x}-\mathbf{q}\|^4}. \end{cases}$$

Proof. To prove (36.16) we first use (36.9) to compute

$$(36.17) \quad \mathbf{y}_{uu} = \frac{\rho^2 \mathbf{x}_{uu}}{\|\mathbf{x}-\mathbf{q}\|^2} - \frac{4\rho^2\big(\mathbf{x}_u\cdot(\mathbf{x}-\mathbf{q})\big)\mathbf{x}_u}{\|\mathbf{x}-\mathbf{q}\|^4} - \frac{2\rho^2\big(\mathbf{x}_{uu}\cdot(\mathbf{x}-\mathbf{q})\big)(\mathbf{x}-\mathbf{q})}{\|\mathbf{x}-\mathbf{q}\|^4}$$

$$- \frac{2\rho^2 \|\mathbf{x}_u\|^2(\mathbf{x}-\mathbf{q})}{\|\mathbf{x}-\mathbf{q}\|^4} + \frac{8\rho^2\big(\mathbf{x}_u\cdot(\mathbf{x}-\mathbf{q})\big)^2(\mathbf{x}-\mathbf{q})}{\|\mathbf{x}-\mathbf{q}\|^6}.$$

Then the first equation of (36.16) follows from Corollary 36.6 and (36.17); the proofs of the other equations of (36.16) are similar. $\blacksquare$

From Lemmas 36.4 and 36.7 we see that an inversion of $\mathbb{R}^3$ multiplies the first fundamental form by a function, but the relation between the second fundamental forms is more complicated. Nevertheless, we have

Corollary 36.8. *Let* **x** *be a patch in* $\mathbb{R}^3$, *and let* **y** *be the inverse patch of* **x** *with respect to* $\mathbf{q} \in \mathbb{R}^3$ *with radius of inversion* ρ. *Then*

(36.18)
$$\begin{cases} f(\mathbf{y})E(\mathbf{y}) - e(\mathbf{y})F(\mathbf{y}) = \dfrac{-\rho^6}{\|\mathbf{x} - \mathbf{q}\|^6}\big(f(\mathbf{x})E(\mathbf{x}) - e(\mathbf{x})F(\mathbf{x})\big), \\ g(\mathbf{y})E(\mathbf{y}) - e(\mathbf{y})G(\mathbf{y}) = \dfrac{-\rho^6}{\|\mathbf{x} - \mathbf{q}\|^6}\big(g(\mathbf{x})E(\mathbf{x}) - e(\mathbf{x})G(\mathbf{x})\big), \\ g(\mathbf{y})F(\mathbf{y}) - f(\mathbf{y})G(\mathbf{y}) = \dfrac{-\rho^6}{\|\mathbf{x} - \mathbf{q}\|^6}\big(g(\mathbf{x})F(\mathbf{x}) - f(\mathbf{x})G(\mathbf{x})\big). \end{cases}$$

Making use of these preparatory lemmas, we can now prove the following important result.

Theorem 36.9. *Let* $\Phi \colon \mathbb{R}^3 \setminus \mathbf{q} \longrightarrow \mathbb{R}^3 \setminus \mathbf{q}$ *be an inversion which maps a surface* $\mathcal{M}$ *onto a surface* $\Phi(\mathcal{M})$. *Then* Φ *maps the principal curves of* $\mathcal{M}$ *onto the principal curves of* $\Phi(\mathcal{M})$.

Proof. Let $\alpha \colon (a,b) \longrightarrow \mathcal{M}$ be a principal curve on $\mathcal{M}$ and let $\mathbf{x} \colon \mathcal{U} \longrightarrow \mathcal{M}$ be a patch. On the portion of the trace of α that is contained in the trace of $\mathbf{x}$ we can write $\alpha(t) = \mathbf{x}(u(t), v(t))$. Let E, F, G, e, f, g denote the coefficients of the first and second fundamental forms of $\mathbf{x}$. Corollary 28.2, page 643, states that u and v satisfy the differential equation

(36.19) $(fE - eF)(\alpha(t))u'(t)^2 + (gE - eG)(\alpha(t))u'(t)v'(t)$
$$+ (gF - fG)(\alpha(t))v'(t)^2 = 0.$$

Let $\mathbf{y} = \Phi \circ \mathbf{x}$ and let $\tilde{E}, \tilde{F}, \tilde{G}, \tilde{e}, \tilde{f}, \tilde{g}$ denote the coefficients of the first and second fundamental forms of $\mathbf{y}$. Then $(\Phi \circ \alpha)(t) = \mathbf{y}(u(t), v(t))$. It follows from Corollary 36.8 that

(36.20) $(\tilde{f}\tilde{E} - \tilde{e}\tilde{F})((\Phi \circ \alpha)(t))u'(t)^2 + (\tilde{g}\tilde{E} - \tilde{e}\tilde{G})((\Phi \circ \alpha)(t))u'(t)v'(t)$
$$+ (\tilde{g}\tilde{F} - \tilde{f}\tilde{G})((\Phi \circ \alpha)(t))v'(t)^2 = 0.$$

Now Corollary 28.2 implies that $\Phi \circ \alpha$ is a principal curve on $\Phi(\mathcal{M})$ in the trace of $\mathbf{y} = \Phi \circ \mathbf{x}$. Since $\mathbf{x}$ is arbitrary, the result follows. ∎

Asimpler proof of Theorem 36.9 will be given on page 833.

Next, we determine the relation between the principal curvatures of a surface $\mathcal{M}$ and the image of $\mathcal{M}$ under an inversion Φ. Let $\mathbf{x}$ be a principal patch on $\mathcal{M}$; then the principal curvatures on the trace of $\mathbf{x}$ are given by

$$\mathbf{k}_\pi(\mathbf{x}) = \frac{e(\mathbf{x})}{E(\mathbf{x})}, \quad \text{and} \quad \mathbf{k}_\mu(\mathbf{x}) = \frac{g(\mathbf{x})}{G(\mathbf{x})}.$$

36.3 Inversion of Surfaces

Theorem 36.9 implies that $\Phi \circ \mathbf{y}$ is a principal patch, and so

$$\mathbf{k}_\pi(\mathbf{y}) = \frac{e(\mathbf{y})}{E(\mathbf{y})}, \quad \text{and} \quad \mathbf{k}_\mu(\mathbf{y}) = \frac{g(\mathbf{y})}{G(\mathbf{y})}.$$

From Lemmas 36.4 and 36.7 we get

Lemma 36.10. *Let $\mathbf{x}$ be a principal patch in $\mathbb{R}^3$, and let $\mathbf{y}$ be the inverse patch of $\mathbf{x}$ with respect to $\mathbf{q} \in \mathbb{R}^3$ with radius of inversion ρ. Then*

(36.21)
$$\begin{cases} \mathbf{k}_\pi(\mathbf{y}) = \dfrac{-\|\mathbf{x}-\mathbf{q}\|^2}{\rho^2}\mathbf{k}_\pi(\mathbf{x}) - \dfrac{2\mathbf{U}_\mathbf{x} \cdot (\mathbf{x}-\mathbf{q})}{\rho^2}, \\[1em] \mathbf{k}_\mu(\mathbf{x}) = \dfrac{-\|\mathbf{x}-\mathbf{q}\|^2}{\rho^2}\mathbf{k}_\mu(\mathbf{x}) - \dfrac{2\mathbf{U}_\mathbf{x} \cdot (\mathbf{x}-\mathbf{q})}{\rho^2}, \end{cases}$$

so that

(36.22) $$\mathbf{k}_\pi(\mathbf{y}) - \mathbf{k}_\mu(\mathbf{x}) = \frac{-\|\mathbf{x}-\mathbf{q}\|^2}{\rho^2}\left(\mathbf{k}_\pi(\mathbf{x}) - \mathbf{k}_\mu(\mathbf{x})\right).$$

Furthermore, we have the following relations between the Gaussian and mean curvatures of $\mathbf{x}$ and $\mathbf{y}$:

(36.23) $$K(\mathbf{y}) = \frac{\|\mathbf{x}-\mathbf{q}\|^2}{\rho^4} + \frac{4\|\mathbf{x}-\mathbf{q}\|^4 \mathbf{U}_\mathbf{x} \cdot (\mathbf{x}-\mathbf{q})H(\mathbf{x})}{\rho^4} + \frac{4\left(\mathbf{U}_\mathbf{x} \cdot (\mathbf{x}-\mathbf{q})\right)^2}{\rho^4},$$

(36.24) $$H(\mathbf{y}) = -\frac{\|\mathbf{x}-\mathbf{q}\|^2 H(\mathbf{x})}{\rho^2} - \frac{2\mathbf{U}_\mathbf{x} \cdot (\mathbf{x}-\mathbf{q})}{\rho^2}.$$

Proof. (36.21) is a consequence of (36.16). Then (36.22)–(36.24) follow from (36.21). ∎

Corollary 36.11. *An inversion of $\mathbb{R}^3$ maps umbilic points into umbilic points.*

Proof. This is an immediate consequence of (36.22). ∎

The relevant *Mathematica* definitions corresponding to an inverse patch are:

```
inversionsurf[q_List,rho_][surf_][u_,v_]:=
            inversion[q,rho][surf[u,v]]

inversionsurf[rho_][surf_][u_,v_]:=
            inversion[rho][surf[u,v]]
```

The fact that an inversion preserves principal curvatures can be used to construct many interesting parametrizations of surfaces. For example, the plot of the image of an ellipsoid under an inversion showing the lines of curvature is:

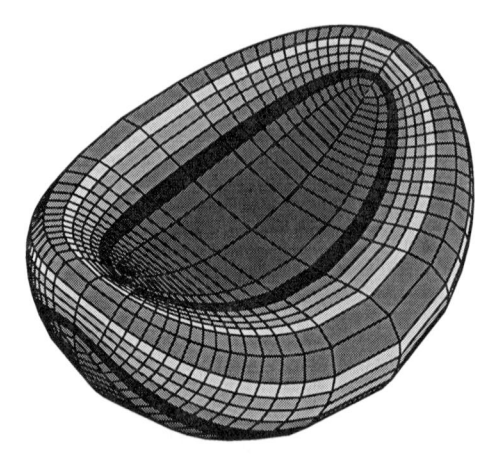

Inversion of an ellipsoid

The plotting command for this picture is the rather complicated

```
ParametricPlot3D[{
Append[inversion[{0,0,5},1][ellipticcoor[1,1,1][16,7,1][0][u,v]],
FaceForm[Hue[(u-1)/6],Hue[(v-7)/9]]],
Append[inversion[{0,0,5},1][ellipticcoor[-1,1,1][16,7,1][0][u,v]],
FaceForm[Hue[(v-7)/9],Hue[(u-1)/6]]],
Append[inversion[{0,0,5},1][ellipticcoor[1,-1,1][16,7,1][0][u,v]],
FaceForm[Hue[(v-7)/9],Hue[(u-1)/6]]],
Append[inversion[{0,0,5},1][ellipticcoor[-1,-1,1][16,7,1][0][u,v]],
FaceForm[Hue[(u-1)/6],Hue[(v-7)/9]]],
Append[inversion[{0,0,5},1][ellipticcoor[1,1,-1][16,7,1][0][u,v]],
FaceForm[Hue[(v-7)/9],Hue[(u-1)/6]]],
Append[inversion[{0,0,5},1][ellipticcoor[-1,1,-1][16,7,1][0][u,v]],
FaceForm[Hue[(u-1)/6],Hue[(v-7)/9]]],
Append[inversion[{0,0,5},1][ellipticcoor[1,-1,-1][16,7,1][0][u,v]],
FaceForm[Hue[(u-1)/6],Hue[(v-7)/9]]],
Append[inversion[{0,0,5},1][ellipticcoor[-1,-1,-1][16,7,1][0][u,v]],
FaceForm[Hue[(v-7)/9],Hue[(u-1)/6]]]
}//Evaluate,
{u,1,7},{v,7,16},Lighting->False,PlotPoints->{9,16},
Axes->None,Boxed->False,ViewPoint->{2.7,2.7,4.2}]
```

36.4 Cyclides of Dupin via Inversions

One definition of a cyclide of Dupin is the image under the inversion of a torus. This leads to the following parametrization in *Mathematica*:

36.4 Cyclides of Dupin via Inversions

```
invertedtorus[{p1_,p2_,p3_},rho_,a_,b_][u_,v_]:=
    {p1 + rho^2(-p1 + Cos[u](a + b Cos[v]))/
    ((-p1 + Cos[u](a + b Cos[v]))^2 +
    (-p2 + (a + b Cos[v])Sin[u])^2 +
    (-p3 + b Sin[v])^2),
    p2 + rho^2(-p2 + Sin[u](a + b Cos[v]))/
    ((-p1 + Cos[u](a + b Cos[v]))^2 +
    (-p2 + (a + b Cos[v])Sin[u])^2 +
    (-p3 + b Sin[v])^2),
    p3 + rho^2(-p3 + b Sin[v])/
    ((-p1 + Cos[u](a + b Cos[v]))^2 +
    (-p2 + Sin[u](a + b Cos[v]))^2 +
    (-p3 + b Sin[v])^2)}
```

Here {p1,p2,p3} is the center of inversion, **rho** is the radius of the inversion, **a** is the wheel radius of the torus and **b** is the tube radius of the torus.

To plot the image of a torus of wheel radius 8 and tube radius 3 under an inversion of radius 2 and center $(0, 2, 0)$, we use

```
ParametricPlot3D[Evaluate[
invertedtorus[{0,2,0},2,8,3][u,v]],
{u,0,2Pi},{v,0,2Pi},Boxed->False,Axes->None,
PlotPoints->{40,40}];
```

to obtain

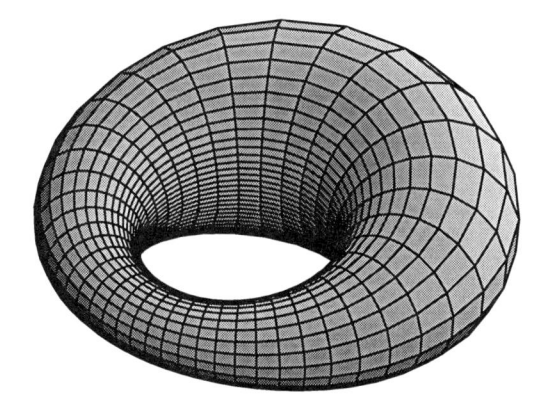

Ring cyclide

Here is the inside-out torus (see page 306) inverted with respect to the point $(0,5,0)$ using an inversion of radius 3; it is plotted with the command

```
ParametricPlot3D[Evaluate[
invertedtorus[{0,5,0},3,8,8][u,v]],
{u,0,2Pi},{v,0,2Pi},Boxed->False,Axes->None,
PlotPoints->{40,40}];
```

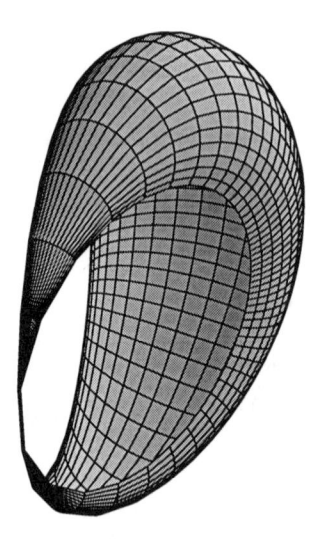

The inside-out torus
$$(u,v) \longmapsto$$
$$\bigl(\cos u(8 + 8\cos v), \sin u(8 + 8\cos v), 8\sin v\bigr)$$
inverted with respect to $(0,5,0)$

36.5 Liouville's Theorem

Conformal maps of a Euclidean space of dimension greater than 2 differ markedly from conformal maps of the plane. In fact, any analytic function F with nonzero derivative is a conformal diffeomorphism from its domain of definition to its range. Moreover, the Riemann mapping theorem (see, for example, [Ahlf, pages 229-335]) implies that for any two simply connected open subsets $\mathcal{U}_1$ and $\mathcal{U}_2$ of the plane, there is an analytic function F with nonvanishing derivative that maps $\mathcal{U}_1$ onto $\mathcal{U}_2$ (provided the complements of $\mathcal{U}_1$ and $\mathcal{U}_2$ contain at least two points). Thus the number of conformal diffeomorphisms between subsets of the plane is vast.

On the other hand, we now prove that every conformal map between open subsets of a Euclidean space of dimension 3 is the composition of inversions and Euclidean motions. First, we observe

Lemma 36.12. *Any conformal diffeomorphism* $\Phi\colon \mathcal{U} \longrightarrow \Phi(\mathcal{U}) \subset \mathbb{R}^n$ *maps a triply orthogonal system of surfaces onto another triply orthogonal system of surfaces.*

Proof. Since a conformal diffeomorphism preserves angles, it preserves orthogonality of surfaces. ∎

Theorem 36.9 admits the following generalization.

Corollary 36.13. *A conformal diffeomorphism maps principal curves onto principal curves.*

Proof. The corollary is a consequence of Lemma 36.12 and Dupin's Theorem (Theorem 29.4, page 668). ∎

A key ingredient in the proof of Liouville's Theorem is

Theorem 36.14. *Let* $\Phi\colon \mathcal{U} \longrightarrow \Phi(\mathcal{U}) \subset \mathbb{R}^3$ *be a conformal diffeomorphism, where* $\mathcal{U}$ *is an open subset of* $\mathbb{R}^3$, *and let* $S^2(\mathbf{p}, a) \subset \mathcal{U}$ *be a sphere of radius* a *centered at* $\mathbf{p} \in \mathcal{U}$. *Then* $\Phi(S^2(\mathbf{p}, a))$ *is isometric to a plane or sphere.*

Proof. For each great circle C of $S^2(\mathbf{p}, a)$ there is a plane $\Pi(C)$ that intersects $S^2(\mathbf{p}, a)$ in C. Consider the family $\mathfrak{A}(\mathbf{p})$ of all spheres centered at $\mathbf{p}$. There is a family $\mathfrak{B}(\mathbf{p})$ of planes, one of whose members is $\Pi(C)$, and a family $\mathfrak{C}(\mathbf{p})$ of cones such that $\{\mathfrak{A}(\mathbf{p}), \mathfrak{B}(\mathbf{p}), \mathfrak{C}(\mathbf{p})\}$ is a triply orthogonal system. Lemma 36.12 implies that the image

$$\{\Phi(\mathfrak{A}(\mathbf{p})), \Phi(\mathfrak{B}(\mathbf{p})), \Phi(\mathfrak{C}(\mathbf{p}))\}$$

is also a triply orthogonal system. Then $\Phi(\Pi(C))$ is a principal curve in $\Phi(S^2(\mathbf{p}, a))$ by Dupin's Theorem (Theorem 29.4). Thus all great circles through a point $\mathbf{q} \in S^2(\mathbf{p}, a)$ are mapped onto principal curves on $\Phi(S^2(\mathbf{p}, a))$. Since Φ is a diffeomorphism, distinct great circles are mapped onto distinct principal curves. Thus there is a principal curve of $\Phi(S^2(\mathbf{p},a))$ through $\Phi(\mathbf{q})$ in every direction. It follows that $\Phi(\mathbf{q})$ is an umbilic point of $\Phi(S^2(\mathbf{p},a))$. Since $\mathbf{q}$ is an arbitrary point of $\Phi(S^2(\mathbf{p},a))$, we have shown that all points of $\Phi(S^2(\mathbf{p},a))$ are umbilic points. Now Theorem 28.5 on page 646 implies that $\Phi(S^2(\mathbf{p},a))$ is isometric to a plane or sphere. ∎

Next, we prove Liouville's Theorem.

Theorem 36.15. *Let* $\Phi:\mathcal{U} \longrightarrow \Phi(\mathcal{U}) \subset \mathbb{R}^3$ *be a conformal diffeomorphism, where* $\mathcal{U}$ *is an open subset of* $\mathbb{R}^3$. *Then* Φ *is a composition of inversions and affine maps.*

Proof. Let $\mathbf{p} \in \mathcal{U}$, and let Ψ be an inversion with pole $\mathbf{p}$. Then Ψ maps any plane Π not passing through $\mathbf{p}$ into a sphere $\Psi(\Pi)$ passing through $\mathbf{p}$, and vice versa. There exists a translation T of $\mathbb{R}^3$ such that $T^{-1}(\Phi(\mathbf{p})) = \mathbf{p}$. Theorem 36.14 implies that $T^{-1} \circ \Phi$ maps $\Psi(\Pi)$ onto a plane or sphere passing through $\mathbf{p}$. Then the transformation
$$A = \Psi \circ T^{-1} \circ \Phi \circ \Psi$$
maps Π into another plane not passing through $\mathbf{p}$. Hence A is an affine map (it is also conformal). ∎

36.6 Exercises

1. A circular cylinder is parametrized in *Mathematica* by

```
circularcylinder[a_][u_,v_] := {a Cos[u], a Sin[u], v}
```

36.6 Exercises

Show that the image of a circular cylinder under an inversion is given by

```
invertedcircularcylinder[{p1_,p2_,p3_},rho_,a_][u_,v_]:=
    {p1 + (rho^2(-p1 + a Cos[u]))/
    ((-p3 + v)^2 + (-p1 + a Cos[u])^2 +
        (-p2 + a Sin[u])^2),
    p2 + (rho^2(-p2 + a Sin[u]))/
    ((-p3 + v)^2 + (-p1 + a Cos[u])^2 +
        (-p2 + a Sin[u])^2),
    p3 + (rho^2(-p3 + v))/
    ((-p3 + v)^2 + (-p1 + a Cos[u])^2 +
        (-p2 + a Sin[u])^2)}
```

Plot several inverted circular cylinders.

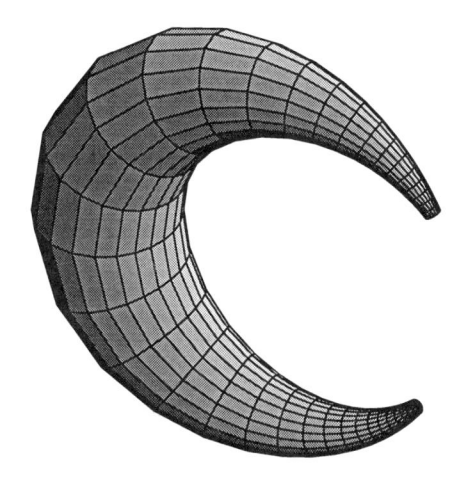

Inverted circular cylinder

2. A circular cone is parametrized in *Mathematica* by

```
circularcone[a_,b_][u_,v_]:= {a v Cos[u],
                              a v Sin[u],b v}
```

Show that the image of a circular cone under an inversion is given by

```
invertedcircularcone[{p1_,p2_,p3_},rho_,a_][u_,v_]:=
    {p1 + (rho^2(-p1 + a v Cos[u]))/
      ((-p3 + b v)^2 + (-p1 + a v Cos[u])^2 +
       (-p2 + a v Sin[u])^2),
     p2 + (rho^2(-p2 + a v Sin[u]))/
      ((-p3 + b v)^2 + (-p1 + a v Cos[u])^2 +
       (-p2 + a v Sin[u])^2),
     p3 + (rho^2 (-p3 + b v))/
      ((-p3 + b v)^2 + (-p1 + a v Cos[u])^2 +
       (-p2 + a v Sin[u])^2)}
```

Plot several inverted circular cones.

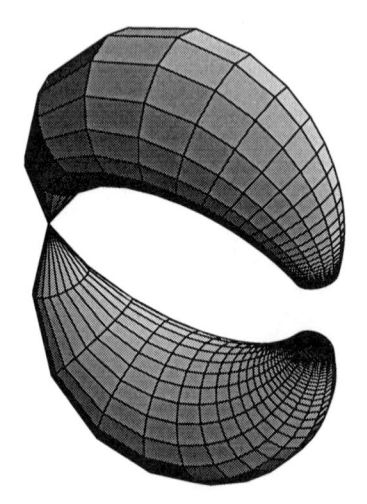

Inverted circular cone

3. Show that the inverse curve with respect to the origin of an equilateral hyperbola is a lemniscate. The parametrization of an equilateral hyperbola that corresponds to the lemniscate is $t \longmapsto (a\sec(t), a\tan(t))$.

4. Plot the inverse curves of an ellipse with respect to several points.

APPENDIX A: GENERAL PROGRAMS

Arc Length Functions

A.1 arclength

t⟼**arclength[alpha][t]** is the arc length function of the curve **alpha**:$(a,b) \longrightarrow \mathbb{R}^n$ computed using **Integrate**.

```
arclength[alpha_][t_]:=
    Integrate[arclengthprime[alpha][tt],tt]   /. tt->t
```

A.2 arclengthea

t⟼**arclength[alpha][t]** is the equiaffine arc length function of the curve **alpha**:$(a,b) \longrightarrow \mathbb{R}^n$ computed using **Integrate**.

```
arclengthea[alpha_][t_]:=
    Integrate[arclengthprimeea[alpha][tt],tt]   /. tt->t
```

A.3 arclengtheaprime

t⟼**arclengtheaprime[alpha][t]** is the derivative of the equiaffine arc length function of the curve **alpha**:$(a,b) \longrightarrow \mathbb{R}^n$.

```
arclengtheaprime[alpha_][t_]:=
    Simplify[Det[Table[D[alpha[tt],{tt,k}],
        {k,Length[alpha[tt]]}]]]^(2/(Length[alpha[tt]]*
        (Length[alpha[tt]] + 1)))  /. tt->t
```

A.4 arclengthprime

t⟼arclengthprime[alpha][t] is the derivative of the arc length function of the curve alpha:$(a,b) \longrightarrow \mathbb{R}^n$.

```
arclengthprime[alpha_][t_]:=
    Sqrt[Simplify[D[alpha[tt],tt].
        D[alpha[tt],tt]]]  /. tt->t
```

A.5 descenttime

descenttime[a,b][g,alpha] is the time required for a bead on a plane curve alpha to descend from alpha[a] to alpha[b]. g is the constant of gravity. The time is computed symbolically.

```
descenttime[a_,b_][g_,alpha_]:=
    Integrate[descenttimeprime[a,g,alpha][u],{u,a,b}]
```

A.6 descenttimen

descenttimen[a,b][g,alpha] is the time required for a bead on a plane curve alpha to descend from alpha[a] to alpha[b]. g is the constant of gravity. The time is computed using **NIntegrate**.

```
descenttimen[a_,b_][g_,alpha_]:=
    NIntegrate[descenttimeprime[a,g,alpha][u],{u,a,b}]
```

A.7 descenttimends

t⟼descenttimends[a,b][g,alpha][t] is the time required for a bead on a plane curve alpha to descend from alpha[a] to alpha[t]. g is the constant of gravity. The time is computed using **NDSolve** over the interval $[a,b]$.

Appendix A: General Programs

```
descenttimends[a_,b_][g_,alpha_][t_]:=
    Chop[N[Module[{aaa,bbb,s,tt},
        aaa = NDSolve[{s'[tt]==
            descenttimeprime[a,g,alpha][tt],
            s[a]==0},s,{tt,a,b}];
        bbb = s[tt] /. aaa;
        bbb /. tt->t][[1]]]]
```

A.8 descenttimeprime

$t \longmapsto$ **descenttimeprime[a,g,alpha][t]** is the derivative of descenttime.

```
descenttimeprime[a_,g_,alpha_][t_]:=
    Sqrt[Simplify[
    arclengthprime[alpha][t]^2/
    (2*g*(alpha[a][[2]] - alpha[t][[2]]))]]
```

A.9 length

length[a,b][alpha] is the length of the curve **alpha**:$(a,b) \longrightarrow \mathbb{R}^n$ computed symbolically.

```
length[a_,b_][alpha_]:=
    Integrate[arclengthprime[alpha][u],{u,a,b}]
```

A.10 lengthea

lengthea[a,b][alpha] is the equiaffine length of the curve **alpha**:$(a,b) \longrightarrow \mathbb{R}^n$ computed symbolically.

```
lengthea[a_,b_][alpha_]:=
    Integrate[arclengtheaprime[alpha][u],{u,a,b}]
```

A.11 lengthean

lengthean[a,b][alpha] is the equiaffine length of the curve **alpha**:$(a,b) \longrightarrow \mathbb{R}^n$ computed using **NIntegrate**.

```
lengthean[a_,b_][alpha_]:= NIntegrate[Evaluate[
    arclengtheaprime[alpha][u]],{u,a,b}]
```

A.12 lengtheands

t⟼**lengtheands[a,b][alpha][t]** is the equiaffine length of the curve
alpha:$(a,b) \longrightarrow \mathbb{R}^n$ over the interval $[a,t]$ computed using **NDSolve** over the interval $[a,b]$.

```
lengtheands[a_,b_][alpha_][t_]:=
    Module[{aaa,bbb,s,tt},
        aaa = NDSolve[{s'[tt]==
            arclengtheaprime[alpha][tt],
            s[a]==0},s,{tt,a,b}];
        bbb = s[tt] /. aaa;
        bbb /. tt->t][[1]]
```

A.13 lengthn

lengthn[a,b][alpha] is the length of the curve **alpha**:$(a,b) \longrightarrow \mathbb{R}^n$ computed using **NIntegrate**.

```
lengthn[a_,b_][alpha_]:= NIntegrate[Evaluate[
    arclengthprime[alpha][u]],{u,a,b}]
```

A.14 lengthnds

t⟼**lengthnds[a,b][alpha][t]** is the length of the curve
alpha:$(a,b) \longrightarrow \mathbb{R}^n$ over the interval $[a,t]$ computed using **NDSolve** over the interval $[a,b]$.

```
lengthnds[a_,b_][alpha_][t_]:=
    Module[{aaa,bbb,s,tt},
        aaa = NDSolve[{s'[tt]==
            arclengthprime[alpha][tt],
            s[a]==0},s,{tt,a,b}];
        bbb = s[tt] /. aaa;
        bbb /. tt->t][[1]]
```

Brioschi Curvature

A.15 brioschicurvature
(Patch Version)

{u,v}⟼brioschicurvature[x][u,v] is the Gaussian curvature of the patch x: $\mathcal{U} \longrightarrow \mathbb{R}^n$ computed using Brioschi's formula.

```
brioschicurvature[x_][u_,v_]:=
    (Det[{{-D[ee[x][uu,vv],{vv,2}]/2 +
        D[ff[x][uu,vv],uu,vv] -
        D[gg[x][uu,vv],{uu,2}]/2,
        D[ee[x][uu,vv],uu]/2,
        D[ff[x][uu,vv],uu] -
        D[ee[x][uu,vv],vv]/2},
       {D[ff[x][uu,vv],vv] -
        D[gg[x][uu,vv],uu]/2,
        ee[x][uu,vv],ff[x][uu,vv]},
       {D[gg[x][uu,vv],vv]/2,
        ff[x][uu,vv],gg[x][uu,vv]}}]
    -Det[{{0,D[ee[x][uu,vv],vv]/2,
        D[gg[x][uu,vv],uu]/2},
       {D[ee[x][uu,vv],vv]/2,
        ee[x][uu,vv],ff[x][uu,vv]},
       {D[gg[x][uu,vv],uu]/2,
        ff[x][uu,vv],gg[x][uu,vv]}}])/
    (ee[x][uu,vv]*gg[x][uu,vv] - ff[x][uu,vv]^2)^2 /.
        {uu->u,vv->v}
```

A.16 brioschicurvature
(Abstract Metric Version)

{u,v}⟼brioschicurvature[ee,ff,gg][u,v] is the Gaussian curvature of the metric whose coefficients are {ee,ff,gg}.

```
brioschicurvature[ee_,ff_,gg_][u_,v_]:=
    (Det[{{-D[ee[uu,vv],{vv,2}]/2 + D[ff[uu,vv],uu,vv] -
        D[gg[uu,vv],{uu,2}]/2,
        D[ee[uu,vv],uu]/2,
        D[ff[uu,vv],uu] - D[ee[uu,vv],vv]/2},
       {D[ff[uu,vv],vv] - D[gg[uu,vv],uu]/2,ee[uu,vv],ff[uu,vv]},
       {D[gg[uu,vv],vv]/2,ff[uu,vv],gg[uu,vv]}}] -
     Det[{{0,D[ee[uu,vv],vv]/2,D[gg[uu,vv],uu]/2},
          {D[ee[uu,vv],vv]/2,ee[uu,vv],ff[uu,vv]},
          {D[gg[uu,vv],uu]/2,ff[uu,vv],gg[uu,vv]}}])/
    (ee[uu,vv]*gg[uu,vv] - ff[uu,vv]^2)^2    /. {uu->u,vv->v}
```

Christoffel Symbols

A.17 christoffel
(Patch Version)

$\{u,v\} \longmapsto$ christoffel[i,j,k][x][u,v] is the Christoffel symbol Γ^i_{jk} of the metric tensor of the patch $\mathbf{x}: \mathcal{U} \longrightarrow \mathbb{R}^n$.

```
christoffel[1,1,1][x_][u_,v_]:=
    Simplify[(D[ee[x][uu,vv],vv]*ff[x][uu,vv] -
    2*ff[x][uu,vv]*D[ff[x][uu,vv],uu] +
    D[ee[x][uu,vv],uu]*gg[x][uu,vv])/
    (2*(ee[x][uu,vv]*gg[x][uu,vv] -
    ff[x][uu,vv]^2))]  /. {uu->u,vv->v}
```

```
christoffel[2,1,1][x_][u_,v_]:=
    Simplify[-(D[ee[x][uu,vv],vv]*ee[x][uu,vv] -
    2*ee[x][uu,vv]*D[ff[x][uu,vv],uu] +
    D[ee[x][uu,vv],uu]*ff[x][uu,vv])/
    (2*(ee[x][uu,vv]*gg[x][uu,vv] -
    ff[x][uu,vv]^2))]  /. {uu->u,vv->v}
```

Appendix A: General Programs

```
christoffel[1,1,2][x_][u_,v_]:=
    Simplify[(D[ee[x][uu,vv],vv]*gg[x][uu,vv] -
    ff[x][uu,vv]*D[gg[x][uu,vv],uu])/
    (2*(ee[x][uu,vv]*gg[x][uu,vv] -
    ff[x][uu,vv]^2))] /. {uu->u,vv->v}
```

```
christoffel[1,2,1][x_][u_,v_]:=
    christoffel[1,1,2][x][u,v]
```

```
christoffel[2,1,2][x_][u_,v_]:=
    Simplify[-(D[ee[x][uu,vv],vv]*ff[x][uu,vv] -
    ee[x][uu,vv]*D[gg[x][uu,vv],uu])/
    (2*(ee[x][uu,vv]*gg[x][uu,vv] -
    ff[x][uu,vv]^2))] /. {uu->u,vv->v}
```

```
christoffel[2,2,1][x_][u_,v_]:=
    christoffel[2,1,2][x][u,v]
```

```
christoffel[1,2,2][x_][u_,v_]:=
    Simplify[(-D[gg[x][uu,vv],vv]*ff[x][uu,vv] +
    2*gg[x][uu,vv]*D[ff[x][uu,vv],vv] -
    D[gg[x][uu,vv],uu]*gg[x][uu,vv])/
    (2*(ee[x][uu,vv]*gg[x][uu,vv] -
    ff[x][uu,vv]^2))] /. {uu->u,vv->v}
```

```
christoffel[2,2,2][x_][u_,v_]:=
    Simplify[(D[gg[x][uu,vv],vv]*ee[x][uu,vv] -
    2*ff[x][uu,vv]*D[ff[x][uu,vv],vv] +
    D[gg[x][uu,vv],uu]*ff[x][uu,vv])/
    (2*(ee[x][uu,vv]*gg[x][uu,vv] -
    ff[x][uu,vv]^2))] /. {uu->u,vv->v}
```

A.18 christoffel
(Abstract Metric Version)

{u,v}⟼christoffel[i,j,k][ee,ff,gg][u,v] is the Christoffel symbol Γ^i_{jk} of the metric whose coefficients are {ee,ff,gg}.

```
christoffel[1,1,1][ee_,ff_,gg_][u_,v_]:=
    Simplify[(D[ee[u,v],v]*ff[u,v] -
    2*ff[u,v]*D[ff[u,v],u] +
    D[ee[u,v],u]*gg[u,v])/
    (2*(ee[u,v]*gg[u,v] - ff[u,v]^2))]
```

```
christoffel[2,1,1][ee_,ff_,gg_][u_,v_]:=
    Simplify[-(D[ee[u,v],v]*ee[u,v] -
    2ee[u,v]*D[ff[u,v],u] +
    D[ee[u,v],u]*ff[u,v])/
    (2*(ee[u,v]*gg[u,v] - ff[u,v]^2))]
```

```
christoffel[1,1,2][ee_,ff_,gg_][u_,v_]:=
    Simplify[(D[ee[u,v],v]*gg[u,v] -
    ff[u,v]*D[gg[u,v],u])/
    (2*(ee[u,v]*gg[u,v] - ff[u,v]^2))]
```

```
christoffel[1,2,1][ee_,ff_,gg_][u_,v_]:=
    christoffel[1,1,2][ee,ff,gg][u,v]
```

```
christoffel[2,1,2][ee_,ff_,gg_][u_,v_]:=
    Simplify[-(D[ee[u,v],v]*ff[u,v] -
    ee[u,v]*D[gg[u,v],u])/
    (2*(ee[u,v]*gg[u,v] - ff[u,v]^2))]
```

```
christoffel[2,2,1][ee_,ff_,gg_][u_,v_]:=
    christoffel[2,1,2][ee,ff,gg][u,v]
```

Appendix A: General Programs 845

```
christoffel[1,2,2][ee_,ff_,gg_][u_,v_]:=
    Simplify[(-D[gg[u,v],v]*ff[u,v] +
    2*gg[u,v]*D[ff[u,v],v] -
    D[gg[u,v],u]*gg[u,v])/
    (2*(ee[u,v]*gg[u,v] - ff[u,v]^2))]
```

```
christoffel[2,2,2][ee_,ff_,gg_][u_,v_]:=
    Simplify[(D[gg[u,v],v]*ee[u,v] -
    2*ff[u,v]*D[ff[u,v],v] +
    D[gg[u,v],u]*ff[u,v])/
    (2*(ee[u,v]*gg[u,v] - ff[u,v]^2))]
```

Complex to Real to Complex

A.19 ctor

ctor[z] converts a complex number **z** to an ordered pair of real numbers whose components are the real and imaginary parts of **z**; it also works for lists of complex numbers.

```
ctor[z_]:= ComplexExpand[{Re[z],Im[z]},{z}]
```

A.20 rtoc

rtoc[{x,y}] converts an ordered pair **{x,y}** of real numbers to a complex number whose real part is **x** and whose imaginary part is **y**; it also works for lists of ordered pairs.

```
rtoc[{x_,y_}]:= x + I*y
```

Complex Structure

A.21 J

J is the linear map $J\colon\mathbb{R}^2 \longrightarrow \mathbb{R}^2$ that maps **{p1,p2}** onto **{-p2,p1}**.

```
J[{p1_,p2_}]:= {-p2,p1}
```

Cross Product

A.22 cross

{v,w}⟼**cross[v,w]** is the cross product of the vectors **v={v1,v2,v3}** and **w={w1,w2,w3}**.

```
cross[v_List,w_List]:=
    RotateLeft[v*RotateLeft[w] -
    RotateLeft[v]*w] /;
    Length[v] == Length[w] == 3
```

Construction of Curves in $\mathbb{R}^n$ from Curves in $\mathbb{R}^n$

A.23 combescure

t⟼**combescure[alpha][f][t]** is the curve **t**⟼**beta[t]** such that **beta'[t]=f[t]*alpha'[t]**.

```
combescure[alpha_][f_][t_]:=
    Integrate[f[tt]*D[alpha[tt],tt],tt] /. tt->t
```

A.24 involute

t⟼**involute[alpha][t]** is the curve traced by a point on a thread kept taut as it is unwound from the curve **alpha**.

Appendix A: General Programs

```
involute[alpha_,c_][t_] := alpha[tt] +
                (arclength[alpha][c] -
                arclength[alpha][tt])*D[alpha[tt],tt]/
                Sqrt[Factor[D[alpha[tt],tt].
                D[alpha[tt],tt]]] /. tt->t
```

```
involute[alpha_][t_] := involute[alpha,0][t]
```

A.25 involutends

$t \mapsto$ **involutends[a,b][alpha][t]** is the involute of **alpha** starting at **alpha[a]** computed over the interval $[a,b]$ using **NDSolve**.

```
involutends[a_,b_][alpha_][t_] :=
    Module[{aaa,bbb,s,tt},
        aaa = NDSolve[{s'[tt]==
            arclengthprime[alpha][tt],
            s[a]==0},s,{tt,a,b}];
        bbb = alpha[tt] +
            (s[a] - s[tt])*D[alpha[tt],tt]/
            Sqrt[D[alpha[tt],tt].
            D[alpha[tt],tt]] /.aaa;
        bbb /. tt->t]
```

A.26 tangentline

$t \mapsto$ **tangentline[alpha][s,t]** is the tangent line of length **s** emanating from the plane curve **alpha** at **alpha[t]**.

```
tangentline[alpha_][s_,t_] := Line[{alpha[tt],alpha[tt] +
    s*D[alpha[tt],tt]/Sqrt[Factor[D[alpha[tt],tt].
        D[alpha[tt],tt]]]}] /. tt->t
```

Construction of Plane Curves from Plane Curves

A.27 contrapedal

t⟼contrapedal[{a,b},alpha][t] is the contrapedal curve of the plane curve **alpha** with respect to the point **{a,b}**. It is the base of the perpendicular line let down from **{a,b}** to the normal line to alpha at **alpha[t]**.

```
contrapedal[{a_,b_},alpha_][t_]:= {a,b} +
    ((alpha[tt] - {a,b}).J[D[alpha[tt],tt]]/
    (D[alpha[tt],tt].D[alpha[tt],tt]))*J[
    D[alpha[tt],tt]] /. tt->t
```

t⟼contrapedal[alpha][t] is the pedal curve of the plane curve **alpha** with respect to the origin.

```
contrapedal[alpha_][t_]:= contrapedal[{0,0},alpha][t]
```

A.28 evolute

t⟼evolute[alpha][t] is the evolute of the plane curve **alpha**. It is the locus of the centers of the osculating circles of **alpha**.

```
evolute[alpha_][t_]:= Simplify[alpha[tt] +
    D[alpha[tt],tt].D[alpha[tt],tt]/
    (D[alpha[tt],{tt,2}].J[D[alpha[tt],tt]])*
    J[D[alpha[tt],tt]] /. tt->t]
```

t⟼evolute[n,alpha][t] is the nth evolute of the plane curve **alpha**.

```
evolute[1,alpha_][t_]:= evolute[alpha][t]
```

```
evolute[n_,alpha_][t_]:= evolute[n,alpha][t] =
    evolute[1,evolute[n - 1,alpha]][t]//Simplify
```

A.29 normalline

$t \mapsto$ `normalline[alpha][s,t]` the normal line of length **s** emanating from the plane curve **alpha** at **alpha[t]**.

```
normalline[alpha_][s_,t_]:=
    Line[{alpha[tt],alpha[tt] + s*J[D[alpha[tt],tt]]/
    Sqrt[D[alpha[tt],tt].D[alpha[tt],tt]]}] /. tt->t
```

A.30 osculatingcircle

$t \mapsto$ `osculatingcircle[alpha][t]` is the osculating circle to the plane curve **alpha** at **alpha[t]**.

```
osculatingcircle[alpha_][t_]:=
    Circle[alpha[tt] + D[alpha[tt],tt].D[alpha[tt],tt]/
    (D[alpha[tt],{tt,2}].
    J[D[alpha[tt],tt]])*J[D[alpha[tt],tt]],
    Abs[(D[alpha[tt],tt].D[alpha[tt],tt])^(3/2)/
    (D[alpha[tt],{tt,2}].J[D[alpha[tt],
        tt]])]] /. tt->t
```

A.31 parcurve

$t \mapsto$ `parcurve[alpha][s][t]` is one component of the curve parallel to a plane curve **alpha** at a distance **Abs[s]**, where **Abs[s]** is small. The other component is $t \mapsto$ `parcurve[alpha][-s][t]`.

```
parcurve[alpha_][s_][t_]:= alpha[tt] +
    s*J[D[alpha[tt],tt]]/Sqrt[Simplify[
        D[alpha[tt],tt].D[alpha[tt],tt]]] /. tt->t
```

A.32 pedal

$t \mapsto$ `pedal[{a,b},alpha][t]` is the pedal curve of the plane curve **alpha** with respect to the point **{a,b}**. It is the base of the perpendicular line let down from **{a,b}** to the tangent line to alpha at **alpha[t]**.

```
pedal[{a_,b_},alpha_][t_] := {a,b} +
    ((alpha[tt] - {a,b}).J[D[alpha[tt],tt]]/
    (D[alpha[tt],tt].D[alpha[tt],tt]))*J[
    D[alpha[tt],tt]] /. tt->t
```

t⟼**pedal[alpha][t]** is the pedal curve of the plane curve **alpha** with respect to the origin.

```
pedal[alpha_][t_] := pedal[{0,0},alpha][t]
```

A.33 rollcurve

t⟼**rollcurve[alpha,b,h,c][t]** is the parametrized curve traced out by a point P attached to a circle of radius **b** rolling along a plane curve **alpha**. The distance of P to the center of the rolling circle is **b**, and **alpha[c]** is the point at which the circle starts rolling.

```
rollcurve[alpha_,b_,h_,c_][t_] := (alpha[tt] -
    h*Sin[(arclength[alpha][tt] - arclength[alpha][c])/b]*
    D[alpha[tt],tt]/Sqrt[Factor[D[alpha[tt],tt].
    D[alpha[tt],tt]]] +
    (b - h*Cos[(arclength[alpha][tt] - arclength[alpha][c])/b])*
    J[D[alpha[tt],tt]]/Sqrt[Factor[D[alpha[tt],tt].
    D[alpha[tt],tt]]]) /. tt->t
```

A.34 plotl

plotl[alpha,t0,tmin,tmax,opts] plots the tangent line emanating from the plane curve **alpha** at **alpha[t0]**.

```
plotl[curve_,t0_,{tmin_,tmax_},opts___] :=
    ParametricPlot[curve[t0 + t]//Evaluate,{t,tmin,tmax},
    Epilog->{tangentline[curve][t0][tmax]//Evaluate},
    AspectRatio->Automatic,opts];
```

Construction of Space Curves from Plane Curves

A.35 `curveonsurf`

t⟼`curveonsurf[surf,alpha][t]` is the composition of the patch **surf**:$\mathcal{U} \longrightarrow \mathbb{R}^3$ with the curve **alpha**:$(a,b) \longrightarrow \mathcal{U}$.

```
curveonsurf[surf_,alpha_][t_]:=
    surf[alpha[t][[1]],alpha[t][[2]]]
```

A.36 `helical`

t⟼`helical[alpha,a][t]` is the helical curve in $\mathbb{R}^3$ with slant **a** over the plane curve **alpha**.

```
helical[alpha_,a_][t_]:= {alpha[t][[1]],alpha[t][[2]],a*t}
```

t⟼`helical[alpha][f][t]` is the helical curve in $\mathbb{R}^3$ with slant function **f** over the plane curve **alpha**.

```
helical[alpha_][f_][t_]:= {alpha[t][[1]],alpha[t][[2]],f[t]}
```

A.37 `inscurve`

t⟼`inscurve[n,a,alpha][t]` is the curve that results when the value **a** is inserted as the nth coordinate into the parametrization of the curve **alpha**.

```
inscurve[n_,a_,alpha_][t_]:= Insert[alpha[t],a,n]
```

A.38 `kappa23d`

t⟼`kappa23d[alpha][t]` is the space curve whose first two components are those of the plane curve t⟼**alpha** and whose third component is the signed curvature of **alpha**.

```
kappa23d[alpha_][t_]:= Append[alpha[t],kappa2[alpha][t]]
```

t⟼**kappa23d[mag,alpha][t]** is the space curve whose first two components are those of the plane curve t⟼**alpha** and whose third component is the signed curvature of **alpha** multiplied by **mag**.

```
kappa23d[mag_,alpha_][t_]:=
     Append[alpha[t],mag*kappa2[alpha][t]]
```

A.39 writhe

t⟼**writhe[alpha][t]** is a curve in $\mathbb{R}^3$ that is a twisted generalization of the plane curve t⟼**alpha[t]**.

```
writhe[alpha_][t_]:= {alpha[t][[1]],
     Cos[t]*alpha[t][[2]],Sin[t]*alpha[t][[2]]}
```

Construction of Space Curves from Space Curves

A.40 evolute3d

t⟼**evolute3d[gamma][t]** is the locus of the centers of the osculating spheres of the space curve **gamma**: $(a,b) \longrightarrow \mathbb{R}^3$.

```
evolute3d[gamma_][t_]:= gamma[tt] +
     (1/kappa[gamma][tt])*normal[gamma][tt] -
     (D[kappa[gamma][tt],tt]/(arclengthprime[gamma][tt]*
     kappa[gamma][tt]^2*tau[gamma][tt]))*
     binormal[gamma][tt] /. tt->t
```

A.41 filament

t⟼**filament[gamma][f,r][t]** is a curve in $\mathbb{R}^3$ that winds around the space curve **gamma**: $(a,b) \longrightarrow \mathbb{R}^3$, resembling an electrical filament of radius **r** and twisting factor **f**.

Appendix A: General Programs 853

```
filament[gamma_][f_,r_][t_] := Simplify[gamma[t] +
    r*(-Cos[f*t]*normal[gamma][t] +
        Sin[f*t]*binormal[gamma][t])]
```

Construction of Surfaces from Functions

A.42 monge

{u,v}⟼monge[h][u,v] is the parametrization {u,v}⟼{u,v,h[u,v]} of the graph of **h**:$\mathcal{U} \longrightarrow \mathbb{R}$, where $\mathcal{U}$ is an open subset of $\mathbb{R}^2$.

```
monge[h_][u_,v_] := {u,v,h[u,v]}
```

A.43 mongepolar

{r,theta}⟼mongepolar[g][r,theta] is the parametrization
{r,theta}⟼{r,theta,g[r,theta]} of the graph of **g**:$\mathcal{U} \longrightarrow \mathbb{R}$, where $\mathcal{U}$ is an open subset of $\mathbb{R}^2$.

```
mongepolar[g_][r_,theta_] :=
    {r*Cos[theta],r*Sin[theta],g[r,theta]}
```

A.44 singmonge

{u,v}⟼singmonge[n,h][u,v] is the parametrization
{u,v}⟼{u^n,v^n,h[u,v]} generalizing a monge patch.

```
singmonge[n_,h_][u_,v_] := {u^n,v^n,h[u,v]}
```

A.45 singmongepolar

{r,theta}⟼singmongepolar[g][r,theta] is the parametrization
{r,theta}⟼{r^n*Cos[theta],r^n*Cos[theta],g[r,theta]}
generalizing a polar monge patch.

```
singmongepolar[n_,g_][r_,theta_] :=
    {r^n*Cos[theta],r^n*Sin[theta],g[r,theta]}
```

Construction of Surfaces from Plane Curves

A.46 cone

{u,v}⟼cone[{p1,p2,p3},gamma][u,v] is the cone with vertex {p1,p2,p3} over the curve **gamma**:$(a,b) \longrightarrow \mathbb{R}^3$.

```
cone[{p1_,p2_,p3_},gamma_][u_,v_]:=
    {p1,p2,p3} + v*gamma[u]
```

A.47 cylinder

{u,v}⟼cylinder[q1,q2,q3,gamma][u,v] is the cylinder in the direction {q1,q2,q3} over the curve **gamma**:$(a,b) \longrightarrow \mathbb{R}^3$.

```
cylinder[{q1_,q2_,q3_},gamma_][u_,v_]:=
    gamma[u] + {v*q1,v*q2,v*q3}
```

A.48 genhel

{u,v}⟼genhel[alpha][sl][u,v] is the generalized helicoid with slant **sl** generated by the plane curve **alpha**.

```
genhel[alpha_][sl_][u_,v_]:= {Cos[u]*alpha[v][[1]],
                              Sin[u]*alpha[v][[1]],
                              sl*u + alpha[v][[2]]}
```

A.49 gensurfrev

{u,v}⟼gensurfrev[alpha,gamma][u,v] is the generalized surface of revolution generated by the plane curves **alpha** and **gamma**.

```
gensurfrev[alpha_,gamma_][u_,v_]:=
        {alpha[u][[1]]*gamma[v][[1]],
         alpha[u][[2]]*gamma[v][[1]],gamma[v][[2]]}
```

Appendix A: General Programs

A.50 ruled

{u,v} ⟼ ruled[alpha,gamma][u,v] is the ruled patch generated by the curves **alpha** and **gamma**.

```
ruled[alpha_,gamma_][u_,v_]:= alpha[u] + v*gamma[u]
```

A.51 surfast

{u,v} ⟼ surfast[n,alpha][u,v] is the astroidial patch of degree **n** generated by the plane curve **alpha**.

```
surfast[n_,alpha_][u_,v_]:= {Cos[u]^n*alpha[v][[1]],
                             Sin[u]^n*alpha[v][[1]],
                             alpha[v][[2]]}
```

A.52 surfrev

{u,v} ⟼ surfrev[alpha][u,v] is the surface of revolution generated by the plane curve **alpha**.

```
surfrev[alpha_][u_,v_]:= {Cos[u]*alpha[v][[1]],
                          Sin[u]*alpha[v][[1]],alpha[v][[2]]}
```

A.53 tubeplanecurve

{r,theta} ⟼ tubeplanecurve[gamma][r][t,theta] is the tube in $\mathbb{R}^3$ of radius **r** about the curve **gamma**:$(a,b) \longrightarrow \mathbb{R}^2$.

```
tubeplanecurve[gamma_][r_][t_,theta_]:= Simplify[
    Append[gamma[tt] - r*Cos[theta]*J[D[gamma[tt],tt]]/
    Sqrt[D[gamma[tt],tt].D[gamma[tt],tt]],
    r*Sin[theta]]] /. tt->t
```

A.54 twist

{u,v} ⟼ twist[a,b,gamma][u,v] is the twisted patch generated by a curve gamma:$(a,b) \longrightarrow \mathbb{R}^2$. The Möbius strip and Klein bottle are special cases.

```
twist[a_,b_,gamma_][u_,v_]:=
    (a + Cos[b*u]*gamma[v][[1]] - Sin[b*u]*gamma[v][[2]])*
    {Cos[u],Sin[u],0} +
    (Sin[b*u]*gamma[v][[1]] + Cos[b*u]*gamma[v][[2]])*
    {0,0,1}
```

Construction of Surfaces from Space Curves

A.55 binormaldev

{u,v} ⟼ binormaldev[alpha][u,v] is the developable surface generated curve alpha:$(a,b) \longrightarrow \mathbb{R}^3$ and its binormal vector field.

```
binormaldev[alpha_][u_,v_]:= alpha[u] + v*binormal[alpha][u]
```

A.56 canalsurf

{r,theta} ⟼ canalsurf[gamma][pm,r][t,theta] is the canal surface of variable radius r[t] about the curve gamma:$(a,b) \longrightarrow \mathbb{R}^3$. The parameter pm is ±1.

```
canalsurf[gamma_][r_][t_,theta_]:= Simplify[gamma[tt] -
    r[tt]*D[r[tt],tt]*tangent[gamma][tt] +
    r[tt]*Sqrt[r[tt]^2 - D[r[tt],tt]^2]*
    (-Cos[theta]*normal[gamma][tt] +
        Sin[theta]*binormal[gamma][tt])] /. tt->t
```

A.57 normaldev

{u,v} ⟼ normaldev[alpha][u,v] is the developable surface generated from the curve alpha:$(a,b) \longrightarrow \mathbb{R}^3$ and its normal vector field.

Appendix A: General Programs

```
normaldev[alpha_][u_,v_]:= alpha[u] + v*normal[alpha][u]
```

A.58 perpdev

{u,v}⟼perpdev[phi,alpha][u,v] is a 1-parameter family of developable surfaces generated by the curve alpha:$(a,b) \longrightarrow \mathbb{R}^3$ and a normal vector field to alpha and some normal vector field to alpha. Special cases are {u,v}⟼perpdev[0,alpha][u,v], which is the normal developable of alpha, and {u,v}⟼perpdev[Pi/2,alpha][u,v], which is the binormal developable of alpha.

```
perpdev[phi_,alpha_][u_, v_]:=
    alpha[u] + v*(Cos[phi]*normal[alpha][u] +
        Sin[phi]*binormal[alpha][u])
```

A.59 seashell

{u,v}⟼seashell[gamma][r][t,theta] is the tube of variable radius r*t about the curve gamma:$(a,b) \longrightarrow \mathbb{R}^3$.

```
seashell[gamma_][r_][t_,theta_]:= gamma[t] +
    r*t*(Cos[theta]*normal[gamma][t] +
        Sin[theta]*binormal[gamma][t])
```

A.60 tandev

{u,v}⟼tandev[alpha][u,v] is the developable surface generated by the curve alpha:$(a,b) \longrightarrow \mathbb{R}^n$ and its tangent vector field.

```
tandev[alpha_][u_,v_]:= alpha[uu] + v*D[alpha[uu],uu]/
    Sqrt[D[alpha[uu],uu].D[alpha[uu],uu]] /. uu->u
```

A.61 transsurf

{u,v}⟼transsurf[alpha,gamma][u,v] is the translation surface {u,v}⟼alpha[u,v] + gamma[u,v] generated by the curves alpha and gamma.

```
transsurf[alpha_,gamma_][u_,v_] := alpha[u] + gamma[v]
```

A.62 tubecurve

`{r,theta} ⟼ tubecurve[gamma][r][t,theta]` is the tube of radius **r** about the curve **gamma**:$(a,b) \longrightarrow \mathbb{R}^3$.

```
tubecurve[gamma_][r_][t_,theta_] := Simplify[gamma[t] +
    r*(Cos[theta]*normal[gamma][t] +
      Sin[theta]*binormal[gamma][t])]
```

Construction of Surfaces from Surfaces

A.63 parsurf

`{u,v} ⟼ parsurf[x][r][u,v]` is the parallel patch at a distance **r** from a patch x:$\mathcal{U} \longrightarrow \mathbb{R}^3$.

```
parsurf[x_][r_][u_,v_] := Module[{xu,xv,n1,n2,n3,n4},
    xu = D[x[uu,vv],uu];
    xv = D[x[uu,vv],vv];
    n1 = cross[xu,xv]//Simplify;
    n2 = Simplify[Factor[n1.n1]];
    n3 = PowerExpand[Sqrt[n2]]//Simplify;
    n4 = Simplify[n1/n3];
    Simplify[Together[x[uu,vv] +
         r*n4]]] /. {uu->u,vv->v}
```

A.64 pedsurf

`{u,v} ⟼ pedsurf[{a,b,c},x][u,v]` is the pedal patch of the patch **x** with respect to the point **{a,b,c}**.

Appendix A: General Programs

```
pedsurf[{a_,b_,c_},x_][u_,v_]:=
    Module[{xu,xv,n1,s2,s3,n4},
    xu = D[x[uu,vv],uu];
    xv = D[x[uu,vv],vv];
    n1 = cross[xu,xv]//Simplify;
    s2 = Simplify[Factor[n1.n1]];
    s3 = PowerExpand[Sqrt[s2]]//Simplify;
    n4 = Simplify[n1/s3];
    {a,b,c} + Simplify[Together[((x[uu,vv] -
    {a,b,c}).n4)*n4]]  /. {uu->u,vv->v}
```

{u,v}⟼pedsurf[x][u,v] is the pedal patch of the patch x with respect to the origin.

```
pedsurf[x_][u_,v_]:= Module[{xu,xv,n1,s2,s3,n4},
    xu = D[x[uu,vv],uu];
    xv = D[x[uu,vv],vv];
    n1 = cross[xu, xv]//Simplify;
    s2 = Simplify[Factor[n1.n1]];
    s3 = PowerExpand[Sqrt[s2]]//Simplify;
    n4 = Simplify[n1/s3];
    Simplify[Together[(x[uu,vv].n4)*n4]]   /.
        {uu->u,vv->v}
```

A.65 stretch

{u,v}⟼stretch[a,b,c,surf][u,v] is the patch **surf**, but stretched by **a,b,c**.

```
stretch[a_,b_,c_,surf_][u_,v_]:=
    {a*surf[u,v][[1]],b*surf[u,v][[2]],c*surf[u,v][[3]]}
```

A.66 tangentplane

{u,v}⟼tangentplane[x][u0,v0][u,v] is the tangent plane to the patch **x** at **x[u0,v0]**.

```
tangentplane[x_][u0_,v0_][u_,v_]:= x[u0,v0] +
    D[x[uu0,vv0],uu0]*u + D[x[uu0,vv0],vv0]*v /.
    {uu0->u0,vv0->v0}
```

A.67 plottp

`plottp[x,u0,v0,umin,umax,vmin,vmax,opts]` plots a patch **x** and its tangent plane at **x[u0,v0]**.

```
plottp[surf_,{u0_,v0_},{umin_,umax_},
    {vmin_,vmax_},opts___]:=
    ParametricPlot3D[{surf[u0 + u,v0 + v],
        tangentplane[surf][u0,v0][u,v]}//Evaluate,
        {u,umin,umax},{v,vmin,vmax},
        Boxed->False,Axes->None,opts];
```

Curvature of Surfaces (Fast)

A.68 gcurvature

$\{u,v\} \mapsto$ `gcurvature[x][u,v]` is the Gaussian curvature of the patch $x: \mathcal{U} \longrightarrow \mathbb{R}^3$.

```
gcurvature[x_][u_,v_]:= Simplify[
    (Det[{D[x[uu,vv],uu,uu],D[x[uu,vv],uu],
        D[x[uu,vv],vv]}]*
    Det[{D[x[uu,vv],vv,vv],D[x[uu,vv],uu],
        D[x[uu,vv],vv]}] -
    Det[{D[x[uu,vv],uu,vv],D[x[uu,vv],uu],
        D[x[uu,vv],vv]}]^2)/
    (D[x[uu,vv],uu].D[x[uu,vv],uu]*
        D[x[uu,vv],vv].D[x[uu,vv],vv] -
    D[x[uu,vv],uu].
        D[x[uu,vv],vv]^2)^2] /. {uu->u,vv->v}
```

A.69 mcurvature

$\{u,v\} \longmapsto$ mcurvature[x][u,v] is the mean curvature of the patch $x: \mathcal{U} \longrightarrow \mathbb{R}^3$.

```
mcurvature[x_][u_,v_]:= Simplify[
    (Det[{D[x[uu,vv],uu,uu],D[x[uu,vv],uu],
        D[x[uu,vv],vv]}]*
        D[x[uu,vv],vv].D[x[uu,vv],vv] -
    2*Det[{D[x[uu,vv],uu,vv],D[x[uu,vv],uu],
        D[x[uu,vv],vv]}]*
        D[x[uu,vv],uu].D[x[uu,vv],vv] +
    Det[{D[x[uu,vv],vv,vv],D[x[uu,vv],uu],
        D[x[uu,vv],vv]}]*
        D[x[uu,vv],uu].D[x[uu,vv],uu])/
    (2*(D[x[uu,vv],uu].D[x[uu,vv],uu]*
        D[x[uu,vv],vv].D[x[uu,vv],vv] -
        D[x[uu,vv],uu].
        D[x[uu,vv],vv]^2)^(3/2))] /. {uu->u,vv->v}
```

Curvature of Surfaces (Slow)

A.70 gaussiancurvature

$\{u,v\} \longmapsto$ gaussiancurvature[x][u,v] is the Gaussian curvature of the patch $x: \mathcal{U} \longrightarrow \mathbb{R}^3$.

```
gaussiancurvature[x_][u_,v_]:= Simplify[
    (eee[x][uu,vv]*ggg[x][uu,vv] - fff[x][uu,vv]^2)/
    (ee[x][uu,vv]*gg[x][uu,vv] -
        ff[x][uu,vv]^2)] /. {uu->u,vv->v}
```

A.71 meancurvature

$\{u,v\} \longmapsto$ meancurvature[x][u,v] is the mean curvature of the patch $x: \mathcal{U} \longrightarrow \mathbb{R}^3$.

```
meancurvature[x_][u_,v_]:= Simplify[
    (eee[x][uu,vv]*gg[x][uu,vv] -
        2*fff[x][uu,vv]*ff[x][uu,vv] +
    ggg[x][uu,vv]*ee[x][uu,vv])/
    (2*(ee[x][uu,vv]*gg[x][uu,vv] -
        ff[x][uu,vv]^2))] /. {uu->u,vv->v}
```

A.72 meancurvaturevf

$\{u,v\} \mapsto$ **meancurvaturevf[x][u,v]** is the mean curvature vector field of the patch $\mathbf{x}: \mathcal{U} \longrightarrow \mathbb{R}^n$.

```
meancurvaturevf[x_][u_,v_]:=
    Module[{h0,ar,h1,h2,h3},
    h0=gg[x][uu,vv]*D[x[uu,vv],uu,uu] -
        2*ff[x][uu,vv]*D[x[uu,vv],uu,vv] +
        ee[x][uu,vv]*D[x[uu,vv],vv,vv];
    ar=Simplify[(ee[x][uu,vv]*gg[x][uu,vv] -
            ff[x][uu,vv]^2)];
    h1 = (h0.D[x[uu,vv],uu])*gg[x][uu,vv] -
        (h0.D[x[uu,vv],vv])*ff[x][uu,vv];
    h2 = (h0.D[x[uu,vv],vv])*ee[x][uu,vv] -
        (h0.D[x[uu,vv],uu])*ff[x][uu,vv];
    Simplify[h0/(2*ar) - h1*D[x[uu,vv],uu]/(2*ar^2) -
        h2*D[x[uu,vv],vv]/(2*ar^2) /. {uu->u,vv->v}]]
```

Curvature and Torsion of Curves

A.73 geodesiccurvature

$t \mapsto$ **geodesiccurvature[surf,alpha][t]** calculates the geodesic curvature of the curve $\mapsto$ **surf[alpha[t]]**, contained in the parametrized surface **surf**: $\mathcal{U} \longrightarrow \mathbb{R}^3$, where **alpha** is a plane curve whose trace is contained in the subset $\mathcal{U}$ of $\mathbb{R}^2$.

Appendix A: General Programs

```
geodesiccurvature[surf_,alpha_][t_]:=
    Module[{unc,tmp,crv1,crv2},
    unc=unitnormal[surf][alpha[tt][[1]],alpha[tt][[2]]];
    tmp=surf[alpha[tt][[1]],alpha[tt][[2]]];
    crv1=D[tmp,tt];crv2=D[tmp,{tt,2}];
    Simplify[PowerExpand[Det[
    {crv2,unc,crv1}]/(crv1.crv1)^(3/2) /. tt->t]]]
```

A.74 geodesictorsion

t ⟼ **geodesictorsion[surf,alpha][t]** calculates the geodesic torsion of the curve ⟼ **surf[alpha[t]]**, contained in the parametrized surface **surf**:$\mathcal{U} \longrightarrow \mathbb{R}^3$, where **alpha** is a plane curve whose trace is contained in the subset $\mathcal{U}$ of $\mathbb{R}^2$.

```
geodesictorsion[surf_,alpha_][t_]:=
    Module[{unc,tmp,crv1,unc1},
    unc=unitnormal[surf][alpha[tt][[1]],alpha[tt][[2]]];
    tmp=surf[alpha[tt][[1]],alpha[tt][[2]]];
    crv1=D[tmp,tt];unc1=D[unc,tt];
    Simplify[PowerExpand[Det[
    {unc,unc1,crv1}]/(crv1.crv1)^(3/2) /. tt->t]]]
```

A.75 kappa

t ⟼ **kappa[alpha][t]** is the unsigned curvature of the curve
alpha:$(a,b) \longrightarrow \mathbb{R}^3$.

```
kappa[alpha_][t_]:=
    Simplify[Sqrt[Factor[cross[D[alpha[tt],tt],
    D[alpha[tt],{tt,2}]].
    cross[D[alpha[tt],tt],D[alpha[tt],{tt,2}]]]]]/
    Simplify[Factor[D[alpha[tt],tt].
    D[alpha[tt],tt]]]^(3/2) /. tt->t
```

A.76 kappan

t⟼**kappan[alpha][t]** is the unsigned curvature of the curve
alpha:$(a,b) \longrightarrow \mathbb{R}^n$.

```
kappan[alpha_][t_]:=
    Simplify[Sqrt[Factor[
    (Sqrt[Factor[D[alpha[tt],tt].
                D[alpha[tt],tt]]]*D[alpha[tt],{tt,2}] -
    D[Sqrt[Factor[D[alpha[tt],tt].
                D[alpha[tt],tt]]],tt]*D[alpha[tt],tt]).
    (Sqrt[Factor[D[alpha[tt],tt].
                D[alpha[tt],tt]]]*D[alpha[tt],{tt,2}] -
    D[Sqrt[Factor[D[alpha[tt],tt].
                D[alpha[tt],tt]]],tt]*
    D[alpha[tt],tt])]]]/
    Simplify[Factor[D[alpha[tt],tt].
                D[alpha[tt],tt]]]^(3/2) /. tt->t
```

A.77 kappa2

t⟼**kappa2[alpha][t]** is the signed curvature of the plane curve
alpha:$(a,b) \longrightarrow \mathbb{R}^2$.

```
kappa2[alpha_][t_]:=
    D[alpha[tt],{tt,2}].J[D[alpha[tt],tt]]/
    Simplify[D[alpha[tt],tt].
            D[alpha[tt],tt]]^(3/2) /. tt->t
```

A.78 kappa2c

t⟼**kappa2c[alpha][t]** is the signed curvature of the complex plane curve
alpha:$(a,b) \longrightarrow \mathbb{R}^2$.

Appendix A: General Programs 865

```
kappa2c[alpha_][t_]:= Simplify[
    ComplexExpand[Re[D[alpha[t],{t,2}]*
    Conjugate[I*D[alpha[t],t]]]]/
    ComplexExpand[Abs[D[alpha[t],t]],
    TargetFunctions->{Re,Im}]^3]
```

A.79 kappa2ea

t ⟼ **kappa2ea[alpha][t]** is the equiaffine curvature of the plane curve
alpha:$(a,b) \longrightarrow \mathbb{R}^2$.

```
kappa2ea[alpha_][t_]:=
    (3*Det[{D[alpha[tt],tt],D[alpha[tt],{tt,2}]}]*
    Det[{D[alpha[tt],tt],D[alpha[tt],{tt,4}]}] +
    12*Det[{D[alpha[tt],tt],D[alpha[tt],{tt,2}]}]*
    Det[{D[alpha[tt],{tt,2}],D[alpha[tt],{tt,3}]}] -
    Det[{D[alpha[tt],tt],D[alpha[tt],{tt,3}]}]^3)/
    (9*Det[{D[alpha[tt],tt],D[alpha[tt],{tt,2}]}])^(8/3) /. tt->t
```

A.80 kappa2implicit

kappa2implicit[g][x,y] is the signed curvature of the plane curve
g[x,y]==0.

```
kappa2implicit[g_][x_,y_]:=
    (D[g[xx,yy],{xx,2}]*D[g[xx,yy],yy]^2 -
    2D[g[xx,yy],xx,yy]*D[g[xx,yy],xx]*D[g[xx,yy],yy] +
    D[g[xx,yy],{yy,2}]*D[g[xx,yy],xx]^2)/
    (D[g[xx,yy],xx]^2 + D[g[xx,yy],yy]^2)^(3/2) /.
        {xx->x,yy->y}
```

A.81 kappa2polar

t ⟼ **kappa2polar[rho][t]** is the signed curvature of the plane curve whose radius function is **rho**.

```
kappa2polar[rho_][t_]:= (-D[rho[tt],{tt,2}]*rho[tt]
    + 2*D[rho[tt],tt]^2 + rho[tt]^2)/
    (D[rho[tt],tt]^2 + rho[tt]^2)^(3/2) /. tt->t
```

A.82 tau

$t \mapsto$ **tau[alpha][t]** is the torsion of the curve **alpha**:$(a,b) \longrightarrow \mathbb{R}^3$.

```
tau[alpha_][t_]:= Simplify[Det[
    {D[alpha[tt],tt],D[alpha[tt],{tt,2}],
    D[alpha[tt],{tt,3}]}]]/
    Simplify[Factor[cross[D[alpha[tt],tt],
    D[alpha[tt],{tt,2}]].cross[D[alpha[tt],tt],
    D[alpha[tt],{tt,2}]]]] /. tt->t
```

A.83 totalscalarcurvature

totalscalarcurvature[a,b,alpha] is the total curvature of the curve **alpha**:$(a,b) \longrightarrow \mathbb{R}^2$.

```
totalscalarcurvature[a_,b_,alpha_]:=
    Integrate[kappa2[alpha][u]*
    Sqrt[Simplify[D[alpha[u],u].
    D[alpha[u],u]]],{u,a,b}]
```

A.84 turningangle

$t \mapsto$ **turningangle[alpha][t]** is the turning angle of the curve **alpha**:$(a,b) \longrightarrow \mathbb{R}^2$.

```
turningangle[alpha_][t_]:= ArcTan[D[alpha[tt],tt][[1]],
    D[alpha[tt],tt][[2]]] /. tt->t
```

A.85 turningnumber

turningnumber[alpha,{a,b},opts] is the turning number of the closed curve **alpha**:$(a,b) \longrightarrow \mathbb{R}^2$.

Appendix A: General Programs 867

```
turningnumber[alpha_,{a_,b_},opts___]:=
    Module[{tt,tmp1,tmp2},
        tmp1 = kappa2[alpha][tt]*Sqrt[alpha'[tt].
            alpha'[tt]];
        tmp2 = NIntegrate[Evaluate[tmp1],{tt,a,b},
            AccuracyGoal->2,opts];
        Round[tmp2/N[2Pi]]]
```

Euclidean Motions Acting on Plane Curves

A.86 tl

t⟼tl[p,alpha][t] is the curve **alpha**:$(a,b) \longrightarrow \mathbb{R}^n$ translated by **p**.

```
tl[p_,alpha_][t_]:= p + alpha[t]
```

A.87 rt

t⟼rt[theta,alpha][t] is the curve **alpha**:$(a,b) \longrightarrow \mathbb{R}^2$ rotated by **theta**.

```
rt[theta_,alpha_][t_]:=
    {Cos[theta]*alpha[t][[1]] - Sin[theta]*alpha[t][[2]],
     Sin[theta]*alpha[t][[1]] + Cos[theta]*alpha[t][[2]]}
```

A.88 rf

t⟼rf[alpha][t] is the reflection about the horizontal axis of the curve **alpha**:$(a,b) \longrightarrow \mathbb{R}^2$.

```
rf[alpha_][t_]:= {alpha[t][[1]],-alpha[t][[2]]}
```

Frames for Plane Curves

A.89 planeframe

t⟼**planeframe[alpha][t]** is the frame **{T,JT}** of the curve
alpha:$(a,b) \longrightarrow \mathbb{R}^2$.

```
planeframe[alpha_][t_]:=
    {Line[{alpha[tt],alpha[tt] +
    D[alpha[tt],tt]/Sqrt[D[alpha[tt],tt].D[alpha[tt],tt]]}],
    Line[{alpha[tt],alpha[tt] +
    J[D[alpha[tt],tt]]/Sqrt[D[alpha[tt],tt].
      D[alpha[tt],tt]]}]} /. tt->t
```

t⟼**planeframe[alpha][t]** is the frame **{T,JT}** of the curve
alpha:$(a,b) \longrightarrow \mathbb{R}^2$ with offset text labels.

```
planeframe[alpha_,offset_][t_]:=
  {Line[{alpha[tt],alpha[tt] +
   D[alpha[tt],tt]/Sqrt[D[alpha[tt],tt].D[alpha[tt],tt]]}],
   Line[{alpha[tt],alpha[tt] +
   J[D[alpha[tt],tt]]/Sqrt[D[alpha[tt],tt].D[alpha[tt],tt]]}],
   Text["T",alpha[tt] +
     offset*D[alpha[tt],tt]/Sqrt[D[alpha[tt],tt].
         D[alpha[tt],tt]]],
   Text["JT",alpha[tt] +
     offset*J[D[alpha[tt],tt]]/Sqrt[D[alpha[tt],tt].
       D[alpha[tt],tt]]]} /. tt->t
```

t⟼**planeframe[alpha][t]** is the frame **{T,JT}** of the curve
alpha:$(a,b) \longrightarrow \mathbb{R}^2$ with offset text labels in a given font and size.

```
planeframe[alpha_,offset_,font_,size_][t_]:=
  {Line[{alpha[tt],alpha[tt] +
    D[alpha[tt],tt]/Sqrt[D[alpha[tt],tt].D[alpha[tt],tt]]}],
  Line[{alpha[tt],alpha[tt] +
    J[D[alpha[tt],tt]]/Sqrt[D[alpha[tt],tt].D[alpha[tt],tt]]}],
  Text[FontForm["T",{font,size}],alpha[tt] +
    offset*D[alpha[tt],tt]/Sqrt[D[alpha[tt],tt].
        D[alpha[tt],tt]]],
  Text[FontForm["JT",{font,size}],alpha[tt] +
    offset*J[D[alpha[tt],tt]]/Sqrt[D[alpha[tt],tt].
    D[alpha[tt],tt]]]} /. tt->t
```

Focal Set Programs

A.90 focalkmu

$\{u,v\} \longmapsto$ **focalkmu[surf][u,v]** is a parametrization of the focal set of a principal patch **surf** corresponding to the principal curvature **kmu** for which **ff[surf]** = **fff[surf]** = 0.

```
focalkmu[surf_][u_,v_]:= (surf[uu,vv] +
    (1/kmu[surf][uu,vv])*
    unitnormal[surf][uu,vv] /. {uu->u,vv->v})
```

A.91 focalkpi

$\{u,v\} \longmapsto$ **focalkpi[surf][u,v]** is a parametrization of the focal set of a principal patch **surf** corresponding to the principal curvature **kpi** for which **ff[surf]** = **fff[surf]** = 0.

```
focalkpi[surf_][u_,v_]:= (surf[uu,vv] +
    (1/kpi[surf][uu,vv])*
    unitnormal[surf][uu,vv] /. {uu->u,vv->v})
```

A.92 focalk1

{u,v}⟼focalk1[surf][u,v] is a parametrization of the focal set of a patch **surf** corresponding to the principal curvature **k1**.

```
focalk1[surf_][u_,v_]:= surf[uu,vv] +
    (1/k1[surf][uu,vv])*
    unitnormal[surf][uu,vv] /. {uu->u,vv->v}
```

A.93 focalk2

{u,v}⟼focalk2[surf][u,v] is a parametrization of the focal set of a patch **surf** corresponding to the principal curvature **k2**.

```
focalk2[surf_][u_,v_]:= surf[uu,vv] +
    (1/k2[surf][uu,vv])*
    unitnormal[surf][uu,vv] /. {uu->u,vv->v}
```

A.94 tangentkmu

{u,v}⟼tangentkmu[x][u,v] is an eigenvector of unit length of a patch **x** corresponding to the principal curvature **kmu**.

```
tangentkmu[x_][u_,v_]:= (D[x[u,vv],vv]/
    Sqrt[gg[x][u,vv]] /. vv->v) /;
    ff[x][u,v]==fff[x][u,v]==0
```

A.95 tangentkpi

{u,v}⟼tangentkpi[x][u,v] is an eigenvector of unit length of a patch **x** corresponding to the principal curvature **kpi**.

```
tangentkpi[x_][u_,v_]:= (D[x[uu,v],uu]/
    Sqrt[gg[x][uu,v]] /. uu->u) /;
    ff[x][u,v]==fff[x][u,v]==0
```

Appendix A: General Programs 871

Frenet Frame Programs

A.96 `binormal`

t⟼**binormal[alpha][t]** is the binormal of the curve **alpha**:$(a,b) \longrightarrow \mathbb{R}^3$.

```
binormal[alpha_][t_]:= Simplify[cross[
    D[alpha[tt],tt],D[alpha[tt],{tt,2}]]]/
    Simplify[Sqrt[Factor[cross[D[alpha[tt],tt],
    D[alpha[tt],{tt,2}]].
    cross[D[alpha[tt],tt],D[alpha[tt],
    {tt,2}]]]]]    /. tt->t
```

A.97 `darbouxvector`

t⟼**darbouxvector[alpha][t]** is the Darboux vector of the curve **alpha**:$(a,b) \longrightarrow \mathbb{R}^3$.

```
darbouxvector[alpha_][t_]:=
    Sqrt[D[alpha[tt],tt].D[alpha[tt],tt]]*(
        tau[alpha][tt]*tangent[alpha][tt] +
        kappa[alpha][tt]*binormal[alpha][tt]) /. tt->t
```

A.98 `frenetframe`

t⟼**frenetframe[alpha][t]** is the Frenet frame of the curve **alpha**:$(a,b) \longrightarrow \mathbb{R}^3$.

```
frenetframe[alpha_][t_]:=
    {Line[{alpha[t],alpha[t] + tangent[alpha][t]}],
     Line[{alpha[t],alpha[t] + normal[alpha][t]}],
     Line[{alpha[t],alpha[t] + binormal[alpha][t]}]}
```

t⟼**frenetframe[alpha][t]** is the Frenet frame of the curve **alpha**:$(a,b) \longrightarrow \mathbb{R}^3$ with offset text labels.

```
frenetframe[alpha_,offset_][t_]:=
    {Line[{alpha[t],alpha[t] + tangent[alpha][t]}],
     Line[{alpha[t],alpha[t] + normal[alpha][t]}],
     Line[{alpha[t],alpha[t] + binormal[alpha][t]}],
     Text["T",alpha[t] + offset*tangent[alpha][t]],
     Text["N",alpha[t] + offset*normal[alpha][t]],
     Text["B",alpha[t] + offset*binormal[alpha][t]]}
```

t⟼**frenetframe[alpha][t]** is the Frenet frame of the curve **alpha**:$(a,b) \longrightarrow \mathbb{R}^3$ with offset text labels in a given font and size.

```
frenetframe[alpha_,offset_,font_,size_][t_]:=
    {Line[{alpha[t],alpha[t] + tangent[alpha][t]}],
     Line[{alpha[t],alpha[t] + normal[alpha][t]}],
     Line[{alpha[t],alpha[t] + binormal[alpha][t]}],
     Text[FontForm["T",{font,size}],
          alpha[t] + offset*tangent[alpha][t]],
     Text[FontForm["N",{font,size}],
          alpha[t] + offset*normal[alpha][t]],
     Text[FontForm["B",{font,size}],
          alpha[t] + offset*binormal[alpha][t]]}
```

A.99 normal

t⟼**normal[alpha][t]** is the normal of the curve **alpha**:$(a,b) \longrightarrow \mathbb{R}^3$.

```
normal[alpha_][t_]:= cross[binormal[alpha][t],
                           tangent[alpha][t]]
```

A.100 tangent

t⟼**tangent[alpha][t]** is the tangent of the curve **alpha**:$(a,b) \longrightarrow \mathbb{R}^n$.

```
tangent[alpha_][t_]:= D[alpha[tt],tt]/
    Simplify[Factor[[D[alpha[tt],tt].
        D[alpha[tt],tt]]]   /. tt->t
```

Appendix A: General Programs   873

Geodesics

A.101 clairautu

{t,v} ⟼ clairautv[pm,c,u0][x][t,v] gives the first integral of the geodesic equations for a *u*-Clairaut patch. The parameter **pm** is ±1.

```
clairautu[pm_,c_,u0_][x_][t_,v_]:=
    PowerExpand[u0 + Integrate[pm*Sqrt[c^2*ee[x][tt,v]/
    (gg[x][tt,v]*(gg[x][tt,v] - c^2))],tt]] /. tt->t
```

A.102 clairautv

{u,t} ⟼ clairautv[pm,c,u0][x][u,t] gives the first integral of the geodesic equations for a *v*-Clairaut patch. The parameter **pm** is ±1.

```
clairautv[pm_,c_,u0_][x_][u_,t_]:=
    PowerExpand[u0 + Integrate[pm*Sqrt[c^2*gg[x][u,tt]/
    (ee[x][u,tt]*(ee[x][u,tt] - c^2))],tt]] /. tt->t
```

A.103 geo

geo[x][p,q,u,v] gives the system of first-order differential equations for a patch $\mathbf{x}: \mathcal{U} \longrightarrow \mathbb{R}^n$.

```
geo[x_][p_,q_,u_,v_]:=
    {p,q,-christoffel[1,1,1][x][u,v]*p^2 -
    2*christoffel[1,1,2][x][u,v]*p*q -
    christoffel[1,2,2][x][u,v]*q^2,
    -christoffel[2,1,1][x][u,v]*p^2 -
    2*christoffel[2,1,2][x][u,v]*p*q -
    christoffel[2,2,2][x][u,v]*q^2}
```

geo[ee,ff,gg][p,q,u,v] gives the system of first-order differential equations for the geodesics of a metric with coefficients **ee**, **ff**, **gg**.

```
geo[ee_,ff_,gg_][p_,q_,u_,v_]:=
    {p,q,-christoffel[1,1,1][ee,ff,gg][u,v]*p^2 -
    2*christoffel[1,1,2][ee,ff,gg][u,v]*p*q -
    christoffel[1,2,2][ee,ff,gg][u,v]*q^2,
    -christoffel[2,1,1][ee,ff,gg][u,v]*p^2 -
    2*christoffel[2,1,2][ee,ff,gg][u,v]*p*q -
    christoffel[2,2,2][ee,ff,gg][u,v]*q^2}
```

A.104 solvegeoeqs

solvegeoeqs solves the geodesic equations of the patch **x** using **NDSolve**.

```
solvegeoeqs[x_,a_:0,{u0_:0,v0_:0},tfin_,div_,optsnd___]:=
    Flatten[Map[NDSolve[#,{u,v},{ss,0,tfin}]&,
    Table[Join[MapThread[Equal,{{u''[ss],v''[ss]},
    Take[geo[x][u'[ss],v'[ss],u[ss],v[ss]],{3,4}]}],
    {u[a]==u0,v[a]==v0,
    u'[a]==Cos[theta],v'[a]==Sin[theta]}],
    {theta,2Pi/div,2Pi,2Pi/div}]],1]
```

Intrinsic Curvature of Curves

A.105 intrinsic

intrinsic[fun,a,{c,d,theta0},optsnd,{smin,smax}] is the plane curve **alpha** starting at **{c,d}** and initial turning angle **theta0** such that the signed curvature of **alpha** is **fun**.

```
intrinsic[fun_,a:_0,{c_:0,d_:0,theta0_:0},optsnd___,
    {smin_:-10,smax_:10}][t_]:=
    Flatten[Module[{x,y,theta},{x[t],y[t]} /.
    NDSolve[{x'[ss]==Cos[theta[ss]],
        y'[ss]==Sin[theta[ss]],
        theta'[ss]==fun[ss],
        x[a]==c,y[a]==d,theta[a]==theta0},
        {x,y,theta},{ss,smin,smax},optsnd]]]
```

Appendix A: General Programs 875

A.106 intrinsicea

`intrinsicea[fun,a,{c1,d1,c2,d2,c3,d3},optsnd,{smin,smax}]` is the plane curve **alpha** starting at **{c1,d1}** with initial velocity **{c2,d2}** and acceleration **{c3,d3}** such that the signed equiaffine curvature of **alpha** is **fun**.

```
intrinsicea[fun_,a_,{c1_,d1_,c2_,d2_,c3_,d3_},optsnd___,
    {smin_:-10,smax_:10}][t_]:=
  Flatten[Module[{x,y},
  {x[t],y[t]} /.
  NDSolve[{x'''[ss]== -fun[ss]*x'[ss],
          y'''[ss]== -fun[ss]*y'[ss],
   x[a]==c1,y[a]==d1,x'[a]==c2,y'[a]==d2,
   x''[a]==c3,y''[a]==d3},
   {x,y},{ss,smin,smax},optsnd]]]
```

A.107 plotintrinsic

`plotintrinsic[fun,a,{c,d,theta0},optsnd,{smin,smax},optspp]` plots the curve **alpha** starting at **{c,d}** and initial turning angle **theta0**, such that the signed curvature of **alpha** is **fun**.

```
plotintrinsic[fun_,a_:0,{c_:0,d_:0,theta0_:0},optsnd___,
        {smin_:-10,smax_:10},optspp___]:=
    ParametricPlot[Evaluate[
    intrinsic[fun,a,{c,d,theta0},optsnd,{smin,smax}][t]],
    {t,smin,smax},AspectRatio->Automatic,optspp];
```

```
plotintrinsic[fun_,optsnd___,
    {smin_:-10,smax_:10},optspp___]:=
  plotintrinsic[fun,0,{0,0,0},{smin,smax},optspp]
```

A.108 plotintrinsicea

`plotintrinsicea[fun,a,{c1,d1,c2,d2,c3,d3},`
`optsnd,{smin,smax},optspp]`
plots the plane curve **alpha** starting at **{c1,d1}** with initial velocity **{c2,d2}**

and acceleration `{c3,d3}` such that the signed equiaffine curvature of **alpha** is **fun**.

```
plotintrinsicea[fun_,a_,{c1_,d1_,c2_,d2_,c3_,d3_},optsnd___,
        {smin_:-10,smax_:10},optspp___]:=
    ParametricPlot[Evaluate[
    intrinsicea[fun,a,{c1,d1,c2,d2,c3,d3},optsnd,{smin,smax}][t]],
        {t,smin,smax},AspectRatio->Automatic,optspp];
```

A.109 plotintrinsicsurfrev

`plotintrinsicsurfrev[fun,rad,a,{c,d,theta0},optsnd,`
`{umin,umax},{vmin,vmax},optspp3d]`
plots a surface of revolution of radius a whose profile curve has signed curvature **fun**.

```
plotintrinsicsurfrev[fun_,rad_,a_:0,{c_:0,d_:0,theta0_:0},
        optsnd___,{umin_:0,umax_:2*Pi},{vmin_:-10,vmax_:10},
        optspp3d___]:=
    Module[{tmpitr,tmpsr},
        tmpitr=intrinsic[fun,a,{c,d,theta0},optsnd,
            {vmin,vmax}][v];
        tmpsr={Cos[u]*(rad + tmpitr[[1]]),
            Sin[u]*(rad + tmpitr[[1]]),tmpitr[[2]]};
        ParametricPlot3D[Evaluate[tmpsr],
            {u,umin,umax},{v,vmin,vmax},optspp3d]]
```

A.110 plotintrinsic3d

`plotintrinsic3d[kk,tt,a,{p1,p2,p3},{q1,q2,q3},{r1,r2,r3},`
`{smin,smax},opts]` plots a unit-speed space curve **alpha** with curvature **kk** and torsion **tt** such that `alpha[a]={p1,p2,p3}`,
`alpha'[a]={q1,q2,q3}` and `alpha''[a]=kk[a]*{r1,r2,r3}`.

```
plotintrinsic3d[{kk_,tt_},{a_:0,{p1_:0,p2_:0,p3_:0},
      {q1_:1,q2_:0,q3_:0},{r1_:0,r2_:1,r3_:0}},
            {smin_:-10,smax_:10},opts___]:=
    ParametricPlot3D[Module[
       {x1,x2,x3,t1,t2,t3,n1,n2,n3,b1,b2,b3},
       {x1[s],x2[s],x3[s]} /.
       NDSolve[{x1'[ss]==t1[ss],
               x2'[ss]==t2[ss],
               x3'[ss]==t3[ss],
            t1'[ss]==kk[ss]*n1[ss],
            t2'[ss]==kk[ss]*n2[ss],
            t3'[ss]==kk[ss]*n3[ss],
               n1'[ss]==-kk[ss]*t1[ss] + tt[ss]*b1[ss],
               n2'[ss]==-kk[ss]*t2[ss] + tt[ss]*b2[ss],
               n3'[ss]==-kk[ss]*t3[ss] + tt[ss]*b3[ss],
                b1'[ss]==-tt[ss]*n1[ss],
                b2'[ss]==-tt[ss]*n2[ss],
                b3'[ss]==-tt[ss]*n3[ss],
         x1[a]==p1,x2[a]==p2,x3[a]==p3,
         t1[a]==q1,t2[a]==q2,t3[a]==q3,
         n1[a]==r1,n2[a]==r2,n3[a]==r3,
             b1[a]==q2*r3 - q3*r2,
             b2[a]==q3*r1 - q1*r3,
             b3[a]==q1*r2 - q2*r1},
       {x1,x2,x3,t1,t2,t3,n1,n2,n3,b1,b2,b3},
       {ss,smin,smax}]]//Evaluate,
       {s,smin,smax},opts];
```

Inversion Programs

A.111 inversion

p$\mapsto$`inversion[q,rho][p]` is the inversion of a point **p** with respect to a sphere in $\mathbb{R}^n$ of radius **rho** centered at **q**.

```
inversion[q_,rho_][p_]:= q + (rho^2/((p - q).(p - q)))*(p - q)
```

$p \longmapsto$ **inversion[rho][p]** is the inversion of a point **p** with respect to a sphere in $\mathbb{R}^n$ of radius **rho** centered at the origin.

```
inversion[rho_][p_] := (rho^2/p.p)*p
```

A.112 inversioncurve

$t \longmapsto$ **inversioncurve[q,rho][alpha][t]** is the inversion of a curve **alpha** with respect to a sphere in $\mathbb{R}^n$ of radius **rho** centered at **q**.

```
inversioncurve[q_List,rho_][curve_][t_] :=
                inversion[q,rho][curve[t]]
```

$t \longmapsto$ **inversioncurve[rho][alpha][t]** is the inversion of a curve **alpha** with respect to a sphere in $\mathbb{R}^n$ of radius **rho** centered at the origin.

```
inversioncurve[rho_][curve_][t_] :=
                inversion[rho][curve[t]]
```

A.113 inversionsurf

$t \longmapsto$ **inversionsurf[q,rho][surf][u,v]** is the inversion of a patch **surf** with respect to a sphere in $\mathbb{R}^n$ of radius **rho** centered at **q**.

```
inversionsurf[q_List,rho_][surf_][u_,v_] :=
                inversion[q,rho][surf[u,v]]
```

$t \longmapsto$ **inversionsurf[rho][surf][u,v]** is the inversion of a patch **surf** with respect to a sphere in $\mathbb{R}^n$ of radius **rho** centered at the origin.

```
inversionsurf[rho_][surf_][u_,v_] :=
                inversion[rho][surf[u,v]]
```

Jacobians and Normals

A.114 `jacobian2`

`{u,v}`⟼`jacobian2[x][u,v]` is the square of the determinant of the Jacobian matrix of the patch $x: \mathcal{U} \longrightarrow \mathbb{R}^3$.

```
jacobian2[x_][u_,v_]:= Module[{xu,xv,N1},
    xu = D[x[uu,vv],uu];
    xv = D[x[uu,vv],vv];
    N1 = cross[xu,xv];
    Simplify[N1.N1]] /. {uu->u,vv->v}
```

A.115 `surfacejacobianmatrix`

`{u,v}`⟼`surfacejacobianmatrix[x][u,v]` is the Jacobian matrix of the patch $x: \mathcal{U} \longrightarrow \mathbb{R}^n$.

```
surfacejacobianmatrix[x_][u_,v_]:= {D[x[uu,vv],uu],
    D[x[uu,vv],vv]}   /. {uu->u,vv->v}
```

A.116 `surfacejacobian`

`{u,v}`⟼`surfacejacobian[x][u,v]` is the determinant of the product of the Jacobian matrix of the patch $x: \mathcal{U} \longrightarrow \mathbb{R}^n$ with its transpose.

```
surfacejacobian[x_][u_,v_]:= Det[
    surfacejacobianmatrix[x][u,v].
    Transpose[surfacejacobianmatrix[x][u,v]]]
```

A.117 `surfacenormal`

`{u,v}`⟼`surfacenormal[x][u,v]` is the cross product of the tangent vectors x_u and x_v of the patch $x: \mathcal{U} \longrightarrow \mathbb{R}^3$.

```
surfacenormal[x_][u_,v_]:= Module[{xu,xv},
    xu = D[x[uu,vv],uu];
    xv = D[x[uu,vv],vv];
    Simplify[cross[xu,xv]] /. {uu->u,vv->v}]
```

A.118 unitnormal

{u,v} ⟼ unitnormal[x][u,v] is the unit normal to the patch $x:\mathcal{U} \longrightarrow \mathbb{R}^3$. It is the Gauss map of x.

```
unitnormal[x_][u_,v_]:= Module[{xu,xv,N1},
    xu = D[x[uu,vv],uu];
    xv = D[x[uu,vv],vv];
    N1 = cross[xu,xv];
    Simplify[N1.N1]^(-1/2)*N1] /. {uu->u,vv->v}
```

Metric (First Fundamental Form) Programs

A.119 ee

{u,v} ⟼ ee[x][u,v] is the first coefficient of the metric tensor (or first fundamental form) of the patch $x:\mathcal{U} \longrightarrow \mathbb{R}^n$. The usual mathematics symbol is E.

```
ee[x_][u_,v_]:= Simplify[D[x[uu,vv],uu].
            D[x[uu,vv],uu]] /. {uu->u,vv->v}
```

A.120 ff

{u,v} ⟼ ff[x][u,v] is the second coefficient of the metric tensor (or first fundamental form) of the patch $x:\mathcal{U} \longrightarrow \mathbb{R}^n$. The usual mathematics symbol is F.

```
ff[x_][u_,v_]:= Simplify[D[x[uu,vv],uu].
            D[x[uu,vv],vv]] /. {uu->u,vv->v}
```

A.121 gg

{u,v} ⟼ gg[x][u,v] is the third coefficient of the metric tensor (or first fundamental form) of the patch $x:\mathcal{U} \longrightarrow \mathbb{R}^n$. The usual mathematics symbol is G.

```
gg[x_][u_,v_]:= Simplify[D[x[uu,vv],vv].
            D[x[uu,vv],vv]] /. {uu->u,vv->v}
```

A.122 infarea

{u,v} ⟼ infarea[x][u,v] is the infinitesimal area of the metric tensor (or first fundamental form) of the patch x: $\mathcal{U} \longrightarrow \mathbb{R}^n$.

```
infarea[x_][u_,v_]:= Sqrt[Simplify[ee[x][u,v]*gg[x][u,v] -
                     ff[x][u,v]^2]]
```

A.123 ds2

{u,v} ⟼ ds2[x][u,v][du,dv] is the metric tensor (or first fundamental form) of the patch x: $\mathcal{U} \longrightarrow \mathbb{R}^n$.

```
ds2[x_][u_,v_][du_,dv_]:= ee[x][u,v]**du^2 +
    ff[x][u,v]**(du*dv) + gg[x][u,v]**dv^2 /.
    {0**_->0,_**0->0}
```

Minimal Surface Programs

A.124 analytictominimal

analytictominimal[Psi][t][u,v] is the minimal patch corresponding to the analytic n-tuple z ⟼ E^(-I*t)*Psi[z].

```
analytictominimal[Psi_][t_][u_,v_]:=
    ComplexExpand[Re[E^(-I*t)*Psi[zz]]
    /. zz->u + I*v]
```

A.125 geomincurve

z ⟼ geomincurve[alpha][z] is the minimal curve in $\mathbb{C}^3$ that contains a plane curve alpha: $(a,b) \longrightarrow \mathbb{R}^2$ as a geodesic.

```
geomincurve[alpha_][z_]:= Append[alpha[z],
    I*Simplify[PowerExpand[arclength[alpha][z]]]]
```

A.126 `harmonictoanalytic`

$z \longmapsto$ `harmonictoanalytic[x][z]` is the analytic function whose real part is the harmonic n-tuple $\{u,v\} \longmapsto x[u,v]$ and whose imaginary part vanishes at `{0,0}`.

```
harmonictoanalytic[x_][z_]:=
    Simplify[2x[z/2,z/(2*I)] - x[0,0]]
```

$z \longmapsto$ `harmonictoanalytic[x][x0,y0][z]` is the analytic function whose real part is $\{u,v\} \longmapsto x[u,v]$ and whose imaginary part vanishes at `{x0,y0}`.

```
harmonictoanalytic[x_][x0_,y0_][z_]:=
    Simplify[2x[(z + x0 - I*y0)/2,
    (z - x0 + I*y0)/(2*I)] - x[x0,y0]]
```

A.127 `wei`

$z \longmapsto$ `wei[f,g][z]` is the triple of meromorphic functions constructed from meromorphic functions f and g via the Weierstrass representation.

```
wei[f_,g_][z_]:= Module[{zz,tmp},
    tmp = Integrate[{(1 - g[zz]^2)*f[zz]/2,
    I*(1 + g[zz]^2)*f[zz]/2,f[zz]g[zz]},zz];
    Simplify[tmp /. zz->z]]
```

A.128 `weicurvature`

$z \longmapsto$ `weicurvature[f,g][z]` is the Gaussian curvature of the Weierstrass patch determined by meromorphic functions **f** and **g**.

```
weicurvature[f_,g_][z_]:= -16*Abs[D[g[zz],zz]]^2/
    (Abs[f[zz]]^2*(1 + Abs[g[zz]]^2)^4) /. zz->z
```

A.129 `weimetric`

$z \longmapsto$ `weimetric[f,g][z]` is the scaling function of the metric of the Weierstrass patch determined by meromorphic functions **f** and **g**.

```
weimetric[f_,g_][z_]:=
    (1/4)*Abs[f[zz]]^2*(1 + Abs[g[zz]]^2)^2 /. zz->z
```

Principal Curvatures

A.130 kmu

$\{u,v\} \mapsto$ kmu[x][u,v] is a principal curvature of a patch **x** for which ff[x] = fff[x] = 0.

```
kmu[x_][u_,v_]:= ggg[x][u,v]/gg[x][u,v] /;
                 ff[x][u,v]==fff[x][u,v]==0
```

A.131 kpi

$\{u,v\} \mapsto$ kpi[x][u,v] is a principal curvature of a patch **x** for which ff[x] = fff[x] = 0.

```
kpi[x_][u_,v_]:= eee[x][u,v]/ee[x][u,v] /;
                 ff[x][u,v]==fff[x][u,v]==0
```

A.132 k1

$\{u,v\} \mapsto$ k1[x][u,v] is a principal curvature of a patch **x**.

```
k1[x_][u_,v_]:= mcurvature[x][u,v] +
    Sqrt[Simplify[mcurvature[x][u,v]^2 -
        gcurvature[x][u,v]]]
```

A.133 k2

$\{u,v\} \mapsto$ k2[x][u,v] is a principal curvature of a patch **x**. (**k1** is greater than or equal to **k2**).

```
k2[x_][u_,v_]:= mcurvature[x][u,v] -
    Sqrt[Simplify[mcurvature[x][u,v]^2 -
        gcurvature[x][u,v]]]
```

Second Fundamental Form Programs

A.134 eee

$\{u,v\} \mapsto$ eee[x][u,v] is the first coefficient of the second fundamental form of the patch $x: \mathcal{U} \longrightarrow \mathbb{R}^3$. The usual mathematics symbol is e.

```
eee[x_][u_,v_]:= Simplify[Det[{
    D[x[uu,vv],uu,uu],D[x[uu,vv],uu],D[x[uu,vv],vv]}]/
    Sqrt[Simplify[D[x[uu,vv],uu].D[x[uu,vv],uu]*
        D[x[uu,vv],vv].D[x[uu,vv],vv] - D[x[uu,vv],uu].
        D[x[uu,vv],vv]^2]]] /. {uu->u,vv->v}
```

A.135 eeee

$\{u,v\} \mapsto$ eeee[x][u,v] is the first coefficient of the second fundamental form of the patch $x: \mathcal{U} \longrightarrow \mathbb{R}^3$ multiplied by the infinitesimal area. The Gauss symbol is D.

```
eeee[x_][u_,v_]:= Simplify[Det[{
    D[x[uu,vv],uu,uu],D[x[uu,vv],uu],
    D[x[uu,vv],vv]}]] /. {uu->u,vv->v}
```

A.136 fff

$\{u,v\} \mapsto$ fff[x][u,v] is the second coefficient of the second fundamental form of the patch $x: \mathcal{U} \longrightarrow \mathbb{R}^3$. The usual mathematics symbol is f.

```
fff[x_][u_,v_]:= Simplify[Det[{
    D[x[uu,vv],uu,vv],D[x[uu,vv],uu],D[x[uu,vv],vv]}]/
    Sqrt[Simplify[D[x[uu,vv],uu].D[x[uu,vv],uu]*
        D[x[uu,vv],vv].D[x[uu,vv],vv] - D[x[uu,vv],uu].
        D[x[uu,vv],vv]^2]]] /. {uu->u,vv->v}
```

A.137 ffff

$\{u,v\} \mapsto$ ffff[x][u,v] is the second coefficient of the second fundamental form of the patch $x: \mathcal{U} \longrightarrow \mathbb{R}^3$ multiplied by the infinitesimal area. The Gauss symbol is D'.

Appendix A: General Programs

```
ffff[x_][u_,v_]:= Simplify[Det[{
    D[x[uu,vv],uu,vv],D[x[uu,vv],uu],
    D[x[uu,vv],vv]}]] /. {uu->u,vv->v}
```

A.138 ggg

{u,v} ⟼ ggg[x][u,v] is the third coefficient of the second fundamental form of the patch $\mathbf{x}:\mathcal{U} \longrightarrow \mathbb{R}^3$. The usual mathematics symbol is g.

```
ggg[x_][u_,v_]:= Simplify[Det[{
    D[x[uu,vv],vv,vv],D[x[uu,vv],uu],D[x[uu,vv],vv]}]/
    Sqrt[Simplify[D[x[uu,vv],uu].D[x[uu,vv],uu]*
        D[x[uu,vv],vv].D[x[uu,vv],vv] - D[x[uu,vv],uu].
        D[x[uu,vv],vv]^2]]] /. {uu->u,vv->v}
```

A.139 gggg

{u,v} ⟼ gggg[x][u,v] is the third coefficient of the second fundamental form of the patch $\mathbf{x}:\mathcal{U} \longrightarrow \mathbb{R}^3$ multiplied by the infinitesimal area. The Gauss symbol is D''.

```
gggg[x_][u_,v_]:= Simplify[Det[{
    D[x[uu,vv],vv,vv],D[x[uu,vv],uu],
    D[x[uu,vv],vv]}]] /. {uu->u,vv->v}
```

A.140 sff

{u,v} ⟼ sff[x][u,v][du,dv] is the second fundamental form of the patch $\mathbf{x}:\mathcal{U} \longrightarrow \mathbb{R}^n$.

```
sff[x_][u_,v_][du_,dv_]:= eee[x][u,v]**du^2 +
    fff[x][u,v]**(du*dv) + ggg[x][u,v]**dv^2 /.
    {0**_->0,_**0->0}
```

A.141 weingarten

{u,v} ⟼ weingarten[x][u,v] is the matrix of the shape operator of the patch $\mathbf{x}:\mathcal{U} \longrightarrow \mathbb{R}^3$ with respect to $\mathbf{x}_u$ and $\mathbf{x}_v$.

```
weingarten[x_][u_,v_]:=
    {{eee[x][u,v]*gg[x][u,v] - fff[x][u,v]*ff[x][u,v],
      fff[x][u,v]*ee[x][u,v] - eee[x][u,v]*ff[x][u,v]},
     {fff[x][u,v]*gg[x][u,v] - ggg[x][u,v]*ff[x][u,v],
      ggg[x][u,v]*ee[x][u,v] - fff[x][u,v]*ff[x][u,v]}}/
    (ee[x][u,v]*gg[x][u,v] - ff[x][u,v]^2)
```

Vector Analysis Programs

A.142 der

der[v][ww][x,y] is the derivative of the vector field **ww** on $\mathbb{R}^2$ in the direction **v**.

```
der[v_List][ww_][x_,y_]:= {v[[1]]*D[ww[xx,yy],xx],
                          v[[2]]*D[ww[xx,yy],yy]} /. {xx->x,yy->y}
```

der[v][ww][x,y,z] is the derivative of the vector field **ww** on $\mathbb{R}^3$ in the direction **v**.

```
der[v_List][ww_][x_,y_,z_]:= {v[[1]]*D[ww[xx,yy,zz],xx],
                              v[[2]]*D[ww[xx,yy,zz],yy],
                              v[[3]]*D[ww[xx,yy,zz],zz]} /. {xx->x,yy->y,zz->z}
```

A.143 divergence

divergence[ww][x,y] is the divergence of a vector field **ww** on $\mathbb{R}^2$.

```
divergence[ww_][x_,y_]:= D[ww[xx,yy],xx][[1]] +
        D[ww[xx,yy],yy][[2]] /. {xx->x,yy->y}
```

divergence[ww][x,y,z] is the divergence of the vector field **ww** on $\mathbb{R}^3$.

```
divergence[ww_][x_,y_,z_]:= D[ww[xx,yy,zz],xx][[1]] +
                            D[ww[xx,yy,zz],yy][[2]] +
        D[ww[xx,yy,zz],zz][[3]] /. {xx->x,yy->y,zz->z}
```

Appendix A: General Programs

A.144 gmap

gmap[g][x,y] is the Gauss map for a level curve of a function $g: \mathbb{R}^2 \longrightarrow \mathbb{R}$.

```
gmap[g_][x_,y_]:= grad[g][x,y]/Sqrt[grad[g][x,y].grad[g][x,y]]
```

gmap[g][x,y,z] is the Gauss map for a level surface of a function $g: \mathbb{R}^3 \longrightarrow \mathbb{R}$.

```
gmap[g_][x_,y_,z_]:=grad[g][x,y,z]/
    Sqrt[grad[g][x,y,z].grad[g][x,y,z]]
```

A.145 grad

grad[g][x,y] is the gradient of a function $g: \mathbb{R}^2 \longrightarrow \mathbb{R}$.

```
grad[g_][x_,y_]:= {D[g[xx,yy],xx],
                   D[g[xx,yy],yy]} /. {xx->x,yy->y}
```

grad[g][x,y,z] is the gradient of a function $g: \mathbb{R}^3 \longrightarrow \mathbb{R}$.

```
grad[g_][x_,y_,z_]:=
    {D[g[xx,yy,zz],xx],D[g[xx,yy,zz],yy],
     D[g[xx,yy,zz],zz]} /. {xx->x,yy->y,zz->z}
```

A.146 laplace

laplace[g][x,y] is the Laplacian of a function $g: \mathbb{R}^2 \longrightarrow \mathbb{R}$.

```
laplace[g_][x_,y_]:= D[g[xx,yy],{xx,2}] +
    D[g[xx,yy],{yy,2}] /. {xx->x,yy->y}
```

laplace[g][x,y,z] is the Laplacian of a function $g: \mathbb{R}^3 \longrightarrow \mathbb{R}$.

```
laplace[g_][x_,y_,z_]:= D[g[xx,yy,zz],{xx,2}] +
    D[g[xx,yy,zz],{yy,2}] + D[g[xx,yy,zz],{zz,2}] /.
    {xx->x,yy->y,zz->z}
```

Miscellaneous Programs

A.147 `asymeq1`

{u,v} ⟼ `asymeq1[x][u,v]` is the differential equation

```
eee[x][u,v]*u'[v]^2 + 2*fff[x][u,v]*u'[v] + ggg[x][u,v]==0
```

of the patch x: $\mathcal{U} \longrightarrow \mathbb{R}^3$.

```
asymeq1[x_][u_,v_]:=
    Module[{uuu,a,xxx,eqa1,eqa2},
    xxx = Solve[eeee[x][uuu,v]*a^2 + 2*ffff[x][uuu,v]*a +
    gggg[x][uuu,v]==0,a];
    eqa1 = u'[v]== a /. xxx[[1]];
    eqa2 = u'[v]== a /. xxx[[2]];
    {eqa1,eqa2} /. {a->u'[v],uuu->u[v]}]
```

A.148 `asymeq2`

{u,v} ⟼ `asymeq2[x][u,v]` is the differential equation

```
eee[x][u,v] + 2*fff[x][u,v]*v'[u] + ggg[x][u,v]*v'[u]^2==0
```

of the patch x: $\mathcal{U} \longrightarrow \mathbb{R}^3$.

```
asymeq2[x_][u_,v_]:=
    Module[{vvv,b,xxx,eqb1,eqb2},
    xxx = Solve[eeee[x][u,vvv] + 2*ffff[x][u,vvv]*b +
    gggg[x][u,vvv]*b^2==0,b];
    eqb1 = v'[u]== b /. xxx[[1]];
    eqb2 = v'[u]== b /. xxx[[2]];
    {eqb1,eqb2} /. {b->v'[u],vvv->v[u]}]
```

A.149 `circle3point`

`circle3point[p1,q1,p2,q2,p3,q3]` is the circle in the plane passing through the points {p1,q1}, {p2,q2}, {p3,q3}.

Appendix A: General Programs

```
circle3point[{p1_,q1_},{p2_,q2_},{p3_,q3_}]:=
    Circle[{((p2^2*q1 - p3^2*q1 - p1^2*q2 + p3^2*q2 -
    q1^2*q2 + q1*q2^2 + p1^2*q3 - p2^2*q3 + q1^2*q3 -
    q2^2*q3 - q1*q3^2 + q2*q3^2)/
    (2*(p2*q1 - p3*q1 - p1*q2 + p3*q2 + p1*q3 - p2*q3)),
    (p1^2*p2 - p1*p2^2 - p1^2*p3 + p2^2*p3 + p1*p3^2 -
    p2*p3^2 + p2*q1^2 - p3*q1^2 - p1*q2^2 + p3*q2^2 +
    p1*q3^2 - p2*q3^2)/
    (2*(p2*q1 - p3*q1 - p1*q2 + p3*q2 + p1*q3 - p2*q3))},
    Sqrt[(p1^2 - 2*p1*p2 + p2^2 + q1^2 - 2*q1*q2 + q2^2)*
    (p1^2 - 2*p1*p3 + p3^2 + q1^2 - 2*q1*q3 + q3^2)*
    (p2^2 - 2*p2*p3 + p3^2 + q2^2 - 2*q2*q3 + q3^2)]/Abs[
    2*(-p2*q1 + p3*q1 + p1*q2 - p3*q2 - p1*q3 + p2*q3)]]
```

A.150 nateq

nateq[alpha][k2,s] sometimes finds the natural equations of a plane curve **alpha**.

```
nateq[alpha_][k2_,s_]:= ToRules[Eliminate[
    {s==arclength[alpha][t]//PowerExpand,
    k2==kappa2[alpha][t]//PowerExpand//Simplify},t]]
```

A.151 paramtoimp

paramtoimp[alpha,x,y] attempts to convert the parametric form of a curve **alpha** to the implicit form. The result is a number of equations, one of which is the implicit form of the curve. The command will usually not succeed if the components of **alpha** are not rational functions.

```
paramtoimp[alpha_,{x_,y_}]:= Eliminate[{x,y}==alpha[t],t]
```

A.152 polarize

{r,theta}⟼polarize[func][r,theta] is the composition of the maps **{u,v}⟼func[u,v]** and **{r,theta}⟼{r*Cos[theta],r*Sin[theta]}**, where **{u,v}⟼func[u,v]** is an n-tuple of functions.

```
polarize[func_][r_,theta_]:= func[r*Cos[theta],r*Sin[theta]]
```

A.153 stereoinj

p⟼**stereoinj[a][p]** is the stereographic injection from $\mathbb{R}^n$ to a sphere of radius **a** centered at the origin in $\mathbb{R}^{n+1}$.

```
stereoinj[a_][p_]:=Append[2*a^2*p,a*(p.p - a^2)]/(p.p + a^2)
```

A.154 stereoproj

q⟼**stereoproj[a][q]** is the stereographic projection from the sphere of radius **a** centered at the origin in $\mathbb{R}^{n+1}$ to $\mathbb{R}^n$.

```
stereoproj[a_][q_]:=a*Drop[q,-1]/(a - Last[q])
```

A.155 umeqs

umeqs[k][x][u,v] attempts to find the umbilic points of the patch $\mathbf{x}:\mathcal{U} \longrightarrow \mathbb{R}^3$.

```
umeqs[k_][x_][u_,v_]:=
    {ee[x][u,v],ff[x][u,v],gg[x][u,v]} -
    {k*eee[x][u,v],k*fff[x][u,v],k*ggg[x][u,v]}
```

APPENDIX B: CURVES

Parametrically Defined Plane Curves

B.1 agnesi

t⟼**agnesi[a][t]** is the parametrized curve whose implicit equation is

$$x^2 y = 4a^2(2a - y).$$

```
agnesi[a_][t_] := {2*a*Tan[t],2*a*Cos[t]^2}
```

B.2 alysoid

s⟼**alysoid[a,b,c][s]** is a unit-speed curve which is a generalization of a catenary. Its curvature is $-c/(a b + s^2)$. s⟼**alysoid[a,a,a][s]** is the same as s⟼**catenaryunitspeed[a][s]**.

```
alysoid[a_,b_,c_][s_] :=
    Integrate[alysoidprime[a,b,c][ss],{ss,0,s}] + {a,0}
```

t⟼**alysoidprime[a,b,c][t]** is the velocity of t⟼**alysoid[a,b,c][t]**.

```
alysoidprime[a_,b_,c_][t_] :=
    {Sin[(c/Sqrt[a*b])*ArcTan[t/Sqrt[a*b]]],
     Cos[(c/Sqrt[a*b])*ArcTan[t/Sqrt[a*b]]]}
```

B.3 archimedesspiral

$t \mapsto$ **archimedesspiral[n,a][t]** is the spiral of Archimedes of radius **a** and index **n**.

```
archimedesspiral[n_,a_][t_]:=
    {a*t^(1/n)*Cos[t],a*t^(1/n)*Sin[t]}
```

B.4 astroid

$t \mapsto$ **astroid[a][t]** is the parametrized curve whose implicit equation is

$$x^{2/3} + y^{2/3} = a^{2/3}.$$

```
astroid[a_][t_]:= {a*Cos[t]^3,a*Sin[t]^3}
```

$t \mapsto$ **astroid[a,b][t]** is the parametrized curve whose implicit equation is

$$\left(\frac{x}{a}\right)^{2/3} + \left(\frac{y}{b}\right)^{2/3} = 1.$$

```
astroid[a_,b_][t_]:= {a*Cos[t]^3,b*Sin[t]^3}
```

$t \mapsto$ **astroid[n,a,b][t]** is the parametrized curve whose implicit equation is

$$\left(\frac{x}{a}\right)^n + \left(\frac{y}{b}\right)^n = 1.$$

```
astroid[n_,a_,b_][t_]:= {a*Cos[t]^n,b*Sin[t]^n}
```

B.5 bellcurve

$t \mapsto$ **bellcurve[mean,variance][t]** is the bell curve of statistics.

```
bellcurve[mean_,variance_][t_]:=
    {t,(1/Sqrt[2*Pi*variance^2])*
        E^(-(t - mean)^2/(2*variance^2))}
```

B.6 bicorn

t⟼**bicorn[a][t]** is a curve with two horns.

```
bicorn[a_][t_]:= {a*Sin[t],
                  a*Cos[t]^2*(2 + Cos[t])/(3 + Sin[t]^2)}
```

B.7 bow

t⟼**bow[a][t]** is a parametrized curve whose shape is a bow. The polar equation is $r = a(1 - \tan^2 \theta)$.

```
bow[a_][t_]:= {a*(1 - Tan[t]^2)*Cos[t],
               a*(1 - Tan[t]^2)*Sin[t]}
```

B.8 bowtie

t⟼**bowtie[a,b][t]** is a parametrized curve whose shape is a bowtie.

```
bowtie[a_, b_][t_]:= {a*(1 + Cos[t]^2)*Sin[t],
                      (b + Sin[t]^2)*Cos[t]}
```

B.9 bulletnose

t⟼**bulletnose[a,b][t]** is the parametric form of the curve whose implicit equation is

$$\frac{a^2}{x^2} - \frac{b^2}{y^2} = 1.$$

```
bulletnose[a_,b_][t_]:= {a*Cos[t],b*Cot[t]}
```

B.10 cardioid

t⟼**cardioid[a][t]** is a cardioid or heart-shaped curve that is traced by a point on the circumference of a circle of radius $2a$ rolling around a fixed circle of the same radius. It has the polar equation

$$r = 2a(1 + \cos\theta).$$

The implicit equation is

$$(x^2 + y^2 - 4a^2)^3 - 108\,a^4 y^2 = 0.$$

A cardioid is a special case of an epicycloid.

```
cardioid[a_][t_] := {2*a*Cos[t]*(1 + Cos[t]),
                    2*a*Sin[t]*(1 + Cos[t])}
```

B.11 cardioidunitspeed

s⟼cardioidunitspeed[a][s] is a unit-speed parametrization of
t⟼cardioid[a][t].

```
cardioidunitspeed[a_][s_] :=
    {2*a*Cos[2*ArcSin[s/(8*a)]]*(1 + Cos[2*ArcSin[s/(8*a)]]),
     2*a*(1 + Cos[2*ArcSin[s/(8*a)]])*Sin[2*ArcSin[s/(8*a)]]}
```

B.12 cartesianoval

u⟼cartesianoval[a,b][u,v] and v⟼cartesianoval[a,b][u,v] are orthogonal Cartesian ovals.

```
cartesianoval[a_,b_][u,v] := {a*Re[WeierstrassP[u + I*v,b,0]],
                              a*Im[WeierstrassP[u + I*v,b,0]]}
```

B.13 cassini

t⟼cassini[a,b,pm1,pm2][t] is a parametrization of an oval of Cassini. The parameters **pm1** and **pm2** are ±1. The implicit equation is

$$(x^2 + y^2 + a^2)^2 - b^4 - 4a^2 x^2 = 0.$$

```
cassini[a_,b_,pm1_,pm2_][t_] :=
   {pm1*Sqrt[a^2*Cos[2*t] +
       pm2*Sqrt[b^4 - a^4*Sin[2*t]^2]]*Cos[t],
    pm1*Sqrt[a^2*Cos[2*t] +
       pm2*Sqrt[b^4 - a^4*Sin[2*t]^2]]*Sin[t]}
```

Appendix B: Curves

B.14 catenary

t⟼**catenary[a][t]** is the curve formed by a perfectly flexible inextensible chain of uniform density hanging from two supports.

```
catenary[a_][t_]:= {a*Cosh[t/a],t}
```

B.15 catenaryunitspeed

s⟼**catenaryunitspeed[a][s]** is a unit-speed parametrization of t⟼**catenary[a][t]**.

```
catenaryunitspeed[a_][s_]:=
    {a*Sqrt[1 + s^2/a^2],a*ArcSinh[s/a]}
```

B.16 cayleysextic

t⟼**cayleysextic[a][t]** is Cayley's sextic curve.

```
cayleysextic[a_][t_]:=
    {4*a*Cos[t/2]^4*(-1 + 2*Cos[t]),
    4*a*Cos[t/2]^3*Sin[3*t/2]}
```

B.17 circle

t⟼**circle[a][t]** is a circle centered at the origin of radius **a**. The nonparametric form is $x^2 + y^2 = a^2$.

```
circle[a_][t_]:= {a*Cos[t],a*Sin[t]}
```

B.18 circleinvolute

t⟼**circleinvolute[n,a][t]** is the nth involute starting at **{a,0}** of the circle t⟼**{a*Cos[t],a*Sin[t]}**. t⟼**circleinvolute[a][t]** is the same as t⟼**circleinvolute[1,a][t]**.

```
circleinvolute[n_,a_][t_]:= Module[{tmp1,tmp2},
    tmp1=E^(I*tt)Normal[Series[E^(-I*tt),{tt,0,n}]];
    tmp2=a*ComplexExpand[{Re[tmp1],Im[tmp1]}];
    tmp2 /. tt->t]
```

B.19 circleinvoluteunitspeed

s⟼circleinvoluteunitspeed[n,a][s] is a unit-speed parametrization of t⟼circleinvolute[n,a][t]. s⟼circleinvoluteunitspeed[a][s] is the same as s⟼circleinvoluteunitspeed[1,a][s].

```
circleinvoluteunitspeed[n_,a_][s_]:=
    circleinvolute[n,a][((n + 1)! s/a)^(1/(n +1))]
```

B.20 circleunitspeed

s⟼circleunitspeed[a][s] is a unit-speed parametrization of $t \longmapsto$ circle[a][t].

```
circleunitspeed[a_][s_]:= {a*Cos[s/a],a*Sin[s/a]}
```

B.21 cissoid

t⟼cissoid[a][t] is an ivy-shaped curve whose implicit equation is

$$x^3 + x y^2 - 2a y^2 = 0.$$

```
cissoid[a_][t_]:= {2*a*t^2/(1 + t^2),2*a*t^3/(1 + t^2)}
```

B.22 clothoid

s⟼clothoid[n,a][s] is the nth clothoid. It is a unit-speed curve. The curvature is proportional to the nth power of the arc length function.
s⟼clothoid[1,a][s] is the ordinary clothoid, also known as Euler's spiral.
s⟼clothoid[a][s] is the same as s⟼clothoid[1,a][s].

Appendix B: Curves

```
clothoid[n_,a_][s_]:=
    Integrate[clothoidprime[n,a][ss],{ss,0,s}]
```

s⟼clothoidnds[smin,smax][n,a][s] is an approximation over the interval smin<s<smax found by NDSolve to s⟼clothoid[n,a][s].

```
clothoidnds[smin_,smax_][n_,a_][s_]:=
    Module[{x,y,tmp},
        tmp=NDSolve[{x'[ss]==clothoidprime[n,a][ss][[1]],
            y'[ss]==clothoidprime[n,a][ss][[2]],
            x[0]==0,y[0]==0},{x,y},{ss,smin,smax}];
        First[{x[s],y[s]} /. tmp]]
```

t⟼clothoidprime[n,a][t] is the velocity of the nth clothoid. It is used in the definitions of s⟼clothoid[n,a][s] and s⟼clothoid[n,a][s].

```
clothoidprime[n_,a_][t_]:=
    {a*Sin[t^(n + 1)/(n + 1)],a*Cos[t^(n + 1)/(n + 1)]}
```

B.23 cnccoprofile

s⟼cnccoprofile[pm,a,b][s] is a profile curve for a surface of revolution of constant negative curvature $-a^{-2}$ of conic type. It is a unit-speed curve. The parameter pm is +1 or -1.

```
cnccoprofile[pm_,a_,b_][s_]:=
    {pm*b*Sinh[s/a],-I*Sqrt[a^2 - b^2]*EllipticE[I*s/a,
        -b^2/(a^2 - b^2)]}
```

B.24 cnchyprofile

s⟼cnchyprofile[pm,a,b][s] is a profile curve for a surface of revolution of constant negative curvature $-a^{-2}$ of hyperboloid type. It is a unit-speed curve. The parameter pm is +1 or -1.

```
cnchyprofile[pm_,a_,b_][s_]:=
    {pm*b*Cosh[s/a],-I*a*EllipticE[I*s/a,-b^2/a^2]}
```

B.25 cochleoid

t⟼**cochleoid[n,a][t]** is the parametrized curve whose polar equation is

$$r = \left(\frac{a\sin\theta}{\theta}\right)^n.$$

```
cochleoid[n_,a_][t_]:=
    {a*(Sin[t]/t)^n*Cos[t],a*(Sin[t]/t)^n*Sin[t]}
```

B.26 conchoid

t⟼**conchoid[a,b][t]** is the parametrized curve whose implicit equation is

$$(x^2 + y^2)(x - b)^2 = a^2 x^2.$$

```
conchoid[a_,b_][t_]:= {b + a*Cos[t],b*Tan[t] + a*Sin[t]}
```

B.27 coshspiral

t⟼**coshspiral[n,a][t]** is a hyperbolic cosine spiral or Poinsot cosh spiral of radius **a**.

```
coshspiral[n_,a_][t_]:={a*Cos[t]/Cosh[n*t],
                       a*Sin[t]/Cosh[n*t]}
```

B.28 cothspiral

t⟼**cothspiral[a][t]** is the hyperbolic cotangent spiral of radius **a**.

```
cothspiral[a_][t_]:= {-Sinh[2*t]/(Cos[2*a*t] - Cosh[2*t]),
                     Sin[2*a*t]/(Cos[2*a*t] - Cosh[2*t])}
```

Appendix B: Curves

B.29 cpcprofile

$s \mapsto$ `cpcprofile[pm,a,b][s]` is a profile curve for a surface of revolution of constant positive curvature a^{-2}. It is a unit-speed curve. The parameter **pm** is +1 or −1. There are 3 cases: football type ($a > b$), sphere ($a = b$), barrel type ($a < b$).

```
cpcprofile[pm_,a_,b_][s_]:=
    {pm*b*Cos[s/a],a*EllipticE[s/a,b^2/a^2]}
```

B.30 crosscurve

$t \mapsto$ `crosscurve[a,b][t]` is the parametric form of the curve whose implicit equation is

$$\frac{a^2}{x^2} + \frac{b^2}{y^2} = 1.$$

```
crosscurve[a_,b_][t_]:= {a*Sec[t],b*Csc[t]}
```

B.31 cubicparabola

$t \mapsto$ `cubicparabola[a,b,c,d][t]` is the general cubic parabola whose implicit equation is

$$y = a x^3 + b x^2 + c x + d.$$

```
cubicparabola[a_,b_,c_,d_][t_]:= {t,a*t^3 + b*t^2 + c*t + d}
```

B.32 cycloid

$t \mapsto$ `cycloid[a,b][t]` is the locus of points traced out by a point on a circle of radius **b** when a concentric circle of radius **a** rolls without slipping along a straight line. The cycloid is called prolate if **b > a** and curtate if **a > b**. $t \mapsto$ `cycloid[a][t]` is the same as $t \mapsto$ `cycloid[a,a][t]`.

```
cycloid[a_,b_][t_]:= {a*t - b*Sin[t],a - b*Cos[t]}
```

B.33 cycloidunitspeed

s⟼cycloidunitspeed[a,a][s] is a unit-speed parametrization of
t⟼cycloid[a,a][t].

```
cycloidunitspeed[a_,a_][s_] := {4*a*ArcSin[Sqrt[s/(8*a)]] -
    (4*a - s)*Sqrt[(s/a)*(8 - s/a)]/8, s - s^2/(8*a)}
```

B.34 delaunay

s⟼delaunaynds[smin,smax][a,e][s] is an approximation over the interval smin<s<smax found by NDSolve to a unit-speed parametrization of a Delaunay curve. A **Delaunay curve** is the path of a focus of a conic with long axis **a** and eccentricity **e** as the conic rolls on a straight line. If $0 < e < 1$, the conic is an ellipse; if $e > 1$, the conic is a hyperbola.

```
delaunaynds[smin_,smax_][a_,e_][s_] :=
    intrinsic[delaunaykappa2[a,e],0,{0,0,0},
    MaxSteps->2000,{smin,smax}][s]
```

s⟼delaunaykappa2[a,e][s] is the signed curvature of a Delaunay curve. It is used in the definition of s⟼delaunaynds[smin,smax][a,e][s].

```
delaunaykappa2[a_,e_][t_] :=
    e*(e - Cos[t/a])/(a*(1 - 2*e*Cos[t/a] + e^2))
```

B.35 deltoid

t⟼deltoid[a][t] is a parametrized curve which resembles a Δ. A deltoid is a special case of an hypocycloid.

```
deltoid[a_][t_] := {2*a*Cos[t]*(1 + Cos[t]) - a,
                    2*a*Sin[t]*(1 - Cos[t])}
```

B.36 deltoidinvolute

t⟼deltoidinvolute[a][t] is the involute of t⟼deltoid[a][t]

Appendix B: Curves

```
deltoidinvolute[a_][t_]:=
    {(a/3)*(8*Cos[t/2] + 2*Cos[t] - Cos[2*t]),
     (a/3)*(-8*Sin[t/2] + 2*Sin[t] + Sin[2*t])}
```

B.37 deltoidunitspeed

s⟼deltoidunitspeed[a][s] is a unit-speed parametrization of t⟼deltoid[a][t].

```
deltoidunitspeed[a_][s_]:=
    {a*(2*Cos[2*ArcCot[3*s/Sqrt[64*a^2 - 9*s^2]]/3] +
     Cos[4*ArcCot[3*s/Sqrt[64*a^2 - 9*s^2]]/3]),
     -8*a*Cos[ArcCot[3*s/Sqrt[64*a^2 - 9*s^2]]/3]*
     Sin[ArcCot[3*s/Sqrt[64*a^2 - 9*s^2]]/3]^3}
```

B.38 diamond

t⟼diamond[a][t] is a parametrized curve whose shape is a diamond.

```
diamond[a_][t_]:=
    {a*Cos[t]*Sqrt[Cos[t]^2],a*Sin[t]*Sqrt[Sin[t]^2]}
```

t⟼diamond[n,a,b][t] is a generalization of t⟼diamond[a][t] that is similar to t⟼astroid[n,a,b][t].

```
diamond[a_][t_]:=
    {a*Cos[t]*Sqrt[Cos[t]^2],a*Sin[t]*Sqrt[Sin[t]^2]}
```

B.39 eight

t⟼eight[t] is a parametrized curve which resembles a figure eight.

```
eight[t_]:= {Sin[t],Sin[t]*Cos[t]}
```

B.40 ellipse

t⟼**ellipse[a,b][t]** is an ellipse centered at the origin. The nonparametric form is

$$\frac{x^2}{a^2} + \frac{y^2}{b^2} = 1.$$

```
ellipse[a_,b_][t_] := {a*Cos[t],b*Sin[t]}
```

B.41 ellipseinc

t⟼**ellipseinc[a,b,c,psi][t]** is the locus of points at a distance **c** along a line meeting an ellipse t⟼**ellipse[a,b][t]** at the point **ellipse[a,b][t]** with an angle **psi**.

```
ellipseinc[a_,b_,c_,psi_][t_]:=
    Module[{k},k=1 - b^2/a^2;
      {a*JacobiSN[t,k] + (c/a)(a*Cos[psi]*JacobiCN[t,k] +
              b*Sin[psi]*JacobiSN[t,k])/JacobiDN[t,k],
      b*JacobiCN[t,k] + (c/a)(a*Sin[psi]*JacobiCN[t,k] -
              b*Cos[psi]*JacobiSN[t,k])/JacobiDN[t,k]}]
```

B.42 epicycloid

t⟼**epicycloid[a,b][t]** is a parametrized curve traced out by a point P on the circumference of a circle (of radius **b**) rolling outside another circle (of radius **a**). Special cases are: cardioid (**b=a**), nephroid (**b=a/2**).

```
epicycloid[a_,b_][t_]:=
    {(a + b)*Cos[t] - b*Cos[((a + b)/b)*t],
     (a + b)*Sin[t] - b*Sin[((a + b)/b)*t]}
```

B.43 epicycloidunitspeed

s⟼**epicycloidunitspeed[a,b][s]** is a unit-speed parametrization of t⟼**epicycloid[a,b][t]**.

```
epicycloidunitspeed[a_,b_][s_]:=
{(a + b)*Cos[4*b*ArcSin[Sqrt[a*s/(b*(8*a + 8*b))]]/a] -
  b*Cos[4*(a + b)*ArcSin[Sqrt[a*s/(b*(8*a + 8*b))]]/a],
 (a + b)*Sin[4*b*ArcSin[Sqrt[a*s/(b*(8*a + 8*b))]]/a] -
  b*Sin[4*(a + b)*ArcSin[Sqrt[a*s/(b*(8*a + 8*b))]]/a]}
```

B.44 epispiral

s⟼**epispiral[a,n][t]** is an epispiral. The curve has n branches if n is odd and $2n$ branches if n is even. Each branch has two asymptotes radiating out from the center. The polar equation is $r = a/\sin(n\theta)$.

```
epispiral[a_,n_][t_]:= {a*Cos[t]/Sin[n*t],a*Sin[t]/Sin[n*t]}
```

B.45 epitrochoid

t⟼**epitrochoid[a,b,h][t]** is a parametrized curve traced out by a point P attached to a circle (of radius **b**) rolling outside another circle (of radius **a**). Let **h** be the distance from P to the center of the moving circle **h**. Special cases are: circle (**a=0**), limacon (**a=b**), epicycloid (**h=b**), cardioid (**h=b=a**), nephroid (**h=b=a/2**). There are **a/b** loops. For an epicycloid the loops degenerate to points.

```
epitrochoid[a_,b_,h_][t_]:=
      {(a + b)*Cos[t] - h*Cos[((a + b)/b)*t],
       (a + b)*Sin[t] - h*Sin[((a + b)/b)*t]}
```

B.46 fermatspiral

t⟼**fermatspiral[a][t]** is a spiral of Fermat of radius **a**, where **t** is positive.
t⟼**fermatspiral[n,a][t]** is a spiral of Fermat of radius **a** and index **n**, where **t** may be positive or negative. Special cases are: fermatspiral (**n=1/4**), lituus (**n=3/4**), archimedesspiral (**n=(m-1)/(2m)**).

```
fermatspiral[a_][t_]:= {a*Sqrt[t]*Cos[t],a*Sqrt[t]*Sin[t]}
```

t⟼**fermatspiral[n,a][t]** is the spiral of Fermat of radius **a** and index **n**.

```
fermatspiral[n_,a_][t_]:= {a*t/(t^2)^n*Cos[Sqrt[t^2]],
                          a*t/(t^2)^n*Sin[Sqrt[t^2]]}
```

B.47 folium

t ⟼ **folium[t]** is a generalized folium of Descartes. The implicit equation is

$$x^3 + y^3 - 3xy = 0.$$

```
folium[t_]:= {3*t/(1 + t^3),3*t^2/(1 + t^3)}
```

t ⟼ **folium[n,a,b,b,t]** is a generalized folium of Descartes. The implicit equation is

$$a x^{2n+1} + b y^{2n+1} - c x^{2n} y^{2n} = 0.$$

```
folium[n_,a_,b_,c_][t_]:= {c*t^n/(a + b*t^(2*n + 1)),
                           c*t^(n + 1)/(a + b*t^(2*n + 1))}
```

B.48 genparabola

t ⟼ **genparabola[n][t]** is the curve $t \longmapsto t^n$.

```
genparabola[n_][t_]:= {t,t^n}
```

B.49 hippopede

t ⟼ **hippopede[a,b][t]** is a curve whose shape resembles a horse fetter.

```
hippopede[a_,b_][t_]:=
        {2*Cos[t]*Sqrt[a*b - b^2*Sin[t]^2],
         2*Sin[t]*Sqrt[a*b - b^2*Sin[t]^2]}
```

Appendix B: Curves

B.50 hyperbola

t⟼**hyperbola[a,b][t]** is a parametrization of a hyperbola centered at the origin using cosh and sinh. The nonparametric form is

$$\frac{x^2}{a^2} - \frac{y^2}{b^2} = 1.$$

```
hyperbola[a_,b_][t_]:= {a*Cosh[t],b*Sinh[t]}
```

B.51 hyperbolabis

t⟼**hyperbolabis[a,b][t]** is a parametrization of a hyperbola centered at the origin using sec and tan. The nonparametric form is

$$\frac{x^2}{a^2} - \frac{y^2}{b^2} = 1.$$

```
hyperbolabis[a_,b_][t_]:= {a*Sec[t],b*Tan[t]}
```

B.52 hyperbolicspiral

t⟼**hyperbolicspiral[a][t]** is the hyperbolic spiral of radius **a**. It is a special case of a spiral of Archimedes.

```
hyperbolicspiral[a_][t_]:= {a*Cos[t]/t,a*Sin[t]/t}
```

B.53 hypocycloid

t⟼**hypocycloid[a,b][t]** is a parametrized curve traced out by a point P on the circumference of a circle (of radius **b**) rolling inside another circle (of radius **a**). Special cases are: deltoid (**b=a/3**), astroid (**b=a/4**).

```
hypocycloid[a_,b_][t_]:=
    {(a - b)*Cos[t] + b*Cos[((a - b)/b)*t],
    (a - b)*Sin[t] - b*Sin[((a - b)/b)*t]}
```

B.54 hypocycloidinvolute

t⟼hypocycloidinvolute[a,b][t] is the involute of
t⟼hypocycloid[a,b][t]

```
hypocycloidinvolute[a_,b_][t_]:=
    {(a^2*Cos[t] - 3*a*b*Cos[t] + 2*b^2*Cos[t] -
     a*b*Cos[t - a*t/b] + 2*b^2*Cos[t - a*t/b] +
     4*a*b*Cos[t - a*t/(2*b)] -
     4*b^2*Cos[t - a*t/(2*b)])/a,
     (a^2*Sin[t] - 3*a*b*Sin[t] + 2*b^2*Sin[t] -
     a*b*Sin[t - a*t/b] + 2*b^2*Sin[t - a*t/b] +
     4*a*b*Sin[t - a*t/(2*b)] -
     4*b^2*Sin[t - a*t/(2*b)])/a}
```

B.55 hypocycloidunitspeed

s⟼hypocycloidunitspeed[a,b][s] is a unit-speed parametrization of
t⟼hypocycloid[a,b][t].

```
hypocycloidunitspeed[a_,b_][s_]:=
    {(a - b)*Cos[4*b*ArcSin[Sqrt[a*s/((8*a - 8*b)*b)]]/a] +
     b*Cos[4*(a - b)*ArcSin[Sqrt[a*s/((8*a - 8*b)*b)]]/a],
     (a - b)*Sin[4*b*ArcSin[Sqrt[a*s/((8*a - 8*b)*b)]]/a]
     -b*Sin[4*(a - b)*ArcSin[Sqrt[a*s/((8*a - 8*b)*b)]]/a]}
```

B.56 hypotrochoid

t⟼hypotrochoid[a,b,h][t] is a parametrized curve traced out by a point P on a moving circle (of radius **b**) rolling inside another circle (of radius **a**). Let **h** be the distance from P to the center of the moving circle. Special cases are: ellipse (**a=2b**), hypocycloid (**h=b**), deltoid (**h=b=a/3**), astroid (**h=b=a/4**). There are **a/b** loops. For a hypocycloid the loops degenerate to points.

```
hypotrochoid[a_,b_,h_][t_]:=
    {(a - b)*Cos[t] + h*Cos[((a - b)/b)*t],
     (a - b)*Sin[t] - h*Sin[((a - b)/b)*t]}
```

Appendix B: Curves

B.57 kampyle

t⟼kampyle[a][t] is the kampyle or crooked staff of Eudoxus. The polar equation is $r = a(\sec\theta)^2$.

```
kampyle[a_][t_]:= {a*Sec[t],a*Sin[t]*Sec[t]^2}
```

B.58 keplerorbit

t⟼keplerorbit[a,e][t] is a Kepler orbit, where e is the eccentricity. The orbit is part of an ellipse ($0 \leq e < 1$), parabola ($e = 1$) or hyperbola ($e > 1$).

```
keplerorbit[a_,e_][t_]:={a*Cos[t]/(1 - e*Cos[t]),
                         a*Sin[t]/(1 - e*Cos[t])}
```

B.59 lemniscate

t⟼lemniscate[a][t] is the parametrized curve whose implicit equation is

$$(x^2 + y^2)^2 = a^2(x^2 - y^2).$$

```
lemniscate[a_][t_]:=
    {a*Cos[t]/(1 + Sin[t]^2),
     a*Sin[t]*Cos[t]/(1 + Sin[t]^2)}
```

B.60 lemniscateunitspeed

s⟼lemniscateunitspeed[a][s] is a unit-speed parametrization of
t⟼lemniscate[t].

```
lemniscateunitspeed[a_][s_]:=
    {a*JacobiCN[s/a,-1]/JacobiDN[s/a,-1]^2,
     a*JacobiCN[s/a,-1]JacobiSN[s/a,-1]/JacobiDN[s/a,-1]^2}
```

B.61 limacon

t⟼limacon[a,b][t] is the parametrized curve whose implicit equation is

$$(x^2 + y^2 - 2ax)^2 = b^2(x^2 + y^2).$$

t⟼limacon[a,2a][t] coincides with t⟼cardioid[a][t].

```
limacon[a_,b_][t_]:= {(2*a*Cos[t] + b)*Cos[t],
                     (2*a*Cos[t] + b)*Sin[t]}
```

B.62 line

t⟼line[a1,b1,a2,b2][t] is a parametrized line in $\mathbb{R}^2$ with
line[a1,b1,a2,b2][0]={a2,b2} and line[a1,b1,a2,b2]'[0]={a1,b1}.

```
line[a1_,b1_,a2_,b2_][t_]:= {a1*t + a2,b1*t + b2}
```

B.63 lissajous

t⟼lissajous[n,d,a,b][t] is a Lissajous or Bowditch curve.

```
lissajous[n_,d_,a_,b_][t_]:= {a*Sin[n*t + d],b*Sin[t]}
```

B.64 lituus

t⟼lituus[a][t] is the lituus of radius **a**. It is a trumpet-shaped curve that is the locus of points P such that the square of distance of P from the origin is inversely proportional to the angle theta that P makes with the horizontal axis. The polar equation is

$$r^2 = \frac{a^2}{\theta}.$$

```
lituus[a_][t_]:= {a*t/(t^2)^(3/4)*Cos[Sqrt[t^2]],
                  a*t/(t^2)^(3/4)*Sin[Sqrt[t^2]]}
```

Appendix B: Curves 909

B.65 logistic

t ⟼ **logistic[k,a,b][t]** is the logistic curve used to model growth in biological populations for which saturation occurs.

```
logistic[k_,a_,b_][t_]:= {t,k/(1 + E^(a + b*t))}
```

B.66 logspiral

t ⟼ **logspiral[a,b][t]** is the logarithmic spiral of radius **a** and slant **b**. The polar equation is
$$r = a\, e^{bt}.$$

```
logspiral[a_,b_][t_]:= {a*E^(b*t)*Cos[t],a*E^(b*t)*Sin[t]}
```

B.67 logspiralunitspeed

s ⟼ **logspiralunitspeed[a,b][s]** is a unit-speed parametrization of t ⟼ **logspiral[a,b][t]**.

```
logspiralunitspeed[a_,b_][s_]:=
   {b*s*Cos[Log[b*s/(a*Sqrt[1 + b^2])]/b]/Sqrt[1 + b^2],
    b*s*Sin[Log[b*s/(a*Sqrt[1 + b^2])]/b]/Sqrt[1 + b^2]}
```

B.68 nephroid

t ⟼ **nephroid[a][t]** is a nephroid or kidney-shaped curve that is traced by a point on the circumference of a circle of radius a rolling around a fixed circle of radius $2a$. It has the polar equation
$$r = 2a \left(\sin(\theta/2)^{2/3} + \cos(\theta/2)^{2/3} \right)^{3/2}.$$

The implicit equation is
$$(x^2 + y^2 - 4a^2)^3 - 108\, a^4 y^2 = 0.$$

A nephroid is a special case of an epicycloid.

```
nephroid[a_][t_]:= {a*(3*Cos[t] - Cos[3*t]),
                    a*(3*Sin[t] - Sin[3*t])}
```

B.69 nephroidunitspeed

s⟼**nephroidunitspeed[a][s]** is a unit-speed parametrization of
t⟼**nephroid[a][t]**.

```
nephroidunitspeed[a_][t_]:=
   {s/2 - a*Cos[3*ArcCos[s/(6*a)]],
    a*(3*Sqrt[1 - s^2/(36*a^2)] - Sin[3*ArcCos[s/(6*a)]])}
```

B.70 ngon

t⟼**ngon[n,a][t]** is a parametrization of a regular polygon with n sides, where a is the length of the apothem.

```
ngon[n_,a_][t_]:=
    Which@@Flatten[Table[{(k - 1)/n<=t<=k/n,
    a*{Cos[2*Pi*t],Sin[2*Pi*t]}/Cos[2*Pi*t - (2*k - 1)*Pi/n]},
    {k,1,n}],1]
```

B.71 nielsenspiral

t⟼**nielsenspiral[a][t]** is Nielsen's spiral of radius **a**.

```
nielsenspiral[a_][t_]:= {a*CosIntegral[t],a*SinIntegral[t]}
```

B.72 parabola

t⟼**parabola[a][t]** is the vertical parabola with vertex at the origin, focus at **{0,a}** and directrix t⟼**{t,-a}**. The nonparametric form is $x^2 = 4ay$.

```
parabola[a_][t_]:= {2*a*t,a*t^2}
```

B.73 parabolicspiral

t⟼**parabolicspiral[n,a,b][t]** is a parabolic spiral of radius **a**, offset **b** and index **n**, where **t** may be positive or negative. Special cases are: fermatspiral (**b=0**), lituus (**n=3/4, b=0**), archimedesspiral (**n=(m - 1)/(2*m), b=0**).

```
parabolicspiral[n_,a_,b_][t_]:=
    {(a*t/(t^2)^n + b)*Cos[Sqrt[t^2]],
    (a*t/(t^2)^n + b)*Sin[Sqrt[t^2]]}
```

B.74 piriform

t⟼**piriform[a,b][t]** is a pear-shaped curve whose implicit equation is

$$a^4 y^2 - b^2 x^3 (2a - x) = 0.$$

```
piriform[a_,b_][t_]:=
    {a*(1 + Sin[t]),b*Cos[t]*(1 + Sin[t])}
```

B.75 pseudocatenary

s⟼**pseudocatenary[a,b][s]** is a unit-speed curve which is a generalization of a catenary. s⟼**pseudocatenary[a,a][s]** is the same as s⟼**catenaryunitspeed[c,b][s]**

```
pseudocatenary[a_,b_][s_]:=
    Integrate[pseudocatenaryprime[a,b][ss],{ss,0,s}] + {a,0}
```

t⟼**pseudocatenaryprime[a,b][t]** is the velocity of the pseudocatenary.

```
pseudocatenaryprime[a_,b_][t_]:=
    {Sin[Sqrt[a/b]*ArcTan[t/Sqrt[a*b]]],
    Cos[Sqrt[a/b]*ArcTan[t/Sqrt[a*b]]]}
```

B.76 rectangle

t⟼**rectangle[n,a,b][t]** is a generalized rectangle.
t⟼**rectangle[1,a,b][t]** is an ellipse.

t⟼**rectangle[2,a,b][t]** is a parametrization of an ordinary rectangle.
t⟼**rectangle[3,a,b][t]** astroid rotated by $-\pi/4$.

```
rectangle[n_,a_,b_][t_]:=
        {a*(Cos[t]*Sqrt[Cos[t]^2]^(n - 1) +
           Sin[t]*Sqrt[Sin[t]^2]^(n - 1)),
         b*(-Cos[t]*Sqrt[Cos[t]^2]^(n - 1) +
           Sin[t]*Sqrt[Sin[t]^2]^(n - 1))}
```

B.77 reuleauxpolygon

t⟼**reuleauxpolygon[n,a][t]** is a parametrization of a Reuleaux constant width curve with $2n+1$ sides, where a is the distance between any vertex and the center.

```
reuleauxpolygon[n_,a_][t_]:=Which@Flatten[
    Table[{k/(2n + 1)<=t<=(k + 1)/(2n + 1),
        a{Re[E^(2Pi I k/(2n + 1)) +
           2I Sin[Pi n/(2n + 1)]*
           E^(Pi I((k + n)/(2n + 1) + t))],
          Im[E^(2Pi I k/(2n + 1)) +
           2I Sin[Pi n/(2n + 1)]*
           E^(Pi I ((k + n)/(2n + 1) + t))]}},
        {k,0,2n}],1]
```

B.78 rose

t⟼**rose[n,a][t]** is a rose of radius **a** with **n** petals if **n** is odd and **2*n** petals if **n** is even. The polar equation is $r = a\cos n\theta$.

```
rose[n_,a_][t_]:= {a*Cos[n*t]*Cos[t],a*Cos[n*t]*Sin[t]}
```

B.79 sc

t⟼**sc[t]** is the semicubical parabola. The nonparametric form is $y^2 = x^3$.

```
sc[t_]:= {t^2,t^3}
```

B.80 scarab

t ⟼ **scarab[a,b][t]** is a parametrization of a scarab.

```
scarab[a_, b_][t_]:= {(a*Cos[2*t] - b*Cos[t])*Cos[t],
                     (a*Cos[2*t] - b*Cos[t])*Sin[t]}
```

B.81 scunitspeed

s ⟼ **scunitspeed[s]** is a unit-speed parametrization of t ⟼ **sc[t]**

```
scunitspeed[s_]:={s^(2/3),(s^(2/3) - 4/9)^(3/2)}
```

B.82 serpentine

t ⟼ **serpentine[a,b][t]** is a serpentine curve.

```
serpentine[a_,b_][t_]:= {t,a*t/(1 + b*t^2)}
```

B.83 sinoval

t ⟼ **sinoval[n,a][t]** is a hyperbolic sine spiral or Poinsot sinh spiral of radius **a**.

```
sinoval[n_,a_][t_]:= {a*Cos[t],a*Nest[Sin,t,n]}
```

B.84 sinhspiral

t ⟼ **sinhspiral[n,a][t]** is a hyperbolic sine spiral or Poinsot sinh spiral of radius **a**.

```
sinhspiral[n_,a_][t_]:={a*Cos[t]/Sinh[n*t],
                       a*Sin[t]/Sinh[n*t]}
```

B.85 spring

t⟼**spring[a,b,v][t]** is a plane curve which resembles a spring. It is the projection of t⟼**helix[a,b][t]** onto a plane whose normal makes an angle **v** with the z-axis.

```
spring[a_,b_,v_][t_]:={a*Cos[t],a*Cos[v]*Sin[t] + b*t*Sin[v]}
```

B.86 strophoid

t⟼**strophoid[a][t]** is the parametrized curve whose implicit equation is $y^2(a-x) = x^3$.

```
strophoid[a_][t_]:= {a*(t^2 - 1)/(t^2 + 1),
                    t*a*(t^2 - 1)/(t^2 + 1)}
```

t->**strophoid[m,a][t]** is the parametrized curve whose implicit equation is

$$y^2(a-x) = x^2(am+x).$$

The case **m=0** is the cissoid $t \mapsto$ **cissoid[a/2][t]**, and the case **m=1** is the strophoid $t \mapsto$ **strophoid[a][t]**.

```
strophoid[m_,a_][t_]:= {a*(t^2 - m)/(t^2 + 1),
                        t*a*(t^2 - m)/(t^2 + 1)}
```

B.87 tanhspiral

t⟼**tanhspiral[a][t]** is the hyperbolic tangent spiral of radius **a**.

```
tanhspiral[a_][t_]:= {Sinh[2*t]/(Cos[2*a*t] + Cosh[2*t]),
                      Sin[2*a*t]/(Cos[2*a*t] + Cosh[2*t])}
```

B.88 teardrop

t⟼**teardrop[a,b,c,d,n][t]** is a curve resembling a teardrop that has high order contact with the horizontal axis.

```
teardrop[a_,b_,c_,d_,n_][t_]:=
    {a*Sin[t],d*(b*Sin[t] + c*Sin[2*t])^n}
```

B.89 tractrix

$t \mapsto$ **tractrix[a][t]** is a tractrix. It represents the path of a particle P pulled by an inextensible string whose end moves along a line. The evolute of a catenary is a tractrix. A tractrix is the profile curve of a surface of revolution of constant Gaussian curvature $-a^{-2}$.

```
tractrix[a_][t_]:= {a*Sin[t],a*(Cos[t] + Log[Tan[t/2]])}
```

$t \mapsto$ **tractrix[a,b][t]** is a generalization of a tractrix in the same way that an ellipse is a generalization of a circle. In the terminology of Bianchi, $t \mapsto$ **tractrix[a,b][t]** is a shortened tractrix for $b < a$ or extended tractrix for $b > a$.

```
tractrix[a_,b_][t_]:= {a*Sin[t],b*(Cos[t] + Log[Tan[t/2]])}
```

B.90 tractrixunitspeed

$s \mapsto$ **tractrixunitspeed[a][s]** is a unit-speed parametrization of $t \mapsto$ **tractrix[a][t]**.

```
tractrixunitspeed[a_][s_]:= {a*E^(-Sqrt[s^2]/a),
    a*ArcTanh[Sqrt[1 - E^(-2*Sqrt[s^2]/a)]] -
    a*Sqrt[1 - E^(-2*Sqrt[s^2]/a)]}
```

B.91 triangle

$t \mapsto$ **triangle[t]** is a parametrization of an equilateral triangle.

```
triangle[t_]:= If[t<1,t*{1,0},
    If[t<2,{1,0} + (t - 1)*{-1/2,Sqrt[3]/2},
    {1/2,Sqrt[3]/2} + (t - 2)*{-1/2,-Sqrt[3]/2}]]
```

B.92 trisectrix

$t \longmapsto$ **trisectrix[a][t]** is the trisectrix of Maclaurin.

```
trisectrix[a_][t_]:=
    {a*(1 - 4*Cos[t]^2),a*(1 - 4*Cos[t]^2)*Tan[t]}
```

B.93 tschirnhausen

$t \longmapsto$ **tschirnhausen[n,a][t]** is a parametrization of Tschirnhausen's cubic.

```
tschirnhausen[n_,a_][t_]:={a*Cos[t]/Cos[t/3]^n,
                          a*Sin[t]/Cos[t/3]^n}
```

B.94 zcprofile

$t \longmapsto$ **zcprofile[pm,a,c][t]** is the profile curve of a generalized helicoid of zero Gaussian curvature. The parameter **pm** is $+1$ or -1.

```
zcprofile[pm_,a_,c_][t_]:=
    {t,pm*Sqrt[a^2*t^2 - c^2] + c*ArcSin[c/(a*t)]}
```

Implicitly Defined Plane Curves

B.95 agnesiimplicit

agnesiimplicit[a][x,y]==0 is the nonparametric form of the parametrized curve $t \longmapsto (2a \tan t, 2a(\cos t)^2)$.

```
agnesiimplicit[a_][x_,y_]:= x^2*y - 4*a^2*(2*a - y)
```

B.96 astroidimplicit

astroidimplicit[n,a,b][x,y]==0 is the nonparametric form of the parametrized curve $t \longmapsto \bigl(a(\cos t)^n, b(\sin t)^n\bigr)$.

```
astroidimplicit[n_,a_,b_][x_,y_]:=
    (x/a)^(2/n) + (y/b)^(2/n) - 1
```

B.97 bulletnoseimplicit

bulletnoseimplicit[a,b][x,y]==0 is the nonparametric form of the parametrized curve $t \longmapsto (a\cos t, b\cot t)$.

```
bulletnoseimplicit[a_,b_][x_,y_]:= (a/x)^2 - (b/y)^2 - 1
```

B.98 cardioidimplicit

cardioidimplicit[a][x,y]==0 is the nonparametric form of the parametrized curve $t \longmapsto a(1+\cos t)(\cos t, \sin t)$.

```
cardioidimplicit[a_][x_,y_]:=
    (x^2 + y^2 - 2*a*x)^2 - 4*a^2*(x^2 + y^2)
```

B.99 cartesianimplicit

cartesianimplicit[a,c,m][x,y]==0 is an oval of Descartes.

```
cartesianimplicit[a_,c_,m_][x_,y_]:=
    ((1 - m^2)*(x^2 + y^2) + 2*m^2*c*x + a^2 - m^2*c^2)^2 -
        4a^2*(x^2 + y^2)
```

B.100 cassiniimplicit

cassiniimplicit[a,b][x,y]==0 is an oval of Cassini.

```
cassiniimplicit[a_,b_][x_,y_]:=
    (x^2 + y^2 + a^2)^2 - b^4 - 4*a^2*x^2
```

B.101 circleimplicit

`circleimplicit[a][x,y]==0` is the nonparametric form of the parametrized circle $t \longmapsto a(\cos t, \sin t)$.

```
circleimplicit[a_][x_,y_]:= x^2 + y^2 - a^2
```

B.102 cissoidimplicit

`cissoidimplicit[a][x,y]==0` is the nonparametric form of the parametrized curve
$$t \longrightarrow \left(\frac{2at^2}{1+t^2}, \frac{2at^3}{1+t^2}\right).$$

```
cissoidimplicit[a_][x_,y_]:= x^3 + x*y^2 - 2*a*y^2
```

B.103 conchoidimplicit

`conchoidimplicit[a,b][x,y]==0` is the nonparametric form of the parametrized curve $t \longmapsto (b + a\cos t, b\tan t + a\sin t)$.

```
conchoidimplicit[a_,b_][x_,y_]:=
    (x^2 + y^2)*(x - b)^2 - a^2*x^2
```

B.104 crosscurveimplicit

`crosscurveimplicit[a,b][x,y]==0` is the nonparametric form of the parametrized curve $t \longmapsto (a\sec t, b\csc t)$.

```
crosscurveimplicit[a_,b_][x_,y_]:= (a/x)^2 + (b/y)^2 - 1
```

B.105 cubicparabolaimplicit

`cubicparabolaimplicit[a,b,c,d][x,y]==0` is the nonparametric form of the parametrized curve $t \longmapsto (t, at^3 + bt^2 + ct + d)$.

```
cubicparabolaimplicit[a_,b_,c_,d_][x_,y_]:=
    a*x^3 + b*x^2 + c*x + d - y
```

Appendix B: Curves 919

B.106 deltoidimplicit

`deltoidimplicit[a][x,y]==0` is the nonparametric form of the parametrized curve $t \longmapsto (2a\cos t(1+\cos t) - a, 2a\sin t(1-\cos t))$.

```
deltoidimplicit[a_][x_,y_] := (x^2 + y^2)^2 -
    8*a*x*(x^2 - 3*y^2) + 18*a^2*(x^2 + y^2) - 27*a^4
```

B.107 devilimplicit

`devilimplicit[a,b][x,y]==0` is the devil's curve.

```
devilimplicit[a_,b_][x_,y_] :=
    y^2*(y^2 - b^2) - x^2*(x^2 - a^2)
```

B.108 ellipseimplicit

`ellipseimplicit[a,b][x,y]==0` is the nonparametric form of the parametrized curve $t \longmapsto (a\cos t, b\sin t)$.

```
ellipseimplicit[a_,b_][x_,y_] := x^2/a^2 + y^2/b^2 - 1
```

B.109 foliumimplicit

`foliumimplicit[a][x,y]==0` is the nonparametric form of the parametrized curve
$$t \longmapsto \frac{(3t, 3t^2)}{(1+t^3)}.$$

```
foliumimplicit[x_,y_] := x^3 + y^3 - 3*x*y
```

B.110 hyperbolaimplicit

`hyperbolaimplicit[a,b][x,y]==0` is the nonparametric form of the parametrized curve $t \longmapsto (a\cosh t, b\sinh t)$.

```
hyperbolaimplicit[a_,b_][x_,y_] := x^2/a^2 - y^2/b^2 - 1
```

B.111 keplerimplicit

`keplerimplicit[a,b][x,y]==0` is the folium of Kepler.

```
keplerimplicit[a_,b_][x_,y_]:= ((x - b)^2 + y^2)*
    (x*(x - b) + y^2) - 4*a*(x - b)*y^2
```

B.112 lemniscateimplicit

`lemniscateimplicit[a][x,y]==0` is the nonparametric form of the parametrized curve
$$t \longmapsto \frac{(a\cos t, a\sin t \cos t)}{1+\sin^2 t}.$$

```
lemniscateimplicit[a_][x_,y_]:=
    (x^2 + y^2)^2 - a^2*(x^2 - y^2)
```

B.113 limaconimplicit

`limaconimplicit[a,b][x,y]==0` is the nonparametric form of the parametrized curve $t \longmapsto (2a\cos t + b)(\cos t, \sin t)$.

```
limaconimplicit[a_,b_][x_,y_]:=
    (x^2 + y^2 - 2a*x)^2 - b^2*(x^2 + y^2)
```

B.114 nephroidimplicit

`nephroidimplicit[a][x,y]==0` is the nonparametric form of the parametrized curve $t \longmapsto a(3\cos t - \cos 3t, 3\sin t - \sin 3t)$.

```
nephroidimplicit[a_][x_,y_]:=
    (x^2 + y^2 - 4*a^2)^3 - 108*a^4*y^2
```

B.115 parabolaimplicit

`parabolaimplicit[a][x,y]==0` is the nonparametric form of the parametrized curve $t \longmapsto (2at, at^2)$.

Appendix B: Curves

```
parabolaimplicit[a_][x_,y_]:= x^2 - 4*a*y
```

B.116 perseusimplicit

perseusimplicit[a,b,c][x,y]==0 is the nonparametric form of a spiric of Perseus. **perseusimplicit[a,b^2/(2a),b^2/(2a)][x,y]==0** coincides with **cassiniimplicit[a,b][x,y]==0**. A section of a torus by a plane parallel to the axis of revolution is a spiric of Perseus.

```
perseusimplicit[a_,b_,c_][x_,y_]:=
        (x^2 + y^2 + a^2 -b^2 +c^2)^2 - 4a^2*(c^2 + x^2)
```

B.117 predatorpreyimplicit

predatorpreyimplicit[a,b,p,q,c][x,y]==0 is the implicit solution of the predator-prey system

$$\begin{cases} x' = x(a - b\,y), \\ y' = y(-p + q\,x) \end{cases}$$

centered at $(p/q, a/b)$ with pseudoradius c.

```
predatorpreyimplicit[a_,b_,p_,q_,c_][x_,y_]:=
        y^a*x^p*E^(-b*y - q*x) -
        c*(a/b)^a*(p/q)^p*E^(-a - p)
```

B.118 piriformimplicit

piriformimplicit[a,b][x,y]==0 is the nonparametric form of the parametrized curve $t \longmapsto (1 + \sin t)(a, b\cos t)$.

```
piriformimplicit[a_,b_][x_,y_]:= a^4*y^2 - b^2*x^3*(2*a - x)
```

B.119 scimplicit

`scimplicit[x,y]==0` is the nonparametric form of the semicubical parabola $t \longmapsto (t^2, t^3)$.

```
scimplicit[x_,y_]:= y^2 - x^3
```

B.120 semicubicimplicit

`semicubicimplicit[a,b,c,d][x,y]==0` is the nonparametric form of the curve $y^2 = a x^3 + b x^2 + c x + d$.

```
semicubicimplicit[a_,b_,c_,d_][x_,y_]:=
               a*x^3 + b*x^2 + c*x + d - y^2
```

B.121 serpentineimplicit

`serpentineimplicit[a,b][x,y]==0` is the nonparametric form of the parametrized curve $t \longmapsto \dfrac{(t, a t)}{1 + b t^2}$.

```
serpentineimplicit[a_,b_][x_,y_]:= y - a*x + b*x^2*y
```

B.122 strophoidimplicit

`strophoidimplicit[a][x,y]==0` is the nonparametric form of the parametrized curve
$$t \longmapsto \frac{a(t^2 - 1)}{t^2 + 1}(1, t).$$

```
strophoidimplicit[a_][x_,y_]:= y^2*(a - x) - x^2*(a + x)
```

`strophoidimplicit[m,a][x,y]==0` is the nonparametric form of the parametrized curve
$$t \longmapsto \frac{a(t^2 - m)}{t^2 + 1}(1, t).$$

```
strophoidimplicit[m_,a_][x_,y_]:= y^2*(a - x) - x^2*(a*m + x)
```

Polar Defined Plane Curves

B.123 `archimedesspiralpolar`

`theta` ⟼ `archimedesspiralpolar[n,a][theta]` is the polar form of a spiral of Archimedes of radius **a** and order **n**.

```
archimedesspiralpolar[n_,a_][theta_] := a*theta^(1/n)
```

B.124 `bowpolar`

`theta` ⟼ `bowpolar[a][theta]` is the polar form of a parametrized curve whose shape is a bow.

```
bowpolar[a_][theta_] := a*(1 - Tan[theta]^2)
```

B.125 `cardioidpolar`

`theta` ⟼ `cardioidpolar[a][theta]` is the polar form of a heart-shaped curve that is traced out by a point on the circumference of a circle rolling around a fixed circle of the same radius.

```
cardioidpolar[a_][theta_] := 2*a*(1 + Cos[theta])
```

B.126 `cassinipolar`

`theta` ⟼ `fermatspiralpolar[a][theta]` is the polar form of an oval of Cassini. The parameters **pm1** and **pm2** are ±1. The implicit equation is

$$(x^2 + y^2 + a^2)^2 - b^4 - 4a^2 x^2 = 0.$$

```
cassinipolar[a_,b_,pm1_,pm2_][theta_] :=
    pm1*Sqrt[a^2*Cos[2*theta] +
    pm2*Sqrt[b^4 - a^4*Sin[2*theta]^2]]
```

B.127 cochleoidpolar

theta⟼cochleoidpolar[n,a][theta] is the polar form of the curve whose parametrized form is

$$t \longmapsto a\left(\frac{\sin t}{t}\right)^n (\cos t, \sin t).$$

```
cochleoidpolar[n_,a_][theta_]:= a*(Sin[theta]/theta)^n
```

B.128 epispiralpolar

theta⟼epispiralpolar[n,a][theta] is the polar form of a curve that has n branches if n is odd and $2n$ branches if n is even. Each branch has two asymptotes radiating out from the center.

```
epispiralpolar[n_,a_][theta_]:= a/Sin[n*theta]
```

B.129 fermatspiralpolar

theta⟼fermatspiralpolar[a][theta] is the polar form of the spiral of Fermat of radius **a**, which may be positive or negative.

```
fermatspiralpolar[a_][theta_]:= a*Sqrt[theta]
```

B.130 foliumpolar

theta⟼foliumpolar[theta] is the polar form of the folium of Descartes. The implicit equation is

$$x^3 + y^3 - 3xy = 0.$$

```
foliumpolar[theta_]:=
    3*Cos[theta]*Sin[theta]/(Cos[theta]^3 + Sin[theta]^3)
```

B.131 hyperbolicspiralpolar

theta⟼hyperbolicspiralpolar[a][theta] is the hyperbolic spiral of radius **a**. It is a special case of a spiral of Archimedes.

Appendix B: Curves 925

```
hyperbolicspiralpolar[a_][theta_]:= a/theta
```

B.132 kampylepolar

theta⟼**kampylepolar[a][theta]** is the polar form of the kampyle or crooked staff of Eudoxus.

```
kampylepolar[a_][theta_]:= a*Sec[theta]^2
```

B.133 keplerorbitpolar

theta⟼**keplerorbitpolar[a,e][theta]** is the polar form of a Kepler orbit, where e is the eccentricity. The orbit is part of an ellipse ($0 \leq e < 1$), parabola ($e = 1$) or hyperbola ($e > 1$).

```
keplerorbitpolar[a_,e_][theta_]:=a/(1 - e Cos[theta])
```

B.134 lemniscatepolar

theta⟼**lemniscatepolar[pm,a][theta]** is the polar form of the curve whose implicit equation is

$$(x^2 + y^2)^2 = a^2(x^2 - y^2).$$

The parameter **a** is positve or negative.

```
lemniscatepolar[a_][theta_]:= a*Sqrt[Cos[2*theta]]
```

B.135 limaconpolar

theta⟼**limaconpolar[a,b][theta]** is the polar form of the curve whose implicit equation is

$$(x^2 + y^2 - 2a\,x)^2 = b^2(x^2 + y^2).$$

```
limaconpolar[n_,a_,b_][theta_]:= 2*a*Cos[n*theta] + b
```

B.136 lituuspolar

`theta ⟼ lituuspolar[a][theta]` the polar form of the lituus spiral of radius **a**, which may be positive or negative.

```
lituuspolar[a_][theta_] := a/Sqrt[theta]
```

B.137 logspiralpolar

`theta ⟼ logspiralpolar[a,b][theta]` is the polar form of the log spiral of radius **a** and slant **b**.

```
logspiralpolar[a_,b_][theta_] := a*E^(b*theta)
```

B.138 nephroidpolar

`theta ⟼ nephroidpolar[a][theta]` is the polar form of a kidney-shaped curve that is traced by a point on the circumference of a circle of radius a rolling around a fixed circle of radius $2a$.

```
nephroidpolar[a_][theta_] :=
    2*a*(Sin[theta/2]^(2/3) + Cos[theta/2]^(2/3))^(3/2)
```

B.139 pacman

`t ⟼ pacman[n][theta]` is the polar form of a pacman-shaped curve.

```
pacman[n_][theta_] := Cos[theta]^n + 1
```

B.140 rosepolar

`theta ⟼ rosepolar[n,a][theta]` is the polar form of a rose of radius **a** with **n** petals if **n** is odd and **2*n** petals if **n** is even.

```
rosepolar[n_,a_][theta_] := a*Cos[n*theta]
```

Parametrically Defined Space Curves

B.141 ast3d

t ⟼ **ast3d[n,a,b][t]** is a space curve analog of the plane curve astroid.

```
ast3d[n_,a_,b_][t_]:= {a*Cos[t]^n,b*Sin[t]^n,Cos[2*t]}
```

B.142 baseballseam

t ⟼ **baseballseam[a,b,c][t]** is an approximation to the seam on a baseball of radius **a**.

```
baseballseam[a_,b_,c_][t_]:=
    {a*Sin[Pi/2 - (Pi/2 - c)*Cos[t]]*Cos[t/2 + c*Sin[2*t]],
     a*Sin[Pi/2 - (Pi/2 - c)*Cos[t]]*Sin[t/2 + c*Sin[2*t]],
     a*Cos[Pi/2 - (Pi/2 - c)*Cos[t]]}
```

B.143 besselcurve

t ⟼ **besselcurve[a,b,c][t]** is a space curve whose components are Bessel functions.

```
besselcurve[a_,b_,c_][t_]:=
    {BesselJ[a,t],BesselJ[b,t],BesselJ[c,t]}
```

B.144 bicylinder

t ⟼ **bicylinder[a,b,pm][t]** is a parametrization of the intersection of the circular cylinders $x^2 + y^2 = a^2$ and $y^2 + z^2 = b^2$. One component is obtained by taking **pm = 1**, the other by taking **pm = -1**.

```
bicylinder[a_,b_,pm_][t_]:= {a*Cos[t],a*Sin[t],
                             pm*Sqrt[b^2 - a^2*Sin[t]^2]}
```

B.145 biquadratic

t⟼**biquadratic[a,k][t]** is a parametrization of the intersection of circular cylinders $x^2 + y^2 = a^2$ and the elliptic cylinder $k^2x^2 + z^2 = a^2$.

```
biquadratic[a_,k_][t_]:=
    {a*JacobiSN[t,k],a*JacobiCN[t,k],a*JacobiDN[t,k]}
```

B.146 clelia

t⟼**clelia[a][m,n][t]** is a clelia of type **{m,n}** on a sphere of radius **a**. The number of petals is **n**.

```
clelia[a_][m_,n_][t_]:=
    {a*m*Cos[t]*Sin[n*t],a*m*Sin[t]*Sin[n*t],
    a*Sqrt[1 - (m*Sin[n*t])^2]}
```

B.147 conicalhelix

t⟼**conicalhelix[a,b,c,d][t]** is a parametrization of a helix on a cone over t⟼**ellipse[a,b][t]**. The slant of the cone is determined by **c** and the amount of twisting is determined by **d**.

```
conicalhelix[a_,b_,c_,d_][t_]:={a*t*Cos[d*t],b*t*Sin[d*t],c*t}
```

B.148 cubicalellipse

t⟼**cubicalellipse[a,b,c,d,e,f,pm][t]** is a cubical ellipse for **pm = 1** and a cubical hyperbola for **pm = -1**.

```
cubicalellipse[a_,b_,c_,d_,e_,f_,pm_][t_]:=
    {a*t/(1 + pm*t^2),
    (b*t + c*t^2)/(1 + pm*t^2),
    (d*t + e*t^2 + f*t^3)/(1 + pm*t^2)}
```

B.149 eightknot

t⟼**eightknot[t]** is a parametrization of an eight knot.

Appendix B: Curves 929

```
eightknot[t_]:=
    {10*(Cos[t] + Cos[3*t]) + Cos[2*t] + Cos[4*t],
     6*Sin[t] + 10*Sin[3*t],
     4*Sin[3*t]*Sin[5*t/2] + 4*Sin[4*t] - 2*Sin[6*t]}
```

B.150 epitrochoid3d

t⟼**epitrochoid3d[a,b,h,omega][t]** is the parametrized curve traced out by a point P attached to a circle of radius **b** rolling along a fixed circle of radius **a**. The rolling circle lies in a plane whose inclination with the plane of the fixed circle is **omega**, and **h** is the distance of P to the center of the rolling circle. The curve t⟼**epitrochoid3d[a,b,h,0][t]** is a (planar) epitrochoid and the curve t⟼**epitrochoid3d[a,b,h,Pi][t]** is a (planar) hypotrochoid.

```
epitrochoid3d[a_,b_,h_,omega_][t_]:=
    {Cos[t]*(a + b*Cos[omega]) -
      h*(Cos[t]*Cos[a*t/b]*Cos[omega] - Sin[t]*Sin[a*t/b]),
     (a + b*Cos[omega])*Sin[t] -
      h*(Cos[a*t/b]*Cos[omega]*Sin[t] + Cos[t]*Sin[a*t/b]),
     (b - h*Cos[a*t/b])*Sin[omega]}
```

B.151 genhelix

t⟼**genhelix[a,b][f][t]** is a generalized elliptical helix with slant function **f**.

```
genhelix[a_,b_][f_][t_]:= {a*Cos[t],b*Sin[t],f[t]}
```

B.152 helix

t⟼**helix[a,c][t]** is a circular helix of radius **a** and slant **c**.

```
helix[a_,c_][t_]:= {a*Cos[t],a*Sin[t],c*t}
```

t⟼**helix[a,b,c][t]** is an elliptical helix of slant **c**.

```
helix[a_,b_,c_][t_]:= {a*Cos[t],b*Sin[t],c*t}
```

B.153 helixunitspeed

$t \longmapsto$ `helixunitspeed[a,b][t]` is a unit-speed parametrization of circular helix of radius **a** and slant **b**.

```
helixunitspeed[a_,b_][s_]:=
    {a*Cos[s/Sqrt[a^2 + b^2]], a*Sin[s/Sqrt[a^2 + b^2]],
    b*s/Sqrt[a^2 + b^2]}
```

B.154 horopter

$t \longmapsto$ `horopter[a,b][t]` is a horopter.

```
horopter[a_,b_][t_]:=
    {2*a/(1 + b^2*t^2),2*a*b*t/(1 + b^2 t^2),t}
```

B.155 line3d

$t \longmapsto$ `line3d[a1,b1,c1,a2,b2,c2][t]` is a parametrized line in $\mathbb{R}^3$ with `line3d[a1,b1,c1,a2,b2,c2][0]= {a2,b2,c2}` and `line3d[a1,b1,c1,a2,b2,c2]'[0]={a1,b1,c1}`.

```
line3d[a1_,b1_,c1_,a2_,b2_,c2_][t_]:=
    {a1*t + a2,b1*t + b2,c1*t + c2}
```

B.156 lissajous3d

$t \longmapsto$ `lissajous3d[n,d,a,b,c][t]` is a space curve analog of lissajous curve.

```
lissajous3d[n_,d_,a_,b_,c_][t_]:=
    {a*Sin[n*t + d],b*Sin[t],c*Cos[t]}
```

B.157 oneparametersubgroup

$t \longmapsto$ `oneparametersubgroup[mat,vec][t]` is the curve in $\mathbb{R}^n$ formed by the one parameter subgroup $t \longmapsto \exp(t * \mathbf{mat})$ acting on a vector **v** in $\mathbb{R}^n$, where **mat** is an $n \times n$ matrix.

Appendix B: Curves

```
oneparametersubgroup[mat_,vec_][t_]:=
    ComplexExpand[Normal[MatrixExp[Rationalize[t*mat]]].vec]
```

B.158 pseudospheregeodesic

t⟼pseudospheregeodesic[a,c,u0,pm][t] is a geodesic of slant c on a pseudosphere of constant negative curvature $-a^{-2}$. The parameter pm is +1 or −1.

```
pseudospheregeodesic[a_,c_,u0_,pm_][t_]:=
    {a*Sin[t]*Cos[u0 + pm*Sqrt[a^2 - 2*c^2 - a^2*Cos[2*t]]/
     (Sqrt[2]*c*Sin[t])],
     a*Sin[t]*Sin[u0 + pm*Sqrt[a^2 - 2*c^2 - a^2*Cos[2*t]]/
     (Sqrt[2]*c*Sin[t])],a*(Cos[t] + Log[Tan[t/2]])}
```

B.159 seiffertspiral

t⟼seiffertspiral[a,k][t] is Seiffert's spiral of slant **k** on a sphere of radius **a**.

```
seiffertspiral[a_,k_][t_]:=
    {a*Cos[t]*JacobiSN[t,k],a*Sin[t]*JacobiSN[t,k],
     a*JacobiCN[t,k]}
```

B.160 sphericalcardioid

t⟼sphericalcardioid[a][t] is a spherical cardioid on a sphere of radius $3|a|$. It is a curve of constant slope that projects to t⟼cardioid[-a][t + Pi].

```
sphericalcardioid[a_][t_]:=
    {2*a*Cos[t] - a*Cos[2*t],2*a*Sin[t] - a*Sin[2*t],
     Sqrt[8]*a*Cos[t/2]}
```

B.161 sphericalhelix

t⟼sphericalhelix[a,b][t] is a spherical helix on a sphere of radius $|a+2b|$. It is a curve of constant slope that projects to t⟼epicycloid[a,b][t].

```
sphericalhelix[a_,b_][t_]:=
    {(a + b)*Cos[t] - b*Cos[(a + b)*t/b],
     (a + b)*Sin[t] - b*Sin[(a + b)*t/b],
     2*Sqrt[a*b + b^2]*Cos[a*t/(2*b)]}
```

B.162 sphericalloxodrome

t⟼**sphericalloxodrome[a,b,c][t]** is a loxodrome on a sphere of radius **a**. It is the image under the stereographic injection {u,v}⟼**stereographicsphere[a][u,v]** of the logarithmic spiral t⟼**logspiral[b,c][t]**. A sphericalloxodrome meets each meridian of the sphere at the same angle.

```
sphericalloxodrome[a_,b_,c_][t_]:=
    {2*a*b*E^(c*t)*Cos[t]/(1 + b^2*E^(2*c*t)),
     2*a*b*E^(c*t)*Sin[t]/(1 + b^2*E^(2*c*t)),
     a*(-1 + b^2*E^(2*c*t))/(1 + b^2*E^(2*c*t))}
```

B.163 sphericalnephroid

t⟼**sphericalnephroid[a][t]** is a spherical nephroid on a sphere of radius $4|a|$. It is a curve of constant slope that projects to t⟼**nephroid[a][t]**.

```
sphericalnephroid[a_][t_]:=
    {3*a*Cos[t] - a*Cos[3*t],3*a*Sin[t] - a*Sin[3*t],
     Sqrt[12]*a*Cos[t]}
```

B.164 sphericalspiral

t⟼**sphericalspiral[a][m,n][t]** is a spherical spiral of type **{m,n}** on a sphere of radius **a**.

```
sphericalspiral[a_][m_,n_][t_]:=
         {a*Cos[m*t]*Cos[n*t],
          a*Sin[m*t]*Cos[n*t],
          a*Sin[n*t]}
```

B.165 torusknot

$t \mapsto$ **torusknot[a,b,c][p,q][t]** is a torus knot of type **{p,q}** on an elliptical torus of type **{a,b,c}**.

```
torusknot[a_,b_,c_][p_,q_][t_]:=
    {(a + b*Cos[q*t])*Cos[p*t],
     (a + b*Cos[q*t])*Sin[p*t],c*Sin[q*t]}
```

B.166 twicubic

$t \mapsto$ **twicubic[t]** is a twisted cubic.

```
twicubic[t_]:= {t,t^2,t^3}
```

B.167 veryflat

$t \mapsto$ **veryflat[1][t]** and $t \mapsto$ **veryflat[2][t]** are space curves with the same curvature and torsion, but are not congruent by a Euclidean motion.

```
veryflat[1][t_]:= {t,0,5*E^(-1/t^2)}
```

```
veryflat[2][t_]:= If[t<0,{t,5*E^(-1/t^2),0},
                   If[t==0,{0,0,0},{t,0,5*E^(-1/t^2)}]]
```

B.168 viviani

$t \mapsto$ **viviani[a][t]** is a Viviani curve on a sphere of radius $2a$.

```
viviani[a_][t_]:= {a*(1 + Cos[t]),a*Sin[t],2*a*Sin[t/2]}
```

Parametrically Defined Curves in $\mathbb{R}^n$

B.169 besseln

t⟼**besseln[n,a][t]** is the curve $t \longmapsto a(J_0(t), \ldots, J_{n-1}(t))$.

```
besseln[n_,a_][t_]:=Table[a*BesselJ[k,t],{k,0,n - 1}]
```

B.170 cosn

t⟼**cosn[n,a][t]** is the curve $t \longmapsto a(\cos(t), \ldots, \cos(n\,t))$.

```
cosn[n_,a_][t_]:=Table[a*Cos[k*t],{k,1,n}]
```

B.171 sinn

t⟼**sinn[n,a][t]** is the curve $t \longmapsto a(\sin(t), \ldots, \sin(n\,t))$.

```
sinn[n_,a_][t_]:=Table[a*Sin[k*t],{k,1,n}]
```

B.172 twistedn

t⟼**twistedn[n,a][t]** is the curve $t \longmapsto a(t, \ldots, t^n)$.

```
twistedn[n_,a_][t_]:=Table[a*t^k,{k,1,n}]
```

APPENDIX C: SURFACES

Parametrically Defined Surfaces

C.1 `agnesirev`

{u,v}⟼`agnesirev[a][u,v]` is a parametrization of the surface of revolution formed from a rotated witch of Agnesi.

```
agnesirev[a_][u_,v_]:=
 {-2*a*Cos[u]*Cos[v]^2,-2*a*Cos[v]^2*Sin[u],2*a*Tan[v]}
```

C.2 `astell`

{u,v}⟼`astell[a,b,c][u,v]` is a parametrization of an astroidial ellipsoid with axes of lengths **a**, **b** and **c**.

```
astell[a_,b_,c_][u_,v_]:= {a*Cos[u]*Cos[v],
                          b*Sin[u]*Cos[v],c*Sin[v]}^3
```

C.3 `bohdom`

{u,v}⟼`bohdom[a,b,c][u,v]` is a parametrization of a Bohemian dome. It is formed by moving an ellipse along a circle in a perpendicular plane so that the ellipse remains parallel to a plane.

```
bohdom[a_,b_,c_][u_,v_] := {a*Cos[u],
                           a*Sin[u] + b*Cos[v],c*Sin[v]}
```

C.4 bour

{u,v} ⟼ bour[u,v] is a parametrization of Bour's minimal surface.

```
bour[u_,v_] := {(2*u^2 - u^4 - 2*v^2 + 6*u^2*v^2 - v^4)/4,
                u*v*(-1 - u^2 + v^2),(2/3)*(u^3 - 3*u*v^2)}
```

C.5 bourpolar

{r,theta} ⟼ bourpolar[m][t][r,theta] is the polar parametrization of a 1-parameter family of minimal surfaces such that

{r,theta} ⟼ bourpolar[m][0][r,theta] is Bour's minimal surface of degree **m**.

```
bourpolar[m_][t_][r_,theta_] :=
    {r^(m - 1)*Cos[t - (m - 1)*theta]/(m - 1) -
     r^(m + 1)*Cos[t - (m + 1)*theta]/(m + 1),
     r^(m - 1)*Sin[t - (m - 1)*theta]/(m - 1) +
     r^(m + 1)*Sin[t - (m + 1)*theta]/(m + 1),
     2*r^m*Cos[t - m*theta]/m}
```

{r,theta} ⟼ bourpolar[m,n][t][r,theta] is the polar parametrization of a 1-parameter family of minimal surfaces such that

{r,theta} ⟼ bourpolar[m,n][0][r,theta] is Bour's minimal surface of degree **(m,n)**.

```
bourpolar[m_,n_][t_][r_,theta_] :=
    {r^(m - 1)*Cos[t - (m - 1)*theta]/(m - 1) -
     r^(m + 2*n - 1)*Cos[t - (m + 2*n - 1)*theta]/(m + 2*n - 1),
     r^(m - 1)*Sin[t - (m - 1)*theta]/(m - 1) +
     r^(m + 2*n - 1)*Sin[t - (m + 2*n - 1)*theta]/(m + 2*n - 1),
     2*r^(m + n - 1)*Cos[t - (m + n - 1)*theta]/(m + n - 1)}
```

Appendix C: Surfaces

C.6 boy

$\{u,v\} \mapsto$ `boy[a,b,c][u,v]` is a parametrization of Boy's surface.

```
boy[a_,b_,c_][u_,v_]:=
  {(a/2)*(-Cos[v]^2 + 2*Cos[u]^2*Sin[v]^2 +
    Cos[v]*Cos[u]*Sin[v]*(-Cos[v]^2 + Cos[u]^2*Sin[v]^2) -
    Sin[v]^2*Sin[u]^2 + 2*Cos[v]*Sin[v]*Sin[u]*
    (-Cos[v]^2 + Sin[v]^2*Sin[u]^2) - Sin[v]^4*Sin[4*u]/4),
   (Sqrt[3]/2)*b*(-Cos[v]^2 +
    Cos[v]*Cos[u]*Sin[v]*(Cos[v]^2 - Cos[u]^2*Sin[v]^2) +
    Sin[v]^2*Sin[u]^2 - (Sin[v]^4*Sin[4*u])/4),
   c*(Cos[v] + Cos[u]*Sin[v] + Sin[u]*Sin[v])*
    (-4*Sin[v]*(Cos[u] - Sin[u])*(-Cos[v] + Cos[u]*Sin[v])*
    (Cos[v] - Sin[u]*Sin[v]) +
    (Cos[v] + Cos[u]*Sin[v] + Sin[u]*Sin[v])^3)}
```

C.7 catalan

$\{u,v\} \mapsto$ `catalan[a][u,v]` is a parametrization of Catalan's minimal surface.

```
catalan[a_][u_,v_]:= {a*(u - Sin[u]*Cosh[v]),
                     a*(1 - Cos[u]*Cosh[v]),
                     -4*a*Sin[u/2]*Sinh[v/2]}
```

C.8 catalandef

$\{u,v\} \mapsto$ `catalandef[a][t][u,v]` is a 1-parameter family of minimal surfaces connecting Catalan's minimal surface to its conjugate.

```
catalandef[a_][t_][u_,v_]:=
  {a*Cos[t]*(u - Cosh[v]*Sin[u]) +
   a*Sin[t]*(v - Cos[u]*Sinh[v]),
   a*Cos[t]*(1 - Cos[u]*Cosh[v]) +
   a*Sin[t]*Sin[u]*Sinh[v],
   4*a*(1 - Cos[u/2]*Cosh[v/2])*Sin[t] -
   4*a*Cos[t]*Sin[u/2]*Sinh[v/2]}
```

C.9 catenoid

`{u,v} ⟼ catenoid[c][u,v]` is the minimal surface of revolution generated by `t ⟼ catenary[a][t]`.

```
catenoid[c_][u_,v_] := {c*Cos[u]*Cosh[v/c],
                       c*Sin[u]*Cosh[v/c],v}
```

C.10 catenoidasym

`{p,q} ⟼ catenoidasym[c][p,q]` is a parametrization of a catenoid by asymptotic curves.

```
catenoidasym[a_][p_,q_] :=
    {a*Cos[(p + q)/2]*Cosh[(q - p)/2],
     a*Sin[(p + q)/2]*Cosh[(q - p)/2],
     a*(q - p)/2}
```

C.11 catenoidiso

`{p,q} ⟼ catenoidiso[a][u,v]` is an isothermal parametrization of a catenoid.

```
catenoidiso[a_][u_,v_] := {a*Cos[u/a]*Cosh[v/a],
                          a*Sin[u/a]*Cosh[v/a],v}
```

C.12 circleinvolutemin

`{u,v} ⟼ circleinvolutemin[1,a][t][u,v]` is a 1-parameter family of minimal surfaces such that `{u,v} ⟼ circleinvolutemin[1,a][0][u,v]` is a minimal surface containing the involute of a circle as a geodesic.

Appendix C: Surfaces

```
circleinvolutemin[1,a_][t_][u_,v_]:=
  {a*Cos[t]*(Cos[u]*Cosh[v] +
      u*Cosh[v]*Sin[u] - v*Cos[u]*Sinh[v]) +
    a*Sin[t]*(v*Cosh[v]*Sin[u] +
      u*Cos[u]*Sinh[v] - Sin[u]*Sinh[v]),
  a*Sin[t]*(-v*Cos[u]*Cosh[v] + Cos[u]*Sinh[v] +
      u*Sin[u]*Sinh[v]) + a*Cos[t]*(-u*Cos[u]*Cosh[v] +
      Cosh[v]*Sin[u] - v*Sin[u]*Sinh[v]),
  -a*u*v*Cos[t] + a*(u^2 - v^2)*Sin[t]/2}
```

C.13 circularcone

$\{u,v\} \mapsto$ `circularcone[a,b][u,v]` is a parametrization of a circular cone of radius **a** and slope **b/a**.

```
circularcone[a_,b_][u_,v_]:= {a*v*Cos[u],a*v*Sin[u],b*v}
```

C.14 circularcylinder

$\{u,v\} \mapsto$ `circularcylinder[a][u,v]` is a parametrization of a circular cylinder of radius **a**.

```
circularcylinder[a_][u_,v_]:= {a*Cos[u],a*Sin[u],v}
```

C.15 clothoidmin

$\{u,v\} \mapsto$ `clothoidmin[a][t][u,v]` is a 1-parameter family of minimal surfaces such that $\{u,v\} \mapsto$ `clothoidmin[a][0][u,v]` is a minimal surface containing a clothoid as a geodesic.

```
clothoidmin[a_][t_][u_,v_]:=
  {a*Sqrt[Pi]*Cos[t]*Re[FresnelS[(u + I*v)/Sqrt[Pi]]] +
   a*Sqrt[Pi]*Im[FresnelS[(u + I*v)/Sqrt[Pi]]]*Sin[t],
   a*Sqrt[Pi]*Cos[t]*Re[FresnelC[(u + I*v)/Sqrt[Pi]]] +
   a*Sqrt[Pi]*Im[FresnelC[(u + I*v)/Sqrt[Pi]]]*Sin[t],
   -a*v*Cos[t] + a*u*Sin[t]}
```

C.16 cnccosurfrev

{u,v} ⟼ cnccosurfrev[a,b][u,v] is a parametrization of a surface of revolution of constant negative curvature $-a^{-2}$ of conic type.

```
cnccosurfrev[a_,b_][u_,v_]:= {b*Cos[u]*Sinh[v/a],
                              b*Sin[u]*Sinh[v/a],
              -I*Sqrt[a^2 - b^2]*EllipticE[I*v/a,
              -b^2/(a^2 - b^2)]}
```

C.17 cnchysurfrev

{u,v} ⟼ cnchysurfrev[a,b][u,v] is a parametrization of a surface of revolution of constant negative curvature $-a^{-2}$ of hyperboloid type.

```
cnchysurfrev[a_,b_][u_,v_]:= {b*Cos[u]*Cosh[v/a],
                              b*Sin[u]*Cosh[v/a],
               - I*a*EllipticE[I*v/a,-b^2/a^2]}
```

C.18 cossurface

{u,v} ⟼ cossurface[a][u,v] is the surface
{u,v} ⟼ a*{Cos[u],Cos[v],Cos[u + v]}. The implicit equation is
$4x^2y^2(a^2 - z^2) = a^2(x^2 + y^2 - z^2)^2$.

```
cossurface[a_][u_,v_]:= {a*Cos[u],a*Cos[v],a*Cos[u + v]}
```

C.19 costa

{u,v} ⟼ costa[u,v] is a parametrization of Costa's minimal surface.

```
costa[u_,v_]:=
  {((1/2)*Re[-WeierstrassZeta[u + I*v,189.07272,0] +
   3.14159*u + 0.35889 +
   0.228473*(WeierstrassZeta[u + I*v - 1/2,189.07272,0] -
       WeierstrassZeta[u + I*v - I/2,189.07272,0])],
   (1/2)*Re[-I*WeierstrassZeta[u + I*v,189.07272,0] +
   3.14159*v + 0.35889 -
   0.228473*(I*WeierstrassZeta[u + I*v - 1/2,189.07272,0] -
       I*WeierstrassZeta[u + I*v - I/2,189.07272,0])],
   0.626657*Log[Abs[(WeierstrassP[u + I*v,189.07272,0] -
       6.87519)/
       (WeierstrassP[u + I*v,189.07272,0] + 6.87519)]]}
```

C.20 cpcsurfrev

{u,v} ⟼ cpcsurfrev[a,b][u,v] is a parametrization of a surface of revolution of constant positive curvature a^{-2}. There are 3 cases: football type $(a > b)$, sphere $(a = b)$, barrel type $(a < b)$.

```
cpcsurfrev[a_,b_][u_,v_]:= {b*Cos[u]*Cos[v/a],
                           b*Sin[u]*Cos[v/a],
                           a*EllipticE[v/a,b^2/a^2]}
```

C.21 crosscap

{u,v} ⟼ crosscap[a][u,v] is a parametrization of a cross cap in $\mathbb{R}^3$.

```
crosscap[a_][u_,v_]:= {a^2*Sin[u]*Sin[2*v]/2,
                      a^2*Sin[2*u]*Cos[v]^2,a^2*Cos[2*u]*Cos[v]^2}
```

C.22 deltoidinvolutemin

{u,v} ⟼ deltoidinvolutemin[a][t][u,v] is a 1-parameter family of minimal surfaces such that {u,v} ⟼ deltoidinvolutemin[a][0][u,v] is a minimal surface containing the involute of a deltoid as a geodesic.

```
deltoidinvolutemin[a_][t_][u_,v_]:=
    {(a/3)*Cos[t]*(8*Cos[u/2]*Cosh[v/2] + 2*Cos[u]*Cosh[v] -
      Cos[2*u]*Cosh[2*v]) +
     (a/3)*Sin[t]*(-8*Sin[u/2]*Sinh[v/2] - 2*Sin[u]*Sinh[v] +
      Sin[2*u]*Sinh[2*v]),
     (Cos[t]*(-8*a*Cosh[v/2]*Sin[u/2] + 2*a*Cosh[v]*Sin[u] +
      a*Cosh[2*v]*Sin[2*u]))/3 +
     (Sin[t]*(-8*a*Cos[u/2]*Sinh[v/2] + 2*a*Cos[u]*Sinh[v] +
      a*Cos[2*u]*Sinh[2*v]))/3,
     (4*a*Sin[t]*(3*u - 2*Cosh[3*v/2]*Sin[3*u/2]))/9 -
     (4*a*Cos[t]*(3*v - 2*Cos[3*u/2]*Sinh[3*v/2]))/9}
```

C.23 deltoidmin

{u,v} ⟼ deltoidmin[a][t][u,v] is a 1-parameter family of minimal surfaces such that {u,v} ⟼ deltoidmin[a][0][u,v] is a minimal surface containing a deltoid as a geodesic.

```
deltoidmin[a_][t_][u_,v_]:=
    {a*Cos[t]*(2*Cos[u]*Cosh[v] + Cos[2*u]*Cosh[2*v]) +
     2*a*Sin[t]*Sin[u]*(-Sinh[v] - Cos[u]*Sinh[2*v]),
     -2*a*(Cos[u]*Sinh[v]*((-1 + Cos[u]*Cosh[v])*Sin[t] +
      2*a*Cos[t]*Sin[u]*Sinh[v])) +
     Cosh[v]*Sin[u]*(2*a*Cos[t]*(1 - Cos[u]*Cosh[v]) +
     2*a*Sin[t]*Sin[u]*Sinh[v]),
     -(8/3)*a*Cos[3*u/2]*Cosh[3*v/2]*Sin[t] -
     (8/3)*a*Cos[t]*Sin[3*u/2]*Sinh[3*v/2]}
```

C.24 dini

{u,v} ⟼ dini[a,b][u,v] is the standard parametrization of Dini's surface of constant negative curvature $-a^{-2}$. It is the generalized helicoid of slant b generated by a tractrix. The case $b = 0$ is the standard parametrization of a pseudosphere.

Appendix C: Surfaces

```
dini[a_,b_][u_,v_]:= {a*Cos[u]*Sin[v],
                     a*Sin[u]*Sin[v],
                     a*(Cos[v] + Log[Tan[v/2]]) + b*u}
```

C.25 dinitcheb

{u,v}⟼dinitcheb[a,b][u,v] is a Tchebyshef principal patch and
{p,q}⟼dinitcheb[a,b][p + q,p - q] is a Tchebyshef asymptotic patch
for Dini's surface of constant negative curvature $-a^{-2}$.

```
dinitcheb[a_,b_][u_,v_]:=
    {a*Cos[b]*Cos[v]/Cosh[(u - v*Sin[b])/Cos[b]],
     a*Cos[b]*Sin[v]/Cosh[(u - v*Sin[b])/Cos[b]],
     a*(u - Cos[b]*Tanh[(u - v*Sin[b])/Cos[b]])}
```

C.26 eighteight

{u,v}⟼eighteight[u,v] is a parametrization of the generalized surface of revolution formed by moving a figure eight about a figure eight.

```
eighteight[u_,v_]:= {Sin[u]*Sin[v],
                    Cos[u]*Sin[u]*Sin[v],Cos[v]*Sin[v]}
```

C.27 eightsurface

{u,v}⟼eightsurface[u,v] is a surface of revolution generated by
t⟼eight[t].

```
eightsurface[u_,v_]:=
    {Cos[u]*Sin[2*v],Sin[u]*Sin[2*v],Sin[v]}
```

C.28 ellhypcyclide

{u,v}⟼ellhypcyclide[a,c,k][u,v] is a parametrization of the cyclide of Dupin of radius **k** whose focal sets are the ellipse $u \longmapsto \left(a\cos u, \sqrt{a^2 - c^2}\sin u, 0\right)$ and the hyperbola $v \longmapsto \left(c\sec v, 0, \sqrt{a^2 - c^2}\tan v\right)$.

```
ellhypcyclide[a_,c_,k_][u_,v_]:=
    {(c*(k - c*Cos[u]) + a*Cos[u]*(a - k*Cos[v]))/
            (a - c*Cos[u]*Cos[v]),
     (Sqrt[a^2 - c^2]*(a - k*Cos[v])*Sin[u])/
            (a - c*Cos[u]*Cos[v]),
     (Sqrt[a^2 - c^2]*(k - c*Cos[u])*Sin[v])/
            (a - c*Cos[u]*Cos[v])}
```

C.29 ellipsemin

{u,v}⟼ellipsemin[a,b][t][u,v] is a 1-parameter family of minimal surfaces such that {u,v}⟼ellipsemin[a,b][0][u,v] is an elliptical catenoid and {u,v}⟼ellipsemin[a,b][Pi/2][u,v] is an elliptical helicoid. The patch {u,v}⟼ellipsemin[a,b][0][u,v] contains the ellipse t⟼ellipse[a,b][t] as a geodesic.

```
ellipsemin[a_,b_][t_][u_,v_]:=
    {a*Cos[t]*Cos[u]*Cosh[v] - a*Sin[t]*Sin[u]*Sinh[v],
     b*Cos[t]*Sin[u]*Cosh[v] + b*Sin[t]*Cos[u]*Sinh[v],
     -b*(Cos[t]*Im[EllipticE[u + I*v,1 - a^2/b^2]] -
         Sin[t]*Re[EllipticE[u + I*v,1 - a^2/b^2]])}
```

C.30 ellipsoid

{u,v}⟼ellipsoid[a,c][u,v] is a parametrization of an ellipsoid with axes lengths **a, a** and **c**.

```
ellipsoid[a_,c_][u_,v_]:= {a*Cos[v]*Cos[u],
                          a*Cos[v]*Sin[u],c*Sin[v]}
```

{u,v}->ellipsoid[a,b,c][u,v] is a parametrization of an ellipsoid with axes of lengths **a, b** and **c**.

```
ellipsoid[a_,b_,c_][u_,v_]:= {a*Cos[v]*Cos[u],
                              b*Cos[v]*Sin[u],c*Sin[v]}
```

Appendix C: Surfaces

C.31 ellipticalhyperboloidplus ellipticalhyperboloidminus

{u,v} ⟼ ellipticalhyperboloidplus[a,b,c][u,v] is a parametrization of a hyperboloid of one sheet as a ruled surface.

```
ellipticalhyperboloidplus[a_,b_,c_][u_,v_]:=
    {a*Cos[u] - a*v*Sin[u],b*v*Cos[u] + b*Sin[u],c*v}
```

{u,v} ⟼ ellipticalhyperboloidminus[a,b,c][u,v] is a parametrization of a hyperboloid of one sheet as a ruled surface.

```
ellipticalhyperboloidminus[a_,b_,c_][u_,v_]:=
    {a*Cos[u] + a*v*Sin[u],-b*v*Cos[u] + b*Sin[u],c*v}
```

C.32 ellipticcoor

{u,v} ⟼ ellipticcoor[d,e,f][a,b,c][w][u,v] is an elliptic coordinate patch on $\mathbb{R}^3$. The parameters d, e, f are +1 or −1.

```
ellipticcoor[d_,e_,f_][a_,b_,c_][w_][u_,v_]:=
    {d*Sqrt[(a - u)*(a - v)*(a - w)/((a - b)*(a - c))],
     e*Sqrt[(b - u)*(b - v)*(b - w)/((b - a)*(b - c))],
     f*Sqrt[(c - u)*(c - v)*(c - w)/((c - a)*(c - b))]}
```

C.33 enneper

{u,v} ⟼ enneper[u,v] is a parametrization of Enneper's minimal surface.

```
enneper[u_,v_]:= {u - u^3/3 + u*v^2,
                  -v + v^3/3 - v*u^2,u^2 - v^2}
```

C.34 ennepercatenoidpolar

{r,theta} ⟼ ennepercatenoidpolar[t][r,theta] is the polar parametrization of a 1-parameter family of minimal surfaces constructed from the minimal curve whose Weierstrass representation is given by $f(z) = 1$ and $g(z) = 1/z + z^3$.

```
ennepercatenoidpolar[t_][r_,theta_]:=
    {-r^7*Cos[t - 7*theta]/7 - (2*r^3*Cos[t - 3*theta])/3 +
     r*Cos[t - theta] + Cos[t + theta]/r,
     (r^7*Sin[t - 7*theta])/7 + (2*r^3*Sin[t - 3*theta])/3 +
     r*Sin[t - theta] - Sin[t + theta]/r,
    (r^4*Cos[t - 4*theta])/2 + Cos[t]*Log[r^2] + 2*theta*Sin[t]}
```

C.35 enneperpolar

`{r,theta}⟼enneperpolar[n][t][r,theta]` is the polar parametrization of a 1-parameter family of minimal surfaces such that
`{r,theta}⟼enneperpolar[n][0][r,theta]` is Enneper's minimal surface of degree **n**.

```
enneperpolar[n_][t_][r_,theta_]:=
    {r*Cos[t - theta] - r^(2*n + 1)*
      Cos[t - (2*n + 1)*theta]/(2*n + 1),
     r*Sin[t - theta] + r^(2*n + 1)*
      Sin[t - (2*n + 1)*theta]/(2*n + 1),
     2*r^(n + 1)*Cos[t - (n + 1)*theta]/(n + 1)}
```

C.36 exptwist

`{u,v}⟼exptwist[a,c][u,v]` is a parametrization of a helicoidial-like surface whose twisting varies exponentially.

```
exptwist[a_,c_][u_,v_]:={a*v*Cos[u],a*v*Sin[u],a*E^(c*u)}
```

C.37 exptwistasym

`{p,q}⟼exptwistasym[a,c][p,q]` is the parametrization by asymptotic curves of `{u,v}⟼exptwist[a,c][u,v]`.

```
exptwistasym[a_,c_][p_,q_]:= {a*E^(c*(p - q)/2)*Cos[p],
                              a*E^(c*(p - q)/2)*Sin[p],
                              a*E^(c*p)}
```

C.38 flattorus

$\{u,v\} \longmapsto$ `flattorus[a,b][u,v]` is a parametrization of a torus into $\mathbb{R}^4$ as the product of two circles. The Gaussian curvature is zero.

```
flattorus[a_,b_][u_,v_] := {a*Cos[u],a*Sin[u],
                            b*Cos[v],b*Sin[v]}
```

$\{u,v\} \longmapsto$ `flattorus[a,b,c,d][u,v]` is a parametrization of an elliptical torus into $\mathbb{R}^4$ as the product of two ellipses. The Gaussian curvature is zero.

```
flattorus[a_,b_,c_,d_][u_,v_] := {a*Cos[u],b*Sin[u],
                                  c*Cos[v],d*Sin[v]}
```

C.39 funnel

$\{u,v\} \longmapsto$ `funnel[1,1,c][u,v]` is the surface of revolution formed by rotating the graph of $t \longmapsto c*Log[t]$.

```
funnel[a_,b_,c_][u_,v_] := {a*v*Cos[u],
                            b*v*Sin[u],c*Log[v]}
```

C.40 funnelasym

$\{p,q\} \longmapsto$ `funelasym[c][p,q]` is the parametrization by asymptotic curves of the surface of revolution formed by rotating the graph of $t \longmapsto c*Log[t]$.

```
funnelasym[c_][p_,q_] := {c*E^(p - q)*Cos[p + q],
                          -c*E^(p - q)*Sin[p + q],c*(p - q)}
```

C.41 gencube

$\{u,v\} \longmapsto$ `gencube[n,a,b,c][u,v]` is a generalized cube.
$\{u,v\} \longmapsto$ `gencube[1,a/Sqrt[2],a/Sqrt[2],a][u,v]` is a sphere.
$\{u,v\} \longmapsto$ `gencube[2,a,a,a][u,v]` is a parametrization of an ordinary cube.
$\{u,v\} \longmapsto$ `genoctahedron[3,a,a,a][u,v]` is an astroidial cube.

```
gencube[n_,a_,b_,c_][u_,v_]:=
    {a*(Cos[u]*Sqrt[Cos[u]^2]^(n - 1) +
        Sin[u]*Sqrt[Sin[u]^2]^(n - 1))*
      (Cos[v]*Sqrt[Cos[v]^2]^(n - 1) +
        Sin[v]*Sqrt[Sin[v]^2]^(n - 1)),
     b*(-Cos[u]*Sqrt[Cos[u]^2]^(n - 1) +
        Sin[u]*Sqrt[Sin[u]^2]^(n - 1))*
      (Cos[v]*Sqrt[Cos[v]^2]^(n - 1) +
        Sin[v]*Sqrt[Sin[v]^2]^(n - 1)),
     c*(-Cos[v]*Sqrt[Cos[v]^2]^(n - 1) +
        Sin[v]*Sqrt[Sin[v]^2]^(n - 1))}
```

C.42 genhypparab

{u,v} ⟼ genhypparab[a,b][u,v] is a parametrization of a hyperbolic paraboloid as a ruled surface.

```
genhypparab[a_,b_][u_,v_] := {a*(u + v),b*v,u^2 + 2*u*v}
```

C.43 genoctahedron

{u,v} ⟼ genoctahedron[n,a,b,c][u,v] is a generalized octahedron.
{u,v} ⟼ genoctahedron[1,a,b,c][u,v] is an ellipsoid.
{u,v} ⟼ genoctahedron[2,a,a,a][u,v] is a parametrization of an ordinary octahedron. {u,v} ⟼ genoctahedron[3,a,b,c][u,v] is an astroidial ellipsoid.

```
genoctahedron[n_,a_,b_,c_][u_,v_]:=
    {a*Cos[u]*Cos[v]*Sqrt[Cos[u]^2*Cos[v]^2]^(n - 1),
     a*Sin[u]*Cos[v]*Sqrt[Sin[u]^2*Cos[v]^2]^(n - 1),
     a*Sin[v]*Sqrt[Sin[v]^2]^(n - 1)}
```

C.44 handkerchief

{u,v} ⟼ handkerchief[a][u,v] is a parametrization of a handkerchief-shaped surface.

Appendix C: Surfaces 949

```
handkerchief[a_][u_,v_]:= {u,v,
                (1/3)*u^3 + u*v^2 +a*(u^2 - v^2)}
```

C.45 heart

{u,v}⟼heart[a][u,v] is a surface of revolution generated by t⟼cardioid[a][t].

```
heart[a_][u_,v_]:= {2*a*Cos[u]*(1 + Cos[v])*Sin[v],
                2*a*(1 + Cos[v])*Sin[u]*Sin[v],
                -2*a*Cos[v]*(1 + Cos[v])}
```

C.46 helicoid

{u,v}⟼helicoid[a,c][u,v] is the standard parametrization of a circular helicoid of radius **a** and slant **c**.

```
helicoid[a_,c_][u_,v_]:= {a*v*Cos[u],a*v*Sin[u],c*u}
```

{u,v}⟼helicoid[a,b,c][u,v] is the standard parametrization of an elliptical helicoid of slant **c**.

```
helicoid[a_,b_,c_][u_,v_]:= {a*v*Cos[u],b*v*Sin[u],c*u}
```

C.47 helicoidprin

{p,q}⟼helicoidprin[b][p,q] is a parametrization of a circular helicoid of radius **b** and slant **b** by principal curves. The parametrization is also isothermal.

```
helicoidprin[b_][p_,q_]:=  {b*Cos[p + q]*Sinh[-p + q],
       b*Sin[p + q]*Sinh[-p + q],b*(p + q)}
```

C.48 heltocat

{u,v}⟼heltocat[t][u,v] is a 1-parameter family of minimal surfaces connecting a helicoid to a catenoid.

```
heltocat[t_][u_,v_]:=
    {Cos[u]*Cosh[v]*Sin[t] + Cos[t]*Sin[u]*Sinh[v],
     Cosh[v]*Sin[t]*Sin[u] - Cos[t]*Cos[u]*Sinh[v],
     u*Cos[t] + v*Sin[t]}
```

C.49 henneberg

{u,v} ⟼ henneberg[u,v] is a parametrization of Henneberg's minimal surface.

```
henneberg[u_,v_]:=
    {2*Sinh[u]*Cos[v] - (2/3)*Sinh[3*u]*Cos[3*v],
     2*Sinh[u]*Sin[v] + (2/3)*Sinh[3*u]*Sin[3*v],
     2*Cosh[2*u]*Cos[2*v]}
```

C.50 hyperbolicparaboloid

{u,v} ⟼ hyperbolicparaboloid[u,v] is a parametrization of a hyperbolic paraboloid.

```
hyperbolicparaboloid[u_,v_]:= {u,v,u*v}
```

C.51 hyperboloid

{u,v} ⟼ hyperboloid[a,c][u,v] is the standard parametrization of a circular hyperboloid of one sheet.

```
hyperboloid[a_,c_][u_,v_]:= {a*Cosh[v]*Cos[u],
                             a*Cosh[v]*Sin[u],
                             c*Sinh[v]}
```

{u,v} ⟼ hyperboloid[a,b,c][u,v] is the standard parametrization of an elliptical hyperboloid of one sheet.

```
hyperboloid[a_,b_,c_][u_,v_]:= {a*Cosh[v]*Cos[u],
                                b*Cosh[v]*Sin[u],
                                c*Sinh[v]}
```

Appendix C: Surfaces

C.52 hy2sheet

{u,v} ⟼ hy2sheet[a,c][u,v] is the standard parametrization of a circular hyperboloid of two sheets.

```
hy2sheet[a_,c_][u_,v_]:= {a*Cosh[u]*Cosh[v],
                         a*Sinh[u]*Cosh[v],
                         c*Sinh[v]}
```

{u,v} ⟼ hy2sheet[a,b,c][u,v] is the standard parametrization of an elliptical hyperboloid of two sheets.

```
hy2sheet[a_,b_,c_][u_,v_]:= {a*Cosh[u]*Cosh[v],
                             b*Sinh[u]*Cosh[v],
                             c*Sinh[v]}
```

C.53 invertedcircularcone

{u,v} ⟼ invertedcircularcone[{p1,p2,p3},rho,a,b][u,v] is a parametrization of {u,v} ⟼ circularcone[a,b][u,v] inverted with respect to a sphere of radius rho centered at {p1,p2,p3}.

```
invertedcircularcone[{p1_,p2_,p3_},rho_,a_,b_][u_,v_]:=
    {p1 + rho^2*(-p1 + a*v*Cos[u])/
     ((-p3 + b*v)^2 + (-p1 + a*v*Cos[u])^2 +
     (-p2 + a*v*Sin[u])^2),
     p2 + rho^2*(-p2 + a*v*Sin[u])/
     ((-p3 + b*v)^2 + (-p1 + a*v*Cos[u])^2 +
     (-p2 + a*v*Sin[u])^2),
     p3 + rho^2*(-p3 + b*v)/
     ((-p3 + b*v)^2 + (-p1 + a*v*Cos[u])^2 +
     (-p2 + a*v*Sin[u])^2)}
```

{u,v} ⟼ invertedcircularcone[rho,a,b][u,v] is a parametrization of {u,v} ⟼ circularcone[a,b][u,v] with respect to a sphere of radius rho centered at the origin.

```
invertedcircularcone[rho_,a_,b_][u_,v_]:=
    {a*rho^2*Cos[u]/(a^2*v + b^2*v),
     a*rho^2*Sin[u]/(a^2*v + b^2*v),
     b*rho^2/(a^2*v + b^2*v)}
```

C.54 invertedcircularcylinder

{u,v}⟼invertedcircularcylinder[{p1,p2,p3},rho,a][u,v] is a parametrization of {u,v}⟼invertedcircularcylinder[a][u,v] inverted with respect to a sphere of radius **rho** centered at {p1,p2,p3}.

```
invertedcircularcylinder[{p1_,p2_,p3_},rho_,a_][u_,v_]:=
    {p1 + rho^2*(-p1 + a*Cos[u])/
     ((-p3 + v)^2 + (-p1 + a*Cos[u])^2 +
      (-p2 + a*Sin[u])^2),
     p2 + rho^2*(-p2 + a*Sin[u])/
     ((-p3 + v)^2 + (-p1 + a*Cos[u])^2 +
      (-p2 + a*Sin[u])^2),
     p3 + rho^2*(-p3 + v)/
     ((-p3 + v)^2 + (-p1 + a*Cos[u])^2 +
      (-p2 + a*Sin[u])^2)}
```

{u,v}⟼invertedcircularcylinder[rho,a][u,v] is a parametrization of {u,v}⟼circularcylinder[a][u,v] with respect to a sphere of radius **rho** centered at the origin.

```
invertedcircularcylinder[rho_,a_][u_,v_]:=
    {a*rho^2*Cos[u]/(a^2 + v^2),
     a*rho^2*Sin[u]/(a^2 + v^2),
     rho^2*v/(a^2 + v^2)}
```

C.55 invertedtorus

{u,v}⟼invertedtorus[{p1,p2,p3},rho,a,b][u,v] is a parametrization of {u,v}⟼torus[a,b][u,v] inverted with respect to a sphere of radius **rho** centered at {p1,p2,p3}.

Appendix C: Surfaces

```
invertedtorus[{p1_,p2_,p3_},rho_,a_,b_][u_,v_]:=
    {p1 + (rho^2*(-p1 + Cos[u]*(a + b*Cos[v])))/
    ((-p1 + Cos[u]*(a + b*Cos[v]))^2 +
    (-p2 + (a + b*Cos[v])*Sin[u])^2 + (-p3 + b*Sin[v])^2),
    p2 + (rho^2*(-p2 + (a + b*Cos[v])*Sin[u]))/
    ((-p1 + Cos[u]*(a + b*Cos[v]))^2 +
    (-p2 + (a + b*Cos[v])*Sin[u])^2 + (-p3 + b*Sin[v])^2),
    p3 + (rho^2*(-p3 + b*Sin[v]))/
    ((-p1 + Cos[u]*(a + b*Cos[v]))^2 +
    (-p2 + (a + b*Cos[v])*Sin[u])^2 + (-p3 + b*Sin[v])^2)}
```

{u,v}⟼invertedtorus[rho,a,b][u,v] is a parametrization of
{u,v}⟼torus[a,b][u,v] with respect to a sphere of radius **rho** centered at the origin.

```
invertedtorus[rho_,a_,b_][u_,v_]:=
    {rho^2*Cos[u]*(a + b*Cos[v])/(a^2 + b^2 + 2*a*b*Cos[v]),
    rho^2*(a + b*Cos[v])*Sin[u]/(a^2 + b^2 + 2*a*b*Cos[v]),
    b*rho^2*Sin[v]/(a^2 + b^2 + 2*a*b*Cos[v])}
```

C.56 kidney

{u,v}⟼kidney[a][u,v] is a surface of revolution generated by
t ⟼nephroid[a][t].

```
kidney[a_][u_,v_]:=
    {a*Cos[u]*(3*Cos[v] - Cos[3*v]),
    a*Sin[u]*(3*Cos[v] - Cos[3*v]),
    a*(3*Sin[v] - Sin[3*v])}
```

C.57 kleinbottle

{u,v}⟼kleinbottle[a][u,v] is a parametrization of the surface formed by moving and twisting a figure eight along a circle of radius **a**. The surface is nonorientable and a neighborhood of the self-intersection curve is nonorientable.

```
kleinbottle[a_][u_,v_]:=
    {(a + Cos[u/2]*Sin[v] - Sin[u/2]*Sin[2*v])*Cos[u],
     (a + Cos[u/2]*Sin[v] - Sin[u/2]*Sin[2*v])*Sin[u],
     Sin[u/2]*Sin[v] + Cos[u/2]*Sin[2*v]}
```

C.58 kleinbottlebis

$\{u,v\} \mapsto$ kleinbottlebis[e,f][u,v] is a Klein bottle in $\mathbb{R}^3$. A neighborhood of the self-intersection curve is orientable.

```
kleinbottlebis[e_,f_][u_,v_]:=If[u<0,
    {f*Sin[u]/2,(-e - f/2 + f*Cos[u]/2)*Cos[v],
     (e + f/2 - f*Cos[u]/2)*Sin[v]},
    {10*Sin[0.105*u] + 2*(e + 0.033*f*u)*
     (Cos[0.105*u] + 4*Cos[0.21*u])*Cos[v]*
     (Sin[0.105*u] + 2*Sin[0.21*u])/
     Sqrt[100*Cos[0.105*u]^2 + 4*(Cos[0.105*u] +
       4*Cos[0.21*u])^2*(Sin[0.105*u] + 2*Sin[0.21*u])^2],
     (Sin[0.105*u] + 2*Sin[0.21*u])^2 -
     (10*(e + 0.033*f*u)*Cos[0.105*u]*Cos[v])/
     Sqrt[100*Cos[0.105*u]^2 + 4*(Cos[0.105*u] +
       4*Cos[0.21*u])^2*(Sin[0.105*u] + 2*Sin[0.21*u])^2],
     (e + 0.033*f*u)*Sin[v]}]
```

C.59 klein4

$\{u,v\} \mapsto$ klein4[a,b][u,v] is a parametrization of a Klein bottle as a surface in $\mathbb{R}^4$.

```
klein4[a_,b_][u_,v_]:=
    {(a + b*Cos[v])*Cos[u],(a + b*Cos[v])*Sin[u],
     b*Sin[v]*Cos[u/2],b*Sin[v]*Sin[u/2]}
```

Appendix C: Surfaces

C.60 kuen

$\{u,v\} \longmapsto$ `kuen[a][u,v]` is a parametrization of Kuen's surface. It has constant negative curvature $-a^{-2}$.

```
kuen[a_][u_,v_]:=
    {2*a*(Cos[u] + u*Sin[u])*Sin[v]/(1 + u^2*Sin[v]^2),
     2*a*(Sin[u] - u*Cos[u])*Sin[v]/(1 + u^2*Sin[v]^2),
     a*(Log[Tan[v/2]] + 2*Cos[v]/(1 + u^2*Sin[v]^2))}
```

C.61 kuentcheb

$\{u,v\} \longmapsto$ `kuentcheb[a][u,v][u,v]` is a Tchebyshef principal patch and $\{p,q\} \longmapsto$ `kuentcheb[a][p + q,p - q]` is a Tchebyshef asymptotic patch for Kuen's surface of constant negative curvature $-a^{-2}$.

```
kuentcheb[a_][u_,v_]:=
    {2*a*Cosh[u]*(Cos[v] + v*Sin[v])/(v^2 + Cosh[u]^2),
     2*a*Cosh[u]*(-v*Cos[v] + Sin[v])/(v^2 + Cosh[u]^2),
     a*(u - Sinh[2*u])/(v^2 + Cosh[u]^2)}
```

C.62 lemniscatemin

$\{u,v\} \longmapsto$ `lemniscatemin[a][t][u,v]` is a 1-parameter family of minimal surfaces such that $\{u,v\} \longmapsto$ `lemniscatemin[a][0][u,v]` is a minimal surface containing a lemniscate as a geodesic.

```
lemniscatemin[a_][t_][u_,v_]:=
    {(-16*a*Cos[u]*Cosh[v]*Sin[u]*Sinh[v]*
    (Cos[u]*Cosh[v]*Sin[t] + Cos[t]*Sin[u]*Sinh[v]))/
    (18 + Cos[4*u] - 12*Cos[2*u]*Cosh[2*v] + Cosh[4*v]) +
    (4*a*(3 - Cos[2*u]*Cosh[2*v])*
    (Cos[t]*Cos[u]*Cosh[v] - Sin[t]*Sin[u]*Sinh[v]))/
    (18 + Cos[4*u] - 12*Cos[2*u]*Cosh[2*v] + Cosh[4*v]),
    (a*(-Sin[t - 4*u] + 6*Cosh[2*v]*Sin[t - 2*u] -
    6*Cosh[2*v]*Sin[t + 2*u] + Sin[t + 4*u] -
    6*Sin[t - 2*u]*Sinh[2*v] - 6*Sin[t + 2*u]*Sinh[2*v] +
    2*Sin[t]*Sinh[4*v]))/(2*(-18 - Cos[4*u] +
    12*Cos[2*u]*Cosh[2*v] - Cosh[4*v])),
    a*(-(Cos[t]*Im[EllipticF[u + I*v,-1]]) +
    Re[EllipticF[u + I*v,-1]]*Sin[t])}
```

C.63 menn

{u,v} ⟼ menn[u,v] is a parametrization of Menn's surface.

```
menn[a_][u_,v_]:= {u,v,a*u^4 + u^2*v - v^2}
```

C.64 mercatorellipsoid

{u,v} ⟼ mercatorellipsoid[a,b,c][u,v] is the Mercator injection of an ellipsoid with axes of lengths **a**, **b** and **c** into $\mathbb{R}^3$.

```
mercatorellipsoid[a_,b_,c_][u_,v_]:=
    {a*Sech[v]*Cos[u],b*Sech[v]*Sin[u],c*Tanh[v]}
```

C.65 moebcirc

{u,v} ⟼ moebcirc[a][u,v] is a parametrization of a Möbius strip that has a circle as boundary.

Appendix C: Surfaces

```
moebcirc[a_][u_,v_]:={a*(-2*Cos[2*v]*Sin[u])/
  (-2 + Sqrt[2]*Cos[u]*Sin[v] + Sqrt[2]*Sin[u]*Sin[2*v]),
   a*(Sqrt[2]*(Cos[u]*Sin[v] - Sin[u]*Sin[2*v]))/
  (-2 + Sqrt[2]*Cos[u]*Sin[v] + Sqrt[2]*Sin[u]*Sin[2*v]),
   a*(-2*Cos[u]*Cos[v])/
  (-2 + Sqrt[2]*Cos[u]*Sin[v] + Sqrt[2]*Sin[u]*Sin[2*v])}
```

C.66 moebiusstrip

{u,v} ⟼ moebiusstrip[a][u,v] is the standard parametrization of a Möbius strip.

```
moebiusstrip[a_][u_,v_]:= {a*(Cos[u] + v*Cos[u/2]*Cos[u]),
   a*(Sin[u] + v*Cos[u/2]*Sin[u]),a*v*Sin[u/2]}
```

C.67 monkeysaddle

{u,v} ⟼ monkeysaddle[u,v] is a parametrization of a monkey saddle.

```
monkeysaddle[u_,v_]:= {u,v,u^3 - 3*u*v^2}
```

{u,v} ⟼ monkeysaddle[n][u,v] is a parametrization of a monkey saddle for a monkey with $n-2$ tails.

```
monkeysaddle[n_][u_,v_]:={u,v,
   Simplify[ComplexExpand[Re[(u + I*v)^n],
     TargetFunctions->{Re,Im}]]}
```

C.68 monkeysaddlepolar

{r,theta} ⟼ monkeysaddlepolar[n][r,theta] is a polar parametrization of a monkey saddle with $n-2$ tails.

```
monkeysaddlepolar[n_][r_,theta_]:=
   {r*Cos[theta],r*Sin[theta],r^n*Sin[n*theta]}
```

C.69 nodoid

{u,v} ⟼ nodoidnds[vmin,vmax][a,b,c,e][u,v] is an approximation over the interval vmin<v<vmax found by NDSolve to a nodoid. It is a surface of revolution of a curve of Delaunay of eccentricity e and has constant mean curvature.

```
nodoidnds[vmin_,vmax_][a_,b_,c_,e_][u_,v_]:=
    Module[{tmpintr},
    tmpintr= intrinsic[delaunaykappa2[c,e],0,{0,0,-Pi/2},
        MaxSteps->2000,{vmin,vmax}][v];
    {Cos[u]*(a + b*tmpintr[[1]]), Sin[u]*(a + b*tmpintr[[1]]),
        b*tmpintr[[2]]}]
```

C.70 paraboliccoor

{u,v,w} ⟼ paraboliccoor[d,e,f][a,b,c][w][u,v] is a parabolic coordinate patch on $\mathbb{R}^3$. The parameters d, e, f are +1 or −1.

```
paraboliccoor[d_,e_,f_][a_,b_,c_][w_][u_,v_]:=
    {d*Sqrt[(a - u)*(a - v)*(a - w)/(b - a)],
    e*Sqrt[(b - u)*(b - v)*(b - w)/(a - b)],
    f*(c - a - b + u + v + w)/2}
```

C.71 paraboliccyclide

{u,v} ⟼ paraboliccyclide[a,k][u,v] is a parametrization of the cyclide of Dupin of radius k whose focal sets are the parabolas $u \longmapsto (u, 0, -u^2/(8a) + a)$ and $v \longmapsto (0, v, v^2/(8a) - a)$.

```
paraboliccyclide[a_,k_][u_,v_]:=
    {(u*(8*a^2 + k + v^2))/(16*a^2 + u^2 + v^2),
    (8*a^2 - k + u^2)*v/(16*a^2 + u^2 + v^2),
    (16*a^2*k - 16*a^2*u^2 - k*u^2 + 16*a^2*v^2 - k*v^2)/
    (8*a*(16*a^2 + u^2 + v^2))}
```

Appendix C: Surfaces

C.72 paraboloid ellipticparaboloid

{u,v} ⟼ ellipticparaboloid[a,b][u,v] is the surface
{u,v} ⟼ {u,v,u^2/a^2 + v^2/b^2}.

```
ellipticparaboloid[a_,b_][u_,v_]:=
                {u,v,u^2/a^2 + v^2/b^2}
```

{u,v} ⟼ paraboloid[a][u,v] is a circular paraboloid of radius **a**.

```
paraboloid[a_][u_,v_]:= {u,v,a*(u^2 + v^2)}
```

{u,v} ⟼ paraboloid[a,b][u,v] and {u,v} ⟼ paraboloid[a,b,c][u,v] is either an elliptic or a hyperbolic paraboloid.

```
paraboloid[a_,b_][u_,v_]:= {u,v,a*u^2 + b*v^2}
```

```
paraboloid[a_,b_,c_][u_,v_]:= {u,v,a*u^2 + b*v^2 + c*u*v}
```

C.73 paraboloidpolar

{r,theta} ⟼ paraboloidpolar[n,a,b][r,theta] is a polar parametrization of a generalized paraboloid..

```
paraboloidpolar[n_,a_,b_][u_,v_]:=
                a*r*Cos[theta],a*r*Cos[theta],b*r^n}
```

C.74 pear

{u,v} ⟼ pear[a,b][u,v] is a surface of revolution generated by
t ⟼ piriform[a,b][t] with the axis of revolution passing through the cusp.

```
pear[a_,b_][u_,v_]:= {b*Cos[u]*Cos[v]*(1 + Sin[v]),
                     b*Cos[v]*Sin[u]*(1 + Sin[v]),
                     -a*(1 + Sin[v])}
```

C.75 perturbedms

$\{u,v\} \mapsto$ `perturbedms[a][u,v]` is a parametrization of a monkey saddle perturbed by a circular paraboloid.

```
perturbedms[a_][u_,v_]:=
    {u,v,u^3 - 3*u*v^2 + a*(u^2 + v^2)}
```

C.76 pillow

$\{u,v\} \mapsto$ `pillow[a][u,v]` is a pillow-shaped surface.

```
pillow[a_][u_,v_]:= {a*Cos[u],a*Cos[v],a*Sin[u]*Sin[v]}
```

C.77 plane

$\{u,v\} \mapsto$ `plane[a1,a2,b1,b2,c2,c3][u,v]` is a parametrization of a plane in $\mathbb{R}^3$.

```
plane[a1_,a2_,b1_,b2_,c1_,c2_][u_,v_]:=
    {a1*u + a2*v,b1*u + b2*v,c1*u + c2*v}
```

C.78 plucker

$\{u,v\} \mapsto$ `plucker[u,v]` is the standard parametrization of Plucker's surface.

```
plucker[u_,v_]:= {u,v,2*u*v/(u^2 + v^2)}
```

$\{u,v\} \mapsto$ `plucker[n,u,v]` is a parametrization of Plucker's surface with n folds.

```
plucker[n_][u_,v_]:= {u,v,Sin[n*ArcTan[u,v]]}
```

$\{u,v\} \mapsto$ `plucker[m,n,a][u,v]` is a generalization of the monkey saddle and Plucker's surface.

```
plucker[m_,n_,a_][u_,v_]:= {a*u,a*v,
        a*(u + v)^(m/2)*Sin[n*ArcTan[u,v]]}
```

C.79 pluckerpolar

{r,theta} ⟼ pluckerpolar[n][r,theta] is polar parametrization of Plucker's surface with n folds.

```
pluckerpolar[n_][r_,theta_]:=
            {r*Cos[theta],r*Sin[theta],Sin[n*theta]}
```

{r,theta} ⟼ pluckerpolar[m,n,a][r,theta] is a generalization of the polar parametrizations of the monkey saddle and Plucker's surface.

```
pluckerpolar[m_,n_,a_][r_,theta_]:=
       {a*r*Cos[theta],a*r*Sin[theta],a*r^m*Sin[n*theta]}
```

C.80 polarplane

{r,theta} ⟼ polarplane[r,theta] is the polar parametrization of the plane.

```
polarplane[u_,v_]:= {v*Cos[u],v*Sin[u],0}
```

C.81 pseudocrosscap

{u,v} ⟼ pseudocrosscap[u,v] is a parametrization of a pseudocross cap in $\mathbb{R}^3$.

```
pseudocrosscap[u_,v_]:=
            {(1 - u^2)*Sin[v],(1 - u^2)*Sin[2*v],u}
```

C.82 pseudosphere

{u,v} ⟼ pseudosphere[a][u,v] is the standard parametrization of a pseudosphere of constant negative curvature $-a^{-2}$.

```
pseudosphere[a_][u_,v_]:= {a*Cos[u]*Sin[v],
                           a*Sin[u]*Sin[v],
                           a*(Cos[v] + Log[Tan[v/2]])}
```

C.83 pseudospherebis

{u,v}⟼pseudospherebis[a][u,v] is a parametrization of a pseudosphere of constant negative curvature $-a^{-2}$.

```
pseudospherebis[a_][u_,v_] := {a*Cos[u]*Tanh[v],
                               a*Sin[u]*Tanh[v],
                               a*(Sech[v] + Log[Tanh[v/2]])}
```

C.84 pseudospheretcheb

{u,v}⟼pseudospheretcheb[a][u,v] is a Tchebyshef principal patch and {p,q}⟼pseudospheretcheb[a][p + q,p - q] is a Tchebyshef asymptotic patch for a pseudosphere of constant negative curvature $-a^{-2}$.

```
pseudospheretcheb[a_][u_,v_] := {a*Cos[v]/Cosh[u],
                                 a*Sin[v]/Cosh[u],
                                 a*(u - Tanh[u])}
```

C.85 rconoid

{u,v}⟼rconoid[u,v] is a right conoid. It coincides with
{u,v}⟼2*plucker[1][v/2,u].

```
rconoid[u_,v_] := {v*Cos[u],v*Sin[u],2*Sin[u]}
```

C.86 richmond

{u,v}⟼richmond[u,v] is a minimal surface with one planar end.

```
richmond[u_,v_] :=
    {(-3*u - u^5 + 2*u^3*v^2 + 3*u*v^4)/(6*(u^2 + v^2)),
     (-3*v - 3*u^4*v - 2*u^2*v^3 + v^5)/(6*(u^2 + v^2)),u}
```

C.87 richmondpolar

{r,theta}⟼richmondpolar[n][t][r,theta] is the polar parametrization of a 1-parameter family of minimal surfaces such that
{r,theta}⟼richmondpolar[n][0][r,theta] is a minimal surface with one planar end of degree **n**.

```
richmondpolar[n_][t_][r_,theta_]:=
    {-Cos[t + theta]/(2*r) -
        r^(2*n + 1)*Cos[t - (2*n + 1)*theta]/(4*n + 2),
    -Sin[t + theta]/(2*r) +
        r^(2*n + 1)*Sin[t - (2*n + 1)*theta]/(4*n + 2),
    r^n*Cos[t - n*theta]/n}
```

C.88 roman

{u,v}⟼roman[a][u,v] is a parametrization of Steiner's Roman surface of radius **a** in $\mathbb{R}^3$.

```
roman[a_][u_,v_]:= {(a^2/2)*Sin[2*u]*Cos[v]^2,
                    (a^2/2)*Sin[u]*Sin[2*v],
                    (a^2/2)*Cos[u]*Sin[2*v]}
```

C.89 rosemin

{u,v}⟼rosemin[n,a][t][u,v] is a 1-parameter family of minimal surfaces such that {u,v}⟼rosemin[n,a][0][u,v] contains the rose
t⟼$(a\cos(t)\cos(nt), a\cos(nt)\sin(t))$ as a geodesic.

```
rosemin[n_,a_][t_][u_,v_]:=
{Cos[n*u]*Cosh[n*v]*(a*Cos[t]*Cos[u]*Cosh[v] -
   a*Sin[t]*Sin[u]*Sinh[v]) +
   Sin[n*u]*(-a*Cos[u]*Cosh[v]*Sin[t] -
   a*Cos[t]*Sin[u]*Sinh[v])*Sinh[n*v],
  -Cos[u]*Sinh[v]*(-a*Cos[n*u]*Cosh[n*v]*Sin[t] -
   a*Cos[t]*Sin[n*u]*Sinh[n*v]) +
   Cosh[v]*Sin[u]*(a*Cos[t]*Cos[n*u]*Cosh[n*v] -
   a*Sin[t]*Sin[n*u]*Sinh[n*v]),
  -a*Cos[t]*Im[EllipticE[n*(u + I*v),1 - n^2]]/n +
   a*Re[EllipticE[n*(u + I*v),1 - n^2]]*Sin[t]/n}
```

C.90 scherk

{u,v} ⟼ scherk[a][u,v] is a parametrization of Scherk's minimal surface.

```
scherk[a_][u_,v_]:= {u,v,(1/a)*Log[Cos[a*v]/Cos[a*u]]}
```

{u,v} ⟼ scherk[a,theta][u,v] is a parametrization of twisted Scherk's minimal surface of angle **theta**.

```
scherk[a_,theta_][u_,v_]:=
    {Sec[2*theta]*(Cos[theta]*u + Sin[theta]*v),
     Sec[2*theta]*(Sin[theta]*u + Cos[theta]*v),
     (1/a)*Log[
     Cos[a*Sec[2*theta]*(Sin[theta]*u + Cos[theta]*v)]/
     Cos[a*Sec[2*theta]*(Cos[theta]*u + Sin[theta]*v)]]}
```

C.91 scherk5

{u,v} ⟼ scherk5[a,b,c][u,v] is a parametrization of Scherk's fifth minimal surface, where the parameters **a, b, c** are +1 or −1.

```
scherk5[a_,b_,c_][u_,v_]:= {a*ArcSinh[u],
                            b*ArcSinh[v],
                            c*ArcSin[u*v]}
```

Appendix C: Surfaces 965

C.92 shoe

{u,v} ⟼ shoe[u,v] is a shoe-shaped surface.

```
shoe[u_,v_]:= {u,v,(1/3)*u^3 - (1/2)*v^2}
```

C.93 sievert

{u,v} ⟼ sievert[a][u,v] is a parametrization of Sievert's surface of constant positive curvature a^2.

```
sievert[a_][u_,v_]:=
  {2/(a + 1 - a*Sin[v]^2*Cos[u]^2)*
    (Sqrt[(a + 1)*(1 + a*Sin[u]^2)]*Sin[v]/Sqrt[a])*
    Cos[-u/Sqrt[a + 1] + ArcTan[Sqrt[a + 1]*Tan[u]]],
   2/(a + 1 - a*Sin[v]^2*Cos[u]^2)*
    (Sqrt[(a + 1)*(1 + a*Sin[u]^2)]*Sin[v]/Sqrt[a])*
    Sin[-u/Sqrt[a + 1] + ArcTan[Sqrt[a + 1]*Tan[u]]],
   (Log[Tan[v/2]] +
    2*(a + 1)/(a + 1 - a*Sin[v]^2*Cos[u]^2)*
    Cos[v])/Sqrt[a]}
```

C.94 sinovaloid

{u,v} ⟼ sinovaloid[a][u,v] is the surface of revolution generated by
t ⟼ sinoval[n,a][t].

```
sinovaloid[n_,a_][u_,v_]:= {a*Cos[u]*Cos[v],
                            a*Sin[u]*Cos[v],
                            a*Nest[Sin,v,n]}
```

C.95 sinsurface

{u,v} ⟼ sinsurface[a][u,v] is the surface
{u,v} ⟼ a*{Sin[u],Sin[v],Sin[u + v]}. The implicit equation is
$4x^2y^2(a^2 - z^2) = a^2(x^2 + y^2 - z^2)^2$.

```
sinsurface[a_][u_,v_] := {a*Sin[u],a*Sin[v],a*Sin[u + v]}
```

C.96 sphere

{u,v} ⟼ sphere[a][u,v] is the standard parametrization of a sphere of radius **a** centered at the origin.

```
sphere[a_][u_,v_] := {a*Cos[v]*Cos[u],a*Cos[v]*Sin[u],a*Sin[v]}
```

C.97 stereographicellipsoid

{u,v} ⟼ stereographicellipsoid[a,b,c][u,v] is a stereographic parametrization of an ellipsoid with axes of lengths **a**, **b** and **c**.

```
stereographicellipsoid[a_,b_,c_][u_,v_] :=
    {2*a*u/(u^2 + v^2 + 1),2*b*v/(u^2 + v^2 + 1),
    c*(u^2 + v^2 - 1)/(u^2 + v^2 + 1)}
```

C.98 stereographicsphere

{u,v} ⟼ stereographicsphere[a][u,v] is a stereographic parametrization of a sphere of radius **a**.

```
stereographicsphere[a_][u_,v_] :=
    {2*a*u/(u^2 + v^2 + 1),2*a*v/(u^2 + v^2 + 1),
    a*(u^2 + v^2 - 1)/(u^2 + v^2 + 1)}
```

C.99 stereographicspherepolar

{r,theta} ⟼ stereographicsphere[a][r,theta] is a polar stereographic parametrization of a sphere of radius **a**.

```
stereographicspherepolar[a_][r_,theta_] :=
    {2*a*r*Cos[theta]/(1 + r^2),2*a*r*Sin[theta]/(1 + r^2),
    2*a*(-1 + r^2)/(1 + r^2)}
```

Appendix C: Surfaces 967

C.100 swallowtail

{u,v} ⟼ swallowtail[u,v] is a parametrization of a swallow tail-shaped surface.

```
swallowtail[u_,v_]:={3*v^4 + u*v^2,-4*v^3 - 2*u*v,u}
```

C.101 tetrahedral

{u,v} ⟼ tetrahedral[m,n][A,B,C][a,b,c][u,v] is a parametrization of a tetrahedral surface.

```
tetrahedral[m_,n_][A_,B_,C_][a_,b_,c_][u_,v_]:=
        {A*(u - a)^m*(v - a)^n,
         B*(u - b)^m*(v - b)^n,
         C*(u - c)^m*(v - c)^n}
```

C.102 thomsen

{u,v} ⟼ thomsen[a,b][u,v] is a parametrization of Thomsen's minimal surface.

```
thomsen[a_,b_][u_,v_]:=
    {b*u/a + Sqrt[1+b^2]*Sinh[a*u]*Cos[a*v]/a^2,
     Sqrt[1 + b^2]*v/a + b*Cosh[a*u]*Sin[a*v]/a^2,
     Sinh[a*u]*Sin[a*v]/a^2}
```

C.103 torus

{u,v} ⟼ torus[a,b][u,v] is a parametrization of the torus formed by revolving a circle of radius **b** in the xz-plane about the z-axis along a circle of radius **a** in the xy-plane.

```
torus[a_,b_][u_,v_]:= {(a + b*Cos[v])*Cos[u],
                       (a + b*Cos[v])*Sin[u],
                       b*Sin[v]}
```

$\{u,v\} \mapsto$ **torus[a,b,c][u,v]** is a parametrization of the torus formed by revolving an ellipse with axes of lengths **b** and **c** in the xz-plane about the z-axis along a circle of radius **a** in the xy-plane.

```
torus[a_,b_,c_][u_,v_]:= {(a + b*Cos[v])*Cos[u],
                         (a + b*Cos[v])*Sin[u],
                         c*Sin[v]}
```

C.104 twiflat

$\{u,v\} \mapsto$ **twiflat[pm,a,b][sl][u,v]** is a parametrization of a generalized helicoid of zero Gaussian curvature of slant **sl**. The parameter **pm** is $+1$ or -1.

```
twiflat[pm_,a_,c_][sl_][u_,v_]:= {v*Cos[u],v*Sin[u],
    sl*u + ArcSin[c/(v*a)]*c + Sqrt[v^2*a^2 - c^2]*pm}
```

C.105 twisphere

$\{u,v\} \mapsto$ **twisphere[a,b][u,v]** is a parametrization of a generalized helicoid formed by rotating and twisting a semicircle. It is also called a corkscrew surface.

```
twisphere[a_,b_][u_,v_]:= {a*Cos[u]*Cos[v],
                           a*Sin[u]*Cos[v],
                           a*Sin[v] + b*u}
```

C.106 veronese

$\{u,v\} \mapsto$ **veronese[a][u,v]** is a parametrization of a Veronese surface in $\mathbb{R}^5$.

```
veronese[a_][u_,v_]:= a*{u,v,u^2,u*v,v^2}
```

C.107 wallis

$\{u,v\} \mapsto$ **wallis[a,b,c][u,v]** is a parametrization of a conical edge of Wallis.

Appendix C: Surfaces 969

```
wallis[a_,b_,c_][u_,v_]:= {v*Cos[u],v*Sin[u],
                          c*Sqrt[a^2 - b^2*Cos[u]^2]}
```

C.108 whitneyumbrella

{u,v} ⟼ whitneyumbrella[u,v] is a parametrization of a Whitney umbrella.

```
whitneyumbrella[u_,v_]:= {u*v,u,v^2}
```

C.109 wrinkles

{u,v} ⟼ wrinkles[a][u,v] is the Monge patch

$$(u,v) \longrightarrow \left(u, v, a\left(\frac{u}{v} + \frac{v}{u}\right)\right).$$

```
wrinkles[a_][u_,v_]:= {u,v,a*(u/v + v/u)}
```

Implicitly Defined Surfaces

C.110 cossurfaceimplicit

cossurfaceimplicit[a][x,y,z]==0 is the nonparametric form of the surface {u,v} ⟼ {a*Cos[u],a*Cos[v],a*Cos[u + v]}.

```
cossurfaceimplicit[a_][x_,y_,z_]:=
        (z - a*x*y)^2 - a^2(1 - x^2)(1 - y^2)
```

C.111 crosscapimplicit

crosscapimplicit[a,b][x,y,z]==0 is the nonparametric form of a crosscap.

```
crosscapimplicit[a_,b_][x_,y_,z_]:=
    (a*x^2 + b*y^2)(x^2 + y^2 + z^2) - 2*z*(x^2 + y^2)
```

C.112 ellipsoidimplicit

ellipsoidimplicit[a,b,c][x,y,z]==0 is the nonparametric form of an ellipsoid.

```
ellipsoidimplicit[a_,b_,c_][x_,y_,z_]:=
        (x/a)^2 + (y/b)^2 + (z/c)^2 - 1
```

C.113 ellipticparaboloidimplicit

ellipticparaboloidimplicit[a,b,c][x,y,z]==0 is the nonparametric form of an elliptic paraboloid.

```
ellipticparaboloidimplicit[a_,b_,c_][x_,y_,z_]:=
        (x/a)^2 + (y/b)^2 - c*z
```

C.114 equihom1implicit

equihom1implicit[x,y,z]==0 is the nonparametric form of of the first equi-affinely homogeneous surface.

```
equihom1implicit[x_,y_,z_]:= x*y*z - 1
```

C.115 equihom2implicit

equihom2implicit[x,y,z]==0 is the nonparametric form of of the second equi-affinely homogeneous surface.

```
equihom2implicit[x_,y_,z_]:= (x^2 + y^2)*z - 1
```

C.116 equihom3implicit

equihom3implicit[x,y,z]==0 is the nonparametric form of of the third equi-affinely homogeneous surface.

```
equihom3implicit[x_,y_,z_]:= x^2*(z - y^2)^3 - 1
```

Appendix C: Surfaces 971

C.117 `equihom4implicit`

`equihom4implicit[x,y,z]==0` is the nonparametric form of of the fourth equi-affinely homogeneous surface.

```
equihom4implicit[x_,y_,z_]:= x^2*(z - y^2)^3 + 1
```

C.118 `equihom5implicit`

`equihom5implicit[x,y,z]==0` is the nonparametric form of of the fifth equi-affinely homogeneous surface.

```
equihom5implicit[x_,y_,z_]:= z - x*y - x^3/3
```

C.119 `equihom6implicit`

`equihom6implicit[x,y,z]==0` is the nonparametric form of of the sixth equi-affinely homogeneous surface.

```
equihom6implicit[x_,y_,z_]:= z - x*y + Log[x]
```

C.120 `goursatimplicit`

`goursatimplicit[a,b,c][x,y,z]==0` is the nonparametric form of Goursat's surface.

```
goursatimplicit[a_,b_,c_][x_,y_,z_]:= x^4 + y^4 + z^4 +
        a*(x^2 + y^2 + z^2)^2 + b*(x^2 + y^2 + z^2) + c
```

C.121 `hyperbolicparaboloidimplicit`

`hyperbolicparaboloidimplicit[a,b,c][x,y,z]==0` is the nonparametric form of a hyperbolic paraboloid.

```
hyperbolicparaboloidimplicit[a_,b_,c_][x_,y_,z_]:=
        (x/a)^2 - (y/b)^2 - c*z
```

C.122 `hyperboloidimplicit`

`hyperboloidimplicit[a,b,c][x,y,z]==0` is the nonparametric form of a hyperboloid of one sheet.

```
hyperboloidimplicit[a_,b_,c_][x_,y_,z_]:=
        (x/a)^2 + (y/b)^2 - (z/c)^2 - 1
```

C.123 `hy2sheetimplicit`

`hy2sheetimplicit[a,b,c][x,y,z]==0` is the nonparametric form of a hyperboloid of two sheets.

```
hy2sheetimplicit[a_,b_,c_][x_,y_,z_]:=
        (x/a)^2 - (y/b)^2 - (z/c)^2 - 1
```

C.124 `kummerimplicit`

`kummerimplicit[x,y,z]==0` is the nonparametric form of Kummer's surface.

```
kummerimplicit[x_,y_,z_]:= x^4 + y^4 + z^4 -
        (y^2*z^2 + z^2*x^2 + x^2*y^2) - (x^2 + y^2 + z^2) + 1
```

C.125 `scherkimplicit`

`scherkimplicit[a][x,y,z]==0` is the nonparametric form of Scherk's minimal surface.

```
scherkimplicit[a_][x_,y_,z_]:= E^(a*z)*Cos[a*x] - Cos[a*y]
```

`scherkimplicit[a,phi][x,y,z]==0` is the nonparametric form of a twisted Scherk's minimal surface.

```
scherkimplicit[a_,phi_][x_,y_,z_]:=
        Tan[x/(a*Cos[phi])]*Tan[y/(a*Sin[phi])] - Tanh[z/a]
```

C.126 scherk5implicit

scherkimplicit[a][x,y,z]==0 is the nonparametric form of a Scherk's fifth minimal surface.

```
scherk5implicit[x_,y_,z_]:=
        Sinh[x]*Sinh[y] - Sin[z]
```

C.127 sinsurfaceimplicit

sinsurfaceimplicit[a][x,y,z]==0 is the nonparametric form of the surface $\{u,v\} \mapsto \{a*Sin[u], a*Sin[v], a*Sin[u + v]\}$.

```
sinsurfaceimplicit[a_][x_,y_,z_]:=
        4x^2y^2(a^2 - z^2) - a^2(x^2+y^2 - z^2)^2
```

C.128 sphereimplicit

sphereimplicit[a][x,y,z]==0 is the nonparametric form of a sphere.

```
sphereimplicit[a_][x_,y_,z_]:=
        x^2 + y^2 + z^2 - a^2
```

sphereimplicit[n,a][x,y,z]==0 is the nonparametric form of the surface $x^{2n} + y^{2n} + z^{2n} = a^{2n}$.

```
sphereimplicit[n_,a_][x_,y_,z_]:=
        x^(2*n) + y^(2*n) + z^(2*n) - a^(2*n)
```

C.129 torusimplicit

torusimplicit[a,b][x,y,z]==0 is the nonparametric form of a torus.

```
torusimplicit[a_,b_][x_,y_,z_]:=
        z^2 + (Sqrt[x^2 + y^2] - a)^2 - b^2
```

C.130 twocuspsimplicit

`twocuspsimplicit[x,y,z]==0` is the nonparametric form of a surface with two cusps.

```
twocuspsimplicit[x_,y_,z_]:= (z - 1)^2*(x^2 - z^2) -
    (x^2 - z)^2 - y^4 - y^2*(2x^2 + z^2 + 2*z - 1)
```

Drum Plots

C.131 circularmembrane

`{r,theta}⟼circularmembrane[m,n,t][r,theta]` is a parametrization of a circular membrane of degree (m,n) at time t.

```
circularmembrane[m_,n_,t_][r_,theta_]:=
    Flatten[{r*Cos[theta],r*Sin[theta],
        BesselJ[m,r*BesselJZeros[m,{n,n}]]*
        Cos[m*theta]*
        Cos[t*BesselJZeros[m,{n,n}]]}]
```

`circularmembraneplot[m,n,t]` plots a circular membrane of degree (m,n) at time t.

```
circularmembraneplot[m_,n_,t_,opts___]:=
    Module[{zzz},
        zzz=circularmembrane[m,n,t][r,theta];
        ParametricPlot3D[Evaluate[zzz],
            {r,0,1},{theta,0,2Pi},opts]]
```

`circularmembranemovie[m,n,t,tmin,tmax,numf]` animates a circular membrane of degree (m,n) at time t.

Appendix C: Surfaces 975

```
circularmembranemovie[m_,n_,{t_,tmin_,tmax_,numf_},
    opts___]:=
    Do[Module[{zzz},
        zzz=circularmembrane[m,n,t][r,theta];
        ParametricPlot3D[Evaluate[zzz],
            {r,0,1},{theta,0,2Pi},opts]],
        {t,tmin,tmax,numf - 1}]
```

C.132 squaremembrane

{u,v}⟼squaremembrane[m,n,t][u,v] is a parametrization of a square membrane of degree (m, n) at time t.

```
squaremembrane[m_,n_,t_][u_,v_]:=
    {u,v,Sin[m*u]*Sin[n*v]*Cos[Sqrt[m^2 + n^2]*t]}
```

squaremembraneplot[m,n,t] plots a square membrane of degree (m, n) at time t.

```
squaremembraneplot[m_,n_,t_,opts___]:=
    ParametricPlot3D[Evaluate[
        squaremembrane[m,n,t][u,v]],
        {u,0,Pi},{v,0,Pi},opts]
```

squaremembranemovie[m,n,t,tmin,tmax,numf] animates a square membrane of degree (m, n) at time t.

```
squaremembranemovie[m_,n_,{t_,tmin_,tmax_,numf_},
    opts___]:=
    Do[ParametricPlot3D[Evaluate[
        squaremembrane[m,n,t][u,v]],
        {u,0,Pi},{v,0,Pi},opts],
        {t,tmin,tmax,numf - 1}]
```

Minimal Curves

C.133 astroidmincurve

z⟼astroidmincurve[a][z] is the minimal curve generated by the astroid t⟼astroid[a][t].

```
astroidmincurve[a_][z_] := {a*Cos[z]^3,a*Sin[z]^3,
                            (-3*I)/4*a*Cos[2*z]}
```

C.134 bourmincurve

z⟼bourmincurve[m][z] is the minimal curve corresponding to Bour's minimal surface of order m.

```
bourmincurve[m_][z_] :=
    {z^(m - 1)/(m - 1) - z^(m + 1)/(m + 1),
     I*(z^(m - 1)/(m - 1) + z^(m + 1)/(m + 1)),2*z^m/m}
```

z⟼bourmincurve[m,n][z] is the minimal curve corresponding to Bour's minimal surface of order (m, n). Note that bourmincurve[m]=bourmincurve[m,1] and enneper[n]=bourmincurve[2,n].

```
bourmincurve[m_,n_][z_] :=
    {z^(m - 1)/(m - 1) - z^(m + 2*n - 1)/(m + 2*n - 1),
     I*(z^(m - 1)/(m - 1) + z^(m + 2*n - 1)/(m + 2*n - 1)),
     2*z^(m + n - 1)/(m + n - 1)}
```

C.135 cardioidmincurve

z⟼cardioidmincurve[a][z] is the minimal curve generated by the cardioid t⟼cardioid[a][t].

```
cardioidmincurve[a_][z_] := {2*a*Cos[z]*(1 + Cos[z]),
                2*a*(1 + Cos[z])*Sin[z],8*I*a*Sin[z/2]}
```

Appendix C: Surfaces 977

C.136 catalanmincurve

z⟼catalanmincurve[a][z] is the minimal curve corresponding to Catalan's minimal surface. It contains a cycloid as a geodesic.

```
catalanmincurve[a_][z_]:= {a*z - a*Sin[z],a - a*Cos[z],
                                -4*I*a*Cos[z/2] + 4*I*a}
```

C.137 catenoidmincurve

z⟼catenoidmincurve[a][z] is the minimal curve corresponding to a catenoid. It contains a circle as a geodesic.

```
catenoidmincurve[a_][z_]:= {a*Cosh[z/a],z,-I*a*Sinh[z/a]}
```

C.138 cayleysexticmincurve

z⟼cayleysexticmincurve[a][z] is the minimal curve generated by Cayley's sextic t⟼cayleysextic[a][t].

```
cayleysexticmincurve[a_][z_]:=
    {4*a*Cos[z/2]^4*(-1 + 2*Cos[z]),
     4*a*Cos[z/2]^3*Sin[3*z/2],3*I*a*(z + Sin[z])}
```

C.139 circleinvolutemincurve

z⟼circleinvolutemincurve[n,a][z] is the minimal curve generated by the nth involute starting at $(a,0)$ of the circle t⟼{a*Cos[t],a*Sin[t]}.

```
circleinvolutemincurve[n_,a_][z_]:=
    Module[{tmp1,tmp2,tmp3},
        tmp1=E^(I*tt)Normal[Series[E^(-I*tt),{tt,0,n}]];
        tmp2=a*ComplexExpand[{Re[tmp1],Im[tmp1]}];
        tmp3=tmp2 /. tt->z;
        Append[tmp3,I*a*z^(n + 1)/(n + 1)!]]
```

C.140 `cissoidmincurve`

`z⟼cissoidmincurve[a][z]` is the minimal curve generated by the cissoid `t⟼cissoid[a][t]`.

```
cissoidmincurve[a_][z_]:= {2*a*z^2/(1 + z^2),
    2*a*z^3/(1 + z^2),2*I*a*(Sqrt[4 + z^2] -
    Sqrt[3]*ArcTanh[Sqrt[4 + z^2]/Sqrt[3]])}
```

C.141 `clothoidmincurve`

`z⟼clothoidmincurve[a][z]` is the minimal curve generated by the clothoid `t⟼clothoid[1,a][t]`.

```
clothoidmincurve[a_][z_]:=
    {a*Sqrt[Pi]*FresnelS[z/Sqrt[Pi]],
     a*Sqrt[Pi]*FresnelC[z/Sqrt[Pi]],I*a*z}
```

C.142 `cnchyprofilemincurve`

`z⟼cnchyprofilemincurve[pm,a,b][z]` is the minimal curve generated by the curve `t⟼cnchyprofile[pm,a,b][t]`. The parameter `pm` is plus 1 or −1.

```
cnchyprofilemincurve[pm_,a_,b_][z_]:=
    {pm*b*Cosh[z/a],-I*a*EllipticE[I*z/a,-b^2/a^2],I*z}
```

C.143 `costamincurve`

`z⟼costamincurve[z]` is the minimal curve that determines Costa's minimal surface.

Appendix C: Surfaces

```
costamincurve[z_]:=
  {(1/2)*(-WeierstrassZeta[z,189.07272,0] + Pi*(z - I) +
  (1 + I)*0.35889 +
  0.228473*(WeierstrassZeta[z - 1/2,189.07272,0] -
      WeierstrassZeta[z - I/2,189.07272,0])),
  (I/2)*(-WeierstrassZeta[z,189.07272,0] + Pi*(-z + 1) -
  (1 + I)*0.35889 -
  0.228473*(WeierstrassZeta[z - 1/2,189.07272,0] -
      WeierstrassZeta[z - I/2,189.07272,0])),
  Sqrt[2*Pi]/4*(Log[(WeierstrassP[z,189.07272,0] - 6.87519)/
      (WeierstrassP[z,189.07272,0] + 6.87519)] - Pi*I)}
```

C.144 cpcprofilemincurve

z⟼cpcprofilemincurve[pm,a,b][z] is the minimal curve generated by the curve t⟼cpcprofilemincurve[pm,a,b][t]. The parameter pm is plus 1 or −1.

```
cpcprofilemincurve[pm_,a_,b_][z_]:= {b*pm*Cos[z/a],
                    a*EllipticE[z/a,b^2/a^2],I*z}
```

C.145 deltoidinvolutemincurve

z⟼deltoidinvolutemincurve[a][z] is the minimal curve generated by the involute starting at $t = 0$ of the deltoid t⟼deltoid[a][t].

```
deltoidinvolutemincurve[a_][z_]:=
    {(a/3)*(8*Cos[z/2] + 2*Cos[z] - Cos[2*z]),
    (a/3)*(-8*Sin[z/2] + 2*Sin[z] + Sin[2*z]),
    (4*a*I/9)*(3*z - 2*Sin[3*z/2])}
```

C.146 deltoidmincurve

z⟼deltoidmincurve[a][z] is the minimal curve generated by the deltoid t⟼deltoid[a][t].

```
deltoidmincurve[a_][z_]:= {-a + 2*a*Cos[z]*(1 + Cos[z]),
        2*a*(1 - Cos[z])*Sin[z],-8*(I/3)*a*Cos[3*z/2]}
```

C.147 ellipsemincurve

z⟼ellipsemincurve[a,b][z] is the minimal curve generated by the ellipse t⟼ellipse[a,b][t].

```
ellipsemincurve[a_,b_][z_]:= {a*Cos[z],b*Sin[z],
                I*b*EllipticE[z,-a^2/b^2]}
```

C.148 ennepermincurve

z⟼ennepermincurve[n][z] is the minimal curve corresponding to Enneper's minimal surface of order n.

```
ennepermincurve[n_][z_]:= {z - z^(2*n + 1)/(2*n + 1),
    I*(z + z^(2*n + 1)/(2*n + 1)),2*z^(n + 1)/(n + 1)}
```

C.149 ennepercatenoidmincurve

z⟼ennepercatenoidmincurve[z] is the minimal curve whose Weierstrass representation is given by $f(z) = 1$ and $g(z) = 1/z + z^3$.

```
ennepercatenoidmincurve[z_]:= {z^(-1) + z - 2*z^3/3 - z^7/7,
        -I/z + I*z + 2*I/3*z^3 + I/7*z^7,z^4/2 + 2*Log[z]}
```

C.150 epicycloidmincurve

z⟼epicycloidmincurve[a,b][z] is the minimal curve generated by the epicycloid t⟼epicycloid[a,b][t].

```
epicycloidmincurve[a_,b_][z_]:=
        {(a + b)*Cos[z] - b*Cos[(a + b)*z/b],
         (a + b)*Sin[z] - b*Sin[(a + b)*z/b],
         -4*I*b*(a + b)*Cos[a*z/(2*b)]/a}
```

Appendix C: Surfaces

C.151 hennebergmincurve

z⟼hennebergmincurve[z] is the minimal curve corresponding to Henneberg's minimal surface.

```
hennebergmincurve[z_]:= {-8*Sinh[z]^3/3,
    8*I/3 - 2*I*Cosh[z] - 2*I/3*Cosh[3*z],2*Cosh[2*z]}
```

C.152 hyperbolamincurve

z⟼hyperbolamincurve[a,b][z] is the minimal curve generated by the hyperbola t⟼hyperbola[a,b][t].

```
hyperbolamincurve[a_,b_][z_]:= {a*Cos[z],b*Sin[z],
                               b*EllipticE[I*z + a^2/b^2]}
```

C.153 hypocycloidinvolutemincurve

z⟼hypocycloidinvolutemincurve[a,b][z] is the minimal curve generated by the involute starting at $t = 0$ of the hypocycloid t⟼hypocycloid[a,b][t].

```
hypocycloidinvolutemincurve[a_,b_][z_]:=
    {(a^2*Cos[z] - 3*a*b*Cos[z] + 2*b^2*Cos[z] -
      a*b*Cos[z - a*z/b] + 2*b^2*Cos[z - a*z/b] +
      4*a*b*Cos[z - a*z/(2*b)] - 4*b^2*Cos[z - a*z/(2*b)])/a,
     (a^2*Sin[z] - 3*a*b*Sin[z] + 2*b^2*Sin[z] -
      a*b*Sin[z - a*z/b] + 2*b^2*Sin[z - a*z/b] +
      4*a*b*Sin[z - a*z/(2*b)] - 4*b^2*Sin[z - a*z/(2*b)])/a,
     2*I*(a^2 - 3*a*b + 2*b^2)*(a*z - 2*b*Sin[a*z/(2*b)])/a^2}
```

C.154 hypocycloidmincurve

z⟼hypocycloidmincurve[a,b][z] is the minimal curve generated by the hypocycloid t⟼hypocycloid[a,b][t].

```
hypocycloidmincurve[a_,b_][z_]:=
    {(a - b)*Cos[z] + b*Cos[(a - b)*z/b],
     (a - b)*Sin[z] - b*Sin[(a - b)*z/b],
     -4*I*(a - b)*b*Cos[a*z/(2*b)]/a}
```

C.155 lemniscatemincurve

$z \mapsto$ `lemniscatemincurve[a][z]` is the minimal curve generated by the lemniscate $t \mapsto$ `lemniscate[a][t]`.

```
lemniscatemincurve[a_][z_]:=
    {(a*Cos[z])/(1 + Sin[z]^2),
     (a*Cos[z]*Sin[z])/(1 + Sin[z]^2),I*a*EllipticF[z,- 1]}
```

C.156 logspiralmincurve

$z \mapsto$ `logspiralmincurve[a,b][z]` is the minimal curve generated by the log spiral $t \mapsto$ `logspiral[a][t]`.

```
logspiralmincurve[a_,b_][z_]:= {a*E^(b*z)*Cos[z],
    a*E^(b*z)*Sin[z],(I*a*Sqrt[1 + b^2]*E^(b*z))/b}
```

C.157 nielsenspiralmincurve

$z \mapsto$ `nielsenspiralmincurve[a][z]` is the minimal curve generated by Nielsen's spiral $t \mapsto$ `nielsenspiral[a][t]`.

```
nielsenspiralmincurve[a_][z_]:=
    {a*CosIntegral[z],a*SinIntegral[z],I*a*Log[z]}
```

C.158 nephroidmincurve

$z \mapsto$ `nephroidmincurve[a][z]` is the minimal curve generated by the nephroid $t \mapsto$ `nephroid[a][t]`.

```
nephroidmincurve[a_][z_]:= {a*(3*Cos[z] - Cos[3*z]),
            a*(3*Sin[z] - Sin[3*z]),-6*I*a*Cos[z]}
```

Appendix C: Surfaces 983

C.159 parabolamincurve

z⟼**parabolamincurve[a][z]** is the minimal curve generated by the parabola t⟼**parabola[a][t]**.

```
parabolamincurve[a_][z_]:= {2*a*z,a*z^2,
    I*a*(z*Sqrt[1 + z^2] + Log[z + Sqrt[1 + z^2]])}
```

C.160 richmondmincurve

z⟼**richmondmincurve[n][z]** is the minimal curve corresponding to Richmond's minimal surface of order n.

```
richmondmincurve[n_][z_]:=
        {-1/(2*z) - z^(2*n + 1)/(4*n + 2),
        -I/(2*z) + I*z^(2*n + 1)/(4*n + 2),z^n/n}
```

C.161 rosemincurve

z⟼**rosemincurve[n,a][z]** is the minimal curve generated by the rose t⟼**rose[n,a][t]**.

```
rosemincurve[n_,a_][z_]:= {a*Cos[z]*Cos[n*z],
        a*Cos[n*z]*Sin[z],I*a*EllipticE[n*z,1 - n^2]/n}
```

C.162 scmincurve

z⟼**scmincurve[z]** is the minimal curve generated by the semicubical parabola t⟼**sc[t]**.

```
scmincurve[z_]:= {z^2,z^3,
    I*(4/(27*z) + z/3)*Sqrt[4*z^2 + 9*z^4]}
```

C.163 tractrixmincurve

z⟼**tractrixmincurve[a][z]** is the minimal curve generated by the tractrix t⟼**tractrix[a][t]**.

```
tractrixmincurve[a_][z_]:= {a*Sin[z],
                           a*(Cos[z] + Log[Tan[z/2]]),
                   I*a*(Log[Cos[z/2]] + Log[Sin[z/2]])}
```

Surface Metrics

C.164 `geopolar`

$$ds^2 = du^2 + f(u,v)^2 dv^2$$

is the geodesic polar coordinate metric relative to the function f.

```
ee[geopolar[f_]][u_,v_]:= 1
```

```
ff[geopolar[f_]][u_,v_]:= 0
```

```
gg[geopolar[f_]][u_,v_]:= f[u,v]^2
```

C.165 `liouville`

$$ds^2 = \bigl(f(u) + g(v)\bigr)(du^2 + dv^2)$$

is the metric of a Liouville patch corresponding to the functions f and g.

```
ee[liouville[f_,g_]][u_,v_]:= f[u] + g[v]
```

```
ff[liouville[f_,g_]][u_,v_]:= 0
```

```
gg[liouville[f_,g_]][u_,v_]:= f[u] + g[v]
```

Appendix C: Surfaces

C.166 poincare

$$ds^2 = \frac{du^2 + dv^2}{\lambda^2 v^2}$$

is the metric on the Poincaré upper half-plane with constant curvature $-1/\lambda^2$.

```
ee[poincare[lambda_]][u_,v_]:= 1/(lambda*v)^2
```

```
ff[poincare[lambda_]][u_,v_]:= 0
```

```
gg[poincare[lambda_]][u_,v_]:= 1/(lambda*v)^2
```

$$ds^2 = \frac{du^2}{(\lambda v)^m} + \frac{dv^2}{(\lambda v)^n}$$

is a generalization of the metric on the Poincaré upper half-plane with constant curvature $-1/\lambda^2$.

```
ee[poincare[m_,n_,lambda_]][u_,v_]:= 1/(lambda*v)^m
```

```
ff[poincare[m_,n_,lambda_]][u_,v_]:= 0
```

```
gg[poincare[m_,n_,lambda_]][u_,v_]:= 1/(lambda*v)^n
```

C.167 poindisk

$$ds^2 = \frac{du^2 + dv^2}{1 - \lambda^2(u^2 + v^2)/4}$$

is the metric on the Poincaré disk with constant curvature $-1/\lambda^2$.

```
ee[poindisk[lambda_]][u_,v_]:=
            1/(1 - lambda^2*(u^2 + v^2)/4)^2
```

```
ff[poindisk[lambda_]][u_,v_]:= 0
```

```
gg[poindisk[lambda_]][u_,v_]:=
        1/(1 - lambda^2*(u^2 + v^2)/4)^2
```

C.168 schwarzschildplane

$$ds^2 = \frac{dv^2}{\left(1 - \dfrac{2m}{v}\right)} - \left(1 - \frac{2m}{v}\right)du^2$$

is the 2-dimensional Schwarzschild metric.

```
ee[schwarzschildplane[m_]][u_,v_]:=  -(1 - 2*m/v)
```

```
ff[schwarzschildplane[m_]][u_,v_]:= 0
```

```
gg[schwarzschildplane[m_]][u_,v_]:= (1 - 2*m/v)^-1
```

$$ds^2 = \frac{dv^2}{\left(1 - \dfrac{2m}{v} + f(v)\right)} - \left(1 - \frac{2m}{v} + f(v)\right)du^2$$

is the 2-dimensional generalized Schwarzschild metric relative to the function f.

```
ee[schwarzschildplane[m_,f_]][u_,v_]:=  -(1 - 2*m/v + f[v])
```

```
ff[schwarzschildplane[m_,f_]][u_,v_]:= 0
```

```
gg[schwarzschildplane[m_,f_]][u_,v_]:= (1 - 2*m/v + f[v])^-1
```

APPENDIX D: PLOTTING PROGRAMS

Miscellaneous Plotting Programs

D.1 centerlines

centerlines has the same syntax as **ParametricPlot**. It plots a parametric patch in $\mathbb{R}^3$, but adds lines connecting the center of each quadrilateral to its vertices.

```
centerlines[surface_,{u_,u0_,u1_},{v_,v0_,v1_},opts___]:=
    Module[{plottmp,grtmp},
            plottmp = ParametricPlot3D[
            surface//Evaluate,{u,u0,u1},{v,v0,v1},
            DisplayFunction->Identity,opts];
            grtmp = plottmp /.
  (Polygon[pts_]:>
  {Polygon[Append[pts[[{1,2}]],Plus@pts/4]],
   Polygon[Append[pts[[{2,3}]],Plus@pts/4]],
   Polygon[Append[pts[[{3,4}]],Plus@pts/4]],
   Polygon[Append[pts[[{4,1}]],Plus@pts/4]]});
    Show[grtmp,DisplayFunction->$DisplayFunction]]
```

D.2 dip

DisplayFunction->dip[ins] inserts a **Graphics3D** primitive at the beginning of a **Graphics3D** list when the list is displayed. For example, **DisplayFunction->dip[EdgeForm[]]** displays polygons with no edges.

```
dip[ins_][g_]:= $DisplayFunction[Insert[g,ins,{1,1}]]
```

D.3 elrot

elrot[p,q][theta][aa] is the elementary rotation by an angle **theta** of pth and q^{th} elements of the list **aa**.

```
elrot[p_,q_][theta_][aa_List]:=ReplacePart[ReplacePart[
    aa,aa[[p]]*Cos[theta] + aa[[q]]*Sin[theta],p],
    -aa[[p]]*Sin[theta] + aa[[q]]*Cos[theta],q]
```

D.4 filledcurve

filledcurve has the same syntax as **ParametricPlot**. It fills a curve in $\mathbb{R}^2$.

```
filledcurve[curve_,{u_,u0_,u1_},opts___]:=
    Module[{plottmp,grtmp},
        plottmp = ParametricPlot[
        curve//Evaluate,{u,u0,u1},AspectRatio->Automatic,
        DisplayFunction->Identity,opts];
        grtmp = plottmp /.(Line[pts_]:> Polygon[pts]);
    Show[grtmp,DisplayFunction->$DisplayFunction]]
```

D.5 graphics2dtosurfrev

graphics2dtosurfrev constructs a surface of revolution of radius r from a **Graphics** list of lines. Each circle of latitude is approximated by a regular polygon with n vertices.

```
graphics2dtosurfrev[r_,n_][Line[{{a_,b_},{c_,d_}}]]:=
    Table[Polygon[{
    {r*a*Cos[2*Pi*(k - 1)/n],r*a*Sin[2*Pi*(k - 1)/n],b},
    {r*a*Cos[2*Pi*k/n],r*a*Sin[2*Pi*k/n],b},
    {r*c*Cos[2*Pi*k/n],r*c*Sin[2*Pi*k/n],d},
    {r*c*Cos[2*Pi*(k - 1)/n],
        r*c*Sin[2*Pi*(k - 1)/n],d}}],{k,n}]
```

Appendix D: Plotting Programs

D.6 lathe

lathe[dots] creates a surface of revolution with no edges from a list of points in the plane.

```
lathe[dots_]:=Show[Graphics3D[
  Map[graphics2dtosurfrev[0.5,60],
  linetolines[Line[dots]]]],Boxed->False,
  DisplayFunction->dip[EdgeForm[]]]
```

D.7 linetolines

linetolines converts a broken line segment **Line[{pt1,...,ptn}]** into a list of line segments connecting consecutive pairs of the points **pt1,...,ptn**.

```
linetolines[Line[pts_]]:= Table[Line[{pts[[i]],
  pts[[i + 1]]}],{i,Length[pts] - 1}]
```

D.8 orth4d

orth4d[surf,{u,umin,umax},{v,vmin,vmax},opts] uses **GraphicsArray** to plot a surface in $\mathbb{R}^4$ via the four orthogonal projections from $\mathbb{R}^4$ to $\mathbb{R}^3$.

```
    orth4d[surf_,{u_,umin_,umax_},{v_,vmin_,vmax_},opts___]:=
        Module[{g1,g2,g3,g4},
        g1=ParametricPlot3D[Evaluate[
                Drop[surf,{1}]],{u,umin,umax},{v,vmin,vmax},
                DisplayFunction->Identity,opts];
        g2=ParametricPlot3D[Evaluate[
                Drop[surf,{2}]],{u,umin,umax},{v,vmin,vmax},
                DisplayFunction->Identity,opts];
        g3=ParametricPlot3D[Evaluate[
                Drop[surf,{3}]],{u,umin,umax},{v,vmin,vmax},
                DisplayFunction->Identity,opts];
        g4=ParametricPlot3D[Evaluate[
                Drop[surf,{4}]],{u,umin,umax},{v,vmin,vmax},
                DisplayFunction->Identity,opts];
        Show[GraphicsArray[{{g1,g2},{g3,g4}}],
                GraphicsSpacing->{0.2,0},
                DisplayFunction->$DisplayFunction]]
```

D.9 paramcurveplot1

paramcurveplot1 has the same syntax as **ParametricPlot3D**, except that a **PlotPoints** specification is obligatory. It plots the parameter lines $\mathbf{u} \mapsto \mathbf{x[u,v0]}$ of a parametric patch **surface**:$\mathcal{U} \longrightarrow \mathbb{R}^3$.

```
    paramcurveplot1[surface_,{u_,u0_,u1_},{v_,v0_,v1_},
        PlotPoints->{upoints_,vpoints_},opts___]:=
        ParametricPlot3D[Table[
        surface,{v,v0,v1,(v1-v0)/(vpoints - 1)}]//Evaluate,
        {u,u0,u1},PlotPoints->upoints,opts]
```

D.10 paramcurveplot2

paramcurveplot2 has the same syntax as **ParametricPlot3D**, except that a **PlotPoints** specification is obligatory. It plots the parameter lines $\mathbf{v} \mapsto \mathbf{x[u0,v]}$ of a parametric patch **surface**:$\mathcal{U} \longrightarrow \mathbb{R}^3$.

Appendix D: Plotting Programs 991

```
paramcurveplot2[surface_,{u_,u0_,u1_},{v_,v0_,v1_},
    PlotPoints->{upoints_,vpoints_},opts___]:=
ParametricPlot3D[Table[
surface,{u,u0,u1,(u1-u0)/(upoints - 1)}]//Evaluate,
{v,v0,v1},PlotPoints->vpoints,opts]
```

D.11 plotpatch2d

plotpatch2d has the same syntax as **ParametricPlot**. It plots the images of horizontal and vertical lines under the function $f : \mathcal{U} \longrightarrow \mathbb{R}^2$, where $\mathcal{U}$ is an open subset of $\mathbb{R}^2$. The miniprogram **plotpatch2d** has the same functionality as Maeder's **CartesianMap** and **PolarMap**, but draws curved instead of straight lines.

```
plotpatch2d[expr_,{x_,xmin_,xmax_},{y_,ymin_,ymax_},
    PlotPoints->{xpoints_,ypoints_},opts___]:=
Show[ParametricPlot[Evaluate[
Table[ComplexExpand[
{Re[expr],Im[expr]}],
{y,ymin,ymax,(ymax - ymin)/(ypoints - 1)}]],
{x,xmin,xmax},PlotPoints->xpoints,
DisplayFunction->Identity],
ParametricPlot[Evaluate[
Table[ComplexExpand[
{Re[expr],Im[expr]}],
{x,xmin,xmax,(xmax - xmin)/(xpoints -1)}]],
{y,ymin,ymax},PlotPoints->ypoints,
DisplayFunction->Identity],
AspectRatio->Automatic,
DisplayFunction->$DisplayFunction,opts]
```

D.12 selectgraphics3d

selectgraphics3d[graphics3dobj,bound,opts] selects from a list of polygons those whose vertices have their coordinates bounded by bound, and then displays the results.

```
selectgraphics3d[graphics3dobj_,bound_,opts___]:=
    Show[Graphics3D[Select[graphics3dobj,
    (Abs[#[[1,1,1]]]<bound && Abs[#[[1,1,2]]]<bound &&
    Abs[#[[1,1,3]]]<bound && Abs[#[[1,2,1]]]<bound &&
    Abs[#[[1,2,2]]]<bound && Abs[#[[1,2,3]]]<bound &&
    Abs[#[[1,3,1]]]<bound && Abs[#[[1,3,2]]]<bound &&
    Abs[#[[1,3,3]]]<bound && Abs[#[[1,4,1]]]<bound &&
    Abs[#[[1,4,2]]]<bound && Abs[#[[1,4,2]]]<bound
    )&]],opts]
```

D.13 selectcolgraphics3d

selectgraphics3d[graphics3dobj,bound,opts] selects from a list of colored polygons those whose vertices have their coordinates bounded by bound, and then displays the results.

```
selectcolgraphics3d[graphics3dobj_,bound_,opts___]:=
    Show[Graphics3D[Select[graphics3dobj,
    (Abs[#[[2,1,1,1]]]<bound && Abs[#[[2,1,1,2]]]<bound &&
    Abs[#[[2,1,1,3]]]<bound && Abs[#[[2,1,2,1]]]<bound &&
    Abs[#[[2,1,2,2]]]<bound && Abs[#[[2,1,2,3]]]<bound &&
    Abs[#[[2,1,3,1]]]<bound && Abs[#[[2,1,3,2]]]<bound &&
    Abs[#[[2,1,3,3]]]<bound && Abs[#[[2,1,4,1]]]<bound &&
    Abs[#[[2,1,4,2]]]<bound && Abs[#[[2,1,4,2]]]<bound
    )&]],opts]
```

D.14 stereo4d

stereo4d[{a,b,c,d},surf,{u,umin,umax},{v,vmin,vmax},opts] uses **GraphicsArray** to plot a surface in $\mathbb{R}^4$ via the four stereographic projections from $\mathbb{R}^4$ to $\mathbb{R}^3$ with respect to the point $(a,b,c,d) \in \mathbb{R}^4$.

```
stereo4d[{a_,b_,c_,d_},surf_,
    {u_,umin_,umax_},{v_,vmin_,vmax_},opts___]:=
  Module[{g1,g2,g3,g4},
    g1=ParametricPlot3D[Evaluate[
          Drop[surf,{1}]/(a-surf[[1]])],
          {u,umin,umax},{v,vmin,vmax},
          DisplayFunction->Identity,opts];
    g2=ParametricPlot3D[Evaluate[
          Drop[surf,{2}]/(b-surf[[2]])],
          {u,umin,umax},{v,vmin,vmax},
          DisplayFunction->Identity,opts];
    g3=ParametricPlot3D[Evaluate[
          Drop[surf,{3}]/(c-surf[[3]])],
          {u,umin,umax},{v,vmin,vmax},
          DisplayFunction->Identity,opts];
    g4=ParametricPlot3D[Evaluate[
          Drop[surf,{4}]/(d-surf[[4]])],
          {u,umin,umax},{v,vmin,vmax},
          DisplayFunction->Identity,opts];
    Show[GraphicsArray[{{g1,g2},{g3,g4}}],
          GraphicsSpacing->{0.2,0},
          DisplayFunction->$DisplayFunction]]
```

D.15 surfacecurveplot

surfacecurveplot[surface,{u,u0,u1},{v,v0,v1},PlotLines->
{ulines,vlines}, PlotPoints->{upoints,vpoints},opts] plots both
the u- and v-parameter curves of a patch **surface**:$\mathcal{U} \longrightarrow \mathbb{R}^3$. The number of
curves is specified with **PlotLines**, and the resolution of the lines is specified with
PlotPoints.

```
surfacecurveplot[surface_,{u_,u0_,u1_},{v_,v0_,v1_},
    PlotLines->{ulines_,vlines_},
    PlotPoints->{upoints_,vpoints_},opts___]:=
        Show[paramcurveplot1[surface,{u,u0,u1},{v,v0,v1},
            PlotPoints->{upoints,ulines},
            DisplayFunction->Identity,opts],
            paramcurveplot2[surface,{u,u0,u1},{v,v0,v1},
            PlotPoints->{vlines,vpoints},
            DisplayFunction->Identity,opts],
            DisplayFunction->$DisplayFunction]
```

D.16 wireframe

wireframe has the same syntax as **ParametricPlot3D**. It plots a wireframe of a parametric surface in $\mathbb{R}^3$.

```
wireframe[surface_,{u_,u0_,u1_},{v_,v0_,v1_},opts___]:=
    Module[{plottmp,grtmp},
        plottmp = ParametricPlot3D[
        surface,{u,u0,u1},{v,v0,v1},
        DisplayFunction->Identity,opts];
        grtmp = plottmp /.(Polygon[pts_]:>
        Line[Append[pts,First[pts]]]);
    Show[grtmp,DisplayFunction->$DisplayFunction]]
```

Mathematica to Acrospin Programs

Acrospin is a program for PCs that animates diagrams consisting of points or lines. It is very fast because it is written in assembly language. The following miniprograms write a 3-dimensional *Mathematica* graphic to an acrospin file, which by convention has a suffix **acd**. The file can then be animated with **Acrospin**.

Acrospin is written by David B. Parker and is available from:

> MathWare PO Box 3025
> Urbana IL 61801
> USA

Telephone: (800) 255-2468 Fax: (217) 384-7043
email: info@mathware.com
http:// www.mathware.com

D.17 primitiveplot3d

primitiveplot3d[{fx,fy,fz},{t,tmin,tmax}] produces a list of points of a space curve parametrized by a variable **t** which runs from **tmin** to **tmax**.
primitiveplot3d[{fx,fy,fz},{t,tmin,tmax},{u,umin,umax}]
produces a list of points by polygons of a surface parametrized by **t** and **u**.

```
primitiveplot3d[exp_,{t_,tmin_,tmax_},opts___]:=
    First[ParametricPlot3D[exp,{t,tmin,tmax},opts,
    DisplayFunction->Identity]] /.
    (Line[pts_] :> pts)
```

```
primitiveplot3d[exp_,{u_,umin_,umax_},{v_,vmin_,vmax_},
    opts___]:=
    First[ParametricPlot3D[exp,
    {u,umin,umax},{v,vmin,vmax},opts,
    DisplayFunction->Identity]] /.
    (Polygon[pts_] :> pts)
```

D.18 WriteAcroPoints

WriteAcroPoints[{color},file,{fx,fy,fz},{t,tmin,tmax}]
outputs to a file in acrospin point format the plot of a space curve.
**WriteAcroPoints[{color},file,{fx,fy,fz},{t,tmin,tmax},
{u,umin,umax}]** outputs to a file in acrospin point format the plot of a surface.

```
WriteAcroPoints[{color_},file_String,curve_,
    {t_,tmin_,tmax_},opts___]:=
    Module[{tmp1,tmp2,tmp3,tmp4},
    OpenWrite[file];
    WriteString[file,"Set Color "];
    WriteString[file,color];
    WriteString[file,"\n"];
    WriteString[file,"PointList X Y Z\n"];
    tmp1=primitiveplot3d[curve//Evaluate,
    {t,tmin,tmax},opts];
    tmp2=Union[Flatten[tmp1,1]];
    tmp3=Chop[tmp2,10^-3];
    tmp4=Map[PaddedForm[#,{4,4}]&,tmp3,{2}];
    Scan[(WriteString[file,First[#]];
    Scan[WriteString[file,"   ",#]&,Rest[#]];
    WriteString[file,"\n"])&,
    tmp4];Close[file]]
```

```
WriteAcroPoints[{color_},file_String,surface_,
    {u_,umin_,umax_},{v_,vmin_,vmax_},opts___]:=
    Module[{tmp1,tmp2,tmp3,tmp4},
    OpenWrite[file];
    WriteString[file,"Set Color "];
    WriteString[file,color];
    WriteString[file,"\n"];
    WriteString[file,"PointList X Y Z\n"];
    tmp1=primitiveplot3d[surface//Evaluate,
    {u,umin,umax},{v,vmin,vmax},opts];
    tmp2=Union[Flatten[tmp1,1]];
    tmp3=Chop[tmp2,10^-3];
    tmp4=Map[PaddedForm[#,{4,4}]&,tmp3,{2}];
    Scan[(WriteString[file,First[#]];
    Scan[WriteString[file,"   ",#]&,Rest[#]];
    WriteString[file,"\n"])&,
    tmp4];Close[file]]
```

D.19 WriteAcroLines

`WriteAcroLines[{color},file,{fx,fy,fz},{t,tmin,tmax}]`
outputs to a file in acrospin point format the plot of a space curve.
`WriteAcroLines[{color1},{color2},file,{fx,fy,fz},`
`{t,tmin,tmax},{u,umin,umax}]` outputs to a file in acrospin line format the plot of a surface.

```
WriteAcroLines[{color_},file_String,curve_,
    {t_,tmin_,tmax_},opts___]:=
  Module[{tmp1,tmp2,tmp3,tmp4,tmp5,tmp6,tmp7,tmp8},
  OpenWrite[file];
  WriteString[file,"Set Color "];
  WriteString[file,color];
  WriteString[file,"\n"];
  WriteString[file,"EndPointList name X Y Z\n"];
  tmp1=primitiveplot3d[curve//Evaluate,
    {t,tmin,tmax},opts];
  tmp2=Flatten[tmp1,1];
  tmp3=Chop[tmp2,10^-3];
  tmp4=Map[PaddedForm[#,{4,4}]&,tmp3,{2}];
  tmp5=Table[
    {StringForm["VERTEX`'",j],tmp4[[j]]},
    {j,Length[tmp3]}];
  tmp6=Map[Flatten[#,1]&,tmp5,{1}];
  Scan[
    (WriteString[file,First[#]];
    Scan[WriteString[file,"  ",#]&,Rest[#]];
    WriteString[file,"\n"])&,
    tmp6];
  WriteString[file,"\nLineList from to\n"];
  tmp7=Table[
    StringForm["VERTEX`' VERTEX`'\n",j,j+1],
    {j,Length[tmp3]-1}];
  Scan[WriteString[file,#,"\n"]&,tmp7];
  Close[file]]
```

```
WriteAcroLines[{color1_:12,color2_:14},
    file_String,surface_,
    {u_,umin_,umax_},{v_,vmin_,vmax_},opts___]:=
    Module[{tmp1,tmp2,tmp3,tmp4,tmp5,tmp6,
    tmp7,tmp8,tmp9,tmp10,tmp11,tmp12},
    OpenWrite[file];
    WriteString[file,"EndPointList name X Y Z\n\n"];
    tmp1=primitiveplot3d[surface//Evaluate,
    {u,umin,umax},{v,vmin,vmax},opts];
    tmp2=Chop[tmp1,10^-3];
    tmp3=Map[PaddedForm[#,{4,4}]&,tmp2,{3}];
    tmp4=Table[
    {StringForm["POLY``VERTEX``",j,k],tmp3[[j,k]]},
    {j,Length[tmp3]},{k,4}];
    tmp5=Map[Flatten[#,1]&,tmp4,{2}];
    tmp6=Flatten[tmp5,1];
    Scan[WriteString[file,"   ",#]&,Rest[#]];
    Scan[
    (WriteString[file,First[#]];
    Scan[WriteString[file,"   ",#]&,Rest[#]];
    WriteString[file,"\n"])&,
    tmp6];
    WriteString[file,"\nLineList from to Color Layer\n\n"];
    tmp7=Table[
    StringForm["POLY``VERTEX`` POLY``VERTEX`` `` ``\n",
    j,1,j,2,color1,1],{j,Length[tmp3]}];
    tmp8=Table[
    StringForm["POLY``VERTEX`` POLY``VERTEX`` `` ``\n",
    j,4,j,3,color1,1],{j,Length[tmp3]}];
    tmp9=Join[tmp7,tmp8];
    Scan[WriteString[file,#,"\n"]&,tmp9];
    tmp10=Table[
    StringForm["POLY``VERTEX`` POLY``VERTEX`` `` ``\n",
    j,1,j,4,color2,2],{j,Length[tmp3]}];
    tmp11=Table[
    StringForm["POLY``VERTEX`` POLY``VERTEX`` `` ``\n",
    j,3,j,2,color2,2],{j,Length[tmp3]}];
    tmp12=Join[tmp10,tmp11];
    Scan[WriteString[file,#,"\n"]&,tmp12];
    Close[file]]
```

Mathematica to Geomview Programs

Geomview is an interactive program for Iris, NeXT and Sun workstations that can be used for viewing and manipulating geometric objects. It was written at the Geometry Center of the University of Minnesota by Tamara Munzner, Stuart Levy and Mark Phillips. It can be obtained free via anonymous ftp on Internet from the host **geom.umn.edu**.

Geomview is capable of reading many graphics file formats, one of which is the mesh file format. The following miniprograms are quite similar to those of **ParametricPlot3D**, but instead of outputting an object for immediate viewing by *Mathematica*, they write the parametric plot of a surface to a file in mesh format. The file can then be used as input to Geomview.

D.20 Write3DMESH

`Write3DMESH[file,expression,{u,umin,umax},{v,vmin,vmax}, PlotPoints->{udiv,vdiv},opts]` outputs to a file in mesh format the parametric plot of a surface.

```
Write3DMESH[file_String,expression_,
    {u_,umin_,umax_},{v_,vmin_,vmax_},
    PlotPoints->{udiv_,vdiv_},otmp___]:=
Module[{tmp1,tmp2,tmp3,tmp4},
OpenWrite[file];
    WriteString[file,"MESH\n"];
    WriteString[file,ToString[vdiv+1]];
    WriteString[file," "];
    WriteString[file,ToString[udiv+1]];
    WriteString[file,"\n"];
    tmp1=N[Table[expression,
        {u,umin,umax,(umax-umin)/udiv},
        {v,vmin,vmax,(vmax-vmin)/vdiv}]];
    tmp2=Flatten[tmp1,1];
    tmp3=Chop[tmp2,10^-4];
    tmp4=Map[PaddedForm[#,{4,4}]&,tmp3,{2}];
Scan[
    (WriteString[file,First[#]];
    Scan[WriteString[file,"   ",#]&,Rest[#]];
    WriteString[file,"\n"])&,tmp4];
Close[file]]
```

D.21 Write3DCMESH

Write3DCMESH[file,{expression,color},{u,umin,umax}, {v,vmin,vmax},PlotPoints->{udiv,vdiv},opts] outputs to a file in mesh format the parametric plot of expression colored by **color**.

```
Write3DCMESH[file_String,{expression_,color_},
    {u_,umin_,umax_},{v_,vmin_,vmax_},
    PlotPoints->{udiv_,vdiv_},otmp___]:=
Module[{tmp1,tmp2,tmp3,tmp4,tmp5,tmp6,tmp7},
    OpenWrite[file];
    WriteString[file,"CMESH\n"];
    WriteString[file,ToString[vdiv+1]];
    WriteString[file," "];
    WriteString[file,ToString[udiv+1]];
    WriteString[file,"\n"];
    tmp1=N[Table[{expression,HSBColor[color,1,1]},
        {u,umin,umax,(umax-umin)/udiv},
        {v,vmin,vmax,(vmax-vmin)/vdiv}]];
    tmp2=Apply[List,tmp1,{3}];
    tmp3=Map[Flatten[#]&,tmp2,{2}];
    tmp4=Flatten[tmp3,1];
    tmp5=Map[Append[#,1]&,tmp4];
    tmp6=Chop[tmp5,10^-4];
    tmp7=Map[PaddedForm[#,{4,4}]&,tmp6,{2}];
    Scan[
        (WriteString[file,First[#]];
        Scan[WriteString[file,"   ",#]&,Rest[#]];
        WriteString[file,"\n"])&,tmp7];
    Close[file]]
```

D.22 Write3DCNMESH

Write3DCNMESH[file,{expression,unitn,color},{u,umin,umax}, {v,vmin,vmax},PlotPoints->{udiv,vdiv},opts] outputs to a file in mesh format the parametric plot of expression colored by **color** with normal information **unitn**.

```
Write3DCNMESH[file_String,{expression_,unitn_,color_},
    {u_,umin_,umax_},{v_,vmin_,vmax_},
    PlotPoints->{udiv_,vdiv_},otmp___]:=
Module[{tmp1,tmp2,tmp3,tmp4,tmp5,tmp6,tmp7},
    OpenWrite[file];
    WriteString[file,"CNMESH\n"];
    WriteString[file,ToString[vdiv+1]];
    WriteString[file," "];
    WriteString[file,ToString[udiv+1]];
    WriteString[file,"\n"];
    tmp1=N[Table[{expression,unitn,HSBColor[color,1,1]},
        {u,umin,umax,(umax-umin)/udiv},
        {v,vmin,vmax,(vmax-vmin)/vdiv}]];
    tmp2=Apply[List,tmp1,{3}];
    tmp3=Map[Flatten[#]&,tmp2,{2}];
    tmp4=Flatten[tmp3,1];
    tmp5=Map[Append[#,1]&,tmp4];
    tmp6=Chop[tmp5,10^-4];
    tmp7=Map[PaddedForm[#,{4,4}]&,tmp6,{2}];
    Scan[
        (WriteString[file,First[#]];
        Scan[WriteString[file,"  ",#]&,Rest[#]];
        WriteString[file,"\n"])&,tmp7];
    Close[file]]
```

D.23 Write2COLMESH

Write2COLMESH[{color1,color2},file,expression,{u,umin,umax}, {v,vmin,vmax},PlotPoints->{udiv,vdiv},opts] outputs to a file in mesh format the parametric plot of a surface with one side **color1** and the other **color2**.

```
Write2COLMESH[{color1_,color2_},file_String,expression_,
    {u_,umin_,umax_},{v_,vmin_,vmax_},
    PlotPoints->{udiv_,vdiv_},otmp___]:=
  Module[{cc1,cc2,cc3,dd1,dd2,dd3,df,em,tmp1,tmp2,tmp3,tmp4},
    OpenWrite[file];
    WriteString[file,"appearance {\n"];
    WriteString[file,"material {\n"];
    {cc1,cc2,cc3}=List@color1;
    em=StringForm["emission `` `` ``\n",cc1,cc2,cc3];
    WriteString[file,em];
    {dd1,dd2,dd3}=List@color2;
    df=StringForm["diffuse `` `` ``}}\n",dd1,dd2,dd3];
    WriteString[file,df];
    WriteString[file,"MESH\n"];
    WriteString[file,ToString[vdiv+1]];
    WriteString[file," "];
    WriteString[file,ToString[udiv+1]];
    WriteString[file,"\n"];
    tmp1=N[Table[expression,
        {u,umin,umax,(umax-umin)/udiv},
        {v,vmin,vmax,(vmax-vmin)/vdiv}]];
    tmp2=Flatten[tmp1,1];
    tmp3=Chop[tmp2,10^-4];
    tmp4=Map[PaddedForm[#,{4,4}]&,tmp3,{2}];
    Scan[
       (WriteString[file,First[#]];
        Scan[WriteString[file,"   ",#]&,Rest[#]];
        WriteString[file,"\n"])&,tmp4];
    Close[file]]
```

BIBLIOGRAPHY

[AbSt] M. Abramowitz and I.A. Stegun, **Handbook of Mathematical Functions**, Dover Publications, New York, 1965.

[Agnesi] M.G. Agnesi, **Instituzioni analitiche ad uso della gioventú italiana**, Nella Regia-ducal corte, Milan, 1748. English by John Colson, Taylor and Wilks, London, 1801.

[Ahlf] L.V. Ahlfors, **Complex Analysis**, Third Edition, McGraw-Hill, New York, 1979.

[AhSa] L.V. Ahlfors and L. Sario, **Riemann Surfaces**, Princeton University Press, Princeton, NJ, 1960.

[Alek] A.D. Aleksandrov, Curves and Surfaces, **Mathematics–Its Content, Methods and Meaning**, Volume 2, M.I.T. Press, Cambridge, MA, 1962, 57-117.

[Aoust] L'abbé L. Aoust, **Analyse Infinitésimale des Courbes**, Gauthier-Villars, Paris, 1869.

[Apéry] F. Apéry, **Models of the Real Projective Plane**, Friedr. Vieweg & Sohn, Braunschweig, 1987.

[Arch] R.C. Archibald, Curve, **Encyclopedia Britannica**, 14th Edition, Encyclopedia Britannica, New York, 1929.

[Arg] D.E. Arganbright, **Practical Handbook of Spreadsheet Curves and Geometric Constructions**, CRC Press, Boca Raton, FL, 1993.

[Arm] M.A. Armstrong, **Basic Topology**, *Undergraduate Texts in Mathematics*, Springer-Verlag, Berlin-New York, 1983.

[Bck] A.V. Bäcklund, Om ytori med konstant negativ Krökning, *Lunds Universitets Arsskrift* **19**, 1883. English translation by E.D. Coddington: *Concerning Surfaces with Constant Negative Curvature*, New Era Printing Co., Lancaster, PA, 1905.

[Bahd] T. Bahder, **Mathematica for Scientists and Engineers**, Addison-Wesley, Reading, MA, 1995.

[BGM] T. Banchoff, T. Gaffney and C. McCrory, **Cusps of Gauss Mappings**, *Res. Notes in Math.* **55**, Pitman, Boston-London, 1982.

[Bar] J.É. Barbier, Note sur le probème de l'aiguille et le jeu di joint couvert, *J. Math. pures et appl.* (**2**) **5** (1860), 273–386.

[BaCo] J.L.M. Barbosa and A.G. Colares, **Minimal Surfaces in** $\mathbb{R}^3$, *Lecture Notes in Mathematics* **1195**, Springer-Verlag, Berlin-New York, 1986.

[Baud] P. Baudoin, **Les ovales de Descartes et le limaçon de Pascal**, Librairie Vuibert, Paris, 1938.

[Beltr] E. Beltrami, Saggio di interpretazione della geometria non-euclidea, *Giorn. di Mat. Napoli*, **6**, 284-312. Also, **Opere I**, 406-429.

[BeGo] M. Berger and B. Gostiaux, **Differential Geometry: Manifolds, Curves, and Surfaces**, *Graduate Texts in Mathematics* **115**, Springer-Verlag, Berlin-New York, 1988.

[Bers] L. Bers, **Riemann Surfaces**, New York University, Institute of Mathematical Sciences, 1958.

[BDP] J. Bertrand, Diguet and V. Puiseux, Démonstration d'un théorème de Gauss, *Journal de Mathématiques* **13** (1848), 80–90.

[Bian] L. Bianchi, **Lezioni di Geometria Differenziale**, Terza edizione interamente rifatta, Nicola Zanichelli, Bologna, 1927. German translation by M. Lukat: **Vorlesungen über Differentialgeometrie**, Teubner, Leipzig, 1899, 1910.

[Bieb] L. Bieberbach, **Differentialgeometrie**, *Teubners Mathematische Leitfäden* **31**, Teubner, Leipzig, 1932.

[Bj] E.G. Björling, In integrationem aequationis derivatarum partialum superficiei, cujus in puncto unoquoque principales ambo radii curvedinis aequales sunt signoque contrario, *Arch. Math. Phys.* (1) **4** (1844), 290–315.

[Blach] N. Blachman, **Mathematica Quick Reference, Version 2**, Variable Symbols, Berkeley, CA, 1992.

[Blas2] W. Blaschke, **Differentialgeometrie**, Volumes 1 and 2, Chelsea Publishing Co., New York, 1967, Volume 3, Springer-Verlag, Berlin, 1929.

[Boas] R.P. Boas, **Invitation to Complex Analysis**, Random House, New York, 1987.

[Bolt] V.C. Boltyanski, **Envelopes**, Pergamon Press, Oxford, 1964.

[Bomp] E. Bompiani, Rappresentazione Grafica delle Ciclidi di Dupin e delle loro Lossodromiche, *Memorie del R. Istituto Lombardo di Scienze e Lettere*, (**3**) **21-2** (1915), 205–241.

[Bonn1] P.O. Bonnet, Mémoire sur la théorie général des surfaces, *J. Éc. Polytech.* **19** (1848), 129–168.

[Bonn2] P.O. Bonnet, Mémoire sur la théorie des surfaces applicables sur une surface donée, *J. Éc. Polytech.* **24** (1865), 209–230.

[BrKn] E. Brieskorn and H. Knörrer, **Plane Algebraic Curves**, Birkhäuser Verlag, Basel-Boston, 1986.

[Brio] F. Brioschi, Intorno ad alcuni punti della teorica delle superficie, *Annali di Scienze Matematiche e Fisiche* **3** (1852), 293–321. Also, **Opere 3** 11–33.

[BrLa] Th. Bröcker and L. Lander, **Differentiable Germs and Catastrophes**, *London Math. Soc. Lecture Notes* **17**, Cambridge University Press, Cambridge, 1984.

[BrGi] J.W. Bruce and P.J. Gibbin, **Curves and Singularities**, Cambridge University Press, Cambridge, 1982.

[Buck] R.C. Buck, **Advanced Calculus**, Second Edition, McGraw-Hill, New York, 1965.

[BuBu]   C. Burali-Forti and P. Burgatti, **Analisi vettorale generale e applicazioni**, Volume 2, Geometria differenziale, Nicola Zanichelli, Bologna, 1930.

[BuZi]   G. Burde and H. Zieschang, **Knots**, *de Gruyter Studies in Mathematics* **5**, Walter de Gruyter, Berlin-New York, 1985.

[Cajori] F. Cajori, **A History of Mathematics**, Third Edition, Chelsea Publishing Co., New York, 1980.

[dC1]   M.P. do Carmo, **Differential Geometry of Curves and Surfaces**, Prentice-Hall, Englewood Cliffs, NJ, 1976.

[dC2]   M.P. do Carmo, **Superfícies Mínimas**, 16º *Colóquio Brasileiro de Matemática*, IMPA, Rio de Janeiro, 1987.

[Cartan] É. Cartan, **La méthode du repère mobile, la théorie des groupes continus et les espaces généralisés**, *Exposés de Géométrie*, V, Hermann, Paris, 1935.

[Cata]   E. Catalan, Mémoire sur les surfaces dont les rayons de courbure, en chaque point, sont égaux et de signes contraires, *J. Éc. Polytech.* **21** (1858), 129–168.

[CeRy]   T.E. Cecil and P.J. Ryan, **Tight and Taut Immersions of Manifolds**, *Res. Notes in Math.* **131**, Pitman, Boston-London, 1985.

[Ces]   E. Cesàro, **Lezioni di Geometria Intrinseca**, Tipografia della R. Accademia delle Scienze, Napoli, 1896. German translation by G. Kowalewski: **Vorlesungen über natürliche Geometrie**, Teubner, Leipzig, 1901.

[Chand] K. Chandrasekharan, **Elliptic Functions**, *Grundlehren der Mathematischen Wissenschaften* **281**, Springer-Verlag, Berlin-New York, 1985.

[Ch]   M. Chasles, Les lignes géodésiques et les lignes de courbure des surfaces du second degré, *Journ. de Math.* **11** (1846), 5–20.

[Chern1] S.S. Chern, **Introduction to Differential Geometry**, Notes, University of California, Berkeley, CA, 1963.

[Chern2] S.S. Chern, Studies in Global Analysis and Geometry, **MAA Studies in Mathematics**, **27**, The Mathematical Association of America, Washington, DC, 1989.

[Chern3] S.S. Chern, Surface Theory with Darboux and Bianchi, **Miscellanea Mathematica**, 59–69, Springer-Verlag, Berlin-New York, 1991.

[Chris] E.B. Christoffel, Über die Transformation der homogenen Differentialausdrücke zweiten Grades, *J. reine angew. Math. (Crelle's Journal)* **70** (1869), 46–70.

[Codaz] D. Codazzi, Sulle coordinate curvilinee d'una superficie dello spazio, *Annali di matematica pura ed applicata* (2) **2** (1868-1869), 101–119.

[CV] S. Cohn-Vossen, Kürzste Wege und Totalkrümmung auf Flächen, *Composito Math.* **2** (1935), 69–133.

[Cool2] J.L. Coolidge, **A History of Geometrical Methods**, Oxford University Press, London, 1940. Reprint by Dover Publications, New York, 1963.

[Costa1] C.J. Costa, Example of a complete minimal immersion in $\mathbb{R}^3$ of genus one and three embedded ends, *Bol. Soc. Brasil Math.* **15** (1984), 47–54.

[Costa2] S.I.R. Costa, On closed twisted curves, *Proc. Amer. Math. Soc.* **109** (1990), 205–214.

[Crowe] M.J. Crowe, **A History of Vector Analysis**, Dover Publications, New York, 1994.

[CF] R.H. Crowell and R.H. Fox, **Introduction to Knot Theory**, Ginn and Company, Boston, 1963. Reprint as *Graduate Texts in Mathematics* **57**, Springer-Verlag, Berlin-New York, 1978.

[Darb1] G. Darboux, **Sur une Classe Remarquable de Courbes et de Surfaces Algebriques et sur la Theorie des Imaginaires**, Gauthier-Villars, Paris, 1873. Reprint by University Microfilms International, Ann Arbor, MI, 1992.

[Darb2] G. Darboux, **Leçons sur la Théorie Générale des Surfaces**, 4 volumes, Gauthier-Villars, Paris, 1914, 1915, 1894, 1896. Reprint by Chelsea Publishing Co., New York, 1972.

[Darb3] G. Darboux, **Leçons sur les Systèmes Orthogonaux et les Coordonnées**, Gauthier-Villars, Paris, 1898.

[Darb4] G. Darboux, **Principes de Géométrie Analytique**, Gauthier-Villars, Paris, 1917.

[Dark]　H.E. Dark, **The Wankel Rotary Engine: Introduction and Guide**, Indiana University Press, Bloomington, IN, 1974.

[De]　A. Deprit, Letter to the editor, *Mathematica J.*, **1**, Issue 2, (1990), 4.

[DHKW]　U. Dierkes, S. Hildebrandt, A. Küster and O. Wohlrab, **Minimal Surfaces**, 2 volumes, Springer-Verlag, Berlin-New York, 1992.

[Dom]　P. Dombrowski, **150 Years after Gauss'** *"Disquisitiones Generales Circa Superficies Curvas"*, *Astérisque* **62**, Société Mathématique de France, Paris, 1979.

[Dupin1]　C. Dupin, **Développements de Géométrie**, Courcier, Paris, 1813.

[Dupin2]　C. Dupin, **Applications de Géométrie et de Méchanique**, Bachelier, Paris, 1822.

[Eisen1]　L.P. Eisenhart, **A Treatise on the Differential Geometry of Curves and Surfaces**, Ginn and Company, Boston, 1909. Reprint by Dover Publications, New York, 1960.

[Eisen2]　L.P. Eisenhart, **Transformations of Surfaces**, Princeton University Press, Princeton, NJ, 1923.

[Enn1]　A. Enneper, Analytisch-geometrische Untersuchungen, *Z. Math. Phys.* **9** (1864), 96–125.

[Euler1]　L. Euler, De constructione aequationum ope motus tractorii aliisque ad methodum tangentium inversam pertinentibusm, *Comment. Acad. Petropolit.* **8** (1736), 66–85. Reprinted in *Opera omnia* **(1) 22**, Basel 1926, 83-107.

[Euler2]　L. Euler, Methodus inveniendi lineas curvas maximi minimive proprietate gaudentes sive solutio problematis isoperimetrici latissimo sensu accepti, *Opera omnia* **(1) 24**, Füssli Turici, Lausanne, 1952. English translation (in part) by D.J. Struik [Stru3, pages 399–406].

[Euler3]　L. Euler, Recherches sur la courbure des surfaces, *Mém. de l'Acad. des Sciences, Berlin* **16** (1760), 119–143. Reprinted in *Opera omnia* **(1) 28**, Füssli Turici, Lausanne, 1955, 1-22.

[Euler4]　L. Euler, De curvis triangularibus, *Acta Acad. Petropolit.* **2** (1778), 3-30. *Opera omnia* **(1) 28**, Füssli Turici, Lausanne 1955, 298-321.

[Fava] F. Fava, **Geometria Differenziale**, Levrotto & Bella, Torino, 1991.

[Fed] A.S. Fedenko, **A Collection of Problems in Differential Geometry**, Nauka, Moscow, 1979. Spanish translation, Mir, 1981. French translation, Mir, 1982.

[FeGrMa] H. Fergason, A. Gray and S. Markvorsen, Costa's minimal surface via Mathematica *Mathematica in Research and Education* **5**, (1996) 5-10.

[Fischer] G. Fischer, **Mathematical Models**, 2 volumes, Friedr. Vieweg & Sohn, Braunschweig, 1986.

[FoTu] A.T. Fomenko and A.A. Tuzhilin, **Elements of the Geometry and Topology of Minimal Surfaces in Three-Dimensional Space**, *Translations of Mathematical Monographs* **93**, American Mathematical Society, Providence, RI, 1991.

[Francis] G.K. Francis, **A Topological Picture Book**, Springer-Verlag, Berlin-New York, 1987.

[Frenet] F. Frenet, Sur les courbes à double courbure, Thèse, Toulouse, 1847. Abstract in *Journ. de Math.* **17**, 1852.

[Frost] P. Frost, **Curve Tracing**, MacMillan, London, 1926.

[Gauss] C.F. Gauss, Disquisitiones Generales Circa Superficies Curvas, *Comm. Soc. Reg. Sc. Gott. Rec.*, **6** (1828).

[Gomes] F. Gomes Teixeira, **Traité des courbes spéciales remarquables planes et gauches**, Coimbra, 1908–1915. Reprint by Chelsea Publishing Co., New York, 1971.

[Gott] D.H. Gottlieb, All the Way with Gauss-Bonnet and the Sociology of Mathematics, *Amer. Math. Monthly* **103** (1996), 457–469, **104** (1997), 35.

[Gour] É. Goursat, Sur un mode de transformation des surfaces minima, *Acta Math.* **11** (1887–1888), 135–186 and 257–264.

[Gray1] A. Gray, **Tubes**, The Advanced Book Program, Addison-Wesley, Redwood City, CA, 1990.

[Henn] L. Henneberg, Über diejenige Minimalfläche, welche die Neil'sche Parabel zur ebenen geodätischen Linie hat, *Vierteljahresschr. Naturforsch. Ges. Zürich* **21** (1876), 66–70.

[Hicks] N.J. Hicks, **Notes on Differential Geometry**, Van Nostrand Reinhold, Princeton, NJ, 1965.

[Higg] A.L. Higgins, **The Transition Spiral and Its Introduction to Railway Curves with Field Exercises in Construction and Alignment**, Constable, London, 1921.

[Hil] D. Hilbert, Über Flächen von konstanter Gausscher Krümmung, *Trans. Amer. Math. Soc.* **2** (1901), 87-99.

[HC-V] D. Hilbert and S. Cohn-Vossen, **Geometry and the Imagination**, Chelsea Publishing Co., New York, 1952.

[Hille] E. Hille, **Analytic Function Theory**, Second Edition, 2 volumes, Chelsea Publishing Co., New York, 1973.

[Hoff] D. Hoffman, The Computer-Aided Discovery of New Embedded Minimal Surfaces, *Mathematical Intelligencer* **9** (1987), 8–21.

[HoMe] D. Hoffman and W. Meeks, A complete minimal surface in $\mathbb{R}^3$ with genus one and three ends, *J. Differential Geometry* **21** (1985), 109–127.

[Hopf1] H. Hopf, Über die Drehung der Tangenten und Sehen ebener Kurven, *Composito Math.* **2** (1935), 50–62.

[Hopf2] H. Hopf, **Differential Geometry in the Large**, Seminar Lectures, New York University 1946 and Stanford University 1956, *Lecture Notes in Mathematics* **1000**, Springer-Verlag, Berlin-New York, 1983.

[HR] H. Hopf and W. Rinow, Über den Begriff der vollständigen differentialgeometrischen Flächen, *Comm. Math. Helv.* **3** (1931), 209–225.

[Jacobi] C.G.J. Jacobi, Ueber einige merkwürdige Curventheoreme, *Astr. Nachr.* **20** (1843), 115-120. **Math. Werke 7**, 34-39.

[Joach1] M.F. Joachimsthal, Demostrationes theorematum ad superficies curvas spectantium, *J. reine angew. Math. (Crelle's Journal)* **30** (1846), 347–350.

[Joach2] M.F. Joachimsthal, **Anwendung der Differential- und Integralrechnung auf die Allegemeine Theorie der Flächen und der Linien Doppelter Krümmung**, Third Edition, Teubner, Leipzig, 1890.

[Kauf1] L.H. Kauffman, **On Knots**, *Ann. Math. Stud.* **115**, Princeton University Press, Princeton, NJ, 1987.

[Kauf2] L.H. Kauffman, **Knots and Physics**, World Scientific, Singapore, 1991.

[Klein] F. Klein, Bemerkungen über den Zusammenhang der Flächen, *Math. Ann.* **7** (1874), and **9** (1875/76), **Math. Werke 2**, 63-77.

[Kosz] J.L. Koszul, Homologie et cohomologie des algèbres de Lie, *Bull. Soc. Math. France* **78** (1950) 65–127.

[Krey1] E. Kreyszig, **Introduction to Differential Geometry and Riemannian Geometry**, *Mathematical Expositions No. 16*, University of Toronto Press, 1968. Reprint by Dover Publications, New York, 1991.

[Krey2] E. Kreyszig, On the theory of minimal surfaces, **The Problem of Plateau**, 138–164, World Scientific, Singapore, 1993.

[Kuen] Th. Kuen, Fläche von constantem negativen Krümmungsmass nach L. Bianchi, *Sitzungsber. d. Bayer. Akad.*, 1884.

[Lag] J.L. Lagrange, Essai d'une nouvelle méthode pour déterminer les maxima et les minima des formules intégrales indéfinies. *Miscellanea Taurinensia* **2** (1760-1761), 173–195. Œuvres de Lagrange, Volume 1, Gauthier-Villars, Paris, 1867, 335–362. English translation (in part) by D.J. Struik [Stru3, 406–413].

[Lait] E.V. Laitone, Relation of the conjugate harmonic function to $f(z)$, *Amer. Math. Monthly* **84** (1977), 281–283. Review: *Math. Rev.* **55**, # 10646.

[Lamé] G. Lamé, **Leçons sur les Coordonnées Curvilignes**, Mallet-Bachelier, Paris, 1859. Reprint by University Microfilms International, Ann Arbor, MI, 1992.

[Law] J.D. Lawrence, **A Catalog of Special Plane Curves**, Dover Publications, New York, 1972.

[Lie] S. Lie, Beiträge zur Theorie der Minimalflächen I, *Math Ann.* **14** (1879), 331–416.

[Lem] J. Lemaire, **Hypocycloïdes et Épicycloïdes**, Librairie Vuibert, Paris, 1938. Reprint by Blanchard, Paris, 1967.

[LeviC]   T. Levi-Civita, **Lezioni di Calcolo Differenziale Assoluto**, Roma, 1925. Reprinted by Dover Publications, New York, 1977 as **The Absolute Differential Calculus**.

[LCRC]   T. Levi-Civita and G. Ricci-Curbastro, Méthodes de calcul différentiel absolu et leurs applications. *Math. Ann.* **54** (1901), 125–201.

[Liou]   J. Liouville, Note au sujet de l'article précédent, *J. Math. pures et appl.* **12** (1847), 265–290.

[Lip]   M.M. Lipschutz, **Theory and Problems of Differential Geometry**, *Schaum's Outline Series*, McGraw-Hill, New York, 1969.

[Lock]   E.H. Lockwood, **A Book of Curves**, Cambridge University Press, Cambridge, 1961.

[Loria1]   G. Loria, **Curve Piane Speciali**, Ulrico Hoepli, Milano, 1930. German translation: **Spezielle algebraische und transzendente Kurven**, Teubner, Leipzig, 1910.

[Loria2]   G. Loria, **Curve Sghembe Speciali**, 2 volumes, Nicola Zanichelli, Bologna, 1910.

[Ma]   R. Maeder, **Programming in Mathematica**, Second Edition, The Advanced Book Program, Addison-Wesley, Redwood City, CA, 1991.

[Main]   G. Mainardi, Sulle coordinate curvilinee d'una superficie dello spazio, *Giornale del R. Istituto Lombardo* **9** (1856), 385-398.

[Max]   J.C. Maxwell, On the cyclide, *Quarterly J. of Pure and Applied Math.* **34** (1867), 144–159. Also, Collected Works **2**, 144–159.

[Massey]   W.S. Massey, **Algebraic Topology: An Introduction**, Springer-Verlag, Berlin-New York, 1977.

[McCl]   J. McCleary, **Geometry from a Differentiable Viewpoint**, Cambridge University Press, Cambridge, 1993.

[Mell]   A.P. Mellish, Notes on Differential Geometry, *Ann. of Math.* **(2) 32** (1931), 181–190.

[Melz]   Z.A. Melzak, **Companion to Concrete Mathematics**, 2 volumes, John Wiley and Sons, New York, 1973, 1976.

[Meu] J.B. Meusnier, Mémoire sur la courbure des surfaces, *Mémoires des savans étrangers* **10** (lu 1776) (1785), 477–510.

[MiPa] R.S. Millman and G.D. Parker, **Elements of Differential Geometry**, Prentice-Hall, Englewood Cliffs, NJ, 1977.

[Mind] F.A. Minding, Wie sich unterscheiden läßt, ob zwei gegebene krumme Fächen aufeinander abwickelbar sind oder nicht; nebst Bemerkungen über Fächen von unveränderlichem Krümmungsmaß, *J. reine angew. Math. (Crelle's Journal)* **19** (1839), 370–387.

[Monge] G. Monge, **Application de l'Analyse à la Géométrie**, cinquième édition, revue, corrigée et annotée par M. Liouville, Bachelier, Paris, 1850. Reprint by University Microfilms International, Ann Arbor, MI, 1979.

[MoRo] S. Montiel and A. Ros, **Curvas y Superficies**, Granada, 1966.

[MoSp] P. Moon and D. Spencer, **Field Theory Handbook**, Springer-Verlag, Berlin-New York, 1961.

[MS] G. Mostow and J.H. Sampson, **Linear Algebra**, McGraw-Hill, New York, 1969.

[Muk] Mukhopadhyaya, New methods in the geometry of a plane arc, *Bull. Calcutta Math. Soc.* **1** (1909), 31–37.

[Newton] I. Newton, **Curves, The Mathematical Works of Issak Newton**, Volume 2, 135–161, Johnson Reprint Corporation, New York-London, 1962.

[NiSh] V.V. Nikulin and I.R. Shafarevich, **Geometries and Groups**, Springer-Verlag, Berlin-New York, 1987.

[Nits] J.C.C. Nitsche, **Lectures on Minimal Surfaces**, Volume 1, Cambridge University Press, Cambridge, 1989. English translation (with additions) of the first part of: **Vorlesungen über Minimalflächen**, *Grundlehren der Mathematischen Wissenschaften* **199**, Springer-Verlag, Berlin-New York, 1975.

[Og1] A. Ogawa, The trinoid revisited, *Mathematica J.* **2**, Issue 1, (1992), 58–60.

[Og2] A. Ogawa, Helicatenoid, *Mathematica J.* **2**, Issue 2, (1992), 21–23.

[ON1] B. O'Neill, **Elementary Differential Geometry**, Academic Press, New York-London, 1966.

[ON2]   B. O'Neill, **Semi-Riemannian Geometry**, Academic Press (Harcourt Brace Jovanovich), New York-London, 1983.

[Oprea]   J. Oprea, **Differential Geometry and Its Applications**, Prentice-Hall, Englewood Cliffs, NJ, 1997.

[Os1]   R. Osserman, **A Survey of Minimal Surfaces**, Dover Publications, New York, 1986.

[Os2]   R. Osserman, The Geometry Renaissance in America: 1938-1988, **A Century of Mathematics in America**, volume 2, American Mathematical Society, Providence, 1988.

[Pet]   K.M. Peterson, **Ueber die Biegung der Flächen**, Kandidatenschrift, Dorpat, 1853.

[RaTo1]   H. Rademacher and O. Töplitz, **Von Zahlen und Figuren**, Second Edition, Springer-Verlag, Berlin-New York, 1933.

[RaTo2]   H. Rademacher and O. Töplitz, **The Enjoyment of Mathematics; Selections from Mathematics for the Amateur**, A translation from [RaTo1] with Chapters 15 and 28 added by Herbert Zuckerman. Princeton University Press, Princeton, NJ, 1957. Reprint by Dover Publications, New York, 1990.

[Reich]   K. Reich, Die Geschichte der Differentialgeometrie von Gauß bis Riemann (1828-1868), *Arch. for the Hist. of Exact Sci.* **11** (1973), 273-382.

[Reid]   K. Reidemeister, **Knotentheorie**, Springer-Verlag, Berlin-New York, 1933. Reprint, 1974. English translation: **Knot theory**, BCS Associates, Moscow, ID, USA, 1983.

[Reul]   F. Reuleaux, **Lehrbuch der Kinematik**, 2 volumes, Friedr. Vieweg & Sohn, Braunschweig, 1875.

[Ricci]   G. Ricci-Curbastro, Résumé de quelques travaux sur les systémes variables de fonctions, *Bulletin des Sciences Mathématiques* **16** (1892), 167–189.

[Riem]   B. Riemann, Über die Hypothesen, welche der Geometrie zugrunde liegen, *Abhandlungen der Königlichen Gesellschaft der Wissenschaften zu Göttingen.* **13** (1867).

[Rolf]   D. Rolfsen, **Knots and Links**, Publish or Perish, Wilmington, DE, 1976.

[Roth]   E.H. Roth, **Revised Railway Transition Spiral: Text and Tables**, J.P. Bell, Lynchburg, VA, 1955.

[Row]   T.S. Row, **Geometric Exercises in Paper Folding**, Dover Publications, New York, 1963.

[Ryan]   P.J. Ryan, Homogeneity and some curvature conditions for hypersurfaces, *Tôhoku Math. J.* (2) **21** (1969), 363–388.

[Salmon1]   G. Salmon, **A Treatise on the Analytic Geometry of Three Dimensions**, Longmans, New York, 1928. Reprint by Chelsea Publishing Co., New York, 1965.

[Salmon2]   G. Salmon, **A Treatise on the Higher Plane Curves**, G.E. Stechert, 1934. Reprint by Chelsea Publishing Co., New York, 1960.

[Sav]   A.A. Savelov, **Ploskie Krivye** (Plane Curves), Fizmatgiz, Moscow, 1960.

[Scherk]   H.F. Scherk, Bemerkung über die kleinste Fläche innerhalb gegebener Grenzen, *J. reine angew. Math. (Crelle's Journal)* **13** (1835), 185–208.

[Schill]   F. Schilling, **Die Pseudosphäre und die nichteuklidische Geometrie**, Teubner, Leipzig, 1935.

[Schwa]   H.A. Schwarz, **Gesammelte Mathematische Abhandlungen**, Volumes 1 and 2, Springer-Verlag, Berlin, 1890. Reprint by Chelsea Publishing Co., New York, 1972.

[Scott]   W.R. Scott, **Theory of Groups**, Prentice-Hall, Englewood Cliffs, NJ, 1964.

[Serret1]   J.A. Serret, Sur quelques formules relatives à la théorie des courbes à double courbure, *Journ. de Math.* **16** (1851), 193–207.

[Serret2]   P.J. Serret, **Théorie nouvelle géométrique et mécanique des lignes à double courbure**, Mallet-Bachelier, Paris, 1860.

[Serret3]   J.A. Serret, **Lehrbuch der Differential- und Integralrechnung**, 3 volumes, German Translation by A. Harnacks, adapted by G. Scheffers, Fourth and Fifth Editions, Teubner, Leipzig, 1912.

[ShTi]   W.T. Shaw and T. Tigg, **Applied Mathematica, Getting Started, Getting it Done**, The Advanced Book Program, Addison-Wesley, Redwood City, CA, 1993.

[Shikin] E.V. Shikin, **Handbook and Atlas of Curves**, CRC Press, Boca Raton, FL, 1995.

[Siev] H. Sievert, Über die Zentralflächen der Enneperschen Flächen konstanten Krümmungsmaßes, Dissertation, Tübingen, 1886, Kap VI.

[SmBl] C. Smith with N. Blachman, **The Mathematica Graphics Guide Book**, Addison-Wesley, Reading, MA, 1994.

[Sny] V. Snyder, Criteria for Nodes in Dupin's Cyclides, with a Corresponding Classification, *Ann. of Math.* (Series 1) **11** (1898), 137–147.

[Spivak] M. Spivak, **A Comprehensive Introduction to Differential Geometry**, Second Edition, 5 volumes, Publish or Perish, Wilmington, DE, 1979.

[Springer] G. Springer, **An Introduction to Riemann Surfaces**, Chelsea Publishing Co., New York, 1981.

[Still1] J. Stillwell, **Mathematics and Its History**, Springer-Verlag, Berlin-New York, 1989.

[Still2] J. Stillwell, **Geometry of Surfaces**, Springer-Verlag, Berlin-New York, 1990.

[Stoker] J.J. Stoker, **Differential Geometry**, John Wiley and Sons, New York, 1969.

[Strub] Karl Strubecker, **Differentialgeometrie**, Second edition, 3 volumes, *Sammlung Göschen* 1113/1113a, 1179/1179a, 1180/1180a, Walter de Gruyter, Berlin, 1964.

[Stru1] D.J. Struik, Outline of a history of differential geometry, *Isis* **19** (1933), 92–120, **20** (1933), 161–191.

[Stru2] D.J. Struik, **Lectures on Classical Differential Geometry**, Second Edition, Addison-Wesley, Reading, MA, 1961. Reprint by Dover Publications, New York, 1988.

[Stru3] D.J. Struik, **A Source Book in Mathematics, 1200–1800**, Harvard University Press, Cambridge, MA, 1969.

[Study] E. Study, Über einige imaginäre Minimalflächen, *Leipz. Akad.-Ber.* **63** (1911), 14–26.

[Thom]   D'A.W. Thompson, **Growth and Form**, Cambridge University Press, Cambridge, 1917.

[Thor]   J.A. Thorpe, **Elementary Topics in Differential Geometry**, *Undergraduate Texts in Mathematics*, Springer-Verlag, Berlin-New York, 1979.

[vonSeg] D.H. von Seggern, **CRC Handbook of Mathematical Curves and Surfaces**, CRC Press, Boca Raton, FL, 1990.

[Visc]   P. Viscensini, Differential geometry in the nineteenth century, *Scientia* **107**, 661-696. Translation of La géométrie différentiele au XIXeme siécle, *Scientia* **107**, 617-660.

[Wagon] S. Wagon, **Mathematica in Action**, W.H. Freeman and Company, New York, 1991.

[Walk]   R.J. Walker, **Algebraic Curves**, Princeton University Press, Princeton, NJ, 1950. Reprint by Springer-Verlag, Berlin-New York, 1978.

[Wea]   C.E. Weatherburn, **Differential Geometry of Three Dimensions**, 2 volumes, Cambridge University Press, Cambridge, 1927, 1930.

[Wei1]   K. Weierstrass, Über die geodätischen Linen auf dem dreiachsigen Ellipsoid, **Math. Werke I**, 257–266.

[Wei2]   K. Weierstrass, Untersuchungen über die Flächen, deren mittlere Krümmung überall gleich Null ist, *Moatsber. Berliner Akad.* (1866), 612–625. **Math. Werke III**, 39–52.

[Weyl]   H. Weyl, **The Concept of a Riemann Surface**, Addison-Wesley, Reading, MA, 1964. (English translation of **Die Idee der Riemannschen Flächen**, Third Edition, Teubner, Stuttgart, 1955.)

[Wick]   T. Wickham-Jones, **Graphics with Mathematica**, Springer-Verlag, TELOS, Berlin-New York, 1994.

[Wiel1] H. Wieleitner, **Theorie der ebene algebraische Kurven höherer Ordnung**, *Sammlung Schubert* **13**, G.J. Göschen'sche Verlagshandlung, Leipzig, 1908.

[Wiel2] H. Wieleitner, **Spezielle ebene Kurven**, *Sammlung Schubert* **56**, G.J. Göschen'sche Verlagshandlung, Leipzig, 1908.

[Will]   T.J. Willmore, **An Introduction to Differential Geometry**, Oxford University Press, London, 1959.

[Wm]   S. Wolfram, **Mathematica**, Second Edition, The Advanced Book Program, Addison-Wesley, Redwood City, CA, 1990.

[Wood]   R. Woodhouse, **Mathematica, A History of the Calculus of Variations in the Eighteenth Century**, Chelsea Publishing Co., New York, 1964. Reprint of **A Treatise on Isoperimetrical Problems and the Calculus of Variations**, originally published in 1810.

[Yates]   R.C. Yates, **Curves and Their Properties**, The National Council of Teachers of Mathematics, 1974. Reprint by University Microfilms International, Ann Arbor, MI, 1992.

[Zwi]   C. Zwikker, **The Advanced Geometry of Plane Curves and their Applications**, Dover Publications, New York, 1963.

INDEX

Absolute value, 4
Abstract surface, 574
Acceleration of a curve, 6, 191, 265, 418
   normal component of the, 362
   with respect to a connection, 563
Acrospin, 994
Addition of vectors, 2
Agnesi, curve of, 72, 891, 916
Airy functions, 148
Algebra
   Lie, 546
   of real-valued differentiable
      functions, 527
Algebraic curve, 146
Analytic
   function, 521, 705, 832
   manifold, 524
Angle
   between
      tangent vectors to $\mathbb{R}^n$, 247
      tangent vectors to a surface, 349
      vectors in $\mathbb{R}^n$, 3
   exterior, 629
   function between curves in $\mathbb{R}^2$, 18
   interior, 629
   of a rotation in $\mathbb{R}^2$, 134
   oriented, 4, 372, 520
   tangent-radius angle, 23
   turning, 19, 585
Anticomplex map, 694

Antipodal map, 290, 330
Antisymmetric bilinear form, 564
Arc length, 11
   of a polar parametrization, 89
Archimedes, spiral of, 90, 892
Area
   element of, 351, 352, 354, 881
   of a subset of a patch, 353
   of a surface, 351
Associated family
   of a minimal curve, 717
   of conjugate harmonic minimal
      isothermal patches, 707
Associative, 130
Astroid, 23, 46, 241, 479, 766, 974
   3-dimensional, 203, 927
Astroidial
   ellipsoid, 396, 414
   sphere, 397
Asymptotic
   curve, 418
      differential equation for an, 418, 419
   direction, 364, 418
   patch, 421
   vector, 364
Atlas, 523
   complete, 524
Axis
   of a generalized helicoid, 475
   of revolution, 457

Barrel type, 485, 498
Base curve, 431, 603
Baseball seam, 927
Basis
   orthonormal, 132
   Theorem, 536
Beltrami-Enneper Theorem, 421
Bernoulli, lemniscate of, 52, 53, 82, 770, 771, 907, 920, 980
Bessel function, 143, 206, 927
Bicylinder, 204, 205
   tangent developable to, 456
Binormal vector field to a curve in $\mathbb{R}^3$, 185
Björling
   curve, 762
   formula of, 762, 763
Bohemian dome, 389, 935
Bonnet, Theorem of, 781
Boundary of a set, 629
Bounded set, 353
Bour
   minimal curve of, 974
   minimal surface of, 732, 936
Bowditch curve, 70, 787, 908, 930
Bracket, 546, 547
Brioschi, formula of, 504, 506, 574, 580, 593, 841

C^∞-functions, 521
C^k-manifold, 524
Canal surface, 799
Capture point, 67
Cardioid, 54, 88, 203, 241, 479, 772, 893, 975
   curvature of a, 54
   evolute of a, 120
   parallel curve to a, 120
   pedal curve to a, 119, 120
   spherical, 932
Cassini, oval of, 82, 917, 923

Catalan
   minimal curve of, 975
   minimal surface of, 693, 724, 766, 937
      isometric deformation of, 725, 937
Catenary, 23, 55, 56, 103, 467, 895
   involute of a, 121
   natural equation of a, 138, 139
   pedal curve to a, 121
Catenoid, 467, 470, 519, 685, 686, 741, 760, 777, 938, 975
   elliptic, 769, 944, 978
Cauchy-Riemann equations, 705, 706
Cauchy-Schwarz inequality, 3, 231, 247, 762
Cayley, sextic of, 119, 895, 975
Cayley-Hamilton Theorem, 381
Center
   curve of a canal surface, 799
   of curvature, 98
Chain rule, 7, 249, 250, 265, 538
   for curves on a manifold, 545
Change of coordinates, 284, 523
Christoffel symbols, 510, 512, 513, 516, 519, 570, 576, 577, 597, 598, 703, 842, 844
Circle, 6, 36, 50, 54, 59, 82, 105, 107, 371, 465, 895, 918
   central, 325, 328
   involute of a, 23, 46, 105–107, 121
   natural equation of a, 138
   osculating, 98, 111, 113, 121, 228
      to a cubic parabola, 112, 121
      to a logarithmic spiral, 111, 112
      to a parabola, 115, 120
   pedal curve to a, 119
Cissoid, 57, 479, 976
   curvature of a, 60
   evolute of a, 101, 102
   pedal curve to a, 119
Clairaut
   patch, 604

relation of, 605
Closed curve, 155
   rotation index of a, 160
   simple, 160
Clothoid, 64, 141, 145, 471, 771, 896, 939, 976
   natural equation of a, 138
Coefficients of the
   first fundamental form, 342, 368, 381
   second fundamental form, 368, 381, 709, 720
Coherently oriented, 321
Compact set, 353
Complete
   atlas, 524
   elliptic integral of the second kind, 482
   geodesically, 600
   metric space, 624
Complex
   conjugate, 4
   derivative of a patch, 711, 716
   Euclidean n-space, 714
   manifold, 524
   map, 694
   structure, 4, 29, 132, 247, 261, 317, 318, 582, 846
Complexification, 713
Composition of maps, 131
Cone
   circular, 387, 497
   generalized, 439
   vertex of a, 439
Confocal surfaces, 669, 676
Conformal
   diffeomorphism, 349, 820
   map, 349, 702
Conic type, 487, 498
Conical edge of Wallis, 454, 968
Conjugate
   harmonic, 706
   Weierstrass patch, 737, 738

Conjugation, 715
Connected topological space, 320
Connection
   natural, on $\mathbb{R}^n$, 559
   on a differentiable manifold, 558
   Riemannian, 566
   torsion of a, 572
   with torsion zero, 567
Contact of order k, 225
Contrapedal, 124
Convex plane curve, 163
Coordinate
   function, 522
   neighborhood of a patch, 522
Coordinates
   change of, 284, 523
      rectangular to polar, 307
   elliptic, 670, 945
   natural, of $\mathbb{R}^n$, 286
   parabolic, 676, 958
   polar, 307
Coradial tangent vector, 620
Corkscrew surface, 477, 968
Cornu, spiral of, 64
Costa, minimal
   curve of, 977
   surface of, 747, 757, 940
Covariant
   derivative, 558
   tensor, 551
   tensor field, 550
Critical point, 267
Cross
   cap, 333, 334, 415, 969
   product, 182, 194, 318
Cubic, twisted, 189, 455, 933
Curtate cycloid, 51, 70
Curvature
   center of, 98
   Gaussian, 373, 374, 502, 574, 580, 703
   of a minimal curve, 720

geodesic, 514, 519, 583, 862
mean, 373, 374, 388, 704
    vector field, 589
meaning of the sign of, 15
normal
    of a curve, 513
    of a surface, 359, 363, 373
of a cardioid, 54
of a circular helix, 189, 199
of a cissoid, 60
of a curve in $\mathbb{R}^2$, 14, 30, 136
of a curve in $\mathbb{R}^3$, 191, 195
of a curve in $\mathbb{R}^n$, 184
of a cycloid, 51
of a figure eight, 32
of a lemniscate, 53
of a parallel curve, 116
of a polar parametrization, 89
of Viviani's curve, 201, 232
principal, 364, 369, 373
radius of, 98, 139
signed, of a curve in $\mathbb{R}^2$, 14, 30, 184
total, geodesic, 588
total, signed, 154
unsigned geodesic, 513
Curve
    acceleration of a, 6, 191, 265, 362, 418
    algebraic, 146
    arc length function of a, 11
    asymptotic, 418, 597
    base, of a ruled surface, 431
    bicylinder, 204, 456
    Bowditch, 70, 787
    center, of a canal surface, 799
    closed, 155
        length of a, 155
        period of a, 155
        rotation index of a, 160
        simple, 160
    convex, 163
    devil's, 92, 919

differentiable, 5
director, of a ruled surface, 431
family, 172
geodesic, 596–598, 606, 608, 609, 611, 613, 615
    curvature of a, 514
    maximal, 600
    torsion of a, 517
helicoidial over
    a cardioid, 203
    a cycloid, 203
    a figure of eight, 203
    a logarithmic spiral, 200
    a plane curve, 199
implicitly defined in $\mathbb{R}^2$, 76
in $\mathbb{R}^n$, 5
integral, of a vector field, 560
isotropic, 716
length of a, 7
Lissajous, 70, 787
minimal, 716, 762
    Bour's, 732, 974
    Catalan's, 975
    catenoid, 975
    Costa's, 977
    ellipsemincurve, 978
    Enneper's, 978
    Henneberg's, 979
    regular, 716
    Richmond's, 981
negative reparametrization of a, 7
nonparametric form of a, 6, 76
normal
    curvature of a, 513
    line to a, 100
of constant slope, 234
of pursuit, 67
on a differentiable manifold, 543
on a surface, 286
on one side of a line, 163
parallel to a, 116

parametric form of a, 5, 6, 76
pedal to a, 118, 121
piecewise-differentiable, 5
positive reparametrization of a, 7
pregeodesic, 600, 693, 696
principal, 459, 517, 643, 668
radius of curvature of a, 14
regular, 6
   at a point, 22
signed curvature of a, 14, 30
simple closed, 160
speed of a, 6
striction, 445, 446
tangent line to a, 100
total
   geodesic curvature of a, 588
   signed curvature of a, 154
trace of a, 6
tube around a, 207
u-parameter, 275, 296
unit-speed, 6
unsigned geodesic
   curvature of a, 513
v-parameter, 275, 296
vector field along a, 12
velocity of a, 6, 30, 191
Viviani's, 201, 202, 205, 232, 453, 456
   curvature of, 201, 232
   torsion of, 201, 232
Curvilinear patch, 665, 783
Cusp, 57, 102, 124
Cyclide of Dupin, 809, 812, 951, 952
Cycloid, 50, 51, 70, 120, 203, 241, 479, 693, 696, 766, 899
   curtate, 51, 70
   curvature of a, 51
   evolute of a, 120
   prolate, 51, 70
Cylinder, 322, 387
   circular, 273, 497, 508, 603
   elliptical, 603

generalized, 439
over a figure of eight, 441

Darboux
   frame field, 518
   vector field, 205, 871
Deformation
   from a helicoid to a catenoid, 685, 742
   isometric, 707, 717
   of Catalan's minimal surface, 725
Degree
   of a continuous function of a circle onto itself, 159
   of a tensor field, 550, 551
Deltoid, 70, 479, 977, 978
   parallel curve to a, 120
Derivation, 545
Derivative
   complex, of a patch, 711, 716
   covariant, 558
   directional, 248, 249
   of a vector field, 259, 261
Descartes
   folium of, 78, 80, 904, 919, 924
      perturbed, 78, 79
Developable surface, 439
Devil's curve, 92, 919
Diffeomorphism
   between
      manifolds, 531
      open sets of $\mathbb{R}^n$, 250
      regular surfaces, 289
   conformal, 349, 820
   local, 531
Differentiable
   curve
      in $\mathbb{R}^n$, 5
      on manifold, 543
      on a regular surface, 286
   function, 285
      between $\mathbb{R}^n$ and $\mathbb{R}^m$, 250

between regular surfaces in $\mathbb{R}^n$, 288
manifold, 521, 523
map between manifolds, 530, 543
real-valued function, 521
vector field on $\mathbb{R}^n$, 255
Differential
  1-form, 552
  of a function, 344, 553
Dimension, 523
Dini, surface of, 493, 494, 942, 943
Diocles, cissoid of, 57, 896, 918
Direction
  asymptotic, 364, 418
  on a surface, 363
  principal, 364
Directional derivative, 248
Directrix, 431
  of a parabola, 45
  of a ruled surface, 431
Distance
  between points in $\mathbb{R}^n$, 3
  function, 348, 623
  infinitesimal, 343
  intrinsic, 348, 349
Distribution parameter, 447
Divergence, 263, 267
Dot product, 3, 714
Doubly ruled surface, 431
Dual 1-form of a vector field, 566
Dupin
  cyclide of, 951, 952
  Theorem of, 668

Edge, 631
  of regression, 441
Eigenvalues and eigenvectors, 371
Eight
  figure of, 31, 35, 53, 105, 203, 241, 480, 787, 901
    curvature of a, 32
    cylinder over, 441

involute of a, 105, 106, 122, 124
surface, 310
Element of area, 351, 352
Ellipse, 37, 47, 108–110, 135, 210, 241, 767, 902, 944, 978
  cubical, 928
  curve parallel to an, 116
  evolute of an, 99, 100
  focus of an, 38
  length of an, 46, 499
  pedal curve to an, 121
  rotated, 135
Ellipsoid, 358, 387, 626, 641, 671, 675, 679, 944, 969
  astroidial, 396, 414
    Gaussian curvature of an, 415
  Gauss map of an, 407
  Gaussian curvature of an, 414
  Mercator, 312, 956
  nonparametric form of an, 301
  of revolution, 301, 680
  parametric form of an, 302
  pedal surface to an, 785
  stereographic, 302, 966
  support function of an, 414
Elliptic
  coordinate patch, 670, 945
  integral of the second kind, 482
  point, 375, 418
Elliptical
  catenoid, 769, 944
  helicoid, 422
  torus, 210, 304
Enneper
  minimal curve of, 978
  minimal surface of, 358, 388, 415, 519, 684, 726, 728, 760, 945
Envelope, 812
  of curves, 173
  of surfaces, 798
Epicycloid, 150, 158, 978

natural equation of an, 151
Epitrochoid, 157
Equation
   differential, of a geodesic, 597
   natural, 138
      of a catenary, 138, 139
      of a circle, 138
      of a clothoid, 138
      of a curve in $\mathbb{R}^2$, 147
      of a hypocycloid, 151
      of a logarithmic spiral, 138
      of an epicycloid, 151
      of an involute of a circle, 138, 146
Equations
   Cauchy-Riemann, 705, 706
   Gauss, 511, 704
      for abstract surfaces, 593
   Peterson-Mainardi-Codazzi, 641, 649, 651, 660, 705
   Weingarten, 369, 392, 461, 650, 705
Euclidean
   motion of $\mathbb{R}^n$, 130, 217, 362
   n-space, 2, 246, 521, 523
      complex, 714
Euler
   characteristic, 635
   constant of, 147
   spiral of, 896
   Theorem of, 372
Evolute
   of a circle, 99
   of a cissoid, 101, 102
   of a cycloid, 120
   of a logarithmic spiral, 120
   of a plane curve, 98–100, 111
   of a space curve, 230
   of a tractrix, 102, 103
   of an ellipse, 99
Exponential
   graph, length of an, 46
   map, 617

Exterior angle, 629

Face, 632
Fermat, spiral of, 90, 903, 924
Figure of eight, 328, 338
Filling of plane curves, 34
First fundamental form, 342, 380, 381, 880, 881
   coefficients of the, 342, 368, 381
Flat, 374
   generalized helicoid, 490
   ruled surface, 439
   torus, 579, 947
Focal
   interval, 790
   patch, 791
   point, 790
   set, 790
Focus
   of a lemniscate, 52
   of a parabola, 45
   of an ellipse, 38
Folium
   of Descartes, 77, 78, 80, 904, 919, 924
      perturbed, 78, 79
   of Kepler, 93, 920
Football type, 484, 498
Form
   bilinear
      antisymmetric, 564
      signature of a, 565
      symmetric, 564
      symmetric nondegenerate, 564
      symmetric positive definite, 565
   differential, 552
   first fundamental, 380, 381, 880
   second fundamental, 380, 381, 884
   third fundamental, 381
Formula
   Björling's, 762
   Brioschi's, 504, 506, 574, 580, 593, 841

Frame field
   Darboux, 518
   Frenet, 185, 420, 871
   natural, of $\mathbb{R}^n$, 256
   plane, 868
Freeway exit, 64
Frenet
   formulas, 186, 187, 189, 191, 192, 214, 223, 420, 421, 517
   frame field, 185, 420, 871
Fresnel integral, 65
Function, 26
   Airy, 148
   analytic, 521, 705, 832
   arc length, 11, 343
   Bessel, 143
   coordinate, of a patch, 522
   differentiable, 285, 521, 526
   differential of a, 344, 553
   distance, 348, 623
      square of, 286
   elliptic, 768
   Gamma Γ, 148
   harmonic, 706
      conjugate, 706
   height, 286
   holomorphic, 524, 679
   in *Mathematica*, 27, 107, 741
   Kronecker delta, 537
   meromorphic, 736
   natural coordinate, of $\mathbb{R}^n$, 255, 286, 522
   sawtooth, 149
   scaling, 702
   square wave, 150
   support
      of a hyperboloid of one sheet, 414
      of a hyperboloid of two sheets, 414
      of a plane curve, 175
      of a surface in $\mathbb{R}^3$, 410
      of an ellipsoid, 414

Fundamental
   form
      first, 380, 381, 880
      second, 380, 381, 884
   period parallelogram, 748
   theorem of plane curves, 136, 137
   theorem of space curves, 219
   theorem of surfaces, 654, 656

Gamma function Γ, 148
Gauss
   equations, 511, 704
      for abstract surfaces, 593
   Lemma, 620
   map, 279, 293, 319, 362, 470, 760, 884, 887
      of a minimal curve, 718
      of a minimal surface, 694
   notation, 342, 368
   Theorema Egregium, 507–509, 577
Gaussian curvature, 373, 374, 384, 502, 574, 580, 703
   of a minimal curve, 720
General relativity, 579
Generalized
   cone, 439
   cylinder, 439
   parabola, 226, 904
Geodesic, 563, 596, 598, 693, 873
   curvature, 514, 519, 583, 862
   differential equation of a, 597, 613
   maximal, 600
   on a monkey saddle, 615
   on a surface of revolution, 611
   on the plane, 608
   on the Poincaré upper half-plane, 609
   polar metric, 579, 619
   pseudosphere, 931
   radial, 620
   slant on a Clairaut patch, 606
   torsion, 517, 863

Geodesically complete surface, 600
Geometric genus, 635
Geomview, 999
Gibbs-Heaviside notation, 360
Globalization, 530
Golden ratio, 33
Gradient, 257, 267, 291, 566
Great circle, 238, 458
Group, 130
Grouping, 26

Harmonic
  function, 706
  patch, 706
Height function, 286
Helicoid, 388, 415, 456, 508, 643, 685, 741, 760, 949
  circular, 449
    Gaussian curvature of a, 430
    mean curvature of a, 430
  elliptical, 422, 769
  general, 724
  generalized, 475, 479, 495
    flat, 490
  geometric definition of a, 357
  principal patch on a, 644
Helix, 198, 929
  circular, 188, 198, 199, 208, 309, 456
    curvature of a, 189
    tangent developable to a, 444
    torsion of a, 189
  elliptical, 205
  generalized, 929
  radius of a, 198
  slope of a, 198
  spherical, 237, 932
  tube around a, 208
Henneberg
  minimal curve of, 979
  minimal surface of, 691, 743, 767, 950
Hessian, 267

Holomorphic function, 524
Homothety, 146
Homotopy, 161
Horn torus, 306
Hundekurve, 61
Hyperbola, 772, 905, 979
  cubical, 928
Hyperbolic
  paraboloid, 297, 298, 449
  point, 375, 418
Hyperboloid, 945
  elliptic, 470
  of one sheet, 387, 404, 416, 432, 671, 675, 679, 950, 970
    Gaussian curvature of a, 414
    support function of a, 414
  of revolution, 470, 680
  of two sheets, 387, 406, 671, 675, 679, 951, 970
    Gaussian curvature of a, 414
    support function of a, 414
  type, 487, 498
Hyperplane, 129, 225
Hypersphere, 225
  radius of a, 225
Hypocycloid, 150, 158
  natural equation of a, 151
Hypotrochoid, 157

Identity
  element, 130
  Jacobi, 546
  Lagrange, 182, 183, 715
  linear map, 4
Implicitly defined hypersurface, 225
Inequality
  Cauchy-Schwarz, 3, 231, 247, 762
  triangle, 3, 9
Infinitesimal
  arc length, 343
  distance, 343

volume element, 351
Inner product, 565
Integral
   curve of a vector field, 560
   elliptic of the second kind, 482
      complete, 482
Interior angle, 629
Intrinsic distance, 348
Inverse, 131
   function theorem, 542
   patch, 825
Inversion, 820
   center, 820, 831
   pole of an, 820
   radius, 820, 831
Involute, 103, 104
   iterated, 107
   of a catenary, 121
   of a circle, 23, 46, 105–107, 121, 895, 896
      length of a, 46
      natural equation of an, 138, 146
   of a cubic parabola, 122, 124
   of a figure of eight, 105, 106, 122, 124
Isometric deformation, 707, 717
Isometry, 130, 346
   local, 346
Isothermal patch, 702, 737, 760
Isotropic curve, 716

Jacobi identity, 546
Jacobian matrix, 254, 271, 275, 280, 320, 331, 353, 542, 671, 677, 879
Joachimsthal, Theorem of, 460, 461
   converse to, 479, 495

Kepler, folium of, 93, 920
Klein bottle, 323, 327, 335, 338, 339, 358, 415, 592, 787, 856, 953, 954
Knot
   eight, 215, 928
   torus, 209, 210, 213, 215, 933
   trefoil, 210
Kronecker delta function, 537
Kuen's surface, 496, 955

L'Hôpital's rule, 529
Lagrange identity, 182, 183, 715
Laplacian, 263, 703
Lathed surface, 474
Legendre's relation, 749
Leibnizian property, 533, 547
Lemma, Gauss, 620
Lemniscate
   curvature of a, 53
   of Bernoulli, 52, 53, 82, 120, 479, 770, 771, 907, 920, 980
      parallel curve to a, 120
   of Gerono, 31
Length, 46, 839, 840
   of a closed curve, 155
   of a curve, 7, 136
   of the graph of the exponential function, 46
   of the graph of the logarithm function, 46
   of the involute of a circle, 46
Level surface, 291
Lie
   algebra, 546
   bracket, 546
Limaçon, 71, 88, 119, 908, 920, 925
Line
   in $\mathbb{R}^n$, 6
   normal to a curve in $\mathbb{R}^2$, 108
   tangent to a curve in $\mathbb{R}^2$, 108
Linear map, 3
Linearity property, 533, 547
Lissajous curve, 70, 787, 908, 930
Local
   diffeomorphism, 289, 531
   homeomorphism of a patch, 522
   isometry, 346

surface in $\mathbb{R}^n$, 269
system of coordinates
   of a surface, 281
Logarithm graph, length of a, 46
Logarithmic spiral, 40, 200, 479
Lower g-ij's, 569
Loxodrome, 238

Maclaurin, trisectrix of, 94, 916
Manifold, 521
   analytic, 524
   complex, 524
   differentiable, 521, 523
   product, 525
   pseudo-Riemannian, 565
   Riemannian, 565
   topological, 523
Map
   affine, 130
   anticomplex, 694
   antipodal, 290, 330
   complex, 694
   conformal, 349
   differentiable, 250
      between manifolds, 530, 543
   exponential, 617
   Gauss, 279, 293, 319, 362, 470, 760, 884, 887
      of a minimal curve, 718
      of a minimal surface, 694
   identity, 4
   linear, 3
   of surfaces, 288
   orientation-preserving, 128, 130
   orientation-reversing, 128, 130
   orthogonal, 132
   projection, of the tangent bundle, 548
   secant, 161
   tangent, 251, 271, 289, 362, 540
Matrix
   associated with a linear
   transformation, 253

Jacobian, 254, 271, 275, 280, 320, 331, 353, 542, 671, 677, 879
   of maximum rank, 254
   rank of a, 254, 272
   transpose of a, 275, 879
Maximal geodesic, 600
Mean curvature, 373, 374, 384, 388, 704
   vector field, 589
Membrane
   circular, 972
   square, 973
Menn's surface, 956
Mercator ellipsoid, 312, 956
Meridian, 238, 355
   of a generalized helicoid, 476, 495
   of a sphere, 238
   of a surface of revolution, 458
Meromorphic function, 736
Metric, 565, 881
   components of a, 568
   geodesic polar, 579
   of $\mathbb{R}^2$
      in Cartesian coordinates, 344
      in polar coordinates, 344
   Poincaré, 578, 580, 604, 609
      generalized, 581
   pseudo-Riemannian, 564
   Riemannian, 342, 347, 565
   Schwarzschild, 579, 581
   space, 623
      complete, 624
Minimal
   curve, 716
   isothermal patch, 706
   surface, definition of, 374
Möbius strip, 322, 325, 326, 338, 358, 407, 415, 437, 786, 856, 956, 957
   as a ruled surface, 437
   Gaussian curvature of a, 438
   mean curvature of a, 438
   parallel surfaces to a, 777

Monge patch, 398, 415, 429, 688
   Gaussian curvature of a, 399
   mean curvature of a, 399
Monkey saddle, 299, 300, 312, 366, 382, 615, 957, 959
   area of a, 355
   asymptotic curves of a, 423
   Gauss map of a, 408
   Gaussian curvature of a, 383, 507
   generalized, 300, 301
   geodesics on a, 614, 615
   mean curvature of a, 383, 507
   metric of a, 355
   normal sections of a, 366
   principal curvatures of a, 403
Monotone, 163

Natural
   connection on $\mathbb{R}^n$, 559
   coordinate functions of $\mathbb{R}^n$, 255, 522
   equation, 138
      a hypocycloid, 151
      an epicycloid, 151
      of a catenary, 138, 139
      of a circle, 138
      of a clothoid, 138
      of a curve in $\mathbb{R}^2$, 147
      of a logarithmic spiral, 138
      of an involute of a circle, 138, 146
   frame field of $\mathbb{R}^n$, 256
Nephroid, 479, 772, 981
   spherical, 932
Nielsen, spiral of, 146, 910, 980
Noncylindrical ruled surface, 445
Nondegenerate symmetric bilinear form, 564
Nonparametric form of a curve, 6, 76
Norm, 715
   in $\mathbb{R}^n$, 3
Normal
   curvature

      of a curve, 513
      of a surface, 359, 363, 373
   neighborhood, 618
   patch, 619
   section, 365
   to a patch, 277
   to a regular surface in $\mathbb{R}^n$, 288
   variation of a surface, 681, 682
North pole, 526
Notation
   Gauss, 342, 368
   Gibbs-Heaviside, 360

Open
   set of $\mathbb{R}^n$, 250
   submanifold, 525
   subset of a regular surface in $\mathbb{R}^n$, 281
Opposite point, 169
Order
   of an algebraic curve, 146
   of contact, 225
Orientable, 318
   abstract surface, 582
Orientation, 320
   preserving curvilinear patch, 665
Oriented, 318, 582
   angle, 4, 372, 520
   coherently, 321
Orthogonal system of curves, 679
Orthonormal basis, 132
Osculating
   circle, 98, 111, 228
   sphere, 229
Oval, 168
   of Cassini, 82, 85, 917, 923
   of constant width, 169
   spread of an, 169
   width of an, 169

Parabola, 15, 45, 115, 910, 981
   cubic, 46, 98, 112, 122, 899, 918
      involute of a, 122, 124

osculating circle to a, 121
directrix of a, 45
focus of a, 45
generalized, 226, 904
osculating circle to a, 115, 120
pedal curve to a, 119
semicubical, 21, 34, 95, 912, 913, 922, 981
vertex of a, 45
Parabolic
  coordinate patch, 676, 958
  point, 375, 418
Paraboloid, 307, 387, 645, 948, 959, 969, 970
  elliptic, 358, 415, 647, 676, 678
  hyperbolic, 287, 297, 298, 300, 312, 358, 365, 387, 415, 422, 449, 676, 678, 950
    asymptotic lines of a, 366
    general, 434
    generalized, 430
    principal curves of a, 366
    unit normal to a, 280
  nonparametric form of a, 307
  perturbed, 959
Parallel
  curve
    curvature of a, 116
    to a cardioid, 120
    to a curve in $\mathbb{R}^2$, 116
    to a deltoid, 120
    to a lemniscate, 120
    to an ellipse, 116
  helical curve of a
    generalized helicoid, 476
  of a sphere, 238
  of a surface of revolution, 458
  patch, 774
  surface, 773
    shape operator of a, 779
    to a Möbius strip, 777

Parametric form of a curve, 6
Parametrization
  by arc length, 12
  pedal, 175
Pascal's snail, 71, 908
Patch, 269, 277, 523
  Clairaut, 604
  coordinate
    function of a, 522
    neighborhood of a, 522
  curvilinear, 665, 783
    orientation preserving, 665
    regular, 665
  elliptic coordinate, 670, 945
  geodesic polar, 619
  harmonic, 706
    conjugate, 706
  injective, 273
  inverse, 825
  isothermal, 702, 737, 760
  minimal isothermal, 706
  Monge, 398, 415, 429, 688
    Gaussian curvature of a, 399
    mean curvature of a, 399
  normal, 619
  on a topological space, 522
  parabolic coordinate, 676, 958
  parallel to a patch, 774
  principal, 461, 644
  regular, 273
    at a point, 274
  ruled, 431
  trace of a, 270
  triply orthogonal, 666
    unit normal of a, 667
  vector perpendicular to, 277
  Weierstrass, 737, 738
    conjugate, 737, 738
Pedal
  curve, 118, 121
    to a cardioid, 120

to a catenary, 121
to a circle, 119
to a cissoid, 119
to a parabola, 119
to an ellipse, 121
parametrization, 175
surface, 785
to an ellipsoid, 785
Pendulum, 123
Period
of a closed curve, 155
parallelogram, fundamental, 748
Peterson-Mainardi-Codazzi
equations, 641, 649, 651, 660, 705
Piecewise-differentiable curve in $\mathbb{R}^n$, 5
Pinch point, 310
Piriform, 178
Pitch of a loxodrome, 238
Planar point, 375, 403, 418
Plane, 2, 497, 508
frame field, 868
geodesic on the plane, 608
Plücker, conoid of, 435
generalization of, 436
Poincaré metric, 578
Point
data, 474
elliptic, 375, 418
focal, 790
hyperbolic, 375, 418, 419
of application of a tangent vector
to $\mathbb{R}^n$, 246
of inflection, 123
opposite, 169
parabolic, 375, 418
pinch, 310
planar, 375, 403, 418
umbilic, 645, 790
Polar parametrization, 86
arc length of a, 89
curvature of a, 89

radius function of a, 86
Pole
north, 526
of an inversion, 820
south, 526
Polygon, regular, 47
Polygonal region, 632
Positive definite symmetric
bilinear form, 565
Pregeodesic, 600, 693, 696
Principal
curvature, 364, 369, 373
curve, 459, 517, 668
differential equation for a, 643
direction, 364
normal vector field to a curve in $\mathbb{R}^3$,
185
patch, 461, 644
radius of curvature, 789
vector, 364
Product
by scalars, 2, 714
dot, 3
interior, 565
manifold, 525
scalar, 3, 27, 247, 342, 714
symmetric, 351, 568
vector cross, 182, 194, 247, 261, 318,
715, 846
vector triple, 183, 392
wedge, 351
Profile curve
of a generalized helicoid, 475
of a surface of revolution, 457
Projection
map of the tangent bundle, 548
stereographic, 719
Projective space, real, 554
Prolate cycloid, 51, 70
Pseudo-Riemannian
manifold, 565

metric, 564
Pseudocrosscap, 337, 358
Pseudosphere, 487, 489, 498, 961, 962
   geodesic of a, 612, 931
Pull back, 552, 575
Pursuit curve, 67
Pythagorean Theorem, 59, 341, 342

Quaternions, 181

Radial
   geodesic, 620
   tangent vector, 620
Radius
   function of a polar parametrization, 86
   inversion, 820, 831
   of a canal surface, 799
   of a helix, 198
   of a hypersphere, 225
   of curvature of a curve in $\mathbb{R}^2$, 14, 98, 139
   tube, of a torus, 304, 831
   wheel, of a torus, 304, 831
Railroad construction, 64
Rank of a matrix, 254, 272
Real projective
   plane, 324, 330
   space, 554
Reflection
   of $\mathbb{R}^n$, 128
Region, 629
   polygonal, 632
   regular, 631
   triangular, 632
Regression, edge of, 441
Regular
   curve, 6
      at a point, 22
   curvilinear patch, 665
   isothermal patch, 713
   map between manifolds, 543
   minimal curve, 716

patch in $\mathbb{R}^n$, 273
polygon, 47
region, 631
surface in $\mathbb{R}^n$, 281
Reparametrization, 7
   negative, 7
   positive, 7
Rhumb line, 238
Richmond
   minimal curve of, 981
   minimal surface of, 744, 962
Riemann
   mapping theorem, 833
   sum approximation, 10
Riemannian
   connection, 566
   manifold, 565
   metric, 342, 565
Right conoid, 450, 451, 456
Ring cyclide, 813
Rose, 981
Rotation
   angle, 134
   index, 160
   of $\mathbb{R}^n$, 128
Ruled
   patch, 431
   surface, 431
Rulings, 431, 603

Saddle surface, 299
Sawtooth
   function, 149
   wave, 149
Sc, 922
Scalar
   multiplication, 2
   product, 3
      in $\mathbb{C}^n$, 714
Scale factor, 349
Scaling function, 702

Scarab, 913
Scherk
   fifth minimal surface of, 697, 698, 964, 971
   minimal surface of, 688, 689, 760, 964, 971
      implicit plot of, 690
      twisted generalization of, 696
Schwarzschild
   metric, 579, 581
   surface, 580, 581
Screw motion, 459, 475
Sea shell, 309
Secant map, 161
Second fundamental form, 380, 381, 884, 885
   coefficients of the, 368, 381, 709, 720
Section
   normal, 365
   of the tangent bundle, 548
Semicubical parabola, 21, 95
Serpentine, 913
Set
   bounded, 353
   compact, 353
   of zeros, 75
Shape operator, 360–362
   definition of the, 360
   eigenvalues and eigenvectors of the, 371
Sign, of a number, 15
Signature, 565
Signed curvature of a curve in $\mathbb{R}^2$, 14, 30
Simple curve, 160
Sine surface, 315
Singular curve at a point, 22
Slant
   of a generalized helicoid, 475
   of a geodesic in a v-Clairaut patch, 606
Slope, 234
   of a helix, 198

South pole, 526
Space
   Euclidean, 2, 246, 521, 523
      complex, 714
   metric, 623
      complete, 624
   tangent, 582
      to a differentiable manifold, 534, 564
Speed
   of a curve in $\mathbb{R}^n$, 6
   ratio, 67
Sphere, 270, 275, 282, 292, 294, 312, 387, 407, 482, 484, 508, 519, 525, 626, 971
   area of a, 354
   astroidial, 397
   Gaussian curvature of a, 395
   mean curvature of a, 395
   metric of a, 354
   osculating, 229
   shape operator of a, 395
   stereographic, 966
   twisted, 477, 968
Spherical
   cardioid, 932
   helix, 237, 932
   nephroid, 932
   spiral, 216
Spindle torus, 306
Spiral
   hyperbolic, 91, 905, 924
      cotangent, 898
      tangent, 914
   logarithmic, 40, 87, 111, 120, 135, 479, 909, 926, 980
      evolute of a, 120
      natural equation of a, 138
   Nielsen's, 146, 910, 980
   of Archimedes, 90, 892, 923
   of Cornu, 64
   of Euler, 896

of Fermat, 90, 903, 924
parabolic, 911
Seiffert, 931
spherical, 216, 932
Spread of an oval, 169
Spring, 914
Square
of the distance function, 286
wave, 150
Weierstrass functions for a, 750
Steiner, Roman surface of, 331–333, 336, 963
Stereographic
ellipsoid, 302
injection, 526
map, 238, 350
projection, 719
Striction curve, 445, 446
Subgroup, 131
Sum of vectors, 247
Super quadric, 414
Support function
of a plane curve, 175
of a surface in $\mathbb{R}^3$, 410
Surface
abstract, 574
orientable, 582
Boy's, 937
canal, 799
corkscrew, 477, 478
developable, 439
Dini's, 493, 494, 942, 943
eight, 310, 358, 943
ennepercatenoidpolar, 945
exptwist, 946
funnel, 424, 508, 947
funnelasym, 947
generalized helicoid
axis of a, 475
meridian of a, 476
profile curve of a, 475

standard parametrization of a, 475
geodesically complete, 600
Kuen's, 496, 955
lathed, 474
level, 291, 320
local, 269
mapping, 288
Menn's, 956
minimal
Bour's, 732, 936, 974
Catalan's, 693, 724, 766, 937, 975
Catalan's, isometric deformation of, 725
catenoid, 467, 470, 519, 685, 686, 777, 938, 975
Costa's, 747, 757, 940, 977
definition of, 374
ellipsemincurve, 978
Enneper's, 358, 388, 415, 519, 684, 726, 728, 945, 978
helicoid, 357, 685, 949
Henneberg's, 691, 743, 767, 950, 979
lemniscatemin, 955
Richmond's, 744, 962, 981
Scherk's, 688–690, 696, 760, 964, 971
Scherk's fifth, 697, 698, 964, 971
Thomsen's, 967
trinoid, 760
with one planar end, 744
monkey saddle, 299, 300, 312, 366, 382, 509, 614, 615, 957, 959
nonorientable, 322, 324, 327, 407
nonparametric form of a, 291, 311, 409
normal
curvature of a, 359, 363, 373
to a patch in $\mathbb{R}^3$, 278
variation of a, 681, 682
of revolution, 457
axis of revolution of a, 457
barrel type, 485

conic type, 487, 498
flat, 497
football type, 484
generalized, 479
geodesic on a, 610
hyperboloid type, 487, 498
meridian of a, 458
parallel of a, 458
principal curves of a, 460
profile curve of a, 457
standard parametrization of a, 457
orientable, 318
oriented, 318, 582
parallel, 773
shape operator of a, 779
to a Möbius strip, 777
pedal, 785, 788
pseudosphere, 961, 962
regular, 526
in $\mathbb{R}^n$, 281, 574
parametric representation of a, 291
ruled, 431
base curve of a, 431
director curve of a, 431
directrix of a, 431
doubly, 431
flat, 439
Gaussian curvature of a, 431
noncylindrical, 445
rulings of a, 431, 603
Schwarzschild, 580, 581
shoe, 964
Sievert's, 964
Steiner's Roman, 331–333, 336, 963
tangent developable, 439, 594
torus, 294, 304, 323, 356, 384, 388, 415, 430, 466, 519, 788, 967
translation, 388, 416
twisted, 786
vector perpendicular to, 288
Symmetric

bilinear form, 564
product, 351, 568
System
of local coordinates, 281, 522
orthogonal, of curves, 679
triply orthogonal, of surfaces, 661

Tangent
bundle, 547, 556
projection of the, 548
developable, 439, 594
map, 251, 289, 362, 540
space
of $\mathbb{R}^n$, 246
to a differentiable manifold, 534
to a surface, 287
vector
coradial, 620
radial, 620
to $\mathbb{R}^n$, 246
to a manifold, 533
to a patch, 276
Tangent-radius angle, 23
Tchebyshef net, 593
Tensor, 551
covariant, 551
field, 550, 551
vectorvariant, 551
Theorem
basis, 536
Beltrami-Enneper, 421
Bonnet's, 781
Cayley-Hamilton, 381
Dupin's, 668
Euler's, 372
fundamental
of plane curves, 136, 137
of space curves, 219
of surfaces, 654, 656
inverse function, 274, 290, 542

Joachimsthal's, 460, 461
   converse to, 479, 495
mean value, 9
Pythagorean, 59, 341, 342
Riemann mapping, 833
Theorema Egregium, 507–509, 577
Third fundamental form, 381
Thomsen, minimal surface of, 967
Topological manifold, 523
Torsion
   geodesic, 517, 863
   of a circular helix, 189, 199
   of a connection, 572
   of a curve in $\mathbb{R}^3$, 186, 191, 195, 421, 866
   of Viviani's curve, 201, 232
   zero for a connection, 567
Torus, 294, 312, 323, 356, 384, 388, 415, 430, 466, 519, 788, 967, 972
   crossed, 339
   elliptical, 210, 304
   flat, 579, 591, 947
   Gaussian curvature of a, 384
   horn, 306
   knot, 209, 210, 213
      tube around a, 212, 214, 215
   mean curvature of a, 384
   ring, 304
   spindle, 306
Total
   geodesic curvature of a curve on a surface, 588
   signed curvature of a curve in $\mathbb{R}^2$, 154
Trace
   of a curve in $\mathbb{R}^n$, 6
   of a patch, 270
Tractoid, 489
Tractrix, 61, 63, 64, 73, 102, 121, 493, 915, 982
   differential equation of a, 62
   evolute of a, 102, 103

Transformation
   affine, 130, 262
   by reciprocal radii, 820
   orthogonal, 128, 132
Translation
   of $\mathbb{R}^n$, 130
   surface, 388, 416
Trefoil knot, 210
Triangle
   equilateral, 42, 479
   inequality, 3, 9
Triangular region, 632
Triangulation, 632
Trinoid, 760
Triple product, 183
Triply orthogonal
   patch, 666
   system of surfaces, 661
Trisectrix of Maclaurin, 94, 916
Tube
   around
      a circular helix, 208
      a curve, 207
      a torus knot, 212, 214
      an eight knot, 215
      Viviani's curve, 216
   as a canal surface, 801
   radius of a torus, 304, 831
Turning angle
   of a curve in $\mathbb{R}^2$, 19
   of a curve on a surface with respect to a vector field, 585
Twisted
   cubic, 189, 190, 455, 933
   surface, 786

u-parameter curve, 275, 296
Umbilic point, 645, 790
Unit
   normal
      to a paraboloid hyperbolic, 280

to a patch in $\mathbb{R}^3$, 278
normals of a triply orthogonal patch, 667
speed curve in $\mathbb{R}^n$, 6
tangent vector field to a curve in $\mathbb{R}^n$, 185

Unsigned geodesic curvature, 513
Upper g-ij's, 569

v-parameter curve, 275, 296
Vector
  cross product, 182, 318, 715, 846
  field
    along a curve, 12, 185, 265, 561
    binormal, 185
    differentiable, 255
    dual to a 1-form, 566
    image by a diffeomorphism, 261
    mean curvature, 589
    normal to a surface, 277
    on $\mathbb{R}^n$, 255
    on a differentiable manifold, 547
    on a patch, 277
    on a surface, 277, 288
    principal normal, 185
    tangent to a surface, 277
    unit tangent to a curve, 185
    velocity of a
    curve in $\mathbb{R}^n$, 265
  in $\mathbb{C}^n$, 714
  in $\mathbb{R}^n$, 2, 245
  normal
    to a patch, 277
    to a surface, 288
  part of a tangent vector to $\mathbb{R}^n$, 246
  perpendicular
    to a patch, 277
  principal, 364
  tangent, 533
    point of application of a, 246
    to a patch, 276

vector part of a, 246
triple product, 183
velocity, 543
Vectorvariant
  tensor, 551
  tensor field, 551
Velocity, 6, 30, 191, 265
  initial, 544
  of a curve on a differentiable manifold, 543
Versiera, 72
Vertex
  of a cone, 439
  of a parabola, 45, 115
  of a piecewise-differentiable simple closed curve, 628
  of a plane curve, 166
Viviani
  curve of, 201, 202, 205, 232, 933
    binormal surface to, 453
    curvature of, 201, 232
    evolute of, 242
    normal surface to, 453
    tangent developable to, 456
    torsion of, 201, 232
    tube around, 216

Wallis, conical edge of, 454, 968
Wedge product, 351
Weierstrass
  $\wp$ function, 748
  ζ function, 748
  functions for a square, 750
  patch, 737, 738
Weingarten
  equations, 369, 392, 461, 650, 705
  surface, 369
Wheel radius of a torus, 304, 831
Whitney umbrella, 310, 311, 358, 401, 968
Width of an oval, 169

NAME INDEX

Agnesi, Maria Gaetana (1718-1799), 72, 1007
Ahlfors, Lars Valerian (1907-1996), 1007
Airy, Sir George Biddel (1801-1892), 148
Aleksandrov, Aleksandr Danilovich (1912-), 1007
Aoust, L'abbé Louis (1811-1885), 1007
Apollonius of Perga (262-180 B.C.), 14, 37
Archibald, Raymond Claire (1875-1955), 1007
Archimedes of Syracuse (287-212 B.C.), 91

Barbier, Joseph Émile (1839-1889), 169, 1008
Bartels, Karl Eduard (1769-1836), 187
Beltrami, Eugenio (1835-1900), 420, 610
Berkeley, Bishop George (1685-1753), 94
Bernoulli, Daniel (1700-1782), 595
Bernoulli, Jakob (1654-1705), 52, 64
Bernoulli, Johann (1667-1748), 595
Bernoulli, Johann (II) (1710-1790), 595
Bernoulli, Nicolaus (1695-1726), 595
Bertrand, Joseph Louis Francois (1822-1900), 1008
Bessel, Friedrich Wilhelm (1784-1846), 143
Bianchi, Luigi (1856-1928), 360, 494, 1008
Bieberbach, Ludwig (1886-1982), 1008

Björling, Emanuel Gabriel (1808-1872), 761, 1009
Blaschke, Wilhelm Johann Eugen (1885-1962), 360, 1009
Bolyai, Farkas Wolfgang (1775-1856), 610
Bolyai, János (1802-1860), 610
Bompiani, Enrico (1889-1975), 1009
Bonnet, Pierre Ossian (1819-1892), 627, 638, 654, 781, 1009
Bour, Edmond (1832-1866), 732
Bowditch, Nathaniel (1773-1838), 70
Brioschi, Francesco (1824-1897), 504, 1009
Burali-Forti, Cesare (1861-1931), 360, 1010

Cajori, Florian (1859-1930), 1010
Cartan, Élie (1869-1951), 518, 574, 1010
Cassini, Gian Domenico (1625-1712), 82
Catalan, Eugène Charles (1814-1894), 693, 1010
Cauchy, Augustin Louis (1789-1857), 624, 705
Cayley, Arthur (1821-1895), 119
Cesàro, Ernesto (1859-1906), 50, 139, 1010
Chasles, Michel (1793-1880), 133, 1010
Chern, Shiing-Shen (1911-), 781, 1010
Christoffel, Elwin Bruno (1829-1900), 510, 1011

Clairaut, Alexis Claude (1713-1765), 187, 604
Codazzi, Delfino (1824-1873), 649, 1011
Cohn-Vossen, Stephan (1902-1936), 639, 1011, 1014
Coolidge, Julian Lowell (1873-1954), 1011
Cornu, Marie Alfred (1841-1902), 64

Darboux, Jean Gaston (1842-1917), 205, 360, 809, 1011
Descartes, René du Perron (1596-1650), 1, 41, 77, 80, 1008
Dini, Ulisse (1845-1918), 493
Diocles (180 B.C.), 49
Dupin, Baron Pierre Charles Francois (1784-1873), 668, 809, 1012

Eisenhart, Luther Pfahler (1876-1965), 380, 1012
Enneper, Alfred (1830-1885), 358, 420, 684, 726, 1012
Euclid of Alexandria (365-300 B.C.), 2, 609
Euler, Leonhard (1707-1783), 64, 139, 168, 372, 507, 635, 735, 1012

Fermat, Pierre de (1601-1665), 92
Frenet, Jean Frédéric (1816-1900), 185, 1013
Fresnel, Augustin Jean (1788-1827), 65
Frost, Percival (1817-1898), 25, 1013

Gauss, Carl Friedrich (1777-1855), 143, 279, 319, 342, 362, 368, 373, 374, 507, 573, 574, 587, 610, 620, 638, 649, 694, 720, 1013
Germain, Sophie (1776-1831), 374
Gerono, Camille Christophe (1799-1891), 31
Gibbs, Josiah Willard (1839-1903), 181

Gomes Teixeira, Francisco (1851-1933), 50, 1013
Goursat, Édouard Jean Babtiste (1858-1936), 314, 1013
Grassmann, Hermann Günther (1809-1877), 181
Graustein, William Caspar (1888-1941), 162
Green, George, (1793-1841), 629

Halley, Edmond (1656-1742), 187
Hamilton, Sir William Rowan (1805-1865), 181, 381
Harriot, Thomas (1560-1621), 627
Henneberg, Ernst Lebrecht (1850-1922), 691, 1013
Hesse, Ludwig Otto (1811-1874), 267
Hilbert, David (1862-1943), 652, 1014
Hill, Thomas (1818-1891), 139
Hopf, Heinz (1894-1971), 161, 600, 625
Huygens, Christiaan (1629-1695), 77, 123

Jacobi, Carl Gustav Jacob (1804-1851), 254, 1014
Joachimsthal, Ferdinand (1818-1861), 460, 1014

Kästner, Abraham Gotthelf (1719-1800), 20
Kepler, Johannes (1571-1630), 93
Klein, Christian Felix (1849-1925), 323, 610, 1015
Kowalewski, Gehrhard (1876-1950), 1010
Kronecker, Leopold (1823-1891), 537
Kummer, Ernst Eduard (1810-1893), 313

L'Hôpital, Guillaume Francois Antoine Marquis de (1661-1704), 529
Lacroix, Sylvestre Francois (1765-1843), 139

Name Index

Lagrange, Joseph Louis (1736-1813), 183, 681
Lamé, Gabriel (1795-1870), 661, 702, 1015
Laplace, Pierre Simon, Marquis de (1749-1827), 70
Legendre, Adrien Marie (1752-1833), 481, 749
Leibniz, Gottfried Wilhelm (1646-1716), 532
Leonardo da Vinci (1452-1519), 66
Levi-Civita, Tullio (1873-1941), 559, 1016
Lie, Marius Sophus (1842-1899), 716
Liebmann, Heinrich (1874-1939), 483, 653
Liouville, Joseph (1809-1882), 587, 592, 595, 809, 1016
Lissajous, Jules Antoine (1822-1880), 70
Lobachevsky, Nikolai Ivanovich (1792-1856), 610
Loria, Gino (1862-1954), 50, 1016

Maclaurin, Colin (1698-1746), 94
Mainardi, Gaspare (1800-1879), 649, 1016
Maupertuis, Pierre Louis Moreau de (1698-1759), 187
Maxwell, James Clerk (1831-1879), 809, 1016
Mellish, Arthur Preston (1905-1930), 171, 1016
Mercator, Gerardus (1512-1594), 312
Meusnier, Jean Babtiste (1754-1793), 735, 1017
Minding, Ernst Ferdinand Adolf (1806-1885), 475, 1017
Möbius, August Ferdinand (1790-1868), 322
Monge, Gaspard (1746-1818), 398, 507, 1017

Newton, Sir Isaac (1642-1727), 49, 57, 1017
Nielsen, Niels (1865-1931), 146
Nuñez Salaciense, Pedro (1502-1578), 238

Pascal, Étienne (1588-1651), 71
Pascal, Blaise (1623-1662), 71
Peterson, Karl Mikhailovich (1828-1881), 649
Plateau, Joseph Antoine Ferdinand (1801-1883), 681
Playfair, John (1748-1819), 610
Plücker, Julius (1801-1868), 435
Poincaré, Jules Henri (1854-1912), 578, 610

Rademacher, Hans (1892-1969), 1018
Reidemeister, Kurt Werner Friedrich (1893-1971), 1018
Reuleaux, Franz (1829-1905), 176, 1018
Ricci-Curbastro, Georgorio (1853-1925), 550, 1016, 1018
Richmond, Herbert William (1863-1949), 744
Riemann, Georg Friedrich Bernhard (1826-1866), 573, 1018
Rinow, Willi (1907-1978), 600, 625
Rolle, Michel (1652-1719), 113

Saccheri, Giovanni Girolamo (1667-1733), 420
Salmon, George (1819-1904), 1019
Scheffers, Georg (1866-1945), 1019
Scherk, Heinrich Ferdinand (1798-1885), 688, 1019
Schwarz, Herman Amandus (1843-1921), 762, 1019
Schwarzschild, Karl (1873-1916), 579
Senff, Karl Eduard (1810-1849), 187
Serret, Joseph Alfred (1819-1885), 187, 1019

Serret, Paul Joseph (1827-1898), 187, 1019
Snyder, Virgil Snyder (1869-1950), 1020
Steiner, Jakob (1796-1863), 331
Stoker, James Johnston (1905-1988), 1020
Strubecker, Karl (1904-?), 1020
Struik, Dirk Jan (1894-), 50, 1020
Study, Eduard (1862-1930), 733, 1020

Töplitz, Hans (1881-1940), 1018
Tchebyshef, Pafnuty Lvovich (1821-1894), 593
Thompson, D'Arcy Wentworth (1860-1948), 1021
Tschirnhausen, Ehrenfried Walter (1651-1708), 73

Veronese, Giuseppe (1854-1917), 594
Viviani, Vincenzo (1622-1703), 201

Walker, Robert John (1909-?), 1021
Wallis, John (1616-1703), 454
Wankel, Felix Heinrch (1902-1988), 177
Weatherburn, Charles Ernest (1884-1974), 1021
Weber, Wilhelm Eduard (1804-1891), 279
Weierstrass, Karl Theodor Wilhelm (1815-1897), 736, 1021
Weingarten, Julius (1836-1910), 369
Weyl, Hermann Klaus Hugo (1885-1955), 574, 1021
Whitney, Hassler (1907-1989), 76, 162, 310
Wieleitner, Heinrich (1874-1933), 50, 1021
Woodhouse, Robert (1773-1827), 1022

Yates, Robert Carl (1904-1963), 1022

Zwikker, Cornelius (1900-?), 1022

MINIPROGRAM AND MATHEMATICA COMMAND INDEX

(), 26
->, 30
/., 30
<=, 47, 176
===, 180
[], 26
{}, 26

agnesi, 72, 891
agnesiimplicit, 916
agnesirev, 935
AiryAi, 148
AiryBi, 148
alysoid, 891
analytictominimal, 723, 741, 742, 765–767, 881
Append, 335
archimedesspiral, 892
archimedesspiralpolar, 90, 923
arclength, 33, 768, 837
arclengthea, 837
arclengtheaprime, 837
arclengthprime, 33, 838
ArcSin, 27

ArcTan, 149
AspectRatio, 32, 33
AspectRatio->Automatic, 33
ast3d, 203, 927
astell, 396, 935
astroid, 46, 99, 767, 892
astroidimplicit, 916
astroidmincurve, 974
asymeq1, 427, 888
asymeq2, 427, 888
Axes, 297

baseballseam, 927
bellcurve, 892
besselcurve, 206, 927
BesselJ, 144, 145, 206, 472, 927
besseln, 934
betaconoid, 450
bicorn, 893
bicylinder, 204, 927
binormal, 195, 196, 871
binormalsurf, 452
biquadratic, 928
bjorling, 763

Block, 280
bohdom, 389, 935
bour, 936
bourmincurve, 733, 974
bourpolar, 733, 936
bow, 893
bowpolar, 923
bowtie, 893
Boxed, 297
boy, 937
brioschicurvature, 506, 519, 580, 593, 841
bulletnose, 893
bulletnoseimplicit, 917

Cancel, 197
cardioid, 54, 71, 88, 119, 120, 893
cardioidimplicit, 917
cardioidmincurve, 975
cardioidpolar, 923
cardioidunitspeed, 894
cartesianimplicit, 917
cartesianoval, 894
cassini, 894
cassiniimplicit, 82, 917
cassinipolar, 923
catalan, 693, 724, 725, 937
catalandef, 937
catalanmincurve, 975
catenary, 56, 103, 121, 122, 479, 895
catenaryunitspeed, 895
catenoid, 467, 686, 938
catenoidasym, 938
catenoidiso, 938
catenoidmincurve, 975
cayleysextic, 119, 895
cayleysexticmincurve, 975
centerlines, 987
christoffel, 512, 842, 844
Circle, 37
circle, 36, 895

circle3point, 114, 888, 889
circleimplicit, 918
circleinvolute, 46, 895
circleinvolutemin, 938
circleinvolutemincurve, 975
circleinvoluteunitspeed, 896
circleunitspeed, 896
circularcone, 939
circularcylinder, 939
circularmembrane, 972
cissoid, 57–60, 73, 77, 101, 119, 896
cissoidimplicit, 918
cissoidmincurve, 976
clairaut, 873
clelia, 928
clothoid, 65, 66, 896
clothoidmin, 939
clothoidmincurve, 771, 772, 976
clothoidprime, 65
cnccoprofile, 498, 897
cnccosurfrev, 499, 940
cnchyprofile, 498, 897
cnchyprofilemincurve, 976
cnchysurfrev, 498, 499, 940
cochleoid, 898
cochleoidpolar, 924
ColorFunction, 400
combescure, 846
Composition, 332
conchoid, 47, 898
conchoidimplicit, 918
cone, 441, 854
conicalhelix, 928
ContourGraphics, 79
ContourPlot, 300
contrapedal, 848
Cos, 27
coshspiral, 898
CosIntegral, 147
cosn, 934
cossurface, 940

cossurfaceimplicit, 969
costa, 757, 940
costamincurve, 977
cothspiral, 898
cpcprofile, 497, 899
cpcprofilemincurve, 977
cpcsurfrev, 498, 941
cross, 194–196, 279, 846
cross1, 194
crosscap, 334, 941
crosscapimplicit, 969
crosscapmap, 333, 339
crosscurve, 899
crosscurveimplicit, 918
ctor, 845
cubicalellipse, 928
cubicparabola, 899
cubicparabolaimplicit, 918
curveonsurf, 851
cycloid, 50, 766, 899
cycloidunitspeed, 900
cylinder, 440, 854

D, 29, 30
darbouxvector, 205, 871
delaunay, 900
deltoid, 70, 120, 900
deltoidimplicit, 919
deltoidinvolute, 178, 900
deltoidinvolutemin, 941
deltoidinvolutemincurve, 977
deltoidmin, 942
deltoidmincurve, 978
deltoidunitspeed, 901
der, 886
descenttime, 838
descenttimen, 838
descenttimends, 838
descenttimeprime, 839
devilimplicit, 92, 919
diamond, 43, 901

dini, 494, 942
dinitcheb, 943
dip, 474, 757, 772, 987
DisplayFunction, 109, 404, 644
divergence, 264, 886
ds2, 881
DSolve, 688

E, 353
ee, 354, 880
eee, 392, 884
eeee, 426, 884
eight, 31, 901
eighteight, 943
eightknot, 215, 928
eightsurface, 294, 296, 310, 943
Eliminate, 77
ellhypcyclide, 943
ellipse, 38, 99, 100, 108, 110, 116, 768, 902
ellipseimplicit, 919
ellipseinc, 902
ellipsemin, 768, 769, 944
ellipsemincurve, 768, 978
ellipsoid, 302, 944
ellipsoidimplicit, 969
ellipticalhyperboloidminus, 433, 945
ellipticalhyperboloidplus, 433, 945
ellipticcoor, 670–673, 945
EllipticE, 482
EllipticIntegrate, 768
ellipticparaboloid, 647
ellipticparaboloidimplicit, 969
elrot, 988
enneper, 358, 684, 726, 727, 945
ennepercatenoidmincurve, 760, 978
ennepercatenoidpolar, 760, 945

ennepermincurve, 728, 978
enneperpolar, 729–731, 946
epicycloid, 902
epicycloidmincurve, 978
epicycloidunitspeed, 902
Epilog, 46
epispiral, 903
epispiralpolar, 924
epitrochoid, 47, 903
epitrochoid3d, 929
Evaluate, 37
evolute, 99, 101, 848
evolute3d, 852
Exp, 27
exptwist, 946

FaceForm, 335
Factor, 197
fermatspiral, 903
fermatspiralpolar, 924
ff, 354, 880
fff, 392, 884
ffff, 427, 884
filament, 852
filledcurve, 34, 988
Flatten, 47, 176
flattorus, 579, 591, 947
focalk1, 870
focalk2, 870
focalkmu, 869
focalkpi, 869
folium, 80, 904
foliumimplicit, 919
foliumpolar, 924
frenetframe, 871
FresnelC, 65
FresnelS, 65
Function, 27
funnel, 424, 947
funnelasym, 947

Gamma, 148

gammaconoid, 450
gaussiancurvature, 393, 394, 861
gcursq, 413
gcurvature, 394, 466, 477, 492, 494, 507, 519, 745, 746, 860
gencube, 947
genhel, 475–477, 492, 493, 854
genhelix, 929
genhypparab, 434, 948
genoctahedron, 948
genparabola, 904
gensurfrev, 480, 854
geo, 613, 873
geodesiccurvature, 519, 862
geodesictorsion, 863
geomincurve, 764–767, 881
geopolar, 982
gg, 354, 880
ggg, 392, 885
gggg, 427, 885
gmap, 887
goursatimplicit, 970
grad, 258, 887
graph, 478
Graphics, 79, 87, 108, 472
graphics2dtosurfrev, 473, 988
Graphics3D, 79, 85, 472, 685
GrayLevel, 385

handkerchief, 313, 948
harmonictoanalytic, 723, 882
hcursq, 413
heart, 949
helical, 199, 851
helicoid, 357, 422, 449, 643, 742, 949
helicoid1, 724
helicoidprin, 644, 949
helix, 198, 929
helixunitspeed, 930
heltocat, 685, 686, 696, 949
henneberg, 691, 692, 743, 950

hennebergmincurve, 979
hippopede, 904
horopter, 930
Hue, 746
hy2sheet, 406, 951
hy2sheetimplicit, 970
hyperbola, 45, 905
hyperbolabis, 905
hyperbolaimplicit, 919
hyperbolamincurve, 979
hyperbolicparaboloid, 280, 297, 449, 950
hyperbolicparaboloidimplicit, 970
hyperbolicspiral, 905
hyperbolicspiralpolar, 924
hyperboloid, 404, 950
hyperboloidimplicit, 970
hypocycloid, 905, 980
hypocycloidinvolute, 906
hypocycloidinvolutemincurve, 979
hypocycloidmincurve, 980
hypocycloidunitspeed, 906
hypotrochoid, 47, 906

IdentityMatrix, 194
If, 43
ImplicitPlot, 27, 75, 79, 82, 92, 93
ImplicitPlot3D, 292, 311
infarea, 354, 881
inscurve, 851
Integrate, 33, 355
InterpolatingFunction, 471
intrinsic, 140, 471, 874
intrinsicea, 874
inversion, 877
inversioncurve, 878
inversionsurf, 878
invertedcircularcone, 951
invertedcircularcylinder, 952

invertedtorus, 952
involute, 104, 846
involutends, 122, 847

J, 29, 846
jacobian2, 280, 879

k1, 402, 883
k2, 403, 883
kampyle, 907
kampylepolar, 925
kappa, 195, 196, 863
kappa2, 30, 583, 864
kappa23d, 240, 851
kappa2c, 864
kappa2ea, 865
kappa2implicit, 865
kappa2polar, 89, 865
kappan, 203, 864
keplerimplicit, 93, 920
keplerorbit, 907
keplerorbitpolar, 925
kidney, 953
klein4, 592, 954
kleinbottle, 327, 787, 953
kleinbottlebis, 330, 954
kmu, 467, 883
kpi, 467, 883
kuen, 496, 955
kuentcheb, 955
kummerimplicit, 970

laplace, 887
laplacian, 264
lathe, 989
lemniscate, 52, 907
lemniscateimplicit, 920
lemniscatemin, 770, 955
lemniscatemincurve, 770, 980
lemniscatepolar, 925
lemniscateunitspeed, 907
length, 33, 839

lengthea, 839
lengthean, 839
lengtheands, 840
lengthn, 33, 840
lengthnds, 840
limacon, 71, 908
limaconimplicit, 920
limaconpolar, 88, 925
Line, 35, 46, 108, 472
line, 908
line3d, 930
linetolines, 472, 989
liouville, 982
lissajous, 70, 908
lissajous3d, 930
lituus, 91, 908
lituuspolar, 926
logistic, 909
logspiral, 40, 909
logspiralmincurve, 980
logspiralpolar, 87, 926
logspiralunitspeed, 909
loxodrome, 238, 930

Map, 473
mcurvature, 394, 466, 861
meancurvature, 393, 394, 861
meancurvaturevf, 593, 862
menn, 313, 956
mercatorellipsoid, 312, 956
Module, 35, 140, 223, 280
moebcirc, 956
moebiusstrip, 325, 437, 786, 957
monge, 399, 853
mongepolar, 853
monkeysaddle, 299, 957
monkeysaddlepolar, 957

N, 34, 196
nateq, 147, 889
NDSolve, 122, 140, 222, 223

Needs, 79, 85, 87, 335, 477, 685, 727, 768
Negative, 81, 148
nephroid, 47, 909
nephroidimplicit, 920
nephroidmincurve, 981
nephroidpolar, 926
nephroidunitspeed, 910
Nest, 107, 121, 165
ngon, 47, 910
nielsenspiral, 146, 910
nielsenspiralmincurve, 980
NIntegrate, 33, 157, 356
nodoidnds, 958
normal, 196, 872
normalcurvature, 519
normalline, 108, 120, 849
normalsurf, 452
NSolve, 82

oneparametersubgroup, 931
orth4d, 989
osculatingcircle, 111, 121, 849
oval, 175

pacman, 94, 926
parabola, 45, 119, 910
parabolaimplicit, 920
parabolamincurve, 981
paraboliccoor, 676, 677, 958
paraboliccyclide, 958
parabola, 911
paraboloid, 307, 959
paramcurveplot1, 990
paramcurveplot2, 990
ParametricPlot, 27, 33, 34, 53, 75, 87
ParametricPlot3D, 197, 198, 241, 296, 298, 335, 385, 403, 404, 422, 458, 695
paramtoimp, 77, 889
parcurve, 116, 849

parsurf, 776, 858
pear, 959
pedal, 118, 849
pedsurf, 785, 858
perpsurf, 456
perseusimplicit, 921
perturbedms, 313, 959
pillow, 960
piriform, 47, 178, 911
piriformimplicit, 921
plane, 960
planeframe, 868
Plot, 27, 33
Plot3D, 296, 298, 300
plotintrinsic, 141, 875
plotintrinsic3d, 223, 876
plotintrinsicea, 875
plotintrinsicsurfrev, 876
Plotpatch2d, 991
PlotPoints, 40, 197
PlotRange, 334
plottl, 850
plottp, 294, 860
plucker, 435, 960
pluckerpolar, 436, 960
poincare, 580, 581, 983
poindisk, 984
polarize, 308, 889
polarplane, 961
PolarPlot, 27, 87, 92, 94
Polygon, 35
PowerExpand, 34, 60, 61, 105, 197, 434, 467
predatorpreyimplicit, 921
primitiveplot3d, 995
Prolog, 110
pseudocatenary, 911
pseudocrosscap, 337, 415, 961
pseudosphere, 489, 961
pseudospherebis, 961
pseudospheregeodesic, 931

pseudospheretcheb, 962
Purple, 335

rconoid, 962
rectangle, 911
Release, 37, 197
reuleauxpolygon, 176, 912
rf, 134, 867
RGBColor, 335
richmond, 962
richmondmincurve, 744, 981
richmondpolar, 745, 746, 962
rollcurve, 850
roman, 332, 963
romanmap, 331
rose, 912
rosemin, 963
rosemincurve, 981
rosepolar, 926
Rotate3D, 727
RotateLeft, 194
Rotations, 727
Round, 157
rt, 134, 867
rtoc, 845
ruled, 450, 855

sawtooth, 149
sc, 34, 912
scarab, 913
scherk, 689, 697, 964
scherk5, 698, 964
scherk5implicit, 971
schwarzschildplane, 581, 984
scimplicit, 922
scmincurve, 981
scunitspeed, 913
seashell, 309
seiffertspiral, 931
Select, 82, 425
selectcolgraphics3d, 992
selectgraphics3d, 425, 757, 991

semicubicimplicit, 922
serpentine, 913
serpentineimplicit, 922
sff, 885
Shadow, 685
shoe, 313, 964
Show, 108
sievert, 499, 964
FullSimplify, 31
Simplify, 31, 105, 196, 197
singmonge, 853
singmongepolar, 853
Sinh, 27
sinhspiral, 913
SinIntegral, 147
sinn, 934
sinoval, 165, 913
sinovaloid, 965
sinsurface, 965
sinsurfaceimplicit, 971
Solve, 60, 79, 84, 113, 115, 648
solvegeoeqs, 613, 614, 873
SpaceCurve, 197
sphere, 270, 965
sphereimplicit, 971
sphericalcardioid, 932
sphericalhelix, 932
sphericalnephroid, 932
sphericalspiral, 216, 932
spring, 914
squaremembrane, 973
squarewave, 150
StackGraphics, 85
stereo4d, 992
stereographicellipsoid, 303, 966
stereographicsphere, 966
stereoinj, 556, 890
stereoproj, 556, 890
stretch, 859
strophoid, 47, 92, 121, 914
strophoidimplicit, 922

surfacecurveplot, 993
SurfaceGraphics, 79
surfacejacobian, 275, 692, 879
surfacejacobianmatrix, 275, 879
surfacenormal, 879
surfast, 855
surfrev, 457, 465, 466, 471, 855
swallowtail, 966

Table, 47, 53, 54, 176, 497
tandev, 444
tangent, 195, 196, 872
tangentkmu, 870
tangentkpi, 870
tangentline, 108, 120, 847
tangentplane, 293, 859
tanhspiral, 914
tau, 195, 196, 866
teardrop, 914
tetrahedral, 966
Thickness, 110
thomsen, 967
Ticks, 40
Timing, 212, 519
tl, 134, 867
torus, 304, 356, 788, 967
torusimplicit, 972
torusknot, 210, 933
totalscalarcurvature, 866
tractrix, 62, 102, 121, 915
tractrixmincurve, 982
tractrixunitspeed, 915
Transpose, 275, 879
triangle, 43, 915
Trig, 197
TrigCanonical, 197
TrigReduce, 477
trisectrix, 916
trisectrixpolar, 94
tschirnhausen, 916
tubecurve, 208

tubeplanecurve, 855
turningangle, 866
turningnumber, 866
twicubic, 190, 193, 933
twiflat, 967
twisphere, 477, 968
twist, 786, 856
twistedn, 206, 934
twocuspsimplicit, 972

umeqs, 647, 890
unitnormal, 280, 880

veronese, 968
veryflat, 933
viviani, 201, 933

wallis, 454, 968
wei, 741, 742, 744, 760, 882
weicurvature, 741, 882
weimetric, 882
weingarten, 392, 885
Which, 47, 176
whitneyumbrella, 310, 968
wireframe, 35, 404, 994
wrinkles, 968
Write2COLMESH, 1003
Write3DCMESH, 1001
Write3DCNMESH, 1002
Write3DMESH, 1000
WriteAcroLines, 997
WriteAcroPoints, 995
writhe, 241, 852

zcprofile, 491, 492, 916